CRC Series in
COMPUTATIONAL MECHANICS and APPLIED ANALYSIS

COMBUSTION SCIENCE AND ENGINEERING

CRC Series in

COMPUTATIONAL MECHANICS and APPLIED ANALYSIS

Series Editor: J.N. Reddy
Texas A&M University

Published Titles

ADVANCED THERMODYNAMICS ENGINEERING
Kalyan Annamalai and Ishwar K. Puri

APPLIED FUNCTIONAL ANALYSIS
J. Tinsley Oden and Leszek F. Demkowicz

COMBUSTION SCIENCE AND ENGINEERING
Kalyan Annamalai and Ishwar K. Puri

EXACT SOLUTIONS FOR BUCKLING OF STRUCTURAL MEMBERS
C.M. Wang, C.Y. Wang, and J.N. Reddy

THE FINITE ELEMENT METHOD IN HEAT TRANSFER AND FLUID DYNAMICS, Second Edition
J.N. Reddy and D.K. Gartling

MECHANICS OF LAMINATED COMPOSITE PLATES AND SHELLS: THEORY AND ANALYSIS, Second Edition
J.N. Reddy

PRACTICAL ANALYSIS OF COMPOSITE LAMINATES
J.N. Reddy and Antonio Miravete

SOLVING ORDINARY and PARTIAL BOUNDARY VALUE PROBLEMS in SCIENCE and ENGINEERING
Karel Rektorys

CRC Series in
COMPUTATIONAL MECHANICS and APPLIED ANALYSIS

COMBUSTION SCIENCE AND ENGINEERING

Kalyan Annamalai
Ishwar K. Puri

CRC Press
Taylor & Francis Group
Boca Raton London New York

CRC Press is an imprint of the
Taylor & Francis Group, an informa business

CRC Press
Taylor & Francis Group
6000 Broken Sound Parkway NW, Suite 300
Boca Raton, FL 33487-2742

Printed in the United States of America on acid-free paper
10 9 8 7 6 5 4 3 2 1

International Standard Book Number-10: 0-8493-2071-2 (Hardcover)
International Standard Book Number-13: 978-0-8493-2071-2 (Hardcover)

Library of Congress Cataloging-in-Publication Data

Annamalai, Kalyan.
Combustion science and engineering / Kalyan Annamalai & Ishwar K. Puri.
p. cm. -- (CRC series in computational mechanics and applied analysis ; 7)
Includes bibliographical references and index.
ISBN 0-8493-2071-2 (alk. paper)
1. Combustion engineering. I. Puri, Ishwar Kanwar, 1959- II. Title. III. Series.

TJ254.5.A56 2005
621.402'3--dc22 2005050219

Visit the Taylor & Francis Web site at
http://www.taylorandfrancis.com

and the CRC Press Web site at
http://www.crcpress.com

Dedication

KA dedicates this text to his mother, Kancheepuram Pattammal Sundaram, who could not read or write, for educating him in all aspects of his life and his late brother-in-law Mr. Masilamani, for without him KA would not have been where he is today. He also thanks his children, Shankar, Sundhar, and Jothi, and grandchildren, Rohen and Ravi, for providing a constant source of energy in his career, and his wife, Vasantha, for providing "bioenergy" and keeping him healthy. Finally, he wishes to thank the forefathers of the United States for developing a system that provides opportunities to anyone from any corner of the world who wishes to succeed on U.S. soil.

IKP thanks his wife, Beth, for her friendship and unstinting support, and acknowledges his debt to his sons, Shivesh, Sunil, and Krishan ("Great kids!") for allowing him to take time off from other pressing responsibilities, such as playing catch and chess. He calls his career "a very fortunate journey," which would have been impossible without the vital involvement of his entire family, including his parents, Krishan and Sushila Puri.

Table of Contents

Preface

We have written this text for engineers in mechanical, chemical, aeronautical, energy, and environmental engineering who wish to understand combustion science and apply it to engineering problems. Whereas the basic concepts in any discipline are relatively invariant, the problems keep changing. In several instances we have included physical explanations along with the mathematical relations and equations so that the principles can be applied to solve problems ranging from simple candle burning to real-world combustion and pollution problems.

The authors have a rich experience teaching combustion fundamentals and engineering for more than two decades. We were instructed, trained, and mentored by eminent professors, for which we are grateful. We modeled this book on our previous effort, *Advanced Thermodynamics Engineering*, by including ample examples as self-instruction tools. We acknowledge the debt that we owe to our students who asked the questions that clarified each chapter as we wrote it. The text uses a down-to-earth and, perhaps, unconventional approach in teaching advanced concepts in combustion. However, the mathematics of combustion cannot be avoided, because concerns about emissions of pollutants require information on the detailed differential structure of a flame. Mathematics, in particular, is kept to a minimum in the first seven chapters. A background on molecular concepts is given, followed by engineering details. As engineers, we have stressed applications throughout the text. There are at least 100 figures and 150 engineering examples in 18 chapters. Many of the examples and exercise problems are solved using Microsoft Excel spreadsheets. Attempts have been made to use a common nomenclature for all chapters; further dimensionless numbers relevant to combustion are summarized.

Chapter 1 contains an outline of the corpuscular aspects of thermodynamics. An overview of microscopic thermodynamics illustrates the physical principles governing the macroscopic behavior of substances. Fundamental concepts related to matter, energy, entropy, and property in a mixture are discussed along with presentation of the first law of thermodynamics (energy conservation), the entropy balance equation (which dictates the direction of reaction), and definitions of Gibbs and Helmhotz functions.

Chapter 2 discusses the stoichiometry, thermochemistry, and the first law for closed and open systems in mass and molar forms. It includes several problems related to gas analyses, which can be used to understand experiments for determining the A:F ratio, concentration of pollutants, and oxygen-based normalization. It also explains the role of reactive mixtures in determining complete combustion, flame temperatures, and entropy generation in reactive systems. In Chapter 3, criteria for the direction that chemical reactions proceed in are developed. This is followed by a description of equilibrium in single-phase and multiphase systems and a discussion of the Gibbs minimization method.

Chapter 4 introduces the background related to properties of solid, liquid, and gaseous fuels. It discusses the Wobbe number, proximate and ultimate analyses, biomass fuels, gas yield in digestion processes, and solid and liquid fuel particle size distribution along with presentation of empirical equations for determining heating values. Chapter 5 presents the chemical kinetics for homogeneous and heterogeneous reactions and the pyrolysis of solid fuels, and a discussion on second law when multistep and global kinetics are used.

Chapter 6 briefly discusses mass transfer processes and transport properties relevant to combustion. Chapter 7 deals with applications of the global form of energy and mass conservation equations for simple ignition problems, combustion of gaseous fuels in closed systems (e.g., automobile engines), plug flow reactors, and perfectly stirred reactors; the resistance concept is introduced for the combustion of solid and liquid fuels. Chapter 1 to Chapter 7 may be used as introductory material for a senior elective course in combustion along with parts of Chapter 9, Chapter 10, and finally Chapter 17, dealing with pollutant formation and destruction.

Chapter 8 to Chapter 16 and Chapter 18 involve a more detailed mathematical treatment of a wider range of combustion problems. Chapter 8 presents the conservation relations in differential form for mass, momentum, energy, and species, and the simplified Shvab–Zeldovich formulation for single and two component fuels. Chapter 9 to Chapter 12 deal mostly with diffusion flames (nonpremixed). Chapter 9 applies the conservation equations to a simple spherically symmetric problem of carbon combustion, including multiple heterogeneous reactions. Here, the single- and double-film models are discussed. Chapter 10 deals with liquid droplet combustion including evaporation and combustion of multicomponent drops. Chapter 11 presents the solutions for the combustion within boundary layers for forced, free, and stagnation flows. Many of the derivations for transformations and similarity conditions are left as exercise problems. Integral and "nonconventional integral" techniques suitable for senior courses are presented. Chapter 12 extends the solutions obtained in Chapter 11 to the combustion of gaseous fuels involving 2-D and circular laminar jets. The classical literature of Burke and Schumann is also presented. The results from Chapter 11 and Chapter 12 could be presented to seniors without details of derivations.

Chapter 13 introduces ignition and extinction during gaseous and solid fuel combustion. Chapter 14 describes limiting flame propagation speeds at subsonic (deflagrations) and minimum supersonic velocities (detonations) of premixed fuels, whereas Chapter 15 discusses the concept of a kinetically limited laminar burning velocity and flammability limits for combustible mixtures. Chapter 16 deals with interactive array and cluster combustion, whereas Chapter 17 describes pollutant formation and control methods and describes various methods of reporting emissions. Chapter 18 discusses turbulent combustion and various computational approaches.

Exercise problems and a list of formulae are placed at the end of the text. This is followed by more than 45 tables and charts containing thermodynamic and combustion properties and other useful information. Because of interest in Hg pollution, a small section has been added in Chapter 17, along with an equilibrium chart for Hg compounds. Spreadsheet-based software containing many programs covering many chapters and unit conversion software can be downloaded from the Web at the site http://www.crcpress.com/e_products/downloads by following the links for this book.

The combustion field is vast and all topics cannot be covered in a single book. Readers who discover errors, conceptual conflicts, or have any comments are encouraged to e-mail these to the authors (kannamalai@tamu.edu and ikpuri@vt.edu). The assistance of Charlotte Sims in typing a part of the manuscript is gratefully acknowledged. We wish to acknowledge helpful suggestions and critical comments from several students and faculty. IKP wishes to thank numerous graduate students in his research group who have helped considerably to advance his understanding of combustion.

We wish to acknowledge helpful suggestions and critical comments from several students and faculty and many of our colleagues who educated the authors by sharing their research work. Years of fruitful research collaboration with Prof. John M. Sweeten of Texas A&M on biomass is greatly appreciated by KA. We wish to acknowledge helpful inputs, suggestions, project reports, and critical comments from several students in our combustion classes. We specially thank P. Soyuz, S. Senthil, V. Subramanian, and many other graduate students for reading appropriate chapters; Dr. Tillman of ITT for Chapter 12; our colleagues Prof. Caton of Texas A&M and Prof. Lucht (currently at Purdue University); and the following reviewers: Prof. Blasiak (Royal Institute of Technology, Sweden); Prof. S. Gollahalli (University of Oklahoma), whose classroom combustion lectures motivated KA to select combustion as his field of research; Prof. Hernandez (Guanajuato, Mexico); Prof. X. Li (Waterloo, Ohio); Profs. Nagarajan and Ramaprabhu of Anna University, Chennai, India; Prof. M. Sibulkin of Brown University, who mentored Annamalai in experimental combustion research; and Prof. Krishnan (Indian Institute of Technology, Chennai, India, and Kyungpook National University, Korea); Margaret Wooldridge of the University of Michigan, Ann Arbor, and Mich Wittneben of TAMU Computer Services.

Kalyan Annamalai

Ishwar K. Puri

The Authors

Kalyan Annamalai, Ph.D., is the Paul Pepper professor of mechanical engineering at Texas A&M University. He received his B.S. degree from Anna University, Chennai, M.E. from Indian Institute of Science, Bangalore and Ph.D. from the Georgia Institute of Technology, Atlanta. After obtaining his doctoral degree, he worked as a research associate in the Division of Engineering, Brown University, and at AVCO-Everett Research Laboratory in Massachusetts. He has taught several courses at Texas A&M University, including advanced thermodynamics, and combustion science and engineering at the graduate level; and thermodynamics, heat transfer, combustion, and fluid mechanics at the undergraduate level. He is the recipient of the Senior TEES Fellow Award and Dow Chemical Fellowship from the College of Engineering for excellence in research, a teaching award from the Mechanical Engineering Department, and a service award from the American Society of Mechanical Engineers (ASME). He is a Fellow of the ASME, and a member of the Combustion Institute and the Texas Renewable Industry Association. He has served on several federal panels. He was coauthor of the textbook *Advanced Thermodynamics Engineering* (CRC Press). His research funded by the DOE-UCR, DOE-NETL, DOE-Golden, EPRI, DOE, TCEG, and ATP of the State of Texas ranges from basic research on coal combustion, group combustion of oil drops and coal, etc., to applied interdisciplinary research on the cofiring of coal with animal waste in a boiler burner and NO_x and Hg reduction using biomass fuels. He has published more than 145 journal and conference articles on the results of this research. He participated in the Student Transatlantic Exchange Program (STEP; U.S. DO Ed.–FIPSE).

Ishwar K. Puri, Ph.D., has served as professor and department head of Engineering Science and Mechanics at Virginia Tech since 2004. He obtained his Ph.D. (1987), and M.S. (1984) degrees in engineering science (applied mechanics) from the University of California–San Diego after obtaining a B.Sc. (1982) in mechanical engineering from the University of Delhi. He served as an assistant research engineer at the University of California–San Diego from

1987 to 1990. He was appointed as an assistant professor in the Mechanical Engineering Department at the University of Illinois at Chicago in 1990, was promoted to the rank of associate professor with tenure in 1994 and to the rank of professor in 1999. He served as director of graduate mechanical and industrial engineering programs from 1994 to 1997, and 1999 to 2000. He served as associate dean for research and graduate studies (2000 to 2001), and as executive associate dean of engineering (2001 to 2004). He served on the steering committee of the UIC Institute for Environmental Studies and facilitated UIC's micro- and nanotechnology initiatives. Puri has conducted research through grants from NASA, NSF, DOE, U.S. EPA, State of Illinois, the natural gas industry (GTI, IGT, GRI), and other industries. He established a European–U.S. consortium to conduct engineering student exchanges at the undergraduate and graduate levels that was funded through the U.S. Department of Education's FIPSE program. His students are placed in major corporations and universities worldwide. Puri is the author of more than 200 archival and conference publications in the fields of combustion and transport phenomena (e.g., related to emissions, self-assembly and magnetic drug targeting, and hydrogen storage). He has edited a book on the environmental implications of combustion processes and the textbook *Advanced Thermodynamics Engineering* (with Kalyan Annamalai). Puri is a Fellow of the American Society of Mechanical Engineers (ASME) and of the American Association for the Advancement of Science (AAAS). He was a 1993 American Association for the Advancement of Science-Environmental Protection Agency (AAAS-EPA) Environmental Fellow, and a 1992 NASA/Stanford University Center for Turbulence Research Fellow. He is an editor of the journal *Experimental Heat Transfer.*

Relations/Accounting Equations

The following relations/accounting equations are useful in the engineering analysis of combustion systems:

Accumulation rate of an extensive property B: $\frac{dB}{dt}$ = {rate of B entering a volume ($\dot{B}_i$) – rate of B leaving a volume ($\dot{B}_e$) + rate of B generated or consumed in a volume}, ($\dot{B}_{gen}$) > 0 for generation, < 0 for consumption

Mass conservation: $\frac{dm_{cv}}{dt} = \dot{m}_i - \dot{m}_e,\quad \frac{dm_{k,cv}}{dt} = \dot{m}_{k,i} - \dot{m}_{k,e} + \dot{w}_{k,m}$

Molar balance: $\frac{dN_{cv}}{dt} = \dot{N}_{k,i} - \dot{N}_{k,e} + \dot{w}_k$

First law of energy conservation: $\frac{dE_{cv}}{dt} = \dot{Q} - \dot{W} + \dot{m}_i\, e_{T,i} - \dot{m}_e\, e_{T,e},$

where $e_T = h + ke + pe,\quad h = E_{cv} = U + KE + PE,$

$$\delta w_{rev,\ open} = -v\ dP,\quad \delta w_{rev,\ closed} = P\ dv$$

Molar form of energy equation:

$$\frac{dE_{cv}}{dt} = \dot{Q}_{c.v.} - \dot{W}_{c.v.} + \sum_{k,\,i} \dot{N}_{k,i}\ \hat{e}_{k,T,i} - \sum_{k,\,e} \dot{N}_{k.e}\ \hat{e}_{k,T,e}$$

where

$$E = U + KE + PE,\quad \hat{e}_{k,T,e} \approx \bar{e}_k = \bar{h}_k + \overline{ke}_k + \overline{pe}_k - \bar{h}_k \approx \bar{h}^\circ_{f,k} + (\bar{h}^\circ_{t,T} - \bar{h}^\circ_{t,298})_k$$

$$\overline{ke}_k = (1/2) M_k V^2,\quad \overline{pe} = M_k gZ$$

Second law or entropy balance equation:

$$\frac{dS_{cv}}{dt} = \frac{\dot{Q}_{cv}}{T_b} + \dot{m}_i s_i - \dot{m}_e s_e + \dot{\sigma}_{cv} = \frac{\dot{Q}_{cv}}{T_b}$$

$$+\sum \dot{N}_{k,i}\, \hat{s}_{k,T,i} \sum - \dot{N}_{k.e}\, \hat{s}_{k,T,e} + \dot{\sigma}_{cv}$$

where $\dot{\sigma}_{cv} > 0$ for all irreversible processes and is equal to zero for a reversible process.

Availability balance: $\frac{dE_{cv} - T_o S_{cv}}{dt} = \dot{Q}\left(1 - \frac{T_0}{T_R}\right) + \dot{m}_i \psi_i - \dot{m}_e \psi_e - \dot{W}_{cv} - T_o \dot{\sigma}_{cv}$,

$\psi = (e_T - T_0\, s) = h + ke + pe - T_0 s$, and $E_{cv} = U + KE + PE$

Third law: $S \rightarrow 0$ *as* $T \rightarrow 0$

Law of stoichiometry: *e.g.,* $CO + 1/2\ O2 \rightarrow CO2$

$$\frac{\dot{w}'''_{CO}}{-1} = \frac{\dot{w}'''_{O2}}{-1/2} = \frac{\dot{w}'''_{CO2}}{+1} = \dot{w}''', \quad \dot{w}''' \geq 0$$

Law of mass action and Arrhenius law:

$\dot{w}''' = k\,[CO]^{\alpha}[O2]^{\beta}[M]^{\gamma}$, $\dot{w}''' \geq 0$, $k = A\,T^n \exp\left(-\frac{E}{\bar{R}T}\right)$, E: activation energy, M: third body

Dimensionless Numbers in Combustion, Heat, and Mass Transfer

Group	*Definition*	*Interpretation*
Biot number (*Bi*)	$\frac{hL}{k_s}$	Ratio of the internal thermal resistance of a solid to the boundary layer thermal resistance
Mass transfer Biot number (Bi_m), porous media	$\frac{h_m L}{D_{ABS}}$	Ratio of the internal species transfer resistance to the boundary layer species transfer resistance
Bond number (*Bo*)	$\frac{g(\rho_l - \rho_v)L^2}{\sigma}$	Ratio of gravitational and surface tension forces
Cauchy number (*Ca*)	$\frac{F}{\rho V^2 L^2}$	The pressure coefficient
Coefficient of friction (C_f)	$\frac{\tau_s}{(\rho V^2/2)}$	Dimensionless surface shear stress
Damkohler number (*Da*) (I)	$\frac{\dot{w}'''L}{\rho V}$	Ratio of chemical reaction rate to mass flow rate
(II)	$\frac{q^\circ \dot{w}'''L}{\rho V c_p T}$	Ratio of chemical heat liberation rate to advection q^0, heating value
(III)	$\frac{q^\circ \dot{w}'''L}{\rho D}$	Ratio of chemical reaction rate to diffusion
(IV)	$\frac{q^\circ \dot{w}'''L^2}{\lambda T}$	Ratio of chemical heat to diffusive heat flow
(V)	$\frac{\dot{w}'''R}{\rho \phi}$	Ratio of chemical rate to elutriation rate, used in fluidized bed combustion
Eckert number (*Ec*)	$\frac{V^2}{c_p(T_s - T_\infty)}$	Kinetic energy of the flow relative to the boundary layer enthalpy difference
Euler number (*Eu*)	$\frac{\Delta P}{\rho V^2}$	Pressure drop to dynamic pressure
Fourier number (*Fo*)	$\frac{\alpha t}{L^2}$	Ratio of the diffusive conduction rate to the rate of thermal energy storage

Mass transfer Fourier number (*Fo*)	$\frac{D_{AB}t}{L^2}$	Ratio of the species diffusion rate to the rate of species storage; dimensionless time
Froude number (*Fr*)	$\frac{V}{\sqrt{gd}}$ or $\frac{V^2}{gd}$	Ratio of inertial force to buoyancy forces
Friction factor (*f*)	$\frac{\Delta p}{(L/D)(\rho u_m{}^2/2)}$	Dimensionless pressure drop for internal flow
Graetz number (*Gr*)	$\left(\frac{L}{D}\right) \times$ Peclet No.	See Peclet number
Grashof number (Gr_L)	$\frac{g\beta(T_s - T_\infty)L^3}{v^2}$	Ratio of buoyancy to viscous forces
Colburn *j* factor (j_H)	$StPr^{2/3}$	Dimensionless heat transfer coefficient
Colburn *j* factor (j_m)	$St_m Sc^{2/3}$	Dimensionless mass transfer coefficient
Jakob number (*Ja*)	$\frac{c_p(T_S - T_{sat})}{h_{fg}}$	Ratio of sensible to latent energy absorbed during liquid–vapor phase change
Karlovitz number (Ka)	$\frac{\sigma\delta_f}{S_L}$	Reaction of flame stretch to rate flame velocity
Knudson number (*Kn*)	$\frac{\ell}{L}$	Ratio of mean free path to characteristic dimension
Lewis number (*Le* = *Sc*/*Pr*)	$\frac{\alpha}{D_{AB}}$	Ratio of the thermal and mass diffusivities
Mach number (*Ma*)	$\frac{V}{c}$	Ratio of velocity to sound speed
Markstein number (Ma)	L'/δ_f	L′, Markstein length; $L'\delta$, effect on S_L
Nusselt number (Nu_L)	$\frac{hL}{k_f}$	Ratio of convective to diffusive rate of heat transfer
Peclet number (Pe_L)	$\frac{VL}{\alpha} = Re_L\,Pr$	Ratio of convective velocity to thermal diffusion velocity
Prandtl number (*Pr*)	$\frac{c_p\mu}{k} = \frac{v}{\alpha}$	Ratio of the momentum and thermal diffusivities
Rayleigh number (*Ra*)	$Ra = Gr\,Pr$	Buoyancy diffusivity to thermal diffusivity

Reynolds number (Re_L)	$\frac{VL}{v}$	Ratio of the inertial and viscous forces
Schmidt number (Sc)	$\frac{v}{D_{AB}}$	Ratio of the momentum and mass diffusivities
Sherwood number (Sh_L)	$\frac{h_m L}{D_{AB}}$	Ratio of convective to diffusive mass transfer rates
Stanton number (St)	$\frac{h}{\rho V c_p} = \frac{Nu_L}{Re_L Sc}$	Ratio of convective heat transfer to sensible energy of flow medium
Stanton number (St_m): mass transfer	$\frac{h_m}{V} = \frac{Sh_L}{Re_L Sc}$	Ratio of convective mass transfer to bulk species flow
Stokes number (St)	τ*velocity/d, velocity $(\rho_p d_p^2/18\mu)*$ (v/dp)	Ratio of stopping distance, S to characteristic dimension, d
Strouhal number (Str)	$\frac{nd}{u_o}$	Characteristic frequency
Weber number (We)	$\frac{\rho V^2 L}{\sigma}$	Ratio of inertia to surface tension forces

Nomenclature*

(Some of the symbols are used for more than one parameter; their interpretation may be according to the context.)

Symbol	Description	SI	English	Conversion SI to English
A	Helmholtz free energy	kJ	Btu	0.9478
A	Area	m^2	ft^2	10.764
A,	Preexponential constant			
$A_{[\,]}$	Preexponential constant, mole concentration			
$A_{(\,)}$	Preexponential constant, mass concentration			
A_X	Preexponential constant, mole fraction based			
A_Y	Preexponential constant, mass fraction based			

* Lowercase (lc) symbols denote values per unit mass, lc symbols with a bar (e.g., $\bar{h}$) denote values on mole basis, lc symbols with a caret and tilde (respectively, $\hat{h}$ and $\tilde{h}$) denote values on partial molal basis based on moles and mass, and symbols with a dot (e.g., $\dot{Q}$) denote rates.

A:F	Air to fuel ratio			
a	Radius	m	ft	3.281
a_o	Initial radius	m	ft	3.281
a	Specific Helmholtz free energy, mass	kJ kg^{-1}	Btu lb_m^{-1}	0.4299
a	Strain rate	sec^{-1}	sec^{-1}	
$\bar{a}$	Specific Helmholtz free energy, mole	kJ $kmol^{-1}$	Btu $lb\text{-}mol^{-1}$	0.4299
a1, a2, etc. (see Chapter 11)				
B	Transfer number			
B	Generalized variable (v_x, Y_k, h_t, etc.)			
B1, b2, etc. (see Chapter 11)				
C_μ =	Constant used in turbulence model			
C_1 =	Constant used in turbulence model; also Chapter 11			
C_2 =	Constant used in turbulence model; also Chapter 11			
C_1 =	Also see Chapter 11 for laminar flow			
C_2 =	Also see Chapter 11 for laminar flow			
C_{g1} =	Constant used in combustion model			
C_{g2} =	Constant used in combustion model			
C_m =	Also see Chapter 11 for laminar flow			

c	Molal concentration	kmol/m³	lb-mol/ft³	
c	Specific heat	$kJ\ kg^{-1}\ K^{-1}$	Btu/lb R	0.2388
$\bar{c}$	Molal specific heat	$kJ\ kmol^{-1}K^{-1}$	$Btu\ lb\text{-}mol^{-1}R^{-1}$	
c_p	Specific heat at constant p	$kJkmol^{-1}K^{-1}$	$Btu\ lb\text{-}mol^{-1}R^{-1}$	
D	Diffusion coefficient	m^2sec^{-1}	$ft^2\ sec^{-1}$	10.764
D_{ij}	Binary diffusion coefficient	m^2sec^{-1}	$ft^2\ sec^{-1}$	10.764
Da	Damköhler number			
d	Diameter of particle	m	ft	3.281
dref	Reference length	m	ft	3.281
E	Energy, (U + KE + PE)	kJ	Btu	0.9478
E	Activation energy	kJ/kmol	$Btu/lb\text{-}mol^{-1}$	0.4299
E_I	Ignition energy	joules		
E_T	Total energy (H + KE + PE)	kJ	Btu	0.9478
E(X)	Exponential integrals			
e	Specific energy	$kJ\ kg^{-1}$	$Btu\ lb_m^{-1}$	0.4299
e_T	Methalpy = h + ke + pe	$kJ\ kg^{-1}$	$Btu\ lb_m^{-1}$	0.4299
F	Force	kN	lb_f	224.81
F	Chemical force			
F	Interface function for B #			
F_β	Interface function, Chapters 8, 10, 11, 16			

f′	Velocity in similarity coordinate			
f_m	Mixture fraction variable			
$\vec{f}$	Body force per unit mass			
G	Gibbs free energy	kJ	Btu	0.9478
G	Group combustion number, Chapter 16			
g	Specific Gibbs free energy	kJ kg^{-1}	Btu lb_m^{-1}	0.4299
G_1	Modified group combustion, Chapter 16 (mass basis)			
g	Gravitational acceleration	m sec^{-2}	ft sec^{-2}	3.281
g	Variance of mixture fractions			
g_c	Gravitational constant			
$\bar{g}$	Gibbs free energy (mole basis)	kJ $kmol^{-1}$	Btu lb-mol^{-1}	0.4299
$\hat{g}$	Partial molal Gibbs function	kJ $kmol^{-1}$	Btu lb-mol^{-1}	0.4299
H	Flame height	m	ft	
H	Enthalpy, extensive	kJ	Btu	0.9478
HHV	Higher or gross heating value	kJ kg^{-1}	Btu lb^{-1}	0.4299
HV	Heating value	kJ/kg	Btu/lb	0.4299
HV_{FC}	Heating value of fixed carbon	kJ/kg	Btu/lb	0.4299
HV_V	Heating value of volatiles	kJ/kg	Btu/lb	0.4299
h	Specific enthalpy or enthalpy	kJ kg^{-1}	Btu lb_m^{-1}	0.4299

$\bar{h}_f$	Enthalpy of formation	kJ $kmol^{-1}$	Btu lb-mol^{-1}	0.4299
h_H	Heat transfer coefficient	kW/(m^2sec)	Btu/(ft^2sec)	
h_m	Mass transfer coefficient	kg/m^2sec	lb_m/(ft^2sec)	0.205
h_c, LHV	Lower heating value	kJ/kg	Btu/lb	0.4299
h_p	Planck's constant			
ΔH_R^0	Enthalpy of reaction			
h_t	Thermal enthalpy	kJ/kg	Btu/lb	0.4299
$h_{t,\infty}$	$c_p(T_\infty - T_w)$ or $c_p(T_\infty - T_{ref})$	kJ/kg	Btu/lb	0.4299
J	Joules' work equivalent of heat	(1 Btu = 778.14 ft lb_f)		
J′	Total scalar flux per unit width 2-D jet			
J	Total scalar flux, circular jet			
j_k	Fluxes for species, heat, etc.	kg sec^{-1}, kW	Btu sec^{-1}	0.9478
j_k''	Diffusive mass flux of species *k*	kg/m^2 sec	$lb_m ft^{-2} sec^{-1}$	0.205
K	Karlovitz number, Chapter 15			
K^0	Equilibrium constant, 1 bar			
K_c	Equilibrium constant based on concentrations			
KE	Kinetic energy	kJ	Btu	0.9478
ke	Specific kinetic energy	kJ kg^{-1}	Btu lb_m^{-1}	0.4299
k	Ratio of specific heats, c_p/c_v			

k	Specific reaction rate constant			
[k]	Concentration of species k	$kmol/m^3$	$lbmol/ft^3$	0.06243
(k)	Concentration of species k	kg/m^3	lb/ft^3	0.06243
k_B	Boltzmann constant			
L	Latent heat or h_{fg}	kJ/kg	Btu/lb_m	0.429
LHV	Lower heating value	kJ/kg	Btu/lb_m	0.429
Le	Lewis number			
l	Intermolecular spacing	m	ft	3.281
l_m	Mean free path	m	ft	3.281
ℓ	Interdrop spacing			
M	Molecular weight, molal mass	$kg\ kmol^{-1}$	$lb_m\ lb\text{-}mo^{-1}$	
M′	Total momentum flux per unit width, 2-D jet	Nm^{-1}	$lb_f ft^{-1}$	0.06852
M	Total momentum flux, circular jet	N	lb_f	0.2248
m	Mass	kg	lb_m	2.2046
m_0	Initial mass	kg	lb_m	2.2046
$\dot{m}$	Mass rate	kg/sec	lb_m/sec	2.2046
$\dot{m}_C$	Carbon loss rate	kg/sec	lb_m/sec	2.2046
$\dot{m}_V$	Volatile loss rate	kg/sec	lb_m/sec	2.2046
$\dot{m}''$	Mass flow rate per area	kg/m^2sec	lb_m/ft^2sec	0.205
$\dot{q}''$	Heat flux	kW/m^2	Btu/ft^2sec	317.1

Y_k	Mass fraction of species k			
N	Number of moles	kmol	lb-mol	2.2046
$\dot{N}$	Mole flow rate	kmol sec^{-1}	lb mole sec^{-1}	2.2046
N_{Avag}	Avogadro number	molecules $kmol^{-1}$	molecules $lb\text{-}mol^{-1}$	0.4536
NO_X	Nitric oxide	kg/GJ	lb/mmBtu	2.326
Nu_0	Nusselt number under zero mass Tr			
n	Temperature exponent in Arrhenius law			
n	Number density	per m^3	per ft^3	0.0283
n′	Number of molecules per unit volume	per m^3	per ft^3	0.0283
P	Pressure	kN m^{-2}	kPa lb_f in^{-2}	0.1450
P	Pressure	bar	atm	0.987
$\vec{\vec{P}}$	Stress tensor			
p_k	Partial pressure, (X_kP)			
ppm	Parts per million			
PE	Potential energy	kJ	Btu	0.9478
pe	Specific potential energy	kJ	Btu/lb	0.4299
Q	Heat transfer	kJ	Btu	0.9478
$\dot{Q}$	Heat transfer rate	kW	Btu/sec	0.9478
$\dot{Q}_G$	Heat generation rate	kW	Btu/sec	0.9478
$\dot{Q}_L$	Heat loss rate	kW	Btu/sec	0.9478
Q1, Q2, Q3, and Q4	heat liberated or absorbed in carbon reactions (I) to (IV)			
q	Energy released per kg mixture	kJ/kg mix	Btu/lbmix	0.4299

I

$\dot{q}''$	Heat flux rate per unit area	kW/m^2	Btu/ft^{2}sec	0.4299
$\dot{q}_R$	Radiation heat loss rate	kW	Btu/sec	
$\dot{q}_{loss}$	Heat loss per unit volume	kW	Btu/sec	0.9478
R	Gas constant	kJ kg^{-1} K^{-1}	Btu lb^{-1} R^{-1}	0.2388
Rc	Cloud radius	m	Ft	3.281
Re	Reynolds number, $Vd\rho/\mu = Vd/\nu$			
$\overline{R}$	Universal gas constant	kJ kmol^{-1}	Btu lb-mol^{-1}	0.2388
r,	Radial distance	m	ft	3.281
S	Entropy	kJ K^{-1}	Btu R^{-1}	0.5266
S	Source term			
S	Steric factor			
S, S_L	Laminar burn velocity, Chapter 14 and Chapter 15	m sec^{-1}	ft sec^{-1}	3.281
Sc	Schmidt number			
Sh	Sherwood number $(h_m d_p)/(\rho D)$			
SR	Stoichiometric ratio $(1/\varphi)$			
SO_X	Sulfur oxide	kg/GJ	lb/mm Btu	2.326
s	Specific entropy (mass basis)	kJ kg^{-1} K^{-1}	Btu lb^{-1} R^{-1}	0.2388
s	≈ Inverse of A:F	$Y_{O2\infty}/\{\nu_{O2}\, Y_{F,i}\}$ or $Y_{O2\infty}/\{\nu_{O2}\, Y_{F,c}\}$		
$\bar{s}$	Specific entropy (mole basis)	kJ kmol^{-1} K^{-1}	Btu lb-mol^{-1} R^{-1}	0.2388
T	Temperature	°K	°R	(9/5)* T in K
t	Time	sec	Sec	
U	Internal energy	kJ	Btu	0.9478

$\vec{\vec{U}}$	Unit stress tensor			
u	Specific internal energy	kJ kg^{-1}	Btu lb^{-1}	0.4299
$\bar{u}$	Internal energy (mole basis)	kJ $kmol^{-1}$	Btu lb-mol^{-1}	0.4299
V	Volume	m^3	ft^3	35.315
V	Volume	m^3	gallon	264.2
VM	Volatile matter	m^3	ft^3	35.245
V	Velocity	m sec^{-1}	ft sec^{-1}	3.281
V*	Ultimate-volatile yield			
V_k	Diffusion velocity (mass) of k	m sec^{-1}	ft sec^{-1}	3.281
$\bar{V}_k$	Diffusion velocity (mole) of species k	m sec^{-1}	ft sec^{-1}	3.281
v_x	Velocity in x direction	m sec^{-1}	ft sec^{-1}	3.281
v_y	Velocity in y direction	m sec^{-1}	ft sec^{-1}	3.281
v_r	Radial velocity (m/sec)	m sec^{-1}	ft sec^{-1}	3.281
v	Mass-averaged velocity	m sec^{-1}	ft sec^{-1}	3.281
$\bar{v}$	Mole-averaged or bulk velocity	m sec^{-1}	ft sec^{-1}	3.281
v_k	Absolute velocity of species k	m sec^{-1}	ft sec^{-1}	3.281
v_0	Laminar-burn velocity	m sec^{-1}	ft sec^{-1}	3.281
v	Specific volume (mass basis)	m^3 kg^{-1}	ft^3 lb_m^{-1}	16.018
$\bar{v}$	Specific volume (mole basis)	m^3 $kmol^{-1}$	ft^3 lb-mol^{-1}	16.018
W	Work	kJ	Btu	0.9478
w	Work per unit mass	kJ kg^{-1}	Btu lb^{-1}	0.4299

w	Specific humidity	kg kg^{-1}	$1b_m$ lb_m^{-1}	
$\dot{w}$	Reaction rate	kg sec^{-1}	lb sec^{-1}	
$\dot{w}_k$	Reaction rate of species k	kg sec^{-1}	lb sec^{-1}	2.205
$\lvert\dot{w}'''\rvert$	Reaction rate per unit volume	kgm^{-3} sec^{-1}	lb ft^3 sec^{-1}	0.0624
$\lvert\dot{w}'''\rvert$	Reaction rate per unit volume	$kmolm^{-3}$ sec^{-1}	lb-mol ft^{-3} sec^{-1}	0.0624
x	Coordinate	m	ft	3.281
x′	Incompressible coordinate, Chapter 11			
X_k	Mole fraction of species k			
Y_k	Mass fraction of species k			
Y_{ash}	Ash fraction in solid fuel			
$Y_{F,c}$	Fuel fraction in condensate phase			
y	Coordinate	m	ft	3.281
y′	Incompressible coordinate, Chapter 11			
$Z_{k'}$	Element mass fraction			
Z	Collision number			
Z	Elevation	m	ft	3.281
$\delta()$	Differential of a nonproperty, e.g., δQ, δW, etc.			
d ( )	Differential of property, e.g., du, dh, dU, etc.			
D	Change in value			
ΔH_R	Enthalpy of reaction			

$\vec{\nabla} = \vec{i}\frac{\partial}{\partial x} + \vec{j}\frac{\partial}{\partial y} + \vec{k}\frac{\partial}{\partial z}$, Laplacian operator

Greek Symbols

α	Nondimensional mass flow $= \dot{m} / \dot{m}_c$			
α_T	Thermal diffusivity, $\alpha_T = \frac{\rho c_P}{\lambda}$ (m^2/sec)	m^2/s	ft^2/s	10.764
α_j	kg element/kg species			
β_{ij}, β_{ht-j},	Shvab–Zeldovich (SZ) variable, β_s			
χ	Generalized transport property (ν, α, D, etc.)			
δ	Boundary-layer thickness	m	ft	3.281
$\vec{\delta}$	Unit vector			
δ'	Stretched boundary-layer thickness			
δ_f	Flame thickness			
δ_s	Species boundary-layer thickness	m	ft	3.281
δ_m	Velocity boundary-layer thickness	m	ft	3.281
δ_t	Thermal boundary-layer thickness	m	ft	3.281
ε	Emissivity, burned fraction			
ε	Dissipation of kinetic energy			
ε_k	Species k flux/average mass flux			
ε_T	Turbulent diffusivity	m^2/s	ft^2/s	10.764
γ	Ratio of specific heats cp/cv			
η	Similarity coordinate			
η	Correction factor			

η	Efficiency			
φ	Equivalence ratio			
φ	Angular variable in spherical coordinates			
ϕ	Normalized SZ variable			
Φ	Potential			
κ	Bulk viscosity; also Chapter 15, fuel loading Chapter 16	kg/(m sec)	lb/(ft sec)	0.672
λ	Thermal conductivity	kW m^{-1} K^{-1}	Btu ft^{-1} R^{-1}	0.1605
λ	Lagrange multiplier	kW m^{-1} K^{-1}	Btu ft–^{1} R^{-1}	0.1605
μ	Reduced mass	kg/(m sec)	lb_m/(ft sec)	0.672
μ	Absolute viscosity	kg/(m sec)	lb_m/(ft sec)	0.672
μ	Chemical potential	kJ $kmol^{-1}$	Btu lb-mol^{-1}	0.4299
θ	Angle			
ρ	Density	kgm^{-3} sec^{-1}	lbm ft^{-3} sec^{-1}	00624
$\dot{\sigma}$	Entropy generation	kJ $sec^{-1}K^{-1}$	Btu $sec^{-1}R^{-1}$	0.2388
ν	Stoichiometric coefficient			
ν	Dynamic, kinematic viscosity	m^2/sec	ft^2/sec	10.764
ξ ,	Nondimensional distance x, Chapter 11 and Chapter 12			
ξ	Normalized inverse radius Chapter 10			
ξ	Extent of reaction, Chapter 3			
ξ_w	Normalized burn rate, $\dot{m}$ /(4*πρDa)			
Ψ	Stream function , see Chapter 13 and Chapter 16			

Ω_D	Collision integral, diffusion			
Ω_μ	Collision integral, viscosity			
ρ_{ref}	Reference density	kg m^{-3}	lb ft^{-3}	0.0624
σ	Stefan–Boltzmann constant; strain rate			
σ	Entropy generation	kJ K^{-1}	Btu R^{-1}	0.2388
τ	Shear stress, characteristic time	N m^{-2}, s	lb ft^{-2}, s	0.0209
ν	Kinematic viscosity of gas	m^2/s	ft^2/s	10.764
$\nu_1, \nu_2, \nu_3, \nu_4$	Stoichiometric coefficients for Reaction I to Reaction IV			
υ	Stoichiometric coefficients			

Chapter 1

Introduction and Review of Thermodynamics

1.1 Introduction

The mention of fire evokes familiar images and sensations: the warmth of a hearth fire, the holocaust of a forest fire, the flickering of a candle flame, the flaming exhaust of a jet engine, etc. [Blackshear, 1961]. For historians, fire is related to momentous events. First used by prehistoric humans about half a million years ago, fire was responsible for the survival of many tribes through periods of glaciation. Fire was the means by which humans emerged from the Stone Age, through pyrometallurgy. Engines that use fire have been sources of power and propulsion only in the past several centuries, and have succeeded in liberating humankind from drudgery. Finally, it appears that fire will play a key role in our first adventures away from this planet.

Fire (or combustion) also helps create technologies for industrial use. Combustion is the most common method of energy conversion currently used. The major portion of power is generated from coal- and natural-gas-burning power plants that produce steam, which in turn drives electric generators. These are known as external combustion (EC) systems. Internal combustion (IC) engines or systems are so interwoven with the global economy that should a shortage of fossil fuels suddenly occur, crises and chaos would inevitably follow. In IC engines chemical energy is converted into thermal energy and then directly to mechanical energy, unlike coal-fired plants. Metallurgical and chemical processes involving "fire" are global businesses. To the conservationist, it is obvious that this activity cannot continue indefinitely. There is enough oil and coal on earth to carry us forward another 20 to 200 years depending on what data one accepts, but clearly there is a finite supply that cannot be replenished. For many environmentalists combustion is an anathema, because it is the cause of global warming (through CO_2 emission), smog (NO_x and O_3 generation), and acid rain (through SO_x). The

U.S. consumes about 100 PkJ per year, whereas world energy usage is 300 PkJ (Table A.1A of Appendix A for units). The U.S. emits 5.5 Gt of CO_2 per year, whereas the world CO_2 emissions is about 21 Gt per year. The amount of atmospheric CO_2 is expected to triple before 2100, and it is widely recognized that global warming is now an engineering problem.

For the engineer who is enough of a historian to appreciate the past and enough of an economist, conservationist, and environmentalist to be interested in the future, a technical description of combustion involves many disciplines and presents many interesting challenges. A typical combustion problem will call upon a student or practicing engineer to display a grasp of thermodynamics, fluid mechanics, transport processes, chemical kinetics, as well as a proficiency in mathematics. Combustion is a very active field of technology and an excellent proving ground for advanced engineers to employ the tools that they have acquired.

The primary energy sources in North America in 1975 included fossil fuels such as coal (22%), oil (40%), natural gas (32%), hydroelectric energy (4%), and conventional and breeder nuclear reactors (8%). Clearly, the major energy source is fossil fuels. In 1993 alone, almost 1 Gt of coal was mined in the U.S. The projected consumption of fossil fuels in 2010 can be approximately divided as shown in Table 1.1.

The applications of combustion include home heating, cooking with gas stoves, welding, transportation, forest fires, electric power, material, and agricultural processing. The chemical energy of fossil fuels is converted into thermal energy using combustors, and this energy is then used to generate steam and run steam turbines as in steam-based power plants (EC systems). Alternatively, the hot gases can be directly used to run a gas turbine as in gas turbine plants or to deliver work through a piston–cylinder assembly as in automobiles (IC systems). In coal-fired utilities, coal is transported to silos by conveyors, and then ground and fed to burners (Figure 1.1a). The heat from combustion is used to produce steam at a temperature of 815 K and pressure of 170 bar and hence electrical power using steam turbines running at about 3000 rpm. The exhaust steam is then condensed to water and pumped at a pressure of 170 bar back to the boiler. The

Table 1.1 Projected 2010 Energy Consumption

	Total Energy Consumption	*Biomass Energy Consumption*	*Biomass % Total*	*Total Energy Consumption*	*Biomass Energy Consumption*	*Biomass % Total*
Industrial	37.5	2.8	8.0	41.0	3.4	8.7
Electric generation	45.1	0.2	0.5	51.0	0.2	0.4
Subtotal power	82.6	3.0	8.5	92.0	3.6	9.1
Transportation	35.5	0.3	0.7	41.7	0.3	0.7

Note: Energy in PkJ/yr; (1 PkJ = 10^{15} kJ ≈ 1 quad).

Source: DOE/EIA, *Annual Energy Outlook*, 2002.

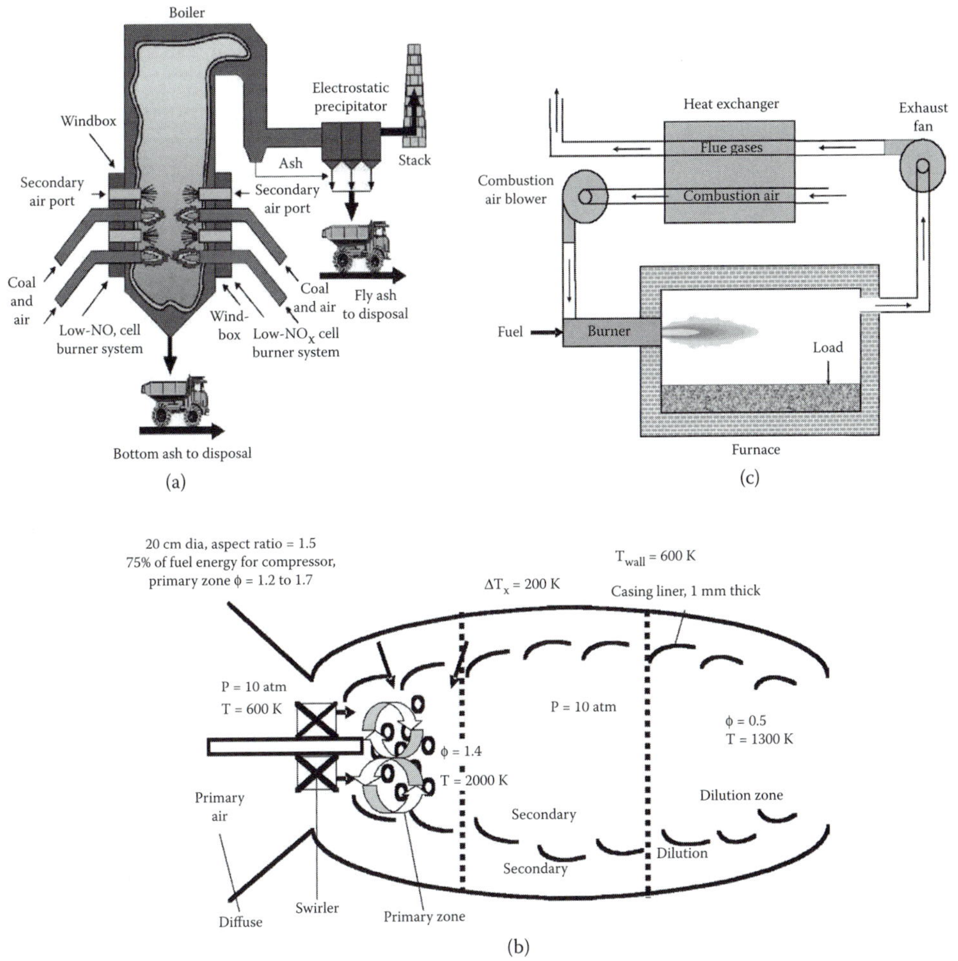

Figure 1.1(a) A low-NOx coal-fired boiler burner [Adapted from http://www.lanl.gov/projects/cctc/factsheets/clbrn/cellburnerdemo.html, McDermott Technology, Inc.]; (b) schematic of gas turbine combustor, average velocity of 25–40 m/sec (aircraft), 15–25 m/sec (industrial gas turbine); (c) a typical industrial combustion system: load could be water tubes in boiler, HC tubes in a refinery, metal melting in furnaces, brick in kilns, etc. (Adapted from Baukal.)

products of combustion consist of bottom ash (larger particles) and fly ash (smaller particles), which are then collected for disposal.

In the case of the gas turbine, the combustor is fired with either natural gas or atomized liquid fuels with swirling air (about 45° swirl angle for imparting tangential motion to air) in order to vaporize the fuel; ignition is started with a spark plug, and the flame is stabilized with a rich mixture near the injector (called the *primary zone*); and secondary air and a large amount of diluted air are used to limit the temperature to about 1300 K at the exit, reduce the temperature gradient, and minimize hot spots prior to entry into the turbine (Figure 1.1b). In other industrial applications, the combustion process is used to melt metals (Figure 1.1c), or make bricks in kilns.

Combustion in practical systems as shown in Figure 1.1 is very complex. It can involve:

1. Two-phase flows (liquid and air as in the gas turbine of Figure 1.1b, or coal and air as in the boiler of Figure 1.1a)
2. Turbulence and variable densities within phases
3. Conduction, convection, and radiation heat transfer to heat exchangers as in boilers and gas turbine liners
4. Interactions between liquid and solid particles in dense flows for problems related to items 1 to 3
5. High pressures
6. The thermochemistry of fuel and reactants that can involve problems with multiple components
7. Complex chemical kinetic mechanisms
8. Property changes of multiple components with temperature

These complexities of combustion must be understood to reduce air pollution, smog, fire and explosion hazards, and to conserve energy.

There are power plants using combined steam and gas turbine cycles with an efficiency of 60% for generating power. In contrast to 250-MW turbines weighing about 300 t, microturbines weighing only about 45 kg that produce 25 kW electric power and heat are being developed. Combustion in such microsystems requires a fundamental understanding of complex processes governing the rate phenomena.

1.2 Combustion Terminology

A brief overview of terminology used in combustion engineering is presented in the following text:

Combustion: It generally involves the oxidation of a fuel (i.e., a chemical reaction of a fuel with oxygen). This is accompanied by release of chemical energy that usually produces a large temperature increase. For instance, from the methane-burning reaction

$$CH_4 + 2O_2 \rightarrow CO_2 + 2H_2O,$$

it is possible to release as high as 55,000 kJ per kg of fuel that is burned. Recently the knowledge of combustion science has been expanded to include "fluid dynamics combined with exothermic reactions and everything it implies" [Oran, 2004], which seems to include even nuclear reactions occurring on earth and in the universe.

Continuum: A fluid is said to be a continuum if the dimensions of the region of interest are large compared to the mean free path of the molecules contained within the region.

Density is given by:

$$\rho = limit\ \Delta V \rightarrow 0 \quad for \quad \frac{\Delta m}{\Delta V}$$

The limit $\Delta V \rightarrow 0$ should be interpreted as $\Delta V \rightarrow$ multiple mean free paths.
Radiation: Energy from the sun reaches the earth via electromagnetic waves (or via photons of energy $h_p{}^*\nu$, where h_p is Planck's constant and ν, the frequency of light; Table A.1A of Appendix A). The electromagnetic spectrum extends from microwaves to amplitude-modulated (AM) and frequency-modulated (FM) waves. When the wave is in the visible wavelength region, we can see the wave as light. The wave is reflected from black surfaces (about 3%), and 97% is absorbed. The absorbed energy results in heating the surfaces. For silvery surfaces, the reflection could be as high as 90%, whereas absorbed energy is only 10%. At steady state, the emitted radiation is equal to the absorbed amount. In a perfect black body, no energy is reflected, and hence one will not see the object (as a result, it appears black). For example, when humans emit radiation at a low temperature (310 K), the waves are in the nonvisible or infrared region and we cannot see them, particularly at night. During the day time we can see them because they reflect the light of the sun's radiation. However, electrical coils at 1000 K emit light in the visible wavelength region, and we see the coil colored red (the wavelength lies in the red portion). For a tungsten lamp at 3000 K, one sees white light because the wavelengths span the entire visible range, similar to the sun. When combustion occurs, there is significant radiation from hot gases that lies within the visible wavelength range, and thus one can observe the flames.
Ignition: Initiation of combustion, e.g., igniting the candle with a lighter.
Flame: The appearance of light, i.e., radiation in the visible wavelength region because of rise in thermal energy associated with the combustion phenomenon. There are two kinds of flames: (1) premixed flame and (2) diffusion flame.
Diffusion: Just as heat is transferred (i.e., thermal energy transfer) because of temperature gradient, species such as fuel vapor or O_2 are transferred or diffused because of concentration gradient, e.g., fuel evaporation from wicks, droplets, etc., (chapters 6, 7, 10, 11).
Flame Colors:

Deep Violet: A deep violet color of flame indicate radiation from CH species produced during reaction (Chapter 5).

Blue: Typically in excess air combustion due to radiation from excited H radicals in high T zone.

Blue-Green: Occurs when air is slightly less than stoichiometric due to radiation from C_2 radicals.

Visible radiation: OH radicals.

Yellow: At richer mixtures soot is produced, radiation from soot is yellow. Indicating continuum radiation, with maximum intensity in "infra red" region.
Diffusion flame (Figure 1.2a): A nonpremixed flame is a flame due to combustion of initially unmixed fuel and air, e.g., candle flame. The candle melts, permeates through the cotton wick, vaporizes, diffuses into the ambience, meets O_2, and then forms the flame. More rigorously, a nonpremixed flame may be

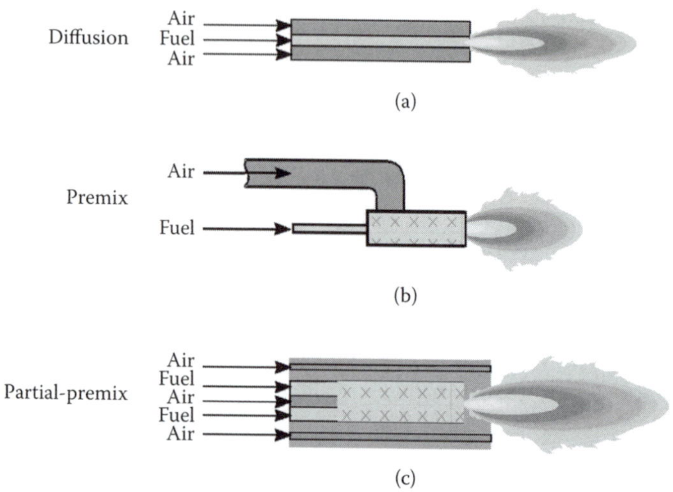

Figure 1.2 Schematic illustrations of: (a) diffusion flame, (b) premixed flame, and (c) partially premixed flame. (Adapted from Baukal, C.E., Gershtein, V.Y., and Li, X., Eds., *Industrial Combustion*, CRC Press, Boca Raton, FL, 2000.)

defined as any flame resulting from the diffusion of fuel and oxidizer to the *flame zone* (where they are mixed and burnt) from zones that contain only oxygen or fuel. In biological systems (e.g., humans), fuel (food) and air enter separately, are processed separately, food via the digestion system and air via lungs, then mixed in the bloodstream as glucose (fuel) and O_2 (carried by hemoglobin), and burned in cells (which are similar to microcombustors distributed throughout the body) but at a slower rate (without any flame!).

Premixed flame (Figure 1.2b): Flame due to combustion of an already premixed fuel and oxidizer is called a premixed flame, (e.g., oxyacetylene [C_2H_2] torch, premixed gasoline vapor and air in IC engines, and gas-fired furnaces and stoves).

Partially premixed (Figure 1.2c): Here, fuel and a small amount of air are premixed (e.g., domestic gas burners).

Laminar flame: This is a flame in which viscous forces are dominant compared to inertial forces, e.g., a smooth flame at low velocity.

Turbulent flame: This is a flame in which inertial forces are dominant compared to viscous forces, e.g., a wrinkled flame at high velocity.

Stationary flame: Here, the flame does not move, e.g., candle flame, flame from a gas stove, etc.

Propagating flame: This kind of a flame propagates into a premixed combustible mixture, e.g., igniting premixed gasoline vapor and air in an automobile engine. Flames travel from the spark plug toward the rest of the mixture (Figure 1.3).

Flame velocity: The rate of flame propagation into an unburned combustible mixture is expressed in m/sec. Under laminar conditions it is of the order of 0.4 m/sec.

Deflagration: Here, flame propagation speed is less than the speed of sound, which is about 700 m/sec in an IC engine.

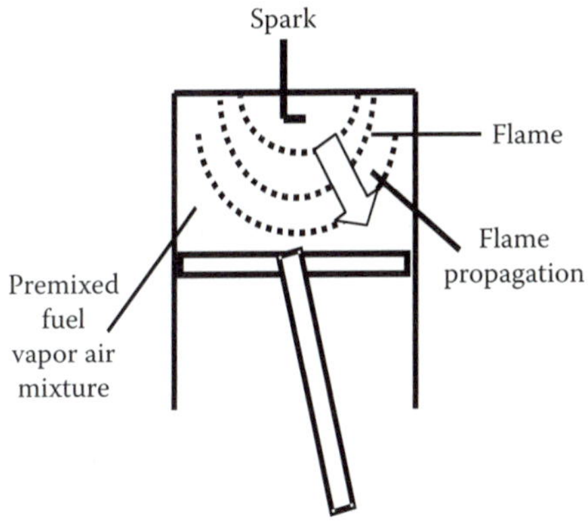

Figure 1.3 Schematic illustration of flame propagation in a gasoline engine after sparking.

Detonation: The flame propagation speed is greater than the speed of sound.
Chemical kinetics: A subject concerned with rates of reactions and factors on which the rates depend. For example, how fast is the fuel consumed per second per unit volume (kg/m^3 sec) in an automobile engine? For example, one can mount a candle on a balance, ignite it, and weigh the candle as a function of time. The mass vs. time curve yields the overall reaction rate of the candle fuel.
Heterogeneous combustion: These are combustion reactions that involve two phases, e.g., charcoal combustion in a grill in which the fuel is solid carbon and the oxidant is in the gas phase.
Homogeneous combustion: These are combustion reactions that involve same phases within the system of interest.
Extinction: It is the disappearance of the flame (i.e., the phenomenon of apparent cessation of chemical action), e.g., blowing out a candle flame.
Flammability: The ignitability of the mixture. For example, take an oxyacetylene torch, and let the mixture have a large amount of oxygen, but very little acetylene. The mixture will not ignite. On gradually increasing the amount of acetylene, the mixture ignites at a particular point. Such a mixture is said to flammable.
Bunsen burner: [http://www.wikipedia.org/wiki/Bunsen_burner]. The Bunsen burner, named after Robert Wilhelm Bunsen, its inventor, is a device used in many combustion experiments to study premixed and diffusion flames and flame propagation velocity. Typically, natural gas or any selected pure hydrocarbon (HC) gas (such as propane or butane) is used as fuel. The burner has a weighted base and a tube rising vertically from it, through which fuel is supplied. Just at the inlet to the tube, a metal collar is turned to gradually open or close holes made on the side of the tube. If all holes are open, the flame is almost premixed and is blue, and if all the holes are closed, the burner produces a diffusion flame colored yellow.

1.3 Matter and Its Properties

Sections 1.3 to 1.6 review some basic principles of thermodynamics as applied to combustion.

1.3.1 Matter

The amount of matter contained within a system is specified either by a molecular number count or by the total mass. An alternative to using the number count is a mole unit. Matter consisting of 6.023×10^{26} molecules (or Avogadro number of molecules) of a species is called 1 kmol of that substance. The total mass of those molecules (i.e., the mass of 1 kmol of the matter) equals the molecular mass of the species in kilograms. Likewise, 1 lb-mole of a species contains its molecular mass in pounds. For instance, 18.02 kg of water corresponds to 1 kmol, 18.02 g of water contains 1 g-mol, and 18.02 lb mass of water has 1 lb-mol of the substance. Unless otherwise stated, throughout the text the term mole refers to the unit kmol.

1.3.2 Mixture

A system that consists of more than one component (or species) is called a mixture. Consider a mixture of 18×10^{26} molecules of H_2 and 18×10^{26} molecules of He (Figure 1.4), i.e., the total number of molecules is 36×10^{26}. Then the hydrogen fraction

$$X_{H_2} = (18 \times 10^{26} \text{ molecules of } H_2)/(36 \times 10^{26} \text{ total molecules}) = 0.5.$$

Instead of using molecules, we can employ moles, as 1 kmol = 6×10^{26} molecules. In this case the number of kmoles of hydrogen

$$N_{H_2} = (18 \times 10^{26}/6 \times 10^{26}) = 3 \text{ kmol}.$$

Similarly $N_{He} = 3$ kmol, i.e., the hydrogen mole fraction

$$X_{H_2} = N_{H_2}/(N_{He} + N_{H_2}) = 0.5.$$

The composition determination based on the molecule number fraction or on the mole fraction is sometimes referred to as *molar analysis*.

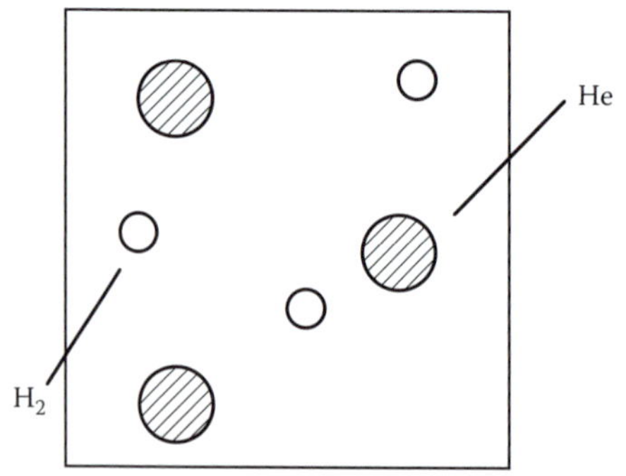

Figure 1.4 Illustration of mole and mass fractions.

In this hydrogen–helium system, the mass of hydrogen $(m_{H_2})$is 3 kmol × 2 kg kmol[1] and that of helium is 3 kmol × 4 kg kmol[1]. The hydrogen mass fraction

$$Y_{H_2} = m_{H_2}/(m_{H_2} + m_{He}) = 3 \times 2/(3 \times 2 + 3 \times 4) = 0.333.$$

Likewise, $Y_{He} = 0.67$.

If N_k denotes the number of moles of the k-th species in a mixture, the mole fraction of that species X_k is given by the relation

$$X_k = N_k/N, \tag{1.1}$$

where $N = \Sigma N_k$ is the total number of moles contained in the mixture. A mixture can also be described in terms of the species mass fractions Y_k as

$$Y_k = m_k/m, \tag{1.2}$$

where m_k denotes the mass of species k and m the total mass. Note that $m_k = N_k M_k$, with the symbol M_k representing the molecular weight of any species k. Therefore, the mass of a mixture

$$m = \sum N_k M_k$$

The molecular weight of species k, M_k, can be calculated using atomic weights. See Table A.1B for a partial listing of atomic weights.

The molecular weight of a mixture M is defined as the average mass contained in a kmol of the mixture, i.e.,

$$M = m/N = \left(\sum N_k M_k\right)/N = \sum X_k M_k. \tag{1.3}$$

It should be apparent from Equation 1.1 and Equation 1.2 that

$$\sum_k X_k = 1, \quad \text{and that} \quad \sum_k Y_k = 1.$$

Air is an example of a mixture containing molecular nitrogen, and oxygen, and argon.

Example 1

Assume that a vessel contains 3.12 kmol of N_2, 0.84 kmol of O_2, and 0.04 kmol of Ar. Determine the constituent mole fractions, the mixture molecular weight, and the species mass fractions.

Solution

Total number of moles N = 3.12 + 0.84 + 0.04 = 4.0 kmol

$$X_{N_2} = N_{N_2}/N = 3.12/4 = 0.78.$$

$$\text{Similarly,}\quad X_{N_2} = 0.21 \quad \text{and} \quad X_{Ar} = 0.01.$$

The mixture molecular weight can be calculated using Equation 1.3, i.e., M = 0.78 × 28 + 0.21 × 32 + 0.01 × 39.95 = 28.975 kg per kmol of mixture.

The total mass m = 3.12 × 28.02 + 0.84 × 32 + 0.04 × 39.95 = 115.9 kg, and the mass fractions are:

$$Y_{N_2} = m_{N_2}/m = 3.12 \times 28.02/115.9 = 0.754.$$

$$\text{Similarly,}\quad Y_{O_2} = 0.232 \quad \text{and} \quad Y_{Ar} = 0.0138.$$

Remark

A mixture of N_2, O_2, and Ar in the molal proportion of 78:21:1 (mol or volume percent) is representative of the composition of air. Many times Ar is combined with N_2 and air composition is reported as 79% N_2 and 21% O_2.

1.3.3 Property

A *property* is a characteristic of a system in equilibrium. Properties can be classified as follows:

- *Primitive* properties are those that appeal to human senses, e.g., T, P, V, and m.
- *Derived* properties are derived from primitive properties; e.g., force derived from Newton's law, enthalpy H, entropy S, and internal energy U.
- *Intensive* properties are independent of the extent or size of a system, e.g., P (kN m^2), v (m^3 kg^1), specific enthalpy h (kJ kg^1), and T (K).
- *Extensive* properties depend upon system extent or size, e.g., m (kg), V (m^3), total enthalpy H (kJ), and total internal energy U (kJ). The extensive properties are additive.

An extensive property can be converted into an intensive property provided it is distributed uniformly throughout the system, by determining its value per unit mass, unit mole, or unit volume. For example, the specific volume v = V/m (in units of m^3kg^{-1}) or V/N (in terms of m^3kmol^{-1}). The density ρ = m/V is the inverse of the mass-based specific volume. We will use lowercase symbols to denote specific properties (e.g., v, $\bar{v}$, u, and $\bar{u}$, etc.) and the bars denote mole-based specific properties. The exceptions to the lowercase rule are temperature T and pressure P. Furthermore, we will represent the differential of a property as d (property), e.g., dT, dP, dV, dv, dH, dh, dU, and du and those variables crossing the boundary as δ (variables); e.g., δQ, δW, where Q and W are heat and work crossing the boundary.

1.3.4 State

The condition of a system is its *state*, which is normally identified and described by the observable primitive properties of the system. Generally, a set of properties, such as T, V, P, N_1, N_2, etc., representing system characteristics define the state of a given system.

Mechanical equilibrium prevails if there are no changes in pressure. *Thermal equilibrium* exists if the system temperature is unchanged. *Phase equilibrium* occurs if, at a given temperature and pressure, there is no change in the mass distribution of the phases of a substance, i.e., if the physical composition of the system is unaltered. For instance, if a vessel containing liquid water is placed in a room with both the liquid water and room air being at the same temperature, and if the liquid water level in the vessel is unchanged, then the water vapor in the room and liquid water in the vessel are in phase equilibrium.

Chemical equilibrium exists if the chemical composition of a system does not change. For example, if a mixture of H_2, O_2, and H_2O of arbitrary composition is enclosed in a vessel at a prescribed temperature and pressure, and there is no subsequent change in chemical composition, the system is in chemical equilibrium. The term *thermodynamic state* refers only to equilibrium states.

1.3.5 Equation of State

Having described systems, and the type and state of matter contained within them in terms of properties, we now explore whether all of the properties describing a state are independent or related.

A thermodynamic state is characterized by macroscopic properties called *state variables* denoted by x_1, x_2, …, x_n, and F. Examples of state variables include T, P, V, U, H, etc. It has been experimentally determined that, in general, at least one state variable, say F, is not independent of x_1, x_2, …, x_n, so that

$$F = F\,(x_1, x_2, \ldots, x_n). \quad (1.4)$$

Equation 1.4 is referred to as a *state postulate* or *state equation*. The number of independent variables, x_1, x_2, …, x_n (in this case there are n variables), is governed by the laws of thermodynamics. For example, if $x_1 = T$, $x_2 = V$, $x_3 = N$, and F = P, then

$$P = P\,(T, V, N).$$

For an ideal gas, the functional form of this relationship is given by the ideal gas law, i.e.,

$$P = N\bar{R}\,T/V, \quad (1.5)$$

where $\bar{R}$ is known as the universal gas constant, the value of which is 8.314 kJ kmol^{-1} K^{-1}. The universal gas constant can also be deduced from Boltzmann's constant, which is the universal constant for one molecule of matter (defined as $k_B = \bar{R}/N_{Avog} = 1.38 \times 10^{28}$ kJ molecule^{-1} K^{-1}; see Table A.1A). Defining the molar specific volume, $\bar{v} = V/N$, we can rewrite Equation 1.5 as

$$P = \bar{R}\,T/\bar{v}. \tag{1.6}$$

Equation 1.6 (stated by J. Charles and J. Gay Lussac in 1802) can be called *an intensive equation of state*, as all the variables contained in it are intensive. The ideal gas equation of state may be also expressed in terms of mass units after rewriting Equation 1.5 in the form

$$P = (m/M)\,\bar{R}\,T/V = mRT/V, \tag{1.7}$$

where $R = \bar{R}/M$. Similarly,

$$P = RT/v. \tag{1.8}$$

Equation 1.8 demonstrates that $P = P\ (T,\ v)$ for an ideal gas and is known once T and v are specified.

For a nonequilibrium system, state equations for the entire system are meaningless. However, the system can be divided into smaller subsystems A and B, with each assumed to be in a state of internal equilibrium. State equations are applicable to subsystems that are in local equilibrium.

1.3.6 Standard Temperature and Pressure

Using Equation 1.6, it can be verified that the volume of 1 kmol of an ideal gas at *standard* (scientific) *ambient temperature and pressure* (SATP), given by the conditions T = 25°C (77°F) and P = 1.013 bar (1 atm), is 24.5 m^3 kmol^{-1} (392 ft^3 lbmol^{-1}). This volume is known as a "standard" cubic meter (SCM) or a standard cubic foot (SCF).

Interestingly, there is no general agreement among engineers and chemists regarding standard temperature and pressure (STP). Engineers often define STP (called US standard or International Standard Atmosphere, ISA) at 15.6°C (60°F) and 101.3 kPa (1 atm). For ideal gases, 1 kmol of any fuel at STP, 15.6°C (288.6K) and 101.3 kPa, occupies a volume of 23.68 m^3 ($\bar{R}T/P$, where $\bar{R}$ = 8.314 atm m^3 K^{-1}). Chemists use STP at 0°C, 1 atm (chemist standard atmosphere, CSA) with STP volume of 22.4 m^3/kmol. See Table A.1A in Appendix A for the values of volume at various STP conditions. Thus, mole ratio for gaseous species is the same as the corresponding ratio on a volume basis. This equality is very useful for gas-fired burners, because if the volume flow of gases in cubic meters is known, then the mole-based flows can be readily determined. Using STP volume, one can determine the density of any species k at STP as M/(STP volume) where M is the molecular weight of species k.

Specific gravity (Sg) is the relative density of a given gas to air at the same T and P; thus,

$$Sg = \rho/\rho_a = \{PM/\bar{R}\,T\}/\{PM_a/\bar{R}\,T\} = M/M_a$$

Thus, for methane, Sg = 16.05/28.97 = 0.55. For Sg of liquids, see Section 4.3.9.1.

1.3.7 Partial Pressure

The equation of state for a mixture of ideal gases can be generalized if the number of moles in Equation 1.5 is replaced by

$$N = N_1 + N_2 + N_3 + \ldots = \sum N_k, \tag{1.9}$$

so that Equation 1.5 transforms into

$$P = N_1\,\bar{R}\,T/V + N_2\,\bar{R}\,T/V + \ldots \tag{1.10}$$

The first term on the right-hand side of Equation 1.10 is to be interpreted as the partial pressure (p_1) for an ideal gas mixture. This is the pressure that would have been exerted by component 1 if it alone had occupied the entire volume. Therefore,

$$p_1 = N_1\,\bar{R}\,T/P = X_1 N\bar{R}\,T/V = X_1 P. \tag{1.11}$$

Assuming air at a standard pressure of 101 kPa to consist of 21 mole percent of molecular oxygen, the pressure exerted by O_2 molecules alone p_{O_2} is 0.21 × 101 = 21.21 kPa. The partial pressures are extremely important in combustion calculations because we deal with gas mixtures.

1.3.8 Phase Equilibrium

When liquid fuels are analyzed for combustion, a knowledge of phase equilibrium is required in order to predict the rate of evaporation or combustion. Here, a brief overview of phase equilibrium is presented. Consider a small quantity of liquid contained in a piston–cylinder assembly with a weight at the top, so that P is constant during the process of heating. If T of the fluid is less than $T_{boiling}$, at given P the state is called *subcooled* or *compressed liquid*. If a small bubble is formed at $T = T_{boiling}$ (or more appropriately called T^{sat} at a given P), the liquid is said to be a *saturated liquid* (state F; Figure 1.5a), and the specific volume of the liquid at this saturated state is denoted by v_f. As more heat is added, the liquid and vapor phases coexist at state W (in the two-phase or wet region, $v > v_f$). The ratio of the mass of vapor (m_g) to total mass m is termed *quality* ($x = m_g/m$). As more heat is added, the liquid completely converts to vapor at state G and is called the *saturated vapor* state or the *dew point* (state G; Figure 1.5a), and the specific volume of the vapor at this saturated state is denoted by v_g. The heat

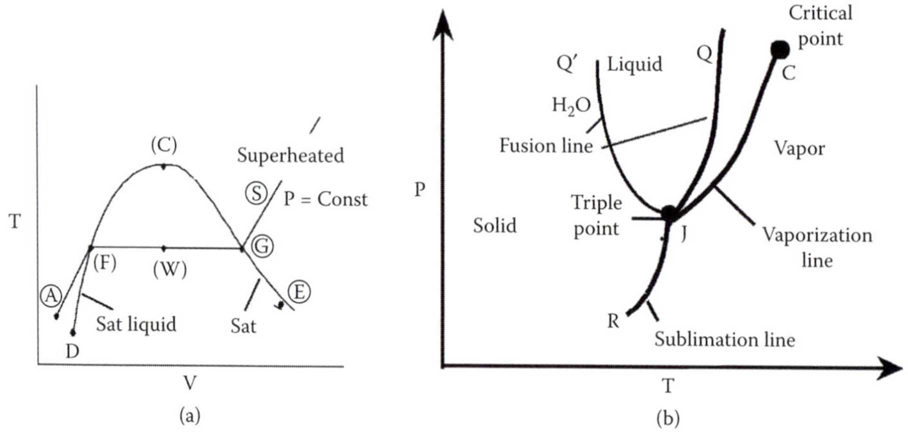

Figure 1.5 (a) T-v-P and (b) P-T diagrams of a pure substance.

required to vaporize from saturated liquid to saturated vapor is called *heat of vaporization* or *latent heat* (L). Upon further heat addition at the specified pressure, the system temperature becomes larger than the saturation temperature, and the vapor enters state *S*, which is known as the *superheated vapor state*. The curve AFWGS in Figure 1.5a describes an isobaric process on the T-v diagram. Experiments could be repeated at different pressures P and variation of T_{sat} with P can be obtained, or vice versa. Often, the following empirical fit, known as Cox–Antoine relation can be obtained:

$$\ln P^{sat} = A + \frac{B}{T + C}.$$

When C = 0, the relation is known as the Clausius–Clapeyron relation (see Chapter 10).

Curve DFC is called saturated liquid, and CGE is called saturated curve (Figure 1.5a). The two curves intersect at the critical point C that corresponds to a distinct critical temperature and pressure, T_c and P_c. The latent heat is L = 0 at the critical point. Table A.2A of Appendix A contains critical data for many substances. Substances at $P > P_c$ and $T > T_c$ are generally referred to as fluids that exist in a *supercritical state*. Both liquid and vapor are contained in the *two-phase dome* where $P < P_c$, $T < T_c$, and $v_f < v < v_g$.

On plotting the pressure with respect to the saturation temperature T_{sat} along the saturated curve, the phase diagram of Figure 1.5b is obtained. In this figure, the vaporization curve is represented by JC, JQ is the melting curve for most solids but JQ′ is the melting curve for ice, and JR represents the sublimation curve. The intersection J of the curves JC and JQ at which all three phases coexist is called the *triple point*. For water, this point is characterized by P = 0.0061 bar (0.006 atm) and T = 273.16 K (491.7°R), whereas for carbon, the analogous conditions are T ≈ 3800 K and P ≈ 1 bar. Knowledge of liquid–vapor equilibrium is essential in understanding liquid fuel combustion as well as condensation of H_2O when product gases are cooled.

1.4 Microscopic Overview of Thermodynamics

In order to understand the processes governing combustion, it is useful to understand the microscopic behavior of molecules constituting the matter of those systems. A brief overview of the subject is presented in the following subsections.

1.4.1 Matter

Atoms are of the order of 1 to 2 Å (i.e., 1×10^{-10} 2×10^{-10} m) in radius. Because g and kg are too large to denote the mass of atom, we define atomic mass unit (amu). One amu = $(1/6.023 \times 10^{23})$ g. If Z is the number of protons and N the number of neutrons, then

$$A = Z + N = \text{atomic mass number or total mass number.}$$

For example, A for He = 4; it has 2 protons, 2 neutrons, and 2 electrons; He is written as ${}^4_2\text{He}$, where 4 is the atomic mass number, and 2 is the number of protons. H has no neutrons and is written as ${}^1_1\text{H}$. Many atoms do not exist alone and combine with other atoms forming a molecule. Oxygen is a homonuclear molecular compound, whereas the water molecule, H_2O, is a heteronuclear molecule consisting of two atoms of H (separated by 105°) 1 Å distant from one atom of O (Figure 1.6). Water is an inorganic compound (i.e., it does not have a C atom).

Electrons orbit at different energy levels ($2n^2$, where n is the orbit number; 2 [n = 1], 8 [n = 2], 18 [n = 3]). Electrons get attracted to protons, and this is greatest

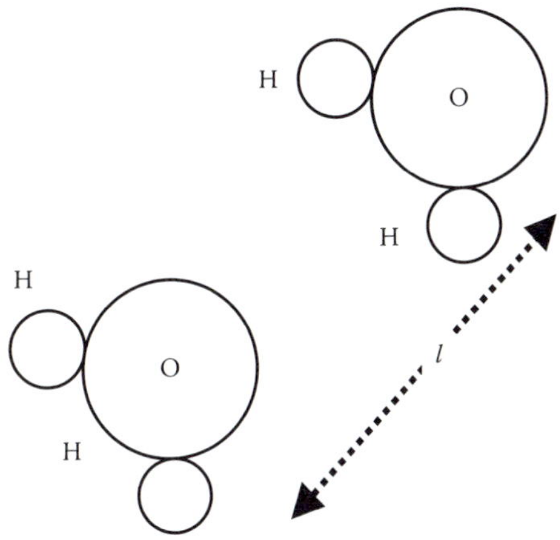

Figure 1.6 Schematic illustration of a water molecule.

for electrons at closest orbits and least for those at farthest orbits (called valence electrons). Electrons in the farthest orbit have high interatomic potential energy, and they are more likely to get involved with chemical interactions because they are least tightly held. An atom is in a stable state if its outermost orbit is filled, or has 8 electrons. Such an atom is said to be chemically inert, such as He with 2 electrons. And Ne with 10 (2 + 8) electrons. On the other hand, the H atom has 1 electron and hence combines with another H to form H_2. Strong attractive forces called *chemical forces* hold the atoms together in a molecule (e.g., two H atoms bound together in H_2), and the corresponding potential energy is called a *chemical bond (intramolecular forces)*. Another example is, two atoms of H and one atom of O held together as a molecule of H_2O by a chemical bond (Figure 1.6). These bonds are created by pulling electrons away from H atoms to O atoms, forming H^+ and O, which results in chemical bonds. On the other hand, if a second H_2O molecule is brought near the first molecule, there exist *intermolecular* forces. Chemical forces (strengths within molecules) are always much higher than the attractive forces between the molecules, and this keeps the molecules separated from one another.

1.4.2 Intermolecular Forces and Potential Energy

Consider the N_2 molecule. Adjacent N_2 molecules are separated by an intermolecular distance ℓ. The variation of the intermolecular force F between molecules as a function of ℓ is illustrated in Figure 1.7. In a piston–cylinder–weight assembly, this distance can be varied by changing the volume by the addition or removal of weights. The intermolecular force is negative when attractive, i.e., it attempts to draw molecules closer together, whereas positive forces correspond to closer

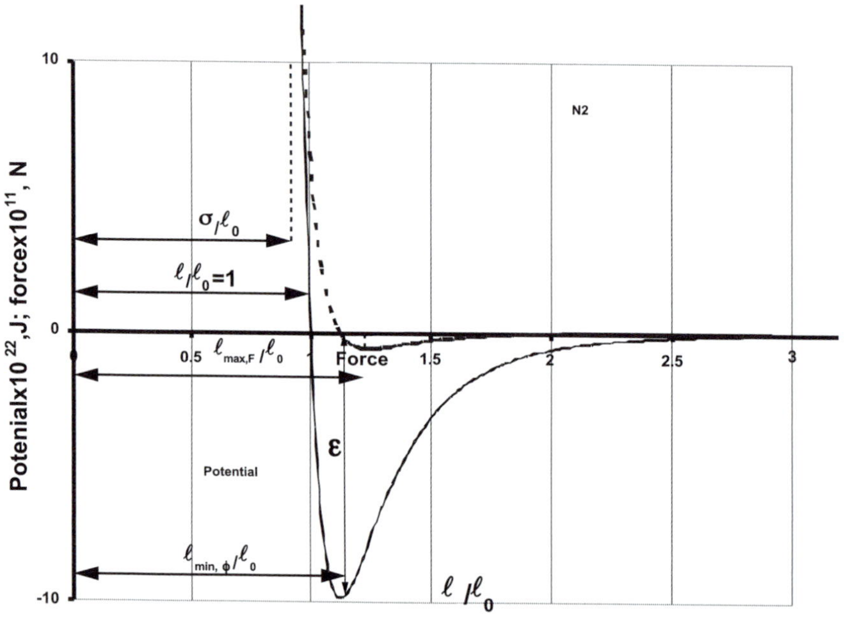

Figure 1.7 Calculated LJ potential and force field for nitrogen molecules. $\ell_{min}/\ell_o = 1.1225$, $\ell_{max}/\ell_o = 1.2445$.

intermolecular spacing and are repulsive, i.e., they attempt to move the molecules away from each other. The distances ℓ, $\ell_{max,F}$ $\ell_{min,\phi}$, ℓ_0, and σ that are illustrated in Figure 1.7 will be described later. The Lennard–Jones (LJ) (6–12 law) empirical approach for like molecular pairs, such as the homonuclear molecular pair N_2 – N_2, furnishes the intermolecular potential energy (IPE) in the form

$$\Phi(\ell) = 4\varepsilon\,((\ell_0/\ell)^{12} - (\ell_0/\ell)^6), \tag{1.12}$$

where ε represents the characteristic interaction energy between molecules, i.e., the minimum potential energy

$$\Phi_{min} = \varepsilon \approx 0.77\ k_B\ T_c, \tag{1.13}$$

where k_B denotes Boltzmann's constant ($= \bar{R}/N_{Avog} = 1.33 \times 10^{-26}$ (kJ/molecule K)), T_c the critical temperature, and ℓ_0 represents the distance at which the potential is zero (cf. Figure 1.7) and is approximately equal to the characteristic or collision diameter σ of a molecule at which the potential curve shown in Figure 1.7 is almost vertical. Table A.3 of Appendix A tabulates σ and ε/k_B (in K) for many substances. In order to determine $\ell_{min,\phi}$, Equation 1.12 can be differentiated with respect to ℓ and set equal to zero. From this exercise,

$$\ell_{min,\phi}/\ell_0 = 2^{1/6} = 1.1225, \tag{1.14}$$

and the corresponding value of $\Phi_{min} = \varepsilon$. Hence,

$$\Phi(\ell)/\Phi_{min} = 4((\ell_0/\ell)^{12} - (\ell_0/\ell)^6). \tag{1.15}$$

The interaction force between the molecules is given by the relation $F(\ell) = d\Phi/d\ell$, so that

$$F(\ell)/\Phi_{min} = (4/\ell_0)(12(\ell_0/\ell)^{13} - 6(\ell_0/\ell)^7). \tag{1.16}$$

The maximum attractive force occurs at $\ell_{max,F}/\ell_0 = 1.2445$, and the corresponding force $|F_{max}| = 2.3964\ |\Phi_{min}|\,\ell_0$.

Assuming ℓ_0 to equal σ, for molecular nitrogen, $\sigma = \ell_0 = 3.681$ Å and $\varepsilon/k_B = 91.5$ K (Table A.3 of Appendix A). Using the value of $k_B = 1.38 \times 10^{23}$ J molecule^{-1} K^{-1}, Φ and F can be determined for given values of ℓ_0/ℓ. Results are presented for molecular nitrogen in Figure 1.7. Note that the molecular diameter is σ (also called the collisional diameter), which is the closest distance at which another molecule can approach it.

If the molecules are spaced relatively far apart, the attractive force is negligible. Ideal gases fall into this regime. As the molecules are brought closer together, although the attractive forces increase, the momentum of the moving molecules is high enough to keep them apart. As the intermolecular distance is further decreased, the attractive forces become so strong that the matter changes phase from gas to liquid. Upon decreasing this distance further, the forces experienced by the molecules become smaller (Figure 1.7). Any further compression results in strong repulsive forces, and the matter is now a solid in which the molecules are well positioned.

1.4.3 Molecular Motion

At low pressures and high temperatures, the intermolecular spacing in gases is usually large, and the molecules move incessantly over a wide range of velocities. The molecules also vibrate, rotate (Figure 1.8), and constantly collide with each other, particularly in gases and liquids.

1.4.3.1 Collision Number and Mean Free Path

Molecules contained in matter travel a distance ℓ_{mean} before colliding with another molecule. Collision number between molecular A and B per unit volume is given as

$$Z_{coll,AB} = n_A\, n_B\, N_{Avog}^2\, \pi\sigma_{AB}^2\, V_{avg,AB} \tag{1.17}$$

and average molecular velocity and molecular diameter are given as

$$V_{avg,AB} = \{8k_BT/(\pi\mu_{AB})\}^{1/2} \quad \text{and} \quad \sigma_{AB} = (\sigma_A + \sigma_B)/2 \tag{1.18a}$$

$$\mu_{AB} = \frac{(m_A m_B)}{(m_A + m_B)}, \quad m = \frac{M}{N_{Avog}} \tag{1.18b}$$

where n is the number of kmol/m^3.

Reverting to similar molecules, the average distance traveled by a molecule during this time is called its mean free path ℓ_{mean}, where

$$\ell_{mean} = 1/(2^{1/2}\pi n'\sigma^2), \quad n' = nN_{Avog} \tag{1.19}$$

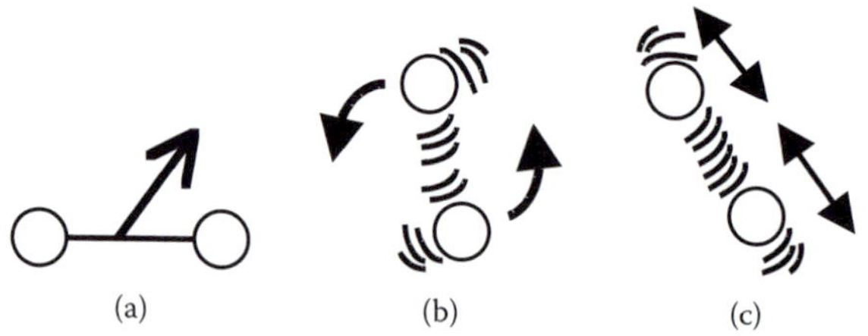

Figure 1.8 Illustration of the energy modes associated with a diatomic molecule. (a) translational energy (TE), (b) rotational energy (RE), (c) vibrational energy (VE).

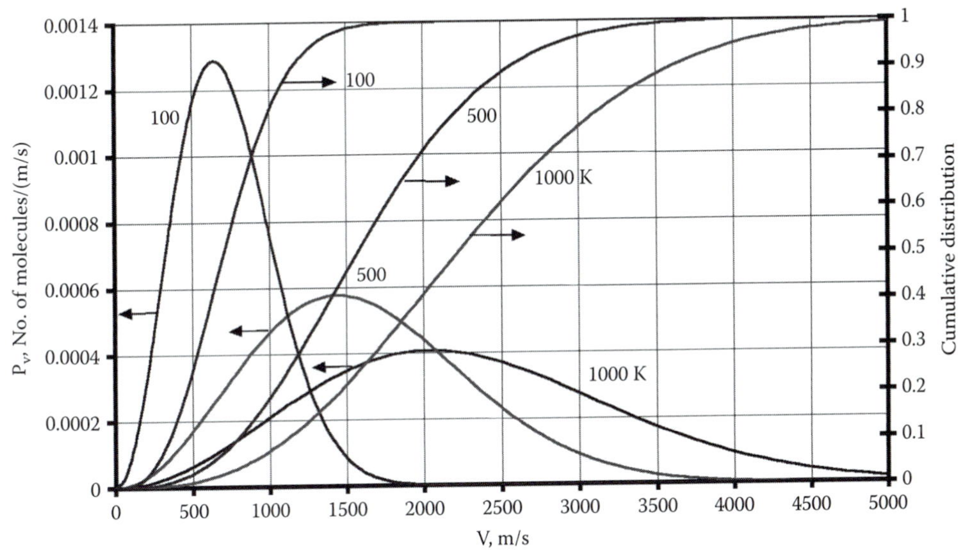

Figure 1.9 Maxwellian distribution of the absolute velocity in helium, which is a perfect gas (Helium with m = 6.65 × 10^{-23}g).

Typically, the number of collisions is of the order of 10^{39}m^{-3} sec^{-1}. The mean free path is not the same as the average distance between adjacent molecules, because molecules' paths may not be collinear.

1.4.3.2 Molecular Velocity Distribution

All of the molecules do not travel at the average velocity. The typical velocity distributions (also called the Maxwellian distributions) of Helium molecules at different temperatures are illustrated in Figure 1.9. The typical velocity distributions can be determined from the expression

$$(1/\mathrm{N})(\mathrm{dN}'_V/\mathrm{d}V) = (4\pi)^{-1/2}\,[\mathrm{m}/(2\mathrm{k_B T})]^{3/2} V^2 \exp[(1/2)\mathrm{m}V^2/(\mathrm{k_B T})] \quad (1.20)$$

where N_V represents the number of molecules moving with a velocity in the range V and $V + dV$, N is the total number of molecules, and m, the molecular mass (M/N_{Avog}), with M denoting the molecular weight.

Equation 1.20 can be rewritten in terms of the energy $e' = mV^2/2$ and integrated to obtain the fraction of molecules possessing energy in the range E to ∞, i.e.,

$$\mathrm{N}'_E/\mathrm{N} = 2\pi^{-1/2}((\mathrm{E}/(\bar{R}\ \mathrm{T}))^{1/2}\exp(-(\mathrm{E}/(\bar{R}\ \mathrm{T})) + (1 - \mathrm{erf}((\mathrm{E}/\bar{R}\ \mathrm{T})^{0.5})), \quad (1.21)$$

where $E = M\,V^2/2$, M denotes the molecular weight (or the mass of 1 kmol), and $\bar{R} = k_B\,N_{Avog}$ is the universal gas constant. Because $E/\bar{R}T$ is typically large, the value of the last term on the right-hand side of Equation 1.21

is negligibly small. Hence, the fraction of molecules with a velocity in the range V to ∞ (or E ≤ Energy ≤ ∞) may be expressed as

$$N'_V/N = 2\pi^{-1/2}(E/(\bar{R}T))^{1/2}\exp\{-(E/(\bar{R}T)\}. \qquad (1.22)$$

Equation 1.22 indicates that the fraction of molecules associated with an energy of value E and greater is proportional to exp $\{-(E/(\bar{R}T)\}$. Note that E generally includes all forms of energies of molecules. Chemical reactions between reactant molecules occur when the energy E exceeds the minimum activation value that is required to overcome the molecular bond energies, thereby allowing the atoms to be rearranged in the form of products. The lower is E, the higher is the fraction of molecules having an energy of E and higher.

1.4.3.3 Average, RMS, and Most Probable Molecular Speeds

In order to determine the average and RMS speeds, etc., from Equation 1.20, the following integrals are useful:

$$\int_0^\infty x^n \exp(-ax^2) = \frac{1}{2}\sqrt{\frac{\pi}{a}}, \frac{1}{2a}, \frac{1}{4}\sqrt{\frac{\pi}{a^3}}, \frac{1}{2a^2}, \frac{3}{8}\sqrt{\frac{\pi}{a^5}}, \frac{1}{2a^3},$$

$$\textit{for} \quad n = 0, 1, 2, 3, 4, 5\,\text{etc.}$$

The average molecular speed V_{avg} is

$$V_{avg} = [8/(3\pi)]^{1/2}\, V_{RMS} = (8\, k_B T/(\pi m))^{1/2} = (8\bar{R}\, T/(M\pi))^{1/2}, \qquad (1.23)$$

The expression for the most probable speed (where dN_V/dV is a maximum) is

$$V_{mps} = (2/3)^{1/2}\, V_{rms} = (2k_B T/m)^{1/2} = (2\bar{R}\, T/M)^{1/2}. \qquad (1.24)$$

The rms speed V_{RMS} can be expressed as

$$V_{RMS} = (3k_B T/m)^{1/2} = (3\bar{R}\, T/M)^{1/2}, \qquad (1.25)$$

where $V_{RMS}^2 = V_x^2 + V_y^2 + V_z^2$ is based on the three velocity components. The average translational energy (TE) is given by

$$\frac{1}{2}mV_{RMS}^2 = \frac{1}{2}m\left(V_x^2 + V_y^2 + V_z^2\right) = \frac{3}{2}k_B T \qquad (1.26)$$

From Equation 1.24 note that average transnational energy per molecule is $3k_B\, T/2$ where $k_B = \bar{R}/N_{Avog}$, whereas for 1 kmol, the average TE = (1/2) M V^2_{RMS}.

It is customary to assume that $|V_x| = |V_y| = |V_z|$, i.e., each degree of freedom contributes energy equivalent to $(1/2)k_BT$ to the molecule. At standard conditions $V_{RMS} \approx$ 1,918, 515, and 481 m sec^{-1}, respectively, for H_2, N_2, and O_2, and is typically of the same magnitude as the speed of sound in those gases. For an ideal gas, the speed of sound $c = (k\bar{R}\,T/M)^{1/2}$, where $1 \le k \le 5/3$. For gaseous N_2 and H_2, respectively, at standard conditions $V_{avg} \approx$ 475 and 1770 m sec^{-1}; m = 4.7 × 10^{-26} kg and 0.34 × 10^{-26} kg; σ = 3.74 Å and 2.73 Å; ℓ_{mean} = 650 Å and 1230 Å. At 27°C the average TE per molecule is (3/2) * 8314 * 300 (J/kmol)/6.023 × 10^{26} = 6.2 × 10e^{-21} J per molecule; TE per kmol = 1/2 MV_{RMS}^2= (3/2) * 8314 * 300 (J/kmol) = 3734 and 260 J/kmol, respectively.

1.4.4 Temperature

One can define temperature only when there is equilibrium between the various internal degrees of freedom. For example, if one increases the "temperature" instantaneously, say, from 300 K to 3000 K, the molecular speed increases. According to Newton's law, the increase requires a finite time called *relaxation time*, whose timescale is of the order of 10^{-14} sec. Similarly, the molecule may increase its rotational speed and vibrational frequency with corresponding relaxation timescales that are of the order of 10^{-8} sec and 10^{-4} sec. Therefore, unless the timescale for temperature change is much higher than 0.1 msec, one may not assume internal molecular equilibration. For example, in deterioration waves (Chapter 14), "temperature" rises very rapidly and internal equilibration may not exist.

1.4.5 Knudsen Number

State equations and many heat and mass transfer relations (e.g., Fourier Law, $PV = N\bar{R}T$) are valid only when the surface on which a molecule impinges has a dimension much larger than the mean free path ℓ_{mean}. Consider a small particle of the order of, say, 0.01 μm surrounded by N_2 gas. We wish to determine the pressure exerted on this small particle. Let the mean free path of the N_2 molecule be 0.1 μm. It is, therefore, possible that molecules located 0.1 μm apart may not collide at all on the surface of the particle. For such cases the pressure cannot be calculated through state equations like Equation 1.5. The Knudsen number is defined as

$$Kn = \ell_{mean}/d, \tag{1.27}$$

where d denotes the particle diameter. This number is useful in defining continuum properties such as pressure, thermal conductivity, etc. If Kn << 1, the continuum approximation is valid. Particularly, when carbon or liquid droplets burn or evaporate, the size of the particle becomes extremely small and continuum approximation may not be valid. Kn is also useful in dealing with combustion analyses of nanotubes.

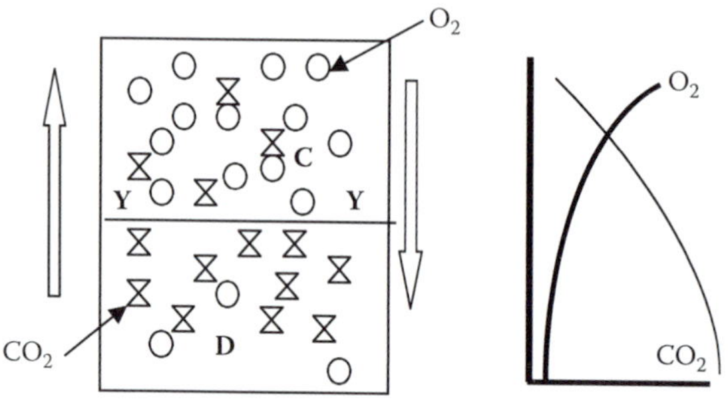

Figure 1.10 Illustration of species transfer. Oxygen molecules are denoted by O, and CO_2 molecules by ⧖.

1.4.6 Chemical Potential and Diffusion

The chemical potential drives mass (or species) transfer in a manner similar to the thermal potential that drives heat transfer $\dot{Q}$ from higher to lower temperatures.

Consider a vessel divided into two sections C and D separated by a permeable membrane (as shown in Figure 1.10) that initially contain oxygen throughout, and in which charcoal is spread over the floor of section D. Because the charcoal burns, sections C and D consist of two components: oxygen and CO_2. Figure 1.10 shows the variation of mole fractions of O_2 and CO_2. Further, consider a specific time at which the mole fraction of O_2 in section C (say, X_{O_2} = 80%) is larger compared to that in section D (say, X_{O_2} = 30%).

Because molecules move randomly, for every 1000 molecules that migrate from C into D through the section Y–Y, 1000 molecules will move from D into C. Consequently, 800 molecules of O_2 will move into D, whereas only 300 molecules of this species will migrate to C from D, so that there is net transfer of 500 molecules of O_2 from section C to D. Simultaneously, there is a net transfer of 500 molecules of CO_2 across the Y–Y plane from section D into C. The oxygen transfer enables continued combustion of the charcoal. This mass transfer (or species transfer) due to random molecular motion is called *diffusion* (also see Chapter 6).

The chemical potential μ, alternatively known as Gibbs function (g) for ideal gases, is related to the species concentrations (hence, their mole fractions). Macroscopically speaking, a higher species mole fraction of O_2 in section C implies a higher chemical potential for that species, thereby inducing oxygen transfer from C to D. If the charcoal is extinguished, CO_2 production (therefore, O_2 consumption) ceases, and eventually a state of species equilibrium is reached. At this state the chemical potential of each species or its concentration is uniform in the system.

1.4.7 Entropy (S)

1.4.7.1 Overview

Molecules undergo random motion. The energy of random motion is indicated by the temperature (e.g., $T \propto mV^2$, where V denotes the molecular velocity; a random velocity distribution is provided by Maxwell's law), whereas "ipe" depends on the volume V of the system. In addition to "te," molecules contain energy in the form of "ve" and "re" at various rotational speeds.

Entropy is a measure of the number of random states in which molecules store energy, just as there are several ways to store physical items in a cabinet (depending on the number of items and the shelves in the cabinet). Assuming all of these states to be equally probable, the entropy S is defined as a quantity proportional to the logarithm of the total number of microstates Ω_{tot} in which energy is stored, i.e., from statistical thermodynamics, $S = k_B \ln \Omega_{tot}$ (known as Boltzmann's law).

1.4.7.2 *Entropy S = S (U, V)*

Thus, for a given gas within a room, the probable macroscopic state is one having various velocity magnitudes, different directions, different positions, etc. If U is increased at the same volume, there are more possible energy states and hence more entropy. S increases with U. Further, in smaller volumes, they collide more frequently per unit time, and the distribution becomes more uniform, hence, entropy is lesser for the same U. Thus, at a given U, S is also $S(V)$; if both U and V vary, then S = S(U, V).

1.5 Conservation of Mass and Energy and the First Law of Thermodynamics

Section 1.2 to Section 1.4 presented basic concepts on macroscopic and microscopic thermodynamics and the physical meanings of pressure P, temperature T, and entropy S. In this section, conservation equations of energy (called *the first law* in mathematical form) and mass are presented for both closed and open systems.

1.5.1 Closed System

1.5.1.1 Mass Conservation

For closed systems the mass conservation equation is simply that the total mass

$$m = \text{constant}, \tag{1.28}$$

1.5.1.2 Energy Conservation

An informal statement regarding energy conservation is as follows: "Although energy assumes various forms, the total quantity of energy is constant, with the consequence that when energy disappears in one form, it appears simultaneously in other forms."

1.5.1.2.1 Elemental Process

For a closed system undergoing an infinitesimal process, during which the only allowed interactions with its environment are those involving heat and work, the first law can be expressed quantitatively as follows

$$\delta Q - \delta W = dE(\text{kJ or Btu}) \tag{1.29}$$

where δQ denotes the elemental (heat) energy transfer across the system boundaries because of temperature differences, δW the elemental (work) energy in transit across the boundaries in all forms, such as Pdv, shaft work, electrical equivalent of work (e.g., the piston weight lifted because of the expansion of the system, Figure 1.11), and dE the energy change in the system. The term E = U + KE + PE, where U is the non-nuclear internal energy U (= transitional energy (TE) + vibrational energy (VE) + rotational energy (RE) + IPE + chemical or bond energy, etc.) that resides in matter, the kinetic energy (KE) and the potential energy (PE). Note that Q and W are transitory forms of energy, and their differentials are written in the inexact forms δQ and δW, whereas differentials of resident energy E are written as exact differentials. Dividing Equation 1.29 by m,

$$\delta q - \delta w = de, \text{ kJ/kg} \quad \text{or} \quad \text{Btu/lb} \tag{1.30}$$

where q denotes the heat transfer per unit mass Q/m, w the analogous work transfer W/m, and, likewise, e = E/m.

A positive sign is assigned for heat transfer into a system, while the work done by the system is assigned a positive sign ($W > 0$), and work done on the system is assigned a negative sign (so that $W < 0$).

Equation 1.30 may be rewritten for a static system in the form

$$\delta Q - \delta W = dU. \tag{1.31}$$

1.5.1.2.2 Energy

At a microscopic level, the internal energy U is due to the molecular energy, which is the sum of the (1) molecular translational, vibrational, and rotational energies (also called the thermal portion of the energy), (2) the molecular bond energy (also called the chemical energy), and (3) the intermolecular potential energy, IPE. At a specified temperature, the energy depends upon the nature of the substance, and, hence, is known as an *intrinsic* form of energy. Note that in chemical reactions, the atomic structure changes (even though atoms are conserved) producing new species, and hence bond or chemical energies are different for different species. For nonreacting systems the bond or chemical energy can be ignored.

The first law is

$$\delta Q - \delta W = dU + d(PE) + d(KE), \tag{1.32}$$

where PE and KE denote the potential and kinetic energies, which are called *extrinsic* forms of energy.

1.5.1.2.3 Integrated Form

Integrating Equation 1.29 between any two thermodynamic states (1) and (2) we have

$$Q_{12} - W_{12} = E_2 - E_1 = \Delta E \text{ kJ or Btu.} \quad (1.33)$$

The heat and work transfers are energy forms in transit and, hence, do not belong to the matter within the system, the implication being that neither Q nor W is a property of matter.

1.5.1.2.4 Cyclic Form

Because for a cycle the initial and final states are identical (Figure 1.12),

$$\oint \delta Q = \oint \delta W = 0 \quad \text{or} \quad Q_{cycle} = Q_{in} - Q_{out} = W_{cycle}. \quad (1.34)$$

Example 2

Using the first law for a closed system, obtain an expression for the escape velocity of mass m if it has to overcome the earth's gravitational field with "g" given by $\frac{Cm_E}{r^2}$. Compute the speed in m/sec.

Solution

$Q_{12} - W_{12} = E_2 - E_1$, $Q = 0$, $W = 0$; $E_2 = E_1$; $KE_2 + PE_2 + U_2 = KE_1 + PE_1 + U_1$; $U_2 = U_1$; $KE_2 + PE_2 = KE_1 + PE_1$.
If 1 is a location on the earth's surface (say, Cape Canaveral, Florida) and 2 is at a location where the earth's gravity is almost zero, then

$$KE_2 + PE_2 = 0 + (-C\ m\ m_E/r) \text{ with } r \to \infty = KE_1 + PE_1 = 1/2\ m\ V_1^2 + (-C\ m\ m_E/r_E)$$

Then $1/2\ m\ V_1^2 = (C\ m\ m_E/r_E)$, $V_1 = (2\ C\ m_E/r_E)^{1/2}$ or $(2\ C\ \rho_E\ (4/3)\ \pi\ r_E^2)^{1/2}$
Average density of earth, $\rho_E \approx 5{,}500$ kg/m^3, $C = 6.67 \times 10^8$ Nm^2 kg^{-2}, r_E, radius of earth = 6,378,137 m, $V_1 = (2 \times 6.67 \times 10^{-11} \times 5.97 \times 10^{24}/6{,}378{,}137)^{1/2}$ = 11,200 m/sec = 25,000 mi per h.

1.5.1.2.5 Rate Form

Equation 1.33 can be used to express the change in state over a short time period δt (i.e., $\delta Q = \dot{Q}\delta t$ and $\delta W = \dot{W}\delta t$) to obtain the first law in rate form, namely,

$$\dot{Q} - \dot{W} = dE/dt. \quad (1.35)$$

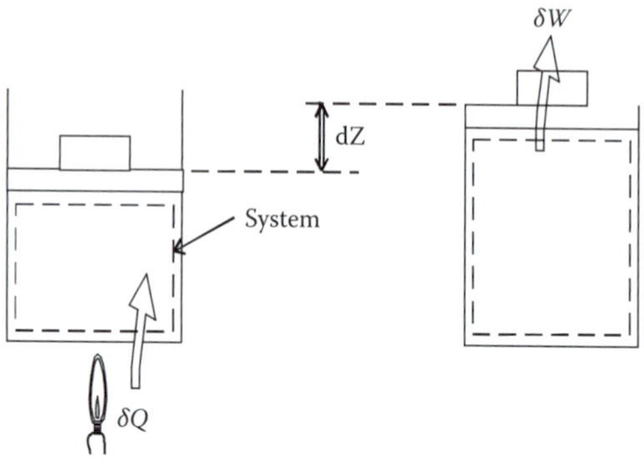

Figure 1.11 Illustration of the first law for a single process.

The rate of work $\dot{W}$ is the energy flux crossing the boundary in the form of work through distance dz as illustrated in Figure 1.11. The heat flux $\dot{Q}$ is a consequence of a temperature differential, and does not itself move the boundary but alters molecular velocity that manifests itself in the form of temperature. Equation 1.35 is useful for the analysis of compression, combustion, and expansion in automobile engines.

The laws of thermodynamics are constitutive equation independent, while calculations of $\dot{Q}$ and $\dot{W}$ may require constitutive relations. For example, Fourier heat conduction relation $\dot{Q} = -\lambda \nabla T$, a constitutive equation is employed for heat transfer.

1.5.1.2.6 First Law in Enthalpy Form

The enthalpy can replace the internal energy in Equation (1.31). The enthalpy of any substance is defined as

$$H = U + PV, \; h = u + Pv. \tag{1.36a,b}$$

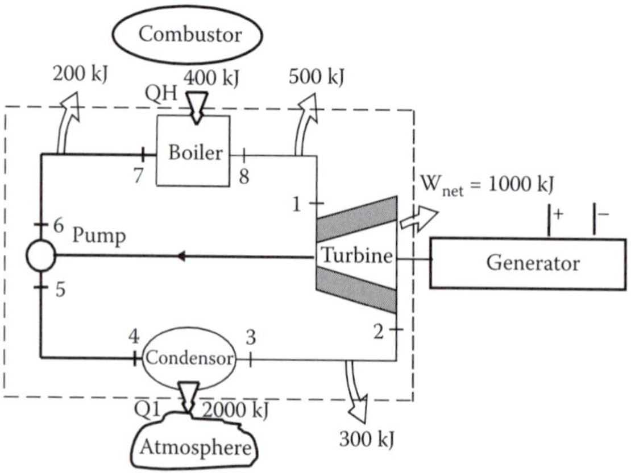

Figure 1.12 Illustration of the first law for a cyclic process.

For ideal gases PV = mRT and, hence, H = U + mRT.
For a quasi-equilibrium process $\delta W = PdV + \delta W_{other}$.
If $\delta W_{other} = 0$, Equation (1.31) becomes

$$\delta Q + VdP = dH. \tag{1.37}$$

For a quasi-equilibrium process at constant pressure,

$$\delta Q_P = dH. \tag{1.38}$$

Note that the first law is valid whether a process is reversible or not. However, once the equality $\delta W = P\ dV$ is accepted, a quasi-equilibrium process is also assumed.

1.5.1.2.7 Internal Energy and Enthalpy

The internal energy is the aggregate energy contained in the various molecular energy modes (translational, rotational, vibrational, and bond energy), which depend upon both the temperature and the intermolecular potential energy, the latter being a function of intermolecular spacing or volume. Therefore, u = u(T, v) or u = u(T, P), because the specific volume is a function of pressure.

Enthalpy, h, is a sum of internal energy and additional Pv energy which has the potential to cause a flow or perform flow work.

$$h = u + RT, \text{ ideal gas} \tag{1.39}$$

Once h(T) is known, u can be derived. Denoting the enthalpy of an ideal gas by h(T),

$$u = h(T) - Pv = h(T) - RT = u(T), \text{ ideal gas.} \tag{1.40}$$

1.5.1.2.8 Specific Heats at Constant Pressure and Volume

The specific heat at constant volume c_v is defined as,

$$c_v = (\partial u/\partial T)_v = \delta q_v/dT_v. \tag{1.41}$$

For an ideal gas, u = u(T) only and hence,

$$c_v = \frac{du}{dT}, \quad du = c_v dT. \tag{1.42}$$

If the matter contained in a piston–cylinder–weight assembly that ensures isobaric processes is likewise heated, the constant-pressure specific heat c_p is defined as

$$c_p = (\partial h/\partial T)_P = (\delta q/dT)_P \tag{1.43}$$

For an ideal gas, h = u(T) + RT = h(T) only and hence,

$$c_p = dh/dT, \quad dh = c_p dT \tag{1.44}$$

For any substance, the values of the properties c_p and c_v can be experimentally measured.

The ratio of the two specific heats $k = c_p/c_v$ is an important thermodynamic parameter. Typically, the value of k is 1.6 for monatomic gases (such as Ar, He, and Ne), 1.4 for diatomic gases (such as CO, H_2, N_2, and O_2), and 1.3 for triatomic gases (CO_2, SO_2, and H_2O).

For ideal gases, the subscript "v" could be interpreted as differentiation of u with respect to T, whereas the subscript "p" may be interpreted as differentiation of h with respect to T. Because $dh - du = (c_p - c_v)\, dT$ and $h = u + RT$, then $d(h - u) = (RT) = (c_p - c_v)\, dT$

$$c_{po}(T) = c_{vo}(T) + R. \tag{1.45}$$

For solids and liquids $c_p \approx c_v = c$. Table A.4A and Table A.4B, as well as Table A.5, in Appendix A present density and specific heats at specified temperature for solid, liquids, and gases, whereas Table A.6A and Table A.6B of Appendix A present relations for $c_{po}(T)$ for liquids, solids, and many gaseous fuels. The internal energy and enthalpy of an ideal gas can be calculated by integrating Equation 1.44, i.e.,

$$h = \int_{T_{ref}}^{T} c_{po}(T)\, dT, \quad \text{and} \quad u = \int_{T_{ref}}^{T} c_{vo}(T)\, dT,$$

where $h_{ref} = u_{ref} = 0$. Once either the enthalpy or internal energy is known, the other property can be calculated from the ideal gas relation $u = h - RT$. For instance, if $c_{po}(T)$ is specified, one can generate h and u tables for ideal gases. Table A.9 and Table A.11 make use of known c_{p0} (T) relations in evaluating h and u values of various ideal gases with the exception of carbon(s) (Table A.10). The term $h_T - h_{298}$ is called *thermal enthalpy* and is equal to $\int_{298}^{T} c_{po}\, dt$ in Table A.10 to Table A.30 of Appendix A with T_{ref} selected at 298K. Thermal enthalpy will be discussed further in Chapter 2.

1.5.2 Open Systems

In open systems (also known as a control volume (c.v.) system, based on the presumption that the volume is rigid), mass crosses the system boundary. In addition to heat and work interactions with the environment, interactions also occur through an exchange of constituent species between the system enclosure and its environment. Consequently, the mass contained within the system may change. Examples of open systems include combustors, turbines, etc., which have a rigid boundary, thereby implying a fixed c.v.

(as in Figure 1.13), or automobile engine cylinders in which the c.v. deforms during the various strokes.

An open-system energy conservation equation is equivalent to that for a closed system if the energy content of an appropriate fixed mass in the open system is temporally characterized using the Lagrangian method of analysis. However, the problem becomes complicated if the matter contains multiple components. It is customary to employ an Eulerian approach that fixes the c.v., and analyzes the mass entering and leaving it. We now present the Eulerian mass and energy conservation equations, and illustrate their use by analyzing various flow problems. At the end of the chapter, we will also develop simplified differential forms of these equations that are useful in problems involving chemically reacting flows.

1.5.2.1 Mass

1.5.2.1.1 Integrated Form

The mass conservation for a c.v. system (i.e., volume of system is fixed) with multiple inlets and exits is given as,

$$dm_{cv}/dt = \sum \dot{m}_i - \sum \dot{m}_e. \tag{1.46}$$

If the rigid boundary around a turbine is demarcated, the properties within the c.v. vary spatially from inlet to exit. Unlike closed systems, property gradients inherently exist as shown in Figure 1.13 because $P_A > P_B > P_C$ and $T_A > T_B > T_C$. Therefore, the mass within the c.v. can only be determined by considering a small elemental volume *dV*, and integrating therefrom over the entire turbine volume. Because the velocity distribution also varies spatially, the inlet and exit mass flow rates can be represented as

$$\dot{m} = -\int_{Ai} \rho \vec{V} \cdot d\vec{A}\,. \tag{1.47}$$

The negative sign in Equation 1.47 is due to the velocity vector of the entering mass that points toward the elemental area *dA*, whereas the area vector always points outward normal to the c.v. According to our previously stated convention, the mass entering the turbine must carry a positive sign. This is satisfied by providing a negative sign to the equation. Therefore, Equation (1.46) becomes

$$d/dt\left(\int_{cv} \rho dV\right) = -\oint \rho \vec{V} \cdot d\vec{A}. \tag{1.48}$$

The cyclic integration implies that the mass is tracked both in and out throughout the c.v. of the system.

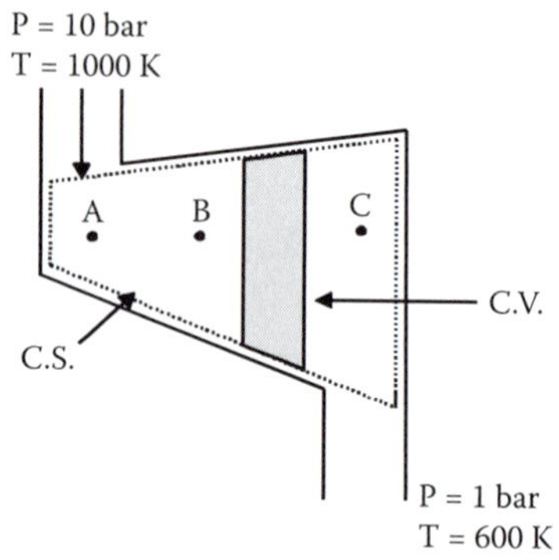

Figure 1.13 Nonuniform property within a control volume.

1.5.2.1.2 Differential Form

Applying the Gauss divergence theorem to the right-hand side of Equation 1.48,

$$-\oint \rho\vec{v}\cdot d\vec{A} = -\oint \vec{\nabla}\cdot\rho\vec{v}\, dV. \tag{1.49}$$

(The x-wise component of the right-hand side is $\{\partial(\rho v_x)/\partial y)\ dy(dxdz)\}$). If the control volume is time independent, i.e., it has rigid boundaries, Equation 1.48 may be written in the form

$$\int_{cv} (\partial\rho/\partial t + \vec{\nabla}\cdot\rho\vec{v})dV = 0. \tag{1.50}$$

Because the c.v. is arbitrarily defined, the mass conservation becomes,

$$\partial\rho/\partial t + \vec{\nabla}\cdot\rho\vec{v} = 0. \tag{1.51a}$$

For multicomponents we replace $\rho\vec{v}$ by $\Sigma\rho_K v_K$ and hence

$$\partial\rho/\partial t + \nabla\cdot\sum\rho_K v_K = 0 \tag{1.51b}$$

where v_K, absolute velocity of species k. (More details in Chapter 6.)

1.5.2.2 First Law of Thermodynamics or Energy Conservation Equation

1.5.2.2.1 Integrated Form

For an open system, we modify the first law of closed system (Equation 1.35), accounting for energy with inlet and exit flows:

$$\mathrm{dE_{cv}/dt} = \dot{Q}_{cv} - \dot{W}_{cv} + \dot{m}_i(\mathrm{h+ke+pe})_i - \dot{m}_e\,(\mathrm{h+ke+pe})_e \tag{1.52}$$

Compared to Equation 1.35, there are two modifications: (1) addition of energy transport via mass entering and leaving, and (2) the energy per unit mass of transport contains h (= u + Pv) rather than u because each unit mass has to perform Pv (flow or pumping work) to push the a mass of volume v in and out against the resisting pressures. Hence, flow work is added with u in the enthalpy term. The sum h + ke + pe is called *methalpy*. (*Note:* ke (kJ/kg) = $V^2/2000$, pe (*kJ/kg*) = g Z/1000 (SI); h_{st} = h + ke = h + $V^2/2$, called *stagnation enthalpy*.) Equation 1.52 simplified to Equation 1.35 with $\dot{m}_i = 0$, $\dot{m}_e = 0$.

The following example illustrates an application of the first law.

Example 3

Consider a spherical tank of radius 0.38 m (radius R). We wish to pack electric bulbs each of radius 0.01 m (radius a). The power to each bulb is adjusted such that the surface temperature of the tank is maintained at 37°C. The heat transfer coefficient is about 4.63 W/m² K. Assume steady state. Determine (a) heat loss from the tank ($\dot{Q}_L = h_H A\,(T - T_0)$) for T_0 = 25°C, (b) number of bulbs you can pack in the tank, (c) amount of electrical power required for each bulb so that the tank surface is always at 37°C, and (d) what are the answers to (a) through (c) if the tank radius is reduced to 0.19 m but the surface temperature is still maintained at 37°C and the bulb size is fixed?

Solution

(a) $\dot{Q} = h_H\ A\,(T - T_0) = 4.63 \times 4\,\pi \times 0.38^2 \times (37 - 25)\ 101$ W.

(b) Because the volume of each bulb V = 4/3 × π × a³, the number of bulbs R^3/a^3 = 54,872.

(c) Assuming steady state, $dE_{c.v}/dt = \dot{Q} - \dot{W} + \dot{m}_i\, e_{T,i} - \dot{m}_e e_{T,e} = 0$, $\dot{m}_i = \dot{m}_e = 0$ and $\dot{Q} - \dot{W} = 0$. Therefore, $\dot{W} = \dot{Q}$, and $\dot{W}/N = \dot{Q}/N$ = 101/54,872 = 0.00184 W per bulb.

(d) $\dot{Q}/N = h_H\ A(T - T_0)/(R^3/a^3) = h_H\ 4\,\pi\ (T - T_0)\,a^3/R = 0.0007/R$.

Therefore, by reducing the radius R, the electrical work per bulb increases, and the power per bulb doubles to 0.00368 W.

Remarks

(i) As an analogy, the cells in a human (cell dia ~ 10μm and cell wall thickness ~ 10 nm) or animal body can be thought of as replacing the bulbs in the above example. The electrical power can then be replaced by the slow metabolism or slow combustion of fuel (glucose, fat, and proteins). As the size of a species decreases (Figure 1.14a), there is a smaller number of cells (which decrease in proportion to the length scale R³), and the surface area decreases more slowly (proportional to R²). Thus, a larger amount of fuel metabolism is required in each cell. We can now obtain scaling groups. The heat loss from an organism $\dot{Q}_L = \dot{Q}_L''\ A = h_H A(T_b - T)$, where A denotes the organism body area. Assuming the heat transfer rate h_H to be constant, we note that $\dot{Q}_L \propto m_b^{2/3}$

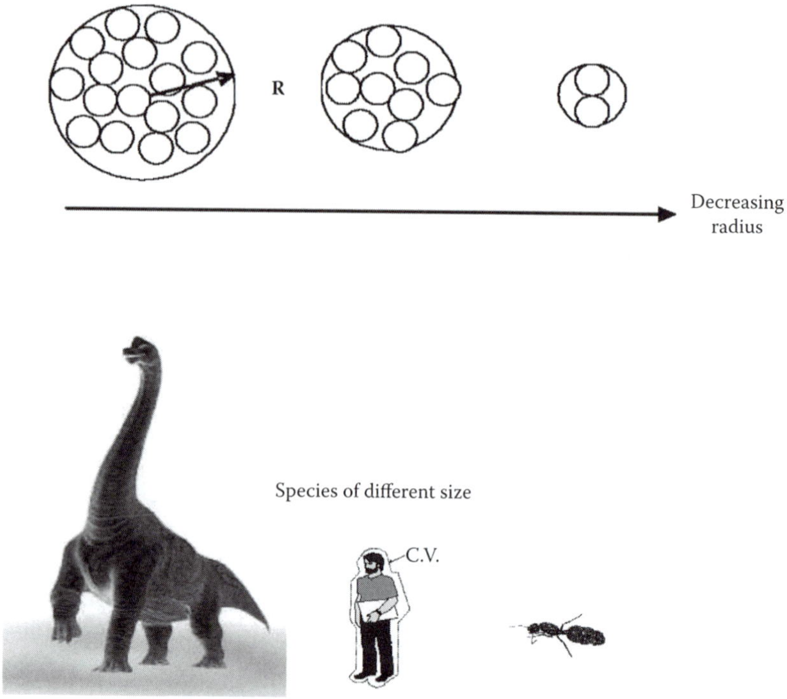

(a)

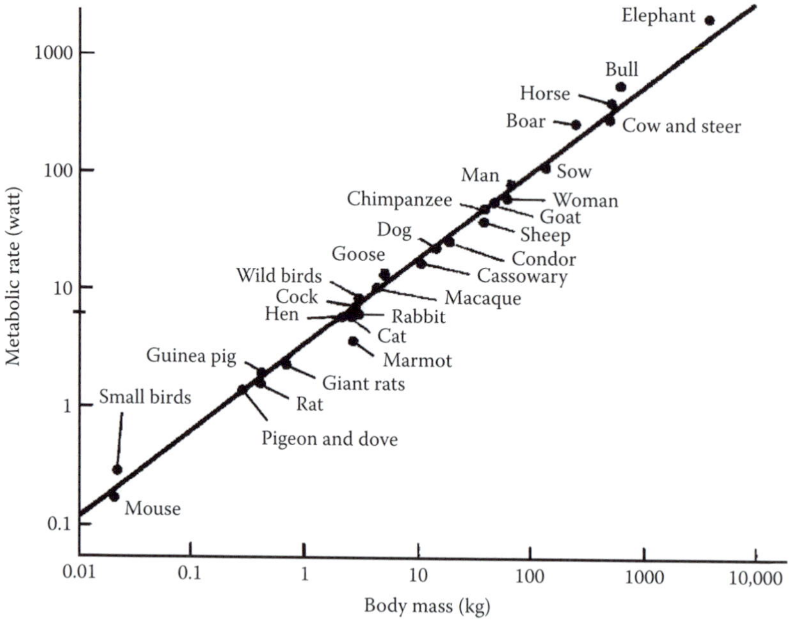

(b)

Figure 1.14 **(a) Metabolic rates and sizes of species, (b) metabolic rates of different species. (Adapted from Scaling: *Why Is Animal Size So Important*. K.S. Nielsen, Cambridge University Press, p 57, 1984. With permission.)**

(Euclidean geometrical scaling). Experiments yield that $\dot{Q}_L = 3.552\ m_b^{0.74}$. The metabolic rates of various species in relation to their body mass are illustrated in Figure 1.14b.

(ii) The metabolic rate during the human lifetime keeps varying, with the highest metabolic rate being that for a baby and the lowest being that for a relatively senior citizen.

(iii) The minimum metabolic rate required for maintaining bodily functions is of the order of 1 W. The open-system energy balance under steady state provides the relation $\dot{Q}_{c.v.} = \dot{m}_e h_e - \dot{m}_i h_i = \dot{m}(h_e - h_i)$.

(iv) If the body temperature rises (e.g., during fever), then $dE_{c.v.}/dt \neq 0$.

■ ■ ■

Energy accounting and conservation (single inlet and exit) with multiple components as in chemical reactions) is rewritten as,

$$dE_{cv}/dt = \dot{Q}_{cv} - \dot{W}_{cv} + \sum_k \dot{m}_{k,i}\ (h + ke + pe)_{k,i} - \sum_k \dot{m}_{k,e}\ (h + ke + pe)_{k,e} \tag{1.53}$$

where k refers to species k. In a combustor (e.g., boilers, in which fuel burns over a large volume), the terms m_{cv} and $E_{c.v.}$ must be evaluated for the entire combustor in which the properties are spatially nonuniform. Therefore, the specific energy and specific volume vary within the c.v. (e.g., regions A, B, C, etc., in Figure 1.13 are at different T and P)

$$m_{cv} = \int m dV/v, \quad E_{cv} = \int (u + ke + pe)\, dV/v. \tag{1.54}$$

1.5.2.2.2 Integral Form

The methalpy or total enthalpy $e_T = h + ke + pe$ crossing the c.v. is

$$\dot{m}_k\ e_T = -\int_K \rho_k e_T \vec{v}_k \cdot d\vec{A}, \tag{1.55}$$

where $\vec{v}_k$ is the absolute velocity of each species k, $\vec{v}_k \cdot d\vec{A} < 0$ for the incoming flow, and $\vec{v}_k \cdot d\vec{A} > 0$ for the exiting flow. A negative sign is added to the value for consistency with our sign convention for the mass inflow and outflow,

$$\dot{W}_{cv} = \int_{cv} \dot{w}'''_{cv} dV, \tag{1.56}$$

where $\dot{w}'''_{cv}$ denotes the work done per unit volume per unit time. The heat crossing the system boundary $\dot{Q}_{cv}$ is given by

$$\dot{Q}_{cv} = -\int_{cv} \dot{q}'' \cdot d\vec{A}, \tag{1.57}$$

where $\vec{q}''$ is the energy flux solely due to temperature gradients (e.g., $\vec{q}'' = -\lambda \vec{\nabla} T$). The negative sign associated with Equation 1.57 is due to heat input being positive.

Using the relationships of Equation 1.54 to Equation 1.57, the integral form of the energy conservation equation for a rigid c.v. assumes the form,

$$d/dt\left(\int_{cv} \rho e dV\right) = \oint \vec{q}'' \cdot d\vec{A} - \int \dot{w}_{cv}''' dV - \oint \rho_k e_T \vec{v}_k \cdot d\vec{A} \qquad (1.58)$$

where the term d/dt(∫ρedV) in Equation 1.37 denotes the rate of change of energy in the entire c.v.

1.5.2.2.3 Differential Form

Applying the Gauss divergence theorem to Equation 1.37 and converting the surface integral into a volume integral, and simplifying the result,

$$\partial(\rho e)/\partial t + \vec{\nabla}\cdot(\rho e_T \vec{v}) = -\vec{\nabla}\cdot\dot{\vec{Q}}'' - \dot{w}_{cv}'''. \qquad (1.59)$$

If the heat transfer in the c.v. occurs purely through conduction, and the Fourier law is applied, no work is delivered, and the kinetic and potential energies are negligible (namely, e_T = h, and e = u), Equation 1.59 may be expressed in the form

$$\partial(\rho u)/\partial t + \vec{\nabla}\cdot(\rho\vec{v}h) = \vec{\nabla}\cdot(\lambda\vec{\nabla}T). \qquad (1.60)$$

Using the relation u = h − P/ρ, Equation 1.60 may be written as

$$\partial(\rho h)/\partial t + \vec{\nabla}\cdot(\rho\vec{v}h) = \partial P/\partial t + \vec{\nabla}\cdot(\lambda\vec{\nabla}T). \qquad (1.61)$$

These differential forms of the energy conservation equation are commonly employed in analyses involving heat transfer, combustion, and fluid mechanics.

For multicomponent systems, ρ h = Σ ρ_k h_k, $\rho v h = \Sigma\, \rho_k v_k\, h_k$

$$\partial\left(\sum \rho_k h_k\right)/\partial t + \vec{\nabla}\cdot\left(\sum \rho_k \vec{v}_k h_k\right) = \partial P/\partial t + \vec{\nabla}\cdot(\lambda\vec{\nabla}T) \qquad (1.62)$$

where λ is for a mixture of fluids and v_k, the velocity of species k, which is different from the average velocity of flow v.

1.6 The Second Law of Thermodynamics

1.6.1 Introduction

The simplest statement of the second law is that *the heat input for network delivery in a cyclic process requires heat to be rejected.* It can also be stated informally as: *the efficiency of a heat engine is less than unity.*

The efficiency η of a thermodynamic cycle is defined as the ratio of the work output to the thermal input, i.e.,

$$\eta = \text{sought/bought} = W_{cycle}/Q_{in}. \tag{1.63}$$

For a cyclical process, the first law states that

$$\oint \delta Q = Q_{in} - Q_{out} = W_{cycle} = \oint \delta W. \tag{1.64}$$

Using Equation 1.63 and Equation 1.64 the informal statement follows,

$$\eta = (1 - Q_{out}/Q_{in}) < 1. \tag{1.65}$$

The first law conserves energy, and the second law prohibits the complete conversion of heat into work during a cyclical process (e.g., automobiles).

1.6.2 *Entropy and Second Law*

Entropy is a measure of the number of states in which energy is stored. The calculation of entropy requires knowledge of energy states of molecules. Now, using classical thermodynamics, a mathematical definition will be given for estimating the entropy in terms of macroscopic properties.

1.6.2.1 Mathematical Definition

If we take the first law of thermodynamics given by Equation 1.31, apply it to a reversible process ($\delta W = PdV$), divide by T, and integrate for a cyclic process involving ideal gases ($du = c_v\, dT$, $Pv = RT$), then $\oint (\delta Q/T)_{int,rev} = 0$. It can be shown that the same statement is valid for any substance used as a medium in a reversible cyclic process. Because the cyclic integral for any property is also zero, e.g., $\oint du = \oint dh = \oint dT = \oint dP = 0$, we can define $(\delta Q/T)_{rev}$ in terms of the entropy, which is a property. The subscript "int" will be omitted here onward for the sake of convenience. Therefore,

$$(\delta Q/T)_{rev} = dS, \quad \text{kJ/kg} \quad \text{or} \quad \text{Btu/R} \tag{1.66}$$

The absolute entropy can be expressed in units of kJ/K or Btu/°R. On a unit-mass basis $(\delta q/T)_{rev}$ ds (in units of {kJ/(Kg K)} or {Btu/(lb°R)}). Similarly, on a mole basis, $(\delta \bar{q}/T)_{rev}$ $d\bar{s}$ (expressed in units of kJ/{kmol K} or Btu{lb-mole°R}).

1.6.2.2 Relation between dS, Q, and T during an Irreversible Process

For an infinitesimal irreversible process, the Clausis Inequality modified Equation (1.66) as

$$dS > \delta Q/T \tag{1.67}$$

and if temperature is not uniform within the system, the system boundary temperature T_b is used.

$$dS > \delta Q/T_b. \tag{1.68}$$

Thus, the entropy change between two equilibrium states 1 and 2 exceeds the entropy change induced by the heat transfer process or the transit entropy alone because of irreversibility. The entropy transfer due to heat transfer across a boundary $\delta Q/T$ is termed as the *transit entropy* (abbreviated as tentropy). It is a not a property. The transit entropy, i.e., $\delta Q/T$, equals 0 only if $Q = 0$.

1.6.2.3 Entropy Balance Equation for a Closed System

1.6.2.3.1 Infinitesimal Form

For a closed system, the differential relation of Equation 1.68 may be rewritten in balanced form by explicitly including the infinitesimal entropy generation $\delta\sigma$, i.e.,

$$dS = \delta Q/T_b + \delta\sigma \tag{1.69}$$

or

$$T_b\, dS = \delta Q + T_b\, \delta\sigma, \quad \delta\sigma \geq 0. \tag{1.70}$$

This relation is also known as the Gibbs equation. The entropy generation $\delta\sigma$ is greater than zero for internally irreversible processes and is zero for internally reversible processes.

Even if the boundary temperature is uniform and at the same temperature as the system constituents, thereby indicating thermal reversibility, other irreversibilities, such as those due to chemical reactions (e.g., CO and air mixture in a room), can contribute to σ.

1.6.2.3.2 Integrated Form

The integrated form of Equation 1.69 is

$$S_2 - S_1 = \int_1^2 \delta Q / T_b + \sigma\,. \tag{1.71}$$

If a process satisfies Equation 1.71 (e.g., $S_2 > S_1$, in an adiabalic process with $\sigma_{12} > 0$), there is no assurance that the end state 2 is realized. As will be seen later, for a process to occur, the condition $\delta\sigma \geq 0$ must be satisfied during each elemental part of the process. Therefore, Equation 1.69 is more meaningful than Equation 1.71.

In order to evaluate $\delta\sigma$, dS and $\delta Q/T_b$ must be evaluated during every elemental step.

1.6.2.3.3 Rate Form

The time derivative of Equation 1.69 returns the rate form of that relation, i.e.,

$$dS/dt = \dot{Q}/T_b + \dot{\sigma}. \quad (1.72)$$

1.6.2.3.4 Entropy Evaluation

For any given state change (u_1, v_1) to (u_2, v_2) of a reversible or irreversible process, the first law yields $\delta q - \delta w = du$; for the same du, one can connect a reversible path and hence get $\delta q_{rev} - \delta w_{rev} = du$. For this path the second law yields $\delta q_{rev} = Tds$ and $\delta w_{rev} = Pdv$. Thus, using the first and second laws along a reversible path for unit mass, $T\,ds - Pdv = du$, and solving for ds,

$$ds = du/T + Pdv/T. \quad (1.73)$$

Once integrated, $s_2 - s_1$ will depend only on u_1, u_2, v_1, and v_2, but not on the reversible path used in reaching state 2 from state 1. Thus, s is a property and is independent of path. For a closed system of fixed mass, Equation (1.73) states that

$$s = s\,(u, v). \quad (1.74)$$

Because $du = dh - d\,(Pv)$, Equation 1.73 assumes the form

$$ds = dh/T - vdP/T. \quad (1.75)$$

It is apparent from Equation 1.75 that $s = s(h, P)$.

1.6.2.3.4.1 Ideal Gases

— Substituting for the enthalpy $dh = c_{p0}\,(T)\,dT$, Equation 1.75 may be written in the form

$$ds = c_{po}\,dT/T - R\,dP/P. \quad (1.76)$$

Constant specific heats. Integrating Equation 1.77 from (T_{ref}, P_{ref}) to (T,P)

$$s(T, P) - s(T_{ref}, P_{ref}) = c_{po}\,\ln(T/T_{ref}) - R\,\ln(P/P_{ref}). \quad (1.77)$$

Selecting $P_{ref} = 1$ atm and letting $s(T_{ref}, 1) = 0$, we have

$$s(T, P) = c_{p0}\,\ln\,(T/T_{ref}) - R\,\ln\,(P(atm)/1(atm)). \quad (1.78)$$

Selecting an arbitrary value for T_{ref}, and applying Equation 1.78 at states 1 and 2,

$$s(T_2, P_2) - s(T_1, P_1) = c_{po}\,\ln(T_2/T_1) - R\,\ln(P_2/P_1). \quad (1.79)$$

Properties for c_{p0} are tabulated in Table A.4A, Table A.4B, and Table A.5 of Appendix A. In isentropic processes (e.g., adiabatic reversible compression of air within the cylinder of an automobile engine), s remains constant and hence Equation 1.79 yields

$$P_2/P_1 = \{T_2/T_1\}^{(k/(k-1))} \quad \text{or} \quad P/T^{(k/(k-1))} = \text{constant} \tag{1.80a}$$

where $k = c_p/c_v$. Using the ideal gas law $Pv = RT$, one can eliminate T and obtain

$$P_2/P_1 = \{v_1/v_2\}^k \quad \text{or} \quad P\,v^k = \text{constant} \tag{1.80b}$$

Variable specific heats. Consider an ideal gas that changes state from (T_{ref}, P_{ref}) to (T, P). Integrating Equation 1.76 and setting $s(T_{ref}, P_{ref}) = 0$, we have,

$$s(T, P) = \int_{T_{ref}}^{T} (c_{p0}(T)/T)dT - R\ln(P/P_{ref}).$$

For ideal gases, the first term on the right is a function of temperature alone. Let

$$s^0(T) = \int_{T_{ref}}^{T} (c_{p0}(T)/T)dT. \tag{1.81}$$

With known data for the specific heat $c_{p0}(T)$ (Table A.6B of Appendix A), Equation 1.81 can be readily integrated. In general, tables listing $s^0(T)$ assume that $T_{ref} = 0$ K (Table A.9 to Table A.30 of Appendix A). Therefore,

$$s(T, P) = s^0(T) - R\ln(P/1), \tag{1.82}$$

where the pressure is expressed in units of atm or bar. If P = 1 atm (≈1 bar), s(T, 1) s°(T), which is the entropy of an ideal gas at a pressure of 1 bar (≈ 1 atm) and a temperature *T*. The second term on the right-hand side of Equation 1.82 is a pressure correction because P is different from 1 bar. Applying Equation 1.82 to states 1 and 2,

$$s(T_2, P_2) - s(T_1, P_1) = s^0(T_2) - s^0(T_1) - R\ln(P_2/P_1). \tag{1.83}$$

In isentropic processes, Equation 1.83 yields $(P_2/P_1) = \exp[\{s^0(T_2) - s^0(T_1)\}/R] = \exp\{s^0(T_2)/R\}/\exp(s^0(T_1)\}/R] = p_{r2}(T_2)/p_{r1}(T_1)$, where $p_r(T)$, which is proportional to $\exp(s^0/R)$, is called *relative pressure*. Using the ideal gas law, one can eliminate *P* and obtain $v_2/v_1 = \{v_{r2}(T_2)/v_{r1}(T_1)\}$, where $\{v_{r2}/v_{r1}\} = \{(p_{r2}/T_2)/(p_{r1}/T_1)\}$; v_r is called the *relative volume*. Relative volumes are useful for closed systems, whereas relative pressures are more useful in open systems; however, they can be applied to any system. Table A.9 of Appendix A tabulates the properties $p_r(T)$ and $v_r(T)$ for air.

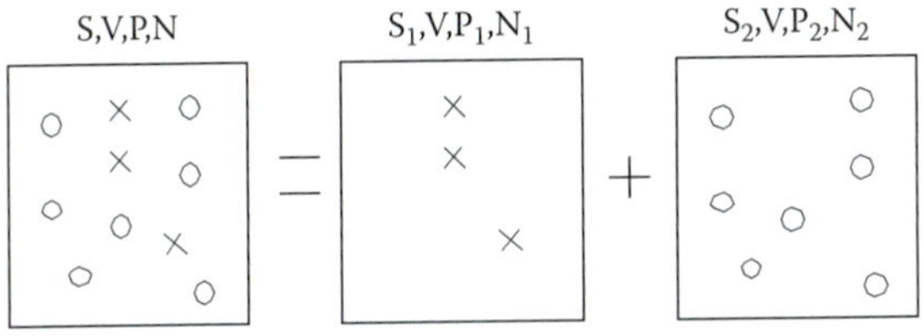

Figure 1.15 Illustration of the Gibbs–Dalton law.

1.6.2.3.4.2 Solids and Liquids — If solids and liquids are assumed to be incompressible, then

$$ds = du/T + Pdv/T = c\ dT/T + 0.$$

Integrating,

$$s_2 - s_1 = c \ln (T_2/T_1). \tag{1.84}$$

For example, if 1 kg of coffee is cooling from 340 K (T_1) to 330 K (T_2), $s_2 - s_1 = 4.184 \ln (330/340) = -0.125$ kJ/kg.

1.6.2.3.5 Properties of a Mixture of Ideal Gases

Gibbs–Dalton law. The application of the Gibbs–Dalton law to characterize a multicomponent gaseous mixture is illustrated in Figure 1.15, in which the mixture contains two kinds of species (x, o). Two components species (x, o) are hypothetically separated, and the component pressures p_1 and p_2 are obtained. The component pressure p_k is then determined as though component k alone occupies the entire volume at the mixture temperature. Thereafter, using the component pressures, the entropy is evaluated, i.e.,

$$S\ (T, P, N) = \sum S_k\ (T, p_k, N_k) = N_k \bar{s}_k(T, p_k), \tag{1.85}$$

where p_k, the component pressure, is the same as the conventional partial pressure for ideal gases, i.e., $p_k = X_k P$. However, for real gas mixtures $p_k \neq X_k P$.

Similarly for ideal gas mixture

$$H = \sum N_K \bar{h}_k(T)$$

$$G = H - TS$$

$$= \sum N_K \bar{h}_k(T) - T \sum N_K s_K(T, p_k)$$

$$= \sum N_k \bar{g}_k(T, p_k)$$

where

$$\bar{g}_k(T, p_k) = \bar{h}_k(T) - T\bar{s}(T, p_k)$$

Example 4

An automobile engine contains a 0.1-kmol mixture consisting of 40% CO_2 and 60% N_2 at 10 bar and 1000 K (state 1). The mixture is heated to 11 bar and 1200 K (state 2) using a thermal energy reservoir (TER) maintained at 1300 K. The work output from the assembly is 65.3 kJ. Evaluate the entropy change $S_2 - S_1$ and σ_{12} for the following cases: (a) the boundary temperature T_b equals that of the gas mixture and (b) T_b is fixed and equals 1300 K (TER temperature) during heat up.

Solution

$$S = \sum_k N_k \bar{s}_k(T, p_k), \tag{A}$$

$$S_1 = \{N_{CO_2}\, \bar{s}_{CO_2}(T, P_{CO_2}) + N_{N_2}\, \bar{s}_{N_2}(T, p_{N_2})\}_1, \tag{B}$$

$$S_2 = \{N_{CO_2}\bar{s}_{CO_2}(T, P_{CO_2}) + N_{N_2}\, \bar{s}_{N_2}(T, p_{N_2})\}_2, \tag{C}$$

where $N_{CO_2} = 0.4 \times 0.1 = 0.04$ kmol, and $N_{N_2} = 0.6 \times 0.1 = 0.06$ kmol.

$$\text{Now, } \bar{s}_{CO2}(T, p_k) = \bar{s}^0_{CO_2}(T) - \bar{R}\ln(P_{CO_2}/1), \quad \text{where} \tag{D}$$

$$(p_{CO_2})_1 = 0.4 \times 10 = 4 \text{ bar}, \quad (p_{CO_2})_2 = 0.4 \times 11 = 4.4 \text{ bar}$$

$$\text{and } (p_{N_2})_1 = 0.6 \times 10 = 6 \text{ bar}, \quad (p_{N_2})_2 = 0.6 \times 11 = 6.6 \text{ bar}.$$

Therefore, at conditions 1 and 2, respectively,
$\bar{s}_{CO_2}(1200\text{K, } 4.4 \text{ bar}) = \bar{s}^0_{CO_2}(1200 \text{ K}) - \bar{R}\ln(4.4 \div 1) = 234.1 - 8.314 \times \ln(4.4 \div 1) = 221.8$ kJ kmol^{-1} K^{-1}, and

$$\bar{s}_{CO_2}(1000\text{K, } 4 \text{ bar}) = \bar{s}^0_{CO_2}(1000 \text{ K}) - \bar{R}\ln(4 \div 1) = 216.6 \text{ kJ kmol}^{-1} \text{ K}^{-1}.$$

Likewise,

$$\bar{s}_{N_2}(1200\text{K, } 6.6 \text{ bar})$$

$$= 263.6 \text{ kJ kmol}^{-1} K^{-1}, \quad \text{and} \quad \bar{s}_{N_2}(1000\text{K, } 6 \text{ bar})$$

$$= 254.3 \text{ kJ kmol}^{-1} \text{ K}^{-1}.$$

Using Equation B and Equation C

$S_1 = 0.04 \times 216.6 + 0.06 \times 254.3 = 23.92$ kJ K^{-1}, $S_2 = 0.04 \times 221.8 + 0.06 \times 263.6 = 24.69$ kJ K^{-1},

(a) $$S_2 - S_1 = 24.69 - 23.92 = 0.77 \text{ kJ K}^{-1},$$

$$S_2 - S_1 - Q_{12}/T_b = \sigma_{12}. \quad \text{(E)}$$

For this case $\sigma_{12} = 0$ because there is no thermal gradient; $Q_{12}/T_b = S_2 - S_1 = 0.77$ kJ K^{-1}.

(b) Applying the first law, $Q_{12} = U_2 - U_1 + W_{12}$. Therefore,

$U_2 = 0.04 \times 43871 + 0.06 \times 26799 = 3362.8$ kJ, $U_1 = 0.04 \times 34455 + 0.06 \times 21815 = 2687.1$ kJ,

and

$$Q_{12} = 3362.8 - 2687.1 + 65.3 = 741 \text{ kJ}. \quad \text{(F)}$$

Using Equation F in Equation E

$$\sigma_{12} = 0.77 - 741/1300 = 0.2 \text{ kJ K}^{-1}.$$

1.6.2.4 Entropy Balance Equation for an Open System

We will now formulate the entropy balance for an open system in an Eulerian reference frame.

1.6.2.4.1 General Expression

Heat transfer can occur across the c.v. boundary. In general, the boundary temperature at its inlet is different from the corresponding temperature at the exit, and the heat transfer rate δQ may vary from the inlet to exit. For the sake of analysis, we divide the boundary into sections such that at any section j the boundary temperature is $T_{b,j}$ and the heat transfer rate across the boundary is $\dot{Q}_j$. The term $\dot{Q}/T_b$ in Equation 1.72 is given as

$$\dot{Q}/T_b = \sum_j \dot{Q}_j/T_{b,j}. \quad (1.86)$$

The entropy balance equation for an open system is almost the same as for a closed system except that entropy added and removed with mass flows must be accounted for. Modifying Equation 1.72,

$$dS_{c.v.}/dt = \dot{m}_i s_i - \dot{m}_e s_e + \sum_j \dot{Q}_j/T_{b,j} + \dot{\sigma}_{cv}. \quad (1.87)$$

Equation 1.87 may be interpreted as follows: The rate of entropy accumulation in the cv entropy inflow through advection entropy outflow + change in the transit entropy through heat transfer + entropy generated due to irreversible processes (e.g., entropy generation due to gradients within the system). All chemical reactions are irreversible, and hence entropy is generated (i.e.,$\dot{\sigma}_{cv} > 0$). This concept is extremely important in that the direction of reaction i.e., whether CO + 1/2 O $\rightarrow CO_2$ or $CO_2 \rightarrow$ CO + 1/2 O_2 is determined by the condition that $\dot{\sigma}_{cv} > 0$. The applications of the entropy generation concept range from predicting the life spans of biological species [Annamalai and Puri, 2000] to estimating the damping coefficient of materials when subjected to bending [Bishop and Kinra, 1995, 1996].

Equation 1.87 may be rewritten in the infinitesimal form (over a period "δt")

$$dS_{c.v.} = dm_i\ s_i - dm_e\ s_e + \sum_j \delta Q_j/T_{b,j} + d\sigma. \tag{1.88}$$

For a closed system, Equation 1.88 reduces to Equation 1.69. If the process is reversible, Equation 1.87 becomes

$$dS_{c.v.}/dt = \dot{m}_i s_i - \dot{m}_e s_e + \sum_j \dot{Q}_j/T_{b,j}. \tag{1.89}$$

Under steady state $\dot{m}_i = \dot{m}_e = \dot{m}$ and dividing by $\dot{m}$,

$$s_i - s_e + \sum_{q_j} /T_{bj} + \sigma_m = 0 \quad or \quad s_e - s_i = \sum_{q_j} /T_{bj} + \sigma_m, \tag{1.90}$$

where $\sigma_m = \dot{\sigma}_{cv}/\dot{m}$ is the entropy generated per unit mass. In the case of a single inlet and exit but for a substance containing multiple components, the relevant form of Equation 1.87 is

$$dS_{c.v.}/dt = \sum_k \dot{m}_{k,i}\ \bar{s}_{k,i} - \sum_k \dot{m}_{k,e}\ \bar{s}_{k,e} + \sum_j \dot{Q}_j/T_{b,j} + \dot{\sigma}_{cv}. \tag{1.91}$$

where$\bar{s}_k$ denotes the entropy of the k-th component in the mixture. For ideal gas mixtures $\bar{s}_k(T, P, X_k) = \bar{s}_k^0(T) - \bar{R}\ \ln(PX_k/1)$ if P is in bar. Equation 1.91 is extremely useful in evaluating entropy generation in reacting systems.

1.6.3 Entropy Balance in Integral and Differential Form

1.6.3.1 Integral Form

Following same procedure outlined for mass and energy conservation equations in differential form, Equation 1.87 may be expressed in integral form as

$$d/dt\left(\int_{c.v.} \rho s dV\right) + \oint \rho \vec{V} s \cdot d\vec{A} = -\oint (\vec{\dot{Q}}''/T) \cdot d\vec{A} + \int_{c.v.} \dot{\sigma}''' dV. \tag{1.92}$$

1.6.3.2 Differential Form

The entropy balance equation can be written in a differential form to evaluate the entropy generation in a c.v. Applying the Gauss divergence theorem to Equation 1.92, we obtain

$$d/dt\,(\rho s) + \vec{\nabla} \cdot \rho \vec{V} \vec{s} = -\vec{\nabla} \cdot (\dot{\vec{Q}}''/T) + \dot{\sigma}'''_{cv}. \tag{1.93}$$

Employing the mass conservation equation, Equation 1.93 can be simplified as

$$\rho ds/dt + \rho \vec{V} \cdot \vec{\nabla} s = -\vec{\nabla} \cdot (\dot{\vec{Q}}''/T) + \dot{\sigma}'''_{cv}. \tag{1.94}$$

Recall that the Fourier law is represented by the relation $\dot{\vec{Q}}'' = -\lambda \vec{\nabla} T$.

The heat flux vector has a positive direction with respect to x when heat flows from a higher temperature to a lower temperature, applying the Fourier law to Equation 1.94,

$$\rho \partial s/\partial t + \rho \vec{V} . \nabla s = \vec{\nabla} \cdot (\lambda \vec{\nabla} T/T) + \dot{\sigma}'''_{cv}. \tag{1.95}$$

1.6.4 Combined First and Second Laws

1.6.4.1 Fixed-Mass System

Using the first law, $\delta Q - \delta W = dU$, to determine δQ and inserting in the second law (Equation 1.69) for a fixed-mass system, one obtains

$$dU - T_b\, dS + \delta W = -T_b\, \delta\sigma \tag{1.96}$$

If $T_b = T$, $\delta W = P\, dV$

$$dU - T\, dS + P\, dV = -T\, \delta\sigma, \quad \delta\sigma > 0 \tag{1.97}$$

Defining $H = U + PV$ the Equation 1.99 can be written in the following form

$$dH - T\, dS - V\, dP = -T\, \delta\sigma \tag{1.98}$$

Using the definition of Helmholtz (A) and Gibbs (G) functions

$$A = U - TS \tag{1.99}$$

$$G = H - TS \tag{1.100}$$

and differentiating Equation 1.99 and Equation 1.100 and using Equation 1.97 and Equation 1.98 in the result, one obtains the following

$$dA + S\,dT + PdV = -T\,\delta\sigma \tag{1.101}$$

$$da + s\,dT + Pdv = -T\,\delta\sigma_m \tag{1.102}$$

$$dG + S\,dT - VdP = -T\,\delta\sigma \tag{1.103}$$

$$dg + s\,dT - vdP = -T\,\delta\sigma_m \tag{1.104}$$

where A denotes the Helmholtz function or free energy (A = U − TS, kJ) and "a" denotes the specific Helmholtz function or free energy (a = u− Ts, kJ/kg). The Helmholtz function "a" of a unit mass is a measure of the potential of a system having internal energy u and entropy s to perform optimum work in a closed system. G denotes the Gibbs function or Gibbs free energy. On a unit-mass basis, g is also referred to as the chemical potential of a single component and is commonly used during discussions of chemical reactions. The Gibbs function g of a unit mass is a measure of the potential of a system having enthalpy h and entropy s to perform optimum work in a steady-state steady-flow reactor. It is seen that if irreversible chemical reactions occur within a closed system, then $\delta\sigma > 0$, and if it occurs at constant T and P, then $dG_{T,P} < 0$ for a fixed mass undergoing an irreversible process (e.g., chemical reactions). For multicomponent systems, $H = \Sigma N_K \bar{h}_K(T)$, $S = \Sigma N_K \bar{s}_k(T, p_K)$, $G = \Sigma N_K \bar{g}_K(T, p_K)$. The differential of G at fixed T and P must ratify

$$d\left\{\sum N_K \bar{g}_K(T, p_K)\right\}_{T,P} < 0, \tag{1.105}$$

when chemical reactions occur irreversibly.

Example 5

(a) Determine the chemical potential of pure O_2 ($\bar{g}_{O2} = \mu_{O2}$) at T = 2000 K and P = 6 bar, and (b) O_2 present in a gaseous mixture at T = 2000 K and P = 6 bar, and $X_{O2} = 0.3$, assuming the mixture to behave as an ideal gas.

Solution

(a) $\mu = \bar{g}^0$ $(\bar{h} - T\bar{s})$, where for ideal gases $\bar{s} = \bar{s}^0 - R \ln(P/Pref)$. Using Table A.26 of Appendix A,

$\bar{g}_{O2} = \mu_{O_2} = 67881$ kJ kmol^{-1} − 2000 K × (268.655 kJ kmol^{-1} K^{-1} − 8.314 kJ kmol^{-1} K^{-1}× ln(6 bar ÷1 bar)) = −439,636 kJ kmol^{-1}.

(b) For an ideal gas mixture, according to the Gibbs–Dalton law,
$\bar{s}_k(T, p_k) = \bar{s}_k^0\ (T) - \ln(p_k/Pref) = (268.655\ kJ\ kmol^{-1}\ K^{-1} - 8.314\ kJ\ kmol^{-1}\ K^{-1}) \times \ln((6\ bar \times 0.3) \div 1\ bar) = 272.747\ kJ\ kmol^{-1}\ K^{-1}$.
Because $\bar{g}_k(T, p_K) = \mu_k = (\bar{h}_k - T\bar{s}_k), \mu_{O_2} = -477{,}613$ kJ (kmol O_2 in the mixture)$^{-1}$.

Remarks

The chemical potential of O_2 decreases as its concentration is reduced from 100% (pure gas) to 30%.

You will see later that chemical potential plays a major role in determining the direction of chemical reactions (Chapter 2 and Chapter 3) and of mass transfer (Chapter 6), just as temperature determines the direction of heat transfer.

1.7 Summary

An introduction to combustion and a brief review of other thermodynamic concepts have been presented in Chapter 1. Species, mass, and energy conservation equations were briefly summarized and were be used to solve simple combustion problems. We were apply these thermodynamic concepts in Chapter 2 and Chapter 3 to learn about directions of reactions.

Chapter 2

Stoichiometry and Thermochemistry of Reacting Systems

2.1 Introduction

Chemical reactions result in a rearrangement of atoms among molecules. Examples of chemical reactions include charcoal burning in barbecue grills, photosynthesis, metabolism, combustion, petroleum refining, and plastics manufacture. Combustion occurs because of rapid exothermic oxidation reactions. In this chapter, we will answer questions on topics such as how much air is required to burn a given quantity of fuel either on volume basis or mass basis, what the equivalence or stoichiometric ratio is, why a cloud of vapor might appear near a chimney stack, and the maximum temperature of an oxyacetylene welding torch, of the flame in a barbecue grill, or of gasoline burning in an automobile engine. Energy cannot be extracted from a system that is constrained unless the thermodynamic constraint is removed. Fossil fuels such as coal, oil, and natural gas are stored below the earth's surface and are in thermal, mechanical, and chemical equilibrium. If the fuels are allowed physical contact with air, i.e., if the chemical constraint is removed, it is possible to release chemical energy. Thus, reactions at finite rate are always irreversible, and entropy is generated.

2.2 Overall Reactions

2.2.1 Stoichiometric Equation with O_2

A stoichiometric or a theoretical reaction results in the complete combustion of fuel and without leaving any oxygen in products as illustrated below. Consider the reaction

$$CH_4 + aO_2 \rightarrow bCO_2 + cH_2O \tag{2.1}$$

involving 1 kmol of fuel and "a" kmol of oxygen. The species on the left-hand side of Equation 2.1 are usually called *reactants* (which react and are consumed during the overall chemical reaction), and those on the right-hand side are termed *products* (which are produced as a result of the chemical reaction). In the case of a steady-flow reactor, the input and output streams contain the following quantities in kilomoles of the elements (C, H, and O).

	Input	=	*Output*
C	1	=	b
H	4	=	2c
O	2a	=	2b + c

We deduce from the carbon atom balance that b = 1, from the H atom balance that c = 2, and from the O atom balance that a = (2b + c)/2. The coefficients a, b, and c are called stoichiometric coefficients. Consequently, Equation 2.1 can be written in the form of the stoichiometric reaction

$$CH_4 + 2O_2 \rightarrow CO_2 + 2H_2O \tag{2.2}$$

More generally, chemical reactions can be expressed in the form

$$\upsilon_1' \; S_1' + \upsilon_2' \; S_2 + \ldots \rightarrow \upsilon_1'' \; S_3 + \upsilon_2'' \; S_4 + \upsilon_3'' \; S_5 + \upsilon_4'' \; S_6 + \ldots \tag{2.3}$$

where υ_1', υ_2', etc., represent the stoichiometric coefficients of the reactant species and υ_1'', υ_2'', etc., are the corresponding coefficients for the product species. The symbols S_1, S_2, ..., denote the species symbols. In the context of Equation 2.3, for Equation 2.2, $\upsilon_1' = 1$, $S_1 = CH_4$, $\upsilon_2' = 2$, $S_2 = O_2$, $\upsilon_1'' = \upsilon_2'' = 0$, $\upsilon_3'' = 1$, $S_3 = CO_2$, $\upsilon_4'' = 2$, $S_4 = H_2O$, and $\upsilon_3' = \upsilon_4' = \upsilon_3'' = \upsilon_4'' = \ldots = 0$.

The oxygen source could be liquid (e.g., hydrogen peroxide), solid (e.g., ammonium perchlorate used in booster rockets), or gaseous (pure O_2 or air).

Example 1

(a) Determine the stoichiometric oxygen coefficient for the combustion of 1 kmol of CO, and (b) also express it in mass units.

Solution

(a) In the equation

$$CO + a\ O_2 \rightarrow b\ CO_2, \tag{A}$$

the terms a and b are called stoichiometric coefficients, which are determined using mass or atom conservation. For instance, for C atoms

$$b = 1,$$

and, similarly, for O atoms

$$1 + a \times 2 = b \times 2, \quad \text{i.e., } a = 1/2.$$

Therefore,

$$CO + (1/2)\ O_2 \rightarrow CO_2. \quad (B)$$

(b) Rewriting Equation B

28.01 kg of CO + 16 kg of oxygen → 44.01 kg of CO_2

1kg of CO + 0.57 kg of oxygen → 1.571 kg of CO_2 (C)

where $\nu_{O2,st}$ = 0.57 kg of stoichiometric oxygen per kg of CO.

Remarks

(i) The number of product moles leaving the reactor does not equal the number of moles of reactants entering the reactor (1 + 0.5 = 1.5). There is a net reduction in the number of moles within the reactor for this case.

(ii) However, on a mass basis, the mass of the products leaving (44 kg per kmol of CO consumed) is the same as that of the reactants (28 + 16 = 44 kg per kmol of CO consumed), because

$$28 \text{ kg of CO} + 16 \text{ kg of } O_2 \rightarrow 44 \text{ kg of } CO_2.$$

Mass is conserved, whereas moles are not.

Similarly, for methane,

$$CH_4 + 2\ O_2 \rightarrow CO_2 + 2\ H_2O.$$

In this special case, the molar flow leaving the reactor is the same as the molar flow entering it.

2.2.2 Stoichiometric Equation with Air

When air is used to supply the oxidizer O_2, each kilomole of O_2 is associated with 3.76 kmol of molecular nitrogen and Equation 2.2 can be rewritten in the form

$$CH_4 + 2\ (O_2 + 3.76\ N_2) \rightarrow CO_2 + 2\ H_2O + 7.52\ N_2, \quad (2.4)$$

where N_2 is unused during the combustion of methane in air. The stoichiometric air-to-fuel (A:F) ratio on a mole basis is

$$(A{:}F)_{st,mole} = (2 + 2 \times 3.76) \div 1 = 9.52 \text{ kmol of air to 1 kmol of fuel.} \quad (2.5a)$$

(*Note*: Air is an abbreviated term to indicate a mixture of O_2 and N_2 with the restriction that the N_2/O_2 ratio is 3.76; thus, 9.52 mol of air implies that it is a mixture of 2 mol of O_2 and 7.52 mol of N_2). Alternately,

$$(F{:}A)_{st,mole} = 1/9.52 = 0.105 \text{ kmol of stoichiometric fuel is supplied.} \quad (2.5b)$$

On a mass basis, Equation 2.5a could be converted into

$$(A{:}F)_{st,mass} = (2 \times 32 + 2 \times 3.76 \times 28) \div (1 \times 16)$$

$$= 17.16 \text{ kg of air kg}^{-1} \text{ of fuel.} \quad (2.6)$$

More accurately, air is a mixture containing 78% N_2, 1% Ar, and 21% O_2. If the argon is included, Equation (2.4) is written as,

$$CH_4 + 2\,(O_2 + 3.71\,N_2 + 0.05\,Ar) \rightarrow CO_2 + 2\,H_2O + (7.42\,N_2 + 0.1\,Ar). \quad (2.7)$$

In this case, $(A{:}F)_{st,\,mole} = 9.52$, but on a mass basis it changes to

$$(A{:}F)_{st,\,mass} = (2 \times 32 + 7.42 \times 28 + 0.1 \times 40) \div 1617.24 = 17.24. \quad (2.8)$$

Generally, the contribution of Ar is lumped with N_2. For above example, in the case of methane, lumping 0.05Ar with 3.71 N_2, we get the same equation or Equation (2.4).

Therefore, $(A{:}F)_{st} = 9.52$ kmol of air per kmol of CH_4, or $(A{:}F)_{st} = 17.16$ kg of air per kg of fuel CH_4. The stoichiometric A:F ratio in terms of the free-stream oxygen mass fraction can also be determined. Assume that 1 kg of fuel burns with ν_{O_2} kg of stoichiometric oxygen to produce $(1 + \nu_{O_2})$ kg of CO_2 and H_2O. If the free-stream oxygen mass fraction is $Y_{O_2,f}$ (expressed in kg of O_2 per kg of air in the free stream), the stoichiometric amount of air in kg per kg of fuel is

$$(A{:}F)_{st} = \nu_{O_2}/Y_{O_2,\infty} = 1/s \quad (2.9)$$

where s is the stoichiometric fuel per kg of air. (See Table A.2A for ν_{O2} values of many fuels.) As the inert mass per kg air is $1 - Y_{O_2,\infty}$, then

$$(\text{inert mass:F}) = (A{:}F)_{st}\,(1 - Y_{O2,\infty}) = (1/s)\,(1 - Y_{O_2,\infty}) \quad (2.10)$$

The mass of products, including inerts, is

$$1 + (A{:}F)_{st} = (1 + s)/s. \quad (2.11)$$

Example 2

(a) Determine the stoichiometric air-to-fuel ratio for CH_4 on kilomole and mass basis assuming that air consists of 21% O_2 and 79% N_2 (mole basis).
(b) Repeat (a) if air consists of 21% O_2, 78% N_2, and 1% Ar.
(c) Using the s values, verify the stoichiometric (A:F) mass obtained in part (b) if the oxygen mass fraction in air is 0.232.

Solution

(a) The reaction is

$$CH_4 + 2\,(O_2 + 3.76\,N_2) \rightarrow CO_2 + 2\,H_2O + 7.52\,N_2.$$

If air is used as the oxygen source.

$$(A{:}F)_{st} = (2 + 7.52)/1$$

$$= 9.52 \text{ kmol of air per kmol}^{-1} \text{ of } CH_4.$$

We can also express A:F on a mass basis, i.e.,

$$(A{:}F)_{st} = (2 \times 32 + 7.52 \times 28)/(1 \times 16) = 17.2 \text{ kg of air per kg of } CH_4.$$

(b) If Ar is included,

$$CH_4 + 2\,(O_2 + 78/21\ N_2 + 1/21\ Ar) \rightarrow CO_2 + 2\ H_2O$$
$$+ 2 \times (78/21)\ N_2 + 2 \times (1/21)\ Ar.$$

Again,

$$(A{:}F)_{st} = 2 + 2 \times (78/21) + 2 \times (1/21)$$

$$= 9.52 \text{ kmol of air per kmol of } CH_4.$$

The molecular weight of air including Ar is

$$0.21 \times 32 + 0.78 \times 28.02 + 0.01 \times 39.95 = 28.97 \text{ kg per kmol.}$$

On a mass basis,

$$(A{:}F)_{st} = 9.54 \times 28.97/16.05 = 17.24 \text{ kg}$$

(c) $s = 0.232/4 = 0.058$, $(A{:}F)_{st} = \dfrac{1}{0.058} = 17.24$ kg air per kg of fuel.

Because a 1 kg mass of fuel burns with 17.24 kg of air, the mass of products (including CO_2, H_2O, and N_2) is 18.24 kg, by mass conservation.

Remarks

Typically, Ar is lumped with N_2. In order to get the same molecular weight of air (28.97) for the N_2 and O_2 mixture, the molecular weight of N_2 must be adjusted to $M_{N2,\ equiv} = \{28.02*78 + 39.95*1)/79 = 28.17$ for better accuracy.

2.2.3 Reaction with Excess Air (Lean Combustion)

Fuel and air are often introduced separately (i.e., without premixing) into a combustor, e.g., in a boiler. Owing to the large flow rates and short residence times, there is no assurance that each molecule of fuel is surrounded by the appropriate number of oxygen molecules required for stoichiometric combustion. Therefore, it is customary to supply excess air in order to facilitate better mixing and thereby ensuring complete combustion. The excess oxygen remains unused and appears in the products. For example, if one uses 3 kmol of O_2 for CH_4,

$$CH_4 + 3\,(O_2 + 3.76\,N_2) \rightarrow CO_2 + 2\,H_2O + O_2 + 11.28\,N_2. \quad (2.12)$$

In this case, the air-to-fuel ratios are

$$(A{:}F)_{mole} = (3 \times 3.76 + 3) \div 1 = 14.28, \quad \text{and} \quad (A{:}F)_{mass}$$
$$= (3 \times 3.76 \times 28 + 3 \times 32) \div 16 = 25.74.$$

The excess-air percentage (mole percentage basis, which is equally valid for mass percentage basis) is

$$((A{:}F) - (A{:}F)_{st}) \times 100 \div (A{:}F)_{st} = (3 - 2) \times 100 \div 2 = 50\%.$$

2.2.4 Reaction with Excess Fuel (Rich Combustion) or Deficient Air

Incomplete combustion occurs when the air supplied is less than the stoichiometric amount required or when there is poor mixing of fuel and air. For this condition, the products of incomplete oxidation may contain a mixture of CO, CO_2, H_2, H_2O, and unburned fuel.

For example, if all of the C atoms in CH_4 are transformed into CO and H atoms into H_2O, then

$$CH_4 + 1.5\,(O_2 + 3.76\,N_2) \rightarrow CO + 2\,H_2O + 5.64\,N_2. \quad (2.13)$$

In this case, the minimum value of A:F (so that no carbon is released) is 7.14 kmol per kmol of fuel. Often, CO poisoning occurs when domestic gas burners are used in a tightly sealed house with a partially blocked air supply (so that combustion occurs with a deficient amount of air and CO is produced).

2.2.5 Equivalence Ratio ϕ and Stoichiometric Ratio (SR)

The equivalence ratio

$$\phi = (F{:}A)/(F{:}A)_{st} = (A{:}F)_{st}/(A{:}F) = (O_2{:}F)_{st}/(O_2{:}F). \quad (2.14a)$$

Here (F:A) denotes the fuel-to-air ratio, which is the inverse of (A:F). As an example, if $\phi = 0.5$ for methane air combustion, this implies that only 50% of stoichiometric fuel is supplied for given air flow or twice the stoichiometric air is supplied for given fuel flow. As (F:A) for an infinitely lean mixture tends to 0, $\phi \rightarrow 0$. Likewise, as (F:A) for an infinitely rich mixture tends to, $\phi \rightarrow \infty$. For a fuel-rich mixture, $\phi > 1$, and for a lean mixture, $\phi < 1$.

Another term used to represent the fuel–air mixture composition is the stoichiometric ratio (SR), which is the inverse of the equivalence ratio, i.e.,

$$SR = (\text{actual air supplied})/(\text{stoichiometric air demand of fuel})$$
$$= 1/\phi. \quad (2.15)$$

$$\text{excess air \%} = \text{excess } O_2\% = 100^* \{A{:}F - (A{:}F)_{st}\}/(A{:}F)_{st}$$
$$= 100^* (1 - \phi)/\phi = 100[SR - 1]. \quad (2.16a)$$

The parameter "γ" is defined as

$$\gamma = (F{:}A)/\{F{:}A + (F{:}A)_{st}\} = (F{:}O_2)/\{(F{:}O_2) + (F{:}O_2)_{st}\} \quad (2.16b)$$

For very lean mixtures [as (F:A) $\rightarrow$ 0], $\gamma \rightarrow 0$. The stoichiometric value of γ is 0.5, and for an infinitely rich mixture, $\gamma = 1$. Thus, γ is bounded.

In general, for the combustion of a fuel $C_cH_h\ N_n\ O_o\ S_s$

$$C_cH_h\ N_n\ O_oS_s + \{(c + h/4 + s - o/2)SR\}(O_2 + 3.76\ N_2) \rightarrow cCO_2$$
$$+ (h/2)\ H_2O + s\ SO_2 + (c + h/4 + s - o/2)\ \{(SR) - 1\}\ O_2$$
$$+ [\{3.76\ (c + h/4 + s - o/2)SR\} + n/2]\ N_2,\ SR > 1 \quad (2.17)$$

$$C_cH_h\ N_n\ O_oS_s + (c + h/4 + s - o/2)\ SR\ (O_2 + 3.76\ N_2) \rightarrow \{2\ c$$
$$- o - (c + h/4 - o/2 + s) * SR*2 + h/2 + 2s\}CO + \{o +$$
$$(c + h/4 - o/2 + s) * SR * 2 - h/2 - 2\ s - c\}CO_2 + (h/2)\ H_2O$$
$$+ s\ SO_2 + (c + h/4 - o/2 + s) * SR* 3.76* N2,\ SR < 1. \quad (2.18)$$

If all the c goes to CO, then the number of moles of CO_2, $N_{CO2} = 0$ and, hence, $\{SR\}_{min} = \{c + h/2 + 2\ s - o\}/\{(c + h/4 - o/2 + s)*2\}$.

For ethanol with SR = 0.9,

$$C_2H_5\ OH + 3*0.9\ (O2 + 3.76N_2) \rightarrow 0.6\ CO + 1.4\ CO_2 + 3H_2O + 10.152\ N_2.$$

Example 3

Consider the metabolism of glucose in the human body. As we breathe in air, we transfer oxygen from our lungs into our bloodstream. This oxygen is transported to the cells of our tissues, where it oxidizes glucose.

(a) Write down the stoichiometric reaction for the consumption of glucose(s) $C_6H_{12}O_6$ by pure oxygen and by air.
(b) Determine the amount of air required if 400% excess air is involved.
(c) Express (A:F) in terms of the percentage of theoretical air and the equivalence ratio.
(d) Write the associated reaction equation.
(e) If the breathing rate in humans ≈360 l(STP) h^{-1}, how much glucose is consumed per minute?

Solution

(a) The stoichiometric or theoretical reaction equation for this case is

$$C_6H_{12}O_6 + a\ O_2 \rightarrow b\ CO_2 + c\ H_2O \quad \text{(A)}$$

Applying an atom balance,

$$\text{Carbon C: } 6 = b, \quad \text{(B)}$$

$$\text{Hydrogen H: } 12 = 2\ c, \quad \text{i.e., } c = 6, \quad \text{(C)}$$

and

$$\text{Oxygen O: } 6 + 2a = 2\ b + c, \quad \text{i.e., } a = 6. \quad \text{(D)}$$

Therefore, the stoichiometric relation assumes the form

$$C_6H_{12}O_6 + 6\ O_2 \rightarrow 6\ CO_2 + 6\ H_2O. \quad \text{(E)}$$

The corresponding stoichiometric or theoretical reaction equation in the case of air is

$$C_6H_{12}O_6 + 6\ (O_2 + 3.76\ N_2) \rightarrow 6\ CO_2 + 6\ H_2O + 22.56\ N_2. \quad \text{(F)}$$

(b) $(A{:}F)_{st}$ = 28.56, 400% excess air = (supplied air – stoichiometric air) × 100/(stoichiometric air) = $(A{:}F)_{sub,mol}$ = 142.8 kmol of air per kmol of glucose. (G)

(c) Therefore, air as a percentage of theoretical air equals

$$(A{:}F/(A{:}F)_{st}) \times 100 = 142.8 \div 28.56 = 500\%. \quad \text{(H)}$$

$$\phi = ((A{:}F)_{st}/(A{:}F)) = 28.56 \div 142.8 = 0.2 \quad \text{or} \quad SR = 5.0. \quad \text{(I)}$$

(d) The actual reaction equation five times the theoretical air is

$$C_6H_{12}O_6 + 5 \times 6(O_2 + 3.76\ N_2) \rightarrow 6CO_2 + 6\ H_2O + (30 - 6)\ O_2 + 112.8\ N_2. \quad \text{(J)}$$

(e) The air consumption is specified as 360 l h^{-1} or using SATP volume 0.0147 kmol h^{-1}. Therefore, the glucose consumption per hour is

$$\omega = 0.0147 \div 142.8 = 0.0001029\ \text{kmol h}^{-1}. \quad \text{(K)}$$

The glucose consumption per hour in terms of mass is

$$\dot{m}_F = 00.0001029\ \text{kmol h}^{-1} \times 180.2\ \text{kg kmol}^{-1}$$

$$= 18.6\ \text{g h}^{-1} \text{ or } 0.31\ \text{g min}^{-1}. \quad \text{(L)}$$

Remarks

(i) Typically, fats (e.g., palmitic acid) are burned after glucose falls below a certain level in the bloodstream, which occurs after 20 to 25 min of vigorous exercise. If fat is represented by $C_{15}H_{31}COOH$, then one can obtain the fat-burn rate for a given breathing rate. The stoichiometric reaction for fat is represented by the relation

$$C_{15}H_{31}COOH + 23O_2 \rightarrow 16CO_2 + 16H_2O. \quad \text{(M)}$$

(ii) The molar ratio of CO_2 to O_2 is referred to as the respiratory quotient (RQ) in the medical literature. For glucose, RQ = 1; for fat RQ = 16/23 = 0.7. Because older people have problems excreting CO_2 from the bloodstream, fats are preferable to glucose owing to their lower RQ values.

(iii) Human body is mostly water; the chemical formula for the 70 kg body is appoximately given as $\mathbf{C}_{1.442}\mathbf{H}_{6.931}\mathbf{N}_{0.128}\mathbf{O}_{2.688}\mathbf{S}_{0.00437}\mathbf{P}_{0.0052}\mathbf{Ca}_{0.025}\ \mathbf{Na}_{0.0044}\ \mathbf{Cl}_{0.00268}$ [Emsley J, 1998].

2.3 Gas Analyses

The *in situ* measurements (e.g., CO_2, NO, CO, etc.) involve direct measurements of flue gas, usually on a wet basis. Sometimes, the sample line must be heated (for example, if there is unburned diesel in the diesel engine exhaust), or it may be diluted with inert gas to minimize to condensation for wet-basis measurements. In other cases, the sample is withdrawn from flue gas through a line and dried to prevent condensation in the sampling line in order to prevent interference with other measurements. Hence, concentrations are typically measured on a dry basis. The composition of gases is reported either on wet basis (H_2O included in products) or on dry basis (H_2O excluded in products) and the percentage is given either on a volumetric or mass basis.

2.3.1 Dew Point Temperature of Product Streams

Although it is generally reasonable to believe that water is present as vapor at elevated combustion temperatures, there is no assurance that it always exists in a gaseous phase in the products. Condensation can occur if the product temperature is lower than the dew point temperature T_{DP} (Figure 2.1). For example, combustion products in a boiler are cooled because of heat transfer to the water in the boiler tubes and, consequently, the temperature of the products decreases. The partial pressure of water vapor in the products

$$p_{H_2O} = X_{H_2O}\ P = \{N_{H_2O}/N\}\ P \quad (2.19)$$

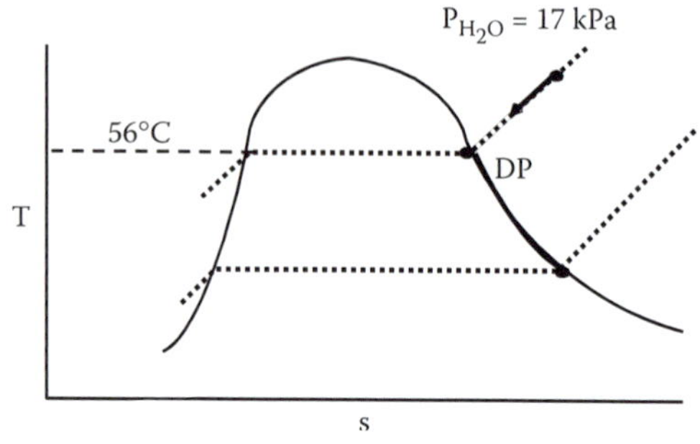

Figure 2.1 Schematic illustration of the determination of the dew point.

where N_{H2O} is the number of moles of gaseous H_2O in products, N is the total number of product moles, and P is the pressure. For stoichiometric combustion of CH_4 in air

$$CH_4 + 2\ (O_2 + 3.76\ N_2) \rightarrow CO_2 + 2\ H_2O + 7.52\ N_2$$

$$p_{H2O} = \{2/(1 + 2 + 7.52)\}\ P = 0.19\ P$$

If P = 100 kPa, then p_{H2O} = 19 kPa and the corresponding $T^{sat} = T_{DP}$ = 56°C (Table A.7 of Appendix A or use $\ell_n\ p^{sat}$ (bars) = A − B/T where A = 13.09, B = 5209 K, Tin K, 273 < T < 373 K); i.e., if the product temperature is lowered below 56°C, condensation occurs and the gas-phase mole fraction of H_2O starts decreasing. It is possible to have both liquid water (say, on the walls of a chimney or as fog from the exhaust duct from an automobile) and water vapor in the product stream. It is also possible to cool the gases to a low enough temperature to condense almost the entire amount of water vapor.

The percentage of CO_2 in the products equals {100/(1 + 2 + 7.52)} = 9.51% and that of H_2O is 19.01%. This is an example of a wet analysis, as the products also include water vapor. If the combustion products are analyzed with water removed as a constituent (by cooling to a temperature much below the dew point), the analysis is called a dry gas analysis. The CO_2 percentage on a dry basis is 100/(1 + 7.52) = 11.74%. Note that the volume percentages of the constituents of an ideal gas mixture are identical to their molar percentages.

Example 4

(a) Determine the wet and dry gas composition if CH_4 is fired with 50% excess air.
(b) Calculate the molecular weight of wet and dry products.
(c) Determine the mass fraction of CO_2 on a dry basis.

Solution

(a) $CH_4 + 2 * 1.5\ (O_2 + 3.76\ N_2) \rightarrow CO_2 + 2\ H_2O + O_2 + 11.28\ N_2$

Mole fraction of CO_2, i.e., X_{CO_2} = 1/(1 + 2 + 1 + 11.28) = 0.06545.

Similarly, X_{H_2O} = 0.1309, X_{O_2} = 0.065445, and x_{N2} = 0.74.

The dry gas composition is determined by condensing H_2O from the products. Based on dry analysis,

X_{CO_2} = 1/(1 + 1 + 11.28) = 0.075 moles of CO_2 for each kilomole of dry mixture, similarly, X_{O_2} = 0.075 and X_{N2} = 0.85.

(b) Molecular weight of gases:

M = 0.065 × 44.01 + 0.131 × 18.02 + 0.065 × 32 + 0.74 × 28.02 = 28.04 kg $kmol^{-1}$ of wet mixture

Molecular weight of dry gases: M = 0.075 × 44.01 + 0.075 × 32 + 0.85 × 28.02 = 29.52 kg $kmol^{-1}$ of dry mixture

(c) Mass fraction of CO_2, Y_{CO2} = 0.075 × 44.01/29.52 = 0.112.

Remarks

(i) Because each kilomole of gas at a specified value of T and P occupies the same volume, the mole fractions of gases equal their volume fractions.

(ii) If all molecular weights are equal, then $Y_k = X_k$.

(iii) Inversely if the measured dry or wet gas compositions are known, one can determine (A:F) as illustrated in the following example.

Example 5

Consider the combustion of natural gas (which is assumed to be almost CH_4). Determine (A:F) if the products contain 3% O_2 (on a dry basis) and the composition on both a wet and dry basis.

Solution

The reaction equation is

$$CH_4 + a\,(O_2 + 3.76\,N_2) \rightarrow CO_2 + 2H_2O + b\,N_2 + d\,O_2. \qquad (A)$$

From the O atom balance,

$$2\,a = 2 + 2 + 2\,d, \quad \text{or} \quad a = 2 + d. \qquad (B)$$

The exhaust contains 3% O_2 on a dry basis, i.e.,

$$d/(1 + 3.76\,a + d) = 0.03. \qquad (C)$$

Using Equation B and Equation C and solving

$$d = 0.2982, \qquad (D)$$

$$a = 2.2982. \qquad (E)$$

Also, from an N atom balance

$$b = 3.76\ a. \qquad (F)$$

Therefore,

$$(A:F) = 2.2982 \times 4.76 \div 1 = 10.9, \text{ and} \qquad (G)$$

Equation A assumes the form

$$CH_4 + 2.3\ (O_2 + 3.76\ N_2) \rightarrow CO_2 + 2\ H_2O + 8.6\ N_2 + 0.3\ O_2. \qquad (H)$$

The composition is as follows:

$$CO_2\%\ \text{in exhaust (wet)} = 100 \times (1 \div (1 + 2 + 3.76 \times 2.2982 + 0.2982)) = 8.4,$$

$$CO_2\%\ \text{in exhaust (dry)} = 100 \times (1 \div (1 + 3.76 \times 2.2982 + 0.2982)) = 10.1,$$

$$N_2\%\ \text{in exhaust (wet)} = 72.4\%,\ N_2\ \text{in exhaust (dry)} = 86.9.$$

Remarks

Using Equation H, the mole fraction of H_2O on wet basis has a value equal to $2 \div (1 + 2 + 8.6 + 0.3) = 0.17$. Consequently, if it exists as vapor, the partial pressure of H_2O is 0.17 bar if P = 1 bar. At this pressure $T^{sat} = T_{DP} = 56°C$, where T_{DP} denotes the dew point temperature. Inversely, if T_{DP} is known, the water percentage and (A:F) can be determined (cf. Figure 2.1).

2.3.2 Generalized Dry Gas Analysis for Air with Ar

2.3.2.1 Excess Air % from Measured CO_2% and O_2%

$$\text{Fuel + air supplied} \rightarrow CO_2,\ H_2O,\ O_2,\ N_2,\ Ar$$

The % excess $O_2 = 100^* N_{O_2,P}/\{N_{O_2,R} - N_{O_2,P}\}$
where $N_{O_2,R}$ moles of O_2 are supplied with reactants and $N_{O_2,P}$ moles of O_2 leave with products. Dividing denominator and numerator by $N_{Dry,P}$ of products,

$$\%\ \text{excess}\ O_2 = 100^*\ \{N_{O_2,P}/N_{dry,P}\}/\{(N_{O2,R} - N_{O2,P})/N_{Dry,P}\}$$

Noting that $N_{O2,R} = N_{N2,R}\ (O_2/N_2)_R = N_{N2,P}(O_2/N_2)_R$

$$\%\ \text{excess}\ O_2 = 100^*\ X_{O2}/\{[X_{N2}\ (O_2/N_2)_R] - X_{O_2}\}$$

$$= 100^*\ O_2\%/\{[(N_2\%)(O_2/N_2)_R] - O_2\%\} \qquad (2.20)$$

where $N_2\%$ (dry) = 100 − CO_2% − O_2%; typically, $(O_2/N_2)_R = 0.27$

Thus, knowing $CO_2\%$, $O_2\%$, one can get $N_2\%$ and, hence, excess air % is calculated from Equation 2.20 if the fuel is pure HC and hence get A:F or ϕ. If the actual air-to-fuel ratio is known from measurements, then the theoretical A:F ratio can be determined from the gas analyses. Here,

$$X_{N2} = N_{N2,P}/N_{Dry,P} = \text{supplied}N_2/N_{dry,P} = (\text{air supplied}^* \; X_{N_2,a})/N_{dry,P}$$

where $X_{N2,a}$:N_2 mole fraction in ambient is

$$N_{dry,P}/N_{air} = \{X_{N2,a}/X_{N2}\}. \tag{2.21}$$

Example 6

Consider the combustion of natural gas. Determine (A:F) if the products contain 3% O_2 (on a dry basis) and the CO_2 % is 10.1.

Solution

With $CO_2\% = 10.1\ \%$, $O_2\% = 3.0\%$, then $N_2\% = 100 - 10.1 - 3 = 86.9$, excess air % = $100^*\ 3/(86.9/3.76 - 3) = 14.92\%$ and the equivalence ratio is $1/(1 + 0.149) = 0.87$. Note that $CO_2\%$ and $O_2\%$ depend on the type of fuel fired. SR $= 1/\phi = 1.152$.

Dry moles of products per kilomole of dry air = 0.79/0.87 = 0.91

2.3.2.2 Generalized Analysis for Fuel $CH_hO_oN_n\,S_s$

$$CH_hO_oN_n\,S_s + a(O_2 + B\,N_2 + F\,H_2O(g) + G\,Ar) \rightarrow CO_2 + (h/2 + aF)$$

$$H_2O + \{a - (1 + s + h/4 - o/2)\}\,O_2 + (aB + n/2)N_2 + aG\,Ar \tag{2.22}$$

where the ratios B, F, and G in air are defined as (in molar or volumetric amounts of N_2, O_2, H_2O, and Ar) and all fuel N goes to N_2 (see Problem 2.43).

$$B = N_{N_2}/N_{O2}, \quad F = N_{H2O}/N_{O2}, \quad \text{and} \quad G = N_{Ar}/N_{O2}, \quad \text{i.e.,} \tag{2.23}$$

$$B + G + 1 = \{(N_{N_2} + N_{O_2} + N_{Ar})/N_{O_2}\} = 1/X_{O_2,a} \tag{2.24}$$

where $X_{O_2,a}$ is the oxygen mole fraction in ambient on a dry basis.

The air supplied

$$(A{:}F)_{dry} = a/X_{O_2,a} = \{(-h/4 + o/2 + n/2)\,X_{O_2}$$

$$+ (1 + h/4 + s - o/2)\}/(X_{O_2,a} - X_{O_2}) \tag{2.25}$$

The total dry moles of products could be derived as

$$N_{P,dry} = \{a/X_{O_2,a}\} - h/4 + o/2 + n/2 = (A{:}F)_{dry} - h/4 + o/2 + n/2 \tag{2.26}$$

On a wet basis,

$$(A{:}F)_{wet} = \text{(dry air moles + water moles) per fuel mole}$$
$$= (A{:}F)_{dry} + b, \quad (2.27)$$

where

$$b = (w\ M_a/M_{H_2O}), \quad (2.28)$$

where w denotes the specific humidity in kg of vapor per kg dry air and M_a, the molecular weight of dry air. The expression for w in terms of the relative humidity RH is

$$w = 0.622\ p_v/p_a = 0.622 \times RH \times p_g^{sat}\ (T_{DB})/$$
$$\{P - RH \times P_g^{sat}\ (T_{DB})\}. \quad (2.29)$$

The amount of CO_2 can be derived using $X_{CO2} = 1/N_{p,dry}$

$$X_{CO_2} = (1 - X_{O_2}/X_{O_2,a})/\{(-h/4 + o/2 + n/2)$$
$$+ (1/X_{O_2,a})(1 + h/4 + s - o/2)\} \quad (2.30)$$

Because $X_{CO_2,max}$ occurs when $X_{O_2} = 0$ (i.e., stoichiometric conditions for given fuel),

$$X_{CO_2,max} = 1/\{(-h/4 + o/2 + n/2) + (1/X_{O_2,a})(1 + h/4 - o/2)\} \quad (2.31)$$

and

$$\{X_{CO_2}/X_{CO_2,max}\} = (1 - X_{O_2}/X_{O_2,a}) = (X_{O_2,a} - X_{O_2})/X_{O_2,a} \quad (2.32a)$$

If X_{O_2} is set at $X_{O_2,std}$ then $X_{CO_2} \rightarrow X_{CO_2,std}$ and hence

$$X_{CO_2,std}/X_{CO_2} = \frac{1-(X_{O_2,std}/X_{O_2,a})}{1-(X_{O_2,}/X_{O_2,a})} \quad (2.32b)$$

In the case of CH_4, $X_{CO_2,max} = 1/10.52$ when $X_{O_2,a} = 0.21$. If the amount of CO_2 is measured, we can determine the values of X_{O_2} from Equation 2.30, that of $(A{:}F)_{dry}$ from Equation 2.25, and $(A{:}F)_{wet}$ from Equation 2.27.

2.3.3 *Emissions of NO_x and Other Pollutants*

Gas analyses are used in reporting NO in products as ppm (parts per million or molecules per million molecules for gases = $X_k * 10^6$, where X_K is the mole fraction of pollutant species). As $X_k = N_k/N_p$, where N_p is the number of product moles affected by the amount of excess air used, X_k will vary depending on the excess O_2%, even though N_k may remain unaffected and the emission on mass basis (= $N_k\ M_K$) may remain the same. Thus, the emission in ppm is corrected to standard O_2% in the exhaust. For example, if k = NO, and X_{NO} is measured at X_{O2}, then following Equation 2.32b the NO at standard condition is given by

$$X_{NO,std}/X_{NO} = NO_{ppm,std}/NO \text{ in ppm}$$

$$= (X_{O_2,a} - X_{O_2,std})/(X_{O_2,a} - X_{O_2}), \tag{2.33}$$

where $X_{O2,a}$ is the ambient O_2 mole fraction (typically, 0.21 for dry air). More details, examples, and other methods of emission reporting are given in Chapter 17.

2.4 Global Conservation Equations For Reacting Systems

2.4.1 *Mass Conservation and Mole Balance Equations*

Reacting systems consume fuel and oxygen species and produce CO_2 and H_2O, and convert chemical energy into thermal energy. The energy and mass conversions can be analyzed using mass and energy conservation equations.

2.4.1.1 Closed System

The overall mass conservation equation for a closed system is

$$dm/dt = 0 \quad \text{or} \quad m = \text{constant} \tag{2.34}$$

(For systems involving nuclear reactions in a closed system, $dm_{sys}/dt = -dm_{nuclear}/dt$, where $dm_{nuclear}/dt$ is the mass destruction rate; e.g., the universe.) Chemical reactions involve consumption of species such as fuel and oxygen and the production of species such as CO_2, H_2O, etc. If CO, O_2, and CO_2 are placed in a piston–cylinder assembly, the total mass is conserved but not the total moles. If the generation rate of the k-th species $\dot{m}_{k,chem} < 0$ (e.g., CO is consumed as reaction proceeds), then it is consumed, but if $\dot{m}_{k,chem} > 0$ (e.g., as CO_2 is produced), this implies production. For a closed system

$$dm_k/dt = \dot{m}_{k,chem}. \tag{2.35}$$

Summing over all species k,

$$\sum dm_k/dt = \sum \dot{m}_{k,chem} = dm/dt = 0, \tag{2.36}$$

as there is no net production of mass. Dividing Equation 2.34 by M_k on a mole basis

$$dN_k/dt = \dot{N}_{k,chem}, \tag{2.37}$$

and summing over all species k,

$$dN/dt = \sum dN_k/dt = \sum \dot{N}_{k,chem} \neq 0. \tag{2.38}$$

2.4.1.2 Open System

2.4.1.2.1 Integral Form

Recall that the overall mass conservation equation for an open system is

$$dm/dt = \dot{m}_i - \dot{m}_e, \tag{2.39}$$

and for a specified species k

$$dm_k/dt = \dot{m}_{k,i} - \dot{m}_{k,e} + \dot{m}_{k,chem}. \tag{2.40}$$

Summing over all species k for the single inlet and exit case

$$dm/dt = \sum dm_k/dt = \sum \dot{m}_{k,i} - \sum \dot{m}_{k,e} + \sum \dot{m}_{k,chem} = \dot{m}_i - \dot{m}_e \tag{2.41}$$

since

$$\sum \dot{m}_{k,chem} = 0.$$

Dividing Equation 2.40 by m_k, Equation 2.40 in molar form is,

$$dN_k/dt = \dot{N}_{k,i} - \dot{N}_{k,e} + \dot{N}_{k,gen}, \tag{2.42}$$

where $\dot{N}_{k,gen}$ denotes the molar production of species k by the chemical reaction. For instance, consider the flow of 5 kmol sec^{-1} of CO, 3 kmol sec^{-1} of O_2, and 4 kmol sec^{-1} of CO_2 into a reactor. If all of the CO (fuel) is completely burned, then $\dot{N}_{CO,e} = 0$. Assuming steady state, $dN_{CO}/dt = 0$, Equation 2.42 with k = CO assumes the form

$$dN_{CO}/dt = 0 = 5 - 0 + \dot{N}_{CO,gen} \quad \text{or} \quad \dot{N}_{CO,gen} = -5 \text{ kmol sec}^{-1}.$$

The negative sign associated with $\dot{N}_{CO,gen}$ implies consumption of CO.

2.4.1.2.2 Differential Form

Because the fluid of interest is a mixture consisting of several components, the mass flow rate for each component is determined from the mass conservation equation for the individual species and then added for all species to obtain the overall mass balance. If the velocity vector of species k is represented by $\vec{v}_k$, then because the velocity distribution can spatially vary, the inlet and exit mass flow rates of species can be expressed as

$$\dot{m}_{k,i} = -\int_{A_i} \rho\vec{v}_k \cdot d\vec{A}, \quad \dot{m}_{k,e} = \int_{A_e} \rho\vec{v}_k \cdot d\vec{A},$$

and, more generally, for inlet or exit,

$$\dot{m} = -\oint_A \rho_k\, \vec{v}_k \bullet d\vec{A} \tag{2.43}$$

where $\vec{v}_k$ is the absolute velocity of species k with respect to a stationary observer. It will be seen in Chapter 6 that different species travel with different absolute velocities.

$$\dot{m}_{k,chem} = \int_V \dot{m}'''_{k,chem} dV \tag{2.44}$$

Therefore, using Equation 2.43 and Equation 2.44 in Equation 2.40

$$d/dt(\textstyle\int_{cv} \rho_k dV) = -\oint \rho_k \vec{v}_k \cdot d\vec{A} + \int_{cv} \dot{m}_{k,chem}''' \; dV\,. \tag{2.45}$$

For any c.v., Equation 2.45 still holds, and hence

$$\partial\rho_k/\partial t + \vec{\nabla}\cdot \rho_k \vec{v}_k = \dot{m}'''_{k,chem}\,. \tag{2.46}$$

2.4.2 Energy Conservation Equation in Molar Form

2.4.2.1 Open System

The energy conservation equation in mass and molar forms are (Figure 2.2):

$$dE_{c.v.}/dt = \dot{Q}_{c.v.} - \dot{W}_{c.v.} + \sum_K \dot{m}_{k,i}\, e_{T,k,i} - \sum_K \dot{m}_k\, e_{T,k,e} \tag{2.47}$$

$$(dE_{c.v.}/dt) = \dot{Q}_{c.v.} - \dot{W}_{c.v.} + \sum \dot{N}_{k,i}\, \bar{e}_{k,T,i} - \sum \dot{N}_{k,e}\, \bar{e}_{k,T,e} \tag{2.48}$$

$$e_{T,k} = h_k + ke_k + pe_k \,\frac{kJ}{kg} \tag{2.49}$$

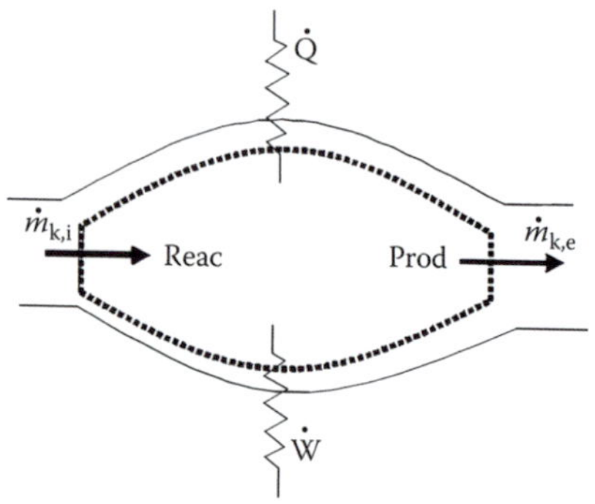

Figure 2.2 Illustration of the first law for a chemically reacting system.

$\bar{h}_k$ is the enthalpy of component k in its pure form at the same temperature and pressure as the mixture.

$$\bar{e}_{T,k} = \bar{h}_k + \overline{ke}_k + \overline{pe}_k, \text{ kJ/kmol} \tag{2.50}$$

2.4.2.2 Differential Form

For multicomponent systems, $\rho h = \Sigma \rho_k h_k$, $\rho u = \rho(h - P/\rho) = \rho h - P = \Sigma\, \rho_k h_k - P$; let $E_{c.v.} = U$. Using the following substitutions in Equation 2.47, neglecting ke and pe, and using the Gauss divergence theorem, we have

$$\frac{dU}{dt} = \frac{d}{dt}\int_{cv} \rho u\, dV, \quad \dot{Q}_{c.v.} = -\int_{cs} \dot{\vec{Q}}'' \bullet d\vec{A}, \quad \dot{\vec{Q}}_{c.v.}'' = -\lambda\, \nabla T, \quad \dot{W}_{c.v.} = 0$$

$$\partial\left(\sum \rho_k h_k\right)/\partial t + \vec{\nabla}\cdot\left(\sum \rho_k \vec{v}_k h_k\right) = \partial P/\partial t + \vec{\nabla}\cdot(\lambda\vec{\nabla}T), \tag{2.51}$$

where λ is calculated for the mixture. As $\bar{h}_k = h_k M_k$, $n_k = \rho_k/M_k = \text{kmol/m}^3$, then $\rho_k\, h_k = n_k\, \bar{h}_k$

$$\partial\left(\sum n_k \bar{h}_k\right)/\partial t + \vec{\nabla}\cdot\left(\sum_k n_k \vec{v}_k \bar{h}_k\right) = \partial P/\partial t + \vec{\nabla}\cdot(\lambda\vec{\nabla}T). \tag{2.52}$$

2.4.2.3 Unit Fuel-Flow Basis

Dividing Equation 2.48 by the fuel flow (or assuming 1 kmol of fuel flow) and ignoring the ke and pe,

$$(dE_{c.v.}/dt)/\dot{N}_F = \bar{q} - \bar{w} + \bar{h}_I - \bar{h}_e, \tag{2.53}$$

where

$\bar{h}_i = \Sigma_k (\dot{N}_k/\dot{N}_F)\bar{h}_k)_i = \Sigma \nu'_{k,i} \bar{h}_{k,i}$, in units of kJ for the mixture per kmol of fuel (2.54)

$\nu'_k = (\dot{N}_{k,i}/\dot{N}_F)$, in units of kmol of species k per kmol of fuel (2.55)

$\bar{h}_e = \Sigma_k (\dot{N}_k/\dot{N}_F)\bar{h}_k)_e = \Sigma \nu''_{k,e} \bar{h}_{ke}$, for the mixture exiting per kmol of fuel

$\nu_k'' = (\dot{N}_{k,e}/\dot{N}_F)$ kmol of species k per kmol of fuel

$$\bar{q} = \dot{Q}_{c.v.}/\dot{N}_F,\ \bar{w} = \dot{W}_{c.v.}/\dot{N}_F,\ \frac{\text{kJ/sec}}{\text{kmol of fuel/sec}} = \frac{\text{kJ}}{\text{kmol of fuel}} \quad (2.56)$$

Under steady-state conditions, dE/dt = 0. On a unit molar fuel-flow basis,

$$0 = \bar{q} - \bar{w} + \bar{h}_i - \bar{h}_e. \quad (2.57)$$

2.5 Thermochemistry

2.5.1 *Enthalpy of Formation*

Elemental C(s) can be burned in O_2 at 25°C and 1 atm to form CO_2. The combustion reaction process is exothermic and results in temperature rise, but the product temperature can be reduced to the same temperature of 25°C through heat transfer. The enthalpy of formation or chemical enthalpy of CO_2, $\bar{h}^o_{f,CO_2}$ in this context equals the amount of heat added to or removed from the compound at 25°C and 1 atm from its elemental constituents (which themselves exist in their natural state at that temperature and pressure). The subscript f denotes formation, and the superscript o denotes that the value corresponds to a pressure of 1 atm.

Another example involves the formation of NO from elemental nitrogen and oxygen in their natural forms at STP (under STP, nitrogen exists as N_2 and oxygen in the form of O_2). In this case,

$$1/2\ N_2 + 1/2\ O_2 \rightarrow NO + 90{,}297 \text{ kJ kmol}^{-1} \text{ of NO.}$$

This chemical reaction requires a heat input of 90,297 kJ kmol^{-1}. Using the first law of thermodynamics on mole basis,

$$Q = \bar{h}^0_{f,NO} = 90{,}297 \text{ kJ kmol}^{-1} = \bar{h}_{NO} - (1/2\,\bar{h}_{N_2} + 1/2\,\bar{h}_{O_2}).$$

It is apparent that $\bar{h}_{NO} > (1/2\,\bar{h}_{N_2} + 1/2\,\bar{h}_{O_2})$. If we arbitrarily set $\bar{h}_{N_2} = \bar{h}_{O_2} = 0$, then $\bar{h}_{NO} = \bar{h}^o_{f,NO} = 90{,}297$ kJ kmol^{-1}, which implies that the energy contained in NO at 298 K and 1 atm is 90,297 kJ more than the energy associated with N_2 and O_2. Conventionally, all elements in their natural form at STP (scientific) are set at zero enthalpy. This energy can be interpreted as that which is absorbed or released when the atoms form the compound. See Table A.10 to Table A.30 and Table A.32 of Appendix A for a tabulation of enthalpy of formation of various substances. Figure 2.3 illustrates the enthalpies of formation of several compounds from elements in natural form.

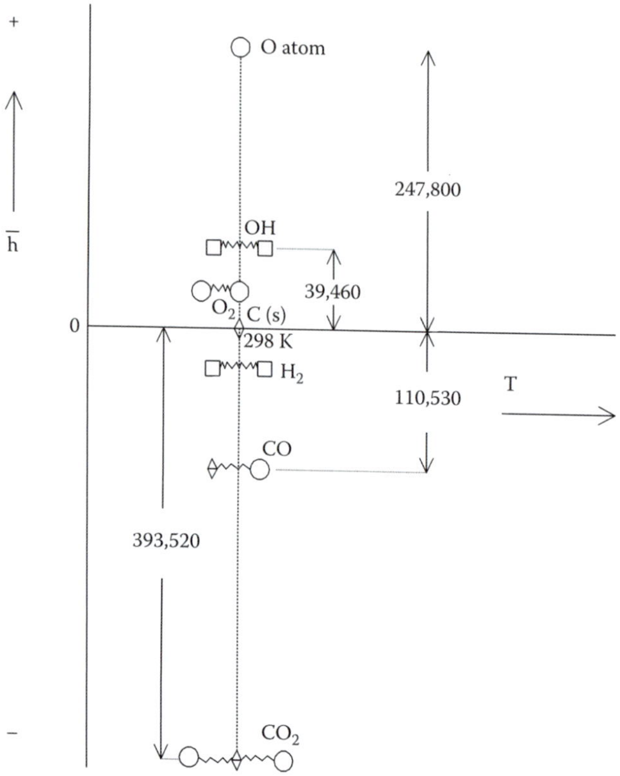

Figure 2.3 Illustration of enthalpies of formation; all units are in 1 J k mole^{-1}.

2.5.1.1 *Enthalpy of Formation from Measurements*

For the reactions

$$C(s) + (1/2)\ O_2 \rightarrow CO - 110{,}530 \text{ kJ } (Q_1) \text{ at 298 K and 1 atm,} \quad (A)$$

$$C(s) + O_2 \rightarrow CO_2 - 393{,}520 \text{ kJ } (Q_2) \text{ at 298 K and 1 atm,} \quad (B)$$

$$CO + 1/2\ O_2 \rightarrow CO_2 + Q_3 \text{ at 298 K and 1 atm,} \quad (C)$$

where Q_3 denotes the heat rejected when CO is burned with 1/2 kmol of O_2 at 298 K.

Equation 2.57 yields $Q_1 = -110{,}530$ kJ $h_{f,CO}$, with the values $\bar{h}_{C(s)} = 0$, $\bar{h}_{O_2} = 0$, $\bar{h}_{f,CO} = -110{,}530$ kJ. This implies that the enthalpy of CO at 298 K is 110,530 kJ below the "energy" level of the stable elements producing the compound (see Figure 2.3).

2.5.1.2 *Bond Energy and Enthalpy of Formation*

If calorimetric data is not available, h_f can still be determined using the bond energy, which is the energy required to break the bonds between the atoms in a molecule (see Section 2.7 for an example of such a procedure). Bond energies are tabulated in Table A.31A and Table A.31B of Appendix A.

2.5.2 Thermal or Sensible Enthalpy

Once a product is formed from elements at 298 K and 1 atm, its temperature can be raised to T by adding heat isobarically, and then raising the pressure to P isothermally. Consequently, the enthalpy of that compound increases. For any product species, the difference between the enthalpy at T and P and that at 298 K and 1 atm of the same product is called *sensible enthalpy*. For ideal gases, the sensible enthalpy is the same as the thermal enthalpy.

$$h_{t,T} - h_{t,298} = \int_{298}^{T} c_p^0 \, dT, \tag{2.58}$$

where c_p^0 refers to the specific heat of the species at 1 atm. For an ideal gas, the superscript ^{0} is redundant as c_p is not a function of pressure. As an example, NO, which is an ideal gas, requires 22,230 kJ kmol^{-1} of energy to raise its temperature from 298 K to 1000 K (cf. Figure 2.4).

For an ideal gas, the superscript 0 can be omitted as c_p and h are independent of P. See Table A.10 to Table A.30 in Appendix A for a tabulation of values of $\Delta h_{t,T} = h_{t,T} - h_{t,298}$ for several ideal gases.

2.5.3 Total Enthalpy

The total enthalpy of a species at any state (T, P) equals the value of its chemical enthalpy at 298 K at 1 atm (when species are formed from elements)

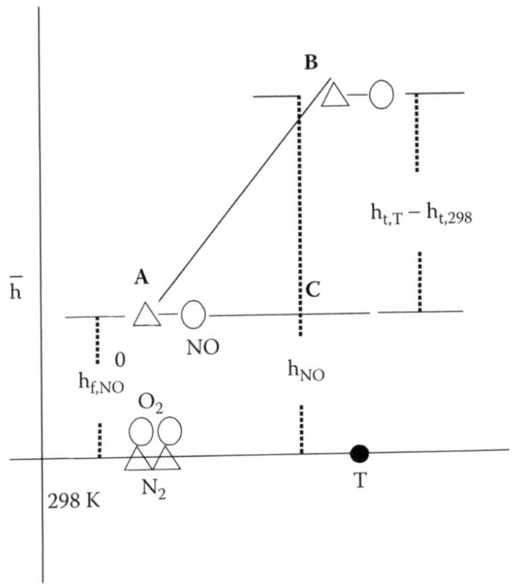

Figure 2.4   Illustration of chemical and thermal enthalpies.

plus the additional enthalpy (called sensible enthalpy) required to raise the temperature of the same species from 298 K to T and pressure from 1 atm to P. At P = 1.013 bar ≈ (100 kPa)

$$\bar{h}_k^o(\mathrm{T}) = \bar{h}_{k,f}^o + \{\bar{h}_{t,298\to\mathrm{T}}\}_\mathrm{k} = \bar{h}_{k,f}^o + \int_{298}^{T} \bar{c}_{p0,k}\, dT. \tag{2.59}$$

Example 7

Determine the total enthalpy of NO at 500 K and 200 kPa assuming ideal gas behavior.

Solution

Consider two hypothetical reactors in series. The first reactor produces NO at 298 K and 1 atm (state 2) from the elements that enter at state 1 (298 K, 1 atm). The second reactor heats the NO to 500 K and 1 atm (state 3) by adding heat (i.e., the thermal portion of the enthalpy $h_t^o(500) - h_{t,298}^o = \int_{298}^{500} \bar{c}_{p,\mathrm{NO}}^o dT$).

The total enthalpy (i.e., the enthalpy of NO at 500 K and 100 kPa) with reference to the enthalpies of the elements constituting the compound at that state are

$$\bar{h}_{\mathrm{NO}}\ (500\ \mathrm{K}, 1\ \mathrm{atm}) = \bar{h}_{\mathrm{f,NO}}\ (298\ \mathrm{K},\ 101.3\ \mathrm{kPa}) + (\bar{h}_{t,500} - \bar{h}_{t,298})$$

$$= 90{,}297 + 6079 = 96{,}776\ \mathrm{kJ}.$$

As the gas is ideal,

$$\bar{h}_{\mathrm{NO}}\ (500\ \mathrm{K},\ 200\ \mathrm{kPa}) = \bar{h}_{\mathrm{NO}}\ (500\ \mathrm{K},\ 101.3\ \mathrm{kPa}).$$

Now consider the pressure effect. The enthalpy of a species at an arbitrary pressure P and temperature T equals its enthalpy at that temperature and 1-atm pressure, plus the additional enthalpy required to change the pressure to P, i.e.,

$$\bar{h}\ (\mathrm{T, P}) = \bar{h}^\circ(\mathrm{T}) + \Delta\bar{h}_{\mathrm{T},1\to\mathrm{P}}, \text{ where} \tag{2.60}$$

$$\bar{h}^\circ(\mathrm{T}) = \bar{h}\mathrm{f}^\circ + \bar{h}_{t,T}^\circ - \bar{h}_{t,298}^\circ. \tag{2.61}$$

The term $\Delta\bar{h}_{\mathrm{T},1\to\mathrm{P}}$ in Equation 2.60, called *enthalpy correction*, to account for real gas effect. For ideal gases, $\Delta h_{\mathrm{T},1\to\mathrm{P}} = 0$. The correction term $\Delta h_{\mathrm{T},1\to\mathrm{P}}$ is obtained from enthalpy correction charts shown in Figure B.2 of Appendix B. If internal energy is required, then $\bar{u} = \bar{h} - p\bar{v} = \bar{h} - Z\bar{R}\,\mathrm{T}$, where Z for real fluids is obtained from Figure B.1 of Appendix B; Z = 1 for ideal gases.

2.5.4 Enthalpy of Reaction and Combustion

The enthalpy of a reaction refers to the difference between the enthalpy of products and reactants at a specified temperature and standard pressure of 1 atm for that reaction. The amount of heat added ($\Delta H_R > 0$) or removed ($\Delta H_R < 0$) for any reaction when the reactants enter and the products leave an isothermal system is called the *enthalpy of reaction*, i.e.,

$$\Delta H^o_{R,T} = H^o_{Prod,T} - H^o_{Reac,T} \cdot \tag{2.62}$$

$\Delta H^o_R < 0$ for exothermic reactions and > 0 for endothermic reactions. If a reaction involves either oxidation or combustion, then the enthalpy of reaction ΔH^o_R is termed as the enthalpy of combustion ΔH^o_C. For instance, consider the oxidation of CO to CO_2 at 298 K and 1 atm, i.e.,

$$CO + 1/2\ O_2 \rightarrow CO_2,$$

for which $H_{prod} = \bar{h}_{CO2} = -393{,}520$ kJ, and $H_{Reac} = 1\ \bar{h}_{CO} + 1/2\ \bar{h}_{O2} = -110{,}530$ kJ. Therefore,

$$\Delta H^o_{R,T} = \Delta H^o_C = -282{,}990 \text{ kJ (kmol of CO)}^{-1},$$

i.e., when a kmol of CO is burned, 282,990 kJ of heat must be removed for the products to leave at the same temperature as the reactants (298 K). Because the values of the enthalpy of reaction are normally tabulated at a temperature of 298 K, the subscript T = 298 K is omitted.

Example 8

Determine the enthalpy of reaction for CO at 500 K (Figure 2.5).

Solution

Using Figure 2.5, reactants enter at (1) at 500 K and products leave at (2) at 500 K. Assume that combustion proceeds according to

$$CO + (1/2)\ O_2 = CO_2. \tag{A}$$

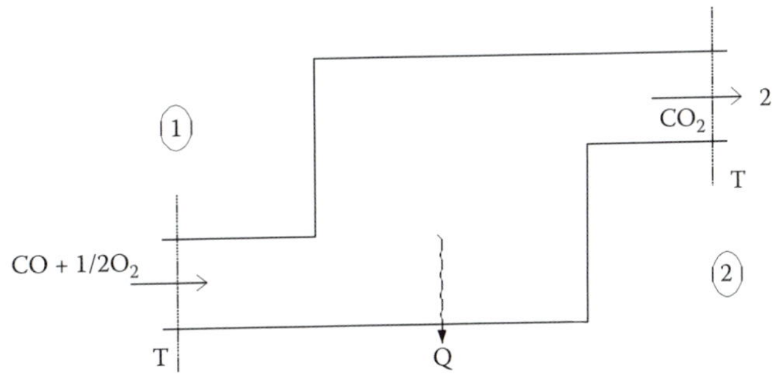

Figure 2.5 Illustration of enthalpy of reaction, Q = $\Delta h°R$.

By definition,

$$\Delta H_R(T) = H_{prod} - H_{reac} = \bar{h}_{CO_2}(500\ K) - \bar{h}_{CO}(500\ K) + (1/2)\,\bar{h}_{O2}(500\ K) \quad (B)$$

$$\Delta H_R(T) = \{-393520 + 8301\} - \{-110530 + 5943\} + (1/2)\,\{0 + 6097\}$$

$$= -277600,\ kJ\ \Delta H_C = \Delta H_R = -277600\ kJ\ kmol^{-1}.$$

Remarks

(i) If calculations are repeated at 298 K, at $\Delta H^0{}_C = \Delta H^0{}_R = -283005$ kJ kmol^{-1}. We find that $\Delta H^0{}_C = \Delta H^0{}_R$ does not change very much with T and is assumed to be constant in most combustion problems.

(ii) The specific enthalpies could be either on mole basis, $\bar{h}_k$, or mass basis, h_k, where $h_k = \bar{h}_k/M_k$; for mass basis.

(iii) H_{prod} denotes the enthalpy of all species in products per kmol of fuel.

2.5.5 Heating Value (HV), Higher HV (HHV) or Gross HV (GHV), and Lower HV (LHV)

2.5.5.1 Heating Values

The heating value of a fuel is the negative of enthalpy of combustion, i.e.,

$$HV = -\Delta H^o_C = H_{Reac} - H_{prod}. \quad (2.63)$$

When CH_4 is burnt, the products are H_2O and CO_2. Depending on the state of H_2O (i.e., liquid or gas) in products, the heating values (HVs) differ. The lower heating value is based on the assumption that the water is in gas form in the products, whereas the higher heating value is based on liquid-phase water in the products. The lower heating value

$$LHV = H_{Reac} - H_{Prod,H_2O_{(g)}}, \quad (2.64)$$

and

$$HHV = H_{Reac} - H_{Prod,H_2O_{(\ell)}}. \quad (2.65)$$

Generally, HV, ΔH^o_R, and ΔH^o_C are tabulated at 298 K and 1 atm. The difference between HHV (or gross heating value) and LHV (or net heating value) is illustrated in Figure 2.6.

For gaseous fuels, the heating values are typically stated on a volumetric basis. If the $\overline{HV}$ is known on a mole basis, one can obtain HV′ based on a volumetric basis using the relation HV′ = HV in kJ/24.5 m^3 or HV′ = HV in (Btu/lb-mol)/392 ft^3 with STP of 25°C (77°F) and 1 atm. The heating values are determined using "bomb" calorimeter. A small mass of solid or liquid fuel is placed in a bomb calorimeter (Figure 2.6), then pumped with O_2 at about

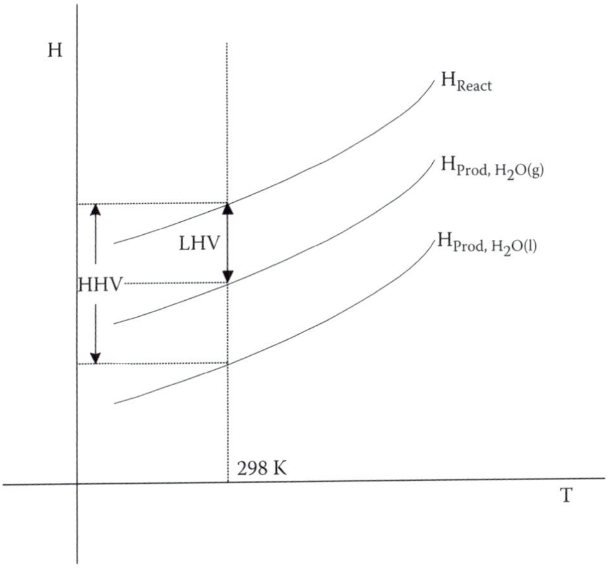

Figure 2.6 Difference between LHV and HHV.

30 atm, burned, and heat is then transferred to the cooling water and container and the temperature rise is measured. From the energy balance ($Q_V = \Delta U = \Delta H - \Delta(n\,\bar{R}\,T) = \Delta H - \Delta n\,\bar{R}\,T$ for constant T system), one can solve for change in enthalpy and, hence, the heating value is determined. For developing HHV relations, one can use closed-system energy conservation equations. Typically, the calorimeter is standardized by combusting benzoic acid (C_6H_5COOH). For gaseous fuels, a constant-pressure flow calorimeter is used in which gaseous fuel flows through a wet gas test meter (or rotameter to measure the flow rate), fuel is continuously burned and the heat is transferred to flowing water. The temperature rise of water indicates the HV of the gaseous fuel. For developing HHV relations, the open-system energy conservation equations are used. If HHV is known, then it is possible to determine the enthalpy of formation of species using Equation (2.62).

Heating values of food (e.g., table sugar — $C_{12}H_{22}O_{11}(s)$), which are the fuels used by humans, are reported in "food" Calories with 1 food Calorie = 1 kCal (e.g., table sugar: 3940 calories per kg = 3940 "conventional" kCal per kg). Food carbohydrates typically release 4150 food calories per kg, which is close to the value for sugar; food fats release 9450 and proteins, 5650. Our digestion system absorbs 97% of carbohydrates, 95% of fats, and 92% of proteins [Cengel and Boles, 2002]. Almost all absorbed carbohydrates and fats are converted into CO_2 and H_2O, etc., but only 77% of proteins are converted into CO_2, H_2O, etc. Thus, the net heat produced is 0.97* 4.15 = 4000 food calories from carbohydrates, 9000 from fats, and 4000 from proteins. Thus, if 0.227 kg of yogurt is consumed, from its composition, i.e., carbohydrates (0.045 kg), fats (0.003 kg), proteins (0.009 kg), one can compute the calories as 0.045* 4000 + 0.003* 9000 + 0.009* 4000 = 243 food calories or 243 conventional kCal [Oxtoby et al., 1998].

Example 9

(a) Determine HV and HV′ for CO. Typically, they are reported at 298 K.
(b) Determine HHV and LHV for CH_4.
(c) If manure is mixed with water in an anaerobic digester that releases biogas with 60% CH_4 and 40% CO_2 (mole %), what is the HHV of the mixture in kJ/m³?

Solution

(a) The heat of reaction of CO from Example 8 is −283005 kJ kmol⁻¹ at 298 K. The HV per kg of CO is given as:

$$HV = 283005/28 = 10{,}100 \text{ kJ kg}^{-1}.$$

Because a kmol of any ideal gas occupies 24.5 m³ at STP, the HV′ of CO is 283,000/24.5 or 11,550 kJ m³.

(b) Unlike CO oxidation, CH_4 oxidation produces both CO_2 and H_2O.

$$CH_4 + 2\ O_2 \rightarrow CO_2 + 2\ H_2O, \qquad \text{(A)}$$

$$\Delta H^\circ_R = H^\circ_P - H^\circ_R = h^0_{f,CO_2} + 2\, h^0_{f,H_2O} - h^0_{f,CH_4}. \qquad \text{(B)}$$

If H_2O is in the liquid state,
HHV in kJ/kmol of fuel = {393520 + 2*285830 − 74850} = 890330 kJ/kmol of fuel

$$HHV = 890330/24.5 = 36340 \text{ kJ m}^3 (= 982 \text{ Btu/SCF}). \qquad \text{(C)}$$

The corresponding LHV is given as 32750 kJ m⁻³ (= 885 Btu/SCF).

(c) The mole % is the same as the volume % for ideal gases. For 60% CH_4 and 40% CO_2, HHV = 0.60*36340 + 0.40*0 = 21492 kJ m⁻³ (= 589 Btu/SCF).

Remarks

In Equation C, we made use of the knowledge of $\bar{h}_{f,CH4}$ to calculate the value of HHV as 890,355 kJ/kmol or 890355/16.05 = 55475 kJ/kg (23850 Btu/lb); conversely, if HHV of CH_4 is given as 55,475 kJ/kg, we can determine the enthalpy of formation of the fuel $\bar{h}_{f,CH4}$.

2.5.6 Heating Value Based on Stoichiometric Oxygen

The heating value based on stoichiometric oxygen

$$HV_{O_2} = HV/\upsilon_{O_2} = (H_R - H_P)/\upsilon_{O_2} \text{ constant.} \qquad (2.66)$$

For instance, the HHV of methane is 36,340 kJ m⁻³ of fuel (or 982 Btu/SCF; see Example 9), whereas that based on unit m⁻³ of stoichiometric air is 36.34/9.52 = 3.82 MJ m⁻³ of air (or 3.82/0.21 = 18.2 MJ m⁻³ of oxygen or 18.2 (MJ/m³) * 24.5 (m³/kmol)/32 (kg/kmol) = 13.9 MJ/kg of O_2). The corresponding value in

Btu is 103 Btu ft^{-3} based on stoichiometric air (and 490 Btu ft^{-3} based on oxygen). The heating value of a low-Btu gas (which is typically a mixture of about 20% CO, 12% H_2, 6% CO_2, and the remainder being N_2) is 100 Btu ft^{-3} based on the fuel. This value is about 10% of the heating value of methane, but the stoichiometric $(A:F)_{st}$ for a low-Btu gas is also about a tenth of that required for methane. Therefore, the heating value per unit mass of stoichiometric oxygen for the low-Btu fuel is 15 MJ m^{-3} of oxygen, similar to that for methane.

The HHV per unit mass of O_2 burned is approximately the same for most hydrocarbon fuels. For methane, the HV per unit O_2 is 13.9 MJ per kg of O_2, whereas for gasoline ($CH_{2.46}$), it is 13.6 MJ per kg of O_2 and for *n*-octane, 13.6 MJ per kg of O_2.

Example 10

(a) Determine the amount of carbon dioxide (which is a greenhouse gas) emitted in kg per kWh if coal is used as a fuel. Assume that the chemical formula for coal is $CH_{0.8}\,O_{0.3}$, its gross heating value is 30,000 kJ kg^{-1}, and the power plant efficiency is 30%.

(b) Compare your answer with the result for natural gas (CH_4), which has a gross heating value of 55,475 kJ kg^{-1} and for the same plant efficiency.

Solution

(a) The chemical reaction governing the burning of coal is

$$CH_{0.8}\,O_{0.3} + 1.05\,O_2 \rightarrow CO_2 + 0.4\,H_2O.$$

The amount of CO_2 produced per kg of coal consumed is

$$44.01/(12.01 + 0.8 \times 1.01 + 16 \times 0.3) = 2.498 \text{ kg}$$

Note that 1 kWh = 1 kJ $sec^{-1} \times 3600$ sec $h^{-1} \times 1$ h^{-1} = 3,600 kJ.

Therefore, the gross heat value released by combustion per kWh = 3600 ÷ 0.3 = 12,000 kJ.

Consequently, the amount of fuel burned per kWh = 12,000 ÷ 30,000 = 0.4 kg, and the amount of CO_2 released = 2.498 × 0.4 = 0.999 kg per kWh of power or 0.208 kg/MJ of electrical energy.

(b) In the case of methane, the CO_2 produced per kg of fuel consumed = 44.01 ÷ 16.05 = 2.74, and the fuel consumed per kWh = 12,000 ÷ 55,475 = 0.22 kg.

Therefore, the CO_2 produced = 0.22 × 2.74 = 0.60 kg kWh^{-1} or 0.17 kg MJ^{-1}.

Remarks

(i) Global human activity releases 6 Gt of carbon every year in the form of CO_2 from various sources. Roughly 1.5 Gt of this amount is sequestered in the oceans, and an additional 1.5 Gt utilized by flora or vegetation worldwide. The ocean and vegetation act as carbon sinks. The remainder of the CO_2 remains in the atmosphere.

(ii) As CO_2 is a greenhouse gas (i.e., it can absorb heat), the larger the amount of CO_2 in the atmosphere, the higher the average global atmospheric temperature. This is a simplification of an accepted process widely known as global climate warming due to fossil fuels predicted by Swedish scientist Svante Arrhenius in 1896. About 18,000 years ago, the amount of CO_2 in the atmosphere was 180 to 200 ppm (on a volume or mole basis). In the 1900s, it rose to 285 ppm in 1958, 358 in 1994, and is now 365 ppm, and climbing. An elevated atmospheric temperature can lead to higher oceanic temperatures. Roger Revelle, an American geophysicist, was the first person to make modern high-level predictions of global warming (1965).

(iii) If the ocean temperature rises, the amount of dissolved gases in it decreases, with detrimental consequences for many forms of marine life.

(iv) The CO_2 emitted when natural gas is used as a fuel is reduced owing to the higher heating value of methane. The Boie equation is an empirical relation that can be used to determine the HHV of many CHNO fuels, including solid and liquid fuels. The relation is

$$\text{HHV kJ kg}^{-1} = 35{,}160 \times Y_C + 116{,}225 \times Y_H - 11{,}090 \times Y_O$$
$$+ 6{,}280 \times Y_N + 10{,}465 \times Y_S$$

where Y_C, Y_H, Y_O, Y_N, and Y_S denote the mass fractions of carbon, hydrogen, oxygen, nitrogen, and sulfur in the fuel. Using this relation, the formula for the mass of CO_2 emitted

CO_2 in kg per GJ of input $= 10^6 \div \{9595 + 31717\,(Y_H/Y_C) - 3026\,(Y_O/Y_C) + 1714\,(Y_N/Y_C) + 2856\,(Y_S/Y_C)\}$.

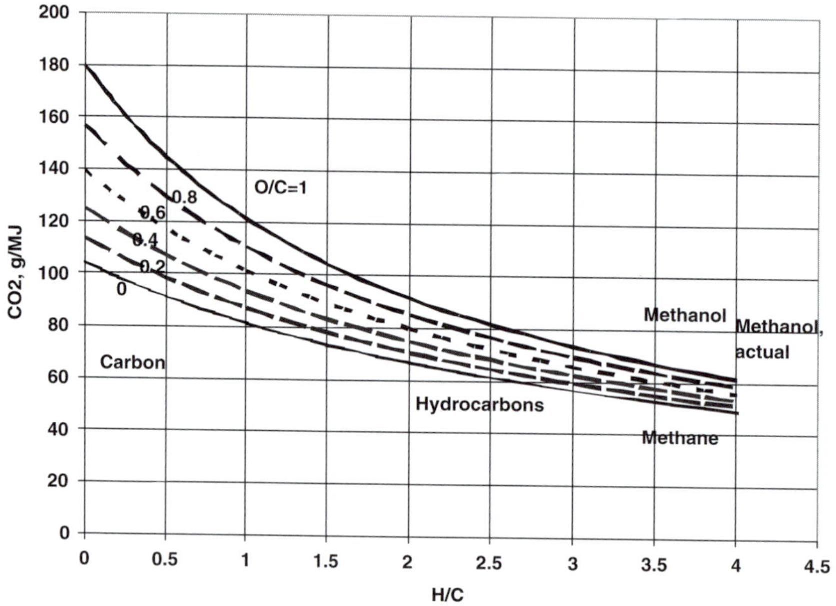

Figure 2.7 Emission of CO_2 as a function of H/C and O/C atom ratios in C–H–O fuels (g/MJ = kg/GJ; multiply by 2.326 to get lb/mmBTU).

In terms of atom ratios, it can be shown that kg of CO_2 per GJ of heat input $= 10^6 \div \{9595 + 2667\,(H/C) - 4032(O/C) + 1999\,(N/C) + 7623\,(S/C)\}$.

(v) In general, CO_2 emissions can be reduced by burning fuels with higher H/C ratios. Therefore, the higher the H/C ratio and the lower the O/C ratio, the lower the CO_2 emission to the atmosphere. Figure 2.7 illustrates the CO_2 emission for different H/C and O/C ratios of C-H-O fuels.

Example 11

Dry-ash-free (DAF) cattle waste can be represented empirically as $CH_{1.253}N_{0.0745}O_{0.516}S_{0.00813}$. The higher heating value of dry waste with 53% ash (e.g., sand) is known to be 9,215 kJ kg^{-1} of dry fuel. Determine (a) the molecular weight of 1 kmol of DAF fuel, (b) the stoichiometric air required in kg per kg of dry fuel, and (c) the enthalpy of formation in kJ kmol^{-1} for the DAF fuel.

Solution

The heating value of dry waste with 53% ash is 9215 kJ kg^{-1}. This contains 47% of pure DAF fuel. The HHV value of 100% DAF fuel is

$$HV_{DAF} = 9215/0.47 = 19606.38 \text{ kJ kg}^{-1} \text{ of DAF fuel.}$$

Using the chemical formula $CH_{1.253}N_{0.0745}O_{0.516}S_{0.00813}$,

$$M_{Fuel} = (1 \times 12.01) + (1.253 \times 1.01) + (0.0745 \times 14.01) + (0.516 \times 16) + (0.00813 \times 32) = 22.84 \text{ kg kmol}^{-1} \text{ of DAF fuel.}$$

Using atom balance the stoichiometric reaction may be written as

$$CH_{1.253}N_{0.0745}O_{0.516}S_{0.00813} + 1.063\,(O_2 + 3.76\,N_2) \rightarrow CO_2 + 0.6265H_2O\,(1) + 4.034N_2 + 0.00813\,SO_2.$$

A:F on a mass basis is

$$A{:}F = \{1.063 \times 32 + 1.063 \times 3.76 \times 28.01\}/\{1 \times 22.812\} = 6.40.$$

The stoichiometric air per kg of dry fuel (with ash)

$$6.40 \times 0.47/1 = 3.0074 \text{ kg of air per kg dry fuel}$$

$\overline{HHV} = 19606.38$ kJ kg^{-1} × 22.84 kg kmol^{-1} = 447261 kJ kmol^{-1}, where

$$447261 = h^{o}_{f,DAF} - h^{o}_{f\,CO_2}\,(298\text{ K}) - 0.6265\,h^{o}_{fH_2O}\,(\ell)\,(298\text{ K}) - 0.00813\,h^{o}_{fSO_2}\,(298\text{ K})$$

$$= h^{o}_{f\,dry\,fuel} - (-393546) - 0.6265\,(-285{,}830) - 0.00813\,(-296842),$$

or

$$h^{O}_{DAF}\,(298\text{ K}) = -122771\text{ kJ kmol}^{-1}.$$

Example 12

A burner facility is operated with fuel 1, $CH_{h1}\,N_{n1}O_{o1}S_{s1}$ which has a rated thermal output of $\dot{P}_t$ at an equivalence ratio of ϕ. The fuel is then switched to fuel 2, $CH_{h2}\,N_{n2}O_{o2}S_{s2}$ with the flow rate of fuel adjusted so that the same equivalence ratio is maintained. Typically, the HHV per stoichiometric airflow is approximately constant.

(a) Show that the equivalence ratio is same the as long as the O_2 percentage on a dry basis is kept the same for both the fuels
(b) Show that the airflow rate must be kept the same for the same rated thermal output.

Solution

(a) Let $\bar{\nu}_{O2,1}$ be the stoichiometric O_2 oxygen-to-fuel mole ratio for fuel 1 and $\bar{\nu}_{O2,2}$, the corresponding ratio for fuel 2. Then, O_2 percentage for fuel $CH_{h1}\,N_{n1}O_{o1}S_{s1}$ is given as

$$X_{O2,1} = \bar{\nu}_{O2,1}\,((1/\phi) - 1)/\{\bar{\nu}_{O2,1}\,((1/\phi) - 1) + \bar{\nu}_{O2,1}\,(3.76/\phi) + 1 + s_1 + n_1/2\}$$

Rewriting

$$\bar{\nu}_{O2,1}\,((1/\phi)-1)\,X_{O2,1} + X_{O2,1}\,\bar{\nu}_{O2,1}\,(3.76/\phi)$$

$$+ \{1 + s_1 + n_1/2\}\,X_{O2,1} = \{\bar{\nu}_{O2,1}\,((1/\phi) - 1)$$

where $\{(1/\phi) - 1\}$ = excess O_2 fraction = excess air fraction.
Simplifying and solving

$$(1/\phi) = [1 - X_{O2,1}\,\{1 - (1+ s_1+n_1/2)/\bar{\nu}_{O2,1}\}]/[1 - 4.76X_{O2,1}]$$

Typically $\bar{\nu}_{O2,1} \approx 1 - 2$, $X_{O2} \approx 0.03 - 0.05$, $s_1 = 0.01$, $n_1 = 0.1$;

$$(1/\phi) \approx 1/[1 - 4.76X_{O2,1}]$$

Thus, as long as X_{O2} is the same in both cases, $(1/\phi)$ remains the same.
(b) Fuel is switched in boilers many times; but thermal output must be maintained in order to produce the same electric power

$$\text{Thermal output} = \dot{m}_{air,st.} * h_{c,air} = \dot{P}_t$$

where $h_{c,air}$, heat released per unit stoichiometric mass of air; $\dot{P}_t$, thermal (not electric) power output

$$\dot{m}_{air,st.} = \dot{P}/h_{c,air},\ \dot{m}_{air} = \dot{P}_t/(\phi h_{c,air})$$

As $h_{c,air}$ for most fuels has the same value, then for the same throughput and equivalence ratio, the air supplied must be maintained at the same rate when switching fuels.

2.5.7 Applications of Thermochemistry

2.5.7.1 First Law Analysis of Reacting Systems

Example 13

Glucose, i.e., $C_6H_{12}O_6$, is oxidized in the human body in its cells. Air is inhaled at 298 K, and a mixture of air and metabolism products (CO_2, H_2O, O_2, and N_2) is exhaled at 310 K. Assume that glucose is supplied steadily to the cells at 25°C and that 400% excess air is utilized during its consumption. If the breathing rate of humans is 360 l/h, determine the amount of heat loss from the human body given that $\bar{h}_{f,glucose} = -1.26 \times 10^6$ kJ kmol^{-1} (Table A.32 of Appendix A), and the higher heating value of glucose is 15,628 kJ kg^{-1}.

Solution

The reaction equation with 400% of excess air (or 5 × theoretical air) is

$$C_6H_{12}O_6 + 5 \times 6\,(O_2 + 3.76\,N_2) \rightarrow 6\,CO_2 + 6\,H_2O + 24\,O_2 + 112.8\,N_2.$$

We now conduct an energy balance, i.e.,

$$dE_{c.v.}/dt = \dot{Q}_{c.v.} - \dot{W}_{c.v.} + \sum_{k,i} \dot{N}_k \bar{h}_k - \sum_{k,e} \dot{N}_k \bar{h}_k$$

Ignoring the breathing (PdV) and other forms of work and using appropriate tables (Table A.10 to Table A.30 of Appendix A)

$$0 = \dot{Q}_{c.v.} + 1 \times (-1.26 \times 10^6) - 6 \times (-393520 + 443) - 6 \times (-241820)$$
$$+ (398) - 24 \times (0 + 348) - 112.8 \times (0 + 345).$$

Hence,

$$\bar{q}_{c.v.} = -2.5 \times 10^6 \text{ kJ kmol}^{-1} \text{ of glucose.}$$

$M_F = 6 \times 12.01 + 12 \times 1.01 + 6 \times 16 = 180.2$ kg kmol^{-1}.
Therefore, $q_{c.v.} = 13872$ kJ per kg^{-1} of glucose.
The air consumption is 360 l h^{-1} or 0.360 m^3 h^{-1}/(24.5 m^3 kmol^{-1}) = 0.0147 kmol h^{-1} so that the glucose consumption is 0.0147 ÷ 142.8 = 0.0001031 kmol h^{-1}, or (0.0001031 kmol × 180.2 kg kmol^{-1} × 1000 g kg^{-1} h^{-1}) ÷ 60 min h^{-1} = 0.31 g min^{-1}.
The heat loss rate

$$\dot{Q}_{c.v.} = \dot{m}_F\, q_{c.v.} = 0.31 \text{ g min}^{-1} \times 13872 \text{ J g}^{-1} \div 60 \text{ sec min}^{-1} = 72 \text{ W}.$$

Remarks

(i) We can interpret from this result that the hypothetical human body considered in this example has the same energy consumption as a 70-W light bulb. However, during fever, there is increased metabolic activity to maintain the fever (higher temperature) [Annamalai and Morris, 1997].

(ii) Our typical food consists of carbohydrates (HHV = 18,000 kJ dry kg^{-1}), fatty acids (HHV = 40,000 kJ dry kg^{-1}), and proteins (HHV = 22,000 kJ dry kg^{-1}), of which almost 96%, 98%, and 78% are metabolized, respectively. In general, the larger the amount of moisture in food, the lower its heating value. Fats have a lower moisture content and, therefore, have a larger heating value, whereas the carbohydrates, having relatively more moisture, have a smaller heating value.

(iii) If glucose is not transported to cells or is not metabolized, its concentration will increase in the bloodstream, which is characteristic of the diabetic condition.

(iv) The body burns 0.7 kJ min^{-1} kg^{-1} of energy during aerobic dancing, 0.8 kJ min^{-1} kg^{-1} while running a mile in 9 min, 1.2 kJ min^{-1} kg^{-1} for a 6-min mile, and 0.3 kJ min^{-1} kg^{-1} while walking.

2.5.7.2 Adiabatic Flame Temperature

Consider a flammable premixed mixture of methane and air jet that emerges from a burner. The contour of the hottest location is often called the flame (cf. Figure 2.8). The maximum possible flame temperature is called the *adiabatic flame temperature*, which is attained in the absence of heat losses and work transfer.

2.5.7.2.1 Analysis of Steady-State Steady-Flow Processes in Open Systems

We will apply the energy balance equation in the context of steady state (Figure 2.8), in the absence of work transfer, complete combustion and neglect

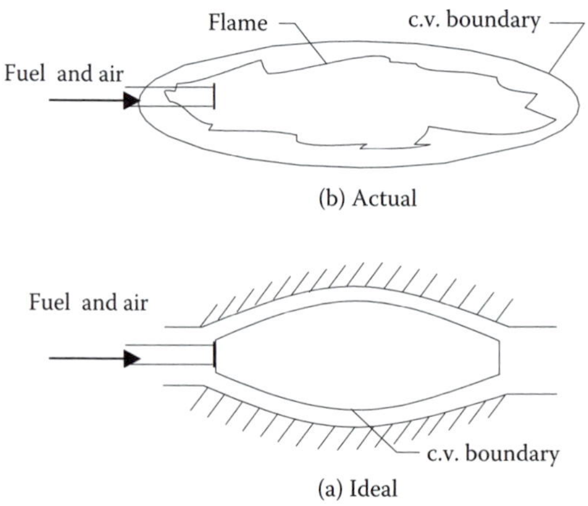

Figure 2.8 Schematic diagram of a flame and its idealization.

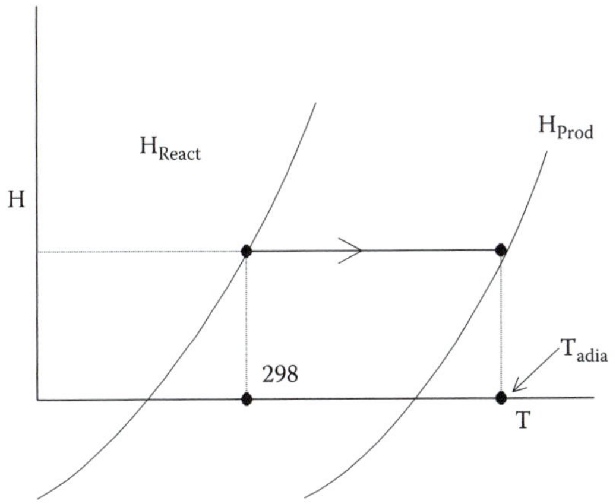

Figure 2.9 Illustration of adiabatic flame temperature.

the kinetic and potential energies so as to obtain the maximum possible flame temperature

$$dE_{c.v.}/dt = \dot{Q}_{c.v.} - \dot{W}_{c.v.} + \sum_{k,i} \dot{N}_k \bar{e}_{TK} - \sum_{k,e} \dot{N}_k \bar{e}_{TK}$$

Dividing by the molar flow of fuel, for an adiabatic combustor, and for steady state

$$0 = H_{prod} - H_{reac}, \tag{2.67}$$

where $H_{prod} = \Sigma_{k,e} (\dot{N}_k/\dot{N}_F)\, \bar{h}_k$ and $H_{Reac} = \Sigma_{k,i} (\dot{N}_k/\dot{N}_F)\, \bar{h}_k$. This implies that the total enthalpy of the products leaving the combustor equals that of the entering reactants, which allows us to calculate the adiabatic flame temperature (Figure 2.9).

2.5.7.2.2 Analysis of Closed Systems

The energy balance for a closed system results in the relation

$$\bar{q} - \bar{w} = U_P - U_R = \sum_{k,P} (\dot{N}_{k,P}/\dot{N}_F)\bar{u}_{k,p} - \sum_{k,R} (\dot{N}_{k,R}/\dot{N}_F)\bar{u}_{k,R} \tag{2.68}$$

where $\bar{q}$, $\bar{w}$ in kJ/kmol of fuel

$$\bar{u}_k = \bar{h}_k - (P\bar{v})_k$$

For ideal gases,

$$\bar{u}_k = \bar{h}_k - RT, \text{ i.e.,}$$

Example 14

Liquid octane burns in a reactor with 40% excess air. Determine the adiabatic flame temperature in the case of (a) a steady-state steady-flow device (Figure 2.10a)

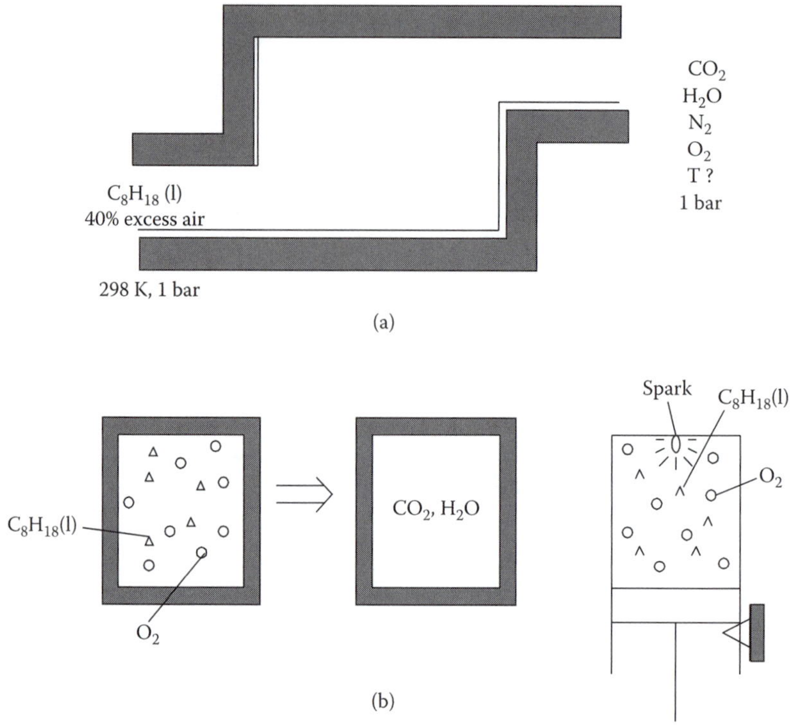

Figure 2.10 Adiabatic flame temperature for liquid octane in (a) an open system, and (b) a closed system.

with both the fuel and air entering at 298 K and 1 bar, and (b) a rigid closed system (Figure 2.10b) with both the fuel and air being initially at 298 K and 1 bar. Also determine the final pressure for case (b). Assume ideal gas behavior and that $\rho_{octane(l)} = 703$ kg m^{-3}.

Solution

For a stoichiometric reaction,

$$C_8H_{18}(\ell) + 12.5\ O_2 \rightarrow 8CO_2 + 9H_2O. \quad \text{(A)}$$

In this example, when 40% excess air is supplied,

supplied $O_2 = 12.5 \times 1.4 = 17.5$ mol per mole of fuel, and

supplied $N_2 = 17.5 \times (79 \div 21) = 65.83$ mol per mole of fuel.

Therefore, the reaction equation assumes the form

$$C_8H_{18}(\ell) + 17.5\ O_2 + 65.83\ N_2 \rightarrow 8CO_2 + 9\ H_2O + 5\ O_2 + 65.83\ N_2. \quad \text{(B)}$$

(a) The energy balance equation for a steady-state steady-flow device is

$$dE_{c.v.}/dt = \dot{Q}_{c.v.} - \dot{W}_{c.v.} + \sum\{\dot{N}_k\bar{h}_k\}_i - \sum\{\dot{N}_k\bar{h}_k\}_e \qquad (C)$$

As the system is adiabatic, steady, and does not involve work,

$$0 = (\Sigma\dot{N}_k\bar{h}_k)_i - (\Sigma\dot{N}_k\bar{h}_k)_e = H_{reac} - H_{prod} \qquad (D)$$

$$\text{where } H_{prod} = \{8 * (-393520 + \bar{h}_{t,T} - \bar{h}_{t,298})CO_2 + 9 * (-241845$$

$$+ \bar{h}_{t,T} - \bar{h}_{t,298})H_2O + 5\ (0 + \bar{h}_{t,T} - \bar{h}_{t,298})\ O_2$$

$$+ 65.83 * (0 + \bar{h}_{t,T} - \bar{h}_{t,298})N_2\} \qquad (E)$$

$$\text{And } H_{reac} = \{1 * (-249950)_{C_8H_{18(l)}} + 17.5 * 0 + 65.83 * 0\}$$

$$= -249950 \text{ kJ} \qquad (F)$$

We use Equations E and F in D and solve T by iteration (see the following table).

Assume T = 2000 K; then select $\bar{h}$ from Table A.13, Table A.18, Table A.23, and Table A.26 of Appendix A, for CO_2, H_2O, N_2, and O_2 for the assumed T and Table A.32 of Appendix A for $\bar{h}^{\circ}_{f,C8H18(l)}$.

T	$(\bar{h}_{t,T} - \bar{h}_{t,298})_{CO2}$ at *T*	h_{H2O}	h_{N2}	h_{O2}	*Hp*	$H_R - Hp$
1900	85,392	67,725	52,541	55,402	−296,000	−46,050
2000	91,420	72,805	56,130	59,169	52,720	302,670

Look at the data for 1900 and 2000 K. As $H_{reac} - H_{prod} = 0$, then interpolating T = 1913 K ≈ 1900 K.

(*Note:* $h_{t2} - h_{t1} \neq c_{p2}(T_2)\ T_2 - c_{p1}(T_1)\ T_1$)

(b) The corresponding energy balance equation for a rigid closed system is

$$Q - W = U_{prod} - U_{reac}.$$

The work $W = \int PdV = 0$, as $dV = 0$, and $Q = 0$ as we assume that the system is adiabatic. Therefore,

$$U_{prod} = U_{reac} \text{ or } \bar{u}_{prod} = \bar{u}_{reac}$$

In this context,

$$\bar{u}_{octane(l)} = \bar{h}_{octane(l)} - P\bar{v} = -249950 \text{ kJ} - 1 \text{ bar} \times 100 \text{ kN m}^{-2} \text{ bar}^{-1}$$

$$\div (703 \text{ m}^3 \text{ kg}^{-1}) \times 114 \text{ kg kmol}^{-1} \approx -249950 \text{ kJ kmol}^{-1}.$$

For an ideal gas, $P\bar{v} = \bar{R}T$, i.e.,

$$\bar{u}_{O_2,298} = \bar{h}_{O_2,298\text{ K}} - 8.314 \times 298 = -2478 \text{ kJ kmol}^{-1}.$$

$$\bar{u}_{N_2,298} = h_{N_2,298\text{K}} - 8.314 \times 298 = -2478 \text{ kJ kmol}^{-1}.$$

Therefore,

$$U_{Reac} = (1 \times (-249950) + 17.5 \times (-2478) + 65.83 \times (-2478))$$

$$= -456406 \text{ kJ kmol}^{-1}.$$

At 2400 K,

$$u_{CO_2} = \bar{h}_{CO_2} - 8.314 \times 2400 = -393520 + (115, 798) - 8.314 \times 2400$$

$$= -297685 \text{ kJ kmol}^{-1}.$$

Hence,

$$\begin{aligned} U_{prod} - U_{reac} = \{&(8 \times (-393520 + 115798) - 8.314 \times 2400)\} \\ &+ 9 \times (-241820 + 93744) - 8.314 \times 2400) \\ &+ 5 \times (0 + 74467) - 8.314 \times 2400) \\ &+ 65.83 \times (0 + 70645) - 8.314 \times 2400)) - (-456406) \\ &\approx 174000 \text{ kJ kmol}^{-1}. \end{aligned}$$

Assume that T = 2200 K,

$$U_{prod} - U_{reac} \approx -389{,}000 \text{ kJ kmol}^{-1}.$$

Interpolating, we obtain that the adiabatic flame temperature

$$T \approx 2340 \text{ K}.$$

The increase in pressure is

$$P_2/P_1 = N_2RT_2/(N_1RT_1)$$

$$= 87.83 \times 2300 \div (84.33 \times 298) = 8.03 \text{ bar}.$$

Remarks

(i) CO_2 and H_2O are known to be radiant species in the products. Thus, if 10% of the heat value (HHV: 48275 kJ/kg or 5515900 kJ/kmol, LHV = 44790 kJ/kg or 5117700 kJ/kmol) is lost as radiation, then the total loss is 511770 kJ, which is essentially because of CO_2 and H_2O. The total thermal energy carried by CO_2 and H_2O is 8 * 85392+ 9 * 67725 = 1292660 kJ.

Thus, 511770/12926600 = 39.6, i.e., 40% of the thermal energy of CO_2 and H_2O is lost at 40% excess air; thus, as CO_2 and H_2O get cooled by radiation, the O_2 and N_2 must supply the necessary thermal energy to CO_2 and H_2O.

(ii) You can repeat these calculations for air that is preheated to 700 K. The open-system adiabatic flame temperature is roughly 2200 K, which is about 300 K higher than that when air at 298 K is used (1913 K). If the preheat required for the secondary air is supplied by the combustion products at 2200 K, then the combustion products' temperature drops to 1913 K. In other words, if one tracks the temperature in such a system, it will rise gradually from 298 K to 2200 K (with an enthalpy above the conventional value without preheating) during an adiabatic process and decrease to 1913 K once heat is removed for preheating the combustion air. Such a scheme is called excess enthalpy combustion (EEC), which is useful for fuels, particularly of low heating values, or combustion involving a large amount of excess air in order to stabilize the flame. On the other hand, if one wants to maintain a temperature of 1913 K without preheating, then the oxygen concentration must be reduced as is done in high temperature air combustion (HiTAC) systems [Tsuji et al.].

2.5.8 Second Law Analysis of Chemically Reacting Systems

2.5.8.1 Entropy

We will now introduce methodologies for determining the entropy, Gibbs function, and other species properties, which are useful in the application of the second law to combustion problems.

Because for an ideal gas,

$$ds = c_{p0}\ dT/T - R\ dP/P,$$

then for a pure component,

$$s_k(T,P) - s_k(T_{ref},P^O) = s_k^0(T) - R\ \ln\ (P/P^O), \tag{2.76}$$

where

$$s_k^0 = \int_{T_{ref}}^{T} (c_{p.k}/T)dT \tag{2.77}$$

It is usual to select the conditions T_{ref} = 298 K, P^O = 1 atm, and $s_k(T_{ref}, P^O)$ = 0 for an ideal gas.

2.5.8.2 Entropy Generated during Any Chemical Reaction

A chemical reaction is irreversible and hence $\delta\sigma > 0$; the second law yields ds = $\delta q/Tb + \delta\sigma$ and under adiabatic conditions, ds (= $\delta\sigma$) > 0. Why does this happen? Combustion is usually a process of oxidation of fossil fuels to CO_2

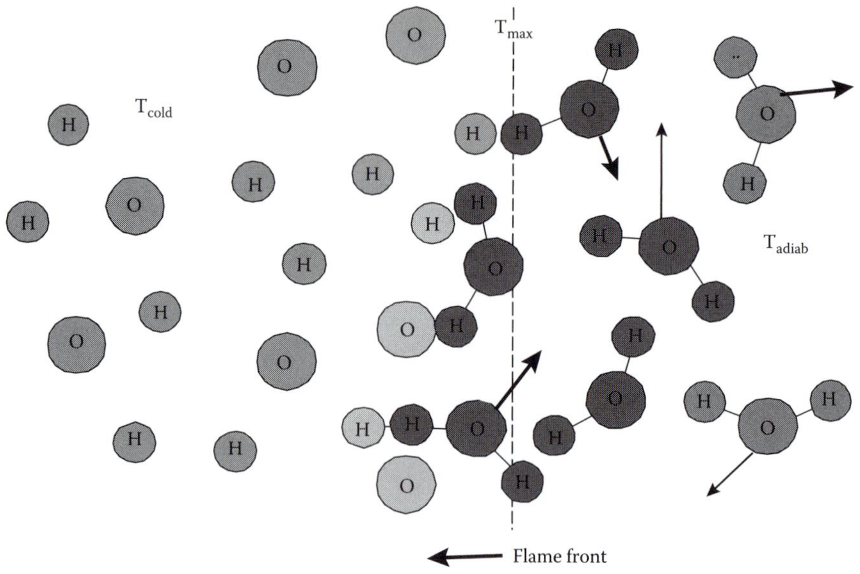

Figure 2.11 Entropy generation during flame propagation. (Adapted from Daw, C.S., Proceedings of Central States Section of the Combustion Institute, Austin, Texas, March 21–23, 2004.)

and H_2O. Fuel and oxygen are called reactants in this case, whereas CO_2 and H_2O are called products. The chemical bonds (which are more organized forms of energy) between the atoms in fuel are broken during combustion and energy is converted into thermal energy, which is eventually stored as te, ve, and re (random or disorganized energy) in a large number of quantum states, or random energy states. Hence, the entropy of the products is higher than that of the reactants. Figure 2.11 shows a schematic view of entropy generation between burned products (random energy) and an unburned mixture of H_2 and O_2 (better-organized energy). The entropy generation results in lower work capability of the fluid (Chapter 1). The work loss capability due to entropy generation is given as $T_0\ \sigma$ [Annamalai and Puri, 2001, Bejan, 1988]. For example, the exergy or loss of work capability in the gas turbine combustor could be as high as 20% to 30% of heat value and forms the largest of all losses in a gas turbine system [Nishida et al., 2003].

2.5.8.3 Entropy Balance Equation

2.5.8.3.1 Entropy Generated during an Adiabatic Reaction

Example 15

One kmol of CO, 0.5 kmol of O_2, and 1.88 kmol of N_2 enter an adiabatic reactor and produce CO_2 because of the exothermic reaction of the CO and O_2. (a) Determine the temperature and entropy of the products assuming that CO reacts in varying amounts, i.e., 0.1, 0.2, … , 0.9, 1 kmol., (b) Obtain an expression for $S_p - S_R$ in kJ/(kg mix k) if $C_{pmix} = Y_k\, C_{pk}$ and M of mix are constant.

Solution

(a) The overall reaction equation for each kmol of CO oxidized is

$$CO + 1/2\ O_2 + 1.88\ N_2 \rightarrow CO_2 + 1.88\ N_2 \quad \text{(A)}$$

For a steady adiabatic process that involves no work, we have shown before that

$$H_{Prod} - H_{Reac} = \sum_{k,e} \dot{N}_k \bar{h}_k - \sum_{k,i} \dot{N}_k \bar{h}_k = 0. \quad \text{(B)}$$

For every kmol of CO consumed, a half kmol of oxygen is consumed and a kmol of carbon dioxide is produced. For instance, in the case when 0.2 mol of CO react:

$$CO + 1/2\ O_2 + 1.88\ N_2 \rightarrow 0.8\ CO,\ 0.2\ CO_2,\ 0.4\ O_2,\ 1.88\ N_2$$

$$H_{Reac} = 1 \times (-110530) + 0 + 0 \times (-393520) + 1.88 \times 0 = -110530 \text{ kJ}, \quad \text{(C)}$$

and

$$H_{Prod} = 0.8 \times (-110530 + (h_T - h_{298})) + 0.4 \times (0 + (h_T - h_{298}))$$
$$+ 0.200 \times (-393520 + (h_T - h_{298})) + 1.88 \times (0 + (h_T - h_{298})). \quad \text{(D)}$$

Equating Equation C and Equation D, one can iteratively solve for T, which in this case is 848 K. Figure 2.12 presents the variation of T and entropy generated vs. the moles of CO that are burned. The corresponding entropy balance equation for steady-state and adiabatic conditions assumes the form

$$0 = \dot{Q}_{c.v.}/T_b + \sum_{k,i} \dot{N}_K \bar{s}_k - \sum_{k,e} \dot{N}_K \bar{s}_k + \dot{\sigma}, \quad \text{(E)}$$

$$\dot{\sigma} = \sum_{k,e} \dot{N}_K \bar{s}_k - \sum_{k,i} \dot{N}_K \bar{s}_k \quad \text{(F)}$$

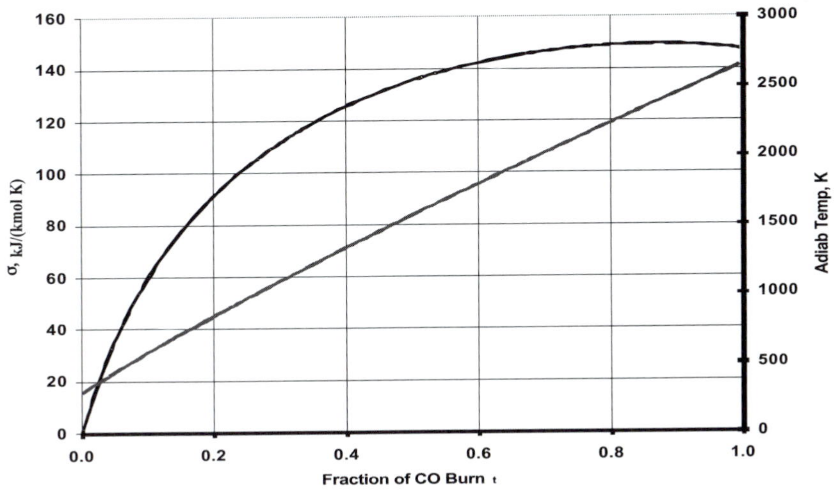

Figure 2.12 Variation of the adiabatic temperature and entropy generation with the CO burned fraction.

Denoting $S_{reac} = (\Sigma N_k \bar{s}_k)_i$, $S_{prod} = (\Sigma N_k \bar{s}_k)_e$, and $\hat{s}_k(T,P,X_k) = \bar{s}_k(T,p_k) = \bar{s}_k^o - \bar{R}\ln(p_k/p^o)$, we obtain

$$\bar{\sigma} = \bar{S}_{prod} - \bar{S}_{reac}. \quad (G)$$

As S_{reac} is specified (687 kJ/kmol) and $\sigma > 0$, S_{prod} must increase with the burned fraction. For instance, if 0.2 mol of CO react, the product stream contains 0.8 mol of CO, 0.4 mol of oxygen, 0.2 mol of the dioxide, and 1.88 mol of nitrogen, which is inert. In that case,

$$p_{CO} = N_{CO}P/N = 0.8 \times 1/(0.8 + 0.4 + 0.2 + 1.88) = 0.244 \text{ atm.}$$

Similarly, $p_{O_2} = 0.122$, $p_{CO_2} = 0.0.061$, and $p_{N_2} = 0.573$. Hence,

$$S_{prod} = 0.8 \times \bar{s}_{CO}(T, p_{co}) + 0.40 \times \bar{s}_{O_2}(T, p_{O_2}) + 0.20 \times \bar{s}_{CO_2}(T, p_{CO_2})$$

$$+ 1.88 \times \bar{s}_{N_2}(T, p_{N_2}) = 0.8 \times (\bar{s}^o_{CO}(848\text{ K}) - 8.314 \times \ln 0.57/1)$$

$$+ 0.4 \times (\bar{s}^o_{O_2}(848\text{ K}) - 8.314 \times \ln 0.122/1) + 0.2 \times (\bar{s}^o_{CO_2}(848\text{ K})$$

$$- 8.314 \times \ln 0.061/1) + 1.88 \times (\bar{s}^o_{N_2}(848\text{ K}) - 8.314 \times \ln 0.573/1) = 779 \text{ kJ/K.}$$

Figure 2.12 plots the variation in S_{prod} with respect to the CO burn fraction. The S_p reaches a maximum at CO burn fraction = 0.87 with T = 2377 K and S_{prod} = 837 kJ/kmol of CO. Then $\sigma_{max} = S_{p,max} - S_{reac} = 837 - 687 = 150$ kJ/kmol CO.

(b) $s_P - s_R = \Sigma_{k,e} Y_{k,e}\, s_{k,e} - \Sigma_{k,i}\, Y_{k,e}\, s_{k,e}$

Consider $s_p = \underline{s_e} = \Sigma_{k,e} Y_{k,e} \{c_{pk} \ln(T_e/T_{ref}) - (\bar{R}/M_{k,e}) \ln\{X_{k,e}\, P_e/P^0\} = c_{pmix} \ln(T_e/T_{ref}) - \Sigma_{k,e}(\bar{R}\, X_{k,e}/M) \ln\{X_{k,e}\, P_e/P^0\}$ where c_p of mixture $= \Sigma Y_k\, \underline{c}_{pk}$

σ_m {kJ/(kg mix K)} $= s_P - s_R = c_{pmix} \ln(T_e/T_i) - (\bar{R}/M) \ln\{P_e/P_i\} - (\bar{R}/M) \ln\{\Pi_k X_{k,e}{}^{Xk,e}/(\Pi_k X_{k,i}{}^{Xk,I})\}$

Entropy generated in adiabatic irreversible chemical reaction = entropy change = change due to increased random energy + change due to pressure change (which changes intermolecular spacing) + change due to irreversible mixing of different components. Typically the mixing part is negligible. Thus random energy contributes to major part of entropy increase.

2.5.8.3.2 Entropy Generated during Any Chemical Reaction

Example 16

The following species are introduced into a reactor at 3000 K and 101 kPa: 5 kmol sec^{-1} of CO, 3 kmol sec^{-1} of O_2, and 4 kmol sec^{-1} of CO_2. One percent of the inlet CO oxidizes to form CO_2. Determine $\delta\sigma$ for a fixed mass of the reactants. Assume instantaneous mixing at entry and ideal gas behavior.

Solution

Consider a reactant mixture consisting of 5 kmol of CO, 3 kmol of O_2, and 4 kmol of CO that has a total mass of $5 \times 28 + 3 \times 32 + 4 \times 44 = 412$ kg. We will follow this mass as the reaction proceeds. The volume of their mass may change but P remains constant (Figure 2.13). Assume that CO is oxidized according to the chemical reaction

$$CO + 1/2\ O_2 \rightarrow CO_2.$$

As only 1% of the inlet or initial CO undergoes this reaction, then the number of moles of CO will be

$$N_{CO} = 5 - 0.01 \times 5 = 4.95 \text{ kmol}.$$

Hence, $dN_{CO} = -0.05$ kmol. Likewise, $N_{CO_2} = 4 + 0.05 = 4.05$ kmol.

$$N_{O_2} = 3 - 0.05 \times 0.5 = 2.975 \text{ kmol}.$$

At constant pressure, the first law for a fixed mass system yields

$$\delta Q_R = \Delta H.$$

From Tables A.12, A.13, and A.26 at 3000 K

$$\bar{h}_{CO} = -110{,}530 + 93562 = -16979 \text{ kJ kmol}^{-1}.$$

$$\bar{h}_{CO_2} = -393520 + 152891 = -240629 \text{ kJ kmol}^{-1}, \text{ and}$$

$$\bar{h}_{O2} = 0 + 98036 = 98036 \text{ kJ kmol}^{-1}.$$

Note that at 3000 K,

$$\Delta H_R^o \text{ (3000 K)} = -272370 \text{ kJ kmol}^{-1} \text{ of CO}.$$

Enthalpy after reaction:

$H_{Prod} = 4.95 \times (-16979) + 2.975 \times (98036) + 4.05 \times (-240{,}629) = -766936$ kJ (or -1862 kJ kg^{-1} of mixture).

The reactant enthalpy or enthalpy before reaction:

$H_{Reac} = 5 \times (-16979) + 3.0 \times (98036) + 4 \times (-240{,}629) = -753303$ kJ (or -1828 kJ kg^{-1} of mixture).

Therefore,

$$\delta Q_R = H_{Prod} - H_{Reac} = -13633 \text{ kJ (or } -33.1 \text{ kJ kg}^{-1} \text{ of mixture)}.$$

This is the amount of heat that has to be removed from 412 kg of mixture so that the products can be maintained at 3000 K.

The entropies for ideal gases before reaction are:

$$X_{CO} = 5/(5 + 4 + 3) = 0.417,\ X_{O_2} = 0.25, \text{ and } X_{CO_2} = 0.3323, \text{ i.e.,}$$

$$\hat{s}_{CO}(T,P,X_{CO}) = \bar{s}_{CO}(T,P) - \bar{R} \ln X_{CO}$$

$$= 273.58 - 8.314 \times \ln 0.417 = 280.85 \text{ kJ K}^{-1} \text{ kmol}^{-1},$$

$$\hat{s}_{O_2}(T,P,X_{O_2}) = 284.4 - 8.314 \times \ln 0.25 = 295.9 \text{ kJ K}^{-1} \text{ kmol}^{-1}, \text{ and}$$

$$\hat{s}_{CO_2}(T,P,X_{CO_2}) = 334.08 - 8.314 \ln 0.33 = 343.3 \text{ kJ K}^{-1} \text{ kmol}^{-1}.$$

The total reactants entropy is

$$S_{Reac} = 5.00 \times 280.85 + 3 \times 295.9 + 4 \times 343.3 = 3665.1 \text{ kJ K}^{-1}, \text{ i.e.,}$$

After reaction,

$X_{CO} = 0.413$, $X_{O_2} = 0.248$, and $X_{CO_2} = 0.338$, so that

$$\hat{s}_{CO} (T, P, X_{CO,e}) = 273.58 - 8.314 \times \ln (0.413) = 280.93 \text{ kJ K}^{-1} \text{ kmol}^{-1},$$

$$\hat{s}_{O_2} = 284.4 - 8.4314 \times \ln 0.248 = 296.0 \text{ kJ K}^{-1} \text{ kmol}^{-1}, \text{ and}$$

$$\hat{s}_{CO_2} = 334.08 - 8.314 \times \ln 0.338 = 343.1 \text{ kJ K}^{-1} \text{ kmol}^{-1}.$$

The product entropy or S after reaction

$$S_{Prod} = 4.95 \times 280.93 + 2.975 \times 296 + 4.05 \times 343.1 = 3660.8 \text{ kJ}.$$

$$dS = S_{Prod} - S_{Reac} = 3665.1 - 3660.8 = 4.3 \text{ kJ K}^{-1} \text{ kmol}^{-1}.$$

Therefore,

$$\delta\sigma = dS - \delta Q_P/T = 4.3 - (-13{,}633) \div 3000 = 0.244 \text{ kJ K}^{-1}$$

$$\text{or} \quad 0.0006 \text{ kJ kg}^{-1} \text{ of mixture}$$

Remarks

(i) For an elemental reaction, $\delta\sigma > 0$.

(ii) Because no entropy is generated owing to nonuniform temperature or pressure gradients within the fixed mass, the finite positive value of $\delta\sigma$ arises on account of the irreversible chemical reaction.

(iii) In such a system, there are no thermal or mechanical irreversibilities. Generalizing,

$$dS - \delta Q_R/T = \delta\sigma, \quad \text{i.e.,}$$

$$T\ \delta\sigma = T\ dS - \delta Q = T(ds) - (dh) = -d\ (H - Ts) = -dG_{T,P} = T\ \delta\sigma$$

$$= 3000 \times 0.935 = (2806 \text{ kJ K}^{-1} \text{ kmol}^{-1} > 0), \text{ and}$$

$$dG_{T,P} = -\ \delta\sigma < 0.$$

Here,

$$G = H - TS = \sum \hat{h}_k N_k - T \sum N_k \hat{s}_k = \sum N_k(\hat{h}_k - T\hat{s}_k) = \sum N_k \hat{g}_k.$$

The Gibbs energy of a fixed mass at a specified temperature and pressure decreases and the entropy generated increases as the chemical reaction proceeds (see Figure 2.13). As G is a measure of availability, the availability decreases during this irreversible process.

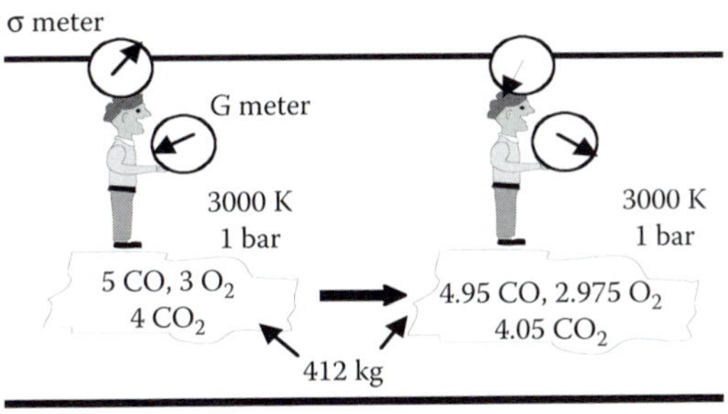

Figure 2.13 Illustration of entropy generation and decrease in G of a fixed mass with reaction. The value of G keeps decreasing as a reaction proceeds.

In Chapter 3, we will see that the direction of reaction (i.e., from the reactant CO to the product CO_2) must be such that $\delta\sigma > 0$ or $dG_{T,P} < 0$. If this is not satisfied, then the assumed reaction is impossible.

2.5.8.4 Gibbs Function, and Gibbs Function of Formation

The Gibbs function for a pure component k is

$$\bar{g}_k = \bar{h}_k - T\bar{s}_k \tag{2.78}$$

For instance, the Gibbs function of H_2O at 298 K and 1 atm is

$$\bar{g} = \bar{h} - T\,\bar{s} = -285830 - 298 \times 69.95 = -306675 \text{ kJ kmol}^{-1} \text{ of } H_2O.$$

Because combustion problems typically involve mixtures, the entropy of the k-th component in a mixture must be first determined, e.g., following the relation

$$\hat{s}_k(T,P,X_k) = \bar{s}_k(T,P) - \bar{R}\ln(X_k) = \bar{s}^0(T) - \bar{R}\ln(p_k/1), \text{ where } p_k = X_k\,P.$$

The Gibbs function of the k-th mixture component $\hat{g}_k$ can be obtained with

$$\hat{g}_k = \hat{h}_k - T\hat{s}_k. \tag{2.79}$$

For ideal gases, $\hat{h}_k = \bar{h}_k$, and $\bar{h}_k = \bar{h}_{f,k}{}^0 + \Delta\bar{h}_k$.

Alternatively, the values of $\bar{g}_k$(298 K, 1 atm) for any compound can also be determined by ascertaining the Gibbs function of formation $\bar{g}_{k,f}$ under those conditions. The value of $\bar{g}_{k,f}$ is identical to the Gibbs energy of formation ΔG for a reaction that forms the compound from its elements, which exist in a natural form. For instance, in the case of water,

$$\bar{g}_{f,H_2O} = \Delta G^o_{R,H_2O} = \bar{g}_{H_2O}(298\text{ K},1\text{ atm}) - \bar{g}_{H_2}(298\text{ K, }1\text{ atm})$$

$$-0.5\,\bar{g}_{O_2}(298\text{ K, }1\text{ atm}). \tag{2.80}$$

The value of $\bar{g}_{k,f}$(298 K, 1 atm) is assigned as zero for elements that exist in their natural form. Knowing $\Delta G^o_{R,H_2O}$, and with $\bar{g}_{H_2}$ (298 K,1 atm) = 0, $\bar{g}_{O_2}$ (298 K,1 atm) = 0, $\bar{g}_{H_2O}$ (298 K,1 atm) $=\bar{g}_{H_2O,f}=\Delta G^o_{R,H_2O}$. Table A.32 of Appendix A contains values of $\bar{g}_{k,f}$ (298 K, 1 atm) of many substances. Unless otherwise stated, $\bar{g}$ values will be evaluated using Equation 2.78 for pure components. Particularly at P = 1 atm, $\bar{g}^0 = \bar{h}$ (T) – T$\bar{s}^0$ are tabulated for various ideal gases in Table A.10 to Table A.30 of Appendix A. The Gibbs function during a reaction may increase or decrease depending on energy input or output. For example, consider the process of photosynthesis.

$$6\ CO_2 + 6\ H_2O\ (liq) \rightarrow C_6H_{12}O_6\ (\text{aq, called glucose, one form of sugar}) + 6\ O_2, \quad (2.81)$$

where $\Delta S^0 = -207$ kJ/K per kmol of sugar, $\Delta G^0 > 0$ and $\Delta G^0 = +2872$ kJ per kmol of glucose at 298.15 K. The right-hand side has high G, the left-hand side has low G, and sunlight provides the energy (it is like heating water, in which you increase G of water) and the reaction is not spontaneous. Once we produce high-G glucose, then we can burn it or decompose it spontaneously resulting in decrease of G during reactions. For example, if the leaves of trees were made of C_3H_8 instead of glucose, then C_3H_8 could be spontaneously burned without any outside intervention!

Example 17

(a) Determine the Gibbs energy for water at 25°C and 1 atm using $\bar{g} = \bar{h} - T\bar{s}$ conditions, (b) determine the change in the Gibbs energy when H_2O (ℓ) is formed from its elements, and (c) determine the Gibbs energies of formation for H_2O (ℓ).

Solution

(a) $$\bar{g}_{H_2O(\ell)} = \bar{h}_{H_2O(\ell)} - T\,\bar{s}_{H_2O(\ell)} = -285{,}830 - 298 \times 69.95$$

$$= -306{,}675\ \text{kJ kmol}^{-1}. \quad (A)$$

(b) For the chemical reaction

$$H_2 + 1/2\ O_2 \rightarrow H_2O(l), \quad (B)$$

$$\Delta G = \bar{g}_{H_2O(\ell)} - \bar{g}_{H_2} - (1/2)\,\bar{g}_{O_2}, \quad (C)$$

where

$$\bar{g}_{H_2} = \bar{h}_{H_2} - T\bar{s}_{H_2} = 0 - 298 \times 130.57 = -38{,}910\ \text{kJ kmol}^{-1}, \quad (D)$$

and

$$\bar{g}_{O_2} = 0 - 298 \times 205.04 = -61{,}102 \text{ kJ kmol}^{-1}. \quad \text{(E)}$$

Applying Equation A, Equation D, and Equation E in Equation C,

$$\Delta G = -306{,}675 - (-38{,}910) - 0.5 \times (-61{,}102) = -237{,}214 \text{ kJ kmol}^{-1}. \quad \text{(F)}$$

(c) Therefore, the Gibbs function of a kmol of H_2O (ℓ) is 237,214 kJ lower than that of its elements (when they exist in a natural form at 25°C and 1 atm). As in the case of the enthalpy of formation, the Gibbs function of all elements in their natural states can be arbitrarily set to zero (i.e., $\bar{g}_{f,H_2} = 0$, $\bar{g}_{f,O_2} = 0$). Thereafter, the Gibbs energy of formation is

$$\bar{g}_{f,\, H_2O(\ell)} = \Delta G = -237214 \text{ kJ kmol}^{-1}. \quad \text{(G)}$$

Example 18

Consider Example 3. Calculate the amount of energy available as work in kJ/day if diet (fuel) is only glucose and if metabolic efficiency defined as {work/ $|\Delta G_{glucose}|$} = 0.38; other data same as in Example 3.

Solution

For example 3, the glucose burn rate per day = 18.6 g/hr × (24 hr/day) = 446 g/day

$\Delta G_{C,310} = 6\,\bar{g}_{CO2}\,(310\,K,\ 1\,atm) + 6\,\bar{g}_{H2O}\,(310\,K,\ 1\,atm) - \{\,\bar{g}_{C6H12O6}\,(310\,K,\ 1\,atm)$
$+\ 6\,\bar{g}_{O2}\,(310K, 1atm)\} \approx \Delta G_{c,298} = -2870$ MJ/kmole of glucose
Energy available as work/day = (0.446 kg/day) (kmole/180.2kg) (2,870,000 kj/kmole)(0.38) = 2.7 kJ/day or 30 W.

Remarks:
If fat and proteins are diet, the metabolic efficiencies are 0.32 and 0.10

2.6 Summary

Chemical reactions occur when species rearrange their atoms, and different compounds with different bond energies are produced. Dry and wet gas analyses were presented in this chapter, which are analytical tools to measure species transformations. Examples are presented for determining (A:F) from dry gas analyses. The enthalpy of formation or chemical enthalpy, thermal enthalpy, and the total enthalpy are defined. Energy conservation (first law) and entropy balance (second law) of reacting systems are introduced both in mass and molar forms, and illustrative examples are provided. Finally, entropy generation and decrease in G at a given T and P were illustrated for reacting systems.

2.7 Appendix

2.7.1 Determination of h_f from Bond Energies

If it is not feasible to obtain calorimetric data using a bomb calorimeter, $h^o_{f,k}$ can still be estimated using bond energy. Bond energy is the energy required to break up a bond (e.g., the O-H bond) between atoms so that atoms are released (O(g) atom and H(g) atom); the bonds can be single bond (e.g., O-H), single-electron sharing, double bond (e.g., CO: C=O), and two-electron sharing. They are tabulated in Table A.31 in Appendix A. The chemical bond strength varies according to the atomic number. The chemical bond becomes weaker as the atomic number increases. For diatomic gases, N_2 (atomic mass = 14, chemical bond strength = 945 MJ per kmol) > O_2 (16; 498 MJ) > F_2 (19; 158 MJ). Similarly, HF > HCl > HBr > HI, (note F:19, Cl:35.5, Br:79.9, I:126.9). One may classify chemical bonds as: a weak chemical bond, <200 MJ per kmol; an average bond, around 500 MJ; and a strong bond, >800 MJ. A reaction occurs by the breaking of a bond and its recombination. For example,

$$CH_4(g) \rightarrow CH_3\ (g) + H\ (g)$$

CH_4 is a stable molecule; however, CH_3 is a highly unstable reaction intermediary as it does not have a stable electron structure [Oxtoby, et al., 1998; Chapter 10]. Here, one C-H bond (sharing one electron) is broken, and H is released as gas. The energy required to initiate this reaction is 438 MJ per kmol of CH_4 or per $6.023 * 10^{26}$ molecules; if we break CH_4 into CH_2 and 2 H(g), then we break two C-H bonds. Similarly, breaking H from $CHCl_3$

$$CHCl_3(g) \rightarrow CCl_3\ (g) + H\ (g) + 380\ \text{MJ/kmol of}\ CHCl_3.$$

However, this value is slightly different (but within a small percentage error) compared to the first reaction. Sometimes, one may have double bonds C=C bonds as in C_2H_2 (sharing two electrons); the C atoms share two electrons (two from an atom of C are shared with two electrons of another C atom) rather than a single electron from each C atom. Hence, energy required to break such a C=C bond (615 MJ/kmol) is higher than to break a C-C bond (348 MJ per kmol).

Example 19

Determine the enthalpy of reaction of CH_4 from its bond energies.

Solution

$$CH_4 + 2O_2 \rightarrow CO_2 + 2H_2O \quad (A)$$

Figure 2.14 illustrates the procedure.

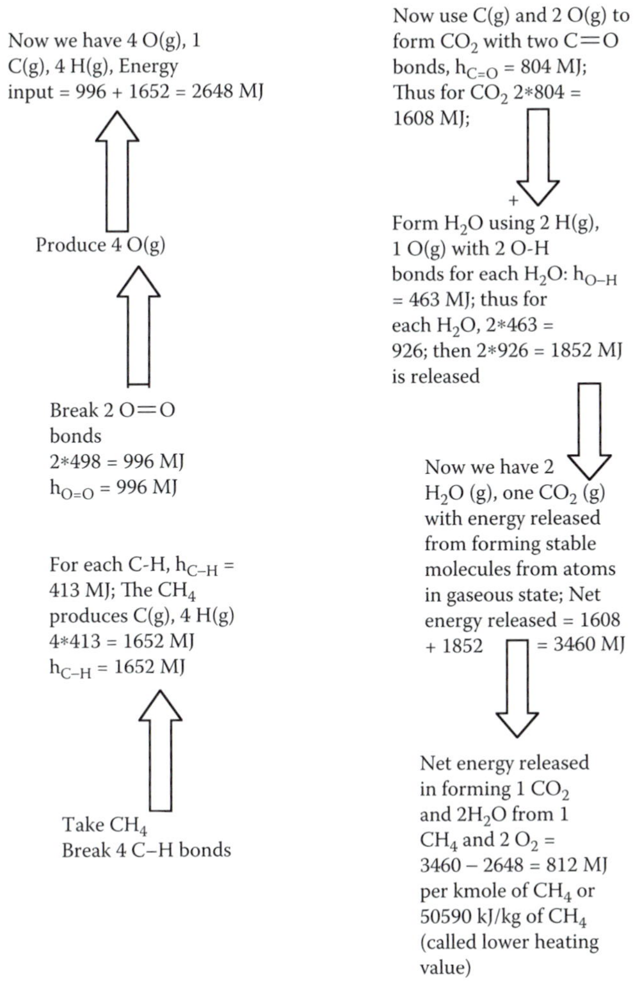
Now we have 4 O(g), 1 C(g), 4 H(g), Energy input = 996 + 1652 = 2648 MJ
Produce 4 O(g)
Break 2 O=O bonds 2*498 = 996 MJ $h_{O=O}$ = 996 MJ
For each C-H, h_{C-H} = 413 MJ; The CH_4 produces C(g), 4 H(g) 4*413 = 1652 MJ h_{C-H} = 1652 MJ
Take CH_4 Break 4 C–H bonds
Now use C(g) and 2 O(g) to form CO_2 with two C=O bonds, $h_{C=O}$ = 804 MJ; Thus for CO_2 2*804 = 1608 MJ;
+
Form H_2O using 2 H(g), 1 O(g) with 2 O-H bonds for each H_2O: h_{O-H} = 463 MJ; thus for each H_2O, 2*463 = 926; then 2*926 = 1852 MJ is released
Now we have 2 H_2O (g), one CO_2 (g) with energy released from forming stable molecules from atoms in gaseous state; Net energy released = 1608 + 1852 = 3460 MJ
Net energy released in forming 1 CO_2 and $2H_2O$ from 1 CH_4 and 2 O_2 = 3460 – 2648 = 812 MJ per kmole of CH_4 or 50590 kJ/kg of CH_4 (called lower heating value)

Figure 2.14 Illustration of the computation of heat of reaction for $CH_4 + 2O_2 \rightarrow CO_2(g) + 2H_2O(g)$.

There are four C-H bonds in CH_4 with an energy of 4 C-H bonds × 413,000 kJ per C-H bond = 1,652,000 kJ and one double bond O=O (with a value of 498,000 kJ) in O_2 (see Table A.31 of Appendix A for bond energies). Because we use 2 kmol of O_2, this energy is twice the latter value, i.e., 996,000 kJ. Therefore, 1,652,000 + 996,000 = 2,648,000 kJ are required to break the six bonds (four for methane and one each for the two oxygen molecules) and produce C, H, and O, i.e.,

$$CH_4 \rightarrow C\,(g) + 4\,H(g) + 1{,}652{,}000 \text{ kJ kmol}^{-1},$$

$$2O_2 \rightarrow 4\,O(g) + 996{,}000 \text{ kJ kmol}^{-1}, \text{ i.e.,}$$

$$CH_4 + 2\,O_2 \rightarrow C(g) + 4\,H(g) + 4\,O(g) + 2{,}648{,}000 \text{ kJ kmol}^{-1}. \qquad \text{(B)}$$

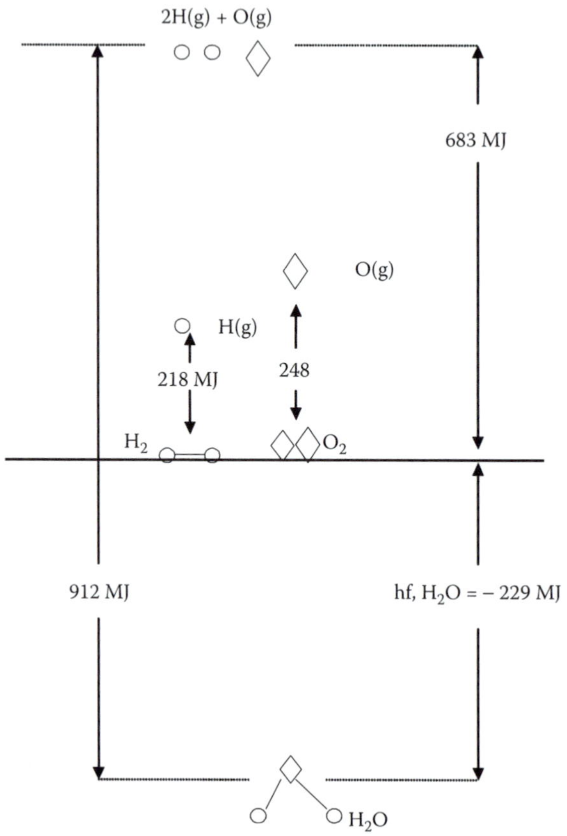

Figure 2.15 Energy levels of various species.

We supply 2,648,000 kJ to a mixture of CH_4 + 2 O_2 to produce C(g), 4 H(g), and 4 O(g). Then CH_4 and 2 O_2 are at an energy level 2,648,000 kJ below the energy level of C(g) + 4 H(g) + 4 O(g). Similarly, to break CO_2 (O=C=O), 2 × 804,000 (twice the energy of each O=C bond) = 1,608,000 kJ is required to produce C(g) and 2 O(g). To break H_2O(g) into 2H(g) and O(g), 2 × (2 × 463,000) or 1,852,000 kJ of energy is required. Therefore, the energy level of 4 H(g) + 2O(g) + 2 O(g) + 1 C(g) is higher by 3,308,000 kJ as compared to that of a mixture containing CO_2 and 2 H_2O, i.e.,

$$CO_2 + 2H_2O \rightarrow C(g) + 4\ H(g) + 4\ O(g) + 3{,}460{,}000\ \text{kJ}, \quad \text{or}$$

$$C(g) + 4\ H(g) + 4\ O(g) \rightarrow CO_2 + 2H_2O(g) - 3{,}460{,}000\ \text{kJ}. \qquad (C)$$

Alternatively, CO_2 + $2H_2O$ are at an energy level of 3,308,000 kJ below the energy level of C(g) + 4 H(g) + 4 O(g). While CH_4 and $2O_2$ are 2,648,000 kJ below energy level of C(g) + 4 H(g) + 4 O(g). Thus 812,000 kJ of energy is released. This is evident from the addition of Equation C and Equation B:

$$CH_4 + 2O_2 \rightarrow CO_2 + 2H_2O - 812{,}000\ \text{kJ}. \qquad (D)$$

Here, ΔH_R = –812,000 kJ is the enthalpy of reaction based on differences in the bond energies. Developing a formula, $LHV_{CH4(g)} = 2 * h_{C=O} + 2 * 2 * h_{O\text{-}H} - 4 * h_{C\text{-}H} - 2 * h_{O=O} = 2 * 804000 + 2 * 2 * 463000 - 4 * 413000 - 2 * 498000 = 812000$ kJ/kmol or 812000/16.05 = 50590 kJ/kg of CH_4 which is almost the same as 49990 kJ/kg from the h_f° method.

Example 20

Determine the enthalpy of formation of $H_2O(g)$ from its bond energies (see Figure 2.15).

Solution

H_2O may be represented as H–O–H. It has two O–H bonds. The energy required to break H_2O into the elements 2H(g) + O(g) is equal to: 2 × 456,100 = 912,200 kJ $kmol^{-1}$ of H_2O. This is represented as $H_2O(g) \rightarrow 2H(g) + O(g) +$ 912,200 kJ $kmol^{-1}$.

This is also the energy removed from 2H(g) and O(g) to form $H_2O(g)$, i.e.,

$$2H(g) + O(g) \rightarrow H_2O(g) - 912{,}200 \text{ kJ kmol}^{-1}. \quad \text{(A)}$$

Likewise,

$$H_2(g) \rightarrow 2H(g) + 435{,}900 \text{ kJ kmol}^{-1}, \text{ and} \quad \text{(B)}$$

$$1/2\ O2(g) \rightarrow O(g) + 247{,}525 \text{ kJ kmol}^{-1}. \quad \text{(C)}$$

Using Hess's law (adding Equation A through Equation C),

$$H_2 + 1/2\ O_2(g) \rightarrow H_2O\ (g) - 228{,}775 \text{ kJ kmol}^{-1}.$$

From Table A.32 of Appendix A, we note that the heat of formation of $H_2O(g)$ is −241,820 kJ $kmol^{-1}$. This discrepancy is due to a precise identification of the minimum bond energy (which occurs at a specific interatomic distance).

Chapter 3

Reaction Direction and Equilibrium

3.1 Introduction

Nature is inherently heterogeneous, and consequently, natural processes occur in a direction that lead to homogeneity and equilibrium. This is essentially a restatement of the second law of thermodynamics. In the previous chapter we assumed that fuels react with oxygen to produce CO_2, H_2O, and other products. We now ask the question whether these products, e.g., CO_2 and H_2O, can react among themselves to give back the fuel and molecular oxygen. If they cannot, then we must answer the questions, "Why not? What governs the direction of a chemical reaction? Why do reactions stop prior to completion, i.e., complete consumption of the reactants? Can we predict the composition of the products if reactions are not completed?" The direction of heat flow is determined by a temperature gradient, or thermal potential, and that of fluid flow by a pressure gradient, or mechanical potential. In this chapter we will characterize the predominant parameters that govern the direction of a chemical reaction. We will learn that the gradient of the Gibbs energy, or chemical potential, determines the direction in which a chemical reaction proceeds. We will also discuss the composition of reaction products under equilibrium conditions.

3.2 Reaction Direction and Chemical Equilibrium

3.2.1 Direction of Heat Transfer

Prior to discussing the direction of a chemical reaction, we will consider the direction of heat transfer. Heat transfer occurs because of a thermal potential (temperature, T), and heat flows from a higher to a lower temperature. Thermal equilibrium is reached when the temperatures of the two systems (one that is

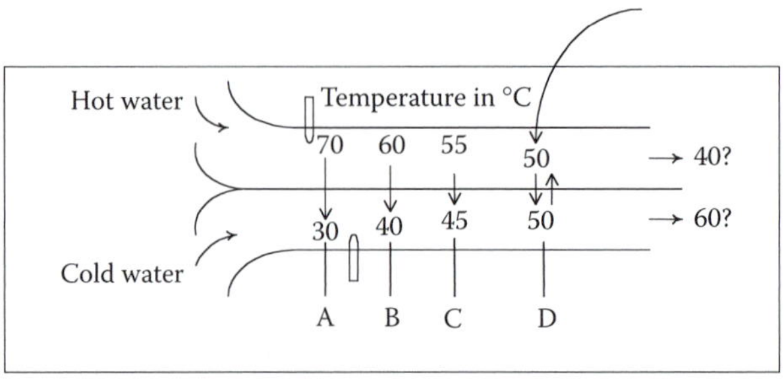

Figure 3.1 Illustration of thermal equilibrium.

transferring and the other that is receiving heat) become equal (Figure 3.1, Location D). Because of the irreversible heat transfer from a warmer to a cooler system, $\delta\sigma > 0$. For example, we never observe warmer water cooling down to 40°C and colder water heating to 60°C (location beyond D in Figure 3.1) even though the first law may be satisfied for such an event, because we find that $\delta\sigma < 0$ in the region beyond the equilibrium point for such a hypothetical event.

3.2.2 Direction of Reaction

The direction of heat transfer is governed by a thermal potential, T. For any infinitesimal irreversible process, $\delta\sigma > 0$. Similarly, the direction of a chemical reaction within a fixed mass under specified conditions is also irreversible and occurs in a direction such that $\delta\sigma > 0$. For instance, consider the combustion of gaseous CO at low temperatures and high pressures, i.e.,

$$CO + 1/2\ O_2 \rightarrow CO_2. \tag{3.1a}$$

At high temperatures and at relatively low pressures,

$$CO_2 \rightarrow CO + 1/2\ O_2. \tag{3.1b}$$

Reaction 3.1a is called an *oxidation* or *combustion reaction*, and Reaction 3.1b is termed a *dissociation reaction*. The direction in which a reaction proceeds depends on the temperature and pressure. We will show that the chemical force potentials, $F_R = \hat{g}_{CO} + (1/2)\hat{g}_{O_2}$ for the reactants (left-hand side of Reaction 3.1a) and $F_P = \hat{g}_{CO_2}$ for the products (right-hand side of Reaction 3.1a), govern the direction of chemical reaction at specified values of T and P. If $F_R > F_P$, Reaction 3.1a dominates, and 3.1b dominates if $F_R < F_P$, just as the thermal potential T governs the direction of heat transfer.

Consider a premixed gaseous mixture that contains 5 kmol of CO, 3 kmol of O_2, and 4 kmol of CO_2 placed in a piston–cylinder–weight (PCW) assembly at specified constant temperature and pressure. It is possible that the oxidation of CO within the cylinder releases heat, in which case heat must be transferred from the system to an ambient thermal reservoir in order to maintain T. Assume that the observer keeps an experimental log, which is reproduced in Table 3.1. An observer will notice that after some time the oxidation reaction (Reaction 3.1a) ceases when chemical equilibrium is reached (at the specified temperature and pressure).

In the context of Reaction 1a, if 0.002 kmol of CO (dN_{CO}) are consumed, then 1/2 × (0.002) = 0.001 kmol of $O_2(dN_{O_2})$ are also consumed and 0.002 kmol of $CO_2(dN_{CO_2})$ are produced. Assigning a negative sign to the species that are consumed and using the associated stoichiometric coefficients,

$$\frac{dN_{CO}}{-1} = \frac{-0.002}{-1} = +0.002; \quad \frac{dN_{O_2}}{-1/2} = +0.002; \quad \frac{dN_{CO_2}}{+1} = +0.002,$$

yielding a constant 0.002. One can now define the extent of the progress of reaction ξ by the relation

$$d\xi = dN_{CO_2}/\upsilon_{CO_2}, \tag{3.2}$$

where dN_{CO_2} denotes the increase in the number of moles of CO_2 as a result of reaction and υ_{CO_2}, the stoichiometric coefficient of CO_2 in the reaction equation. During times $0 < t < t_A$,

$$d\xi = (4.002 - 4.0)/1 = 0.002.$$

Generalizing, we obtain the relation

$$d\xi = dN_k/\upsilon_k, \tag{3.2b}$$

where υ_k assumes a negative sign for reactants and a positive sign for products. The production of CO_2 ceases at a certain mixture composition when chemical equilibrium is attained (time t_E in Table 3.1).

Table 3.1 Experimental Log Regarding the Oxidation of CO at Specified Conditions

Time (sec)	*CO (kmol)*	*O_2 (kmol)*	*CO_2 (kmol)*
0	5	3	4
t_A	4.998	2.999	4.002
t_B	4.5	2.75	4.5
t_C	4.25	2.625	4.75
t_D	3.75	2.375	5.25
t_E	3.75	2.375	5.25

3.2.3 Mathematical Criteria for a Closed System

3.2.3.1 Specified Values of U, V, and m

For a closed or fixed-mass system (operating at specified values of U, V, and m) undergoing an irreversible process (Chapter 1 and Chapter 2),

$$(dS - \delta Q/T_b = \delta\sigma) > 0. \tag{3.3}$$

We wish to determine irreversibility due to reaction alone and eliminate other irreversibilities due to temperature and pressure gradients within the system; hence we set $T_b = T$. Thus, there is no internal thermal irreversibility.

$$(dS - \delta Q/T = \delta\sigma) > 0 \text{ for irreversible processes;}$$
$$(dS - \delta Q/T = \delta\sigma) = \text{“0” for a reversible processes.} \tag{3.4a}$$

If $\delta W_{other} = 0$, then with $\delta Q = \delta W + dU$, $T_b = T$, and $\delta W = PdV$ in Equation 3.3,

$$dU = TdS - PdV - T\,\delta\sigma. \tag{3.4b}$$

Note that Equation 3.4b is valid for irreversible ($\delta\sigma > 0$) or reversible processes ($\delta\sigma = 0$).

For an adiabatic reactor, at fixed, U, V, m

$$(dS_{U,V} = \delta\sigma) \geq 0. \tag{3.4c}$$

Thus, for adiabatic reactions within a rigid vessel, the entropy S reaches a maximum.

3.2.3.2 Specified Values of S, V, and m

For a system operating at specified values of S, V, and m, Equation (3.4b) yields

$$(dU_{S,V} = -T\,\delta\sigma) \leq 0. \tag{3.5}$$

For a reaction occurring at constant S and V, internal energy is minimized at equilibrium.

3.2.3.3 Specified Values of S, P, and m

As in the preceding section, we obtain

$$dH = dU + d(PV) = TdS + VdP - T\delta\sigma. \tag{3.6a}$$

For specified values of S, P, and m,

$$(dH_{S,P} = T\,\delta\sigma) \leq 0. \tag{3.6b}$$

For a reaction occurring at constant S and P, enthalpy is minimized.

3.2.3.4 Specified Values of H, P, and m

From Equation 3.6a,

$$(dS_{H,P} = \delta\sigma) \geq 0, \tag{3.6c}$$

which is similar to Equation 3.4c. Note that for adiabatic reactions in a constant-pressure closed system involving constant enthalpy, the entropy reaches a maximum value at equilibrium.

3.2.3.5 Specified Values of T, V, and m

Similarly, with A = U − TS, dA = dU − d(TS)

$$dA = -SdT - PdV - T\,\delta\sigma, \text{ i.e., } (dA_{T,V} = -T\,\delta\sigma) \leq 0. \tag{3.7}$$

If CO and O_2 are suddenly mixed at temperature T and charged in a rigid vessel immersed in a bath maintained at temperature T, then the Helmholtz function A keeps decreasing as the reaction gradually produces CO_2, and eventually A reaches a minimum when chemical equilibrium is attained.

3.2.3.6 Specified Values of T, P, and m

Similarly, for any state change from G to G + dG, with G = H − TS, dG = dH − d(TS)

$$dG = -SdT + VdP - T\,\delta\sigma, \tag{3.8a}$$

i.e.,

$$(dG_{T,P} = -T\,\delta\sigma) \leq 0. \tag{3.8b}$$

If CO and O_2 are suddenly mixed at temperature T and charged in a PCW assembly, then the Gibbs function G keeps decreasing as reaction gradually produces CO_2, and eventually G reaches a minimum when chemical equilibrium is attained. Note that Equation 3.8a is valid for irreversible ($\delta\sigma > 0$) or reversible processes ($\delta\sigma = 0$). However, if state change occurs reversibly between the same two equilibrium states G and G + dG,

$$dG = -SdT + VdP.$$

In order to validate Equation 3.8b for an irreversible process of a fixed mass, we must determine the value of G, which is a property of the substance of interest, and then use Equation 3.8b for determining $\delta\sigma$. We must determine the value of G for the reacting system as the reaction proceeds.

3.2.4 Evaluation of Properties during Irreversible Reactions

For a closed system, dU = TdS − PdV. However, if mass crosses the system boundary and the system is no longer closed (e.g., pumping of air into a tire). Chemical work is performed for the species crossing the boundary (= $-\Sigma\mu_k\, dN_k$), where $\mu_k = \hat{g}_k(T, P_k) = \hat{h}_k - T\hat{s}_k$, chemical potential of species k. Hence, the change in energy between two equilibrium states for an open system (U and U + dU) is represented by the relation

$$dU = T\,dS - PdV + \sum \mu_k dN_k. \tag{3.9a}$$

The change in entropy between two equilibrium states for an open system (S and S + dS) is obtained from Equation 3.9a as

$$dS = dU/T + PdV/T - \sum \mu_k dN_k/T. \tag{3.9b}$$

Also, adding d(PV) to Equation (3.9a)

$$dH = TdS + VdP + \sum \mu_k dN_k, \tag{3.9c}$$

$$dA = -SdT - PdV + \sum \mu_k dN_k, \tag{3.9d}$$

$$dG = -SdT + VdP + \sum \mu_k\, dN_k. \tag{3.9e}$$

It is apparent from Equation 3.9a to Equation 3.9e that

$$\begin{aligned} -T(\partial S/\partial N_k)_{U,V} &= -T(\partial S/\partial N_k)_{H,P} = (\partial U/\partial N_k)_{S,V} \\ &= (\partial A/\partial N_k)_{T,V} = (\partial G/\partial N_k)_{T,P} = \hat{g}_k = \mu_k, \end{aligned} \tag{3.9f}$$

$$\begin{aligned} -TdS_{U,V,,m} &= -TdS_{H,,P,m} = dU_{S,V,m} = dH_{S,P,m} = dA_{T,V,,m} = dG_{T,P,,m} \\ &= \sum \mu_k dN_k. \end{aligned} \tag{3.9g}$$

Note that Equation 3.9a to Equation 3.9e could be integrated between the initial (1) and final (2) states irrespective of whether state change from (1) to (2) is reached reversibly or irreversibly because U, S, H, A, and G are all properties of the states.

3.2.4.1 Nonreacting Closed System

In a closed, nonreacting system in which no mass crosses the system boundary, $dN_k = 0$. Therefore, for a change in state along a reversible path,

$$\begin{aligned} &dS = dU/T + PdV/T, \\ &dU = TdS - PdV, \quad dH = TdS + VdP, \quad dA = -SdT - PdV, \\ &dG = -SdT + VdP. \end{aligned} \tag{3.10}$$

It is apparent that for a closed, nonreacting system,

$$\sum \mu_k\, dN_k = 0. \tag{3.11}$$

3.2.4.2 Reacting Closed System

Assume that 5 kmol CO, 3 kmol of O_2, and 4 kmol of CO_2 (with a total mass of $5 \times 28 + 3 \times 32 + 4 \times 44 = 412$ kg) are introduced into two identical PCW assemblies A and B. System A has provisions for exchange of mass and acts as an open system. We will assume that system A contains anticatalysts or inhibitors that suppress any reaction, whereas system B can have chemical reactions that result in the final presence of 4.998 kmol of CO, 2.999 kmol of O_2, and 4.002 kmol of CO_2. The species changes in system B are $dN_{CO} = 0.002$, $dN_{O2} = 0.001$, and $dN_{CO2} = 0.002$ kmol, respectively. But the total mass is fixed. The Gibbs energy change dG of system B is now determined by hypothetically injecting 0.002 kmol of CO_2 into system A and withdrawing 0.002 kmol of CO and 0.001 kmol of O_2 from it (so that total mass is still 412 kg) so as to simulate the final conditions in system A same as B. The Gibbs energy after the injection and withdrawal processes, $G_A = G + dG_{T,P}$. System A is open even though its mass has been fixed. The change $dG_{T,P}$ during this process is provided by Equation 3.9e at constant T and P. Thus,

$$\begin{aligned} dG_{T,P;A} &= (-0.002)_A\ \mu_{CO} + (-0.001)_A\ \mu_{O_2} + (+0.002)_A\ \mu_{CO_2} \\ &= \sum \{\mu_k dN_k\}_A \end{aligned} \tag{3.12}$$

where $\{dN_k\}_A$ are the number of moles crossing the boundaries. Because the final states are identical in both systems A and B, $(dN_k)_A = (dN_k)_B$; thus, Gibbs energy change $dG_{T,P;B}$ during this process must equal to $dG_{T,P;A}$. Therefore,

$$dG_{T,P;B} = \left(\sum \mu_k\ dN_k\right)_B = dG_{T,P;A} = \left(\sum \mu_k\ dN_k\right)_A = -T\delta\sigma,$$

i.e.,

$$\left(\sum \mu_k\ dN_k\right)_B < 0. \tag{3.13}$$

Omitting the subscript B, the sum of the changes in the Gibbs energy associated with the three species CO, CO_2, and O_2 are

$$dG_{T,P} = \left(\sum \mu_k\ dN_k\right) = -T\delta\sigma = (\mu_{CO} dN_{CO} + \mu_{O_2} dN_{O_2} + \mu_{CO_2} dN_{CO_2}) < 0, \tag{3.14a}$$

i.e.,

$$\begin{aligned} dG_{T,P} &= \left(\sum \mu_k\ dN_k\right) = (-0.002)\mu_{CO} + (-0.001)\ \mu_{O_2} + (+0.002)\ \mu_{CO_2} \\ &= -T\delta\sigma < 0 \end{aligned} \tag{3.14b}$$

where $dG_{T,P}$ is the change in Gibbs function of a chemically reacting system of fixed mass. Thus, as reaction proceeds at fixed T and P, G decreases and reaches minimum at chemical equilibrium. Now consider Equation 3.9g. For all chemically reacting systems operating with different set of properties (U, V), (H, P), (S, V), (S, P), (T, V), and (T, P), the term $(\Sigma\mu_k\ dN_k) < 0$. For example, at fixed U and V, involving irreversible process within an adiabatic rigid closed systems (Equation 3.9g), $-TdS_{U,V,m} < 0$ or $dS_{U,V,m} > 0$. S increases to a maximum at constant U, V, and m. For constant (T, V) process, A is minimized, whereas at constant (T, P) processes, G is minimized. The inequality represented by Equation 3.13 is a powerful tool for determining the reaction direction and equilibrium condition for any process.

3.2.4.3 Reacting Open System

If one follows a mixture of 5 kmol of CO, 3 kmol of O_2, and 4 kmol of CO_2 as it flows into a chemical reactor in Legrangian frame of reference and then the same criteria that are listed in Equation 3.4 through Equation 3.8 can be used as the system is a fixed-mass one. As reaction proceeds inside the fixed-mass system, the value of G should decrease at specified (T, P) so that $dG_{T,P} \leq 0$.

3.2.5 Criteria in Terms of Chemical Force Potential

3.2.5.1 Single Reaction

The reaction $CO + 1/2O_2 \rightarrow CO_2$ must satisfy the criterion provided by Equation 3.14b to proceed. Dividing Equation 3.14b by 0.002, or the degree of reaction, we get

$$\{dG_{T,P}/0.002\} = (-1)\,\mu_{CO} + (-1/2)\,\mu_{O_2} + \mu_{CO_2} < 0. \tag{3.14c}$$

More generally,

$$dG_{T,P} = \sum \nu_k\,\hat{g}_k = \sum \nu_k \mu_k \leq 0, \tag{3.15}$$

where k and ν_k represent the reacting species and its stoichiometric coefficient for a reaction. In case of the reaction $CO + 1/2O_2 \rightarrow CO_2$, $\nu_{CO} = -1$, $\nu_{O2} = -1/2$, and $\nu_{CO2} = 1$. Equation 3.14c can be alternatively expressed in the form $\mu_{CO} + 1/2\mu_{O_2} > \mu_{CO_2}$. We define the chemical force for the reactants and products as

$$F_R = \mu_{CO} + (1/2)\mu_{O_2} \quad \text{and} \quad F_P = \mu_{CO_2} \quad \text{for reaction} \quad CO + 1/2\,O_2 \rightarrow CO_2 \tag{3.16}$$

The criterion $dG_{T,P} < 0$ leads to the relation $F_R > F_P$. This criterion is equally valid for an adiabatic closed rigid system (U, V, and m specified), adiabatic and isobaric system (H, P, and m specified), isentropic rigid closed system (S, V, and m specified), isentropic and isobaric system (S, P, and m specified), and, finally, isothermal and isovolume system (T, V, and m specified).

For reaction to proceed,

$$F_R > F_P \quad \text{or} \quad F_P - F_R < 0$$

which is similar to the inequality $T_{hot} > T_{cold}$ that allows heat transfer to occur from a hotter to a colder body (Figure 3.1). Analogous to temperature (thermal potential), F_R and F_P are intensive properties called *chemical force potentials*. The chemical potential μ_k is the same quantity as the partial molal Gibbs function

$$\mu_k = \hat{g}_k = \hat{h}_k - T\hat{s}_k = \bar{h}_k(T) - T\bar{s}_k(T,P_k) \tag{3.17}$$

which is a species property. Each species has a unique way of distributing its energy and thus fixing the entropy. A species distributing energy to a larger number of states has a low chemical potential and is relatively more stable. During chemical reactions, the reacting species proceed in a direction that leads to the formation of more stable products (i.e., toward lower chemical potentials).

The physical meaning of the reaction potential is as follows: For a specified temperature, if the population of the reacting species (e.g., CO and O_2) is greater (i.e., higher value of F_R) than that of the product molecules (i.e., CO_2 at lower F_P), then there is a high probability of collisions between CO and O_2, resulting in a reaction that produces CO_2. On the other hand, if the population of the product molecules (e.g., CO_2) is greater (higher F_P value) than that of the reactant molecules CO and O_2 (i.e., lower F_R), there is a higher probability of collisions among CO_2 molecules, which will breakup into CO and O_2. In addition, the reaction potential is also a function of T. If the temperature is lowered, the molecular velocities are reduced, and the translational energy may be insufficient to overcome the bond energy among the atoms in the molecules and cause a reaction, i.e., the reaction potential $F(T, P, X_k)$ is changed by T, P, and X_k.

Example 1

Five kmol of CO, 3 kmol of O_2, and 4 kmol of CO_2 are instantaneously mixed at 3000 K and 101 kPa at the entrance to a reactor. (a) Determine the reaction direction and the values of F_R, F_P, and G. (b) Discuss the change in F_R and F_P as reaction continues to occur. (c) What is the equilibrium composition of the gas leaving the reactor? (d) How is the process altered if 7 kmol of inert N_2 are injected into the reactor?

Solution

(a) We assume that if the following reaction occurs in the reactor

$$CO + 1/2\ O_2 \rightarrow CO_2, \tag{A}$$

so that the criterion $dG_{T,P} < 0$ is satisfied. The reaction potential for this reaction is

$$F_R = (1)\ \mu_{CO} + (1/2)\ \mu_{O_2}, \tag{B}$$

and

$$F_P = (1)\ \mu_{CO2} \tag{C}$$

For ideal gas mixtures, using Equation (3.17),

$$\mu_{CO} = \bar{g}_{CO} = \bar{h}_k(T) - T\left\{\bar{s}_k^{o}(T) - \bar{R}\ \ln\ \frac{X_k\ P}{P^o}\right\} = \bar{h}_k - T\bar{s}_k(T, P) + \bar{R}\ \ln\ X_k$$

where

$$\bar{s}_k(T, P) = s_k^{o}(T) - \bar{R}\ \ln\ (P/P^o) \tag{D}$$

$$\hat{g}_{CO} = \bar{g}_{CO}(T, P) + \bar{R}T\ \ln\ X_{CO} = \bar{g}_{CO}(T, p_{CO}). \tag{E}$$

It is apparent from Equation E that the larger the CO mole fraction, the higher the value of g_{CO} or k_{CO}.

$$\bar{g}_{CO}(T, P) = (\bar{h}_{f,CO}^{0} + (\bar{h}_{t,3000K} - \bar{h}_{t,298K})_{CO}) - 3{,}000 \times (\bar{s}_{CO}^{\circ}(3{,}000) - 8.314(\ln \times P/1))$$

$$= (-110{,}530 + 93{,}541) - 3{,}000 \times 273.508 - 8.314 \times \ln 1)$$

$$\bar{g}_{CO} = -837{,}513 \text{ kJ per kmole}^{-1} \quad \text{(F)}$$

Similarly, at 3000 K and 1 bar,

$$\bar{g}_{O_2}(T, P) = 755{,}099 \text{ kJ kmol}^{-1}, \quad \bar{g}_{CO_2}(T, P) = -1{,}242{,}910 \text{ kJ kmol}^{-1}. \quad \text{(G)}$$

The species mole fractions

$$X_{CO} = 5 \div (5 + 3 + 4) = 0.417, \; X_{O_2} = 0.25, \text{ and } X_{CO_2} = 0.333. \quad \text{(H)}$$

Further,

$$\begin{aligned} \mu_{CO} &= \bar{g}_{CO}(3{,}000 \text{ K}, 1 \text{ bar}, X_{CO} = 0.417) \\ &= \bar{g}_{CO}(3{,}000 \text{ K}, 1 \text{ bar}) + 8.314 \times 3{,}000 \times \ln (0.417) \\ &= -837{,}513 \; + \; 8.314 \times 3{,}000 \times \ln \; 0.467 \\ &= -856{,}504 \text{ kJ kmol}^{-1} \text{ of CO in the mixture.} \end{aligned} \quad \text{(I)}$$

Similarly,

$$\mu_{O_2} = (3000 \text{ K}, 1 \text{ bar}, X_{O_2} = 0.25) = -789675 \text{ kJ per kmol}^{1} \text{ of } O_2. \quad \text{(J)}$$

$$\mu_{CO_2} = (3000 \text{ K}, 1 \text{ bar}, X_{CO_2} = 0.333) = -1270312 \text{ kJ per kmol}^{1} \text{ of } CO_2. \quad \text{(K)}$$

$$F_R = -856504 + 1/2(-789675) = -1254190 \text{ kJ}, \quad \text{(L)}$$

$$F_P = -1270312 \text{ kJ}, \quad \text{(M)}$$

i.e.,

$$F_R > F_P, \quad \text{(N)}$$

This implies that assumed direction is correct, and hence CO will oxidize to CO_2.

The Gibbs energy at any section

$$G = \sum \mu_k N_k = \mu_{CO} N_{CO} + \mu_{O_2} N_{O_2} + \mu_{CO_2} N_{CO_2}$$

At entry to reactor

$G = -856504 \times 5 - 789675 \times 3 - 1270312 \times 4 = -11{,}732{,}793$ kJ.

(b) The oxidation of CO occurs gradually. As more and more moles of CO_2 are produced, its molecular population increases, increasing the potential F_P. Simultaneously, the CO and O_2 populations decrease, thereby decreasing the reaction potential F_R until the reaction ceases when chemical equilibrium is attained. Thus, chemical equilibrium is achieved when $F_R = F_P$, i.e., $dG_{T,P} = 0$. This is illustrated in Figure 3.2 with a plot of F_R and F_P vs. N_{CO2}.

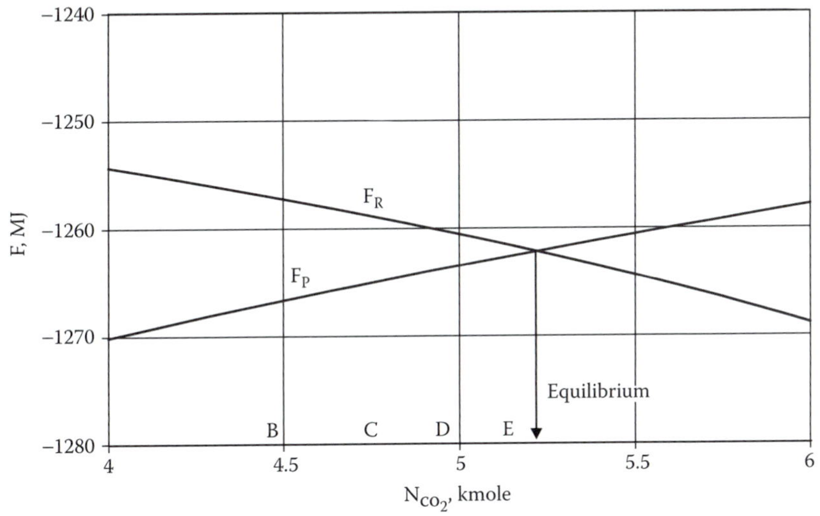

Figure 3.2 The reaction potential with respect to the number of moles of CO_2 produced.

The plot in Figure 3.3 shows that G reaches a minimum value of E when $F_R = F_P$.

(c) The species concentrations at equilibrium ($F_R = F_P$)are

$$N_{CO_2} = 5.25 \text{ kmol}, N_{CO} = 3.75 \text{ kmol}, \quad \text{and} \quad N_{O_2} = 2.375 \text{ kmol}.$$

(d) Nitrogen does not participate in the reaction. Therefore, $dN_{N_2} = 0$, and so the mathematical expressions for F_R and F_P are unaffected. However, the

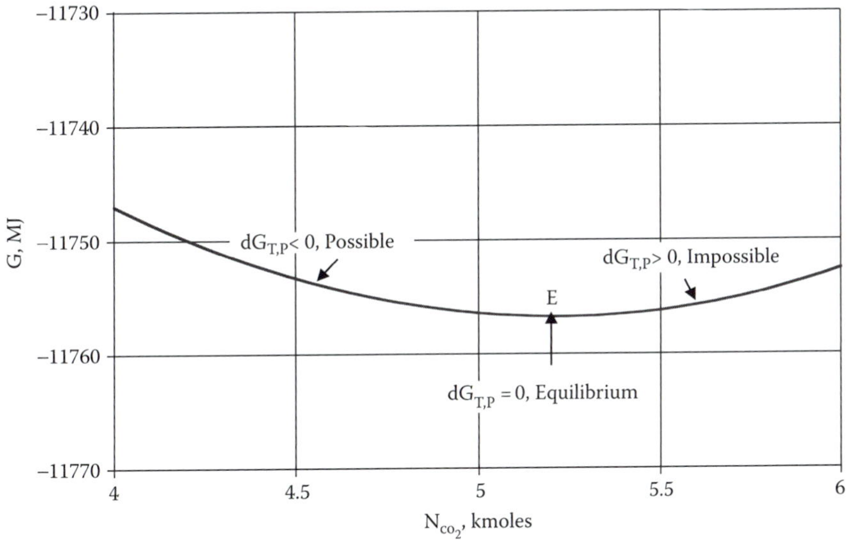

Figure 3.3 Illustration of the minimization of the Gibbs energy at equilibrium with respect to the number of moles of carbon dioxide produced.

mole fractions of the reactants and products change so that the quantitative values of F_R and F_P are different, as is the equilibrium composition. Further, the G expression for this case is

$$G = \sum \mu_k \, N_k = \mu_{CO} N_{CO} + \mu_{O_2} N_{O_2} + \mu_{CO_2} N_{CO_2} + \mu_{N_2} \, N_{N_2}$$

Remarks

(i) The overall reaction has the form

$$5\ CO + 3\ O_2 + 4\ CO_2 \rightarrow 3.75\ CO + 2.375\ O_2 + 5.25\ CO_2.$$

(ii) The assumed direction (i.e., $CO + 1/2\ O_2 \rightarrow CO_2$) is possible if $dG_{T,P} < 0$ or $F_R > F_P$. The mixture is at equilibrium if $dG_{T,P} = 0$ (as illustrated in Figure 3.2) or $F_R = F_P$. If $F_R < F_P$, the reverse reaction $CO_2 \rightarrow CO + 1/2\ O_2$ becomes possible.

3.2.5.2 Multiple Reactions

Suppose we have

$$CO_2 \rightarrow CO + 1/2\ O_2 \qquad \text{(A)}$$

$$CO_2 + H_2 \rightarrow CO + H_2O \qquad \text{(B)}$$

Using $\{\Sigma\mu_k dN_k\}_A < 0$ and $\{\Sigma\mu_k dN_k\}_B < 0$, one can show

$$= d\xi_A\ \{\mu_{CO} + \mu_{O_2}(1/2) - \mu_{CO_2}\} + d\xi_B\{\mu_{CO} + \mu_{H_2O} - \mu_{H_2} - \mu_{CO_2}\} < 0.$$

Because degree of progress of reactions are independent, for each reaction,

$$\{\mu_{CO} + \mu_{O_2}(1/2) - \mu_{CO_2}\} < 0$$

$$\{\mu_{CO} + \mu_{H_2O} - \mu_{H_2} - \mu_{CO_2}\} < 0$$

Generalizing for all multiple reactions, each of the reactions "j" must satisfy $F_{R,j} > F_{P,j}$ for the reaction "j" to move forward, and $F_{R,j} = F_{P,j}$ at chemical equilibrium for each reaction. If one is interested in a mixture of ideal gas, Sections 3.2.6 to 3.3.2 may be skipped.

3.2.6 Generalized Relation for the Chemical Potential

For any species k of a real or ideal gas, solid, or liquid, the partial molal Gibbs function, or chemical potential $\mu_k = \hat{g}_k$, when the species k is inside the mixture given by

$$\mu_k = \hat{g}_k(T,P,X_k) = \bar{g}_k(T,P) + \bar{R}T \ln \hat{\alpha}_k = \bar{g}_k(T,P^0) + \bar{R}T \ln \hat{\alpha}_k{}', \qquad (3.18)$$

where $\bar{g}_k$ (T,P) is the Gibbs function when the species k exists in pure state, the activity $\hat{\alpha}_k = \hat{f}_k(T,P,X_k)/f_k(T,P)$ is the ratio of the fugacity $\{\hat{f}_k(T,P,X_k)\}$ of species k in a mixture to the fugacity $\{f_k(T,P)\}$ of the same species in its pure state at same T and P as of mixture, and $\hat{\alpha}'_k = \{\hat{f}_k(T,P,X_k)/f_k(T,P^0)\}$ where $f_k(T,P^0)$ is the pure fugacity and P^0 is the standard pressure typically selected to be 1 atm. If the species k exists as an ideal gas at state T and P, f_k = P, $\hat{f}_k(T,P,X_k) = p_k = X_k P$, $\hat{\alpha}_k = X_k$, and $\hat{\alpha}'_k = X_k P/P^0$; further, for any ideal mix of real gases or ideal mix of substances, $\hat{\alpha}_k = X_k$, but $f_k \neq P$ and $\hat{f}_k(T,P,X_k) \neq p_k \neq X_k P$. Equation 3.13 can be expanded for any reaction in the form

$$\sum \hat{g}_k\, dN_k = \sum \{\bar{g}_k(T,P) + \bar{R}T \ln \hat{\alpha}_k\} \quad dN_k \le 0, \tag{3.19}$$

$$\sum \hat{g}_k\, dN_k = \sum \{\bar{g}_k(T,P) + \bar{R}T \ln \hat{\alpha}_k\} \quad dN_k \le 0 \text{ (for any mixture)} \tag{3.20}$$

$$\sum \hat{g}_k\, dN_k = \sum \{\bar{g}_k(T,P) + \bar{R}T \ln X_k\} \quad dN_k \le 0$$

(for an ideal mix of substances including an ideal gas) (3.21)

Under ideal mix conditions,

$$\hat{g}_k(T,P,X_k) = \{\bar{g}_k(T,P) + \bar{R}T \ln X_k\}. \tag{3.22}$$

For pure solids and liquids, $\bar{g}_k(T,P) \approx \bar{g}_k^o(T)$.

Example 2

Consider the reactions

$$C(s) + 1/2\ O_2 \rightarrow CO, \tag{I}$$

and

$$C(s) + O_2 \rightarrow CO_2 \tag{II}$$

Which of the two reactions is more likely when 1 kmol of C reacts with 50 kmol of O_2 in a reactor at 1 bar and 298 K? Assume that $\bar{c}_{p,C}/\bar{R} = 1.771 + 0.000877\ T - 86700/T^2$ in SI units and T is in K. Assume ideal gas mixtures.

Solution

If $|(F_R - F_P)|_I > |(F_R - F_P)|_{II}$, then the first reaction dominates and vice versa. Using Equation 3.22,

$$\hat{g}_C = \bar{g}_C(T,P) + \bar{R}T \ln \hat{\alpha}_k. \tag{A}$$

Because solid carbon is a pure component, the activity $\hat{\alpha}_{C(S)} = 1$. Further, at 298 K

$$\bar{h}_C = \bar{h}^o_{f,C} + \int_{298K}^{T} \bar{c}_{p,C} dT = 0 + 0. \quad \text{(B)}$$

$$\bar{s}_C(298,1) = \bar{s}^0_C(298\ \text{K}) + \int_{298\ K}^{T} (\bar{c}_{p,c}/\text{T})d\text{T} = (5.74 + 0) = 5.74\ \text{kJ kmol}^{-1}\ \text{K}^{-1} \quad \text{(C)}$$

Hence, using Equation B and Equation C,

$$\bar{g}^o_C = \bar{h}_{298K} - 298 \times \bar{s}_C(298\ \text{K} = 0 - 298 \times 5.74 = -1710\ \text{kJ kmol}^{-1}. \quad \text{(D)}$$

Assume that 0.001 mol of C(s) reacts with 0.0005 mol of O_2 to produce 0.001 mol of CO. Hence,

$$p_{O_2} = X_{O_2}\ P = (50 - 0.0005) \div (0.001 + (50 - 0.0005)) = 0.9999\ P = 0.9999\ \text{bar}.$$

Therefore,

$$\bar{s}_{O_2} = 205.03 - 8.314 \times \ln 0.9999 = 205.03\ \text{kJ K}^{-1}\ \text{kmol}^{-1}$$

$$\bar{g}_{O_2}\ (298\ \text{K}, 1\ \text{bar}) = 0 - 298 \times 205.03 = -61099\ \text{kJ kmol}^{-1}. \quad \text{(E)}$$

Similarly,

$$X_{CO} = 0.001 \div (0.001 + 49.9995) \approx 0.00002,$$

and

$$\bar{s}_{CO}(\text{T}, p_{CO}) = 197.54 - 8.314 \times \ln(0.00002)$$
$$= 287.5\ \text{kJ K}^{-1}\ \text{kmol}^{-1},$$

so that

$$\bar{g}_{CO}(298\ \text{K}, 1\ \text{bar}) = -110530 - 298 \times 287.5 = -196205\ \text{kJ kmol}^{-1}. \quad \text{(F)}$$

Employing Equation D and Equation E for Reaction I,

$$F_{R,I} = \bar{g}_C + 1/2\ \bar{g}_{O_2} = -1710 + 0.5 \times (61099) = -32260\ \text{kJ}$$

and

$$F_{P,I} = \bar{g}_{CO} = -196205\ \text{kJ kmol}^{-1},$$

i.e.,

$$F_{R,I} - F_{P,I} = 32260 + 196205 = 163945\ \text{kj}$$

Figure 3.4 shows a plot of $F_{R,I}$, $F_{P,I}$ versus number of carbon moles consumed.

$$G_1 = N_{C(S)} * \bar{g}_{C(S)} + N_{O_2} * \bar{g}_{O_2} + N_{CO_2} * \bar{g}_{CO_2}$$

$$G_I = 1710 * (1-0.001) + 61099 * 49.9995 + 196205 * 0.001 = 3056825 \text{ kJ}$$

Figure 3.5 shows a plot of G_I versus N_C.

Similarly proceeding for Reaction II,

$$X_{O_2} = 0.999,\ X_{CO_2} \approx 0.00002\ \bar{s}_{O2} \approx 205.3 \text{ kJ K}^{-1} \text{ kmol}^{-1}, \bar{s}_{CO_2} \approx 303.70 \text{ kJ K}^{-1} \text{ kmol}^{-1},$$

$$\bar{g}_{O_2} = -61{,}099 \text{ kJ kmol}^{-1},$$

and

$$\bar{g}_{CO_2} = -393{,}546 - 298 \times 303.70 = -484{,}048 \text{ kJ kmol}^{-1}.$$

For this Reaction II

$$F_{R,II} = -1710 + (-61{,}099) = -62{,}810 \text{ kJ, and } F_{P,II} = -484{,}048 \text{ kJ kmol}^{-1}, \quad \text{(G)}$$

i.e.,

$$F_{R,II} - F_{P,II} = -62{,}810 + 484{,}048 = 421{,}238 \text{ kJ}.$$

The variations in the reaction potentials for Reaction I and Reaction II with respect to the number of moles of carbon that are consumed at a reactant temperature of 298 K are presented in Figure 3.4, and the corresponding

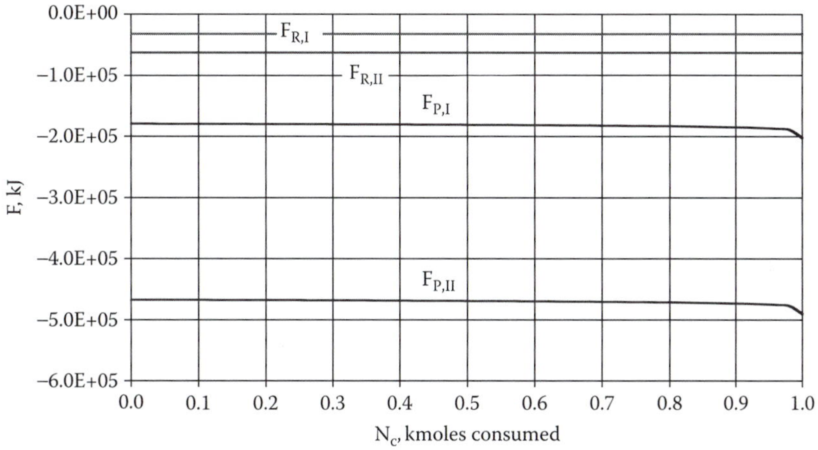

Figure 3.4 The reaction potentials for Reaction I and Reaction II with respect to the number of moles of carbon that are consumed.

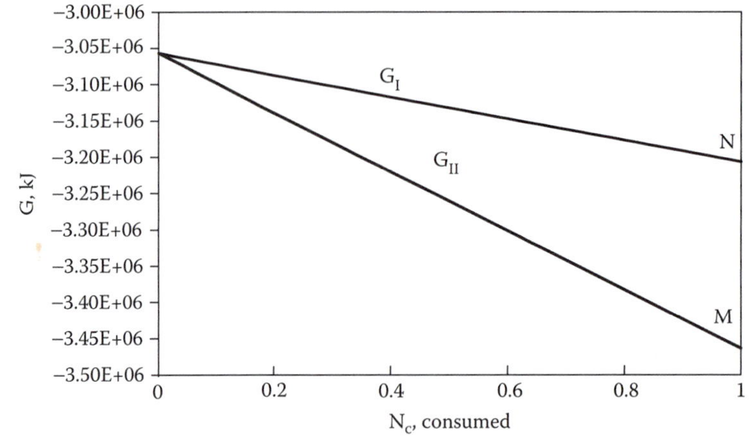

Figure 3.5 Variation in G_I and G_{II} with respect to the number of moles of carbon consumed for Reaction I and Reaction II at 298 K.

variation in G_I and G_{II} in Figure 3.5. At 298 K, CO_2 production dominates. The analogous variations in G_I and G_{II} at 3500 K are presented in Figure 3.6. At higher temperatures, CO formation is favored.

3.2.7 Approximate Method for Determining Direction of Reaction, ΔG⁰, and Gibbs Function of Formation

In general, the values of $\bar{g}_k(T, P, X_k)$ are functions of the species mole fractions (cf. Equation 3.22). If we assume that $|\bar{g}_k(T, P)| >> |\bar{R}T \ln X_k|$, then

$$\bar{g}_k(T, P, X_k) \approx \bar{g}_k(T, P). \tag{3.23}$$

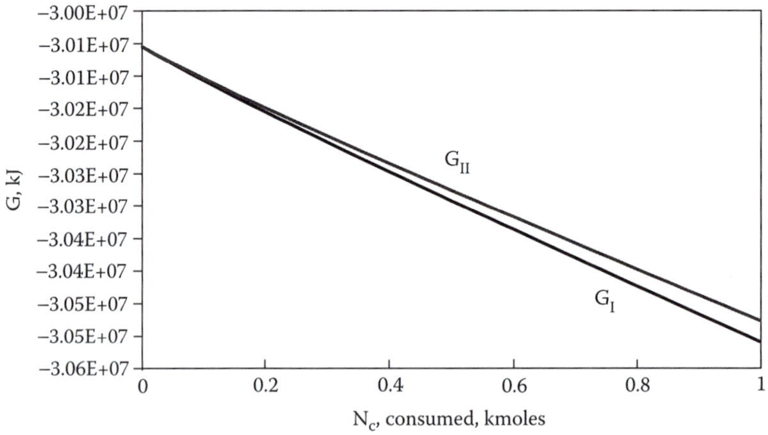

Figure 3.6 Variation in G_I and G_{II} with respect to the number of moles of carbon consumed for Reaction I and Reaction II at 3500 K.

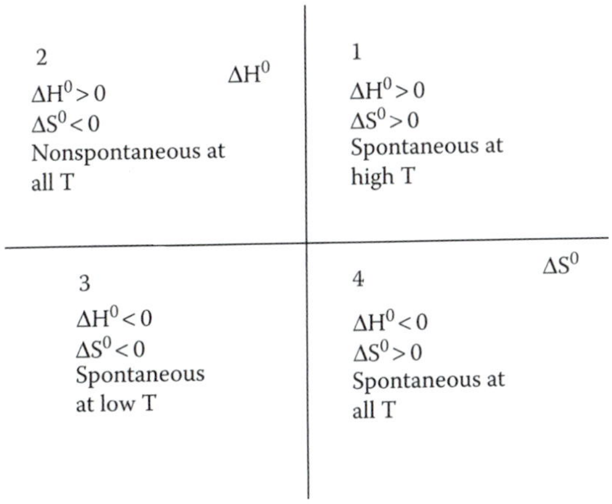

Figure 3.7 The regimes of reaction.

This offers an approximate method of determining whether Reaction I or Reaction II is favored in Example 2. For instance, if Reaction I proceeds in the indicated direction, then $F_{P,I} - F_{R,I} = \Delta G_I\,(T, P) = \bar{g}_{CO}\,(T, P) - (\bar{g}_C\,(T, P) + 1/2\ \bar{g}_{O_2}$ $(T, P)) < 0$. Similarly, we can evaluate $\Delta G_{II}\,(T, P) = \{\bar{g}_{CO_2} - (\bar{g}_C + \bar{g}_{O_2})\} < 0$. Then, we determine whether $|\Delta G_{II}\ (T, P)| > |\Delta G_I\ (T, P)|$; if so, the CO_2 production reaction is favored. If P = 1 bar, $\Delta G^0 < 0$, then reaction has strong possibility of proceeding in the indicated direction. The requirement $\Delta G^0 < 0$ indicates the likelihood of $\Delta H^0 < 0$ (but not necessarily) or exothermicity of reaction. Exothermicity drives the reaction toward the spontaneity. Similarly, if a particular fuel is burnt, say $C_2H_2 + 2.5\ O_2 \rightarrow 2\ CO_2 + H_2O(g)$ at 298 K, then $\Delta h_C{}^0 = -48{,}300$ kJ/kg of fuel, $\Delta s^0 = [2 * \bar{s}^{\,0}_{CO_2} + 1\,\bar{s}^{\,0}_{H_2O} - 2.5\,\bar{s}^{\,0}_{O_2} - \bar{s}_{C2H2}{}^0]/M_{C2H2} = -3.74$ kJ/kg of fuel K; $\Delta G^0(298) = \Delta h_C{}^0 - T_0\Delta s^0 = -48{,}300 - 298 * (-3.74) = -47{,}190$ kJ/kg of fuel < 0. See Table A.33 for tabulation of these values.

Consider $\Delta G^0 = \Delta H^0 - T\Delta S^0$. The value and sign of ΔG^0 depends on ΔH^0, ΔS^0, and T > 0. The four possibilities are indicated in Figure 3.7. Reactions in quadrant 4 are always spontaneous ($\Delta H^0 < 0$, $\Delta S^0 > 0$), and those in quadrant 2 are always nonspontaneous ($\Delta H^0 > 0$, $\Delta S^0 < 0$), e.g., production of glucose in plants driven by sunlight. If T is low, the second term is negligible, the reaction becomes spontaneous even though $\Delta S^0 < 0$, it is in quadrant 3, and $\Delta H^0 > 0$ and $\Delta S^0 > 0$; if T is high, the second term is significant, the reaction becomes spontaneous, it is in quadrant 1. As an example, determine which quadrant the following reaction belongs to: $C(s) + H_2O\ (g)$ at 298.15 K, P = 1 atm, $\Delta H^0 = 131{,}300$ kJ kmol^{-1} of C(s), $\Delta S^0 = 133.67$ kJ kmol^{-1} K^{-1}? It belongs to quadrant 1; if T > 980 K, it is spontaneous because ΔH^0 and ΔS^0 change with temperature T.

At 298 K, it is seen from Table A.12 and Table A.13 that $\bar{g}^{\,0}_{CO_2} < \bar{g}^{\,0}_{CO}$; thus, CO_2 formation is favored.

In addition to Reaction I and Reaction II in Example 2, consider the following reactions:

$$C(s) + CO_2 \rightarrow 2\ CO, \qquad \text{(III)}$$

$$CO + 1/2\ O_2 \rightarrow CO_2, \qquad \text{(IV)}$$

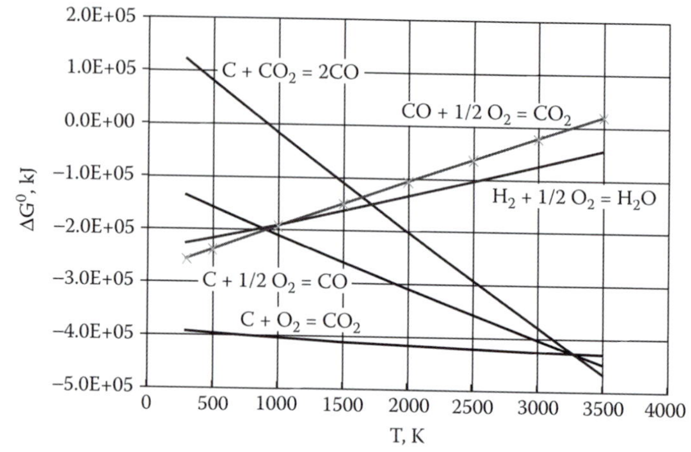

Figure 3.8 The variation in the value of ΔG° with respect to the temperature for several reactions.

and

$$H_2 + 1/2\ O_2 \rightarrow H_2O. \qquad \text{(V)}$$

Figure 3.8 plots values of ΔG° with respect to the temperature for these five reactions. For instance, for Reaction III, $\Delta G^o(298\ K) = 2\bar{g}_{CO} - (\bar{g}_C + \bar{g}_{CO_2}) = 120080$ kJ, which a positive number. This implies that the reaction cannot proceed in the indicated direction.

A relatively large negative value of ΔG° implies that $F_R >> F_P$, and this requires the largest decrease in the reactant population (or extent of completion of reaction) before chemical equilibrium is reached. A positive value for ΔG implies that the reaction will produce an insignificant amount of products (Reaction III).

One can show that the value of ΔG° for Reaction IV can be obtained in terms of the corresponding values for Reaction I and Reaction II.

$$\Delta G^\circ_{IV}(298^\circ\ K) = \Delta G^\circ_{II} - \Delta G^\circ_{I}$$

$$\Delta G^\circ_{IV} = \bar{g}^\circ_{CO_2}(298\ K) - \bar{g}^\circ_{CO}(298\ K) = -394390 + 137137 = -257253\ kJ. \qquad (3.24)$$

Note (from Chapter 2) that

$$\Delta G^\circ_{I}(298\ K) = \bar{g}^\circ_{f,CO}(298\ K),\ \Delta G^\circ_{II}(298\ K) = \bar{g}^\circ_{f,,CO_2}(298\ K),$$

where $\bar{g}^0_{f,,k}$ is called the Gibbs function of formation of species k from elements in natural form.

3.3 Chemical Equilibrium Relations

For the reaction $CO_2 \rightarrow CO + 1/2\ O_2$ to occur, $F_R(=\hat{g}_{CO_2}) > F_P(=\hat{g}_{CO} + 1/2\hat{g}_{O_2})$. In general, Equation 3.13 becomes:

$$dG_{T,P} = \sum \mu_k dN_k = \sum \hat{g}_k(T,P,X_k)\ dN_k \leq 0. \tag{3.25}$$

Ideal gas mixtures are discussed in Subsection 3.3.3. The change in the mole numbers is related to the stoichiometric coefficients; at specified values of T and P,

$$\sum v_k \{\bar{g}_k(T,P) + \bar{R}T \ln \hat{\alpha}_k\} \leq 0, \tag{3.26}$$

where $\hat{\alpha}_k = X_k$ for ideal gas mixtures. For example, for the reaction $CO_2 \rightarrow CO + 1/2O_2$, $\upsilon_{CO2} = -1$, $v_{CO} = +1$, $v_{O2} = +1/2$.

In the following text, equilibrium constant K°(T) will be defined and the criteria for the direction of a reaction given in terms of K° (T).

3.3.1 Real Mix of Substances

For any reaction, $v_A A + v_B B \rightarrow v_C C + v_D D$

$$\Delta G(T,P) = \Delta G(T,P^0) + \bar{R}T \ln \{\hat{\alpha}'_C\ \hat{\alpha}'_D\}/(\hat{\alpha}'_A\ \hat{\alpha}'_B) \leq 0 \tag{3.27a}$$

$$\Delta G(T,\ P) = \Delta G\ (T,\ P)\ +\ \bar{R}T\ \ln \{\hat{\alpha}_C\ \hat{\alpha}_D\}/\{\hat{\alpha}_A\ \alpha_B\} \leq 0, \tag{3.27b}$$

where

$$\Delta G(T,P^0) = \sum \upsilon_k \{\bar{g}_k(T,P^0)\} = \{\upsilon_C \bar{g}_C(T,P^0) + \upsilon_D \{\bar{g}_D(T,P^0)\} - \{\upsilon_A \bar{g}_A(T,P^0) + \upsilon_B \bar{g}_B(T,P^0)\} \tag{3.28a}$$

$$\Delta G(T,P) = \Sigma \upsilon_k \{\bar{g}_k(T,P)\} = \{\upsilon_C\ \bar{g}_C(T,P) + \upsilon_D \bar{g}_D(T,P)\} - \{\upsilon_A \{\bar{g}_A(T,P) + \upsilon_B\ \bar{g}_B(T,P)\} \tag{3.28b}$$

$$\hat{\alpha}_k = \{\hat{f}_k(T,P,X_k)/f_k(T,P)\} \quad \text{and} \quad \hat{\alpha}'_k = \{\hat{f}_k(T,P,X_k)/f_k(T,P^0)\}. \tag{3.29a,b}$$

The fugacities of species k in mixtures and as pure substance are topics in advanced thermodynamics, and readers are referred to Smith and Van Ness [1987], Modell and Reid [1983], and Annamalai and Puri [2001].

3.3.2 Ideal Mix of Liquids and Solids

Henceforth, all chapters deal only with the ideal mix of substances. In this case, $\hat{\alpha}_k = \{\hat{f}_k(T,P,X_k)/f_k(T,P)\} = X_k$. Dividing Equation 3.27b by $\bar{R}T$ and rewriting the inequality,

$$\sum_k \ln X_k^{\nu_k} \le -\sum_k \upsilon_k\ \bar{g}_k(T,P)/(\bar{R}T). \tag{3.30}$$

We define the term

$$K(T,P) = \exp\left\{-\sum_k \upsilon_k \bar{g}_k(T,P)/(\bar{R}T)\right\} \tag{3.31}$$

Equation 3.30 assumes the form

$$\Pi\ X_k^{\upsilon_k} \le K\,(T,P), \tag{3.32}$$

where X_k denotes the mole fraction of a gas, solid, or liquid species k (e.g., CO gas in exhaust of an automobile engine, $CaSO_4$(s) in a mixture of Fe_2O_3(s), $CaSO_4$(s), and CaO(s)). Specifically for a solid or a liquid,

$$\bar{g}_{k(s)}(T, P) \approx \bar{g}_{k(s)}(T, 1\ \text{bar}) \quad \text{or} \quad \bar{g}_{k(\ell)}(T, P) \approx \bar{g}_{k(\ell)}(T, 1\ \text{bar})$$

If K(T, P) in Equation 3.32 is evaluated using $\bar{g}_{k(s)}(T, P)(\approx \bar{g}_{k(s)}(T, 1\ \text{bar}))$, then for solid and liquid mixtures,

$$\Pi\ X_k^{\upsilon_k} \le K\,(T, P) \approx K^\circ(T), \tag{3.33}$$

where K°(T) is equilibrium constant, but it is a constant only at specified value of T. The physical meaning of this relation is as follows. In solid or liquid mixtures, X_k changes as the reaction proceeds. The term $\Pi\ X_k^{\nu_k}$ keeps increasing toward K°(T) and becomes equal to K°(T) at chemical equilibrium.

3.3.3 Ideal Gases

3.3.3.1 Equilibrium Constant and Gibbs Free Energy

Unlike in the case of superheated steam, for which properties are tabulated as functions of (T, P), in the case of ideal gases K(T, P) is tabulated typically only at P^0, the standard pressure because simple relations are available that relate K(T, P) to K(T, P^0). For ideal gases,

$$\bar{g}_k(T, P, X_k) = \bar{g}_k(T, p_k) = \bar{h}_k^0(T) - T\ \bar{s}_k(T, p_k) = \bar{h}_k^0(T) - T\ \bar{s}_k^{\ 0}(T) + \bar{R}T \ln(p_k/P^0)$$

where $\bar{s}_k^{\ 0}$ (T) is at standard pressure P^0 and $p_k = X_k P$. Typically, P^0 is selected as 1 bar or 1 atm. Rewriting, we get

$$\bar{g}_k(T, p_k) = \bar{g}_k(T,p_k) = \bar{g}_k^0(T) + \bar{R}T\ \ln(p_k/P^0) \tag{3.34}$$

$$\bar{g}_k^0(T) = \bar{h}_k^0(T) - T\ \bar{s}_k^{\ 0}(T) \tag{3.35}$$

where the term $\bar{h}_k^{\,0}(T)$ includes chemical and thermal enthalpies. Using Table A.10 to Table A.30, $\bar{g}_k^{\,0}(T)$ can be estimated at 1 bar for any species from the known values of $\bar{h}_k^{\,0}(T)$ and $\bar{s}_k^{\,0}(T)$. For convenience, the $\bar{g}_k^{\circ}$ values are also tabulated. The values of $\bar{g}_k^{\,0}(T)$ can also be estimated using Gibbs function of formation; unless otherwise stated, Equation 3.35 will be used for determining the values of $\bar{g}_k^{\,0}(T)$. Using this relation in Equation 3.13 and replacing dN_k with $\nu_k\, d\xi$ for any specified reaction, the criterion for reaction is given as

$$\sum \nu_k \bar{g}_k(T,p_k)/(\bar{R}T) = \sum \nu_k \{\bar{g}_k^0(T)/(\bar{R}T)\} + \sum \nu_k \ln(p_k/P^0) \le 0. \quad (3.36)$$

Now, let us define

$$K^\circ(T) = \exp\left\{\left(-\sum \nu_k \bar{g}_k^0(T)/(\bar{R}T)\right)\right\} = \exp\{-\Delta G^\circ/(\bar{R}T)\} \quad (3.37)$$

For the reaction $CO_2 \rightarrow CO + 1/2\ O_2$,

$$\Delta G^0(T)/\bar{R}\,T = \sum \nu_k \bar{g}_k^0(T)/(\bar{R}\,T) = \left\{\bar{g}_{CO}^0 + \frac{1}{2}\, g_{O2}^0 - \bar{g}_{CO2}^0\right\}/\{\bar{R}\,T\}$$

The term K°(T) is conventionally called the *equilibrium constant* and is tabulated for many standard reactions, as shown in Table A.36A and Table A.36B of Appendix A. If reactants are elements in their natural forms producing compounds or elements in unnatural form, then the equilibrium constants for the "elementary" reactions can be easily obtained. Using these elemental equilibrium constants, the equilibrium constants for other reactions involving compounds can be calculated (see Section 3.8) as in the thermochemical heat calculations (Chapter 2).

3.3.3.2 Criteria for Direction of Reaction and Chemical Equilibrium in Various Forms

3.3.3.2.1 Partial Pressure Form

Using Equation 3.37 in Equation 3.36, and simplifying the criterion for a reaction to proceed in a specified direction is given as

$$\Pi\, \{p_k/P^0\}^{\nu_k} \le K^\circ(T)\,. \quad (3.38a)$$

and at chemical equilibrium, the relation in terms of partial pressures is

$$\Pi\, \{p_k/P^0\}^{\nu_k} = K^\circ(T)\,. \quad (3.38b)$$

The left-hand side may be called a *reaction quotient.* If the reaction occurs in the assumed direction at a given T, K°(T) is constant and Equation 3.38 stipulates that nondimensional partial pressures must approach the value of K°(T) as the reaction proceeds and eventually become equal to K°(T) at chemical equilibrium.

3.3.3.2.2 Mole Fraction Form

Replacing the partial pressure with mole fraction and total pressure in Equation 3.38, i.e., $p_k = X_k P$, we get

$$\Pi \left\{ X_k \left(\frac{P}{P^0} \right) \right\}^{\nu_k} \leq K^\circ(T) \quad (3.39)$$

3.3.3.2.3 Concentration Form

If we represent molar concentration as $[k] = p_k/(\bar{R}T)$ in units of kmol $\bar{m}^3$, then Equation 3.38a becomes

$$\Pi\ [k]^{\nu_k} / [P^0/\bar{R}\,T]^{\Sigma\nu k}\} \leq K^\circ(T) \quad (3.40)$$

Equation 3.40 is useful in chemical kinetics (discussed in Chapter 5).

3.3.3.2.4 Mole Form

With $X_k = N_k/N$, where N is total number of moles in products, Equation 3.39 assumes the form

$$\left\{ \left(\frac{P}{N P^0} \right)^{\Sigma \nu_k} \right\} \Pi\ N_k^{\nu_k} \leq K^\circ(T) \quad (3.41)$$

Example 3

Consider the following reaction at equilibrium

$$CO_2 \rightarrow CO + 1/2\ O_2 \quad (A)$$

Write down the criteria in terms of (a) partial pressure, (b) mole fraction, (c) molal concentration [k], and (d) number of moles.

Solution

$$\text{(a)}\ \sum \nu_k = 1 + \tfrac{1}{2} - 1 = \tfrac{1}{2} \quad (B)$$

$$\{(P/P^0)^{1/2}\}(p_{CO})^1 (p_{O_2})^{1/2} (p_{CO_2})^{-1} \leq K^\circ(T) \quad (C)$$

$$\text{(b)}\ ((P/P^0)^{1/2} (X_{CO})^1 (X_{O_2})^{1/2} (X_{CO_2})^1) \leq K^\circ(T) \quad (D)$$

$$\text{(c)}\ \{[CO]^1 [O_2]^{1/2} / [CO_2]^1\} / \{P^0/(\bar{R}T)\}^{\frac{1}{2}} \leq K^\circ(T) \quad (E)$$

$$\text{(d)}\ ((P/N\,P^0)^{1/2} (N_{CO})^1 (N_{O_2})^{1/2} / (N_{CO_2})^1) \leq K^\circ(T) \quad (F)$$

Example 4

Consider the following reaction at equilibrium

$$CO_2 \rightarrow CO + 1/2\ O_2 \quad (A)$$

(a) Determine the equilibrium constant $K^0(1800)$ using ΔG^0 and compare with values from Table A.36A and Table A.36B of Appendix A.
(b) Consider a mixture with the following composition at 1800 K and 2 MPa, i.e., $N_{CO_2} = 1.2$ kmol, $N_{O_2} = 0.6$ kmol, $N_{CO} = 3.6$ kmol, and $N_{N_2} = 6.6$ kmol. In which direction will the following reaction proceed,

$$CO + 1/2\ O_2 \rightarrow CO_2 \text{ or } CO_2 \rightarrow CO + 1/2\ O_2,$$

if we maintain T and P to be constants ? Use the equilibrium constant approach.

Solution

(a) $$\Delta G^0(T)/\bar{R}T = \sum \upsilon_k \bar{g}_k^0(T)/(\bar{R}T) = \{\bar{g}_{CO}^o + (1/2)\bar{g}_{O_2}^0 - \bar{g}^0_{CO_2}\}/\{\bar{R}T\} \quad (B)$$

Using the g values listed in Table A.12, Table A.13, and Table A.26, ΔG^o = (1) × (–269,164) + (1/2) × (0) + (–1) × (–396,425) = 127,261 kJ,

$$K^o(T) = \exp(-127260 \div (8.314 \times 1800)) = 0.00020.$$

From Table A.36A and Table A.36B of Appendix A, log 10 ($K^0(T)$) = –3.696, i.e., $K^0(T) = 0.0002$.

(b) For the reaction $CO_2 \rightarrow CO + 1/2\ O_2$ to occur, Equation 3.38a must be satisfied. For the specified composition,

$$X_{CO} = 3.6/12 = 0.3,\ p_{CO} = 0.3 \times 20 = 6 \text{ bar}.$$

Similarly, $p_{O2} = 1$ bar, $p_{CO2} = 2$ bar. The ratios of the partial pressures are

$$(p_{CO}/P^0)^1\ (p_{O_2}/P^0)^{1/2}/(p_{CO_2}/P^0)^1 = (6/1)^1\ (1/1)^{1/2}/(2/1)^1 = 3.$$

The criterion

$$(p_{CO}/P^0)^1(p_{O_2}/p^0)^{1/2}(p_{CO_2}/P^0)^1 \le \{K^o(T) = 0.0002\}$$

is violated. Therefore, CO will oxidize to CO_2, i.e., the reverse path is favored.

Example 5

One kmol of air is in a closed PCW assembly placed at 298 K and 1 bar. Trace amounts of NO and NO_2 are generated according to the overall reaction

$$0.79\ N_2 + 0.21\ O_2 \rightarrow a\ NO + b\ NO_2 + c\ N_2 + d\ O_2. \quad (A)$$

Assume the following equilibrium reactions

$$N_2 + O_2 \Leftrightarrow 2\ NO, \text{ and} \tag{B}$$

$$N_2 + 2\ O_2 \Leftrightarrow 2\ NO_2. \tag{C}$$

The values of Gibbs function of formation $\bar{g}_f^{\circ}(298\text{ K})$ for NO and NO_2 are, respectively, 86,550 and 51,310 kJ kmol^{-1}, and for the elements j in their natural forms $\bar{g}_j^{\circ}(298\text{ K}) = 0$ kJ kmol^{-1}. Determine (a) equilibrium constants for Reactions B and Reaction C using $\bar{g}_f^{\circ}$ and (b) the NO and NO_2 concentrations at chemical equilibrium.

Solution

(a) For Reaction B and Reaction C

$$\Delta G^{\circ}_{NO} = 2\bar{g}^{\circ}_{NO} - \bar{g}^{\circ}_{N_2} - \bar{g}^{\circ}_{O_2} = 2\times 86{,}550 = 173{,}100 \text{ kJ per kmol of } N_2.$$

$$\Delta G^{\circ}_{NO_2} = 2\bar{g}^{\circ}_{NO_2} - \bar{g}^{\circ}_{N_2} - 2\bar{g}^{\circ}_{O_2} = 2\times 51{,}310 = 102{,}620 \text{ kJ per kmol of } N_2.$$

Therefore,

$$K^{\circ}_{NO} = \text{Exp}\{-\Delta G^{\circ}_{NO}/(\bar{R}T)\} = 14.54\times 10^{-31},$$

$$K^{\circ}_{NO_2} = \text{Exp}\{-\Delta G^{\circ}_{NO_2}/(\bar{R}T)\} = 1.03\ \times 10^{-18},$$

(b)

$$K^{\circ}_{NO} = (p_{NO}/P^0)^2/((p_{N_2}/P^0)(p_{O_2}/P^0)), \tag{D}$$

$$K^{\circ}_{NO_2} = (p_{NO_2}/P^0)^2/((p_{N_2}/P^0)(p_{O_2}/P^0)^2. \tag{E}$$

Because NO exists in trace quantities the partial pressures of N_2 and O_2 in Equation D are virtually unaffected by Reaction A. Hence,

$$(p_{NO}/P^0) = (4.54\times 10^{-31}\times (0.79/1)\times (0.21/1))^{1/2} = 2.75\times 10^{-18},$$

$$X_{NO} = 2.75\times 10^{-18} \quad \text{or} \quad NO = 2.75\times 10^{-12} \text{ ppm},$$

$$K^{\circ}_{NO_2} = \exp(-\Delta G^{\circ}_{NO_2}/\bar{R}T) = (p_{NO_2}/P^0)^2/((p_{N_2}/1)(p_{O_2}/P^0)^2),$$

where

$$(p_{NO_2}/P^0) = (1.03\times 10^{-18}\times (0.79/1)\times (0.21/1)^2)^{1/2} = 1.89\times 10^{-10}, \text{ and}$$

$$X_{NO_2} = 1.89\times 10^{-4} \text{ ppm}.$$

Remarks

It is seen that equilibrium favors NO_2, rather than NO, formation at 298K. Thus, most of the NO emitted into the atmosphere eventually gets converted into NO_2.

Example 6

A reactor is fired with propane, and the amounts of O_2 and N_2 are measured to be 10% and 76%, respectively, at its exit. At 2200 K, determine the maximum amount of NO that can be present at equilibrium. Assume P = 1 bar.

Solution

Even though the percentages of O_2 and N_2 are specified in this example, you can determine these values if A:F is specified assuming complete combustion of propane and that NO exists as a trace species.
Assume chemical equilibrium only for the reaction

$$(1/2)N_2 + (1/2)\ O_2 \Leftrightarrow NO.$$

At 2200 K, K = 0.033 = $\{p_{NO}/P^0\}/\{(p_{O2}/P^0)^{1/2}\ (p_{N2}/P^0)^{1/2}\} = X_{NO}/\{(X_{O2})^{1/2}\ (X_{N2})^{1/2}\} = X_{NO}/\{(0.21)^{1/2}(0.79)^{1/2}\}$; X_{NO} = 0.00897 or 8970 ppm, where ppm is an expression for parts per million or molecules per million molecules = mole fraction × 10^6.

At 1500 K, X_{NO} = 905 ppm.

Remarks

(i) When the temperature decreases, NO is reduced.
(ii) Note that we assumed equilibrium only between NO, N_2, and O_2 but not for other species, i.e., we assumed partial equilibrium in the products.

Example 7

Five kmol of CO, 3 kmol of O_2, 4 kmol of CO_2, and 7 kmol of N_2 are introduced into a reactor at 3000 K and 2000 kPa. (a) Determine the equilibrium composition of the gas leaving the reactor, assuming that the outlet (product) stream contains CO, O_2, N_2, and CO_2. (b) Will the equilibrium composition change if the feed is altered to 6 kmol of CO, 3 kmol of CO_2, 3.5 kmol of O_2, and 7 kmol of N_2? (c) Will the CO concentration at the outlet change if the pressure changes, say, to 101 kPa?

Solution

(a) We have four unknowns N_{CO}, N_{O2}, N_{N2}, and N_{CO2}; hence, we need four equations.

The overall balance equation in terms of the four unknown concentrations is

$$5\ CO + 3O_2 + 4\ CO_2 + 7N_2 \rightarrow N_{CO}CO + N_{O_2}O_2 + N_{CO_2}CO_2 + N_{N_2}N_2 \quad (A)$$

The conservation of C, O, and N atoms provide three atom balance equations.

$$\text{Carbon atoms: } N_C = 5 + 4 = N_{CO} + N_{CO_2} \tag{B}$$

$$\text{Oxygen atoms: } N_O = 5\times1+3\times2+4\times2 = N_{CO}\times1+N_{CO_2}\times2+N_{O_2}\times2 \tag{C}$$

$$\text{Nitrogen atoms: } N_N = 7\times2 = N_{N_2}\times2 \tag{D}$$

The fourth equation is provided by the equilibrium condition at 3000 K for the $CO_2 \Leftrightarrow CO + 1/2\ O_2$ reaction for which $\log_{10}K^0$ (3000) = −0.48.

$$10^{-0.487} = 0.327 = (N_{CO}N_{O_2}^{1/2})(P^0/(P^0\times N))^{1/2}/N_{CO_2} \tag{E}$$

Using Equation B

$$N_{CO} = N_C - N_{CO_2}, \quad \text{i.e.,} \quad N_{CO} = 9 - N_{CO_2}. \tag{F}$$

Further, using Equation F in Equation C and solving for N_{O_2},

$$N_{O_2} = (19-9-N_{CO_2})/2. \tag{G}$$

Therefore, using Equations G and F, the total number of moles at the exit

$$N = N_{CO} + N_{O_2} + N_{CO_2} + N_{N_2} = 21 - N_{CO_2}/2\,. \tag{H}$$

Applying Equation F to Equation H in Equation E, with P = 20 bar, and P^0 = 1 bar, we solve for N_{CO2} at the exit

$N_{CO_2} = 6.96$ kmol, and hence using Equations F and G

N_{CO} = 2.04 kmol, N_{O_2} = 1.52 kmol, and N = 17.52 kmol.

(b) When the feed stream is altered to react 6 kmol of CO, 3 kmol of CO_2, 3.5 kmol of O_2, and 7 kmol of N_2, the respective inputs of C, O, and N atoms remain unaltered at 9, 19, and 14, respectively. Therefore, the equilibrium composition is unchanged. This indicates that it does not matter in which form the atoms of the reacting species enter the reactor. The same composition, for instance, could be achieved by reacting a feed stream containing 9 kmol of C(s) (solid carbon, such as charcoal), 9.5 kmol of O_2, and 7 kmol of N_2 (which is treated as an inert in this problem).

(c) From Equation E we note that for a specified temperature, the value of $K^o(T)$ is unique. Therefore, if the pressure changes, the temperature does not. Equation E dictates that the composition is altered and more CO_2 is produced as the pressure is increased.

Example 8

Consider a PCW assembly that is immersed in an isothermal bath at 3000 K. It initially consists of 9 kmol of C atoms and 19 kmol of O atoms (total mass = 9 × 12.01 + 19 × 16 = 412 kg). This is allowed to reach chemical equilibrium at 3000 K and 1 bar. What is the equilibrium composition? What is the value of the Gibbs energy? If we keep placing sand particles one at a time on the piston, let it equilibrate, and then keep proceeding to a final pressure of 4 bar, we have allowed sufficient time for chemical equilibrium to be reached at that pressure. What is the resulting equilibrium composition and Gibbs energy?

Solution

$$9\,C + 19\,O \rightarrow N_{CO_2}\;CO_2 + N_{CO}\;CO + N_{O_2}\;O_2$$

For the three unknowns, we have two atom balance equations and one equilibrium relation. We leave it to the reader to show that at equilibrium (Example 1),

$$N_{CO_2} = 5.25\text{ kmol},\ N_{CO} = 3.7\text{ kmol},\quad\text{and}\quad N_{O_2} = 2.37\text{ kmol}. \tag{A}$$

Therefore,

$$N = \sum N_k = 11.37\text{ kmol}. \tag{B}$$

The Gibbs energy is

$$G = N_{CO_2}\;\hat{g}_{CO_2} + N_{CO}\;\hat{g}_{CO} + N_{O_2}\;\hat{g}_{O_2}, \tag{C}$$

where

$$X_{CO_2} = N_{CO_2}/N = 0.462. \tag{D}$$

At 3000 K and 1 bar, $\hat{g}_{CO_2} = -1262000\text{ kJ kmol}^{-1}$, $\hat{g}_{CO} = -865200\text{ kJ kmol}^{-1}$ and $\hat{g}_{O_2} = -794400\text{ kJ kmol}^{-1}$. Hence, at T = 3000 K, P = 1 bar (state 1)

$$G = -11{,}753{,}000\text{ kJ},$$

which at this equilibrium state must be at a minimum value.

At a temperature of 3000 K and a pressure of 4 bar (state 2), the equilibrium composition changes to

$$N_{CO_2} = 6.4\text{ kmol},\ N_{CO} = 2.6\text{ kmol},\ \text{and}\ N_{O_2} = 1.8\text{ kmol}, \tag{E}$$

and

$$N = \sum N_k = 10\text{ kmol and } G = -11374000\text{ kJ}. \tag{F}$$

Remarks

(i) The difference in the minimum Gibbs free energies (i.e., at the equilibrium states) between the two states (3000 K, 1 bar) to (3000 K, 4 bar) is

$$dG = -SdT + VdP,\ dG_T = VdP = (N\,\bar{R}\,T/P)\ dP \quad \text{or}$$
$$G_2\ (3000,\ 1) - G_1(3000,\ 4) = (-11374000) - (-11757000) = 383000\ \text{kJ}.$$

(ii) If N ≈ constant ≈ (11.37 + 10)/2 = 10.69 kmol, then $dG_T = (N\,\bar{R}\,T/P)\ dP$. Integrating, $G_2 - G_1 \approx N\,\bar{R}\,T\ \ln\ (P_2/P_1) = 369455$ kJ, which is slightly different from 383000 kJ.

(iii) The relations dU = T dS − P dV, dH = T dS + V dP, dG = −S dT + V dP, etc., for closed systems can be applied even for chemical reactions as long as we connect a reversible path between the two equilibrium states for which we compute $U_2 - U_1$, $S_2 - S_1$, and $G_2 - G_1$. However, these equations cannot be applied during the process of irreversible chemical reaction or process.

(iv) If we compress the products very slowly from 1 to 4 bar isothermally at 3000 K, the reaction tends to produce more CO_2, i.e., N_{CO2} increases from 5.25 to 6.4 kmol. Instead, rapid compression to 4 bar (i.e., allowing a very short time for CO to react with O_2 and produce additional CO_2) produces an insignificant change from the composition at 1 bar, i.e., the products will be almost frozen at N_{CO2} = 5.25 kmol, N_{CO} = 3.75 kmol, and N_{O2} = 2.37 kmol even though the state is now at 3000 K and 4 bar. The products during this initial period are in a nonequilibrium state. The value of G at this state is $G_{Frozen} = 5.25 \times \hat{g}_{CO_2}$ (3000, 4 bar, X_{CO2} = 5.25/11.37) + 3.75 × $\hat{g}_{CO}$(3000, 4 bar, X_{CO} = 3.75/11.37) + 2.37× $\hat{g}_{O_2}$ (3000, 4 bar, X_{O2} = 3.75/ 11.37) = 5.25 × (1,227,400) + 3.75 × −(830,600) + 2.37 × −(759,800) = −11,360,000 kJ, which is higher than G = 11,374,000 kJ at the equilibrium composition corresponding to 3000 K, 4 bar. If we allow more time at the state (3000 K, 4 bar), G will decrease. After a long time, chemical equilibrium will have been reached and G will have approached its minimum value of −11,374,000 kJ from −11,360,000 kJ.

(v) A similar phenomenon occurs when these reacting gases flow at the slowest possible velocity through a diffuser, where the pressure at the inlet to the diffuser is 1 bar and at exit it is 4 bar. The 412-kg mass, when it flows through the diffuser, will reach its equilibrium composition given by Equation E. However, if the same mass flows at high velocity, the composition at the exit of the diffuser can be almost the same as that at the inlet.

3.3.4 Gas, Liquid, and Solid Mixtures

Consider the following chemical reactions at equilibrium

$$CaCO_3(s) \Leftrightarrow CaO(s) + CO_2(g), \tag{3.42}$$

$$H_2O(\ell) + CO(g) \Leftrightarrow CO_2(g) + H_2(g), \tag{3.43}$$

$$CaO(s) + SO_2(g) + 1/2\ O_2(g) \Leftrightarrow CaSO_4(s). \tag{3.44}$$

Equilibrium relations can be written as follows. Consider Equation 3.44. Assume ideal gas behavior for SO_2 and O_2. Therefore, Equation 3.44 at equilibrium assumes the form

$$K°(T) = \{(X_{CaSO_4(s)}/(X_{SO_2} X_{O_2}^{1/2} X_{CaO(s)})\}(P/P^0)^{-3/2} \quad (3.45)$$

Example 9

Consider the water gas shift reaction $H_2O(\ell) + CO(g) \Leftrightarrow H_2(g) + CO_2(g)$. Determine the equilibrium constant for the reaction at 298 K, treating the gaseous species as ideal.

Solution

Because CO, H_2, and CO_2 are treated as ideal gases, $\hat{\alpha}_k = X_k$, $\hat{\alpha}_{H_2O(\ell)} = 1$ Equation 3.38b transforms to

$$\begin{aligned} K°(T) &= (p_{H_2}/P^0)(p_{CO_2}/P^0)/(p_{CO}/P^0) \\ &= \exp(-(\bar{g}^0_{H_2} + \bar{g}^0_{CO_2} - \bar{g}^0_{CO} - \bar{g}^0_{H_2O(\ell)})/\bar{R}\,T). \end{aligned} \quad (A)$$

Using Table A.18, Table A.12, Table A.13, and Table A.17

$\bar{g}^0_{H_2O(\ell)} = \bar{h}_{H_2O(\ell)} - T\,\bar{s}_{H_2O(\ell)} = -285830 - 298 \times 69.95 = -306675$ kJ $kmole^{-1}$ of $H_2O(\ell)$,

$\bar{g}^{\circ}_{CO} = -169403$ kJ $kmol^{-1}$ $\bar{g}^{\circ}_{CO_2} = -457203$ kJ $kmol^{-1}$ and $\bar{g}^{\circ}_{H_2} = 0 - 298 \times 130.57 = -38910$ kJ $kmol^{-1}$.

Therefore,

$$\begin{aligned} \Delta G° &= -38910 - 457203 - (-169403 - 306675) \\ &= -20035 \text{ kJ kmol}^{-1} \end{aligned}$$

and $$K°(298\text{ K}) = \exp(20{,}035 \div (8.314 \times 298)) = 3250.4.$$

Remarks

Because K°(298 K) is extremely large, $X_{CO_2}X_{H_2}/X_{CO}$ is also large, and consequently the value of X_{CO} at chemical equilibrium is extremely small. Therefore, if CO gas is bubbled through a vast reservoir of $H_2O(\ell)$ at 298 K, very little unreacted CO is left over. Note that the results pertain only to an equilibrium condition. However, the timescale required to reach it may be inordinately large.

Example 10

Consider the reaction of $SO_3(g)$ with CaO(s), a process that is used to capture the SO_3 released during the combustion of coal, i.e., $CaO(s) + SO_3(g) \rightarrow CaSO_4(s)$. Determine the equilibrium relation assuming that the sulfates and CaO are mixed at the molecular level (i.e., they are mutually soluble) and when they are unmixed. What is the partial pressure of SO_3 at 1200 K if $K°(1200\text{ K}) = 2.93 \times 10^7$ for $CaO(s) + SO_3(g) \Leftrightarrow CaSO_4(s)$?

Solution

Assume the standard for solids to be in solid state and for gases to be ideal gases is P = 1 bar, and use the approximation that $f_k(s)(T,P) \approx f_k(s)(T, 1\text{ bar})$.

The solid phase contains both $CaSO_4$ and CaO. Assume the ideal solution model for the solid phase. The equilibrium relation for the reaction is

$$K^o(T) = X_{CaSO_4}/(X_{CaO(s)}\, P\, X_{SO_3}/P^0).$$

If the solids are not mixed at the molecular level, they exist separately. Therefore, $X_{CaSO_4(s)} = X_{CaO(s)} = 1$, and

$$K^o(T) = 1/\{(P\, X_{SO_3})/P^0\}.$$

In the unmixed case with $P^0 = 1$ bar,

$$2.93\times 10^7 = 1/(p_{SO_3}/1), \quad \text{i.e.,} \quad p_{SO_3} = 0.41\times 10^{-8}\ \text{bar}.$$

Remarks

(i) If the pressure P = 1 bar, then for the unmixed case $X_{SO_3} = 0.41\times 10^{-8} = p_{SO_3}/P$ or 0.0041 ppm (parts per million). If the pressure is increased isothermally, the value of p_{SO_3} remains unchanged, but X_{SO_3} decreases, i.e., more of the sulfate will be formed.

(ii) In many instances in power plants, SO_2 released because of coal combustion is allowed to react with lime (CaO (s)) in order to produce sulfates according to the reaction $CaO(s) + SO_2 + 1/2O_2 \rightarrow CaSO_4$ (s), favored for 1000 K < T < 1200 K.

Here, CaO is produced from limestone $CaCO_3(s)$, and a temperature range of 1400 < T < 1600 is preferred to reduce the time of reaction. At T = 1700 K, lime is dead burned (with less surface area for SO_2 reaction). The equilibrium relation for this reaction is

$$K^o(T) = 1/\{(p_{SO_2}/P^0)(p_{O_2}/P^0)^{1/2}\} = (P/P^0)^{-3/2}/X_{SO_2}X_{O_2}^{\ 1/2}.$$

Increasing the pressure at constant T causes X_{SO_2} to decrease so that a lesser amount of SO_2 will be emitted, i.e., more SO_2 is captured from the combustion gases.

3.3.5 Dissociation Temperatures

Molecular dissociation is a very significant effect for hydrogen and hydrocarbon flames with equivalence ratios near 1. One can compute the 1% dissociation temperatures for P = 1 bar for several reactions [Lucht, 2006]. For example, consider 1 kmol of CO_2 entering a reactor at T. Let 1% of CO2 dissociate resulting in $N_{CO2} = 0.99$, $N_{CO} = 0.01$, and $N_{O2} = 0.005$, with the total product moles being 0.99 + 0.01 + 0.005 = 1.015 kmol. Then, applying Equation 3.41 to $CO_2 \Leftrightarrow CO + 1/2\ O_2$, we get $\{N_{CO}\ N_{O2}^{1/2}/N_{CO2}\}\{P/(P^0 *N)\} = \{0.01*0.005^{1/2}/0.99\}\{1/(1*1.015)\} = 7.09*10^{-4} = K^o(T)$.

From Table A.36A and Table A.36B of Appendix A, $\log_{10} K^o(T) = 3.15$ and $T = 1930$ K. The following table lists a few dissociation temperatures at $P = 1$ bar:

Reaction	*Temperature (K)*
$CO_2 \Leftrightarrow CO + 1/1\ O_2$	1930
$H_2O \Leftrightarrow OH + 1/2\ H_2$	2080
$H_2O \Leftrightarrow H_2 + 1/2\ O_2$	2120
$H_2 \Leftrightarrow 2H$	2430
$O_2 \Leftrightarrow 2O$	2570
$N_2 \Leftrightarrow 2N$	3590

3.3.6 Equilibrium for Multiple Reactions

This concept will be illustrated through the following example.

Example 11

Let the number of kilomoles of O_2 entering a reactor be "a," while the number of moles of O_2, CO, CO_2, and C(s) leaving that reactor at equilibrium be N_{O_2}, N_{CO}, N_{CO_2} and $N_{C(s)}$. Assume the following reactions (taking place during entry and exit, respectively) to be in equilibrium:

$$C(s) + 1/2\ O_2 \Leftrightarrow CO \tag{A}$$

and

$$CO_2 \Leftrightarrow CO + 1/2\ O_2 \tag{B}$$

(a) Determine $N_{CO}(T)$ in the products and (b) minimum amount of carbon(s) that should enter the reactor so as to maintain equilibrium at the exit.

Solution

$$\text{(a) O atom balance: } 2a = 2\,N_{O_2} + N_{CO} + 2N_{CO_2}, \tag{C}$$

$$K_A^O = \{N_{CO}/N_{O_2}{}^{1/2}\}\{P/(P^0N)\}^{1/2}, \tag{D}$$

$$K_B^O = \{N_{CO}\ N_{O_2}{}^{1/2}/N_{CO_2}\}\{P/(P^0N)\}^{1/2}. \tag{E}$$

Hence,

$$K_B^O/K_A^O = N_{O_2}/N_{CO_2}. \tag{F}$$

Using Equation F in C and solving for NO_2

$$N_{O_2} = (2a - N_{CO})/\{2(K_A/K_B + 1)\}. \tag{H}$$

Using Equation H in F to solve for N_{CO_2} and calculating total moles leaving reactor as

$$N = (2\ a + N_{CO})/2. \tag{I}$$

Using Equation I and H in D and solving for N_{co}

$$N_{CO}(T) = 2\ a/\{(1 + 4(\{P/P^0\}/K_A)\ (1/K_A+1/K_B))^{1/2}\}, \tag{J}$$

$$X_{CO}(T) = N_{CO}/N = 2/(1 + (1 + 4\ (\{P/P^0\}/K_A)\ (1/K_A + 1/K_B))^{1/2})). \tag{K}$$

(b) Once N_{CO} is solved from Equation J, N_{O2} is solved from Equation H, and N_{CO2} can be obtained from Equation F. X_{CO_2} can be similarly expressed. The required carbon atom input must be such that

$$N_{C,\ in} \geq N_{CO} + N_{CO_2} \tag{L}$$

where the equality applies to the minimum carbon input required for achieving chemical equilibrium. A mathematical expression for the minimum carbon input is

$$N_{C,in,min}/a = \left[\left\{\left(K_A^O/K_B^O\right)+1\right\}/\left\{1+4\left(1/K_B^O+1/K_A\right)P'/\ K_A^O\right\}^{/2}\right.$$

$$\left. +2\left(K_A^O/K_B^O\right)\right]/\left(2+K_B^O/K_A^O\right),$$

where $P^1 = P/P^O$

Remarks

Now consider 2 kmol of O_2 entering the reactor, and assume that the carbon flow is always above minimum value. Thus, there are always a few carbon moles present at the exit. Thus, equilibrium exists between C(s), O_2, CO, and CO_2. For Reaction A, $\ln K_A = 4.3478 + 5874.1/T$ and $\ln K_B = 4.383 - 14604/T$ for Reaction B.

Figure 3.9 shows a plot of CO, CO_2, and minimum number of moles of C/O ratio required for equilibrium at exit vs. temperature at pressures 0.1 to 10 bar. Thus, it can be seen that there is more CO_2 at high pressure.

■ ■ ■

General Procedure for Solving Equilibrium Problems

1. Specify the fuel (say CH_3OH) and amount of air supplied (called reactants); Now specify product species, say CO2, CO, H2, H2O, O2, N2, OH (Total number of unknown product species $N_s = 7$).
2. Make atom conservation between reactants and products; we have 4 different atom types: C, H, N, O ($N_A = 4$) and hence 4 atom conservation equations.

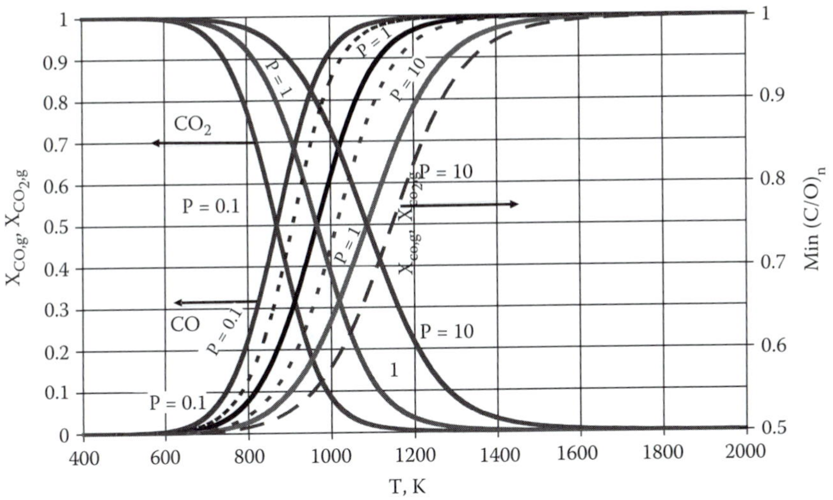

Figure 3.9 CO and CO_2 products under equilibrium when 2 kmol of O_2 is supplied with enough carbon for producing CO and CO_2. P in bars.

3. Now ignore the reactants, i.e., air and fuel at inlet; assume equilibrium *among the exit product species* only. Since $N_s > N_A$ we need $N_s - N_A = N_E$ equilibrium relations. For the example there are three equilibrium relations: (I) CO2, CO O2; (II) H2O, H2 O2; (III) H2O, H2, OH for which data for K^0 are available.
4. When selecting relations for each equilibrium relation, *write exactly the same way they are written in tables* for which K^0 were tabulated (e.g., CO2 ⇔ CO + 1/2 O2 in tables but not CO + 1/2 O2 ⇔ CO2) and then write in terms of partial pressures, e.g., CO2 ⇔ CO + 1/2 O2, $K^0(T) = \left(N_{CO} N_{O_2}^{1/2}\right) \{P/(P^0 \times N)\}^{1/2}/N_{CO2}$. Note that P is the exit pressure of products.
5. Now you have N_s equations and N_s unknowns (for this example 4 atom balance eqs and 3 equilibrium relations for 7 unknown product species). Solve. The procedure is illustrated in the following example.

Example 12

Consider the stoichiometric combustion of 1 kmol of CH_4 with air. The products are at 2250 K and 1 bar and contain CO_2, CO, H_2O, H_2, O_2, N_2, and OH. (a) Determine the equilibrium composition. (b) Compute total enthalpy of products, H_P.

Solution

(a) The overall chemical reaction is

$$CH_4 + 2(O_2 + 3.76\, N_2) \rightarrow N_{CO_2}\, CO_2 + N_{CO}\, CO + N_{H_2O}\, H_2O + N_{H_2}\, H_2 + N_{O_2}\, O_2 + N_{N_2} N_2 + N_{OH}\, OH. \quad (A)$$

There are seven species of unknown composition. The four atom conservation equations for C, H, N, and O atoms are:

$$\text{C atoms: } 1 = N_{CO_2} + N_{CO}, \tag{B}$$

$$\text{H atoms: } 4 = 2 \times N_{H_2O} + 2 \times N_{H_2} + N_{OH}, \tag{C}$$

$$\text{N atoms: } 7.52 \times 2 = 2 \times N_{N2}, \tag{D}$$

$$\text{O atoms: } 2 \times 2 = 2 \times N_{CO_2} + N_{CO} + N_{H_2O} + 2 \times NO_2 + N_{OH}. \tag{E}$$

We, therefore, require three additional relations. At equilibrium, we have the reactions

$$CO_2 \Leftrightarrow CO + 1/2\ O_2,\ K_{CO2}{}^{o} = \{p_{CO}/P°\}\,(p_{O_2}/P°)^{0.5}/\,(p_{CO_2}/P°), \tag{F}$$

$$H_2O \Leftrightarrow H_2 + 1/2\ O_2,\ K_{H_2O} = (p_{H_2}/P°)(p_{O_2}/P°)^{0.5}/(p_{H_2O}/P°), \tag{G}$$

$$OH \Leftrightarrow 1/2\ H_2 + 1/2\ O_2,\ K_{OH} = (p_{H_2}/P°)^{0.5}(p_{O_2}/P°)^{0.5}/(p_{OH}/P°). \tag{H}$$

Assume, for example, the values N_{CO}, N_{O2}, and N_{CO2}. Solve for other species from Equation 3.B to Equation 3.E. Then, check whether Equation 3.F to Equation 3.H are satisfied. If not, iterate. One of the present authors has developed a spreadsheet-based program called THERMOLAB, which presents solution for all species at any given T and P and any given air composition. The results (in terms of 1 kmol of methane consumed) are $N_{CO_2} = 0.910$, $N_{CO} = 0.09$, $N_{H_2O} = 1.96$, $N_{H_2} = 0.04$, $N_{O_2} = 0.064$, and $N_{N_2} = 7.52$.

(b) Total enthalpy of products $H_P = N_{CO_2}\ \bar{h}_{O_2} + N_{CO} + \bar{h}_{CO} + N_{H_2O}\ \bar{h}_{H_2O} + N_{H_2}\bar{h}_{H_2} + N_{O_2}\bar{h}_{O_2} + N_{N_2}\bar{h}_{N_2} + N_{OH}\ \bar{h}_{OH}$. Using tables $H_P = -60{,}632$ kJ.

Remarks

If NO exists in trace amounts, its concentration at equilibrium can be determined during combustion by considering the following reactions

$$NO \Leftrightarrow 1/2\ N_2 + 1/2\ O_2 \tag{I}$$

$$K^{O}_{NO} = (p_{N_2}/P°)^{0.5}(p_{O_2}/P°)^{0.5}/(p_{NO}/P°). \tag{J}$$

Figure 17.4 in Chapter 17 presents Equilibrium NO vs. T.

3.4 Vant Hoff Equation

The vant Hoff equation, due to Jacobus Henricus vant Hoff (1852–1911), presents a relation between the equilibrium constant K°(T) and the enthalpy of reaction ΔH_R.

3.4.1 Effect of Temperature on K°(T)

From thermodynamics, $\bar{g}_k(T,P) = \bar{h}_k(T,P) - T\,\bar{s}_k(T,P)$, and the relation for any given pure component k is

$$T\,d\bar{s}_k + \bar{v}_k dP = d\bar{h}_k,$$

$$\frac{\partial}{\partial T}\left(\frac{\bar{g}_k}{T}\right) = \frac{1}{T}\left(\frac{\partial \bar{h}_k}{\partial T}\right) - \frac{\bar{h}_k}{T^2} - \left(\frac{\partial \bar{s}_k}{\partial T}\right) = \frac{\bar{c}_{pk}}{T} - \frac{\bar{h}_k}{T^2} - \frac{\bar{c}_{pk}}{T} = -\frac{\bar{h}_k}{T^2}. \quad (3.46)$$

Rewriting,

$$(\partial(\bar{g}_k/T)/\partial(1/T))_P = \bar{h}_k. \quad (3.47)$$

Consider the equilibrium reaction $CO_2 \Leftrightarrow CO + 1/2\ O_2$, for which $\Delta G(T, P) = \bar{g}_{CO}(T, P) + 1/2\,\bar{g}_{O_2}(T, P) - \bar{g}_{CO_2}(T, P)$.

Differentiating,

$$\partial(\Delta G/T)/\partial(1/T) = \partial\{\bar{g}_{CO}(T, P)\}/\partial(1/T) + (1/2)\partial\{\bar{g}_{O_2}(T, P)\}/\partial(1/T) - \partial\{\bar{g}_{CO_2}(T, P)\}/\partial(1/T) = \Delta H_R(T, P) \quad (3.48)$$

where the enthalpy of reaction is

$$\Delta H_R\ (T,\ P) = \bar{h}_{CO}\ (T,\ P) + 1/2\,\bar{h}_{O_2}(T, P) - \bar{h}_{CO_2}\ (T,\ P). \quad (3.49)$$

At 1 atm, Equation 3.48 becomes

$$d(\Delta G^o/T)/d(1/T) = \Delta H_R^o\,(T) \quad \text{or} \quad d(\Delta G^o/T)/dT = -\Delta H_R^o\,(T)/T^2. \quad (3.50)$$

Because $\ln K^o(T) = -\Delta G^o/\bar{R}T$ (see Equation 3.37), Equation 3.50 becomes

$$d\ \ln K^o(T)/dT = \Delta H_R^o\,(T)/\bar{R}\,T^2, \quad (3.51)$$

which is known as the *vant Hoff equation.* If $\Delta H_R^0 \approx$ constant, Equation 3.51 is integrated to yield $\ln K^o(T) = -\Delta H_R^o/\bar{R}\,T +$ constant, which is a linear relationship between $\ln K^o(T)$ and $1/T$. If $K^o = K_{ref}^o$ at $T = T_{ref}$, the constant can be eliminated, i.e.,

$$\ln(K^o/K_{ref}^o) = -(\Delta H_R^0/\bar{R})(1/T - 1/T_{ref}). \quad (3.52)$$

This relation can be written in the form

$$\ln K^o(T) = A - B/T \quad (3.53a)$$

where

$$A = \ln K_{ref}^o + (\Delta H_R^o/\bar{R})(1/T_{ref}) = \frac{-\Delta G_{ref}^O}{\bar{R}\,T_{ref}} + \frac{\Delta H_R^O}{\bar{R}\,T_{ref}} \quad (3.53b)$$

and

$$B = (\Delta H_R^o/\bar{R}). \tag{3.53c}$$

Figure B.2 in Appendix B presents plots of $\log_{10} K^o(T)$ vs. $10^4/T$ for various reactions. The approximate relation in Equation 3.53a appears to be valid, i.e., assumption of $\Delta H_R^0 \approx$ constant seems to be reasonable.

The constants A and B for any given reaction can be estimated using values given in Table A.36A and Table A.36B of Appendix A. Since $\Delta G^o_{ref} = \Delta H^o_{R,ref} - T_o \Delta S^o_{R,ref}$

$$A = \Delta S_R^o(T_{ref})/\bar{R}. \tag{3.54}$$

If $\Delta H_R^o > 0$ (e.g., endothermic decomposition reactions), B > 0 and $K^o(T)$ increases with temperature. Applying Equation 3.54 and Equation 3.53c in Equation 3.53a, we get $K^o(T) = \exp(\Delta S^o/\bar{R}) \exp(-\Delta H_R^o/\bar{R}\, T_{trans})$. At the transition temperature $T = T_{trans}$, the value of K^0 equals unity. Thus $1 = \exp\left(-(\Delta H_R^o - T_{trans}\Delta S_R^o)/\bar{R}\, T_{trans}\right)$,

i.e.,

$$T_{trans} = (\Delta H_R^o/\Delta S_R^o). \tag{3.55}$$

The transition temperature is the temperature at which significant amounts of products start to be formed.

Example 13

Determine the equilibrium constant for the reaction $NH_4HSO_4(l) \rightarrow NH_3(g) + H_2O(g) + SO_3(g)$. At 298 K, ΔH^o = 336500 kJ kmol^{-1} and ΔS^o = 455.8 kJ kmol^{-1} K^{-1}. Determine the transition temperature (i.e., at which $K^0(T)$ =1).

Solution

Recall from Equation 3.52 with $T_{ref} = T_0$,

$$\ln K^o(T) = \ln K^o(T_{ref}) - \{\Delta H^o/\bar{R}\}\{1/T - 1/T_{ref}\} \tag{A}$$

where $K(T_{ref}) = \exp(-\Delta G^o/\bar{R}\, T_{ref})$ and $\Delta G^o = \Delta H^o - T\Delta S^o = 336500 - 298 \times 455.8 = 200672$ kJ kmol^{-1}, i.e., at T_{ref} = 298K,

$$K^o(T_{ref}) = \exp(-\Delta G^o/\bar{R}\, T_{ref}) = 6.67 \times 10^{-36}, \tag{B}$$

or

$$K^o(T) = 6.67 \times 10^{-36} \exp\{-(336500/8.314)(1/T - 1/298)\}. \tag{C}$$

By setting $K^o(T) = 1$, the transition temperature $T_{trans} = 336500 \div 456 = 738$ K.

Remarks

Because

$$K^{\circ}(T) = (p_{NH_3}/P^0)(p_{H_2O}/P^0)(p_{SO_3}/P^0), \quad \text{(D)}$$

and K° increases with temperature (enthodermic reaction), decomposition is favored at higher temperatures and particularly at T > 738K.

Using the vant Hoff equation (3.52) one can actually derive the Clausius–Clapeyron relation, which presents a relation between saturation temperature and pressure for phase equilibrium such as $H_2O(\ell) \Leftrightarrow H_2O(g)$:

$$\ln(P/P_{ref}) = -\{(\bar{h}_g - \bar{h}_f)/\bar{R}\}(1/T - 1/T_{ref}). \quad (3.56)$$

where T_{ref} is saturation temperature at reference pressure P_{ref} and T is saturation temperature at P. Equation 3.56 is useful in deriving results for evaporation and combustion rate of droplets.

For the vaporization process,

$$\Delta H_R^o = \bar{h}_g - \bar{h}_f, \quad \text{or} \quad \Delta H_R^o = \bar{h}_{fg} \quad \text{(E)}$$

i.e.,

$$\ln(p_{H_2O(g)}/p_{H_2O(g),ref}) = -(\bar{h}_{fg}/\bar{R})(1/T - 1/T_{ref}). \quad \text{(F)}$$

This is a relation for the change in the saturation pressure of the H_2O as the temperature changes. This is also known as the Clausius–Clapeyron relation, given by

$$\ln(P/P_{ref}) = -((\bar{h}_g - \bar{h}_f)/\bar{R})(1/T - 1/T_{ref}). \quad \text{(G)}$$

3.4.2 Effect of Pressure

Because $d\hat{g}_{k,T} = \hat{v}_k dP$ and $d\hat{g}/\partial P = \hat{v}_k$,

$$\{\partial(\Delta G)/\partial P\}_T = \sum \nu_k \hat{v}_k = \Delta V_R, \quad (3.57)$$

where ΔV_R denotes the volume change between the products and the reactants. In this context, because $\ln K(T,P) = -\Delta G(T,P)/\bar{R}\,T$, and using Equation 3.57, we get

$$\{\partial(\ln K(T,P))/\partial P\}_T = -\{\partial(\Delta G/\bar{R}T)/\partial P\}_T = -(1/\bar{R}\,T)\Delta V_R. \quad (3.58)$$

For an incompressible species (e.g., during the reaction Na(s) + Cl(s) → NaCl (s)), $\Delta V_R \approx$ constant and $\ln\{K(T,P)\} = -(1/\bar{R}\,T)\,P\Delta V_R^o +$ constant.

If the value of $K(T,P_{ref})$ is known, then

$$\ln(K(T,P)/K_{ref}(T,P_{ref})) = -(1/\bar{R}\,T)\Delta V_R^o(P - P_{ref}). \quad (3.59)$$

For ideal gases, with each species at T and P,

$$\Delta V_R = \sum \nu_k (\bar{R}T/P) \tag{3.60}$$

$$\ln (K(T,\ P)) = \left(\sum \nu_k\right) \ln P + \text{Constant.} \tag{3.61}$$

Example 14

Consider the combustion of CH_4 in air at 1 bar. Determine the number of moles of products produced (and, in particular, the CO concentration) per kmol of fuel consumed if the oxygen mole fraction in the products is 3% on a dry basis. Assume equilibrium reaction $CO_2 \Leftrightarrow CO + 1/2O_2$,

$$\ln K^o = A - B/T \tag{A}$$

where A = 9.868 and B = 33742.4 K. Assume that CO is present in trace amounts.

Solution

Assume that the overall combustion reaction is represented by the equation

$$CH_4 + a\ (O_2 + 3.76\ N_2) \rightarrow CO_2 + 2H_2O + b\ N_2 + d\ O_2. \tag{B}$$

It is now straightforward to determine that a = 2.2982, b = 8.641, and d = 0.2982.

The carbon dioxide concentration in the exhaust on a wet basis is

$$100 \times (1 \div (1 + 2 + 3.76 \times 2.2982 + 0.2982))\% = 8.4\%$$
$$p_{CO2} = X_{CO2}{*}P = 0.084{*}1 = 0.084 \text{ bar.}$$

Similarly, the nitrogen concentration is 72.4%. Note that CO is not produced during combustion according to our model, but is instead formed because of the dissociation of CO_2 through the reaction $CO_2 \Leftrightarrow CO + 1/2\ O_2$. The equilibrium constant for that reaction is

$$K^\circ(T) = \{p_{CO}/P^\circ\}(p_{O_2}/P^\circ)^{0.5}/(p_{CO2}/P^\circ),\ \text{i.e.,}\ p_{CO}/P^\circ = K^\circ(T)\{p_{CO_2}/P^\circ\}/\ (p_{O_2}/P^\circ)^{0.5}. \tag{C}$$

Because CO exists in trace amounts, $P^\circ = 1$ bar, $p_{CO_2} \approx 0.084$ bar, and $p_{O_2} \approx 0.025$ bar (in proportion to their concentrations). Applying Equation A at 1500 K, $K^o(T) = 3.28 \times 10^{-6}$, i.e., $p_{CO}/P^\circ = 3.28 \times 10^{-6}\ (0.084/1)/(0.025/1)^{0.5} = 1.743 \times 10^{-6}$. At one bar, $X_{CO} = p_{CO}/P = p_{CO}/1 \text{ bar} = 1.743 \times 10^{-6}$.

Remarks

Figure 3.10 (left) presents the CO concentration in ppm vs. 1/T for methane and C fuels with 20% excess air using THERMOLAB software. Runs could be made for rich mixtures and plots similar to those shown in Figure 3.10 (right) can be obtained. The CO concentration does not significantly alter when the fuel changes from methane to carbon. Similarly, one can determine the equilibrium concentrations of other trace species such as O and H_2.

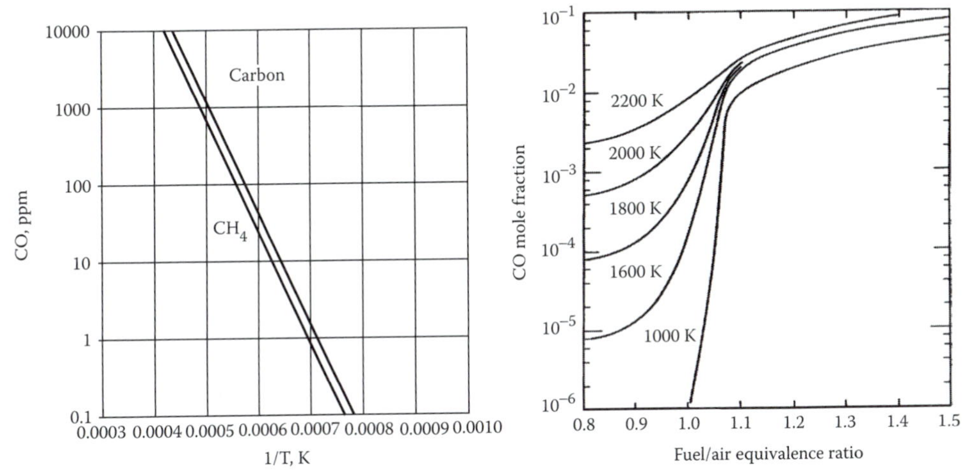

Figure 3.10 (Left) The CO emission due to the combustion of carbon and CH_4 at 3% O_2, using THERMOLAB software; (Right) CO vs. equivalence ratio for CH_4. (Adapted from Bartok, W. and Sarofim, A., *Fossil Fuel Combustion*, John Wiley & Sons, New York, 1991.)

3.5 Adiabatic Flame Temperature with Chemical Equilibrium

The energy balance equation applicable during combustion in a reactor is

$$dE_{cv}/dt = \dot{Q}_{cv} - \dot{W}_{cv} + \sum\nolimits_k \dot{N}_{ik}\hat{h}_{i,k} - \sum\nolimits_k \dot{N}_{e,k}\hat{h}_{e,k}. \quad (3.62)$$

3.5.1 Steady-State, Steady-Flow (SSSF) Process

At steady state, $dE_{cv}/dt = 0$. For an adiabatic reactor, $\dot{Q}_{cv} = 0$. Because $\dot{W}_{cv} = 0$,

$$\sum\nolimits_k \dot{N}_{ik}\hat{h}_{i,k} - \sum\nolimits_k \dot{N}_{e,k}\hat{h}_{e,k} = 0 \quad \text{or} \quad \dot{H}_i = H_e = 0 = \dot{H} = \sum\nolimits_k \dot{N}_k\breve{h}_k . \quad (3.63)$$

For ideal gas mixtures or ideal solutions, $\hat{h}_k = \bar{h}_k = \bar{h}^o_{f,k} + (\bar{h}_{t,T} - \bar{h}_{t,298})_k$. Table A.34 tabulates adiabatic flame temperatures and product composition at chemical equilibrium for a few fuels under SSSF process.

3.5.2 Closed Systems

The energy balance for an adiabatic closed system is $dE_{cv}/dt = \dot{W}_{cv}$. If the boundary work is neglected (e.g., for a fixed volume), $dE_{cv}/dt = 0$. If other forms of the energy are neglected and we assume that E = U,

$$\left(\sum N_k\ \hat{u}_k\right)_{products} - \left(\sum N_k\ \hat{u}_k\right)_{reactants} = 0 \quad \text{or} \quad U_{reactants} = U_{products} \quad (3.64)$$

where

$$\hat{u}_k = \hat{h}_k - P\,\hat{v}_k \tag{3.65}$$

and for ideal gas mixtures,

$$\hat{u}_k = \bar{u}_k = \bar{h}_k - \bar{R}T \tag{3.66}$$

Example 15

Consider 5 mol of CO, 3 mol of O_2, 4 mol of CO_2, and 7 mol of N_2 entering a reactor at 2000 kPa. If the inlet enthalpy (H_{in}) is equal to 90184.4 kJ. (a) Determine the adiabatic flame temperature and equilibrium composition of the gas leaving the reactor. Assume that CO, O_2, N_2, and CO_2, are products at equilibrium. (b) Will the results change if we supply 9 mol of C, 19 mol of O atoms, and 14 mol of N atoms but with the same input enthalpy?

Solution

(a) The three atom balance equations are:

Input C balance: $5+4=9=N_{CO}+N_{CO_2}$.
O balance: $5\times 1+3\times 2+4\times 2=19=N_{CO}\times 1+N_{CO_2}\times 2+N_{O_2}\times 2$.
N balance: $2\times 7=14=2N_{N_2}$.

The five unknowns are N_{CO}, N_{O2}, N_{CO2}, N_{N2}, and T. If we assume T, then the equilibrium solution provides the 4th relation, and one can solve for N values at a given T. Assume T = 3010 and $K^0_{CO_2}$ = 0.343 for $CO_2 \Leftrightarrow CO + 1/2\ O_2$.
Solving,
N_{CO} = 1.893, N_{CO_2} = 7.107, N_{N_2} = 7, N_{O_2} = 1.446, and N = 17.446 (all in units of kmol). For the assumed value of T of 3010 K,
$\bar{h}_{CO} = -16579.9$, $\bar{h}_{CO_2} = -240107.94$, $\bar{h}_{N_2} = 93072.75$, and $\bar{h}_{O_2} = 98483.15$ (all units in kJ kmol^{-1}). The assumed value of T must satisfy the energy balance equation

$$Q = H_{out} - H_{in} = N_{O_2}\,\bar{h}_{O_2} + N_{CO_2}\,\bar{h}_{CO_2} + N_{CO}\,\bar{h}_{CO} + N_{N_2}\,\bar{h}_{N_2} - H_{in}$$
$$= -35659.784 \text{ kJ}.$$

But Q = 0. Assuming T = 3030 K and repeating the procedure, we get $\bar{h}_{CO}$ = −15834.8, $\bar{h}_{CO_2} = -238864.3$, $\bar{h}_{N_2} = 93815$, and $\bar{h}_{O_2} = 99282.2$ kJ kmol^{-1}. $K^o_{CO_2}$ = 0.369, i.e., N_{CO} = 1.981, N_{CO_2} = 7.019, N_{N_2} = 7, and N_{O_2} = 1.491 (all in units of mol). N = 17.491 and Q = 5042.34 kJ.
Interpolating, T = 20(0 − (−35660))/(5042 − (−35660)) + 3010 ≈ 3028 K.

(b) If the atom inputs are unchanged (C = 9, O = 19, N = 14), enthalpy is unchanged, because T is the same.

Remarks

Computational codes use similar equilibrium approach in determining the value of local T and species concentrations from specified total enthalpy and atom composition.

Example 16

Consider the stoichiometric combustion of 1 kmol of CH_4 with air at 1 bar. The species enter an adiabatic reactor at 298 K (state 1), and the products leaving it are the mix of species CO_2, CO, H_2O, H_2, O_2, N_2, and OH at equilibrium. Determine the product temperature and composition.

Solution

First, let us assume a value of T. The method of solving for the composition is similar to example 12. The overall chemical reaction is the same as Equation A of example 12.

Let us assume T is 2250K. Using example 12, $H_p = -60{,}632$kJ. The energy conservation requires H_R or $H_i = H_p$ or H_e. The input enthalpy at 298 K is $-74{,}855$ kJ $\neq H_p$.

This problem is solved iteratively. The adiabatic temperature is found to be 2230 K.

The equilibrium composition is

$$N_{CO_2} = 0.91,\ N_{CO} = 0.09,\ N_{H_2O} = 1.94,\ N_{H_2} = 0.04,\ N_{O_2} = 0.052,$$

$$N_{N_2} = 7.52 \text{ kmol},\ N_{OH} = 0.041.$$

Remarks

Table A.34 tabulates the equilibrium flame temperature and entropy generated at 298 K in an adiabatic reactor for selected fuels. One can download the Microsoft-Excel-based THERMOLAB software from the CRC Web site to evaluate the flame temperatures at equilibrium for both open and closed systems for any specified C-H-N-O-S fuels.

3.6 Gibbs Minimization Method

3.6.1 General Criteria for Equilibrium

At a specified temperature and pressure, for any reacting system, $dG_{T,P} = \Sigma\mu_k dN_k \leq 0$, where the equality holds at equilibrium. The Gibbs energy for a chemical reaction decreases as the reaction progresses (cf. $dG_{T,P} < 0$) until it reaches a minimum value. Because $G = G(T, P, N_1, N_2, \ldots)$, the criteria for G to be a minima are

$$dG_{T,P} = 0 \tag{3.67a}$$

$$d^2G_{T,P} > 0 \tag{3.67b}$$

Consider methane–air combustion at a specified temperature and pressure according to the overall reaction

$$\begin{aligned} CH_4 + a\,O_2 + b\,N_2 \rightarrow\ & c\,CO + d\,CO_2 + e\,O + f\,O_2 \\ & + g\,NO + h\,H_2O + i\,H_2 + j\,N_2 + k\,C(s), \end{aligned} \tag{3.68}$$

where a and b are known for specified equivalence ratios. There are nine unknown product species moles (c, d, e, …, k) and four atom balance equations for C, H, N, and O atoms. Therefore, five equilibrium relations are required, which may include the reactions

$$CH_4 \rightarrow C(s) + 2\ H_2, \tag{I}$$

$$CO_2 \rightarrow CO + 1/2\ O_2, \tag{II}$$

$$H_2O \rightarrow 1/2\ O_2 + H_2, \tag{III}$$

$$O_2 \rightarrow 2\ O, \tag{IV}$$

$$N_2 + O_2 \rightarrow 2\ NO. \tag{V}$$

We may also select a linear combination of reactions. For instance, subtracting Reaction B from Reaction C, we obtain the reaction

$$H_2O - CO_2 \rightarrow 1/2\ O_2 + H_2 - CO - 1/2\ O_2,$$

i.e.,

$$H_2O + CO \rightarrow CO_2 + H_2,$$

which is the familiar water gas shift reaction. Note that Reaction V′ does not provide an independent equilibrium relation that can be selected in addition to Reaction I to Reaction V. The procedure becomes far more complex in a realistic applications in which there are hundreds of species. Therefore, it is useful to adopt a more general procedure, during which the species moles are adjusted until the Gibbs energy reaches a minimum value at equilibrium, i.e., $dG_{T,P} = 0$ and $d^2G_{T,P} > 0$ subject to the atom balance constraints and the specified temperature and pressure. The Lagrange multiplier method is useful in this regard.

3.6.2 G Minimization Using Lagrange Multiplier Method

We will now generalize the methodology for multicomponent systems. The procedure to minimize $G = G(T, P, N_1, N_2, \ldots, N_K)$ subject to the atom balance equations is as follows.

Formulate the atom balance equations for each element j

$$\sum_k d_{jk} N_k = A_j,\ j = 1, \ldots, J,\ k = 1, \ldots, K, \tag{3.69}$$

where d_{jk} denotes the number of atoms of an element j in species k (e.g., for the element j = O in species k = CO_2, d = 2) and A_j is the number of atoms of type j entering the reactor. This relation can be expressed using the Lagrange multiplier method. Rewrite Equation 3.69 as

$$\lambda_j\left(\sum_k d_{jk} N_k - A_j\right),\ j = 1, \ldots, J,\ k = 1, \ldots, K \tag{3.70}$$

where λ_j is the Lagrange multiplier for element j.

The Gibbs energy at equilibrium must be minimized subject to this condition. We create a function

$$F = G + \sum_j \lambda_j \left(\sum_k d_{jk} N_k - A_j \right),\ j = 1, \ldots, J,\ k = 1, \ldots, K. \quad (3.71)$$

We minimize F with respect to N_k, k = 1, …, K. The minimization of F requires

$$\partial F/\partial N_k = (\partial G/\partial N_k)_{T,P} + \left(\sum_j \lambda_j d_{jk} \right)_{T,P} = 0,\ j = 1, \ldots, J,\ k = 1, \ldots, K \quad (3.72)$$

Simplifying the above equation,

$$\mu_k + \sum_j \lambda_j d_{jk} = 0,\ j = 1, \ldots, J,\ k = 1, \ldots, K. \quad (3.73)$$

We can interpret λ_j as the element potential of element j in species k, $\lambda_j\ d_{jk}$ as contribution by element j to the k-th species potential, and the term $\mu_{E,k} = \Sigma_j \lambda_j d_{jk}$ is the combined element potential of species k due to all elements within K. Unknown are K species and J multipliers. Equation 3.73 provides K equations while Equation 3.69 yields J equations and hence the problem can be solved.

Example 17

A steady-flow reactor is fired with 1 kmol of $C_{10}H_{20}$, with 5% excess air. The species (1 to 8) leaving are CO, CO_2, H_2, H_2O, OH, O_2, NO, and N_2 at T = 2500 K and 1 bar. Determine equilibrium composition of species leaving the reactor. Assume ideal gas behavior.

Solution

The stoichiometric reaction is rewritten as:

$$(C_{10}H_{20}) + 15(O_2 + 3.76N_2) = 10\ CO_2 + 10\ H_2O + 59.4\ N_2.$$

With 5% excess air, the O_2 supplied is 15.8 kmol and the N_2 supplied is 59.4 kmol.

$$(C_{10}H_{20}) + 15.\ 8(O_2 + 3.76\ N_2) \rightarrow \text{Products}.$$

This system is an open system. Now, we follow a fixed mass (140 + 528 + 1737 = 2405 kg) as it travels through the reactor. We will assume that this mixture is instantaneously heated to 2500 K at 1 bar and then calculate G of the mixture assuming various values for the moles of the 8 species subject to the four atom conservation relations for C, H, N, and O. We will then alter the composition subject to the four atom conservation relations (C, H, N, O, Equation 69), calculate G, and then determine the composition at which G has a minimum value. The Lagrange multiplier method is used to minimize G. The four elements C, H, N, and O are denoted by the subscript j (j = 1, …, 4) and the eight species denoted by subscript k (k = 1, …, 8). The coefficients d_{jk}, i.e., $d_{11} = 1$, $d_{12} = 1$, …, are provided in the following table:

Table of Coefficients d_{jk}

	Element j			
Species k	C	H	N	O
CO	1	—	—	1
CO_2	1	—	—	2
H_2	—	2	—	—
H_2O	—	2	—	1
NO	—	—	1	1
N_2	—	—	2	—
OH	—	1	—	1
O_2	—	—	—	2

The atom conservation equations (Equation 69) yield the relations:

$$j=1\ (\text{C atoms}):$$
$$1N_{CO}+1N_{CO_2}+0N_{H_2}+0N_{H_2O}+0N_{NO}+0N_{N_2}+0N_{OH}+0N_{O_2}-10=0, \tag{A}$$

$$j=2\ (\text{H atoms}):$$
$$0\,N_{CO}+0\,N_{CO_2}+2\,N_{H_2}+2\,N_{H_2O}+0N_{NO}+0N_{N_2}+1N_{OH}+0N_{O_2}-20=0, \tag{B}$$

$$j=3\ (\text{O atoms}):$$
$$1N_{CO}+2N_{CO_2}+0\,N_{H_2O}+1N_{NO}+0N_{N_2}+1N_{OH}+2N_{O_2}-31.6=0, \tag{C}$$

$$j=4\ (\text{N atoms}):$$
$$0\,N_{CO}+0\,N_{CO_2}+0N_{H_2}+0N_{H_2O}+1N_{NO}+2N_{N_2}+0N_{OH}+0N_{O_2}-118.6=0. \tag{D}$$

Dividing these equations by the total moles of products ($N = \Sigma N_k$), we obtain the relations

$$1X_{CO}+1X_{CO_2}+0X_{H_2}+0X_{H_2O}+0X_{XO}+0X_{N_2}+0X_{OH}+0X_{O_2}-10/N=0, \tag{E}$$

$$0X_{CO}+0\,X_{CO_2}+2\,X_{H_2}+2\,X_{H_2O}+0X_{NO}+0X_{N_2}+1X_{OH}+0X_{O_2}-20/N=0, \tag{F}$$

$$1X_{CO}+2\,X_{CO_2}+0\,X_{H_2}+1\,X_{H_2O}+1X_{NO}+0X_{N_2}+1X_{OH}+2X_{O_2}-31.6/N=0, \tag{G}$$

$$0X_{CO}+0\,X_{CO_2}+0X_{H_2}+0\,X_{H_2O}+1X_{NO}+2X_{N_2}+0X_{OH}+0X_{O_2}-118.6/N=0. \tag{H}$$

N is solved from the identity

$$\sum X_k = 1. \tag{I}$$

$$G = G(T, P, N_1, N_2, \ldots, N_K). \tag{J}$$

Multiplying Equation A to Equation D, respectively, by λ_C, λ_H, λ_N, and λ_O (4 multipliers for 4 atoms) and adding with Equation J, we form a function

$$F = G + \lambda_C\,(N_{CO} + N_{CO_2} - 10) + \lambda_H(2N_{H_2} + 2N_{H_2O} + N_{OH} - 20)$$
$$+ \lambda_O(N_{CO} + 2N_{CO_2} + N_{H_2O} + N_{OH} + N_{NO} + 2\,N_{O_2} - 31.6) \tag{K}$$
$$+ \lambda N(N_{NO} + 2\,N_{N_2} - 118.6)$$

that is minimized at equilibrium, i.e.,

$$\partial F/\partial N_{CO} = (\partial G/\partial N_{CO})_{T,P} + \lambda_C + \lambda_O = 0, \tag{L}$$

$$\partial F/\partial N_{CO2} = (\partial G/\partial N_{CO2})_{T,P} + \lambda_C + 2\lambda_O = 0, \tag{M}$$

$$\partial F/\partial N_{H_2} = (\partial G/\partial N_{H_2})_{T,P} + 2\lambda_H = 0. \tag{N}$$

Similarly, $\partial F/\partial N_{H2O} = 0$, $\partial F/\partial N_{NO} = 0$, $\partial F/\partial N_{N2} = 0$, $\partial F/\partial N_{OH} = 0$, and $\partial F/\partial N_{O2} = 0$, labeled as Equations O, P, Q, R, and S.

Rewrite Equation L to Equation S in the form

$$\mu_{CO} + \lambda_C + \lambda_O = 0, \tag{T}$$

$$\mu_{CO_2} + \lambda_C + 2\,\lambda_H = 0, \tag{U}$$

$$\mu_{H_2} + 2\lambda_H = 0, \tag{V}$$

where $\mu_k = (\partial G/\partial N_k)_{T,P} = \hat{g}_k$. Similarly, we get five more equations labeled as W, X, Y, Z, and a.

Recall that for an ideal mixture of gases or ideal mix of liquids or solids

$$\mu_k = \hat{g}_k = \bar{g}_k\,(T,\ P) + \bar{R}T\,\ln\,(X_k). \tag{b}$$

Dividing by $\bar{R}$T,

$$1_k^{\ell}/(\bar{R}T) = \{\bar{g}_k\,(T,\ P)/\,\bar{R}\,T\} + \ln\,(X_k). \tag{c}$$

Divide Equation L to Equation S by $\bar{R}$ T and using Equation C,

$$(\bar{g}_{CO}\,(T,\ P)/\bar{R}T) + \ln\,(X_{CO}).+ \lambda_C' + \lambda_O' = 0, \tag{d}$$

$$(\bar{g}_{CO_2}\,(T,\ P)/\,\bar{R}\,T) + \ln\,(X_{CO_2}) + \lambda_C' + 2\,\lambda_H' = 0, \tag{e}$$

$$(\bar{g}_{H_2}\,(T,\ P)/\,\bar{R}\,T) + \ln\,(X_{H_2}) + 2\lambda_H' = 0, \tag{f}$$

where $\lambda_j' = \lambda_j/\,\bar{R}\,T$. Similarly, we get five more equations labeled as g, h, i, j, and k. The values of ln (X_k) can be determined from Equation d to Equation k for the assumed set of values of the modified Lagrange multipliers of λ_C', λ_H', λ_N', and λ_O'. Once solved for X_k's, we check whether these four multipliers satisfy the four element conservation equations, Equation E to Equation H. Thus, knowing the values of μ_k, those of X_k can be determined from equations such as Equation c. Irrespective of the number of unknown species, the number of assumptions made for the Lagrange multipliers are equal to number of atom balance equations. Good starting values can be obtained by assuming complete combustion and determining those mole fractions of four significant species to obtain initial guesses for four multipliers λ_C', λ_H', λ_N', and λ_O'.

$$\text{For } T = 2500 \text{ K and } P = 1 \text{ bar}, \ \bar{g}_k = \bar{h}_k - T\,\bar{s}_k,$$

$$\bar{g}_{CO} = 701455 \text{ kJ kmol}^{-1},$$

$$\bar{g}_{H_2} = 419840 \text{ kJ kmol}^{-1},$$

$$\bar{g}_{CO_2} = 1078464 \text{ kJ kmol}^{-1}, \text{ and so on.} \quad \text{(S)}$$

Good starting values could be λ_C', λ_H', λ_N', and λ_O' = 19.7, 12.5, 14.0, and 17.1, respectively. The converged solutions are:

$$N_{CO} = 2.04 \text{ kmol}, N_{CO_2} = 7.96, N_{H_2} = 0.42, N_{H_2O} = 9.24,$$

$$N_{NO} = 0.57, N_{N_2} = 59, N_{OH} = 0.69, N_{O2} = 1.52, \text{ all in units of kmol.}$$

Hence $N = \Sigma N_k = 81.4$ kmol.

The NASA equilibrium code uses a steepest descent Newton–Raphson method with typically 8 to 12 iterations.

3.7 Summary

The chemical force potentials of reactant $(\Sigma \nu_k \hat{g}_{k,\text{reactants}})$ and product species $(\Sigma \nu_k \hat{g}_{k,\text{products}})$ can be used to determine the direction of a chemical reaction just as the temperature (or thermal potential) can be used to determine the direction of heat transfer. The first law and second law can be used to determine the equilibrium temperature and chemical equilibrium composition. THERMOLAB spreadsheet software, available as a free download, may be used to solve for equilibrium flame temperature and composition of any given fuel.

3.8 Appendix

3.8.1 Equilibrium Constant in Terms of Elements

Combustion reactions customarily involve compounds consisting of elemental C, H, N, and O. Consider the reaction

$$H_2O \rightarrow H + OH. \quad \text{(A)}$$

For which we are asked to determine the value of K°. If the K° value is available for the reactions involving elements in their natural forms,

$$H_2 + 1/2\ O_2 \Leftrightarrow H_2O, \tag{B}$$

$$1/2\ H_2 \Leftrightarrow H, \tag{C}$$

$$1/2\ H_2 + 1/2\ O_2 \Leftrightarrow OH, \tag{D}$$

then for Reaction B to Reaction D

$$-\bar{R}\,T\ \ln K^{\circ}_{H_2O} = \bar{g}_{H_2O} - \bar{g}_{H_2} - 1/2\ \bar{g}_{O_2}, \tag{E}$$

$$-\bar{R}\,T\ \ln K^{\circ}_{H} = \bar{g}^{\circ}_{H} - 1/2\ \bar{g}^{\circ}_{H_2}, \tag{F}$$

$$-\bar{R}\,T\ \ln K^{\circ}_{OH} = \bar{g}^{\circ}_{OH} - 1/2\ \bar{g}^{\circ}_{H_2} - 1/2\ \bar{g}^{\circ}_{O_2}, \tag{G}$$

where K^0_{H2O}, K^0_{H}, and K^0_{OH} are the equilibrium constants when products are formed from elements in their natural states. Likewise, for Reaction A,

$$-\bar{R}\,T\ \ln K^{\circ} = = \bar{g}^{\circ}_{H} + \bar{g}^{\circ}_{OH} - \bar{g}^{\circ}_{H_2O}, \tag{H}$$

Substituting $\bar{g}^{\circ}_{H}$ from Equation F, $\bar{g}^{\circ}_{OH}$ from Equation G, and $\bar{g}^{\circ}_{H_2O}$ from Equation 3.E into Equation 3.H, we obtain

$$K^{o} = K^{\circ}_{OH} K^{\circ}_{H} / K^{\circ}_{H_2O}. \tag{I}$$

Similarly, for the equilibrium constant for the reaction $CO_2 \rightarrow CO + 1/2\ O_2$, one can write, by observation, $K^{\circ} = K^{\circ}_{CO}/K^{\circ}_{CO_2}$. Note that K^o (T) will not appear for elements in natural form. Typically, $\ln K_k^0 = A_k - B_k/T$. Table A.35 presents the values of A_k and B_k of selected species formed from elements in natural states. For example, $K^0 = K^{\circ}_{CO}/K^{\circ}_{CO_2} = \exp\{A_{CO} - B_{CO}/T\}/\exp\{A_{CO2} - B_{CO2}/T\} = \exp[\{A_{CO} - A_{CO2}\} - \{B_{CO} - B_{CO2}\}/T] = \exp\{\Delta A - \Delta B/T\}$, where $\Delta A = A_{CO} - A_{CO2} = 10.108$, $\Delta B = \{B_{CO} - B_{CO2}\} = 33770$. Thus, $K^0 = \exp\{10.108 - 33770/T\}$.

Example 18

Determine the equilibrium constant for the reaction CO2 $\Leftrightarrow$ CO + 1/2 O_2 at 3000 K using atomic equilibrium constants.

Solution

$K = K^{\circ}_{CO}/K^{\circ}_{CO2}$, $\exp\{A_{co} - B_{co}/T\}/\exp\{A_{CO_2} - B_{CO_2}/T\} = \exp\{\Delta A - \Delta B/T\}$, $\Delta A = A_{CO} - A_{CO_2} = 10.108$, $\Delta B = 33770$, $K^{\circ} = \exp(10.108 - 33770/T)$ from the data in the Table A.35, $K^{\circ} = \mathrm{Exp}(10.108 - 33770/3000) = 0.317$.

Chapter 4

Fuels

4.1 Introduction

This chapter presents an overview of fuels used and their properties. Solar energy is stored as chemical energy in fossil and biomass fuels (wood residues, sawdust, paper, cow manure and poultry litter, crop straw, switch grass, etc.) containing C, H, N, O, and S elements. Many devices that harness the energy released during combustion of these fuels and convert it into a useful form have been developed. Examples of devices that utilize combustion to harness fossil fuel energy are power plants, automobiles, aircrafts and ships, and boilers for the production of steam.

Figure 4.1 shows the energy and percentage of use for various fuels in the U.S. The U.S. consumed 29 quads in 1949 and 83 quads in 2002 (1 quad = 10^{15} Btu), indicating an 86% increase. As of June 2006 the shares by energy source are: coal, 49.3%; natural gas, 17.8%; nuclear, 30%; petroleum, 1.5%; hydroelectric, 8.3%; others, 3.1%. The current worldwide energy consumption is 400 quads. By 2050, world consumption will reach 1000 quads, with that of the U.S. about 260 quads (a 300% increase). The energy consumption was 215 mmBtu in 1949 but 337 mmBtu per person in 2005.

Fuels can be classified according to the phase or state in which they exist, i.e., as gaseous (e.g., propane), liquid (e.g., gasoline), or solid (e.g., coal). The selection of a particular fuel for energy conversion depends on the total cost including the sourcing cost (fuel cost, Table 4.1), and the operational and environmental compliance costs. The fuel cost depends on its physical (e.g., moisture and ash contents) and chemical properties (heating values). Thus, it is essential to describe the various classes of fuels and determine their physical and chemical characteristics in order to determine the required air-to-fuel ratios for combustion, flame temperatures, combustion characteristics, and economics of the use of fuel. Fuels can either occur naturally (e.g., fossil fuels) or be synthesized (e.g., synthetic fuels). The important fuel properties are composition of fuel, heating values, density, etc.

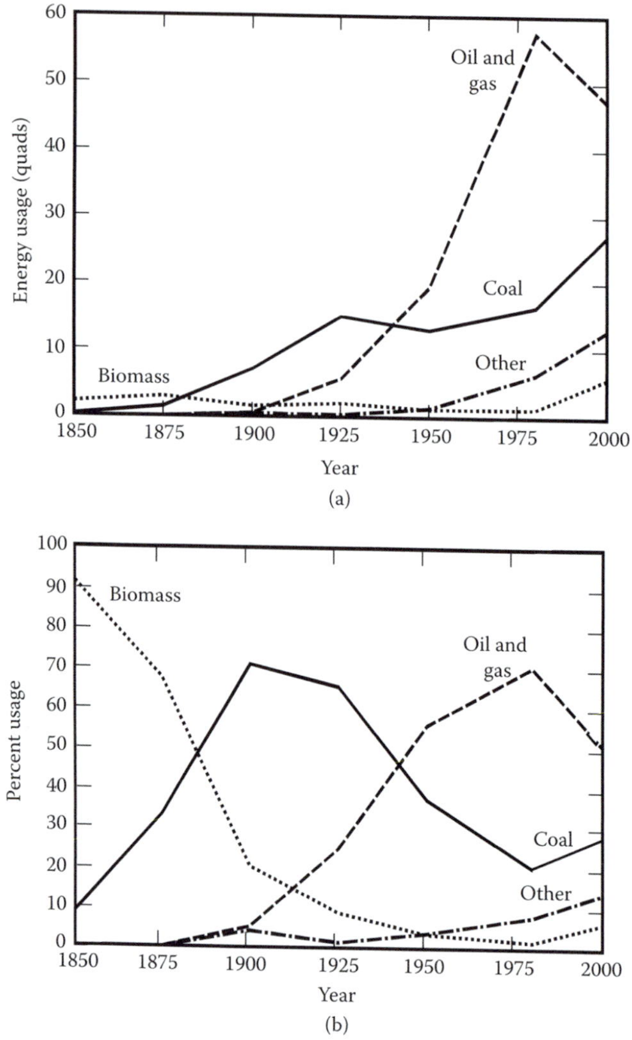

Figure 4.1 Energy consumption.

4.2 Gaseous Fuels

Gaseous fuels are the most clean burning fuels compared with liquid and solid fuels. These fuels are a mixture of hydrocarbons (HC) with highly volatile CH_4 as the dominant component and with very little S and N. The volumetric fraction of each chemical constituent in a gaseous mixture characterizes the mixture composition. There are various kinds of gaseous fuels: Natural gas is found in compressed form below rock strata and is extracted from gas wells, unassociated gas is found in petroleum reservoirs but contains no oil, and associated gas is found in crude oil reservoirs. Natural gas is liquefied (LNG stands for liquefied natural gas) by cooling below –164°C. LNG has the advantage of a larger heating capacity in a smaller volume,

Table 4.1 Cost of Various Fuels

Fuel	*$*	*$ per 10^6 Btu*	*$ per GJ*
Natural gas (residence)	6.3/1000 SCF	6.3	5.97
Natural gas (industrial)	3.34/1000 SCF	3.3	3.13
Gasoline	1.23/gal	10	9.48
Kerosene	0.72	5.2	4.93
No. 2 fuel oil	0.67	5.22	4.95
No. 2 Diesel	0.68	4.9	4.65
Coal	29/ton	1.3	1.23
Wood (15% Moist)	40/cord	2.1	1.99
Electricity (residence)	0.084/kwh	24.6	23.32
Electricity (industrial)	0.047/kwh	13.8	13.08

and hence transportation cost is reduced. Liquefied petroleum gas (LPG) consists of higher HCs — ethane, propane, butane, ethylene, butylenes, and propylene. It is stored as liquid under high pressure but is a gas at atmospheric pressure. Typically, LPG is a by-product of petroleum refining and it consists of 90% propane. CNG is compressed natural gas stored at pressures of 150 to 250 atm.

4.2.1 Low- and High-BTU Gas

A low-Btu gas contains 3,400 kJ/SCM (100 Btu/SCF), a medium-Btu gas 17,000 kJ/SCM (500 Btu/SCF), whereas a high-Btu gas (e.g., methane) can provide 34,000 kJ/SCM (1,000 Btu/SCF). Hydrogen is another gaseous fuel with a heat value of 11,525 kJ/SCM. Although extremely flammable, H_2 has the advantage of not producing any CO_2, a greenhouse gas, during combustion.

4.2.2 Wobbe Number

The ultimate objective is to obtain a fixed thermal output rate when fuels are changed. *Wobbe number* provides a relation between the volume-based heating value and the density of fuel, and is a measure of the interchangeability of gaseous fuels. Similar Wobbe number assures similar thermal output rate. The volume flow rate of a gas through an orifice is given,

$$\dot{Q} = V \times A \tag{4.1}$$

$$\dot{Q} = A(2\Delta P/\rho)^{1/2} = A(2P_{gage}/\rho)^{1/2} \tag{4.2}$$

where P_{gage} denotes the gage pressure drop across the orifice; ρ, the gas density; and A, the cross-sectional area. The thermal output rate is

$$\dot{H} = \dot{Q}\,(HHV') = A(HHV')(2P_{gage}/\rho_a)^{1/2}(\rho_a/\rho)^{1/2} \quad (4.3)$$

where HHV′ denotes the higher heating value per unit volume and ρ_a, the air density. Hence,

$$\dot{H} = A(2P_{gage}/\rho_a)^{1/2}\,Wo \quad (4.4)$$

$$Wo\ (\text{Wobbe number}) = \{HHV'/(\rho/\rho_a)^{1/2}\} = kJ/m^3 \quad (4.5)$$

where the term ρ/ρ_a is called the specific gravity (sg) of the gas.

If the gas line pressure is constant (typically 6 inches or 15 cm of water for residential burners), the heat throughput $\dot{H}$ is also the same for identical values of Wo when the gaseous fuel is changed. The lower the value of Wo, the lower is the heat release rate. If the Wobbe number changes, then the burner velocity must be altered (and the burner dimensions must sometimes be changed) to obtain a fixed value of the heat release, which may have consequences for flame stability. Note that similar Wo does not imply similar volume flow rates; thus, if HHV′ is higher, the volume flow rate through same burner will be lower for fixed Wo. Thus, if a gas company measures the fuel flow in SCM, the price per unit SCM will be adjusted depending on HHV′.

Example 1

Natural gas (NG) with a 37925 kJ/m³ (1025 Btu/ft³) higher heating value and a specific gravity of 0.592 is to be replaced with a propane–air (PA) mixture (HHV′ = 96755 kJ/m³ (2615 Btu/ft³) and specific gravity = 1.552). (a) What is the volume fraction x of propane for identical thermal output and upstream pressure? (b) What will be the ratio of volume flow rate ($\dot{V}$) of the replacement gas (PA) to that of the original gas (NG)?

Solution

$$HHV'_{PA}/(\rho_{PA}/\rho_A)^{1/2} = HHV_{NG}/(\rho_{NG}/\rho_A)^{1/2} \quad (A)$$

$$\rho_{PA} = x\,\rho_P + (1-x)\,\rho_A, \quad \text{or} \quad (\rho_{PA}/\rho_A) = x\,\rho_P/\rho_A + (1-x) \quad (B)$$

$$HHV_{PA} = x\,HHV_P + (1-x)\,HHV_A = HHV_{PA} = x\,HHV_P + 0 \quad (C)$$

From Equation A to Equation C and for similar Wo,

$$x\,HHV_P/\{(x\,\rho_P/\rho_A + (1-x))^{1/2}\} = HHV_{NG}/(\rho_{NG}/\rho_A)^{1/2} \quad (D)$$

Solving with $\rho_{NG}/\rho_A = 0.592$ and $\rho_P/\rho_A = 1.552$, HHV′ = 96755 kJ/m^3, HHV′$_{NG}$ = 37925 kJ/m^3, i.e., $(96755x)^2/(1.552\ x + (1 - x)) = 37925^2/0.592$, or x = 0.586. (b) $\dot{V}_{NG} = A(2P_{gage}/\rho_{NG})^{1/2}$ and $\dot{V}_{PA} = A(2P_{gage}/\rho_{PA})^{1/2}$; using Equation B, $(\rho_{PA}/\rho_A) = x\ \rho_P/\rho_A + (1 - x) = 0.586 \times 1.552 + 0.414 = 1.323$, we get $\dot{V}_{PA}/\dot{V}_{NG} = (\rho_{NG}/\rho_{PA})^{1/2} = (0.592/1.323)^{1/2} = 0.669$. Hence, the propane:air volume flow rate through the line is only 66.9% of natural gas flow for same heat throughput rate and same upstream pressure. Even though the volume or velocity of gas through the kitchen range may be less, the heat throughput per unit time with PA will be same as NG.

4.3 Liquid Fuels

The world's petroleum reserves are about 4×10^{15} MJ, and the U.S. oil consumption is approximately 4×10^{13} MJ/yr. Crude oil occurs naturally as a free-flowing liquid with a density $\rho \approx 780$ to 1000 kg/m^3. Commonly used liquid fuels are generally derived from petroleum and are not pure substances. Typically, the ash quantity in petroleum is 0.1%, whereas in coal the ash can be as high as 15 to 25%. Crude oil contains 0.15 to 0.5% nitrogen by weight, whereas in a heavy distillate this fraction is in the range of 0.6 to 2.15%.

4.3.1 Oil Fuel Composition

The crude oil normally contains a mixture of hydrocarbons and the "boiling" temperature keeps increasing as the oil is distilled (Figure 4.2 and Figure 4.3) and lighter components are evaporated. Thus, oil in a refinery is broken into chemical fractions by distillation in a fractionating column. Crude oil yields 3.7% LPG, 2.9% refinery gas, 1.3% naptha, 39% motor gasoline ($C_4 - C_{10}$), 0.2% aviation gas, 5.6% jet fuels ($C_{10} - C_{13}$) (Jet A in aircraft gas turbine, $C_{12}H_{23}$), 0.8% kerosene (industrial gas turbine), 18.2% heating fuel (C_{13}–C_{18}) and diesel, 16.6% residual fuel oil (C_{13}–C_{18}), 5.9% petrochemical feed, 0.9% lube oil and greases (C_{18}–C_{45}), 2.3% road oil, asphalt (>C_{45}), and 2.5% coke and wax, with 0.2% losses [Borman

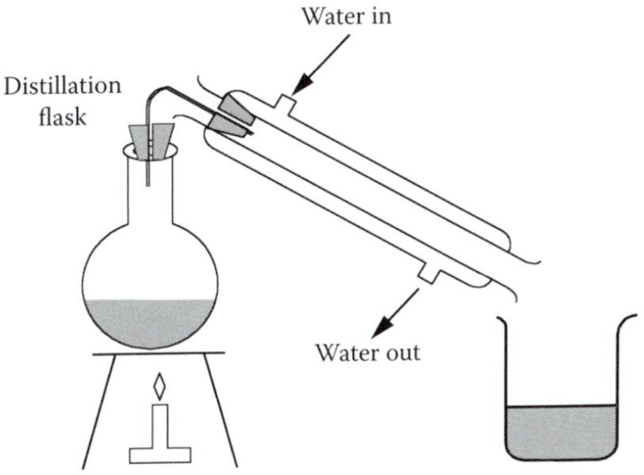

Figure 4.2 Distillation experiment for determining variation of boiling point versus percentage of mass evaporated.

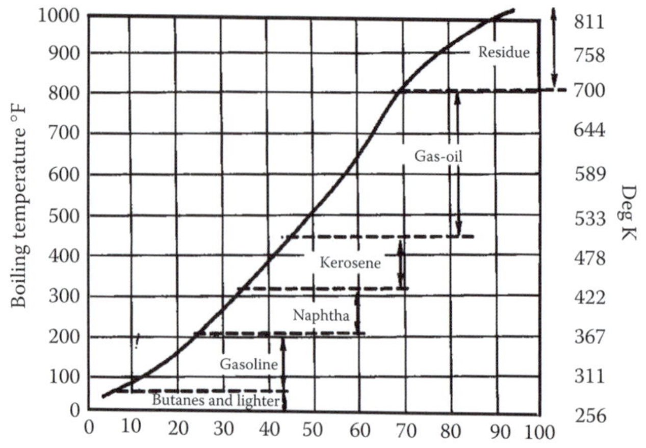

Figure 4.3 Crude oil distillation curves. (Adapted from Baukal C.E., Gershtein, V.Y., and Li, X., Eds., *Computational Fluid Dynamics in Industrial Combustion*, CRC Press, Boca Raton, FL., 2000.)

and Ragland, 1998]. The type and amount of yield varies depending on local requirements.

Most fuel oils contain 83 to 88% carbon and 6 to 12% (by mass) hydrogen [Reed, 1997]. The specific gravity is typically defined at 15.6°C (60°F) with water as the reference fluid. Distillation tests as shown in Figure 4.2 and Figure 4.3 are used to define the terms 10% and 90% (volume loss) boiling temperature. If the 10% temperature is very low, then the fuel is easily vaporizable. Similarly, a high 90% temperature will indicate difficulty in vaporization and formation of droplets. Liquid fuels contain several groups of well-defined organic chemicals, and the major constituents are described in the following subsections.

4.3.2 Paraffins

Paraffins (alkanes) are saturated HC (i.e., impossible to add any more H atom into HC) and chemically represented as C_nH_{2n+2}. The structure of straight- or normal-chain alkanes are shown in Figure 4.4a. They are denoted

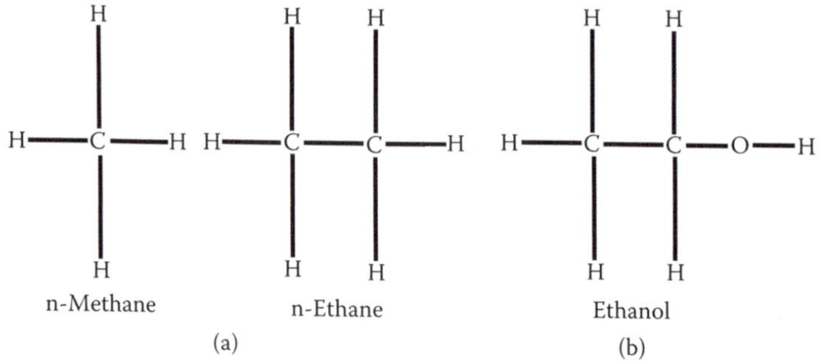

Figure 4.4 (a) Straight chain (normal n) paraffins C_nH_{2n+2}, (b) Paraffinic alcohols $C_nH_{2n+1}OH$.

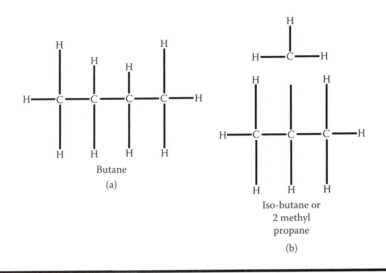

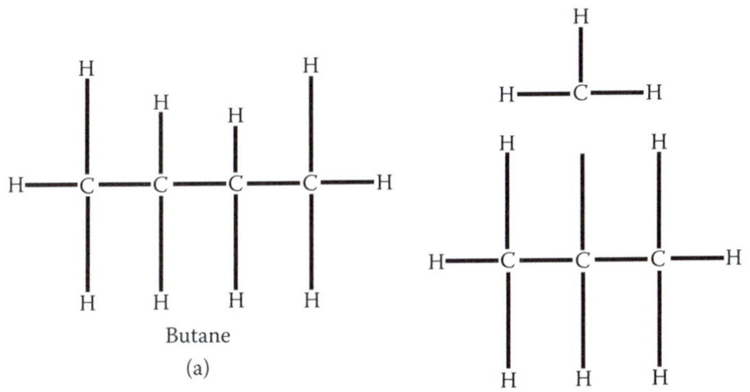

Figure 4.5 (a) Straight- and (b) branched-chain paraffins, C_nH_{2n+2}.

by a prefix "*n*" (e.g., *n*-Octane). Examples of alkanes include methane (CH_4), ethane (C_2H_6), propane C_3H_8), and butane (C_4H_{10}). The naming sequence is based on number of C atoms — 1: meth, 2: etha, 3: prop, 4: but, 5: pent, 6: hex, 7: hept, 8: oct, 9: non, and 10: dec. These names all end in "ane." The number of bonds in each alkane is as follows: methane: 4 C–H and *n*-ethane: 6 C–H and 1 C–C. Typically, as we increase the number of hydrogen and carbon atoms, we produce heavier gases, increase the bond energy, and increase molar heating value because of increase in C and H atoms. Hence, volumetric heat content (1 kmol at STP = 24.6 m^3) will gradually increase with increase in the number of C and H atoms. For example, CH_4 has heat content of 37000 kJ/SCM (standard cubic meter), propane 92500 kJ/SCM, and butane 118000 kJ/SCM. Further, the ignition temperature (at which the fuel ignites) decreases with increase in the number of C atoms in the paraffin. A few paraffinic components in natural gas are CH_4 and C_2H_6; in liquid petroleum gas, C_3H_8 and C_4H_{10}; and in gasoline, C_8H_{18}.

Branched-chain alkanes (isomers) also occur, e.g., isobutane C_4H_{10} (also known as 2-methyl propane), in which a methyl radical CH_3 is attached to the second carbon atom in the C–C–C chain of propane (Figure 4.5b) and isooctane C_8H_{18} or 2,2,4-trimethylpentane, in which two methyl (or alkyl) radicals are attached with one to the second and the other to the fourth carbon atoms in a pentane molecule. Paraffins have maximum H/C ratio, highest HV in kJ/kmol, lowest HV (kJ/m^3), and clean combustion. The greater the C in paraffins, the greater the number of isomers.

4.3.3 *Olefins*

Olefins or alkenes are HCs having open unsaturated (i.e., H atom <2n+2 in C_nH_{2n}) chain structures and are represented by the chemical formula C_nH_{2n} as shown in Figure 4.6. They contain both single and double bonds. The names of alkenes generally end in "ene." They can also be either straight (Figure 4.6)

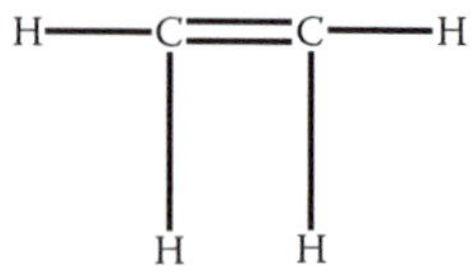

Ethylene, C_2H_4

Figure 4.6 Olefins, C_nH_{2n} (straight chain).

or branched chain (Figure 4.7). They are clean burning and have a higher octane rating than alkanes. Mono-olefins have an open-chain structure. Olefins are unstable (i.e., more prone to oxidation during storage) and can form residues.

4.3.4 Diolefins

Diolefins $C_nH_{(2n-2)}$ have two double bonds, and their names end in "diene," e.g., heptadiene contains 7 carbon and 12 hydrogen atoms. Use of the expanded term 1,5-heptadiene implies that first and fifth carbon atoms have double bonds.

4.3.5 Naphthenes or Cycloparaffin

The naphthenes or cycloparaffin family is denoted by the formula C_nH_{2n}. These fuels are similar to paraffins except for their ring saturated (instead of straight chain) structure (Figure 4.8a and Figure 4.8b).

4.3.6 Aromatics

Double-bonded HCs can also be arranged in cyclic forms. Aromatics are represented by $C_6H_{(2n-6)}$ and include benzene C_6H_6 (Figure 4.9a), toluene C_7H_8, ethyl benzene C_8H_{10} (which is identical to benzene but with a H atom replaced by an ethyl radical C_2H_5). The double bonds can alternate in aromatic rings.

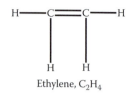

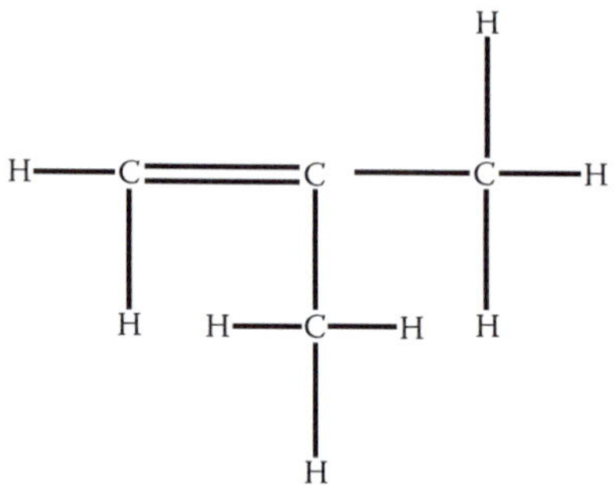

Figure 4.7 2-Methyl propylene, C_4H_8 (branched chain).

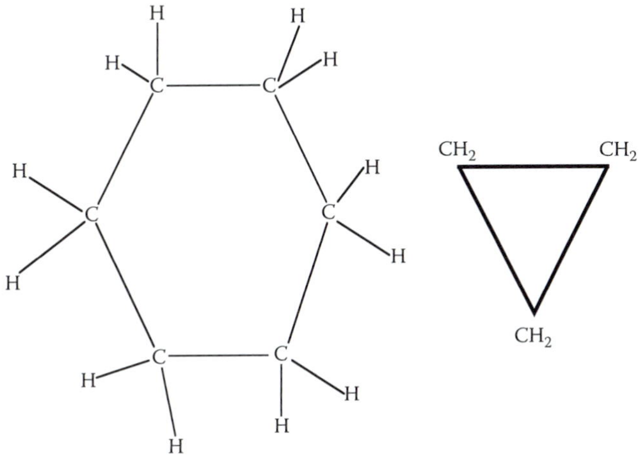

Figure 4.8 (a) Cyclohexane C_6H_{12}, (b) Cyclopropane C_3H_6.

Aromatics are chemically more stable, carcinogenic, and have smoky combustion, highest HV (volume based), lowest HV on mass basis. Rings can be connected in polycyclic form. For example, when two rings are united it forms naphthalene $C_{10}H_8$ (Figure 4.9b), when 3 rings are united it forms anthracene $C_{14}H_{10}$, and benzopyrene has 5 rings united.

4.3.7 Alcohols

Alcohols are partially oxidized HCs having saturated structure. In the case of alcohols, an H atom of a paraffin is replaced by an OH radical (C_nH_{2n+1} OH), e.g., ethyl alcohol C_2H_5OH (see Figure 4.4b) and pentyl alcohol $C_5H_{11}OH$.

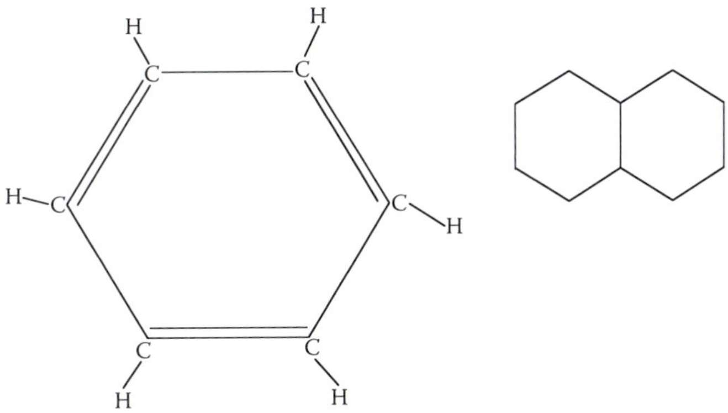

Figure 4.9 (a) Benzene C_6H_6 (b) Naphthalene $C_{10}H_8$.

The number of bonds in ethanol is: 5 C–H, 1 C–C, 1 C–O, and 1 O–H. If OH group is attached to benzene, it forms phenol.

Ethanol for beverages and almost 50% of industrial ethanol are made by fermentation, which is a sequence of reactions in the absence of oxygen. Corn is converted into sugar or is cooked with enzymes, and yeast (fungi) is added for fermentation. Energy is released when sugar is fermented into ethanol and carbon dioxide:

$$C_6H_{12}O_6 \text{ (sugar)} \rightarrow 2CH_3CH_2OH + 2CO_2$$

4.3.8 Common Liquid Fuels

Gasoline (C_8H_{16}): Gasoline is a mixture that results from the fractionation of crude oil. It consists mostly of C_5–C_{10} hydrocarbon groups; it can be represented as $C_nH_{1.97n}$ and has a molecular weight of 110, HHV = 47300 kJ/kg, LHV = 44000 kJ/kg, T_{BP} = 30 to 200°C, LHV per unit stoichiometric mixture = 2830 kJ/kg of mix, A:F = 14.6, density of 0.72–0.78 g/cm³, $h_{fg} \approx 305$ kJ/kg, c = 2.4 kJ/kg K, and $c_{p,vapor}$ = 1.7 kJ/kg K. In order to stabilize gasoline during storage tetracyano ethylene $(NC_2)_2 = C(CN)_2$ is added.

An octane number of N implies that for every 100 m³ of a surrogate mixture of octane and *n*-heptane, N m³ is occupied by isooctane C_8H_{18} and (100–N) m³ by *n*-heptane C_7H_{16}. If gasoline is rated with an octane number of 93, then its knock characteristics are similar to an isooctane and *n*-heptane surrogate mixture that contains 93% isooctane by volume. They are determined using ASTM-D2699 for RON (research octane number) and D2700 for MON (motar octane number). Aromatics lower the octane quality.

Diesel fuel ($C_{11.6}H_{23.2}$): This is a mixture boiling in the range of 240 to 370°C; A:F ≈ 14.7 a typical surrogate fuel is *n*-heptane. Sometimes diesel fuel is treated as dodecane $C_{12}H_{26}$.

Kerosenes: These are crude oil derivatives (12 to 16 carbon atoms per molecule) boiling in the range of 140 to 250°C. Surrogate fuels are tridecane $C_{13}H_{28}$, $C_{12}H_{26}$, etc. JP5 is a kerosene distillate fuel. Aircraft fuels are essentially kerosene fuels in the C_{10} to C_{18} hydrocarbon groups. Some military and commercial aircraft use JP–XX (with XX denoting an integer) fuels that have relatively lower boiling points. A higher C/H ratio in a fuel results in enhanced soot formation.

Fuel oil (12 to 16 carbon atoms per molecule): Typically, they are used in burners for heating and power generation. They boil in the range of 340 to 420°C.

Naphthas: Typically, they used as feedstocks in manufacturing solvents, and boil in the range of 95 to 315°C.

4.3.9 API Gravity, Chemical Formulae, Soot, and Flash and Fire Points

4.3.9.1 API Gravity

For liquids, the specific gravity is defined as SG = ρ (T)/ρ_w (T) with T at 15.8°C (60°F), where ρ_w (T) is the density of water. Alternatively, the American Petroleum Institute (API) uses API Gravity = 141.5/SG(15.8°C) – 131.5.

Table 4.2 A, B, and C Values in API Gravity Relations

Fuel	*A*	*B*	*C*
Fuel oil	43380	93	10
Kerosene	42890	93	10
Gasoline	42612	93	10
Heavy cracked fuel	41042	126	0

HHV (kJ/kg) = A + B (API–C). See Table 4.2 for A, B, and C values.

Naphthenes have higher C/H ratio compared with paraffins and hence higher SG values. Higher the SG, lower is the API gravity and lower the heating value.

4.3.9.2 Empirical Formulae and Formulae Unit for Complex Fuels

Gasoline, diesel, etc., are mixtures of pure hydrocarbons such as C_7H_{16}, C_8H_{18}, etc., that are represented by molecular formulae. For example, California gasoline is a mixture of 15% (by volume) isobutane, 20% hexane, 45% isooctane, and 20% decane. If the overall carbon content of gasoline is 82.6% (by mass) and that of hydrogen is 17.4%, then the empirical formula is $C_{6.88}H_{17.23}$ (C = (82.6/12.01) = 6.88 and H = 17.23) or, after normalization with C atoms, $CH_{2.5}$. The empirical formulae are more suitable for gasoline, diesel, etc. When intramolecular forces (bonds within atoms of a molecule) are strong compared to intermolecular forces (between molecules), the molecules maintain their integrity. However, when intermolecular forces are comparable to chemical bond forces, then one cannot differentiate one molecule from another one. Consider table salt, NaCl. We have Na^+ and Cl^- ions. Each ion is surrounded by 6 oppositely charged ions (e.g., Na^+ surrounded 6 Cl^- ions or Cl^- surrounded by 6 Na^+ ions). There exist no definable molecules of NaCl; thus, the empirical formula is NaCl, which implies that the atom ratio between Na and Cl is unity; other examples include $LiCl_2$ (called *one formulae unit* of $LiCl_2$), which is not a molecular formula. Similarly, $CoCl_2{\bullet}6H_2O$ means six molecules of water attached to one formulae unit of $CoCl_2$. An empirical chemical formula for a fuel is analogous to the aforementioned formula unit and can also be derived from the ultimate analysis illustrated in Section 4.4.

4.3.9.3 Soot

One commonly observes soot in the exhaust of diesel trucks. Soot and smoke consist of carbonaceous solids produced during combustion through the formation of polycyclic aromatic hydrocarbons (PAHs), i.e., gas-phase condensation reactions. These solids are known carcinogens (cancer causing chemical) and are produced by the cracking of HCs. Cracking is a process of breaking larger molecules into smaller ones, with 1 to 14 C atoms per molecule, such as alkanes

C_nH_{2n+2} and alkenes C_nH_{2n}. When cracking takes place in the presence of hydrogen, it helps in picking up H atoms to form saturated carbons, a process called *hydrogenation*. *Polymerization* is the inverse of the cracking process, where smaller HC molecules combine to form a larger HC. It is also called *alkylation*, for example, $CH_4 + bC_2H_4 + cC_2H_6 + dC_3H_6 \leftrightarrow C_4H_{10}$.

Aromatics tend to form more soot than paraffins and multirings (anthracene as compared with benzene) form more soot than single rings. Luminometer number (L_N) is a measure of color temperature of gas temperature at which radiation is in green–yellow regime. Higher the luminometer number, lesser is the tendency to smoke. Smoke point (S_P), measured in mm, is the height of flame in a standard lamp without causing smoking. Higher the volatility, greater is the premixing of fuel with air and lesser is the smoking tendency. Higher the smoke point, lesser is the tendency to smoke. Smoke point is related to L_N as S_P in mm $= 4.16 + 0.331\ L_N + 6.48 \times 10^{-4}\ L_N^2$ [Bartok and Sarofim, 1991].

4.3.9.4 Flash, Fire, Pour, and Cloud Points

When a liquid fuel contained in a pan is heated, its temperature rises and it produces vapor. One can pass a flame or any other ignition source or create a spark at a specified distance from the liquid surface and check whether the A:F ratio or vapor concentration is favorable for ignition. A flame cannot be sustained if the mixture is too weak (i.e., too lean). If the amount of vapor is limited or if the heat released by a flame or spark is incapable of vaporizing enough additional fuel, the flame will consume the entire vapor around the spark plug or ignition source and dissipate. The lowest liquid temperature at which the fuel:air mixture formed above the liquid is capable of flaming is called the *initial flaming temperature* or *flash point temperature*. The lowest liquid temperature at which a sustained flame (at least 5 sec) occurs is called the *fire point*. A sustained flame will propagate toward the liquid surface to react with more fuel vapor. It is apparent that sustainment depends on the fuel volatility or vapor pressure and flammability limit. Thus, fire point denotes the fuel temperature at which one may form sufficient vapor to sustain a flame. Table 4.3 contains the flash and fire points of various liquid fuels. Flash and fire points serve to indicate the fire hazard of fuels. Lower the flash point, higher is the hazard in storing the fuels. Another property of interest is the *pour point*, which indicates the lowest temperature at which the liquid will flow. Cloud point is the temperature at which wax crystals start forming (e.g., NO_2 fuel oil). ASTM provides specifications of various liquid fuels, including flash and fire points. The *autoignition temperature* is the lowest furnace temperature required to initiate sustained flaming in atmospheric air without an ignition source.

4.4 Solid Fuels

4.4.1 Coal

Coal is a compact, aged form of biomass (plant debris) containing combustibles, moisture, intrinsic mineral matter originating from salts dissolved in water, and extrinsic ash. Coal is used to generate about 52% of U.S. electricity, which

Table 4.3 Flash and Fire Points of Various Liquid Fuels

Fuel	*SG*	T_{flash} °C	T_{fire} °C	T_{BP} °C	$(T_{fire}-T_{flash})$ °C	*Tign* C	h_{fg} kJ/kg
Ethyl alcohol	0.796	10	69–76	78	59–66	423	920
N-Butyl alcohol	0.595	34	105	117	71	—	—
Acetone	0.790	−20	55	56	75	—	—
Petroleum	—	30	—	—	—	—	—
Solar oil	—	148	—	—	—	—	—
Gasoline	0.72–0.78	10	16	100–150	6	370	380
Cylinder oil	—	215	—	—	—	—	—
Diesel	0.85	38–49–52	—	149–300	—	254	375
Biodiesel	0.88	100–170	—	182–338	—	—	—
Synthetic diesel	0.77	160–382	246	160–382	—	257	—
Lube oil	0.8–1.01	204	344	—	59	—	320
Machine oil	—	196	—	—	—	—	—
Methanol	0.794	11	—	65	—	385;464	1185
Kerosene	0.806	38	—	160–280	—	230	425

Note: T_{BP}: boiling point

Source: The data is partly from Blinov and Khudyakov, *Diffusive Burning of Liquids*, Pergamon, London, 1960.

is expected to continue through the next 20 years. Coal is a very abundant fuel accounting for 75% of the U.S. energy reserves, which are estimated to be 240 Gt. Because of large coal reserves, the U.S. relies heavily on coal for electrical power generation. The U.S. electrical power utilities consumed about 87% of the nearly 1.1 Gt of coal produced in 2003 and the open market price is \$17.85 per ton while the average delivered price is \$25.30 per ton [www.EIA.DOE.gov]. The demand for electricity grows at an average of 3.2% a year. Coal provided 51% of total U.S. electricity generation in 1995.

Coal is usually found interspersed with smaller amounts of inorganic matter beneath sedimentary rocks and is believed to have been formed by ancient vegetation collected over a million years.

Plant materials have a high cellulosic ($\approx CH_2O$, Cellulose: $C_{24}H_{40}O_{20}$) content and a molecular weight of approximately 500000 kg kmol^{-1}. Anaerobic microorganisms or bacteria in dense swampy conditions assisted in converting the plant debris into peat-like deposits. For example, cellulose a component of plant debris decomposes to peat as follows:

$$3C_6H_{10}O_5 \rightarrow C_8H_{10}O_5\ (\text{peat}) + 5CH_4 + 5CO_2$$

It is a slow process, with deposition taking place at the rate of 300 m per million years. The peat bed was then covered with sediment and folded in the earth's crust, increasing the pressure and temperature. The large thermal gradient due to the geological depth at which the coal was buried lead to chemical decomposition, which drive off volatile gases and water.

This lowered the O and H content in the coal as it was being formed. It is believed that during the "coalification" process, lignite was produced first (40 million years under heated conditions, brown colored wood-like structure with almost same heat value as cellulose), followed by subbituminous (black lignite, with typically low sulfur content, and noncaking), bituminous (soft coal, tends to stick when heated, and with typically high S content), and finally, anthracite (dense coal, has the highest carbon content [>90%], low volatility [<15%]) lasting over 200 million years under heat and pressure with a gradual increase in the coal C/O ratio. The transformation of plant debris to coal is influenced by (1) bacteria — bacterial action determines the extent of decay that takes place before the deposit is covered by an impervious sedimentary layer, (2) temperature and time — these are important after bacterial action has ceased, and (3) pressure — this increasing with depth and accentuated by severe earth movements such as folding or buckling of strata. The chemical properties of coal depend on: (1) the proportions of the different chemical constituents present in the parent plant debris, (2) the nature and extent of the changes that these constituents have undergone since their deposition, and (3) the nature and quantity of the inorganic matter present.

4.4.2 Solid Fuel Analyses

Coal, a polymer, has a higher C/H ratio compared to HCs and is largely aromatic. In order to classify coals and ascertain the quality of coal, it is essential to perform proximate and ultimate analyses in accordance with the ASTM standards codes specified in Table 4.4.

4.4.2.1 Proximate Analysis (ASTM D3172)

A solid fuel consists of combustibles, ash, and moisture, as shown in Figure 4.10. The combustibles contain fixed carbon and volatiles (for instance, dry-ash-free coal contains 63% carbon with the remaining being

Table 4.4 ASTM Analysis Standards for Coal and Coke

Drying	D3173
Total moisture	D3302
C, H, N	D5373
S	D4239
ASH	D3174
Volatile matter	D3175
HHV	D5865
Ash analysis	Measured with atomic emission spectroscopy

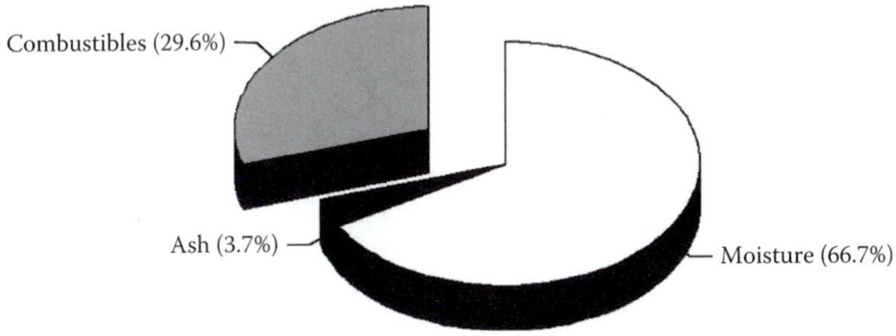

Figure 4.10 Typical solid waste fuel composition.

volatiles). Proximate analysis provides the following information [PETC Review, 1990]: surface moisture (SM), i.e., moisture in air dried coal, the inherent moisture in the coal (M), volatile matter (VM, a thermal decomposition process resulting in release of water, gases, oil, and tar), fixed carbon (FC) and mineral matter (MM), and heating value (HV). The higher or gross heating value of solid fuel is determined according to ASTM D5865 using an isothermal-jacket bomb calorimeter. The surface moisture percentage is determined as follows. A coal of mass m_1 (that is approximately 1 to 2 g) is exposed to the atmosphere at 297 K and 75% humidity over 24 hr. The moisture evaporates and the remaining mass reduces to m_2. Consequently,

$$SM\% = (m_1 - m_2) \times 100/m_1. \quad (4.6)$$

Note that sometimes H_2O may be chemically bound. The air-dried (i.e., surface moisture removed) coal of mass m_2 is then dried in an oven in a nitrogen atmosphere at 380 K for 1 hr. The coal mass further reduces to m_3 (following ASTM D3173) and the inherent moisture on an as-received basis

$$M = (m_2 - m_3) \times 100/m_1. \quad (4.7)$$

The moisture loss is also called dry loss (DL). During the procedure, part of the air-dried coal (about 1 g) is kept in a crucible that is closed with a lid to prevent oxidation. The crucible is placed in a furnace at 1200 K for 7 min and the mass reduces to m_4 so that

$$(VM + M)\% = (m_2 - m_4) \times 100/m_1. \quad (4.8)$$

Knowing M%, VM% can be determined from Equation 4.8. The ash content is determined as follows. The air-dried coal is placed in a furnace at 700 K for 30 min in air, and then the furnace is heated to 1050 K. The temperature is maintained until the sample mass remains constant. This mass m_5 is used in the calculation of ash quantities. The following relations are applied

$$VM + M + FC = (m_2 - m_5), \quad (4.9)$$

and

$$A\% = m_5 \times 100/m_1. \quad (4.10a)$$

Table 4.5a Comparative Proximate Analyses for Different Fuels, Percent and Heating Values

Fuel	DL	VM	FC	Ash	HHV, MJ/kg
Coal	21.2	32.6	41.9	4.3	21.8
AFB	10.9	57.0	17.3	14.8	15.0
LB	7.5	40.3	8.4	43.8	9.2

Ash is mainly MM (mineral matter) and may include sodium carbonates formed during combustion. The following formula is used for MM.

$$MM\% = 1.10A + 0.74CO_2 + 0.53S_T - 0.32 \tag{4.10b}$$

where S_T = total sulfur percent.

It is possible that some ash may be lost during the char combustion. It is noted that FC% and hence VM% under actual combustion conditions could be different compared with those from proximate analyses. For biomass fuels, proximate analysis is typically limited to 880 K. Table 4.5a shows comparative analyses of coal, advanced feedlot biomass (AFB, low-ash cattle manure), and litter biomass (LB, chicken manure) [Priyadarsan et al, 2004].

The ash mass is almost invariant (except for slight evaporation) during coal combustion, but its percentage increases as combustion proceeds. The ash fraction is, therefore, sometimes useful for determining the extent of combustion or gasification (Section 4.8). The free swelling index (FSI) of coal is determined by placing 1 g of sample in a crucible and then heating it at 820°C for 25 min. The ratio of increase in surface area to the original surface surface area is called FSI.

A microscopic examination of coal reveals its heterogeneous physical characteristics. At 40 to 60 times magnification, organic matter (yellow, orange, red, and brown) and inorganic matter (aluminosilicate clays and pyrites that do not transmit light) can be observed. Ash includes gold-colored particles of pyrite (FeS_2). Organic matter is classified into three different constituents that are called macerals: (1) vitrinite, which accounts for 85% of the organic matter, and it arises because of the decomposition of cellulose that is destroyed or degraded by microorganisms, leaving the more resistant lignin, (2) exinite or liptinite, derived from pores, resins, and waxes, and (3) inertinite. Vitrinite is a major component of low-rank coals.

4.4.2.2 *Ultimate or Elemental Analysis (ASTM D3176)*

Solid fuels contain different amounts of C, H, N, O, S, moisture, and MM (mineral matter, which includes pyrite, quartz, calcite, clays, etc., and trace elements). Many solid fuels do not have simple chemical formulae specifying the number of atoms. The chemical composition of these fuels is specified in terms of the mass percentage of its various elements. An elemental-mass-based composition is called *ultimate analysis*. It can be expressed either on an as-received basis, a dry basis (with the moisture in the solid fuel removed), or a dry-ash-free basis (DAF, also known as the moisture ash-free basis, MAF). The ultimate analysis uses the following procedure. The tests and analyses are conducted using the ASTM standards. When m_F kg of a hydrocarbon fuel is completely burned to produce CO_2 and H_2O, the carbon dioxide can be absorbed in a bed of NaOH. The CO_2 mass can be determined by measuring the weight gain of the bed, which, in turn, can provide information about carbon mass. Note that while most of the carbon originates from organic C, there may be traces of C from mineral constituents. Water can be simultaneously absorbed in $Mg(ClO_4)_2$ and its mass measured. For an oxygenated fuel, $C_cH_hO_o$, the amount of oxygen added to the reactor to burn the fuel must also be measured in order to determine the oxygen content in the fuel. For "as received" basis the oxygen percentage is calculated as the difference between 100 and the sum of C, H, N, S, Cl, F, Br, ash and water percentage. Thus, the C, H, and O atom masses can be calculated.

Using laser spark spectroscopy for ultimate analysis [Bilger, 1980], one may pulse a small volume of gas with a high-intensity laser and convert it to plasma. Spectroscopic analysis of light emission from this plasma gives the concentration of elements, because each element shows atomic and ionic lines of intensity proportional to mass fractions Y_k.

The nitrogen percentage in fuels is important in the context of NOx formation. It is not present normally in natural gas, but crude oil contains about 0.15 to 0.5% by mass, heavy distillate 0.6 to 2.15%, and coal 1 to 1.5%. Table 4.5b shows a sample of ultimate analyses: Table A.2B and Table A.2D present ultimate analyses of coal and woody biomasses.

4.4.2.3 Coal Classification, Composition, and Rank

The composition of coal can be given on an as-received (AR), dry (D), and DAF basis. The heating values are measured with a bomb calorimeter using ASTM D2105. Solid fuels can be characterized by the elemental composition and heating values.

Table 4.5b Ultimate Analyses

Fuel	*%C*	*%H*	*%O*	*%N*	*%S*	*AF*
Coal	58.9	6.2	33.8	0.9	0.3	8.1
Cattle manure	43.7	6.2	44.9	4.0	0.8	4.4
Chicken litter	39.1	6.7	48.3	4.7	1.2	3.9

Source: Priyadarsan, S., Annamalai, K., Sweeten, J.M., Holtzapple, M., and Mukhtar, S., Co-gasification of Blended Coal with Feedlot and Chicken Litter Biomass, 30th International Symposium on Combustion, 2004.

Table 4.5c Mass Percentage of Elements in Various Coals

Coal	*C*	*H*	*N*	*O*	*S*	$(C/H)_{atom}$	$(C/O)_{atom}$
Lignite	71	4	1	23	1	1.62	4.11
Subbituminous (Cl < 0.011%)	77	5	1	16	1	1.30	6.41
Bituminous	80	6	1	8	1	1.12	13.33
Anthracite	92	3	1	3	1	2.58	40.85

ASTM has developed a system for classifying coal by rank, which is a measure of the degree of change of chemical composition during the transition from cellulose to graphite. The highest-ranked coal has a graphitic structure, and older the coal, the higher is its rank. An ascending order in rank corresponds to an increase in the carbon content of coal (or a corresponding decrease in the hydrogen and oxygen contents; see Table 4.5c), the coal heating value, vitrinite content, aromatic complexity, and a decrease in functional group content. Thus, a low-rank coal has a lower carbon content (or a small degree of change from the original cellulose) as in peat (produced during the initial stage of formation of coal), whereas a higher rank indicates a higher carbon content (or a larger degree of change) as in anthracite (black coal). Under the action of heat and pressure, water and oxides of carbon are expelled. The volatile matter content varies with rank — the lower the rank, the higher is the volatile matter content. Figure 4.11 shows the HHV coal estimated for HV of volatiles (see Example 6) and VM% of various coals. Oxygen levels are generally low for the higher-ranks coal. Coal is said to have caking properties when, as a result of heating, it tends to adhere or stick and forms a solid mass. Caking is measured by the capability of the coal to swell and agglutinate. Table 4.5c shows the mass percentage of C, H, N, O, and S for various coals. Table 4.6 shows the ASTM classification of coal. The ASTM classification is based on FC, whereas for FC < 69%, it is based on VM. Coals with FC > 86% are called as anthracite, and those with 69% < FC < 86% as bituminous. Bituminous coals produce more liquids and tars during pyrolysis; lignites produce more gases, involving more water and CO_2. The British classification method is based on the volatile content of a coal. The international method classifies coal in 10 classes from 0 through 9. Classes 0–5 are based on VM (which, for FC > 67, is similar to the British method), and classes 6–9 are similar to the ASTM method (see Table 4.4).

Example 2

Wood is a biomass fuel with the following composition: carbon: 50%, hydrogen: 6%, and oxygen 44%. Its HHV is 20000 kJ/kg or 8700 Btu/lb; its bulk density 640 kg/m^3 (40 lb/ft^3). (a) Determine its chemical formula empirically. (b) What is the molecular weight? (c) If fuel formulae are normalized with C atom (i.e., expressed as H/C and O/C), what is the molecular weight?

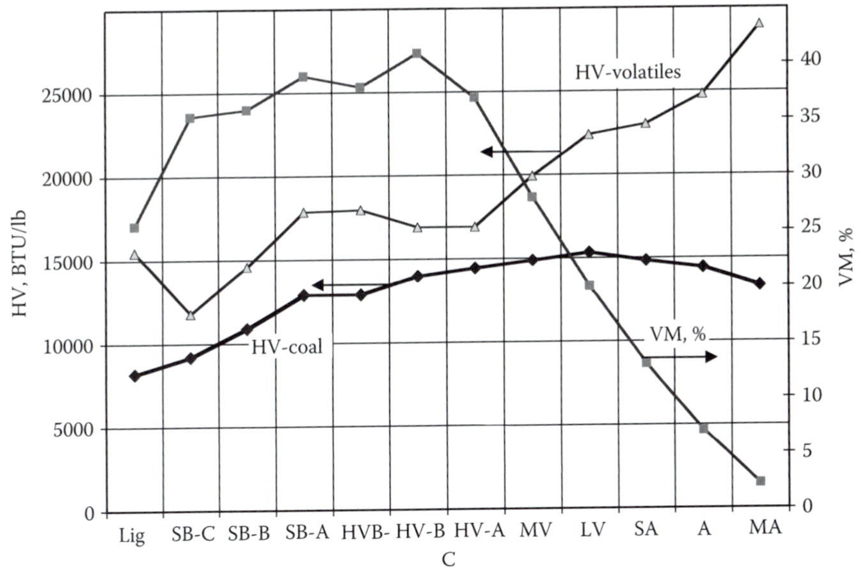

Figure 4.11 HV of volatiles versus type of coal. (Coal HV and VM data from Smoot, L. and Pratt, D.T., *Pulverized Coal Combustion and Gasification*, Plenum Press, New York, 1979.)

Table 4.6 ASTM Classification of Coal

Class	*Group*	*FC%*	*HHV MJ/Kg*	*Physical Properties*
Anthracite	Metaanthracite	>98		Nonagglomerating;
FC > 86%	Anthracite	92–98		brighter black in
	Semianthracite	86–92		appearance.
Bituminous	Low volatile	78–86		Agglomerating and
FC 69–86%	Medium volatile	69–78		nonweathering;
	High-volatile A			bituminous dark black.
	High-volatile B	<69	>32.7	
HHV 26–33%	High-volatile C		30.3–32.7	
			25.7–30.3	
Subbituminous (SB)	SB-A		25.7–30.3	Weathering and
HHV 19–26%	SB-B		22.2–25.7	nonagglomerating;
	SB-C		19.4–22.2	subbituminous dull black.
Lignite	Lignite		<19.4	Consolidated; brownish
HHV< 19%	Brown		<19.4	black.
				Unconsolidated.

Note: Apart from rank, another classification is based on "grade of coal," but it is only qualitative. Higher the ash, lower is the ash fusion temperature (AFT, see Table A.2F for values), higher the S, lower is the grade.

Source: ASTM Specifications for Petroleum Products, American Society for Testing Materials, Philadelphia, 1996.

Solution

(a) Neglecting the ash, 100 kg of dry wood contains 50 kg carbon, 6 kg hydrogen, and 44 kg oxygen. Therefore, a representative sample contains 50/12.01 = 4.16 atoms of carbon, 6/1.01 = 5.94 atoms of hydrogen, and 44/16 = 2.75 atoms of oxygen in a composite wood molecule. The empirical chemical formula for such a wood is $C_{4.16}H_{5.94}O_{2.75}$.

(b) The above empirical formulae has a molecular weight of 100 kg kmol^{-1}.

(c) It can also be expressed as ratio of H/C, O/C, etc. For example, $C_{4.16}H_{5.94}O_{2.75}$ can be expressed as $CH_{5.94/4.16}O_{2.75/4.16} = CH_{1.43}O_{.66}$, with a molecular weight of dry-ash-free normalized fuel as $M_{DAF,n}$ = 100/4.16 = 24.04. Note that for the normalized fuel $M_{DAF,n}$ = 12.01/{fraction of carbon in DAF} = 24.04. If ash and water are present, then $M_{DAF,n}$ = 12.01 × (100–ash%–water%)/(% of carbon in as-received fuel).

Remarks

(i) Typically, solid (such as coal) liquid and gaseous fuels are represented crudely as: $CH_{0.9}$, CH_2, and CH_4.

(ii) Wood usually contains ash. Assuming 5% ash, the empirical chemical formula is $C_4H_5O_{2.6}$. It has a molecular weight of 95 kg kmol^{-1} for the ash-free portion. The chemical reaction equation for stoichiometric combustion is

$$C_4H_5O_{2.6} + \text{ash} + 5.2O_2 \rightarrow 4CO_2 + 5H_2O + \text{ash}.$$

(iii) For dry air, empirical formula $N_{1.562}\ O_{0.420}\ Ar_{0.00932}\ C_{0.0003}$

(iv) Typically, the elemental composition of C, H, N, etc., is reported on an as-received basis. Sometimes the total moisture, ash, and ultimate analysis percentage may not add up to 100. Therefore, it is possible that H = 5.3% includes the hydrogen both in the fuel and the accompanying moisture. If the moisture is 9%, the hydrogen content due to the moisture is 1 kg for every 100 kg of the as-received fuel, whereas the oxygen mass in the moisture is 8 kg. Thus, (5.3 − 1) = 4.3 kg must have originated from combustible content of fuel. Hence, there is 4.3 kg of H for 91 kg of dry fuel, and the aforementioned percentages can be corrected to a dry basis.

(v) Empirical formulae do not provide information on molecular structure [Oxtoby et al., 1998]. For example, acetic acid has a formula $C_2H_4O_2$, whereas glucose has a formula $C_6H_{12}O_6$. The normalized empirical formula (CH_2O) is the same for both acetic acid and glucose.

Example 3

(a) Determine the empirical formulae for a gaseous fuel containing carbon (81.68% by mass), hydrogen (18.32%), and a negligible amount of oxygen. (b) Determine the A:F ratio on mole, volume, and mass basis. (c) Comment on the volume-based A:F ratio for the empirical fuels.

Solution

(a) Consider 100 kg of the fuel. The carbon mass is 81.68 kg so that its kmol in the fuel is 81.68/12.01 = 6.807. Similarly, hydrogen kmol in the fuel is

18.32/1.01 = 18.14. Therefore, the empirical formula is $C_{6.80}\,H_{18.14}$, with an empirical molecular weight of 100 kg kmol^{-1}. The empirical formula normalized with respect to the carbon content is $CH_{(8/3)} = CH_{2.667}$, with an empirical molecular weight of 14.70 kg kmol^{-1}.

(b) The stoichiometric equation for complete combustion is

$$CH_{2.667} + (1 + 2.667/4)\ O_2 \rightarrow CO_2 + 1.333\ H_2O$$

The stoichiometric A:F (mole) ratio is (1 + 2.667/4)/0.21 = 7.94 if the fuel is represented as $CH_{2.667}$. The A:F mole ratio is 53.98 if the fuel is considered to be $C_{6.80}\,H_{18.14}$. However, the stoichiometric air-to-fuel ratio in terms of kg of air per kg of fuel is identical in both cases (15.64).

(c) Results show that because the fuel is a gas, the value of the A:F ratio on a volumetric basis could be either 7.94 for fuel $CH_{2.667}$ or 53.98 for fuel $C_{6.8}H_{18.14}$, which is an absurd answer. The A:F ratio on a volumetric basis for gaseous fuels requires the exact specification of the number of carbon, hydrogen, and other atoms in the fuel. Consider propane, for instance, which has 3 carbon atoms and 8 hydrogen atoms. The molecular formula of propane is specified on this basis, and we do not consider multiples of the C and H atoms, such as C_6H_{16} or $C_{12}H_{32}$, which are entirely different chemical species. The A:F ratio on a molar basis for propane is 3 × 7.94 = 23.82 kmol of air per kmol of fuel or 23.82 m^3/m^3 of propane (which has molecular weight of 44.11 kg).

Example 4

A low-Btu gas has the following composition:

CH_4: 3 (mole % or volume %)
CO: 13
CO_2: 10
H_2: 5
N_2: 69
O_2: 0

(a) Determine (i) υ_{O2} in kg of O_2 per kg of fuel, (ii) $Y_{F,i}$ mass fraction of combustibles and (iii) $\upsilon_{O2}Y_{F,i}$.

(b) Determine the overall empirical chemical formulae and using these repeat part (a).

Solution

(a) (i) The low-Btu gas mixture consists of fuel and inert components. The combustible components are CH_4, CO, and H_2. For CH_4, CO, and H_2,

$$CH_4 + 2O_2 \rightarrow CO_2 + 2H_2O,$$

$$CO + 1/2O_2 \rightarrow CO_2,$$

and

$$H_2 + 1/2O_2 \rightarrow H_2O$$

Then, oxygen required by the fuel part of the mixture (CH_4, CO, and H_2) is given as 0.03 × 2 + 0.13 × 1/2 + 0.05 × 1/2 = 0.150 kmol.

Thus, υ_{O2} (based on fuel within the mixture only) = 0.150 × 32/(0.03 × 16.05 + 0.13 × 28.01 + 0.05 × 2.02) = 1.1364 kg of O_2 per kg of fuel in the mixture, with the fuel being a mixture of CH_4, CO, and H_2.

(ii) The combustible mass fraction $Y_{F,i}$ in the gas mixture is given by

$$Y_{F,i} = (0.03 \times 16.05 + 0.13 \times 28.01 + 0.05 \times 2.02)/(0.03 \times 16.05 + 0.13 \times 28.01 + 0.05 \times 2.02 + 0.10 \times 44.01 + 0.69 \times 28.02) = 0.1511.$$

(iii) Then, $\upsilon_{O2} Y_{F,i}$ = (kg of O_2/kg of fuel) × (kg of fuel/kg of gas mixture) = kg of O_2/kg of gas mixture = 1.1364 × 0.1511 = 0.17.

(b) (i) We can get an overall chemical formulae for the low-Btu gas including inerts in the fuel as follows

Number of C atoms = 0.03 × 1 + 0.13 × 1 + 0.1 × 1 = 0.26.

Similarly, the number of H atoms = 0.22, N atoms = 1.36, and O atoms = 0.33 Thus, the empirical chemical formulae for the low Btu gas is given as $C_{0.26}H_{0.22}N_{1.36}O_{0.33}$ or $CH_{0.046}N_{5.31}O_{1.27}$.

For such a gas including inerts,

$$C_{0.26}H_{0.22}N_{1.36}O_{0.33} + (0.26 + 0.055 - 0.33/2)\ O_2\ 6\ 0.26\ CO_2 + 0.11\ H_2O + \ldots$$

Then, υ_{O2} = 0.15 × 32/(12.01 × .26 + 0.22 × 1.01 + 1.36 × 14.01 + 0.33 × 16) = 4.8/27.678 = 0.17 kg/kg of gas fired through the injector.

(ii) Note that we treat the fuel mass fraction in the inlet to be 1 because we treated the whole low-Btu gas with an empirical formulae. Thus, $Y_{F,I} = 1$

(iii) $Y_{F,i}\upsilon_{O2} = \upsilon_{O2}$ = 0.17 kg of O_2 per kg of mixture.

4.4.3 Coal Pyrolysis

Coal is a complex polymer consisting of C, H, N, and O. The heating of bituminous coal results in the evolution of gases during *pyrolysis*, which is the process of volatile evolution, and the formation of a solid, mostly carbon skeleton. With increasing temperature, coal begins to soften like a plastic. The individual coal particles fuse, stick together, and form a porous mass that expands and solidifies. *Coking* is the phenomenon of polymerization in which molecules repeat themselves in a chain-like structure. As volatiles are driven off, long-chain molecules try to reattach to each other. The resulting product is called *coke* (essentially char produced from coal). *Coking coal* is a term used for the manufacture of metallurgical cokes.

The process of evolution of combustible gases due to the thermal decomposition of the solid is referred to as *pyrolysis*. Biomass typically consists of 70 to 80% volatile matter (VM) (which is the fraction of solid fuel that can be released in the form of combustible gases), and coal consists of 10 to 50% VM depending on its age or rank (Figure 4.11). As the aging period increases, VM decreases as a result of the gradual release of gases. Coal rank increases with decreasing VM. Typically, a medium rank coal consists of 40% VM with volatile species and tar (which is a macromolecule), and about

60% fixed carbon (FC). If the heat of pyrolysis is neglected, the heat of combustion of the coal can be represented as a combination of the contribution from the VM and the contribution from the FC in relation to their mass percentages, which will be seen later.

4.4.3.1 Chemical Formulae for Volatiles

When a coal particle is placed in a hot furnace, it is heated and pyrolysis occurs. Typically, bituminous coal pyrolyzes at about 700 K (as is also the case for most plastics). Pyrolysis products range from lighter volatiles such as CH_4, C_2H_4, C_2H_6, CO, CO_2, H_2, and H_2O to heavier tars. Tars thermally crack to form CH_4, C_2H_4, C_2H_6, C_3H_6, C_2H_2, and CO above 1000 K and are responsible for soot formation.

Tars produced from coal may occur in the form of liquids that vaporize if sufficient heat is supplied. If vaporization does not occur, tars can undergo repolymerization (or primary cracking) to produce lighter volatiles and char. The mixture of gases or volatiles can usually be represented by an overall empirical chemical formulae $C_{cv}H_{hv}N_{nv}O_{ov}S_{sv}$ where the subscripts cv, hv, etc., represent the number of atoms in volatiles for carbon, hydrogen, etc. If the ultimate and proximate analyses are known, it is possible to obtain an overall empirical chemical formula for the volatiles.

Pyrolysis is a relatively slow chemical process as compared with the rapid physical vaporization of oil drops. Coal can swell upon heating, resulting in a larger particle size (whereas the drop size of a liquid remains virtually constant during heating). This effect is stronger in an inert environment or under reducing conditions. The swelling factor, that is the ratio of the larger swollen coal particle to its original dimension, can range from 1.3 (under oxidative conditions) to 4 in an inert environment. Swelling introduces thin-walled cenospheres, which can produce a sudden decrease in the particle size when burned and a density increase during burn off.

4.4.4 Ash and Loss on Ignition (LOI)

Ash is produced during incomplete coal combustion or because of poor combustion efficiency of coal. Accumulation of ash tends to insulate boiler tubes, lowering the heat transfer rate. The ash is swirled in the tubes, the impact of which can cause erosion leading to holes in the tubes through which steam can escape. Seven to twelve percent ash is desirable because it protects boiler tubes, and up to 25% ash is tolerable. The accumulated ash is periodically blown off using the soot blowers. The carbon content in coal ash can range from 1 to 20%, and it is also called the *loss on ignition* (LOI). It can be detected as follows. A dry coal sample can be collected and irradiated with infrared light (e.g., with a diode laser) that oxidizes the carbon in a closed chamber. The heat produced causes a minute pressure increase in the chamber, and the resulting sound wave is sensed by a microphone. This sound pulse intensity approximately doubles for each percentage increase in carbon content. Low-NOx burners operating at low-O_2 concentrations can cause the combustion efficiency to decrease, which promotes LOI. Ash disposal costs are about \$5 to \$10 per ton.

The ash fouling tendencies of coal in boilers are related to ash fusion temperature (AFT) [Lawn, 1987; Annamalai et al, 2003]. Thus, it is important to determine ash fusion characteristics of coal ash. Coal ash begins to soften at about 1450 K and becomes a fluid at 1580 K. AFT (ASTM 1857) is determined as follows. The fuel is first heated in air at 1123 K and then in O_2 in order to produce ash, and it is then mixed with a solution of dextrin to form a paste and is made into a conical shape using a mold. The conical ash is placed in a furnace heated by a controlled heating rate dT/dt, the change in shape is observed, and the corresponding T is recorded. Acid in oxides (SiO_2, Al_2O_3, TiO_2, and Fe_2O_3) increase T_{melt}, whereas basic oxides (CaO, MgO, Na_2O, and K_2O) reduce T_{melt}. Table A.2E and Table A.2F tabulate mineral matter and phase-change properties of ash.

4.4.4.1 Physical Properties

The *specific heat* of graphite lies in the range of 0.67 to 1.6 kJ/kg K depending on its form, whereas for coal with 50% VM, it has a value from 0.95 to 1.5 kJ/kg K at 350 K. For coal with 25% VM, its value is 2.2 kJ/kg K at 700 K.

The *thermal conductivity* λ of coal is listed in the literature within a range of 125×10^{-6} to 340×10^{-6} kW/m K. The emissivity of spherical carbon is roughly 0.85, whereas that of partially reacted char is approximately 0.7.

Density: The *density* of coal could be measured using Hg as a displacing medium. Thus, a known mass of coal could be placed in Hg and the displaced volume measured. Thus, apparent density is estimated as mass/volume. If water is used instead of Hg, an impervious film must be applied to the coal surface to prevent penetration of water into pores. Typically, the densities of coal and carbon range from 1100 kg/m^3 for low-rank coals to 2330 kg/m^3 for high-density pyrolytic graphite (Figure 4.12).

Pore Distribution can be described as follows: Coal consists of pores of various sizes. The pore sizes vary from micropores (2 nm or less) to macropores (more than 20 nm). Approximately 10 to 30% of the coal volume consists of pores. The volume percent of micropores increases with the rank of the coal. If helium is used as displaced media, the density values will be different because helium can penetrate the pores. Thus, the density difference between He and Hg methods could be used to infer the porosity of coal. Figure 4.12 shows the variation of actual density and porosity of coal with C content [Chigier, 1981].

The *porosity* of coal compares the volume of its pores with its total volume. Chars are very porous with a surface area ranging from 1,000 m^2/kg to 200,000–400,000 m^2/kg.

4.4.5 Heating Value (ASTM D5865)

The heating value (HV) of a fuel is the amount of heat released when a unit (mass or volume) of the fuel is burned. Typically, bomb calorimeters are used to determine the HV (Chapter 2). A relationship between HHV and the

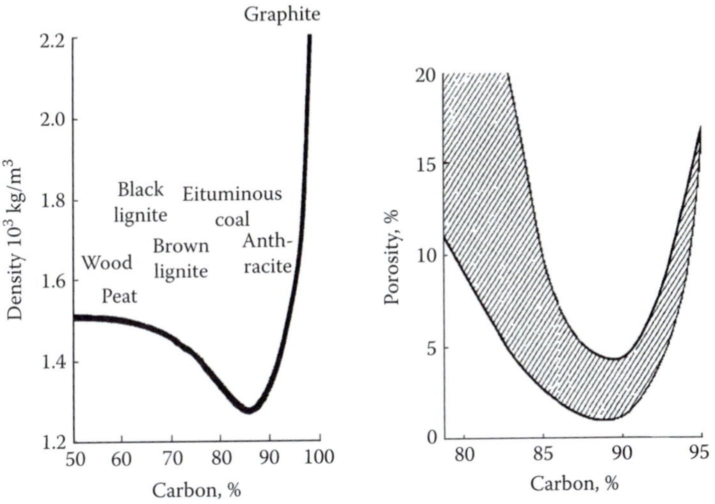

Figure 4.12 Variation of density and porosity with carbon content percentage. (Adapted from Chigier, N., *Energy, Combustion, and Environment*, McGraw-Hill, New York, 1981.)

elemental composition was noticed by Boie, Dulong, and others in the 19th century. The Boie equation is given as [Annamalai et al, 1987.]

$$HHV_F,\ kJ/kg_{fuel} = 35160Y_C + 116225Y_H - 11090Y_O + 6280Y_N + 10465Y_S, \quad (4.11)$$

$$LHV_F,\ kJ/kg_{fuel} = 35160Y_C + 94438Y_H - 11090Y_O + 6280Y_N + 10465Y_S, \quad (4.12)$$

$$HHV_F,\ kJ/kmol_{fuel} = 422270N_C + 117385N_H - 177440N_O + 87985N_N + 335510N_S, \quad (4.13)$$

where Y_k denoted the mass fraction of an element with k = C, H, O, N, and S in the fuel, and N_k, the number of atoms of the respective species in a fuel molecule. Note that the sign for the terms corresponding to O in Boie equation are negative, indicating that higher the O content, lower is the heat value because O is bound to carbon and is already in an oxidized state. For example, if C is bound as CO_2, it cannot contribute to the heat value. The N and S contributions also indicate conversion of fuel N (at least a part of it) to NO and S to SO_2. Table A.2D compares the heating values of several woody biomasses computed from Boie equation with measured values. Based on the Boie equation, the enthalpy of formation can be derived as

$$\bar{h}_{F,f}^{0} = 28752 \times \{N_C - 0.888 \times N_H - 6.168 \times N_O + 6.199\ N_N + 1.337\ N_S\},\ kJ/kmol$$

$$HHV'_{O_2} = 17.24 \times (N_C + 0.278\ N_H - 0.4202\ N_O + 0.2084\ N_N + 0.7945\ N_S)/(N_C + 0.25\ N_H - 0.5\ N_O + N_S),\ kJ/liter \quad (4.14)$$

where HHV'_{O_2} is heating value on volume basis in kJ per unit volume of stoichiometric oxygen at STP, i.e., in kJ/liter of oxygen (STP corresponds to 25°C and 1 atm). The value of $HHVo_2$ is roughly constant for many common fuels. For glucose ($C_6H_{12}O_6$) and fat (palmitic acid, $C_{16}H_{32}O_2$), these values are 19.2 and 17.8 kJ/liter of oxygen (SATP) even though on a mass basis the value of HHV_F for fat is about 2.5 times larger than that for glucose. The exception is H_2 which has 23.33 kJ/liter of O_2.

The heating values for coals can also be empirically obtained by using the Dulong equation, namely,

$$\text{HHV (kJ/kg)} = 33800\ N_C + 144153\ N_H - 18019\ N_O + 9412\ N_S. \quad (4.15)$$

Channiwala [1992] considered over 200 species of biomass and fitted the following equation to the data:

$$\text{HHV (kJ/kg)} = 349100Y_C + 1178300\ Y_C - 103400\ Y_O - 211100\ Y_A + 100500\ Y_S - 15100\ Y_N \quad (4.16a)$$

The experimental data has an error of about 1.5%. For biomass fuels like onion waste HHV $\approx 0.63 + 0.39 \times$ C, MJ/kg, C percentage dry basis.

The flue gas volume can also be estimated given the ultimate-analysis- and Boie-equation-based heating values. Hence, flue gas volume per GJ can be estimated and plotted against the H/C and O/C atom ratio for CHO fuels at 6% O_2, as shown in Figure 4.13. The flue gas volume is almost independent of O/C ratios. The fit gives the following empirical equation (Problem 4.24) at 6% O_2 in flue gas.

Flue gas STP volume (m^3/GJ) = $4.9564\ (H/C)^2 - 38.628\ (H/C) + 389.72$, which has a correlation coefficient of 0.9938. STP corresponds to 25°C and 1 bar. For any arbitray 0% < O_2 <9% (volume), a fit is given as

Flue gas STP volume (m^3/GJ) = $\{3.55 + 0.131\ O_2\% + 0.018 \times (O_2\%)^2\}\ (H/C)^2 - \{27.664 + 1.019\ O_2\% + 0.140 \times (O_2\%)^2\}\ (H/C) + \{279.12 + 10.285\ O_2\% + 1.416 \times (O_2\%)^2\}$. (4.16b)

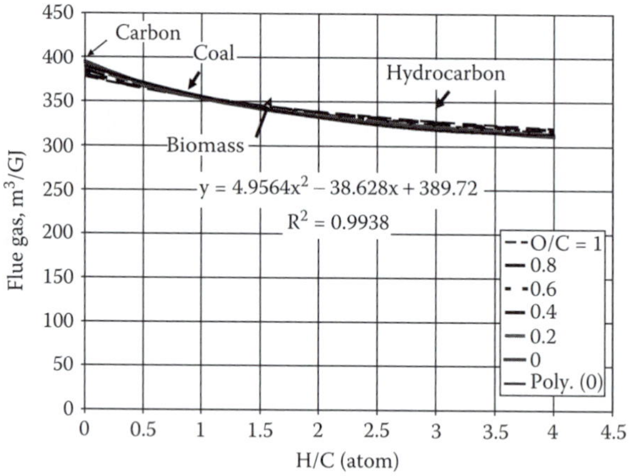

Figure 4.13 Flue gas at STP (25°C and 1 bar) volume for various fuels at 6% O_2. Multiply ordinate by 37.26 to get fr^3/mmBtu.

Example 5

For a coal undergoing pyrolysis according to the chemical reaction $CH_{0.8}$ (s) → a CH_n (g) + b C(s) and for which VM (on a DAF mass basis) = 0.4, determine the value of a, n, and b.

Solution

With C and H atom balance, 1 = a + b, 0.8 = n × a, M_{Coal} = (12.01 + 1.01 × 0.8), and because VM = 0.4, FC = 0.6 = b × 12.01/M_{Coal}. Hence, b = 0.640, and the carbon atom balance yields a = 1 − b = 0.36. The H atom balance yields n = 0.8/a = 0.8/0.36 = 2.2.

Remarks

(i) At a higher heating rate of coal, the amount of VM that evolves increases. If, for a hypothetical case, the value of VM = 0.8, then b = 0.21, a = 0.79, and n = 1.1.
(ii) The H/C ratio in volatiles decreases as the value of VM increases. H/C is 2.2 when VM = 0.4, and 1.1 when VM = 0.8.
(iii) Note that CH_n represents overall H/C ratio. It is possible that volatiles may consist of say 20% C_2H_4, 70% CH_4, and 10% C_2H_6, but with overall H/C = {0.2 × 4 + 0.7 × 4 + 0.1 × 6}/{0.2 × 2 + 0.7 × 1 + 0.1 × 2} = 3.2.

Example 6

(a) Determine the heating value of volatiles from coals if the heat of pyrolysis is 400 kJ/kg of volatiles. Assume the coal to have an empirical composition $CH_{0.8}$, VM = 0.4, and a heating value of 35000 kJ/kg. (b) If HV coal ≈ VM × HV_{VM} + FC × HV_{FC}, estimate HV_{VM}. (c) Estimate HV_{VM} in kJ/kmol of the empirical volatile.

Solution

(a) The enthalpy of formation of coal can be determined by examining its chemical reaction, i.e.,

$$CH_{0.8} + 1.2\ O_2 \rightarrow CO_2 + 0.4\ H_2O, \tag{A}$$

or

$$HV \times (12.8) = \bar{h}_{f,coal} - \left(\bar{h}_{f,co} + 0.4\bar{h}_{f,H_2O}\right), \tag{B}$$

so that

$$\bar{h}_{f,\ coal} = 35000 \times 12.8 - 285830 \times 0.4 - 393520 = -59852\ \text{kJ/kmol}. \tag{C}$$

Because each kg of coal produces 0.4 kg of volatiles, the heat required per kg of coal = −400 × 0.4 = −160 kJ/kg of coal. For the pyrolysis process

$$CH_{0.8} \rightarrow 0.36\ CH_{2.2} + 0.64\ C, \tag{D}$$

if (H_{prod} − H_{reac}) = 160 × (12.8) kJ/kmol of coal = 2048 kJ/kmol of coal, then

$$0.36\,\bar{h}_{f,CH_{2.2}} + 0.64\,\bar{h}_{f,C} - \bar{h}_{f,\ coal} = 2048\ \text{kJ/kmol of coal}. \tag{E}$$

Because $\bar{h}_{f,C} = 0$, using Equation C in Equation E, we get

$$\bar{h}_{f,CH_{2.2}} = (2048 - 59852)/0.36 = -160567 \text{ kJ/kmol of volatiles.} \quad \text{(F)}$$

Therefore,

$$CH_{2.2} + 1.55O_2 \rightarrow CO_2 + 1.1\ H_2O, \quad \text{(G)}$$

and

$$HV_{VM} = (H_{prod} - H_{reac})/(12 + 2.2) = (-160{,}567 + 285{,}830 \times 1.1 + 393{,}520)/14.2$$
$$= 38{,}547 \text{ kJ/kg.} \quad \text{(H)}$$

(b) If the HV of coal is approximated by the relation

$$HV \approx VM \times HV_{VM} + HV_{FC} \times FC, \quad \text{(I)}$$

it becomes possible to determine the value of HV_{VM}. Applying the chemical reaction $C + 1/2\ O_2 \rightarrow CO_2$ yields HV_{FC} = 32601 kJ/kg of C. Using the relation FC = 1 – VM,

$$HV_{VM} \approx HV - (1 - VM) \times HV_{FC}/VM,$$

or

$$HV_{FC} = 393520/12.01 = 32766 \text{ kJ/kg,}$$

and

$$HV_{VM} = (35000 - 0.6 \times 32766)/0.4 = 38351 \text{ kJ/kg.} \quad \text{(J)}$$

This is a slightly lower value than one obtained using the more exact method of Equation H. Both methods agree if the heat of pyrolysis is zero. When heat is supplied for pyrolysis, the enthalpy of the volatiles increases. Because HV_{VM} = (H_{reac}(including volatiles) – H_{prod}), the more exact method yields a slightly larger value. The difference between the two values increases when HV for the parent coal is low.

(c) HHV for volatiles can also be computed on mole basis if the volatiles are a mix of ideal gases. From Equation D, M_{VM} = 12.01 + 2.2 × 1.01 = 14.23 kg/kmol. Thus, HHV = 38351 kJ/kg × 14.23 kg/kmol = 545735 kJ/kmol of empirical volatile.

Remarks

(i) Anthracite coal has a lower VM value as compared to bituminous coals. At lower values of VM, the H/C ratio is larger and, therefore, the HV of the volatile matter is higher. Thus, HV differs depending on the coal type. Figure 4.12 presents a plot of the average HV of volatiles with respect to the type of coal.

(ii) In gasification processes, fuel is completely pyrolyzed, and fixed carbon is preferentially oxidized to CO so that the heating value of gases can be maximized. If the O_2 mass fraction in oxidant stream at the inlet of the gasifier is $Y_{O2,in}$, then the maximum gross heat content per unit mass of gases produced can be derived. Note that any moisture in the fuel is condensed in the exit gases.

1 kg solid DAF fuel → VM kg of volatiles + (1 − VM) kg of fixed carbon. If heat of pyrolysis is neglected, HV_{VM} = {HV_{fuel} − (1 − VM) × HV_{FC}}/VM, where HV_{FC} = 32766 kJ/kg of C.
(1 − VM) kg of C + (16/12) × (1 − VM) kg of oxygen → (28/12) × (1 − VM) kg of CO.
The heat value of the mixture containing volatiles and is CO given as HV_{VM} × VM + HV_{CO} × (1 − VM) × (28/12), where HV_{CO} = 10103 kJ/kg of CO.
{HV_{fuel} − (1 − VM) × HV_{FC}} + HV_{CO} × (1 − VM) × (28/12) = {HV_{fuel} − (1 − VM) × {HV_{FC} − HV_{CO} × (28/12)}.
Then, the gross HV per unit mass combustible gas produced from the gasifier is given as $\{HV\}_{gas}$ = {HV_{fuel} − FC × {HV_{FC} − HV_{CO} × (28/12)}/1 + (1/$Y_{O2,in}$) × (16/12)FC} (in units of kJ/kg of gas) where FC = 1 − VM.
Simplifying with $Y_{O2,in}$ = 0.23 for air as oxidant stream and noting that HV and FC are on DAF basis, $\{HV\}_{gas,max}$ = {$HV_{fuel,DAF}$ − FC_{DAF} × 9196}/{1 + 5.8 × FC_{DAF}} (in units of kJ/kg).
For coal, HV_{fuel} = 29786 kJ/kg, FC = 0.533, $Y_{O2,in}$ = 0.23, and HV_{gas} = 6090 kJ/kg, which is about 11% of HV of CH_4.

Example 7

The mass-based composition of Wyoming low-sulfur coal is as follows: moisture: 10.65%, ash: 5.7%, C: 54.9%, H: 4.33%, N: 0.76%, O: 23.32%, and S: 0.34%. A proximate analysis provides the following values: VM: 30.72%, FC: 52.8%, and HHV: 11410 Btu/lb. Determine the DAF composition and heating value.

Solution

The DAF carbon fraction is 54.9 × 100/(100 – ash% – moisture%) = 54.9 × 100/(100 – 10.65 – 5.7) = 65.63%. Hence, C: 65.6%, H: 5.17%, N: 0.9%, O: 27.9%, S: 0.4%, VM: 36.7%, FC: 63.1%, and the heating value is 13640.1 Btu/lb.

Remarks

Typically, 95% of S is converted into SOx, while the rest goes as sulfates in ash; about 30% of N in fuel is converted into NOx called fuel NO. The NO released is usually converted into NO_2 in the atmosphere. Hence, for reporting NO emission, the molecular weight of NO_2 is used. With 30% N conversion, NO will be 0.76 × 0.3 × 46.02/14.01 = 0.75 kg per 100 kg of fuel.

4.5 Other Fuels

4.5.1 Industrial Gaseous Fuels

These include the following:

Blast furnace gas: a low-Btu gas produced from blast furnaces
Catalytically cracked gas: gas produced from cracking natural gas or HC using nickel oxide catalysts
Coal gas: gas mixture consisting of pyrolysate and water gas (CO and H_2) produced by steam carbon reaction

Coke oven gas: the pyrolysate from coal when coke is made from coal
H_2 gas: produced by reforming natural gas by a reaction with steam using a catalyst (1100–1200 K); e.g., $CH_4 + H_2O \rightarrow H_2$, CO, CO_2 and shift reaction $CO + H_2O \rightarrow CO_2 + H_2$. The CO_2 is removed by scrubbing.
Natural gas (crude): natural gas as obtained from wells consisting of He, H_2S, etc.
Oil gas: made from cracking of oils with medium-to-high-Btu gas, has HHV 550 to 1000 Btu/ft^3
Producer gas: mainly a mixture of CO, H_2, N_2 (<55%), and small amount of CO_2, gases produced by passing air (< stoichiometric) over coal and biomass bed
Refinery gas: gas produced as a by-product from refineries with HHV 1400 to 2000 Btu/ft^3
Reformed natural gas: gas obtained by thermal cracking of natural gas or other petroleum gases using water gas generator
Synthesis gas: made from various processes in coal, coke, or HC, has $H_2/CO \approx 2$
Water gas: gas made by blowing a mix of air and steam for the reaction $C + H_2O \rightarrow CO + H_2$, is better than producer gas because H_2 production is enhanced by steam (See Bartok and Sarofim for details.)

4.5.2 Synthetic Liquids

Coal has an average H/C ratio of 0.75, and liquids derived directly from coal by liquefaction and separation also have the same, relatively low, H/C ratio. The conversion of a low-H/C-ratio liquid fuel to that with a larger value of, say, 2 as in petroleum, requires hydrogen addition through hydrogenation processes. About 1780 m^3 of gaseous hydrogen is required to produce 1 m^3 of petroleum. This hydrogen can be produced from coal or, alternatively, electrochemically from other power sources. An energy balance of the overall process of producing a liquid fuel from coal that has similar chemical properties to petroleum shows that the combination of hydrogenation, conversion to a lower molecular weight, and hydrocracking results in an overall efficiency of 43% based on the amount of hydrogen used.

Methanol is produced from CO and H_2 using the overall reaction $CO + 2H_2(g) = CH_3OH(g)$, steam reforming reactions such as $CH_4 + H_2O(g) = CO(g) + 3H_2(g)$, and $C_3H_8 + 3H_2O = 3CO + 7H_2(g)$ are employed in the presence of catalysts at 425°C. The product mixture $CO + 3H_2$ may be called a *synthesis gas*, which is used to produce methanol. More hydrogenation is possible with $CO + H_2O = CO_2 + H_2$, which is called the *shift reaction* (typically at T = 350°C) and is also exothermic. The shift reaction is used in commercial production of CO_2 or dry ice. The reaction $C(s) + H_2O = CO + H_2$ is also used for H_2 production. Synthetic diesel is produced by gasifying wood, corn, etc., and using the Fisher-Tropsch process.

4.5.3 Biomass

Biomass is defined as "any organic material from living organisms that contain stored sunlight in the form of chemical energy," [US DOE, www.eere.energy.gov/

Table 4.7 Projected 2010 Total Energy Consumption and Biomass Energy Consumption (BEC)

	Total Energy Consumption[a]	*Biomass Energy Consumption*[a]	*Biomass % Total*
Industrial	38.9	3.2	8.2
Electric Generation	48.3	0.2	0.4
Subtotal power	87.2	3.4	3.9
Transportation	39.5	0.28	0.7

[a]Energy in quadrillion Btu/yr (10^{15} Btu/yr)

Source: DOE/EIA, Annual Energy Outlook, 2002.

biopower] which include agrobased (vegetation, trees, and plants) and waste biomass from municipal solid waste, animal waste, etc. Biomass now supplies 3% of the U.S. energy (see Table 4.7 for projected biomass consumption) and it could be as high as 20% in the future. The U.S. alone produces 1 Gt annually (which is the largest tonnage in the world), of which it consumes 0.8 Gt for power generation. Biomass resources can be estimated to be 14 quads (55 quads if energy crops are included) in U.S.; but only 3 quads are used, whereas U.S. energy consumption is 84 quads. Biomass can be woody (soft and hard wood, poplar, bark, and saw dust) and nonwoody (switchgrass, straw, bagasse, husks, manure, etc.). Bark contains more ash (1 to 3%) than wood (0.2 to 1%). Charcoal is made from wood in the absence of air. Drywood is made of cellulose ($C_6H_{10}O_5$, 40 to 50%), hemicellulose lignin (20 to 35%), resins ($C_{40}H_{44}O_6$, 15 to 30%), etc. The ultimate analyses and heating values of several woody biomasses are presented in Table A.2D. Table 4.8 shows analyses on animal wastes.

Biomass fuels such as agricultural crops (sugar cane, 83–90% sugar; 10–17% cellulose on DAF; and bagasse, which is a wasteful product after extracting sugar from sugar cane), biowastes (municipal solid waste [MSW], feedlot manure [nitrogen in manure exists mostly in the form of urea from 0.82 to 1.29 lb N per mm Btu and chlorine 2.73 lb per mm Btu], and chicken litter), etc., are renewable fuels and provide an alternative energy source. Biowastes can undergo anaerobic digestion to produce 50% CH_4 and 50% CO_2 gases. About 185 Mt of MSW and 200 million waste tires are generated per yr. About 25% of these can be recycled. The MSW that can be used for power generation (about 11,000 MW) amounts to 140 Mt per yr, equivalent to 58 Mt of coal obtained either through synthetic solid, liquid, and gaseous fuels or direct firing [McGowin and Highes]. The fuels derived from solid refuse (via shredding, screening, and separation of metals by magnets) is called refuse-derived fuel (RDF). RDF can also be palletized or pulverized.

Biomass can be pyrolyzed and gasified to yield low-Btu gases, or they can also be converted to transportation fuels like biodiesel (from soybeans) methanol, ethanol, etc.; they may be either used alone or blended with gasoline.

Table 4.8 Coal, Feedlot Biomass (FB), and Litter Bimass (LB)

Parameter	Wyoming Coal	Cattle Manure FB	Chicken Manure LB[a]
Dry Loss	22.8	6.8	7.5
Ash	5.4	42.3	43.8
FC	37.25	40.4	8.4
VM	34.5	10.5	40.3
C	54.1	23.9	39.1
H	3.4	3.6	6.7
N	0.81	2.3	4.7
O	13.1	20.3	48.3
S	0.39	0.9	1.2
Cl	<0.01%	1.2	
HHV as-received	21385 kJ/kg	9560 kJ/kg	9250
$T_{adiab,Equil}$[b]	2200 K (3500°F)	2012 K (3161°F)	
DAF formula	$CH_{0.76}O_{0.18}N_{0.013}S_{0.0027}$	$CH_{1.78}O_{.64}N_{.083}S_{.014}$	$CH_{2.04}O_{0.93}N_{0.10}S_{0.012}$
HHV-DAF	29785	18785	18995

[a] From Priyadarsan, S., Annamalai, K., Sweeten, J.M., Holtzapple, M., and Mukhtar, S., Co-gasification of Blended Coal with Feedlot and Chicken Litter Biomass, 30th International Symposium on Combustion, 2004.

[b] Equilibrium temperature for stoichiometric mixture, from THERMOLAB spreadsheet software (available from CRC Web site) for any given fuel of known composition. (From Annamalai, K. and Puri, I., *Advanced Thermodynamics Engineering*, CRC Press, Boca Raton, FL, 2001.)

Biomass fuels (switchgrass, willow, energy cane, and corn stoves) can also be burned directly either alone or cofired with coal. Many biomass fuels contain alkali metal species (Na, K, and Ca); products of these species *viz.*, metal chlorides and silicates cause fouling problems (forming deposits, causing corrosion and erosion) in heat exchangers and reduce the heat transfer rates. When biomass is cofired with coal, fouling problem can become more severe. Some cofiring can reduce NOx because of smaller fuel nitrogen (Chapter 5).

Example 8

Anaerobic digestion involves microbial action on the organic components of animal waste or sludge, which results in release of biogas, containing a mixture of CH_4 and CO_2. If the chemical formulae for DAF animal waste is given as $CH_hO_oN_nS_s$,

(a) Obtain an expression for the number of kmol of CH_4 and CO_2 per kmol of DAF and the fraction of CH_4 in the biogas.
(b) If h = 1.52, o = 0.6, n = 0.086, and s = 0.0084 for dry animal waste, determine the fraction of CH_4 in the biogas.
(c) If h = 1.98, o = 0.83, n = 0.086, and s = 0.0084 for wet animal waste, determine the fraction of CH_4 in the biogas.

(d) Ignoring the presence of N and S in the solid waste after digestion, estimate heat of reaction if the heating value of DAF fuel is 20532 kJ/kg.
(e) Obtain the volume of biogas and CH_4 produced in SCM (assume T = 15°C and P = 1 bar) per kg of fuel input and the heating value of biogas in kJ/m³ of mixture.

Solution

(a) $CH_hO_oN_nS_s$ (s) + N_{H2O} $H_2O \rightarrow N_{CH4}$ $CH_4(g)$ + N_{CO2} CO_2 (g) + N_nS_s (s) (A.1)

Using atom balance for C, H, and O,

$$C\text{: } 1 = N_{CH4} + N_{CO2} \quad (A.2)$$

$$H\text{: } h + 2\, N_{H2O} = 4N_{CH4} \quad (B)$$

$$O\text{: } o + N_{H2O} = 2\, N_{CO2} \quad (C)$$

From Equation C, $N_{CO2} = \{o + N_{H2O}\}/2$ (D)

From Equation B, $N_{CH4} = \{h + 2N_{H2O}\}/4$ (E)

Using Equation D and Equation E in Equation A.2,

$$1 = N_{CH4} + N_{CO2} = \{h + 2\, N_{H2O}\}/4 + \{o + N_{H2O}\}/2 \quad (F)$$

Solving for N_{H2O},

$$8 - 2\,h - 4\,o = 8N_{H2O},\ N_{H2O} = \{8 - 2h - 4\,o\}/8 = 1 - h/4 - o/2 \quad (G)$$

Using Equation D and Equation E,

$$N_{CO2} = \{o + (8 - 2\,h - 4\,o)/8\}/2 = \{8o + 8 - 2\,h - 4\,o\}/16$$
$$= \{8 + 4o - 2\,h\}/16 \quad (H)$$

$$N_{CH4} = \{h + 2\,(8 - 2\,h - 4\,o)/8\}/4 = \{8h + 16 - 4\,h - 8\,o)\}/32$$
$$= \{4 + h - 2o\}/8 \quad (I)$$

Using Equation A.1

$$CH_hO_oN_nS_s\,(s) + \{1 - h/4 - o/2\}H_2O \rightarrow \{1/2 + h/8 - o/4\}CH_4 + \{1/2 + o/4 - h/8\}CO_2 + N_nS_s(s) \quad (J)$$

$$X_{CH4} = \{1/2 + h/8 - o/4\} \quad (K)$$

(b) $X_{CH4} = 0.5 + 1.52/8 - 0.6/4 = 0.54$, i.e., a theoretical maximum of 54%; if o = 0, then $X_{CH4} = 0.69$.

$N_{H2O} = N_{H2O} = 1 - h/4 - o/2 = 1 - 1.53/4 - 0.6/2 = 1 - 0.38 - 0.3 = 0.32$.

$$CH_{1.53}O_{0.6}N_{0.086}S_{0.0084}\,(s) + 0.32H_2O \rightarrow 0.54CH_4(g) + 0.46CO_2(g) + N_{0.086}S_{0.0084}(s) \quad (L)$$

(c) $X_{CH4,max} = 0.54$
$N_{H2O} = 1 - h/4 - o/2 = 1 - 1.98/4 - 0.83/2 = 1 - 0.0.495 - 0.41 = 0.0.09$.
Here, less water is required because water is already present in manure.
$CH_{1.98}O_{0.83}N_{0.086}S_{0.0084}$ (s) + $0.09H_2O \rightarrow 0.54CH_4(g) + 0.46CO_2(g) + N_{0.086}$ $S_{0.0084}(s)$ (M)
This is the same result as before except that external water required is reduced.

(d) Heat of reaction: $\{1/2 + h/8 - o/4\}\ h_{f,\ CH4} + \{1/2 + o/4 - h/8\}\ h_{f,\ CO2} + hf_{\ NnSs\ (s)}$ $-\ hf_{\ CH1.98O0.83N0.086S0.0084\ (s)}$
Using Chapter 2, $h_{f\ CH1.98O0.83N0.086S0.0084\ (s)} = -106,357$ kJ/kmol of fuel.
$\Delta H_R^0 = \{1/2 + h/8 - o/4\}\ h_{f,CH4} + \{1/2 + o/4 - h/8\}\ h_{f,\ CO2} + h_{f,NnSs\ (s)}$ $-\ h_{f,CH1.98O0.83N0.086S0.0084\ (s)} = 0.541 \times (-74850) + 0.459 \times (-393546) - (-106357)$ $-\ 0.318 \times (-285830) = -23944$ kJ/kmol fuel (exothermic).
Molecular weight of fuel = 24.64 kg/kmol of fuel.
$\Delta H_R^0 = -23944/24.64 = -972$ kJ/kg of fuel (exothermic).
Anaerobic digestion is exothermic.

(e) Standard volume at 15°C and 1 bar is 23.7 m³/kmol.
Maximum gas produced per kmol of DAF fuel = 23.7 m³.
Maximum gas produced per kg of DAF fuel = 23.7/24.64 = 0.962 m³/kg of DAF fuel.
Maximum CH_4 produced = 0.962 × 0.541 = 0.52 m³ of CH4 per kg of DAF fuel.
Heat value of biogas = 0.541 × (−74850) − 0.541 × (−393546) − 0.541 × 2 × (−285830) = 481905 kJ/kmol of gas mixture. Heat value of biogas = 481905 kJ/kmol of gas mixture/(23.7 m³/kmol mixture) = 20334 kJ/m³ of mixture or 529 Btu/ft³ of mixture.

■ ■ ■

The arguments in favor of use of biomass energy are as follows. Unlike other nonrenewable forms of energy, biomass energy can be produced and consumed in a sustainable fashion, and there is no net contribution of carbon dioxide (which is a greenhouse gas) to global warming. In an ideal closed-loop system, carbon dioxide would be absorbed by new plant at the same rate at which it is released by harvested biomass used as fuel. Such bioenergy crops would make a negligible, if any, contribution to atmospheric carbon dioxide, in contrast to fossil fuels (which consist of carbon that has been geologically stored for millions of years) that make a net contribution to atmospheric greenhouse gases.

Oxygenating additives such as methyl tertiary butyl ether [Gayle, 2003] are used to raise the octane rating of gasoline. But these were banned after 2004. Ethanol will, therefore, find more use as additive. Further, ethanol is blended with denaturant so it cannot be consumed as an alcohol. The chemical industry used ethylene as feedstock for ethanol. Ethanol fuel can also be derived using sugars from biomass. Corn starch (plant fibers $(C_6H_{10}O_5)_n$) is used to produce ethanol by feeding starch into digesters and adding yeasts. Almost 1.5 billion gallons of ethanol are produced from corn every year; it can also be blended with gasoline. Switch grass and wood chips can also be used as feedstock in the fermentation process.

$C_{12}H_{22}O_{11} + H_2O$ with enzyme invertase $\rightarrow C_6H_{12}O_6$ (glucose also called dextrose) + $C_6H_{12}O_6$ (fructose also called levulose), where glucose and fructose are called invert sugars.

$$C_6H_{12}O_6 \text{ with enzyme zymase} \rightarrow 2CH_3CH_2OH \text{ (ethanol)} + 2CO_2$$

In dry milling, one uses enzymes to convert the starch in corn into sugar and then adds yeast to ferment sugar. It is noted that for starch sources [Keating, 1993].

$$2(C_6H_{10}O_5)_n + H_2O \text{ with malt enzyme diastase} \rightarrow C_{12}H_{22}O_{11} \text{ (maltose sugar)},$$

$$C_{12}H_{22}O_{11} \text{ (maltose sugar) with yeast enzyme} \rightarrow C_6H_{12}O_6 \text{ (dextrose)}.$$

During this process, one obtains ethanol and CO_2. The nonfermentable part is used for other applications. New processes use new enzymes to convert cellulose and hemicellulose (in wood) into sugar; leftover lignin is burned to produce steam for generating electrical power.

Currently, ethanol:gasoline blends are marketed: E10 means 10% ethanol in gasoline, whereas E-diesel is a blend of 10 to 15% ethanol, 80% diesel, and 5% additive for keeping it as a blended fuel. However, one gallon of ethanol has 28% less energy than 1 gallon of gasoline. The 2003 Ford Taurus and DaimlerChrysler cars can run on gasoline or E85 blends (i.e., 85% of volume is due to ethanol). A car running on E10 gasoline emits 30% less CO, 12% less VOC (volatile organic compounds), 30% less benzene and toluene, and 3% less NOx. Similarly, methanol can be made from biomass through gasification into synthesis gas; however, most methanol (1.2 billion gallons/yr) is made from natural gas and used as a solvent, an antifreeze, etc. Only about 38% is used as blended fuel with gasoline

4.5.4 Municipal Solid Waste (MSW)

Municipal refuse includes residential, commercial, and industrial wastes, which could be as used as fuel for the production of steam and, hence, electric power. About 20,000 t/d is disposed in a city of 5 million people. Major components of typical MSW: paper 43%, yard waste including grass clippings 10%, food: 10%, glass or ceramics 9%, ferrous materials 6%, and plastics and rubber 5%. Typically MSW is fired with 80–100% excess air as opposed to 16–20% excess air used in pf fired burners. When the waste is processed to remove glass, metals, etc., it is called RDF (refused derived fuel). These can decompose in two ways: aerobic and anaerobic (without air). Aerobic decomposition (also called composting) occurs when O_2 is present. Composting produces CO and water but no usable energy products. Anaerobic decomposition occurs in the absence of O_2. It produces landfill gas of CH_4 (55%) and CO_2 (45%) [Solid Waste and Power, 1993]. Gas from perimeter wells contain less than 4% methane, whereas those at center contain 50 to 60% methane. Sometimes natural gas is used along with landfill gas because of the fluctuating quality of landfill gas. Note that commercial pipeline gas has more than 95% CH_4. Over 20 to 30 yr, 20 to 60% of landfill is converted into landfill gas by anaerobic decomposition process. One kilogram of refuse with a HHV of 11600 kJ (5000 Btu/lb or 9700 Btu/ DAF lb) will generate 0.029 to 0.087 m^3 (1 to 3 ft^3) of CH_4. Burning landfill gas (LFG) is similar to burning dilute natural gas. Landfill gas has a HV of 11100 to 20350 kJ per m^3 (300 to 550 Btu per ft^3, which is about 50% of that of natural gas; cost of electricity is 4.5 to 5 cents per kWh). Greater the penetration of air into landfill, greater is the amount of CO_2. Normally air can penetrate 6 to 12 m (20 to 40 ft) into land refill. The typical production rate is on the order of 210,00 ft^3/h (HV = 430 Btu per SCF) [Distributed energy, Jan 2005, pp 30–35]. Thus, shallow landfills are not recommended. Gas normally escapes at 100 to 110°F with 3 to 7% moisture. Gas is used as boiler fuel in the Rankine cycle for producing steam. More than 100 power plants that burn LFG are distributed in 31 states.

4.6 Size Distributions of Liquid and Solid Fuels

The combustion intensity of a gas burner is essentially controlled by the molecular mixing rate of fuel with oxygen. However, many practical combustor systems such as boilers, gas turbines, diesel engines, rockets, etc., use condensate fuels (liquid and solid fuels) as the energy source. The condensate must be atomized, or pulverized, into smaller droplets, or particles, to increase the surface area per unit volume of the combustor space in order to facilitate rapid heating, gasification, and mixing with the oxygen-rich ambience. Thus, atomization leads to improved ignition and combustion characteristics. A liquid fuel is atomized into finer droplets with nonuniform sizes and different velocities, and is usually introduced into combustors in the form of a spray. Atomization is a process that transforms a liquid into a dispersion of small droplets by a "shattering" process between the liquid fuel and the surrounding gas. This process is employed in several industrial processes and has many applications in agriculture, meteorology, and medicine. A liquid when sprayed is stretched to a thin sheet because of the drag, then it ripples because of surface tension forces and breaks into ligaments; the ligaments form drops, which further break up. The drag on a larger drop shears the drops; if the drag exceeds the surface tension forces holding them together, then drop breaks into smaller drops. The Weber number (We) characterizes the break up.

$$\mathrm{We} = \rho D V^2/\sigma = \text{disruption drag force/holding surface tension force},$$

where ρ is the density; σ, the surface tension; and $V = \{V_{fuel} - V_{air}\}$. If We >10, the drop breaks up into smaller drops. Thus, drop size depends on injection velocity or Re, and surface tension forces. For breakup in absence of ambient gas, $Z = (\mathrm{We}^{1/2}/\mathrm{Re}) = (\mu/(\rho\sigma\, d_o)^{1/2})$, where d_o atomizer dia the Z. The Z and We provide a measurement of quality for atomization. The fuel injectors consist of nozzles with multiple holes (4 to 10), which could be of pressure type to create high velocities and to make thin sheets of liquids.

Twin-blast and air-blast atomizers (Figure 4.14a and Figure 4.14b) are used when the penetration of jet is large. Prefill atomizers (another form of air-blast atomization; Figure 4.15) and spill-type swirl atomizers (Figure 4.16) are used when the penetration is small. In swirl atomizers, fuel enters tangentially and exits as a spray; the fuel flow is controlled by the spill return. The pressure of

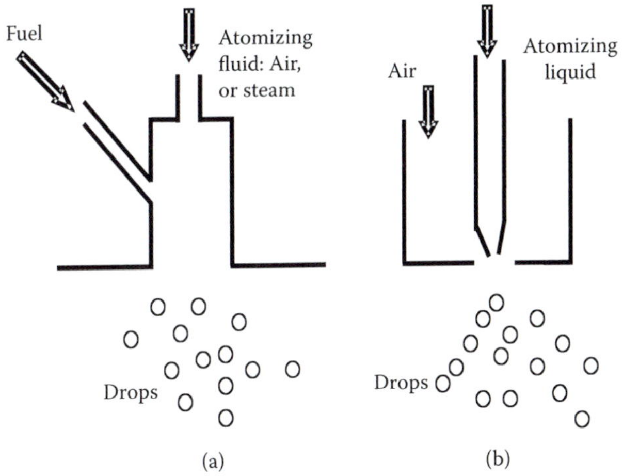

Figure 4.14 (a) Twin-blast calorimeter, (b) Air-blast calorimeter.

injection and atomization characteristics (e.g., spray angle) may remain the same for all flows. For air-blast atomizers, there is less smoke because air is almost premixed with fuel. For gas turbine air blast, there can be a plain jet or a prefilming jet in which liquid is spread out into a continuous sheet and then exposed to high-velocity air. Wider spray angles are preferred if there is a high degree of air movement around the spray (e.g., wall-fired boiler burners). A narrow angle is preferred if air movement is restricted (e.g., diesel engines).

The vapors released from drops may or may not surround the drop depending on relative Reynolds number. Further, the local A:F ratio may vary from rich conditions near a burner to lean mixtures far from the burner, depending on the extent of mixing. Rich mixtures lead to soot and unburned HC and CO formation.

Solid fuels are typically pulverized to finer ones in suspension-fired boilers prior to firing. When $d_p < 10\ \mu m$, the coal is called ultrafine and can be used as fuel in gas turbines provided the ash content is reduced below 1%. The Hardgrove grindability (HG) test (ASTM D409) is used to determine the grindability

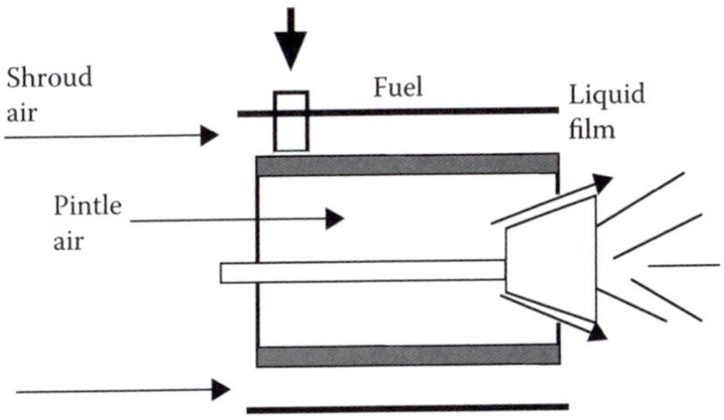

Figure 4.15 Prefill atomizer.

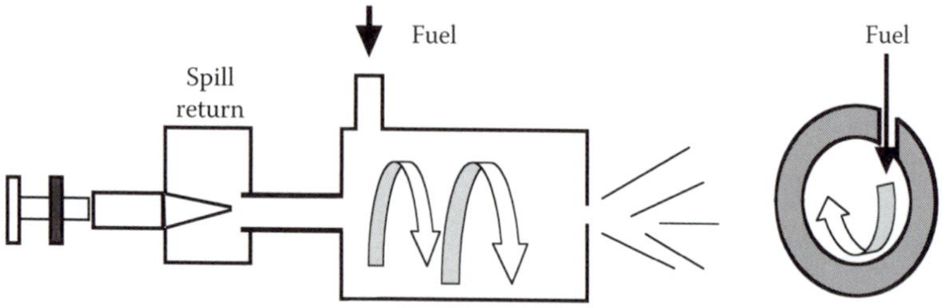

Figure 4.16 Spill-type swirl atomizer.

of coal. Fifty grams of sample is loaded into a bowl containing steel balls each of 2.5-cm diameter and is driven by a spindle at 20 rpm for over 3 min exerting a force of 285 N. The amount of coal in grams passing through 20 mesh is called the *HG index*.

4.6.1 Size Distribution

The experimental measurements for drop size distribution include local number density (called *spatial distribution*) and liquid number flux (called *temporal distribution*). Consider a spray having smaller and larger drops of diameters d_s and d_L, respectively. If local number density of smaller drops with velocities v_s is n_s and that of larger drops with velocities v_L is n_L, then the number fluxes at a given surface are $\dot{N}_s = n_s v_s$ and $\dot{N}_L = n_L v_L$. Thus, measurement of number flux alone does not yield information on local number density unless velocity is measured simultaneously. Spray characteristics are obtained using phase doppler analyzer, which measures drop size and liquid volume fluxes [Dodge and Schwalb, 1988]. From drop size and liquid volume fluxes, $\dot{V}_s = (\pi\, d_s^3/6)\, n_s v_s$, and $\dot{V}_L = (\pi\, d_L^3/6)\, n_L v_L$, the number density and spacing may be obtained.

4.6.1.1 Lognormal Distribution

For sprays, drop size distribution is typically represented by a number fraction N_F. Let dN_F represent the fraction of the drop having sizes between d_p and $d + dd_p$. A lognormal size distribution can be expressed in the form

$$dN_F/dd_p = f = \frac{1}{\sqrt{2\pi}\, d_p \sigma_d} \exp\left\{ -\frac{1}{2\sigma_d^2} \left(\ln \frac{d_p}{\bar{d}_p} \right)^2 \right\} \tag{4.17}$$

where d_p is the particle or drop size; $\bar{d}_p$, the mean drop size; and σ_d, the geometric standard deviation. For a lognormal distribution, the mean could be $\bar{d}_p = \bar{d}_{p,n}$, AMD (area mean diameter), and VMD (volume mean diameter) as defined in the following text.

The mean diameters (mean, AMD, and VMD) are defined as

$$\bar{d}_{p,n} = \left\{ \int_0^\infty f\, d_p \, dd_p \right\}, \tag{4.18a}$$

$$AMD = \left[\int_0^\infty f\, d_p^{\,2} \, dd_p \right]^{1/2}, \tag{4.18b}$$

$$VMD = \left[\int_0^\infty f\, d_p^{\,3} \, dd_p \right]^{1/3} \tag{4.18c}$$

If tabulated data are available for number distribution, say, $\Delta N_{F,I}$ (number fraction) for size between d_p and dd_p,

$$\bar{d}_{p,n} = \left\{ \sum_i \Delta N_{F,i}\, d_{p,i} \right\}, \tag{4.19a}$$

$$AMD = \left\{ \sum_i \Delta N_{F,i}\, d_{p,i}^{\,2} \right\}^{1/2}, \tag{4.19b}$$

$$VMD = \left\{ \sum_i \Delta N_{F,i}\, d_{p,i}^{\,3} \right\}^{1/3} \tag{4.19c}$$

VMD represents a diameter that provides the average volume per drop (but the corresponding area does not provide the total surface area). In general, mean diameter can be expressed as

$$(d_{ab})^{a-b} = \frac{\int_{d_o}^{d_m} d^a (dN_F/dd_p) dd_p}{\int_{d_o}^{d_m} d_p^{\,b} (dN_F/dd_p) dd_p} = \frac{\sum_i d_{p,i}^{\,a-3} \Delta N_{F,i}}{\sum_i d_{p,i}^{\,b} \Delta N_{F,i}} = \frac{\sum_i d_{p,i}^{\,a-3} \Delta m_{F,i}}{\sum_i d_{p,i}^{\,b-3} \Delta m_{F,i}}, \tag{4.20}$$

where a and b can take any values corresponding to the effect investigated (Table 4.9). The sum (a + b) is called the *order of the mean diameter*, and $\Delta m_{F,I}$ is the mass fraction of practices of sizes between dp and ddp. Other choices of representative diameter can be made, i.e., $d_{0.1}$ may represent a drop diameter for a case in which 10% of the total liquid volume is contained in drops diameters less than $d_{0.1}$; similarly, $d_{0.5}$ or $d_{0.99}$.

The *Sauter mean diameter* (SMD) is defined as

$$SMD = \frac{\left[\int_0^\infty f\, d^3 \, d(d_p) \right]}{\left[\int_0^\infty f\, d^2 \, d(d_p) \right]} = \frac{\left\{ \sum_i \Delta N_{F,i}\, d_p^{\,3} \right\}}{\left\{ \sum_i \Delta N_{F,i}\, d_p^{\,2} \right\}} = \frac{1}{\sum \frac{\Delta m_{F,i}}{d_{pi}}} \tag{4.21}$$

Table 4.9 Mean Diameter and Their Applications

a	*b*	*Symbol*	*Name of mean diameter*	*Expression*[a]	*Application*
1	0	d_{10}	Length	$\frac{\sum N_i d_{pi}}{\sum N_i}$	Comparisons
2	0	d_{20}	Surface area	$\left(\frac{\sum N_i d_{pi}^2}{\sum N_i}\right)^{1/2}$	Surface area controlling, also known as average mean diameter (AMD)
3	0	d_{30}	Volume	$\left(\frac{\sum N_i d_{pi}^3}{\sum N_i}\right)^{1/3}$	Volume controlling also known as volume mean diameter (VMD)
2	1	d_{21}	Surface area length	$\frac{\sum N_i d_{pi}^2}{\sum N_i d_{pi}}$	Absorption
3	1	d_{31}	Volume length	$\left(\frac{\sum N_i d_{pi}^3}{\sum N_i d_{pi}}\right)^{1/2}$	Evaporation, molecular diffusion
3	2	d_{32}	Sauter (SMD)	$\frac{\sum N_i d_{pi}^3}{\sum N_i d_{pi}^2}$	Mass transfer, reaction
4	3	d_{43}	De Brouckere or Herdan	$\frac{\sum N_i d_{pi}^4}{\sum N_i d_{pi}^3}$	Combustion equilibrium

[a] d_p = d diameter of particle or drop.

which is the volume-to-surface-area ratio for all droplets. Generally, in problems on combustion, SMD is widely used. In Chapter 10, we will see that the evaporation and combustion rates of sprays are controlled by drop size, and SMD may be used to estimate spray evaporation time. The most probable diameter (MPD) is the one for which f is a maximum. The relation between the various mean diameters can be expressed in terms of the number geometric mean diameter d_p, i.e.,

$$\ln(AMD) = \ln \bar{d}_{p,n} + 2\sigma_d^2 \quad \text{and} \quad \ln(VMD) = \ln \bar{d}_p + 3\sigma_d^2. \tag{4.22}$$

For any of the aforementioned distributions, the fraction of drops (or particles) below a size d_p is

$$F = \int_0^{d_p} f \, dd_p$$

The cumulative volume fraction (CVF) of drops (or particles) with a dimension lesser than d_p is

$$CVF = \frac{\int_0^{d_p} f \, d_p^3 \, dd_p}{\int_0^{\infty} f \, d_p^3 \, dd_p}$$

In the next section an expression for CVF will be obtained. Note that CVF is the same as cumulative mass production if d_p is the same for all drops/particles.

4.6.1.2 Rosin–Rammler Relation

This distribution is used widely for pulverized solid and liquid fuels. Note that typically for solid fuels, the particles, particularly biomass, may not be spherical. Biomass particles have aspect ratios ranging from 3 to 12.

The fraction of drops (or particles) below a dimension x is

$$F(x, \alpha, \beta) = 1 - \exp-\{(x/\beta)^{\alpha}\} \tag{4.23}$$

where α is a measure of the system uniformity, and β represents a characteristic dimension. Therefore,

$$dF/dx = (\alpha/\beta^{\alpha})\ x^{(\alpha-1)} \exp-\{(x/\beta)^{\alpha}\}. \tag{4.24}$$

Denoting $1/\beta^{\alpha}$ by b, α by n, and the particle size x by d_p, all three of which are generally constant for a system, one form of the Rosin–Rammler relation is obtained as

$$d(Y_p)/d(d_p) = \{\text{Fractional mass with size } d_p \text{ and } d_p + d(d_p)\}/\{d(d_p)\}$$

$$d(Y_p)/d(d_p) = n\ b\ d_p^{n-1} \exp(-bd_p^{n}). \tag{4.25}$$

Then,

$$d(Y_p) = \int \exp(-x)\, dx, \quad \text{where } x = bd_p^{n}. \tag{4.26}$$

Integrating from x = 0 to x (with $Y_p = 0$ at x = 0), CVF of drops (or particles) with a dimension lesser than d_p is

$$CVF = CMF = (1 - \exp(-bd_p^{\,n})), \tag{4.27}$$

where b is the size constant and n the distribution constant (2 < n < 5).

In terms of the $d_{p,Charac}$ size,

$$CVF = 1 - \exp\{-(d_p/d_{p,Charac})^n\} \tag{4.28}$$

where $d_{p,Charac}$ denotes the characteristic drop or particle size for which CVF = 1 − exp(−1) = 63.2%, and n is a measure of spread of the drop sizes. Higher the value of n, greater is the uniformity of drop size. The fraction R of drops having sizes greater than d_p is

$$R = 1 - CVF = 1 - CMF = \exp\{-(d_p/d_{p,Charac})^n\}. \tag{4.29}$$

The plot of ln R versus d_p must be linear if n = 1.

CVF and the probability density function for these values is presented in Figure 4.17 for n = 4.16 and b = 0.001634 (typical for coal-fired plants). Typically, for coal-fired boilers, about 70% of mass passes through 75-μm (200 mesh) sieves.

The Nukiyama–Tanasawa relation is based on a number of particles and can be directly measured from a sample of the drops or particles. It has the form

$$d(N)/d(d_p) = a d_p^{\alpha} \exp(-b d_p^{\beta}),$$

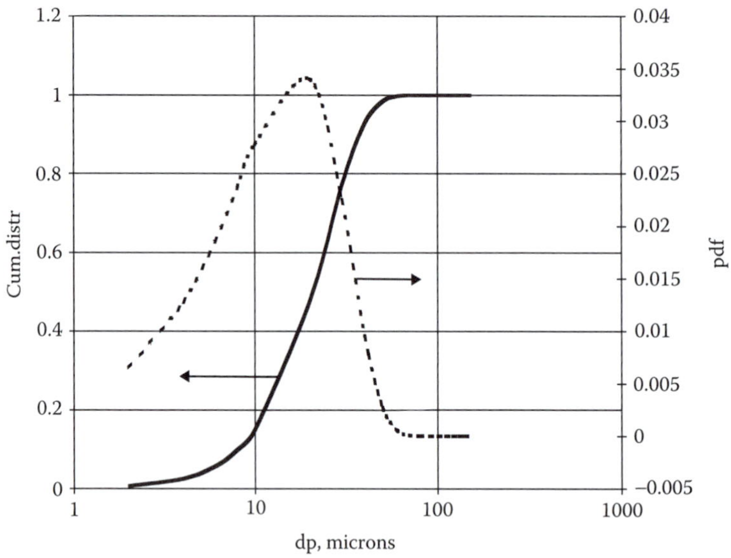

Figure 4.17 Cumulative particle distribution and probability density function for characteristic power plant coal particles.

Table 4.10
Characteristic Values of α and β for Various Flows

Flow	α	β
Swirl	5	1
Impinging jet	1	1
Pressurized Jet	2	1
Twin fluid jet	1	1

where α, a, and β are constants (Table 4.10), and dN represents the number of drops or particles within a range of dimension d_p and $(d_p + d(d_p))$.

4.6.2 Some Empirical Relations

For prefilming air-blast atomizers used in gas turbines, [Lefevbre, 1989]:

$$SMD = 0.073\left(\frac{\sigma_\ell}{u_{rel}^{\;2}\rho_a}\right)^{0.6}\left(\frac{\rho_\ell}{\rho_a}\right)^{0.1} D_p^{0.4}\left(1+\frac{\dot m_\ell}{\dot m_a}\right)+0.015\left(\frac{\mu_\ell^{\;2} D_p}{\sigma_\ell \rho_\ell}\right)^{0.5}\left(1+\frac{\dot m_\ell}{\dot m_a}\right)$$

where ν is the kinematic viscosity in m^2/sec; the mass flow is in kg/sec, a stands for air, ℓ for liquid, and h is the height of annulus in mm. The SMD (or d_{32}) for diesel fuel spray is empirically given by various authors as:

1. By Haywood [1998]: SMD (in μm) = A (ΔP in MPa)$^{-0.135}$ (ρ_a in kg/m^3) $^{0.121}$ (V_f in mm^3)$^{0.131}$
   where ΔP is the pressure drop; ρ_a, the air density; V_f, the amount of fuel injected per stroke; and A = 25.1 for pintle nozzles and 23.9 for hole nozzles, and 22.4 for throttling pintle nozzles.
2. By Hiroyasu and Kadota [1974]: SMD (μm) = 2.33 × 10^3 (ΔP)$^{-0.135}$ $\rho_a^{0.121}$ {V}$^{0.131}$ (Pa, kg m^2 sec)
   where V is the injection fluid volume; for diesel sprays using heavy duty diesels, typical diameters of openings is 0.3 mm, ΔP = 36 MPa, ρ_l = 850 kg/m^3, and σ = 0.03 N/m.
   [Hiroyasu et al. 1989]

$$\frac{d_{32}}{d_{\text{nozzle}}} = \mathrm{Re}_i^m\ \mathrm{We}_i^n\left(\frac{\mu_i}{\mu_a}\right)^{p}\left(\frac{\rho_\ell}{\rho_a}\right)^{q}$$

3. By Knight [1955]: SMD (μm) = 1.605 × 10^6 (ΔP)$^{-0.458}$ ($\dot m$)$^{0.209}$ $\nu^{0.215}$ {A/A(t)}$^{0.916}$ (Pa kg m^2 sec)
   where A(t) is the actual opening of the nozzle where the pressure difference is ΔP.
4. By El-Kotb [1982]: SMD (μm) = 3.08x10^6 $\nu^{0.185}$ $\sigma^{0.737}\rho_\ell^{\;0.737}\rho_a^{\;0.06}\Delta P^{-0.054}$ (Pa kg m^2 sec N m).

Example 9

The SMD and v_f (volume flow of liquid/volume flow of air) for swirl jet atomizer are measured to be 35 μm and 8×10^{-05}, and 80 μm and 1.2×10^{-4} at r = −2 cm and +4 cm, respectively, and x = 50 mm from injector. (a) Compute number density and interdrop spacing ℓ. (b) If combustor pressure is 6.9 bar, T = 600 K, compute local A:F at r = −2 cm. Assume ρ_ℓ = 900 kg/m³.

Solution

(a) If number density is n (drops per m³ of air), then liquid volume/air volume $= n\ \pi d^3/6 = 8 \times 10^{-5}$. Thus $n = 8 \times 10^{-5}/\pi\{(35 \times 10^{-6})^3/6\} = 3.56 \times 10^9$ drops per m³, $\ell \approx 1/n^{1/3} = 7200$ μm $\ell/a = 411$.

Similarly, n = 4.48 × 109, ℓ = 14380 μm, ℓ/a = 360 at r = +4 mm.

(b) $A{:}F = \rho_a/\{n\ (\pi d^3/6) \times \rho_\ell\} = \rho_a/\{v_f \times \rho_\ell\}$

$$\rho_a = P_a/(R_aT) = 690\text{ kPa}/(0.286\text{ kJ/(kg K)} \times 600\text{ K}\} = 4.02\text{ kg/m}^3$$

$$A{:}F = \{4.02\text{ kg/m}^3\text{ of air})/\{1.2 \times 10^{-4}\ (\text{m}^3/\text{m}^3\text{ air}) \times 900\text{ kg/m}^3\} = 37.22.$$

Remarks

The interdrop spacing is important in estimating drop combustion rate (Chapter 10) and studying the effect of interactions (Chapter 16).

4.7 Summary

Fuels can be classified according to the state in which they exist, i.e., gas, liquid, and solid. They are naturally occurring (e.g., fossil fuels), but could also be synthetic fuels in gaseous, liquid, and solid phases. Proximate and ultimate analyses characterize a solid fuel, whereas basic components characterize a liquid fuel, which is a mixture of pure HCs. Empirical equations for HV are presented along with relations for standard flue gas volumes for any CHO fuel. Formulae for maximum possible heating value from gasification of solid fuels are presented. The size distributions for solid and liquid fuels are briefly described.

4.8 Appendix

The ash percentage increases as a solid fuel burns. For instance, if 50% of a solid fuel with a 5% initial ash content is burned, the ash percentage in the remaining 50% of the fuel is 0.05/(0.05 + 0.5 × 0.95) = 9.52%. The ash percentage is a measure of how much combustible material has been burned.

4.8.1 Ash Tracer Method for Coal Analysis

4.8.1.1 Ash Fraction

For a dry solid fuel with a combustible fraction F and an ash fraction A,

$$A_o = m_{A,o}/m = m_{A,o}/(m_{A,o} + m_{F,o}), \tag{A}$$

where A_o denotes the initial fraction of ash on a dry basis; m_o, the initial mass of dry solid fuel; $m_{A,o}$, the initial mass of ash in the fuel; m_o, the mass of dry solid fuel after gasification; and $m_{F,o}$, the initial combustible mass in the fuel. After combustion, the ash mass remains constant, but the amount of combustibles decreases. Therefore, the ash fraction increases. This is conveyed through the expression

$$A = m_A/m = m_A/(m_{A,o} + m_F), \tag{B}$$

where A represents the fraction of ash in a dry sample after gasification.

4.8.1.2 *Burned or Gasification Fraction*

The initial mass is

$$m_o = m_A/A_0, \quad \text{i.e., because } m = m_A/A, m/m_o = A_o/A, \tag{B'}$$

and

$$m_o - m = m_0(1 - A/A_0) = m_{F,o} - m_F. \tag{B''}$$

Using Equation A and Equation B,

$$m_{F,o} = (m_A/A_0) - m_A, \tag{C}$$

and

$$m_F = (m_A/A) - m_A. \tag{D}$$

Then,

$$m_F/m_{F0} = \text{combustibles/initial/combustibles} = 1 - BF = ((1/A) - 1)/(1/A_0 - 1), \tag{E}$$

where the burned fraction, η_g = combustibles gasified/initial combustibles and

$$\eta_g = ((1/A_0) - 1/A)/(1/A_0 - 1) = (A - A_0)/\{A(1 - A_0)\}. \tag{F}$$

4.8.1.3 *Fraction of S or Element Conversion*

The above analysis can be extended to determine the fraction of sulfur that is captured by the ash, i.e.,

$$1 - \eta_{s,g} = m_S/m_{S,o} = m\, Y_S/m_o\, Y_{S,o} \tag{G}$$

or

$$1 - \eta_{s,g} = F_{S,ret} = m_S/m_{S,o} = (1 - \eta_g)\, Y_S/Y_{S,o}, \tag{H}$$

where Y_S and $Y_{S,o}$ denote the sulfur fraction in the dry fuel and that was originally present in the fuel on a dry basis, respectively; $F_{S,ret}$, the sulfur fraction retained with the ash in the fuel; and $\eta_{s,g}$, the sulfur fraction that is gasified. In terms of ash fractions,

$$1 - \eta_{s,g} = (Y_S/Y_{S,o})(1/A)(A_o) = (Y_S/Y_{S,o})\,(A_o/A). \tag{I}$$

4.8.1.4 Ideal Conversion

If sulfur is removed in proportion to its burned mass fraction, then Equation H yields the relation

$$1 - \eta_{s,g,\ ideal} = F_{S,ret,ideal} = (1 - \eta_g). \tag{J}$$

This relation is valid for any combustible element in its general form, e.g.,

$$F_{C,ret,ideal} = F_{S,ret,ideal} = m_{C,ideal}/m_{C0} = m_{S,ideal}/m_{S0} = (1 - \eta_g). \tag{K}$$

Typically, about 95% of the sulfur in the combustible (organic) part of the fuel is burned into SO_2. However, about 5% of SO_2 can be captured by the ash and the refractory walls. Therefore, the amount of retained sulfur does not usually correspond to its ideal value.

Chapter 5

Chemical Kinetics

5.1 Introduction

Chapter 2 and Chapter 3 discussed the overall chemical reaction with complete combustion and products at chemical equilibrium. However, the rate of reaction and time required to achieve either complete combustion or chemical equilibrium have not been dealt with. In this chapter, we consider the change in the chemical structure (i.e., molecular rearrangement of atoms) that results in desired chemical products and the rate at which the products are formed. Consider a premixed mixture of two species: fuel CH_4 and oxidizer O_2. The fuel species reacts with the oxidizer and produces CO_2 and H_2O.

$$CH_4 + 2O_2 \rightarrow CO_2 + 2H_2O. \tag{5.1}$$

Here, the C-H bonds are broken (Table A.31A and Table A.31B of Appendix A), and C and H are separated and attached with O atoms, resulting in an atomic rearrangement called a *chemical reaction*. This chapter deals with types of reactions, reaction rate theory, Arrhenius law, intermediate steps involved in formation of product species, equilibrium reaction and its relation to reaction rates and, finally, pyrolysis and heterogeneous reactions involving solid fuels.

5.2 Reaction Rates: Closed and Open Systems

Consider a general reaction involving species A, B, etc., resulting in products E, F, etc.

$$\mathrm{aA + bB + \ldots \rightarrow eE + fF + \ldots +}\ \Delta h_R^0, \tag{5.2}$$

where a, b, etc., are called *stoichiometric coefficients*, e.g., for reaction 5.1, a = 1, A = CH_4, b = 2, B = O_2, e = 1, E = CO_2, f = 2, F = H_2O. This reaction may occur either in a closed system (e.g., automobile engines) or in an open system (e.g., gas turbines). From Chapter 2, the species conservation equation for species E in molar form for any system is written as

$$dN_E/dt = \dot{N}_{E,i} - \dot{N}_{E,e} + \dot{N}_{E,gen,} = \dot{N}_{E,i} - \dot{N}_{E,e} + \dot{w}_E''' V \quad (5.3)$$

where V is the volume of reaction zone; $\dot{w}_E'''$, is the production rate of species E per unit volume (assumed to be uniform throughout volume of interest); and dN_E/dt, the accumulation rate of moles of species E. Applying Equation 5.3 for a closed system (Figure 5a for reaction 5.1),

$$\frac{dN_E}{dt} = \frac{d}{dt}\int_V [E]dV = \int_V \dot{w}_E''' dV \quad (5.4)$$

where dN_E/dt represents the rate of accumulation of species E per unit volume owing to Reaction 2, and [E], the molal concentration of species E kmole/m^3. In a closed system, the interior may be hot (e.g., an automobile engine), whereas the regions near the walls may be cold. In this case, reaction rates are faster near the interior and slower near the walls. Hence, $\dot{w}_E'''$ is not uniform and is a function of spatial coordinates. In a closed system or fixed-mass system, the accumulation rate is the same as the production rate of species E because of the chemical reaction rate. If we consider species A in Equation 2, $dN_A/dt = \dot{w}_A'''$, where A is consumed and N_A decreases with time; hence $\dot{w}_A'''$ or $dN_A/dt < 0$. Negative values imply consumption (e.g., $d[CH_4]/dt < 0$), and positive values are associated with species production. If the control volume is fixed, Equation 5.4 becomes

$$\int_V \left\{ \frac{\partial [E]}{\partial t} - \dot{w}_E''' \right\} dV = 0 \quad \text{or} \quad \frac{\partial [E]}{\partial t} = \dot{w}_E''' \text{ kmol/(m}^3\text{ sec)} \quad (5.5a)$$

If the reaction rate is uniform within the whole volume,

$$\frac{d[E]}{dt} = \dot{w}_E''' \quad (5.5b)$$

Note that $d[E]/dt \neq c\,\{dX_E/dt\}$, where c represents molal concentration of the mixture in kmol/m^3 as the number of moles of the mixture may change during the reaction. If molal-concentration change of fuel F (e.g., CH_4) with respect to time is known (Figure 5.1), the rate at which the fuel is burned, $d[F]/dt$, and that at which oxygen is consumed, $d[O_2]/dt$, are also known.

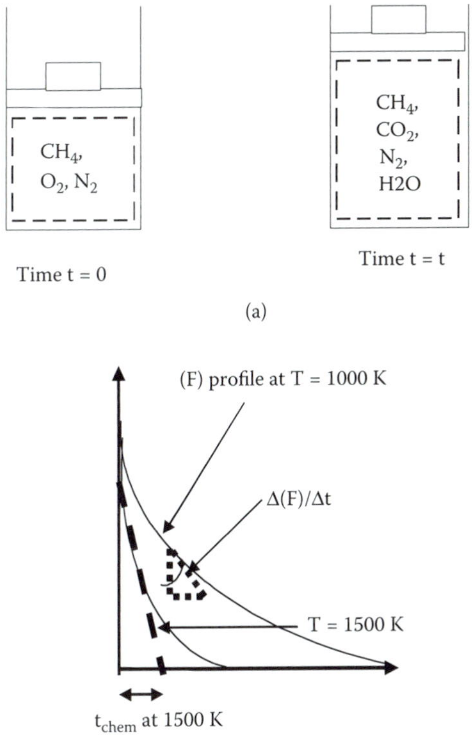

Figure 5.1 Illustration of chemical reaction at constant P: (a) piston-cylinder-weight assembly, (b) change in [F] with time.

Now consider a closed system in which the volume may vary but the mass is still fixed. If the chemical production rate is written as $\dot{w}'''_{E,m}$ in kg/m^3 sec (mass basis), then one can apply the species equation in mass form (Chapter 2):

$$\frac{dm_E}{dt} = \frac{d}{dt}\int_V (E)dV = \frac{d}{dt}\int_V dm_E = \int_V \dot{w}'''_{E,m}dV$$

where (E) represents the concentration of E in kg/m^3. For uniform reaction

$$\frac{d\{m_E\}}{dt} = \frac{d\{mY_E\}}{dt} = m\frac{dY_E}{dt} = V\dot{w}'''_{E,m}$$

Simplifying,

$$\rho\frac{dY_E}{dt} = \dot{w}'''_{E,m} \qquad (5.5c)$$

where $Y_E = m_E/m$, the mass fraction of species E.

Now consider the system in which we have an opening with a valve (e.g. automobile engine). Suppose the CO_2 formation rate is 0.05 kmol/m³ sec and that once the level CO_2 reaches, say, 0.1 kmol/m³, we open the valve and start removing the CO_2 species at the rate of 0.05 kmol/m³ sec; then the CO_2 concentration in the system remains fixed at 0.1 kmole/m³. Note that our system is no longer a closed system, and hence from Equation 5.3,

$$dN_E/dt = \dot{N}_{E,i} - \dot{N}_{E,e} + \dot{w}'''_E\, V = -\dot{N}_{E,e} + \dot{w}'''_E\, V$$

and if steady state is reached, then $dN_E/dt = 0$ and hence,

$$\dot{N}_{E,e} = \int_V \dot{w}'''_E dV$$

Thus, the exit flow of species E in kmol/sec must be same as the chemical production rate. For example, consider a gas turbine combustor in which fuel enters with air and burns to CO_2 and H_2O. The concentrations are steady in the combustor at any given location. As such, the exiting flow rate of CO_2 must be equal to the production rate of CO_2 within the whole volume of the combustor. Using local time derivatives may yield misleading results.

On the other hand, if we use the Lagrangian frame (following a fixed mass of premixed fuel CH_4 and air) in an open system, then $d[CH_4]/dt < 0$ and $d[CO_2]/dt > 0$; this implies that $[CH_4]$ will decrease and $[CO_2]$ will increase with time as we follow the fixed mass. Thus, the differentials here represent material derivative as we follow a fixed mass. Unless otherwise mentioned, we will imply system of fixed mass when we use time derivatives in this chapter.

The enthalpy of reactionΔh_R^0 is positive for endothermic and negative for exothermic reactions. For the methane-burning reaction of Equation 1,Δh_R^0, −49,988 kJ/kg (lower heat of reaction), and is exothermic.

5.2.1 Law of Stoichiometry

The law of stoichiometry for the reaction given in Equation 5.1 is written as

$$(d[O_2]/dt)/(-2) = (d[CH_4]/dt)/(-1) = (d[CO_2]/dt)/1 = (d[H_2O]/dt)/2 = \dot{w}'''. \tag{5.6}$$

Equation 5.6 is termed the *law of stoichiometry*. The stoichiometric coefficients in the denominators are molar coefficients in the reaction equation. Negative coefficients are assigned to reacting species and positive ones to products in the law of mass stoichiometry. The law of stoichiometry for the reaction in Equation 5.2 is written as

$$\{1/(-a)\}d[A]/dt = \{1/(-b)\}d[B]/dt = \ldots = \{1/e\}d[E]/dt$$
$$= \{1/f\}d[F]/dt = \dot{w}''', \text{ i.e.,} \tag{5.7}$$

Rewriting,

$$\{1/(\text{-a})\}\,\dot{w}_A''' = \{1/(-b)\}\,\dot{w}_B''' = \ldots = \{1/e\}\,\dot{w}_E''' = \{1/f\}\,\dot{w}_F''' = \dot{w}''', \quad (5.8)$$

where $\dot{w}'''$ denotes the reaction rate per unit volume (kmol/m^3 sec), which is the production rate for any species for which the molal stoichiometric coefficient is unity.

5.2.2 *Reaction Rate Expression, Law of Mass Action, and the Arrhenius Law*

Obviously, the reaction rate must depend upon the concentrations of its participating species. The functional relation of the reaction rate with the concentration and temperature is called the *reaction rate expression*. If A + b B → e E, then the phenomenological law of mass action relates $\dot{w}'''$ in terms of molal (or mass) concentration as

$$\dot{w}_A''' = (-a)\dot{w}''', \dot{w}_E''' = e\dot{w}''' \quad (5.9a)$$

$$\dot{w}'' = k[A]^{\alpha}[B]^{\beta}, \quad (5.9b)$$

where α and β are indicative of the reaction orders with respect to the species A and B, respectively, and the overall order of the reaction is $\alpha + \beta$. Because $\dot{w}''' = k(T)\,[A]^{\alpha}\,[B]^{\beta}$ is represented in units of kmol/m^3 sec, Equation 5.9b dictates that the specific reaction rate constant must be in units of

$$k = \text{kmol}^{(1-\alpha-\beta)}/(\text{sec m}^{3(1-\alpha-\beta)}).$$

It is seen that the reaction rate $\dot{w}'''$ increases with increased concentration of reacting species or increased mass of reacting species (when $\alpha > 0$ $\beta > 0$), and the equation $\dot{w}''' = k(T)\ [A]^{\alpha}\,[B]^{\beta}$ is hence called *law of mass action.* Generally, k is a strong function of temperature as given by the Arrhenius law

$$k = A\ T^n \exp\{-E/(\bar{R}\,T)\}, \quad (5.9b)$$

where A is the preexponential factor; E, the activation energy; and T, the temperature in K. The value n is an empirical constant typically varying from 0 to 1/2. The theory behind the Arrhenius law will be presented later in the chapter.

5.3 Elementary Reactions and Molecularity

An elementary reaction is a reaction in which products are formed by direct collision or contact. We have discussed global reactions for multiple reactants, e.g., A and B, and assumed that as soon as A and B are mixed, they directly react to form the product AB. However, reactions do not proceed in this manner and, instead, form intermediate species that react to form other intermediates, and so on, until the products are formed. For instance, during methane oxidation $CH_4 + 2\ O_2 \rightarrow CO_2 + 2\ H_2O$, methane first forms CO through a complex pathway, which then oxidizes to CO_2. Likewise, both atomic and molecular hydrogen are formed, which produce OH, and then H_2O. Therefore, there could be multiple steps involved in an actual reaction. Further reactions could involve one or more molecules of a species. The reaction mechanism for reactants-to-products conversion is a group of elementary reactions resulting in an overall reaction. The following illustrates the various types of single-step reactions.

5.3.1 Unimolecular Reaction

The molecularity of a reaction is defined by the number of reactant molecules that must be present for a reaction to proceed. A unimolecular reaction involves a single reactant molecule and can be represented in the form

$$A \rightarrow B \tag{5.10a}$$

$$d[A]/dt = -k(T)[A]. \tag{5.10b}$$

The first step in the reaction can involve an activation step that produces activated A^*. The activation

$$A + M \xrightarrow{k_{1f}} A^* + M \tag{5.11}$$

can occur when a molecule of A collides with a third-body molecule M and the reaction has a specific reaction rate constant k_{1f}. The third body represents one or more species (e.g., inert species such as N_2) in the system or the container walls. In the reaction, the translational energy (te) of the molecule M is transferred to A. Consequently, the vibrational energy (ve) and rotational energy (re) of A increase, and it transforms into the excited state A^*. The excited state can be deenergized either by collision with another third-body molecule ($A^* + M \xrightarrow{k_{1,b}} A + M$, which increases the te of the third body) or by the emission of radiant energy ($A^* \xrightarrow{k_{1,b'}} A + h\nu$). Thus, the third body is similar to a flywheel, storing and transferring energy when necessary.

The energization of A into A^* can also occur by the reaction $A + A \rightarrow A^* + A$. This reaction depends upon the concentration of A and, because species A is consumed as the reaction proceeds, its concentration also decreases. This reduces the reaction rate faster than in the third-body reaction given by Equation 5.11. The third-body efficiency compares the effectiveness of M with A in producing A^*. Therefore, the specific reaction rate k_1 assumes different values, depending on the type of third body.

The reaction rate constant $k_{i,b}$ for the backward deactivation reaction

$$A^* + M \xrightarrow{k_{1,b}} A + M \tag{5.12}$$

has a smaller value than k_{1f}. The forward step of Equation 5.11 can also be followed by another deactivation reaction occurring in product B:

$$A^{*} + M \rightarrow B + M. \quad (5.13)$$

When M is a solid, the reaction is of first order. An example of a unimolecular reaction is

$$Br_2 \rightarrow 2\ Br \quad (5.14)$$

where third body could be Br_2. Applying the law of stoichiometry,

$$\dot{w}'''_{Br_2}/(-1) = \dot{w}'''_{Br}/(2) = \dot{w}'''. \quad (5.15)$$

Using the law of mass action, $\dot{w}''' = k_{uni}\,[Br_2]$. Then

$$\dot{w}'''_{Br_2} = 2k_{uni}\,[Br_2]^1,\ \text{kmol/m}^3\ \text{sec} \quad (5.16)$$

Likewise, $\dot{w}'''_{Br_2} = -k_{uni}\,[Br_2]$ and the overall order m = 1. The units for k are determined as follows. $[Br_2]$ is expressed in kmol/m^3, and the units of k_{uni} are (kmol/m^3 sec)/(kmol/m^3) = 1/sec. This is a *unimolecular reaction* because one molecule of Br_2 decomposes into two Br atoms. Other examples are:

$$C_2H_6 \rightarrow 2\ CH_3,$$

and ozone decomposition in the upper atmosphere,

$$O_3\ (g) \rightarrow O_2\ (g) + O(g).$$

5.3.1.1 Characteristic Reaction Times (t_{char})

Integrating Equation 5.10b for an isothermal reaction,

$$[A] = [A]_0 \exp\{-k_{uni}\,(T)\ t\}$$

where $[A]_0$ is the concentration at t = 0. Letting $[A]/[A]_0 = 1/e$ (note that e = 2.718, 1/e = 0.368), t_{char} is given as

$$t_{char} = 1/k_{uni}(T)$$

Example 1

Consider a closed rigid vessel initially consisting of H_2. The reactor is rapidly heated to a high temperature T, and after heating and before decomposition the pressure becomes P_0. The hydrogen then undergoes a unimolecular reaction

$$H_2 \rightarrow 2\ H \quad (A)$$

During decomposition, T is maintained constant. (a) Show that the pressure history is given as

$$P = [H_2]_0\{2 - \exp(-k_{uni}(T)\ t)\}\ \bar{R}\ T.$$

(b) Develop a method for obtaining a solution for k (T) if P vs. t is measured.

Solution

As H_2 decomposes, each molecule of H_2 produces two H atoms and, hence, the total number of molecules increases with time resulting in a pressure rise. Thus, we track $[H_2]$ and [H] with time. For Reaction A,

$$1/(-1)\ d[H_2]/dt = \{1/(2)\}\ \{d[H]/dt\} = \dot{w}''' = k_{uni}\ (T)\ [H_2]. \quad (B)$$

Hence,

$$d[H_2]/dt = -k_{uni}\ (T)\ [H_2] \quad (C)$$

$$d[H]/dt = 2\ k_{uni}\ (T)\ [H_2]. \quad (D)$$

Dividing Equation D by Equation C,

$$d[H]/d[H_2] = -2 \quad (E)$$

Integrating Equation E and using the initial condition $[H_2] = [H_2]_0$ at $[H] = 0$, we have

$$[H] = 2\ ([H_2]_0 - [H_2])$$

Integrating Equation C, $[H_2] = [H_2]_0$ at $t = 0$

$$[H_2] = [H_2]_0\ \exp(-k_{uni}\ (T)\ t)$$

Total moles per unit volume, $n = [H] + [H_2] = 2\{[H_2]_0 - [H_2]\} + [H_2] = \{2 - \exp(-k_{uni}\ t)\}$

$$\text{At } t = 0,\ n = [H_2]_0 \text{ and as } t \to \infty,\ n = 2[H_2]_0$$

$$P = n\ \bar{R}\ T = [H_2]_0\ \{2 - \exp(-k_{uni}\ (T)\ t)\}\ \bar{R}\ \ T \quad (F)$$

Thus, pressure increases linearly with time initially. Rearranging Equation F

$$\{P(t)\ n/(\ \bar{R}\ T\ [H_2]_0)\} = \{2 - \exp(-k_{uni}\ t)\}$$

Thus, measuring P vs. t and plotting $\ln\ \{2 - P(t)\ n/(\ \bar{R}\ T\ [H_2]_0)\}$ vs. t, one can determine the slope as $(-k_{uni})$.

Remarks

One can alter T, and get k_{uni} vs. T for the elementary reaction; with a plot of ln k_{uni} vs. 1/T the E can be obtained.

5.3.2 Bimolecular Reaction

Two molecules are involved in the reaction, say A + B → C. For example,

$$H_2 + I_2 \rightarrow 2\ HI, \quad \text{i.e.,} \tag{5.17}$$

$$\dot{w}'''_{H_2}/(-1) = \dot{w}'''_{I_2}/(-1) = \dot{w}'''_{HI}/(1) = \dot{w}''' \tag{5.18}$$

where

$$\dot{w}''' = k_{bi}\ [H_2]^1\ [I_2]^1. \tag{5.19a}$$

Using Equation 5.19 in Equation 5.18,

$$\dot{w}'''_{H_2} = d[H_2]/dt = -\ k_{bi}\ [H_2]^1\ [I_2]^1 \tag{5.19b}$$

The units of k are now $m^3/(kmol\ sec)$. The molecularity of reaction 5.17 is m = 2, where the orders with respect to both H_2 and I_2 are unity. In both examples (i.e., in Equation 5.10 and Equation 5.19), the reaction order equals the molecularity of the reaction.

Other examples of second-order reactions are as follows:

$$O + O_3 = 2O_2,\ CH_4 + OH \rightarrow H_2O + CH_3,\ CO + O_2 \rightarrow CO_2 + O,\ H + O_2 \rightarrow OH + O,$$
$$O + H_2 \rightarrow OH + H,\ H_2 + OH \rightarrow H_2O + H$$

5.3.2.1 Characteristic Reaction Times (t_{char})

Integration of Equation 5.19b requires solution of $[I_2]$ in terms of $[H_2]$, which is provided by Equation 5.18. From the law of stoichiometry,

$$(d[H_2]/dt)/(-1) = (d[I_2]/dt)/(-1) \tag{5.20a}$$

Integrating Equation 5.20a,

$$[H_2]_0 - [H_2] = [I_2]_0 - [I_2] \tag{5.20b}$$

Using Equation 5.20b in Equation 5.19b and integrating, one can obtain

$$\frac{[H_2]}{[H_2]_0} = \frac{\dfrac{[I_2]_0}{[H_2]_0} - 1}{\left(\dfrac{[I_2]_0}{[H_2]_0}\exp\{k_{bi}\ t([I_2]_0 - [H_2]_0)\} - 1\right)} = \frac{1 - \dfrac{[H_2]_0}{[I_2]_0}}{\left(\exp\{k_{bi}\ t([I_2]_0 - [H_2]_0)\} - \dfrac{[H_2]_0}{[I_2]_0}\right)}$$

Letting $[H_2]/[H_2]_0 = 1/e$, t_{char} is given as

$$t_{char} = \frac{\ell n\left(e+(1-e)\frac{[H_2]_0}{[I_2]_0}\right)}{\left([I_2]_0 - [H_2]_0\right)k_{bi}(T)}. \tag{5.21}$$

The relation can be generalized for any bimolecular reaction $A + B \rightarrow C + D$.

5.3.3 Trimolecular Reactions

Three molecules are involved in a trimolecular reaction (three-body reactions), which is relatively less frequent in combustion applications. The third body, M, is required to carry the energy liberated when a stable species is formed, which explains why the translation energy of M increases.

$$A + B + M \rightarrow C + M \tag{5.22}$$

$$d[A]/dt = -k_{tri} [A] [B][M] \tag{5.23}$$

Three-body recombination reactions are very important in hydrogen-oxygen chemistry. The H_2–O_2 reaction schemes were summarized by Westbrook and Dryer [1984]. They become more important relative to bimolecular reactions as the pressure increases. A few reactions do not occur unless a third body is present, e.g.,

$$H + H + M \rightarrow H_2 + M. \tag{5.24a}$$

For the reaction in Equation 5.24a,

$$d[H]/dt = -2k_{tri} [H]^2[M].$$

Other examples of a trimolecular reaction are:

$$H + OH + Cl \rightarrow H_2O + Cl \tag{5.24b}$$

$$2\ NO + O_2 \rightarrow 2\ NO_2$$

$$H + H + H \rightarrow H_2 + H$$

In the last reaction, three H atoms collide, produce H_2, and leave one H atom as third body.

$$d[H]/dt = k_{tri} [H]^3 \{-3 + 1\}$$

5.3.3.1 Characteristic Reaction Times (t_{char})

In trimolecular reactions involving third bodies, the concentration of the third body remains unaffected. For example, consider Equation 5.24b. Cl acts as the third body whose concentration remains unaffected by Reaction 5.24b.

$$d[H]/dt = -k_{tri}\,[Cl]\,[H][OH] \tag{5.24c}$$

Because [Cl] remains unaffected, k_{tri} [Cl] is a constant; we then find that Equation 5.24b remains similar to Equation 5.19b. Thus, t_{char} is the same as Equation 5.21 with H replacing H_2 and OH replacing I_2.

5.4 Multiple Reaction Types

Earlier, we dealt with the types of molecules involved in a reaction. Many times, a series of reactions is involved in the formation of final products. The intermediate reactions are summarized in the following subsections.

5.4.1 Consecutive or Series Reactions

In consecutive or series reactions, intermediary species are produced, which subsequently form the products

$$A+B \xrightarrow{k_1} AB \xrightarrow{k_2} E+F, \tag{5.25}$$

where AB is an intermediary product, the net production of which is

$$d[AB]/dt = k_1\,[A][B] - k_2\,[AB] = \text{AB produced by } k_1 - \text{AB consumed by } k_2 \tag{5.26}$$

5.4.2 Competitive Parallel Reactions

The species A and B can form either the species AB or the products E and F directly through competitive parallel reactions, e.g.,

$$A+B \xrightarrow{k_1} AB, \text{ and} \tag{5.27}$$

$$A+B \xrightarrow{k_2} E + F. \tag{5.28}$$

$$d[A]/dt = -k_1\,[A]\,[B] - k_2\,[A]\,[B].$$

5.4.3 Opposing or Backward Reactions

We have assumed thus far that the reaction A + B proceeds in the forward direction to form the composite species AB. It is also possible for AB to disintegrate into the species A and B through a backward reaction. Consider CO oxidation with O_2

$$CO + 1/2\ O_2 \xrightarrow{k_f} CO_2, \tag{A}$$

$d[CO_2]/dt = \dot{w}'''_{CO_2} = k_f[CO][O_2]^{1/2}$, indicating that $d[CO_2]/dt > 0$. Let us call this direction as *forward direction.* Now consider

$$CO_2 \xrightarrow{k_b} CO + 1/2\ O_2, \tag{B}$$

for which $d[CO_2]/dt = \dot{w}'''_{CO_2} = -k_b[CO_2] < 0$. Denoting Reaction B as backward reaction, we can combine both reactions in the form

$$CO + 1/2\ O_2 \underset{k_b}{\overset{k_f}{\rightleftarrows}} CO_2, \tag{C}$$

Then the rate of production of CO_2 is written as

$$\dot{w}'''_{CO_2} = d[CO_2]/dt = k_f[CO][O_2]^{1/2} - k_b\ [CO_2] \tag{5.29}$$

Although the Fourier law appropriately accounts for the rate of heat transfer in terms of gradient of T, an equivalent law for chemical reaction rate could not be written in terms of G by including both forward and backward reaction rates, but the second law still dictates whether $d[CO_2]/dt > 0$ or $d[CO_2]/dt < 0$ at any given T and P. Thus, Equation 5.29 could be hypothesized as a kind of Fourier law analogy for chemical reactions. If $k_f[CO][O_2]^{1/2} > k_b\ [CO_2]$, then in an overall sense CO_2 is produced and vice versa. It will be shown later that k_b is related to k_f through the equilibrium constant.

5.4.4 Exchange or Shuffle Reactions

$$O + N_2 \rightarrow NO + N \tag{A}$$

$$N + O_2 \rightarrow NO + O \tag{B}$$

where O and N are shuffled between reactant and product sides. It is almost the same as the bimolecular atom exchange reaction, $AB + C \rightarrow BC + A$. Other examples include

$$H + O_2 \rightleftarrows OH + O \tag{C}$$

$$O + H_2 \rightleftarrows OH + H \tag{D}$$

$$H_2 + OH \rightleftarrows H_2O + H \tag{E}$$

$$2OH \rightleftarrows H_2O + O \tag{F}$$

The OH production rate for the multiple reactions Reaction C to Reaction F is given as

$\dot{w}'''_{OH} = k_{G,f}[H][O_2] - k_{G,b}\,[OH]\,[O] + k_{H,f}\,[O][H_2\,] - k_{H,b}\,[OH][H] - k_{I,f}[H_2][OH] + k_{I,b}\,[H_2O]\,[H] - 2\,k_{J,f}\,[OH]^2 + 2\,k_{H,b}\,[H_2O][O]$

Example 2

Nitrogen is released in the form of HCN from coal, which undergoes oxidation to NO through the reaction

$$HCN + O_2 \underset{k_{A,b}}{\overset{k_{A,f}}{\rightleftarrows}} NO + HCO. \qquad (A)$$

HCN also combines with NO to produce N_2 through the reaction

$$HCN + NO \underset{k_{B,b}}{\overset{k_{B,f}}{\rightleftarrows}} N_2 + HCO. \qquad (B)$$

These two reactions are also sometimes referred to as *De Soete's kinetics*. Reaction A is favored if oxygen is abundant, and Reaction B if O_2 is relatively deficient. Note that Reaction A and Reaction B proceed in both directions.

(a) Write an expression for the production rate of NO, neglecting the backward paths of the two reactions.
(b) Determine the production rate of NO if the molar concentration of HCN is 1%, O_2: 5%, and NO: 0.2% and the pressure and temperature are 1 bar and 1500 K. The specific reaction rate constants are $k_{A,f} = 1 \times 10^{11} \exp(-280{,}328/\bar{R}T)$, $k_{B,f} = 3 \times 10^{12} \exp(-251{,}040/\bar{R}T)$, where $\bar{R}$ is expressed in units of kJ/kmol K. Assume for Reaction A, its order with respect to HCN is unity, as is its order with respect to O_2. The order of Reaction B with respect to HCN is zero, whereas its order with respect to NO is unity.
(c) If the temperature is maintained constant, what is the steady-state concentration of NO? Assume that NO is the only trace species in the system.

Solution

(a)

$d[NO]/dt = k_{A,f}\,[HCN]\,[O_2] - k_{A,b}\,[NO]\,[HCO] - k_{B,f}\,[HCN]\,[NO] + k_{B,b}\,[N_2]\,[HCO].$

Neglecting the backward paths,

$$d[NO]/dt = k_{A,f}\,[HCN]\,[O_2] - k_{B,f}\,[HCN]\,[NO]$$

(b) The species considered are HCN, O_2, NO, and HCO with molecular weights of 27.02, 32, 30.01, and 29.02 kg $kmol^{-1}$ respectively.
The mixture concentration c is 0.0080 $kmol/m^3$ (= P/RT).

Therefore, the species mole fractions and concentrations are as shown in the following table:

	HCN	O_2	NO	HCO
X_k	0.01	0.05	0.002	0
[k] = cX_k, kmol/m^3	8.0186E–05	0.00040093	1.60372E–05	0

$k_{A,f} = 1 \times 10^{11} \exp(-280{,}328/\bar{R}\,T) = 1.0 \times 10^{11} \exp(-280{,}328/8.314 \times 1500) = 17.3$ m^3/(kmol sec), and

$k_{B,f} = 3 \times 10^{12} \exp(-251{,}040/\bar{R}\,T) = 5430.2$ m^3/(kmol sec).

Hence,

$k_{A,f}$ [HCN] [O_2] = 5.55×10^{-7} kmol/m^{3}sec,

and

$k_{B,f}$ [NO] = 0.0871 kmol/m^{3}sec.

The net production rate d[NO]/dt = is −0.087 kmol/m^{3}sec. The negative sign indicates net consumption of NO.

(c) At steady state, d[NO]/dt = 0, i.e.,

$0 = k_{A,f}$ [HCN] [O_2] − $k_{B,f}$ [NO],

i.e.,

[NO] = $k_{A,f}$ [HCN] [O_2]/$k_{B,f}$,

or

[NO] = 1.02×10^{-10} kmol/m^3,

$X_{NO} = 1.02 \times 10^{-10}/0.008 = 1.28 \times 10^{-7}$ or 0.0128 ppm.

5.4.5 Synthesis

It is the reaction in which molecules form a larger molecule out of simpler molecules, typically energy absorbing; e.g., amino acids combining to form a protein molecule.

$$A + B \rightarrow AB$$

5.4.6 Decomposition

It is the reverse of the synthesis reaction; here, molecules form a simpler molecule out of larger molecules, typically energy releasing; e.g., larger glycogen molecules decomposing into glucose (sugar) molecule.

$$AB \rightarrow A + B$$

$$AB + M \rightarrow A + B + M$$

Cells break down glucose, react with oxygen carriers, form CO_2 and water, and release energy.

5.5 Chain Reactions and Reaction Mechanisms

Section 5.4 discussed the various types of reactions. In this section, reaction mechanisms are classified based on the chain-reaction mechanism through radical production or consumption. Noble gases like He, Ne, and Ar with outermost electrons being 2, 8, etc., are stable. Radicals are atoms or molecules with unpaired electrons. For example, O_2 (which shares two electrons in the two O atoms) is stable, but O atom has six electrons that are unpaired. O, OH, CH_3, N, etc., are called *radicals* and typically have positive h_f^o and, hence, absorb energy from collisions. Oxidation begins with O_2 molecules acquiring sufficient energy to break the C-H bond (weaker compared to the C-C bond; see Table A.31A and Table A.31B of Appendix A for bond energies) and removing the H atom (say from C_2H_6), which joins with O_2 to form, say, R + HO_2, where R = C_2H_5 in the present example. Both are radicals. On the other hand, if the temperature is high and there is no O_2 present, then breakup is thermally induced by dissociation with a third body, M. Thus, $C_3H_8 + M \rightarrow C_2H_5 + CH3 + M$, where M absorbs or loses energy. The radicals build up, and product species could be different in the absence of O_2; for example, the HC species may form soot. On the other hand, if O_2 is present, radicals are extremely reactive with O_2, and reaction results in CO_2, H_2O, releasing energy, etc. The overall reaction $A_2 + B_2 \rightarrow 2\ AB$ is used as an illustration in explaining different reaction mechanisms.

5.5.1 Chain Initiation Reactions

Chain initiation reactions involve the initiation of radicals R from stable species S; e.g., if S = A_2,

$$A_2 + M \rightarrow A + A + M, \text{ with a specific reaction rate } k_1. \quad (5.30)$$

Here, two radicals (A, A) are produced from one stable species, A_2.

5.5.2 Chain Propagating Reactions

Chain propagating reactions involve same number of radical species on both reactant and product sides such that the radical pool is propagated, e.g.,

$$A + B_2 \xrightarrow{k_2} AB + B, \quad (5.31)$$

and

$$B + A_2 \xrightarrow{k_3} AB + A. \quad (5.32)$$

For example, $O + NO \rightarrow O_2 + N$.

5.5.3 Chain Branching Reactions

In chain branching reactions, more radicals are produced than consumed, e.g.,

$$A + B_2 + M \xrightarrow{k_4} A + 2\ B + M. \quad (5.33)$$

For example, $CH_4 + O \rightarrow CH_3 + OH$, $O + H_2 \rightarrow OH + H$.

5.5.4 Chain Breaking Reactions

In chain breaking reactions, unlike branching reactions, more radicals are consumed than are formed, e.g.,

$$A + 2B + M \xrightarrow{k_5} A + B_2 + M. \qquad (5.34)$$

5.5.5 Chain Terminating Reactions

In chain terminating reactions, the radicals are completely consumed either forming stable product or by wall collisions, e.g.,

$$A + B + M \xrightarrow{k_6} AB + M. \qquad (5.35)$$

For example, $H + OH + M \rightarrow H_2O + M$.

Reaction rates can be slowed down by decreasing the temperature (so that molecules have insufficient energy during collisions for breaking the bonds) or the radical concentrations. A thermal inhibitor (e.g., N_2) reduces the reaction rate by cooling the medium (for instance, upon the addition of nitrogen to air, a greater inert mass is present to absorb the sensible heat in the system). A chemical inhibitor can involve an endothermic reaction that reduces the temperature or acts as a radical scavenger and reduces the radical pool that is available for chemical reactions to proceed. Chemical inhibitors can also burn and produce heat, e.g., in the case of chloromethane, CH_3Cl.

The sequence of methane oxidation, which dominates the combustion of natural gas, can involve 207 reactions and 40 species [Hunter et al., 1994], whereas propane oxidation can involve 493 reactions [Hoffman et al., 1991].

5.5.6 Overall Reaction Rate Expression

Suppose we initially charge A_2 and B_2 into a system. The overall reaction still produces 2AB. We will neglect the chain branching and breaking reactions. The rates for consumption and production of the reactants (species A_2 and B_2) and product (i.e., AB) are [Turns, 1996]. Using Equations (5.30) to (5.32), and (5.35)

$$d[A_2]/dt = -k_1[A_2][M] - k_3[A_2][B], \qquad (5.36)$$

$$d[B_2]/dt = -k_2[B_2][A], \qquad (5.37)$$

and

$$d[AB]/dt = k_2[A][B_2] + k_3[A_2][B] + k_6[A][B][M]. \qquad (5.38)$$

Likewise, the consumption and production rates for the radical species A and B are (again neglecting Reaction 5.33 and Reaction 5.34)

$$d[A]/dt = 2k_1[A_2][M] - k_2[A][B_2] + k_3[A_2][B] - k_6[A][B][M], \qquad (5.39)$$

and

$$d[B]/dt = k_2[A][B_2] - k_3[A_2][B] - k_6[A][B][M]. \quad (5.40)$$

One can solve five differential equations (5.36) to (5.40) and obtain d[AB]/dt. The radical species involved in the law of mass action are very reactive and difficult to measure during experiments. Thus, they could be related to concentration of stable species using certain approximations.

5.5.7 Steady-State Radical Hypothesis

Suppose we charge reacting species in a reactor maintained at T and P. Suppose radical A is produced, as well as consumed, by two reactions (Equation 5.31 and Equation 5.35). Initially [A] increases rapidly with time, then flattens out, and by the end of the reaction, decreases rapidly to zero. For most of the reaction time, the chain propagation steps dominate compared to chain initiation, branching, and termination; thus, [A] remains flat, indicating that as soon it is produced or it is consumed (d[A]/dt = 0). This is called the *steady-state radical hypothesis* (Figure 5.2); i.e., the production rate is equal to the radical destruction rate most of the time. Using steady-state approximations for radicals, we set d[A]/dt = 0, d[B]/dt = 0 (Equation 5.39 and Equation 5.40) and solve for [A] and [B]. The solution for [A] is

$$[A] = \left(\frac{k_1[A_2][M]}{2k_2[B_2]} + \left(\frac{k_1k_3}{k_6k_2}\right)^{1/2} \frac{[A_2]}{[B_2]^{1/2}} \left\{ 1 + \frac{k_6k_1}{4k_2k_3} \left\langle \frac{[M]^2}{[B_2]} \right\rangle \right\}^{1/2} \right)$$

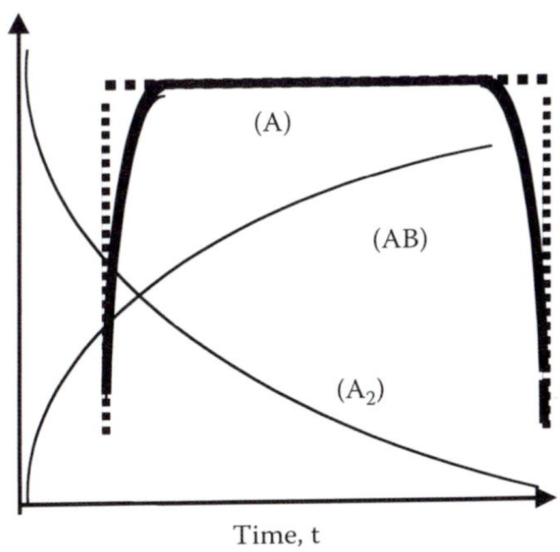

Figure 5.2 Illustration of concentration with time: reactants [A_2], products [AB], radical [A].

Then, substituting the expressions for [A] in Equation 5.37, we obtain the relation

$$\frac{d[B_2]}{dt} = -\frac{k_1[A_2][M]}{2} - \left\{\frac{k_2k_1k_3}{k_6}\right\}^{1/2} [A_2][B_2]^{1/2}$$

$$\left\{1+\frac{1}{4}\left\langle\frac{k_1^2\,k_6^2\,[A_2]^2\,[M]^4}{k_6k_2[B_2][M]k_1\,k_3[A_2]^2[M]}\right\rangle\right\}^{1/2} \qquad (5.41a)$$

The term $k_1k_6/(k_2\ k_3) << 1$, because the radical propagation reactions (involving the specific rates k_2 and k_3) are generally very fast. With that additional assumption, we obtain the relation

$$d[B_2]/dt = -(k_1/2)[A_2][M] - (k_1^2k_4[A_2][M]^3)/(4\ k_1k_3[B_2]) \qquad (5.41b)$$

At very low pressures, the concentrations are also low, so that $[A_2][M]^3/[B_2] << 1$. Therefore,

$$d[B_2]/dt \approx -(k_1/2)\ [A_2][M], \quad \text{i.e.,} \quad d[B_2]/dt \propto P^2. \qquad (5.41c)$$

At high pressures, $[A_2][M]^3/[B_2] >> 1$; the first term in Equation 5.41b can be ignored. Thus, $d[B_2]/dt \propto P^3$ if M is a gaseous species, but $d[B_2]/dt \propto P^0$ if M is a solid. This hypothesis illustrates complexities on simplified reaction rate expressions.

5.5.8 Catalytic Reactions

A substance is a catalyst if it can increase the speed of reaction rate at a given temperature without getting consumed and without altering its own overall concentration. Further, the energy difference Δh_R is not altered. It provides a new pathway in which the activation energy is reduced. Catalytic additives, such as calcium, potassium, and lithium acetate, are added in small amounts (1 to 2% by weight) to improve the combustion intensity of coal-fired boilers [Seshan and Kumar, 1982].

Other catalysts that promote oxidation of VOCs (volatile organic compounds) are from wood ash (mullite) ($3Al_2O_3 \bullet 2SiO_2$; sometimes Al replaced by $Fe^{3}+$ or Ti), magnetite (Fe_3O_4), and hematite (Fe_2O_3, with small amounts of V, Mn, Cu, and Co substituting for Fe) and from coal ash (e.g., mullite). CaO, MgO, and Fe_2O_3 have been used as catalysts to enhance H_2 yields in biomass gasification [Sutton et al., 2001]. Platinum is used as a catalyst in the exhaust of internal combustion engines to reduce the emissions of CO and unburned hydrocarbons. Protein molecules called *enzymes* act as catalysts enhancing reaction rates. For example, two nitrogenase enzymes provide low-activation-energy pathways for bacteria to convert atmospheric N_2 to NH_3, which plants can assimilate.

Catalysts involve both heterogeneous and homogeneous reactions. Consider ethylene molecule, which is an unsaturated HC. We wish to saturate it with more hydrogen by the reaction $C_2H_4 + H_2 \rightarrow C_2H_6$. A solid catalyst such as Pt, Pa, or Ni can adsorb H_2 (collection of molecules on the surface in attached mode), as well as C_2H_4. The adsorbed H can migrate toward adsorbed C_2H_4, form solid C_2H_6 (adsorbed C_2H_6), and get desorbed as gas [Zumdahl, 1986]. An example of a homogeneous catalyst is NO, which catalyzes O_3 production in the lower atmosphere, whereas in the upper atmosphere, it depletes O_3. Similarly, Freon 12 acts as a catalyst in O_3 depletion.

Heterogeneous catalysis occurs in catalytic converters in which the oxidation catalysts convert CO and unburned C_nH_m into CO_2 and H_2O at low temperature. There are other catalysts that can convert NO and NO_2 into N_2 and O_2. Pt (oxidation catalysts for CO to CO_2) and Rh (reduction catalysis, NO to N_2) are deposited on the honeycomb of Al_2O_3 in automobile engines. Other catalysts are Cu-Zn catalysts used in synthetic gas reactions ($CO + 2H_2 \rightarrow CH_3OH$) and the nickel oxide catalyst used in $CH_4 + H_2O \rightarrow CO + 3H_2$. Inhibitors are negative catalysts, which suppress reactions, whereas enzymes are positive catalysts.

Auto-catalytic reactions are those reactions in which the reaction rate increases with increased product concentrations [P]. For example, consider the n-th-order reaction:

$$d[A]/dt = -k\ [A]^{(n-j)}\ [P]^j.$$

5.6 Global Mechanisms for Reactions

5.6.1 Zeldovich Mechanism for NOx from N_2

The extended Zeldovich mechanism for NO formation involves the following simplified reaction scheme:

$$O + N_2 \rightarrow NO + N, \quad k_1, \tag{5.42a}$$

$$N + O_2 \rightarrow NO + O, \quad k_2, \tag{5.42b}$$

and

$$N + OH \rightarrow NO + H. \tag{5.42c}$$

The simpler Zeldovich reaction scheme for NO production is represented by the first two reactions, i.e., Reaction 5.42a and Reaction 5.42b only. In the context of the simpler scheme, if N atoms are assumed to be in steady state,

$$d[N]/dt = k_{1f}\ [N_2]\ [O] - k_{2f}\ [N][O_2] = 0$$

or

$$[N] = k_{1f}\ [N_2]\ [O]/k_{2f}\ [O_2],$$

so that the relation for NO production is simplified as

$$d[NO]/dt = 2\ k_{1f}\ [N_2][O]. \tag{5.43}$$

The O-atom concentration in Equation 5.43 can be determined from the equilibrium condition for the reaction (see also Subsection 5.8.3).

$$O_2 \Leftrightarrow 2O$$

for which

$$K^o = (p_O/1)^2/(pO_2/1) = (\bar{R}\ T/P^0)[O]^2/[O_2],$$

where $\bar{R} = 0.08314$ bar m^3/(kmol K) and $P^0 = 1$ bar. Therefore, Equation 5.43 becomes

$$d[NO]/dt = 2\ k_{1f}\ [N_2]\ K^{0\ 1/2}\ [O_2]^{1/2}\ \{P^0/(\bar{R}\,T)\}^{1/2}. \tag{5.44}$$

If one adopts the $K^{0\ 1/2}\ \{P^0/(\bar{R}\,T)\}^{1/2} = K_c^0 = 4.1\ \exp(-29{,}150/T)\ (\text{mol/cm}^3)^{1/2}$, and with $k_{1f} = 7 \times 10^{13}\ \exp(-27{,}750/T)$, cm^3/(mol sec) [Williams, 1985] then

$$d[NO]/dt = 5.7 \times 10^{14}\ \exp(-66{,}900/T)\ [O_2]^{0.5}\ [N_2],\ \text{mol cm}^{-3}\ \text{sec}^{-1}. \tag{5.45}$$

The overall nitric oxide formation rate through this mechanism is described by the empirical relation [Heywood, 1988]

$$\frac{d[NO]}{dt} = \frac{6\times10^{16}}{(T)^{0.5}}\exp\left(\frac{-69{,}090}{aT}\right)[O_2]^{0.5}[N_2]\ \text{mol cm}^{-3}\ \text{sec}^{-1}. \tag{5.46}$$

These relations show that the overall or global order of the reaction according to its rate is 1.5 and Equation 5.46 looks similar to Equation 5.44. Also note that Equation 5.45 ignores Reaction 5.42c.

5.6.2 NO_2 Conversion to NO

Consider the global reaction describing the conversion of NO_2 into NO, i.e.,

$$NO_2 + CO \rightarrow NO + CO_2. \tag{5.47}$$

This overall reaction represents the steps

$$NO_2 + NO_2 \rightarrow NO_3 + NO \tag{5.48a}$$

and

$$NO_3 + CO \rightarrow NO_2 + CO_2. \tag{5.48b}$$

In this scheme, NO_3 is neither a reactant nor a product, but is an intermediate species.

5.6.3 Hydrocarbon Global Reactions

This is a black box approach to representing a chemical process that involves several, even thousands, of elementary reactions. For instance, methane oxidation does not follow the simple step of Equation 5.1 but involves numerous reactions, many of which are sequential. The transformation of methane occurs through the formation of several intermediate species, some of which are stable (e.g., CH_2O, CO, and NO), and others are very reactive radical species (molecules with an unpaired electron, e.g., O, OH, N, CH_3).

5.6.3.1 Generic Approaches

The sequence of methane conversion into stable products (e.g., CO_2) occurs through the steps

$$CH_4 \rightarrow CH_3 \rightarrow CH_2O \rightarrow CHO \rightarrow CO \rightarrow CO_2.$$

The pathways involved can be represented in the form

$$CH_4 + \{M, H, O, OH, O_2\} \rightarrow \{H + M, H_2, OH, HO_2\} + CH_3 + \{O, O_2\} \rightarrow$$

$$\{H, OH\} + HCHO + \{M, OH\} \rightarrow \{\ldots, H_2O + CHO\} \text{ and so on.}$$

Initially, methane is oxidized by O, OH, and O_2.

5.6.3.2 Reduced Kinetics

The following are the various types of reduced kinetics mechanisms:

(1) GRI 2.1 Reduced kinetics [Miller and Bowman, 1989]: five-step reactions for methane in air:
   (a) $CH_4 + 1.5\ O_2 \rightarrow CO + 2\ H_2O$
   (b) $CO + 1/2\ O_2 \rightarrow CO_2$
   (c) $CO_2 \rightarrow CO + 1/2\ O_2$
   (d) HCN, NH_3, …, + $O_2 \rightarrow$ NO (nonthermal)
   (e) $N_2 + O_2 \rightarrow$ NO (Thermal)

(2) The four-step paraffin oxidation (C_nH_{2n+2}) [Westbrook and Dryer, 1984]:

$$C_nH_{2n+2} \rightarrow (n/2)\ C_2H_4 + H_2 \quad (5.49)$$

$$C_2H_4 + O_2 \rightarrow 2\ CO + 2\ H_2 \quad (5.50)$$

$$CO + (1/2)\ O_2 \rightarrow CO_2 \quad (5.51)$$

$$H_2(1/2)\ O_2 \rightarrow H_2O \quad (5.52)$$

$$\frac{d[Fuel]}{dt}, \frac{kmole}{m^3\ s} = -A_{homog}[B]^{\alpha}[C]^{\beta}[D]^{\gamma}\exp\left[-\frac{E}{\bar{R}T}\right], \quad \bar{R} = 8.314\frac{kJ}{kmole\ K},$$

kmol,m^3,sec

where A_{homog}, B, C, D, etc., are tabulated for several reactions in Table A.39B in Appendix A. Note that the oxidation of CO by molecular oxygen is facilitated by small amounts of water vapor. Typically, the relation that can represent the overall CO consumption rate is written as

$$d[CO]/dt = -A_{homog}\ [O_2]\ [H_2O]^{\gamma}\ [CO]^{\alpha}\ \exp\{-E/(\bar{R}T)\}. \quad (5.53)$$

Even 20 ppm of H_2 will oxidate CO rapidly.

(3) Global two-step oxidation reaction:

For any fuel, assume that all H is burned to H_2O while C burns to CO. Thus,

$$\text{Fuel} + \nu_{O_2}\ O_2 \rightarrow \nu_{CO}CO + \nu_{H_2O}\ H_2O \quad (5.54)$$

$$d[fuel]/dt \approx -A_{homog}\ [B]^{\alpha}\ [C]\ [D]^{\gamma}\exp(-E/RT),\ [B] = [Fuel],\ [C] = [O_2],\ [D] = 1$$

Then CO burns by Reaction 5.53 mentioned earlier.

(4) Global single-step oxidation schemes: Single-step oxidation occurs thus,

$$\text{Fuel} + \nu_{O_2}\ O_2 \rightarrow \nu_{CO_2}\ CO_2 + \nu_{H_2O}\ H_2O$$

$$d[Fuel]/dt \approx -A_{homog}\ [B]^{\alpha}\ [C]\ [D]^{\gamma}\exp(-E_1/RT),\ [B] = [Fuel],\ [C] = [O_2],\ [D] = 1$$

(5) Other oxidation schemes:

Sometimes the C oxidation is split between CO and CO_2 [Kong et al., 1995]

$$\text{Fuel} + \nu_{O_2}\ O_2 \rightarrow \rightarrow \nu_{CO}CO + \nu_{CO_2}\ CO_2 + \nu_{H_2O}\ H_2O$$

where ν_{CO}/ν_{CO2} (mole ratios) = $\gamma/(1-\gamma)$, $\gamma \approx 2/3$

5.6.4 The H_2–O_2 System

Consider the reactions for the H_2–O_2 system. Typically, one uses the global reaction $2H_2 + O_2 \rightarrow H_2O$. However, the simplified reaction schemes are as follows:

Chain initiation:

$$O_2 \rightarrow 2O \tag{R1}$$

Chain propagation:

$$H_2 + O_2 \rightarrow HO_2 + H, \tag{R2'}$$

$$OH + H_2 \rightarrow H_2O + H \tag{R2}$$

$$HO_2 + H_2 \rightarrow H_2O + OH \tag{R3}$$

$$OH + OH \rightarrow H_2O + O \tag{R4}$$

$$O + H + M \rightarrow OH + M \tag{R5}$$

$$H + O_2 + M \rightarrow HO_2 + M \tag{R6}$$

Chain branching:

$$O + H_2 \rightarrow OH + H \tag{R7}$$

$$H + O_2 \rightarrow OH + O \tag{R8}$$

Chain termination:

$$OH + HO_2 \rightarrow H_2O + O_2 \tag{R9}$$

$$H + H + M \rightarrow H_2 + M \tag{R10}$$

$$O + O + M \rightarrow O_2 + H \tag{R11}$$

$$OH + H + M \rightarrow H_2O + M \tag{R12}$$

The species, H_2, O_2, and H_2O are stable species. H, OH (hydroxyl), O, and HO_2 (hydroperoxyl) are reactive radicals. A single bond of H_2 is broken in Reaction R2' when a hydrogen molecule combines with O_2. The H atoms react with O_2 and form two OH radicals in Reaction R5 and Reaction R8 rather than in the competing reaction $H_2 + O_2 \rightarrow 2\,OH$, which is unlikely, because it requires breaking bonds in both H_2 and O_2. Reaction R6 is a competing pathway for H-atom consumption. Next, OH reacts with H_2 to form H_2O. Approximately 20 elementary reactions are required to adequately describe the $H_2 + O_2$

system. Totally, this system of reactions also involves the seven species O_2, O, H_2, HO_2, H, OH, H_2O, and the third body M.

It is customary to represent the reaction number by a digit, say, i, and each species or third body by another number, say, j. For instance, the index j = 1 could denote O_2, j = 2 could represent H_2, and so on. Likewise, i = 1 could be used to denote Reaction R1. This reaction mechanism can be denoted through the generic relation

$$\sum_{ji} \upsilon'_{ji} \, M_j \rightarrow \sum_{ji} \upsilon'_{ji} \, M_j.$$

Here, Mj denotes the j-th species, and υ'_{ji} and υ'_{ji} the stoichiometric coefficients of that species on the reactant and product sides of the reaction, respectively, for the particular reaction i. Several computer codes have been developed using this methodology to represent chemical reactions (see Subsection 5.13.2).

5.6.5 Carbon Monoxide Oxidation

The oxidation of CO in the absence of any hydrogen-containing material is extremely slow. The mechanism of reactions in the presence of small amounts of water vapor is

$$CO + O_2 \rightarrow CO_2 + O, \quad k_1 \tag{R1}$$

$$O + H_2O \rightarrow OH + OH, \quad k_2 \tag{R2}$$

$$CO + OH \rightarrow CO_2 + H, \quad k_3 \tag{R3}$$

and

$$H + O_2 \rightarrow OH + O, \quad k_4 \tag{R4}$$

The following reaction steps are important when hydrogen is present, namely,

$$O + H_2 \rightarrow H + OH, \quad k_5 \tag{R5}$$

and

$$OH + H_2 \rightarrow H + H_2O, \quad k_6 \tag{R6}$$

At high pressures or during the initial stages of hydrocarbon oxidation, the oxidation of CO by the hydroperoxyl (HO_2) radical may also be important.

$$CO + HO_2 \rightarrow CO_2 + OH, \quad k_7 \tag{R7}$$

Reaction R3 is a very important reaction during hydrocarbon combustion and is usually responsible for nearly all of the oxidation of CO to CO_2. It is a very complex elementary reaction that exhibits some pressure dependence, because its activated complex has a significant lifetime. However, the rate of reaction of OH with CO is lower than that of hydroxyl radicals with various hydrocarbon species. Therefore, the presence of even small amounts of hydrocarbons inhibits CO oxidation. Hence, the CO concentration continues to increase until all of the fuel is consumed, after which CO is oxidized in the presence of high OH concentrations. Consequently, although CO is produced in significant amounts by hydrocarbon oxidation, it is not oxidized until all of the hydrocarbons and hydrocarbon fragments are consumed and the OH concentration increases substantially.

5.7 Reaction Rate Theory and the Arrhenius Law

The previous sections described the types of molecules involved in reactions and the types of reactions. In this section, the theory underlying rate of reaction is presented.

5.7.1 Collision Theory

Collision reaction rate theory assumes that reacting molecules first collide, and hypothesizes that chemical reactions or chemical transformations will occur only in the case of energetic collisions.

5.7.1.1 Collision Number and Mean Free Path — Simple Theory

Molecules contained in matter travel a distance l_{mean} before colliding with another molecule. Consider a moving molecule, A, that first collides with a stationary molecule after traveling a distance l_{mean}, then undergoes another collision with another stationary molecule after moving a distance of $2l_{mean}$, and so on, until colliding for the N-th time with another molecule after having moved along a distance $N^*\ l_{mean}$. If these N collisions occur in 1 sec, the molecule A is said to undergo N collisions per unit time (also known as the *collision number*). Assume that the average molecular velocity of the moving molecule A is

$$V_{avg} = \{8k_B\ T/(\pi\ m_A\ )\}^{1/2}, \tag{5.55}$$

where m_A is the mass of molecule = M_A/N_{Avog}. V_{avg} is the distance through which the molecule travels in 1 sec. Now consider a geometrical space shaped in the form of a cylinder of radius σ and length V_{avg}. There are $n'\pi\sigma^2 V_{avg}$ molecules within this cylinder, where n′ denotes the number of molecules per unit volume. A molecule traveling through the cylinder will collide with all of the molecules contained within it (which are assumed to be stationary), because the cylinder radius equals σ. Therefore, the number of collisions undergone by a single molecule per unit time

$$Z_{coll,1} = n'_A\pi\sigma^2 V_{avg} = n'_A\pi\sigma_A^2\ \{8\ k_B\ T/(\pi\ m_A)\}^{1/2}.$$

If other molecules now move, the number of collisions undergone by a single molecule of A increases by √2; thus, the above equation is now corrected as

$$Z_{coll,1} = n'_A \pi\sigma_A^2 \sqrt{2}\,\{8\,k_B\,T/(\pi\,m_A)\}^{1/2}. \tag{5.56}$$

If n_A is the number of molecules per unit volume, as they move, the number of collisions undergone by all molecules per unit volume is given by

$$Z_{coll} = Z_{AA} = n'^{\,2}_A\ \pi\sigma_A^2\,\{16\,k_B\,T\,/(\pi\,m_A\,)\}^{1/2} \text{ number of collisions }/(\text{sec m}^3).$$

Because $n'_A = n_A\,N_{Avog} = [A]\,N_{Avog}$,

$$Z_{coll} = Z_{AA} = [A]^2\ \pi\sigma_A^2\ N_{Avog}^2\,\{16\,k_B\,T/(\pi\,m_A\,)\}^{1/2} \text{ number of collisions}/(\text{sec m}^3). \tag{5.57}$$

The time taken for a single collision is the inverse of $Z_{coll,1}$, which is given by Equation 5.56. The average distance traveled by the molecule during this time is called its mean free path l_{mean}, where

$$\ell_{mean} = V_{avg}/(n'\pi\sqrt{2}\,\sigma_A^2\,V_{avg}) = 1/(n'\pi\sqrt{2}\,\sigma_A^2)$$

5.7.1.2 Collision Number, Reaction Rate, and Arrhenius Law

Consider a boulder rolling down a hill; as it rolls down, the kinetic energy (KE) increases. Then it can climb another smaller hill if it has sufficient KE. Similarly, molecules with sufficient transnational energy (TE) can increase the distance between the atoms within a molecule and decrease the chemical force (or increase the interatomic potential energy); at some point, they reach the transition state that separates reactants from products. For example, NO_2 and CO require 132 MJ per kmol to reach the transition state, form N-O-O-C-O, and then rearrange to form NO and CO_2. This energy is acquired on impact between NO_2 and CO molecules. Smaller molecules move fast, collide more often, and participate in the reaction, increasing temperature and TE compared to larger molecules.

For bimolecular reactions of the form

$$A + B \rightarrow \text{products},$$

the consumption rate may be modeled as

d[A]/dt ∝ (Number of collisions of A and B per unit volume per unit time Z_{AB})* probability of reaction occurring on collision.

The probability of a reaction depends on various factors, such as bond energy (ε; Figure 5.3), and the species concentrations.

Equation 5.57 presents a relation for the number of collisions between similar molecules [A] and [A]. Now consider the binary species [A] and [B] having molecular diameters σ_A and σ_B, respectively. Following Equation 5.57, the number of A–B collisions per unit volume per unit time is

$$Z_{AB} = (N_{Avog}[A])(N_{Avog}[B])\ \pi\ \sigma_{AB}^2\ V_{rAB}, \tag{5.58}$$

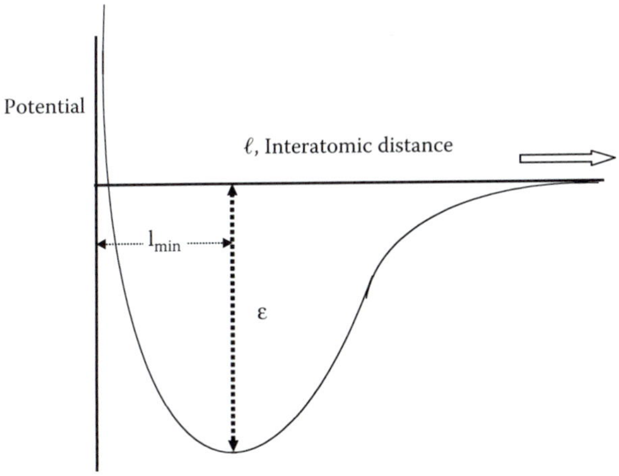

Figure 5.3 Potential field for CO molecules. Here, C and O atoms are separated by an interatomic distance ℓ.

where

$$V_{r,AB} = \{8k_BT/(\pi\mu_{AB})\}^{1/2}, \quad (5.59)$$

$$\sigma_{AB} = (\sigma_A + \sigma_B)/2, \quad (5.60a)$$

$$\mu_{AB} = \frac{m_A m_B}{m_A + m_B} \quad (5.60b)$$

$$m_A = \frac{M_A}{N_{Avog}} \quad (5.60c)$$

The term $\pi\,\sigma_{AB}^2$ denotes the cross-sectional area of collision; V_r, the average relative (but random) velocity between molecules of A and B; N_{Avog}, the Avogadro number of molecules 6.023×10^{26} molecules $kmol^{-1}$; k_B, the Boltzmann constant 1.38×10^{-26} kJ K^{-1} $molecule^{-1}$; and μ_{AB}, the reduced mass. Equation 5.58 provides the number of collisions per cubic meter per second. It reduces to Equation 5.57 for collision between similar molecules. If every collision between A and B results in the consumption of molecules of species A, the number of molecules consumed per unit volume per unit time is Z_{AB}. In this case, the number of moles of A that are consumed is

$$|\dot{W}'''_{hypoth}| = |d[A]/dt|_{hypoth} = Z_{AB}/N_{Avog} = (\pi\,\sigma_{AB}^2\,\bar{V}_r N_{Avog})[A][B] = k_{hypoth}\,[A][B], \quad (5.61)$$

where $k_{hypoth} = \pi\,\sigma_{AB}^2\,\bar{V}_r N_{Avog}$.

Equation 5.61 suggests that in a 1-m^3 volume, about 10^{34} collisions per second will result in the consumption of 10^{34} molecules of species A, which corresponds to a consumption rate of 10^8 kmol/m^3 sec. Experiments for various reactive species involved in combustion chemistry yield realistic values for consumption rates of 10^{-3} to 10^{-8} kmol/m^3sec at room temperature and about

10^3 at temperatures around 2000 K. This implies that not all collisions result in a chemical transformation of species A, because some molecules collide with greater energy, but others contain smaller amounts of energy.

Typically, at room temperature, TE ≈ 1/2 (k_B T), ve = 0.5 * 1.38 × 10^{-26} (kJ/molecule K) * 298 (K), and RE = 2.05 × 10^{-24} kJ/molecule or 1230 kJ/kmol. Compare this with a bond energy of H-H of 436,000 kJ/kmol or 7.24 × 10^{-22} kJ per H-H bond and C-H of 415,000 kJ/kmol or 6.89 × 10^{-22} kJ per C-H bond. Thus, bond energies are higher by a factor of 500 to 1,000. At low temperatures, these energies of collision cannot break the bonds, and chemical reactions may not occur. If T is increased to, say, 3000 K, there are greater numbers of molecules with increased te that thus approach the bond energy, altering the interatomic distance (Figure 5.3) and thus detaching the atom away from the molecule. Further, te, ve, and re have Maxwellian distribution of energy and can also fluctuate with time. However at temperatures of the order of 1,000 K, ve can increase. Hence, the molecule is not stable; it can break up into atoms. From kinetic theory, it is found that the fraction (F) of collisions having relative KE along the line of centers greater than E is given as

$$F = \text{Fraction of A–B collisions with sufficient energy to cause reaction} = \exp\left(-\frac{E}{\bar{R}T}\right) \quad (5.62)$$

where E is the activation energy. The activation energy is related to the bond energy. If A is the species consumed with stoichiometric coefficient "1," then we modify Equation 5.61 as

$$|\dot{w}_A'''| = |d[A]/dt| = \exp(-E/\bar{R}T)\, Z_{AB}/N_{Avog} = k_{hypoth} \exp(-E/\bar{R}T)\,[A][B] \quad (5.63)$$

Equation 5.63 requires one more correction. This correction S is called the orientation or steric factor. For example, if an O atom hits a CO molecule at the C-atom side of a CO molecule, CO_2 is formed because the O atom is symmetrically distributed around the C atom. If the O atom hits the O end of the CO molecule, CO_2 may not form. Thus, one must consider only those collisions with proper orientation. Thus, we rewrite Equation 5.63 as

$$\left|\dot{w}_A'''\right| = \left|d[A]/dt\right| = (-1)\, S\, k\, [A][B] \quad (5.64)$$

$$k = S k_{hypoth} \exp(-E/\bar{R}T) = \exp(-E/\bar{R}T) S\, N_{Avog}\, \pi\sigma_{AB}^2 = A \exp(-E/\bar{R}T) \quad (5.65)$$

A is called the *preexponential factor* and S, the *steric factor* or *orientation factor.* The term $\exp(-E/\bar{R}T)$ is the probability that an energetic collision will lead to reaction. From Equation 5.65, the preexponential factor is given as

$$A = S^* \pi\, \sigma_{AB}^2\, N_{Avog}\, \bar{V}_r \quad (5.66)$$

where $\bar{V}_r \propto T^{1/2}$ and $A \propto T^{1/2} \exp(-E/\bar{R}T)$. The law relating k to temperature is called the Arrhenius law. More generally, one writes

$$k = A\, T^n \exp(-E/\bar{R}T) \tag{5.67}$$

5.7.2 An Application

Let species A be an H atom. Consider the reaction

$$H + H \rightarrow H_2$$

Recall Equation 5.57

$$Z_{HH} = [H]^2 \pi \sigma_H^2 N_{Avog}^2 \{16\, k_B\, T\, N_{Avog}/(\pi\, M_B)\}^{1/2} \text{ number of collision/sec m}^3$$

$$-\frac{d[H]}{dt} = Z_{coll,HH} S^* \exp\left\{-\frac{E}{\bar{R}T}\right\} = 4\pi\sigma_H^2 [H]^2\, N_{Avog}^2 \sqrt{\frac{N_{Avog}\, k_B T}{M_H}}$$

$$S^* \exp\left\{-\frac{E}{\bar{R}T}\right\} \frac{\text{kmol}}{\text{m}^3\text{sec}}.$$

Rewriting,

$$k_{HH} = -\frac{d[H]}{dt[H]^2} = A\, T^n \exp\left\{-\frac{E}{\bar{R}T}\right\} \frac{\text{m}^3}{\text{kmol sec}},$$

$$\text{where } A = 4\pi\sigma_H^2\, N_{Avog}^2 \sqrt{\frac{N_{Avog}\, k_B}{M_H}}\, S \frac{\text{m}^3}{\text{kmol sec K}^{1/2}} \quad \text{and} \quad n = 1/2$$

$$|\dot{w}_H'''| = d[H]/dt = (-2)\, k_{HH}\, [A][B].$$

Although most of the reactions have activation energy E > 0, some do not have barriers at all, i.e., E = 0, e.g., H + NO; the reaction rate does not change between 1900 and 2000 K, i.e., E = 0. Many times, the reaction kinetics are given in units of kmol/m³sec when concentrations are expressed in kmol/m³. Instead, if the rate is in kg/m³sec and mass fractions are used, one needs to convert them (Subsection 5.13.4 shows details). Given,

$$|w_F'''| \frac{\text{kmol}}{\text{m}^3\text{sec}} = k[F]^{nF}\, [O_2]^{nO_2}, \quad k = A\, T^n \exp\left(-\frac{E}{\bar{R}T}\right)$$

one can show that

$$|w_F'''|\frac{kg}{m^3 sec} = k_Y \rho^{nCH4+nO2} Y_F^{nF} Y_{O2}^{nO2}, k_Y = A_Y T^n \exp\left(-\frac{E}{\bar{R}T}\right), A_Y = A\left\{\frac{M_F}{M_F^{nF}}\frac{1}{M_{O2}^{nO2}}\right\}$$

When a multistep reaction is represented with a global reaction rate expression, sometimes the rate can decrease when T increase or the apparent activation energy E < 0 for some global reaction.

Example 3

(a) Estimate the number of collisions of CO molecules with O_2 molecules per unit volume and per unit time at T = 300 K and 3000 K. Assume P = 100 kPa, CO = 40%, and O_2 = 60% (mole basis). The diameters are given as 3.59 Å for CO and 3.433 Å (Table A.3 of Appendix A).

(b) If each collision results in the reaction

$$CO + O_2 \rightarrow CO_2 + O.$$

With CO_2 and O atoms as products, how many kmol of CO will be consumed per unit volume and per unit time?

(c) If E = 120,000 kJ/kmol and steric factor = 1, compute reaction rates.

Solution

(a)

$$M_{co} = \frac{28}{6.023 \times 10^{26}} = 4.65 \times 10^{-26} \text{ kg/molecule} \quad (A)$$

$$M_{O_2} = \frac{32}{6.023 \times 10^{26}} = 5.31 \times 10^{-26} \text{ kg/molecule} \quad (B)$$

Using Equation 5.60b

$$\mu_{AB} = \frac{4.65 \times 10^{-26} \times 5.31 \times 10^{-26}}{4.65 \times 10^{-26} + 5.31 \times 10^{-26}} = 2.47 \times 10^{-26} \text{ kg} \quad (C)$$

At 300 K, using Equation 5.59

$$V_r = \left(\frac{8 \times 1.38 \times 10^{-23} (J/K) \times 300(K)}{3.14 \times 2.47 \times 10^{-26} (kg)}\right)^{\frac{1}{2}} = 652 \text{ m/sec at } 300 \text{ K} \quad (D)$$

Similarly at 3000 K,

$$V_r = \left(\frac{8 \times 1.38 \times 10^{-23} \times 3000}{3.14 \times 2.48 \times 10^{-26}}\right)^{1/2} = 2062 \text{ m/s at } 3000 \text{ K} \quad \text{(E)}$$

$$\sigma_{A+B} \frac{\sigma_A + \sigma_B}{2} = \frac{3.59 + 3.433}{2} = 3.511 \text{ Å}. \quad \text{(F)}$$

With T = 300 K, (N/V) = c = P/RT = 0.04 kmol/m^3 at 300 K

$$Z_{A+B} = \pi\sigma_{AB}^2 V_r \left[[A]N_{Avog}\right]\left[[B]N_{Avog}\right] = 3.14 \times (3.511 \times 10^{-10})^2 \times 652$$

$$\times [0.4 \times 0.04 \times 6.023 \times 10^{26}]\,[0.6 \times 0.04 \times 6.023 \times 10^{26}]$$

$$= 3.54 \times 10^{34} \text{ collisions/m}^3 \text{ sec}.$$

Similarly at 3000 K with c = P/RT = 0.004 kmol/m^3 at 3000 K

$$Z_{A+B} = 3.14 \times (3.511 \times 10^{-10})^2 \times 2061.78 \times [0.4 \times 0.004 \times 6.023 \times 10^{26}]$$

$$[0.6 \times 0.004 \times 6.023 \times 10^{26}] = 1.12 \times 10^{33} \text{ collisions/m}^3 \text{ sec}.$$

(b) If each collision that occurs results in a reaction between the colliding molecules, 3.54 × 10^{34} molecules of CO and O_2 will get consumed per second, i.e., Z_{AB}/N_{Avog} = 5.87 × 10^7 kmol of CO will get consumed per second per unit volume at 300 K and 1.86 × 10^6 of CO at 3000 K. Even though relative velocity is increased, less is consumed at high T due to reduced molal concentrations per unit volume (i.e., less number of molecules per unit volume) and less number of collisions per unit volume.

(c) At 3000 K

$$\dot{\omega}_{AB}''' = S\frac{Z_{AB}}{N_{Avog}} e^{-E/RT} = 1 \times \frac{1.11 \times 10^{33}}{6.023 \times 10^{26}} \times e^{-120000/8.314 \times 3000} = 1.51 \times 10^4 \text{ kmol/m}^3\text{s}.$$

Remarks

When T is increased from 300 to 310 K, reaction rate increases about five times and when it is increased from 300 K to 3000 K, the reaction rate increases 10^{18} times!

5.7.3 Determination of Kinetics Constants in Arrhenius Law

Whereas the preexponential factor A in the Arrhenius law is related to the collision number, E is related to the bond energy of molecules involved in the reactions. It is difficult to select a particular bond energy as many species are involved in producing the final products. Typically, we determine A and E in the Arrhenius law using experimental data.

Uni-molecular Br_2 reaction: $Br_2 \rightarrow 2\ Br$.

We supply Br_2 to a plug flow (uniform velocity V across cross section) quartz reactor maintained at constant T. The Br_2 molecules collide with each other and produce Br atoms. We measure $[Br_2]$ vs. x distance along the reactor. Then dt= dx/V; thus, we can evaluate $d[Br_2]/dt$ along the reactor. Then for the elementary reaction,

$$-d[Br_2]/dt = k\ [Br_2] \qquad (5.68a)$$

where, by the Arrhenius law in simplified form, $k = A \exp\{-E/(\bar{R}T)\}$, we rewrite Equation 5.68a as

$$-d\ (\ln [Br_2])/dt = k = A \exp\{-E/(\bar{R}T)\}. \qquad (5.68b)$$

For various temperatures, determine d (ln $[Br_2]/dt$ and plot

$$\ln \{-d\ (\ln [Br_2]/dt\} = \ln \{k\} \text{ vs. } 1/T.$$

The slope is equal to $E/\bar{R}$ for the elementary reaction.

Bimolecular H_2 and I_2 reaction [Laidler, 1963]: The species H_2 and I_2 react together to form HI

$$H_2 + I_2 \rightarrow 2\ HI$$

The Arrhenius law is written as

$$d[HI]/dt = k\ [H_2]\ [I_2] \text{ and hence,}$$

$$\frac{d[HI]/dt}{[H_2][I_2]} = k = A \exp^{(-E/\bar{R}T)}.$$

The plot of ln k vs. 1/T yields $E/\bar{R}$ for the elementary reaction. An example follows.

Example 4

Given the following experimental set of data, (a) determine the activation energy and preexponential factor A for the bimolecular reaction $H_2 + I_2 \rightarrow 2\ HI$; (b) determine the consumption rate of H_2 at 1000 K.

T	*d[HI]/dt, kmol/(m³ sec)*	$\frac{d[HI]/dt}{[H2][I2]} = k$, *m³/(kmol sec)*
600	1.075×10^{-04}	0.538
800	4.72×10^{-01}	$2.36 \times 10^{+03}$
1000	$7.24 \times 10^{+01}$	$3.62 \times 10^{+05}$
1200	$2.074 \times 10^{+03}$	$1.037 \times 10^{+07}$

Solution

(a) The Arrhenius law is written as

d[HI]/dt = k $[H_2]$ $[I_2]$ and hence

$$\frac{d[HI]/dt}{[H_2][I_2]} = k = A\exp^{(-E/\bar{R}T)}. \quad (5.68')$$

Figure 5.4 shows a plot of $In\left\{\frac{d[HI]/dt}{[H_2][I_2]}\right\}$ = In {k} vs. 1/T. The plot is linear and the slope yields E/$\bar{R}$ = 20,130 K or E = 20,130 K *8.314 (kJ/kmol K) = 167,360 kJ/kmol. The intercept y axis yields the value of A = 2 × 10^{14} m^3/(kmol sec). Thus

$$\frac{d[HI]}{dt} = 2\times 10^{14}[H_2][I_2]\exp^{(-20130/T)}, \frac{kmole}{m^3 \sec}.$$

(b) Using the law of stoichiometry,

$$\frac{d[H_2]/dt}{-1} = \frac{d[I_2]/dt}{-1} = \frac{d[HI]/dt}{2} = k[H_2][I_2].$$

Thus,

d$[H_2]$/dt = −36.2 kmol/m^3sec.

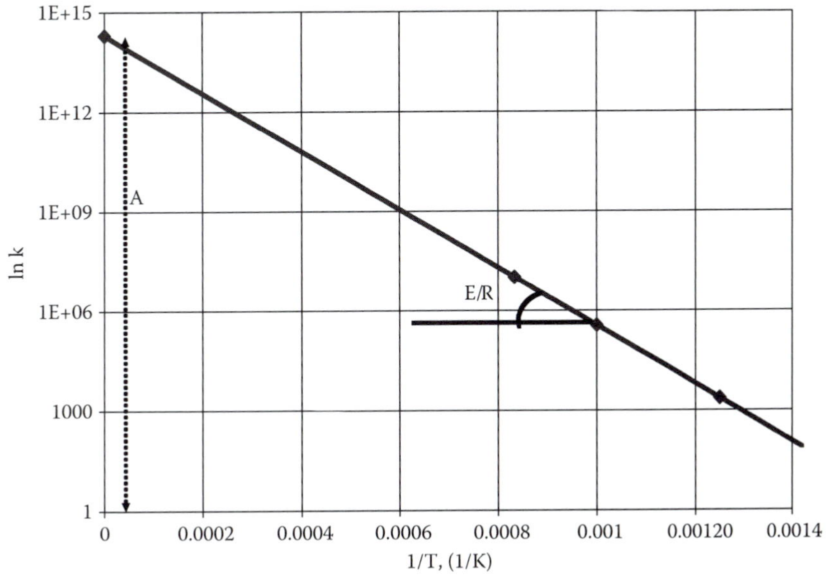

Figure 5.4 Plot of ln k vs. 1/T for HI formation.

Remarks

(i) The negative sign indicates that the species are being consumed. See Glassman for kinetics constants A and E for many elementary reactions.
(ii) There are several other techniques for determining A and E: (1) ignition studies, (2) PFR and PSR (chapter 7), and (3) laminar flame propagation velocity (chapter 15).

5.8 Second Law and Global and Backward Reactions

5.8.1 Backward Reaction Rate and Second Law

Consider the overall reaction at a given T and P

$$CO_2 + H_2 \rightarrow CO + H_2O.$$

From the thermodynamic perspective and the second law (Chapter 3), there should be positive entropy generation at a given T and P. This reaction is possible at a given T and P if and only if

$$\hat{g}\,CO_2 + \hat{g}H_2 > \hat{g}\,CO + \hat{g}\,H_2O$$

where $\hat{g}_k(T, P_k) = \bar{g}_k^0(T) + \bar{R}T \ln (p_k/1) = \bar{g}_k^0(T) + \bar{R}T \ln ([K] \;\; \bar{R}\,T/1)$. Alternatively, this criterion is written as (Chapter 3)

$$\Pi[K]^{\upsilon_k} \le K^0(T)\left(\frac{1}{\bar{R}T}\right)^{\Sigma_{\upsilon_k}} \tag{5.69}$$

where $[K] = p_k/\bar{R}T$, i.e., in kmol/m^3, and $\ln K^o(T) = -\{\Delta G^o/(\bar{R}\,T)\}$, $\Delta G^0(T)/(\bar{R}\,T) = \{\,\bar{g}_{CO}^0\,\}/(\,\bar{g}_{H2O}^0 T)$. Expanding Equation 5.69 for current reaction,

$$[CO][H_2O]/[CO_2][H_2] \le K^o(T)^* \;[1/\bar{R}\,T]^{(1+1-1-1)}.$$

Thus, at any given section, $CO_2 + H_2 \rightarrow CO + H_2O$ proceeds if the preceding criterion is satisfied, and if so, CO_2 is being consumed. From the kinetics point of view, the rate of change in CO_2 at any T and P is written as

$$\frac{d[CO_2]}{dt} = -k_f[CO2][H_2] + k_b[CO][H_2O] \tag{5.70}$$

where the second term on the right implies that $CO + H_2O \rightarrow CO_2 + H_2$, where CO_2 is being produced. How is the backward reaction rate expression possible even though at that location $\hat{g}_{CO2} + \hat{g}_{H2} > \hat{g}_{CO} + \hat{g}_{H2O}$ or $\Delta F_{chem} = \{\,\hat{g}_{CO2} + \hat{g}_{H2} - \hat{g}_{CO} - \hat{g}_{H2O}\} > 0$? Recall that the Maxwell–Boltzmann distribution law (Chapter 1) suggests that there could be a few energetic collisions between CO and H_2O that can result in production of CO_2 and H_2, leading to backward production. However, there is still net production of CO and the second law is still satisfied, because the forward reaction dominates as long as $\Delta F_{chem} > 0$. The literature on heat transfer contains only an expression for the net heat transfer rate, but it does not link the net rate with forward heat transfer, say, from hot water to cold air and backward heat transfer from cold air to hot water. The rate of net heat transfer is governed by the Fourier law $\dot{q}'' = -\lambda \nabla T$, where T is a measure of macroscopic energy. As the gradient reduces to zero, we reach thermal equilibrium. But there is no such law that links the rate of net reaction with the gradient in ΔF_{chem}. Let us postulate that the rate of production of CO_2 within a unit volume depends on the reaction potential difference through f $\{\Delta F_{chem}\}$:

$$d\,[CO_2]/dt = -k_f\,[CO_2]\,[H_2]\,f\,\{\Delta F_{chem}\} \tag{5.71}$$

where f $\{\Delta F_{chem}\}$ is an arbitrary function of $\{\Delta F_{chem}\}$. We have to select the function f $\{\Delta F_{chem}\}$ such that as $\{\Delta F_{chem}\}$ tends to zero, the rate, or f $\{\Delta F_{chem}\}$, approaches zero; as $\{\Delta F_{chem}\}$ tends to a large value, f $\{\Delta F_{chem}\}$ must tend to unity, indicating only forward rate. Therefore, we select f $\{\Delta F_{chem}\}$ as $\{1\text{-exp}\,[-\Delta F_{chem}/(\bar{R}\,T)]\}$. Thus,

$$d[CO_2]/dt = -k_f\,[CO_2][H_2]f\{\Delta F_{chem}\} = k_f\,[CO_2\,]\,[H_2]\,\{1-\exp\,[-\Delta F_{chem}/(\bar{R}\,T)]\}.$$

But

$$\exp\{-\Delta F_{chem}/(\bar{R}\,T)\} = \exp\,\{\Delta G^0/\,\bar{R}\,T + \ln\,(RT/1)^{\Sigma vk}\,[CO][H_2O]/[CO_2][H_2]\} = \{1/K^0\}\,\{[CO][H_2O]/[CO_2][H_2]\}$$

where $K^0 = \exp\{-\Delta G^0/(\bar{R}\,T)\}$ and $\sum \upsilon_k = 1 + 1 - (1 + 1) = 0.$

Hence,

$$d[CO_2]/dt = -k_f\,[CO_2\,]\,[H_2]\,(1-\{1/K^0\}\,\{[CO][H_2O]/[CO_2][H_2]\})$$

$$d[CO_2]/dt = -\{k_f\,[CO_2\,]\,[H_2]\,-\{k_f/K^0\}\,[CO][H_2O]\} = -\{k_f\,[CO_2\,]\,[H_2]\,-k_b\,[CO][H_2O]\} \tag{5.72}$$

$$\text{where } k_b = \{k_f/K^0\}. \tag{5.72$'$}$$

Another way to interpret forward and backward reaction rates is to look at collisions between CO_2 and H_2 that produce CO and H_2O along with collisions between CO and H_2O that result in production of CO_2 and H_2. Net production of CO_2 must conform to the second law for this elementary reaction.

5.8.2 *Equilibrium Constants and Estimation of Backward Reaction Rate Constants*

5.8.2.1 *Equimolecularity of Products and Reactants*

Estimation of k_b using Equation 5.72′ ensures that the second law will be satisfied under all conditions. We will obtain the relation given in Equation 5.72′ using the equilibrium concept. Suppose we have a quartz reactor. We admit large quantities of CO_2 and H_2 and small quantities of CO and H_2O. Consider the chemical reactions:

$$CO_2 + H_2 \rightarrow CO + H_2O, \quad \text{forward rate is } k_f \tag{A}$$

$$CO + H_2O \rightarrow CO_2 + H_2, \quad \text{backward rate is } k_b \tag{B}$$

Let us keep T and P fixed. At any location

$$\frac{d[CO_2]}{dt} = -k_f[CO_2][H_2] + k_b[CO][H_2O] \tag{5.73}$$

$[CO_2]$, $[H_2]$, etc., keep decreasing, CO and H_2O keep increasing, and eventually reach constant values, indicating equilibrium. At the location where equilibrium prevails:

$$\frac{d[CO_2]}{dt} = -k_f[CO_2]_{eq}[H_2]_{eq} + k_b[CO]_{eq}[H_2O]_{eq} = O$$

$$\frac{k_f(T)}{k_b(T)} = \frac{[CO]_{eq}[H_2O]_{eq}}{[CO_2]_{eq}[H_2]_{eq}} = \frac{\left(\frac{P_{CO}}{\bar{R}T}\right)\left(\frac{P_{H2O}}{\bar{R}T}\right)}{\left(\frac{P_{CO}}{\bar{R}T}\right)\left(\frac{P_{H2}}{\bar{R}T}\right)} = \frac{\left(\frac{P_{CO}}{P^0}\right)\left(\frac{P_{H2O}}{P^0}\right)}{\left(\frac{P_{CO2}}{P^0}\right)\left(\frac{P_{H2}}{P^0}\right)} = K^0(T) \tag{5.74}$$

where P^0 is the standard pressure (typically $P^0 = 1$ bar) and K^0 is the equilibrium constant (Table A.36A and Table A.36B of Appendix A). Thus, for the current reaction, in which the overall order of reaction in forward and backward reactions is the same,

$$k_f(T)/k_b(T) = K^0(T). \tag{5.75}$$

Hence, if $k_f(T)$ and $K^0(T)$ are known, one can determine $k_b(T)$. Now we can write Equation 5.73 at any location as

$$\frac{d[CO_2]}{dt} = -k_f[CO2]\,[H_2]\left\{1 - \frac{1}{K^0(T)}\frac{[CO][H_2O]}{[CO_2][H_2]}\right\}. \tag{5.76}$$

Equation 5.76 will yield $\{d[CO_2]/dt\} > 0$ or < 0 depending upon concentrations. According to the second law, $\{d[CO_2]/dt\} < 0$ if the first of the following conditions is satisfied; if the second is satisfied, then $\{d[CO_2]/dt\} > 0$ (Chapter 3).

$$\frac{[CO][H_2O]}{[CO_2][H_2]} < K^0(T) \quad \textit{for} \quad CO_2 + H_2 \rightarrow CO + H_2O$$

$$\frac{[CO][H_2O]}{[CO2][H_2]} > K^0(T) \quad \textit{for} \quad CO + H_2O \rightarrow CO_2 + H_2$$

Thus, the second law is satisfied for each elementary step reaction at every stage of the reaction as it progresses; however, this may not be the case for the global empirical equation describing the overall reaction rate or those using approximations. Now, consider simplified global reactions or reduced mechanisms (e.g., the four-step global reaction for HC oxidation). If the constants for the four steps are evaluated for limited T and P values from detailed chemical kinetics that include forward and backward reactions, then one of the preceding conditions must apply within this range; otherwise you may end up with negative entropy generation. Even the assumption of complete combustion may be a violation of the second law of thermodynamics. Recall the definition of K^0 from Chapter 3:

$$k_f/k_b = K^0(T) = \exp\left(\frac{-\Delta G^o}{\bar{R}T}\right) \quad \text{and} \quad \ln K^o = \frac{-\Delta G^o}{\bar{R}T} = -\left\{\Delta H_R^o(T) - T\Delta S_R^o(T)\right\}$$

Then one can show from the vant Hoff equation (Chapter 3) that

$$\frac{d\ln K^o}{dT} = \frac{\Delta G^o}{\bar{R}T^2} - \left[\frac{1}{\bar{R}T}\frac{d(\Delta G^o)}{dT}\right] = \frac{\Delta H_R^o}{\bar{R}T^2} \tag{5.77}$$

where ΔH_R^0 denotes the enthalpy of reaction in kJ/kmol

$$\Delta H_R^0 = \bar{h}^0_{f,CO} + \bar{h}^0_{H2O} - \{\bar{h}^0_{f,CO2} + \bar{h}^0_{f,H2}\}$$

$$\frac{d\ \text{In}\ k_f}{dT} - \frac{d\ \text{In}\ K_b}{dT} = \frac{\Delta\bar{h}_R^0}{\bar{R}T^2}.$$

Using the Arrhenius law,

$$k = A\ T^n \exp\left(-\frac{E}{\bar{R}T}\right),$$

$$\text{In}\ k = \text{In}\ A - \frac{E}{\bar{R}T} + n\ \ \text{In}\ \ T.$$

Applying this relation for forward and backward directions,

$$\text{In}\ k_f = \text{In}\ A_f - E_f/(\bar{R}T) + n_f\ \ \text{In}\ T$$

$$\text{In}\ k_b = \text{In}\ A_b - E_b/(\bar{R}T) + n_b\ \ \text{In}\ T$$

Assuming $n_f = n_b$

$$\frac{d\ln\left(\frac{k_f}{k_b}\right)}{dT} = \frac{E_f - E_b}{\bar{R}T^2}.$$

But

$$d\ln[K^0]/dT = d\ln[k_f/k_b]/dT = \Delta H_R^0/\bar{R}T^2.$$

Hence

$$(E_f - E_b) = \Delta H_R^0. \tag{5.78}$$

If ΔH_R^0 is constant, then E_b can be estimated if E_f is known; once A_f is known, A_b can also be determined. Figure 5.5 illustrates the relation between activation energies and enthalpy of reaction. For exothermic reactions, $\Delta H_R^0 < 0$ and hence $E_b > E_f$ (Figure 5.5b); for endothermic reactions, $\Delta H_R^0 > 0$ and hence $E_f > E_b$ (Figure 5.5a)

Example 5

Calculate the rate constant k for the reaction $H_2 + O_2 \rightarrow H + HO_2$ at a temperature of 1,500 K. Assume for HO_2, h_f^0 = 20,900 kJ/kmol, s^0 = 297.3 kJ/kmol K and $h_{t,1500} - h_{t,298}$ = 54,511 kJ/kmol.

$$k = A_f T^n \exp\left(\frac{-E_f}{\bar{R}T}\right), n = 0, \log_{10} A_f = 13.4, E_f = 2926\frac{\text{kJ}}{\text{kmole}} \quad \text{[Glassman]}$$

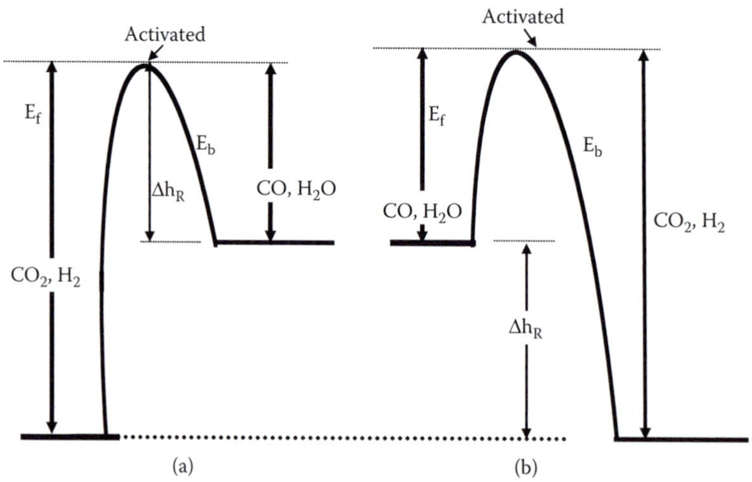

Figure 5.5 Relation between forward and backward activation energies and enthalpy of reactions; (a) E_f-E_b = $\Delta h_R^0 > 0$ for the current scheme ($CO_2 + H_2 \rightarrow CO + H_2O$ or E_f-E_b > 0 for endothermic reactions, (b) E_f-E_b = $\Delta h_R^0 < 0$ for the scheme ($CO + H_2O \rightarrow CO_2 + H_2$ or E_f-E_b < 0 for exothermic reactions).

Solution

For the reaction $H + HO_2 \rightarrow H_2 + O_2$,

$$k_f(T) = 10^{10.4}\,\frac{m^3}{kmol\ sec}\exp\left(-\frac{2926}{8.314\times1500}\right) = 1.986\times10^{10}\,\frac{m^3}{kmol\ sec}$$

$$\bar{g}^0_{HO2} = \bar{h}_{HO_2} - T\bar{s}^0_{HO_2}$$

= [20900 + 54511] kJ/kmol − 1500 K* 297.3 kJ/kmol K = −315940 kJ/kmol.

At a temperature of 1500 K, the change in Gibbs function is

$$\Delta G^0_T = \bar{g}^0_{H2}(1500) + \bar{g}^0_{O2}(1500) - \bar{g}^0_{H}(1500) - \bar{g}^0_{HO2}(1500)$$

$$= -231{,}835 - 346{,}374 - 20679 + 315{,}940 \; = -228350\ \text{kJ/kmol}$$

$$K^0\ (1500\ K) = \exp\left(-\frac{\left(-2.28350\times10^5\,\text{kJ/kmol}\right)}{(8.314\ \text{kJ/kmol}\ K)(1500\ K)}\right) = 8.96\times10^7$$

$$K^0(1510) = \frac{k_f(1510)}{k_b(1510)} \Rightarrow k_b\ (1510) = \frac{k_f(1510)}{K^0(1510)} = \frac{1.986\times10^{10}}{8.96\times10^7} = 222\frac{m^3}{kmol\ s}.$$

5.8.2.2 General Reaction of Any Molecularity

Proceeding as before, consider

$$a\ A + b\ B \Leftrightarrow c\ C + d\ D$$

where a, b, c, and d are stoichiometric coefficients.

Then, the concentration-based equilibrium constant K_c is written as

$$K_c = k_f/k_b = [C]^c\ [D]^d/\{[A]^c\ [B]^b\}$$

and K_c is related to the equilibrium constant K^0 and $[C] = p_c/(\bar{R}T)$.

$$K^0(T) = (p_c/P^0)^c\ (p_D/\ P^0)^{\ d}/\{(p_A/\ P^0)^a\ (p_B/\ P^0)^b\} = K_c\ [\,\bar{R}\ T/\ P^0]^{(c+d-a-b)}$$

where $\bar{R}$ is in units of bar m^3/kmol K, P^0 is the reference pressure in bar and typically $P^0 = 1$ if bar units are adopted for $\bar{R}$; $\Delta G^0 = c\bar{g}^0_C + d\bar{g}^0_D - a\bar{g}^0_A - \bar{g}^0_B$.

$$\text{Thus, } k_f/k_b = K_c = K^0\ (T) * [P^0/\ (\bar{R}T)]^{(c+d-a-b)}. \qquad (5.79)$$

Example 6

Consider a reactor in which we admit O_2 at $[O_2]_0$ and $[O]_0$ at x = 0. As the reaction proceeds, $[O_2]$ decreases and $[O]$ increases.

$$O_2 \underset{k_b}{\overset{k_f}{\rightleftarrows}} 2O \tag{A}$$

(a) Determine $[O_2]$ as a function of t if T is fixed assuming $[O_2]$ is small.
(b) Find the concentrations at equilibrium.

Solution

(a) The $[O_2]$ and [O] are related at any point x in the reactor. We call this the atom balance equation. The rate of formation of [O] is written as

$$\frac{d[O_2]}{dt} = -k_f\,[O_2] + k_b[O]^2. \tag{B}$$

$$\text{O atom balance: } [O] + 2[O_2] = [O]_0 + 2[O_2]_0 \tag{C}$$

Thus, using Equation C in Equation B to eliminate [O] and simplifying

$$d[O_2]/dt \simeq -(k_f + 2\,k_b)[O_2] + k_b\,\{[O]_0 + 2[O_2]_0\}^2. \tag{D}$$

Integrating at constant T, and with known concentration at t = 0

$$[(k_f + 2\,Y\,k_b)[O_2] - k_bY^2]/\{(k_f + 2\,Y\,k_b)[O_2]_0 - k_b\,Y^2\} \simeq \exp\{-(k_f + 2Y\,k_b)t\} \tag{E}$$

where $Y = [O]_0 = 2[O_2]_0$

(b) When equilibrium is reached at given temperatures, the forward reaction rate becomes equal to the backward reaction rate. Thus, from Equation D with $d[O_2]/dt = 0$, and with $[O]_0 = 0$

$$[O_2]_{eq}/[O_2]_0 \simeq 2\,k_b[O_2]_0\,/(k_f + 2[O_2]_0k_b). \tag{F}$$

5.9 The Partial Equilibrium and Reaction Rate Expression

5.9.1 Partial Equilibrium

Most combustion systems entail oxidation mechanisms with numerous individual reaction steps. Under certain circumstances, a group of reactions will proceed rapidly and reach a quasi-equilibrium state. Concurrently, one or more reactions may proceed slowly. If the rate or rate constant of this slow reaction is to be determined and if the reaction contains a species difficult to measure, it is possible, through a partial equilibrium assumption, to express the unknown concentrations in terms of other measurable quantities. Thus, the partial equilibrium assumption is very much like the steady-state approximation discussed earlier. The difference is that in the steady-state approximation one is concerned with "steadiness" of a selected radical species typically present in multiple reactions, and in the partial equilibrium assumption one is concerned with a specific reaction.

5.9.2 Reaction Rate

A specific example can illustrate the use of the partial equilibrium assumption. Consider, for example, CO in an oxidizing medium. By the measurement of the CO, O_2, and H_2O concentrations, it is desired to obtain an estimate of the rate constant of the reaction that governs CO_2 production

$$CO + OH \rightarrow CO_2 + H. \tag{5.80}$$

The rate expression is

$$\frac{d[CO_2]}{dt} = -\frac{d[CO]}{dt} = k[CO][OH]. \tag{5.81}$$

Then, the question is how to estimate the rate constant k without a measurement of the OH concentration in the combustion system. If one assumes there is equilibrium between the H_2–O_2–H_2O chain species, then the following equilibrium reactions could be used to complete the reaction scheme:

$$H_2 + O_2 \rightleftharpoons 2\,OH, \tag{5.82}$$

$$H_2 + \frac{1}{2}O_2 \rightleftharpoons H_2O. \tag{5.83}$$

Note that Reaction 5.82 and Reaction 5.83 involve elements in their natural forms producing a compound. Thus

$$K^0_{OH} = K_{c,OH} = [OH]^2/\{[H_2][O_2]\} \tag{5.84}$$

$$K^0_{H2O} = \frac{[H_2O]}{[H_2][O_2]^{1/2}}\left(\frac{RT}{P^0}\right)^{-1/2} = K_{c,H2O}(\bar{R}\,T/P^0)^{-1/2}. \tag{5.85}$$

$[H_2]$ and [OH] are unknown. Both could be solved from Equation 5.84 and Equation 5.85 in terms of H_2O and O_2. Thus, expressing [OH] in terms of $[H_2O]$ and $[O_2]$, and substituting this result in Equation 5.81,

$$\frac{d[CO_2]}{dt} = -\frac{d[CO]}{dt} = k\left[\frac{K^{0\,2}_{OH}}{K^0_{H_2O}}\right]^{1/2}\left[\frac{P^0}{\bar{R}T}\right]^{1/4}[CO][H_2O]^{1/2}[O_2]^{1/4} \tag{5.86}$$

where P^0 is typically 1 bar. Thus, one observes that the rate expression can be written in terms of readily measurable species. Care must be exercised in applying this assumption. Equilibria do not always exist among the H_2–O_2 reactions in a hydrocarbon combustion system and, indeed, it is questionable if equilibrium exists during CO oxidation in a hydrocarbon system. Nevertheless, it is interesting to note that there is experimental evidence that shows the rate of formation of CO_2 is 1/4 order with respect to O_2, 1/2 order with respect to water, and first order with respect to CO.

Example 7

Thermal NO formation:

$$N_2 + O \underset{}{\overset{k_1}{\rightleftharpoons}} NO + N \tag{1}$$

$$N + O_2 \overset{k_2}{\rightleftharpoons} NO + O \tag{2}$$

$$O_2 \overset{k_3}{\rightleftharpoons} O \tag{3}$$

(a) Use the steady-state approximation for N and obtain an expression for [N]; (b) use the equilibrium assumption for O concentration using $O_2 \Leftrightarrow 2\,O$, and obtain an expression for d[NO]/dt in terms of k_{1f}, $[N_2]$, K^0, $[O_2]$, $\bar{R}$ and T, where K is the equilibrium constant for $O_2 \Leftrightarrow 2\,O$; for forward reaction $E \approx$ 315,000 kJ/kmol.

Solution

$$\text{(a)}\ d[N]/dt = k_{1f}[O][N_2] - k_{1b}[NO][N] - k_{2f}[O_2][N] + k_{2b}[NO][O] \tag{A}$$

$$d[O]/dt = -k_{1f}[O][N_2] + k_{1b}[NO][N] + k_{2f}[O_2][N] - k_{2b}[NO][O] + k_{3f}[O_2] - k_{3b}[O]^2 \tag{B}$$

$$d[N_2]/dt = -k_{1f}[O][N_2] + k_{1b}[NO][N] \tag{C}$$

$$d[O_2]/dt = -k_{3f}[O_2] + k_{3b}[O]^2 - k_{2f}[O_2][N] + k_{2b}[NO][O] \tag{D}$$

$$d[NO]/dt = k_{1f}[O][N_2] - k_{1b}[NO][N] + k_{2f}[O_2][N] - k_{2b}[NO][O] \tag{E}$$

Since d[N]/dt = 0, setting Equation A to zero and solving

$$[N] = \frac{k_{1f}[O][N_2] + k_{2b}[NO][O]}{k_{2f}[O_2] + k_{1b}[NO]} \tag{F}$$

(b) Using Equation F in Equation E,

$$\frac{d[NO]}{dt} = k_{1f}[O][N_2] + \left[\frac{\{k_{2f}[O_2] - k_{1b}[NO]\}\{k_{1f}[O][N_2] + k_{2b}[NO][O]\}}{\{k_{2f}[O_2] + k_{1b}[NO]\}}\right] - k_{2b}[NO][O]. \quad (G)$$

Similarly, with $d[O]/dt = 0$, using Equation F for N, setting Equation B = 0, one can obtain [O]. Alternately adding Equations A and B and setting sum to zero,

$$k_{3f}[O_2] - k_{3b}[O]^2 = 0. \quad (H)$$

That is, $[O_2]$ and [O] are equilibrium concentrations when $d[N]/dt = 0$ and $d[O]/dt = 0$. Simplifying Equation G,

$$\frac{d[NO]}{dt} = \left[\frac{2k_{1f}\,k_{2f}[O_2][N_2][O]}{\{k_{2f}[O_2] + k_{1b}[NO]\}}\right]\left[1 - \frac{k_{1b}[NO]k_{1f}[N_2]}{k_{1f}\,k_{2f}[N_2][O_2]}\}\right] \quad (I)$$

$$[O] \approx [O]_{eq} = \{K_{c,O}\,[O_2]\}^{1/2} = \{K_O^0\,(\bar{R}T/P^0)^{-1}[O_2]\}^{1/2}.$$

Using this result in Equation I and simplifying

$$\frac{d[NO]}{dt} = \left[\frac{2k_{1f}K^0\left(\frac{\bar{R}T}{P^0}\right)^{-1}[O_2]^{1/2}[N_2]}{\left\{1 + \frac{k_{1b}[NO]}{k_{2f}[O_2]}\right\}}\right]\left[\left\{1 - \frac{k_{1b}[NO]\,k_{2b}[NO]}{k_{1f}\,k_{2f}[N_2][O_2]}\right\}\right]. \quad (J)$$

Because NO is in trace amounts, $[k_{1b}/k_{2f}]\,[NO]/[O_2] <<< 1$ and $[k_{2b}/k_{2f}]\,[NO]/[N_2] <<< 1$. Equation J simplifies to

$$\frac{d[NO]}{dt} = 2k_{1f}\,K^0\,[O_2]^{1/2}\,[N_2]\left[\frac{P^0}{\bar{R}T}\right]^{1/2}. \quad (K)$$

Consider an overall reaction

$$N_2 + O_2 \rightarrow 2\,NO \quad (L)$$

$$\{d[NO]/dt\}\,/(2) = \{d[N_2]/dt\}/(-1) = \{d[O_2]/dt\}/(-1) = k_O\,[N_2]^a[O_2]^b) \quad (M)$$

Then,

$$\{d[NO]/dt\} = 2\,k_g\,[N_2]^a[O_2]^b. \quad (N)$$

Comparing Equation S with Equation Q,
$a = 1$, $m = 1/2$; global specific reaction rate, $k_g = k_{1f}\;K_O^0\,\{P^0/(\bar{R}T)\}^{1/2}$

5.10 Timescales for Reaction

5.10.1 Physical Delay

Typically, chemical reactions occur rapidly at temperatures higher than the ignition temperature T_{ign}, the temperature at which reaction rates are high enough to result in appearance of a flame. Chapter 13 deals with ignition in more detail, and Table 39C presents ignition data for fuels. The time required to heat up the mixture to ignition temperature without appreciable chemical reaction (or heat release) is called *physical delay.* Suppose a premixed cold combustible mixture is stored in a rigid container of volume V and surface area A immersed in a bath at T_{wall}. Assume that the mixture is at uniform $T < T_w$ when it heats up.

Then

$$\rho V c_v \frac{dT}{dt} = h_H A (T_w - T) . \tag{5.87a}$$

A similar expression can be written for the mixture at constant P (e.g., a mixture kept in a PCW assembly or a mixture flowing through a duct)

$$\rho V c_p \frac{dT}{dt} = h_H A (T_w - T) . \tag{5.87b}$$

If the wall temperature is low, T will rise from an initial temperature $T_{initial}$ and eventually reach T_w. If T_w is raised beyond a certain value, then T rises, seems to flatten, but again rises rapidly. Figure 5.6 shows a plot of T vs. t (GBCD) and radical concentration A* vs. t (EFG). The lowest gas temperature at which the rapid rise in T is observed is called *ignition temperature.* Physical delay (Figure 5.6) is approximately given by

$$t_{phys} \approx \frac{(T_{ignition} - T_{initial})}{(dT/dt)_{t=0}} . \tag{5.88}$$

5.10.2 Induction Time (t_{ind})

Radicals appear just after a physical delay t_{phys}. A finite time is required to generate enough radicals, which by chain branching reactions multiply to a large number (called an *explosion*), resulting in rapid temperature rise, appearance of flame, or both. The time to reach the critical number of radicals or the explosion level is called *induction time* t_{ind}.

5.10.3 Ignition Time (t_{ign})

$$t_{ign} = t_{phys} + t_{ind} \tag{5.89}$$

Many times, $t_{ind} << t_{phys}$, and thus, $t_{ign} \approx t_{phys}$.

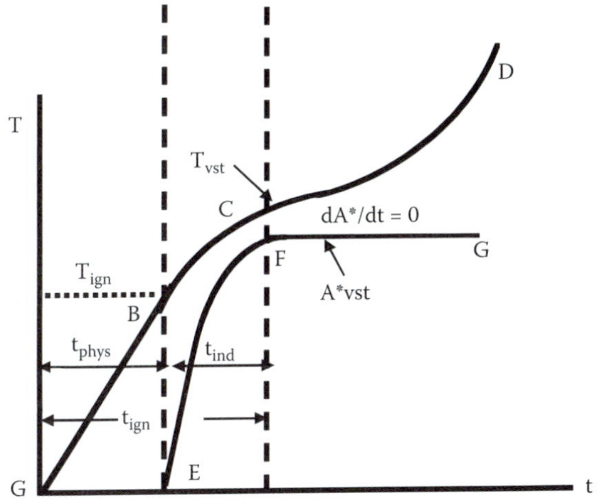

Figure 5.6 Timescales for a reacting system.

5.10.4 *Characteristic Reaction or Chemical/Combustion Times*

Suppose mixture of CH_4 and an excess amount of O_2 are kept in a rigid container at T =1200 K. They start reacting, $[CH_4]$ starts decreasing, and eventually, the $[CH_4]$ reaches a negligible value. One can plot $[CH_4]$ vs. time (Figure 5.1b) and select the reaction rate at t = 0 for estimation, i.e., it is assumed that the reaction takes place at a constant rate at t = 0; then, letting $[CH_4]$ = [F], the time to consume the species is approximately

$$t_{chem} = \frac{[F]_0}{\left|d[F]/dt\right|_{t=0}} \tag{5.90a}$$

where $[F]_0$ is the concentration of fuel at t = 0.

5.10.5 *Half-Life Time and Time Constants ($t_{1/2}$)*

The timescale t_{chem} is unrealistic, because it assumes that concentration remains constant. As the reaction proceeds, concentrations of reactants decrease and those of products increase. Half-life times account for variations in concentration. Half-life time is the time required to reduce the concentration of the main reactant (i.e., fuel) to half of its initial value.

$$t_{1/2} \approx \ln k'(T)/2 \tag{5.90b}$$

where k′(T) is related to k (T) as illustrated in the following example.

Example 8

Suppose the O_2 and H_2O concentrations are large, and the CO concentration is low. When the reaction takes place, the concentration of CO decreases and concentrations of O_2 and H_2O remain approximately constant. Determine the time for CO to reach half the initial concentration when the temperature is fixed.

Solution

Recall Equation 5.86. Rewriting with general order m for CO,

$$d[CO]/dt = -k' \, [CO]^m \qquad (A)$$

$$\text{where } k'(T) = k \, [K^{02}_{OH}/K^{0}_{H_2O}]^{1/2} \, [P^0/\bar{R}\,T]^{1/4} \, [O_2]^{1/4} \, [H_2O]^{1/2}. \qquad (B)$$

Because the temperature is constant and O_2 and H_2O concentrations remain almost constant, k′(T) is fixed. Integrating (m not equal to 1),

$$[CO]^{(1-m)} = [CO]_0^{\;(1-m)} - k'\,t. \qquad (C)$$

With $[CO] = (1/2)\,[CO]_0$ in Equation C

$$t_{1/2,\,m} = \{\{2^{(m-1)} - 1\}/(m-1)\}\,[CO]_0^{\;(1-m)}/k'. \qquad (D)$$

If m = 1, one can integrate Equation A directly

$$\{[CO]/[CO]_0\} = \exp\{-k'(T)\,t\}. \qquad (E)$$

For first order, the half-life time ($t_{(1/2)}$) is defined as the time at which $[CO] = (1/2)\,[CO]_0$ (Figure 5.7). Hence

$$t_{(1/2)} = \ln 2/k'(T). \qquad (F)$$

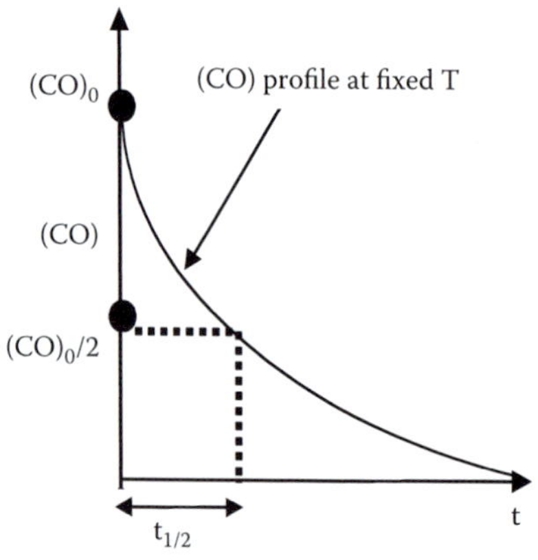

Figure 5.7 Illustration of half-life time.

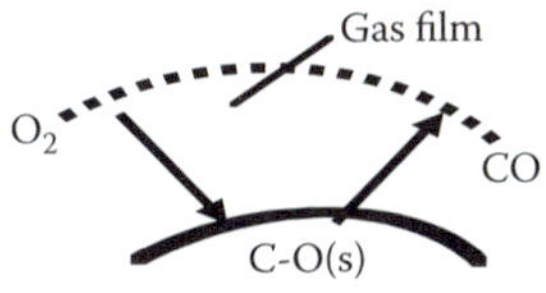

Figure 5.8 Illustration of carbon oxygen reaction.

Remarks

The time constant for the reaction, t_{charac}, is defined as that time at which concentration is e^{-1} of the initial concentration. Thus, from Equation E

$$t_{charac} = 1/k'(T). \tag{5.90c}$$

5.10.6 Total Combustion Time

When a cold fuel-to-air mixture is injected into a furnace, the mixture heats up and starts producing radicals beyond a certain temperature, the radical population rapidly increases resulting in flame, and then an intensive reaction occurs. Hence, combustion time is given by

$$t_{comb} = t_{phys} + t_{ind} + t_{reaction}$$

where $t_{reaction} \approx t_{chem}$ or $t_{1/2}$ or t_{charac}. If the mixture velocity is V, the required combustor length (L) = V × t_{comb}.

5.11 Solid–Gas (Heterogeneous) Reactions and Pyrolysis of Solid Fuels

5.11.1 Solid–Gas (Heterogeneous) Reactions

Consider charcoal, which is mostly carbon. When carbon is heated (e.g., by dousing it with lighter fluid and producing hot gases for heating carbon), it does not vaporize. Hence C(s) cannot diffuse, meet with O_2, and react in the gas phase. Instead, the O_2 molecules from the gas phase collide with the carbon atoms in the charcoal, and if the energy of collision is high, the O atoms get attached to the carbon atoms (this is called *adsorption*) and form C-S(s) (called *solid oxide*); subsequently CO is released as gas (called *desorption*).

5.11.2 Heterogeneous Reactions

Oxygen transfer to the particles can occur via O_2, CO_2, and H_2O, the dominant transfer mechanism being the transfer of diatomic oxygen. Hence, the combustion of carbon/char occurs primarily via one or more of the following reactions (see Chapter 9 for carbon combustion):

$$C(s) + (1/2)\ O_2 \rightarrow CO \tag{I}$$

$$C(s) + O_2 \rightarrow CO_2 \tag{II}$$

$$C(s) + CO_2 \rightarrow 2\ CO \tag{III}$$

$$C(s) + H_2O \rightarrow CO + H_2 \tag{IV}$$

Apart from reaction with stable species, reactions may also occur with O atoms, H atoms, and OH radicals. Reaction kinetics and the enthalpies of reaction are tabulated by Bradley et al. [1984] for carbon reactions involving both stable and radical species.

Chemical reactions between species in two separate phases (e.g., solid and gaseous) are generally governed by the following processes: (1) diffusion of gaseous species (e.g., O_2 in Reaction I) through the gas-phase film surrounding the particle to the particle surface; (2) adsorption of gaseous molecules (e.g., O_2) onto active sites on the surface; (3) transformation of the adsorbed molecules into solid oxides (e.g., C–O (s)) on the surface; (4) desorption of the solid oxides into the gas phase as gas (e.g., CO) through a high-activation-energy reaction; and (5) migration of the gaseous species through the flow boundary layer back into the free stream. If boundary-layer diffusion (items 1 or 5) controls the overall reaction rate, the problem can be easily solved using mass transfer and fluid mechanics principles (Chapter 9 and Chapter 11). In a manner analogous to gas-phase reactions, the heterogeneous reaction rate for the principal species can be expressed in the form [Williams, 1985]

$$\dot{m}'' = k\ \Pi_j\, \gamma_j^{\upsilon_j}, \tag{5.91}$$

where k denotes the specific reaction rate; γ_j, the surface concentration of species j per unit area; and v_j, the stoichiometric coefficient of species j.

Adsorption of gaseous species (e.g., O_2) results in occupation of a fraction θ of the carbon sites on the surface and hence increases the value of θ, whereas desorption decreases it. A fraction of colliding molecules of concentrations c_i and mass m_i is returned to the gas phase. If the collision rate of molecules of species i per unit area of solid surface is denoted by Z_i and the accommodation coefficient is α_i, then the adsorption rate per unit area is $Z_i\alpha_i(1 - \theta)$ and $\theta = \Sigma\theta_i$, where $Z_i = c_i\ (k_BT/(2\pi m_i))^{1/2}$. The rate of return via desorption of species i per unit area is $\mu_i\ \theta_I$, where μ_i is a proportionality constant. For constant or steady-state values of θ_i, the adsorption rate must be equal to the desorption rate. Therefore, for the single-species carbon attack, with i = 1,

$$\theta = Z_1\alpha_1/(\mu_1 + Z_1\ \alpha_1), \tag{5.92}$$

and the reaction rate per unit area is

$$\dot{m}'' = \mu_1\ \theta = \mu_1\ Z_1\ \alpha_1/(\mu_1 + Z_1\ \alpha_1). \tag{5.93}$$

Consider the following limiting cases.

5.11.2.1 *Desorption Control*

Here, the fourth step — desorption of the solid oxides into the gas phase through a high-activation-energy reaction — is limiting. If the desorption rate is slow compared to the adsorption rate ($Z_1\alpha_1 >> \mu_1$), then the reaction rate Equation 93 reduces to

$$\dot{m}'' = \mu_1. \tag{5.94}$$

In this case, all the active sites are covered and, hence, the adsorption-site concentration is independent of the species concentration. Therefore, this is a zeroth-order reaction.

5.11.2.2 *Absorption Control*

In this case, the second step — adsorption of gaseous molecules onto active sites on the surface — is rate limiting. If the collision rate is low (i.e., $Z_1\alpha_1 << \mu_1$), then the reaction rate (Equation 5.93) is

$$\dot{m}'' = \alpha_1\ Z_1. \tag{5.95}$$

The collision rate Z_1 is proportional to the gas-phase concentration of species 1 and, thus, the reaction is first order.

If there is single-species attack of the solid, assuming the reactions to be first order, the species consumption rate per unit area for heterogeneous reactions can be written in terms of the mass fraction in the form

$$\dot{m}'' = \rho_w k\ Y_k,\ \text{kg/m}^2\text{sec} \tag{5.96}$$

where ρ_w is the density of gas at solid surface and Y_k denotes the species mass fraction. Note that even if kinetic data are available for the single-component attack of the carbon sites, these data cannot be used when multiple components are present. For example, if graphite powder is exposed to vitiated gases, including the H atoms, O atoms, OH radicals, and the H_2O, CO_2, and O_2 species, the active sites available for a particular species are reduced and hence there is site interaction [Bradley et al., 1984]. However, these interactions are typically ignored in the formalism of models.

5.11.2.3 Global Reaction

Consider Reaction I. The global reaction rate expression is written as

$$\dot{w}''_c = (-1)\, A\, [O2]^m\, T^n \exp\left[-\frac{E}{\bar{R}T}\right], \quad n = 0 \tag{5.97}$$

where $\dot{w}''_c$, denotes carbon consumption per unit surface area of solid carbon; $\dot{w}''_c = \dot{w}_c/(4\,\pi a^2)$ for a spherical particle of radius a, and m is the order of reaction, typically, $0 < m < 1$. Units for A will vary depending on m. Typical values for carbon are as follows. For fast reactions, $A = 3.3 \times 10^9$ kg/m^2 bar^m sec, E = 184,000 kJ/kmol, whereas for slow reactions, $A = 2.2 \times 10^5$ kg/m^2 bar^m sec, E = 177,000 kJ/kmol. More detailed global reaction kinetics of constants for carbon reactions are given by Mulcahy and Smith [1969] as well as Hamor and Smith [1973].

Reaction II is significant at low temperatures (T < 800 K), whereas Reaction I is significant under typical combustion conditions. Arthur [1951] has given the ratio of CO/CO_2 as

$$[CO]/[CO_2] = 10^{3.4} \exp(-6200/T). \tag{5.98}$$

Alternatively, the relative heterogeneous production rates of CO to CO_2 (kg/kg) [Wicke, 1955; Smith, 1971a, b] can be expressed as

$$\frac{\dot{m}''_{CO}}{\dot{m}''_{CO2}} = 2500 \exp\left(-\frac{E}{\bar{R}T}\right). \tag{5.99}$$

It is apparent that for T > 1100 K, $\{\dot{m}''_{CO}/\dot{m}''_{CO2}\} > 10$. Reaction III and Reaction IV are dominant only at extreme temperatures or when O_2 concentrations are extremely low.

5.11.3 Forward and Backward Reaction Rates

Suppose the following reaction occurs in forward and backward directions

$$C(s) + 1/2\ O_2 \Leftrightarrow CO \tag{I}$$

$$\dot{w}''_{O_2} = -(1/2)\, k_f\, [O_2]^{1/2} + (1/2)\, k_b\, [CO],\ \text{kmol/m}^2\ \text{sec}.$$

Once they reach equilibrium at the same T, then $K_{c,I}(T) = k_{f,I}(T)/k_{b,I}(T) = K^0(T)\,\{P^0/\bar{R}\,T\}^{1/2} = [CO]^1/[O_2]^{1/2}$.

Hence, knowing $k_{f,I}$ $K^0(T)$, $k_{b,I}$ can be computed. Then, at any concentration

$$\dot{w}''_{O_2} = -k_f\,(1/2)\,[O_2]^{1/2} + (1/2)\,k_b\,[CO] = -(1/2)\,k_f\,[O_2]^{1/2} + (1/2)\,[k_f\,\{P^0/\bar{R}\,T\}^{-1/2}/K^0(T)]\,[CO] \tag{5.100}$$

where K_1^0 is the equilibrium constant for Reaction I.

$$d[O_2]/dt = -(1/2)k_{f,I}\ \{[O_2]^{1/2} - \{(\bar{R}\ T/P^0)^{(-1/2)}/K_1^0(T)\}\ [CO]\} \quad (5.101)$$

If K_1^0 (T) is large (for example, O_2 concentration at equilibrium is very low), then

$$\dot{w}''_C \approx -k_f\ (1/2)\ [O_2]^{1/2}. \quad (5.102)$$

Example 9

(a) Determine the carbon burn rate in kg/m²sec if $C + 1/2\ O_2 \rightarrow CO$, A = 5144 m/sec, m = 1, pO_2 = 0.21 bar, E = 85,000 kJ/kmol K and T = 1,500 K.
(b) Estimate combustion time of a particle of 75 μm. Assume kinetics control and ρ_P = 1,200 kg/m³.

Solution

(a) $$\dot{w}''_C/(-1) = \dot{w}''\ O_2/(-1/2) = \dot{w}''$$

$$\dot{w}'' = A\ [O_2]\ \exp(-E/RT)$$

$\{O_2] = 0.21/(0.08314 \times 1500) = 0.00238$ kmol/m³

$\dot{w}''_C = (-1) \times 5144 \times \exp(-85000/(8.314 \times 1500)) \times [0.00238] = 0.0134$ kmol/m²sec;

$$\dot{w}''_C\ (\text{kg/m}^2\text{sec}) = 0.0134 \times 12.01 = 0.16\ \text{kg/m}^2\text{sec}.$$

Surface area = $4\pi\ a^2 = 4 \times 3.1417 \times (37.5 \times 10^{-6})^2 = 1.767 \times 10^{-8}$ m².

(b) $$\dot{w}_c = 0.16 \times 1.767 \times 10^{-8}\ \text{kg/sec}$$

Mass = $(4/3)\ \pi\ a^3\ \rho_c = 1.33 \times 3.1417 \times (37.5 \times 10^{-6})^3 \times 1200 = 2.65 \times 10^{-10}$ kg

The kinetic timescale = $2.65 \times 10^{-10}/1.767 \times 10^{-8} = 0.094$ sec or 94 msec.

Formulae $t_{kinetic} = (4/3)\ \pi\ a^3\ \rho_c/4\pi\ a^2\ wc'' = a\ \rho_c/(3\ \dot{w}''_c) = 94$ msec.

Remarks

(i) If coal is assumed to be entirely made up of carbon, then knowing the number of particles per unit volume, one can estimate the burn rate per unit volume. Knowing the heating value, the space combustion rate per unit volume or the space heating rate can be estimated; typically, it is about 8 Btu/ft³ sec (probably owing to slow burn of solid fuels), whereas in gas turbines burning liquid droplets, it is about 10,000 Btu/ft³ sec.

(ii) For utility boilers, the combustion intensity (based on burner zone volume) ≈180 kJ/m^{3}sec for bituminous pulverized coal, 270 kJ/m^{3}sec for fuel oil, and 310 kJ/m^{3}sec for natural gas.

5.11.4 Pyrolysis of Solids

Recall that energetic collisions result in chemical reactions involving homogeneous and heterogeneous phases. Now consider naphthalene balls, $C_{10}H_{18}(s)$. It simply sublimes at room temperature, and the molecular structure is preserved both in the gas and solid phases.

$$C_{10}H_{18}(s) \rightarrow C_{10}H_{18}(g) \text{ on heating}$$

This process of gasification is called a *physical process*. Consider a polymer, which is similar to a row of balls called *monomers* attached to a string. The monomers get detached from the string during heating. Now consider Plexiglas (e.g., aircraft window), a solid polymer that burns in air. The monomer is represented by $C_5H_8O_2$. On heating the solid, average ve increases and at some energy E, the monomer gets detached (decomposes) and produces gas ($C_5H_8O_2$ (g)) this is called *pyrolysis* (thermal decomposition). The fraction of molecules having energy E and above is given by exp(-E/R T). Thus, the gasification or pyrolysis rate is also subject to the Arrhenius law. For first-order pyrolysis,

$$\dot{m}\ \ (g) = A\ m_V \exp(-E/\bar{R}T) \tag{5.103}$$

where m_V is the amount of volatile matter remaining in the solid. In the case of Plexiglas, a totally pyrolyzing polymer, $m_v = m$. Typically, Plexiglas gasifies when heated to about 600 K.

Now consider coal. Unlike Plexiglas, only about 40% of coal (which may contain polymer molecules that gasify) consists of volatile matter (VM) and 60% remains as fixed carbon (FC). As the polymer part decomposes, the amount of VM changes. Thus, the gasification rate is proportional to the amount of coal and the fraction of the VM having an energy E and above. Thus, for coal

$$\dot{m}\ = -dm_V/dt = km_v = A\ m_v\ \exp(-E/\bar{R}\ T).$$

Generalizing,

$$\dot{m} = -dm_V/dt = A\ m_V^{nv}\ \exp(-E/\ \bar{R}T) \tag{5.104}$$

where n_v is called the *global order of the pyrolysis reaction*.

The thermo-gravimetric analyzer (TGA) is used for conducting pyrolysis studies in N_2. The sample is kept on a crucible and hung from a balance in the furnace. As it is heated at a prescribed heating rate (HR), the temperature increases, the solid pyrolyzes, its weight decreases, and the results are plotted

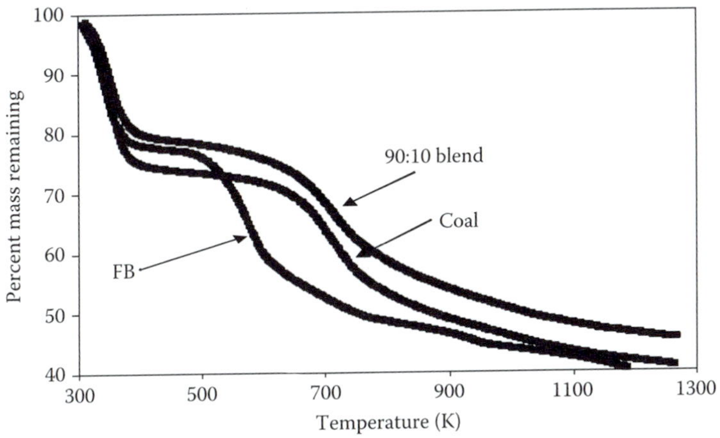

Figure 5.9 TGA of coal and FB. 90:10 blend = 90% coal and 10% FB (Adapted from Thien, B., Annamalai, K., and Bukur, D., Thermogravimetric Analyses of Coal, Feedlot Biomass and Blends in Inert and Oxidizing Atmospheres, Proceedings of IJPGC:IJPGC-2001,JPGC 2001/FACT-19082, New Orleans, LA, June 4-7, 2001.)

as a function of temperature. TGA traces for coal and animal waste (feedlot biomass, FB) are given in Figure 5.9 (HR ~ 10-100° K/min, $T_{peak} \approx 1000$ K). N_2 is used, in order to avoid oxidation. The weight loss can also be predicted for a given A and E by integrating Equation 5.104. Sometimes there could be mass gain. For example, copper could lose H_2O at around 100 °C, indicating weight loss, and there could be gain due to oxidation at around 500 °C. Subsection 5.13.5 summarizes various methods of integrating kinetics relations.

5.11.4.1 Single Reaction Model (at Any Instant of Time)

Many authors have approximated the overall process as a first-order decomposition occurring uniformly throughout the particle. Thus, the rate of devolatilization is expressed as

$$\text{coal} \rightarrow \text{solid residue (S)} + \text{volatile (V)}$$

$$\dot{m} = k(V_\infty - V), \tag{5.105}$$

where V_∞ represents the asymptotic (time $\rightarrow \infty$) volatile yield, V represents the total volatile produced at any instant of time, and $V \rightarrow V_\infty$ as $t \rightarrow \infty$. Badzioch and Hawksley (1970) proposed a single-reaction model for the pyrolysis of coal,

$$dY_V/dt = k_0\ (Y_{V\infty} - Y_V)$$

where k_0 = A exp(−E/RT), $Y_{v\infty}$ = final volatile yield (but not necessarily the same as the initial volatile matter), and Y_V = volatile yield after time t. Another approach has been to use an n-th-order rate expression

$$\dot{m} = k(V_\infty - V)^n. \quad (5.106)$$

5.11.4.2 Competing Reaction Model

When a coal particle is placed in a hot furnace, it is heated and pyrolysis (thermal decomposition of coal) ensues. Typically, bituminous coal pyrolyzes at about 700 K (1% mass loss, for heating rates < 100°C/sec) as in the case of most plastics; this temperature is at or below the autoignition temperature for HC (though not for H_2 or CO). Pyrolysis is presumed to start at around 1500 K at high heating rates (heating rate > 10,000°C/sec). Pyrolysis products range from lighter volatiles, CH_4, C_2H_4, C_2H_6, CO, CO_2, H_2, H_2O, etc., to heavier tars, but the tars invariably crack to form CH_4, C_2H_4, C_2H_6, C_3H_6, C_2H_2, and CO at T > 1000 K [Bradley, 1984] resulting in soot formation. Product gas composition under rapid pyrolysis has been measured by Suuberg et al. [1978]. Tars produced from coal may be in liquid form, which vaporize if sufficient heat is supplied. If not, the tars could undergo repolymerization, producing lighter volatiles and char (primary cracking). If tars are released as gases (primary products), they can crack at temperatures above 900 K. The cracking process producing C_2 and C_1 groups is very rapid.

Pyrolysis is a relatively slow chemical process compared to rapid physical vaporization of oil drops. Coal can swell upon heating; this results in increasing particle size (whereas the drop size remains constant), particularly under pyrolysis in inert environments or reducing conditions. The swelling factor can range from 1.3 (under oxidative) to 4 in an inert environment. A medium value of 1.5 is mostly used for all coals. The swelling increases the size by 60% if pyrolyzed in an O_2-free environment, compared to 10% when heated in 13% O_2 [Street, Weight, and Lightman, 1969]. Swelling introduces thin-walled cenospheres, which when burned can cause sudden decrease in size and increase in density as they burn off [Sergeant and Smith, 1973]. Volatiles are also found to issue like jets from particles of 80 to 100 μm, whereas for d < 40 μm, no trails were found.

Volatile yields from coal can be as high as 80% depending upon the type of coal, final temperature, and rate of heating. [Kimber and Gray, 1967; Badzioch and Hawksley, 1970; Kobayashi, 1972]. For example, Kobayashi found that the volatile yield increased from 30% at 1250 K to 63% at 2100 K for the same coal [Kobayashi, 1972]. Various pyrolysis models ranging from single reaction to multiple reactions are summarized by Ubhayakar et al. [1976, 1977] and Anthony and Howard [1976]. Generally, increased temperature, increased heating rate, decreased particle size, decreased pressure, and reduced bed heights increase the volatile yields. The competing reaction and multiple parallel reaction models have been shown to adequately describe devolatization.

Example 10

Consider coal undergoing pyrolysis. The following competing reactions are hypothesized to occur:

$$CH_{0.8}\ (s) \rightarrow a\ CH_n\ (g) + b\ C(s),\ VM = 0.4 \quad (I)$$

$$CH_{0.8}\ (s) \rightarrow c\ CH_m\ (g) + d\ C(s),\ VM = 0.8 \quad (II)$$

(a) If the coal pyrolyzes by Reaction I yielding volatile matter fraction as 0.4 (α_I) on mass basis, what are a, n, and b? (b) What will be a, n, and b if the volatile matter fraction is 0.8 (α_{II})? (c) During pyrolysis of 1 kg of coal, 0.2 kg of coal pyrolyzes via Reaction I while 0.5 kg pyrolyzes via Reaction II. What is the amount of production of volatiles in kg per kg of coal 40% of the H atoms is exhausted via Reaction I and the remaining 60% by Reaction II. What will be the volatile yield?

Solution

(a) In Example 5 of Chapter 4, we determined a, n, and b as 0.36, 2.2, and 0.540 and M_{coal} = 12.82 kg/kmol so that the atom balance is satisfied.
(b) Following Example 5 of Chapter 4, α_{II} = 0.8, FC = 0.2, c = 0.786, m = 1.017, and d = 0.214.

Summarizing,

$$CH_{0.8}\ (s) \rightarrow 0.36\ CH_{2.2}\ (g) + 0.64\ C(s),\ \text{with VM} = 0.4 \quad (I)$$

$$\text{Or 1 kg of coal} \rightarrow 0.4\ \text{kg of}\ CH_{2.2}\ (g) + 0.6\ \text{kg of}\ C(s)$$

$$CH_{0.8}\ (s) \rightarrow 0.786\ CH_{1.017}\ (g) + 0.214\ C(s),\ \text{with VM} = 0.8. \quad (II)$$

Or

$$1\ \text{kg of coal} \rightarrow 0.8\ \text{kg of}\ CH_{1.017}\ (g) + 0.2\ \text{kg of}\ C(s)$$

(c) $0.2\ \text{kg of}\ CH_{0.8}\ (s) \rightarrow 0.08\ \text{kg of}\ CH_{2.2}\ (g) + 0.12\ \text{kg of}\ C(s)$

$$0.5\ \text{kg of coal} \rightarrow 0.4\ \text{kg of}\ CH_{1.017}\ (g) + 0.1\ \text{kg of}\ C(s).$$

Net volatile yield = (0.08 + 0.4) = 0.48 kg of volatiles for 0.2 + 0.5 coal decomposed or pyrolyzed, i.e., 67% volatile yield from coal decomposed.

Thus, 0.3 kg of coal is undecomposed for each kg of coal.

Remarks

(i) If volatile-mass loss via Reaction I is known to be 0.08 kg, then we can determine coal decomposed as 0.08/0.4 = 0.2 kg and, similarly, mass decomposed via Reaction II as 0.4/0.8 = 0.5 kg. If $\dot{m}_{V,I}$ is the volatile liberation rate via Reaction I, then within dt, volatile-mass loss via Reaction I is $\dot{m}_{V,I}dt$ kg, and hence mass of coal decomposed via Reaction I is $(\dot{m}_{V,I}/\alpha_I)$ dt; similarly, mass of coal decomposed via Reaction II is $(\dot{m}_{V,II}/\alpha_{II})$ dt; Integrating over time t coal decomposed is $\int_0^t \{(\dot{m}_{V,I}/\alpha_I + \dot{m}_{V,II}/\alpha_{II})\}dt$, whereas coal undecomposed is $m_{cu,0} - \int_0^t \{(\dot{m}_{V,I}/\alpha_I + \dot{m}_{V,II}/\alpha_{II})\}dt$.

(ii) If the percentage volatile loss in Reaction I progressively decreases, then more mass is available for Reaction II, which will increase the net volatile yield. The volatiles are liberated at finite rates and, as such, the more the time coal pyrolyzes at a high temperature, the greater the volatile yield; i.e., coal is heated rapidly so that the time available at low T is less. Hence, volatile yield depends on the heating rate of coal.

■ ■ ■

More generally, Kobayashi [1972] and Kobayashi et al. [1976] postulated a model with two competing reactions to explain the increased yield at higher heating rates.

$$\text{coal} \rightarrow (1 - \alpha_1)\, S_I + \alpha_I\, V_I \quad \text{(I)}$$

$$\text{coal} \rightarrow (1 - \alpha_2)\, S_{II} + \alpha_{II}\, V_{II} \quad \text{(II)}$$

where species S_I could have a greater H/C ratio compared to S_{II} (Example 10).

The total volatile evolution rate by both pyrolysis reactions is given as

$$dm_V/dt = \dot{m}_V = \{(\dot{m}_{V,I} + \dot{m}_{V,II}\} = (\alpha_I k_I + \alpha_{II} k_{II})\, m_{cu} \quad (5.107)$$

where m_{cu} denotes the mass still left to be pyrolyzed, or the amount of virgin coal still left. For Reactions I and II respectively,

$$k_{I,II} = A_{I,II} \exp(-E_{I,II}/RT) \quad (5.108)$$

$$m_{cu} = m_{cu,0} - \int_0^t \{(\dot{m}_{V,I}/\alpha_I + \dot{m}_{V,II}/\alpha_{II})\}dt \quad (5.109)$$

where $m_{cu,o} = m_{p0}$. (See Example 10 for rationale of Equation 5.109.) Differentiating Equation 5.109

$$dm_{cu}/dt = -\{(\dot{m}_{V,I}/\alpha_1 + \dot{m}_{V,II}/\alpha_2)\} = \{(k_1 + k_2)\}\, m_{cu}\, dt,\ m_{cu,0} = m_{p,0}. \quad (5.110)$$

Table 5.1 presents the preexponential factor and activation energy for the pyrolysis reactions, Reaction I and Reaction II. Figure 5.10 plots ln k vs. 1/T for Reaction I, Reaction II, and Reaction II′, where $A'_{II} = A_{II} \times 10^7$ is a modified form of Reaction II; it is seen that Reaction II is faster compared to Reaction I at T > 1330 K. Thus, most of the coal will pyrolyze via Reaction II at T > 1330 K.

Table 5.1 Kinetic Parameters, Coal Properties, and Other Data

Density	1300 kg/m^3	
Heat of combustion	30 MJ/kg	
ASTM volatile yield	0.3	
Oxygen mass fraction	0.23	
Heating value of carbon	32810 kJ/kg	
Coal composition	C = 77.12%, H = 6.42%, O = 16.14%	
Devolatilization Kinetics	**Reaction I**	**Reaction II**
Preexponential factor	1.34E5 sec^{-1}	1.46E5 sec^{-1}
Activation energy	74.1 MJ/kmol	251.0 MJ/kmol
Heat of devolatilization	1.0 MJ/kg of volatiles	

Source: From Du, X. and Annamalai, K., Transient ignition of isolated coal particle, *Combustion and Flame*, 97, 339–354, 1994.

5.11.4.3 Parallel Reaction Model

Because coal is a complex polymer having a range of bond energies, one may fit a Gaussian distribution to the bond energy. Because activation energy is related to bond energy, one can assume that the pyrolysis process consists of a series of reactions proceeding in parallel [Anthony et al., 1975, 1976; Anthony, 1976], each reaction having different activation energies. Coal is presumed to consist of a

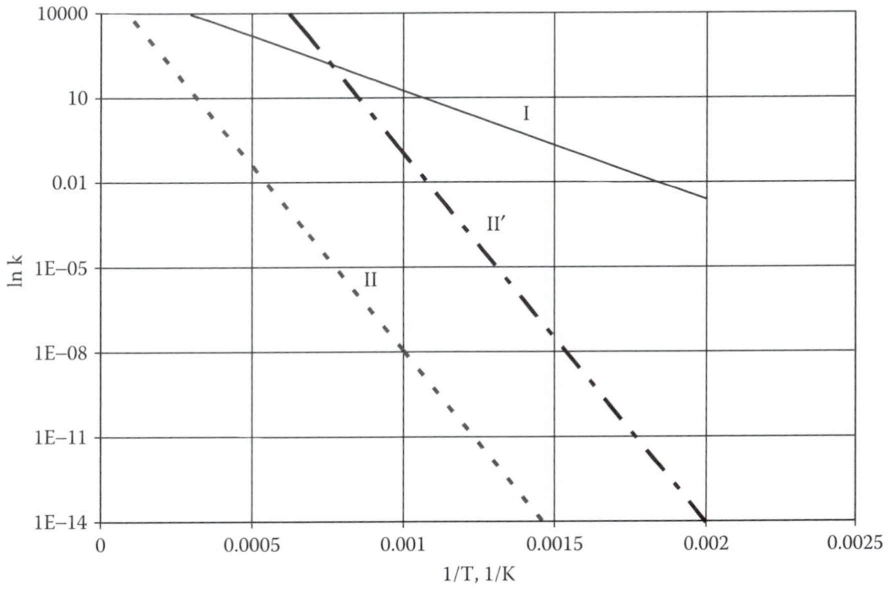

Figure 5.10 Plot of ln k vs. 1/T for competing reactions, Reaction I, Reaction II, and Reaction II′ with $A_{II'} = 10^7 \times A_{II}$.

series of components with activation energy E ranging from 0 to infinity. If dV_i is the mass fraction of the i-th volatile component having an activation energy E_i released over a period dt with a first order reaction, then,

$$dV_i = (V_i^* - V_i)\, k_i(T)\, dt \tag{5.111}$$

where V_i is the mass fraction of the i-th component released up to time t; V_i^*, the initial volatile-mass fraction of the i-th component; and $k_i(T)$, the specific reaction rate constant (1/min) as given by the Arrhenius law,

$$k_i(T) = ko_i \exp\left(-\frac{E_i}{RT}\right) \tag{5.112}$$

Using Equation 5.112 in Equation 5.111,

$$\frac{dV_i}{dt} = \left(V_i^* - V_i\right) ko_i \exp\left(-\frac{E_i}{RT}\right). \tag{5.113}$$

Integrating Equation 5.113 over time t with the initial condition that $V_i = 0$ at $t = 0$, one obtains the mass fraction of the i-th component (i.e., the fraction of mass having activation energy E_i) remaining in coal as

$$V_i^* - V_i = V_i^* \exp\left(-\int_0^t ko_i \exp\left(-\frac{E_i}{RT}\right) dt\right). \tag{5.114}$$

Now it is assumed that the distribution of reactions is continuous and described by the function f(E) such that:

$$dF_i = f(E)dE \tag{5.115}$$

where dF_i is the fraction of mass of the i-th component having energy between E and E + dE. Omitting the subscript i,

$$dV^* = V^* dF = V^* f(E)dE \tag{5.116}$$

$$\int_{-\infty}^{\infty} f(E)dE = 1\,. \tag{5.117}$$

The Gaussian distribution of activation energy is written as

$$f(E) = \frac{1}{\sigma\sqrt{2\pi}} \exp\left(-\frac{\left(E - E_m\right)^2}{2\sigma^2}\right) \tag{5.118}$$

where E_m, average activation energy; σ, standard deviation. Note that f(E) is zero for $E < 0$; further typically $f(E) = 0$ for $E < E_m\ 3\sigma$ and $E > E_m + 3\sigma$.

With $ko_i = ko$, multiply Equation 5.114 by f(E)dE and integrate over all activation energies:

$$V^* - V = V^* \int_{-\infty}^{\infty} \exp\left(-\int_0^t ko \exp\left(-\frac{E}{RT}\right) dt\right) f(E) dE. \quad (5.119)$$

Using heating rate (dT/dt) in Equation (5.114) the following result is obtained:

$$V^* - V = \frac{V^*}{\sigma(2\ )^{1/2}} * \int_{Em-3\sigma}^{Em+3\sigma} \exp\left(-\frac{ko}{\{dT/dt\}} \int_{To}^{T} \exp\left(-\frac{E}{RT}\right) dT - \frac{(E - E_m)^2}{2\sigma^2}\right) dE. \quad (5.120)$$

The limits on outside integrals have been changed to facilitate integration. The limits of $E_m \pm 3\sigma$ include 99% of activation energies. Using experimental data on V*, V vs. T, E_m and σ can be determined.

5.12 Summary

Although Chapter 2 presumes a one-step reaction in estimating flame temperature and product composition, the reaction typically involves many elementary steps. This chapter presents a brief overview of multiple reactions among gaseous species and reactions between gaseous and solid species. The reactions could be simplified to one or two steps using the steady-state radical hypothesis. Further, thermal NOx formation from N_2 and O_2 is briefly discussed using the Zeldovich mechanism. Relations for hetrogeneous reactions and pyrolysis involving solid fuels are presented.

5.13 Appendix

5.13.1 Multistep Reactions

Consider the reactions for an H_2–O_2 system:

$$H_2 + O_2 \rightarrow HO_2 + H, \quad (R1)$$

$$H + O_2 \rightarrow OH + O \quad (R2)$$

$$OH + H_2 \rightarrow H_2O + H \quad (R3)$$

$$H + O_2 + M \rightarrow HO_2 + M \quad (R4)$$

Suppose these are the only four reactions involving H, HO_2, H_2, H_2O, M, O, OH, and O_2 (up to eight species); we represent the reaction number by i and species by j. Thus, j = 1 means O_2, j = 2 means HO_2 .... up to j = 8, which is M. Similarly, i = 1 means Reaction R1, $H_2 + O_2 \Leftrightarrow HO_2 + H$ ..., and so on. We can write all four reactions symbolically as follows:

$$\sum \upsilon_{ji} X_j' \Leftrightarrow \sum \upsilon_{ji}\ X_j',\ i = 1\ \ldots\ 4. \quad (a)$$

Consider Reaction R3. Then $\nu_{13}' = 0$, $\nu_{23}' = 0$, $\nu_{33}' = 1$ …. $\nu_{13}'' = 1$, $\nu_{23}'' = 0$, …. Expanding Equation a for Reaction R3,

$$0^*H + 0^*\ HO_2 + 1^*\ H_2 + 0^*H_2O + 0^*M + 0^*O + 1^*OH + 0^*O_2 + 1^*H_2 + 0^*H \rightarrow$$
$$1^*H + 0^*\ HO_2 + 1^*\ H_2 + 1^*H_2O + 0^*M + 0^*O + 1^*OH + 0^*O_2 + 1\ ^*\ H_2 + 0^*H \tag{b}$$

Similarly, we write the law of mass action for species j in Reaction i as $\dot{w}_j'' = \{k_{ib}\Pi\ [Xj]^{\ \nu ji} - k_{if}\ \Pi\ [Xj]^{\ \nu ji}\}$.

For example, for reaction i = 3 and species j = H, we write the production rate of H as

$$\dot{w}_{H,3}'' = \{k_{3b}\ [O_2]^0 [H_2]^0 [H_2O]^1\ [H]^1\ [HO_2]^0 [O]^0\ [OH]^0\ [M]^0 - k_{3f}\ [O_2]^0\ [H_2]^1\ [H_2O]^0\ [H]^0$$
$$[HO_2]^0\ [O]^0\ [OH]^1\ [M]^0\} \tag{c}$$

$$\dot{w}_H'' = \sum_i \nu_{Hi}\ \dot{w}_{Hi}, \quad i = 1, \ldots, 4 \tag{d}$$

$$\dot{w}_H'' = \dot{w}_{H,1}'' + \dot{w}_{H,3}'' + \dot{w}_{H,2}'' + \dot{w}_{H,4}''. \tag{e}$$

$$\text{Then } d[H]/dt = \dot{w}_H'' = \dot{w}_{H,1}'' + \dot{w}_{H,3}'' + \dot{w}_{H,2}'' + \dot{w}_{H,4}''. \tag{f}$$

Similarly, one can write the expressions for the other seven species and then integrate with time; but reaction timescales should be estimated for each, and the least reaction time should be selected for integration so that the increment or decrement is not too steep. Typically, for the third body, we include all species or [M] = P/RT; sometimes, addition of a species such as H_2O will enhance the third-body efficiency (i.e., efficiency in absorbing heat released by recombination of radicals to stable species, e.g., $H + H = H_2$); in that case, $[M] = P/RT + [H_2O]$.

Combustion codes include many reactions. Say, you specify $[H_2]$, $[O_2]$, $[H_2O]$, and $[N_2]$ at the inlet at a given T. Then various codes will consider all possible reactions, integrate, and give results for [H] and [OH] with the appropriate heat to be added or removed at the given T. Alternatively, you could specify species, say, H, OH, H_2, and OH as participating species.

5.13.2 Simplified CH_4 Reactions

$$CH_4 + OH \rightarrow CH_3 + H_2O \tag{g}$$

Where does OH come from? Consider

$$O_2 \Leftrightarrow 2O \tag{h}$$

$$H_2O + O \Leftrightarrow 2\ OH \tag{i}$$

$$K_O^0 = (pO/P^0)^2\ /(pO_2/P^0) = [\ \bar{R}T/P^0]^{(2-1)}\ [O]^2/[O_2] \tag{j}$$

$$K_{OH}^0 = (p_{OH}/P^0)^2\ /\{(p_{H2O}/P^0)(p_O/P^0)\} = [OH]^2/\{[H_2O][O]\} \tag{k}$$

Use [O] from Equation j in Equation k and solve for [OH]

$$K^0_{OH} = \{ \bar{R}\ T/\ P^0\}^{1/2}\ [OH]^2/\ \{[H_2O]\ \{\ K^0_O\ [O_2]\}^{1/2}\}. \quad (l)$$

Solving for [OH],

$$[OH] = K^{0\ \ 1/2}_{OH}\ [H_2O\ ]^{1/2}\ K^{0\ \ 1/4}_O\ [O_2]^{\ 1/4}/\ \{\ \bar{R}\ T/\ P^0\}^{1/4} \quad (m)$$

$$\text{Now } CH_4 + OH \rightarrow CH_3 + H_2O \quad (n)$$

$$d[CH_4]/dt/(-1) = w''' = k\ [CH_4]\ [OH] \quad (o)$$

Using Equation 5.l for [OH] in Equation o,

$$d[CH_4]/dt = -k\ [CH_4]\ [OH] = -\{k\ \ K^{0\ \ 1/2}_{OH}\ \ K^{0\ 1/4}_O\}\ [CH_4]\ [H_2O\ ]^{1/2}\ [O_2]^{\ 1/4}\}/\{\ \bar{R}\ T/\ P^0\}^{1/4} \quad (p)$$

where $k = A \exp(-E/\ \bar{R}T)$, where E refers to the activation energy required to break bonds of CH_4 and OH.
Order: 1 + (1/2) + 1/4 = 1.75

Note that many times $\ln\ K^0_O = A_o - B_o/T$, $\ln\ K^0_{OH} = A_{OH} - B_{OH}/T$, and as such, $K^0_O = F_O \exp(-B_o/T)$, $K_{OH} = F_{OH} \exp(-B_{OH}/T)$, where $F_O = \exp(A_O)$, $F_{OH} = \exp\ (A_{OH})$. Thus, E will get modified.

$$d[CH_4]/dt = -k_{glob}\ \exp(-E_{glob}/\ \bar{R}T)\ [CH_4]\ [H_2O]^{1/2}\ [O_2]^{1/4}$$

where $k_{glob} = \{k_0\ F^{1/2}_{OH} F^{1/4}_O/(\ \bar{R}T/P^0)^{1/4}\}$, global preexponential factor, and $E_{glob} = E + B_O\ \bar{R}\ + B_{OH}\ \bar{R}$.

5.13.3 Conversion of Reaction Rate Expressions

5.13.3.1 Conversion of Law of Mass Action from Molar Form (kmol/m³) to Mass Form (Concentration in kg/m³)

Consider the oxidation expression for CH_4

$$d[CH_4]/dt = -A_{[\ ]}\ (T/T_{ref})^m\ \exp(-E_O/RT)\ [CH_4]^{\alpha}\ [H_2O] - [O_2]^{\gamma}\ \text{kmol/m}^3\ \text{sec} \quad (A)$$

5.13.3.1.2 Mass Form

The corresponding expression is

$$\{d(CH_4)/dt\},\ \text{kg/m}^3\ \text{sec} = -A_{(\ )}\ (T/T_{ref})^m\ \exp(-E_O/RT)\ (CH_4)^{\alpha}\ (H_2O)^{\beta}(O_2)^{\gamma}\ \text{kg/m}^3\text{sec} \quad (B)$$

If $A_{[\ ]}$ is given, how do you find $A_{(\ )}$ (mass form)?

Multiply (A) by M_{CH4} to get kg/m^{3}sec, i.e.,

$\{d(CH_4)/dt\}$, kg/m^3 sec = $-A_{[\,]}\, M_{CH4}\, (T/T_{ref})^m \exp(-E_O/RT)\, [CH_4]^\alpha\, [H_2O]^\beta\, [O_2]^\gamma$ kg/ m^{3}sec

$\{d(CH_4)/dt\}$, kg/m^3 sec = $-A_{[\,]}\, M_{CH4}\, (T/T_{ref})^m \exp(-E_O/RT)\, (CH_4/M_{CH4})^\alpha\, (H_2O/ M_{H2O})^\beta\, (O_2/M_{O2})^\gamma$ kg/ m^{3}sec.

Simplifying,

$d(CH_4)/dt$ kg/m^3 sec = $-A_{(\,)}\, (T/T_{ref})^m \exp(-E_O/RT)\, (CH4)^\alpha\, (H_2O)^\beta\, (O_2)^\gamma$, kg/ m^{3}sec (C)

where $A_{(\,)} = A_{[\,]}\ M_{CH4}^{(1-\alpha)}\ M_{H2O}^{-\beta}\ -\ M_{O2}^{-\gamma}$.

5.13.3.2 Conversion of Law of Mass Action from Mass Form to Molar Form

Consider the expression

$d(CH_4)/dt$ kg/m^3 sec= $-A_{(\,)}\, (T/T_{ref})^m \exp(-E_O/RT)\, (CH4)^\alpha\, (H_2O)^\beta\, (O_2)^\gamma$ kg/m^{3}sec. (D)

The corresponding expression is

$d[CH_4]/dt = -A_{[\,]}\, (T/T_{ref})^m \exp(-E_O/RT)\, [CH_4]^\alpha\, [H_2O]^\beta\, [O_2]^\gamma$ kmol/ m^{3}sec.

If $A_{(\,)}$ is given how do you find $A_{[\,]}$?

$d(CH_4)/dt$ kg/m^{3}sec = $-A_{(\,)}\, (T/T_{ref})^m \exp(-E_O/RT)\, (CH^4)^\alpha\, (H_2O)^\beta\, (O_2)^\gamma$ kg/m^{3}sec (E)

Divide by M_{CH4}

$d(CH_4)/dt$ kg/m^3 sec = $-A_{[\,]}\, (T/T_{ref})^m \exp(-E_O/RT)\, [CH_4]^\alpha\, [H_2O]^\beta [O_2]^\gamma$ kg/m^{3}sec (F)

where $A_{[\,]} = A_{(\,)}$ and $M_{CH4}^{\alpha-1}\ M_{H2O}^{\beta}\ M_{O2}^{\gamma}$.

5.13.3.3 Summary on Conversions

Conversions

$A_{[\,]}$	$A_{(\,)}$	A_X	A_Y
1	$M_{CH4}{}^{(1-\alpha)}\ M_{H2O}^{\beta}\ M_{O2}^{-\gamma}$	1	$\{M_{CH4}{}^{1-\alpha}\}/ M_{H2O}^{\beta}\ M_{O2}^{\gamma}$
$M_{CH4}{}^{(\alpha-1)}\ M^{\beta}{}_{H2O}\ M_{O2}{}^{\gamma}$	1	$\{M_{H2O}^{\beta}\ M_{O2}{}^{\gamma}\}/M_{CH4}{}^{1-\alpha}$	1

5.13.4 Some Approximations in Kinetics Integrals

Many of the approximations are left as exercise problems. Only results are given.

Case (i)

$$\int_{To}^{T} \exp(-E/\bar{R}T)\, dT = \{E/\bar{R}\}\{e^{x}[1/x^2 + 2/x^3 + 6/x^4 + 24/x^5 + 120/x^6 + \ldots]\}$$

$$-\{E/\bar{R}\}\{e^{x0}[1/x_0^2 + 2/x_0^3 + 6/x_0^4 + 24/x_0^5 + 120/x_0^6 + \ldots]\},\ x = -E/RT$$

Case (ii)

For $T - T_0 = y$, $y << T_0$

$$\int_{To}^{T} \exp(-E/\bar{R}T)\, dT = \int_{To}^{T} \exp(-E/(\bar{R}T_0(1 + y/T_0))\, dT = \int_{To}^{T} \exp\{-E(1 - y/T_0)/(\bar{R}T_0)\}\, dT$$

$$= \exp\{-E/(\bar{R}T_0)\}\left(\int_{0}^{y} \exp\{E\,y/\bar{R}T_0^2\}\right) dy = (\bar{R}T_0^2/E)\exp[-E/(\bar{R}T_0)]\{\exp[E/(\bar{R}T_0^2)]\,y - 1\}$$

Case (iii)

$$\int_{To}^{T} \exp(-E/\bar{R}T)\, dT = ?$$

Let $E/RT = x$, $dT = -E\,dx/(\bar{R}Tx^2)$. Typically, T_0 is very low and hence letting $x_0 \to \infty$

$$\int_{xo}^{x} \exp(-E/\bar{R}T)\, dT = -\{E/\bar{R}\} - \int_{\infty}^{x} e^{-x}\, dx/x^2$$

$$\int_{\infty}^{x} e^{-x}\, dx/x^2 = \int_{\infty}^{x} e^{-x}\, d(1/x) = [-e^{-x}/x^2 - \int_{\infty}^{x} e^{x}\, dx/x^3] = p(x),\ 20 < x < 60$$

$$\log 10\,\{p(x)\} \approx -2.315 - 0.457\,x \quad \text{or}$$

$$p(x) \approx 0.00482\exp(-1.052\,x)$$

$$\int_{To}^{T} \exp(-E/\bar{R}T)\, dT \approx \{0.00482\,E/\bar{R})\exp\{-1.052\,E/\bar{R}T\},\ 20 < E/RT < 60$$

Case (iv)

This deals with exponential integrals.

$$\int_{To}^{T} e^{-(E/\bar{R}T)}\, dT = -(E/\bar{R})\{E_2(X_0)/X_0 - E_2(X)/X\} = (E/\bar{R})\{E_2(X)/X - E_2(X_0)/X_0\}$$

where the exponential integrals E_n are tabulated [Abramovitz and Stegun, 1964 with the following recurrence relation:

$$E_1(X)*X*e^X = (X^2 + a_1X + a_2)/(X^2 + b_1X + b_2)$$

With
$a_1 = 2.334733$, $b_1 = 3.330657$, $a_2 = .250621$, $b_2 = 1.681534$, $1 \leq X < \infty$,
we have:

$$E_2\ (X) = \{\exp(-X) - X\ E_1(X)\}$$

$$E_{n+1}\ (X) = (1/n)\ \{\exp(-X) - X\ E_n(X)\},\ n = 1, 2, 3, \ldots$$

Chapter 6

Mass Transfer

6.1 Introduction

This chapter discusses mass transfer relations and several definitions used in mass transfer theory. In addition, relations for transport properties in terms of molecular properties for pure component and mixtures are also presented.

6.2 Heat Transfer and the Fourier Law

The Fourier law relates the rate of transport of heat (energy flux) via conduction to the temperature gradient in a medium. It has the form

$$\dot{q}'' = -\lambda \, dT/dy = -\alpha_T \rho c_p \frac{dT}{dy} \tag{6.1}$$

as shown in Figure 6.1a, where λ denotes the medium conductivity, α_T, thermal diffusivity, ρ, density, and c_p, specific heat. Similarly, if a cold gas flow with a temperature T_∞ occurs over a hot surface maintained at the temperature T_w, the heat transfer via conduction per unit area is

$$\dot{q}'' = h_H \, (T_w - T_\infty), \tag{6.2}$$

where h_H denotes the *convective heat transfer coefficient* (Figure 6.1b). In this chapter, we are interested in the corresponding relations for rate of mass transport. Assume that boiling water is poured into a thick metal cup (as shown in Figure 6.2a). Whereas the inside surface is almost instantaneously heated, the outside surface takes a relatively longer time to heat up from room temperature because of the thick wall. Consequently, before steady state, the metal molecules adjacent to the inner cup wall vibrate with a higher energy than those near the outer surface. This higher vibration energy is transmitted (i.e., heat is

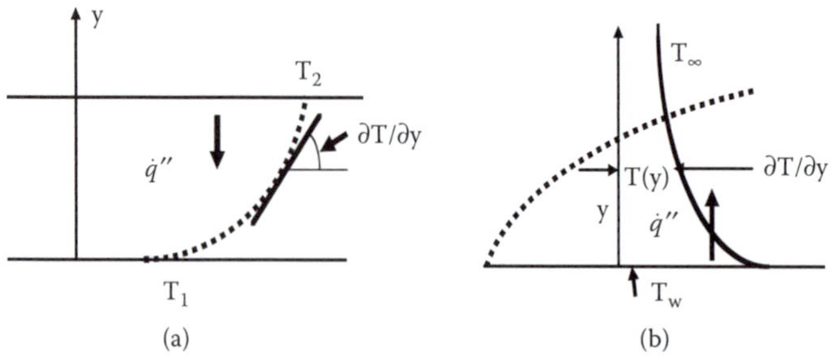

Figure 6.1 Schematic illustration of (a) conduction, (b) convection heat transfer.

conducted) to each adjacent layer over a finite time period, which is known as the *thermal diffusion time*. Typically, this diffusion time is estimated as

$$t_{diff,T} = x^2/\alpha_T. \tag{6.3}$$

In Equation 6.3, x denotes the distance traveled by the diffusing thermal wave within time $t_{diff,T}$ and α_T (= $\lambda/\rho c$) is the thermal diffusivity in m^2/sec, where c represents the specific heat and ρ the density. The energy U_x contained within the layer of thickness x is

$$U_x = x\, A\rho c = A\rho c\ (t\alpha_T)^{1/2} = A(t\lambda/\rho c)^{1/2}, \tag{6.4}$$

where A is the cross sectional area.

If the cup thickness is L, then $t_{diff,T} = L^2/\alpha_T$. The thermal diffusion velocity (a measure of how fast the thermal energy moves from the inside to the outside surface by conduction) is

$$V_{diff,T} = L/t_{diff,T} = \alpha_T/L. \tag{6.5}$$

For instance, the value of α for copper is 117×10^{-6} m^2/sec, whereas for steel it is 18×10^{-6} m^2/sec. If the cup is assumed to be fabricated from a steel plate that has a thickness of 0.003 m, then $t_{diff,T} = 0.003^2/18 \times 10^{-6} = 0.5$ sec and the thermal velocity equals 0.003/0.5 = 0.6 cm/sec (whereas for Cu it is 3.9 cm/sec). A similar analogy exists for mass transfer with mass diffusion velocity.

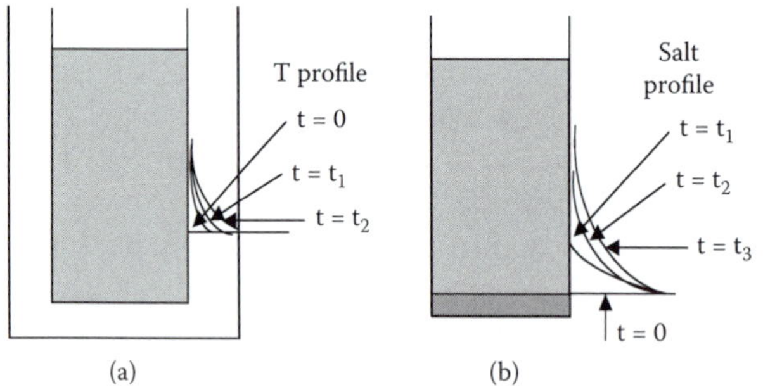

Figure 6.2 Illustration of (a) thermal diffusion, (b) mass diffusion.

6.3 Mass Transfer and Fick's Law

Common examples of mass transfer involve the evaporation of liquid drops, the vaporization of ocean water, and the drying of wet clothes. Consider a beaker containing a specified volume of distilled water (which has a molecular weight of 18 kg/kmol) into which a layer of common salt (NaCl with a molecular weight of 40 kg/kmol) is instantaneously introduced at the bottom (as shown in Figure 6.2b) at time t = 0. Initially, the water is not salty. However, if one were to carefully skim a small portion of water from the surface periodically and taste it, the saltiness due to NaCl would slowly increase. This would be due to the diffusion of the salt from the bottom to the top water surface. Diffusion causes the bulk movement of molecules (of NaCl in this example) from a higher-concentration site to a lower-concentration, and the result is mass transfer. Diffusion involves mixtures, and in this example the mass transfer continues until the salt is uniformly dissolved in the water, i.e., until there are no concentration gradients for either NaCl or H_2O. A microscopic overview is presented in Section 6.4.

6.3.1 *Fick's Law*

Fick's law is a mass transfer rule that is similar to the Fourier law. It has the form

$$j_k'' = -\rho\ D_{kl}\ dY_k/dy,\ kg/(m^2\ sec) \tag{6.6}$$

where j_k'' denotes the diffusive mass transfer rate of a species k (e.g., NaCl) per unit surface area (kg/m² sec); Y_k, the mass fraction of the species k; D_{kl}, the binary diffusion coefficient of species k in species l (e.g., of NaCl in water); and ρ, the mixture density through which species diffuse. Alternatively, the chemical engineering literature uses mole-based flux which is also defined as

$$\dot{N}''_{k,diff} = -c\ D_{k\ell}\ dX_k/dy,\ kmole/(m^2\ sec) \tag{6.7}$$

where c is the molal concentration; X_k, the mole fraction; $\dot{N}''_{k,diff}$, the diffusive mole flow per unit area (kmol/m²sec); and $D_{k\ell}$, the binary diffusion coefficient of k in ℓ. It will be shown later that $\dot{N}''_{k,diff}\ M_k$ is not necessarily equal to j_k''. A relation similar to Equation 6.5 can be developed for mass transfer, i.e.,

$$V_{diff,m} = L/t_{diff,m} = D_{k\ell}/L. \tag{6.8}$$

Typically, for many gases, $D_{k\ell}$ is of the order of 10^{-4} m²/sec at 1000 K. Then, in a gaseous space, if L = 0.003 m, $t_{diff}\ L^2/D_{k\ell} = 0.09$ sec, and $V_{diff,m}$ = 4 cm/sec. The thermal diffusion velocity in typical gas layers is similar to the mass diffusion velocity, because the Lewis number $\alpha_T/D_{k\ell}$ for air (or oxygen species k diffusing – in nitrogen species ℓ) is roughly unity. The number of molecules per unit volume or the mass of species k per unit volume changes in a variable-concentration, or stratified, mixture. Therefore, the mixture has varying species density until diffusion ceases.

6.3.2 Definitions

The *mass concentration* ρ_k of a species k represents the mass of that species per unit volume of a mixture, i.e., its species *density*. For ideal gases

$$\rho_k = p_k/(R_k\ T),\ \text{kg of species k/m}^3, \quad \text{where} \quad R_k = \bar{R}/M_k$$

where p_k denotes the *partial pressure* of species k in a mixture. Likewise, the *molal concentration* c_k is the number of moles of the species per unit volume of a mixture, and

$$c_k = [k] = \rho_k/M_k\ \text{kmol of k/m}^3. \tag{6.9}$$

For ideal gases,

$$c_k = p_k/(\bar{R}T). \tag{6.10}$$

The local *mole fraction* X_k is defined in terms of molal concentration as

$$X_k = c_k/c \quad \text{where} \quad c = \sum c_k \tag{6.11}$$

where c is the molal density. For ideal gases

$$X_k = p_k/P, \quad \text{as} \quad c = P/\bar{R}T. \tag{6.12}$$

The local species *mass fraction* Y_k is defined as the ratio of mass of species k, m_k, to the total mass m within an elemental volume containing the mixture:

$$Y_k = m_k/m = \rho_k/\rho = c_kM_k/\sum c_kM_k. \tag{6.13}$$

Divide numerator and denominator of the fourth term by c

$$Y_k = X_kM_k/M_{mix} \tag{6.14}$$

$$\text{where} \quad M_{mix} = \sum X_k\ M_k \tag{6.15}$$

is the *mixture molecular weight* or *mixture molal mass* representing the mass of a kilomole of the mixture.

Similarly,

$$X_k = (Y_k/M_k)/(1/M_{mix}). \tag{6.16}$$

This can be illustrated using the example of air. In every 100 moles of air collected, on average 21 kmol are of oxygen and 79 kmol are of nitrogen, i.e., (X_{O2} = 0.21). From Equation 6.15, $M_{mix} = 0.21 * 32 + 0.79 * 28 = 28.56$ kg/kmol. Using Equation 6.14 to convert this value to a mass basis, $Y_{O2} = 0.21 \times 32/(0.21 \times 32 + 0.79 \times 28) = 0.23$.

For ideal gases

$$\rho = P/(RT) \quad \text{and} \quad R = \bar{R}/M_{mix}. \tag{6.17}$$

It is also apparent from Equation 6.11 and Equation 6.12 that

$$\sum X_k = \sum p_k/P = P/P = 1 \tag{6.18}$$

and

$$\sum Y_k = \sum \rho_k/\rho = \rho/\rho = 1 \tag{6.19}$$

as $\sum \rho_k = \rho.$

The *mass density* or *mixture density*

$$\rho = \sum \rho_{k.} \tag{6.20}$$

Each species k in a multicomponent mixture moves with its own velocity that is referred to as the *absolute velocity*, v_k, of that species. For example, consider a two-component mixture. Species 1 moves at a slower velocity, whereas species 2 moves at a higher velocity (Figure 6.3). There can be many species in a mixture, each traveling with a different absolute velocity. Consider the example of the salt and water mixture. When the NaCl molecules (M = 58.5; density = 2165 kg/m^3; bulk density = 1154 kg/m^3) move upward, some water molecules (M = 18; density = 1000 kg/m^3) move downward to fill the spaces that are left behind by the migrating salt molecules. The NaCl molecules could be considered to migrate upward with an absolute velocity with respect to a stationary observer standing outside the container (i.e., $v_{NaCl} > 0$). On the other hand, water molecules move downward with a velocity $v_{H_2O}(< 0)$. During the initial period, a larger mass is moved upward than downward owing to the larger molecular weight of NaCl. Thus, there could be a net initial upward mass shift or an upward mass-averaged velocity v.

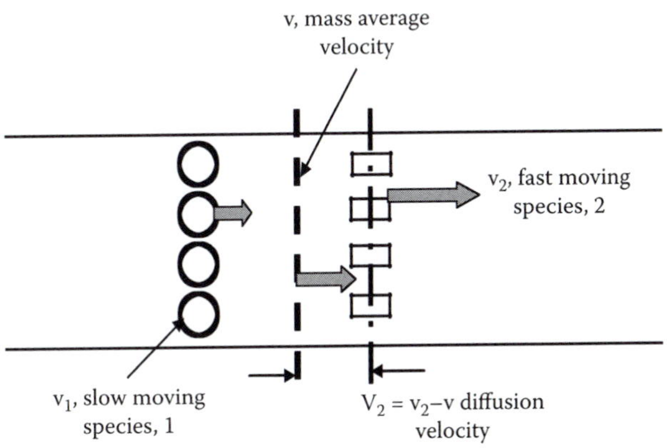

Figure 6.3 Absolute, average, and diffusion velocities.

The mass *average bulk velocity* v (also known as *Stefan velocity*; Figure 6.3) at any section of a flow is obtained from the relation

$$\rho v = \sum \rho_k v_k, \quad \text{or} \quad v = \sum \rho_k v_k/\rho = \sum Y_k v_k \tag{6.22}$$

where v_k could be positive or negative. For the water–salt example, $v_{NaCl} > 0$ and $v_{H_2O} < 0$ initially at the mid plane of the container, so that there is a positive net mass-averaged or average bulk velocity. Likewise, the *mole-averaged velocity* is

$$\bar{v} = \sum X_k\, v_k, \tag{6.23}$$

and is normally different from the mass-averaged velocity. The velocity v_k is the *absolute velocity* of a species k with respect to a stationary observer. The *mass-averaged velocity* is the velocity that is normally measured using a Pitot tube or a hot-wire anemometer. The mass-averaged velocity can also be obtained from solutions of the conservation equations (Chapter 8).

The difference between the absolute species velocity v_k and the mass-averaged velocity v is the mass-based diffusion velocity

$$V_k = v_k - v, \quad \text{or} \quad v_k = v + V_k. \tag{6.24a}$$

Similarly, the mole-based diffusion velocity is given as,

$$\bar{V}_k = v_k - \bar{v}, \quad \text{or} \quad v_k = \bar{v} + \bar{V}_k. \tag{6.24b}$$

The relations in Equation 6.24 enable us to show that

$$\sum \rho_k V_k = \rho \sum Y_k V_k = 0, \quad \sum c_k \bar{V}_k = c \sum X_k \bar{V}_k = 0 \quad \text{and because} \tag{6.25}$$

$$\sum \rho_k (v_k - v) = \sum \rho_k v_k - \sum \rho_k v = \rho\, v - \rho\, v = 0 \quad \text{and} \quad \sum X_k (v_k - \bar{v}) = 0.$$

The following example illustrates the concepts of number and mass-based average velocities and the number and mass-based diffusion velocities of each species k.

Example 1

Consider two horses of type A, one horse of B, three horses of C, two horses of D, and two horses of type E with corresponding masses of 100, 150, 175, 250, and 300 kg running at speeds of 30, 20, 40, 20, and 20 m/sec, respectively. Estimate the number-averaged velocity, mass-averaged velocity, and speed of each type of horse (diffusion velocity) relative to the number-averaged velocity and mass-averaged velocity.

Solution

Number-averaged velocity, v_n = {2 * 30 + 1 * 20 + 3 * 40 + 2 * 20 + 2 * 20}/10 = 28 m/sec

Mass-averaged velocity, v = Σ {Number of horses of type k * m_k v_k}/Σ{Number of horses of type k * m_k}

v = {2 * 30 * 100 + 1 * 20 * 150 + 3 * 40 * 175 + 2 * 20 * 250 + 2 * 20 * 300}/{2 * 100 + 1 * 150 + 3 * 175 + 2 * 250 + 2 * 300} = 52,000/1975 = 26.33 m/sec

Diffusion velocity with respect to number-averaged velocity

$\bar{V}_A = v_A - \bar{v}$ = 30 − 28 = 2 m/sec; similarly, $\bar{V}_B$ = −8 m/sec, $\bar{V}_C$ = 12 m/sec, $\bar{V}_D$ = −8 m/sec, $\bar{V}_E$ = −8 m/sec

$\Sigma N_k \bar{V}_k$ = 2 * 2 + 1 * (−8) + 12 * 3 + 2 * (−8) + 2 * (−8) = 0.

Similarly, diffusion velocity with respect to mass-averaged velocity is

$V_A = v_A - v$ = 30 − 26.3 = 3.67 m/sec; similarly, V_B = −6.33 m/sec, V_C = 13.67 m/sec, V_D = −6.33 m/sec, and V_E = −6.33 m/sec

$\Sigma N_k\, m_k\, V_k$ = 100 * 2 * 3.67 + 150 * 1 * (−6.33) + 175 * 3 * 13.67 + 250 * 2 * (−6.33) + 300 * 2 * (−6.33) = 0

Remarks

(i) Difference between mass-averaged and number-averaged velocity is 26.33 − 28 = −1.67 m/sec.

(ii) $v_A - v_B = \bar{V}_A - \bar{V}_B = V_A - V_B$ = 10 m/sec.

(iii) Diffusive mass flux of A with respect to mass-averaged velocity is 2 * 100 * 3.67 = 734 kg/sec. The average mass flux of A is 26.33 * 2 * 100 = 5266 kg/sec and, hence, absolute mass flux of A is 6000 kg/sec. Number-based diffusive flux is 2 * 100 * 2 = 400 kg/sec, whereas average mass flux with number-based average velocity 2 * 100 * 28 = 5600 kg/sec with absolute flux of A being 5600 + 400 = 6000 kg/sec. Thus, the total absolute flux remains fixed as it should be when one uses number-based fluxes or mass-based fluxes.

(iv) Similarly, we have gas mixtures containing H_2 (A), O_2 (B), and CO_2 (C) etc., with molecular weights of 2, 32, and 44 traveling at different speeds ($v_A = v + V_A$, mass-averaged velocity + diffusion velocity) and their number or mass concentrations vary. The number-based velocity is nothing but the molal velocity = $\bar{v}_A - \bar{v}_B = V_A - V_B$ = 10 m/sec.

■ ■ ■

Most of the conservation equations are formulated using mass-based average velocity and, as such, the emphasis on Fick's law will be mass based. The diffusive velocity can be determined by using relations involving the mass fraction gradients. Considering only diffusive mass transport of species k

$$j_k'' = \rho_k V_k = \rho_k\,(v_k - v). \qquad (6.26)$$

Using Fick's law (Equation 6.6) in Equation 6.28,

$$V_k = j_k''/\rho_k = -D_{kl}(1/Y_k)\,(dY_k/dy) = -D_{kl}\,d(\ln Y_k)/dy. \qquad (6.27)$$

$V_k > 0$ if Y_k decreases with increasing y, and $V_k < 0$ if Y_k increases with increasing y. For a 3-D problem, Fick's law is generalized as

$$\nabla Y_k = -V_k Y_k/D_{kl}. \qquad (6.28)$$

More generally, for a multicomponent species, the relation is given as (Section 6.5)

$$\nabla Y_k = -\sum_1 (V_k - V_l) Y_l Y_k / D_{kl}. \tag{6.29}$$

If all of the D_{kl} values equal D, then

$$\nabla Y_k = -V_k Y_k / D, \tag{6.30}$$

because using Equation 6.25,

$$\sum_1 V_l Y_l = 0. \tag{6.31}$$

The *mass flow rate* of a species k

$$\dot{m}_k'' = \rho_k v_k = \rho_k (v + V_k) = \text{mass flow of k with bulk flow + k with diffusive flow}$$
$$= \rho_k \{v - D_{kl}(d(\ln Y_k)/dy)\}. \tag{6.32}$$

The ratio between species mass flow $\dot{m}_k$ to total mass flow rate $\dot{m}$ is called the flux ratio of species, ε_k

$$\varepsilon_k, \text{Flux ratio} = \frac{\dot{m}_k''}{\dot{m}''} = \frac{\rho_k v_k}{\rho v} = \frac{\rho_k [v + V_k]}{\rho v} = Y_k - \frac{\rho D}{\rho v} \frac{dY_k}{dy} \tag{6.33}$$

where the average, absolute, and diffusion velocities can be illustrated through the next example.

When can one neglect diffusion? Recall from Equation 6.24a that $v_k = v + V_k$. Diffusive velocity is negligible when $V_k << v$, or dY_k/dy is very low; thus $v_k \approx v$ and $\rho_k v_k \approx \rho_k v$. For example, if water evaporates from a vertically oriented tank, then dY_w/dy (y normally to the water surface) near the water surface may be high and we may need both convection and diffusion; however, far above the water surface, $dY_w/dy \approx 0$. Sometimes the average velocity may be very low (i.e., negligible Stefan velocity) compared to diffusion velocity; $v_k = v + V_k$, $v_k \approx V_k$, $\rho_k v_k \approx \rho_k V_k$.

Example 2

Consider a drop of water that has an initial diameter d_o (= 2 a_0) placed in a quiescent ambient consisting of dry air. The drop evaporates and water vapor (k) is transported to the ambient air (Figure 6.4).

(a) Obtain an expression for the diffusive mass transfer rate j_k'', where k denotes water and a, air, if the radial mass fraction profile around an evaporating drop is represented by the relation

$$(Y_k - Y_{k,\infty})/(Y_{k,w} - Y_{k,\infty}) = a/r, \quad r \geq a, \tag{A}$$

where the subscripts w and ∞ refer to the drop surface and far ambient, respectively.

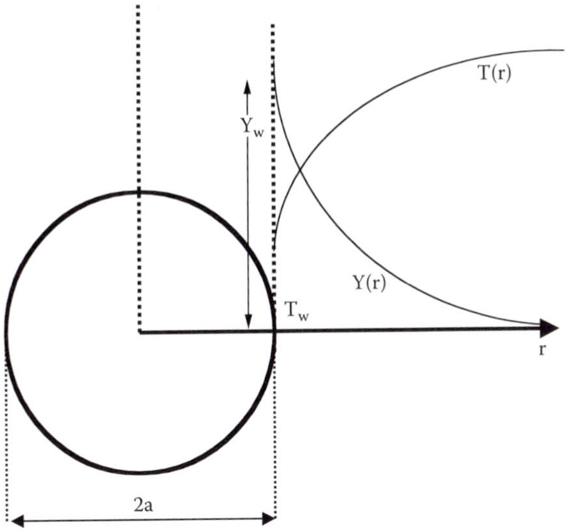

Figure 6.4 Evaporation of a drop in warm air.

(b) Assuming that $V_k >> v$, develop an expression for $\dot{m}''_k(r)$. What is the total evaporation rate $\dot{m}$?
(c) Develop an expression for $\dot{m}_k$ (r) or $\dot{m}_{k,w}$.
(d) In analogy with heat transfer, if $\dot{m}''$ is expressed as $h_m(Y_{k,w} - Y_{k,\infty})$, obtain an expression for the Sherwood number, $Sh = h_m d/(\rho D_{kl})$.
(e) Determine the mass transport of water in air per unit surface area of a 50 micron radius spherical drop at a total pressure of 100 kPa. The temperature of the drop $T_w = 80°C$. At phase equilibrium the partial pressure of water vapor p_{kw} is the same as the saturation pressure at the drop surface:

$$p_{kw} = p_k^{sat} = 1.8 \times 10^8 \exp(-5319/T_w), \text{ kPa} \tag{B}$$

Assume $D_{k1} = 10^{-5}$ m²/sec.
(f) Examine if $V_k >> v$.
(g) The drop surface regresses as a function of time so that the value of d decreases over time. Show that the expression for d^2 with respect to time in terms of evaporation rate constant α_e is provided by the so-called d^2 law,

$$d^2 = d_0^2 - \alpha_e t, \tag{C}$$

where $\alpha_e = 4\, Sh\, (\rho D)(Y_{k,w} - Y_{k,\infty})/\rho_\ell$, where ρ_ℓ denotes the density of liquid water (1000 kg/m³).
(h) Determine the value of α_e and the evaporation time t_e.

Solution

(a) The gradient

$$dY_k/dr = -Y_{k,w}\, a/r^2 \tag{D}$$

where k = $H_2O(g)$. From Fick's law and Equation A,

$$j_k'' = -\rho D_{kl}\, dY_k/dr, \quad \text{i.e.,} \quad j_k'' = \rho\, D_{kl}\, (Y_{k,w} - Y_{k,\infty})(a/r^2). \qquad \text{(E)}$$

(b) Assuming that the evaporation is slow:

$$\dot m_k''(r) = \rho_k v_k \approx \rho_k V_k = j_k''. \qquad \text{(F)}$$

(c) $\dot m_k(r) = \dot m_k''(r)\,(4\,\pi r^2) \approx \rho\, D_{kl}\,(Y_{k,w} - Y_{k,\infty})\, 4\pi a$ = constant (G)
At r = a, $\dot m_k = \dot m_k''(a)\,(4\pi a^2) = \rho D_{kl} Y_{kw}\,(4\pi a)$ = constant as $Y_{k,\infty} = 0$.
(d) In analogy with heat transfer,

$$\dot m_k''(a) = h_m(Y_{k,w} - Y_{k,\infty}) = h_m Y_{k,w} = \rho D_{kl} Y_{k,w}/a = 2\rho D_{kl} Y_{k,w}/d,$$

Simplifying

$$Sh = h_m d/(\rho D_{kl}) = 2 \qquad \text{(H)}$$

where the Sherwood number is analogous to the Nusselt number ($Nu = h_H d/\lambda$) used in heat transfer applications. Typically, Sh = 2 in a quiescent atmosphere as in the case of heat transfer, where Nu = 2. When there is relative velocity, the Ranz–Marshall correlation suggests that $Sh = 2 + 0.6\, Re^{1/2} Sc^{1/3}$, where $Re = v_r d/\nu$, Schmidt number, $(Sc) = D/\nu$, ν denotes the kinematic viscosity, and v_r is the relative velocity of drop with respect to ambient.
(e) Using Equation B at 353 K,
p_w = 51.44 kPa.
As the total pressure is 100 kPa, the partial pressure of air is
p_{air} = 100 − 51.4 = 48.6 kPa.
For ideal gas mixtures,
$p_k/P = X_k$ = 51.44/100 = 0.5144, i.e., X_{air} = 0.486.
The molecular weight of the mixture is
M = 0.514 × 18.02 + 0.486 × 28.97 = 9.26 + 14.08 = 23.34 kg/kmol,
$Y_{k,w} = X_k\, M_k/M$ = 0.5144 × 18.02/23.34 = 0.40.
Assume that the average density in the region $a < r < \infty$ is identical to the mixture density at the surface. Therefore,

$$\rho = PM/(\bar R T) = 100 \times 23.34/(8.314 \times 353) = 0.795\ \text{kg/m}^3.$$

The mass transfer rate

$$\dot m_k'' \approx j_k'' = 1 \times 10^{-5} \times 0.795 \times 0.4/(50 \times 10^{-6}) = 0.0636\ \text{kg/m}^2\ \text{sec}.$$

(f) The density of vapor at the surface

$$\rho_k = p_k\, M_k/(\bar R T) = 51.44 \times 18.02/(8.314 * 353) = 0.315\ \text{kg/m}^3.$$

Therefore,

$$v_k = 0.0636/0.315 = 0.2\ \text{m/sec}.$$

There is no transport of air between drop and ambient.

$$v = (\rho_k v_k + \rho_l\, v_l)/\rho = (0.0636 + 0)/0.795 = 0.08 \text{ m/sec}, \tag{I}$$

and is called the Stefan velocity.

The diffusion velocity $V_k = v_k - v = 0.12$ m/sec is higher than the average velocity by a factor of 1.5. Therefore, the very small Stefan flow assumption that was used to develop Equation G is not strictly valid, because $V_k \approx v$. The Stefan velocity becomes smaller at lower temperatures. Flux ratio $\varepsilon_k = 0.0636/0.0636 = 1$ because no other species are moving.

(g) As the drop evaporates, its diameter decreases over time. As long as the profile given in Equation A is valid even when $a = a(t)$, the time rate of mass decrease must be equal to the evaporation rate, i.e.,

$$-dm/dt = \dot{m} = h_m\,(Y_{k,w} - Y_{k,\infty})(4\pi a^2), \tag{J}$$

where

$$m = \pi d^3 p_p/6 \tag{K}$$

where ρ_p is the particle or liquid density.

Differentiating Equation K, and using Equation J and the definition of Sh,

$$-d((\rho_\ell D)d)/dt = 2Sh(Y_{k,w} - Y_{k,\infty})\rho D. \tag{L}$$

Integrating this equation with the initial condition $t = 0$, $d = d_0$, we obtain

$$d^2 = d_0^2 - \alpha_e\, t, \tag{M}$$

where

$$\alpha_e = 4\ Sh\ \rho D_{kl}(Y_{k,w} - Y_{k,\infty})/\rho_p. \tag{N}$$

(h) The evaporation rate constant for this problem

$$\alpha_e = 4 \times 2 \times 0.795 \text{ kg/m}^3 \times 10^{-5} \text{ m}^2\text{/sec}$$

$$(0.40 - 0)/1000 \text{ kg/m}^3 = 25.4 \times 10^{-9} \text{ m}^2\text{/sec}.$$

When the drop evaporates completely, $d \rightarrow 0$. Therefore, the evaporation time t_e is given by using the value $d^2 = 0$ in Equation M. Here, $t_e = 0.393$ sec or 393 msec.

Remarks

(i) The Sh number concept is particularly useful when there is relative velocity. Using Equation A, $h_{m,kl} = Sh\ (\rho D_{kl})/d$. Thus, when $Re = 4$ and $Sc = 1$,

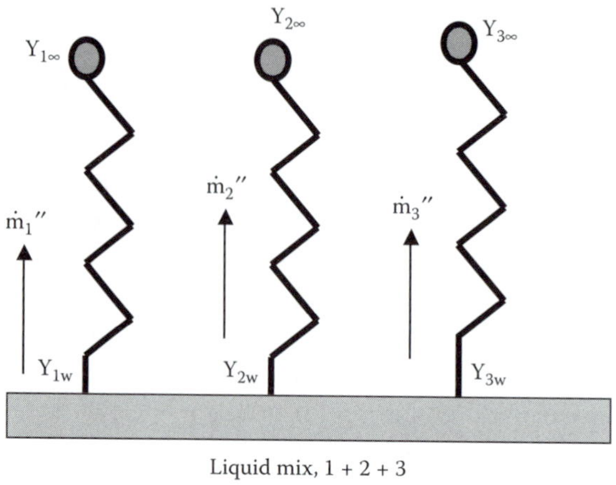

Figure 6.5 Evaporation of multicomponent liquid-resistance concept.

Sh from the Ranz–Marshall correlation = 2 + 0.6 $Re^{1/2}$ Sc $^{1/3}$ = 2 + 0.6 × $4^{1/2}$ = 3.2. Hence, $h_{m,kl}$ and $\dot{m}_k'' = h_{m,k}\,(Y_{k,w} - Y_{k,\infty})$ can be estimated.

(ii) Using the analogy with electrical current, I = voltage/resistance and $\dot{m}_k'' = (Y_{k,w} - Y_{k,\infty})/R_{m,kl}$, where the mass transfer resistance for species k is $R_{m,kl} = 1/h_{m,kl}$. For example, consider a slurry droplet containing methanol, ethanol, and water. Let k = 1 for H_2O, k = 2 for CH_3OH, k = 3 for ethanol diffusing in N_2 (k = 4) (k = 1,2,3 ... diffusing in 4) (Figure 6.5). Then $\dot{m}'' = \dot{m}_1'' + \dot{m}_2'' + \dot{m}_3'' = (Y_{1,w} - Y_{1,\infty})/R_{m,14} + (Y_{2,w} - Y_{2,\infty})/R_{m,24} + (Y_{3,w} - Y_{3,\infty})/R_{m,34}$, where $R_{m,k4} = 1/h_{m,k4}$. Mass transfer rate $\dot{m}_k = h_{m,k}$ A ($Y_{k,w}$ − $Y_{k,\infty}$), is similar to the heat transfer rate Sh = {$h_{m,k}$ d/D_{k1}}. Note that $\dot{m}_k''$ could also be negative. One can show that $1/D_{mix} = 1/D_{14} + 1/D_{24} + 1/D_{34}$. In Chapter 7, these concepts are used to estimate the burn rate of carbon.

Example 3

The elemental mass-based composition of Wyoming low-sulfur coal of diameter 80 µm is as follows: moisture, 10%; ash, 5%; combustibles, 85%; and HHV, 11,410 Btu/lb as received. The particle density is 1300 kg/m³. Assume that water is dispersed throughout the particle such that the pseudo water density is equal to moisture content times density of particle (ρ_p). The particle heats up and vaporizes at T_p. The vapor is transported away from the particle by mass transfer governed by the Sherwood number

$$Sh = h_{m,kl}d/(\rho D_{kl}), \qquad \text{(A)}$$

where d = 2a. Assume that $\rho D_{k1} = \rho D$ has a constant value

The temperature of the particle is T_w = 80°C. Saturation pressure is given by Equation B of Example 2.

Estimate the drying time. Use Ranz–Marshall correlation. Re = 4, Sc = 1. Let $\rho D = 8 \times 10^{-5}$ kg/(msec).

Solution

$\rho_w = Y_{k.w} * \rho_p = 0.1 * 1300 = 130 \text{ kg/m}^3$
$d^2 = d_0^2 - \alpha_e t,$
$t_{dry} = d_0^2/ \alpha_e,$
$\alpha_e \simeq 4 \text{ Sh } (\rho D)(Y_{k,w} - Y_{k,\infty})/\rho_w, \ \rho_w = 130 \text{ kg/m}^3$
Thus, when Re = 4, Sc = 1, Sh = 2 + 0.6 $4^{1/2}$ = 3.2.
From Example 2, $Y_{k,w} = 0.4$

$$\alpha_e \simeq 4 \text{ Sh } (\rho D)(Y_{k,w} - Y_{k,})/\rho_w = 4 * 3.2 * 8 \times 10^{-5} * 0.4/130$$
$$= 3.15 \times 10^{-6} \text{ m}^2\text{/sec.}$$

As $d^2 = d_0^2 - \alpha_e t$, $t_{dry} \simeq d_0^2/\alpha_e = 1000 * (80 \times 10^{-6})^2/3.15 \times 10^{-6} = 2$ msec.

Remarks

As the apparent density of water is low, the liquid layer recedes rapidly within the particle.

6.4 Molecular Theory

6.4.1 *Approximate Method for Transport Properties of Single Component*

6.4.1.1 *Absolute Viscosity*

A simple molecular theory is presented to illustrate the derivation of transport properties μ, ν, D, λ, and α_T. Typically, the molecules move in all six directions: +x, −x, +y, −y, +z, and −z. Consider flow over a plate. Assume that $v_x(y)$ with $v_x = 0$ at y = 0. Consider location B (Figure 6.6). Let $\dot{N}'$ be the transport rate of molecular from B. Suppose $\dot{N}'/6$ molecules move up from B to location A′ (+y direction) through an area dA per second, and they carry an x-momentum of $[\{v_x - (dv_x/dy)\ \ell\}]\ [\dot{N}/6]$, where ℓ is the mean free path. Note that when T's are equal, velocities at A and B have equal random velocities. At the same time, $\dot{N}/6$ molecules move downward from A (−y direction) and carry an x-momentum

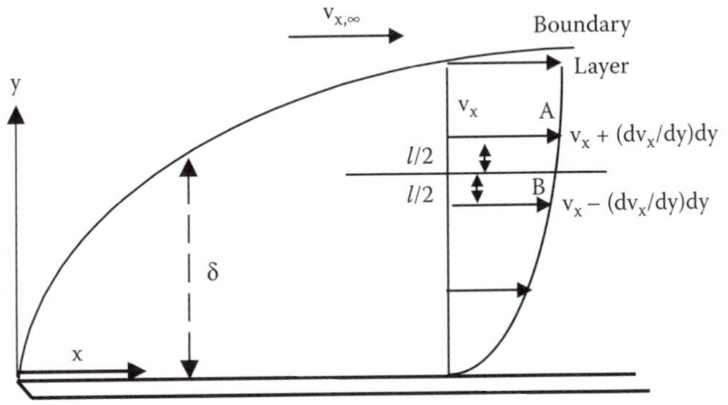

Figure 6.6 Schematic illustration of velocity boundary layer.

of $[\{v_x + (dv_x/dy)\ \ell\}]\ [\dot{N}'/6]$. Thus, the net momentum transported toward B = $[\{(dv_x/dy)\ \ell\}]\ [\{\dot{N}'/3]$. If each molecule has a mass of m′, then the momentum transported downward is given as $[\{(dv_x/dy)\ \ell\}]\ [\dot{N}'m'/3]$.

The molecular flow rate $\dot{N}'$ at, say, B, is given by $\dot{N}' = n'\ v'_{avg}\ dA$, where n′ is the number of molecules per m^3 and $v'_{avg} = \{8\bar{R}T/(\pi M)\}^{1/2}$, the random average velocity. The net momentum transported downward per sec per unit area is the shear stress τ. Equating it to Newton's law of viscosity

$$\{(dv_x/dy)\ \ell\}]\ [\dot{N}'m'/3], = (1/3)\ \{(dv_x/dy)\ \ell\}\ ]\ n'\ m'\ v'_{avg}$$
$$= \mu\{dv_x/dy\} = \tau.$$

Solving for μ,

$$\mu = (1/3)\ n'\ell m'v'_{avg} = (1/3)\ \rho\ \ell\ v'_{avg}$$

and

$$\{\mu/\rho\} = \nu = (1/3)\ \ell\ v'_{avg}$$

As $\ell = 1/\{\pi\sigma^2 n\}$ and $v'_{avg} = \{\ 8\bar{R}T/\pi\ M\}^{1/2}$, we have

$$\nu = (1/3)\ \{1/(\pi\sigma^2\ n'\ N_{Avog})\}\ \{8\bar{R}T/\pi\ M\}^{1/2} = \{2 * 2^{1/2}/(3\pi^{3/2})\}$$

$$\bar{R}^{3/2}\ \{1/M\}^{1/2}\ \{1/(\sigma^2\ P\ N_{Avog})\}\ T^{3/2}$$

$$\mu = \rho\nu = \{2^{1/2}/(3\pi^{3/2})\}\ \{1/(\sigma^2\ N_{Avog})\}\ M^{1/2}\ \bar{R}^{1/2}\ T^{1/2}.$$

If $\bar{R} = 8314$ J/kmol K, $N_{Avog} = 6.023 \times 10^{26}$ molecules per kmol, σ (in m) = σ (in Å) $\times 10^{-10}$ m, then

$$\mu\ (kg/m\ sec) = 2.563 \times 10^{-6}\ M^{1/2}\ T^{1/2}\ (in\ K)/\sigma\ (in\ Å)^2, \quad (6.34)$$

$$\mu/\rho = \nu\ (m^2/sec) = 2.13 \times 10^{-8}\ M^{1/2}\ T^{1/2}\ (in\ K)/\{P\ (bar)\ \sigma\ (in\ Å)^2\}. \quad (6.35)$$

6.4.1.2 Any Property α

The net property transported downward per unit area per sec is $(1/3)\ (d\alpha/dy)\ \ell\ n'\ m'\ v'_{avg}$ where α is property per unit mass.

For example, consider α to be, say, h_t, the thermal enthalpy per mass (open system); then $dh_t/dy = dh_t/dT\ dT/dy = c_p\ dT/dy$, and

$(1/3)\ c_p\ (dT/dy)\ \ell\ n'\ m'\ v'_{avg} = \dot{q}'' = \lambda dT/dy$ is the net transported property downward per unit area per second.

Then $\lambda = (1/3)\ c_p\ \ell\ n'\ m'\ v'_{avg} = (1/3)\ c_p\ \ell\ \rho\ v'_{avg}$; thus, thermal diffusivity

$$\alpha_T = \nu = (1/3)\ \ell v'_{avg} = 2.13 \times 10^{-8}\ M^{1/2}\ T^{1/2}(in\ K)/\{P\ (bar)\ \sigma\ (in\ Å)^2\}. \quad (6.36)$$

Similarly, if $\alpha = Y_k$, mass of the species k per unit mass of mixture is

$$(1/3)\ \{(dY_k/dy)\ \ell\}\ ]\ n'\ m'\ v'_{avg} = \rho\ D\ \{dY_k/dy\}.$$

Then

$$D = (1/3)\ \ell v'_{avg} = 2.13 \times 10^{-8}\ M^{1/2}\ T^{1/2}(\text{in K})/\{P\ (\text{bar})\ \sigma\ (\text{in Å})^2\}. \quad (6.37)$$

Simple molecular theory sugggests that $\nu = \alpha_T = D$, or Sc = 1 and Pr = 1.

6.4.2 Rigorous Method for Transport Properties of Single Component

In the following subsections, a more complex methodology for estimating transport properties is presented. Rigorous computer codes are available for calculating transport properties of gases [Waranatz and Miller, 1983].

6.4.2.1 Absolute Viscosity

For a one-component system, rigorous formulation yields

$$\mu\ \{\text{kg/(m sec) or N sec/m}^2\} = 2.6693 \times 10^{-6}\ M^{1/2}\ T^{1/2}/(\sigma^2 \Omega_u) \quad (6.38)$$

where Ω_u is called the *collision integral for viscosity* and is weakly dependent on the temperature, and T and σ are in units of K and Å, respectively. A fit for Ω_u vs. $\varepsilon/(k_B T)$ yields

$$\Omega_u = -0.104\ \varepsilon/(k_B T)^2 + 0.9663\ \varepsilon/(k_B T) + 0.7273,$$

$$0 < \varepsilon/(k_B T) < 3.5, \quad (6.39)$$

with a correlation coefficient of 0.9996. The $\varepsilon/(k_B T)$ for various substances is tabulated in Table A.3 of Appendix A, whereas the collision integral Ω_u is tabulated in Table A.37 of Appendix A. It is seen that μ is independent pressure but $\mu \propto T^{1/2}$.

6.4.2.2 Thermal Conductivity

6.4.2.2.1 Monatomic Gas

In the case of a monatomic gas,

$$\lambda\ \{\text{kW/(m K)}\} = \lambda' = (15/4)\ R\ \mu, \quad (6.40)$$

R in kJ/kg K and, μ in kg/m sec.

6.4.2.2.2 Polyatomic Gas

For a polyatomic gas,

$$\lambda = \lambda' + \lambda'' \quad (6.41a)$$

$$\lambda'' = 0.88\ \lambda'\ \{0.4\ (c_p/R) - 1\}. \quad (6.41b)$$

Euken formulae are: $\lambda/(\mu R) = c_p/R + 5/4 = c_v/R + 9/4$

$$\lambda/\lambda' = (4/15)\ (c_p/R) + (5/15) = (4/15)\ (c_v/R) + (3/5), \text{ for an ideal gas} \quad (6.42)$$

Using the relation for λ' from Equation 6.40,

$$\{\lambda/(\mu\ c_v)\} = \{9k - 5\}/4. \tag{6.43}$$

Let $f = \{\mu\ c_v/\lambda\} = 4/\{9\ k - 5\}$. The value of $\{\mu\ c_v/\lambda\} = 0.4$ for a monatomic gas, <0.5 for a diatomic gas, and 0.57 for a triatomic gas.

Pr is given as,

$$\mathrm{Pr} = \mu c_p/\lambda = f\ c_p/c_v = \{4k\ /(9k - 5)\} \tag{6.44a}$$

where $f = 4/(9k - 5)$. With $k = 1.4$, $\mathrm{Pr} = 0.75$ and with $k = 1.667$, $\mathrm{Pr} = 0.667$ for a monatomic gas.

According to Hirschfelder, $\lambda/\lambda' = 0.115 + \{0.354k/(k - 1)\}$ (6.44b)

6.4.2.3 Self-Diffusion Coefficient

Consider the evaporation of liquid H_2O into an atmosphere of $H_2O(g)$; the evaporated H_2O molecules must diffuse through $H_2O(g)$. The diffusion coefficient for such a case is given as follows:

$$D_{kk}\ (\mathrm{m^2/sec}) = 2.628 \times 10^{-7}\ [\mathrm{m^2\ bar\ Å^2\ kg^{1/2}}/\{\mathrm{sec\ k\ mol^{1/2}\ K^{3/2}}\}]T^{3/2}/\{P\sigma^2\Omega_D M^{1/2}\} \tag{6.45}$$

where T is in K, σ in Å, P in bar and Ω_D is tabulated in Table A.37. An empirical fit for Ω_D is given as

$$\Omega_D = -0.0759\ (\varepsilon/(k_B T))^2 + 0.8505(\varepsilon/(k_B T)) + 0.6682, \quad 0 < \varepsilon/(k_B T) < 3.5. \tag{6.46}$$

Note that $D \propto^1 \neq M^{1/2}$; lighter gas diffuses faster. Transport properties of air, N_2, and O_2 are tabulated in Table A.8. Using Equations 6.38 and 6.45,

Schmidt #, $\mathrm{Sc} = \nu/D = 0.8439\ \{\Omega_D/\Omega_u\} = 0.8314\{\Omega_D/\Omega_u\}$, a constant almost independent of gas!

6.4.3 Transport Properties of Multiple Components

6.4.3.1 Absolute Viscosity

$$\mu_{mix} = \sum_i X_i \mu_i \Big/ \left(\sum_j X_j \varphi_{ij}\right), \tag{6.47}$$

and

$$\varphi_{ij} = \frac{1}{8^{1/2}}\left[1 + \frac{M_i}{M_j}\right]^{-1/2}\left[1 + \left\{\frac{\mu_i}{\mu_j}\right\}^{1/2}\left\{\frac{M_j}{M_i}\right\}^{1/4}\right]^2$$

$$\varphi_{ji} = \varphi_{ij} \quad \{\mu_j/\mu_i\}\{M_i/M_j\}.$$

The absolute viscosity of a mixture is given by the following empirical relation:

$$\mu_{mix} = \sum_i X_i^2/[\{X_i/\mu_i\} + 1.385 \sum_{j,j\neq i} \{X_i X_j \bar{R}\, T/(p\ M_i\ D_{ij})\}], \tag{6.48}$$

If $\phi_{ij} \approx 1$ $\mu_{mix} = \Sigma_i X_i \mu_i$; $\mu \propto M^{1/2}$; $\nu = \mu/\rho \propto 1/M^{1/2}$; as mixture molecular weight increases, ν decreases. Since $Sc = \nu/D$, then Sc decreases. The Schmidt number, $Sc_{ij} = \mu_{ij}/\{\rho D_{ii}\}$.

6.4.3.2 *Thermal Conductivity*

6.4.3.2.1 Empirical Relations

$$\lambda_{1,mix} = X_A \lambda_A + X_B \lambda_B \tag{6.49a}$$

$$(1/\lambda_{mix}) = X_A/\lambda_A + X_B/\lambda_B \tag{6.49b}$$

$$\lambda_{mix} = a\ \lambda_{1,mix} + (1 - a)\ \lambda_{2,mix} \tag{6.49c}$$

where

$$a = 0.4056\ X_A^2 + 0.0417\ X_A + 0.3327, \quad R^2 = 0.9945. \tag{6.49d}$$

For an N_2–CO_2 mixture, this correlation predicted to within 6% up to 800 K.

6.4.3.2.2 Lindsey and Bromley

For a mixture of gases,

$$\lambda_{mix} = 0.5\left(\sum_k \lambda_k + \frac{1}{\sum_k \frac{X_k}{\lambda_k}}\right). \tag{6.50}$$

6.4.3.2.3 Additive Mixing Rules

$$\lambda_{mix} = \lambda'_{mix} + \lambda''_{mix} \tag{6.51a}$$

$$\lambda'_{mix} = \sum_i X_i\ \lambda'_i \Big/ \left\{\sum_j X_j\ \psi_{ij}\right\} \tag{6.51b}$$

$$\psi_{ij} = \varphi_{ij}\ [1 + \{2.41\ (M_i - M_j)\ (M_i - 0.162\ M_j)/(M_i + M_j)^2\} \tag{6.51c}$$

$$\lambda''_{mix} = \sum_i X_i\ \lambda''_i \Big/ \left\{\sum_j X_j\ \varphi_{ij}\right\} \tag{6.51d}$$

(*Note:* $\varphi_{ji} = \{\varphi_{ij}\ \lambda'_j/\lambda'_i\}$)

6.4.3.2.4 Rule in Terms of Specific Heats

Following Euken formula (Equation 6.42)

$$\{\lambda_{mix}/\lambda'_{mix}\} = (4/15)\ c_v/R + 0.6 \tag{6.52}$$

6.4.3.2.5 Other Mixing Rules

$$\lambda_{mix} = \sum_i X_i \lambda_i \Big/ \left(\sum_j X_i \varphi_{ij} \right), \tag{6.53a}$$

where

$$\varphi_{ij} = (1/8^{1/2})\,(1 + M_i/M_j)^{-1/2}\,\{(1 + (\mu_i/\mu_j)^{1/2}(M_i/M_j)^{1/4})\}^2. \tag{6.53b}$$

Another relation is as follows:

$$\lambda_{mix} = \left(\sum_i \frac{\lambda_i}{\sum_k A_{ik} \frac{X_k}{\lambda_k}} \right) \tag{6.54a}$$

where $A_{ik} = 1$, if $i = k$ and

$$A_{ik} = \frac{1}{4}\left(1 + \sqrt{\frac{\mu_i}{\mu_k}\left(\frac{M_k}{M_i}\right)^{0.75} \frac{1+S_i/T}{1+S_k/T}} \right) 2\,\frac{1+S_{ik}/T}{1+S_k/T}, \quad i \neq k \tag{6.54b}$$

$$S_i = 1.5\,T_{bi}, \quad S_{ik} = 0.735(S_i S_k)^{0.5},$$

where T_b is the boiling temperature.

In evaluating Sc for combustion problems involving fuel one assumes stoichiometric mixture of fuel and air and can determine X_F and X_{air} (= $1 - X_F$). Thus μ_{mix} and λ_{mix} can be evaluated.

6.4.3.3 Diffusion in Multicomponent Systems

6.4.3.3.1 Binary Diffusion Coefficient

If there are only two species in a mixture, the binary diffusion coefficient $D_{k\ell}$ can be determined using principles from statistical thermodynamics [Bird et al., 1960; Welty et al., 1984], i.e.,

$$D_{k\ell} = 2.628 \times 10^{-7}\,T^{3/2}/(P\,\sigma_{k\ell}^{2}\,\Omega_D\,M_{k\ell}^{\ 1/2})\ \text{m}^2/\text{sec}, \tag{6.55}$$

where T is in units of K, P in bar, $\sigma_{kl} = (\sigma_k + \sigma_\ell)/2$ in Å, and $M_{k\ell}$ is the reduced molecular weight for binary species in kg/kmol, and

$$1/M_{k\ell} = (1/2)((1/M_k) + (1/M_\ell)). \tag{6.56}$$

The value of $(\varepsilon/k_B)_{k\ell}$ for the binary mixture

$$(\varepsilon/k_B)_{k\ell} = ((\varepsilon/k_B)_k(\varepsilon/k_B)_\ell)^{1/2}. \tag{6.57}$$

For binary combination of oxygen with N_2, CO_2, and H_2O, $(\varepsilon/k_BT)_{k\ell}$ ranges from about 0.77 to 0.29 for variation of T from 273 K to 350 K. Equation 6.55 seems to yield reasonable values for polyatomic nonpolar gas pairs also. The $D_{k\ell}$ values are at most independent of composition at low pressure.

In general, the diffusivity

$$D \propto T^m/P, \tag{6.58}$$

where m lies in the range from 3/2 to 2. As $\rho = P/(RT)$ for ideal gases, $\rho D \propto T^{m-1}$ is a weak function of temperature and is independent of pressure. Further $\rho^2 D$ is a weak function of T but $\rho^2 D \propto P$.

6.4.3.3.2 Diffusion of Trace Species A in a Mixture

For species A in a multicomponent mixture

$$D_{Ak} = 1/\sum \{X_k/D_{Ak}\}, \tag{6.59}$$

where A is in low concentration.

6.4.3.3.3 Diffusion of Species k into a Mixture of Species

There are two methods to address the diffusion of a species into a mixture of species. In the first method all of the species, b, c, d, etc., except the diffusing species "a" are lumped in into one equivalent species, and the binary diffusivity relation is used. Another method involves Wilke's law, which considers the diffusion of a species into a stagnant mixture; i.e., the species b, c, etc., are stationary or present in large amounts compared to "a" (i.e., mass flows of b, c, d, etc., are almost zero). One can derive the following equation for the diffusion coefficient of moving species k into a stagnant mixture [Wilke, 1950; Welty et al., 1984].

$$D_{k,\text{stag mix}} = \frac{1-Y_k}{\sum\limits_{j\neq k} \frac{Y_j}{D_{kj}}}, \tag{6.60}$$

where k is the moving species.

Example 4

Consider the drying of coal considered in Example 3. Estimate the diffusion coefficient $D_{H2O\text{-stag}}$. Assume the temperature to be 80°C. Use the binary diffusion coefficient formulae for $D_{k\ell}$, and use the relation in Equation 6.60.

Solution

$$Y_{H2O} = 0.4,\ D_{H2O\text{-}N2} = 2.91 \times 10^{-5}\ m^2/sec,\ D_{H2O\text{-}O_2} = 2.93 \times 10^{-5}\ m^2/sec,$$

$$Y_{air} = 0.6,\ Y_{O2} = 0.6 * 0.23 = 0.138,\ Y_{N2} = 0.6 * 0.77 = 0.462,$$

$$D_{k,\text{stag mix}} = \frac{\{1-Y_k\}}{\left\{\sum\limits_{j\neq k} \frac{Y_j}{D_{kj}}\right\}} = \frac{\{1-Y_{H2O}\}}{\left\{\frac{Y_{O2}}{D_{H2O\text{-}O2}} + \frac{Y_{N2}}{D_{H2O\text{-}N2}}\right\}}$$

$$= \frac{\{1-0.4\}}{\left\{\frac{0.138}{2.93\times10^{-5}} + \frac{0.462}{2.93\times10^{-5}}\right\}} = 2.92\times10^{-5} m^2/sec$$

If there are only two components, $D_{k,stag} \to D_{12}$.

6.4.3.3.4 Multicomponent Diffusion Coefficient

From kinetic theory, an expression for multicomponent diffusion coefficient can be given as [Hirschfelder et al., 1954]

$$D_{iK} = \frac{X_i M_i \left\{ \langle L_{ik} \rangle^{-1} - \langle L_{ii} \rangle^{-1} \right\}}{M_k}$$

$$\langle L_{ik} \rangle = \begin{pmatrix} L_{11} & L_{12} & L_{13} & \ldots & L_{1J} \\ L_{21} & L_{22} & L_{23} & \ldots & L_{2J} \\ L_{31} & L_{32} & L_{33} & \ldots & L_{3J} \\ \cdots & \cdots & \cdots & \cdots & \cdots \\ \cdots & \cdots & \cdots & \cdots & \cdots \\ L_{J1} & L_{J2} & L_{J3} & \ldots & L_{JJ} \end{pmatrix} \tag{6.61}$$

$$L_{ik} = \sum_j \frac{X_j}{M_i D_{ij}} \{ M_k X_k (1 - \delta_{ij}) - M_i X_i (\delta_{ik} - \delta_{kj}) \} \tag{6.62}$$

where $<L_{ik}>$ is a matrix and L_{ik} are components of a matrix of the order $J \times J$, where J is the number of species; $L_{ii} = 0$; D_{ij}, binary diffusion coefficient; δ_{ij} is Kronecker delta ($\delta_{ij} = 1$, if I = j and $\delta_{ij} = 0$ if $i \neq j$). In this case, it is noted that $D_{jj} = 0$ because the diagonal element in the matrix is zero.

For more formulae on calculations of ρ, λ, c_p, and D, the reader is advised to consult Ben-Dor et al., [2003a, 2003b]; Reid et al., [1987]; and Vargaftik [1975].

6.5 Generalized Form of Fourier and Fick's Law for a Mixture, with Simplifications

6.5.1 Generalized Law: Multicomponent Heat Flux Vector

The rigorous law for a multicomponent heat flux vector is given as

$$\vec{q} = -\lambda \nabla T + \rho \left[\sum_{\ell=1}^{n} h_\ell Y_\ell \vec{V}_\ell \right] + \left[\sum_{\ell=1}^{n} \sum_{k=1}^{n} \frac{X_k D_{T\ell}}{M_\ell D_{\ell k}} (\vec{V}_\ell - \vec{V}_k) \right] \tag{6.63}$$

where the last term is called Dufour heat flux, $D_{T\ell}$ is the thermal diffusion coefficient of species ℓ, which could be negative or positive, and $D_{T\ell}/\rho D_{\ell k}$ is typically less than 0.1 [Williams, 1985].

6.5.2 Generalized Law: Multicomponent Diffusion

The rigorous law for multicomponent diffusion is presented in the appendix at the end of the chapter.

$$\nabla X_k = \left[\sum_{\ell=1}^{n} \frac{X_k X_\ell}{D_{k\ell}} (\vec{V}_\ell - \vec{V}_k)\right] + \rho \sum Y_k Y_\ell \left[\vec{f}_k - \vec{f}_\ell\right] + (Y_k - X_k) \frac{\Delta p}{p} + \sum_{\ell=1}^{n} \frac{X_k X_\ell}{\rho D_{k\ell}} \left(\frac{D_{T,\ell}}{Y_\ell} - \frac{D_{T,k}}{Y_k}\right) \frac{\Delta T}{T} \quad (6.64)$$

where $D_{k\ell}$ is the binary diffusion coefficient; $D_{T\ell}$, the thermal diffusion coefficient; Δp, the pressure gradient; $\vec{f}_k$, the body force per unit mass (e.g., "g") on species k; ΔT, the temperature gradient; and $V_k - V_\ell$, the relative diffusion velocity between molecules of k and ℓ. Diffusophoresis occurs because of the difference in molecular weights. Heavier vapor molecules will impart greater momentum to the particles of lighter vapor that diffuse from higher vapor concentration. This results in a net force on the particles, for example, cigarette smoke. Smoke particles are heavier, whereas gas molecules are lighter. Apart from the continuum phenomena of transfer of mass, there could be thermophoresis effects. Thermophoresis is a thermal drift or thermal precipitation [MacGibbon et al., 1999]. For example, consider chemical vapor deposition (CVD) occurring in a varying temperature gradient (water at 300 K with gas at 700 K). The particles of interest may have a diameter d_p less than the mean free path ℓ_m within the thermal boundary layer (where a T gradient exists). The hotter gas molecules impinge on the particle with a higher velocity compared to the colder gas and, hence, the asymmetrical interactions result in a net thermal force on the particle from the hotter to colder region (i.e., the particles are repelled from the hotter surface, say, 700 K, or attracted toward the colder water surface at 300 K, as in CVD). The thermal force on the particle is approximately $-\pi\, \ell_m\, d_p^2\, \nabla T/T$.

A molar diffusional gradient arises because of the following:

1. Multiple components and nonuniform composition, with different species having different absolute velocities or diffusion velocities imparting differing momentum flux rates that set up pressure gradients
2. Pressure gradients due to differing molecular weights (e.g., partial pressure variation of k)
3. Nonuniform body forces per unit mass between various components
4. Temperature gradients resulting in the Soret effect as in water tubes in boilers

Simplifications

The diffusion laws can be simplified by using the following approximations:

1. Either all species have the same molecular weight or there is no pressure gradient. Then the third term due to pressure gradients disappears.
2. Body forces for all k are the same, or there is no body force. The second term on the right disappears (e.g., $\vec{f}_k = g$, gravity is the only force acting equally on all species).

3. Typically, $\frac{D_{T,\ell}}{D_{k,\ell}} \approx \frac{1}{10}$; further assume that the effect of T gradient on diffusion is negligible.

The fourth term disappears. Then

$$\nabla X_k = \left[\sum_{\ell=1}^{n} \frac{X_k X_\ell}{D_{k\ell}} (\vec{V}_\ell - \vec{V}_k\right].$$

Now let us assume that all $D_{k\ell}$'s are equal to D; because

$$\frac{1}{D}\left[\sum_{\ell=1}^{n} X_k\ X_\ell\left(\vec{V}_\ell - \vec{V}_k\right)\right] = -\frac{X_k \vec{V}_k}{D}$$

Ignoring body forces, pressure, and temperature gradients, Equation 6.64 yields

$$\nabla X_k = \left[\sum_{\ell=1}^{n} \frac{X_k\ X_\ell}{D_{k\ell}} (\vec{\vec{V}}_\ell - \vec{\vec{V}}_k)\right] = \left[\sum_{\ell=1}^{n} \frac{X_k\ X_\ell}{D_{k\ell}} (\vec{V}_\ell - \vec{V}_k)\right]$$

$$= \left[\sum_{\ell=1}^{n} \frac{X_k\ X_\ell}{D_{k\ell}} (\vec{v}_\ell - \vec{v}_k)\right] \tag{6.65}$$

since $\vec{\vec{V}}_k - \vec{\vec{V}}_\ell = \vec{V}_k - \vec{V}_\ell$, where $\vec{\vec{V}}$ is the mole-based diffusion velocity; $\vec{V}$, the mass-based diffusion velocity; and $\vec{v}_l - \vec{v}_k$ the difference between the absolute velocities of species. For equal binary diffusion coefficients D,

$$\nabla X_k = \sum_{\ell=1}^{n} \frac{X_k X_\ell\left[\vec{\vec{V}}_\ell - \vec{\vec{V}}_k\right]}{D_{k\ell}} = \frac{X_k \sum_{\ell=1}^{n} X_\ell \vec{\vec{V}}_\ell - X_k \vec{\vec{V}}_k}{D} \tag{6.66}$$

$$= \frac{-X_k \vec{\vec{V}}_k}{D}$$

because $\sum X_\ell\, \vec{\vec{V}}_\ell = 0$ and $\vec{v} = \sum X_\ell\, \vec{v}_\ell$. Simplifying Equation 6.66,

$$\frac{D}{X_k} \nabla X_k = -\vec{\vec{V}}_k$$

$$\vec{\vec{V}}_k = -D \nabla \ln\{X_k\}.$$

Or

$$c_k \vec{\vec{V}}_k = \dot{N}''_{k,diff} = -c_k D\ \nabla \ln\{X_k\} = -cD\ \nabla X_k\,. \tag{6.67}$$

Example 5

For a binary system it is known that

$$\vec{\bar{V}}_1 = -D_{12}\nabla \ln X_1 \quad \text{so that} \quad j_1''\left(\frac{\text{kmol}}{\text{m}^2\text{sec}}\right) = cX_1\vec{\bar{V}}_1 = -cX_1 D\nabla X_1.$$

Then show that

$$\vec{\bar{V}}_1 = -D_{12}\nabla \ln Y_1, \quad \vec{\bar{V}}_2 = -D_{12}\nabla \ln Y_2.$$

Solution

$$\vec{\bar{V}}_1 = -D_{12}\nabla \ln X_1, \quad \vec{\bar{V}}_2 = -D_{12}\nabla \ln X_2$$

$$\vec{\bar{V}}_1 - \vec{\bar{V}}_2 = -D_{12}\nabla \ln \frac{X_1}{X_2} = -D_{12}\nabla \ln \frac{Y_1 M_2}{Y_2 M_1} = -D_{12}\nabla \ln \frac{Y_1}{Y_2}$$

As

$$\vec{\bar{V}}_1 - \vec{\bar{V}}_2 = \vec{V}_1 - \vec{V}_2$$

$$\vec{V}_1 - \vec{V}_2 = -D_{12}\nabla \ln \frac{Y_1}{Y_2}$$

More generally,

$$V_k - \vec{V}_\ell = -D_{kl}\nabla \ln \frac{Y_k}{Y_\ell}.$$

Multiplying by Y_ℓ and summing over all ℓ

$$\sum \vec{V}_k Y_\ell - \sum Y_l \vec{V}_\ell = -\sum D_{k\ell}\{\nabla \ln Y_k\} - \sum D_{k\ell}\{Y_\ell \nabla \ln Y_\ell\}.$$

$\vec{V}_k = -D\{\nabla \ln Y_k\}$, if all D's are equal.

Remarks

For a binary system, the following is always true

$$Y_2\vec{V}_1 - Y_2\vec{V}_2 = (1-Y_1)\vec{V}_1 - Y_2\vec{V}_2 = \vec{V}_1 = -D_{12}Y_2\nabla \ln \frac{Y_1}{Y_2}$$

$$\vec{V}_1 = -D_{12}Y_2\nabla \ln \frac{Y_1}{Y_2} = -D_{12}Y_2\left[\nabla \ln Y_1 - \nabla \ln Y_2\right]$$

$$= -D_{12}[(1-Y_1)\nabla \ln Y_1 - \nabla Y_2] = -\frac{D_{12}}{Y_1}[(1-Y_1)\nabla Y_1 - Y_1\nabla Y_2]$$

$$\vec{V}_1 = -D_{12}\nabla \ln Y_1.$$

It can be shown that if one uses mass-averaged velocity [Williams, 1985]

$$\nabla Y_k = -\sum_{l=1}^{n} \frac{Y_k Y_\ell \left[\vec{V}_k - \vec{V}_\ell\right]}{D_{k\ell}} .$$

With $D_{k\ell} = D$ (equal binary diffusion coefficient),

$$\vec{V}_k = -D \nabla \ln X_k .$$

If we define $-D \nabla \ln Y_k = \vec{V}_k$, then the difference between mole diffusion and mass diffusion (vectorial) velocities is given as,

$$\begin{aligned} \vec{\bar{V}}_k - \vec{V}_k &= -D \nabla \ln \frac{X_k}{Y_k} = -D \nabla \ln \left(\frac{M X_k}{X_k M_k} \right) \\ &= -D \nabla \ln \left(\frac{M}{M_k} \right) = -D \nabla \ln \left(\frac{\sum_k X_k M_k}{M_k} \right) \end{aligned} \tag{6.68}$$

then,

$$\vec{V}_k = \vec{\bar{V}}_k + D \nabla \ln \left\{ \frac{M}{M_k} \right\} = \vec{\bar{V}}_k + D \nabla \ln \left\{ \frac{M}{M_k} \right\} .$$

The generalized form of Fick's law for a binary mixture has the form

$$\vec{V}_{k\ell} = -D_{k\ell} \nabla Y_k / Y_k . \tag{6.69}$$

This relation applies to a multicomponent mixture with equal diffusion coefficients, and it is not unusual to use this law even for more realistic multicomponent mixtures. The Lewis number Le (= $\lambda/(\rho D c_p)$) compares the mixture thermal diffusivity α ($\lambda/(\rho c_p)$) with the species diffusivity. It is analogous to the Prandtl number (Pr = ν/α).

Example 6

Consider a liquid fuel consisting of a mixture of *n*-pentane, benzene, and methanol that is stored in a cylindrical tank of diameter d. The distance of the rim of the tank from the liquid surface is L. As evaporation occurs, the fuel vapors diffuse out into the open atmosphere. However if diffusion is slow, these vapors accumulate and form a flammable mixture. If the timescale of methanol diffusion is $t_{methanol}$, obtain expressions for $t_{pentane}/t_{methanol}$ and $t_{benzene}/t_{methanol}$.

Solution

The length L over which diffusion occurs can be determined from the relation

$$D\,t = L^2, \quad \text{i.e.,} \quad t = L^2/D.$$

However, $D \propto 1/M^{1/2}$, so that $t \propto M^{1/2}L^2$.
The timescales are:

$$t_{pentane}/t_{methanol} = (M_{pentane}/M_{methanol})^{1/2} = (72/32)^{1/2} = 1.5,$$

and

$$t_{benzene}/t_{methanol} = (78/32)^{1/2} = 1.56.$$

Remarks

(i) Based on the diffusion timescale, benzene takes longer to diffuse to the tank rim compared to methanol. Therefore, a much steeper gradient is required for benzene to diffuse out of the tank if the evaporation rates for all fuels are to be maintained at the same value.

(ii) This analysis implies that the vapor concentration for benzene should be the highest at the liquid surface for benzene and the lowest for methanol. However, the evaporation rates are not necessarily identical and, typically, methanol will evaporate faster. If the evaporation rate exceeds the diffusion rate, then fuel vapors will accumulate in the tank and create an explosive, combustible mixture.

6.6 Summary

Mass transfer concepts have been introduced in this chapter, including definitions of absolute velocity, average velocity, mass-based diffusion velocity, and mole-based diffusion velocity. Simplified relations are presented for estimation of transport coefficients including μ and λ for single and multiple components, and for the binary diffusion coefficient D_{ij}. Simplified and generalized versions of Fick's law and the Fourier law are presented for mass and heat transport.

6.7 Appendix: Rigorous Derivation for Multicomponent Diffusion

The derivation follows that of Williams [1985]. Consider molecules of species k and ℓ. Let their absolute average velocities be v_k and v_ℓ. Thus, their relative velocity is $v_\ell - v_k$ and reduced mass is $\mu_{k\ell} = m_\ell m_k/(m_k + m_\ell)$, where m_k is mass of a molecule of species k. Because $v_k = v + V_k$, where v is the mass-averaged velocity (e.g., average velocity v gas of a flowing in a pipe) and V_k, the average random velocity of molecules k with respect to the mass-averaged velocity, then $v_\ell - v_k = V_\ell - V_k$. Let the body force be $\vec{f}_k$ on species k. The momentum Mo transferred from ℓ to k per unit volume is given as

$$Mo_k = \sum_\ell \mu_{\ell k} Z_{\ell k}(\vec{v}_\ell - \vec{v}_k) + \rho Y_k \vec{f}_k \qquad (6.70)$$

where $Z_{\ell k}$ is the number of collisions per unit volume per sec between species k and ℓ. Thus,

$$Mo_k = \sum_{\ell} \mu_{\ell k} Z_{\ell k} (\vec{V}_\ell - \vec{V}_k) + \rho Y_k \vec{f}_k \,. \tag{6.71}$$

The momentum transferred to species k can be split into two parts, increase in mass-averaged velocity contributed by species k and change in random velocity of species k. If a unit surface area travels with mass-averaged velocity v, then species k moves with the random diffusive velocity V_k across this surface, resulting in a rate of change of momentum and thus creating a partial pressure gradient ∇p_k, where p_k is the partial pressure ($= X_k$ p) contributed by species k. Thus,

$$\sum_{\ell} \mu_{\ell k} Z_{\ell k} (\bar{V}_\ell - \bar{V}_k) + \rho Y_k \vec{f}_k = \rho Y_k \frac{Dv}{Dt} + \nabla(X_k p). \tag{6.72}$$

Momentum transfer to species k by collisions and body force on k = momentum change of species traveling with average velocity + random momentum exchange. But from the momentum equation

$$\rho \frac{Dv}{Dt} = -\nabla p + \rho Y_k \vec{f}_k. \tag{6.73}$$

Hence using Equation 6.73 in Equation 6.72 and further simplifying, a relation between ∇X_k and the diffusion velocities is obtained. Thus, a rigorous law for multicomponent diffusion is given as

$$\nabla X_k = \left[\sum_{\ell=1}^{n} \frac{\mu_{\ell k} Z_{\ell k}}{p} (\bar{V}_\ell - \bar{V}_k) \right] + (Y_k - X_k) \frac{\Delta p}{p} + \frac{\rho}{P} \sum Y_k Y_\ell \left[\vec{f}_k - \vec{f}_\ell \right]. \tag{6.74}$$

Note that the mole diffusion velocity $\vec{\vec{V}}_k$ (with respect to the mole-based average velocity) is related to mass-based diffusion velocity $\vec{V}_k$ as follows:

$$\begin{aligned} \vec{\vec{V}}_k &= v_k - \sum_j X_j v_j = v_k - \sum_j X_j \{v + \vec{V}_j\} \\ &= v_k - v - \sum_j X_j \vec{V}_j = V_k - \sum_j X_j \vec{V}_j \end{aligned} \tag{6.75}$$

$$\begin{aligned}\vec{\vec{V}}_\ell &= v_\ell - \sum_j X_j v_j = v_\ell - \sum_j X_j \left\{ v + \vec{V}_j \right\} \\ &= v_\ell - v - \sum_j X_j \vec{V}_j = V_k - \sum_j X_j \vec{V}_j\end{aligned} \tag{6.76}$$

Hence,

$$\vec{V}_k - \vec{V}_\ell = \vec{\vec{V}}_k - \vec{\vec{V}}_\ell. \tag{6.77}$$

Simplifying Equation 6.74 for binary components in the absence of body forces (constant-pressure systems),

$$\nabla X_1 = \frac{\mu_{12} Z_{12}}{p} (\vec{V}_2 - \vec{V}_1) = \frac{\mu_{12} Z_{12}}{p} (\vec{\vec{V}}_2 - \vec{\vec{V}}_1). \tag{6.78}$$

Fick's law is given as

$$D_{12} \nabla X_1 = -X_1 \vec{\vec{V}}_1 \quad \text{with} \quad k = 1 \tag{6.79}$$

$$D_{12} \nabla X_2 = -X_2 \vec{\vec{V}}_2 \quad \text{with} \quad k = 2. \tag{6.80}$$

With $X_2 = 1 - X_1$ in Equation 6.80,

$$-D_{12} \nabla X_1 = -(1 - X_1) \vec{\vec{V}}_2 \quad \text{with} \quad k = 2. \tag{6.81}$$

From Equation 6.81 and Equation 6.79, and solving for $\vec{V}_1$ and $\vec{V}_2, \vec{V}_1 = -D_{12} \frac{\nabla X_1}{X_1}, \vec{V}_2 = D_{12} \frac{\nabla X_1}{X_2}$, and using these in Equation 6.78,

$$\begin{aligned}\nabla X_1 &= \frac{\mu_{12} D_{12} Z_{12}}{p} \left(\frac{\nabla X_1}{X_1} + \frac{\nabla X_1}{X_2} \right) \\ &= \frac{\mu_{12} D_{12} Z_{12} \nabla X_1}{p} \left(\frac{1}{X_1 X_2} \right).\end{aligned}$$

Then solving for D_{12},

$$D_{12} = \frac{X_1 X_2 p}{\mu_{12} Z_{12}} \tag{6.82}$$

where the collision number is known from Chapter 5 as

$$Z_{12} = (N_{Avag}[1])(N_{Avag}[2])\ \pi\sigma^2_{12}\,\bar{V}_r, \tag{6.83}$$

$$\bar{V}_r = \{8k_BT/(\pi\mu_{12})\}^{1/2} \quad \text{and} \quad \sigma_{12} = (\sigma_1 + \sigma_2)/2. \tag{6.84a,b}$$

$$D_{12} = \frac{X_1\ X_2\ p}{\mu_{12}\ Z_{12}} = \frac{p}{\mu_{12}\ \bar{V}_r\,\pi\ \sigma_{12}{}^2 N^2{}_{Avog}c^2} = \frac{(k_BT)^{3/2}}{p\,\{8\ \pi\mu_{12}\}^{\ 1/2}\ \ \sigma_{12}{}^{2}}. \tag{6.85}$$

Generalizing for any two species k and ℓ

$$D_{k\ell} = \frac{X_k\ X_\ell\ p}{\mu_{k\ell}\ Z_{k\ell}} = \frac{(k_BT)^{3/2}}{p\,\{8\ \pi\mu_{k\ell}\}^{1/2}\ \sigma_{k\ell}{}^{2}}. \tag{6.86}$$

With $\mu = m/2 = M/N_{avog}$, $k_B = \bar{R}/N_{Avog}$, $k = \ell$, the self-diffusion coefficient is given as,

$$D_{kk}\ (m^2/sec) = 3.55 \times 10^{-7}\ [m^2\ bar\ \text{Å}^2\ kg^{1/2}/\{sec\ k\ mol^{1/2}\ K^{3/2}\}]T^{3/2}\ /\{P(bar)\sigma^2 M^{1/2}\}, \tag{6.87}$$

which is slightly higher than that given by Equation 6.45.
Using Equation 6.86 in Equation 6.74

$$\nabla X_k = \left[\sum_{\ell=1}^{n} \frac{X_k\ X_\ell}{D_{k\ell}}\left(\vec{V}_\ell - \vec{V}_k\right)\right] + (Y_k - X_k)\frac{\Delta p}{p} + \frac{\rho}{p}\sum Y_k\ Y_\ell\left[\vec{f}_k - \vec{f}_\ell\right]. \tag{6.88}$$

Taking thermal diffusion into account,

$$\nabla X_k = \left[\sum_{\ell=1}^{n} \frac{X_k\ X_\ell}{D_{k\ell}}(\vec{V}_\ell - \vec{V}_k)\right] + (Y_k - X_k)\frac{\Delta p}{p} + \frac{\rho}{p}\sum Y_k\ Y_\ell\left[\vec{f}_k - \vec{f}_\ell\right] + \sum_{\ell=1}^{n}\frac{X_k\ X_\ell}{\rho\ D_{k\ell}}\left(\frac{D_{T,\ell}}{Y_\ell} - \frac{D_{T,k}}{Y_k}\right)\frac{\Delta T}{T}. \tag{6.89}$$

Chapter 7

First Law Applications

7.1 Introduction

Suppose diesel is sprayed into a diesel engine. The drops are assumed to instantaneously vaporize and mix with air. If reaction occurs as a homogeneous burn under finite kinetics, how much time does it take to burn? What will be the rate of temperature and pressure rise? Similarly, if natural gas along with air is premixed and injected into a gas turbine combustion chamber of prescribed volume V, how much is burned and what is the temperature? What is the maximum firing rate in a gas turbine before the flame blows out? These questions will be answered in this chapter. We analyze some simple engineering problems first for combustion within closed system at constant volume and pressure, and then for open systems, a plug flow reactor (PFR) or a perfectly stirred reactor (PSR). Finally, resistance concept is introduced for predicting burn rates of carbon particles and droplets.

7.2 Generalized Relations in Molar Form

7.2.1 Mass Conservation and Molar Balance

Recall the overall mass conservation equation for an open system (Chapter 2):

$$dm/dt = \dot{m}_i - \dot{m}_e. \tag{7.1}$$

For a species k,

$$dm_{cv,k}/dt = \dot{m}_{k,i} - \dot{m}_{k,e} + dm_{k,chem}/dt = \dot{m}_{k,i} - \dot{m}_{k,e} + \dot{w}_{k,m}, \ \frac{kg}{sec} \tag{7.2}$$

where $dm_{k,chem}/dt = \dot{w}_{k,m}$ is the change in mass of the species due to chemical reactions occurring over a volume, V, and $\dot{w}'''_{k,m} = \dot{w}_{k,m} / V$.

Since, $\dot{m}_k = \dot{N}_k M_k$, Equation 7.2 assumes the molar form

$$dN_k/dt = \dot{N}_{k,i} - \dot{N}_{k,e} + dN_{k,chem}/dt = \dot{N}_{k,i} - \dot{N}_{k,e} + \dot{w}_k \quad \frac{\text{kmol}}{\text{sec}} \tag{7.3}$$

where $dN_{k,chem}/dt = \dot{w}_k$ denotes the molar production rate of species k due to chemical reactions. Also, $\dot{w}_k = \dot{w}_k''' V$, where V is the volume. The value of $\dot{w}_k$ can be determined by applying the law of mass action (Chapter 5) for any given chemical reaction. For closed system with fixed mass, Equation 7.1 to Equation 7.3 become

$$dm/dt = 0$$

$$dm_{cv,k}/dt = \dot{w}_{k,m}$$

$$dN_k/dt = \dot{w}_k .$$

7.2.2 First Law

Recall the energy conservation equation for a system,

$$dE_{cv}/dt = \dot{Q}_{cv} - \dot{W}_{cv} + \sum \dot{m}_i (h + ke + pe)_i - \sum \dot{m}_e (h + ke + pe)_e. \tag{7.4}$$

In molar form,

$$(dE_{cv}/dt) = \dot{Q}_{cv} - \dot{W}_{cv} + \sum \dot{N}_{k,i} \bar{e}_{k,T,i} - \sum \dot{N}_{k,e} \bar{e}_{k,T,e}. \tag{7.5}$$

Neglecting ke and pe,

$$(dE_{cv}/dt) = \dot{Q}_{cv} - \dot{W}_{cv} + \sum \dot{N}_{k,i} \bar{h}_{k,i} - \sum \dot{N}_{k,e} \bar{h}_{k,e}. \tag{7.6}$$

7.3 Closed-System Combustion

The combustion in an internal combustion (IC) engine occurs typically at almost constant volume for a gasoline engine and at constant pressure for a diesel engine. For the modern diesel engines (operating with a dual cycle, which includes both constant-volume and constant-pressure combustion), combustion can occur both at constant V and P. We will develop a global analysis to predict the change in temperature, pressure, or volume in these engines, which we will model as closed systems. Whereas the mass remains unchanged in a closed system undergoing chemical reactions, this is not necessarily the case for the total number of moles in the system.

7.3.1 Simple Treatment

First, a simple treatment will be presented, which will be followed by a rigorous formulation. The fuel–air mixture reacts volumetrically, and exothermic

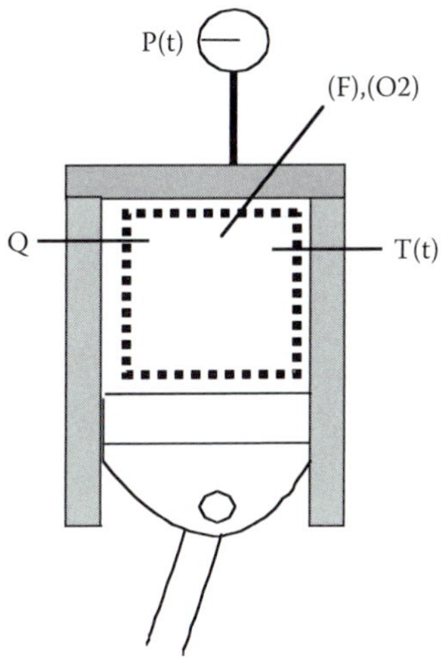

Figure 7.1 Closed-system combustion model.

heat liberation of heat is used to raise the temperature of the products (Figure 7.1). However, work transfer out of the system results in energy transfer, reducing the degree of temperature rise. Heat generated by chemical reactions + heat transfer across boundary work transfer = sensible energy rise

$$mc_v \frac{dT}{dt} = \left|\dot{w}_{F,m}\right| h_c + \dot{Q} - \dot{W}$$

where $\dot{Q} < 0$ if there is heat loss, work transfer rate $\dot{W} = P\ dV/dt$, $\dot{w}_{F,m}$ is the mass-based fuel consumption rate, and h_c is the heat released in kJ per kg of fuel.

$$mc_v dT/dt = \left|\dot{w}_{F,m}\right| V\ h_c + \dot{Q} - PdV/dt. \tag{7.7}$$

From mass conservation for fuel species,

$$\left|\dot{w}'''_{F,m}\right| V = -d\,\{V\,(F)\}/dt = -(F)d\,\{V\}/dt - V\,d\,\{(F)\}/dt \tag{7.8}$$

where (F) is the mass concentration of fuel (kg/m^3).

7.3.1.1 Constant-Volume Reactor

For constant-volume system, dV/dt = 0. With these considerations in Equation 7.7 and Equation 7.8,

$$mc_v \frac{dT}{dt} = \left|\dot{w}_{F,m}\ V\right| h_c + \dot{Q}. \tag{7.9a}$$

The volume V includes cylinder and piston crevice (5 to 10% of the total volume). The fuel consumption by a chemical reaction results in change of fuel concentration. Equation 7.8 simplified to

$$\left|\dot{w}'''_{F,m}\right| = -d\{(F)\}/dt. \tag{7.9b}$$

From Equation 7.9a and Equation 7.9b, then for constant-volume adiabatic reactors only

$$mc_v \frac{dT}{dt} = -V\,h_c \frac{d(F)}{dt} \tag{7.10}$$

which is valid for whatever may be the global order of reaction. Simplifying,

$$\frac{dT}{dt} = -\left(\frac{h_c}{\rho c_v}\right)\frac{d\{(F)\}}{dt} = \frac{-h_c}{c_v}\frac{dY_F}{dt}.$$

Note that (F) = ρY_F and $\rho = m/V$ remains constant. Thus, integrating and with initial condition (F) = $(F)_0$ at $T = T_0$,

$$\frac{T - T_0}{(F) - (F)_0} = -\frac{h_c}{(\rho c_v)}. \tag{7.11}$$

Thus, T and (F) have linear relation if c_v is constant. As (F) = ρY_F, then

$$\frac{T - T_0}{Y_F - Y_{F0}} = -\frac{h_c}{c_v}, \text{ adiabatic, constant-V reactor.} \tag{7.12}$$

When fuel is completely burned, $Y_F = 0$; Equation 7.12 yields

$$T_{f,V} = T_0 + \frac{Y_{F0}\,h_c}{c_v}. \tag{7.13}$$

Using the final condition $Y_F = 0$ at $T = T_{f,V}$, flame temperature for constant-volume combustion, one can rewrite Equation 7.12 as,

$$\frac{T - T_0}{T_{f,V} - T_0} = 1 - \frac{Y_F}{Y_{F0}} \tag{7.14}$$

where the slope of the line of T vs. (F) must be linear as given by

$$\frac{dT}{dY_F} = -\frac{h_c}{c_v} = -\frac{(T_{f,V} - T_0)}{Y_{F,0}}. \tag{7.15}$$

Because $d(F)/dt = \{d(O_2)/dt\}/\nu_{O2}$, and $(O_2) = \rho\, Y_{O2}$, Equation 7.10 can be written as

$$mc_v \frac{dT}{dt} = -\frac{Vh_c}{\nu_{O2}}\frac{d\{(O_2)\}}{dt} = -\frac{\rho V h_c}{\nu_{O2}}\frac{dY_{O_2}}{dt}$$

Integrating

$$\frac{T - T_0}{\left[\frac{Y_{O2}}{\nu_{O2}} - \frac{Y_{O2,0}}{\nu_{O2}}\right]} = -\frac{h_c}{c_v}. \tag{7.16}$$

Consider the global expressions

$$\frac{d[F]}{dt} = -k_{[]}[F]^{nF}\,[O_2]^{nO2}, \qquad k_{[]} = A_{[]}\exp\left(-\frac{E}{\bar{R}T}\right).$$

Because $[F] = (F)/M_F$, $[O_2] = (O_2)/M_{O2}$, converting to mass form and then using $(F) = \rho Y_F$, the reaction rate in terms of mass concentration is written as,

$$\dot{w}'''_{F,m} = \frac{d(F)}{dt} = -k_{()}\,(F)^{nF}\,(O_2)^{nO_2}, \; k_{()} = A_{()}\exp\left(\frac{E}{\bar{R}T}\right), \quad A_{()} = A_{[]}\left\{\frac{M_F}{M_F^{nF}}\frac{1}{M_{O2}^{nO2}}\right\}$$

Writing in terms of mass fractions,

$$\rho\frac{d[Y_F]}{dt} = \dot{w}'''_{F,m} = -k_{()}\rho^{nF+nO_2}\,[Y_F]^{nF}\,[Y_{O_2}]^{nO_2} = -k_Y\,[Y_F]^{nF}\,[Y_{O_2}]^{nO_2}, \tag{7.17}$$

$$k_{()} = A_{()}\,\rho^{nF+nO_2}\exp\left(-\frac{E}{\bar{R}T}\right), \quad A_{()} = A_{[\,]}\left\{\frac{M_F}{M_F^{\;nF}}\frac{1}{M_{O_2}^{\;nO_2}}\right\}.$$

Solution procedure: Given a small increment in T (say $dT = (T_f - T_0)/300$), get dY_F ($= -dT/(h_c/c_v)$ from Equation 7.15), determine dY_{O2} ($= -\nu_{O2}dT/h_c/c_v$), get $P = \rho RT$, where $R = \bar{R}/M_{mix}$, and $M_{mix} = 1/\Sigma\{Y_k/M_k\}$, determine reaction rate from Equation 7.17, use it in Equation 7.9 with $Q = 0$ to get "dt"; for next step, find $Y_{F,new} = Y_{Fprevious} + dY_F$, $Y_{O2,new}$, new $P = \rho R(T + dT)$. Keep proceeding until all fuel is burned. Thus, $T(t)$, $Y_F(t)$, $Y_{O2}(t)$, $P(t)$ can be obtained.

Example 1

Consider combustion of CH_4 with 10% excess air at constant volume. Initial conditions: $T_0 = 900$ K, $P_0 = 10$ bar, Kinetics : $n_F = 1$, $n_{O2} = 1$, $A_{()} = 8.6 \times 10^7$ m³/kg sec, $E = 125500$ kJ/kmol, $h_c = 50{,}000$ kJ/kg $c_v = 0.96$ kJ/kg K. Assume $M_{Mix} = 28.97$ kJ/kmol, a constant: (a) calculate the flame temperature and final pressure, (b) plot T, Y_F vs. t.

Solution

(a) $CH_4 + 2{*}1.1{*}(O_2 + 3.76\ N_2) \rightarrow CO_2 + 2\ H_2O + 0.2\ O_2 + 8.27\ N_2$

$$Y_{F0} = 1{*}16.05/\{16.05 + 2.2{*}32 + 2.2{*}3.76{*}28.02\} = 0.0498$$

$$Y_{O_2,0} = 0.219$$

$$T_f = T_0 + \frac{Y_{F0}\, h_c}{c_v} = 900 + (0.0498*50000/0.96) = 3495 \text{ K}$$

$P_0V_0 = m\ \bar{R}\ T_0/M_{Mix,0}$, $P_fV_f = m\ \bar{R}\ T_f/M_{Mix,f}$. As mixture molecular weights are constant and combustion occurs at constant volume,

$$\{P_f/P_0\} = T_f//T_0 = 3495/900 = 3.88,\ P_f = 38.8 \text{ bar.}$$

(b) Temperature interval is selected as dT = (3495 − 900)/300 = 8.65 K, M_{mix} = 28.97 (assumed), $\rho = P\ M_{mix}/(\bar{R}\ T) = 10*100*28.97/(8.314*900) = 3.87$ kg/m³, a constant because volume and mass are fixed. $(F)_0 = \rho\ Y_{F,0} = 3.87*0.0498 = 0.193$ kg of fuel/m³. Similarly, $(O_2)_0 = \rho\ Y_{O2,0} = 3.87*0.219 = 0.848$ kg of fuel/m³.

$$\frac{d(F)}{dt} = -k_{(\,)}\,(F)^{nF}\,(O2)^{nO2},\ k_{(\,)} = A_{(\,)} \exp\left(\frac{E}{\bar{R}T}\right),$$

T_{avg} = 900 + 8.65/2 = 904.3 K

$k_{(\,)} = 8.6 \times 10^7 \exp\{-125500/[8.314*(904.3)]\} = 4.84$ (m³/kg sec),
d(F)/dt = −4.84 (m³/kg sec)*0.193*0.848 (kg/m³)² = −0.79 kg/(m³ sec).

As ρ = constant, d(F)/dt = d/dt $(\rho Y_F) = \rho dY_F/dt$ = −0.79 kg/(m³ sec);
Similarly, $\rho\ dY_{O_2}/dt$ = −3.16 kg/m³sec

Because

$$\frac{T - T_0}{T_f - T_0} = 1 - \frac{Y_F}{Y_{F0}}.$$

With T = 908.65 K, T_0 = 900 K, and Y_{F0} = 0.498, Y_F = 0.0497; as $\rho dY_F/dt$ = −0.79 kg/(m³ sec), then dt = 3.87*(0.497 − 0.0498)/(−0.79) = 7.53 × 10⁻⁴ sec. One may repeat calculations for all temperature steps. Then, plot T and Y_F vs. t as shown in Figure 7.2. Combustion is complete within about 6 msec.

Remarks

(i) If the compression ratio is increased, pressure and temperature can increase, resulting in increased reaction rate and reduced burn time.
(ii) The final adiabatic flame temperature is extremely high, and as such the final products could be CO, etc., due to dissociation, and there could be significant NO production. The CO will gradually oxidize as the system is expanded during the expansion stroke.
(iii) Instead of using finite chemistry for oxidation reactions in auto engines, characteristic time models are used [Abraham et al., 1985; Reitz and Kuo, 1989].

$$\frac{dY_k}{dt} = -\frac{Y_k - Y_{k,e}}{\tau_c},$$

where $Y_{k,e}$ is the equilibrium mass fraction of species k and τ_c is the characteristic time to reach equilibrium $\tau_c = \tau_\ell + f\tau_t$, where $\tau_c = \tau_\ell$ laminar

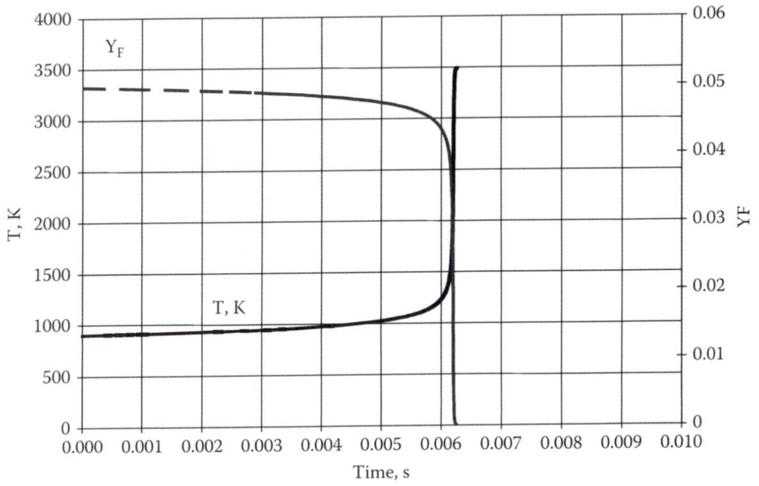

Figure 7.2 Combustion of CH_4 with 10% excess air; $T_0 = 900$ K, $P_0 = 10$ bar, $n_F = 1$, $n_{O2} = 1$, $A_Y = 8.6 \times 10^7$, $E = 125{,}500$ kJ/kmol, $c_v = 0.96$ kJ/kg K.

kinetics time and τ_t turbulent mixing time (= $C_2\kappa/\varepsilon$; see Chapter 18 for $\kappa-\varepsilon$ model in turbulence), f is the delay coefficient given by (1-exp(R))/0.632, where R is local mass fraction of residual gases mixed with fresh air. For example, around 1800 K, $Y_{CO2,e}$ will be close to stoichiometric mass fraction whereas actual value (say) is zero; then

$$\frac{dY_{CO2}}{dt} = \frac{Y_{CO2,e}}{\tau_e}.$$

7.3.1.2 Constant-Pressure Reactor

For a constant-pressure system, Equation 7.7 yields

$$\left|\dot{w}'''_{F,m}\right| V(t)\, h_c + \dot{Q} = P dV/dt + m c_v dT/dt = mP\frac{dv}{dt} + m\frac{du_t}{dt} = m\frac{dh_t}{dt} = mc_p\frac{dT}{dt}.$$

Rewriting,

$$mc_P\frac{dT}{dt} = \left|\dot{w}'''_{F,m}\right| V(\mathrm{t})h_c + \dot{Q} \tag{7.18}$$

where V (t) = m RT(t)/P. Similarly, for constant-pressure adiabatic combustion with flame temperature $T_{f,P}$, Equation 7.15 can be modified to yield

$$\frac{dT}{dY_F} = -\frac{h_c}{c_p} = -\frac{(T_{f,P} - T_0)}{Y_{F,0}} \tag{7.19}$$

$$\frac{T - T_0}{Y_F - Y_{F0}} = -\frac{h_c}{c_p}.$$

Here, fuel concentration varies inversely with increase in temperature. When all fuel is burned (F) = 0 and hence

$$\left\{\frac{T_f}{T_0}\right\} = \frac{Y_{F0} h_c}{c_{p0}\, T_0} + 1 .$$

7.3.2 Rigorous Formulation

For a closed system with negligible KE and PE, omitting the subscript cv,

$$\dot{Q} - \dot{W} = \frac{dU}{dt} = m\frac{du}{dt} + u\frac{dm}{dt} = m\frac{du}{dt}. \tag{7.20}$$

Expressing u in terms of h,

$$\frac{du}{dt} = \frac{dh}{dt} - P\frac{dv}{dt} - v\frac{dP}{dt}. \tag{7.21}$$

If we assume that the only work done is Pdv work (e.g., in an idealized automobile engine), then $\dot{W}$ = P dV/dt, Using Equation 7.21 in Equation 7.20,

$$\dot{Q} - \dot{W} = \dot{Q} - P\frac{dV}{dt} = m\frac{du}{dt} = m\frac{dh}{dt} - P\frac{dV}{dt} - V\frac{dP}{dt} \Rightarrow \dot{Q} + V\frac{dP}{dt} = \frac{dH}{dt}$$

$$\dot{Q} + V\frac{dP}{dt} = \frac{dH}{dt}. \tag{7.22}$$

The first law is written using either Equation 7.20 or Equation 7.22. Consider an automobile engine fired with CO, O_2, and an inert gas I, such as N_2, at the end of the compression stroke. Carbon monoxide and oxygen react to produce CO_2. The enthalpy inside the system changes because of heat losses and the work that it delivers during the expansion stroke. In terms of the system composition,

$$H = \sum_j N_j \bar{h}_j = N_{CO}\,\bar{h}_{CO} + N_{O2}\,\bar{h}_{O2} + N_{CO2}\,\bar{h}_{CO2} + N_I\,\bar{h}_I \tag{7.23}$$

where N_k denotes the number of moles of the species k. The number of moles of the species changes even though the total mass remains constant. Now, from Equation 7.23,

$$\frac{dH}{dt} = \left[\sum_j \left(\bar{h}_j \frac{dN_j}{dt}\right) + \sum_j \left(N_j \frac{d\bar{h}_j}{dt}\right)\right].$$

As $dN_I/dt = 0$,

$$\frac{dH}{dt} = \left(N_{CO} \frac{d\bar{h}_{CO}}{dt} + N_{O2} \frac{d\bar{h}_{O2}}{dt} + N_{CO2} \frac{d\bar{h}_{CO2}}{dt} + N_I \frac{d\bar{h}_I}{dt} \right) + \left(\bar{h}_{CO} \frac{dN_{CO}}{dt} + \bar{h}_{O2} \frac{dN_{O2}}{dt} + \bar{h}_{CO2} \frac{dN_{CO2}}{dt} \right). \tag{7.24}$$

The rate of change of N_k over time is $dN_k/dt = \dot{w}_k$, and applying the stoichiometric reaction relation for a single-step reaction $CO + 1/2\ O_2 \rightarrow CO_2$,

$$\frac{dN_{CO}/dt}{(-1)} = \frac{dN_{O2}/dt}{(-1/2)} = \frac{dN_{CO2}/dt}{(1)} = \dot{\omega} \tag{7.25}$$

where the law of mass action yields

$$\dot{w} = k_{[\,]}\ V\,[CO]^{\alpha}\ [O_2]^{\beta}. \tag{7.26}$$

Using the relation $\bar{h} = \bar{h}_f + \bar{h}_t$ and Equation 7.25 in Equation 7.24, we obtain the expression

$$\frac{dH}{dt} = \left(N_{CO} \frac{d\bar{h}_{CO,t}}{dt} + N_{O2} \frac{d\bar{h}_{O2,t}}{dt} + N_{CO2} \frac{d\bar{h}_{CO2,t}}{dt} + N_I \frac{d\bar{h}_{I,t}}{dt} \right) + \left(\bar{h}_{CO}\, \dot{\omega}(-1) + \bar{h}_{O2}\, \dot{\omega}(-1/2) + \bar{h}_{CO2}\, \dot{\omega}(+1) \right). \tag{7.27}$$

It is noted that

$$\{\bar{h}_{CO}\, \dot{\omega}(-1) + \bar{h}_{O2}\, \dot{\omega}(-1/2) + \bar{h}_{CO2}\, \dot{\omega}(1)\} = \dot{\omega}\Delta\bar{h}_{R,CO,T} \tag{7.28}$$

For ideal gases,

$$\frac{d\bar{h}_j}{dt} = \bar{c}_{p,j} \frac{dT}{dt}. \tag{7.29}$$

Using Equation 7.27 to Equation 7.29 in Equation 7.22,

$$\begin{aligned} \frac{dT}{dt} &= \frac{\dot{Q} + V\frac{dP}{dt} - \dot{\omega}\Delta\bar{h}_{R,CO,T}}{\left(N_{CO}\ \bar{c}_{p,CO} + N_{O2}\ \bar{c}_{p,O2} + N_{CO2}\ \bar{c}_{p,CO2} + N_I\,\bar{c}_{p,I} \right)} \\ &= \frac{\dot{Q} + V\frac{dP}{dt} + \dot{\omega}\ \overline{HV}_{CO,T}}{\left(N_{CO}\ \bar{c}_{p,CO} + N_{O2}\ \bar{c}_{p,O2} + N_{CO2}\ \bar{c}_{p,CO2} + N_I\,\bar{c}_{p,I} \right)} \end{aligned} \tag{7.30}$$

where $\overline{HV}_{CO,T} = -\Delta\bar{h}_{R,T}^{0}$. On a mass basis,

$$\frac{dT}{dt} = \frac{\dot{Q} + V\frac{dP}{dt} - \dot{w}_m h_{c,CO,T}}{\left(m_{CO}\, c_{p,CO} + m_{O2}\, c_{p,O2} + m_{CO2}\, c_{p,CO2} + m_I\, c_{p,I}\right)} = \frac{\dot{Q} + V\frac{dP}{dt} - \dot{w}_m\, h_{c,CO,T}}{\left(m\, c_P\right)} \quad (7.31)$$

where $h_{c,CO,T}$, heating value of CO at temperature T.

$$c_p = (m_{CO}\, c_{pCO} + m_{O2}\, c_{pO2} + m_{CO2}\, c_{p,cO2} + m_I\, c_{p,I})/m$$

$$c_p = Y_{CO}\, c_{pCO} + Y_{O2}\, c_{pO2} + Y_{CO2}\, c_{p,cO2} + Y_I\, c_{p,I}.$$

If P is constant, Equation 7.31 is similar to Equation 7.18

$$\left(\Delta\bar{h}_{R,CO,T}\right) = [\bar{h}_{f,CO}(-1) + \bar{h}_{f,O2}(-1/2) + \bar{h}_{f,CO2}(1)]$$
$$+\{\bar{h}_{t,CO}(-1) + \bar{h}_{t,O2}(-1/2) + \bar{h}_{t,CO2}(1)\}$$

where $\bar{h}_{t,CO} = \int \bar{c}_{pco}\, dT$ at 298 K,

$$\{\bar{h}_{f,CO}\, \dot{\omega}(-1) + \bar{h}_{f,O2}\, \dot{\omega}(-1/2) + \bar{h}_{f,CO2}\, \dot{\omega}(1)\} = \dot{\omega}\Delta\bar{h}^{0}_{R,CO}. \quad (7.32)$$

The enthalpy of reaction $\Delta\bar{h}_{R,T}$ is evaluated at a specified temperature T. It is a very weak function of temperature so that $\Delta\bar{h}_{R,T} \approx \Delta\bar{h}_{R,298} = \Delta\bar{h}^{0}_{R,T}$, i.e., Equation 7.30 is written as

$$\frac{dT}{dt} \approx \frac{\dot{Q} + V\frac{dP}{dt} - \dot{\omega}\Delta\bar{h}^{0}_{R,CO}}{\left(N_{CO}\, \bar{c}_{p,CO} + N_{O2}\, \bar{c}_{p,O2} + N_{CO2}\, \bar{c}_{p,CO2} + N_I \bar{c}_{p,I}\right)}. \quad (7.33)$$

Generalizing Equation 7.30 for a multistep reaction mechanism,

$$\frac{dT}{dt} = \frac{\dot{Q} + V\frac{dP}{dt} - \sum_j \left(\dot{\omega}_j \Delta\bar{h}_{R,T,j}\right)}{\sum_k \left(N_k\, \bar{c}_{p,k}\right)}, \quad (7.34)$$

where k refers to the species in the system and j to the elementary reaction contained in the multistep mechanism. For instance, for the mechanism

$$CO + OH \rightarrow CO_2 + 1/2\ H_2, \quad (A)$$

$$H_2 + 1/2O_2 \rightarrow CO_2, \quad (B)$$

$$\sum_j \left(\dot{\omega}_j \Delta\bar{h}_{R,T,j}\right) = \dot{\omega}_A \Delta\bar{h}_{R,A} + \dot{\omega}_B \Delta\bar{h}_{R,B}, \text{ and} \quad (7.35)$$

$$\sum_k \left(N_k \bar{c}_{p,k} \right) = N_{co} \bar{c}_{p,CO} + N_{OH} \bar{c}_{p,OH} + N_{H2} \bar{c}_{p,H2...,} \quad (7.36)$$

As the fuel burns, the temperature rises and the pressure and volume both increase. Because PV = NRT,

$$(1/P)\, dP/dt = (1/N)(dN/dt) + (1/T)\,(dT/dt) - (1/V)\, dV/dt. \quad (7.37)$$

Using Equation 7.37 in Equation 7.34 to eliminate dP/dt,

$$\frac{dT}{dt} = \frac{\dot{Q} + PV\frac{(dN/dt)}{N} + PV\frac{(dT/dt)}{T} - P\frac{dV}{dt} - \sum_j \left(\dot{\omega}_j \Delta \bar{h}_{R,T,j} \right)}{\left\{ \sum_k \left(N_k\ \bar{c}_{p,k} \right) - N\bar{R} \right\}}.$$

Solving for (dT/dt),

$$\frac{dT}{dt} = \frac{\dot{Q} + \bar{R}T\frac{dN}{dt} - P\frac{dV}{dt} - \sum_j \left(\dot{\omega}_j \Delta \bar{h}_{R,T,j} \right)}{N\left[\bar{c}_p - \bar{R} \right]}$$

where $\bar{c}_p = \sum_k (X_k \bar{c}_{p,k})$. Simplifying and rearranging the relation,

$$N\bar{c}_v \frac{dT}{dt} = -\sum_j \left(\dot{\omega}_j \Delta \bar{h}_{R,T,j} \right) + \dot{Q} - P\frac{dV}{dt} + \bar{R}T\frac{dN}{dt}. \quad (7.38)$$

Here, the term dN/dt can be obtained by considering the chemical reaction rates. For instance, for CO oxidation,

$$dN/dt = dN_{CO}/dt + dN_{O2}/dt + dN_{CO2}/dt = \dot{\omega}\,(-1 - 1/2 + 1) = -(1/2)\,\dot{\omega}. \quad (7.39)$$

The term $\bar{R}$ TdN/dt represents the pumping work due to the changing number of moles. If it is ignored, Equation 7.38 becomes

$$N\bar{c}_v \frac{dT}{dt} = -\sum_j \left(\dot{\omega}_j \Delta \bar{h}_{R,T,j} \right) + \dot{Q} - P\frac{dV}{dt}. \quad (7.40)$$

On a mass basis, Equation 7.38 assumes the form,

$$mc_v \frac{dT}{dt} = -\sum_j \left(\dot{\omega}_{j,m} \Delta h_{R,T,j} \right) + \dot{Q} - P\frac{dV}{dt} + RMT\frac{d(m/M)}{dt}$$

$$mc_v \frac{dT}{dt} \approx -\sum_j \left(\dot{\omega}_{j,m} \Delta h_{R,T,j} \right) + \dot{Q} - P\frac{dV}{dt} = mc_v \frac{dT}{dt} \approx \sum_j \left(\dot{\omega}_{j,m} h_{c,j} \right) + \dot{Q} - P\frac{dV}{dt}$$

(7.41)

where $\dot{w}_{j,m}$ denotes the j-th reaction rate in mass units. We note from this relation that the sensible internal energy rise of a mass within a system = energy released by all chemical reactions + energy added through heat transfer – energy out through work performed. Note the similarity of Equation 7.41 and Equation 7.7.

7.3.3 Applications of Rigorous Treatment

7.3.3.1 Constant Pressure

Consider a piston–cylinder–weight (PCW) arrangement with a perfectly homogeneous mixture of reactants. Assume constant mass and pressure for the system as combustion occurs. T, V, and mole fractions change as the reaction occurs. Appling Equation 7.34 at constant pressure we obtain,

$$\frac{dT}{dt} = \frac{\dot{Q} - \sum_{j}\left(\dot{\omega}_{j}\Delta \bar{h}^{0}_{R,T,j}\right)}{\sum_{k}\left(N_{j}\ \bar{c}_{p,k}\right)} \tag{7.42}$$

Once T vs. t is known, V vs. t can be obtained using the ideal gas law PV = N $\bar{R}$ T, where N could change as reaction proceeds. For the CO oxidation system, the term dN/dt can be obtained from Equation 7.39. Then, using this result in Equation 7.37, the volume change is related to temperature change as

$$(1/V)dV/dt = (1/N)((-1/2)\,\dot{\omega}\,) + (1/T)\ (dT/dt), \tag{7.43}$$

where the value of dT/dt is obtained from Equation 7.42. Thus, one can plot T and V vs. t during combustion at constant pressure as in a diesel engine.

7.3.3.2 Isobaric and Isothermal

In additions, the PCW assembly is now kept in an isothermal bath at T.

In this case Equation 7.42 assumes the form

$$\dot{Q} = \dot{\omega}\Delta\bar{h}_{R,CO,T} = \dot{\omega}\left(\Delta\bar{h}^{0}_{R,CO} + \{\bar{h}_{T,CO}(-1) + \bar{h}_{T,O2}(-1/2) + \bar{h}_{T,CO2}(1)\}\right) \approx \dot{\omega}\Delta\bar{h}^{0}_{R,CO}. \tag{7.44}$$

Using Equation 7.43, the rate of volume change for CO oxidation reaction is given as:

$$(1/V)\ dV/dt = (1/N)\ (-(1/2)\,\dot{\omega}\,). \tag{7.45}$$

7.3.3.3 Constant Volume

Using Equation 7.38, we obtain the relation

$$N\bar{c}_{v}\frac{dT}{dt} = -\sum_{j}\left(\dot{\omega}_{j}\Delta\bar{h}_{R,T,j}\right) + \dot{Q} + \bar{R}T\frac{dN}{dt}. \tag{7.46}$$

For the $CO + 1/2\ O_2 \rightarrow CO_2$ reaction, applying Equation 7.46 for a single reaction,

$$N\bar{c}_v \frac{dT}{dt} = -\left(\dot{\omega}\Delta\bar{h}_{R,T}\right) + \dot{Q} + (-1/2)\dot{\omega}\,\bar{R}T, \quad \text{where} \tag{7.47}$$

where $\bar{c}_v = \Sigma \bar{c}_{vj} X_j$, $Xj = N_j/N$, $\bar{R} = \bar{c}_p - \bar{c}_v$.

Equation 7.47 can also be written as

$$mc_v \frac{dT}{dt} = \left(\dot{\omega}_m h_c\right) + \dot{Q} + (-1/2)\dot{\omega}_m R\,M\,T \approx \left(\dot{\omega}_m HV\right) + \dot{Q}. \tag{7.48}$$

Using Equation 7.39 in Equation 7.37, the rate of pressure rise is given as

$$(1/P)dP/dt = (1/N)\ (-(1/2)\ \dot{\omega}\ ) + (1/T)(dT/dt), \tag{7.49}$$

where the value of dT/dt can be determined from Equation 7.47 or Equation 7.48. Thus, one can plot T and P vs. t during combustion at constant volume as in an Otto engine.

7.3.3.4 Constant Volume and Isothermal

In this case, dT/dt = 0, dV/dt = 0, and Equation 7.47 simplifies to

$$\dot{Q} = \sum_j \left(\dot{\omega}_j \Delta\bar{h}_{R,T,j}\right) + (1/2)\dot{\omega}\bar{R}T\ , \tag{7.50}$$

and the rate of pressure rise is provided by Equation 7.49

$$(1/P)dP/dt = (1/N)\ (-(1/2)\ \dot{\omega}\ ), \tag{7.51}$$

$$\text{where} \quad \dot{\omega} = A_{[\,]}\ \text{Exp}\ \{-E/(\bar{R}\ T)\}\ [F]^{\ nF}\ [O2]^{nO2}. \tag{7.52}$$

Example 2

Consider a fixed-mass, constant-volume, well-insulated reactor filled initially with a stoichiometric mixture of CO and O_2 at a temperature of 1500 K and a pressure of 1 bar (= 10^5 Pa) in a 0.0001–m^3 volume. In the absence of hydrogen or water vapor, CO and O_2 are essentially unreactive. At a certain time, water vapor is injected and instantaneously mixed throughout the chamber to a mole fraction of 0.01. Calculate the initial rate of temperature change dT/dt in the chamber immediately following water injection and mixing. Assume that CO and O_2 react with the following global rate constant: $\dot{\omega}''' = k_{[\,]}\ [CO][H_2O]^{0.5}[O_2]^{0.25}$ kmol $m^3 sec^1$, where $k_{[\,],f} = 10^7\ \exp(20{,}000/T)$, where T is expressed in kelvin.

Solution

From the law of stoichiometry,

$$\frac{d[CO]}{dt(-1)} = \frac{d\left[O_2\right]}{dt(-1/2)} = \frac{d\left[CO_2\right]}{dt} = \dot{\omega}'''$$

Using Equation 7.46,

$$N\bar{c}_v \frac{dT}{dt} = -\sum_j \left(\dot{\omega}_j \Delta \bar{h}_{R,T,j}\right) + \dot{Q} + \bar{R}T\frac{\mathrm{dN}}{\mathrm{dt}} .$$

For the single-step reaction in an adiabatic system,

$$\mathrm{dN/dt} = -(1/2)\,\dot{\omega} = -(1/2)\,\dot{\omega}'''\,\mathrm{V}, \quad \text{or}$$

$$\mathrm{N}\bar{c}_v \frac{dT}{dt} = -\left(\dot{\omega}''' V\, \Delta \bar{h}_{R,T}\right) + \bar{R}T\frac{\mathrm{dN}}{\mathrm{dt}} = -\,\{\dot{\omega}''' V\,(\Delta \bar{h}_{R,T} + (1/2)\bar{R}T)\} .$$

Writing the reaction rate expression in molar form (kmol, m, sec)

$$\dot{\omega}''' = k_{[\,]} \left[\frac{P}{\bar{R}T}\right]^{(1+0.5+0.25)} [X_{CO}][X_{H_2O}]^{0.5}[X_{O_2}]^{0.25} .$$

Expressing in mass form (kg, m, sec)

$$\dot{\omega}'''_m = k_{[\,]} M_{CO} \left[\frac{1}{M_{CO}}\right]\left[\frac{1}{M_{O2}}\right]^{0.25}\left[\frac{1}{M_{H2O}}\right]^{0.5}\left[\frac{PM}{\bar{R}T}\right]^{(1+0.5+0.25)} [Y_{CO}][Y_{H_2O}]^{0.5}[Y_{O_2}]^{0.25} .$$

As $CO + 1/2\ (O_2 + 3.76\ N_2) \rightarrow CO_2 + 1.88\ N_2$. Ignore the small water concentration, assume stoichiometric reactant concentrations, the mixture molecular weight $M_{mix} = 28.809$. Then the initial mole and mass fractions of the various species are:

Species	*Mole fraction*	*Mass fraction*
CO	0.269	0.261
O_2	0.153	0.171
N_2	0.577	0.562
H_2O	0.01	0.006

Therefore,

$P/RT = 1/(0.08314 \times 1500) = 0.008\ \mathrm{kmol/m^3}$,

$N = 0.008 \times 0.0001 = 8 \times 10^{-7}$ kmol

$\dot{\omega}''' = 10^7 \exp(-20000/T) \times 0.008^{1.75} \times 0.269 \times 0.01^{0.5} \times 0.153^{0.25} = 7 \times 10^{-5}$ $\mathrm{kmol/m^3}$ sec.

HV = 10103 kJ/kg or 282990 kJ/kmol of CO, $\bar{c}_v = \Sigma \bar{c}_{vj} X_j = 0.269 * \bar{c}_{v,CO} + 0.153 * \bar{c}_{v,O2} + 0.577 * \bar{c}_{v,N2} + 0.01 * \bar{c}_{v,H2O} \approx 16$ kJ/kmol K, i.e.,

$$N\bar{c}_v \frac{dT}{dt} = -\sum_j \left(\dot{\omega}_j \Delta \bar{h}_{R,T}\right) + \dot{Q} + \bar{R}T\frac{\mathrm{dN}}{\mathrm{dt}}$$

$$8 \times 10^{-7} \times 16 \text{ kJ/(kmol K)} \times dT/dt = 7 \times 10^{-5} \times 0.0001 \times (282990 - 0.08314 \times 0.5 \times 1500)$$

Solving, $dT/dt = 151.35$ K/sec.
Hence,

$(1/P)\, dP/dt = (1/8 \times 10^{-7})\, (-(1/2)7 \times 10^{-5} \times 0.0001) + (1/1500) \times 151.4 = 0.096$ (1/sec).

Therefore,

$$dP/dt = 0.096 \text{ bar/sec.}$$

7.4 Open Systems

Consider a system consisting of a combustible mixture with inlet and exit valves (e.g., automobile engine). Suppose both valves are closed. In such a closed system, the concentrations keep changing with time. When concentrations are uniform within the closed system, the variation of concentration, say [CO], with time is written as $d[CO]/dt = \dot{w}'''_{CO}$. Thus, for chemically reacting closed systems, the problem is inherently unsteady. However, if one opens the inlet valve and admits CO at $\dot{N}_{CO,i}$ at a rate equal to the consumption rate $\dot{w}'''_{CO}$, then [CO] may not change with time within this open system. Such a condition is called *steady state.*

Consider an open system indicated in Figure 7.3. Consider the elementary volume dV = Adx; say, a gaseous fuel, CO, enters the volume at x and then undergoes a change in concentration to [CO] + {d/dx} [CO] dx at x + dx. As the CO consumption will tend to decrease the concentration within this volume Adx because of chemical reactions, fresh CO enters through the inlet and replenishes it. However, CO also is removed as exit flow at x + dx. If the input rate is less than the sum of consumption rate and exit rate, then [CO] within Adx will decrease with time; otherwise, it may increase. The increase or decrease rate is given as $\partial[CO]/\partial t$ ($\neq d[CO]/dt$), whereas the change in concentration with distance is given as $\partial[CO]/\partial x$. If local input rate $\dot{N}_{CO,i}$ is equal to the sum of $\dot{w}'''_{CO}$ and $\dot{N}_{CO,e}$, then $\partial[CO]/\partial t = 0$ and [CO] remains constant at a given location. In this case, $\dot{N}_{CO,i} - \dot{N}_{CO,e} = |\, \dot{w}'''_{C}\, |$.

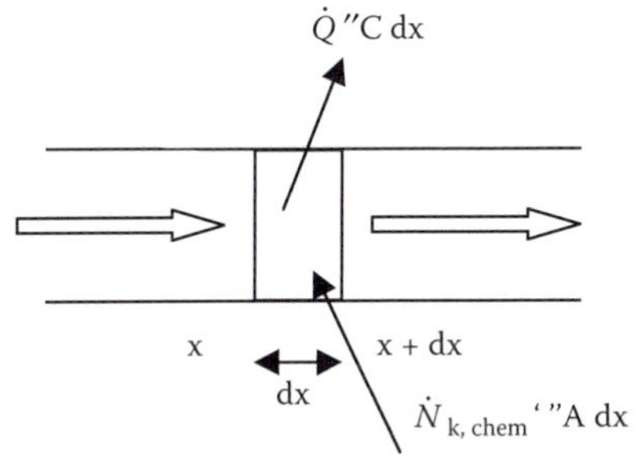

Figure 7.3 Schematic diagram of a plug flow reactor.

7.4.1 Damköhler Numbers

Damköhler numbers characterize the relative magnitude of the timescale or rate of a process. The first Damköhler number is the ratio of the residence time of a packet of fluid in an advective flow to the characteristic chemical reaction time, i.e.,

$$\mathrm{Da_I} = t_{res}/t_{chem} = |\dot{w}'''_{F,m}|\,L_c/(\rho v) = |\dot{w}_{F,m}|V_c/\dot{m} = |\dot{w}_{F,m}|/\dot{m} \qquad (7.53)$$

where $|\dot{w}'''_{F,m}|$ denotes the chemical reaction rate, L_c and, V_c (= AL_c) are the characteristic length and volume, respectively, and v, is a the characteristic velocity. Also, $t_{res} \approx L_c/v = \rho A\, L_c/\{\rho A\, v\} = m_c/\dot{m}$, and $t_{chem} \approx m_c/|\dot{w}_{F,m}| = \rho/|\dot{w}'''_{F,M}|$, where m_c (= $\rho A\, L_c$)) is the characteristic mass. The second Damköhler number can be expressed as

$$\mathrm{Da_{II}} = |\dot{w}'''_{F,m}|\,L_c h_c/(\rho c_p T) = |\dot{w}'''_{F,m}|\,V_c h_c/\{\dot{m}\, c_p\, T\} = \dot{w}_{F,m}|\,h_c/\{\dot{m}\, c_p\, T\}, \qquad (7.54)$$

which is the ratio of the heat generation rate to the convective heat loss rate, where h_c denotes the heating value of fuel. The third Damköhler number is

$$\mathrm{Da_{III}} = t_{res,diff}/t_{chem} = |\dot{w}'''_{F,m} L_c^2|/\rho D = |\dot{w}'''_{F,m} V_c|/\{\rho D L_c^2/L_c\} = |\dot{w}_{F,m}|/\{\dot{m}_{c,diff}\} \qquad (7.55)$$

where $\dot{m}_{c,diff}$ is a characteristic diffusion rate. $\mathrm{Da_{III}}$ represents the ratio of the residence time in a diffusive flow to the chemical reaction time. Now, $t_{res,diff} \approx L_c^2/D$. The fourth such parameter

$$\mathrm{Da_{IV}} = \dot{w}'''\, L_c^2 h_c/(\rho D c p T) = |\dot{w}'''_{F,m}|\, V_c/\{\rho D\, c_p\, T\, L_c^2/L_c\} = |\dot{w}_{F,m}|\, h_c/\{\dot{m}_{c,diff}\, c_p T\} \qquad (7.56)$$

is the ratio of the heat generation rate to the diffusive heat loss rate. In general, $\mathrm{Da_I}$ and $\mathrm{Da_{II}}$ are applied to chemically reacting flow problems (plug flow reactor, perfectly stirred reactor; see Subsection 7.4.2 and Subsection 7.4.3), whereas $\mathrm{Da_{III}}$ and $\mathrm{Da_{IV}}$ are used for diffusive problems (droplet and particle combustion, Chapter 9 and Chapter 10).

7.4.2 Plug Flow Reactor (PFR)

A plug flow reactor is a reactor in which there is no radial variation of the system properties. Recall the overall mass and energy conservation obtained in Chapter 2. Consider the overall mass conservation and molar balance equations for the plug flow reactor that is schematically described in Figure 7.3. Consider any species k. The mole flow in at x is $\dot{N}_k$, whereas the mole flow out at x + dx is $\dot{N}_k$ + dx {d $\dot{N}_k$/dx}. Then, using the molar balance equation (Equation 7.3) at steady state leads to the expression

$$0 = -A\,(d\,\dot{N}''_k/dx)\,dx + \dot{N}'''_{k,chem}\,A dx = -A\,(d\,\dot{N}'''_k/dx)\,dx + \dot{w}'''_k\,.$$

Here, $\dot{N}_k'''$ denotes the molar flux per unit cross-sectional area. Simplifying, we obtain the expression

$$d\,\dot{N}_k'''/dx = \dot{w}_k''' . \quad (7.57)$$

Multiplying the relation by the molecular weight of the k-th species M_k,

$$(d\,\dot{m}_k''/dx) = M_k\,\dot{w}_k''' = \dot{w}_{k,m}'' . \quad (7.58)$$

As $\dot{m} = \rho\ V = \rho_{in} V_{in}$ is constant and

$$\dot{m}_k'' = \rho_k v_k = \rho_k(v + V_k) = \rho_k\ [v - \{D/Y_k)\ dY_k/dx\} = \rho\ Y_k v - (\rho D\ dY_k/dx).$$

Let the flux ratio

$$\varepsilon_k = \rho_k v_k/\rho v = \dot{m}_k''/\dot{m}'' = Y_k - (D/v)(dY_k/dx) = Y_k - (D/v)(dY_k/dx). \quad (7.59)$$

(Note that flux ratio $\varepsilon_k \approx Y_k$ if diffusion terms are ignored or if diffusion velocity $Vk = -(D/Y_k)\ dY_k/dx << v$.)

$$\dot{m}''\ [d\varepsilon_k/dx] = \dot{w}_{k,m}''' \quad \text{or} \quad [\rho v\, d\varepsilon_k/dx] = \dot{w}_{k,m}''' \quad (7.60)$$

Defining $dt = dx/v$

$$[d\varepsilon_k/dt] = \dot{w}_{k,m}'''/\rho \quad (7.61)$$

Note that once ε_k's are known, Y_k is solved from Equation 7.59 and X_k from the relation $X_k = MY_k/M_k$.

The corresponding energy conservation relations in mass and molar forms are

$$(dE_{cv}/dt) = \dot{Q}_{cv} - \dot{W}_{cv} + \sum \dot{m}_{k,i}(h + ke + pe)_{k,i} - \sum \dot{m}_{k,e}(h + ke + pe)_{k,e} \quad (7.62)$$

$$(dE_{cv}/dt) = \dot{Q}_{cv} - \dot{W}_{cv} + \sum \dot{N}_{k,i}\bar{e}_{k,T,i} - \sum \dot{N}_{k,e}\bar{e}_{k,T,e} \quad (7.63)$$

Neglecting ke and pe,

$$(dU_{cv}/dt) = \dot{Q}_{cv} - \dot{W}_{cv} + \sum \dot{N}_{k,i}\bar{h}_i - \sum \dot{N}_{k,e}\bar{h}_e \quad (7.64)$$

Energy inflow at x is $\dot{N}_k\,\bar{h}_k$, and the corresponding exit flow at x + dx is $(\dot{N}_k\,\bar{h}_k + d/dx\ (\dot{N}_k\,\bar{h})dx,)$

$$0 = \dot{Q}_{cv}''C\,dx - 0 + \sum \dot{N}_k\bar{h}_k - \sum \{\dot{N}_k\bar{h}_k + d/dx(\dot{N}_k\bar{h}_k)dx\},$$

where C denotes the circumference, and the axial conducive heat flow due to temperature difference is ignored (Figure 7.3).

From the mole balance equation, the mole flow rate change over distance $(d\,\dot{N}''_k/dx) = \dot{w}'''_k$. Therefore,

$$0 = \dot{Q}''_{cv}\,(C/A) - \sum \dot{w}'''_k\,\bar{h}_k - \sum \dot{N}''_k\,d\bar{h}_k/dx. \tag{7.65}$$

where A represents the reactor cross-sectional area.

As $\bar{h}_k = \bar{h}_{f,k} + \bar{h}_{t,k}$, (hereafter, $\bar{h}_{t,k}$ implies thermal enthalpy) and with simplifying assumptions, this expression has the form

$$0 = \dot{Q}''_{cv}\,(C/A) - \sum \dot{w}'''_k\,\Delta\bar{h}_k - \sum \dot{N}''_k\,\bar{c}_{pk}\,dT/dx. \tag{7.66}$$

We will now demonstrate this concept for a single reaction, e.g., CO + 1/2 O_2 = CO_2. Then, from Equation 7.57,

$$(d\,\dot{N}''_{CO}/dx) = \dot{w}'''_{CO} = -\,\dot{w}''', \tag{7.67}$$

$$(d\,\dot{N}''_{O_2}/dx) = -(1/2)\,\dot{w}''', \tag{7.68}$$

$$(d\,\dot{N}''_{CO_2}/dx) = \dot{w}'''. \tag{7.69}$$

Multiplying these relations by M_k, we obtain mass form of conservation equations:

$$d\,\dot{m}''_{CO}/dx = M_{CO}\,(-1)\,\dot{w}''',$$

as $\dot{m}''_{CO} = \rho Y_{CO} v_{CO} = \rho Y_{CO}\,\{v + V_{CO}\}$. Ignoring diffusive velocity, $\dot{m}''_{CO} \approx \rho v Y_{CO} = \dot{m}''\,Y_{CO}$. Note that ε_k, mass flux ratio $= \dot{m}_k/\dot{m} \approx X_k$, mass fraction when diffusion velocities are small compared to bulk velocity. Then,

$$d\,\dot{m}''_{CO}/dx = \dot{m}''\,dY_{CO}/dx = M_{CO}\,(-1)\,\dot{w}''',\ \text{diffusion ignored} \tag{7.70}$$

$$d\,\dot{m}''_{O2}/dx = M_{O2}\,(-1/2)\,\dot{w}''',\ \text{diffusion ignored}, \tag{7.71}$$

$$\dot{m}''\,/dY_{CO2}/dx = M_{CO2}\,\dot{w}''',\ \text{diffusion ignored}. \tag{7.72}$$

Therefore, using these relations in Equation 7.66,

$$0 = \dot{Q}''_{cv}\,(C/A) - (\dot{w}'''_{CO}\,\bar{h}_{CO} + \dot{w}'''_{O_2}\,\bar{h}_{O2} + \dot{w}'''_{CO_2}\,\bar{h}_{CO2})$$

$$-\,(dT/dx)(\dot{N}''_{CO}\,\bar{c}_{p,CO} + \dot{N}''_{O_2}\,\bar{c}_{p,O2} + \dot{N}''_{CO_2}\,\bar{c}_{p,CO2}). \tag{7.73}$$

Applying the law of mass action and proceeding as before,

$$\dot{w}'''_{CO}/(-1) = \dot{w}'''_{O_2}/(-1/2) = \dot{w}'''_{CO_2}/(1) = \dot{w}'''.$$

Defining,

$$\dot{C}''_p = \dot{N}''(X_{CO}\bar{c}_{p,CO} + X_{O2}\bar{c}_{p,O2} + X_{CO2}\bar{c}_{p,CO2}) = \dot{m}''(Y_{CO}c_{p,CO} + Y_{O2}c_{p,O2} + Y_{CO2}c_{p,CO2}), \quad (7.74)$$

$$0 = \dot{Q}''_{cv}(C/A) - \dot{w}'''((-1)\bar{h}_{CO} + (-1/2)\bar{h}_{O2} + \dot{w}_{CO2}\bar{h}_{CO2}) - \dot{C}''_p(dT/dx). \quad (7.75)$$

Recognizing the terms involving the enthalpy of reaction,

$$0 = \dot{Q}''_{cv}(C/A) - \dot{w}'''\Delta\bar{h}_{R,T} - \dot{C}''_p(dT/dx). \quad (7.76)$$

We can assume the heat loss expression as $\dot{Q}''_{cv}(C/A) = -h_H(T - T_\infty)$ to further simplify the relation so that

$$\dot{C}''_p(dT/dx) = -\dot{w}'''\Delta\bar{h}_{R,T} - h_H(T - T_\infty). \quad (7.77)$$

Furthermore, $\Delta\bar{h}_{R,T} \approx \Delta\bar{h}^0_R = -\bar{h}_c$, which is the molar heating value, i.e.,

$$\dot{C}''_p(dT/dx) = \dot{w}'''\bar{h}_c - h_H(T - T_\infty). \quad (7.78)$$

This relation implies that the sensible enthalpy increase within the incremental length dx = the heat liberation rate within dx less the heat loss rate within dx.

Example 3

Consider a plug flow isothermal reactor in which methane is admitted with excess air at an arbitrary temperature and pressure. Ignore axial diffusion terms. Assume that the overall global oxidation reaction is

$$d[CH_4]/dt\ (kmol/m^3\ sec) = k_{[\,]}[Fuel]^a[O_2]^b, \quad (A)$$

where $k_{[\,]} = A_{[\,]}\exp(-E/\bar{R}T)$, E = 202,600 kJ/kmol, a = −0.3, b = 1.3, and $A_{[\,]} = 1.3 \times 10^8$ 1/sec. Assume that the initial concentrations of the products, namely, CO_2 and H_2O, are zero.

(a) Convert Equation A in terms of mass fraction.
(b) Obtain a differential equation for the fuel mass fraction.
(c) Assuming 10% excess air, a constant 1000 K temperature, a pressure of 2 bar, integrate the relations.
(d) Plot Y_F, Y_{O2}, Y_{CO2}, and Y_{H2O} vs. t.
(e) What is the characteristic reaction time if reaction proceeds at same rate as the initial rate?

Solution

(a) $CH_4 + 2\,O_2 \rightarrow CO_2 + 2\,H_2O$

1 kg of methane + 4 kg of O_2 → 2.75 kg of CO_2 + 2.25 kg of H_2O

$$|w'''_{CH4}|\ (kmol/m^3\ s) = k_{[\,]}[Fuel]^a[O2]^b \quad (A)$$

and the law of stoichiometry is

$$\dot{w}'''_{CH4}/(-1) = \dot{w}'''_{O2}/(-2) = \dot{w}'''_{CO2} / (+1) = \dot{w}'''_{H2O}/(+2) = \dot{w}'''. \quad \text{(B)}$$

Thus, $\dot{w}'''$ represents the magnitude of the fuel being consumed in (kmol per m^3 per sec) for which stoichiometric coefficient is 1 kmol, or CO_2 being produced in kmol per m^3 per sec for which stoichiometric coefficient is again 1 kmol.

$$\dot{w}''' = A_{[\,]} \exp(-E/RT)\,(P/RT)^{\alpha+\beta}\,(X_F)^{\alpha}(X_{O2})^{\beta} \text{ kmol/m}^3 \text{ s},$$

Rewriting,

$$\dot{w} = A_{[\,]}\,(1/M_F)^{\alpha}(1/M_{O2})^{\beta},\exp(-E/RT)\,(\rho)^{\alpha+\beta}\,(Y_F)^{\alpha}(Y_{O2,e})^{\beta}, \quad \text{(C)}$$

$$\dot{w}_{CH4,m}/(-1) = \dot{w}_{O2,m}/(-4) = \dot{w}_{CO2,m}/(+2.75) = \dot{w}_{H2O,m}/(+2.25) = \dot{w}'''_m .$$

Thus, $\dot{w}'''_m$ now represents the magnitude of fuel being consumed for which the stoichiometric coefficient is 1 kg. Equation C becomes

$$\dot{w}'''_m = A_{[\,]}\,(M_F)^{1-\alpha}(1/M_{O2})^{\beta} \exp(-E/RT)\,(\rho)^{\alpha+\beta}\,(Y_F)^{\alpha}(Y_{O2})^{\beta} \text{ kg/m}^3 \text{ sec},$$

$$\dot{w}'''_m = A_Y\,(\rho)^{\alpha+\beta}\exp(-E/RT)\,(Y_F)^{\alpha}(Y_{O2})^{\beta}, \quad \text{(D)}$$

where $A_Y = A_{(\,)} = A_{[\,]}\,(M_F)^{1-\alpha}(1/M_{O2})^{\beta}$.

As $A_{[\,]} = 1.3 \times 10^8$, $A_Y = 1.3 \times 10^{8}$ (1/sec) $\times$ 16.05 (kg/kmol)1 – (–0.3)(32 kg/kmol)$^{(-1.3)}$, i.e.,

$$A_Y = 5.30 \times 10^7 \text{ 1/sec}.$$

(b) Assuming that the value of M_{mix} is everywhere constant in the PFR, ρ is therefore also constant at a specified temperature and pressure.

Hence,

$$(dY_F/dt) = -A_Y \exp(-E/RT)\,(\rho)^{\alpha+\beta}\,(Y_F)^{\alpha}(Y_{O2})^{\beta}/\rho_{in} \quad \text{(E)}$$

(c) $CH_4 + 2{*}1.1\,(O_2 + 3.76\,N_2) \rightarrow CO_2 + 2\,H_2O + 0.1{*}2\,O_2 + 8.27\,N_2$

Every kg of methane requires 4 kg of oxygen for stoichiometric combustion. Therefore, the stoichiometric amount of air is 4/0.23 = 17.4 kg of air. With 10% excess air, the air supplied is 19.13 kg. Therefore,

$Y_{air} = 19.13/(19.13 + 1) = 0.9502$, i.e.,

$Y_{O2} = 0.9502 \times 0.23 = 0.2185$ and $Y_{N2} = 0.9502 \times 0.77 = 0.732$, and

$Y_{CH4} = 1 - 0.9502 = 0.0498$.

The total moles in 1 kg of the mixture are

$$Y_{O2}/32 + Y_{CH4}/16.05 + Y_{N2}/28.02 = 0.036,$$

The mixture molecular weight

$M_{mix} = 1/0.036 = 27.74$ kg/kmol of the mixture, i.e.,

$\rho_{mix} = PM_{mix}/RT = 2 \times 27.74/(0.08314 \times 1000) = 0.667$ kg/m^3.

Using Equation E at t = 0 (or x = 0)

$$dY_F/dt = -5.30 \times 10^7 \exp(-202600/(8.314 \times 1000)) \times 0.667^{(-0.3+1.3)}(Y_F)^{-0.3}(Y_{O2})^{1.3}/\rho_{in}$$

$$= -9.241 \times 10^{-4}\ (Y_F)^{-0.3}(Y_{O2})^{1.3}/\rho_{in},$$

where dt = dx/v. At t = 0 or x = 0,

$$dY_F/dt = -9.241 \times 10^{-4}\ (Y_{F,0})^{-0.3}(Y_{O2,0})^{1.3}/0.667 = 0.000473.$$

Initially,

$$dY_{O2}/dt = 4\ dY_F/dt, \quad \text{and} \quad dY_{CO2}/dt = (44.01/16.05)\ dY_F/dt.$$

(d) Figure 7.4 shows a plot of Y_F, Y_{O2}, Y_{CO2}, and Y_{H2O} vs. t.
(e) The characteristic reaction time is

$$t_{char} = Y_{F,0}/(dY_F/dt) = 0.0498/0.000473 = 105 \text{ sec.}$$

Remarks

(i) If $M_{mix} = M_{air} = 28.97$, there is no significant change in the results.
(ii) If the reactor is adiabatic, then the temperature has to be solved for using the energy equation at each time step.

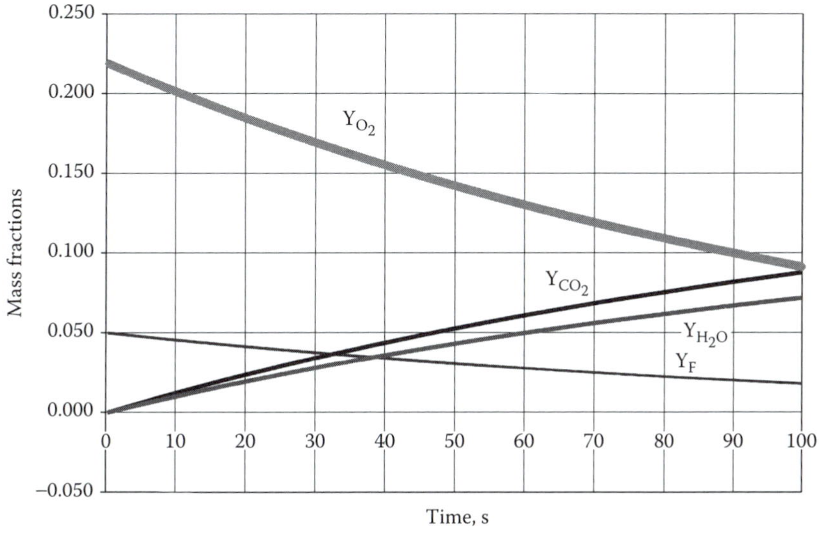

Figure 7.4 Results for a plug flow reactor, 10% excess air, T = 1000 K, Fuel: CH_4.

Example 4

Consider the species CO_2, H_2O, CO, and H_2 entering a PFR. There is no gradient of T within the c.v.; thus, thermal irreversibility is almost zero within the c.v. Furthermore, P will be maintained constant and hence there is no mechanical irreversibility. Hence, the only irreversibility in an isothermal PFR is the chemical reaction

$$CO_2 + H_2 \rightarrow CO + H_2O \tag{A}$$

if this is the assumed direction. Assume steady state. Use the entropy balance equation to prove that $\dot{G}_e < \dot{G}_i$ where $\dot{G} = \Sigma_\kappa \dot{N}_k \bar{g}_k$, if reaction has to proceed in the indicated direction.

$$dS/dt = \dot{Q}_{c.v.}/T_b + \sum \dot{N}_{k,i}\bar{s}_{k,i} - \sum \dot{N}_{k,e}\bar{s}_{k,e} + \dot{\sigma} \tag{B}$$

where $\dot{\sigma}$, entropy generated per sec.

Solution

(a)

$$dN_{CO2}/dt = \dot{N}_{CO2,i} - \dot{N}_{CO2,e} + \dot{N}_{CO2,chem}$$

$$dN_{CO2}/dt = \dot{N}_{CO2,i} - \dot{N}_{CO2,e} + (-1)\dot{w}''', \quad \dot{N}_{CO2,chem} = -\dot{w}''' \tag{C}$$

Similarly,

$$dN_{H2}/dt = N_{2,i} - \dot{N}_{H2,e} + (-1)\dot{w}''', \quad \dot{N}_{H2,chem} = -\dot{w}''' \tag{D}$$

$$dN_{CO}/dt = \dot{N}_{CO,i} - \dot{N}_{CO,e} + (+1)\dot{w}''', \quad \dot{N}_{CO,chem} = \dot{w}''' \tag{E}$$

$$dN_{H2O}/dt = \dot{N}_{H2O,i} - \dot{N}_{H2O,e} + (-1)\dot{w}''', \quad \dot{N}_{H2O,chem} = \dot{w}''' \tag{F}$$

From energy conservation equation,

$$dE/dt = \dot{Q}_{c.v.} - \dot{w}_{c.v} + \sum \dot{N}_{k,i}\bar{h}_{k,i} - \sum \dot{N}_{k,e}\bar{h}_{k,i}$$

and because of steady state and zero work, solving for $\dot{Q}_{c.v.}$,

$$\dot{Q}_{c.v} = \sum \dot{N}_{k,e}\bar{h}_{k,e} - \sum \dot{N}_{k,i}\bar{h}_{k,i} \tag{G}$$

From entropy balance Equation B under steady state

$$0 = \dot{Q}_{c.v.}/T_b + \sum \dot{N}_{k,i}\bar{S}_{k,i} - \sum \dot{N}_{k,e}\bar{S}_{k,e} + \dot{\sigma} \tag{H}$$

Using Equation G in Equation H, and with T_b = T (for c.v containing gas mixture), then under steady state

$$0 = 0 - \sum_{\kappa} \dot{N}_{k,i}\bar{g}_{k,i} + \sum_{\kappa} \dot{N}_{k,e}\bar{g}_{k,e} + T\dot{\sigma}, \quad T\dot{\sigma} > 0 \tag{I}$$

where $\bar{g}_k = \bar{h}_k - T\bar{s}_k$. The requirement $T\dot{\sigma} > 0$ leads to $\dot{G}_e = \Sigma_\kappa \dot{N}_{k,e}\ \bar{g}_{k,e} < \dot{G}_i (= \Sigma_\kappa\ \dot{N}_{k,i}\ \bar{g}_{k,i})$ and $\dot{G}$ keeps decreasing as reaction progresses. Under steady state

$$dN_k/dt = 0 = \dot{N}_{k,i} - \dot{N}_{k,e} + \dot{N}_{k,chem} \tag{J}$$

Now consider an elemental volume with length equal to dx and area of cross section A. If the inlet and exit are located by "dx," hence $\dot{N}_{k,e}$ = d/dx $\{\dot{N}_{k,i}\}$ dx + $\dot{N}_{k,i} + \dot{w}_k'''$ V similarly Equation I reduces to

$$0 = -\sum d/dx\{\dot{N}_k''\ \bar{g}_k\} + T\dot{\sigma}'''$$

Expanding the differential,

$$0 = \sum d/dx\{\dot{N}_k''\}\bar{g}_k + \dot{N}_k'' d/dx\{\bar{g}_k\} + T\dot{\sigma}'''$$

Using elemental mole balance

$0 = -\Sigma\, \dot{w}_k''\ \bar{g}_k - \Sigma\, \dot{N}_k''$ d/dx $\{\bar{g}_k\} - T\dot{\sigma}''' = -\Sigma\, \dot{w}_k'''\ \bar{g}_k - 0 - T\dot{\sigma}'''$ for constant-P systems, as $\Sigma \dot{N}_k'' d\bar{g}_k/dx = 0$ (see Chapter 3). Thus, the requirement that $T\dot{\sigma}''' > 0$ implies

$$-\sum \dot{w}_k'''\ \bar{g}_k = T\dot{\sigma}''' = -\{(-1)\}\dot{w}'''\ \bar{g}_{CO2} + (-1)\bar{g}_{H2}\ \dot{w}''' + \bar{g}_{CO}\dot{w}''' + \bar{g}_{H2O}\dot{w}'''\} > 0 \tag{K}$$

Reducing

$$\bar{g}_{CO2}(-1) + (-1)\bar{g}_{H2} + \bar{g}_{CO} + \bar{g}_{H2O}\}\dot{w}''' = T\dot{\sigma}''' \tag{L}$$

$$\bar{g}_{CO2} + \bar{g}_{H2} > \bar{g}_{CO} + \bar{g}_{H2O} \quad \text{since } \dot{w}''' > 0$$

which is the same result as obtained in Chapter 3.

7.4.3 Nonisothermal Reactor and Ignition

Here, we deal with the variation of temperature and species concentration in a PFR with or without heat loss. The problem is similar to Example 2 except that temperature changes due to heat contribution from reactions are taken into consideration when estimating the reaction rates. Ignition is said to occur when there is rapid temperature rise within a short period. One may determine the ignition time or the distance of flame location and ignition temperature of gaseous mixtures using PFR models.

Example 5

The problem is the same as in Example 2 except that the reactor is an adiabatic PFR. Ignore axial diffusion terms.

(a) Obtain numerical answers for T vs. t and dT/dt vs. t for 10% excess air, T_{in} = 860 K, P = 2 bar. Integrate using time intervals which limit temperature rise to 10 K/sec. Let dt = dx/v.
(b) Ignition time is assumed to be a time at which the temperature rise rate is 10 K/sec. Estimate t_{ign} and the corresponding gas temperature ($T_{g,ign}$), assuming T_{in} = 860 K.
(c) Plot t_{ign} and $T_{g,ign}$ vs. T_{in} with $800 < T_{in} < 950$ K.

Solution

$$dY_F/dt = M_F\,(-1)\,\dot{w}'''/\rho = (-1)\,\dot{w}'''_m/\rho = -A_Y \exp(-E/RT)\,[\rho]^{\alpha+\beta-1}\,[Y_F]^{\alpha}[Y_{O2}]^{\beta} \qquad \text{(A)}$$

where $\rho(t) = P\,M_{mix}/RT(t)$,

Following Equation 7.76 and using the mass form,

$$m''_{cp}\,dT/dx = \dot{Q}''_{cv}\{C/A\} - \sum_j \{\dot{w}'''_j\,\Delta\bar{h}_{R,T,j}\}. \qquad \text{(B)}$$

Neglecting heat loss and for single-step reaction,

$$\dot{m}''_{c_p}\,dT/dx = 0 - \dot{w}'''_m \Delta h_{R,T}.$$

Expressing this in terms of velocity v,

$$v c_p\,dT/dx = 0 - \dot{w}'''_m\,\Delta h_{R,T}/\rho \qquad \text{(C)}$$

$$dT/dt = -A_Y \exp(-E/RT)\,[\rho]^{\alpha+\beta-1}\,[Y_F]^{\alpha}[Y_{O2}]^{\beta}\,\Delta\bar{h}_{R,T}/c_p. \qquad \text{(D)}$$

Equation D is integrated to yield T vs. t_g using Excel spreadsheet. The program determines t_{ign} and T_{gas} at which dT/dt = 10 K/sec. Results are shown in Figure 7.5.

It is seen that T_{gas} at ignition is approximately constant (990K) irrespective of inlet temperature (700–960K).

7.4.4 *Perfectly Stirred Reactor (PSR)*

7.4.4.1 What Is a PSR?

In PFR, the mass fractions and temperature inside the reactor continue to change with distance (or time). Hence, the chemical reaction rate per unit volume changes with distance as we march along the x axis. Suppose we divide the PFR into N sections. Then we let all the sections (except a small volume at inlet) to mix intensively with each other inside the reactor. Then the mass fractions and temperature become almost uniform and the reactor becomes perfectly stirred

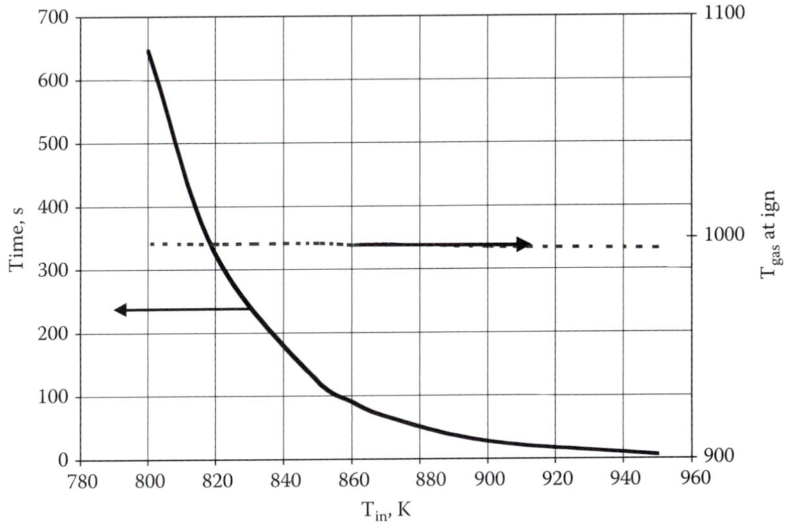

Figure 7.5 Ignition results for methane–air mixture; $T_{gas,\ ign} \approx$ constant.

reactor (PSR). A PSR is a constant-volume (V) vessel in which reactants ($\dot{m}\ Y_{F,in}$, $\dot{m}\ Y_{O2in}$) are injected at multiple inlets and the reactant and product species are stirred so perfectly that its composition is everywhere uniform (except for a small volume adjacent to the walls of the container, Figure 7.6). Furthermore, the stirring process is so efficient that the species concentrations and the temperature in the reactor are identical to those at its outlet through which products are discharged ($\dot{m}\ Y_{P,e}$).

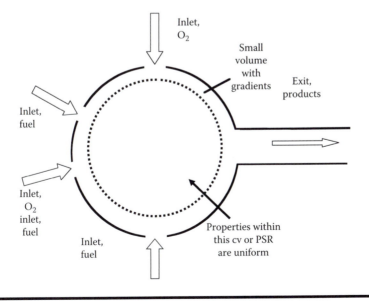

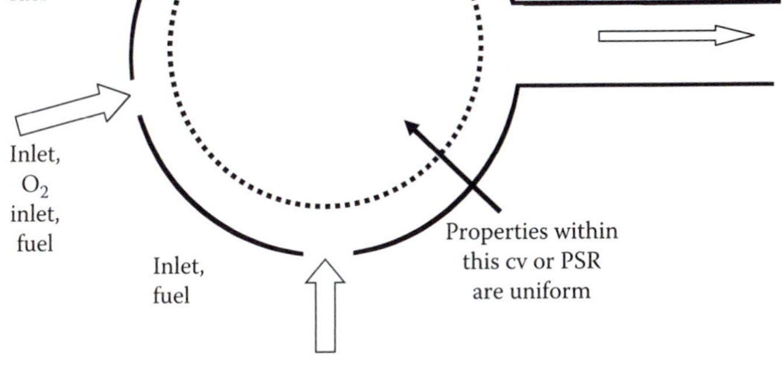

Figure 7.6 Schematic diagram of a perfectly stirred reactor (PSR).

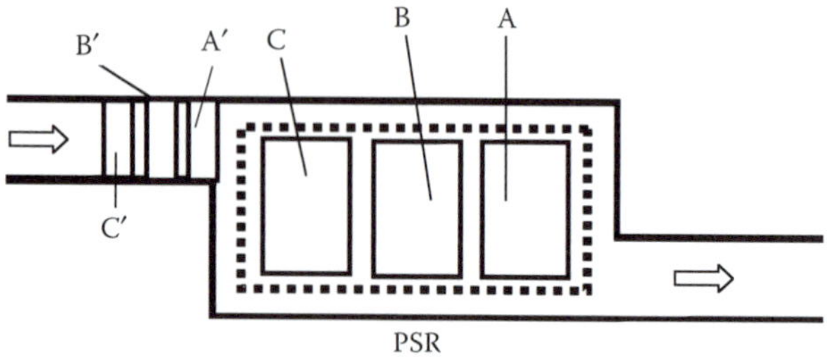

Figure 7.7 Illustration of residence time.

The objective is to determine the combustion rate and predict T and Y_k in PSR. The finite residence time of species is given as $t_{res} = m/\dot{m}$. The residence time concept can be explained using Figure 7.7. Suppose the mass flow rate is 2 kg/sec and mass inside the reactor is 6 kg. Then, according to the residence time equation, $t_{res} = 6/2 = 3$ sec. The mass in reactor is split up into 3 blocks: A, B, and C in Figure 7.7 There are three blocks, A′, B′, and C′, each of mass 2 kg waiting outside the reactor. A′, B′, and C enter at intervals of 1 sec. Within 1 sec, A′ enter and A exits from the reactor within same time interval. Now A′ is inside the reactor. Only when C′ enters the reactor will A′ leave the reactor. Thus, A′ has a residence time of 3 sec. If A′ is now considered to be a block having stoichiometric mass of CO and air, then we find that its mass fraction Y_{CO} decreases from $Y_{CO,in}$ at the inlet almost instantaneously with Y_{CO}, remaining constant within the PSR regime for 3 sec and then leaves the reactor. At the same time, chemical time required for the "block" depends upon the reaction kinetics at Y_{CO}, Y_{O2}, and T.

The chemical reactions inside the reactor occur over a finite time t_{chem}, which is the time required to convert the fuel into products. The first Damköhler number is

$$Da_I = t_{res}/t_{chem}.$$

If $t_{res} >> t_{chem}$ or $Da_I >> 1$, then the fuel is almost completely burned, and the temperature rises should be close to adiabatic flame temperature in an adiabatic reactor. However, when the mass flow rate is very large, $t_{res} \rightarrow t_{chem}$, i.e., $Da_I \rightarrow 1$, and incomplete burning will result in lower temperatures. At even larger flow rates, $Da_I < 1$, $t_{res} <<< t_{chem}$, which quenches the reaction in the reactor. PSR models can predict conditions for ignition, combustion, and extinction. These models can be applied to analyze combustion in boiler burners, gas turbine combustors, gasifiers, and fluidized bed combustors if extensive computations are to be avoided with modern computer codes. We will perform a representative PSR analysis assuming uniform properties within the system. We use energy and species balance to solve for Y_F, Yo_2, and T.

7.4.4.2 *Simplified Method*

7.4.4.2.1 Fuel and Oxygen Mass Balance

$$\text{Mass flow rate of fuel in} = \dot{m}\, Y_{F,in} \quad (7.79a)$$

$$\text{Mass flow rate of fuel out} = \dot{m}\, Y_{F,\,out} \quad (7.79b)$$

$$\dot{m}\,\{Y_{F,in} - Y_{F,\,out}\} = \text{Chemical consumption rate of fuel or fuel burned}$$

$$= |\dot{w}'''_{F,m}|\,V = |\dot{w}_{F,m}| \quad (7.80)$$

where $|\dot{w}'''_{F,m}|$ (kg/m^{3}s) $= A_Y \exp(-E/RT)\, \rho^{\alpha+\beta}\, Y_F^{\alpha} Y_{O2}^{\beta}$ (7.81)

and $\rho = PM_{mix}/RT$,

$$\dot{m}\,\{Y_{O2,in} - Y_{O2,\,out}\} = \text{Chemical consumption rate of oxygen}$$

$$= |\dot{w}'''_{F,m}|\, \nu_{O2} V = \dot{w}'''_{m}\, \nu_{O2} V \quad (7.82)$$

where ν_{O2} is stoichiometric oxygen mass per unit mass of fuel. For example, consider $CO + 1/2\ O_2 \rightarrow CO_2$.

$$1 \text{ kg of } CO + 0.571 \text{ kg of } O_2 \rightarrow 1.571 \text{ kg of } CO_2 \quad (7.82)$$

where $\nu_{O2} = 0.571$.

7.4.4.2.2 Energy Balance

Chemical heat liberated within PSR = sensible energy rise + heat loss from reactor

$$|\dot{w}'''_{F,m}|\,V\, h_c = \dot{m}\, c_p\, \{T - T_{in}\} + |\dot{Q}_{loss}|. \quad (7.83)$$

For known heat loss, inlet conditions, kinetics A_m and E, the four unknowns are $\dot{w}'''_{F,m}$, Y_F, Y_{O2}, and T. With four equations (Equation 7.80 to Equation 7.83), one can solve for four unknowns $\dot{w}'''_{F,m}$, T, Y_F, and Y_{O2}. The fuel burned $|\dot{w}'''_{F,m}|V$ can be expressed in terms of burned fraction ε_b.

7.4.4.2.3 Burned Mass Fraction (ε_b)

$$\varepsilon_b = \text{fuel burned/fuel in} = |\dot{w}'''_{F,m}|\,V/\dot{m}_{F,in} = \dot{w}_{F,m}\{\dot{m}\,Y_{F,in}\} = |\dot{w}'''_{F,m}|\,\{m/\{\rho \dot{m} Y_{F,in}\} \quad (7.84)$$

where $V = m/\rho$, m = mass within PSR, and ρ = density of gas within PSR,

$$\varepsilon_b = \{\dot{m}_{F,in} - \dot{m}_{F,out}\}/\dot{m}_{F,in} == \{Y_{F,in} - Y_{F,out}\}/Y_{F,in} = |\dot{w}'''_{F,m}|\,t_{res}/\{\rho Y_{F,in}\}. \quad (7.85)$$

Similarly, using Equation 7.82,

$$\varepsilon_b\, Y_{Fin}\, \nu_{O2} = \{Y_{O2,in} - Y_{O2}\} \quad (7.86)$$

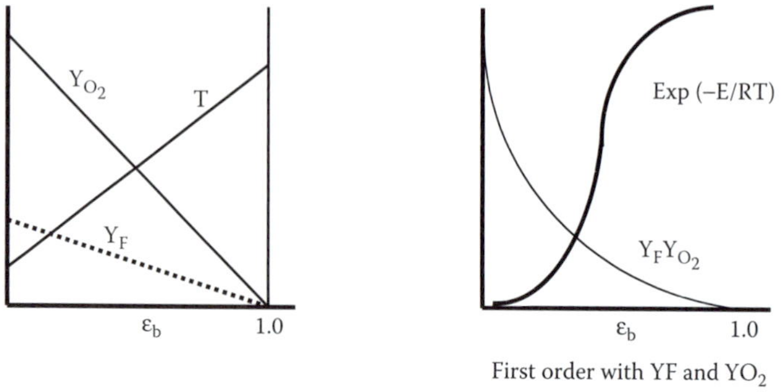

Figure 7.8 (Left) Variation of Y_F, Y_{O2}, and T with burned fraction. (Right) Variation of the product (Y_F Y_{O2}) and exp(E/RT) with burned fraction; illustration for first order with Y_F and Y_{O2}.

Replacing $|\, \dot{w}'''_{F,m} \,|$ in Equation 7.83 in terms of burned fraction and simplifying,

$$\varepsilon_b\ h_c\ Y_{F,in} = c_p\ \{T - T_{in}\} + |\ \dot{Q}_{loss}\ | / \dot{m} \tag{7.87}$$

Knowing the fuel consumption rate $\dot{w}_{F,m}$, the burned fraction can be solved. It is seen that the higher the residence time, the higher the burned fraction and lower the fuel mass fraction leaving (Equation 7.85). The relations between Y_F, Y_{O2}, T, and ε_b are linear (see Equation 7.85 to Equation 7.87, respectively) (Figure 7.8 left side). In the following, a rigorous formulation using standard mass and energy conservation equations are presented with reaction CO + 1/2 $O_2 \rightarrow CO_2$ as an example.

7.4.4.3 Rigorous Formulation from Mass and Energy Equations

7.4.4.3.1 Mass Conservation

(a) Molar Form:

In molar form, the steady-state molar balance equation assumes the form

$$\dot{N}_{k,e} = \dot{N}_{k,in} + \dot{w}_k, \tag{7.88}$$

i.e., for example, with k = CO_2,

$$\dot{N}_{CO2,e} = \dot{N}_{CO2,in} + \dot{N}_{CO2,chem} = \dot{N}_{CO2,in} + \dot{w}_{CO2}. \tag{7.89}$$

From the law of stoichiometry, $\dot{w}_{CO}/(-1) = \dot{w}_{O2}/(-1/2) = \dot{w}_{CO2}/(+1) = V\,\dot{w}''$ Similarly,

$$\dot{N}_{CO,e} = \dot{N}_{CO,in} + \dot{w}(-1) = \dot{N}_{CO,in} - V\ \dot{w}''' \tag{7.90}$$

$$\dot{N}_{O2,e} = \dot{N}_{O2,in} + \dot{w}(-1/2) = \dot{N}_{O2,in} - (1/2)\ V\ \dot{w}''' \quad \text{and} \tag{7.91}$$

$$\dot{N}_{N2,e} = \dot{N}_{N2,in}. \tag{7.92}$$

The reaction rate $\dot{w}'''$ does not appear in Equation 7.92, because nitrogen is assumed to be chemically inert.

(b) Mass Form:

Multiplying Equation 7.89 by the species molecular weight M_{CO2},

$$\dot{m}_{CO2,e} = \dot{m}_{CO2,in} + V \cdot M_{CO2}\, \dot{w}'''_m \tag{7.93}$$

i.e., the mass of CO_2 exiting the PSR = mass of CO_2 in + mass of CO_2 produced,

$$\dot{m}_{CO2,e} = \dot{m}_{CO2,in} + V \cdot M_{CO2}\, \dot{w}'''_m = \dot{m}_{CO2,in} + V\, \nu_{CO2}\, \dot{w}'''_m \tag{7.94}$$

as from law of stoichiometry, $\dot{m}_{CO,m}/(-1) = \dot{w}_{O2,m}/(-0.571) = \dot{w}_{CO2,m}/(1.571) = V\, \dot{w}'''_m$ and $\nu_{CO2} = 1.571$ for the current reaction. Similarly, Equation 7.90 to Equation 7.92 transform to

$$\dot{m}_{CO,e} = \dot{m}_{CO,in} - V\, M_{CO}\, \dot{w}''' = \dot{m}_{CO,in} - V\, \dot{w}'''_m \tag{7.95}$$

$$\dot{m}_{O2,e} = \dot{m}_{O2,in} - (1/2)V\, M_{O2}\, \dot{w}''', \quad \text{and} \tag{7.96}$$

$$\dot{m}_{N2,e} = \dot{m}_{N2,in}. \tag{7.97}$$

Dividing Equation 7.94 to Equation 7.97 by $\dot{m}$, and because diffusional velocities are zero in the reactor due to negligible gradient, the flux ratios become equal to mass fractions. Here,

$$Y_{CO2,e} = Y_{CO2,in} + V\, M_{CO2}\, \dot{w}'''/\dot{m} \tag{7.98}$$

$$Y_{CO,e} = Y_{CO,in} - V\, M_{CO}\, \dot{w}'''/\dot{m} = Y_{CO,in}\,(1 - \varepsilon_b), \tag{7.99a}$$

$$\text{where} \quad \varepsilon_b = |\, \dot{w}'''_{CO,m} \,|V/\{\dot{m}\, Y_{CO,in}\} = |\, \dot{w}'''_{CO,m} \,| t_{res}/\{\rho Y_{CO,in}\} \tag{7.99b}$$

$$Y_{O2,e} = Y_{O2,in} - (1/2)\, V\, M_{O2}\, \dot{w}'''/\dot{m} = Y_{O2,in} - Y_{COin}\, \nu_{O2}\, \varepsilon_b, \quad \text{and} \tag{7.100}$$

where $\nu_{O2} = (1/2)\, M_{O2}/M_{CO}$

$$Y_{N2,e} = Y_{N2,in}. \tag{7.101}$$

Recall that $Y_{CO,e} = Y_{CO}$, $Y_{O2,e} = Y_{O2}$

$$\dot{w}'' = |\, \dot{w}'''_{CO,m} \,| = A_Y \exp(-E/RT)\, (\rho)^{\alpha+\beta}\, (Y_{CO})^{\alpha} (Y_{O2})^{\beta},\ \text{kg/(m}^3\ \text{sec)} \tag{7.102}$$

where $\rho = PM/RT$. $X_k = MY_k/M_k$, $M = 1/(\Sigma Y_k/M_k)$.

Using Equation 7.102 in Equation 7.99

$$\varepsilon_b = \{A_Y V/(Y_{CO,in}\, \dot{m})\} \exp(-E/RT)\, (\rho)^{\alpha+\beta}\, (Y_{CO})^{\alpha} (Y_{O2})^{\beta}. \tag{7.103a}$$

Equation 7.103a provides the relation for ε_b in terms of Y_F, Y_{O2} and T. Because $t_{res} = m/\dot{m} = \rho V/\dot{m}$, generalizing Equation 7.103a,

$$\varepsilon_b = \{A_Y\, t_{res}/Y_{CO,in}\} \exp(-E/RT)\, [\rho]^{\alpha+\beta-1}\, [Y_F]^{\alpha}[Y_{O2}]^{\beta}. \qquad (7.103b)$$

If the reactor temperature is known, using Equation 7.99a, Equation 7.100, and Equation 7.103, the mass fractions of Y_{CO} and Y_{O2} can be determined. The temperature is solved using energy balance equation.

7.4.4.3.2 Energy Equation

(a) Molar Form:

Neglecting the ke and pe and assuming uniform temperature within the system. From First Law:

$$dE_{CV}/dt = 0 = \dot{Q}_{CV} - \dot{W}_{CV} + \sum_{k,i} \dot{N}_k \bar{h}_{,k} - \sum_{k,e} \dot{N}_k \bar{h}_{,k}, \quad \text{i.e.,}$$

$$\dot{Q}_{cv} + \dot{N}_{CO,in}\bar{h}_{CO,in} + \dot{N}_{O2,in}\bar{h}_{O2,in} + \dot{N}_{CO2,in}\bar{h}_{CO_2} + \dot{N}_{N2,in}\bar{h}_{N2,in} - \{\dot{N}_{CO2,in} + V\dot{w}'''(1)\}$$

$$\bar{h}_{CO2} - \{\dot{N}_{CO,in} + V\dot{w}'''(-1)\}\bar{h}_{CO} - \{(\dot{N}_{O_2,in} + V\dot{w}'''(-1/2)\}\bar{h}_{O_2} - \dot{N}_{N_2},\bar{h}_{N_2} = 0. \qquad (7.104)$$

With $\bar{h}_k = \bar{h}_{f,k} + \bar{h}_{t,k}$, and constant specific heats, one obtains

$$V\dot{w}'''\bar{h}_{c,CO} \approx [T - T_{in}]\{\dot{N}_{CO,in}\bar{c}_{pCO} + \dot{N}_{O2,in}\bar{c}_{p,O2} + \dot{N}_{CO2,in}\bar{c}_{p,CO2} + \dot{N}_{n2,in}\bar{c}_{p,N2}\} - \dot{Q}_{cv} \qquad (7.105)$$

$$V\dot{w}'''\bar{h}_{c,CO} \approx [T - T_{in}]\dot{N}_{in}\bar{c}_p - \dot{Q}_{cv} \qquad (7.106)$$

where $\bar{h}_{c,CO} = \bar{h}_{f,CO} + (1/2)\bar{h}_{f,O2} - \bar{h}_{f,CO2}$

$$\bar{c}_p = X_{CO,in}\bar{c}_{p,CO} + X_{O2,in}\bar{c}_{p,O2} + X_{N2,in}\bar{c}_{p,N2} + X_{CO2,in}\bar{c}_{p,CO2}. \qquad (7.107)$$

(b) Mass Form:

Similarly, from Equation 7.105,

$$V\dot{w}_m'''h_{c,CO} \approx [T - T_{in}]\{\dot{m}_{CO,in}c_{pCO} + \dot{m}_{O2,in}c_{p,O2} + \dot{m}_{CO2,in}c_{p,CO2} + \dot{m}_{N2,,in}c_{p,N2}\} + \dot{Q}_{cv}. \qquad (7.108)$$

Or,

$$V|\dot{w}_{CO,m}'''|h_{c,CO} \approx [T - T_{in}]\,c_p\,\dot{m} + \dot{Q}_{cv} \qquad (7.109a)$$

where $c_p = \{Y_{CO,in}\ c_{pcO} + Y_{O2,in}\ c_{p,O2} + Y_{CO2,in}\ c_{p,CO2} + Y_{N2,,in}\ c_{p,N2}\}$

Dividing Equation 7.109 by $\dot{m}$,

$$h_{c,CO}\ \varepsilon_b\ Y_{CO,in} \approx c_p\ [T - T_{in}] + \dot{Q}_{cv}/\dot{m}. \tag{7.109}$$

This equation is same as the one we derived using a simple approach. Having known mass fractions, T can be solved from Equation 7.109. The procedure is as follows. For given t_{res}, assume ε_b, solve Y_{CO} and Y_{O2} from Equation 7.99 and Equation 7.100, T from Equation 7.109, and finally calculate ε_b from Equation 7.103b and verify whether it checks with assumed ε_b. Iterate until convergence is obtained. Thus, solutions for ε_b,Y_{CO}, Y_{O2}, and T for PSR are obtained for a given t_{res}.

Using Equation 7.99, Equation 7.100, and Equation 7.109, one can plot Y_{CO} or Y_F, Y_{O2}, T with ε_b (Figure 7.8a). With $\alpha > 0$, $\beta > 0$, the product $(Y_F^{\alpha}\ Y_{O2}^{\beta})$ varies between a large value at $\varepsilon_b = 0$ and 0 (at $\varepsilon_b = 1$), assuming $\alpha = 1$, and $\beta = 1$ (Figure 7.8 right side). The term exp (–E/RT) varies between 0 (at T_{in} for low ε_b values) and 1 (at very large values of T for higher values of ε_b). Then the product $AY_FY_{O2}Vh_Ce^{-E/RT}$ will vary qualitatively as shown in Figure 7.9.

In Figure 7.9, the curve LIUECH presents the variation of the heat release with burned fraction. The branch LIUEC represents the heat release when the effect of temperature is dominant in the reaction rate, whereas the branch CH occurs when the effect of the product (Y_FY_{O2}) is dominant. The latter branch occurs when the burned fraction is close to unity (which is almost never realized in practical systems). Therefore, the branch LIUEC is more important in analyses of practical systems.

If $\dot{Q}_{cv} = 0$, the right-hand side of Equation 7.109 represents the heat loss $\dot{Q}_L$ through advection, which increases linearly with T of reactor, whereas left-hand side of Equation 7.109 represents the heat generation $\dot{Q}_G$ vs. T, which is exponential with T. Figure 7.10 contains the heat generation curve ($\dot{Q}_G$) LIUEC with respect to temperature and the sensible heat loss ($\dot{Q}_L$) curves IA, LUC, LEB that correspond to various mass flow rates through the reactor.

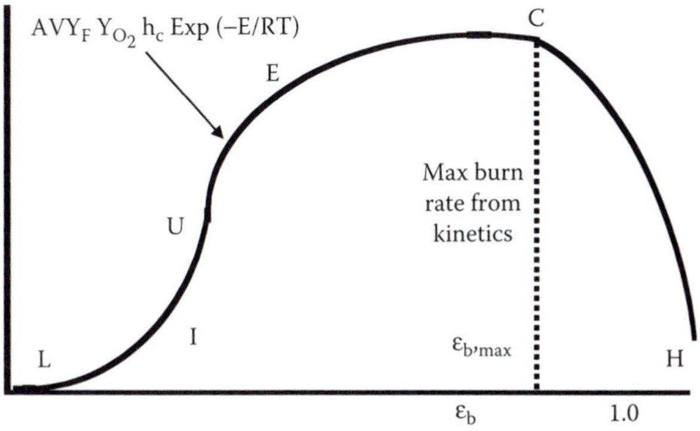

Figure 7.9 Depiction of the variation of heat release with burned fraction (representing the effect of temperature and of the product (YFYO2)).

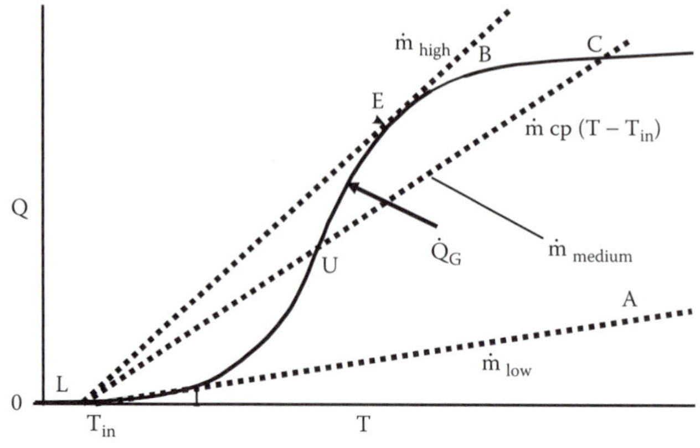

Figure 7.10 Heat generation and sensible heat loss curves.

At medium flow rates there are three solutions, L, U, and C, where C represents high burned fraction and, therefore, high temperature. The point L represents a low temperature where the reaction rate is almost zero. The middle solution U is unstable as a slight positive disturbance in ε_b causes a heat generation rate higher than the heat loss rate, resulting in flaming at point C, whereas a slight negative disturbance causes quenching at point L. As the mass flow is reduced, we continue to obtain three solutions until point I is reached (which is also known as the ignition point), where there are only two solutions. Similarly, as the mass flow rate is increased from a medium to a high value, there are three solutions until point E (also called the extinction point) is reached where we have two solutions. Figure 7.10 can be expressed in terms of $\dot{m}$ or $1/t_{res}$, where $t_{res} = \{V\rho_{in}/\dot{m}\} = \{V\, Y_{COin}\, \rho_{in}/\dot{m}_{CO,in}\}$. Results for ε_b vs. Da_I ($= t_{res}/t_{chem}$) can be obtained using the following example.

Example 6

Obtain a mathematical expression for the burned fraction in terms of the Damköhler number Da_I ($= t_{res}/t_{chem}$, i.e., the ratio of the characteristic residence time to the chemical time) if the overall CO oxidation occurs at the rate

$$|\dot{w}'''_{CO}| = A_Y \exp(-E/RT)\,(\rho)^{\alpha+\beta}\,(Y_{CO})^{\alpha}(Y_{O2})^{\beta},\ \text{kg/(m}^3\text{ s).} \qquad \text{(A)}$$

Solution

For the current example with CO as fuel, the term $\dot{w}'''_{CO,m}V$ represents the mass burn rate of fuel, which is CO $= \varepsilon_b\, \dot{m}_{CO,in}$, and the burned rate of $\dot{m}'''_{co,\,burnt} = |\,w'''_{co,m}\,|$.

One can solve for Y_{CO} from Equation 7.99 and Y_{O2} from Equation 7.100 in terms of burned fraction ε_b and $Y_{CO,in}$, and use the result in Equation 7.103b; then

$$\varepsilon_b = (A_Y t_{res} \nu_{O2}{}^{\beta} Y_{CO,in}{}^{\alpha+\beta-1} \rho_{in}{}^{\alpha+\beta-1}) \exp(-E/RT)\,(\rho/\rho_{in})^{\alpha+\beta}\,(1-\varepsilon_b)^{\alpha}(1/\phi-\varepsilon_b)^{\beta}. \qquad \text{(B)}$$

Let us define the chemical time through the expression,

$$t_{chem} = 1/(A_Y\ Y_{CO,in}{}^{\alpha+\beta-1}\rho_{in}{}^{\alpha+\beta-1}) \quad (C)$$

so that the first Damköhler number for lean mixtures

$$Da_I = t_{res}/t_{chem} = \{A_Y t_{res} \nu_{O2}{}^{\beta} Y_{CO,in}{}^{\alpha+\beta-1}\} = \{A_Y \nu_{O2}{}^{\beta}\ Y_{CO,in}{}^{\alpha+\beta}\rho_{in}{}^{\alpha+\beta-1}\}(V\rho_{in}/\ \dot{m}\ Y_{CO,in}\}. \quad (D)$$

If mixture is rich, the $\nu_{O2}\ Y_{CO,in}$ must be replaced by $Y_{O2,in}$.
Using Equation D in Equation B,

$$\varepsilon_b = Da_I \exp(-E/RT)\ (\rho/\rho_{in})^{\alpha+\beta}\ (1 - \varepsilon_b)^{\alpha}(1/\phi - \varepsilon_b)^{\beta}. \quad (E)$$

where T is solved in terms of ε_b using Equation 7.109:

$$T = T_{in} + \{h_{c,CO}\ \varepsilon_b\ Y_{CO,in} - \dot{Q}_{cv}/\ \dot{m}\ \}/c_p \quad (F)$$

Thus, the burned fraction ε_b is a function of Da_I, which contains the effects of kinetics and residence times, input fuel mass fraction, stoichiometry, etc. If one plots ε_b vs. Da_I, then blowoff or extinction occurs if $Da_I < (Da_I)_{ext}$ or blow off. With Equation C and $(Da_I)_{ext}$ the extinction flow rate is determined. Note that $\alpha = 1$, $\beta = 1$, $\dot{m}''_{ext} \propto VP^2$.

■ ■ ■

Example 6 shows that an ensemble of possible solutions for ε_b vs. Da_I (e.g., various residence times or mass flow rates) yields an S curve shape (Figure 7.11). At medium residence times, there are three solutions (L, U, C). At very low mass flow rates or long residence times there is one upper solution (S)

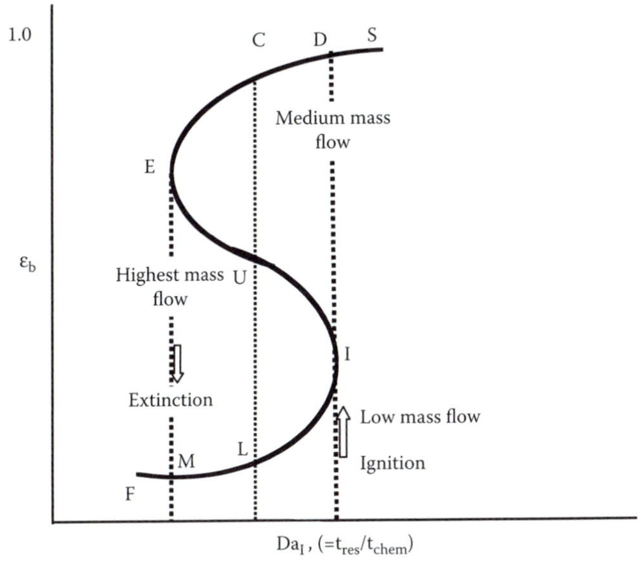

Figure 7.11 The S curve for a PSR.

and one lower solution (F, almost frozen) is obtained at very high mass flow rate corresponding to short times. At particular flow rates called extinction flow rate (or extinction residence time), there are are two solutions, E and M. If the PSR is already burning (point C) and if the mass flow rate is increased, the t_{res} decreases, Da_I decreases, and eventually point E is reached below which there is extinction; i.e., the residence time is much smaller than the chemical reaction time. Conversely, there exists two solutions I and D corresponding to ignition conditions and, hence, ignition of a cold flow requires a low flow rate or long residence time [Strehlow, 1984].

Example 7

$C_{12}H_{26}$ is fired into a quasi-cylindrical gas turbine combustor of 12 cm diameter and 20 cm length. The conditions are $\phi = 1$, $E = 0.15 \times 10^6$ kJ/kmol, $h_c = 44360$ kJ/kg, $T_{in} = 600$ K, $c_p = 1.3$ kJ/kg K, P = 10 bar, and the mixture molecular weight is 28.97 kg/kmol. Assume that $|\dot{w}'''_{C12H26}| = A_Y \exp(-E/RT)$ $(\rho)^{\alpha+\beta} (Y_{CO,e})^{\alpha}(Y_{O2,e})^{\beta}$ kg/m³ sec, where $\alpha = 1$, $\beta = 1$, $A_Y = 10^9$ m³/kg sec, and $t_{res} = (V\ \rho_{in}/\ \dot{m})$. Plot ε_b vs. t_{res} and determine the maximum possible flow without extinction.

Solution

For stoichiometric combustion,

$$C_{12}H_{26}\,(g) + 18.5\ O_2 + 18.5 \times 3.76\ N_2 \rightarrow 12\ CO_2 + 13\ H_2O + 18.5 \times 3.76\ N_2$$

$$\nu_{O_2} = 18.5 \times 32/(144.12 + 26.36) = 3.48 \text{ kg } O_2 \text{ per kg of fuel, and} \qquad (A)$$

$$Y_{F,in} = (1 + (\nu_{O_2}/\phi)(100/23))^{-1} = 0.062. \qquad (B)$$

Now, $Da_I = t_{res}/t_{chem} = (A_Y\ t_{res}\ \nu_{O_2}^{\beta}\ Y_{F,in}^{\alpha+\beta-1}\ \rho_{in}^{\alpha+\beta-1})$, and assuming M = 28.97,

$$\rho_{in} = (PM/RT) = 10 \times 28.97 \times 100/(8.314 \times 600) = 5.81 \text{ kg/m}^3.$$

The variation of the burned fraction CEU with t_{res} is presented in Figure 7.12, whereas the corresponding temperatures CEU are shown in Figure 7.13. The solutions correspond to the branch CEU, where the point E represents the lowest (critical) residence time in the reactor below which extinction occurs.

Similarly, the curve LIU in Figure 7.14 and Figure 7.15 presents the variation of the burned fraction and the corresponding temperatures at very large residence times (i.e., for low mass flow rates). These are the lower branch solutions. The point I represents the ignition condition for the reactor. The branch IU, if extended, will meet with point E, i.e., Figure 7.12 and Figure 7.14 form a composite S curve like the one presented in Figure 7.11.

As the volume

$$V = L\ \pi\ d^2/4 = 0.00226 \text{ m}^3,$$

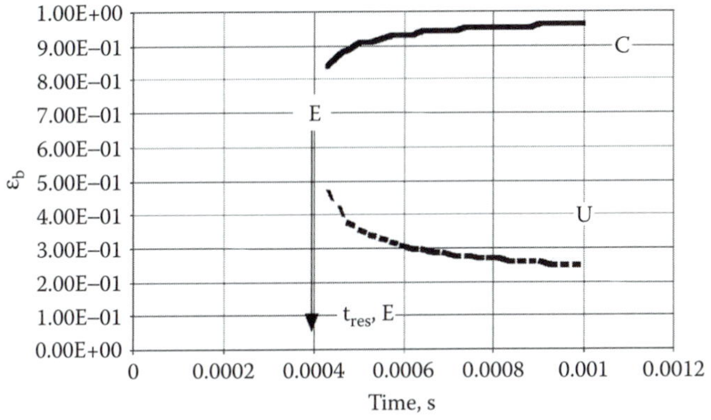

Figure 7.12 The burned fraction with respect to residence time for the extinction through the high-burn regime.

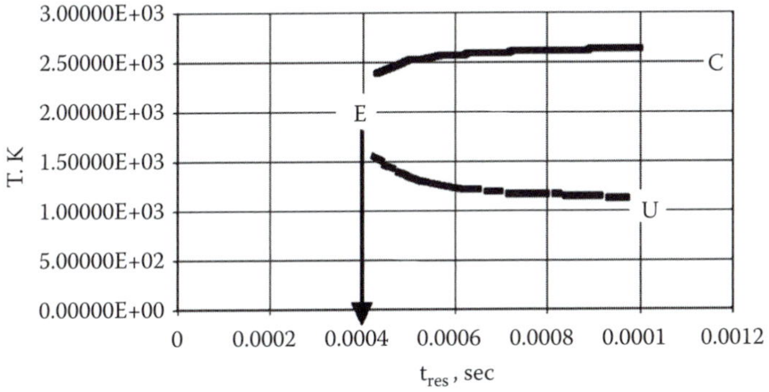

Figure 7.13 Temperature with respect to t_{res} for the extinction through the high-burn regime.

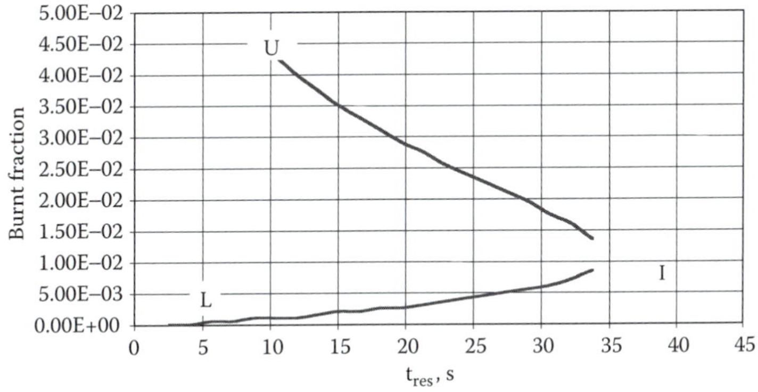

Figure 7.14 The burned fraction with respect to residence time for the ignition through the low-burn regime.

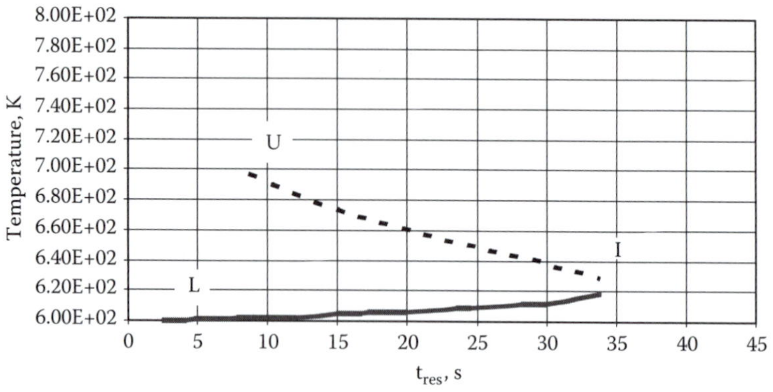

Figure 7.15 Temperature with respect to t_{res} for the ignition through the low-burn regime.

the highest possible flow before extinction occurs (at the lowest possible residence time)

$$t_{res,E} = (V\ \rho_{in}/\ \dot{m}\ ) = 0.0004 \text{ s}, \quad \text{i.e.,}$$

$$\dot{m} = 0.00226/(5.81 \times 0.0004) = 0.92 \text{ kg/sec.}$$

Therefore,

$$t_{chem} = 2.65 \times 10^{-9} \text{ sec.}$$

For this problem, at extinction,

$$Da_I = 0.0004/2.65 \times 10^{-09} = 1.51 \times 10^5.$$

Similarly, the reactor can be ignited when $t_{res,I} = 35$ sec, i.e.,

$$\dot{m} = 0.00226/(5.81 \times 35) = 1.11 \times 10^{-5} \text{ kg/sec}$$

i.e., when at ignition,

$$Da_I = 35/2.65 \times 10^{-9} = 1.32 \times 10^{10}$$

Remarks

(i) The value of ϕ in a gas turbine is very low. Consequently, the value of A:F is very large. This leads to even smaller values of the fuel mass flow rate before extinction occurs. Thus, there is maximum airflow beyond which the combustor cannot operated stably. Experiments are performed at fixed airflow while varying fuel flow; thus, fuel air ratio at blowoff or extinction can be determined. Figure 7.16 shows stability loop of a gas turbine, which is similar to the plots shown for the PSR model.

(ii) It is seen that $Da_I \propto p^2$ if $\alpha = 1$, $\beta = 1$ (Equation D). Thus, jet engines operating at higher altitudes will have lower Da_I. If an extinction occurs during flight, then one must reduce speed, which will increase t_{res} and fly

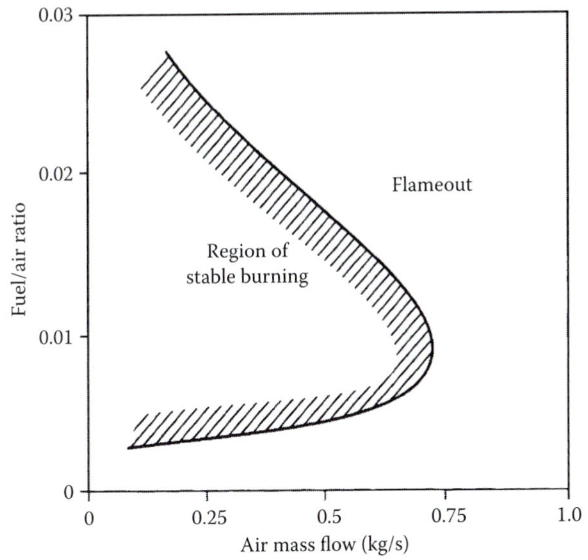

Figure 7.16 Effect of air mass flow rate on stable operating regime at constant pressure. (Adapted from Lefebvre, A., *Gas Turbine Combustion*, Taylor and Francis, London,1983.)

to a lower altitude, which will increase P and enable easier ignition. If Da_I at extinction is fixed, Equation D shows that $\dot{m}/\{VP^2\} \propto \nu_{O2} Y_{CO,in}$ for lean mixtures and $\propto Y_{O2,in}$ for richer mixtures.

(iii) Combustion intensity $\dot{H}'''$ is defined as

$$\dot{H}''' = \frac{\dot{m} h_c}{V} = \frac{\rho_{in} h_c}{t_{res}}$$

For pressure jet atomized gas turbines $\dot{H}''' \approx 10\text{–}100$ Mw/m^3, while for pulverized fuel (pf) fired boilers, $\dot{H}''' \approx 0.01 - 0.1$ Mw/m^3.

7.4.4.3.3 Multiple Reactions

The energy equation must be modified in case multiple reactions are considered, i.e.,

$$V \sum_j \dot{w}_j''' \bar{h}_{c,j} \approx -\dot{Q}_{CV} + \sum_k \dot{N}_{k,in} (\bar{h}_{k,T,in} - \bar{h}_{T,k}) \tag{7.110}$$

where the summation Σ_j is over all reactions and Σ_k is over all species.

7.5 Solid Carbon Combustion

We had applied the combustion concepts for gaseous fuels. However, most of the power plants use coal as fuel for power generation. As shown in

Chapter 4, the dominant component in coal is carbon. Thus, combustion time of coal is controlled by combustion time of carbon. Here simple approximate solutions are presented for combustion rate, change of diameter with time, and combustion time. Rigorous analysis is presented in Chapter 9.

7.5.1 Diffusion Rate of Oxygen

Consider hot carbon or charcoal burning in air (e.g., charcoal in a barbeque grill). A gas film surrounds the carbon surface during combustion, and it supplies oxygen through diffusion as described in Chapter 6. As the reaction between carbon and oxygen proceeds, the O_2 concentration decreases at the particle surface, which establishes an oxygen concentration gradient (Figure 7.17a). Fick's law can be applied to determine the rate of oxygen diffusion to the surface. The diffusion rate can also be expressed using a mass transfer coefficient h_m that is analogous to the heat transfer coefficient. With such an analogy, the transport (or diffusion) rate of O_2 through the gas film is

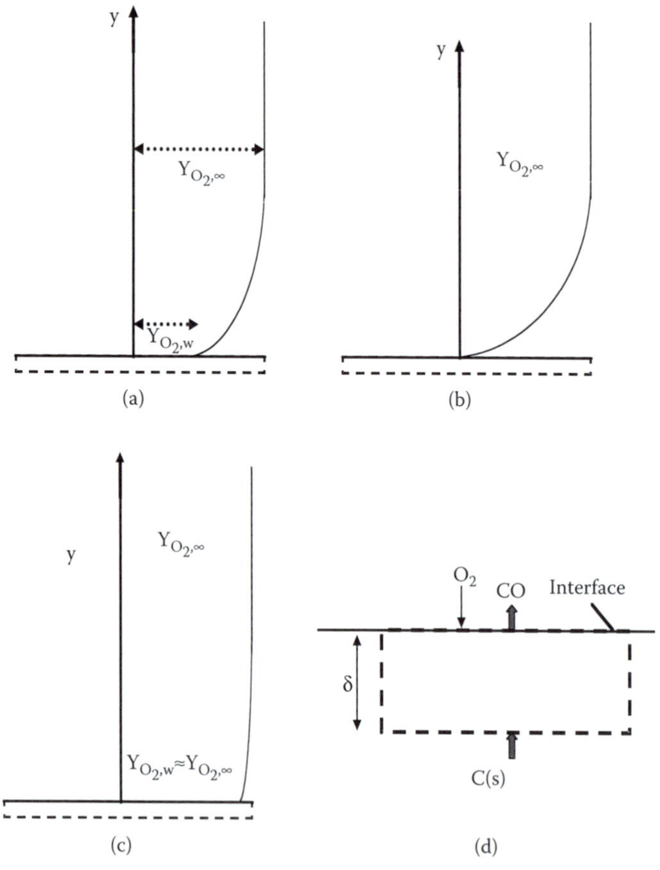

Figure 7.17 Various regimes of burning of carbon: (a) mixed control (diffusion + kinetics), (b) diffusion control, (c) kinetics control.

$$\dot{m}''_{O_2} = h_m(Y_{O2,\infty} - Y_{O2,w}) \tag{7.111}$$

where h_m is given by

$$Sh = h_m\ L/(\rho D). \tag{7.112}$$

L is the characteristic length, L = d for spherical particle, and Sh is the Sherwood number.

7.5.2 Burn Rate

If carbon reaction occurs in the solid within a small depth δ from surface (thus, within the depth δ and over elemental area dA), then carbon consumption rate within $A\delta$ is given as

$$d\,\dot{w} = \dot{w}'''\ dA\ \delta \tag{7.113}$$

$$\dot{w}''\ (\text{kmol/m}^2\ \text{sec}) = d\,\dot{w}\ /dA = \dot{w}'''\ \delta. \tag{7.114}$$

Typically, we measure the regression rate of surface $\dot{w}''$ and, hence, we write Arrhenius equation for $\dot{w}''$ (kmol/m² s) rather than for $\dot{w}'''$. If reaction is $C(s) + 1/2\ O2 \rightarrow CO$, then on a mass basis,

$$\frac{\dot{w}''_{O2,m}}{-1.333} = \frac{\dot{w}''_{C,m}}{-1} = \frac{\dot{w}''_{CO,m}}{+2.333} = \dot{w}''_m\left(\frac{kg}{m^2 s}\right). \tag{7.115}$$

For global first-order reaction, the chemical reaction term $\dot{w}''_m$ is expressed in mass units (e.g., kg/m² sec) as

$$\dot{w}''_m = k_{(\ )}\ \rho\ Y_{O2,w} \tag{7.116}$$

where $k_{(\ )} = A_{(\ ),het}\ \exp\ (-E/RT_w)$ has units of length over time (e.g., m/sec), and the subscript "het" represents the heterogeneous reaction between the gaseous and solid phases. We know at the carbon surface

$$[O_2]_w = p_{O2,w}/(\bar{R}\ T_w) = (p_{O2,w}/P)\ \{P/(\bar{R}\ T_w)\} = X_{O2,w}\ c_w, \tag{7.117}$$

$$\text{where } c_w = P/(\bar{R}\ T_w) = \rho/M_{mix}. \tag{7.118}$$

The following examples illustrate the use of the mass, species, and energy conservation relations to determine the burn rate of solid carbon.

Example 8

A spherical carbon particle burns in air, but a feeding mechanism introduces additional carbon mass so that it is possible to maintain a fixed carbon particle diameter. The carbon mass is introduced at the particle temperature. The oxygen transfer rate to solid carbon surface is expressed as

$$\dot{m}_{O2,i} = h_m\ A\ (Y_{O2,\infty} - Y_{O2,w}) \tag{A}$$

where $Y_{O2,\infty}$ denotes the free-stream oxygen mass fraction and $Y_{O2,w}$,the surface oxygen mass fraction (Figure 7.17a). (a) Derive an expression for the feed rate of carbon to maintain the particle diameter, and (b) the temperature difference ($T_w - T_\infty$) if the heat loss from the particle to the ambient is provided by the relation h_H A ($T_w - T_\infty$). Neglect the influence of thermal radiation.

Solution

(a) Consider any solid carbon surface not necessarily spherical. Oxygen is transported to the surface (Figure 7.17a) which is used by carbon to produce CO with the following overall chemical reaction:

$$C\ (s) + 1/2\ O_2 \rightarrow CO \qquad (I)$$

1 kg of C + $\nu_{O2,I}$ kg of oxygen → (1 + $\nu_{O2,I}$) kg of CO,
where $\nu_{O2,I}$ = 1.333 kg of oxygen per kg of carbon for Reaction (I).

At carbon interface (Figure 7.17d),

$$dm_{cv,C(s),}/dt = \dot{m}_{C(s),i} - \dot{m}_{C(s),e} + \dot{w}_{C(s),} \qquad (B)$$

$$dm_{cv,O2,}/dt = \dot{m}_{O2,i} - \dot{m}_{O2,e} + \dot{w}_{O2} \quad \text{and} \qquad (C)$$

$$dm_{CO,}/dt = \dot{m}_{CO,i} - \dot{m}_{CO,e} + \dot{w}_{CO} \qquad (D)$$

where $\dot{w}_{k,m} = \dot{w}''_{k,m}A$ where A is interface area.

We hypothesize that interface is maintained at the same level by feeding the carbon upward as it gasifies. There is no carbon (s) species leaving interface. At steady state, Equation B to Equation D become

$$0 = \dot{m}_{C(s),i} + \dot{m}_{C(s),chem.} \qquad (E)$$

$$0 = \dot{m}_{O2,i} + \dot{m}_{O2,chem.} \qquad (F)$$

$$0 = -\dot{m}_{CO,e} + \dot{m}_{CO,chem.} \qquad (G)$$

Applying the law of stoichiometry,

$$(dm_{C(s),}/dt)/(-1) = (dm_{O2,}/dt)/(-\nu_{O2,I}) = (dm_{CO,}/dt)/\nu_{CO,I} = \dot{w}_m \qquad (H)$$

where $\nu_{CO,I}$ = (1 + $\nu_{O2,I}$). Using Equation H in Equation E to Equation G,

$$0 = \dot{m}_{C(s),i} - \dot{w}_m \qquad (I)$$

$$0 = \dot{m}_{O2,i} - (-\nu_{O2,I})\,\dot{w}_m \qquad (J)$$

$$0 = \dot{m}_{CO,e} + (1 + \nu_{O2,I})\,\dot{w}_m \qquad (K)$$

From Equation J and Equation A,

$$\dot{w}_m = \dot{w}_{O2,i}/\nu_{O2,I} = h_m\, A\, (Y_{O2,\infty} - Y_{O2,w})/\nu_{O2,I}. \qquad (L)$$

Then, using Equation L in Equation I, the required carbon feed rate or carbon burn rate is given as

$$\dot{m}_{C(s),i} = \dot{w}_m = h_m\, A\, (Y_{O2,\infty} - Y_{O2,w})/\nu_{O2,I} \qquad (M)$$

where Y_{O2w} is still an unknown. Also for a first heterogeneous reaction (Chapter 5),

$$\dot{w}''_m = k\, \rho\, Y_{O2,w}. \qquad (N)$$

(b) We will apply the first law of thermodynamics in simplified form to determine the temperature difference, namely,

$$(dEc,v/dt) = \dot{Q}_{cv} - \dot{W}_{cv} + \sum_i \dot{m}_k h_k - \sum \dot{m}_k h_k$$

where we select small thickness near the interface as c.v. At steady state with no work transfer, $(dE_{c.v.}/dt) = \dot{W}_{cv} = 0$, $\dot{Q}_{cv} = -h_H\, A\, (T_w - T_\infty)$.
Thus, energy conservation simplifies to

$$h_H\, A\, (T_w - T_\infty) = \{\dot{m}_{CO,i} h_{CO,i} + \dot{m}_{O2,i} h_{O2,i} + \dot{m}_{C(s),i} h_{C(s),i}\} - \{\dot{m}_{CO,e} h_{CO,e}$$

$$+ \dot{m}_{O2,e} h_{O2,e}\, \dot{m}_{C(s),e} h_{C(s),e}\}.$$

O_2 enters the cv from the gas phase and CO exits the c.v., whereas C(s) enters from the solid phase.

$$h_H\, A\, (T_w - T_\infty) = \{0 + \dot{m}_{O2,i} h_{O2,i} + \dot{m}_{C(s),i} h_{C(s),ei}\} - \{\dot{m}_{CO,e} h_{CO,e}\}.$$

$\dot{m}_{CO,e}$, $\dot{m}_{O2,i}$ and $\dot{m}_{C(s),I}$ are known from Equation 7I to Equation 7K, then

$$h_H\, A\, (T_w - T_\infty) = \dot{w}_m\{h_{c(s),I} + \nu_{O2,I}\, h_{O2}\} - \dot{w}_m\, \{(1 + \nu_{O2,I})\, h_{CO}\}. \qquad (O)$$

Using $h = h_{f,k} + h_{t,k}$,

$$h_H\, A\, (T_w - T_\infty) = \dot{w}_m\, \{h_{c,I} + \{h_{t,C} + \nu_{O2,I}\, h_{t,O2} - (1 + \nu_{O2,I})\, h_{tCO})\} \qquad (P)$$

where $h_{c,I} = \{\nu_{O2,I}\, h_{f,O2} + h_{f,C(s)} - (1 + \nu_{O2,I})\, h_{f,CO}\}$ is the heating value for Reaction I in kJ/kg of carbon at 298 K. If all specific heats are equal or if $\nu_{O2,I} h_{t,O2} + h_{t,C} - (1 + \nu_{O2,I})\, h_{tCO} << h_{c,I}$, then

$$h_H\, A\, (T_w - T_\infty) \approx \dot{w}_m h_{c,I} = h_m A (Y_{O2,\infty} - Y_{O2,w})\, h_{c,I}/\nu_{O2,I} \qquad (Q)$$

where $h_H L/\lambda = Nu$, $h_m L/\rho D = Sh$ and L, characteristic size. If we assume that the Lewis number is unity, i.e., $D = \lambda/\rho c_p$, or $h_H = h_m$, then the Nusselt and Sherwood numbers are identical. Using this definition and with characteristic size $L = d$,

$$h_H = (Nu\ \lambda/d) = Nu\ \rho D\ c_p/d = Sh\ \rho D\ c_p/d,$$

$$(Sh\ \rho D\ c_p/d)\ A\ (T_w - T_\infty) \approx (Sh\ \rho D/d)\ A\ (Y_{O2,\infty} - Y_{O2,w})\ h_{c,I}/\nu_{O2,I} \qquad (R)$$

Hence,

$$(T_w - T_\infty) \approx \{Y_{O2,\infty} - Y_{O2,w}\}\{h_{c,I}/c_p\}/\nu_{O2,I} \qquad (S)$$

Remarks

For the reaction

$$C + O_2 \rightarrow CO_2, \qquad (II)$$

1 kg of C + ν_{II} kg of oxygen → (1 + ν_{II}) kg of CO_2 so that $\nu_{II} = 2.666$, and

$$\dot{m}_{C,i} = h_m\ A\ (Y_{O2,\infty} - Y_{O2,w})/\nu_{O2,II}$$

i.e.,

$$(T_w - T_\infty) \approx \{Y_{O2,\infty} - Y_{O2,w}\}(h_{c,II}/c_p)/\nu_{O2,II}$$

Other reactions can also be similarly considered, e.g.,

$$C(S) + CO_2 \rightarrow 2\ CO \qquad (III)$$

and

$$C(S) + H_2O \rightarrow CO + H_2 \qquad (IV)$$

If thermal radiation loss is considered, then $\dot{Q}_{cv} = -\{h_H\ A\ (T_w - T_\infty) - \varepsilon\sigma\ A\ (T_w^4 - T_\infty^4)\}$. An iterative procedure can be employed to determine the particle temperature.

Example 9

(a) Determine the value of $\dot{m}_C$ in Example 8 if $Y_{O2,w} \approx 0$ or very small, Sh = 2, d = 100 μm, $\rho D = 2.5 \times 10^{-4}$ m²/sec, and $Y_{O2,\infty} = 0.23$.

(b) Determine the particle temperature if $T_\infty = 1000$ K. Assume that Nu = Sh (i.e., $Le = \lambda/\rho D c_p = 1$), and consider the value of c_p as that for N_2 at 1500 K.

(c) The particle burns in air, but there is no carbon feed. Obtain an expression for d^2 vs. t.

(d) Estimate the burn time. Assume particle density $\rho_p = 700$ kg/m³.

Solution

(a) $Sh = h_m d/\rho D = 2$, i.e., $h_m = 2 \times 2.65 \times 10^{-5}/100 \times 10^{-6} = 0.5$ kg/m² sec. Area $= \pi\, d^2 = 3.1416 \times 10^{-8}$ m². The burn rate is given as

$$\dot{m}_C = -\dot{w}_m = h_m\, A\, (Y_{O2,\infty} - Y_{O2,w})/\nu_{O2,I} =$$

$$\{0.5\ (\text{kg/m}^2\ \text{sec}) \times 3.1416 \times 10^{-8}\ \text{m}^2 \times 0.23\}/1.333 = 0.271 \times 10^{-8}\ \text{kg/sec.} \qquad \text{(A)}$$

(b) At 1500 K, for nitrogen $c_p = 1.242$ kJ/kg K, i.e.,

$$(T_w - T_\infty) \approx (0.23/1.333)\ (9204\ \text{kJ/kg}/1.242\ \text{kJ/kg K}) = 1279\ \text{K.} \qquad \text{(B)}$$

$$T_w = 2279\ \text{K}$$

(c) If there is no feeding, then taking control volume between center and surface of particle $dm_C/dt = \dot{m}_{C,i} - \dot{m}_{C,e} + \dot{m}_{C,chem.} = 0 - \dot{w}_m$.

The mass m_c changes with time. Because the density is constant, a decrease in the mass or consumption of carbon results in a decrease in the diameter.

$$dm_{C,}/dt = -\dot{w}_m = h_m A\ (Y_{O2,\infty}/\nu_{O2,I}) = Sh\ \rho D \pi\ d\ (Y_{O2,\infty}/\nu_{O2,I}).$$

Using the relation $d = 2a$,

$$-4\,\pi\, a^2\, \rho_p\, (da/dt) = Sh\ \rho D\ (2a)\ Y_{O2,\infty}/\nu_{O2,I}, \quad \text{or} \quad -2\pi a\, \rho_p\, (da/dt) = Sh\ \rho D\ Y_{O2,\infty}/\nu_{O2,I}$$

Upon integrating with the initial condition $a = a_0$ at $t = 0$,

$$a_o^2 - a^2 = (Sh\ \rho D B/\rho_p)t \qquad \text{(C)}$$

$$\text{where}\quad B = (Y_{O2,\infty}/\nu_{O2,I}) \qquad \text{(D)}$$

Therefore, the corresponding relation for the particle diameter is

$$d_0^2 - d^2 = \alpha_c\, t \qquad \text{(E)}$$

which is known as the d² law. The parameter $\alpha_c = 4\ Sh\ \rho D\ B/\rho_p$ is known as the burning rate constant and has units of m²/sec.

(d) The burn time t_b is obtained by setting $d = 0$ in Equation E:

$$t_b = d_0^2/\alpha_c \qquad \text{(F)}$$

$$B = 0.17,\ \alpha_c = 4\ Sh\ \rho D\ B/\rho_p = 4.9 \times 10^{-8}\ \text{m}^2/\text{sec}$$

$$t_b = 10^{-8}/4.9 \times 10^{-8} = 0.21\ \text{sec}$$

■ ■ ■

The above example above illustrates that d² law is followed when carbon burns in air and burn time is proportional to d_0^2. Further, if we assume that volatile

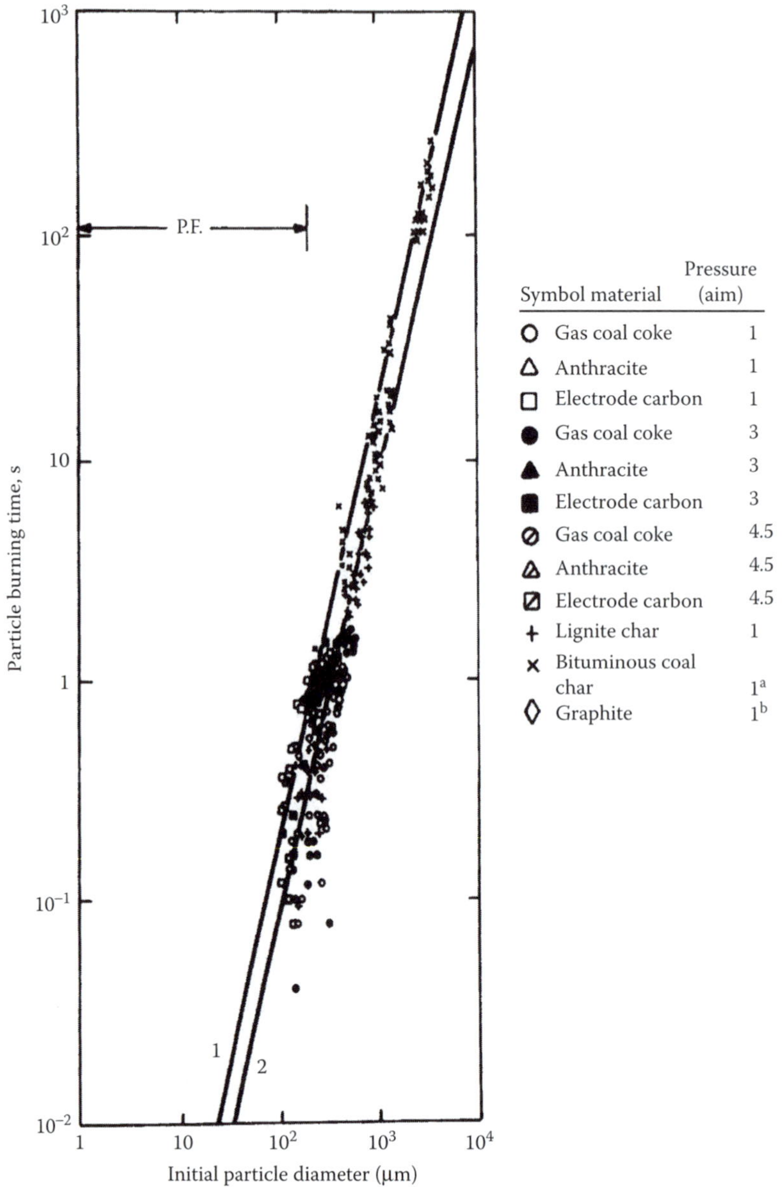

Figure 7.18 Char and coal burnout times in air at 1500 K; curve 1: bulk density: 2000 kg/m^3; curve 2: 1000 kg/m^3. (Adapted from Essenhigh, R.H., *Chemistry Coal Utilization*, M.A. Elliott, Ed., 2nd Suppl., John Wiley & Sons, New York, 1981, p. 1153.)

matter (VM) burns *in situ* along with carbon in the proportion (VM/FC), then we can extend the anlaysis to coal; thus, coal burn rate = carbon burn rate/FC where FC and VM are on DAF basis; again, the d^2 law is followed. The data in Figure 7.18 validates that the burn time $t_b \propto d_0^2$.

Under finite chemistry, carbon combustion can be handled using the resistance concept; the total resistance includes both chemical (R_{chem}) and diffusional resistances and (R_{diff}). See the following example.

Example 10

Consider Example 8. Obtain expressions for the oxygen mass fraction and burn rate when (a) k is finite, (b) $k \to \infty$, and (c) $k \to 0$.

Solution

Oxygen is transported by diffusion from the bulk gas phase. Under steady state, the diffusion rate equals the reaction rate, i.e.,

$$h_m(Y_{O_2,} - Y_{O_2,w}) = \left|\dot{w}_m \nu_{O2}\right| = k\, Y_{O_2,w}\rho_w \nu_{O2} = k_{O_2} Y_{O_2,w}\rho_w \quad \text{(A)}$$

where $k_{O2} = k\ \nu_{O2}$, and $k = A_{het} \exp(-E/RT_w)$. Therefore, solving for $Y_{O2,w}$ from Equation A,

$$Y_{O_2,w} = h_m Y_{O_2,\infty}/(k_{O_2}\rho_w + h_m) = Y_{O_2,}/(k_{O_2}\rho_w/h_m + 1)\,. \quad \text{(B)}$$

Defining the Damkohler number for solid carbon reaction,

$$Da_{III} = k_{O_2}\, \rho_w/h_m, = \rho_w\, \nu_{O_2}\, A_{het} \exp(-E/RT_w)/h_m \quad \text{(C)}$$

$$Y_{O_2,w} = \frac{Y_{O_2,\infty}}{Da_{III} + 1}\,. \quad \text{(D)}$$

The oxygen consumption rate at the wall $|\dot{w}''_{O_2}| = k_{O_2} Y_{O_2,w}\rho_w$. From the O_2 species balance at the interface (Equation A and Equation B):

$$\dot{m}''_{O2} = k_{O2}\, \rho_w\, Y_{O2,w} = \frac{Y_{O_2,\infty}}{\frac{1}{k_{O2}\rho_w} + \frac{1}{h_m}} = \frac{Y_{O_2,\infty}}{R_{chem} + R_{Diff}} \quad \text{(E)}$$

where

$$R_{chem} = \frac{1}{k_{O2}\rho_w},\ R_{Diff} = \frac{1}{h_m}\,. \quad \text{(F)}$$

Equation E looks like current flow in electrical engineering with voltage replaced by free-stream oxygen mass fraction and current flow by oxygen flow. Dividing Equation E by $\nu_{O2,I}$, the carbon burn rate is given as

$$\dot{m}''_c = \frac{B}{R_{chem} + R_{Diff}} = \frac{B}{R} \quad \text{(G)}$$

where $B = Y_{O2,\infty}/\nu_{O2,I}$ is the driving potential or transfer number for carbon burn rate and $R = R_{chem} + R_{diff}$. Figure 7.19 illustrates the resistance concept.

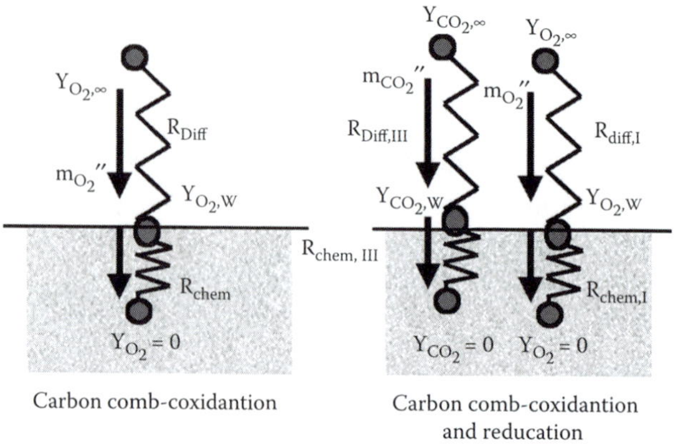

Figure 7.19 Resistance concept: carbon combustion : $R_{diff,j} = 1/h_{m,j}$ $R_{chem,j} = 1/(k_j\ \rho_w)$, j = O_2, CO_2.

Alternatively,

$$\frac{\dot{m}_c''}{h_m} = \frac{B}{\left\{\dfrac{1}{Da_{III}}\right\} + 1} \tag{H}$$

where $Da_{III} = \dfrac{R_{diff}}{R_{chem}}$

Remarks

(i) The burn rate $\dot{m} = \dot{m}''' * A_R$, where A_R is the reaction surface area.

(ii) Diffusion-controlled combustion: When $k_{O_2} \to \infty$, $Da_{III} \to \infty$ (e.g., at high temperatures), and $R_{chem} \to 0$. In this case $Y_{O2,w} \to 0$, as the oxygen is consumed very rapidly at the carbon surface (Figure 7.17b). The oxygen consumption rate is given by,

$$|\ \dot{w}_{O2}''\ | = h_m\ (Y_{O2,\infty} - Y_{O2,w}) \approx h_m\ Y_{O2,\infty}$$

and oxygen consumption is controlled by O_2 diffusion. Under this diffusion-limited condition when $Da_{III} = (\rho_w\ k_{O_2})/h_m = (R_{diff}/R_{chem}) \to \infty$, the carbon consumption is

$$\dot{m}_C'' = h_m\ B,\ kg/m^2sec \tag{I}$$

(iii) Kinetics-controlled combustion: When $k_{O_2} \to 0$ (e.g., at low temperatures), $R_{chem} \to \infty$, $Da_{III} \to 0$, and the oxygen consumption rate is controlled by chemical kinetics. In this case, $Y_{O2,w} \approx Y_{O2,\infty}$ and

$$\dot{m}_{O_2}'' = \dot{w}_{O_2}'' = -k_{O_2} Y_{O_{2,w}} \rho_w = -\ k_{O_2} Y_{O_2,}\ \rho_w.$$

With $B = Y_{O_2,\infty}/\nu_{O2}$, the corresponding carbon consumption rate is

$$\dot{m}_C'' = \rho_w\ k_{O_2}\ B. \tag{J}$$

7.5.3 Sherwood Number Relations for Mass Transfer

The heat transfer literature suggests that Nu = h (2a)/$\lambda \approx 2$ for spheres in a stagnant atmosphere. In analogy, the Sherwood number Sh for a spherical particle transporting mass in a stagnant atmosphere is also 2, namely,

$$Sh = h_m\, d_p/(\rho D) = 2, \quad \text{i.e.,} \quad h_m = 2\,\rho D/d_p = \rho D/a. \tag{7.119}$$

A corresponding relation for a particle burning in a flowing atmosphere is

$$Sh = 2 + 0.6\; Re^{1/2}\; Sc^{1/3}, \tag{7.120}$$

with Re = Vd/ν, V is the relative velocity between gas and particle, and the Schmidt number Sc = ν/D. It is noted that as V $\rightarrow$ 0, Sh $\rightarrow$ 2. Equation 7.119 and Equation 7.120 are strictly valid for low mass transfer conditions. The burn rates of carbon are typically low, and hence one can predict the burning rate of carbon spheres with just the knowledge of the mass transfer coefficient, as long as the diffusion-controlled combustion approximation is valid.

Example 11

100-µm-diameter carbon spheres burn in a quiescent atmosphere under the conditions T_w = 2000 K, P = 1 bar, $D_w = 1 \times 10^{-4}$ m²/sec, and M_{mix} = 25 kg/kmol.

(a) Determine the values of R_{dif}, R_{chem}, and carbon burn rate.
(b) If R_{chem} is ignored compared to R_{diff} (called diffusion-controlled combustion limit), determine the burning time in sec.

Let k (m/sec) = 900 exp (–E/($\bar{R}$ T), where E = 62000 kJ/k mol. Assume C(s) + 1/2 O2 $\rightarrow$ CO and ρ_p = 700 kg/m³ for carbon.

Solution

(a) k_{O_2} = 600 exp (–62000/(2000 × 8.314) = 14.415 m/sec.

$$\rho_w = P\ M/\ \bar{R}\ T = 1 \times 25/(0.08314 \times 2000) = 0.15\ \text{kg/m}^3.$$

$\rho_w D_w = 0.15 \times 10^{-4}$ kg/m sec, and $R_{chem} = 1/(k_{O_2}\,\rho_w) = 0.463$ m² sec/kg.

Assuming Sh = 2, $h_m = 2\,\rho D/d_p = 2 \times 0.15 \times 10^{-4}/(100 \times 10^{-6}) = 0.3$ kg/m² sec, so that

$$R_{dif} = 1/h_m = 3.33\ \text{m}^2\ \text{sec/kg}.$$

Therefore, $\dot{m}''_{O2}$ = 0.23/(0.463 + 3.333) = 0.0605 kg/m² sec, which is analogous to the current flow (Figure 7.19a).

The carbon burn rate is given as 0.0605/1.333 = 0.046 kg/m² sec.

(b) It is clear that this is a diffusion-controlled problem as, $R_{dif} >> R_{chem}$. From Example 9,

α_c = 4 Sh ρD B/ρ_P = 4*2*1.5 × 10^{-5} *0.17/700 = 2.91 × 10^{-8} m²/sec, and t_b $= d_o^2/\alpha_c$ = 100 × 10^{-6}/2.91 × 10^8 = 0.343 sec (or 343 msec).

Example 12

Apply the resistance concept to determine the carbon consumption rate for a particle burning through heterogeneous reactions only, i.e., in a frozen gas phase. The normal heterogeneous reactions are

$$C + 1/2\ O_2 \rightarrow CO, \quad \text{(I)}$$

$$C + O_2 \rightarrow CO_2, \quad \text{(II)}$$

$$CO_2 + C \rightarrow 2\ CO, \quad \text{(III)}$$

$$C + H_2O \rightarrow CO + H_2. \quad \text{(IV)}$$

Ignore Reaction II.

Solution

Consider Reaction I and Reaction III (Figure 7.19b),

$$\dot{m}'' = \dot{m}''_{C,I} + \dot{m}''_{C,III}. \quad \text{(A)}$$

The O_2 and C flows due to reaction I are given by Equation E and Equation G in Example 10:

$$\dot{m}''_{c,I} = \frac{B_I}{R_{chem,I} + R_{Diff,I}} = \frac{B_I}{R_I}. \quad \text{(B)}$$

Similarly, if we consider Reaction III,

$$\dot{m}''_{c,III} = \frac{B_{III}}{R_{chem,III} + R_{Diff,III}} = \frac{B_{III}}{R_{III}}. \quad \text{(C)}$$

If both Reaction I and Reaction III occur simultaneously,

$$\dot{m}''_c = \dot{m}''_{c,I} + \dot{m}''_{c,III} = (\dot{m}''_{O2,I}/\nu_{O2,I}) + (\dot{m}''_{CO2,III}/\nu_{CO2,I}) \quad \text{(D)}$$

$$\dot{m}''_c = \frac{B_I}{R_I} + \frac{B_{III}}{R_{III}}, \quad \text{(E)}$$

Combinig Reaction I, Reaction III, and Reaction IV,

$$\dot{m}''_c = \frac{B_I}{R_I} + \frac{B_{III}}{R_{III}} + \frac{B_{IV}}{R_{VI}}, \quad \text{(F)}$$

where

$R_I = \{(1/(\rho_w k_{O2}))\} + (1/h_m)$, $R_{III} = \{1/(\rho_w k_{CO_2}) + 1/h_m\}$, $R_{IV} = (1/(\rho_w k_{H_2O}) + 1/h_m)$.

For diffusion-controlled combustion,

$$\dot{m}_C'' = h_m\{B_I + B_{III} + B_{IV}\} = h_m\ B$$

where

$$B = \{Y_{O_2\infty}/\nu_{O_2,I}\} + \{Y_{CO_2\infty}/\nu_{O_2,III}\} + \{Y_{H_2O\infty}/\nu_{H_2O,IV}\}.$$

7.5.4 Carbon Temperature during Combustion

The reaction at the particle surface results in the release of heat and subsequent heating of the mass. Consider the energy conservation relation

$$m\ c_p\ dT_p/dt = \dot{Q}_{gen} - \dot{Q}_{loss},\quad \text{where} \tag{7.121}$$

$$\dot{Q}_{gen} = \dot{m}''\ h_c\ A, \tag{7.122}$$

$$\dot{Q}_{loss} = \{h_H\ (T_p - T_\infty) + \varepsilon(T_p^4 - T_{rad}^4)\}A \tag{7.123}$$

where T_{rad}, background temperature for radiation loss,

$$\dot{m}'' = B/R_{chem}\ \text{for chemical control,} \tag{7.124}$$

$$\dot{m}'' = B\ R_{diff} = h_m\ B,\ \text{for diffusion control, and} \tag{7.125}$$

$$\dot{m}'' = B/(R_{chem} + R_{diff})\ \text{for mixed control.} \tag{7.126}$$

Assuming steady state ($dT_p/dt = 0$) and diffusion-controlled burning,

$$h_c h_m\ B = h_H\ (T_p - T_\infty) + \varepsilon\ (T_p^4 - T_{rad}^4). \tag{7.127}$$

If one neglects radiation loss (e.g., for small particles) and with $T_p = T_w$

$$c_p\ (T_w - T_\infty)/h_c = B. \tag{7.128}$$

Equation 7.127 and Equation 7.128 are valid whether a particle burns or a carbon surface burns (as in charcoal in a barbeque grill). For diffusion-controlled combustion, the inclusion of radiative loss requires an iterative solution. An iterative method is also required for mixed and chemical-controlled burning, as the chemical resistance is a function of T_p.

Many of the previous examples illustrate the solutions that could be obtained from integral formulation of species and energy equations for an open system. More detailed treatment involving differential forms of conservation equations and exact solutions to combustion of carbon will be given in Chapter 9.

7.6 Droplet Burning

Droplet burning is addressed in the following example through an analogy with carbon combustion. The example illustrates that pertinent solutions can be obtained from an integral formulation of the species and energy equations considering combustion in an open system. An exact solution of differential forms of the conservation equations is provided in Chapter 10.

Example 13

A gasoline drop could be approximated as an octane droplet. Consider the steady-state burning of a 100-μm liquid octane droplet in the air. As the droplet burns, a liquid feeding mechanism maintains its diameter. The liquid is injected at the droplet bulk temperature. The droplet vaporizes and releases vapors to the gas phase; vapor diffuses to a radius r_f where it reacts with O_2, which diffuses from the ambience. The surface at $r = r_f$ (called flame radius) is analogous to carbon burning in the previous example. A flame is established at a radius r_f, where the gas-phase temperature reaches its highest value and chemical resistance tends to zero at flame surface. Similar to carbon burning, oxygen ($Y_{O2,f}$) and fuel ($Y_{F,f}$) mass fractions at flame surface are assumed to be negligible. Assume that the oxygen transfer rate to the flame surface is described by the expression

$$\dot{m}_{O2,i} = h_m A_f (Y_{O2,\infty} - Y_{O2,f}) \quad \text{(A)}$$

where $Y_{O2,\infty}$ denotes the free-stream oxygen mass fraction; $Y_{O2,f}$, the flame surface oxygen mass fraction; and A_f, the flame surface area. Derive expressions for

(a) The fuel burn rate or the feeding rate of liquid that is required to maintain steady-state conditions.
(b) The temperature difference $(T_f - T_\infty)$ if the heat loss from the flame to the ambient is

$$h_H A_f (T_f - T_\infty). \quad \text{(B)}$$

(c) The burn rate.
(d) Determine the burning rate in kg/sec.

Neglect thermal radiation heat loss. Let $T_\infty = 300$ K, $Y_{O2,\infty} = 0.23$, $L = 362$ kJ/kg K, $h_c = 44350$ kJ/kg, $T_w = T_{BP} = 398$ K, $\upsilon_{O2} = 3.51$, $c_p = 1.1$ kJ/kg K, and $\rho D = 5 \times 10^{-5}$ kg/m sec.

Solution

(a) Oxygen is transported to the flame surface and oxidizes the fuel to CO_2 and H_2O through the chemical reaction occurring around the flame surface

$$C_8H_{18} + 12.5\ O_2 \rightarrow 8\ CO_2 + 9\ H_2O, \text{ i.e.,} \quad \text{(C)}$$

$$1 \text{ kg F} + \nu_{O2} \text{ of } O_2 \rightarrow \nu_{CO2}\, CO_2 + \nu H_2O\, H_2O = (1 + \nu_{O2}) \text{ kg of products},$$

where $\nu_{O2} = 12.5 \times 32/(114) = 3.5$, $\nu_{CO2} = 3.08$, and $\nu_{H2O} = 1.42$.

For an elemental thin volume around flame surface and with finite chemistry within this volume,

$$dm_{k,}/dt = \dot{m}_{k,i} - \dot{m}_{k,e} + \dot{w}_{k}.$$

With k = F, O_2, and P (products),

$$dm_{F,}/dt = \dot{m}_{F,i} - \dot{m}_{F,e} + \dot{w}_{F} \quad \text{(D)}$$

$$dm_{O2,}/dt = \dot{m}_{O2,i} - \dot{m}_{O2,e} + \dot{w}_{O2} \quad \text{and} \quad \text{(E)}$$

$$dm_{P,}/dt = \dot{m}_{P,i} - \dot{m}_{P,e} + \dot{w}_{P}. \quad \text{(F)}$$

At steady state, the mass of fuel, O_2 and products P within the control volume is unchanged, i.e.,

$$0 = \dot{m}_{F,i} + \dot{w}_{F} \quad \text{(G)}$$

where $\dot{m}_{F,i}$ is fuel injection rate and is the same as the burn rate

$$0 = \dot{m}_{O2,i} + \dot{w}_{O2} \quad \text{and} \quad \text{(H)}$$

$$0 = 0 - \dot{m}_{P,e} + \dot{w}_{P}. \quad \text{(I)}$$

Applying the law of stoichiometry,

$$\dot{w}_{F}/(-1) = \dot{w}_{O2}/(-\nu_{O2}) = \dot{w}_{P}/(1 + \nu_{O2}) = \dot{w}_{m}, \quad \text{i.e.,} \quad \text{(J)}$$

where $\dot{w}_{m}$ finite fuel consumption rate

$$0 = \dot{m}_{F,i} - \dot{w}_{m} \quad \text{(K)}$$

$$0 = \dot{m}_{O2,i} - \nu_{O2}\, \dot{w}_{m} \quad \text{and} \quad \text{(L)}$$

$$0 = 0 - \dot{m}_{P,e} + (1 + \nu_{O2})\, \dot{w}_{m}. \quad \text{(M)}$$

Solving for $\dot{w}_{m}$ from Equation L and then using in Equation A,

$$\dot{w}_{m} = \dot{m}_{O2,i}/\nu_{O2} = h_m\, A_f\, (Y_{O2,\infty} - Y_{O2,f})/\nu_{O2}. \quad \text{(N)}$$

Applying this to Equation K, the fuel injection rate,

$$\dot{m}_{F,i} = \text{fuel burn rate } \dot{m}_{F,b} = \dot{m}_{O2,i}/\nu_{O2} = h_{mf}\, A_f\, (Y_{O2,\infty} - Y_{O2,f})/\nu_{O2}, \quad \text{(O)}$$

which is the feeding rate of the liquid fuel.

(b) We will now determine the temperature difference ($T_f - T_\infty$). The energy conservation equation for the c.v. bounded by a surface just below drop surface and flame surface is given as

$$dE_{c.v}/dt = \dot{Q}_{cv} - \dot{W}_{cv} + \sum \dot{m}_{k,i}\, h_{k,t} - \sum \dot{m}_{k,e}\, h_{k,e}. \tag{P}$$

Therefore, at a steady state, using the law of stoichiometry,

$$0 = -h_H A(T_f - T_\infty) + \dot{w}_m\, [h_{F(\ell)} + \nu_{O2} h_{O2} - \nu_{CO2} h_{CO2} - \nu_{H2O} h_{H2O}]. \tag{Q}$$

As before, one can write $h_k = h_{fk} + h_{t,k}$, heating value for gaseous fuel $h_{c,F(g)} = h_{f,F(g)} + \nu_{O2} h_{f,O2} - \nu_{CO2}\, h_{f,CO2} - \nu_{H2O}\, h_{f,H2O}$; hence, using Equation N one can show that [left as exercise] where Sh_{mf} is Sherwood number based on flame diameter d_f and Sh is Sherwood number based on drop diameter d

$$(T_f - T_\infty) \approx (Y_{O2,\infty} - Y_{O2,f})\, \{h_{c,F(g)} - L - C_p\,(T_\infty - T_w)\}/(\nu_{O2}\, c_p) \tag{R}$$

where T_w is drop temperature.

(c) The burn rate cannot be determined unless the flame location r_f is known. Fuel vapor is transported from r = a to $r = r_f$ by mass transfer. From the heat transfer literature for heat transfer between two concentric spheres r = a and $r = r_f$,

$$Nu_{film} = Nu/(1 - a/r_f) \tag{S}$$

where $Nu = h_H\, d/\lambda$. By analogy, for fuel transfer between flame and drop, under low mass transfer condition

$$Sh_{film} = Sh/(1 - a/r_f) \tag{T}$$

where $Sh = h_m\, d/\rho D$. The fuel transfer rate to the flame surface must be such that it is equal to transport rate of oxygen divide by stoichiometric oxygen.

$$\dot{m}_F = h_{m,film}\, A\, (Y_{F,w} - Y_{F,f}) = h_{mf}\, A_f\, (Y_{O2,\infty} - Y_{O2,f})/\nu_{O2} \tag{U}$$

which is the fuel flow rate to flame surface. As $Sh_{mf} = h_{mf}\, d_f/\rho D$, and using Equation T,

$$\{Sh\, d/(1 - a/r_f)\}\, (Y_{F,w} - Y_{F,f}) = Sh_{mf}\, d_f\, (Y_{O2,\infty} - Y_{O2,f})/\nu_{O2}.$$

With $Y_{F,f} = Y_{O2,f} = 0$, $Sh = Sh_{mf}$ and solving

$$r_f/a = (1 + Y_{F,w} \nu_{O2}/Y_{O2,\infty}). \tag{V}$$

Using Equation V in Equation U,

$$\dot{m}_F = Sh_{mf}\, \pi\, \rho D\, d_f\, (Y_{O2,\infty}/\nu_{O2}) = Sh_{mf}\, B'\, \pi\, \rho D\, d \tag{W}$$

where the pseudo transfer number $B' = \{Y_{F,w} + (Y_{O2,\infty}/\nu_{O2})\}$.

If the liquid surface temperature $T_w = T_{BP}$, then $Y_{Fw} = 1$. If the drop surface temperature $T_w < T_{BP}$, then the phase equilibrium requires that the partial pressure of the vapor $p_{F,w}$ is the same as the saturation pressure $p^{sat}(T_w)$ at the surface. Then,

$$X_{F,w} = p_{F,w}/P, \quad \text{and} \quad Y_{F,w} = X_{F,w}\, M_F/M_{mix,w}. \tag{X}$$

If the rest of the gases are inert (e.g., molecular nitrogen) and have a molecular weight equal to M_I,

$M_{mix,w} = X_{F,w}M_F + (1 - X_{F,w})M_I$. Thus given T_w, $Y_{F,w}$ is known and hence $\dot{m}_F$.

(d) Assuming $P^{sat} = 1$ bar, $Y_{Fw} \approx 1.0$, then $B' = 1 + 0.23/3.5 = 1.07$
The B' number estimated by the simplified method is extremely small compared to B from the more rigorous method described in Chapter 10.

$\dot{m} = Sh_{mf}\, B\, \pi\, \rho D\, d = 2*1.07*\, \pi\, \rho D\, d = (2*1.07*\, \pi \times 5 \times 10^{-5} \times 100 \times 10^{-6})$, or

$\dot{m} = 1.07 \times 10^{8}$ kg/sec.

Note that $\dot{m} \propto d$ (i.e., droplet diameter) so that as the drop burns, the value of d should decrease, and hence the burn rate will keep decreasing as the size decreases.

Remarks

(i) If there is no feeding mechanism, the drop size will keep decreasing with time. Proceeding as before (Example 9 in this chapter; also see Chapter 6),

$$d_o^2 - d^2 = \alpha_c\, t, \tag{Y}$$

where $\alpha_c = 4\, Sh\, \rho D\, B'/\rho_l$. Liquid droplets also burn according to the d^2 law.
(ii) If the drop only evaporates and there is no chemical reaction, the mass-loss rate of the droplet can be obtained by imposing the condition $Y_{O2,\infty} = 0$ in definition of B'. Rewriting Equation H

$$\dot{m}_{F,i} \approx Sh\, \pi\, \rho D\, d\, Y_{F,w}. \tag{Z}$$

(iii) A more detailed treatment involving differential forms of conservation equations and exact solutions to the combustion of a drop will be given in Chapter 10.

7.7 Summary

We made use of global mass, species, and energy equations in solving combustion problem in closed (auto engines) and open systems (PFR, PSR). Ignition and extinction conditions are also obtained, Simplified resistance concepts were presented for determining the combustion rates and burning times of carbon and droplets.

Chapter 8

Conservation Relations

8.1 Introduction

This chapter presents conservation equations for species, momentum, and energy for reacting flow processes, as well as explains the transformation of governing equations into a form similar to those in heat transfer and fluid mechanics literature.

8.2 Simple Diffusive Transport Constitutive Relations

Momentum, heat, and mass (species) can be transported through diffusive mechanisms.

8.2.1 Diffusive Momentum Transfer (Newton's Law)

In a 1-D system,

$$\tau = -\mu \; dv/dy \quad \text{or} \quad (-\rho v)/(dv/dy),\ \mathrm{N/m^2},\ \frac{\mathrm{kg}}{\mathrm{m\,sec^2}} \tag{8.1}$$

where τ denotes the shear stress; μ, the absolute viscosity; and v, the kinematic viscosity or momentum diffusivity (m^2/sec). For a 3-D system, the stress tensor

$$\vec{\vec{\tau}} = -\rho v \vec{\nabla} \vec{V} \ ,$$

where the vector $\vec{V} = \vec{i}\,v_x + \vec{j}\,v_y + \vec{k}\,v_z$, and $\vec{\nabla} = \vec{i}\,\partial/\partial x + \vec{j}\,\partial/\partial y + \vec{k}\,\partial/\partial z$. The ∇ operators for Cartesian, cylindrical, and spherical coordinate systems are given in Tables F8.8–F8.10 of the Formulae section. The shear stress on a face normal to y but in the x direction (i.e., acting on the x-z plane),

$$\tau_{yx} = \mu(\partial v_y/\partial x + \partial v_x/\partial y).$$

Similarly,

$$\tau_{yz} = \mu(\partial v_y/\partial z + \partial v_z/\partial y).$$

8.2.2 Diffusive Heat Transfer (Fourier's Law)

The Fourier heat conduction law provides a relation between the heat flux $\vec{q}''$ and temperature gradient. In one dimension,

$$\dot{q}'' = -\lambda \; dT/dy = -\rho\alpha_T c_p \; dT/dy, \; \frac{kw}{m^2K} \tag{8.2}$$

where T, a scalar, λ is the thermal conductivity, α_T is thermal diffusivity, and $\lambda = \rho\alpha_T \; c_p$. In a 3-D system,

$$\vec{\dot{q}}'' = -\rho\alpha_T c_p \vec{\nabla} T. \tag{8.3}$$

8.2.3 Diffusive Species Transfer (Fick's Law)

According to Fick's Law (Chapter 6), the diffusive flux of the k-th species is given as

$$j_k'' = -\rho \; \partial Y_k/\partial y, \; kg/m^2sec. \tag{8.4}$$

In three dimensions,

$$\vec{j}_k'' = \rho Y_k \vec{V}_k = -\rho \; D \vec{\nabla} Y_k \tag{8.5}$$

where D, mass diffusivity, (m^2/sec).

These relations are simplistic and not applicable to all fluids and multicomponent mixtures. However, they are instructive in understanding the behavior of chemically reacting systems. The gradient of a scalar leads to a vector, whereas a vector gradient leads to a tensor. More rigorous constitutive relations are presented in Chapter 6.

8.3 Conservation Equations

This section contains simplified derivations for the overall mass and species conservation equations in order to illustrate the physics involved.

8.3.1 Overall Mass

Overall mass conservation is exemplified by (mass in)–(mass out) = (mass accumulated). In a 2-D system characterized by the coordinates x and y, the

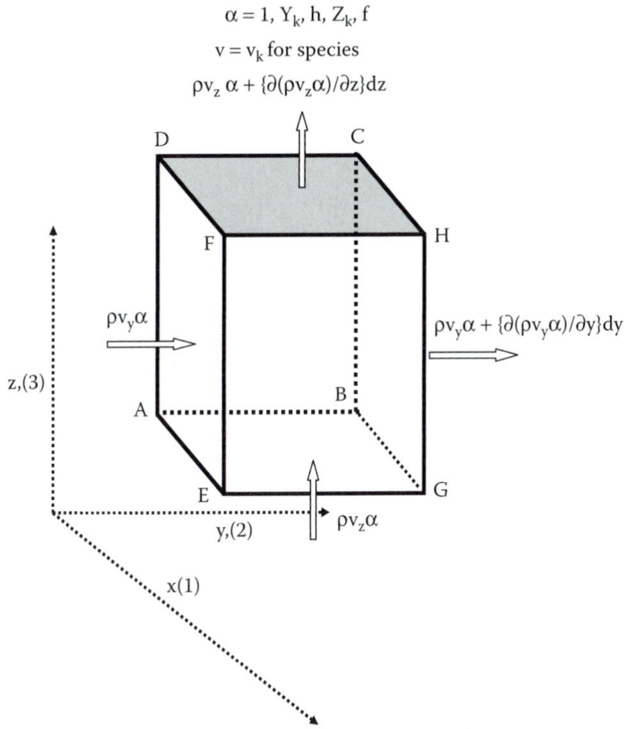

Figure 8.1 Cartesian elemental control volume used to develop the conservation equations.

mass entering a control volume along the x and y directions is

$$[\rho v_x\, dz\, dy + \rho v_y\, dz\, dx]. \tag{8.6}$$

Using Taylor series at x + dx, the mass exiting in the x direction is

$$[\rho v_x]_{x+dx}\, dz\, dy = \left[\rho v_x + \frac{\partial}{\partial x}(\rho v_x)dx\right] dz\, dy, \tag{8.7}$$

and, similarly, along the y direction is (Figure 8.1 with α set to unity)

$$\left[\rho v_y + \frac{\partial}{\partial y}\left(\rho v_y\right)dy\right] dz\, dx. \tag{8.8}$$

The corresponding accumulation rate is

$$\frac{\partial \rho}{\partial t} dx\, dy\, dz. \tag{8.9}$$

Consequently, the mass conservation equation has the form

$$\frac{\partial \rho}{\partial t} = (\rho v_x + \rho v_y) - \left\{ \rho v_x + \frac{\partial(\rho v_x)}{\partial x} \right\} - \left\{ \rho v_y + \frac{\partial(\rho v_y)}{\partial y} \right\}, \tag{8.10}$$

or

$$\frac{\partial \rho}{\partial t} + \frac{\partial(\rho v_x)}{\partial x} + \frac{\partial(\rho v_y)}{\partial y} = 0 . \tag{8.11}$$

The corresponding vectorial form of the relation is,

$$\frac{\partial \rho}{\partial t} + \nabla .\ \rho \vec{v} = 0 \tag{8.12}$$

where the ∇ operators for Cartesian, cylindrical, and spherical coordinate systems are tabulated in Tables F8.8–F8.10 of the Formulae section.

8.3.2 Species Conservation

Consider the elemental volume ($\Delta x \Delta y \Delta z$) illustrated in Figure 8.1 ($\alpha = Y_k$, $v_y = v_{ky}$, $v_x = v_{kx}$) into which a species k enters and leaves with absolute velocity v_k but can be consumed within it (through chemical reactions) at the rate w_k. The amounts of the species entering through the faces ABCD, AEFD, and AEGB are, respectively,

$$\rho_k\ v_{kx}\ \Delta y\ \Delta z,\ \rho_k\ v_{ky}\ \Delta z\ \Delta x, \quad \text{and} \quad \rho_k\ v_{kz}\ \Delta x\ \Delta y, \tag{8.13}$$

where $v_{k,x}$ denotes the absolute velocity of the k-th species in the x-wise direction. Likewise, the amounts leaving the faces EGHF, GHCB, and FHCD are, respectively,

$$\left(\rho_k\ v_{kx}\ \Delta x \Delta y\right)_{x+\Delta x} = \rho_k\ v_{kx}\ \Delta x \Delta y + \frac{\partial}{\partial x}\left(\rho_k\ v_{kx}\ \Delta x \Delta y\right), \tag{8.14}$$

$$\left(\rho_k\ v_{ky}\ \Delta y \Delta z\right)_{y+\Delta y} = \rho_k\ v_{ky}\ \Delta y \Delta z + \frac{\partial}{\partial y}\left(\rho_k\ v_{ky}\ \Delta y \Delta z\right), \tag{8.15}$$

$$\left(\rho_k\ v_{kz}\ \Delta z \Delta x\right)_{z+\Delta z} = \rho_k\ v_{kz}\ \Delta z \Delta x + \frac{\partial}{\partial z}\left(\rho_k\ v_{kz}\ \Delta z \Delta x\right). \tag{8.16}$$

We have expanded the relations using the Taylor series up to the first derivative. The second derivatives vanish in the limit $\Delta x \to 0$, $\Delta y \to 0$, $\Delta z \to 0$. The local density is typically a scalar (unless it is directional like thermal conductivity and density in a rolled carpet) and, hence, is insensitive to the coordinate system. The velocity is a vector and depends upon the local coordinates. The species accumulation rate is

$$\frac{\partial}{\partial t}(\rho_k\ \Delta x \Delta y \Delta z), \tag{8.17}$$

and the species production rate through chemical reactions is

$$\dot{w}'''_{km}\ \Delta x \Delta y \Delta z. \tag{8.18}$$

In subsequent sections and chapters, we may omit the subscript "m" for the chemical reaction rate in mass units. Using the relation (In) – (Out) + (Production) = (Accumulation) and constant values of Δx, Δy, and Δz,

$$\frac{\partial \rho_k}{\partial t} + \frac{\partial}{\partial x}(\rho_k\, v_{kx}) + \frac{\partial}{\partial y}(\rho_k\, v_{ky}) + \frac{\partial}{\partial z}(\rho_k\, v_{kz}) = \dot{w}'''_k. \tag{8.19}$$

The source terms will be written mostly in mass units and as such the notation for specific reaction rate constant $k_{[]}$, k_O, etc., will be omitted. Sometimes we may use a number format for axes and velocities (as in tensor notations),

$$\frac{\partial \rho_k}{\partial t} + \frac{\partial}{\partial x_1}(\rho_k\, v_{k1}) + \frac{\partial}{\partial x_2}(\rho_k\, v_{k2}) + \frac{\partial}{\partial x_3}(\rho_k\, v_{k3}) = \dot{w}'''_k\ , \tag{8.20}$$

where 1, 2, and 3 corresponds to x, y, and z axes. In vectorial form

$$\frac{\partial \rho_k}{\partial t} + \nabla . \rho_k\ \vec{v}_k = \dot{w}'''_k\ , \quad k = 1 \ldots .K \tag{8.21}$$

where the absolute velocity of species k is

$$\vec{v}_k = \vec{v} + \vec{V}_k \tag{8.22}$$

which is the sum of the mass-averaged velocity $\vec{v}$ and the diffusional velocity of species k (Chapter 6), $\vec{V}_k$. One can rewrite Equation 8.21 as

$$\frac{\partial \rho_k}{\partial t} + \vec{\nabla} \cdot \rho_k\ \vec{v} + \vec{\nabla} \cdot \rho_k\ \vec{V}_k = \dot{w}_k\ , \tag{8.23}$$

where

$$\rho_k V_K = \vec{j}''_k,\ \text{kg/m}^2\ \text{sec} \tag{8.24}$$

and $\vec{j}''_k$ denotes the diffusive mass flux of species k.

The overall mass conservation relation Equation 8.12 can be obtained by summing up Equation 8.23 for all species k

$$\frac{\partial \rho}{\partial t} + \vec{\nabla} . \rho\, \vec{v} + \sum_{k=1}^{n} \vec{\nabla} . \rho_k\ \vec{V}_k = \sum_{k=1}^{n} w'''_k\ . \tag{8.25}$$

As the mass-based expression $\Sigma\,\dot{w}_k''' $ and $\Sigma\,\rho_k\vec{V}_k = 0$, Equation 8.25 becomes

$$\frac{\partial\rho}{\partial t}+\nabla\cdot\rho\vec{v}=0. \tag{8.26}$$

Using the relation $\rho_k = \rho Y_k$, and applying the overall mass conservation equation (Equation 8.26), the above equation becomes

$$\rho\frac{\partial Y_k}{\partial t}+\rho\vec{v}\cdot\nabla Y_k=-\nabla\cdot\vec{j}_k''+\dot{w}_k''', \tag{8.27}$$

[accumulation of species k + net convective transport or advection the species into c.v.] = diffusive transport of the species into c.v. + production of species k in c.v. (8.28)

Example 1

Consider the reaction of CO with O_2, where air serves as source of O_2:

(a) Show that $\Sigma\,\dot{w}_k''' = 0$.
(b) Write the mass and species conservation equations for a 2-D flow system. Treat N_2 as an inert gas.
(c) If the density is maintained constant, simplify the mass conservation relation.

Solution

(a) The mass form of the chemical reaction relation is

$$1 \text{ kg CO} + 0.571 \text{ kg of } O_2 \rightarrow 1.571 \text{ kg of } CO_2, \tag{A}$$

$$\frac{\dot{w}_{CO}'''}{-1}=\frac{\dot{w}_{O_2}}{-0.571}=\frac{\dot{w}_{CO_2}'''}{1.571}=\dot{w}'''. \tag{B}$$

Hence $w_{CO}'''=-\dot{w}'''$, $\dot{w}_{O_2}'''=-0.571\,\dot{w}'''$, and $\dot{w}_{CO_2}'''=-1.571\,\dot{w}'''$, i.e.,

$$\sum\dot{w}_k'''=-\dot{w}'''-0.571\,\dot{w}'''+1.571\,\dot{w}'''=0. \tag{C}$$

(b) The mass conservation relation for a 2-D system is

$$\partial\rho/\partial t+\partial(\rho v_x)/\partial x+\partial(\rho v_y)/\partial y=0. \tag{D}$$

Following Equation 8.27, the species conservation relations for CO, O_2, CO_2 (product), and N_2 (inert) for a 2-D system are

$$\rho\frac{\partial Y_{CO}}{\partial t}+\rho\, v_x\frac{\partial Y_{CO}}{\partial x}+\rho v_y\frac{\partial Y_{CO}}{\partial y}=-\frac{\partial j_{CO}''}{\partial x}-\frac{\partial j_{CO}''}{\partial y}+(-1)\dot{w}''', \tag{E}$$

$$\rho \frac{\partial Y_{O2}}{\partial t} + \rho\, v_x \frac{\partial Y_{O2}}{\partial x} + \rho\, v_y \frac{\partial Y_{O2}}{\partial y} = -\frac{\partial j''_{O2}}{\partial x} - \frac{\partial j''_{O2}}{\partial y} + (-0.571)\dot{w}''', \tag{F}$$

$$\rho \frac{\partial Y_{CO2}}{\partial t} + \rho\, v_x \frac{\partial Y_{CO2}}{\partial_x} + \rho\, v_y \frac{\partial Y_{CO2}}{\partial_y} = -\frac{\partial j''_{CO2}}{\partial x} - \frac{\partial j''_{CO2}}{\partial y} + (1.571)\dot{w}''' \tag{G}$$

and

$$\rho \frac{\partial Y_{N_2}}{\partial t} + \rho\, v_x \frac{\partial Y_{N_2}}{\partial x} + \rho\, v_y \frac{\partial Y_{N_2}}{\partial y} = -\frac{\partial j''_{N2}}{\partial x} - \frac{\partial j''_{N2}}{\partial y}\,. \tag{H}$$

(c) If the density is constant (e.g., incompressible flow) Equation D becomes

$$\frac{\partial \rho}{\partial t} + \frac{\partial}{\partial x}(\rho\, v_x) + \frac{\partial}{\partial y}(\rho\, v_y) = 0 + \rho \frac{\partial v_x}{\partial x} + \rho \frac{\partial v_y}{\partial y} = 0\,, \text{ or} \tag{I}$$

$$\frac{\partial v_x}{\partial x} + \frac{\partial v_y}{\partial y} = 0\,. \tag{J}$$

8.4 Generalized Transport

Consider any property α that is described on the basis of per unit mass of a mixture. In that case, Equation 8.27 can be written as

$$(\partial\alpha / \partial t) + \rho\, \vec{v} \cdot \nabla\alpha = -\nabla \cdot \vec{j}_\alpha + \dot{w}'''_\alpha \tag{8.29}$$

where $\vec{j}''_\alpha$ denotes the diffusive flux transfer of α, a property per unit mass and $\dot{w}\ \alpha$ is the production rate of property (e.g., α = Y_k, $\dot{w}'''_{Yk}$ or $\dot{w}'''_k$ refers to production species k).

Defining material derivative (i.e., one travels with a fixed mass and observes changes with time) as

$$\frac{D\{\,\}}{Dt} = \frac{\partial\{\,\}}{\partial t} + \vec{v}.\nabla\{\,\}, \tag{8.30}$$

Thus, for property α,

$$\rho \frac{D\alpha}{Dt} = \rho \frac{\partial \alpha}{\partial t} + \rho\vec{v}.\nabla\alpha\ . \tag{8.31}$$

Species equation in terms of material derivative (α = Y_k, Equation 8.29) is written as

$$\rho \frac{DY_k}{Dt} = \rho \frac{\partial Y_k}{\partial t} + \rho\vec{v}.\nabla Y_k = -\nabla \cdot \vec{j}''_k + \dot{w}'''_k \tag{8.32}$$

(property accumulated + property advected = −property diffused out + property produced).

Using the relation and $\alpha = \vec{v}$, Y_k, h, u, etc. in Equation 8.29, the diffusive and source terms for various transport relations are summarized in Table 8.1. For momentum, we have replaced α with $\vec{V}$, $\dot{w}_\alpha'''$ by $\rho \Sigma Y_k \vec{f}_k$, and $\vec{j}_\alpha''$ by $\vec{\vec{P}}$, stress tensor, and f_k, is the body force on species k per unit mass of species k (e.g., gravitational force "g," which acts equally on all species).

8.4.1 Energy

8.4.1.1 Total Enthalpy Form

Consider a stationary medium. The Fourier law for diffusive flux in a single-component system is $\vec{\dot{q}}'' = -\lambda \vec{\nabla} T$. This law must be modified if there is diffusion of multiple components. The net diffusive energy flux in the x direction is given by

$$\dot{q}'' = j''_{h,x} = -\lambda \partial T/\partial x + \sum \rho_k V_{k,x} h_k. \tag{8.33}$$

The diffusive energy flux includes pure conduction flux and enthalpy flux associated with species diffusion. Thus, for a unit mass traveling in space with a bulk average velocity, $\dot{q}''$ is the flux entering the unit mass due to local temperature and species mass fraction gradients. There can be multiple components that move with diffusive velocities V_k and transport energy. If the species enthalpies are equal, then $h_k = h$, i.e., $\Sigma \rho_k V_k h_k = h \Sigma \rho_k V_k = 0$. In this case there is no mass-based diffusive heat flux because opposing diffusive species fluxes carry equal enthalpies. However, if the species is moving in, say, the "+x" direction has higher enthalpy compared to those moving in "–x" direction, then there will be net diffusive enthalpy flux in the "+x" direction, in addition to the Fourier conduction flux. Figure 8.2 illustrates the diffusive heat flux. Using Equation 8.29 with $\alpha = h$ (total enthalpy of all species, $h = \Sigma Y_k h_k$, $h_k = (h_{f,k} + h_{t,k})$), $\dot{w}_h''' = 0$ and Equation 8.33 for j_h, the 2-D form of the energy equation assumes the form

$$\frac{\partial h}{\partial t} + \left(\rho v_x \frac{\partial h}{\partial x}\right) + \left(\rho v_y \frac{\partial h}{\partial y}\right) = \frac{\partial}{\partial x}\left(\lambda \frac{\partial T}{\partial x} - \sum \rho_k V_{k,x} h_k\right) + \frac{\partial}{\partial y}\left(\lambda \frac{\partial T}{\partial y} - \sum \rho_k V_{k,y} h_k\right). \tag{8.34}$$

Using the relations

$$h = \sum_k h_k Y_k,\ h_k = h_{f,k} + h_{t,k},\ \frac{\partial h_k}{\partial y} = c_{pk} \frac{\partial T}{\partial y},\ c_p = \sum c_{pk} Y_k,\ V_{k,y} = -\frac{\rho D}{Y_k} \frac{\partial Y_k}{\partial y},$$

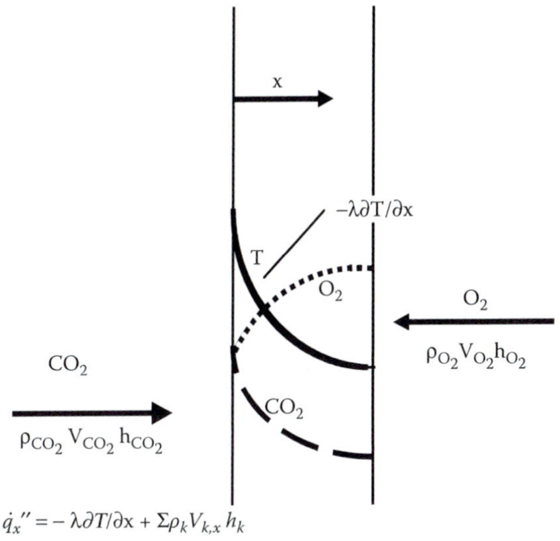

Figure 8.2 Schematic illustration of energy transport by species.

$$\frac{\partial h}{\partial y}=\sum_k \frac{\partial h_k}{\partial y}Y_k+h_k\frac{\partial Y_k}{\partial y}=\sum_k c_{pk}Y_k\frac{\partial T}{\partial y}+h_k\frac{\partial Y_k}{\partial y}$$

$$=c_p\frac{\partial T}{\partial y}-\frac{1}{\rho D}\sum_k h_k Y_k V_{k,y},\ c_p=\sum c_{pk}Y_k$$

where equal binary diffusion coefficients have been assumed.

$$\frac{\lambda}{c_p}\frac{\partial h}{\partial y}=\lambda\frac{\partial T}{\partial y}-\frac{\lambda}{\rho D c_p}\sum_k h_k Y_k V_{k,y}=j''_{h,y}\,. \tag{8.35}$$

Equation 8.35 represents the diffusive energy flux in y direction.

If, $\mathrm{Le}=\dfrac{\lambda}{\rho\,\mathrm{D\,Cp}}=\dfrac{\alpha_{\mathrm{T}}}{\mathrm{D}}=\left(\dfrac{\mathrm{Sc}}{\mathbf{Pr}}\right)=1$, then Equation 8.34 becomes

$$\frac{\partial h}{\partial t}+\left(\rho v_x\frac{\partial h}{\partial x}\right)+\left(\rho v_y\frac{\partial h}{\partial y}\right)=\frac{\partial}{\partial x}\left(\frac{\lambda}{c_p}\frac{\partial h}{\partial x}\right)+\frac{\partial}{\partial y}\left(\frac{\lambda}{c_p}\frac{\partial h}{\partial y}\right) \tag{8.36}$$

Generalizing for all coordinate systems

$$\rho(\partial \mathrm{h}/\partial \mathrm{t})+\rho\,\vec{\mathrm{v}}\cdot\nabla\mathrm{h}=-\left\{\nabla\cdot\frac{\lambda}{\mathrm{c_p}}\nabla\mathrm{h}\right\}. \tag{8.37}$$

Table 8.1 Conservation Equations in Generalized Form

Case	α	$\vec{j}''_{\alpha}$,	$\dot{w}'''_{\alpha}$	*Remarks*
Species	Y_k	$-\rho D V_k Y_k$	$\dot{w}'''_k$	$\dot{w}'''_k$ is chemical source;
Momentum	$\vec{V}$	$\vec{\vec{P}}$	$\rho \sum Y_k \vec{f}_k$	$\vec{\vec{P}}$ is stress tensor
Kinetic energy	$\frac{v^2}{2}$	$-\vec{\vec{P}} : \vec{\nabla}\vec{V} + \vec{\nabla}.$ $[\vec{\vec{P}}.\vec{V}]$	$\rho \sum Y_k \vec{f}_k . \vec{V}$	Mechanical energy
Internal energy	u	$-\lambda \nabla T - \rho_k \vec{V}_k u_k$	$-\vec{\vec{P}} : \vec{\nabla}\vec{V} + \rho \sum Y_k \vec{f}_k . \vec{V}_k$	Conversion of mechanical to thermal energy + pumping work + diffusive potential energy
Enthalpy	h	$-\lambda \nabla T + \sum \rho_k \vec{V}_k h_k$	$(dp/dt) - \vec{\vec{\tau}} : \vec{\nabla}\vec{v} + \rho$ $\sum Y_k \vec{f}_k \cdot \vec{V}_k$	Pumping + viscous dissipation + diffusion of potential energy

where λ and c_p are scalar quantities. There are no source terms in the total enthalpy equation. Equation 8.37 ignores source terms due to pumping work and dissipation due to shear work (see Table 8.1).

8.4.1.2 Thermal Enthalpy Form

Using the relation $h = \Sigma Y_k h_k = \Sigma Y_k (h_{t,k} + h_{f,k})$ and $h_t = \Sigma Y_k h_{t,k}$, Equation 8.36 becomes

$$\rho \frac{\partial h_t}{\partial t} + \left(\rho v_x \frac{\partial h_t}{\partial x}\right) + \left(\rho v_y \frac{\partial h_t}{\partial y}\right) = \frac{\partial}{\partial x}\left(\frac{\lambda}{c_p}\frac{\partial h_t}{\partial x}\right) + \frac{\partial}{\partial y}\left(\frac{\lambda}{c_p}\frac{\partial h_t}{\partial y}\right) + |\dot{w}'''_F| h_c. \tag{8.38}$$

More generally for any coordinate system,

$$\rho(\partial h_t/\partial t) + \rho\, \vec{v} \cdot \nabla h_t = -\{\nabla \cdot \frac{\lambda}{c_p} \nabla h_t\} + |\dot{w}'''_F| h_c. \tag{8.39}$$

8.4.2 Species

Applying the simplified Fick's law for the diffusive term in Equation 8.27

$$\rho \frac{\partial Y_k}{\partial t} + \rho \vec{v} \cdot \nabla Y_k = \nabla \cdot [\rho D \nabla Y_k] + \dot{w}_k''' . \tag{8.40}$$

8.4.3 Momentum

Rigorous equations in tensor forms are presented in the appendix of this chapter. In the context of Equation 8.29 with $\alpha = \vec{v}$ and using the appendix of this chapter for the stress tensor terms, the x and y momentum equations are given as

$$\frac{\partial \rho}{\partial t} + \frac{\partial}{\partial x}\left(\rho v_x \frac{\partial v_x}{\partial x}\right) + \frac{\partial}{\partial y}\left(\rho v_y \frac{\partial v_x}{\partial y}\right) = \frac{\partial}{\partial x}\left(\mu \frac{\partial v_x}{\partial x}\right) + \frac{\partial}{\partial y}\left(\mu \frac{\partial v_x}{\partial y}\right) - \frac{\partial p}{\partial x}, \tag{8.41}$$

$$\frac{\partial \rho}{\partial t} + \frac{\partial}{\partial y}\left(\rho v_x \frac{\partial v_y}{\partial x}\right) + \frac{\partial}{\partial y}\left(\rho v_y \frac{\partial v_y}{\partial y}\right) = \frac{\partial}{\partial x}\left(\mu \frac{\partial v_y}{\partial x}\right) + \frac{\partial}{\partial y}\left(\mu \frac{\partial v_y}{\partial y}\right) - \frac{\partial p}{\partial y}. \tag{8.42}$$

General solution procedure: For more general problems involving 3-D coordinates, the solution procedure is as follows: The unknowns are: (1) 3 velocity components, (v_x, v_y, v_z), (2) K species Y_k's (Y_1, Y_2, ..., Y_K), (3) T or h_t or h, (4) ρ, and (5) P. Thus, 6 + K unknowns. The governing equations are (1) 3 momentum equations in x, y, and z, (2) K − 1 (instead of K) species, with the restriction that $Y_1 + Y_2 + Y_3 \ldots Y_K = 1$, (3) energy equation, (4) state Equation (P, ρ, and T), and (5) the overall mass conservation. Thus, K + 6 equations. Direct numerical simulation (DNS) solves for all the 6 + K unknowns for laminar regime and limited turbulent regime. However, at this time, the DNS programs do not extend to a very high Reynolds number. Thus, simplified turbulent models are used involving turbulent combustion (Chapter 18).

8.4.4 Element

As the species diffuse and convect, elements are transported by the various species. Let the fraction of j-th element in species k be $\alpha_{j,k}$ (e.g., for CO_2, $\alpha_{C,CO2} = 12.01/44.01 = 0.273$ and $\alpha_{O,CO2} = 0.727$). Although within an elemental space, species diffuse, convect, and participate in reactions, the elements are neither produced nor destroyed.

Assuming that all diffusion coefficients are equal in a multicomponent system for a 2-D geometry, then

$$\rho \frac{\partial Y_k}{\partial t} + \rho v_x \frac{\partial Y_k}{\partial x} + \rho v_y \frac{\partial Y_k}{\partial y} = \frac{\partial}{\partial y}\left(\rho D \frac{\partial Y_k}{\partial y}\right) + \frac{\partial}{\partial x}\left(\rho D \frac{\partial Y_k}{\partial x}\right) + \dot{w}_k''' \, .$$

The carbon elemental composition in CO_2 is $\alpha_{C,CO2}$, and in CO it is $\alpha_{C,CO.}$ For example, we multiply Y_k by $\alpha_{c,k}$, and add them to form element balance for "C." Therefore, generalizing for K species,

$$\rho \sum \frac{\partial}{\partial t}(Y_k \alpha_{C,k}) + \rho v_x \frac{\Sigma \partial Y_k \alpha_{C,k}}{\partial x} + \rho v_y \frac{\Sigma \partial Y_k \alpha_{C,k}}{\partial y}$$

$$= \frac{\partial}{\partial x}\left(\rho D \frac{\Sigma \partial Y_k \alpha_{C,k}}{\partial x}\right) + \frac{\partial}{\partial y}\left(\rho D \frac{\Sigma \partial Y_k \alpha_{C,k}}{\partial y}\right)$$

where $\Sigma \, \alpha_{C,k} \, \dot{w}_k'''$ is set to zero as there is no net production of carbon. Thus, for the carbon element,

$$\rho \frac{\partial Z_C}{\partial t} + \rho v_x \frac{\partial Z_c}{\partial x} + \rho v_y \frac{\partial Z_C}{\partial y} = \frac{\partial}{\partial x}\left(\rho D \frac{\partial Z_C}{\partial x}\right) + \frac{\partial}{\partial y}\left(\rho D \frac{\partial Z_C}{\partial y}\right), \quad (8.43)$$

where $Z_c = \sum_k \alpha_{C,k} Y_k$ denotes the carbon element mass in kg per kg of mixture. Generally, the element mass conservation equation for any element j is analogously written as

$$\rho \frac{\partial Z_j}{\partial t} + \rho v_x \frac{\partial Z_j}{\partial x} + \rho v_y \frac{\partial Z_j}{\partial y} = \frac{\partial}{\partial x}\left(\rho D \frac{\partial Z_j}{\partial x}\right) + \frac{\partial}{\partial y}\left(\rho D \frac{\partial Z_j}{\partial y}\right),$$

$$j = C, H, N, O \quad (8.44)$$

where Z_j denotes the mass of element j per unit mass of the mixture, i.e., $Z_j = \Sigma \, Y_k \, \alpha_{j,k}$. Recall that we have assumed that all species diffuse equally to obtain the element conservation equation.

More generally, the element conservation equation is written as

$$\rho\left(\frac{\partial Z_j}{\partial t}\right) + \rho \, \vec{v} \cdot \nabla Z_j = \{\nabla \cdot \rho D \nabla Z_j\} \, . \quad (8.45)$$

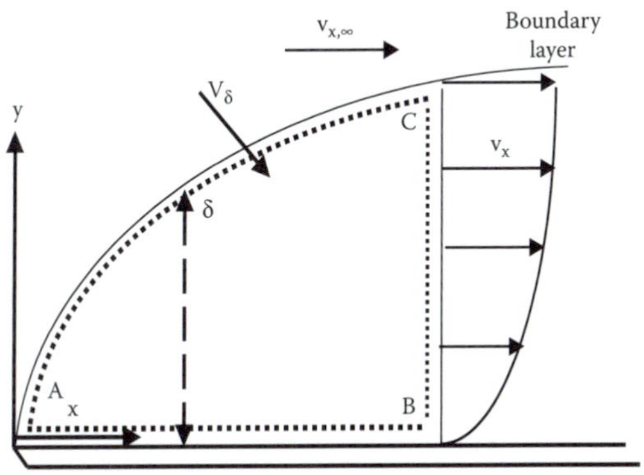

Figure 8.3 Schematic illustration of a velocity boundary layer. $\delta = 4.64\ x/\sqrt{Re_x}$, $Re_x = V\ x\ \rho/\mu$; friction coefficient, $c_f\ (x) = \tau_s/[1/2\rho\ V^2] = 0.646/Re_x^{1/2}$ τ_s = shear stress at wall, Local $Nu_x = h_H x/\lambda = 0.332\ Re_x\ Pr^{1/3}$.

8.5 Simplified Boundary-Layer-Type Problems

Consider flow of a cold combustible mixture over a heated plate. The flow over a horizontal plate is illustrated in Figure 8.3 and the temperature profiles are schematically illustrated in Figure 8.4. The plate heats the mixture, raising its temperature. Consequently, chemical reactions can follow, resulting in ignition and combustion. At y = 0, the velocity $v_x = 0$ and at $y = \delta$, $v_x = v_{x,\infty}$, where δ is the boundary-layer (BL) thickness. If we consider air at 300 K and a free-stream velocity $v_{x,\infty} = 10$ m/sec, then δ has values of 0.19 mm at x = 10 cm and 0.6 mm at 100 cm. The BL thickness is very small compared to the plate length, except at the leading edge. For instance, $\delta/x = 0.019$ at x = 10 cm, but its value is reduced to 0.0006 at x = 100 cm. Other characteristics of the BL are exemplified through the x-wise velocity. As an example, at y = 0.095 mm, $v_x/v_{x,\infty} = 0.69$ (or $v_x = 6.9$ m/sec) at x = 10 cm. However, $v_x/v_{x,\infty} = 0.22$ (or $v_x = 2.2$ m/sec) at the same value of y = 0.095 mm, but at a different location for x = 100 cm, i.e., $v_x = v_x\ (x, y)$ within the BL. The BL can contain velocity, temperature, and species concentration gradients. The gradient $\partial v_x/\partial y \approx v_{x,\infty}/\delta$ is larger than $\partial v_x/\partial x$. For example, at x = 10 cm, $\partial v_x/\partial y \approx 10/(0.19 \times 10^{-3}) = 50000\ \text{sec}^{-1}$ and $\partial v_x/\partial x \approx 10/0.06 = 167\ \text{s}^{-1}$, whereas at x = 100 cm, $\partial v_x/\partial y \approx 10/(1.85 \times 10^{-3}) = 5000\ \text{sec}^{-1}$

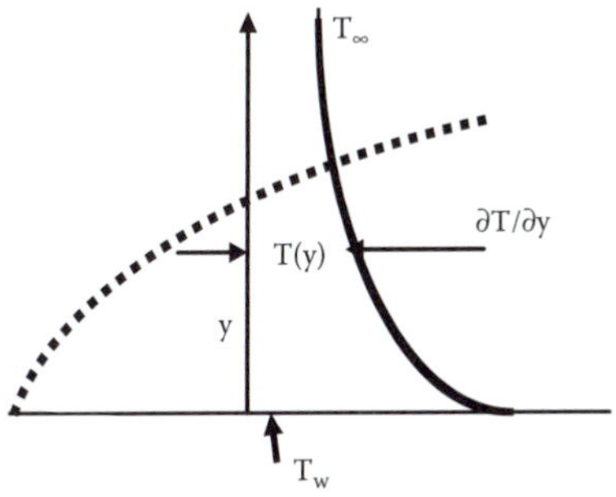

Figure 8.4 Schematic illustration of a thermal boundary layer.

and $\partial v_x/\partial x \approx 10/1 = 10\ \text{sec}^{-1}$. Similarly, $(\partial/\partial x)(\mu\ \partial v_x/\partial x) < (\partial/\partial y)(\mu\ \partial v_x/\partial y)$. Hence, it is possible to neglect gradients in the x direction involving diffusive terms. Similar statements are valid for the temperature or thermal enthalpy gradients, i.e., $(\partial/\partial x)(\lambda\ \partial T/\partial x) < (\partial/\partial y)(\lambda\ \partial T/\partial y)$, and mass fraction gradients, e.g., $(\rho D\ \partial Y_k/\partial x < \rho D\ \partial Y_k/\partial y)$.

8.5.1 Governing Equations

Now we will evaluate the order of magnitude for each term when applied to BLs.

$$\frac{\partial}{\partial y}\left(\mu\frac{\partial v_x}{\partial y}\right)=\frac{\partial}{\partial y}\left(\rho\nu\frac{\partial v_x}{\partial y}\right)\approx\rho\,\nu\frac{v_x}{\delta^2}>>\frac{\partial}{\partial x}\left(\rho\,\nu\frac{\partial v_x}{\partial x}\right) \tag{8.46}$$

$$\approx\rho\,\nu\frac{v_x}{\delta^2},>>\frac{\rho\,\nu\partial v_x}{x^2}\ \text{or}\ \frac{\delta}{x}<<<1 \tag{8.47}$$

where ν denotes the kinematic viscosity. Thus diffusive terms in x direction are ignored.

The 2-D forms of the momentum, species, and energy equations are:

$$\text{x-wise momentum:}\ \frac{\partial\rho}{\partial t}\left(\rho v_x\frac{\partial v_x}{\partial x}\right)+\left(\rho v_y\frac{\partial v_x}{\partial y}\right)=\frac{\partial}{\partial y}\left(\mu\frac{\partial v_x}{\partial y}\right)-\frac{\partial p}{\partial x} \tag{8.48}$$

$$\text{Species:}\ \rho\frac{\partial y_k}{\partial t}+(\rho\,v_x)\frac{\partial Y_k}{\partial x}+\rho\,v_y\frac{\partial Y_k}{\partial y}=\frac{\partial}{\partial y}\left(\rho\ D\frac{\partial Y_k}{\partial y}\right)+\dot{w}_k''' \tag{8.49}$$

$$\text{Energy (total enthalpy):}\ \frac{\partial h}{\partial t}+\rho v_x\frac{\partial h}{\partial x}+\rho v_y\left(\frac{\partial h}{\partial y}\right)=\frac{\partial}{\partial y}\left(\frac{\lambda}{c_p}\frac{\partial h}{\partial y}\right) \tag{8.50}$$

Assuming that $\rho D = \lambda/c_p$, and energy equation (Equation 8.50) reduces to:

$$\frac{\partial h}{\partial t}+\rho v_x\frac{\partial h}{\partial x}+\rho v_y\left(\frac{\partial h}{\partial y}\right)=\frac{\partial}{\partial y}\left(\rho D\frac{\partial h}{\partial y}\right) \tag{8.51}$$

Similarly, the element conservation equation is

$$\rho\,v_x\frac{\partial Z_j}{\partial x}+\rho\,v_y\frac{\partial Z_j}{\partial y}=\frac{\partial}{\partial y}\left(\rho\,D\frac{\partial Z_j}{\partial y}\right),\ j=\text{C, H, N}\ldots \tag{8.52}$$

and finally mixture fraction equation is

$$\rho\ v_x\frac{\partial f_M}{\partial x}+\rho\ v_y\frac{\partial f_M}{\partial y}=\frac{\partial}{\partial y}\left(\rho\ D\frac{\partial f_M}{\partial y}\right) \tag{8.53}$$

where $f_M = Z_C + Z_H$ for a hydrocarbon fuel. See details in Section 8.5.3.

8.5.2 General Solution

We now discuss a method of solution for any property α. Let us define

$$\Phi = \frac{\alpha - \alpha_\infty}{\alpha_w - \alpha_\infty}, \text{ where } \begin{matrix} \alpha = \alpha_w \ at\ y = 0 \\ \alpha = \alpha_\infty \ at\ y \to \infty \end{matrix}. \tag{8.54}$$

Then,

$$\begin{matrix} \phi = 1\ at\ y = 0 \\ \phi = 0\ as\ y \to \infty \end{matrix}, \text{ and } \phi \to 0\ as\ x \to 0\,. \tag{8.55a,b,c}$$

There are two boundary conditions for the y coordinate (Equation 8.55a and Equation 8.55b) and one for the x coordinate (Equation 8.55c) so that differential equations that are first order in x and second order in y can be solved. Once we solve for h_t (x,y) (i.e., T (x,y)) for a reacting system, the heat flux at the wall is simplified as Equation 8.35:

$$\dot{q}''_w \approx -\lambda \frac{\partial T}{\partial y} = -\frac{\lambda}{c_p}\frac{\partial h_t}{\partial y} = h_H (T_w - T_\infty) \tag{8.56}$$

where enthalpy diffusion due to mass flux is ignored or all species have equal enthalpy.

If the gradient $\partial T/\partial y$ is known, the convective heat transfer coefficient $h_H(x)$ and the Nusselt number $Nu(x) = h_H\, x/\lambda$ can also be determined.

Equations that do not have a source term, i.e., $\alpha = h, Z_k$ can be written in the form

$$\rho v_x \frac{\partial \phi}{\partial x} + \rho v_y \frac{\partial \phi}{\partial y} = \frac{\partial}{\partial y}\left(\nu D \frac{\partial \varphi}{\partial y} \right). \tag{8.57}$$

Equation 8.57 presumes that $\alpha_w \neq f(x)$.

8.5.3 Mixture Fraction

8.5.3.1 Definition

Consider an elemental volume containing unit mass of a mixture at an arbitrary location (x,y,z) that arises due to the burning of a fuel C_xH_y (from the primary stream) along with air (introduced in the secondary stream). The mode of introduction of fuel and air, e.g., premixed or nonpremixed, does not matter in this regard. In this context, the mixture fraction (f_M) denotes the fraction of the elemental mass that originates from the fuel or primary stream. For a nonpremixed flame, the local mixedness is also a measure of the mixedness of the fuel. When $f_M = 1$, there is pure fuel and only air when $f_M = 0$. Thus, $f_M = 0$ or 1 indicate unmixedness, whereas the values in between represent degrees of mixedness. Consider a location where products (e.g., CO, O, CO_2, H, H_2, OH, and H_2O), N_2, and unburned fuel are present (Figure 8.5). Here,

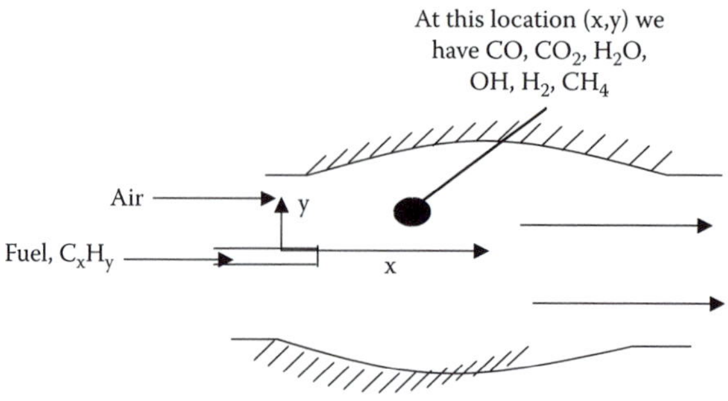

Figure 8.5 Illustration of mixture fraction, fM.

f_M = fractional mass of unburned carbon in the fuel + fractional mass of unburned hydrogen in the fuel + the total carbon elemental mass fraction in the product species + the total hydrogen elemental mass fraction in products, i.e., $f_M = Y_F + \{Y_{CO} \times 12.01/28.01\} + \{Y_H + Y_{H2} + Y_{OH} \times 1.01/17.07 + Y_{H2O} \times 2.02/18.02\}$,

$$f_M = \text{total C elemental mass fraction} + \text{total H elemental mass fraction } (Z_c + Z_H) \tag{8.58}$$

where $1 - f_M$ represent the mass originating from air. The elemental conservation relations can be written in the form

$$\rho v_x \frac{\partial Z_C}{\partial x} + \rho v_y \frac{\partial Z_C}{\partial y} = \frac{\partial}{\partial y}\left(\rho D \frac{\partial Z_C}{\partial y}\right), \quad \text{and} \tag{8.59}$$

$$\rho v_x \frac{\partial Z_H}{\partial x} + \rho v_y \frac{\partial Z_H}{\partial y} = \frac{\partial}{\partial y}\left(\rho D \frac{\partial Z_H}{\partial y}\right). \tag{8.60}$$

Adding

$$\rho v_x \frac{\partial Z_{C+H}}{\partial x} + \rho v_y \frac{\partial Z_{C+H}}{\partial y} = \frac{\partial}{\partial y}\left(\rho D \frac{\partial Z_{C+H}}{\partial y}\right). \tag{8.61}$$

The conservation relation corresponding to Equation 8.61 is

$$\rho v_x \frac{\partial f_M}{\partial x} + \rho v_y \frac{\partial f_M}{\partial y} = \frac{\partial}{\partial y}\left(\rho D \frac{\partial f_M}{\partial y}\right), \tag{8.62}$$

where for HC fuels $f_M = Z_{C+H} = Z_C + Z_H$ where f_M is termed as the mixture fraction.

The mixture fraction equation provides solutions for f_M. If the fuel composition is known, the local atom composition can be determined. The carbon to hydrogen ratio for an HC fuel C_xH_y is

$$Z_C/Z_H = x \times 12.01/(1.01 \times y) = C_m/H_m, \quad \text{i.e.,}$$

$f_M = Z_C + Z_H = Z_C(1 + H_m/C_m) = Z_C \times$ (total fuel mass/carbon mass in the fuel).

8.5.3.2 Local Equivalence Ratio

As f_M denotes the fractional mass of a mixture originating in the fuel, then the fraction $(1 - f_M)$ at the same location must have originated from air. This provides a basis to describe the local fuel-to-air ratio, namely,

$$\text{Local F/A} = f_M/(1 - f_M). \tag{8.63}$$

As the local equivalence ratio $\varphi_{loc} = (F/A)/(F/A)_{st}$,

$$\varphi_{loc} = f_M/\{(1 - f_M)(F/A)_{st}\}. \tag{8.64}$$

If you measure the species concentration near the fuel injector, then $Y_F = 1$, $Y_{O2} = 0$, $Y_{CO2} = 0$, $Y_{H2O} = 0$, $f_M = 1$, and $\varphi_{loc} \to \infty$. If you measure species near the air injection point, $Y_F = 0$, $Y_{CO2} = 0$, $Y_{H2O} = 0$, $Y_{O2} = 0.23$, $Y_{N2} = 0.77$, $f_M = 0$; $\varphi_{loc} = 0$. At any other point, $0 < \varphi_{loc} < \infty$. When the fuel contains oxygen, f_M = carbon elemental mass fraction + hydrogen elemental mass fraction + oxygen element (but pertaining to fuel only) mass fraction, all originating from the fuel stream.

8.5.3.3 Relation between Mixture Fraction, Element Fraction, and Total Enthalpy

For a BL-type problem, if $\Phi = \frac{(f_M - f_{M\infty})}{(f_{Mw} - f_{M\infty})}$, $f_{M\infty} = 0$ and f_{Mw} is known, then the value of $\phi(x,y) = \frac{f_M}{f_{Mw}}$ can be determined. For a BL-type problem, because of the similarity of the conservation equations,

$$\phi = \frac{h - h_\infty}{h_w - h_\infty} = \frac{f_M - f_{M\infty}}{f_{Mw} - f_{M\infty}} = \frac{Z_j - Z_{j\infty}}{Z_{jw} - Z_{j,\infty}}. \tag{8.65}$$

Therefore, h, f_M, and Z_j are linearly related. If solutions to the enthalpy equation can be obtained, the temperature can be determined using the equilibrium assumption (Chapter 3).

Example 2

Consider the reaction of CH_4 with air. Determine the local carbon and hydrogen mass fractions and the equivalence ratio at the locations where (a) $f_M = 0.2$, (b) $f_M = 0.044$, (c) estimate the enthalpy for the mixing problem for $f_M = 0.044$, (d) what is the equilibrium flame temperature for case (b) considering the following species: CO, H_2, CO_2, O_2, N_2, OH, and NO. Assume that there is no heat loss.

Solution

(a) Because $f = Z_C + Z_H = Z_C(1 + H_m/C_m)$, for $f = 0.2$, $0.2 = Z_C(12/16)$, i.e.,

$$Zc = 0.15 \quad \text{and} \quad Z_H = (1/3)\, Z_C = 0.05.$$

$$\text{Local A:F ratio} = (1 - f_M)/\, f_M = 0.8/0.2 = 4.00$$

For the pure mixing case, the local oxygen mass per kg of the mixture is

$$0.8 \times 0.23 = 0.184.$$

The local nitrogen mass fraction is

$$0.8 \times 0.77 = 0.616.$$

Therefore, the C:H:N:O mass ratio is 0.15:0.05:0.616:0.184.

The stoichiometric oxygen requirement for methane combustion is 4 kg of O_2/kg of fuel. Therefore, the local equivalence ratio is

$$\phi = \{O_2 \text{ Stoich}\}/\{O_2 \text{ locally present}\}$$
$$= 4 \text{ kg of } O_2 \text{ per kg fuel}/4.00 \times 0.23 \text{ kg of } O_2 \text{ per kg of fuel} = 4.35.$$

(b) If f = 0.044, on a mass basis

$$A{:}F = 21.72.$$

For CH_4, $A{:}F_{stoich} = 17.4$. Therefore, the equivalence ratio

$$\phi = 17.4/21.72 = 0.8 \text{ or there is 25\% excess air.}$$

(c) The mass-based enthalpy

$$h = (H_{fuel} + H_{air})/(\text{mass of the mixture})$$
$$= (h_{fuel} \times \text{fuel mass} + h_{air} \times \text{air mass})/(\text{total mass})$$
$$= h_{fuel} \times f_M + h_{air} \times (1 - f_M).$$

At 298 K for the pure mixing case,

$$h = 0.044 \text{ kg of fuel/kg of mixture} \times 74850 \text{ kJ/kmol of fuel}/16 \text{ kg/kmol}$$
$$= 205.6 \text{ kJ/kg mix.}$$

(d) In that case, equilibrium calculations with the following ratio C/H/N/O = 0.033:0.011:0.956 × 0.23:0.956 × 0.77 = 0.033:0.011:0.223:0.736 and h = 205.6 kJ/kg mix can be performed. The temperature and composition are predicted as follows (THERMOLAB software):

P	1	Bar
T	1986.15	K
	Mole fraction Wet	Mole fraction Dry
CO	0.00042	0.000498
CO_2	0.07669	0.09034
H_2	0.00021	0.00025
H_2O	0.151	0
NO	0.00317	0.00373
N_2	0.7252	0.85430
OH	0.00576	0.00679
O_2	0.03743	0.04409

Remarks

(a) The aforementioned results for composition are presented on a mole basis.
(b) In arriving at "h" we assumed that there is no heat loss. If there is heat loss, "h" and "f_M" may not be linearly related because the boundary condition for "h" will involve a gradient.
(c) Once the f_M distribution is known, the corresponding contours of h, atom mass ratios, and temperature can be developed for the combustion case.
(d) Methanol is an oxygen-containing fuel for which $Z_C = 0.2 \times (12/32) = 0.075$, $Z_H = 0.2\ 4/32 = 0.025$, and $Z_{O,F} = 0.2 \times (16/32) = 0.1$ (where $Z_{O,F}$ denotes the oxygen mass in the fuel) for $f_M = 0.2$. For the pure mixing case, $Z_{O,air} = 0.8 \times 0.23 = 0.184$ and $Z_{N,air} = 0.616$. Thus, the C:H:N:O mass ratio is 0.075:0.025:0.616:(0.184 + 0.1) = 0.075:0.025:0.616:0.284.

8.6 Shvab–Zeldovich Formulation

The conventional Shvab–Zeldovich (SZ) formulation eliminates the chemical source terms that is typically associated with one-step global chemical reactions and reduces to a form similar to the element and total enthalpy conservation equations. Whereas the element conservation equation can also be applied for multistep reactions, the SZ formualtion is highly useful for single-step or two-step reactions.

8.6.1 Boundary-Layer Problems

8.6.1.1 Single-Step Reaction

As an example of illustration of SZ formulation, consider the species equation in BL-type problem. For species k,

$$\rho\ v_x \frac{\partial Y_k}{\partial x} + \rho\ v_y \frac{\partial Y_k}{\partial y} = \frac{\partial}{\partial y}\left(\rho\ D \frac{\partial Y_k}{\partial y}\right) + \dot{w}_k''' . \tag{8.66}$$

The last term is the source term, which is zero for inert species such as nitrogen. One may generalize the relation in vector form for a 3-D system applicable to Cartesian, cylindrical, and spherical systems. Define a linear operator L as

$$L(\alpha) = \dot{w}_k''' \text{, where for a 3-D problem}$$

$$L(\alpha) = \nabla.(\rho\vec{v}\alpha - \rho D \nabla \alpha) .$$

The thermal enthalpy relation is

$$\rho v_x \frac{\partial h_t}{\partial x} + \rho v_y \frac{\partial h_t}{\partial y} = \frac{\partial}{\partial y}\left(\frac{\lambda}{c_p}\frac{\partial h_t}{\partial y}\right) + \left|\dot{w}_F'''\right| h_c \tag{8.67}$$

where h_c denotes the lower heating value of the fuel in gaseous state. Based on fuel consumption $|\dot{w}_F'''| = -\dot{w}_F'''$, i.e.,

$$\rho\, v_x \frac{\partial h_t}{\partial x} + \rho\, v_y \frac{\partial h_t}{\partial y} = \frac{\partial}{\partial y}\left(\frac{\lambda}{c_p}\frac{\partial h_t}{\partial y}\right) - \dot{w}_F'''\, h_c\,. \tag{8.68}$$

The chemical source terms in Equation 8.66 and Equation 8.68 can be eliminated by combining the two relations, which is the basis of the SZ formulation. Typically, $\lambda/c_p = \rho D$. Let us illustrate the SZ formulation with an example.

Example 3

The aircraft window is made of Plexiglas, which is a polymer. Plexiglas decomposes at about 600 K and produces monomers that can be represented by the chemical formula $C_5H_8O_2$. The monomers mix with air and burn. Assume the following single-step reaction occurring within the BL,

$$C_5H_8O_2 + 6O_2 \rightarrow 5CO_2 + 4H_2O.$$

Convert the governing conservation equations into the SZ forms.

Solution

Writing the reaction in mass form

$$1 \text{ kg } C_5H_8O_2 + 1.92 \text{ kg of } O_2 \rightarrow 2.2 \text{ kg } CO_2 + 0.72 \text{ kg of } H_2O, \tag{A}$$

where $C_5H_8O_2$ or fuel F, O_2, CO_2, H_2O, and N_2 exist in the gas phase. A kg of fuel vapor consumes $6 \times 32/(100.1) = 1.92$ kg of oxygen during combustion produces $5 \times 44.01/100.1 = 2.20$ kg of CO_2 and $4 \times 18.02/100.1 = 0.72$ kg of H_2O. The species conservation equations are

$$\rho v_x \frac{\partial Y_F}{\partial x} + \rho v_y \frac{\partial Y_F}{\partial y} = \frac{\partial}{\partial y}\left(\rho D \frac{\partial Y_F}{\partial y}\right) + \dot{w}_F''' , \tag{B}$$

$$\rho v_x \frac{\partial Y_{O_2}}{\partial x} + \rho v_y \frac{\partial Y_{O_2}}{\partial y} = \frac{\partial}{\partial y}\left(\rho D \frac{\partial Y_{O_2}}{\partial y}\right) + \dot{w}_{O_2}''' , \tag{C}$$

$$\rho v_x \frac{\partial Y_{CO_2}}{\partial x} + \rho v_y \frac{\partial Y_{CO_2}}{\partial y} = \frac{\partial}{\partial y}\left(\rho D \frac{\partial Y_{CO_2}}{\partial y}\right) + \dot{w}_{CO_2}''' , \quad \text{and} \tag{D}$$

$$\rho v_x \frac{\partial Y_{H_2O}}{\partial x} + \rho v_y \frac{\partial Y_{H_2O}}{\partial y} = \frac{\partial}{\partial y}\left(\rho D \frac{\partial Y_{H_2O}}{\partial y}\right) + \dot{w}_{H_2O}''' . \tag{E}$$

Using Equation A and the law of mass action

$$\frac{\dot{w}_F'''}{-1} = \frac{\dot{w}_{O_2}'''}{-1.92} = \frac{\dot{w}_{CO_2}'''}{2.20} = \frac{\dot{w}_{H_2O}'''}{0.72} = \dot{w}''' \tag{F}$$

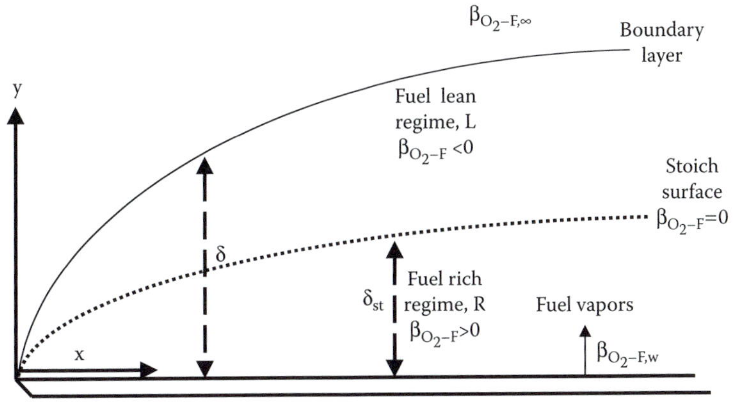

Figure 8.6 Schematic illustration of SZ variable for a boundary-layer problem.

where $\dot{w}'''$ (kg/m^3 sec) = $k_Y\ \rho^n\ Y_F^{nF}\ Y_{O2}^{nO2}$. Therefore, for the BL-type of combustion (Figure 8.6),

$$\rho v_x \frac{\partial Y_F}{\partial x} + \rho v_y \frac{\partial Y_F}{\partial y} = \frac{\partial}{\partial y}\left(\rho D \frac{\partial Y_F}{\partial y}\right) - \dot{w}''' , \tag{G}$$

$$\rho v_x \frac{\partial Y_{O2}}{\partial x} + \rho v_y \frac{\partial Y_{O2}}{\partial y} = \frac{\partial}{\partial y}\left(\rho D \frac{\partial Y_{O2}}{\partial y}\right) - 1.92\dot{w}''' , \tag{H}$$

$$\rho v_x \frac{\partial Y_{CO2}}{\partial x} + \rho v_y \frac{\partial Y_{CO2}}{\partial y} = \frac{\partial}{\partial y}\left(\rho D \frac{\partial Y_{CO2}}{\partial y}\right) + 2.2\,\dot{w}''' , \quad \text{and} \tag{I}$$

$$\rho v_x \frac{\partial Y_{H2O}}{\partial x} + \rho v_y \frac{\partial Y_{H2O}}{\partial y} = \frac{\partial}{\partial y}\left(\rho D \frac{\partial Y_{H2O}}{\partial y}\right) + 0.72\dot{w}''' . \tag{J}$$

Various combinations of these relations are possible. Dividing Equation H by (−1.92) and Equation G by (−1), then subtracting the fuel relation from the oxidizer relation, denoting SZ variables as

$$\beta_{O_2-F} = \frac{Y_{O_2}}{(-1.92)} - \frac{Y_F}{(-1)} = Y_F - \frac{Y_{O_2}}{1.92}, \text{ or } Y_F - \frac{Y_{O_2}}{\nu_{O_2}}, \tag{K}$$

and finally generalizing for any fuel:O_2 combination, we get

$$\rho v_x \frac{\partial \beta_{O_2-F}}{\partial x} + \rho v_y \frac{\partial \beta_{O_2-F}}{\partial y} = \frac{\partial}{\partial y}\left(\rho D \frac{\partial \beta_{O_2-F}}{\partial y}\right), \tag{L}$$

where $\beta_{O_2-F} = Y_F - \frac{Y_{O_2}}{\nu_{O2}}$ for any fuel and ν_{O2} = 1.92 for present example. The variable β is conserved scalar property since there is no source term. Similarly,

$$\rho v_x \frac{\partial \beta_{CO_2-F}}{\partial x} + \rho v_y \frac{\partial \beta_{CO_2-F}}{\partial y} = \frac{\partial}{\partial y}\left(\rho D \frac{\partial \beta_{CO_2-F}}{\partial y}\right), \tag{M}$$

where $\beta_{CO_2-F} = \left(\frac{Y_{CO_2}}{\nu_{CO2}} + Y_F\right)$ and ν_{CO2} = 2 for the present example. The term $\beta_{CO_2\text{-}F}$ denotes the fuel consumed to produce CO_2 plus the left over fuel. Another combination based on $\beta_{H_2O-F} = \frac{Y_{H_2O}}{0.72} - \frac{Y_F}{(-1)} = \frac{Y_{H_2O}}{0.72} + Y_F$ provides the relation

$$\rho v_x \frac{\partial \beta_{H_2O-F}}{\partial x} + \rho v_y \frac{\partial \beta_{H_2O-F}}{\partial y} = \frac{\partial}{\partial y}\left(\rho D \frac{\partial \beta_{H_2O-F}}{\partial y}\right). \tag{N}$$

Generalizing for any fuel, $\beta_{H2O\text{-}F} = Y_F + Y_{H2O}/\nu_{H2O}$.

Using the law of stoichiometry, $\dot{w}_F''' = -\dot{w}'''$ in Equation 8.68 and dividing the result by h_c, thermal enthalpy relation can be written in the form

$$\rho\, v_x \frac{\partial}{\partial x}\left(\frac{h_t}{h_c}\right) + \rho\, v_y \frac{\partial\left(\frac{h_t}{h_c}\right)}{\partial y} = \frac{\partial}{\partial y}\left(\frac{\lambda}{c_p}\frac{\partial\left(\frac{h_t}{h_c}\right)}{\partial y}\right) + \dot{w}'''. \tag{O}$$

Adding Equation G and letting $\beta_{t-f} = \frac{h_t}{h_c} + Y_F$, we obtain

$$\rho\, v_x \frac{\partial \beta_{ht-F}}{\partial x} + \rho\, v_y \frac{\partial \beta_{ht-F}}{\partial y} = \frac{\partial}{\partial y}\left(\rho\, D \frac{\partial \beta_{ht-F}}{\partial y}\right). \tag{P}$$

■ ■ ■

In general form the SZ relation for BL problem is

$$\rho\, v_x \frac{\partial \beta}{\partial x} + \rho\, v_y \frac{\partial \beta}{\partial y} = \frac{\partial}{\partial y}\left(\rho\, D \frac{\partial \beta}{\partial y}\right), \tag{8.69}$$

where $\beta = \beta_{O_2-F}, \beta_{CO_2-F}, \beta_{H_2O-F}$ and β_{h_t-F}. The condition Le $= \frac{\lambda}{\rho D c_p} = 1$ must be invoked for $\beta_{ht\text{-}F}$ in order to couple the conservation relations for the various species and thermal enthalpy. The SZ relation is widely used to solve liquid and solid fuel combustion problems that can be simplified using a one-step global reaction. Generalizing for any coordinate system,

$$\rho \vec{v}.\nabla \beta = \nabla.(\rho D \nabla \beta) \tag{8.70}$$

Defining normalized SZ (NSZ) variable,

$$\phi = \frac{\beta - \beta_\infty}{\beta_w - \beta_\infty}, \tag{8.71}$$

and if $\beta_w - \beta_\infty$ is not a function of x (true for many problems as illustrated in later chapters), then Equation 8.69 has the form,

$$\rho\, v_x \frac{\partial \phi}{\partial x} + \rho\, v_y \frac{\partial \phi}{\partial y} = \frac{\partial}{\partial y}\left(\rho D \frac{\partial \phi}{\partial y}\right), \tag{8.72}$$

which can be solved with the appropriate boundary conditions, e.g., $\phi = 1$ at $y = 0$, $\phi = 0$ as $y \to \infty$, and $\phi = 0$ at $x = 0$. Generalizing for any system of coordinates,

$$\rho \vec{v}.\nabla\phi = \nabla.(\rho D \nabla \phi). \tag{8.73}$$

For the inert (nonreacting) species nitrogen,

$$\rho v_x \frac{\partial Y_{N_2}}{\partial x} + \rho v_y \frac{\partial Y_{N_2}}{\partial y} = \frac{\partial}{\partial y}\left(\rho D \frac{\partial Y_{N_2}}{\partial y}\right), \tag{8.74}$$

which resembles the SZ relations. The NO is sometimes produced from N_2 but only in trace amounts and, hence, N_2 remains almost inert. Any property that is free from sinks and sources and obeys the above relation is called *conserved property*. Note that β is a conserved property or a scalar. The value of $\beta_{CO2\text{-}F}$ is altered by chemical reactions and species transport. Equations such as Equation 8.69 and Equation 8.72 can be solved using principles of fluid mechanics or heat transfer.

The previous example illustrates the SZ formulation for a single-step reaction. For fuel-rich regime R, $\beta_{O2\text{–}F} > 0$ because $Y_F > Y_{O2}/\nu_{O2}$, and for fuel-lean regime L, $\beta_{O2\text{–}F} < 0$ (Figure 8.6). In both the lean and rich regimes, temperatures (T) are typically low. T will be highest when $\beta_{O2\text{–}F} = 0$, called the *stoichiometric surface*, where $Y_F = Y_{O2}/\nu_{O2}$ and $y = \delta_{st}$. Because T is highest at stoichiometric surface, the reaction is rapid at $y = \delta_{st}$. Thus, it will not allow fuel to escape unburned for $y > \delta_f$ and hence $Y_F = 0$ for $y > \delta_f$; similarly, O_2 will not be allowed to seep into rich regimes, i.e., $Y_{O2} = 0$ for $y < \delta_{st}$. The stoichiometric surface acts as a flame surface with the fastest chemical reaction and $\delta_{st} \approx \delta_f$.

8.6.1.2 Fuel Having Two Components

The following example illustrates a procedure for a fuel having two components.

Example 4

When coal is introduced into hot air, it releases combustible gases such as CH_4, C_2H_6, CO, etc. (called *pyrolysate*). We will assume the whole pyrolysate to be represented by a mixture of CH_2 and CO. The gases CH_2 and CO then oxidize in a gas phase:

$$CH_2 + 1.5O_2 \rightarrow CO_2 + H_2O \tag{I}$$

$$CO + 0.5O_2 \rightarrow CO_2 \tag{II}$$

We will assume that a coal plate is subjected to flowing air and treat the problem as for a 2-D boundary-layer problem.

(a) Write down the gas-phase conservation equations for species CO, CH_2, CO_2, O_2, H_2O, N_2 and thermal enthalpy h_t.

(b) For these two reactions taking place in gas phase define suitable variable β_{kij} for species k = CO_2, O_2, H_2O with i = CH_2, j = CO and reduce the conservation equations in Part a in the SZ form.

Solution

$M_{CH2} = 14.03$, $\nu_{O2,I} = 3.42$, $\nu_{CO2,I} = 3.14$, $\nu_{H2O,I} = 1.28$, $M_{CO} = 28.01$, $\nu_{O2,II} = 0.57$, $\nu_{CO2,II} = 1.57$

(a) From law of stoichiometry for Reaction I and Reaction II,

$$\frac{\dot{w}'''_{CH2}}{(-1)} = \frac{\dot{w}'''_{O2}}{-3.42} = \frac{\dot{w}'''_{CO2}}{+3.4} = \frac{\dot{w}'''_{H2O}}{+1.57} = \dot{w}'''_{I} \tag{A}$$

$$\frac{\dot{w}'''_{CO}}{(-1)} = \frac{\dot{w}'''_{O2}}{-0.57} = \frac{\dot{w}'''_{CO2}}{+1.57} = \dot{w}'''_{II} \tag{B}$$

$$\rho v_x \frac{\partial Y_{N2}}{\partial x} + \rho v_y \frac{\partial Y_{N2}}{\partial y} = \frac{\partial}{\partial y}\left(\rho D \frac{\partial Y_{N2}}{\partial y}\right) \tag{C}$$

$$\rho v_x \frac{\partial h_t}{\partial x} + \rho v_y \frac{\partial h_t}{\partial y} = \frac{\partial}{\partial y}\left(\rho D \frac{\partial h_t}{\partial y}\right) + \dot{w}'''_{I}\, h_{c,I} + \dot{w}'''_{II}\, h_{c,II} \tag{D}$$

$$\rho v_x \frac{\partial Y_{O2}}{\partial x} + \rho v_y \frac{\partial Y_{O2}}{\partial y} = \frac{\partial}{\partial y}\left(\rho D \frac{\partial Y_{O2}}{\partial y}\right) + \dot{w}'''_{I}(-3.42) + \dot{w}'''_{II}(-0.57) \tag{E}$$

$$\rho v_x \frac{\partial Y_{CO_2}}{\partial x} + \rho v_y \frac{\partial Y_{CO_2}}{\partial y} = \frac{\partial}{\partial y}\left(\rho D \frac{\partial Y_{CO_2}}{\partial y}\right) + \dot{w}'''_{I}(3.14) + \dot{w}'''_{II}(1.57) \tag{F}$$

$$\rho v_x \frac{\partial Y_{H_2O}}{\partial x} + \rho v_y \frac{\partial Y_{H_2O}}{\partial y} = \frac{\partial}{\partial y}\left(\rho D \frac{\partial Y_{H_2O}}{\partial y}\right) + \dot{w}'''_{I}(1.28) \tag{G}$$

$$\rho v_x \frac{\partial Y_{CH2}}{\partial x} + \rho v_y \frac{\partial Y_{CH2}}{\partial y} = \frac{\partial}{\partial y}\left(\rho D \frac{\partial Y_{CH2}}{\partial y}\right) + \dot{w}'''_{I}(-1) \tag{H}$$

$$\rho v_x \frac{\partial Y_{CO}}{\partial x} + \rho v_y \frac{\partial Y_{CO}}{\partial y} = \frac{\partial}{\partial y}\left(\rho D \frac{\partial Y_{CO}}{\partial y}\right) + \dot{w}'''_{II}(-1)\,. \tag{I}$$

Let us define

$$\beta_{O_2-CH2-CO} = 3.42 Y_{CH2} + 0.57 Y_{CO} - Y_{O_2}. \tag{J}$$

This variable $\beta_{O_2-CH2-CO}$ will be called the modified SZ variable. Then Equation H × 3.42 + Equation I × 0.57 – Equation E

$$\rho v_x \frac{\partial \beta_{O_2-CH2-CO}}{\partial x} + \rho v_y \frac{\partial \beta_{O_2-CH2-CO}}{\partial y} = \frac{\partial}{\partial y}\left(\rho D \frac{\partial \beta_{O_2-CH2-CO}}{\partial y}\right). \tag{K}$$

Similarly

$$\rho v_x \frac{\partial \beta_{\text{CO2-CH2-CO}}}{\partial x} + \rho v_y \frac{\partial \beta_{\text{CO2-CH2-CO}}}{\partial y} = \frac{\partial}{\partial y}\left(\rho D \frac{\partial \beta_{\text{CO2-CH2-CO}}}{\partial y}\right) \quad \text{(L)}$$

where

$$\beta_{\text{CO}_2\text{-CH2-CO}} = (Y_{\text{CO}_2} + 3.14 Y_{\text{CH2}} + 1.57\ Y_{\text{CO}}). \quad \text{(M)}$$

$\beta_{\text{CO}_2\text{-CH2-CO}}$ = CO_2 left over + CO_2 produced by CH_2 + CO_2 produced by CO

$$\rho v_x \frac{\partial \beta_{\text{H}_2\text{O-CH2-CO}}}{\partial x} + \rho v_y \frac{\partial \beta_{\text{H}_2\text{O-CH2-CO}}}{\partial y} = \frac{\partial}{\partial y}\left(\rho D \frac{\partial \beta_{\text{H}_2\text{O-CH2-CO}}}{\partial y}\right), \quad \text{(N)}$$

$$\beta_{\text{H}_2\text{O-CH2-CO}} = Y_{\text{H}_2\text{O}} + 1.28\ Y_{\text{CH2}}\ . \quad \text{(O)}$$

$$\beta_{h_t\text{-CH2-CO}} = h_t + Y_{\text{CH2}}\, h_{c,\text{I}} + Y_{\text{CO}}\, h_{c,II} \quad \text{(P)}$$

$$\rho v_x \frac{\partial \beta_{ht\text{-CH2-CO}}}{\partial x} + \rho v_y \frac{\partial \beta_{ht\text{-CH2-CO}}}{\partial y} = \frac{\partial}{\partial y}\left(\rho D \frac{\partial \beta_{ht\text{-CH2-CO}}}{\partial y}\right).$$

Writing in general form,

$$\rho v_x \frac{\partial \beta}{\partial x} + \rho v_y \frac{\partial \beta}{\partial y} = \frac{\partial}{\partial y}\left(\rho\ D \frac{\partial \beta}{\partial y}\right) \quad \text{(Q)}$$

where β's are suitably defined as before.

Remarks

Because heating-values-based stoichiometric O_2 are approximately constant for most fuels (see Chapter 2 and Chapter 4) Equation D can be rewritten as,

$$\rho v_x \frac{\partial \frac{h_t}{h_{c,\text{O}_2}}}{\partial x} + \rho v_y \frac{\partial \frac{h_t}{h_{c,\text{O}_2}}}{\partial y} = \frac{\partial}{\partial y}\left(\rho D \frac{\partial \frac{h_t}{h_{c,\text{O}_2}}}{\partial y}\right) + \left|\dot{w}'''_{\text{I}}\right| \nu_{\text{O2,I}} + \left|\dot{w}'''_{\text{II}}\right| \nu_{\text{O2,II}} \quad \text{(R)}$$

where $h_{c,O2} = h_{c,I}/\nu_{O2,I} \approx h_{c,II}/\nu_{O2,II}$. Adding Equation E with the preceding equation,

$$\beta_{ht\text{-O2}} = \frac{h_t}{h_{c,\text{O}_2}} + Y_{\text{O2}}. \quad \text{(S)}$$

■ ■ ■

The previous example illustrates a SZ procedure for two-step reactions. Note that for fuel-rich regimes, $\beta = \beta_{\text{O}_2\text{-CH}_2\text{-CO}} > 0$, and fuel-lean regimes $\beta = \beta_{\text{O}_2\text{-CH}_2\text{-CO}} < 0$. Le = 1 is invoked for $\beta_{\text{h}_t\text{-O2}}$ or in case of coupling between species and thermal enthalpy. The surface $\beta = 0$ has the highest temperature. We make use of two component reactions in Chapter 10 when dealing with combustion of multicomponent drops. Unless otherwise mentioned, the variable β implies only a single-step reaction scheme.

8.6.2 Relation between Mixture Fraction (f_M) and an SZ Variable

Consider a hypothetical case of fossil fuel combustion for which the products of combustion consist of only CO_2 and H_2O. Considering the sum of the two species,

$$\rho v_x \frac{\partial \beta_{\text{P-F}}}{\partial x} + \rho v_y \frac{\partial \beta_{\text{P-F}}}{\partial y} = \frac{\partial}{\partial y}\left(\rho D \frac{\partial \beta_{\text{P-F}}}{\partial y}\right) \quad (8.75)$$

where P represents the sum of the products CO_2 and H_2O that are formed with a stoichiometric coefficient ν_P. For instance, on a kg basis

$$1\ C_5H_8O_2 + 1.92\ O_2 \rightarrow (2.2\ CO_2 + 0.72\ H_2O) = 2.92 \text{ kg of products,}$$

i.e., $\nu_P = 2.92 = \nu_{O_2} + \nu_F$. Consequently,

$$\beta_{\text{P-F}} = \frac{Y_P}{2.92} - \frac{Y_F}{(-1)} = \frac{Y_P}{2.92} + Y_F\ .$$

This expression represents a sum of the amount of fuel left unburned at a given location (e.g., Y_F, kg of fuel/kg of the mixture) and the amount of fuel used in the products per unit mass of the mixture (e.g., $Y_P/2.92$). In other words, $\beta_{\text{F-P}}$ = (unburned fuel + mass of the fuel used in products) per unit mass of the mixture. This represents the net mass of material that originates from the fuel in a unit mass of the mixture at location given by (x, y, z), i.e.,

$$\beta_{\text{F-P}} = \text{fuel mass originating in the fuel stream} \div \text{local mass} = f_M. \quad (8.76)$$

As before, the term $(1 - \beta_{\text{F-P}}) = (1 - f_M)$ represents the fraction of the mixture originating from air.

8.6.3 Plug Flow Reactor (PFR)

The definition of PFR is given in Chapter 7. The properties vary only in the "x" direction. There are no lateral property gradients. It can be shown that

$$\rho v_x \frac{d\beta}{dx} = 0 \quad (8.77)$$

and hence β is constant (Exercise problem 8.24).

Thus, a plot of Y_{O2}, Y_{CO2}, and Y_{H2O} with respect to Y_F is linear. Once the fuel conservation equation is solved, the β relation can be used to solve for the other species.

Example 5

A hot, flat, horizontal plate at a uniform temperature T_w is cooled by a cold air flowing past the plate. The boundary-layer temperature profile is approximately

$$T^* = (T_\infty - T)/(T_\infty - T_w) = ((3/2)\,(y/\delta) - 1/2\,(y/\delta)^3), \qquad \text{(A)}$$

which satisfies the boundary conditions $T^* = 1$ at $y = 0$ and $T^* = 0$ at $y = \delta$. Equation A is used to solve the equation for heat transfer (in the absence of chemical reactions), namely,

$$\rho v_x c_p \frac{\partial T}{\partial X} + \rho v_y c_p \frac{\partial T}{\partial y} = \frac{\partial}{\partial y}\left(\lambda \frac{\partial T}{\partial y}\right). \qquad \text{(B)}$$

Further, at $x = 0$, $T = T^* = 0$. Show that if NSZ

$$\Phi = (\beta - \beta_\infty)/(\beta_w - \beta_\infty), \qquad \text{(C)}$$

for a reacting flow, then the NSZ profile Φ is similar to the T^* distribution if $\lambda/c_p = \rho D$ and $\{v\}_{y=0} = 0$.

Solution

For a nonreacting thermal boundary layer, Equation B is written as

$$\rho v_x \frac{\partial T}{\partial x} + \rho v_y \frac{\partial T}{\partial y} = \frac{\partial}{\partial y}\left(\rho D \frac{\partial T}{\partial y}\right) \qquad \text{(D)}$$

Writing Equation D in nondimensional form,

$$\rho v_x \frac{\partial T^*}{\partial x} + \rho v_y \frac{\partial T^*}{\partial y} = \frac{\partial}{\partial y}\left(\rho D \frac{\partial T^*}{\partial y}\right) \qquad \text{(E)}$$

The T^* solution is obtained by satisfying the boundary conditions $T^* = 1$ at $y = 0$ and $T^* = 0$ at $y = \delta$. Further, at $x = 0$, $T^* = 0$. The corresponding NSZ is

$$\rho v_x \frac{\partial \phi}{\partial x} + \rho v_y \frac{\partial \phi}{\partial y} = \frac{\partial}{\partial y}\left(\rho D \frac{\partial \phi}{\partial y}\right). \qquad \text{(F)}$$

At first glance, the corresponding boundary conditions are $\Phi = 1$ at $y = 0$ and $\Phi = 0$ at $y = \delta$. At $x = 0$, $\Phi = 0$. Because of the similarity between the dimensionless equations and the boundary conditions, it is clear that the two solutions are also similar; i.e., $\Phi \approx T^*$; however, in case of combustion $\{v_y\}_{y=0} \neq 0$. Thus, $\Phi_{exact} \neq T^*$ (see Subsection 8.6.4); however under low mass transfer $v_y \approx 0$ and hence $\Phi \simeq T^*$.

Remarks

(i) At $y = 0$,

$$\frac{\partial \phi}{\partial y} = \frac{\partial T^*}{\partial y}. \qquad \text{(G)}$$

(ii) For a flat plate,

$$\rho v_x \frac{\partial v_x}{\partial x} + \rho v_y \frac{\partial v_x}{\partial y} = \frac{\partial}{\partial y}\left(\rho D \frac{\partial v_x}{\partial y}\right). \tag{H}$$

Defining

$$v_x^* = (v_{x\infty} - v_x)/(v_{x\infty} - v_{x,w}) = (v_{x\infty} - v_x)/(v_{x\infty} - 0)$$
$$= (1 - v_x/v_{x\infty}) = 1 - \{(3/2)\,(y/\delta) - 1/2(y/\delta)^3\}. \tag{I}$$

However, the assumed polynomial profile T* is same as $\Phi = v_x^*$. We see that with the SZ transformation, a conservation equation with a source term can be represented in a form similar to relations used to solve heat transfer and fluid mechanics problems. Consequently, those solutions are applicable to combustion problems.
(iii) Recall that

$$\phi = \frac{f_M - f_{M,\infty}}{f_{M,W} - f_{M,\infty}} \tag{J}$$

so that the f_M profile can also be obtained.
(iv) If the chemistry associated with a process is assumed to be a simple one-step reaction, we are required only to solve the fuel species conservation equation for a chemically reacting flow.

8.6.4 Combustion of Liquids and Solids

The burn rates of liquid and solid fuels in the presence of forced and free convection and stagnation flows can be solved using BL-type equations (Chapter 11). A solid plastic-like Plexiglas ($C_5H_8O_2$) pyrolyzes at about 600 K(T_w). The pyrolysis products react with gas-phase oxygen and burn, establishing a flame at some distance from the solid surface. Similarly, liquid *n*-hexane boils at about T_{BP} = 345 K (exact solution will yield T_w slightly less than T_{BP}) and its vapors burn in oxygen at some distance from the liquid pool or drop surface. In this section, we present general wall (or surface) relations for the fuel mass fraction $Y_{F,w}$, fuel temperature T_w, and the flame temperature T_f for the aforementioned situations. Many flow situations and drop combustions are schematically described in Figure 8.7.

8.6.4.1 Interface Conservation Equations

The SZ conservation equations require boundary conditions in terms of SZ variable. Consider the burning of a fuel over a surface, where y is the normal coordinate.

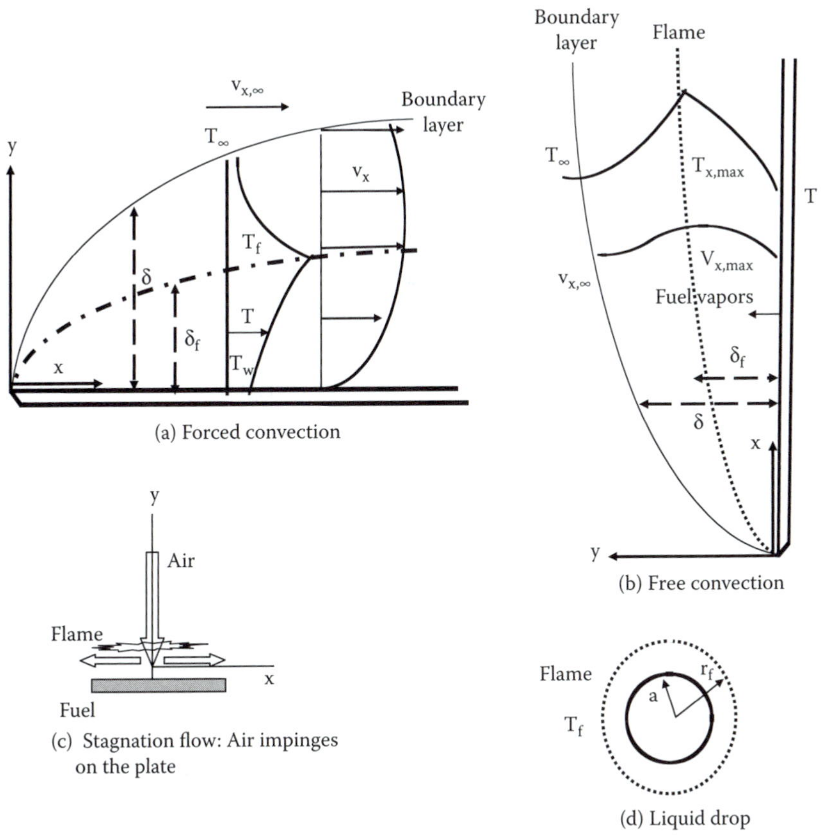

Figure 8.7 Schematic illustration of burning in the presence of a velocity boundary layer.

Species: Assuming that the surface is continuously replenished with fuel (e.g., Figure 8.8), the liquid will move with velocity v_ℓ toward interface the IF, whereas gases move away from IF with velocity v_{yw}. Applying species conservation for a single-component fuel F,

$$\rho_F v_{F,w} = \rho_w v_{y,w} Y_{F,w} - (\rho D\ \partial Y_F/\partial y)_{y=0} = \dot{m}_F'' = \dot{m}''\ Y_{F,c} \quad (8.78)$$

where $\dot{m}''$ is surface-mass-loss rate per unit area $= \rho_w v_{y,w}$. The value of $Y_{F,c}$ (condensate fuel mass fraction) on the liquid or solid side of the surface need not equal unity. Equation 8.78 is rewritten as

$$\dot{m}'' = \rho_w v_{y,w} = -(\rho D\ \partial Y_F/\partial y)_{y=0}/(Y_{F,c} - Y_{F,w}). \quad (8.79)$$

We assume that oxygen does not dissolve into the liquid or solid fuel so that

$$\rho_w v_{y,w}\ Y_{O2,w} - (\rho D\ \partial Y_{O2}/\partial y)_{y=0} = \dot{m}_{O2}'' = 0. \quad (8.80)$$

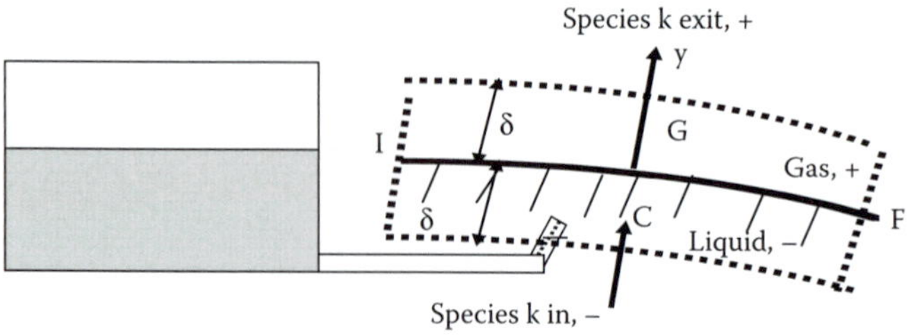

Figure 8.8 Cartesian elemental control volume for conservation equations.

Similar relations are also valid for CO_2 and H_2O. Generalizing for fuel combustion or evaporation,

$$\rho_w v_{y,w}\, Y_{k,w} - (\rho D\, \partial Y_k/\partial y)_{y=0} = \dot{m}''$$
$$= 0,\ k = CO_2, H_2O, N_2, O_2, \ldots \tag{8.81}$$

With $k = O_2$ and dividing the oxygen relation by ν_{O2}, we obtain

$$\rho_w v_{y,w}\, Y_{O2,w}/\nu_{O2} - (\rho D\, \partial(Y_{O2}/\nu_{O2})/\partial y)_{y=0} = \dot{m}''_{\ O2}/\nu_{O2} = 0. \tag{8.82}$$

Subtracting it from Equation 8.78 we have

$$\dot{m}''\, \beta_{O2\text{-}F,w} - \rho D\, (\partial \beta_{O2\text{-}F/}\partial y)_0 = \dot{m}''\, Y_{F,c}, \tag{8.83}$$

More generally,

$$\dot{m}''\, \beta_w - \rho D\, (\partial \beta/\partial y)_0 = \dot{m}''\, F_\beta \tag{8.84}$$

For $\beta = \beta_{O2\text{-}F}$, $F_\beta = F_{O2\text{-}F} = Y_{F,c}$.

$$\dot{m}''\, (F_{O2\text{-}F} - \beta_{O2\text{-}F,w}) = -\rho D\, (\partial \beta_{\ O2\text{-}F}/\partial y)_{y=0} \tag{8.85}$$

For a generic SZ variable β,

$$\dot{m}''\, (F_\beta - \beta_w) = -\rho D\, (\partial \beta/\partial y)_{y=0} \tag{8.86}$$

Dividing the above relation by $(\beta_w - \beta_\infty)$, using NSZ and rearranging

$$\dot{m}'' = -\rho D\ B\ (\partial \phi/\partial y)_w, \tag{8.87}$$

where the transfer number B is defined as

$$B = (\beta_w - \beta_\infty)/(F_\beta - \beta_w). \tag{8.88}$$

Typically, B numbers are defined when $Y_{O2w} = 0$ and $\beta = \beta_{ht\text{-}O2}$ are used for defining transfer number. However, if $\beta = \beta_{O2\text{-}F}$, then

$$\dot{m}'' = -\rho D\ B_{O2\text{-}F}\ (\partial \phi/\partial y)_w, \tag{8.89}$$

$$\text{where} \quad B_{O2\text{-}F} = (\beta_{O2\text{-}F,w} - \beta_{O2\text{-}F,\infty})/(F_{\beta O2\text{-}F} - \beta_{O2\text{-}F,w})$$
$$\approx (Y_{F,w} + Y_{O2,\infty}/\nu_{O2})/(Y_{F,c} - Y_{F,w}). \tag{8.90}$$

Alternatively, if we consider $CO_2 - O_2$ coupling, then

$$\dot{m}'' = -\rho D\; B_{CO2\text{-}O2}\; (\partial\phi/\partial y)_w, \tag{8.91}$$

where $B_{CO2\text{-}O2} = (\beta_{CO2\text{-}O2,w} - \beta_{CO2\text{-}O2,\infty})/(F_{CO2\text{-}O2} - \beta_{CO2\text{-}O2,w})$, and $F_\beta = 0$ so that

$$B_{CO2\text{-}O2} = (Y_{CO2,w} - Y_{O2,\infty}/\nu_{O2})/(0 - Y_{CO2,w}). \tag{8.92}$$

Because $\dot{m}''$ and $(\partial\phi/\partial y)_w$, are the same, Equation 8.87, Equation 8.89, and Equation 8.91 must all be equal to each other. Also, it is apparent that $B_{CO2\text{-}O2} = B_{O2\text{-}F} = B_{ht\text{-}O2} = B_{ij}$ which will lead to solutions for wall mass fraction of various species.

Energy: From Figure 8.8, it can be seen that liquid enters IF from region (C) into region (G). The fuel needs to be heated from ambient temperature (i.e., temperature in tank) to T_w (close to boiling point for combustion case) and, hence, heat must be supplied near the interface, i.e., $\lambda(\partial T/\partial y)_{0-} \neq 0$. Consider the c.v. containing (C + G) of thickness 2δ and area A including conductive and bulk species enthalpy fluxes

$$dE_{cv}/dt = \dot{Q}_{cv} - \dot{W}_{cv} + \sum \dot{m}_{k,i}\,(h + ke + pe)_{k,i} - \sum \dot{m}_{k,e}\,(h + ke + pe)_{k,e}$$

$$\{d\,(\rho_c\, A\delta\, h_{t,-} + \rho_g\, A\,\delta h_{t,+}\}/dt = \{\dot{q}''_-\}\,A + \{\dot{q}''_+\}\,A$$
$$+\left\{\sum_i \dot{m}_{k,-} h_{k,-}\right\} - \left\{\sum_i \dot{m}_{k,+} h_{k,+}\right\}, \tag{8.93}$$

where + denotes the gas side, – denotes the condensate side, *heat fluxes are only due to temperature gradients as mass fluxes of each species are absolute fluxes.* Using Fourier law,

$$\dot{q}'' = -\vec{q}''.\,d\vec{A} \quad \text{where} \quad \vec{q}'' = -\lambda\vec{\nabla}T \tag{8.94}$$

The heat flux enters from the hotter gas phase into the interface and then leaves the interface IF toward the colder condensate phase. The vectorial product of the heat flux with the outer normal vector of surface area is positive in the first law because of sign convention of heat "in."

$$\{d\,(\rho_c\, A\delta\, h_{t,-} + \rho_g\, A\,\delta h_{t,+})\}/dt = -\lambda(\partial T/\partial y)_{y=0-}\, A + \lambda(\partial T/\partial y)_{y=0+}\, A$$
$$+\left\{\sum_i \dot{m}_{k,-} h_{k,-}\right\} - \left\{\sum_i \dot{m}_{k,+} h_{k,+}\right\} \tag{8.95}$$

As $\delta \to 0$ and for nonreacting interface (interface condition for a reacting surface e.g., solid carbon burning in air treated in Chapter 9).

$$0 = -\lambda(\partial T/\partial y)_{y=0-}\} + \lambda(\partial T/\partial y)_{y=0+} + \left\{\sum_i \dot{m}''_{k,-} h_{k,-}\right\} - \left\{\sum_i \dot{m}''_{k,+} h_{k,+}\right\}. \tag{8.96}$$

Because fuel species are the only species entering and leaving, $\dot{m}''_k = \dot{m}''_F = \dot{m}''$ and $h_k = h_F$. Typically, we use $h = h_{f,k} + h_{t,k}$ with $h_{f,k}$ as the enthalpy of formation at 298 K. Because the interface is inert, $h_{F+} - h_{F-} = h_{t,F+} - h_{t,F-} = L$ for nonreacting interface. If the fluid temperature in a tank is at $T_{0,liq}$, then $-\lambda(\partial T/\partial y)_{y=0-} = \dot{m}''\, c_p\, (T_w - T_{0,liq})$.

$$q_w'' = -\lambda(\partial T/\partial y)_{y=0+} = -\dot{m}''\, L_{mod} \tag{8.97}$$

where $L_{mod} = L + c_p(T_w - T_{0,liq})$. If $T_{0,liq} = T_w$, i.e., $[\dot{q}_w'']_-$, the heat flux from the liquid side, is zero or $(dT/dy) = 0$. Then $L_{mod} = L$ and hence

$$q_w'' = -\lambda(\partial T/\partial y)_{y=0+} - \dot{m}''\, L \tag{8.98}$$

which accounts for heat diffusing toward the surface. Hereafter, we will use mainly Equation 8.98. Dividing Equation 8.98 by h_c and adding with Equation 8.82,

$$\dot{m}''\, \beta_w - \rho D\, (\partial\beta/\partial y)_0 = \dot{m}''\, F_\beta \tag{8.99}$$

where $\beta = \beta_{ht\text{-}O2}$, $F_\beta = -L/h_c$, $h_t = \int_{T_w}^{T} c_p dT$.

$$\dot{m}''\, (F_\beta - \beta_{,w}) = -\rho D\, (\partial\beta/\partial y)_{y=0}, \quad \text{i.e.,}$$

$$\dot{m}'' = -\rho D\, B_{ht\text{-}O2}\, (\partial\phi/\partial y)_w, \tag{8.100}$$

where

$$B_{ht\text{-}O2} = (\beta_{ht\text{-}O2,w} - \beta_{ht\text{-}O2,\infty})/(F_{ht\text{-}O2} - \beta_{ht\text{-}O2,w}),\ F_\beta = F_{ht\text{-}O2} = -L/h_c \tag{8.101}$$

$$\beta_{ht\text{-}O2,w} = h_{t,w}/h_c + Y_{O2w}/\nu_{O2} = 0 + Y_{O2w}/\nu_{O2} \approx 0. \tag{8.102}$$

An additional B number could be defined as

$$B_{N2} = (Y_{N2,w} - Y_{N2,\infty})/(F_{N2} - Y_{N2,w}), \tag{8.103}$$

where $F_{N2} = 0$.

Typically the following is used for burn rate within $Y_{O2,w} = 0$, $h_{t,w} = 0$

$$\dot{m}'' = -\rho D\, B\, (\partial\phi/\partial y)_w, \tag{8.104}$$

where

$$B = B_{ht\text{-}O2} = (c_p(T_\infty - T_w)/L + (h_c/L)Y_{O2,\infty}/\nu_{O2})$$
$$= c_p(T_\infty - T_w)/L + s\,(h_c/L) \qquad (8.105)$$

where $s = Y_{O2,\infty}/\nu_{O2}$, (kg of ambient O_2/kg of air)/(kg of stoichiometric O2/kfg of fuel) = $1/\{A{:}F\}_{st}$. The term B is called a transfer number, typically defined with $B_{ht\text{-}O2}$. The B number is used only for problems involving diffusion-controlled combustion (Chapter 9 to Chapter 12). Equating B with B_{O2-F} and letting $Y_{F,c} = 1.0$,

$$B = (Y_{F,w} + Y_{O2,\infty}/\nu_{O2})/(1 - Y_{F,w}) = (c_p(T_\infty - T_w)/h_c + Y_{O2,\infty}/\nu_{O2})/(-L/h_c), \qquad (8.106)$$

Solving Equation 8.106 for $Y_{F,w,}$,

$$Y_{F,w} = 1 - \{(1+s)/(1+B)\}, \quad \text{where} \qquad (8.107)$$

$$s = Y_{O2,\infty}/\nu_{O2}. \qquad (8.108)$$

Typically, T_∞, L, h_c, ν_{O2}, and $Y_{O2,\infty}$ are specified for a particular problem.

8.6.4.2 Numerical Solution

Numerical solution procedure for exact solutions of boundary-layer combustion problems is as follows:

1. Let us first ignore $\partial v_x/\partial x$ at a given location x. The v_x is second order with y in the momentum equation and, hence, with known values of $v_x = 0$ at $y = 0$ and $v_x = v_{x\infty}$ one can solve $v_x = v_x(x, y)$ if v_y is known. v_y is readily determined from mass conservation for assumed v_{yw} or $v_y(0)$ or burn rate because $\dot{m}'' = \rho_w v_{y,w} = \rho_w\, v_y(0)$. For low mass transfer, $v_y(0)$ is extremely low. Now the term $\partial v_x/\partial x$ can be determined if we started from $x = 0$ and marched up to location x.
2. Then, one can solve the β equation with $\beta = \beta_{ht\text{-}O2}$ (Equation 8.69) with known v_x and v_y. For $\beta = \beta_{ht\text{-}O2,}$ surface properties are known because $\beta_w = \beta_{ht\text{-}O2,w} = h_{t,w}/L + Y_{O2,w}/\nu_{O2} = 0$ for diffusion-controlled combustion, and $F_\beta = -L/h_c$. Similarly, $\beta_{ht\text{-}O2,\infty} = h_{t,\infty}/L + Y_{O2,\infty}/\nu_{O2} = c_p\,(T_\infty - T_w)/L + Y_{O2,\infty}/\nu_{O2}$ is also known if one assumes that $T_w = T_{BP}$. Thus, we determine the NSZ profile in Equation 8.71. Then, gradient $\left(\frac{\partial\Phi}{\partial y}\right)_{y=0}$ is evaluated. Using Equation 8.105, one can find find B assuming T_w–T_{Bp} and hence $\dot{m}''$ from Equation 8.104. If assumed value v_{yw} is correct , then the values should match; if not, iterate with different set of v_{yw}.
3. If $T_w \neq T_{BP}$, knowing Y_{Fw} Equation 8.107, we can find $X_{Fw} = Y_{Fw}\,M/M_F$ where M of the mixture is calculated once wall mass fractions of all species are determined, and, hence, $p_{FW} = X_{Fw}\,P$; from known p_{FW}, T_w can be determined from saturation (phase equilibrium) relations of the fuel.

Example 6

Consider the burning of *n*-hexane in air at an ambient temperature $T_\infty = 600$ K. Use the properties L = 335 kJ/kg, h_c = 45105 kJ/kg, T_{BP} = 342.15 K (Table A.2A) c_p = 1.17 kJ/kg K, P = 100 kPa, ln P^{sat} (bar) = A − (B/(T + C)), A = 10.15, B = 3473.3, and C = 0. Obtain the wall or surface properties Y_{Fw}, and T_w.

Solution

The fuel burns according to the chemical relation

$$C_6H_{14} + 9.5\ O_2 = 6\ CO_2 + 7\ H_2O, \quad \text{i.e.,}$$

$$\nu_{O2} = 9.5 \times 32/86 = 3.527,\ s = Y_{O2,\infty}/\nu_{O2} = 0.0652 = 1/(A{:}F)_{st},$$

$$B = 1.17(600 - 342.15)/335 + 0.23 \times 45105/(3.527 \times 335) = 9.68.$$

Using Equation 8.107,

$$Y_{F,w} = 0.9003 \quad \text{and} \quad X_{Fw} = M_{mix} Y_{Fw}/M_F.$$

Assuming that $1 - Y_{Fw} = Y_{N2w,}$ with M_{N2} = 28 kg/kmol,

$$M_{mix} = 1/(0.900/86 + 0.1/28) = 70 \text{ kg/kmol of the mixture.}$$

Then, X_{Fw} = 0.75, p_{FW} = 0.75 × 1 = 0.75 bar, and

$$P^{sat}(\text{bar}) = 0.75 = \exp\,(10.15 - 3473.3/T_w).$$

Determing T_w, reevaluating B, and obtaining $Y_{F,w}$ and $X_{F,w}$ the procedure is repeated until convergence is obtained:

$$T_w = 332.7 \text{ K},\ Y_{Fw} = 0.9006, \quad \text{and} \quad B = 9.7.$$

Remarks

(i) When we assumed $T_w = T_{BP}$ = 342.15 K, then B = 9.68 which is close to exact value of 9.7.

(ii) The value of Y_{CO2w}, Y_{H2Ow}, and Y_{N2w} can also be determined. These properties are valid for any flow system (Chapter 11) or droplet burning (Chapter 10).

8.6.4.3 Thin-Flame or Flame Surface Approximation

Flame Temperature: Recall that $\phi = (\beta - \beta_\infty)/(\beta_w - \beta_\infty)$ for any combination. If $\beta_{O2\text{-}F} = Y_F - Y_{O2}/\nu_{O2}$, then at the flame location the reaction is so fast that $Y_F \approx 0$, $Y_{O2} \approx 0$ assuming that all the fuel and oxidizer are consumed, $\beta_{O2\text{-}F,f} = 0$. Hence, from normalized SZ variable, for $O_2 - F$

$$\phi_f = (\beta_{O2\text{-}F,f} - \beta_{O2\text{-}F,\infty})/(\beta_{O2\text{-}F,w} - \beta_{O2\text{-}F,\infty}) = s/(Y_{F,w} + s), \tag{8.109a}$$

$$\phi_f = (\beta_{ht\text{-}O2,f} - \beta_{ht\text{-}O2,\infty})/(\beta_{ht\text{-}O2,w} - \beta_{ht\text{-}O2,\infty}) \tag{8.109b}$$

With $h_t = \int_{T_w}^{T} c_p dT$, expanding Equation 8.109b, and using Equation 8.107

$$\frac{c_p(T_f - T_\infty)}{h_c} = \frac{s}{(1+s)}\left(1 - \frac{L}{h_c} - \frac{c_p(T_\infty - T_w)}{h_c}\right). \tag{8.110}$$

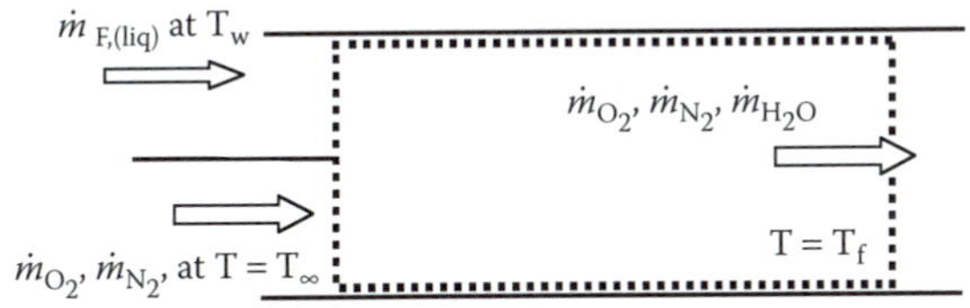

Figure 8.9 Flame temperature around a drop = adiabatic flame temperature.

This relation is often used to determine the flame temperature of droplets, and for forced-, free-, and stagnation-flow combustion problems. This result for T_f is same as the adiabatic flame temperature for a steady-state steady flow (SSSF) reactor operated with fuel (e.g., liquid droplets), entering at a temperature T_w, that burns in stoichiometric air entering the reactor at a temperature T_∞ (Figure 8.9). The stoichiometric relation

$$1 \text{ kg fuel} + \upsilon_{O2} \text{ kg of } O_2 = (1 + \upsilon_{O2}) \text{ products (i.e., } CO_2 \text{ and } H_2O)$$

applies along with any inert diluents. If the diluent is nitrogen in air, the airflow is $\upsilon_{O2}/0.23$ kg per kg of the fuel or $\upsilon_{O2}/Y_{O_2,\infty}$.

Assume a reference temperature of T_w for the gaseous state thermal enthalpy for all species; the air enthalpy at the reactor entrance is $(\upsilon_{O2}/Y_\infty)\, c_p(T_\infty - T_w)$ per kg of fuel. Total mass leaving the reactor is $1 + (\upsilon_{O2}/Y_{O_2,\infty})$. The total thermal energy of the products exiting the reactor per kg of fuel burned is

$$(1 + \upsilon_{O2}/Y_{O_2,\infty})c_p\,(T_f - T_w). \tag{8.111}$$

The heat released by the gaseous fuel h_c is supplemented by the enthalpy of the incoming airflow. The input liquid enthalpy is –L per kg of fuel (the negative sign arises due to the gaseous thermal enthalpy of fuel being zero)

$$h_c + (\upsilon_{O2}/Y_{O2,\infty})\, c_p(T_\infty - T_w) - L = (1 + \upsilon_{O2}/Y_{O2,\infty})\, c_p(T_f - T_w), \quad \text{or}$$

$$c_p\,(T_f - T_\infty) = (h_c - L - c_p\,(T_\infty - T_w))/(1+\upsilon_{O2}/Y_{O2,\infty}). \tag{8.112}$$

Here, radiation loss from the flame to its surroundings is neglected.

Species Mass Fractions at the Flame: If $\beta_{k\text{-}O2} = (Y_k/\upsilon_k) + Y_{O2}$, then at the flame $\beta_{k\text{-}O2,f} = (Y_{kf}/\upsilon_k)$ with k = CO_2 or H_2O. Thus,

$$\Phi_f = \frac{s(1+B)}{B(s+1)} = \frac{\beta_{k\text{-}O2,f} - \beta_{k\text{-}O2,\infty}}{\beta_{k\text{-}O2,w} - \beta_{k\text{-}O2,\infty}} \tag{8.113}$$

Knowing φ_f and expanding β, from Equation 8.113, $Y_{k,f}$ can be determined for k = CO_2, H_2O.

Example 7

Determine the adiabatic flame temperature for the burning of *n*-hexane in air if $T_\infty = 600$ K, L = 335 kJ/kg, $h_c = 45105$ kJ/kg, and $c_p = 1.17$ kJ/kg K. Assume $T_w = 333$ K.

Solution

$$\phi = (\beta - \beta_\infty)/(\beta_w - \beta_\infty).$$

If $\beta_{O2\text{-}F} = Y_F - Y_{O2,}/\nu_{O2}$, $\beta_{O2\text{-}F,f} = 0 - 0 = 0$, then

$$\phi_f = (\beta_{O2\text{-}F,f} - \beta_{O2\text{-}F,,\infty})/(\beta_{O2\text{-}F,w} - \beta_{O2\text{-}F,\infty})$$
$$= (0 + s)/(Y_{F,w} + s) = 0.0652/(0.9006 + 0.0652) = 0.0675.$$

Similarly, for

$$\phi_f = (\beta_{ht\text{-}O2,f} - \beta_{ht\text{-}O2,\infty})/(\beta_{ht\text{-}O2,w} - \beta_{ht\text{-}O2,\infty})$$
$$= \{c_p(T_f - T_w)/h_c - c_p(T_\infty - T_w)/h_c - s\}/\{0 - (c_p\ (T_\infty - T_w)/\ h_c) - s\}$$
$$0.0675 = \{1.17(T_f - 333)/45105 - 1.17(600 - 333)/45105 - 0.0652\}/$$
$$\{0 - (1.17(600 - 332.7)/45105) - 0.0652\} = 0.02664,\ \text{i.e.,}\ T_f = 2926\ \text{K}.$$

8.6.4.4 Burn Rate, Analogy to Heat Transfer and Resistance Concept

Mass transfer: Consider the conservation relations SZ and NSZ given by Equations 8.69 and 8.72.
Recall the interface conservation equation (Equation 8.100). Defining

$$\dot m'' = -\rho D\ B_{ht\text{-}O2}\ (\partial\phi/\partial y)_{y=0} = h_B\ B_{ht\text{-}O2} \qquad (8.114)$$

where $B = B_{ht\text{-}O2} = ((\beta_{ht\text{-}O2,w} - \beta_{ht\text{-}O2,\infty})/(F - \beta_{ht\text{-}O2,w}))$, $F = -L/h_c$, and h_B is defined as mass- or B-transfer coefficient with $h_B = -\rho D\ (\partial\phi/\partial y)_{y=0}$. The mass loss rate has an analogy to the current flow through a resistor. With $R_B = 1/h_\beta$, and potential as B, and the current flow (burn rate) is given by $\dot m''$ (refer to Figure 8.10 for liquid combustion and Figure 8.11 for solid combustion).

Generalizing,

$$\dot m'' = -(\rho D)\, B\, (\partial\phi/\partial y)_{y=0} = h_B\, B,$$

$$-(\partial\phi/\partial y)_{y=0} = h_B/(\rho D) = Sh(x)/x. \qquad (8.115)$$

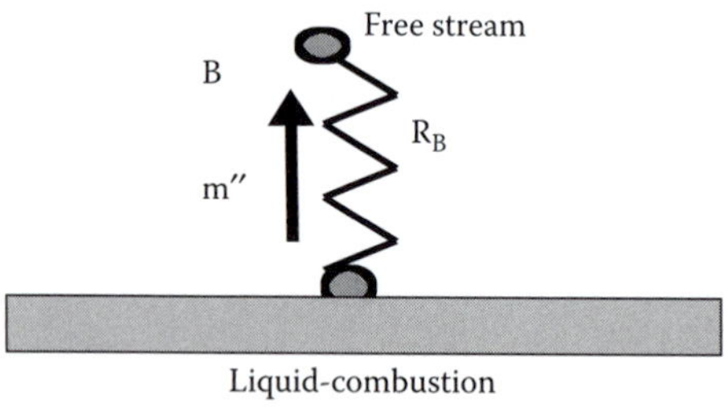

Figure 8.10 Resistance concept for liquid/totally pyrolyzing solid combustion.

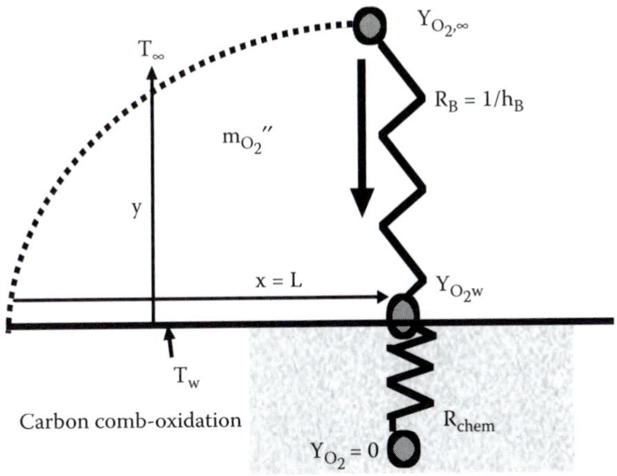

Figure 8.11 Schematic illustration of resistance concept for solid combustion in the boundary layer.

The value of $(\partial\phi/\partial y)_{y=0}$ depends upon the burn rate. For a slow burn rate, $v_y \approx 0$ at $y = 0$, i.e.,

$$-(\partial\phi/\partial y)_{y=0} = h_{B,0}/\rho D = Sh_0(x)/x. \tag{8.116}$$

Analogy to heat transfer: Consider heat transfer in the absence of mass transfer at a liquid or solid surface (or $v_y = 0$ at $y = 0$), e.g., for the flow of a cold gas over a hot plate (Example 5).
The definition of heat transfer coefficient yields,

$$q_w''(x) = -\lambda\,(\partial T/\partial y)_{y=0} = h_H(x)(T_w - T_\infty) = -\lambda\,(T_w - T_\infty)\,(\partial T^*/\partial y)_{y=0},$$

$$\text{where } T^* = \phi_T = (T_\infty - T)/(T_\infty - T_w)$$

$$-(\partial\phi_T/\partial y)_{y=0} = \{h_H(x)/\lambda\} = Nu(x)/x. \tag{8.117}$$

The heat transfer literature contains a relation for forced convection under zero mass transfer.

$$Nu(x) = 0.332\,Rex^{1/2}\,Pr^{\,1/3}. \tag{8.118}$$

The solution for $\phi_T = \phi$ if all boundary conditions are similar, $\lambda/c_p = \rho D$ or $Le = 1$, and $v_y(0) = 0$. Comparing Equation 8.117 with Equation 8.116,

$$Sh_0(x) = Nu(x). \tag{8.119}$$

There are many correlations available for Nu(x) in heat transfer literature for various flow conditions (forced-, free-, and stagnation-flow conditions), which are summarized in the Formulae chapter. If the mass burning rate during combustion is not relatively small, a correction must be applied to this relation of the form (see Section 8.9.6):

$$Sh(x)/Sh_0(x) = \{\ln(1 + B)\}/B. \tag{8.120}$$

Recall that $Sh(x) = h_B\ x/\rho D$ and $Sh_0(x) = h_{B,0}\ x/\rho D$. For relatively high mass transfer rates

$$\dot{m}'' = h_{B,0}\ B\{\ln\ (1 + B)\}/B = h_{B,0}\ \ln(1 + B). \qquad (8.121)$$

Any of the heat transfer correlations can be used and estimations for burn rate can be obtained from Equation 8.121. If one compares with drag coefficient correlations (momentum transfer), the Reynolds' analogy can be used to characterize the gasification of a solid material [Marxmann, 1967].

Combustion under diffusion control for any geometry under luminar or turbulent flow can be approximately estimated using Nu or Sh correlations.

Example 8

A carbon plate burns under forced convection. Assume that $T_p = 1056$ K, $Y_{O2,\infty} = 0.22$, x = 10 cm, $T_\infty = 1030$ K, $v_{x,\infty}$=10m/sec $\rho D = 0.5 \times 10^{-4}$ kg/m sec, $c_p = 1.18$ kJ/kg K, $h_c = 9242$ kJ/kg ($C + 1/2\ O_2 \rightarrow CO$), $Nu_x = S'h_0\ (x) = hx/\lambda = 0.332\ Re_x^{1/2}\ Pr^{1/3}$, Pr = 1, A = 450 m/s, $M_{mix} = 30$, $\rho_{plate} = 650$ kg/m^3, E = 66400 kJ/kmol, and P = 59.9 bar: (a) Determine R_{Diff}, R_{chem}, and the local burning rate of carbon. (b) If the burning is diffusion controlled, what is the local burning rate and average burn rate over a length L?

Solution

This problem is similar to that for carbon particle combustion (analyzed in Chapter 7 with resistance concept) except that the mass transfer coefficient is modified.

At a location x = x, $Nu(x) = Sh_0(x) = h_{m,0}\ x/(\rho D) = 0.332\ Re_x Pr^{1/3}$, $h_{m,0}(x) = Sh_0\ (x)\ \rho D/x$. Therefore,

$R_{Diff}\ (x) = 1/h_{m,0}(x)$ and $R_{chem}\ (x) = 1/\{k(T_w)\ \rho_w\}$ in units of m^2 sec/kg (Chapter 7).

$$\rho = PM/\{\bar{R}\ T\} = 100 * 28.97/(8.314 * 1056) = 0.33\ \text{kg/m}^3$$

$$D = \nu = 1.52 * 10^{-4}\ \text{m}^2/\text{sec}$$

$$Re_x = v_{x,\infty}x/\nu = 10 * 0.10/1.52 \times 10^{-4} = 6600$$

$$Sh_x = 0.332 * 6600^{1/2} = 27$$

$$h_{m,0} = 27 * 1.52 \times 10^{-4}\ /0.1 = 0.0135\ \text{kg/m}^2\ \text{sec}$$

$$R_{diff} = 1/h_{m,0}(x) = 74.2\ \text{m}^2\ \text{sec/kg}$$

$$k(T_w) = A\ \exp\ (-E/\{\bar{R}\ T\}) = 450 * \exp\ \{-66400/(8.314 * 1056)\}$$

$$= 0.284\ \text{m/sec}$$

$$R_{chem}\ (x) = 1/\{k(T_w)\ \rho_w\} = 1/\{0.284 * 0.33\} = 9.73\ \text{m}^2\ \text{sec/kg}$$

$$R = R_{Chem} + R_{diff} = 83.9\ \text{m}^2\ \text{sec/kg}$$

$$B = 0.23/1.333 = 0.173$$

$$\dot{m}_C(s)'' = B/R = 0.173/83.9 = 0.0021\ \text{kg/ m}^2\ \text{sec}$$

(b) For diffusion-controlled combustion,

$$\dot{m}_C''(x) = B/R_{Diff}\,(x) = 0.173/74.2 = 0.0023 \text{ kg/m}^2 \text{ sec}$$

Using the relation $Re_x = v_x x/\nu$, and integrating with respect to x,

$$\dot{m}''_{C,avg}(x) = \left[\frac{\int_0^x \dot{m}''_C(x)dx}{x}\right] = 2\dot{m}''_c(x)$$

$$\dot{m}''_{C,avg} = 0.0046 \text{ kg/m}^2 \text{ sec.}$$

It is seen that combustion rate under diffusion controlled combustion for any geometry under laminar or turbulent flow can be approximately estimated using Nu or Sh correlations.

Example 9

A hexane liquid pool with a flat surface burns in air under forced convection and has a temperature of 600 K. Assume $T_w = T_{BP}$, $c_{pfuel} = 4.39$ kJ/kg K, $\rho_{liq} = 659$ kg/m^3, $T_{BP} = 342.15$ K, L = 335 kJ/kg, $h_c = 45105$ kJ/kg, and the surface dimension L = 0.1 m. In the free stream, $Y_{O2,\infty} = 0.23$, $v_{x,\infty} = 0.1$ m/sec, $T_\infty = 600$ K, and P = 1 bar. The transport properties $\lambda_{gas} = 0.000083471$ kw/mk, Pr = 1, and $c_{p,gas} = 1.17$ kJ/kg K. (a) Estimate B. (b) Use heat transfer analogy $Nu_x = 0.332\, Re_x^{1/2}\, Pr^{1/3}$ and obtain Sh. (c) Obtain burn rate without and with mass-transfer corrections.

Solution

(a) The *n*-hexane molecular weight $M_F = 86.2$.

$$C_6H_{14} + 9.5\ O_2 \rightarrow 6\ CO_2 + 7\ H_2O,$$

$$\nu_{O2} = 9.5 \times 32/86 = 3.527 \text{ kg of } O_2 \text{ per kg of fuel, and}$$

$$s = Y_{O2,\infty}/\nu_{O2=}\ 0.23/3.527 = 0.0652.$$

Assuming $T_w = T_{BP}$, the transfer number B is estimated as

$$B = \{1.17(600 - 342.2)/335\} + (0.23 \times 45105)/(3.527 \times 335) = 9.68.$$

(b) Select the wall properties for air at T_W so that

$$\rho_w = P/RT_w \approx 100 \times 28.97/(8.314 \times 342.2) = 1.018 \text{ kg m}^{-3},$$

$$\nu_w = 2.64 \times 10^{-5} \text{ m}^2/\text{s} = D_w,$$

$$\rho_w D_w = 2.69 \times 10^{-5} \text{ kg/m s.}$$

From heat transfer correlations (see Formulae chapter) we can find that

$$Sh_0 = h_{m,avg,0}\ L/\rho D = 0.664\ Re_\infty^{1/2}\ Pr^{0.333}.$$

$$Re_\infty = (v_{x,\infty}\ d)/\nu_w = 10 \times 0.1/2.64 \times 10^{-5} = 37900,$$

$Sh_0 = 0.664 \times 37900^{1/2}\ 1^{0.333} = 129$, so that

$$h_{m,avg,0} = Sh\ \rho D/L = 129 \times 2.69 \times 10^{-5}/0.1 = 0.0348 \text{ kg/m}^2 \text{ s}.$$

(c) For low mass-transfer rates,

$$\dot{m}'' = h_{B,avg,0}\ B = 0.0348 \times 9.68 = 0.336 \text{ kg/m}^2 \text{ s}.$$

At high mass-transfer rates,

$$\dot{m}''_{avg} = 0.0348 \ln (1 + 9.68) = 0.0823 \text{ kg/m}^2 \text{ s}.$$

Remarks

(a) For flow around a sphere, $Nu_{avg} = 2 + 0.6\ Re^{1/2}\ Pr^{1/3}$. If Re = 0, then $Nu_0 = Sh_0 = 2$ so that the mass-burning rate per unit area is

$$\dot{m}'' = h_{m,0} \ln (1 + B) = Sh_0\ (\rho D/d) \ln (1 + B).$$

(b) Because the sphere surface area is πd^2, the mass-burning rate under stagnant atmosphere.

$$\dot{m} = Sh_0\ \pi\ (\rho Dd) \ln (1 + B) = 2\pi\ (\rho Dd) \ln (1 + B).$$

Example 10

When coal particles are fired in a boiler, a few particles appear like cylinders. If a cylinder is of finite length 5 mm and diameter is 1 mm, estimate the burn rate as though the cylindrical particle burns in a cross flow of gas. The relative gas velocity is 10 m/sec. Assume coal to be carbon, properties of air at 1000 K, diffusion control. $Y_{O2\infty} = 0.1$, C burns to CO, and Le = 1. Estimate the burn rate under diffusion control in kg/sec assuming cylinders in cross flow, $\nu = 1.22 \times 10^{-4}$ m²/s, $\lambda = 6.43 \times 10^{-5}$ kW/mk.

Solution

$$B = Y_{O2\infty}/\nu_{O2} = 0.1/1.33 = 0.075,\ Re = 10\ (m/s)\ 10^{-3}\ m/1.22$$

$$\times\ 10^{-04}(m^2/s) = 82,\ Pr = 0.723$$

Using the table for formulae in the Formulae chapter for a cylinder in cross flow,

$$\overline{Nu_D} = 0.3 + \frac{0.62 Re_D^{1/2} Pr^{1/3}}{[1+(0.4/Pr)^{2/3}]^{1/4}}\left[1+\left(\frac{Re_D}{282{,}000}\right)^{5/8}\right]^{4/5}.$$

These correlations are for infinite length, and assuming that they are valid for finite cylinders in cross flow, Nu = 0.3 + {(0.623 * $82^{(1/2)}$ * $0.723^{(1/3)}$) {1 + (82/ $282000)^{0.625}\}^{0.8}$/{1 + $(0.4/0.723)^{(2/3)}\}^{(1/4)}$ = 4.75

$$\rho D = \lambda/c_p = 6.4310^{-05}/1.141 = 5.64 \times 10^{-05} \text{ kg/(m sec)}.$$

With Sh_0 = Nu = 4.75, $h_{m,0}$ d/ρD = 4.75 * 5.64 × 10^{-05}/1 × 10^{-03} = 0.268 kg/ m^2 sec. Ignoring end areas, burn rate = πD * L * h_m ln (1 + B) = π × 1 × 10^{-03} × 5 × 10^{-03} × 0.268 ln (1 + 0.075) = 1.57 × 10^{-05} kg/sec.

Remarks

The cylinder may tumble around during the flow and hence the burn rate may fluctuate.

8.7 Turbulent Flows

Although laminar flows analyses are amenable to many approximations that greatly reduce the complexity of the solution and yield very simple general solutions, practical problems involve turbulence. Turbulent flows always occur at high Reynolds numbers. Turbulent flows do not lend themselves easily to analysis. Turbulent flows are random and irregular [Tennekes and Lumley, 1990]. This requires the use of statistics to achieve results. The actual velocity u(x, t) for a steady flow is split into an average (u_{avg} (x)) and fluctuating velocity (u(x,t)). Turbulent intensity is normally characterized by root-mean-square fluctuation (rms) of u where $u_{RMS} = \sqrt{\overline{u'^2}}$. As the local u component fluctuates, the mass conservation requires that the other components must also fluctuate. Further, the fluctuations are suppressed through viscous dissipation (i.e., resistance to flow) that involves certain time scales. Eddy is a region of rotating flow instead of flow along a specific line. Within the largest eddy size of a length scale, called integral scale L_I, the turbulent velocities are correlated. The region of rotating flows having kinetic energy breaks (or cascades) into smaller and smaller sizes, passing on the kinetic energy; the smallest size with insufficient kinetic energy to overcome viscous effects is called the Kolmogorov eddy, having a Kolmogorov length scale (L_K) which can be related to L_I as follows:

$$L_K = \sqrt{\frac{\nu^3 L_I}{u'_{RMS}}} \quad . \tag{8.122}$$

The eddy time or turbulent timescale (i.e., the time an eddy retains its identity before being damped and replaced by another eddy) is given by

$$t_{eddy} = \frac{L_I}{u'_{RMS}} \quad . \tag{8.123}$$

The turbulent $Re_{t,I}$ based on an integral scale is defined as

$$Re_{t,I} = \sqrt{\frac{u'_{RMS} L_I}{\nu}} \,. \tag{8.124}$$

When $Re_{t,I} \approx 1$, turbulence is weak, and when $Re_{t,I} >> 1$, turbulence is strong. For weak turbulence, a simple approach is extrapolation of laminar results to turbulent flows with the molecular diffusivity (ν) replaced by turbulent diffusivity (ε_T). Turbulent diffusivity is much greater than the molecular diffusivity in laminar flows. This results in increased mixing and consequently increased momentum, heat, and mass-transfer rates. Vorticity plays a large role in the description of turbulent flows as turbulence is rotational and 3-D. Further, the turbulent flows are always dissipative. Kinetic energy (ke) is converted to internal energy, thus requiring a constant energy supply to maintain the flow (see the appendix of this chapter for kinetic energy balance equation). As such, dissipative terms must be included in the "ke" equations. Turbulence is caused by the type of flow, not the type of fluid flowing. The characteristics of turbulence at a high Reynolds number are not a function of the fluid properties.

Even for nonreacting flows, these turbulent characteristics impose a sizable obstacle to the solution of the problem. When a combusting flow is considered, several effects of turbulence are observed. Unlike a laminar flame, which is stationary and well defined, the turbulent flame consists of random fluctuations, which increase the complexity of the flame and make precisely defining it impossible. This results in a reaction zone, which is rather thick in comparison with the thin flame of the laminar case (see Chapter 7 and Chapter 9 to Chapter 12). Turbulence also causes the flame height to be decreased (e.g., flat plate flame) due to increased diffusivity. In fact, for identical velocities, increased turbulence will cause the flame height to decrease. Many of these differences make turbulent combustion more desirable at times and have prompted the further study of turbulent modeling (Chapter 18).

Mixture fraction is frequently used for turbulent reacting flows. The DNS calculations with 512 × 512 × 1024 grid points for pure fuel and oxidant mixing, assuming one-step irreversible reaction, incompressible flow, and constant fluid properties reveal that the reaction zones are thin, even for isothermal reaction, and becomes thinner for nonisothermal case. Because the reaction zones are so thin, modeling techniques involving averaging, which are common for nonreacting turbulence (Chapter 18) become inaccurate [de Bruyn Kops, 1999, 2001].

8.8 Summary

While Chapter 7 makes use of global conservation equations in solving engineering combustion problems, Chapter 8 presents detailed species, momentum, and energy conservation equations. Detailed conservation equation enables us to obtain the flame structure. Whereas most of the materials deal with 2-D problems, the appendix of this chapter presents detailed conservation equations including tensor notation for momentum equation.

SZ for single and two reaction schemes are illustrated. BL combustion problems are briefly analyzed. The similarity between heat transfer and combustion solutions enable us to make use of many heat-transfer correlations for various flow geometries and estimate the burn rate. Diffusion-controlled combustion of liquid fuels leads to a general solution for fuel mass fraction, liquid surface temperature, flame temperature relations, etc., applicable either to flow problems or liquid drop combustion in quiescent atmospheres.

8.9 Appendix

8.9.1 Vector and Tensors

(i) Dot or Scalar Product of a Vector: Consider a surfboard dragged along a distance ds over water by a power boat that applies a force of F on the board. The board travels along s, while F is applied at an angle to the tangent of s vector. The work done in moving the board through a distance ds is given by

$$\delta W = \vec{F} \cdot d\vec{s} = \{F \cos \theta\}\, ds \tag{8.125}$$

where θ is angle between F vector and ds vector
(ii) Cross or Vector Product of a Vector:

$$\vec{A} = \vec{y}\, X\, \vec{s} = \vec{A} = \vec{k}\, |s|\,|y| \sin\theta \tag{8.126}$$

Thus, the vector or cross product yields an area vector in the direction normal to the plane containing the two vectors.
(iii) Del or Grad of a Scalar: We define grad of T as

$$\nabla T = \left[\vec{i}\,\frac{\partial}{\partial x} + \vec{j}\,\frac{\partial}{\partial y} + \vec{k}\,\frac{\partial}{\partial z}\right] T \quad \text{or} \quad \left\{\sum_{i=1}^{3} \delta_i \frac{\partial}{\partial x_i}\right\} T \tag{8.127}$$

where $\vec{\delta}_1 = \vec{i}, \vec{\delta}_2 = \vec{j}, \vec{\delta}_3 = \vec{k}$, $x_1 = x$, $x_2 = y$, $x_3 = z$, in Cartesian systems; $\vec{\delta}_i$ = unit vector.

The ∇T is a vector normal to surface T = constant; e.g., the Fourier heat flux vector $-\lambda \nabla T$ is normal to surface $T = T_s = C$ (e.g., a spherical surface).
(iv) Curl of a Vector: Now consider cross product of $\nabla \times \vec{F}$

Curl $\vec{F} = \nabla \times \vec{F} = \left[\vec{i}\,\frac{\partial}{\partial x} + \vec{j}\,\frac{\partial}{\partial y} + \vec{k}\,\frac{\partial}{\partial z}\right] x [\vec{i}\, F_x + \vec{j}\, F_y + \vec{k}\, F_z]$

$$\begin{aligned} curl\ F = \nabla \times \vec{F} = \vec{i}\left[\frac{\partial F_z}{\partial y} - \frac{\partial F_y}{\partial z}\right] + \vec{j}\left[\frac{\partial F_x}{\partial z} - \frac{\partial F_z}{\partial x}\right] \\ + \vec{k}\left[\frac{\partial F_y}{\partial x} - \frac{\partial F_x}{\partial y}\right] \end{aligned} \tag{8.128}$$

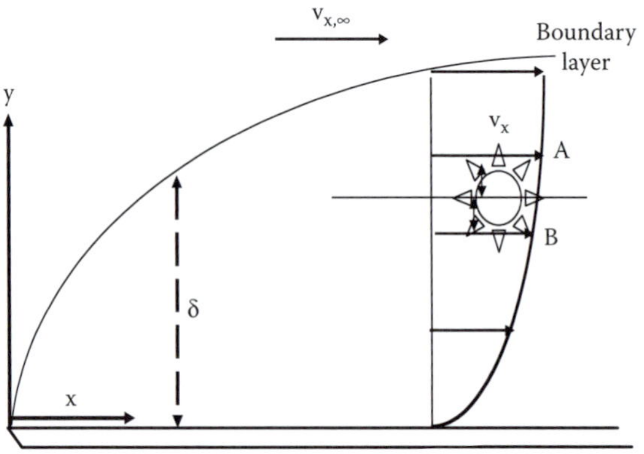

Figure 8.12 Schematic illustration of velocity boundary layer.

where the curl F is a vector whose rotational components are given by preceding relation. What is rotational flow? Consider the velocity profile in a BL (flow of water in a river). The velocity at A is higher compared to B; if we place a paddle wheel into the flow, it can rotate clockwise (Figure 8.12). Such a flow is called rotational. If V is uniform, it becomes irrotational. We know that if $\nabla \times \vec{V} = 0$, then the two vectors ∇ and V must be parallel to each other ($\vec{V}_A = \vec{V}_B$) or the field must be irrotational.

(v) Tensors and Tensor Products: The velocity vector in a general coordinate system is given as

$$\vec{v} = \sum_{j=1}^{3} \vec{\delta}_j v_j \tag{8.129}$$

The product of the gradient and vector is given as

$$\nabla \vec{v} = \sum_{i=1}^{3} \vec{\delta}_i \frac{\partial}{\partial x_i} \sum_{j=1}^{3} \vec{\delta}_j v_j = \sum_{i=1}^{3} \sum_{j=}^{3} \vec{\delta}_i \, \vec{\delta}_j \frac{\partial v_j}{\partial x_i} \tag{8.130}$$

$$\nabla \vec{v} = \begin{pmatrix} \frac{\partial v_1}{\partial x_1} \frac{\partial v_2}{\partial x_1} \frac{\partial v_3}{\partial x_1} \\ \frac{\partial v_1}{\partial x_2} \frac{\partial v_2}{\partial x_2} \frac{\partial v_3}{\partial x_2} \\ \frac{\partial v_1}{\partial x_3} \frac{\partial v_2}{\partial x_3} \frac{\partial v_3}{\partial x_3} \end{pmatrix} \tag{8.131}$$

Note that vector multiplication results in a tensor. When a column in previous matrix is transposed as a row, it is called transpose matrix (e.g., $\nabla \vec{V}^T$).

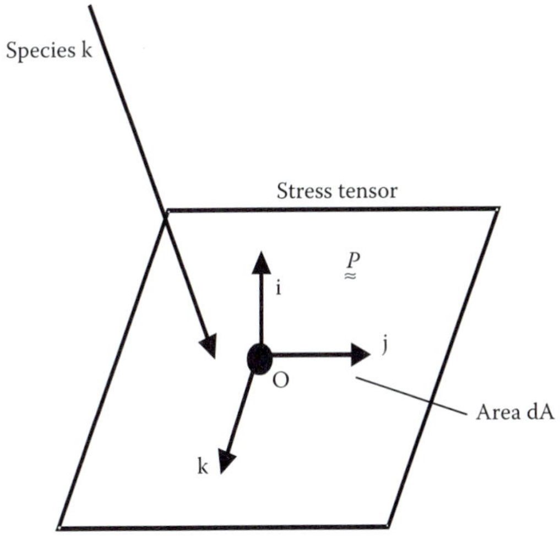

Figure 8.13 Stress tensor acting on an area ΔA.

8.9.2 Modified Constitutive Equations

8.9.2.1 Diffusive Momentum Fluxes and Pressure Tensor

We first consider the pressure or stress as depicted in Figure 8.13.

$$\text{Stress at } O = \frac{\text{Time rate of Change of Momentum due to Moleculer Impacts on surface } dA}{dA} \tag{8.132}$$

As gas molecules collide with a surface, their direction is reversed. According to Newton's law, the momentum change results in pressure or stress. In a multicomponent system the total stress at a point equals the vectorial sum of the stress created by each species. In general, the stress need not be normal to the surface upon which there are molecular collisions because of the presence of viscous forces. The shear and normal stresses are illustrated in Figure 8.14.

As shown in Figure 8.14, on a surface DFHC that is a surface normal to the z axis or axis 3, the stress components are p_{33} (normal), p_{32} (shear on face normal to 3 but in the direction of y axis or axis 2), and its counterpart p_{31}. More generally, p_{ij} denotes the component of stress on the face normal to the i direction but in the direction of j. The stress p_{ij} acting on a surface normal to i is given as

$$p_{ij} = p^0 \delta_{ij} + \tau_{ij} \tag{8.133}$$

where p^o denotes the hydrostatic pressure of the fluid, which is considered to be ideal and in uniform motion (i.e., it is inviscid and has no viscosity).

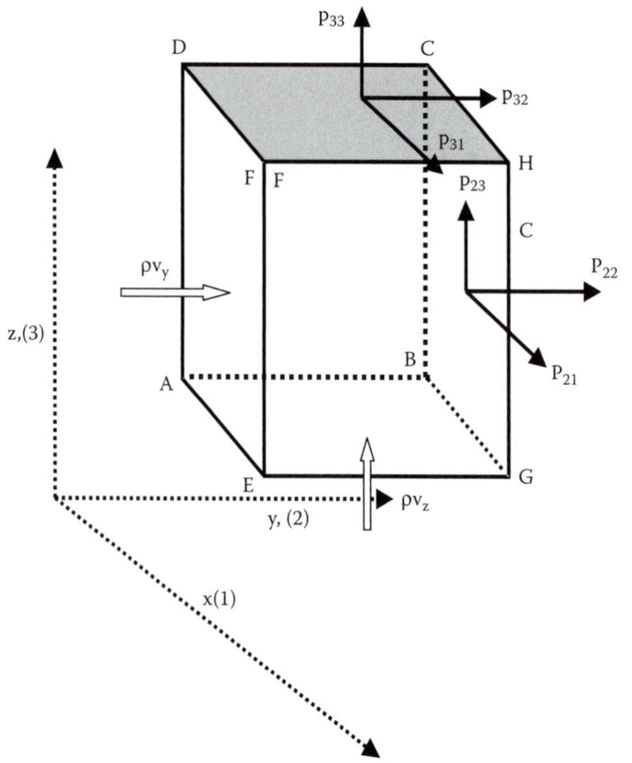

Figure 8.14 Components of stress tensor P_{ij} acting on a surface normal to i but in the direction j.

The shear stress τ_{ij} on a plane normal to i but in the j-th direction is given as (Figure 8.14.),

$$\tau_{ij} = -2\mu\,\gamma_{ij} + \left(\frac{2}{3}\mu - \kappa\right)\left[\sum_{k=1}^{3}\gamma_{kk}\right]\delta_{ij} \tag{8.134}$$

and

$$\delta_{ij},\ Kronecker\ delta = \begin{cases} 1\ when\ i = j \\ 0\ when\ i \neq j \end{cases}. \tag{8.135}$$

Here, μ denotes the absolute viscosity (which is sometimes referred to as the first viscosity) and κ denotes the bulk viscosity or second viscosity. The value of κ is zero for monatomic gases, and is approximately zero for dense gases and liquids. The term $\gamma_{11} = \partial v_1/\partial x_1$ is called the linear strain in the x_1 direction, whereas $\partial v_1/\partial x_2$ is called the shear strain. Generally, the rate of strain is $\partial v_i/\partial x_j$, where if i = j, it is called linear and if i ≠ j it is called shear. The value of summation of linear strain

$$\nabla \cdot \vec{v} = \sum_{k=1}^{3}\gamma_{kk} = 0 \tag{8.136}$$

for an incompressible fluid. In general, strain

$$\gamma_{ij} = \frac{1}{2}\left(\frac{\partial v_i}{\partial x_j} + \frac{\partial v_j}{\partial x_i}\right) \tag{8.137}$$

It is noted that τ_{12} reduces to $= -\mu\partial v_1/\partial x_2 - \mu\partial v_2/\partial x_1$ for 2-D incompressible fluid flow where stress acts in direction x_1. Including all stress components, the pressure tensor $\vec{\vec{P}}$ is given as

$$\vec{\vec{P}} = \left\{p^0 + \left(\frac{2}{3}\mu - \kappa\right)\vec{\nabla}\cdot\vec{v}\right\}\vec{\vec{U}} - \mu\left(\vec{\nabla}\ \vec{v} + \vec{\nabla}\ \vec{v}^T\right) = p^0\vec{\vec{U}} + \vec{\vec{\tau}}\ , \tag{8.138}$$

where $\vec{\vec{U}}$ is unit tensor and $\nabla\vec{v} + \nabla\vec{v}^T$ is the strain tensor γ_{ij}.

$$\vec{\vec{\tau}} = \left[\left(\frac{2}{3}\mu - \kappa\right)\vec{\nabla}\cdot\vec{v}\right]\vec{\vec{U}} - \mu(\vec{\nabla}\ \vec{v} + \vec{\nabla}\,\vec{v}^T)\ . \tag{8.139}$$

$\vec{\vec{P}}$ can also be written as

$$\vec{\vec{P}} = \sum_{i=1}^{3}\sum_{j=1}^{3}\vec{\delta}_i\,\vec{\delta}_j\,p_{ij}$$

The mean hydrostatic (nonequilibrium) pressure is

$$p = \frac{tr(\vec{\vec{P}})}{3} = \frac{(p_{11} + p_{22} + p_{33})}{3}\ , \tag{8.140}$$

where $\frac{tr(\vec{\vec{P}})}{3}$ is the tracer of the tensor (i.e., the sum of the diagonal elements of $\vec{P}$). Actual pressure p under flow conditions $\neq p^0$, hydrostatic pressure which is the same in all directions.

The components of Equation 8.139 are same as Equation 8.134. From Equation 8.133,

$$p_{11} = p^0 + \tau_{11},\ p_{22} = p^0 + \tau_{22},\ p_{33} = p^0 + \tau_{33}, \tag{8.141}$$

where

$$\tau_{11} = (2/3\ \mu - \kappa)(\partial v_1/\partial x_1 + \partial v_2/\partial x_2 + \partial v_3/\partial x_3) - 2\mu\partial v_1/\partial x_1, \tag{8.142}$$

and shear components are given by

$$\tau_{22} = \{(2/3)\mu - \kappa\}(\partial\ v_1/\partial x_1 + \partial v_2/\partial x_2 + \partial v_3/\partial x_3) - 2\mu\partial v_2/\partial x_2, \tag{8.143}$$

and

$$\tau_{33} = (2/3\ \mu - \kappa)(\partial v_1/\partial x_1 + \partial v_2/\partial x_2 + \partial v_3/\partial x_3) - 2\mu\ \partial v_3/\partial x_3. \tag{8.144}$$

$$p_{12} = \tau_{12} = -\mu(\partial v_2/\partial x_1 + \partial v_2/\partial x_1). \tag{8.145}$$

In general,

$$p_{ij} = \tau_{ij} = -\mu(\partial v_j/\partial x_i + \partial v_i/\partial x_j), \quad i \neq j,\ i = 1 \ldots 3,\ j = 1 \ldots 3. \quad (8.146)$$

Using definitions of p_{11}, p_{22}, and p_{33}, Equation 8.140 is written as

$$p = p^0 - \kappa\left(\frac{\partial v_1}{\partial X_1} + \frac{\partial v_2}{\partial X_2} + \frac{\partial v_3}{\partial X_3}\right) = p^0 - \kappa \vec{\nabla} \cdot \vec{v} \quad (8.147)$$

where $p = p^0$ if $\kappa = 0$ or if $\nabla . \vec{v} = 0$ (incompressible).

Often, $\kappa \approx 0$ for monotonic gases and liquids, i.e., $p = p^0$. The divergence of the stress tensor is given as

$$\nabla \cdot \vec{\vec{P}} = \sum_{\ell=1}^{3} \vec{\delta}_\ell \frac{\partial}{\partial x_\ell} \cdot \sum_{i=1}^{3}\sum_{j=1}^{3} \vec{\delta}_i \vec{\delta}_j p_{ij}. \quad (8.148)$$

8.9.2.1.1 Physical Meaning of Stress Tensor

Figure 8.15 illustrates normal and shear stresses. If a person jumps from a bridge on a roller skate, he exerts normal and shear stresses, and hence a skateboard moves in a specified direction. Instead of a person, imagine trillions of molecules striking a surface and exerting stresses. The sum of these stresses is termed *pressure tensor.*

8.9.2.2 Generalized Relation for Heat Transfer

The heat conduction equation has the form

$$\vec{q}'' = -\lambda\, \vec{\nabla} T + \rho \sum Y_k\, \vec{V}_k\, h_k + \bar{R}T \sum\sum \frac{D_{Tj} X_j (\vec{V}_i - \vec{V}_j)}{M_i D_{ij}}. \quad (8.149)$$

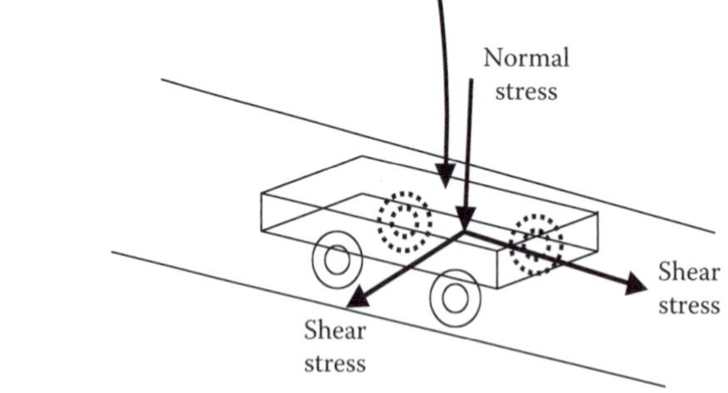

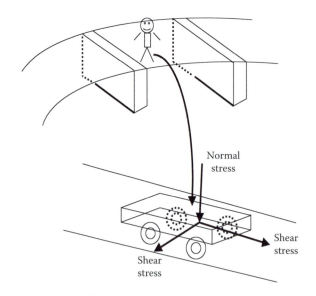

Figure 8.15 Illustration of normal and shear stresses.

where second term is due to mass diffusion and last terms is known as Dufour heat flux (i.e., the thermal diffusion effect).

8.9.3 Rigorous Formulation of the Conservation Relations

8.9.3.1 Momentum

The generalized momentum conservation equation is

$$\rho\,\frac{\partial \vec{V}}{\partial t}+\rho\vec{v}\cdot\vec{\nabla}\,\vec{v}=-\vec{\nabla}\,.\,\vec{\vec{P}}+\rho\sum Y_k\,\vec{f}_k \tag{8.150}$$

where

$$\vec{f}_k=\sum_j \vec{\delta}_j\,\vec{f}_{k,j}\,,\ \text{i.e.,} \tag{8.151}$$

$$\vec{v}\,.\,\vec{\nabla}\,\vec{v}=\sum_{l=1}^{3}\vec{\delta}_\ell\, v_\ell\cdot\sum_{i=1}^{3}\sum_{j=1}^{3}\vec{\delta}_i\vec{\delta}_j\frac{\partial v_j}{\partial x_i}=\sum_{\ell=1}^{3}\sum_{j=1}^{3}\vec{\delta}_j\, v_\ell\frac{\partial v_j}{\partial x_\ell} \tag{8.152}$$

$$\vec{\nabla}\,.\,\vec{\vec{P}}=\sum_{\ell=1}^{3}\vec{\delta}_\ell\frac{\partial}{\partial x_\ell}\cdot\sum_{i=1}^{3}\sum_{j=1}^{3}\vec{\delta}_i\vec{\delta}_j p_{ij}=\sum_{\ell=1}^{3}\sum_{j=1}^{3}\vec{\delta}_j\frac{\partial p_{\ell j}}{\partial x_\ell}. \tag{8.153}$$

Along j = 1 direction, the momentum conservation equation is

$$\rho\frac{\partial v_1}{\partial t}+v_1\frac{\partial v_1}{\partial x_1}+v_2\frac{\partial v_2}{\partial x_2}+v_3\,\frac{\partial v_3}{\partial x_3}=-\left(\frac{\partial p_{11}}{\partial x_1}+\frac{\partial p_{21}}{\partial x_2}+\frac{\partial p_{31}}{\partial x_3}\right)+\rho\sum Y_k\, f_{k,1} \tag{8.154}$$

where p_{11}, p_{21}, and p_{31}, have been defined in Equations 8.141 and 8.146. See formulae list of Chapter 8, Tables 8.2–8.4.

8.9.3.2 Kinetic Energy ($v^2/2$)

Taking the dot product of the momentum equation (Equation 8.150) and velocity vector

$$\rho\,\frac{\partial \vec{v}}{\partial t}\cdot\vec{v}+\rho\vec{v}\cdot\vec{\nabla}\,\vec{v}\cdot\vec{v}=-\vec{v}\cdot\vec{\nabla}\,.\,\vec{\vec{P}}+\rho\,\sum Y_k\,\vec{f}_k\cdot\vec{v} \tag{8.155}$$

Using the identity given in Equation 8.158, Equation 8.155 is written as

$$\rho\,\frac{D\left[\frac{v^2}{2}\right]}{Dt}=\vec{\vec{P}}:\vec{\nabla}\vec{v}+\rho\,\sum Y_k\,\vec{f}_k\,\,.\,\vec{v}-\vec{\nabla}\cdot[\vec{\vec{P}}\cdot\vec{v}]\,. \tag{8.156}$$

8.9.3.3 Internal and Kinetic Energies

Recall from thermodynamics that the energy per unit mass e = u + $v^2/2$ + gz or excluding potential energy, e′ = u + $v^2/2$. Work rate can be expressed in the form (force × distance moved ÷ time) or (stress × elemental area × velocity). Using the Gauss divergence theorem for the work term $\int \nabla.[\vec{\vec{P}}.\vec{v}]$ d(volume),

$$\rho\,\frac{\partial\left[u+\frac{v^2}{2}\right]}{\partial t}+\rho\;\vec{v}\,.\,\vec{\nabla}\left[u+\frac{v^2}{2}\right]$$
$$=-\vec{\nabla}\,.\,\vec{q}''-\vec{\nabla}\,.\,(\vec{\vec{P}}\cdot\vec{v})+\rho\sum Y_k\;\vec{f}_k\;.\vec{v}_k \tag{8.157}$$

Subtracting Equation 8.155 from Equation 8.157 and using identity

$$\vec{v}.\,\vec{\nabla}\,.\,\vec{\vec{P}}=\vec{\nabla}\,.[\vec{\vec{P}}.\vec{v}]-\vec{\vec{P}}.\vec{\nabla}\vec{v}\;, \tag{8.158}$$

we have

$$\rho\,\frac{\partial u}{\partial t}+\rho\;\vec{v}\,.\,\vec{\nabla}u=-\vec{\nabla}\,.\,\vec{q}''-\vec{\vec{P}}\,.\vec{\nabla}\vec{v}+\rho.\sum Y_k\;\vec{f}_k.\vec{V}_k \tag{8.159}$$

where, $\vec{\vec{\text{P}}}$ is given by Equation 8.138.

8.9.4 Enthalpy

$$\rho\frac{\partial h}{\partial t}+\rho\vec{v}\cdot\nabla h=\frac{\partial p}{\partial t}-\frac{P}{\rho}[-\rho\nabla\cdot\vec{v}]-p\nabla.\vec{v}-\bar{\bar{\tau}}\cdot\nabla\vec{v}-\nabla\cdot\vec{q}+\rho$$
$$\sum Y_k\,\vec{f}_k\cdot\vec{V}_{k,diff}=\frac{\partial p}{\partial t}-\bar{\bar{\tau}}\cdot\nabla\vec{v}-\nabla\cdot\vec{q}+\rho\sum Y_k\,\vec{f}_k\cdot\vec{V}_{k,diff} \tag{8.160}$$

8.9.5 Stagnation Enthalpy $\left(h+\frac{v^2}{2}\right)$

$$\rho\frac{\partial\left[h+\frac{v^2}{2}\right]}{\partial t}+\rho\vec{v}\cdot\nabla\left\{\left(h+\frac{v^2}{2}\right)\right\}=\frac{\partial p}{\partial t}-\sum_{j=1}^{3}\sum_{i=1}^{3}\frac{\partial(\tau_{ji}v_i)}{\partial x_j}-\nabla\cdot\vec{q}''$$
$$+\sum\rho Y_k\vec{f}_k\cdot\vec{v}_k \tag{8.161}$$

8.9.6 Film Theory and Mass Transfer

Consider the variation of a property in a film (Figure 8.16). For the property relation,

$$\rho\;v_x\,\frac{\partial\phi}{\partial x}+\rho\,v_y\,\frac{\partial\phi}{\partial y}=\frac{\partial}{\partial y}\left(\frac{\lambda}{c_p}\frac{\partial\phi}{\partial y}\right), \tag{8.162a}$$

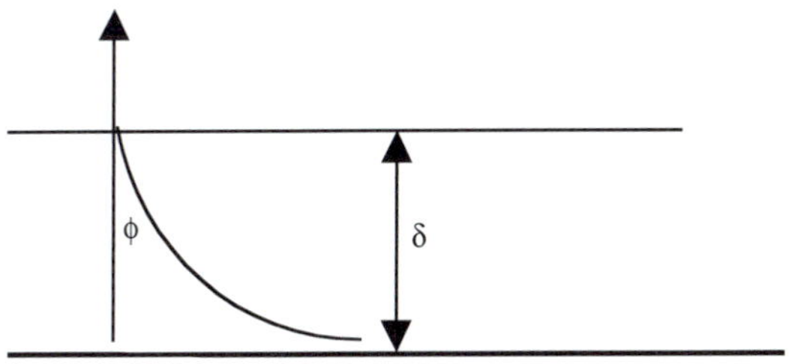

Figure 8.16 The variation of a property in a film.

near the wall $v_x = 0$, i.e.,

$$\rho\ v_y \frac{\partial \phi}{\partial y} \approx \frac{\partial}{\partial y}\left(\frac{\lambda}{c_p}\frac{\partial \phi}{\partial y}\right) \text{ or } \frac{\partial}{\partial y}\left\{\frac{\lambda}{c_p}\frac{\partial \phi}{\partial y} - \rho\ v_y \phi\right\} = 0\,. \tag{8.162b}$$

If $\rho v_y = \dot{m}''$ is constant near the wall,

$$\frac{d\phi}{dy} = \frac{d}{dy}\left(\frac{\rho D}{\dot{m}''}\frac{d\phi}{dy}\right). \tag{8.163}$$

Integrating this expression twice and using the boundary conditions $\phi = 1$ at $y = 0$, $\phi = 0$ at $y = \delta$,

$$1 - \phi = \frac{\exp\left(y\frac{\dot{m}''}{\rho D}\right) - 1}{\exp\left(\delta\frac{\dot{m}''}{\rho D}\right) - 1}\,. \tag{8.164}$$

Differentiating with respect to y and evaluating the resulting expression at $y = 0$,

$$-\left(\frac{d\phi}{dy}\right)_{y=0} = \frac{\frac{\dot{m}''}{\rho D}}{\exp\left(\delta\frac{\dot{m}''}{\rho D}\right) - 1} \tag{8.165}$$

and using in Equation 8.87

$$\dot{m}'' = \frac{\rho D}{\delta}\ln(1 + B)\,. \tag{8.166}$$

If $B \ll 1$, $\dot{m}_0'' = B\,\frac{\rho D}{\delta}$, and hence

$$\frac{\dot{m}''}{\dot{m}_0''} = \frac{\ln(1 + B)}{B} \tag{8.167}$$

Chapter 9

Combustion of Solid Fuels, Carbon, and Char

9.1 Introduction

We have made use of global conservation equations in Chapter 7 for solving the combustion problems in engines and obtained approximate solutions for burn rates of solid and liquid fuels using the resistance concept. Now, we will apply the differential forms of the conservation equations developed in Chapter 8 to solve combustion problems, obtain more exact solutions for combustion of solid fuels, and determine the flame structure around a burning solid surface. The treatment mostly involves the combustion of solid carbon, and the method can be easily extended to the combustion of solid metals.

Most practical combustor systems such as boilers, gas turbines, diesel engines, rockets, etc., use liquid and solid fuels as their energy source. Although gaseous and oil fuels are simple and clean to use, their resources are limited; thus, solid fuels are finding increasing use in electric power generation. The most widely used solid fuel in boilers is coal. A dry-ash-free (DAF) coal consists of almost 60 to 90% carbon (called fixed carbon, FC), whereas the remainder is volatile matter (VM). Thus, the major fraction of heat released is from FC. Further, VM can be burned rapidly (about 10 msec), though FC takes a significant amount of time (on the order of 1 sec for a 0.2-mm particle). It is apparent that the total combustion timescale of coal is dominated by combustion of FC or char. Thus, it is essential to learn the basic principles of carbon combustion.

9.2 Carbon Reactions

9.2.1 Reactions

Consider a hot carbon surface exposed to air. Oxygen, CO_2, and H_2O react with solid carbon and produce various gaseous species. When two phases are involved, the reactions are called *heterogeneous reactions*. There are four significant heterogeneous reactions (Figure 9.1):

$$C(s) + 1/2\ O_2 \rightarrow CO, \qquad h_{c,I} = 9203.2 \text{ kJ/kg of C} \qquad \text{(I)}$$

$$C(s) + O_2 \rightarrow CO_2, \qquad h_{c,II} = 32766 \text{ kJ/kg of C} \qquad \text{(II)}$$

$$C(s) + CO_2 \rightarrow 2\ CO, \qquad h_{c,III} = -14359.7 \text{ kJ/kg of C} \qquad \text{(III)}$$

$$C(s) + H_2O(g) \rightarrow CO + H_2, \qquad h_{c,IV} = -10932 \text{ kJ/kg of C} \qquad \text{(IV)}$$

The first two reactions are exothermic, while the third and fourth reactions are endothermic. The third reaction is known as the *Boudouard reaction*. It is seen that oxidation of carbon can either produce CO or CO_2, or both. Typically, Reaction I dominates for $T > 800$ K, whereas Reaction II dominates below 800 K. Reaction III and Reaction IV are dominant only at extreme temperatures or when O_2 concentrations are extremely low. Generally, the char reaction rate at a given T follows this order: $O_2 >> H_2O >> CO_2$ [Smoot and Smith, 1985]. The CO and H_2 produced from the surface oxidize with O_2 in the gas phase with the following reactions.

$$CO + 1/2\ O_2 \rightarrow CO_2, \qquad h_{c,V} = 10103 \text{ kJ/kg of CO, exo} \qquad \text{(V)}$$

$$H_2 + 1/2\ O_2 \rightarrow H_2O, \qquad h_{c,VII} = 141500 \text{ kJ/kg of } H_2\text{, exo} \qquad \text{(VI)}$$

Reaction V and Reaction VI are called *homogeneous reactions*.

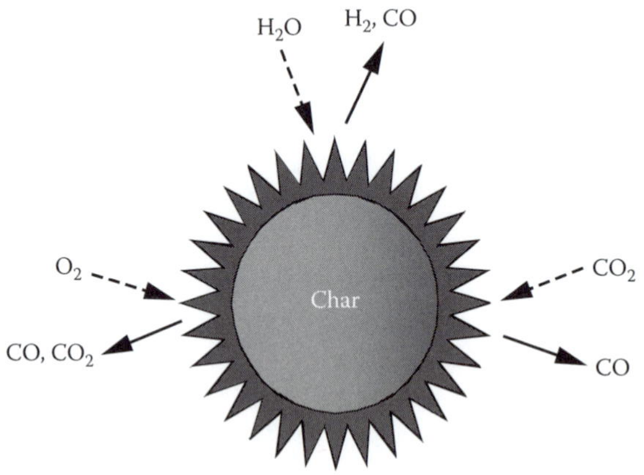

Figure 9.1 Carbon or char reactions.

9.2.2 Identities Involving Carbon Reactions

There are some interesting relations between the heating values (h_c) and stoichiometric coefficients of various reactions, i.e.,

$$\nu_{O2,I}/\nu_{O2,V} = [1/2\ M_{O2}/M_c]/[1/2\ M_{O2}/M_{CO}] = [M_{CO}/M_c] = (1 + \nu_{O2,I}) \quad (9.1)$$

where $\nu_{O2,\ I}$ is the stoichiometric coefficient for Reaction I.

$$\nu_{CO2,III} - \nu_{O2,I} = M_{CO2}/M_C - 1/2\ M_{O2}/M_C = M_{CO}/M_c = (1 + \nu_{O2,I}), \quad (9.2)$$

$$(1 + \nu_{O2,I}) * (1 + \nu_{O2,V}) = [M_{CO}/M_c]\ [M_{CO2}/M_{CO}] = [M_{CO2}/M_c] = \nu_{CO2,III}, \quad (9.2a)$$

$$2(1 + \nu_{O2,I}) = 1 + \nu_{CO2,III}. \quad (9.2b)$$

The difference {Reaction I – Reaction III} yields Reaction V; i.e., C(s) + 1/2 O_2 – C(s) – CO_2 → CO – 2CO leads to CO + 1/2 O_2 → CO_2. Then including the sign for $h_{c,III}$ $\bar{h}_{c,V} = \bar{h}_{c,I} - \bar{h}_{c,III} = 9203.2 \times 12.01 + 14359.7 \times 12.01 = 282991$ kJ/kmol of CO or 10103 kJ/kg of CO. Writing in symbolic form,

$$h_{c,V} = [h_{C,I} \times M_C - h_{C,III} \times M_C]/M_{CO} \quad \text{or} \quad h_{c,V} = [h_{C,I} - h_{C,III}]/(1 + \nu_{O2,I}), \quad (9.3)$$

$$h_{c,V}\ [\nu_{O2,I}/\nu_{O2,V}] = [h_{C,I} - h_{C,III}] \quad \text{or} \quad h_{C,III} = h_{C,I} - h_{c,V}\ [\nu_{O2,I}/\nu_{O2,V}]. \quad (9.4)$$

For example, with $h_{C,I}$ = 9203.2 kJ/kg, $h_{C,V}$ = 10103.2 kJ/kg, $\nu_{O2,I}$ = 1.332, $\nu_{O2,V}$ = 0.571, $\nu_{O2,I}/\nu_{O2,V}$ = 2.332 = (1 + $\nu_{O2,I}$) as it should be; further, $h_{C,III}$ = $h_{C,I} - h_{c,V}\ [\nu_{O2,I}/\nu_{O2,V}]$ = 9203.2 – 10103.2 × 2.332 = –14357 kJ/kg.

9.3 Conservation Equations for a Spherical Particle

9.3.1 Physical Processes

Let us first assume that there is only heterogeneous oxidation involving Reaction I. The carbon reacts locally with oxygen, lowering the oxygen concentration near the surface, whereas the free-stream concentration remains constant. Thus, a concentration gradient in the gas phase is established, which results in diffusion of oxygen toward the particle surface. Under steady conditions, the profiles must adjust in such a way that the oxygen transported via diffusive processes must equal the consumption of oxygen at the particle surface. We are interested in obtaining the profiles under such conditions. The transfer rate of oxygen can then be evaluated by using interface mass conservation equations. Finally, using stoichiometry, the carbon burn rate can then be evaluated.

In an actual situation, CO will react with O_2 in the gas phase and produce CO_2. The CO_2 can in turn react with C(s) and produce CO, involving Reaction III. Finally, when H_2O is present (e.g., humidity, combustion products), Reaction IV can also occur. In the following, we will first formulate the gas-phase conservation equation, including gas-phase oxidation. Later, we will simplify equations for various situations. For the systems with gas-phase oxidation, it will be assumed that Reaction V is the only gas-phase reaction.

9.3.2 Dimensional Form and Boundary Conditions

Assume (i) steady state, (ii) ρD = constant, (iii) spherical geometry, and (iv) property variation in the r direction.

We will use mass, the SZ (β), and the CO species equations. (Chapter 8)

$$\frac{1}{r^2}\frac{d}{dr}(\rho v r^2)=0, \tag{9.5}$$

$$\frac{1}{r^2}\frac{d}{dr}(\rho v r^2 \beta)=\frac{I}{r^2}\frac{d}{dr}\left(\rho D r^2 \frac{d\beta}{dr}\right), \tag{9.6}$$

$$\rho v\frac{dY_{CO}}{dr}=\frac{1}{r^2}\frac{d}{dr}\left[r^2 \rho D\frac{dY_{CO}}{dr}\right]+\dot{w}'''_{CO}. \tag{9.7}$$

When the convection term is included, the problem is known as Stefan flow (SF) problem. Subsection 9.12.2 shows the conditions under which Stefan flow could be ignored. For frozen gas phase (Single Film Model, SFM; Chapter 7), the β's are defined in Table 9.1. For the reacting gas phase, the SZ variable will be illustrated with the gas-phase Reaction V: $CO + (1/2)O_2 \rightarrow CO_2$

Table 9.1 Summary of SZ Variables, Single-Film Model (SFM)

$\dot{m}_{iso}$ = 2Sh $\pi\rho D$ a ln(1 + B) for spherical particles
$h_{c,I}$ = 9203.2 $h_{c,II}$ = 32766.0 kJ/kg of C, $h_{c,III}$ = −14359.7 kJ/kg of C, $h_{c,IV}$ = −10931.7 kJ/kg

$C(s) + (1/2)\,O_2 \rightarrow CO$	(I)
$C(s) + O_2 \rightarrow CO_2$	(II)
$C(s) + CO_2 \rightarrow 2CO$	(III)
$C(s) + H_2O \rightarrow CO + H_2$	(IV)

Single Species Attacking Carbon, Frozen Gas Phase

C(s) + Species j → Prod

j	*Reaction Species*	β_j	β_T	*B*	$-h_{T,\infty}$	*Prod*
I	O_2	$Y_{O2}/\nu_{O2,I}$	$h_T/h_{c,I}$	$Y_{O2,\infty}/\nu_{O2,I}$	$B\,h_{c,I}$	CO
II	O_2	$Y_{O2}/\nu_{O2,II}$	$h_T/h_{c,II}$	$Y_{O2,\infty}/\nu_{O2,II}$	$B\,h_{c,II}$	CO_2
III	CO_2	$Y_{CO2}/\nu_{CO2,III}$	$h_T/h_{c,III}$	$Y_{CO2,\infty}/\nu_{CO2,III}$	$B\,h_{c,III}$	CO
IV	H_2O	$Y_{H2O}/\nu_{H2O,IV}$	$h_T/h_{c,IV}$	$Y_{H2O,\infty}/\nu_{H2O,IV}$	$B\,h_{c,IV}$	CO, H_2
I + III	O_2, CO_2	$\beta_I + \beta_{III}$	—	$B_I + B_{III}$	$\beta_I h_{c,I} + \beta_{III} h_{c,III}$	CO

CO_2 and O_2 Both Reacting Simultaneously Producing CO (SFM)

$C(s) + 1/2O_2 \rightarrow CO$, $C(s) + CO_2 \rightarrow 2CO$

$B = Y_{O2,\infty}/\nu_{O2,I} + Y_{CO2,\infty}/\nu_{CO2,III}$ $-h_{t,\infty} = h_{c,I}\,Y_{O2,\infty}/\nu_{O2,I} + Y_{CO2,\infty}h_{c,III}/\nu_{CO2,III}$

$$\beta_{ij} = \frac{Y_i}{\nu_i} - \frac{Y_j}{\nu_j} \tag{9.8}$$

where i – j combinations are illustrated in column 4, Table 9.2. (If H_2 oxidation is to be included, one must adopt the double SZ formulation, as described in Chapter 8) When i or j is specified, the signs will be included depending on whether species is reactant or product. For example, if $i = O_2$ is selected, $\nu_j = -\nu_{O2}$ and $j = CO$ in Reaction V, $\nu_j = -\nu_{CO}$. For an h_t and CO combination,

$$\beta_{h_t-CO} = \frac{h_t}{h_{c,V}} + Y_{CO} \tag{9.9}$$

$$h_t = \int_{T_w}^{T} c_p dT \tag{9.10}$$

The boundary conditions for Equations 9.6 and 9.7 are:

$$\left.\begin{array}{ll} at\ r = a, & \beta = \beta_w, Y_{CO} = Y_{CO,w} \\ r \to \infty, & \beta = \beta_\infty, Y_{CO} = Y_{CO,\infty} \end{array}\right\} \tag{9.11}$$

The unknowns are mass burn rate $\rho_w v_w$ ($= \dot{m}''$), β_w, and β profiles. If $\rho_w v_w$ ($= \dot{m}''$) and β_w are known, then Equation 9.6 is integrated with two boundary conditions (Equation 9.11); hence, the profile for β(r) is obtained. With the known profile β(r), we can determine $\left(\frac{d\beta}{dr}\right)_{r=a}$ match with the interfacial gradient for β, and determine $\rho_w v_w$ ($= \dot{m}''$) and β_w.

9.4 Nondimensional Conservation Equations and Boundary Conditions

The mass conservation equation (Equation 9.5) can be integrated to obtain

$$4\pi \rho\ v_r\ r^2 = \text{contstant} = \dot{m},\ \text{mass flow} \tag{9.12}$$

One can convert the governing SZ equation (Equation 9.6) in nondimensional form using the following:

$$\xi = \frac{\dot{m}}{4\pi \rho D r} \tag{9.13}$$

and the value of ξ at r = a is written as

$$\xi_w = \dot{m}/(4\pi\rho \text{Da}), \text{ the nondimensional burn rate} \tag{9.14}$$

Table 9.2 Summary of SZ Variables, Double-Film Model (DFM)

Infinite Gas-Phase Oxidation Reactions with Infinite Boudouard Reaction (DFM)

$C(s) + CO_2 \rightarrow 2CO$ (III) $CO + 1/2\ O_2 \rightarrow CO_2$ (V)

The F's for Thin or Thick Flames (DFM)

$$\beta_w + \left(\frac{d\beta}{d\xi}\right)_w = F$$

$\xi_W = \ln(1 + B_F)$, where $B_F = (\beta_w - \beta_\infty)/(F - \beta_w)$; $B_F = B$ for thin flames;

$B = (Y_{O2,\infty}/\nu_{O2,I}) + (Y_{CO2,\infty}/\nu_{CO2,III})$, $s = Y_{O2\infty}/\nu_{O2,V}$

$-h_{t,\infty} = (h_{c,I}Y_{O2,\infty}/\nu_{O2,I}) + (Y_{CO2,\infty}h_{c,III}/\nu_{CO2,III})$ without radiation loss $h_{c,III} < 0$

$h_{t,\infty} = (\tilde{h}_{c,I}Y_{O2,\infty}/\nu_{O2,I}) + (Y_{CO2,\infty}\tilde{h}_{c,III}/\nu_{CO2,III})$ with radiation loss, $\tilde{h}_{C,I} = h_{C,I} - \frac{\dot{q}''_{rad}}{\dot{m}''}$

$$\tilde{h}_{c,III} = h_{c,III} - \frac{\dot{q}''_{rad}}{\dot{m}''},\ h_{c,III} < 0$$

$$\frac{c_p(T_f - T_\infty)}{h_{C,I}} = \frac{Y_{O2,\infty}}{\nu_{O2,V}}\frac{h_{C,V}}{h_{C,I}} + B\frac{s(1+B)}{1+B+s(1+B)-Y_{I,\infty}}\left[\frac{h_{C,III}}{h_{C,I}} - \frac{\dot{q}''_{rad}}{\dot{m}''h_{C,I}}\right]$$

$$\frac{a}{r_f} = 1 - \frac{\ln\left[1 + B\dfrac{B\left(1+\nu_{CO2,III}\right) - \dfrac{Y_{O2,\infty}}{\nu_{O2,V}}}{s + B\left(1+\nu_{CO2,III}+s\right) - \dfrac{Y_{O2,\infty}}{\nu_{O2,V}}}\right]}{\ln[1+B]}$$

#	β_{ij}	F_β	Expanded β
1	$\beta_{ht\text{-}O2}$	$\tilde{h}_{c,III}/h_{c,V}$	$\beta_{hT\text{-}O2} = h_t/h_{c,V} + Y_{O2}/\nu_{O2,V}$
2	$\beta_{ht\text{-}CO}$	$\frac{h_{c,III}}{h_{c,V}} + \left(1+\nu_{CO2,III}\right)$	$\beta_{hT\text{-}F} = h_t/h_{c,CO} + Y_{CO}$
3	$\beta_{O2\text{-}CO}$	$1 + (\nu_{CO2,III})$	$\beta_{O2\text{-}CO} = Y_{CO} - Y_{O2}/\nu_{O2,V}$
4	$\beta_{CO2\text{-}O2}$	$-(1 + \nu_{O_2,I})$	$\beta_{CO_2-O_2} = \frac{Y_{CO2}}{\nu_{CO2,V}} + \frac{Y_{O_2}}{\nu_{O2,V}}$
5	$\beta_{ht\text{-}CO2}$	$\frac{h_{c,III}}{h_{c,V}} + \left(1+\nu_{CO2,III}\right)$	$\beta_{ht\text{-}CO2} = h_t/h_{c,V} - Y_{CO2}/\nu_{CO2,V}$

Note: If $Y_{O2,\infty} = 0.187$, $Y_{CO2,\infty} = 0.171$, dp = 1mm, $\varepsilon = 0.8$, $T_\infty = 1000$ K, then for $C + CO_2 = 2CO$ and $CO^+\ 1/2\ O_2 = CO_2$, $T_w - T_\infty = 514$ K, DFM without radiation correction, 250 K with radiation correction; $T_f - T_\infty = 1730$ K without radiation correction; $T_f - T_\infty = 1475$ K with radiation correction.

Source: Adapted from Annamalai, K. and Ryan, W., Interactive processes in gasification and combustion — II: isolated carbon/coal and porous char particles, *Prog. Energy Combust. Sci.*, 19, 383–446, 1993.

(the suffix w denotes that it is at the wall or surface)

The normalized SZ (NSZ) variable is defined as

$$\phi = \frac{\beta - \beta_\infty}{\beta_w - \beta_\infty} = \frac{Y_I - Y_{I\infty}}{Y_{Iw} - Y_{I\infty}} \tag{9.15}$$

where $\beta = \beta_{ij}, \beta_{b_T - F}$. From Equation 9.15,

$$\beta = \phi\{\beta_w - \beta_\infty\} + \beta_\infty . \tag{9.16a}$$

Differentiating Equation 9.15 twice,

$$\frac{d\beta}{dr} = \frac{d[\phi\{\beta_w - \beta_\infty\}]}{dr} = \{\beta_w - \beta_\infty\}\frac{d\phi}{dr}, \tag{9.16b}$$

$$\frac{d^2\beta}{dr^2} = \{\beta_w - \beta_\infty\}\frac{d^2\phi}{dr^2} . \tag{9.16c}$$

Using Equation 9.13,

$$\frac{d}{dr} = \frac{d}{d\xi}\frac{d\xi}{dr} = \frac{d}{d\xi}\frac{d}{dr}\left[\frac{\dot{m}}{4\pi\rho D r}\right] = -\left[\frac{\dot{m}}{4\pi\rho D r^2}\right]\frac{d}{d\xi} . \tag{9.17}$$

Using Equation 9.16b to Equation 9.17 in Equation 9.6, the SZ equation reduces to

$$\frac{d\phi}{d\xi} + \frac{d^2\phi}{d\xi^2} = 0 \tag{9.18}$$

and Equation 9.7 reduces to,

$$d^2Y_{CO}/d\xi^2 + dY_{CO}/d\xi = -\dot{w}'''_{CO}\,\xi_w{}^2 a^2/(\rho D\xi^4). \tag{9.19}$$

The boundary conditions (BC) given by Equations 9.11 transform to

$$\left.\begin{array}{l} at \quad \xi = \xi_w,\ \phi = 1, Y_{CO} = Y_{CO,w} \\ \xi \to \infty,\ \phi = 0, Y_{CO} = Y_{CO,\infty} \end{array}\right\}. \tag{9.20}$$

The second-order differential equations (Equation 9.18 and Equation 9.19) can be integrated with two boundary conditions for each equation of (Equation 9.20) and the solution for $\phi(\xi)$ and $Y_{CO}(\xi)$ can be obtained.

9.4.1 Gas-Phase Profiles

Integrating Equation 9.18, and using two BCs (Equation 9.20) we obtain the Φ profile

$$\phi = \frac{1-\exp(-\xi)}{1-\exp(-\xi_w)} \tag{9.21}$$

Using Equation 9.21, one can plot Φ vs. ξ with ξ_w as a parameter. We will solve for all unknowns including ξ_w and β_w, using the interfacial BCs that will be dealt with in the following section. Finally, Equation 9.19 needs to be integrated numerically if finite gas-phase chemistry is to be included. For many practical combustion problems, integration of Equation 9.19 is not necessary to obtain burn rates since infinite chemistry will be assumed.

9.5 Interfacial Conservation Equations or BCs

We will use Reaction I, C(s) + 1/2 O_2 → CO, and Reaction V, CO + 1/2 O_2 → CO_2, as an example.

9.5.1 General System with Arbitrary Surface

9.5.1.1 Mass and Species

We will develop a very general procedure for formulating the interface BCs involving all combustion problems involving two phases. Reaction regimes for Reaction I are illustrated in Figure 9.2. For interface conservation, let us select the c.v. near the interface as shown in Figure 9.3.

Overall mass:

$$dm_{cv}/dt = \dot{m}_i - \dot{m}_e. \tag{9.22}$$

Species k:

$$dm_{cv,k}/dt = \dot{m}_{k,i} - \dot{m}_{k,e} + \dot{w}_k. \tag{9.23}$$

We will apply Equation 9.23 to obtain the interfacial species conservation equations for k = C(s), CO, O_2, and CO_2. Species k enters from the solid phase through an area dA (say with a feeding mechanism for carbon, $C_{(s)}$), may undergo chemical reactions within the elemental volume of 2δ, and then exit through the gas phase. Note that δ is extremely small (say, 1 or 2 μm). Because the carbon as solid species (but not the element) does not exit, carbon species conservation yields

$$\{dm_{cv,C}/dt\} = \dot{m}_{c,i} - \dot{m}_{c,e} + \dot{w}_c = \dot{m}_{c,i} + \dot{w}_c. \tag{9.24}$$

Within dt seconds, a mass of dm is lost

$$dm_{cv,C} = d(\rho_-\delta + \rho_+\delta)\ dA,\ \dot{m}_{C,i} = \dot{m}''_{C,i}\ dA, \tag{9.25}$$

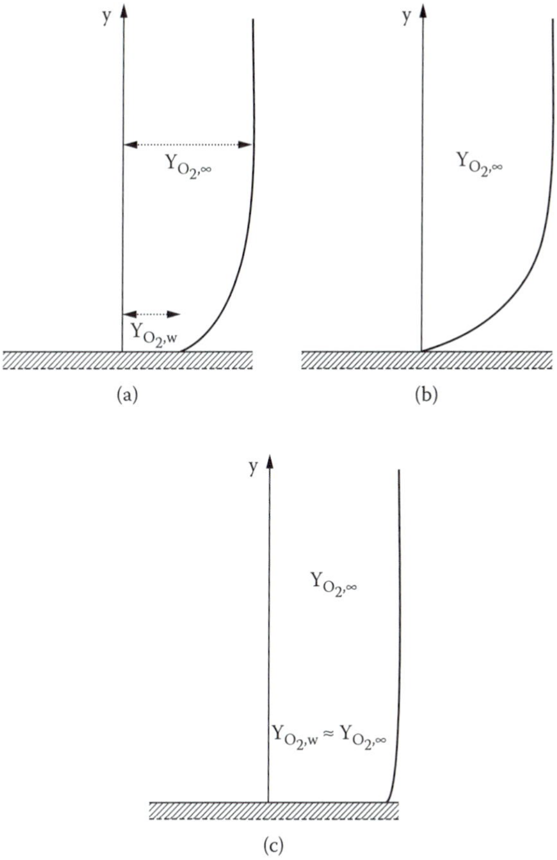

Figure 9.2 Various regimes of burning of carbon: (a) mixed control (diffusion + kinetics), (b) diffusion control, and (c) kinetics control.

where the subscript "+" indicates gas side and "–" indicates solid side.

$$\dot{w}_C = \{\dot{w}'''_{C-}\ \delta dA + \dot{w}'''_{C+}\delta dA\} = \dot{w}'''_{C-}\delta dA + 0\ . \qquad (9.26)$$

Using these results in Equation 9.24

$$\{d(\rho_-\delta + \rho_+\delta)/dt\}dA = \dot{m}''_{C,i}\ \ dA + \dot{w}'''_C\ \ \delta\ dA. \qquad (9.27)$$

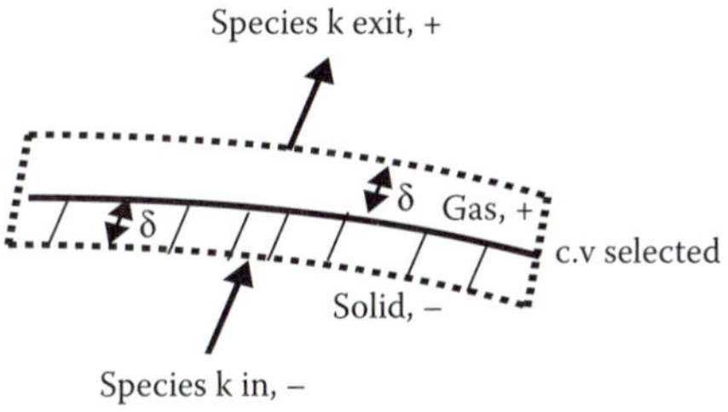

Figure 9.3 Cartesian elemental control volume for conservation equations.

Dividing by dA and letting $\delta \to 0$ with $\dot{w}_c''' \;\; \delta = \dot{w}_c''$ (kg/m²sec) remaining finite,

$$0 = \dot{m}_{c,i}'' + \dot{w}_C'' . \tag{9.28}$$

Similarly, for the oxygen species and CO,

$$0 = \{\dot{m}_{O_2,i}''\}_+ + \dot{w}_{O_2}'', \tag{9.29}$$

$$0 = -\{\dot{m}_{CO,e}''\}_+ + \dot{w}_{CO}'' . \tag{9.30}$$

where species of O_2 enter, and CO exit from gas side. From the law of stoichiometry for Reaction I

$$\frac{\dot{w}_c''}{-1} = \frac{\dot{w}_{O_2}''}{-\nu_{O_2,I}} = \frac{\dot{w}_{CO}''}{-\nu_{CO,I}} = \dot{w}'' . \tag{9.31}$$

From the overall mass balance at the interface, the burn rate or feed rate of carbon into the interface

$$0 = \dot{m}_i'' = \dot{m}_{C,i}'' = \dot{m}_C'' = \dot{m}_e'' .$$

Using Equations 9.29 to 9.31,

$$\left\{\dot{m}_{O2,i}''\right\}_+ = \nu_{O2,I}\,\dot{m}_C'' \tag{9.32}$$

$$\left\{\dot{m}_{CO,e}''\right\}_+ = -\nu_{CO,I}\,\dot{m}_C'' \tag{9.33}$$

Now, we can replace the oxygen flow rate and CO flow rate on the gas side by using mass transfer equations.

Oxygen (O_2)

$$-\left\{\rho_{O2,i} v_{O2,i}\right\}_+ = \nu_{O2,I}\,\dot{m}_C'' \quad \text{or} \quad \left\{\rho_{O2,i} v_{O2,i}\right\}_+ = -\nu_{O2,I}\,\dot{m}_C'' . \tag{9.34}$$

Letting $\rho_+ = \rho_w$, $v_+ = v_w$, $V_{O2+} = V_{O2w}$

$$\rho_w \, Y_{O2,w}[v + V_{O2}] = -\nu_{O2,I}\,\dot{m}_c'' . \tag{9.35}$$

Using Fick's law for diffusion velocity, V_{O2} and $\rho_w v_w = \dot{m}_c''$,

$$\dot{m}_c'' \, Y_{O2,w} - \rho_w D_w \left(\frac{dY_{O2}}{d\,r}\right)_w = -\nu_{O2,I}\,\dot{m}_c'' , \tag{9.36}$$

where O_2 flows toward the surface. Since $(dY_{O2}/dr)_w > 0$, then $\dot{m}_c'' > 0$ when $Y_{O_2,w} = 0$ as it should be.

Mass transfer rate of O_2 from ambient = O_2 consumed by heterogeneous chemical reaction at the interface.

Similarly, for CO

$$\rho_w\, v_w\, Y_{CO,w} - \rho\, D \left(\frac{dY_{CO}}{dr}\right)_w = \dot{m}''_{CO,w} = (1+\nu_{O2,I})\dot{m}''_C \qquad (9.37)$$

where the CO flows away from surface.

For any inert species I (e.g., such as N_2),

$$\rho_w\, v_w\, Y_{I,w} - \rho D \left(\frac{dY_I}{dr}\right)_w = m''_{I,w} = 0. \qquad (9.38)$$

9.5.1.2 Energy

We now consider thermal enthalpy (h_t). The carbon that enters the interface is at T_w. Using the energy conservation equation,

$$dE_{cv}/dt = \dot{Q}_{c.v.} - \dot{W}_{c.v} + \sum \dot{m}_{k,i}\,(h + ke + pe)_{k,i} - \sum \dot{m}_k\,(h + ke + pe)_{k,e} \quad (9.39)$$

where $\dot{Q}_{c.v.}$ is the heat flux solely owing to temperature gradients. As before, dE/dt = 0. The species entering are C(s) and O_2, and the species leaving is CO. Further, the C (s) species entering the area dA is at h_w. There is no shaft or deformation work.

$$0 = [\,\dot{q}''_w\,]_- + [\dot{q}''_w]_+ + [\,\dot{q}''_{ext}] - 0 + \dot{m}_{C(s)}\, h_{c(s)}\,(T_w) + \dot{m}_{O2,w}\, h_{O2}(T_w) - \dot{m}_{CO,w}\, h_{CO}(T_w). \qquad (9.40)$$

$[\dot{q}''_{ext}]$ is the external heat flux (e.g., laser heating, radiative heat transfer) and $\dot{q}''_w$ is solely because of temperature gradients as diffusive and average fluxes have already been accounted for when estimating $\dot{m}_k\, h_k$ in Equation (9.40). Then for a 1-D system with the Fourier law, and recalling that $\{-\vec{q}''.\overrightarrow{dA}\}$ will take care of sign convention in energy equation

$$[\dot{q}''_w]_+ = \left[\lambda \frac{dT}{dr}\right]_+ \qquad (9.41)$$

In Chapter 8, we used $\dot{q}''_w = -\lambda \frac{dT}{dr} + \Sigma \rho_k V_k h_k$ in the differential formulation because the convective energy fluxes in the differential equations are based on average velocity (e.g., convective flux $\rho\vec{v}\cdot\nabla h$ = enthalpy flux carried with average velocity v; see Equation 8.30 of Chapter 8 with α = h).

Since solid is at the assumed to be at uniform T_w, $[\,\dot{q}''_w\,]_- = 0$. The total enthalpy $h = h_{f,k} + h_{t,k}$, where $h_{f,k}$ is the enthalpy of formation at 298 K. In order to minimize the complexities, we will use enthalpy of formation at T_w and hence thermal enthalpy will be defined by Equation 9.10. The energy conservation per unit surface area may be written as

$$0=\left[\lambda\frac{dT}{dr}\right]_{w}+[\dot{q}''_{ext}]+\dot{m}''_{c(s)}[h_{f,c(s)}(T_w)+v_{O2,1}h_{\{f,O_2\}}(T_w)-(1+\nu_{O2,I})h_{f,co}(T_w)] \quad (9.42)$$

$$\left[-\lambda\frac{dT}{dr}\right]=\dot{m}_{C(s)}h_{c,I}+\dot{q}''_{ext}\,. \quad (9.43)$$

Heat conducted away from a hot, heterogeneously reacting surface = heat liberated by the heterogeneous chemical reaction and external heat input. The heating value $h_{c,I}$ is the heat liberated due to the heterogeneous chemical Reaction I and is evaluated at T_w

$$h_{c,I} = [h_{f,c(s)}\,(T_w) + \nu_{O2,I}h_{f,O2}(T_w) - (1 + \nu_{O2,I})\,h_{f,CO}(T_w)]. \quad (9.44)$$

Properties of C(s) are tabulated in Table A.10. Often, $h_{c,I}\,(T_w) \approx h_{c,I}$ (298 K). Assuming $\lambda = \rho D_{Cp}$ and including radiation loss from particle surface to the ambience, then Equation 9.43 is modified as,

$$\left[-\rho D\frac{dh_t}{dr}\right]_{w}=\dot{m}''_{C(s)}h_{c,I}-\dot{q}''_{rad}+\dot{q}''_{ext},\mathrm{kW/m^2} \quad (9.45)$$

where $\dot{q}''_{rad} = \varepsilon_{rad}\sigma\,[T_w^{\,4} - T_\infty^{\,4}]$, ε_{rad} is the emissivity of carbon particle surface ($\approx$ 0.8), and σ = 5.67 × 10^{-11} kW/m^{2}k^4 sec the Stefan Boltzmann constant kW/m^{2}k^4. $[\dot{q}''_{ext}]>0$ if heat enters the surface, and < 0 if heat leaves the surface.

The interface BCs are now complete. Note that all the interfacial conditions involve gradients at the surface that need to be matched with gas-phase gradients.

9.5.2 Spherical Particle

Expressing Equation 9.36 to Equation 9.38 and Equation 9.45 in nondimensional form,

$$Y_{O2,w}+\left(\frac{dY_{O_2}}{d\xi}\right)_{w}=-\nu_{O2,I}=-|\dot{m}''_{O2}|\dot{m}''=-\varepsilon_{O2,w} \quad (9.46)$$

$$Y_{CO,w}+\left(\frac{dY_{CO}}{d\xi}\right)_{w}=\frac{\dot{m}''_{CO,w}}{\dot{m}''}=\varepsilon_{CO,w}=(1+\nu_{O2,I}) \quad (9.47)$$

$$Y_{I,w}+\left(\frac{dY_I}{d\xi}\right)_{w}=0 \quad (9.48)$$

$$\frac{dh_t}{d\xi}=h_{c,I,\mathrm{mod}} \quad (9.49)$$

where ε_{O2}, ε_{CO} are flux ratios, but not necessarily fractions; e.g., $\varepsilon_{CO,w} > 1$ and $\tilde{h}_{c.I} = h_{c.I} - \frac{\dot{q}''_{rad}}{\dot{m}''} + \frac{\dot{q}''_{ext}}{\dot{m}''}$, (kJ/kg), modified heat release per unit mass of carbon after accounting for radiation loss and external heat input.

The gas-phase profiles for ϕ (Equation 9.21) and hence β, Y_I (Equation 9.15) profiles are available; thus, interface BCs given in Equation 9.46 to Equation 9.49 must be given in terms of β or ϕ and should match the gas-phase gradient with interfacial gradients. Dividing Equation 9.46 by $\nu_{O2,V}$, and subtracting the result from the CO Equation 9.47,

$$\beta_{O2\text{-}CO,w} + \{d\beta_{O2\text{-}CO,w}/d\xi\}_w = (1 + \nu_{O2,I}) + (\nu_{O2,I}/\nu_{O2,V}) \tag{9.50}$$

where $\beta_{O2\text{-}CO} = Y_{CO} - Y_{O2}/\nu_{O2,V}$. More generally, let

$$\beta_w + \left(\frac{d\beta}{d\xi}\right)_w = F_\beta \tag{9.51}$$

where for $\beta = \beta_{O2\text{-}CO}$,

$$F_\beta = (1 + \nu_{O2,I}) + (\nu_{O2,I}/\nu_{O2,V}) = 2(1 + \nu_{O2,I}). \tag{9.52}$$

For inerts I, ($\beta = I$)

$$Y_{I,w} + \left(\frac{dY_I}{d\xi}\right)_w = F_\beta = 0 \tag{9.53}$$

Dividing Equation 9.49 by $h_{c,V}$ dividing Equation 9.46 by $\nu_{O2,V}$, and adding, we have

$$\beta_{ht-O2,w} + \left(\frac{d\beta_{ht-O2}}{d\xi}\right)_w = F_\beta = \frac{\tilde{h}_{c,I}}{h_{c,V}} - \frac{\nu_{O2,I}}{\nu_{O2,V}}. \tag{9.54}$$

$$\beta_{ht-O2,w} = \frac{h_{t,w}}{h_{c,V}} + \frac{Y_{O2,w}}{\nu_{O2,V}} = \frac{Y_{O2,w}}{\nu_{O2,V}} \tag{9.55}$$

since $h_{tw} = 0$. We can now obtain explicit solutions for a few specific cases.

9.6 Solutions for Carbon Particle Combustion

9.6.1 Reaction I along with Gas Phase Reaction V

9.6.1.1 Finite Kinetics Gas Phase and Heterogeneous Kinetics

Differentiating Equation 9.21 with ξ, making use of Equation 9.15, and evaluating the differential at $\xi = \xi_w$, we have

$$(d\beta/d\xi)_{\xi=\xi w} = (\beta_w - \beta_\infty)/(\exp[\xi_w] - 1). \tag{9.56}$$

Using Equation 9.56 in the interface condition given by Equation 9.51 and

$$\beta_w + (\beta_w - \beta_\infty)/(\exp[\xi_w] - 1) = F_\beta$$

one can solve for ξ_w

$$\frac{\dot{m}}{4\pi\rho D a} = \xi_w = \ell n\left[1 + B_{kin}\right] \tag{9.57}$$

$$\text{where } B_{kin} = \left(\frac{(\beta_w - \beta_{,\infty})}{(F_\beta - \beta_w)}\right) = \frac{\beta_\infty - \beta_w}{\beta_w - F_\beta}. \tag{9.58}$$

If Stefan flow is ignored (see Equations 9.6 and 9.7) or $B_{kin} << 1$, the burn rate $\xi_w = \ell n\ [1 + B_{kin}] \approx B_{kin}$. Selecting $\beta = \beta_{O2\text{-}CO} = Y_{CO} - Y_{O2}/\nu_{O2,V}$, F_β from Equation 9.52, Equation 9.58 becomes:

$$B_{kin} = \left(\frac{\left(\frac{Y_{O2,\infty}}{\nu_{O2,V}} - Y_{CO,\infty}\right) - \left(\frac{Y_{O2,w}}{\nu_{O2,V}} - Y_{CO,w}\right)}{(1+\nu_{O2,I}) + \frac{\nu_{O2,I}}{\nu_{O2,V}} + \left(\frac{Y_{O2,w}}{\nu_{O2,V}} - Y_{CO,w}\right)}\right). \tag{9.59}$$

Similarly, if $\beta = \beta_{ht\text{-}O2}$, recalling that $h_{t,w} = 0$ and using Equation 9.54,

$$B_{kin} = \left(\frac{\left(\frac{Y_{O2,\infty}}{\nu_{O2,V}} + \frac{h_{t,\infty}}{h_{C,V}}\right) - \left(\frac{Y_{O2,w}}{\nu_{O2,V}}\right)}{\left(\frac{Y_{O2,w}}{\nu_{O2,V}}\right) - \frac{\tilde{h}_{C,I}}{h_{C,V}} + \frac{\nu_{O2,I}}{\nu_{O2,V}}}\right). \tag{9.60}$$

If one uses $\beta_{ht\text{-}CO} = \{h_t/h_{c,V})\} + \{Y_{CO}\}$, and with $F_\beta = \{(1 + \nu_{O2,I})\}$, we have

$$B_{kin} = \left(\frac{Y_{CO,w} - Y_{CO,\infty} - \frac{h_{t,\infty}}{h_{C,V}}}{\frac{\tilde{h}_{C,I}}{\tilde{h}_{C,V}} + (1+\nu_{O2,I}) - Y_{CO,w}}\right). \tag{9.61}$$

Similarly, if $\beta = Y_I$, (for example, N_2), we have

$$B_{kin} = \left(\frac{Y_{I,\infty} - Y_{I,w}}{Y_{I,w}}\right). \tag{9.62}$$

The mass fraction $Y_{O2,w}$ and T_w at the surface are yet to be solved. Suppose the particle is heated with a laser beam to specified T_w; then T_w is known. For the Reaction I chemistry

$$\dot{m}''_C = k_I\ \rho_w\ Y_{O2,w},\ k_I = A_{h,I}\ \exp\ (-E_I/\bar{R}T_w) \tag{9.63}$$

$$\dot{m}_C/(4\ \pi\ \rho D\ a) = \xi_w = \{k_I\ a\ \rho_w/(\rho D)\}\ Y_{O2,w} \tag{9.64}$$

where $\dot{m}_C = \dot{m}''_c\ 4\ \pi a^2$.

Procedure with finite chemistry: The procedure for known T_w is as follows. Assume $Y_{O2,w}$. Get ξ_w from Equation 9.64. Use Equation 9.60 to get B_{kin} and Equation 9.59 to solve for $Y_{CO,w}$ at given T_w (do not use Equation 9.61 as $\tilde{h}_{c,I}$ contains the external heat input, which is not known). Then integrate Equation 9.19 with known $Y_{CO,w}$ and $\varepsilon_{CO,w} = 1 + \nu_{O2,I}$ (Equation 9.47). If the assumed $Y_{O2,w}$ is correct, as $\xi \rightarrow 0$, $Y_{CO,\infty}$ should tend to 0; if not, iterate.

The procedure for unknown T_w is as follows: Follow the procedure for known T_w. Calculate $\tilde{h}_{c,I} = h_{c,I,} - \dot{q}''_{rad}/m'' + \dot{q}''_{ext}/\dot{m}''$, and check whether Equation 9.61 is satisfied for the prescribed T_w; if not, iterate to solve for T_w. For smaller particles, CO diffuses very fast through gas film and hence there is insufficient time for CO oxidation. Then SFM may be adopted.

9.6.1.2 *Finite Kinetics Heterogeneous Chemistry and Frozen Gas Phase (i.e., SFM)*

The gas phase is completely frozen, i.e., no reaction has taken place in the gas phase (Reaction V is absent). In this case, there is a monotonic decrease in CO from the surface and a monotonic increase in O_2 with radius r for Reaction I (Figure 9.4). Similar processes occur for other reactions (Reaction II to Reaction IV), either individually or as combined reactions. It is also possible that the reactions at the surface are infinitely fast (i.e., diffusion-controlled combustion), so that the rate of diffusion controls the combustion rate of carbon (see Chapter 7).

9.6.1.2.1 NSZ Profiles

For this case, Equation 9.6 can be used with the β variables defined in Table 9.1, the solutions being given by Equation 9.21; for the SFM, Equation 9.15 can be simplified as

$$\phi = \frac{Y_{O2} - Y_{O2,\infty}}{Y_{O2,w} - Y_{O2,\infty}} = \frac{Y_{CO} - Y_{CO,\infty}}{Y_{CO,w} - Y_{CO,\infty}} = \frac{Y_I - Y_{I,\infty}}{Y_{I,w} - Y_{I,\infty}} = \frac{h_t - h_{t,\infty}}{h_{t,w} - h_{t,\infty}}. \tag{9.65}$$

Figure 9.4 illustrates the profiles for Y_{O2} and Y_{CO}.

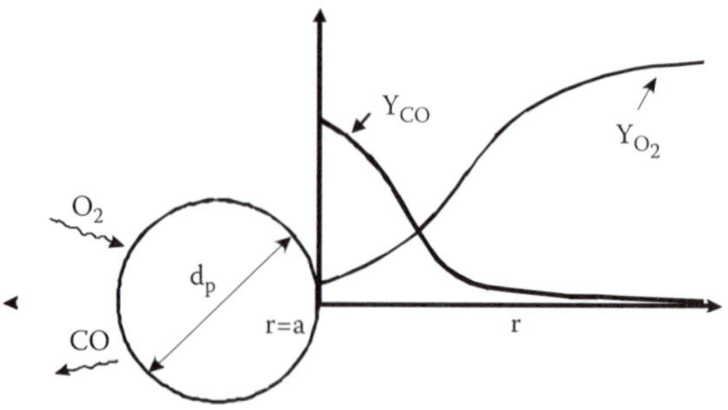

Figure 9.4 Single-film model illustration for Reaction I.

9.6.1.2.2 Transfer Number under Kinetics Control

B_{kin} is obtained by setting $\nu_{O2,V} \rightarrow 0$ in Equation 9.59 and rewriting,

$$B_{kin} = \left(\frac{\left(\frac{Y_{O2,\infty}}{\nu_{O2,I}} - \frac{Y_{O2,w}}{\nu_{O2,I}} \right)}{1 + \left(\frac{Y_{O2,w}}{\nu_{O2,I}} \right)} \right), \qquad C + (1/2)\, O_2 \rightarrow CO \tag{9.66}$$

As $\xi_w = \ln(1 + B_{kin})$, then using Equation 9.66 and equating with Equation 9.64, $Y_{O2,w}$ can be solved for a prescribed T_w. Similarly, using CO and h_t as the variables in frozen gas phase, $Y_{CO,w}$ and T_w can be solved.

$$B_{kin} = \left(\frac{Y_{CO,w} - Y_{CO,\infty}}{1 + \nu_{O2,I} - Y_{CO,w}} \right). \tag{9.67}$$

$$B_{kin} = \left(-\frac{h_{t,\infty}}{\tilde{h}_{C,I}} \right). \tag{9.68}$$

In the absence of radiation loss and external heat input, Equation 9.68 becomes

$$B_{kin} = \left(\frac{c_p (T_w - T_\infty)}{h_{C,I}} \right). \tag{9.69}$$

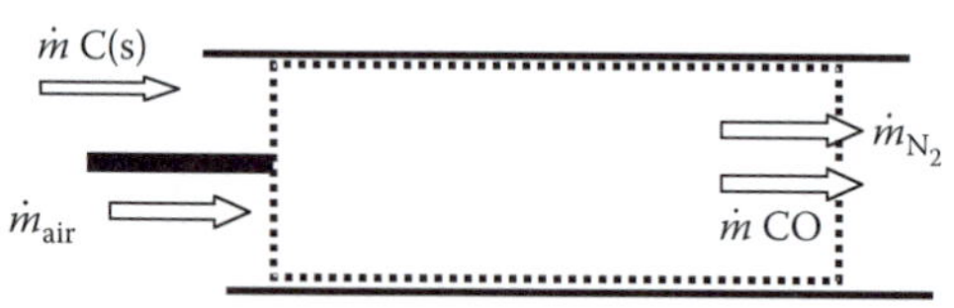

Figure 9.5 $Y_{CO,w}$ at surface is the same as Y_{CO} leaving reactor.

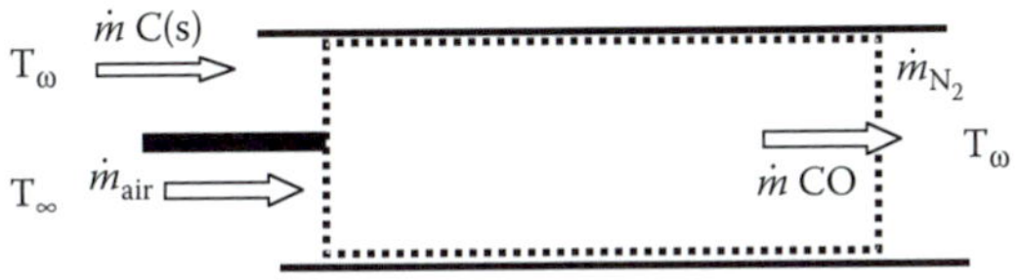

Figure 9.6 T_w at surface is the same as T_w leaving reactor.

9.6.1.3 *Fast Heterogeneous Kinetics and Frozen Gas Phase (i.e., SFM)*

This is an SFM with Reaction I under diffusion-controlled combustion. Setting $Y_{O2,w} = 0$ in Equation 9.66, $B_{kin} = B$, the transfer number

$$B_{kin} = \left[\frac{Y_{O_2,\infty}}{\nu_{O_2,I}}\right] = B. \tag{9.70}$$

Note that B is used typically under diffusion-controlled combustion. The burn rate can be determined from Equation 9.57 as,

$$\xi_w = \frac{\dot{m}}{4\pi\rho D a} = \ell n\left[1+\left(\frac{Y_{O2,\infty}}{\nu_{O2,I}}\right)\right] = \ell n\left[1+B\right]. \tag{9.71}$$

Equating Equation 9.70 with Equation 9.67 and solving for $Y_{CO,w}$

$$Y_{CO,w} = \left(\frac{[1+\nu_{O2,I}][B]+Y_{CO,\infty}}{1+B}\right) = \left(1-\frac{Y_{I,\infty}}{1+B}\right). \tag{9.72}$$

When $Y_{CO,\infty} = 0$, Equation 9.72 becomes,

$$Y_{COw} = \left(\frac{(1+\nu_{O2,I})B}{1+B}\right). \tag{9.73}$$

Similarly, Equation 9.68 becomes

$$-\frac{h_{t\infty}}{\tilde{h}_{C,I}} = \frac{c_p(T_w - T_\infty)}{\tilde{h}_{C,I}} = \left(\frac{Y_{O2,\infty}}{\nu_{O2,I}}\right) = B. \tag{9.74}$$

If the surface radiation loss is neglected and there is no external heat input, $\tilde{h}_{C,I} = h_{C,I}$. If $(T_w - T_\infty) << T_\infty$, then

$$\frac{\tilde{c}_p(T_w - T_\infty)}{h_{C,I}} \simeq B,\ \tilde{c}_p = c_p + \frac{B\varepsilon\nu T_\infty^3}{\dot{m}''} \tag{9.75}$$

The physical meaning of CO mass fraction at the wall, $Y_{CO,w}$ is as follows (Figure 9.5). If stoichiometric air is supplied to combust solid carbon, then O_2

supplied is given by $\dot{m}_{O2} = \nu_{O2,I}\dot{m}_{C(s)}$ and air supplied is given by $\dot{m}_{air} = \dot{m}_{O2}/Y_{O2,\infty} = \nu_{O2,I}\dot{m}_{C(s)}/Y_{O2,\infty} = \dot{m}_{C(s)}/B$; hence, mass flow leaving the reactor is given by $\dot{m} = \dot{m}_{air} + \dot{m}_{C(s)} = \dot{m}_{C(s)}/B + \dot{m}_{C(s)} = \dot{m}_{C(s)}(1 + B)/B$, whereas the mass flow of CO is given by $\dot{m}_{CO} = \dot{m}_{C(s)}\ (1 + \nu_{O2,I})$. As there are no gradients, the ratio of mass flow of CO to the total mass flow is the same as the mass fraction of CO leaving the reactor; $Y_{CO} = \dot{m}_{CO}/\dot{m} = \{(1 + \nu_{O2,I})B\}/(1 + B)$, which is the same as relation Equation 9.73. Similarly, the carbon surface temperature T_w is the same as the adiabatic flame temperature of CO and N_2 products, when C(s) is supplied at T_w, air is supplied at T_∞ (Figure 9.6).

Example 1

Determine the burn rate, $Y_{CO,w}$, and T_w at the particle surface for carbon particle of diameter 100 μm; assume $Y_{O2,\infty} = 0.23$, Reaction I, diffusion-controlled combustion, and $\rho_w{}^*D_w$ (not a function of pressure) = $0.15^* \times 10^{-4}$ kg/m sec. Neglecting radiation loss and assume $T_\infty = 1000$ K.

Solution

We first consider the burn rate:

$d = 100\ \mu m$, $\rho_w{}^*D_w = 0.15 \times 10^{-4}$ kg/msec and $B = 0.23/1.332 = 0.1727$

$$\xi_w = \frac{\dot{m}}{4\pi\rho Da} = \ell n[1+B] = \ln(1 + 0.1727) = 0.1593$$

Burn rate = $\xi_w\ [4^*\ \pi\ \rho D\ (dp/2)] = 0.1593 * [4 * 3.1416 * 0.15 \times 10^{-4} * 50 \times 10^{-6}] = 1.5 \times 10^{-9}$ kg/sec. If $Y_{I,\infty} = 0.77$, $B = 0.23/1.332 = 0.173$.

From Equation 9.73, $Y_{COw} = 0.343$

Using Equation 9.74, $1.2^*\ (T_w - 1000)/9204 = 0.1727$, $T_w = T_p = 2324$ K

Remarks

(i) Now one can include radiation loss and solve for T_w iteratively as 1615 k.

(ii) Figure 9.7 shows a comparison of calculated (with radiation loss) vs. measured temperature ($T_w = T_p$) for burning char particles in a fluidized bed [Annamalai and Caton, 1987; Avedesian and Davidson, 1973; also see Ubhayakar, 1976]. Radiation loss is proportional to a^2, whereas burn rate under diffusion control is proportional to "a" (see Equation 9.71) and, hence, radiation loss per unit burn rate is proportional to "a"; higher radiation loss for larger particles results in lower surface temperatures.

9.6.2 *Other Carbon Reactions*

Similarly, one may consider a combination of Reaction III and Reaction V with finite kinetics. Reaction I to Reaction IV could be considered separately. They are left as exercise problems. Under diffusion-controlled combustion for Reaction II,

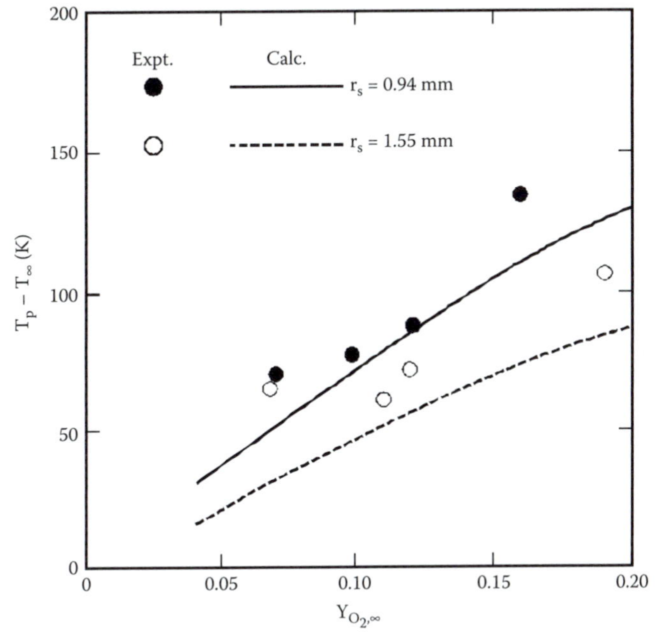

Figure 9.7 Surface temperature of char particles in fluidized bed: measured vs. calculated with radiation loss (adapted from Annamalai, K. and Caton, J., Distinctive burning characteristics of carbon particles, *Can. J. Chem. Eng.*, V. 65, 1027–1032, 1987); r_s = a.

$$\xi_w = \ell n[1+B], \quad B = \frac{Y_{O2,\infty}}{\nu_{O2,II}}, \quad C(s) + O_2 \rightarrow CO_2 \tag{9.76a}$$

Equation 9.76a can be construed as a result with heterogeneous production of CO with infinite CO oxidation at carbon surface.

Following Equation 9.73 for Reaction II

$$Y_{CO2w} = \left(\frac{[1+\nu_{O2,II}]B}{1+B}\right), \quad C(s) + O_2 \rightarrow CO_2 \tag{9.76b}$$

Under diffusion-controlled combustion, for the cases of Reaction III and Reaction IV

$$\xi_w = \ell n\left[1+\frac{Y_{CO2,\infty}}{\nu_{CO2,III}}\right], \quad C(s) + CO_2 \rightarrow 2CO \tag{9.77}$$

$$\xi_w = \ell n\left[1+\frac{Y_{H2O,\infty}}{\nu_{H2O,IV}}\right], \quad C(s) + H_2O \rightarrow 2CO + H_2 \tag{9.78}$$

If Reaction I, Reaction III, and Reaction IV occur together (for example, carbon burning in a boiler in which products are O_2, CO_2, and H_2O), then

$$\xi_w = \ell n\left[1+\frac{Y_{O2,\infty}}{\nu_{O2,I}}+\frac{Y_{CO2,\infty}}{\nu_{CO2,III}}+\frac{Y_{H2O,\infty}}{\nu_{H2O,IV}}\right]. \quad (9.79)$$

9.6.3 Boudouard and Surface Oxidation Reactions with Frozen Gas Phase (i.e., SFM)

There are some interesting results on burn rate, CO mass fraction, and carbon temperature when Boudouard and oxidation reactions are involved. When Reaction I and Reaction III occur simultaneously, Equation 9.47 becomes

$$Y_{CO,w} + \frac{dY_{CO}}{d\xi} = \frac{\dot{m}''_{CO}}{\dot{m}''} = \frac{\dot{m}''_{CO,I}+\dot{m}''_{CO,III}}{\dot{m}''},$$
$$= \frac{\dot{m}''_{C,I}(1+\nu_{O2,I})+\dot{m}''_{C,IIII}(1+\nu_{CO2,III})}{\dot{m}''}. \quad (9.80)$$

The CO_2, CO, and O_2 balance equations are given as

$$Y_{CO_2,w} + \frac{dY_{CO_2}}{d\xi} = -\frac{\dot{m}''_{CO2,III}}{\dot{m}''} = -\frac{\dot{m}''_{C,III}\,\nu_{CO2,III}}{\dot{m}''}, \quad (9.81)$$

$$Y_{O_2,w} + \frac{dY_{O_2}}{d\xi} = -\frac{\dot{m}''_{O2,I}}{\dot{m}''} = -\frac{\dot{m}''_{C,I}\,\nu_{O2,I}}{\dot{m}''}. \quad (9.82)$$

The energy balance equation becomes

$$\left[-\lambda\frac{dT}{dr}\right] + \dot{q}''_{rad} = \dot{m}''_{C,I}h_{C,I,} + \dot{m}''_{C,III,}h_{C,III}, \quad (9.83)$$

where $h_{c,III} < 0$ and $h_{c,I} > 0$.

Normalizing

$$\left(\frac{d\left\{\frac{h_t}{h_{c,I}}\right\}}{d\xi}\right)_w = \frac{\dot{m}''_{C,I}}{\dot{m}''} + \frac{\dot{m}''_{C,III}}{\dot{m}''}\frac{h_{c,III}}{h_{c,I}} - \frac{\dot{q}''_{rad}}{\dot{m}''h_{c,I}}. \quad (9.84)$$

9.6.3.1 Burn Rate

Dividing Equation 9.81 by $\nu_{CO2,III}$, Equation 9.82 by $\nu_{O2,I}$, and adding, we get

$$\left[\frac{Y_{O2,w}}{\nu_{O2,I}}+\frac{Y_{CO2,w}}{\nu_{CO2,III}}\right] + \frac{d\left[\frac{Y_{O2}}{\nu_{O2,I}}+\frac{Y_{CO2}}{\nu_{CO2,III}}\right]}{d\xi} = -\frac{[\dot{m}''_{C,I}+\dot{m}''_{C,III}]}{\dot{m}''} = F_B = -1 \quad (9.85)$$

where $\dot{m}''_{C,I} + \dot{m}_{C,III} = \dot{m}''$. Selecting the SZ variable as $\beta = [Y_{CO2}/\nu_{CO2,III} + Y_{O2}/\nu_{O2,I}]$ for the frozen gas phase, profiles become

$$\phi = \frac{Y_k - Y_{k\infty}}{Y_{kw} - Y_{k\infty}} = \frac{\beta - \beta_\infty}{\beta_w - \beta_\infty} = \frac{h_t - h_{t\infty}}{h_{tw} - h_{t\infty}},\ k = O_2, CO, CO_2, I \tag{9.86}$$

Equation 9.85 assumes the form as in Equation 9.81 with $F_\beta = -11$ and $\beta = \beta_{CO_2-O_2} = \{\frac{Y_{O_2}}{\nu_{O_2,I}} + \frac{Y_{CO_2}}{\nu_{CO_2,III}}\}$. Thus, the burn rate becomes

$$\xi_w = \ell n\left[1 + \left(\frac{\beta_{CO2-O2,w} - \beta_{CO2-O2,\infty}}{F_\beta - \beta_{CO2-O2,w}}\right)\right] = \ln\left[1 + B_{kin}\right]. \tag{9.87}$$

where

$$B_{kin} = \left(\frac{\frac{Y_{O2\infty} - Y_{O2,w}}{\nu_{O2,I}} + \frac{Y_{CO2\infty} - Y_{CO2,w}}{\nu_{CO2,III}}}{1 + \frac{Y_{O2,w}}{\nu_{O2,I}} + \frac{Y_{CO2,w}}{\nu_{CO2,III}}}\right). \tag{9.88}$$

If we assume $Y_{O2,w} \to 0$, $Y_{CO2,w} \to 0$ (diffusion-controlled heterogeneous Reactions I and III), then

$$\xi_w = \ell n[1 + B], \tag{9.89a}$$

where $B = \frac{Y_{O2,\infty}}{\nu_{O2,II}} + \frac{Y_{CO2,\infty}}{\nu_{CO2,III}}$. (9.89b)

9.6.3.2 CO Mass Fraction

Dividing Equation 9.82 by $\nu_{O2,V}$, subtracting it from Equation 9.80, and recalling the identities given by Equation 9.1 and Equation 9.2b,

$$\beta_{CO-O2,w} + \frac{d\beta_{CO-O2}}{d\xi} = F_B = (1 + \nu_{CO2,III}), \tag{9.90}$$

where $\beta_{O2-CO} = Y_{CO} - Y_{O2}/\nu_{O2,V}$; the burn rate is

$$\xi_w = \ell n\left[1 + \left(\frac{\beta_{O2-CO,w} - \beta_{O2-CO,\infty}}{F_\beta - \beta_{O2-CO,w}}\right)\right] \tag{9.91}$$

where $F_\beta = (1 + \nu_{CO2,III})$. Equating Equation 9.91 with Equation 9.87,

$$Y_{CO,w} = \frac{(1 + \nu_{CO2,III})B_{kin} + \frac{Y_{O2w}}{\nu_{O2,V}}(1 + B_{kin}) + \left(Y_{CO,\infty} - \frac{Y_{O2\infty}}{\nu_{O2,V}}\right)}{(1 + B_{kin})}. \tag{9.92}$$

As before, under diffusion-controlled heterogeneous reactions ($Y_{O2,w} = 0$, $Y_{CO2w} = 0$), it can be shown that

$$Y_{CO,w} = 1 - \frac{Y_{I,\infty}}{1+B}, \tag{9.93}$$

which is similar to Equation 9.72 except for a change in the definition of B given by Equation 9.89b (this is left as an exercise problem).

9.6.3.3 Carbon Surface Temperature T_w

One can show that under diffusion-controlled combustion, the particle or carbon surface temperature T_w is given by the relation (left as an exercise problem)

$$\frac{-h_{T,\infty}}{h_{C,I}} = \frac{c_p(T_w - T_\infty)}{h_{C,I}} = \frac{Y_{O2,\infty}}{\nu_{O2,I}} + \frac{Y_{CO2,\infty}}{\nu_{CO2,III}}\frac{h_{C,III}}{h_{C,I}} - \frac{B\dot{q}''_{rad}}{\dot{m}'' h_{C,I}} \tag{9.94}$$

Note that $q''_{rad} = \varepsilon\sigma(T_w^4 - T_\infty^4)$ and, hence, an iterative procedure for T_w is involved. However, if $q''_{rad} = \varepsilon\sigma(T_w^4 - T_\infty^4) \approx 4\,\varepsilon\sigma\,T_\infty^3\,\{T_w - T_\infty\}$ if $T_w - T_\infty << T_\infty$ then the temperature T_w can be explicitly solved with known B. If $Y_{CO2,\infty} = 0$, the B number is governed by Reaction I, which produces CO. Boudouard reaction may still be present owing to CO_2 diffusing from gas-phase oxidation reactions.

If one supplies $\dot{m}_{C(s)}$ along with a gas containing O_2, CO_2, and an inert mixture of $\dot{m}_g$, the O_2 supplied is given by $\dot{m}_g\, Y_{O2,\infty}$ and CO_2 supplied is given by $\dot{m}_g\, Y_{CO2,\infty}$; because all O_2 and CO_2 is consumed, then C consumed by Reaction I is given by $\dot{m}_{C,I} = \dot{m}_g\, Y_{O2,\infty}/\nu_{O2,I}$ and that consumed by Reaction III is given by $\dot{m}_{C,III} = \dot{m}_g\, Y_{CO2,\infty}/\nu_{CO2,III}$. Then, total carbon burned by both Reaction I and Reaction III is given by $\dot{m}_C = \dot{m}_g\, Y_{O2,\infty}/\nu_{O2,I} + \dot{m}_g\, Y_{CO2,\infty}/\nu_{CO2,III} = \dot{m}_g$ B. The CO produced by Reaction I is given by $\dot{m}_{CO,I} = \dot{m}_g\, Y_{O2,\infty}\,(1 + \nu_{O2,I})/\nu_{O2,I}$ and CO produced by Reaction III is given by $\dot{m}_{CO,III} = \dot{m}_g\, Y_{CO2,\infty}\,(1 + \nu_{CO2,III})/\nu_{CO2,III}$. Then total CO leaving the reactor $\dot{m}_{CO} = \dot{m}_g\, YO_{2,\infty}\,(1 + \nu_{O_2,I})/\nu_{O_2,I} + \dot{m}_g\, Y_{CO_2,\infty}\,(1 + \nu_{CO_2,III})/\nu_{CO_2,III}$. CO mass fraction is given by $[\dot{m}_g\, Y_{O2,\infty}\,(1 + \nu_{O2,I})/\nu_{O2,I} + \dot{m}_g\, Y_{CO2,\infty}\,(1 + \nu_{CO2,III})/\nu_{CO2,III}]/[\dot{m}_{C(s)} + \dot{m}_g]$; as $\dot{m}_C = \dot{m}_g B$, $Y_{CO,exit} = [Y_{O2,\infty}\,(1 + \nu_{O2,I})/\nu_{O2,I} + Y_{CO2,\infty}\,(1 + \nu_{CO2,III})/\nu_{CO2,III}]/[1 + B]$, which is the same as Y_{COw}, the mass fraction of CO at the particle surface! Similarly, if the reactor is adiabatic, the temperature of products leaving the reactor is the same as T_w, the particle surface temperature in the absence of radiation loss (see Figure 9.8).

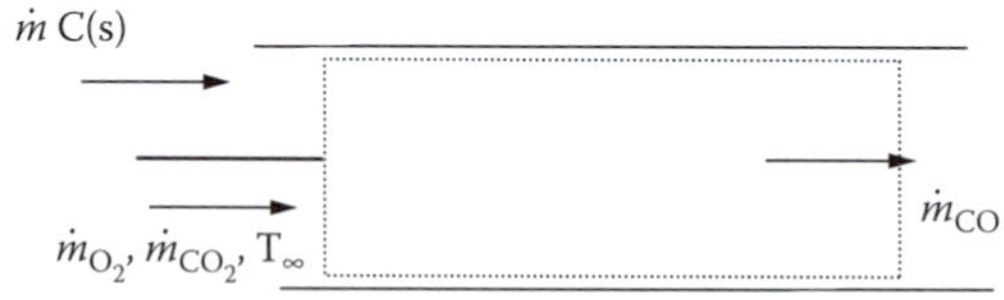

Figure 9.8 T_W **of gas leaving** $\dot{m}_{CO_2} = \dot{m}_g Y_{CO_2,\infty}$ **adiabatic flame temperature,** $\dot{m}_{O_2} = \dot{m}_g Y_{O_2,\infty}$.

9.6.4 Boudouard and Surface Oxidation Reactions along with Gas Phase Oxidation (i.e., DFM))

Heterogeneous reactions (Reaction I and Reaction III) are considered along with the finite gas-phase reaction (Reaction V). The heterogeneous reactions produce CO, which oxidize in the gas phase, raising the gas temperature above the surface temperature. The oxygen concentration decreases in the gas phase, decreasing its contribution to the production of CO via Reaction I. Eventually, a condition is reached in the gas phase in which CO oxidation is so rapid that all the oxygen is consumed by Reaction V at the flame surface (r_f) and, hence, the only CO production at the carbon surface is due to Reaction III. Thus, there exist two films: film I between the carbon surface and flame surface, and film II between the flame surface and ambience. Table 9.2 summarizes the SZ variable for DFM. The gas-phase NSZ profiles are given by Equation 9.21, where NSZ is defined with β, as defined in Table 9.2.

9.6.4.1 Finite Chemistry

The SFM is typically valid for $d_p < 0.05 - 0.1$ mm (due to limited time for co-oxidation) and DFM for $d_p > 1.5$ mm; thus, for $0.1 < d_p < 1$ mm, finite chemistry may be needed [Caram and Amundson]. Selecting $\beta = \beta_{CO2\text{-}O2}$,

$$\beta_{CO2\text{-}O2,} = Y_{CO2}/\{1 + \nu_{O2,V}\} + Y_{O2}/\nu_{O2,V}. \tag{9.95}$$

Dividing Equation 9.82 by $\nu_{O2,V}$, and Equation 9.81 by $(1 + \nu_{O2,V})$ adding, then using the identities given by Equations 9.1 and 9.2a, we obtain

$$\beta_{CO2-O2,w} + \left(\frac{d\beta_{CO2-O2}}{d\xi}\right)_w = F_\beta = -(1+\nu_{O2,I}). \tag{9.96}$$

Thus, the burn rate for the DFM with finite heterogeneous and gas-phase kinetics,

$$\xi_w = \ell n\left[1+\left(\frac{\beta_{CO2-O2,w}-\beta_{CO2-O2,\infty}}{F_\beta-\beta_{CO2-O2,w}}\right)\right] = \ell n\left[1+B_{kin,DFM}\right], \tag{9.97}$$

where

$$B_{kin,DFM} = \left(\frac{\frac{Y_{O2\infty}-Y_{O2,w}}{\nu_{O2,V}}+\frac{Y_{CO2\infty}-Y_{CO2,w}}{(1+\nu_{O2,V})}}{1+\nu_{O2,I}+\frac{Y_{O2,w}}{\nu_{O2,V}}+\frac{Y_{CO2,w}}{(1+\nu_{O2,V})}}\right). \tag{9.98}$$

Dividing numerator and denominator by $(1 + \nu_{O2,I})$, we get another form after simplification:

$$B_{kin,DFM} = \left(\frac{\dfrac{Y_{O2\infty} - Y_{O2,w}}{\nu_{O2,I}} + \dfrac{Y_{CO2\infty} - Y_{CO2,w}}{\nu_{CO2,III}}}{1 + \dfrac{Y_{O2,w}}{\nu_{O2,I}} + \dfrac{Y_{CO2,w}}{\nu_{CO2,III}}} \right) \tag{9.99}$$

for all finite reactions. Equation 9.99 is identical to Equation 9.88. However, due to gas phase oxidation $Y_{O2,w}$ could be different.

9.6.4.1.1 CO Mass Fraction

$$\xi_w = \ell n \left[1 + \left(\frac{\beta_{O2-CO,w} - \beta_{O2-CO,\infty}}{F_\beta - \beta_{O2-CO,w}} \right) \right]$$

Making use of Equation 9.91 and equating with Equation 9.97, Y_{COw} can be solved as

$$Y_{COw} = \left(1 + \frac{Y_{O2,w}}{\nu_{O2,V}} - \frac{\{1 - B_{kin,DFM}\, \nu_{CO2,III}\} + \frac{Y_{O2\infty}}{\nu_{O2,V}} - Y_{CO\infty}}{1 + B_{kin,DFM}} \right) \tag{9.100}$$

9.6.4.1.2 Temperature T_w

$$-\lambda(dT/dr)_{r=a} = -\rho D \frac{dh_t}{dr} = \dot{m}''_{c,I} h_{c,I} + \dot{m}''_{c,III} h_{c,III} + \dot{q}''_{ext} - \dot{q}''_{rad} \tag{9.101}$$

Ignoring external heat flux, dividing by $h_{c,V}$, normalizing it, dividing Equation 9.82 by $\nu_{O2,V}$, adding, and using the identity given by Equation 9.3, we obtain

$$\beta_{ht-O2,w} + \{d\beta_{ht-O2}/d\xi\}_w = F_B = (\tilde{h}_{c,III}/h_{cV}) \tag{9.102}$$

where $\tilde{h}_{c,III} = h_{c,III} - \{\dot{q}''_{rad} / \dot{m}''_c\}$

$$\xi_w = \ell n \left[1 + \left(\frac{\beta_{ht-O_2,w} - \beta_{ht-O2,\infty}}{F_\beta - \beta_{ht-O2,w}} \right) \right] = \ell n \left[1 + \left(\frac{\dfrac{Y_{O2,w}}{\nu_{O2,V}} - \dfrac{h_{t,\infty}}{h_{c,V}} - \dfrac{Y_{O2,\infty}}{\nu_{O2,V}}}{\dfrac{\tilde{h}_{C,III}}{h_{C,V}} - \dfrac{Y_{O2w}}{\nu_{O2,V}}} \right) \right] \tag{9.103}$$

Equating Equation 9.103 with Equation 9.97 and making use of Equation 9.98 for B_{kin}

$$B_{kin,DFM} = \left(\frac{\dfrac{Y_{O2\infty} - Y_{O2,w}}{\nu_{O2,V}} + \dfrac{Y_{CO2\infty} - Y_{CO2,w}}{(1 + \nu_{O2,V})}}{1 + \nu_{O2,I} + \dfrac{Y_{O2,w}}{\nu_{O2,V}} + \dfrac{Y_{CO2,w}}{(1+\nu_{O2,V})}} \right) = \left(\frac{\dfrac{Y_{O2,w}}{\nu_{O2,V}} - \dfrac{h_{t,\infty}}{h_{c,V}} - \dfrac{Y_{O2,\infty}}{\nu_{O2,V}}}{\dfrac{\tilde{h}_{C,III}}{h_{C,V}} - \dfrac{Y_{O_2,w}}{\nu_{O_2,V}}} \right) \tag{9.104}$$

Equation 9.104 provides the solution for T_w.

9.6.5 Fast Chemistry

If the reaction in the gas phase is extremely fast, then O_2 cannot reach the surface and $Y_{O2,w} \to 0$. If the reaction zone or the flame containing both fuel and O_2 becomes extremely thin, then the flame is called a *thin flame*. The reaction (Reaction III) at the surface is infinitely fast. Reaction I should be absent because the homogeneous reaction (Reaction V) is also infinitely fast and, as such, there is no O_2 at the surface. Under fast heterogeneous kinetics, $Y_{O2w} = 0$, $Y_{CO2w} = 0$; Equation 9.104 reduces to

$$B_{DFM} = \left(\frac{\frac{Y_{O2\infty}}{\nu_{O2,V}} + \frac{Y_{CO2\infty}}{(1+\nu_{O2,V})}}{1+\nu_{O2,I}} \right) = \frac{Y_{O2\infty}}{\nu_{O2,I}} + \frac{Y_{CO2\infty}}{\nu_{CO2,III}} = \left(\frac{-\frac{h_{t\infty}}{h_{cV}} - \frac{Y_{O2,\infty}}{\nu_{O2,V}}}{\frac{\tilde{h}_{C,III}}{h_{C,V}}} \right) \tag{9.105}$$

where the identities given by Equation 9.1 and Equation 9.2 have been used. If $Y_{CO2,\infty} = 0$, then the B number is governed by Reaction I, which produces CO. The burn rate $\xi_w = \ln(1 + B)$ is exactly the same as for the case with the frozen gas phase.

9.6.5.1 CO Mass Fraction

Equation 9.100 reduces to

$$Y_{COw} = \left(\frac{B(1+\nu_{CO2,III}) - \frac{Y_{O2\infty}}{\nu_{O2,V}}}{1+B} \right) = \left(1 - \frac{1+\frac{Y_{O2\infty}}{\nu_{O2,V}} - B\nu_{CO2,III}}{1+B} \right) = 1 - \frac{Y_{I,\infty}}{1+B}. \tag{9.106}$$

The same result was obtained for the frozen gas case.

9.6.5.2 Surface Temperature

Using third and fourth terms of Equation 9.105, one can solve for surface temperature as

$$-\frac{h_{t\infty}}{\tilde{h}_{C,I}} = \frac{Y_{O2\infty}}{\nu_{O2,I}} + \frac{Y_{CO2\infty}}{\nu_{CO2,III}} \frac{\tilde{h}_{C,III}}{\tilde{h}_{C,I}}. \tag{9.107a}$$

The same result was obtained for the frozen gas case. Figure 9.9 shows a typical flame structure for DFM

$$\text{where } \tilde{h}_{c,I} = h_{c,I} - \dot{q}''/\dot{m}_c'', \tag{9.107b}$$

$$\tilde{h}_{c,III} = h_{c,III} - \dot{q}''/\dot{m}_c''. \tag{9.107c}$$

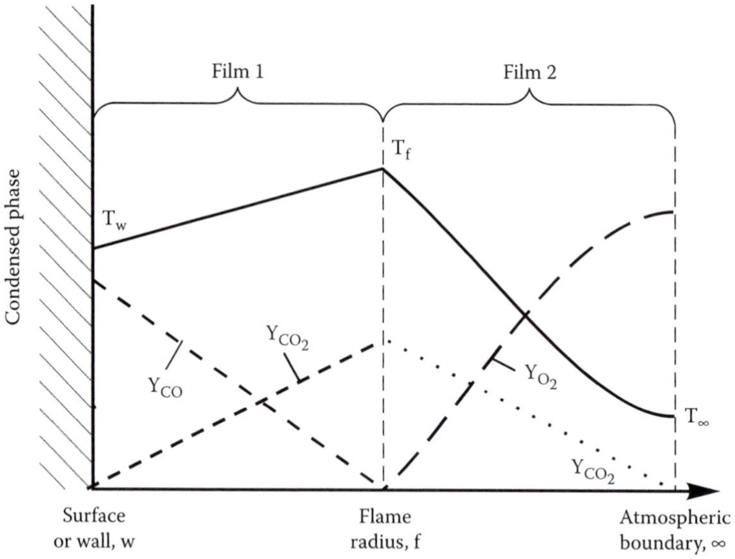

Figure 9.9 Double-film model for C burning: film 1: $Y_{O2} = 0$; film 2: $Y_{CO} = 0$.

9.6.5.3 *Flame Location*

At the flame location, $\beta = \{\beta\}_f$. Using this result in Equation 9.15 and Equation 9.21

$$\phi_f = \frac{\beta_f - \beta_\infty}{\beta_w - \beta_\infty} = \frac{1 - \exp\left(-\xi_f\right)}{1 - \exp\left(-\xi_w\right)}. \tag{9.108}$$

With $\beta_f = \{\beta_{O2\text{-}CO}\}_f = \{Y_{CO} - Y_{O2}/\nu_{O2V}\}_f = 0$, and solving for ξ_f,

$$-\xi_f = \ln\left[1 - \phi_f\left(\frac{B}{1+B}\right)\right] \tag{9.109}$$

where

$$\phi_f = \frac{\beta_f - \beta_\infty}{\beta_w - \beta_\infty} = \frac{0 - \beta_{CO-O2\infty}}{\beta_{CO-O2,w} - \beta_{CO-O2\infty}} = \frac{s}{Y_{CO,w} + s} \tag{9.110}$$

where $s = Y_{O2,\infty}/\nu_{O2,V}$. and $Y_{CO,w}$ is known from Equation 9.106

Using $\xi_w = \ln(1 + B)$ in Equation 9.109,

$$\frac{a}{r_f} = 1 - \frac{\ln\left[1 + B\dfrac{B(1+\nu_{CO2,III}) - \dfrac{Y_{O2,\infty}}{\nu_{O2,V}}}{s + B(1+\nu_{CO2,III} + s) - \dfrac{Y_{O2,\infty}}{\nu_{O2,V}}}\right]}{\ln[1+B]}. \tag{9.111}$$

Letting $Y_{CO2,\infty} = 0$; then $1 - Y_{I,\infty} = Y_{O2,\infty}$; $B = Y_{O2,\infty}/\nu_{O2,I}$ then Equation 9.110 becomes

$$\phi_f = \frac{1+B}{2+B} \tag{9.112}$$

and

$$\frac{r_f}{a} = \frac{\ln[1+B]}{\ln\left[1+\frac{B}{2}\right]}. \tag{9.113}$$

Note:

1. This model is called the double-film model as there are two films: the first film, in which Y_{CO} decreases from a maximum at the particle surface to the thin-flame zone in which the CO mass fraction is zero, and the second film, in which the O_2 mass fraction decreases from a maximum in the ambience to zero at the thin film (see Figure 9.9).
2. By measuring the burn rate, one cannot identify a single- or double-film model if diffusion control is suspected. In the double-film model, the final product is CO_2, whereas in single-film model, the final product is CO. If radiation loss is ignored, the particle temperature T_w for the double-film model with final product CO_2 is exactly the same expression as that obtained for the single-film model, with Reaction I and Reaction III being important and the final product being CO under diffusion-controlled combustion.
3. Figure 9.10a shows a schematic for obtaining Y_{CO} at the exit of an open system in which C(s), CO_2, and O_2 are supplied. Comparing Y_{COw} with SFM, the CO mass fraction at the wall for DFM is the same as before.
4. Note that a/r_f can never be equal to 1. With DFM, $B \to 0$ or as $Y_{O2,\infty} \to 0$, $r_f/a \to 2$, the closest approach when the Boudouard reaction is infinity. At this point, Y_{COw} tends to zero.
5. If finite kinetics are included with prescribed carbon temperatures T_w, the burn rate first increases with T_w and then decreases with further increase in T_w [Strickland–Constable, 1965]. Apart from surface kinetics, a hypothesis postulated was that the carbon produces CO first as T_w is increased; thus, for a mole of O_2 diffusing, two carbon atoms were removed (regime XY); as the surface temperature increases, CO starts oxidizing in the gas phase, reducing the available O_2 concentration at the surface (regime YZ). Thus, as more CO_2 is produced, the CO_2 may increase and for every mole of O_2 diffusion, only one carbon atom mole was removed, thus decreasing the burn rate. The model results are shown in Figure 9.10b; the model does not account for CO_2 reduction reactions at the surface.

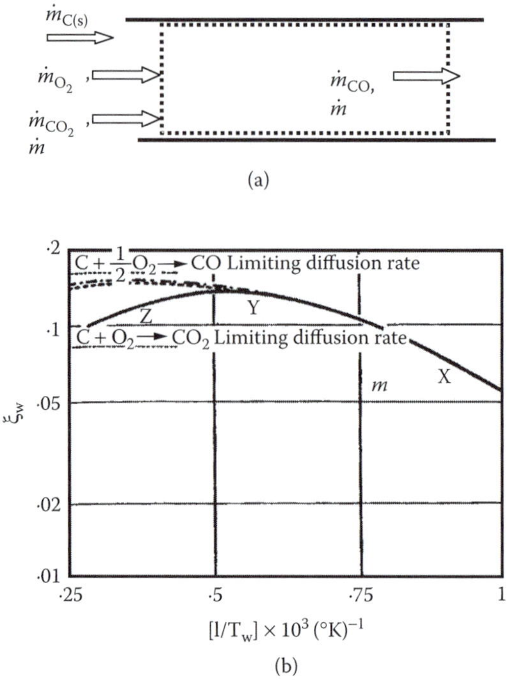

Figure 9.10 (a) $Y_{CO,w}$, at surface = Y_{CO} leaving reactor for DFM; (b) Dimensionless burn rate of carbon vs. surface temperature of particle. (Adapted from Annamalai, K. and Durbetaki, P., Combustion Behavior of Small Char/Graphite Particles, 17th International Symposium on Combustion, University of Leeds, England, Combustion Institute, pp. 169–178, August 20–25, 1979.)

Example 2

Charcoal is burned in pure oxygen. The burning-rate expression for charcoal (pure carbon) in kg of carbon/sec under diffusion-controlled combustion is given by

$$\dot{m} = 4\ \pi\ \rho\ D\ a\ \ln(1 + B)$$

except that $B = Y_{O2\infty}/\nu_{O2}$, where ν_{O2} is the stoichiometric oxygen, say, for the reaction $C(s) + (1/2)\ O_2 \rightarrow CO$. "a" is the particle radius and D is the diffusion coefficient. It is known that {D/a} = 1.68 m/sec and $Y_{O2\infty} = 0.23$. Assume that the gas density of the mixture at the surface is 0.25 kg/m³. Determine the following at the interface, assuming the reaction to be $C(s) + (1/2)\ O_2 \rightarrow CO$.

(a) Oxygen consumption rate per unit surface area of particle in kg/m² sec.
(b) The diffusive flux of oxygen species at surface in kg/m² sec
(c) The CO production rate in kg/m² sec.
(d) The mass fraction of CO at the surface.
(e) The average velocity of gases (v) at the particle surface in m/s and the diffusive velocity of CO in m/sec at the particle surface.

Solution

(a) Burn rate $\dot{m} = 4\,\pi\,\rho$ D a ln(1 + B), $\dot{m}'' = \{\rho$ D/a} ln(1 + B), ρD/a = 0.42, B = 0.173; $\dot{m}''$ = 0.42 ln(1 + 0.173) = 0.0668 kg/m^2 sec

$$\dot{m}''_{O2} = (16/12) \times 0.0668 = 0.0891 \text{kg/m}^2 \text{ sec.}$$

(b) $\rho vY_{O2} - \rho D\partial Y_{O2}/\partial r = -0.0891$; $0 - \rho D\partial Y_{O2}/\partial r = -0.0891$
(c) (28/12) × 0.0668 = 0.156
(d) $Y_{CO} = 1$
(e) $\rho v = 0.0668$, $\rho = 0.25$; v = 0.0668/0.25 = 0.267 m/sec, $V = v_k - v$; $\rho_k v_k = 0.156$; $Y_{CO} + Y_{O2} = 1.0$; $Y_{CO} = 1.0$. Hence, $\rho_k = \rho = 0.25$; $v_k = 0.156/0.25 = 0.624$m/sec; $V_{CO} = 0.624 - 0.267 = 0.357$ m/sec

9.6.5.4 *Flame Temperature — DFM*

The temperature of a flame can be obtained by using $\beta = \beta_{ht\text{-}O2} = h_t/h_{C,V} + Y_{O2}/\nu_{O2,V}$. Evaluating ϕ_f with $\beta > \beta_{ht-O2}$ and using Equation 9.110, one can show that (left as an exercise problem)

$$\frac{c_p(T_f - T_\infty)}{h_{C,I}} = \frac{Y_{O2,\infty}}{\nu_{O2,V}}\frac{h_{C,V}}{h_{C,I}} + B\frac{s(1+B)}{1+B+s(1+B)-Y_{I,\infty}}\left[\frac{h_{C,III}}{h_{C,I}} - \frac{\dot{q}''_{rad}}{\dot{m}''h_{C,I}}\right]. \tag{9.114}$$

9.6.5.5 *CO_2 Mass Fraction in Flame*

$$\text{Using } \phi_f = \frac{s(1+B)}{1+B+s(1+B)-Y_{I,\infty}} = \frac{\beta_{CO2-O2} - \beta_{CO2-O2',\infty}}{\beta_{CO2-O2',w} - \beta_{CO2-O2',\infty}}, \tag{9.115}$$

one can determine that (left as an exercise)

$$Y_{CO2,f} = B[1-\phi_f][1+\nu_{O2,I}][1+\nu_{O2,V}] = B[1-\phi_f]\nu_{CO_2,III}. \tag{9.116}$$

Example 3

Given particle diameter d = 100 μm, ρ_C = 700 kg/m^3, $\lambda = 76.3 \times 10^{-06}$ kW/m K for air at 1200 K, c_p = 1.18 kJ/kg K $\rho D = 3 \times 10^{-5}$ kg/msec, $Y_{O2,\infty} = 0.187$, $Y_{CO2,\infty} = 0.171$, T_∞ = 1000 K; Assume SFM and ignore radiation loss.

C(s) + 1/2 O_2 → CO (I), $h_{c,I}$ = 9203.2 kJ/kg of C, $\nu_{O2,I} = 1.332$,

C(s) + CO_2 → 2 CO (III), $h_{c,III}$ = −14359.7 kJ/kg of C, $\nu_{CO2,III} = 3.664$,

CO + 1/2 O_2 → CO_2, (V), $h_{c,V}$ = 10103.2 kJ/kg of CO, $\nu_{O2,V} = 0.571$

(a) Solve for the burn rate including Reaction I and III only.
(b) Determine the CO mass fraction and particle temperature in K.
(c) Draw CO, CO_2, O_2, N_2, and T profiles.
(d) Determine burn time in msec assuming that the burn rate is given by the same expression as for the steady-state case except that the particle radius shrinks, i.e., a = a(t).
(e) Determine the CO flux rate in kg/sec.

Solution

(a) $B = Y_{O2,\infty}/\nu_{O2,I} + Y_{CO2,\infty}/\nu_{CO2,III} = 0.187/1.332 + 0.171/3.664 = 0.187$, $\xi_w = \ln(1 + B) = 0.1714$.

Assuming Le = 1, $\rho D = \lambda/cp = 76.3 \times 10^{-06}/1.18 = 6.47 \times 10^{-5}$ kg/m sec

$$\xi_w = \frac{\dot{m}}{4\pi\rho Da},\quad \dot{m} = \xi_w[4\pi\rho Da] = 0.1714 * \pi * 6.47\times10^{-5}\left(\frac{100}{2}\right)10^{-6} = 6.97\times10^{-9}\frac{kg}{s}$$

(b) $$Y_{CO,w} = 1 - \frac{Y_{I,\infty}}{[1+B]}.$$

$$Y_{I,\infty} = 1 - Y_{O2,\infty} - Y_{CO2,\infty} = 1 - 0.187 - 0.171 = 0.642.$$

$$Y_{COw} = 1 - 0.642/(1 + 0.187) = 0.46.$$

Ignoring radiation, Equation (9.107a) yields,

$$\frac{-h_{T,\infty}}{h_{C,I}} = \frac{c_p(T_w - T_\infty)}{h_{C,I}} = \frac{Y_{O2,\infty}}{\nu_{O2,I}} + \frac{Y_{CO2,\infty}}{\nu_{CO2,III}}\frac{h_{C,III}}{h_{C,I}} = \frac{0.187}{1.332} + \frac{0.171}{3.664}\frac{(-14360)}{9203} = 0.06757$$

$$T_w = 1427 \text{ K}.$$

If radiation loss is included, then T_w = 1315 K with emissivity of carbon ≈ 1.0 [Hottel and Sarofim, 1978].

(c)

$$\phi = \frac{Y_{CO} - Y_{CO\infty}}{Y_{CO,w} - Y_{CO\infty}} = \frac{Y_{CO2} - Y_{CO2\infty}}{Y_{CO2,w} - Y_{CO2\infty}} = \frac{Y_{O2} - Y_{O2\infty}}{Y_{O2,w} - Y_{O2\infty}} = \frac{h_T - Y_{T\infty}}{h_{T,w} - h_{T,\infty}} = \frac{1 - \exp(-\xi)}{1 - \exp(-\xi_w)}, \tag{A}$$

With $Y_{COw} = 0.46$, $Y_{O2w} = 0$, $h_{t,w} = 0$, $h_t = c_p(T - T_w)$, $\xi/\xi_w = a/r$, $\xi_w = 0.1714$, $Y_{O2\infty} = 0.187$, $Y_{CO2\infty} = 0.171$, and $Y_{CO2w} = 0$, we can get all the profiles. Figure 9.11 shows the profiles.

(d) It was shown in Chapter 7 that when the burning rate of a spherical particle is proportional to the diameter, one obtains the d^2 law.

$$-4\pi a^2 \frac{da}{dt}\rho_C = \dot{m} = \xi_w * 4\pi\rho D\, a, \text{ where } \xi w = \ln(1 + B),$$

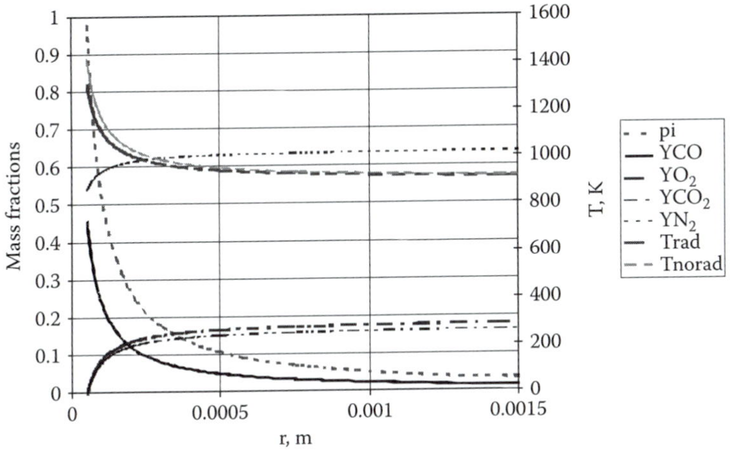

Figure 9.11 Profiles of species mass fractions and T for SFM with Reaction I and Reaction III.

Integrating with initial conditions a = a_0 at t = 0, and simplifying

$$d^2 = d_0^2 - \alpha_b\, t,\ \alpha_b = 8\, \rho\, D\, \ln(1 + B)/\rho_C$$

With the given data, $\alpha_b = 1.270 \times 10^{-7}$ m²/sec, $t_b = (100 \times 10^{-6})^2/1.270 \times 10^{-7}$ = 0.0789 sec.

(e) CO flux from the particle:

$$\text{As } r \to \infty,\ \rho\, v\, 4\, \pi\, r^2\, Y_{CO} - \rho\, D\, 4\, \pi\, r^2\, dY_{CO}/dr = \dot{m}_{CO}.$$

Using nondimensional variables

$$[Y_{CO} + dY_{CO}/d\xi]_{\xi=0} = \dot{m}_{CO}/\dot{m} = \varepsilon_{CO},\ \text{flux ratio}$$

As Y_{CO} at $\xi = 0$ is zero, and using Equation A

$$[dY_{CO}/d\xi]_{\xi=0} = \varepsilon_{CO} = \left[\frac{\exp\ (-\xi)}{1-\exp\ (-\xi_w)}\right]_{\xi=0} \left[Y_{CO,w} - Y_{CO,\infty}\right]$$

$$[dY_{CO}/d\xi]_{\xi=0} = [1/B]\ [(1 - Y_{I,\infty} + B)] = 1 + [1 - Y_{I,\infty}]/B = \varepsilon_{CO}.$$

Thus, flux ratio of CO = $\dot{m}_{CO}/\dot{m}$ = [(1 − 0.642 + 0.187]/0.187 = 2.9144. Note that it is not a fraction i.e., CO flow rate = 2.9144 × $\dot{m}$ = 2.03 × 10^{-06} kg/sec.

9.7 Thermal NO_x from Burning Carbon Particles

It is seen that temperatures can reach as high as 3000 K in the gas phase, which can lead to thermal NO_x formation within the boundary layer. However, the NO_x kinetics are slow and, hence, NO formed is in ppm; thus, it can be treated as a trace species; i.e., the O_2 and N_2 quantities consumed are so small that their concentration profiles are almost unaffected. However, the NO_x concentration in ppm will show a profile in the gas phase.

NO flux from the particle:

$$\text{At any r, } \rho \, v \, 4 \, \pi \, r^2 \, Y_{NO} - \rho \, D \, 4 \, \pi \, r^2 \, dY_{NO}/dr = \dot{m}_{NO}. \quad (9.117)$$

Using nondimensional variables, flux ratio

$$[Y_{NO} + dY_{NO}/d\xi]_{\xi} = \dot{m}_{NO}/\dot{m} = \varepsilon_{NO}. \quad (9.118)$$

Note that ε_{NO} at $\xi = \xi_w$, is zero, as it is assumed that char N content is negligible and NO is assumed to be unreactive with carbon.

From the NO mass conservation equation,

$$d/dr[\rho \, v \, 4 \, \pi \, r^2 \, Y_{NO} - \rho \, D \, 4 \, \pi \, r^2 \, dY_{NO}/dr] = \dot{w}'''_{NO} \, 4\pi r^2.$$

Writing in nondimensional form

$$d/d\xi[Y_{NO} + dY_{NO}/d\xi] = -\frac{\dot{w}'''_{NO} a^2}{\rho D \, \xi_w^2} \frac{r^4}{a^4}. \quad (9.119)$$

Recognizing the term within [ ] as the mass flux ratio ε_{NO}

$$[d\varepsilon_{NO} \, (\xi)/d\xi] = -w'''_{NO} \xi_w{}^2 a^2/(\rho \, D \, \xi^4) \quad (9.120)$$

Integrating

$$[\varepsilon_{NO} \, (\xi) - 0] = -\int_{\xi w}^{\xi} \dot{w}'''_{NO} \, \xi_w{}^2 a^2 \, d\xi/(\rho \, D \, \xi^4). \quad (9.121)$$

Knowing ε_{NO} at any given, ξ and with Equation 9.118, the Y_{NO} profile can be obtained after integration

$$dY_{NO}/d\xi = \dot{m}_{NO}/\dot{m} = \varepsilon_{NO} - Y_{NO}. \quad (9.122)$$

If free stream $Y_{NO,\infty} = 0$, then one can use Equation 9.122 and obtain Y_{NO} vs ξ

$$\text{At } \xi > \xi_f, \text{ there is no } O_2;\ \varepsilon_{NO} = 0; \text{ thus } \ln Y_{NO} = -\xi + C,\ \xi > \xi_f. \quad (9.123)$$

Remarks: We assumd that reaction NO + C → CO + $^1/_2$, and that the N_2 reaction is absent. However the literature suggests [Mitchel et al., 1982]

$$C(s) + NO \rightarrow CO + \frac{1}{2} N_2$$

$$k_{C,XC} = 1.57 \times 10^5 \exp\left(\frac{-34000}{RT_p}\right) \quad (m/s)$$

and, hence, NO can diffuse to the carbon particle surface after which NO can be reduced by this reaction.

9.8 Non-Quasi-Steady Nature of Combustion of Particle

Previously, we assumed that carbon was being fed to maintain the particle size so that the systems were steady. However, practical systems involve decreasing particle size as the particle burns. For example, if the burn rate is $\dot{m}$, the change in particle size within dt is given as

$$\text{Time interval } dt = da \times 4\pi\, a^2\, \rho_C / \dot{m} \text{ or } d\,(a^2)/\alpha_{burn}.$$

Figure 9.12 illustrates the consequence of a decrease in particle size on temperature profiles for SFM. In addition, the oxygen profiles will also be affected. The mass decreases by 20% for a carbon particle of initial size 100 μm within the time interval of 14 msec (according to the d^2 law). Point C is within the particle, whereas points B and A are outside it (Figure 9.12a). The temperature at B is 1995 K, whereas at A it is 1926 K. As the particle shrinks to C, the temperature at B must now decrease. Thus, temperature at B or T and hence density in the gas phase vary as a function of time. Similarly, if O_2 percentages are 0 and 1.2% at B and A (not shown), the new O_2 percentage at B (Figure 9.12b) must be 1.2%. Thus, Y_{O2} at B increases with time. The decrease of temperature requires local heat removal or heat diffusion away from particle, whereas increase of O_2 requires sufficient mass diffusion of O_2. Hence, gas-phase species, energy, and mass conservation equations must have time-dependent terms. So far, we have solved

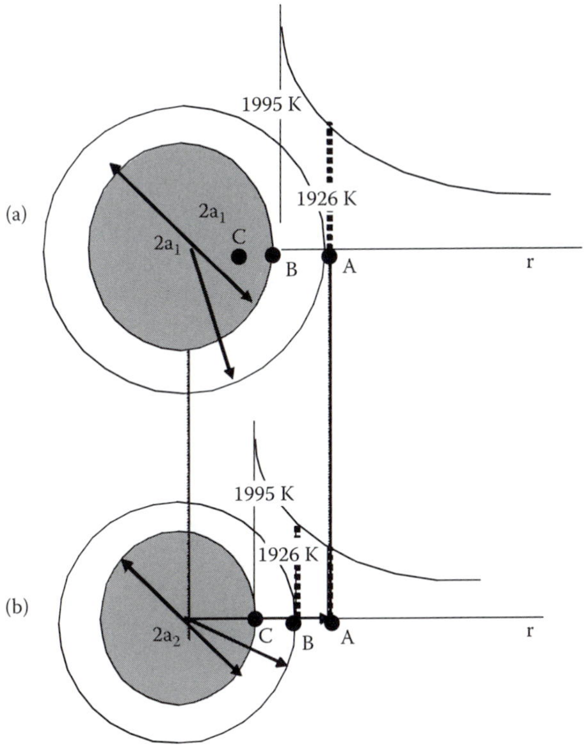

Figure 9.12 Shrinkage of particle and non-quasi-steady nature of problem.

problems dealing with only steady-state combustion; thus, unsteady conservation equations must be used to realistically model the combustion of a particle. However, the steady-state solutions can still be applied to the transient problem using the quasi-steady-state (QS) approximation; e.g., if the gas phase adjusts to new profiles within a timescale much shorter than 14 msec. The diffusion timescale is given as a^2/D, where the characteristic length of diffusion is "a." When the radius changes by da, then incremental diffusion time to adjust to the new profile is given by $dt_{diff} \approx (a\ da/D)$. Then $dt_{diff} << dt$ translates to $(a\ da/D). <<< da \times 4\pi\ a^2\ \rho_C/\dot{m} = da \times 4\pi\ a^2\ \rho_C/\{4\pi\ a\ \rho D \ln(1 + B)\}$

$$\rho_C/\{\rho \ln(1 + B)\} >> 1$$

$$\rho_C/\rho >> 1 \text{ for any given B} \tag{9.143}$$

Thus, the QS approximation can be used for estimating the burn rates as long as $\rho_C/\rho >> 1$. However, when a particle experiences oscillating temperature or concentration fields, as in fluidized beds [Annamalai and Stockbridge, 1984], the d^2 law may not be followed; further, the oscillating fields may cause switching of combustion from kinetics to diffusion-controlled regime and vice versa.

9.9 Element Conservation and Carbon Combustion

So far, we have discussed the solutions with species conservation equations. However, we will now present atom conservation equations, the important difference being the absence of chemical source terms. Consider the combustion of carbon with any number of heterogeneous and homogeneous reactions involving C and O. In Chapter 8, we derived element conservation equations for a QS.

The element and total enthalpy equations for a spherical system reduce to

$$\frac{d\phi}{d\xi} + \frac{d^2\phi}{d\xi^2} = 0 \tag{9.124}$$

$$\text{where } \phi = \frac{Z_k - Z_{k\infty}}{Z_{kw} - Z_{k\infty}} = \frac{h - h_\infty}{h_w - h_\infty}. \tag{9.125}$$

The solution is

$$\phi = \frac{1 - \exp\ (-\xi)}{1 - \exp\ (-\xi_w)}. \tag{9.126}$$

Even if Z_ks are known at the wall and at ambience, ξ_w is still an unknown; thus, we need interfacial conditions at the surface (i.e., between the solid and gaseous interface) in terms of elemental mass fractions. Because the solid contains only C, the burn rate is due to $\dot{m}_C$; then, for QS combustion, interface element conservation for C yields

$$Z_{C,w} + [dZ_C/d\xi]_{\xi\ w} = \dot{m}_C/\dot{m} = \varepsilon_C = F_Z = 1 \tag{9.127}$$

because C atoms are consumed at the interface. Similarly, for O atoms

$$Z_{O,w} + [dZ_O/d\xi]_{\xi\ w} = \dot{m}_O/\dot{m} = \varepsilon_O = F_Z = 0 \tag{9.128}$$

because there is no O atom input from solid material. Then

$$\xi_w = \ell n\left[1+\left(\frac{(Z_{k,w}-Z_{k,\infty})}{(F_{Z,k}-Z_{kw})}\right)\right],\quad k = C, O. \tag{9.129}$$

With k = C and O, we find $F_{Z,k}$ = 1 and 0 respectively, for k = C and O; the burn rate is given as,

$$\xi_w = \ell n\left[1+\left(\frac{Z_{C,w}-Z_{C,\infty}}{1-Z_{Cw}}\right)\right] = \ell n\left[1+\left(\frac{Z_{O,w}-Z_{O,\infty}}{0-Z_{Ow}}\right)\right]. \tag{9.130}$$

Equating the terms in Equation 9.130 within ( ) yields

$$\frac{(1-Z_{C,\infty})}{(1-Z_{C,w})} = \left(\frac{Z_{O,\infty}}{Z_{O,w}}\right) \tag{9.131}$$

The element mass fraction Z_k is related to the species mass fractions. For example, if species are CO_2, O_2, and CO

$$Z_k = Z_O = \frac{Y_{CO}M_c}{M_{CO}} + \frac{Y_{CO2}M_c}{M_{CO2}} + Y_{O_2} \tag{9.132}$$

The element mass fraction at the wall and at ambience can be related to the species mass fractions of CO_2, CO, and O_2. For example, with k = O,

$$Z_{O,\infty} = Y_{CO,\infty}M_O/M_{CO} + Y_{CO2,\infty}M_O/M_{CO2} + Y_{O2,\infty} = Y_{O2,\infty} \tag{9.133}$$

Similarly, at the wall of carbon, $Z_{O,w} = Y_{CO,w}\, M_O/M_{CO} + Y_{CO2,w}\, M_O/M_{CO2} + Y_{O2,w.}$

For example, if Y_{CO2} and Y_{O2} are negligible at the wall, then $Z_{O,w} = Y_{CO,w}M_O/M_{CO} = Y_{CO,w}\ \nu_{O2,I}/(1 + \nu_{O2,I})$.

Similar derivation can be adopted for k = C at surface and ambient.

As $(1 + 2\nu_{O2,I}) = \nu_{CO2,III}$, then

$$Y_{COw} = \left(\frac{B + Y_{O2,\infty} + Y_{CO2,\infty}}{1+B}\right).$$

Simplifying

$$Y_{COw} = \left(1 - \frac{Y_{I,\infty}}{1+B}\right) \tag{9.135}$$

where $Y_{I,\infty} = 1 - Y_{O2,\infty} - Y_{CO2,\infty}$. No assumption has been made regarding the gas-phase oxidation kinetics. Similarly, it can be shown that

$$\xi_w = \ln(1 + B) \tag{9.136}$$

$$\frac{c_p(T_w - T_\infty)}{h_{c,I}} = \frac{Y_{O2,\infty}}{\nu_{O2,I}} + \frac{Y_{CO2,\infty}}{\nu_{CO2,III}}\left[\frac{h_{C,III}}{h_{C,I}}\right]. \tag{9.137}$$

The same results were obtained before. Derivations are left as exercise problem.

9.10 Porous Char

Although it is sometimes reasonable to assume that the reaction area A_R is equal to the external surface area of the particle, i.e., $A_R = 4\pi a^2$, many chars are, however, porous. These pores occur throughout the particle volume. If the pore size is much larger than the diameter of an oxygen molecule, O_2 can penetrate and react inside the particle. In this case, the reaction area must be based on knowledge of the complete pore structure. A sketch of pore structure is shown in Figure 9.13.

Activated carbon (AC) is a carbonaceous material having a highly porous structure (pores occupying 200 cm^3 per kg) and a vast network of internal pores (hence, high internal surface area of about 400,000 m^2/kg), particularly

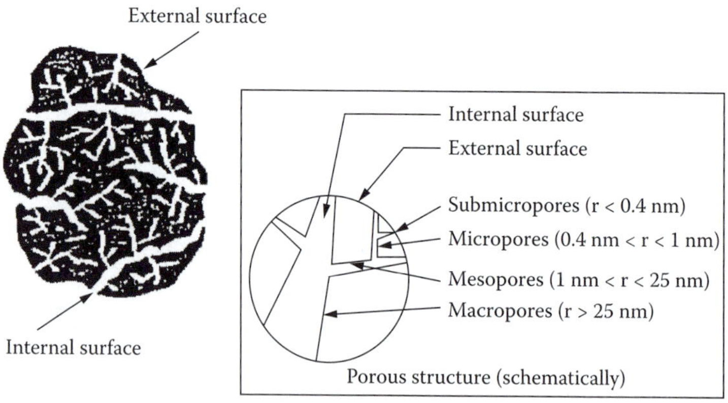

Figure 9.13 Porous char showing macro and micro pores (Fom http://www.netl.doe.gov/publications/proceedings/97/97ps/ps_pdf/PS1-2.PDF).

functioning as adsorbent (attracting molecules to their internal surfaces). The pore sizes range from 0.4 to 25 nm (Figure 9.13; pore sizes are micro for gas-phase applications, meso and macro for liquid-phase applications). Variations of macro, meso, micropores percentages with burn off (η_b) are shown in Figure 9.14. They are also used for removal of impurities from water (atrazine, trichloretane, benzene, etc.), exhaust gas cleanup, food and beverage processing, solvent recovery, etc.; it has been recently used for removal of elemental Hg from coal gases. They are also impregnated with chemicals (S, Cl, I, and Br) to remove particular species; for example, when impregnated with phosphoric acid, AC is used for removal of NH_3: $H_3PO_4 + 3NH_3 = (NH_4)_3\ PO_4$.

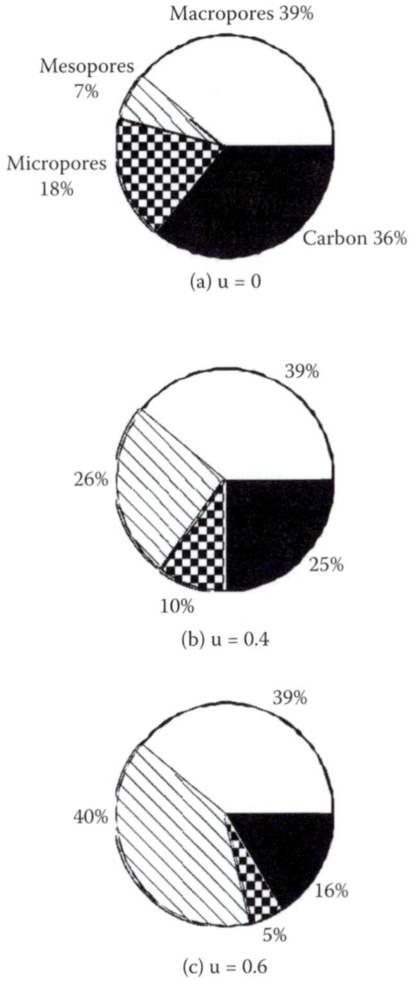

Figure 9.14 Submicro, r < 0.4 nm, micro 0.4 < r < nm, meso 1 < r < 25 nm, macro r > 25 nm. (Adapted from Hurt, R.H., Dudek, D.R., Longwell, J.P., and Sarofim, A.F., The phenomenon of gasification induced carbon densification and its influence on pore structure evolution, *Carbon*, Vol. 26, 4, 433–449, 1988.) u = η_b in text.

They are produced from wood, peat, lignite, and bituminous coals. Some terminology associated with porous char combustion is presented in the following text:

The fraction of mass burned at a particular instant of time can be expressed as

$$\eta_b = (m_0 - m)/m_0. \quad (9.138)$$

For constant density (i.e., no internal burning)

$$d/d_0 = (1 - \eta_b)^{(1/3)}. \quad (9.139)$$

For constant particle size (internal burning, varying density, or porosity)

$$\rho_p/\rho_{po} = (1 - \eta_b). \quad (9.140)$$

Generally, with variable density and size, one may define the change in diameter as

$$d/d_0 = (1 - \eta_b)^s. \quad (9.141)$$

Hamor et al. [1973] determined "s" from experiments as 1/4 and hence

$$d/d_0 = \rho_p/\rho_{p,0} = (1 - \eta_b)^{(1/4)}. \quad (9.142)$$

Why? Since

$$m/m_0 = [\rho_p\ d_p^3/(\rho_{p,0}\ d_{p,0}^3)], \quad (9.143)$$

then, combining Equation 9.142 and Equation 9.143 gives

$$\rho_p/\rho_{p,0} = (1 - \eta_b)^{1-3s} \quad (9.144)$$

and hence at s = 1/4, the normalized density has the same variation with mass burned as the normalized diameter.

Void fraction: The density of a char particle is less than that of the parent coal particle. Representing true solid density as $\rho_{p,t}$ and apparent solid density as ρ_p, the porosity or void fraction (ε_v) for the porous char is given as

$$\varepsilon_v = (1 - \rho_p/\rho_{p,t}). \quad (9.145)$$

It can be shown that

$$(1 - \varepsilon_v)/(1 - \varepsilon_{v,0}) = (1 - \eta_b)V_0/V, \quad (9.146)$$

where V/V_0 is the volume fraction of the solid mass. Porosities range from 0.2 to 0.9.

Internal surface area: As coals and chars are porous, the reaction surface area is higher than the external surface area owing to a vast network of pores. The internal surface area of a porous coal or char (A_{int}) can be expressed either by

$$S_m\text{, the specific internal area per unit mass} = A_{int}/m, \quad (9.147)$$

or alternately

$$S_v\text{, the specific internal area per unit volume} = A_{int}/V = S_m\ \rho_p, \quad (9.148)$$

where ρ_p is the apparent particle density. If only the particle's external surface area is considered, then $S_v = 4\pi a^2/(4/3)\pi a^3 = 3/a$. If the combustion rate is expressed as the combustion rate per unit internal surface area per unit oxygen pressure evaluated at the external surface, it is called the *intrinsic combustion rate.* Internal surface areas range from 100 to 1,000,000 m²/kg for chars and to 10,000,000 m²/kg for wood charcoal. Generally speaking, for a given void fraction, the smaller pores result in larger values of S_m. The external surface area per unit particle mass is given as

$$S_{ext,m} = 6/(d_p\ \rho_p). \tag{9.149}$$

As $S_{ext,m}$ does not include porosity, it provides the lower bound on S_m. For example, with $d_p = 100\ \mu m$, $\rho_p = 1000\ kg/m^3$ and $S_{ext,m} = 60\ m^2/kg$. The internal surface area varies with degree of burn off as shown in Figure 9.15.

Specific combustion rate: The specific combustion rate or mass related rate is defined as the particle combustion rate per unit mass of the particle. It is expressed as

$$\dot{m}_{spec} = (dm/dt)/m. \tag{9.150}$$

Assuming the combustion of the particle is diffusion controlled, then, neglecting buoyancy effects, the specific combustion rate can be expressed as

$$\dot{m}_{spec} \approx 6\ Sh\ \rho DB/(\rho_p\ d^2). \tag{9.151}$$

As dm/dt increases with increasing O_2 concentration, it is often useful to express the specific combustion rate as the rate per unit oxygen mass fraction or per unit oxygen pressure or

$$\dot{m}_{spec,O} = \dot{m}_{spec}/P_{O2} \quad \text{or} \quad \dot{m}_{spec}/Y_{O2} \tag{9.152}$$

Extrinsic combustion rate: The extrinsic combustion rate is defined as the combustion rate per unit external area

$$\dot{m}''_{ext} = \dot{m}/A \tag{9.153}$$

For example, for char particles, Stanmore [1991] presents

$$\dot{m}''_{ext} = 214\ \exp(-119,\ 500/RT). \tag{9.154}$$

Intrinsic combustion rate: The internal surface area of porous char is given by the Brunaver–Emmett–Teller (BET) surface area expressed in m²/kg (S_m) or m²/m³ (S_v). It is seen that as internal combustion occurs, the apparent density decreases and hence voidage increases, increasing S_m (Figure 9.15). Thus, measurement of the apparent density or voidage fraction is an indication of internal reactions. However, sometimes gasification can also induce densification of carbon [Hurt et al., 1988].

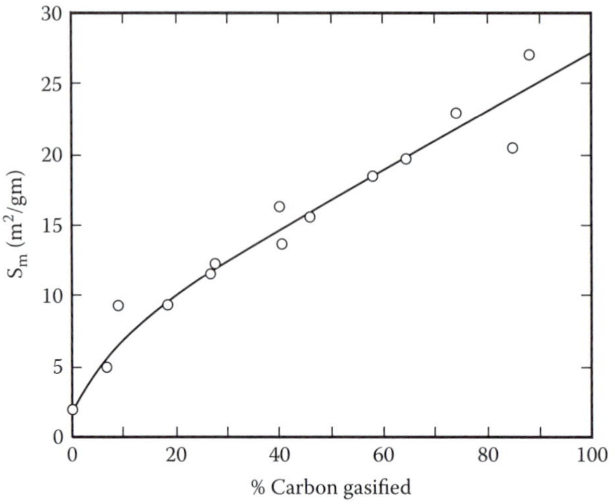

Figure 9.15 Variation of internal surface area with percentage gasified. (Adapted from Wicke, E., Contributions to the Combustion Mechanism of Carbon, 5th International Symposium on Combustion, Reinhold Publishing Co., New York, pp. 245–252, 1955. [See Smith, 1983].)

Defining S_v as the total reaction surface area per unit volume (A_R/V, with units of 1/m), the total reaction area for a particle of radius "a" is

$$A_R = S_v(4/3)\pi a^3. \tag{9.155a}$$

The total reaction area per unit external surface area

$$A''_R = A_R/(4\pi a^2) = S_v(1/3)a \tag{9.155b}$$

The A''_R is a dimensionless number. The kinetic constants based on the accurate reaction area are called *intrinsic kinetic constants* $A_{het,int}$ and E_{int}. In this case, because diffusion to the surface is based on the actual area in contact with the carbon,

$$k_{O_2} = A_{het,int}\, A''_R\, \nu_{O_2} \exp\,(E_{int}/RT) \tag{9.156}$$

If Y_{O2w} is uniform throughout the interior of the particle, then $\dot{w}'''_{02}$ is uniform. In reality, O_2 diffuses from ambience, through the pores, react on the internal surfaces, releases products that diffuse back into the boundary layer via the pores and, hence, Y_{O2} varies internally in the porous particles (Figure 9.16). Hence, $\dot{w}'''_{02}$ will vary inside the pores due to nonuniform Y_{O2}. Therefore, a correction factor called the *effectiveness factor* η_{eff} (representing the ratio of the actual reaction rate to that if the O_2 mass fraction were uniform within the particle) is applied. Charts are available to relate η_{eff} in terms of the Thiele modulus $\psi_{T,sph}$, which, for a spherical porous material of radius "a" is written as

$$\psi_{T,sph} = a\ (S_v\ k_{het}/D_{eff,,pore})^{1/2} \tag{9.157}$$

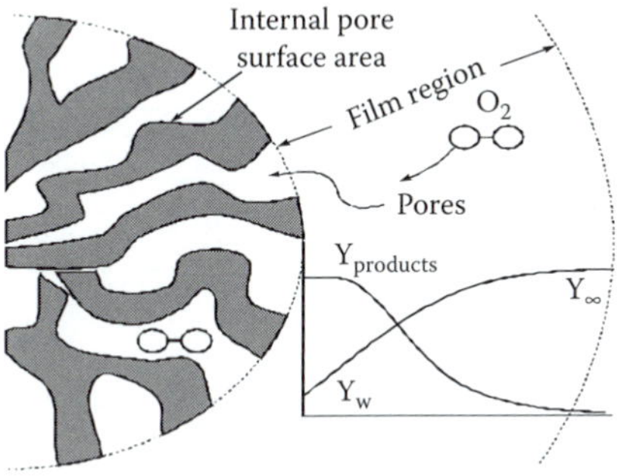
Internal pore
surface area
Film region
O_2
Pores
$Y_{products}$
Y_∞
Y_w
Step 1: O_2 diffuses through the f
to the particle surface.
Step 2: O_2 diffuses into the pores.
Step 3: O_2 under.
Step 4: Products diff toward
the particle surface.
Step 5: Products diff free stream
through the f

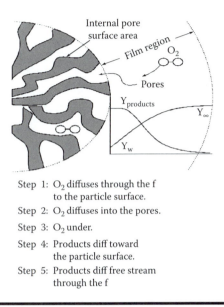

Figure 9.16 Transport and kinetic processes in combustion of porous carbon. (Adapted from Annamalai and Ryan, 2003.)

where $k_{het} = A_{het,int} \exp(-E_{int}/RT)$, m/sec.

$$1/D_{eff,\ pore1} = 1/D_K + 1/D,\ s/m^2,$$

$$D_K = (8/3)r_p\ (RT/\pi\ M)^{1/2} \text{ is the Knudsen diffusion coefficient,}$$

$$D = 1.28 \times 10^{-9}\ T^{1.75}\ m^2/sec, \text{ continuum diffusion coefficient}$$

$$r_p = 2\varepsilon_{Void}/S_v, \text{ the pore radius,} \tag{9.158}$$

and ε_{void}, the void fraction. Figure B.4 of Appendix B presents a chart for the variation in η_{eff} with respect to the Thiele modulus. The effectiveness factors for slab, cylinder, and sphere are shown in the chart. For a semiinfinite cylinder and slab, $\phi = \psi_{T,sph}/3 = L_{charac}\ (S_v\ k_{het}/D_{eff,,pore})^{1/2}$. L_{charac} = Volume/Surface area of the porous material, a/3 for sphere, a/2 for cylinder, and a for a slab of thickness 2a. The Thiele modulus has an analogy to interactive combustion (see Chapter 16; $\psi_{T,sph}{}^2 = G = 9\phi^2$; [Annamalai and Ryan 1993]; [Labowsky and Rosner]). Thus, one can express the intrinsic combustion rate as a ratio of burn rate per unit internal pore area:

$$\dot{m}''_{int} = k_{O_2}\ \eta_{eff}\ A_{het,int}\ A''_R\ \nu_{O_2}\ \exp(-E_{int}/RT)\rho_w Y_{O2,w}. \tag{9.159}$$

If the combustion rate is expressed as the combustion rate per unit internal surface area per unit oxygen pressure evaluated at external surface, it is called the *intrinsic combustion rate.*

$$\dot{m}''_{int} = (dm/dt)/A_{int} = \dot{m}_{spec}/S_m \tag{9.160}$$

The intrinsic combustion rate of a typical char is given by Smith [1982] (also see Smoot and Smith [1985]) (Figure 9.17)

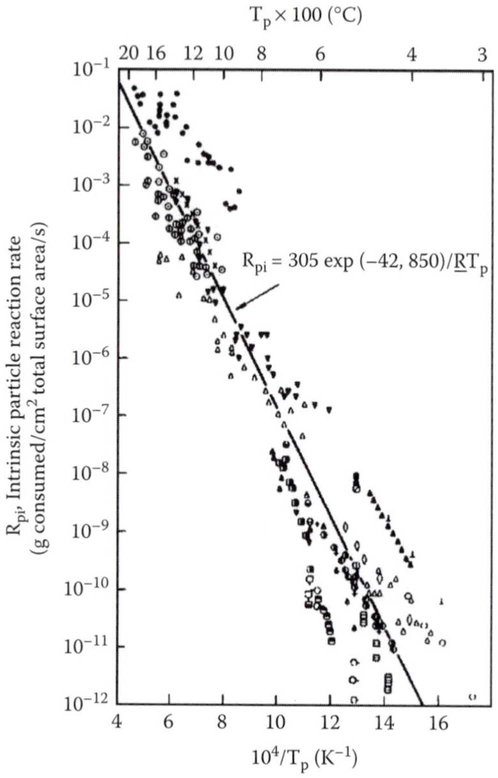

Figure 9.17 Intrinsic kinetics of porous char/carbon. (Adapted from Smith, I.W., 19th International Symposium on Combustion, The Combustion Institute, Pittsburgh, PA, 1045–1065, 1982); $R_{pi} = \dot{m}''_{int}$; R in cal/gmole K.

$$\dot{m}''_{int}\ (\text{g/cm}^2\ \text{atm}) = 305\ \exp\ (-180{,}000/RT). \tag{9.161}$$

Suppose combustion occurs at constant size (variable density). Then

$$\dot{m} = S_v\ \dot{m}''_{int}\ (4/3)\ \pi\ a^3. \tag{9.162}$$

If the particle size and internal surface area are known, then the reaction rate can be expressed per unit external surface area as

$$\dot{m}''_{ext} = S_v\ \ \dot{m}''_{int}\ (a/3). \tag{9.163}$$

Thus, smaller particles have a lower burning rate per unit external area than larger particles for similar S_v values. Note that this assumes $\dot{m}''_{int}$ is uniform throughout the particle and is the same for both the small and large particles (kinetics-controlled combustion).

Reactivity: Reactivity, R, can be expressed as follows. For a given particle diameter with the reaction confined to a thin layer at the surface

$$R = \dot{m}_{spec}\ a_0. \tag{9.164}$$

For a given coke or char, R at a fixed temperature is a fixed constant under kinetics-controlled combustion. High values of R indicate high reactivity. Mulcahy [1969] quotes R at 1667 K as 6800 g μm/cm^2 sec for bituminous coal chars whereas for petroleum coke, R = 2500 g μm/cm^2 sec. The interpretation is that bituminous char is three times more reactive than petroleum coke. Reactivity has also been expressed as the burning rate per unit area per kPa. Thus, care must be taken when interpreting the literature in this area. Note that the reactivity of chars varies as a function of coal rank. Lower C content in the parent coal results in increased char reactivity. Coal impurities (particularly Ca) are known to enhance the reactivity [e.g., lignite chars, Hippo and Walker, 1975].

The rate-limiting step during the oxidation of char can be chemical (such as the absorption of the reactant, reaction at carbon surface, and desorption of products) or gaseous diffusion (in the bulk phase or pore diffusion of reactants or products). Several investigators have postulated the existence of different temperature zones or regimes that determine which resistance is the controlling one. Particularly for a porous char with a vast network of internal pores, there exist three regimes of burning. When a chemical reaction is the rate-determining step (particularly at low T), it is called a zone I reaction (see Figure 9.18), where O_2 concentration is almost uniform within the particle. Zone II is characterized by both chemical reaction and pore diffusion at an intermediate temperature. Zone III combustion occurs at high temperatures that are characterized by bulk mass transfer limitations within the boundary layer, and O_2 concentration is almost zero within the particle.

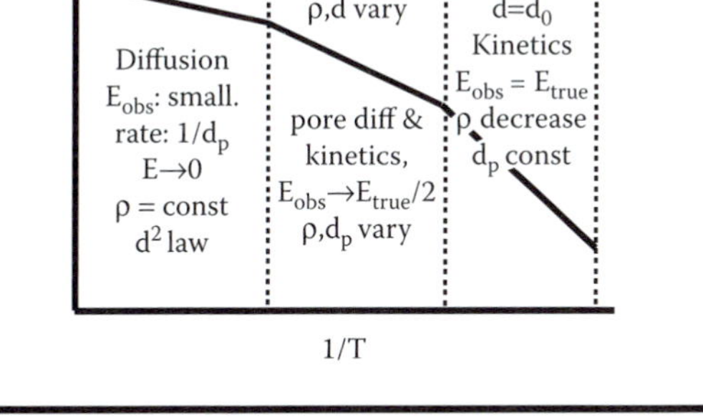

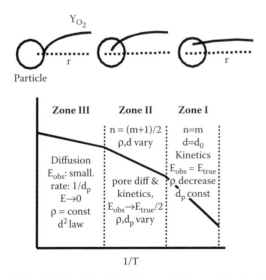

Figure 9.18 Zones of combustion of porous carbon.

9.11 Summary

Rigorous analysis is presented for predicting the burn rate, species mass fraction, and temperature of carbon with four possible reactions: oxidation to CO and CO_2, and reduction reactions with CO_2 and H_2O. Both single-film (C oxidized to CO) and double-film (CO_2 reduction at the surface with CO production and gas-phase oxidation of CO) models are presented, and results are presented for burn rate and particle temperature. Brief overviews on porous char combustion, zone I (kinetics limited), zone II (mixed kinetics), and zone III (diffusion limited) combustion and intrinsic kinetics are presented. Owing to extremely high temperatures surrounding the burning carbon particle, there is a strong likelihood of thermal NO being formed around the particles.

9.12 Appendix: d Law and Stefan Flow Approximation

9.12.1 d Law for Kinetic-Controlled Combustion (i.e., SFM)

The d law for kinetic-controlled combustion is described in the following text. Burning will cause the particle radius to shrink. Hence,

$$-4\pi a^2 \rho_p \frac{da}{dt} = \dot{m}_c'' 4\pi a^2. \qquad \text{(A)}$$

Assuming kinetics control,

$$-4\pi a^2 \frac{da}{dt} \rho_p = \dot{m}_c'' 4\, \pi\, a^2\, . \qquad \text{(B)}$$

As $B = Y_{O2,\infty}/\upsilon_{O2}$,

$$-\frac{da}{dt} = \left\{ \frac{\dot{m}_c''}{\rho_p} \right\}. \qquad \text{(C)}$$

Integrating and using the initial conditions and with $\dot{m}_c''$ = k $\rho_\omega Y_{O_2,\infty}$

$$a = a_0 \text{ at } t = o$$

$$d_0 - d = \alpha_{\text{kint}} \qquad \text{(D)}$$

where

$$\alpha_c = \{k\rho_w Y_{O_2,\infty} / \rho_p\} \text{ burning rate constant under kinetic control.} \qquad \text{(E)}$$

The time for complete combustion is given as

$$t_b = \frac{d_0}{\alpha_{\text{kin}}}\, . \qquad \text{(F)}$$

9.12.2 Stefan Flow Approximation

For slow combustion of carbon, the diffusion velocity V is significant compared to the convective velocity v. Thus, the term $\rho v\, Y_k$, called the *bulk transport term* or *Stefan velocity term,* can be neglected compared to $\rho v V_k\, Y_k$. The neglect of Stefan flow (SF) is justified for slow combustion of carbon as long as $Y_{O2}/\nu_{O2,I}$ is much less than unity, as illustrated in the following text. Consider a carbon particle burning in a quiescent atmosphere, and assume SFM with Reaction I being dominant. The average velocity or SF velocity or bulk velocity v is given as [see Chapter 6]

$$\rho\ v = \rho_{O2}\ v_{O2} + \rho_{CO}\, v_{CO} + \rho_{N2}\ v_{N2} \tag{A}$$

where v_{O2} is the absolute O_2 velocity. As N_2 is inert and transport of each species is related to $\dot{m}''_C$ through the law of mass action

$$\rho\, v = \rho_{O2}\, v_{O2} + \rho_{CO}\, v_{CO} = \dot{m}''_C\ [-\nu_{O2,I} + (1+\nu_{O2,I}\ )] = \dot{m}''_C = \frac{\rho_{O2}\ v_{O2}}{\nu_{O2,I}}. \tag{B}$$

But the absolute velocity is a sum of bulk velocity v and diffusion velocity V. Thus,

$$v_{O2} = v + V_{O2}\ . \tag{C}$$

Using Equation C in Equation B

$$\rho\ v = \frac{\rho_{O2}\ v_{O2}}{\nu_{O2,I}} = \frac{\rho Y_{O2}\ \{v + V_{O2}\}}{\nu_{O2,I}},\quad \frac{v}{V_{O2}} = \frac{\frac{Y_{O2}}{\nu_{O2,I}}}{1-\frac{Y_{O2}}{\nu_{O2,I}}}\ .$$

Then, as long as $\frac{Y_{O2}}{\nu_{O2,I}}$ or B <<< 1, $\frac{v}{V_{O2}}$ <<< 1, Stefan flow can be neglected for carbon combustion. When B << 1, ln(1 + B) ≈ B. The reader can verify this result using the governing equations. Such an approximation is also used for liquid drop evaporation at low temperatures (chapter 10).

Chapter 10

Diffusion Flames — Liquid Fuels

10.1 Introduction

Many practical combustor systems such as boilers, gas turbines, diesel engines, rockets, etc., use condensate fuels (liquid and solid fuels) as the energy source. Combustion intensity is then controlled by gasification, mixing, and chemical kinetics processes. The condensate fuel is atomized or pulverized into smaller droplets or particles to increase the surface area of fuel exposed to the hot gases in order to facilitate rapid gasification and mixing with the oxygen-rich ambience; e.g., rocket engine: $d_{P,0} \approx 300\ \mu m$, diesel engine: 20 μm, gas turbines: 30 to 60 μm (SMD), and oil-fired heaters: 75 μm. The atomization leads to improved ignition and combustion characteristics. Spray ignition and combustion studies are extremely important to determine flame stability behavior at widely varying loads, ensure efficient utilization of fossil energy, and better understand the mechanisms of pollutant formation and destruction. Extensive fundamental and applied research has been conducted on the combustion of oil sprays and pulverized-fuel-particle sprays. Sprays normally involve a large number of drops. Both heterogeneous (drop combustion either in groups or arrays or individually) and homogeneous (small drops less than 10 μm) combustion can occur in a spray. The simplest method of analysis is to treat the fuel spray as an aggregate of fuel droplets. Isolated-drop ignition and combustion studies have been conducted in the last three decades. The reader should refer to the extensive reviews and publications by Law [1976,1978], Law and Law [1982,1980], Law et al. [1980], Faeth [1977], Hedley and Williams, and on moving drops, by Ayyaswamy [1989]. Simple drop or particle models yield parameters controlling the combustion characteristics of individual drops or particles. Thus, the spray combustion rate could be estimated to be the sum of combustion rates of isolated drops if interactions between drops are ignored. However, a drop in the spray does not behave like an isolated drop, because its behavior will be modified owing to the presence of other drops. These interactions will be treated in Chapter 16.

Evaporation assumes special significance, particularly for water drops, in quenching fire and in combustion for boilers or engines fired with micronized droplets. Very small droplets may not even support a flame around the drop [Annamalai and Durbetaki, 1975]. The other applications include diesel engines, gas turbines, and liquid-fuel-fired rockets (shuttle rockets fired with liquid H_2).

10.2 Evaporation, Combustion, and d^2 Law

Evaporation or single-film model (SFM): Examples are water drops in a fire situation and very small oil droplets simply vaporizing. When the gas phase is frozen, the only physical process is evaporation.

Combustion or double-film model (DFM): The evaporation process at the surface is typically heat transfer controlled (e.g., C_8H_{18} (g) from a liquid octane drop). The homogeneous reaction $C_8H_{18} + 12.5O_2 = 8CO_2 + 9H_2O$ is infinitely fast, with a flame sheet generated away from the evaporating surface (Figure 10.1).

We will consider combustion first and then simplify for evaporation. The overall objective is to predict the burning rate (mass-loss rate) of the drop and flame structure of the liquid droplet. By flame structure, we imply determination of T, Y_F, Y_{O_2}, Y_{CO_2}, etc., as a function of spatial coordinates.

Single-drop studies using digital cameras have shown that the drop diameter during QS (quasi-steady) evaporation and combustion follows a diameter square law (or surface area law) with respect to time for drops suspended from fibers and drops in low-Re and low-Gr flow. Typical profiles for temperature and partial pressures of species are shown in Figure 10.1. If d_o, initial diameter of droplet (= 2 a where a is the radius of the drop), and d, the diameter at any time, then the relation is written as

$$d^2 = d_0^2 - \alpha_k t \qquad (10.1)$$

where α_k is the rate constant for an isolated drop with k = e for evaporation and k = c for combustion. The rate constants are evaluated experimentally and compared with the theoretical predictions.

10.3 Model/Physical Processes

10.3.1 Model

Suppose a thermocouple with a bead of very small diameter is dipped in a tank containing octane (≈ gasoline). One observes a droplet of radius a around the bead; the temperature of the drop is T_w (Figure 10.1). Place the drop in a hot furnace at T_∞. From the instant of exposure of the drop to the hot gases, a thermal wave propagates to the drop surface from the ambience and heats the oil drop. T_w increases with time, vapors start being released, T_w reaches a value close to the boiling point T_{BP}, and now vapors start coming off rapidly. The vaporization (gasification) is purely a physical process in that the chemical

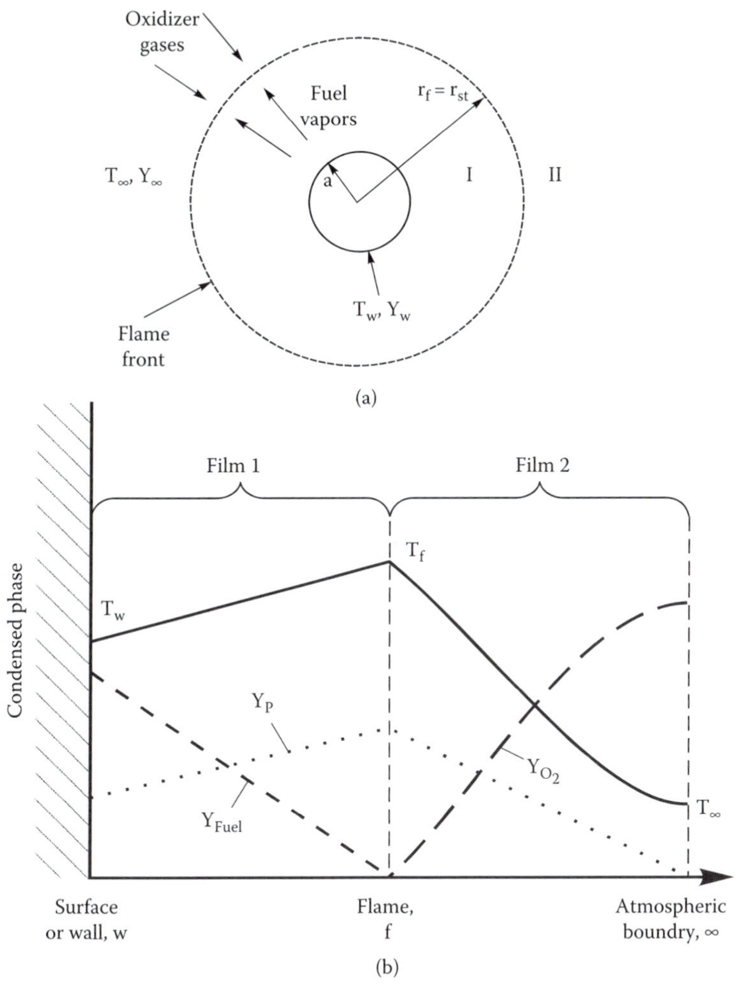

Figure 10.1 Combustion of drop or solid plastic (condensed phase); surface or wall denotes drop surface.

structure of vapor is the same as that of the liquid. Fuel vapors diffuse into the hotter surrounding atmosphere through film I. At the drop surface, O_2 concentration decreases as vapors form; the mixture at the drop surface is relatively cold (close to T_{BP}) and very rich. O_2 then diffuses from the hot ambience, where the mixture is very lean (fuel vapor almost zero but hot) toward the drop surface. At some location away from the drop surface (say, r_{st}), the mixture is near stoichiometric ($Y_{O2}/Y_F = \nu_{O2}$). If the furnace is hot enough, the temperature starts rising rapidly at that location; i.e., a flammable mixture is formed and ignition occurs around r_{st}. Ignition is a process in which there is a very rapid rise in T within a short time (i.e., dT/dt is very large); a flame eventually appears. The reaction rate becomes extremely rapid, resulting in a rapid temperature rise and depletion of oxygen (Y_{O2}) and fuel (Y_F) mass fractions locally and the establishment of a flame at a location $r = r_f$ ($\approx r_{st}$) away from the droplet surface. At the flame, a thermocouple will reveal rapid temperature rise almost to the adiabatic flame temperature (T_f) for a stoichiometric mixture. As the reaction rate α is $e^{-(E/[RT])}$, the rate approaches ∞ (a large value, say, 10^{14}) as T increases. Imagine that the stoichiometric surface is like a char surface; the reaction rate at that surface is very high, kinetic resistance is small, and there is rapid depletion of oxygen and fuel locally. Hence, $Y_{O2} \to 0$ (i.e., extremely low values, say, 4×10^{-7}) and fuel $Y_F \to 0$ (say, 10^{-7}), with $[e^{-(E/[RT])}\, Y_F\, Y_{O2}]$ remaining finite (say, $10^{14} \times 10^{-7}\ 4 \times 10^{-7} = 4$) at the flame surface. Hence, O_2 cannot diffuse past the flame while the fuel cannot diffuse past it. Thus, Y_{O2} is negligible in film 1, whereas Y_F is negligible in film 2. Thus, the flame surface divides the region around the drop into fuel-rich regime (film 1) and fuel-lean regime (film 2). For a quiescent atmosphere, the flame entirely surrounds the drop forming an envelope flame (we will see in Chapter 11 that when flow occurs over a drop or flat liquid surface, the flame is open and does not envelope the drop or flat surface). The flame continues to be supplied with fuel vapors from the oil drop and oxygen from the ambience. Because vaporization is extremely rapid and vapors diffuse out toward the hot oxygen-rich ambience, the combustion rate is extremely high. Once the flame appears, it (at T_f) supplies latent heat to the drop and evaporation is sustained. The drop of diameter d gradually shrinks with time t.

10.3.2 Diffusion-Controlled Combustion

At $r = r_f$, $Y_F = 0$ (extremely low mass fraction), $Y_{O_2} = 0$ (extremely low mass fraction).

$$\text{For} \quad r_w < r < r_f, \quad Y_{O_2} = 0; \quad r_f < r < \infty, \quad Y_F = 0 \tag{10.2}$$

The approximations introduced above are called *diffusion-controlled flame approximations*, also known as *thin-flame approximations*.

10.4 Governing Equations

First, we will analyze the problem for a thick flame (with chemical kinetics) and then reduce it for a thin flame (diffusion controlled). As the drop burns, fuel vapors flow outward from the drop. We use overall mass balance, species, and energy conservation equations and interface energy and species balance equations to solve for mass-loss rate $\dot{m}$, burn rate, flame location (r_f), and flame temperature (T_f). The treatment is somewhat similar to that used in Chapter 9 for carbon combustion.

10.4.1 Assumptions

The analysis is based on the following assumptions:

1. T is uniform in space within the drop (i.e., no temperature gradient).
2. Gas phase is at quasi-steady state.
3. Spherical symmetry.
4. Lewis number = 1 ($\lambda/\{\rho D\ c_p\} = 1$).

The first assumption is justified if conductivity is large or $Bi = h_H d/\lambda_{liq} << 0.1$.

10.4.2 Conservation Equations: Dimensional Form

For mass,

$$(d/dr)(\rho v r^2)=0 \tag{10.3}$$

where v denotes the radial velocity of the vapors. Integrating Equation 10.3 we find that

$$\dot{m} = \rho v 4\pi r^2 = \text{constant}. \tag{10.4}$$

For species,

$$\frac{1}{r^2}\frac{d}{dr}(\rho v r^2 Y_F)=\frac{1}{r^2}\frac{d}{dr}\left(\rho D r^2 \frac{dY_F}{dr}\right)+\dot{w}'''_F .$$

Similarly, one can write equations for reacting species (O_2, CO_2, and H_2O). We have seen before that the species balance for fuel, oxidizer, CO_2, H_2O, etc., can be reduced in terms of Shvab–Zeldovich (SZ) variables (Chapter 8 and Chapter 9).

The energy balance equations for temperature or thermal enthalpy (h_t) can be coupled to the fuel or oxygen conservation equation

$$\frac{1}{r^2}\frac{d}{dr}\left(\rho v r^2 \beta_{ij}\right)=\frac{1}{r^2}\frac{d}{dr}\left(\rho D r^2 \frac{d\beta_{ij}}{dr}\right) \tag{10.5}$$

where β for species variables is

$$\beta_{ij}=\frac{Y_i}{\nu_i}-\frac{Y_j}{\nu_j}. \tag{10.6}$$

For O_2–F,

$$\beta_{O_2-F}=\frac{Y_{O_2}}{(-\nu_{O_2})}-\frac{Y_F}{(-1)}=Y_F-\frac{Y_{O_2}}{\nu_{O_2}} \tag{10.7}$$

where ν_{O2} is the stoichiometric O_2 in kg per kg of fuel.

For CO_2–F,

$$\beta_{CO_2-F}=\frac{Y_{CO_2}}{(+\nu_{CO_2})}-\frac{Y_F}{(-1)}=Y_F+\frac{Y_{CO_2}}{\nu_{CO_2}}. \tag{10.8}$$

When subscripts F and O_2 are used with ν, the sign for stoichiometric coefficients has already been accounted for.

10.4.2.1 Energy

For thermal enthalpy–species variables,

$$\beta_{h_t-F} = \frac{h_t}{h_c} + Y_F, \quad \beta_{h_t-CO_2} = \frac{h_t}{h_c} - \frac{Y_{CO_2}}{\nu_{CO_2}} \tag{10.9}$$

$$h_t = \int_{T_w}^{T} c_p dT. \tag{10.10}$$

Thermal enthalpy–fuel combination,

$$\beta_{h_t-F} = \frac{h_t}{h_c} - \frac{Y_F}{(-1)} = \frac{h_t}{h_c} + Y_F. \tag{10.11a}$$

Thermal enthalpy oxygen and CO_2,

$$\beta_{h_t-O_2} = \frac{h_t}{h_c} - \frac{Y_{O_2}}{(-\nu_{O_2})} = \frac{h_t}{h_c} + \frac{Y_{O_2}}{\nu_{O_2}} \tag{10.11b}$$

$$\beta_{h_t-CO_2} = \frac{h_t}{h_c} - \frac{Y_{O_2}}{(+\nu_{CO_2})} = \frac{h_t}{h_c} - \frac{Y_{O_2}}{\nu_{CO_2}}. \tag{10.12}$$

The lower heating value h_c must be for the gaseous state of fuel and is tabulated in Table A.2A in Appendix A.

10.4.3 Conservation Equations Nondimensional Form

As in Chapter 9 for a steady spherical system, let

$$\xi = \frac{\dot{m}}{4\pi\rho D r}. \tag{10.13}$$

The value of ξ at r = a is written as

$$\xi_w = \dot{m}/(4\pi\rho Da), \text{ the nondimensional mass-loss rate} \tag{10.14}$$

(the suffix w denotes that it is at the wall or surface of the droplet; note that $\xi_w/\xi = r/a$).

Thus, the general governing conservation equation for SZ variable β is written as

$$\frac{d\beta}{d\xi} + \frac{d^2\beta}{d\xi^2} = 0 \tag{10.15a}$$

whereas the fuel species equation becomes

$$\frac{dY_F}{d\xi} + \frac{d^2Y_F}{d\xi^2} = -\frac{\dot{w}_F'''(4\pi)^2\rho D r^4}{\dot{m}^2}. \tag{10.15b}$$

Because the flux ratio is defined as $\varepsilon_k = \dot{m}_k/\dot{m} = \{\rho_k v_k/\rho v\} = Y_k + dY_k/d\xi$ (see Chapter 7), the fuel species equation in terms of flux ratio can be given as

$$\frac{d\varepsilon_F}{d\xi} = -\frac{\dot{w}_F'''(4\pi)^2 \rho D r^4}{\dot{m}^2}. \tag{10.15c}$$

Equation 10.15a along with Equation 10.15b provide the complete solution to droplet combustion. For most practical combustion problems, the solution of Equation 10.15b is unnecessary. Thus, we will concentrate on solving Equation 10.15a.

10.4.4 Boundary Conditions

The solution of the second-order differential equation for β (Equation 10.15a) requires two boundary conditions:

$$\left.\begin{aligned} \text{at} \quad r = a \quad \text{or} \quad & \xi = \xi_w, \beta = \beta_w \\ r \to \infty, \quad \text{or} \quad & \xi \to 0, \beta = \beta_\infty \end{aligned}\right\} \tag{10.16}$$

where ξ_w is yet to be solved.

10.4.5 Solutions

10.4.5.1 SZ Variable

Thus, we solve Equation 10.15a with Equation 10.16 as boundary conditions

$$\phi = \frac{\beta - \beta_\infty}{\beta_w - \beta_\infty} = \frac{1 - \exp(-\xi)}{1 - \exp(-\xi_w)} \tag{10.17}$$

where ϕ is a normalized SZ (NSZ) variable.

Generalizing for all conserved scalars (i.e., without source terms), element k, mass fraction Z, and total enthalpy h

$$\phi = \frac{\beta - \beta_\infty}{\beta_w - \beta_\infty} = \frac{Z_k - Z_{k,\infty}}{Z_{k,w} - Z_{k,\infty}} = \frac{h - h_\infty}{h_w - h_\infty} = \frac{1 - \exp(-\xi)}{1 - \exp(-\xi_w)}$$

where Φ, normalized SZ variable with a value of 1 at the wall and zero far from it.

Note that when ξ_w is extremely small (slow burn or evaporation rate), then Equation 10.17 becomes

$$\phi = \frac{\beta - \beta_\infty}{\beta_w - \beta_\infty} = \frac{\xi}{\xi_w} = \frac{a}{r} \tag{10.18}$$

The β vs. ξ curve from Equation 10.17 can be plotted if ξ_w, β_w, and β_∞ are known. Suppose $\beta = \beta_{ht-O2} = h_t/h_c + Y_{O2}/\nu_{O2}$; at the wall, $\beta_{ht-O2,w} = h_{T,w}/h_c + Y_{O2,w}/\nu_{O2}$. At the wall, $h_{tw} = 0$ because T_{ref} for all enthalpies, including enthalpy of formation, is selected as T_w; $\beta_{ht-O2,\,w} = Y_{O2w}/\nu_{O2}$. If reactions are fast, typically $Y_{O2,w} = 0$; thus, $\beta_{ht-O2,\,w} \approx 0$. As $r \to \infty$, or $\xi \to 0$, $\beta_{hT-O2,\infty} = h_{t,\infty}/h_c + Y_{O2,\infty}/\nu_{O2} = c_p\,(T_\infty - T_w)/h_c + Y_{O2,\infty}/\nu_{O2}$; typically, $T_w \approx T_{BP}$. Thus, $\beta_{ht-O2,\infty} \approx c_p\,(T_\infty - T_{BP})/h_c + Y_{O2,\infty}/\nu_{O2}$. Thus, β vs. ξ can be plotted if we know the burn rate $\dot{m}$.

10.4.5.2 Interface Boundary Conditions

We invoke the interface boundary conditions in order to arrive at the solution of burn rate $\dot{m}$, drop temperature (T_w), and fuel mass fraction ($Y_{F,W}$) at drop surface. In Chapter 8 and Chapter 9, we developed a detailed methodology for developing the interface boundary conditions for species and enthalpy. These details will be omitted now.

Energy (h_T).

As described in Chapter 9, applying energy conservation for a thin film around the interface (Figure 10.2)

$$0 = \left[\dot{q}''_w\right]_- + \left[\dot{q}''_w\right]_+ - 0 + \dot{m}_{F,}\, h_F\,(T_w) - \dot{m}_F\, h_{F(g)}(T_w) \qquad (10.19)$$

where + denotes the gas side and – denotes the condensate side. Heat flux into the control surface is given a positive sign, whereas the heat flux vector times the outer normal vector yields a negative sign. Thus, when writing in terms of the Fourier law

$$q \text{ in First law} = -\vec{q}'' \cdot d\vec{A} \quad \text{where} \quad \vec{q}'' = -\lambda \vec{\nabla} T \qquad (10.20)$$

$$0 = \{-\lambda(\partial T/\partial r)_{r=0-}\} + \{\lambda(\partial T/\partial r)_{r=0+}\} + \left\{\sum_i m_{k,-} h_{k,-}\right\} - \left\{\sum_i m_{k,+} h_{k,+}\right\}. \qquad (10.21)$$

The $\left[\dot{q}''_w\right]_-$, heat flux from liquid side $[dT/dr] = 0$ is zero owing to assumption (Section 10.4.1). On the other hand, dT/dr is finite on the gas side. For a

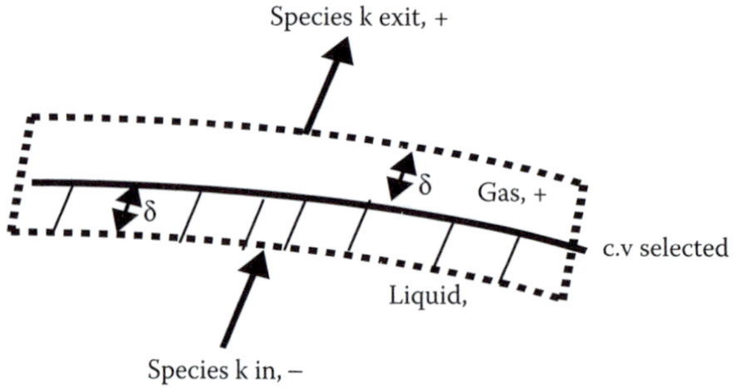

Figure 10.2 Elemental control volume for interface conservation equations.

nonreacting interface, $h_k = h_{t,k}$. In order to minimize the complexities, we will use the reference enthalpy at T_w for fuel in the gaseous state. The energy conservation per unit surface area may be written as

$$0 = \left[\dot{q}''_w\right]_+ + \dot{m}''_F [h_{F(\ell)}(Tw) - h_{F(g)}(T_w)] = [\dot{q}''_w] - \dot{m}''_F \, L. \tag{10.22}$$

Simplifying,

$$\left[\lambda \left(\frac{dT}{dr}\right)_w 4\pi a^2\right] = \dot{m} \;\; L \tag{10.23}$$

where L is the latent heat required for evaporation to change the state of fuel from liquid to vapor at T_w. The left-hand side refers to the condition on the vapor side.

A brief note is appropriate here. If liquid drops are of extremely small size, they get rapidly heated to T_w, and hence during combustion, energy is required to be supplied only for latent heat. If drops are of large size, the drop surface gets heated rapidly, whereas the interior of the drop still remains cold (recall thermal diffusion timescales from Chapter 6). Thus, there exists a temperature gradient within the drop, and hence, $\{\lambda(\partial T/\partial r)_{r=0-}\} \neq 0$. Fourier heat flux on the liquid raises its temperature from $T_{0,liq}$ to T_w, which is used to supply sensible heat of $c_p\,(T_w - T_{0,liq})$ per unit mass of the liquid. If $\{\lambda(\partial T/\partial r)_{r=0}\} \neq 0$, it can be shown that for large liquid drops (Chapter 8), the latent heat L is modified as,

$$\lambda(\partial T/\partial r)_{r=a+} = \dot{m}'' \; \{L + c_p(T_w - T_{0,liq})\} = \dot{m}'' \; L_{mod}$$

where $L_{mod} = L + c_p(T_w - T_{0,liq})$. Now considering only small liquid drops, reverting back to Equation 10.23, rewriting in terms of h_t, and assuming Le = 1,

$$\rho \; D\left(\frac{dh_t}{dr}\right)_w 4\pi a^2 = \dot{m}\, L \tag{10.24}$$

Because the governing equation is in terms of β, we need to get interface balance also in terms of β. As a first step toward that goal, consider $\beta_{hT\text{-}O2} = h_t/h_c + Y_{O2}/\nu_{O2}$. It has two terms, h_t/h_c and Y_{O2}/ν_{O2}. Rewriting Equation 10.24, the energy balance equations in nondimensional form can be presented as follows:

$$\frac{d\;(h_t/h_c)}{d\xi} = -\frac{L}{h_c}. \tag{10.25}$$

Oxygens (O_2).
Since oxygen does not dissolve into the liquid,

$$4\pi \rho_w v_w r^2 Y_{O_2,w} - 4\pi \rho D r^2 \left(\frac{dY_{O2}}{dr}\right)_w = \dot{m}_{O2} = 0. \tag{10.26}$$

The oxygen balance equations in nondimensional form can be written as follows:

$$Y_{O_2,w} + \left(\frac{dY_{O_2}}{d\xi}\right)_w = \varepsilon_{O_2} = 0. \tag{10.27}$$

Generally, for any species not contained in the liquid fuel,

$$4\pi \rho_w v_w a^2 Y_{i,w} - 4\pi \rho D a^2 \left(\frac{dY_i}{dr}\right)_w = \dot{m}_i = 0$$

and

$$Y_{i,w} + \left(\frac{dY_i}{d\xi}\right)_w = \varepsilon_i = 0. \tag{10.28}$$

Dividing Equation 10.27 by ν_{O2} and adding with Equation 10.25

$$\beta_{h_t-O_2,w} + \left(\frac{d\beta_{h_t-O_2}}{d\xi}\right)_w = -\frac{L}{h_c}. \tag{10.29}$$

As in Chapter 9, let us generalize the interface boundary condition as

$$\beta_w + \left(\frac{d\beta}{d\xi}\right)_w = F_\beta \tag{10.30}$$

where for $\beta = \beta_{ht-O2}$, $F_\beta = -L/h_c$. Table 10.1 presents interface conditions involving other β's and lists F_β.

10.5 Solutions

10.5.1 Burn Rate

Using Equation 10.17 and Equation 10.30, we can solve for the burn rate.

10.5.1.1 Thick Flames

Differentiating Equation 10.17 with respect to ξ, and evaluating the differential at $\xi = \xi_w$, we have

$$(d\beta/d\xi)_{\xi=\xi w} = (\beta_w - \beta_\infty)/(\exp[\xi_w] - 1).$$

Table 10.1 The F_β's for Thin or Thick Flames, $\xi_w = \ln\left(1+\frac{\beta_w-\beta_\infty}{F_\beta-\beta_w}\right)$

Number	*SZ β*	F_β	*Definition*	*Thin Flames*
1	$\beta_{ht\text{-}O2}$	$-L/h_c$	$\beta_{ht\text{-}O2} = h_t/h_c + Y_{O2}/\nu_{O2}$	$\beta_w = 0$
2	$\beta_{ht\text{-}F}$	$Y_{F,\ell} - L/h_c$	$\beta_{ht\text{-}F} = h_t/h_c + Y_F$	$\beta_w = Y_{Fw}$
3	$\beta_{O2\text{-}F}$	$Y_{F,\ell}$	$\beta_{O2\text{-}F} = Y_F - Y_{O2}/\nu_{O2}$	$\beta_w = Y_{Fw}$
4	$\beta_{CO2\text{-}F}$	$Y_{F,\ell}$	$\beta_{CO2\text{-}F} = Y_F + Y_{CO2}/\nu_{CO2}$	—
5	$\beta_{H2O\text{-}F}$	$Y_{F,\ell}$	$\beta_{H2O\text{-}F} = Y_F + Y_{H2O}/\nu_{H2O}$	—
6	$\beta_{ht\text{-}CO2}$	L/h_c	$\beta_{ht\text{-}CO2} = h_t/h_c - Y_{CO2}/\nu_{CO2}$	$\beta_w = Y_{CO2w}/\nu_{CO2}$
7	$\beta_{ht\text{-}H2O}$	L/h_c	$\beta_{ht\text{-}H2O} = h_t/h_c - Y_{H2O}/\nu_{H2O}$	$\beta_w = Y_{H2O,w}/\nu_{CO2}$
8	$\beta_{CO2\text{-}O2}$	0	$\beta_{CO2\text{-}O2} = Y_{CO2}/\nu_{CO2} + Y_{O2}/\nu_{O2}$	$\beta_w = Y_{CO2,w}/\nu_{CO2}$
9	$\beta_{H2O\text{-}O2}$	0	$\beta_{ht\text{-}H2O} = Y_{H2O}/\nu_{H2O} + Y_{O2}/\nu_{O2}$	$\beta_w = Y_{H2O,w}/\nu_{H2O}$

Using this result in Equation 10.30, we solve for the nondimensional mass-loss rate ξ_w as

$$\xi_w = \ln\left(1+\frac{\beta_w-\beta_\infty}{F_\beta-\beta_w}\right). \tag{10.31}$$

The "thick flame" transfer number is defined as

$$B_{kin} = \left(\frac{\beta_w-\beta_\infty}{F_\beta-\beta_w)}\right). \tag{10.32}$$

Typically, we evaluate B_{kin} with $\beta_{ht\text{-}O2}$. Thus Equation 10.31 becomes

$$\begin{aligned}\xi_w = \; & n(1+B_{kin}) = n\left[1+\left(\frac{\beta_{ht-O_2,w}-\beta_{ht-O_2,\infty}}{-\frac{L}{h_c}-\beta_{ht-O_2,w}}\right)\right] \\ = \; & n\left[1+\left(\frac{\beta_{ht-O_2,\infty}-\beta_{ht-O_2,w}}{\frac{L}{h_c}+\beta_{ht-O_2,w}}\right)\right].\end{aligned} \tag{10.33}$$

The above solution is for a thick flame (see Figure 10.5). For thick flames the reaction rates are not infinite and $Y_{O2,w} \neq 0$. If O_2 is present everywhere in the bulk phase but with a negligible mass fraction at wall, then

$$B_{kin} = B = \left(\frac{\beta_{ht-O_2,w}-\beta_{ht-O_2,\infty}}{F_\beta-\beta_{ht-O_2,w}}\right) = \frac{\beta_{ht-O_2,\infty}}{(L/h_c)} \quad \text{as} \quad \beta_{ht\text{-}O2,w} = 0.$$

10.5.1.2 Thin Flames

If reaction rates are very fast, then O_2 cannot be present at the drop surface, i.e., $Y_{O2,w} = 0$. The flame is then said to be thin and combustion occurs under diffusion control. From Equation 10.33, the thin-flame solution is given as

$$\xi_w = \frac{\dot{m}}{4\pi\rho Da} = \ell n\left[1+\frac{h_C\beta_{ht-O_2,\infty}}{L}\right] = \ell n\left[1+\left\{\frac{h_{t,\infty}}{L}+\frac{Y_{O_2,\infty}h_c}{\nu_{O_2}L}\right\}\right] = \ell n[1+B] \quad (10.34)$$

where transfer number is defined as

$$\left.\begin{aligned} B &= \frac{h_{t,\infty}}{L}+\frac{Y_{O_2,\infty}}{\nu_{O_2}}\frac{h_c}{L} \\ h_{t,\infty} &= \int_{T_w}^{T_\infty} c_p\, d\,T = c_p\,\{T_\infty - T_w\} \end{aligned}\right\} \quad (10.35)$$

$$B=\{c_p(T_\infty-T_w)/L\}+(Y_{O2,\infty}\,h_c)/(\nu_{O2}\,L)=\{1/L\}\{Y_{O2,\infty}\,h_c/\nu_{O2}+c_p(T_\infty-T_w)\}. \quad (10.36)$$

If $T_w \approx T_{BP}$, then,

$$B \approx \{c_p\,(T_\infty - T_{BP})/L\} + (Y_{O2,\infty}\,h_c)/(\nu_{O2}\,L). \quad (10.37)$$

Then, B is known, ξ_w is solved and, thus, $\dot{m}$ can be determined.

Tables A.2A and A.2C tabulate the B number of many fuels. The higher the B number, the higher the mass loss rate. Tables A.2A and A.2C show that heat values h_c and ν_{O2} vary for plastics, alcohols, and pentane to nonane. But the ratio h_c/ν_{O2} for most fuels is appoximately constant ($\approx$ 13800 kJ/kg of oxygen). Thus B number is mainly affected by L compared to any other terms. B $\approx$ 1 – 2 for plastics (polymers), $\approx$ 2 – 3 for alcohols, and B $\approx$ 8 – 10 for pentane to octane.

10.5.1.3 Physical Meaning of Transfer Number B

Consider an adiabatic open-system reactor supplied 1 kg of air at T_∞ and a stoichiometric amount of liquid fuel (i.e., $1/(A{:}F)_{stoich}$) at T_w:

Sensible energy of liquid drops = 0
Sensible energy of 1 kg of air = $c_p\,(T_\infty - T_w)$, which is relative to energy at T_w
$Y_{O2,\infty}/\nu_{O2}$ = (kg of O_2/kg of air) * kg of fuel/kg of stoichiometric O_2 = kg of fuel burnt/kg of air = $1/(A{:}F)_{stoich}$
$Y_{O2,\infty}\,h_c/\nu_{O2}$ = kJ of heat released per kg of stoichiometric air

The products leave with an energy of $\{c_p\,(T_\infty - T_w) + Y_{O2,\infty}\,h_c/\nu_{O2}\}$. If this thermal energy is supplied to a liquid fuel at T_w (say, noncombustible fuel having same latent heat as the parent fuel), the potential amount of fuel that could be evaporated is $\{c_p\,(T_\infty - T_w) + Y_{O2,\infty}\,h_c/\nu_{O2}\}/L$. The potential capability of evaporating a liquid fuel in kg per kg of stoichiometric air supplied is called the *transfer number B*. The higher the *B* number, the higher the potential for evaporation.

Example 1

Consider a 100-μm octane drop combusting in an environment at $T_\infty = 300$ K, $Y_{O2,\infty} = 0.23$. Other properties of fuel are as follows: L = 362 kJ/kg, h_c = 44350 kJ/kg, T_{BP} = 398 K (Table A.2A in Appendix A), $\nu_{O2} = 3.51$, $c_p = 1.1$ kJ/kg K, and $\rho D = 5 \times 10^{-5}$ kg/m sec. Estimate the burn rate.

Solution

$$B = \{c_p (T_\infty - T_{BP})/L\} + (Y_{O2,\infty}\, h_c)/(\nu_{O2}\, L)$$

$$= 1.1\,(300 - 398)/362 + 0.23 \times 44350/(362 \times 3.51) = 7.73$$

$$\xi_w = \dot{m}/(4\,\pi\,\rho D\,a) = \ln(1 + B) = 2.17$$

$$\dot{m} = 4\,\pi \times 5 \times 10^{-5}\ (\text{kg/msec}) \times 50 \times 10^{-6}\ (\text{m}) \times 2.17 = 6.9 \times 10^{-8}\ \text{kg/sec}$$

Remarks

Note that $\dot{m} \propto$ "a" and as the drop burns, "a" will decrease and, hence, the burn rate in kg/sec will keep decreasing as the size decreases.

10.5.2 D^2 Law

As derived in Chapter 6, Chapter 7, and Chapter 9,

$$-d/dt\,(\rho_\ell (4/3)\,\pi\,a^3) = \dot{m} = 4\,\pi\,\rho D\,a\,\ln(1 + B). \tag{10.38}$$

Simplifying,

$$a\,da/dt = -\rho\,D\,\ln(1 + B)/\rho_\ell. \tag{10.39}$$

Integrating and using the initial conditions that $a = a_0$ at $t = 0$, we have

$$a^2 = a_0^2 - 2\,\rho\,D\,\ln(1 + B)t/\rho_\ell. \tag{10.40}$$

As $a = d/2$, we have

$$d^2 = d_0^2 - [8\,\rho\,D\,\ln(1 + B)/\rho_\ell]t. \tag{10.41}$$

Letting

$$\alpha_c = [8\,\rho\,D\,\ln(1 + B)/\rho_\ell], \tag{10.42}$$

$$d^2 = d_0^2 - \alpha_c t \tag{10.43}$$

where α_c, burning rate constant (m²/sec).

A plot of d^2 vs. t should be linear with a magnitude of slope equal to α_c. Many experiments indicate that the d^2 law is followed. The burning-rate constants for many fuels at higher air temperature are plotted in Figure 10.3. As the ambient temperature increases, there is greater contribution of energy from the ambience (see Subsection 10.5.5.1). Note that $\rho D \propto T^\gamma$, where $0.5 < \gamma < 1$.

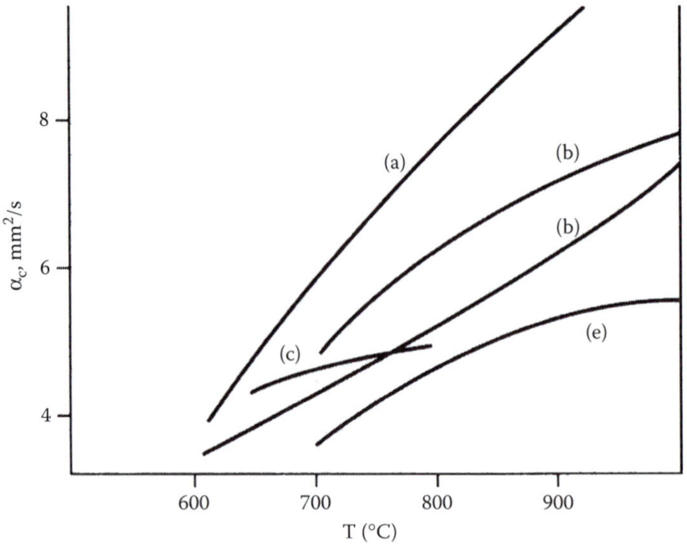

Figure 10.3 Effect of ambient temperatures on burn rate constants: (a) cetane, (b) light diesel, (c) heptane, (d) aviation kerosene, and (e) benzene. (Adapted from Williams, A., *Combustion of Liquid Fuel Sprays*, Butterworth, London, 1990.)

10.5.3 Burning Time

The burning time (t_c) is given by setting d = 0 in Equation 10.43,

$$t_c = d_0^2/\alpha_c. \tag{10.44}$$

For d = 100 μm, $t_b \cong 10$ msec while for char, t_b = 500 to 2000 msec. The higher the B number, the higher the α_c, the lower the burn time.

Example 2

(a) In Example 1, if the diameter is increased to 1 mm, what is the burn rate? (b) If a fuel injector is used to atomize a 1-mm drop into 100-μm droplets, how many droplets are produced and what is the combined burn rate of all those droplets atomized from the 1-mm drop? (c) Compare results from parts (a) and (b) and comment. (d) Determine combustion times of a 1 mm-drop and 100-μm drop of *n*-octane. The liquid density is known to be 707 kg/m³. Use data of Example 1.

Solution

(a) $\dot{m} = 4\ \pi\ 5 \times 10^{-5} \times 500 \times 10^{-6} \times 2.17 = 6.9 \times 10^{-7}$ kg/sec.

(b) $N = (\pi/6) \times 1^3/\{(\pi/6)\ (0.1)^3\} = \{1^3/0.1^3\} = 1000$ drops, $\dot{m}$ of 1000 drops of 0.1 mm diameter $= 6.9 \times 10^{-8} \times 1000 = 6.9 \times 10^{-5}$ kg/sec.

(c) By atomizing from 1 mm (d_{large}) to 0.1 mm (d), the burn rate for the same mass is increased by $6.9 \times 10^{-5}/6.9 \times 10^{-7} = 100$. Burn rate per drop $\dot{m}\ \alpha$

a; then for a large drop $\dot{m}_{large}$ α a_{large}, $\dot{m}$ total α N a, $\dot{m}$ total/ $\dot{m}_{large}$ = N a / a_{large} = $\{/a_{large}{}^3/a^3\}\{a/a_{large}\}$ = $\{a_{large}{}^2/a^2\}$ = 100.

(d) The burn rate constant is evaluated as

$$\alpha_c = [8\ \rho\ D\ \ln(1 + B)/\rho_\ell] = 8\ 5 \times 10^{-5}\ \ln 8.7/707 = 1.22 \times 10^{-6}\ m^2/sec.$$

For a 1-mm drop, $t_b = (1 \times 10^{-3})^2\ m^2/1.22 \times 10^{-6}\ (m^2/sec) = 8.2 \times 10^{-1}$ sec or 820 msec.

For 0.1 mm $t_b = 8.2 \times 10^{-3}$ sec or 8.2 msec.

Remarks

(i) When fuel is atomized, the burn rate is increased, and burn rate of a given mass (or volume) is proportional to the surface area offered by the same mass. The reader can show that the specific burn rate $\dot{m}$ /m = [4 $\pi\rho$ D a ln(1 + B)/{(4/3)π a^3 ρ_ℓ} = 3 ρ D ln(1 + B)/(a^2 ρ_ℓ}, (kg/sec)/kg = 1/sec. The smaller the size, the higher the specific burn rate. One can reduce the burn time by reducing the diameter of the drop.

(ii) The burn time of carbon of similar diameter is on the order of 300 to 500 msec. Thus, carbon burns slowly. The reason is that the surface area where oxygen is consumed is $4\,\pi\,a^2$. This area is much smaller than the flame surface area of the droplet flame where oxygen is consumed.

Example 3

What are the results if T_∞ = 500 K in part (d) of the previous example?

Solution

B = 8.34, ξ_w = 2.23, $\dot{m} = 7.01 \times 10^{-8}$ kg/sec, $\alpha_c = 1.26 \times 10^{-6}$ m²/sec. t_c = 7.93 msec which is a reduction of only 4% as $\{c_p\ (T_\infty - T_{BP})/L\} << (Y_{O2,\infty}\ h_c)/(\nu_{O2}\ L)$ i.e., heat contribution by sensible terms is insignificant compared to the chemical heat release term $(Y_{O2,\infty}\ h_c)/(\nu_{O2}\ L)$.

Example 4

The temperature of gases in rocket engines is about 2800 K. Fuel $C_{12}\ H_{26}$ is injected into the combustion chamber. Assume that the drop travels at the same velocity as the gases. The following data are given:

Drop radius (a) = 50 μm
Gas velocity (v_g) = 30 m/sec
c_p = 1.3 kJ/kg K (same for all species)
ρ D = 1.5×10^{-4} kg/msec
Density of liquid (ρ_ℓ) = 600 kg/m³
Latent heat (L) = 990 kJ/kg
Lower heating value of gaseous fuel (h_c) = 44500 kJ/kg
Temperature of drop (T_w) = boiling point (TBP) = 216°C
Ambient-oxygen mass fraction, $Y_{O2,\infty}$ = 0.1

(a) Estimate required combustor length in cm for complete combustion.
(b) Compute the time in which 80% of the original mass is burned.
(c) What is the effect of decreasing ambient-oxygen mass fraction on the flame radius? Give reasons for your answer using the physics of the problem.

Solution

(a) The Re based on relative velocity $v_{rel}d/\nu = 0$; therefore, combustion is in a quiescent atmosphere. Sh = 2. The results derived so far can be used.

$h_{t\infty} = c_p(T_\infty - T_w) = 1.3 \times (2800 - 489) = 3004$ kJ/kg
$C_{12}H_{26} + 18.5\ O_2 \rightarrow 12\ O_2 + 13\ H_2O$
$\nu_{O2} = 3.475$ kg of O_2 per kg of fuel, $Y_{O2\infty} = 0.1$
$B = 0.1 \times 44500/(3.475 \times 990) + 3004/990 = 4.33$
$\alpha = \{4\ Sh \times \rho D/\rho_\ell\} \ln(1 + B) = 4 \times 2 \times 1.5 \times 10^{-4} \ln(1 + 4.33)/600 = 3.3 \times 10^{-6}$ m²/sec
$t_{burn} = 100 \times 10^{-6} 2/3.35 \times 10^{-6} = 3$ msec
Length = 30 × 0.003 = 9 cm

(b) $d^2 = d_0^2 - \alpha_c t$, $m_l/m_{l0} = (d^3/d_0^3)$; with $m_l/m_{l0} = 0.8$, t = 0.42 msec
(c) As O_2 is decreased, B decreases, and the burning rate decreases, too; the flame moves outward toward O_2 and, hence, r_f/a increases.

10.5.4 Exact Solution for T_w

Recall that we have set $T_w \approx T_{BP}$ in estimating B, but it is only approximate. For example, when water evaporates from lakes and oceans or water drops evaporate in room air, $T_w \neq T_{BP}$ but much less. Similarly, even during combustion, $T_w < T_{BP}$. At phase equilibrium T_w is the same as saturation temperature at a given fuel partial pressure; we need to determine the partial pressure of fuel vapor or the fuel mass fraction at the drop surface.

10.5.4.1 Species: Fuel (F)

The liquid may consist partly of fuel (say, C_8H_{18}) and partly of, say, water (H_2O). Thus, the fuel mass fraction in the liquid phase is denoted by Y_ℓ. If the evaporation rate is $\dot{m}$, then the fuel flow must be set at $\dot{m}_F = \dot{m}\, Y_{F\ell}$ fraction (e.g., suppose we have an oil–water slurry of 80% C_8H_{18} and 20% water; then $Y_\ell = 0.8$).

$$4\pi \rho_w Y_{F,w} v_{F,w} a^2 = \dot{m}_F \tag{10.45}$$

$$4\pi \rho_w v_w r^2 Y_{F,w} - 4\pi \rho D r^2 \left(\frac{dY_F}{dr}\right)_w = \dot{m}_F \tag{10.46}$$

Transforming the r variables in terms of ξ,

$$Y_{F,w} + \left(\frac{dY_F}{d\xi}\right)_w = \varepsilon_F = Y_{F,\ell} \tag{10.47}$$

where

$$\varepsilon_F = \text{flux ratio of fuel} = \dot{m}_F/\dot{m} = Y_{F,\ell}.$$

We need to express the formula in terms of SZ. Selecting $\beta = \beta_{ht} = h_t/h_c + Y_F$ and adding Equation 10.25 to Equation 10.47

$$\beta_{ht-F,w} + \left(\frac{d\beta_{ht-F,w}}{d\xi}\right)_w = F_\beta = Y_{F,\ell} - \frac{L}{h_c}.$$

Using this result in Equation 10.31,

$$\xi_w = n\left[1 + \left(\frac{\beta_{ht-F,w} - \beta_{ht-F,\infty}}{Y_{F,} - \frac{L}{h_c} - \beta_{ht-F,w}}\right)\right]. \quad (10.48)$$

Divide Equation 10.27 by ν_{O2} and subtract it from Equation 10.47, then using Equation 10.31, one obtains ξ_w as

$$\xi_w = \ell n\left[1 + \left(\frac{(\beta_{O_2-F,w} - \beta_{O_2-F,\infty})}{(Y_{F,\ell} - \beta_{O_2-F,w})}\right)\right]. \quad (10.49)$$

Equating Equation 10.48 with Equation 10.49, we obtain the solution for $\beta_{O_2\text{-}F,w}$ from the following equation:

$$\left(\frac{\beta_{h_t-F,w} - \beta_{h_t-F,\infty}}{Y_{F,\ell} - \frac{L}{h_c} - \beta_{h_t-F,w}}\right) = \left(\frac{(\beta_{O_2-F,w} - \beta_{O_2-F,\infty})}{(Y_{F,\ell} - \beta_{O_2-F,w})}\right) \quad (10.50)$$

$$\beta_{O_2-F,w} = \frac{\beta_{O_2-F,\infty}\left(Y_{F,\ell} - \frac{L}{h_c} - \beta_{h_t-F,\infty}\right) + Y_{F,\ell}(\beta_{h_t-F,w} - \beta_{h_t-F,\infty})}{\left(Y_{F,\ell} - \frac{L}{h_c} - \beta_{h_t-F,\infty}\right) + (\beta_{h_t-F,w} - \beta_{h_t-F,\infty})}. \quad (10.51)$$

For any other species not contained in the liquid fuel (there is no O_2, CO_2, etc., in the fuel),

$$4\pi\rho_w v_w a^2 Y_{i,w} - 4\pi\rho D a^2\left(\frac{dY_i}{dr}\right)_w = m_i = 0,\ i = O_2, CO_2, N_2, \text{ etc.} \quad (10.52a)$$

Simplifying in terms of ξ,

$$Y_{i,w} + \left(\frac{dY_i}{d\xi}\right)_w = 0,\ i = O_2, CO_2, N_2, \quad \text{etc.} \quad (10.52b)$$

Then

$$\beta_{i-O_2,w} + \left(\frac{d\beta_{i-O_2}}{d\xi}\right)_w = F_\beta = 0, \quad (i = CO_2, H_2O). \quad (10.53)$$

10.5.4.2 Thin-Flame Results

Comparing Equation 10.49 with Equation 10.34,

$$B=\left(\frac{\beta_{O_2-F,w}-\beta_{O_2-F,\infty}}{Y_{F,\ell}-\beta_{O_2-F,w}}\right)=\left(\frac{Y_{F,w}+\frac{Y_{O_2,\infty}}{\nu_{O_2}}}{Y_{F,\ell}-Y_{F,w})}\right) \tag{10.54}$$

The surface fuel vapor fraction is solved as

$$\frac{Y_{F,w}}{Y_{Fl}}=1-\frac{\left(1+\frac{Y_{O_2,\infty}}{Y_{F,\ell}\nu_{O_w}}\right)}{(1+B)} \tag{10.55}$$

For pure fuel $Y_{F,l}$ = 1.0. The $Y_{F,w}$ and $X_{F,w}$ are given as

$$Y_{F,w}=1-\frac{\left(1+\frac{Y_{O_2,\infty}}{\nu_{O_w}}\right)}{(1+B)},\quad \text{pure fuels} \tag{10.56}$$

$$X_{F,w} = Y_{F,w}\ M_{mix}/M_F \tag{10.56a}$$

where

$$M_{mix}=\sum\{1/(Y_{kw}/M_k)\} \tag{10.56b}$$

where Y_{kw} for k ≠ F is yet to be determined. Species such as CO_2, N_2, and H_2O have molecular weights ranging from 18 to 44. The exact result for M_{mix} requires determination of wall mass fractions of all species. An approximate result for M_{mix} could be obtained by assuming that all species except fuel have an equal molecular weight of M'. Then

$$M = \{Y_{F,w}/M_F + (1 - Y_{F,w})/M'\} \tag{10.57a}$$

X_{Fw} is evaluated using Equation 10.56a. The partial pressure of fuel vapor is given as

$$p_{F,w} = X_{F,w}\ P. \tag{10.57b}$$

The partial pressure of fuel vapor is the same as the saturation pressure of fuel vapor at a given drop temperature, or T_w. Under phase equilibrium, the saturation tables provide the corresponding saturation temperature T_w. Figure 10.4 shows the saturation pressure vs. temperature for selected fuels.

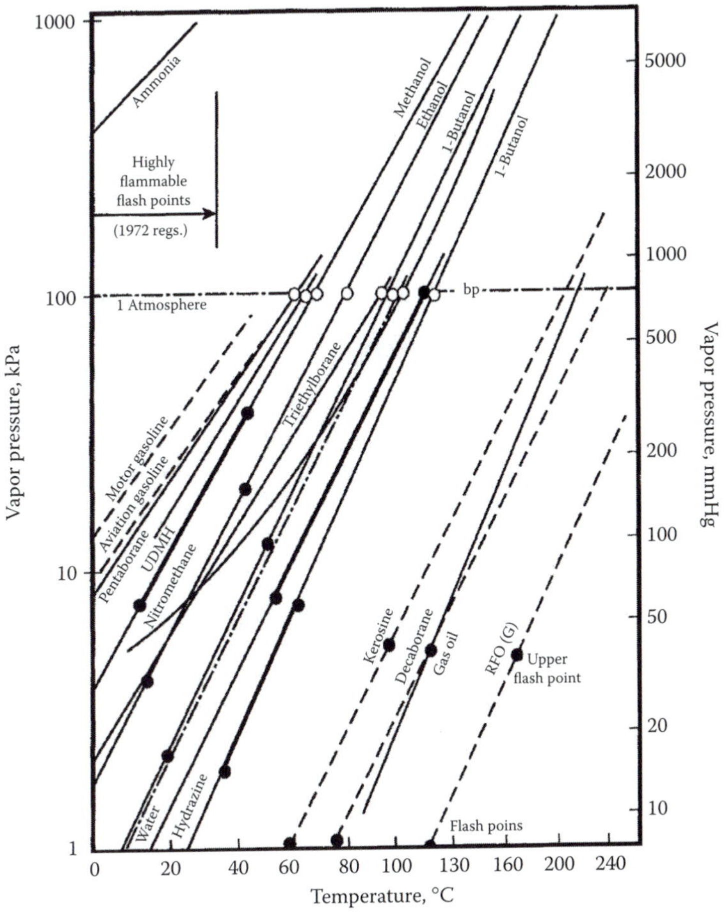

Figure 10.4 Saturation pressures vs. temperature. (Adapted from Goodger, E.M., *Alternate Fuels: Chemical Sources*, Halsted Press, New York, ISBN: 0-470-26953-9, 1980.)

We may also use the Clausius–Cleyperon equation, which relates saturation pressure (i.e., p_F^{sat} = pFW) to the saturation temperature or liquid temperature.

$$\frac{p_{F,w}}{p_{ref}} = \exp\left[\frac{L}{R}\left(\frac{1}{T_{ref}} - \frac{1}{T_w}\right)\right] \tag{10.58a}$$

Rewriting,

$$\ln(p_{F,w}) = A - C/T_w \tag{10.58b}$$

where

$$A = \ln\,(p_{ref}) + (L/RT_{ref}), \quad C = L/R. \tag{10.58c}$$

Normally, P_{ref} = 1 bar and T_{ref} = saturation temperature at 1 bar or normal boiling point (see Table A.2E in Appendix A). Selecting $T_{ref} = T_{BP}$ and p_{ref} = 1 bar,

$$A = \ln(1) + L_{BP}/RT_{BP}, \quad C = L_{BP}/R.$$

Equation 10.58b can also be written as

$$\ln(p_{F,w}) = A - C_1 \{T_{BP}/T_w\} \tag{10.58d}$$

where $C_1 = L_{BP}/(RT_{BP})$. Typically, for many fuels $C_1 = L_{BP}/RT_{BP}$ = 8 to 10 and, hence, A and C_1 are almost constant for many fuels.

Example 5

Determine (a) mass fraction, (b) mole fraction, and (c) partial pressure of octane fuel vapor if B = 7.88, $Y_{O2,\infty} = 0.23$, and other species (e.g., N_2) at the drop surface have a molecular weight of 28. P = 200 kPa. (d) What is the liquid temperature if L = 362 kJ/kg K, T_{BP} = 398 K, and P = 200 kPa?

Solution

(a) $Y_{Fw} = 1 - (1+s)/(1+B)$, $s = Y_{O2,\infty}/\nu_{O2}$, $Y_{Fw} = 1 - \{1 + (0.23/3.51)\}/(1+7.88) = 0.88$

(b) $Y_{I,w} = 1 - 0.88 = 0.12$, $M = \{[0.88/114] + [0.12/28]\}^{-1} = 83.3$

$$X_{F,w} = 0.88 * 83.3/114 = 0.64$$

(c) $p_{F,w} = 0.64 * 200 = 128$ kPa

(d) $R = 8.314/114 = 0.0729$ kJ/kg K, $\frac{128}{100} = \exp\left[\frac{362}{0.0729}\left(\frac{1}{398} - \frac{1}{T_w}\right)\right]$; solving, $T_w = 406$ K

■ ■ ■

In Example 5, we assumed that species other than fuel have a molecular weight of 28; however, the species are CO_2 (M = 44), H_2O (M = 18), and N_2 (M = 28) and M's are not equal. Thus, we need to determine the mass fractions of these species at the wall in order to estimate M_{mix} more accurately.

10.5.4.3 Mass Fractions of CO_2 and H_2O

$$\text{Because } \beta_{CO2\text{-}O2,w} = Y_{CO2,w}/\nu_{CO2} + Y_{O2,w}/\nu_{O2} = Y_{CO2,w}/\nu_{CO2}$$

F_β for $\beta_{CO2\text{-}O2} = 0$. Using Equation 10.32 $B = \{\beta_{CO2\text{-}O2,w} - \beta_{CO2\text{-}O2,\infty}\}/\{F_\beta - \beta_{CO2\text{-}O2,w}\}$ and with $F_\beta = 0$, one solves for the CO_2 wall mass fraction as

$$Y_{CO2,w}/\nu_{CO2} = (Y_{O2,\infty}/\nu_{O2}) * (1 + B).$$

$$\text{In general, } Y_{k,w}/\nu_k = (Y_{O2,\infty}/\nu_{O2}) * (1 + B),\ k = CO_2, H_2O. \tag{10.59}$$

The inert N_2 has a mass fraction of $Y_{N2w} = 1 - Y_{Fw} - Y_{CO2w} - Y_{H2Ow}$. Thus, the mixture molecular weight can be determined using Equation 10.56b.

Example 6

(a) Determine the B number if an *n*-octane drop is burning; T_∞ = 300 K, L = 362 kJ/kg K, h_c = 44350 kJ/kg, and c_p = 1.1 kJ/kg K. Assume the drop temperature to be 350 K.

(b) Determine mass fractions of fuel, CO_2 and H_2O at the drop surface, mixture molecular weight, and fuel mole fraction if the drop temperature is 350 K and R = 0.0729 at P = 2 bar.

Solution

(a) For an octane drop,

$$C_8H_{18} + 12.5\ O_2 \rightarrow 8\ CO_2 + 9\ H_2O$$

$$\nu_{O2} = 3.5,\ \nu_{CO2} = 8 \times 44/114 = 3.1,\ \nu_{H2O} = 1.42,\ \nu_P = 3.5 + 1 = 4.5$$

$$B = 1.1\ (300 - 350)/362 + 0.23 \times 44350/(362 \times 3.5) = 7.787$$

(b) $Y_{Fw} = 1 - (1 + s)/(1 + B) = 1 - (1 + 0.0657)/(1 + 7.79) = 0.88,$

Using Equation 10.59, $Y_{CO2,w} = 0.023$; $Y_{H2O,w} = 0.011$

$$Y_{N2w} = 1 - 0.88 - 0.023 - 0.011 = 0.086$$

$$M_{mix} = 1/(0.88/114 + 0.023/44 + 0.011/18 + 0.086/28) = 83.9$$

$$X_{Fw} = 0.88 * 83.9/114 = 0.65.$$

10.5.4.4 Exact Procedure for Drop Temperature and Burn Rate

The approximate method of estimation of ξ_w uses $T_w = T^{sat}$ at total pressure P. The exact procedure for estimation of ξ_w as follows:

(1) Assume $T_w = T^{sat}$ at the given total pressure P to start with the procedure.
(2) Estimate $h_{t,\infty}$ and B using Equation 10.35.
(3) Evaluate $Y_{F,w}$ using Equation 10.56.
(4) Evaluate $Y_{CO2,w}$, and $Y_{H2O,w}$ using Equation 10.59 and $Y_{N2w} = (1 - Y_{F,w} - Y_{CO2,w}, Y_{H2o,w})$.
(5) Determine $X_{F,w}$ using Equation 10.56a.
(6) Evaluate partial pressure $p_{F,w} = X_{F,w}$ P and use Equation 10.58 to find T_w.
(7) Use the new T_w and check with Step 1 and iterate until convergence occurs.
(8) Then reevaluate h_{∞}, B, and ξ_w.

10.5.5 Flame Structure and Flame Location, r_f

Once ξ_w and the wall properties ($Y_{F,w}$, T_w, etc.) are known, the β profiles and hence profiles of h_T, h, Y_F, Y_{O2}, Y_{CO2}, etc., can be obtained. Recall Equation 10.17. Setting $\beta = \beta_{O2\text{-}F}$ in the Equation

$$[\beta - \beta_\infty]/[\beta_w - \beta_\infty] = [1 - \exp\ (\xi_w\{a/r\})]/[1 - \exp\ (\xi_w)] \qquad (10.60)$$

we can get the relation for flame location by using $\beta = \beta_{O2\text{-}F} = Y_F - (Y_{O2}/\nu_{O2})$ and setting $\beta_{O2\text{-}F,f} = 0$ (because at the flame, $Y_F = Y_{O2} = 0$).

$$\Phi_f = \frac{-\beta_{O_2-F,\infty}}{\beta_{O_2-F,w} - \beta_{O_2-F,\infty}} = \frac{1 - \exp\ (-\xi_f)}{1 - \exp\ (-\xi_w)} = \frac{1 - \exp\left(-\xi_w\left[\frac{a}{r_f}\right]\right)}{1 - \exp\left(-\xi_w\right)} \qquad (10.61)$$

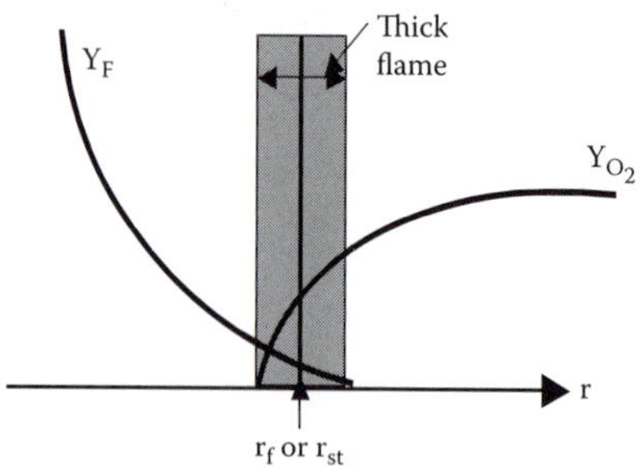

Figure 10.5 Flame location for a thick flame.

One can solve for ξ_f as

$$\xi_f = -\ln\left[1-\Phi_f \frac{B}{1+B}\right] \tag{10.62}$$

$$\beta_{O2\text{-}F,f} = 0 \tag{10.63}$$

$$\phi_f = \frac{\dfrac{Y_{O2,\infty}}{\nu_{O2}}}{\left(\dfrac{Y_{O2,}}{\nu_{O2}} + Y_{F,w}\right)}. \tag{10.64}$$

Using Equation 10.56 in Equation 10.64,

$$\phi_f = \frac{(1+B)s}{B(1+s)} \tag{10.65}$$

$$s = Y_{O,\infty}/\{\nu_{O2}\ Y_{F,l}\}, \tag{10.66}$$

where, s represents stoichiometric fuel consumed for every kg of air.

Using this result in Equation 10.61 and letting $\dot{m} = 4\pi\rho Da$.

$$\xi_f = \frac{\dot{m}}{4\pi\rho D r_f} = \ell n\ (1+s) \tag{10.67}$$

$$\left(\frac{r_f}{a}\right) = \frac{\ell n\ (1+B)}{\ln(1+s)}. \tag{10.68}$$

For a pure mixing problem, this radius r_f is the same as the stoichiometric location r_{st}. As $Y_{O2,\infty}/\nu_{O2}$ is usually small, the denominator in Equation 10.68 can be approximated

$$\left(\frac{r_f}{a}\right) \approx \frac{\ell n(1+B)}{s}. \tag{10.69}$$

Ignoring the weak effects of B, the flame radius is inversely proportional to the oxygen mass fraction

10.5.5.1 Flame Temperature

Because ϕ_f is known from Equation 10.65, $\beta = \beta_{h_t-F}$ is selected as a variable in the NSZ variable ϕ.

At the flame,

$$\phi_f = \frac{s(1+B)}{B(s+1)} = \frac{(\beta_{h_t-F})_f - (\beta_{h_t-F})_\infty}{(\beta_{h_t-F})_w - (\beta_{h_t-F})_\infty}. \quad (10.70)$$

Using Equation 10.56 to eliminate $Y_{F,w}$, and simplifying

$$\frac{c_p(T_f - T_\infty)}{h_c Y_{F,\ell}} = \frac{s}{(1+s)}\left(1 - \frac{L}{h_c Y_{F,\ell}} - \frac{c_p(T_\infty - T_w)}{h_c Y_{F,\ell}}\right). \quad (10.71)$$

With $Y_{F,\ell} = 1$ for pure fuel,

$$\frac{c_p(T_f - T_\infty)}{h_c} = \frac{s}{(1+s)}\left(1 - \frac{L}{h_c} - \frac{c_p(T_\infty - T_w)}{h_c}\right). \quad (10.72)$$

$$s = Y_{O_2,\infty} / \nu_{O_2}, Y_{F,\ell} = 1.0. \quad (10.73)$$

Recall from Chapter 8 that the T_f relation for boundary-layer combustion is exactly the same as T_f for drop combustion.

10.5.5.2 Relation between Flame and Adiabatic Flame Temperatures

The result for T_f in Equation 10.72 is the same as adiabatic flame temperature of an SSSF reactor operated with liquid drops entering at T_w and stoichiometric air entering at T_∞. (See Figure 10.6.)

$$1 + \upsilon_{O2} \text{ kg of } O_2 = (1 + \upsilon_{O2}) \text{ products } CO_2 \text{ and } H_2O$$

Stoichiometric air per kg of fuel = $\upsilon_{O2}/0.23$ or $\upsilon_{O2}/Y_\infty = (A{:}F)_{st}$

Assume the reference to be T_w for thermal enthalpy in the gaseous state.

$$h_c + (\upsilon_{O2}/Y_{O2\infty}) \times c_p (T_\infty - T_w) - L = (1 + \upsilon_{O2}/Y_{O2\infty})\, c_p (T_f - T_w)$$

Simplifying, the relation is found to be the same as Equation 10.72.

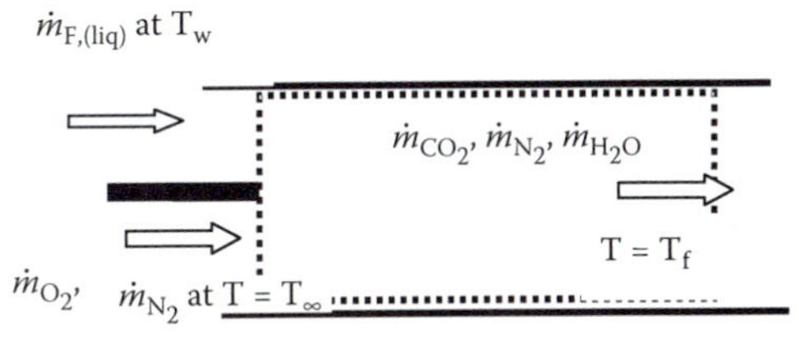

Figure 10.6 Flame temperature around a drop = adiabatic flame temperature.

Table 10.2 Evaluation and Combustion Relations ($s = Y_{O2,\infty}/(\nu_{O2}Y_{F\ell})$, typically $Y_{F\ell} = 1$) $\xi_w = \ln(1 + B)$ where $B = (\beta_w - \beta_\infty)/(F_\beta - \beta_w)$.

Drop	*Evaporation*	*Combustion*
Transfer number, B	$B = \frac{c_p(T_\infty - T_w)}{L}$	$B = \frac{c_p(T_\infty - T_w)}{L} + h_c \frac{s}{L}$
Mass at drop surface, Y_{Fw}	$1 - \frac{1}{1+B}$	$1 - \frac{1+s}{1+B}$, $s = \frac{Y_{O_2\infty}}{\nu_{O_2}}$
Nondimensional stoich/flame temperature	$-\left[\frac{s}{1+s}\right]\left[1 + \frac{c_p(T_\infty - T_w)}{L}\right] = \frac{c_p(T_{St} - T_w)}{L}$	$\left[\frac{s}{1+s}\right]\left[\frac{h_c}{L} - 1 - \frac{c_p(T_\infty - T_w)}{L}\right] = \frac{c_p(T_f - T_\infty)}{L}$
Nondimensional stoich/flame location	$\frac{\ln(1+B)}{\ln(1+s)} = \frac{r_{st}}{a}$	$\frac{\ln(1+B)}{\ln(1+s)} = \frac{r_f}{a}$
Mass-loss rate constant, α	$\alpha_e = \frac{4\ Sh\ \rho D}{\rho_l}\ln(1+B)$	$\alpha_c = \frac{4\ Sh\ \rho D}{\rho_l}\ln(1+B)$
Drop lifetime	$\frac{d_o^2}{\alpha_e}$	$\frac{d_o^2}{\alpha_c}$

10.5.5.2.1 Species Mass Fraction at Flame

Consider $\beta_{k\text{-}O2} = \{Y_k/\upsilon_k\} + Y_{O2}$; at the flame, $\beta_{k\text{-}O2,f} = \{Y_{kf}/\upsilon_k\}$ with $k = CO_2$, H_2O;

$$\Phi_f = \frac{s(1+B)}{B(s+1)} = \frac{\beta_{k-O_2,f} - \beta_{k-O_2,\infty}}{\beta_{k-O_2,w} - \beta_{k-O_2,\infty}}.$$

Knowing ϕ_f and Y_{kw} from Equation 10.55 and Equation 10.64, we can calculate mass fractions at the flame. It can be shown that $Y_{k,f}/\upsilon_k = \{s/(1 + s)\} = 1/\{(A{:}F)_{st} + 1\}$, $k = CO_2$, H_2O. Table 10.2 summarizes combustion relations drop.

Example 7

Determine the flame temperature and partial pressure of *n*-octane fuel drop burning in air at $T_\infty = 900$ K and P = 100 kPa. Other properties: L = 362 kJ/kg, $h_c = 44350$ kJ/kg, $c_p = 1.1$ kJ/kg K, and $\upsilon_{O2} = 3.51$.

Solution

Following the procedure in previous examples, $Y_{F,W} = 0.90$. Assume that the remaining gases at the drop surface have a molecular weight of N_2; then by iteration,

$$T_f = 3332.8 \text{ K}$$

$$s = 0.23/3.5 = 0.0657.$$

We determine adiabatic flame temperature using Equation 10.72, $(T_f - 900) = \{0.0657/1.0657\}\{44350 - 362 - 1.1(900 - 387)\}$; solving, $T_f = 3334$ K.

10.5.5.2.2 Profiles

Now we can determine the profiles of Y_F, Y_{O2}, T, Y_{CO2}, etc., vs. r, using Equation 10.17:

(1) Fuel and oxygen mass fraction with $\beta = \beta_{O2\text{-}F}$

$$Y_F = 0,\ \beta_{O2\text{-}F} = Y_{O2}/\nu_{O2}, \quad \text{for } r > r_f, \text{ fuel-lean regime} \tag{10.74}$$

$$Y_{O2} = 0,\ \beta_{O2\text{-}F} = Y_F, \quad \text{for } r < r_f, \text{ fuel-rich regime} \tag{10.75}$$

Using $\beta_{O2\text{-}F} = Y_F - Y_{O2}/\nu_{O2}$ and Equations 10.74 and 10.75, Y_F, Y_{O_2} profiles are obtained.

(2) With $\beta = \beta_{ht\text{-}O_2} = \{(h_t/h_c) + (Y_{O_2}/\nu_{O_2})\}$, the T profiles are obtained.

Example 8

An *n*-hexane fuel drop burns in air, at $T_\infty = 600$ K and P = 100 kPa. Other properties: L = 335 kJ/kg, $h_c = 45105$ kJ/kg, $c_p = 1.17$ kJ/kg K, $Y_{O2\infty} = 0.23$, $d_0 = 100\ \mu$m, $T_{BP} = 342.15$; $\rho_{liq} = 659$ kg/m^3; $\lambda_{fuel} = 1.53 \times 10^{-4}$ kW/m K, $\lambda_{gas} = 8.34 \times 10^{-5}$ kW/m K, and $c_{pF} = 4.39$ kJ/kg K.

(a) Assume that $T_w = T_{BP}$ at the given P, and obtain wall or surface properties: Y_{Fw}, $Y_{CO2,w}$, Y_{H2Ow}, and Y_{N2w}, and flame properties: T_f, r_f/a, $Y_{CO2,f}$, Y_{H2Of}, and Y_{N2f}.
(b) Determine the initial burn rate in kg/sec, burn rate constant in m^2/sec and burning time for case (a). For burn rate and time, assume λ and c_p for fuel at $(T_w + T_f)/2$ (i.e., $\rho D = \lambda/c_p$); whereas for estimation of B, T_f, etc., assume c_p for gas at $(T_w + T_f)/2$.
(c) For case (a) get T, Y_F, Y_{O2}, and ϕ, f vs. r/a profiles.
(d) How do the results for cases (a) and (b) change if one iteratively calculates T_w?

Solution

A spreadsheet has been developed and used.

$$C_6H_{14} + 9.5\ O_2 + 9.5 \times 3.76\ N_2 = 6\ CO_2 + 7\ H_2O + 9.5 \times 3.76 \times N_2.$$

Thus, $\nu_{CO2} = 3.06$, $\nu_{O2} = 3.53$; $\nu_{H2O} = 1.46$; A:F = 15.33.

$$B = 0.23 \times 45105/(3.53 \times 335) + 1.17 \times (600 - 342.2)/335 = 9.68$$

(a) Wall properties:

$$Y_{Fw}, = 1 - (1 + s)/(1 + B),\ s = (Y_{O2,\infty}/\nu_{O2}) = 0.0652,$$

$$Y_{Fw} = 1 - (1 + 0.0652)/(1 + 9.68) = 0.900$$

$$Y_{CO2,w}/\nu_{CO2} = s/(1 + B),\ Y_{CO2w} = 0.0187,\ Y_{H2Ow} = 0.00893$$

$$Y_{N2w} = 1 - Y_{CO2w} - Y_{H2Ow} = 0.0721$$

Using Equation 10.72, $T_f = 2927$ K

$$\phi_f = s(1 + B)/\{(s + 1)B\} = 0.0675$$

$$\left(\frac{r_f}{a}\right) = \frac{\ell n(1+B)}{\ln(1+s)} = \ln(1 + 9.68)/\ln(1 + 0.0652) = 37.49$$

$$\beta_{k\text{-}O2,w} = \{Y_{CO2,w}/\upsilon_{CO2}\} = 0.006106$$

$$\beta_{k\text{-}O2,\infty} = 0 + Y_{O2,\infty}/\upsilon_{O2} = 0.23/3.52 = 0.0652$$

$$\beta_{k\text{-}O2,f} = \phi_f\,[\beta_{k\text{-}O2,w} - \beta_{k\text{-}O2,\infty}\} + \beta_{k\text{-}O2,\infty} = 0.0675 * \{0.0061 - 0.0652\}$$

$$+ 0.0652 = 0.0612 \text{ with } k = CO_2.$$

$Y_{CO2,f} = \beta_{CO2\text{-}O2,f}\ \nu_{CO2} = 0.0612 \times 3.063 = 0.188$. Similarly,

$$Y_{H2Of} = 0.0896,\ Y_{N2f} = 1 - Y_{CO_2f} - Y_{H_2Of} = 0.723$$

(b) The initial burn rate in kg/sec, $\xi_w = \ln(1 + B) = 2.3685$, $\dot{m}/\{4\pi\rho D\ a] = 2.37$ for $\rho D = \lambda/c_p$, we use for fuel vapor at film temperature: $\lambda = 1.53 \times 10^{-4}$ kW/m K; then $\rho D = \lambda/c_p = 2.53 \times 10^{-5}$ kg/msec, $\dot{m} = \{4\pi \times 2.53 \times 10^{-5} \times 50 \times 10^{-6}] \times 2.36851 = 4.10 \times 10^{-08}$ kg/sec.

$$\alpha_c = [8\ \rho\ D\ \ln(1 + B)/\rho_\ell] = 8 \times 2.53 \times 10^{-5} \times \ln(1 + 9.68)/659 = 7.92 \times 10^{-7}\ m^2/s$$

$$\text{Burning time} = (100 \times 10^{-6})^2\ m^2/7.92 \times 10^{-7} = 0.0126 \text{ sec}$$

(c) Using the procedure in Section 10.5.5.2.2, the Y_F, Y_{O_2}, and T profiles can be plotted as shown in Figure 10.7.

(d) T_w is not equal to T_{BP} as $Y_{Fw} = 0.9 < 1$. Thus, the partial pressure of fuel vapor is less than 1 bar. Using the current wall mass fractions, M_{mix} is calculated and solving exactly $T_w = 332.7$ K, B = 9.71, and $\xi_W = 2.37$.

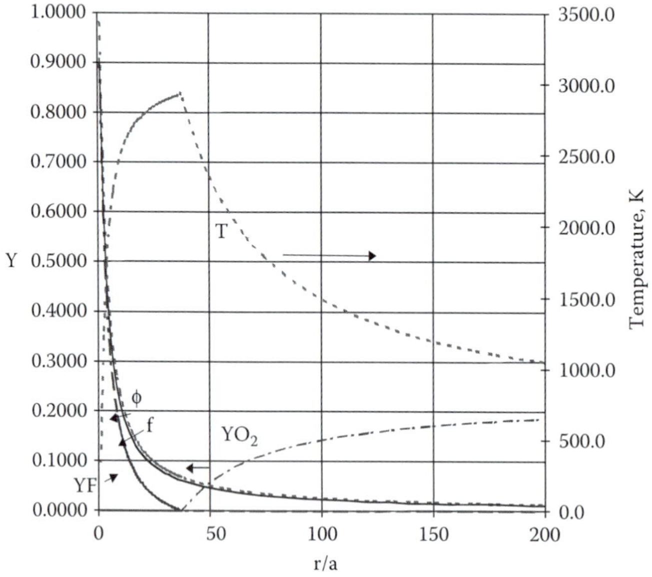

Figure 10.7 Flame structure for a burning *n*-hexane drop.

Remarks

(i) $C_6H_{14} + 9.5\ O_2 + 9.5 \times 3.76\ N_2 = 6\ CO_2 + 7\ H_2O + 9.5 \times 3.76 \times N_2$
Thus, Y_{CO2} in products $= 6 \times 44.01/(6 \times 44.01 + 7 \times 18.02 + 9.5 \times 3.76 \times 28.02) = Y_{CO2f} = 0.189$, which is the same as the mass fraction at the flame.

(ii) Iterative solution for T_w in the combustion problem does not lead to any significant change in results compared to the case of $T_w = T_{BP}$.

10.5.6 Extension to Combustion of Plastics

Plastics on heating pyrolyze (chemical decomposition), releasing combustible gases at temperature T_g, called the pyrolysis/gasification temperature, and the combustible gases oxidize in the gas phase. For example, the polymer PMMA (Plexiglas) pyrolyzes at about 660 K, releasing monomers $C_5H_8O_2$ (Table A.2C in Appendix). Except for direct phase change from solid to gas with heat of gasification (q_g), the combustion is assumed to be similar to that of liquid droplets. Thus, for spherical particles made up of plastics, the B number is defined as

$$B = \{c_p\ (T_\infty - T_g)/q_g\} + (Y_{O2,\infty}\ h_c)/(\nu_{O2}\ q_g) = \{1/q_g\}\ \{Y_{O2,\infty}\ h_c/\nu_{O2} + c_p\ (T_\infty - T_g)\}$$

where L is now replaced by q_g. Plastics are poor conductors; typically, temperature gradients exist within the materials. Hence, q_g needs to be modified as $q_{g,mod} = q_g + c_p\ (T_g - T_{0,s})$.

$$B = \{1/q_{g,mod}\}\ \{Y_{O2,\infty}\ h_c/\nu_{O2} + c_p\ (T_\infty - T_g)\}$$

The B number is typically low owing to higher q_g compared to L and higher T_g for plastics compared to T_w of droplets.

10.5.7 Extension to Combustion of Coal and Biomass

Whereas biomass contains up to 80% mass as volatile matter (VM), bituminous coal contains 40 to 50% VM. If pyrolysis occurs at a fixed T_p (see TGA studies on coal and feedlot biomass in Chapter 5 to determine T_p), then T_w in the definition of B number can be replaced by T_p and h_c, by heating value of volatiles. Thus, the B number for a volatile can be evaluated, i.e., we treat coal VM as though it burns like an oil drop. A more rigorous treatment of VM combustion has been formulated by Timothy et al. [1982, 1986].

10.5.8 Extension of Combustion Analyses to Pure Vaporization/Gasification

For pure evaporation, one can let $h_c = 0$ (i.e., no heat is liberated) and modify B

$$\xi_w = \ln(1 + B). \qquad \text{(From Equation 10.34)}$$

Equation 10.35 defining the B number simplifies to

$$B = (h_{T,\infty}/L). \qquad (10.76)$$

Similarly, with $Y_{O2,\infty} = 0$ in Equation 10.55, we get

$$Y_{F,w} = 1 - \frac{1}{1+B} = \frac{B}{(1+B)}, \text{ pure fuels.} \tag{10.77}$$

The temperature at the stoichiometric location is given by setting $h_c = 0$ in Equation 10.72:

$$c_p(T_{st} - T_\infty) = -\frac{s}{(1+s)}[L + c_p\ (T_\infty - T_w)].$$

Recall that $s = Y_{O2\infty}/\nu_{O2} = 1/\{(A{:}F)_{st}\}$. See Table 10.2 for a summary of relations.

10.5.9 Mass Transfer Correction

The mass transfer correction has been dealt with in Chapter 9. It can also be derived using the current results for evaporation.

$$\xi_w = \ln(1 + B) \qquad \text{(From Equation 10.34)}$$

When B << 1 (typically true of slow evaporation and low mass transfer),

$$\xi_{w,\dot{m}\to 0} \approx B. \tag{10.78}$$

Hence,

$$\dot{m}/\dot{m}_{\dot{m}\to 0} = \xi_w/\xi_{w,\dot{m}\to 0} = \{\ln(1+B)\}/B \tag{10.79}$$

From Equation 10.78 when B <<<1,

$$Y_{F,w} \approx B. \tag{10.80}$$

Although derived for spheres, one can still apply the same correction to any geometry and even to boundary problems (see Chapters 8 and 11).

Using the definition of B given by Equation 10.76 in Equation 10.78,

$$\dot{m}_{\dot{m}\to 0} = 4\ \pi\ \rho\ D\ a\ c_p\ \{T_\infty - T_w\}/L \tag{10.81}$$

$$\dot{m}_{\dot{m}\to 0}\ L = 4\ \pi\ \rho\ D\ a\ c_p\ \{T_\infty - T_w\}, \text{ heat supplied from the gas phase.} \tag{10.82}$$

$$\text{Because heat supplied from the gas phase} = h_H\ \{T_\infty - T_w\} 4\ \pi\ a^2 \tag{10.83}$$

where h_H is the heat transfer coefficient, then it follows from Equation 10.82 and Equation 10.83 that $\rho\ D\ c_p = h_H\ a$. As it was assumed that Le = 1, then $\rho\ D\ c_p = \lambda$; hence,

$$h_H\ a/\lambda = 1 \quad \text{or} \quad h_H\ d/\lambda = Nu = 2$$

Because $Nu = Sh_0 = h_{m0}\ d/(\rho D)$ for Le = 1, under negligible mass transfer rate, $Sh_0 = Nu_0 = 2$ for a heat transfer in a sphere in a quiescent atmosphere with negligible evaporation. Thus, one can use heat transfer correlations for determining slow mass-loss rates and then apply the correction given by Equation 10.79 for mass transfer corrections. For example, if evaporation occurs from a plate under forced convection, the N_{u0} relations deduced without mass transfer effects can now be used for estimating mass-loss rate.

$$\frac{Nu}{Nu_0} \approx \frac{Sh}{Sh_0} = \frac{h_m}{h_{m,\dot{m}\to 0}} = \frac{\ln(1+B)}{B}, \qquad \dot{m} = h_m\, B$$

10.5.10 Evaporation and Combustion inside a Shell of Radius b and the Diameter Law

For a drop burning in a quiescent atmosphere, $\dot{m} = 4\,\pi\rho\,D\,a\,\ln(1 + B)$. One can rewrite it as $\dot{m} = 2\,Sh_0\,\pi\rho\,D\,a\,\ln(1 + B)$, where $Sh_0 = 2$ for a very low mass transfer rate in a quiescent atmosphere. If an isolated drop evaporates or burns within a spherical shell of radius b such that a negligible vapor mass fraction exists at r = b (b > a), then the expression for mass-loss rate can be derived as

$$\dot{m}_k = 2\,Sh_0\,\pi\rho\,D\,a\,\ln(1 + B_k)/\{1 - (a/b)\}, \quad k = e \text{ or } c \tag{10.84}$$

where Sh = 2 [14] and the definition of the B number (Equation 10.35) remains the same with T_∞, and $Y_{O2\infty}$ replaced by T_b and $Y_{O2,b}$, respectively k can represent either the evaporation or combustion case. (See Tishkoff [1979] for evaporation inside a bubble.) The diameter law (for a fixed outer shell radius) under these conditions can be expressed as

$$d^2 = d_0^2 - \alpha_k t - (d_0^3 - d^3)/(3b), \tag{10.85}$$

Thus, the droplet lifetime can be expressed as

$$t_k = d_0^2/\alpha_k - d_0^3/(3\,b\,\alpha_k) \tag{10.86}$$

It is seen that the evaporation time is no longer proportional to d_0^2 and is reduced by an amount $d_0^3/(3\,b\,\alpha_k)$. As $b \to \infty$, the burn time relation reduces to conventional results.

10.6 Convection Effects

10.6.1 Drag Coefficient C_D, Nu and Sh Numbers

The dynamic properties of interest are the drag coefficient and Nu. Drag coefficients and Nusselt numbers for single drops have been summarized by Hedley and A. Williams [1976, 1990]; dynamic properties during combustion have been investigated by Ayyaswamy [1989]. Note that these coefficients are generally lower than those of isothermal impermeable spheres owing to mass transfer, internal circulation, etc. Generally, the Ranz–Marshall correlation for Nusselt number is used to account for convection.

If the time required for a drop to adjust to the local flow field is small compared to the drop evaporation time, then relative velocity effects can be ignored. The *relaxation time* is defined as the time required by a sphere traveling under Stokes flow in a stagnant atmosphere to slow down to 36.8% of its initial velocity (v_0). The distance that the sphere moves before stopping is defined in simple terms as the *stopping, adjustment,* or *relaxation distance.* Assuming the sphere travels under Stokes flow (Re < 1, $C_{D,iso}$ = 24/Re), the relaxation time t_r and stopping distances x_{stop} can be expressed as

$$t_r = \rho_p d_p^2/18\mu \tag{10.87}$$

$$x_{stop} = v_0 t_r = \{\rho_p d_p^2 v_0/18\mu\} = (\rho_p/\rho_g)(d_p/18)Re_0 \tag{10.88}$$

Typically, the initial-drop Reynolds numbers are larger than 1, and Stokes flow is not a valid assumption. The following correlation of C_D with Re for non evaporating spheres has been given by Faeth [1977].

$$C_{D,iso} = 24/Re\ (1 + Re^{(2/3)}/6),\ 3 < Re < 1000 \tag{10.89}$$

$$C_{D,iso} = 24/Re\ (0.82666 + 0.17334Re),\ 1 < Re < 3 \tag{10.90}$$

The following criterion has been put forth to determine whether relative velocity effects are important and need to be considered [Williams, 1985].

$$v_g >> (2/3)\ (\rho_\ell/\rho_g)\ (a/C_D) \tag{10.91}$$

where v_g is the velocity of gas. If this inequality is satisfied, then relative velocity effects should be considered.

Mass transfer from isolated drops is known to decrease the drag coefficient [Hedley et al., 1971; Renksizbulut and Yuen, 1983]:

$$C_D = (24/Re)\ (1 + Re_m^{0.63}/5)/(1 + B_f)^{0.2},\ 10 < Re < 300 \tag{10.92}$$

where all variables are evaluated using the film conditions around the drop (arithmetic average molar concentration and temperature) except for the density in Re_m, which is based on the free-stream conditions.

For ellipsoidal particles with minor axis, y = distortion/radius of sphere (distortion = spherical radius – minor half-axis of ellipse), and $C_D = C_{D,sphere}\{1 + 2.632\ y\}$.

10.6.2 Burn Rates

If combustion of drops occur with a creeping flow, then an envelope flame may still exist. Then one can modify the burn rate relation as follows. If Sh = $h_m d/(\rho\ D)$, then

$$\dot{m} = 4\ \pi\ \rho D\ a\ \ln(1 + B) = 2\ \pi\ \rho D\ d\ \ln(1 + B). \tag{10.93}$$

As before $h_m/h_{m0} = Sh/Sh_0 = \ln(1 + B)/B$,

$$\dot{m} = h_m\ \pi\ h_m\ d^2\ B \quad \text{or} \quad 2\ Sh_0\ \pi\ \rho D\ d\ \ln(1 + B). \tag{10.94}$$

Then, it follows that

$$\alpha_c = 4\ Sh_0\ \rho\ D\ \ln(1 + B)/\rho_\ell. \tag{10.95}$$

Nu_0 and Sh_0 with convection are given in the following text:
(1) Ranz–Marshall:

$$Nu_0 = 2 + 0.572\ Re^{1/2}\ Pr^{1/3},\ 1 < Re < 2000 \quad (10.96)$$

(2) Renksizbulut–Haywood correlation [1988]:

$$Nu_{film} = (2 + 0.57\ Re_m^{\ 1/2}\ Pr_f^{\ 1/3})/(1 + B_f)^{-0.7},\ 10 < Re < 2000,\ B_f < 0.85 \quad (10.97)$$

where all variables are evaluated using the film conditions around the drop (arithmetic average molar concentration and temperature) except for the density in Re_m, which is based on the free-stream conditions.
(3) Faeth's correlation:

$$Nu_0 = 2 + \{0.555Re^{(1/2)}\ Pr^{(1/3)}/[1 + 1.23/\ (Re\ Pr^{(4/3)})^{(1/2)}]\}. \quad (10.98)$$

(4) Free convection [13]:

$$Nu_0 = 2 + 1.066\ Gr^{0.52},\ Gr < 1 \quad (10.99)$$

where $Gr = (g\ d^3\ p^2/\mu^2)\ (T_f - T_{oo})/(T_A)$.

T_A is the average film temperature around the drop.

10.6.3 Wake Flames

Gollahalli et al. [Gollahalli and Brzustowski, 1973, 1974; Gollahalli, 1977] had conducted extensive experiments under convection conditions. Under free convection, for *n*-pentane of 6-mm diameter, near-wake flame is of 5-diameter length, and far-wake flame is 15-diameter length. For Re < 150, envelope flame occurs, whereas for Re > 152, wake flame occurs. The burning rate is very much reduced.

10.7 Transient and Steady-Combustion Results

The d^2 law is based on the QS approximation that the drop size is kept constant by feeding liquid to the drop at the same rate at which it evaporates. Thus, the total mass of vapor between the ambience (or flame) and the drop surface ($m_{F,fw}$) remains constant for the pure-evaporation (or combustion) problem. If there is no feeding mechanism to maintain the drop radius ("a") constant, the drop size along with mass $m_{F,fw}$ keep decreasing with time. The shrinkage leads to an unsteady problem and local T and Y_k at radius r become functions of time. These unsteady effects can be ignored if $\rho_\ell >> \rho$. At combustion near critical conditions as in engines, the gas density approaches the liquid density ρ_ℓ and combustion become transient. Further, the liquid density can decrease as temperature increases; thus, in transient cases, variation of liquid density must be accounted for (see Perry and Green [1997]).

Law et al. [1980] modeled the transient evaporation of droplets. He determined the bubble size b (used in Equation 10.84) by equating the rate of change of fuel vapor accumulated in the shell bounded by the drop surface and the

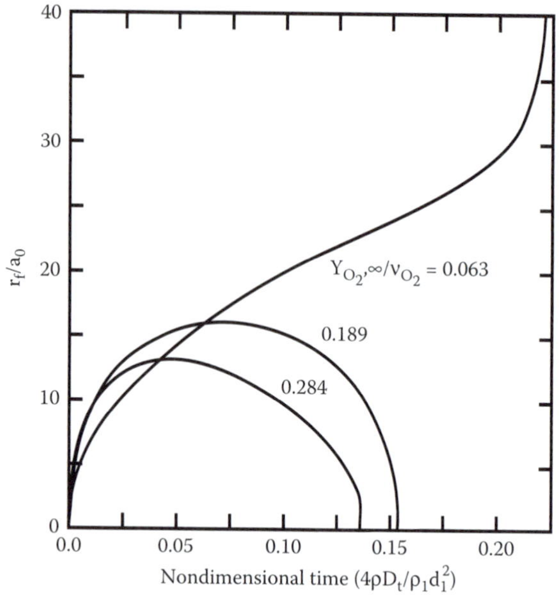

Figure 10.8 Fuel accumulation effects on flame radius, r_f. (Adapted from Law, C.K., Chung, S.H., and Srinivasan, N., Gas phase quasi-steadiness and fuel vapor accumulation effects in droplet burning, *Combust. Flame*, 38, 173–198, 1980.)

bubble radius b (or flame radius in the combustion) to the rate of change of the droplet mass. For pure evaporation, the fuel vapor sphere of radius b propagates (say, a wave with $Y_F = 0.01$) toward the ambience. Equation 10.84 can no longer be used, because b is a function of time. Except for the initial periods, b >> a and, hence, the d² law is essentially followed. The variation of r_f/a_0 with time for the combustion problem with $Y_{O2,\infty}/\nu_{O2}$ as a parameter, is shown in Figure 10.8. In the initial period, the fuel burned at the flame surface (thus less heat is released) is less than the vapor produced and, hence, the flame moves outward because of fuel accumulation; however, as the flame moves outward, the fuel consumption rate increases owing to the increased surface area of oxidation. In the final period, the consumption rate exceeds the evaporation rate, and the flame moves back toward the drop. It is seen that at low oxygen concentrations ($Y_{O2,\infty}/\nu_{O2} = 0.063$), a steady state is not attained, because the flame size is too large to be accommodated by an evaporating drop.

10.8 Multicomponent-Isolated-Drop Evaporation and Combustion

The fuels that are actually used, especially in the field of transportation, are primarily composed of two or more components. For example, gasoline, diesel, and JP 4 are multicomponent fuels [Benaissa et al., 2002; Widman et al.,1997]. When there is an oil spill on water, the process of evaporation and combustion involves multiple components [Walvalker and Kulkarni, 2001]. It is the evaporation of drops that is the precursor of events leading to ignition and combustion and, hence, it is desirable to determine the governing parameters for the multicomponent droplet array evaporation and combustion.

The conservation equations for the evaporation and combustion of multicomponent drops have been presented earlier by Law [1976, 1978], Law and Law [1982] and Annamalai et al., [1993]. It will be assumed that the drop is at uniform temperature (infinite thermal conductivity) and composition (infinite liquid-phase diffusivity), evaporating under QS conditions into a quiescent atmosphere at T_∞. Spherical symmetry and equal binary diffusion coefficients are assumed. The QS solutions will be presented in the following text for the case of Le = 1.0 but with the assumption of uniform chemical composition within the liquid phase at any given time.

10.8.1 Evaporation

10.8.1.1 Governing Equations

The use of conservation equations for each component in a spherical system which includes Stefan flow (SF), yields the species profile for each component, i.

$$\phi = \frac{Y_i - Y_{i,\infty}}{Y_{i,w} - Y_{i,\infty}} = \frac{h_t - h_{t,\infty}}{h_{t,w} - h_{t,\infty}} = \frac{1\text{-}\exp(-\xi)}{1 - \exp(-\xi_w)} \tag{10.100}$$

where $h_{t,\infty} = c_p\,(T_\infty - T_w)$. Equation 10.100 presumes that binary diffusivities of all vapor species are equal.

10.8.1.2 Interface Conservation Equations

Consider binary droplet, where the interface species conservation for each component is given as (Annamalai et al., 1993)

$$4\pi\,\rho_1\,v_{i,w}\,r^2 = \dot{m}_{\text{iso}}\,\varepsilon_i, \quad i = A, B. \tag{10.101}$$

where flux fraction $\varepsilon_i = \dot{m}_i / \dot{m}$.

Equation 10.101 is rewritten as

$$4\pi\rho v r^2 Y_{i,w} - 4\pi\rho D r^2 \left\{\frac{dY_i}{dr}\right\}_w = \dot{m}_{\text{iso}}.\varepsilon_i, \quad i = A, B. \tag{10.102}$$

Normalizing, the interface boundary condition is given as

$$\left[\;Y_i + \left(\frac{dY_i}{d\xi}\right)_w\right] = \varepsilon_i = F_i. \tag{10.103}$$

The interface energy conservation is given as

$$\left[\lambda\left(\frac{dT}{dr}\right)_w 4\,\pi\,a^2\right] = \dot{m}_A\,L_A + \dot{m}_B\,L_B. \tag{10.104}$$

Note: $h_{t,w} = 0$. Rewriting Equation 10.104 as

$$\left[\left(\frac{dh_t}{d\xi}\right)_w\right] = -\{\varepsilon_A \; L_A + \varepsilon_B \; L_B\} = -L_{\text{multi}} = F \qquad (10.105)$$

$$\text{where} \quad L_{\text{multi}} = \{\varepsilon_A \; L_A + \varepsilon_B \; L_B\} \qquad (10.106)$$

and the subscript "multi" denotes the flux-ratio-weighted property for a multi-component system.

10.8.1.3 Solutions

(i) Evaporation Rate

Differentiating Equation 10.100, getting $dh_t/d\xi$, and using the result in Equation 10.112, the evaporation rate in terms of enthalpy is given as

$$\xi_w = \ell n[1+B] \qquad (10.107)$$

$$\text{where} \quad B = \frac{h_{t\infty}}{L_{\text{multi}}} = \frac{c_p(T_\infty - T_w)}{L_{\text{multi}}}. \qquad (10.108)$$

When $B \ll 1$ (slow evaporation),

$$\xi_w \approx B. \qquad (10.109)$$

In slow evaporation, the diffusion velocity is significant compared to convective velocity. Thus, the solution corresponds to the case in which Stefan flow may be neglected in the governing equations. If flux ratios ε_A and ε_B and the drop temperature T_w are known, then one can solve for the burn rate.

(ii) Relation between Surface Fuel Vapor Mass Fraction and Flux Ratios

As before, using the F values in the mass-loss rate relation (Equation 10.103)

$$\xi_W = \ell n\left(1+\frac{\beta_W - \beta_\infty}{F-\beta_W}\right) = \ell n\left(1+\frac{Y_{i,W} - Y_{i,\infty}}{\varepsilon_i - Y_{i,W}}\right)$$

$$\xi_W = \ell n\left(1+\frac{1}{(\varepsilon_i/Y_{i,W})-1}\right) - \ell n\left(1-\frac{Y_{i,W}}{\varepsilon_i}\right), \quad i = A, B, Y_{i,\infty} = 0 \qquad (10.110)$$

if $Y_{iw}/\varepsilon_I \ll 1$, $\xi_W \approx Y_{iw}/\varepsilon_i$.
From Equation 10.110 it is seen that

$$\left\{\frac{\varepsilon_A}{Y_{Aw}}\right\} = \left\{\frac{\varepsilon_B}{Y_{Bw}}\right\} = C. \qquad (10.111)$$

Proceeding similarly for gas phase inert species I,

$$Y_{Iw} + \left\{ \frac{dY_I}{d\xi} \right\}_w = F = 0, \tag{10.112}$$

and with F = 0 in Equation 10.109

$$\xi_w = -\ \ln \left(\frac{Y_{Iw}}{Y_{I\infty}} \right) = -\ \ln\,(Y_{Iw}) \tag{10.113}$$

as $Y_{I,\,\infty} = 1$ for the pure evaporation problem.

Comparing Equation 10.110 with Equation 10.113, we find C as: Y_{Fw}.

$$\frac{1}{1-Y_{Iw}} = \frac{1}{Y_{Fw}} = \left\{ \frac{\varepsilon_A}{Y_{Aw}} \right\} = \left\{ \frac{\varepsilon_B}{Y_{Bw}} \right\} = C \tag{10.114}$$

where $Y_{Fw} = Y_{Aw} + Y_{Bw}$. Because $Y_{Iw} < 1$, $C > 0$ and $C > 1$, the flux ratio is always greater than Y_{Aw} for positive fluxes. This is expected because $\{\varepsilon_k\}_w = Y_{k,w} + \{dY_k/d\xi,\}_w$ and, hence, $\{\varepsilon_k\}_w > Y_{kw}$ as long as $dY_k/d\xi > 0$ (e.g., transport of species A and species B through the film). Expanding Equation 10.122 and generalizing for K components

$$\varepsilon_k = Y_{kw}/\{1 - Y_{Iw}\},\ k = 1 \ldots K, \quad Y_{I\infty} = 1. \tag{10.115}$$

Equating Equation 10.110 with Equation 10.107 and solving for Y_{iw},

$$Y_{iw} = \varepsilon_i \left[1 - \frac{1}{1+B} \right], \quad i = A,\ B \tag{10.116}$$

which is similar to the single-component result (Equation 10.77) except for correction with the flux ratio. Summing up for all fuel components,

$$Y_{Fw} = \left[1 - \frac{1}{1+B} \right] \tag{10.117}$$

which is similar to the result for single-component drop evaporation. The evaluation of B (Equation 10.108) requires that $L_{multi} = L_A\ \varepsilon_A + L_B\ \varepsilon_B = \{\varepsilon_A/Y_{A,w}\}\ L_{avg}$, where $L_{avg} = L = \{L_A\ Y_{A,w} + L_B\ Y_{B,w}\}$, where L is mass-weighted latent heat; one can estimate $Y_{A,w}\ Y_{B,w}$ etc., from phase equilibrium relations.

(iii) Relation between Surface Fuel Mass Fractions and Drop Temperature

The procedure is similar to single component drop except that phase equilibrium must be satisfied for each component. Raoult's law relates the partial pressure of vapor and molar liquid composition for an ideal liquid mixture or ideal solution.

$$p_{iw} = X_{i\ell} p^{sat}{}_i(T_w), \quad i = A, B \tag{10.118}$$

An ideal solution implies that attractive forces between dissimilar molecules (e.g., ethanol:water) are the same as those between similar molecules (e.g., ethanol:ethanol) [Annamalai and Puri, 2002]. Because the solution may not be ideal, the actual pressure p_{iw} may not be the same as that given by Equation 10.118. Thus for real solution one needs a correction factor γ_i; $(p_{iw})_{real} = \gamma_{ii}(p_{iw})$ or one can correct $x_{i\ell}$. $p^{sat}{}_i$ is evaluated using either the Clausius–Clapeyron or the Cox–Antoine relation

$$\ell n\ p^{sat}{}_i = A_i - \frac{C_i}{T + G_i}, \quad i = A,\ B \tag{10.119}$$

where A_i, C_i, and G_i are constants for the given fuel component. The mole and mass fractions at surface (or wall) of drop are given as:

$$X_{iw} = \frac{X_{i\ell} p^{sat}{}_i(T_w)}{P}, \quad i = A, B \tag{10.120}$$

$$Y_{iw} = \frac{X_{iw} M_i}{M} \tag{10.121}$$

Now one can calculate $Y_{F,w}$ and hence ε_i using Equation 10.114.

Finally, the identity given by Equation 10.117 must be satisfied and verify that

$$\sum_i \varepsilon_i = 1.0. \tag{10.122}$$

Procedure for determining drop temperature of a multicomponent drop is as follows:

1. Assume T_w of drop.
2. Obtain Y_{iw} using Raoult's law (Equation 10.118 to Equation 10.121).
3. Evaluate $Y_{Iw} = 1 - Y_{Aw} - Y_{Bw}$ or $Y_{Fw} = Y_{FAw} + Y_{FBw} + \ldots$ then estimate the identity $C = 1/Y_{Fw}$, and evaluate ε_k using $\{\varepsilon_k/Y_{kw}\} = 1/Y_{Fw}$ (Equation 10.114).
4. Determine $L_{multi} = L_A\ \varepsilon_A + L_B\ \varepsilon_B + \ldots = (\varepsilon_A/Y_{A,w})L_{avg}$.
5. Evaluate $B = \{h_{T,\infty}/L_{multi}\}$, where $h_{T,\infty} = c_p(T_\infty - T_w)$.
6. Calculate Y_{Fw} using Equation 10.117.
7. Check the result with step 3. Iterate, if necessary, until convergence is obtained.

Example 9

Consider the evaporation of a multicomponent 100-μm drop containing heptane (component A, $X_{Hep,\ell}$ = 70% by mole) and hexadecane (component B,

$X_{Hexadec,\ell}$ = 30% by mole) in an environment at T_∞ = 1000 K. For heptane, M_A = 100.2, L_A = 316 kJ/kg K, $T_{BP,A}$ = 372 K; for hexadecane $C_{16}H_{34}$, M_B = 226.5, L_B = 232 kJ/kg K, $T_{BP,B}$ = 560 K,

$$c_p = 1.17 \text{ kJ/kg K. } \rho D = 5 \times 10^{-5} \text{ kg/m K} \quad \text{and} \quad P = 1 \text{ bar.}$$

Estimate (a) the temperature of drop under evaporation, (b) flux ratios, and (c) evaporation rate.

Solution

(a) Liquid composition by mass: $Y_{A,\ell}$ = 0.7 × 00.23/{0.7 × 100.23 + 0.3 × 226.5} = 0.508, $Y_{B,\ell}$ = 0.492; using Raoult's law,

$$p_{iw} = X_{i\ell} p^{sat}{}_i (T_w), \quad i = A, B.$$

Using the Clausius–Clapeyron relations, p^{sat} = exp{A − C/T}.

For heptane (See Equation 10.58c), A = ln(1) + (313 × 100.2/8.314) = 10.241, C = 3809.6 K; for hexadecane A = 11.286, C = 632 K.

Assume T_w =315.1 K, p_{Aw} = 0.7 × exp(100.24 − 3809.6/315.1) = 0.110 bar, p_{Bw} = 4.65 × 10^{-5} bar, X_{Aw} = 0.110 bar/1 bar = 0.10, X_{Bw}= 4.62 × 10^{-5}, $X_{air,w}$ = 1 − X_{Aw} − X_{Bw} = 0.889, M_{mix} = 0.110 × 100.23 + 4.62 × 10^{-5} × 226.5 = 36.83, Y_{Aw} = 0.11 × 100.2/36.826 = 0.299, Y_{Bw} = 0.00028, Y_{Fw} = 0.717, $Y_{air,w}$ = 1 − 0.299 = 0.7.

From Equation 10.114,

$$\left\{\frac{\varepsilon_A}{Y_{Aw}}\right\} = \left\{\frac{\varepsilon_B}{Y_{Bw}}\right\} = C = 1/Y_{Fw} = 1/0.3 = 3.333$$

Solving for flux ratios, ε_A = 0.999, ε_B = 0.00095. L_{multi} = $\varepsilon_A L_A + \varepsilon_B L_A$ = 314.96 kJ/kg, B = $h_{T,\infty}/L_{multi}$ = 2.455.

Using Equation 10.117, Y_{Fw} = 1 − (1/3.455) = 0.717, which is different from the value of 0.3 given by phase equilibrium. Iterate. The solution is as follows: T_w = 353.5 K, p_{Aw} = 0.409 bar, p_{Bw} = 0.00041 bar, X_{Aw} = 0.10, X_{Bw} = 4.1 × 10^{-5}, $X_{air,w}$ = 0.59, M_{mix} = 58.2, Y_{Aw} = 0.704, Y_{Bw} = 0.00016, Y_{Fw} = 0.7042, $Y_{air,w}$ = 0.2958, C = 1.42.

ε_A = 0.998, ε_B = 0.00226. L_{multi} = 315.4 kJ/kg. B = $h_{T,\infty}/L_{multi}$ = 2.395, Y_{Fw} = 0.705, which is almost the same as the value of 0.704 given by phase equilibrium.

Using Equation 10.108 and using c_{pgas} for B,

$$\xi_w = \ln\{1 + B\} = \ln\{1 + 2.395\} = 1.222$$

$$\xi_w = \dot{m}/(4\ \pi\ \rho D\ a) = \dot{m}/(4\ \pi\ 5 \times 10^{-5} \times 50 \times 10^{-6})$$

$$\dot{m} = 4\ p\ 5 \times 10^{-5} \times 50 \times 10^{-6} \times 1.222 = 3.84 \times 10^{-7} \text{ kg/sec.}$$

Then, flux rate of heptane = 0.996 × 3.84 × 10^{-7} kg/sec = 3.83 × 10^{-7} kg/sec.

If we use c_p of fuel for B, then T_w, and hence the evaporation rate, could be slightly different.

Remarks

(i) The hexadecane is almost nonvolatile; thus B, and hence $\dot{m}$, is controlled by heptane evaporation. In the initial period, it is heptane vapor that will mix with air, ignite, and burn.
(ii) As time progresses, the heptane concentration will decrease, hexadecane concentration will increase, evaporation will be dominated by hexadecane, and owing to phase equilibrium, T_w will increase gradually along with change in L_{multi}. ε_A will decrease, and ε_B will increase.
(iii) Diameter law: $\dot{m} = d/dt\{\pi\rho_{,\ell} d^3/6) = d\rho_\ell/dt\{\pi d^3/6) + \rho_\ell\{dd/dt\}\{\pi d^2/2)$; because density along with latent heat keeps varying owing to changing composition, d^2 law may not be valid [Wood et al., 1960].

(iv) Isolated Drop of a Volatile and a Nonvolatile Component

Simplified results can be obtained for an isolated drop consisting of volatile (component A) and nonvolatile (component B) components. Because ε_B is zero, $Y_{Bw} = 0$, $Y_{Fw} = Y_{Aw}$, and $\varepsilon_A = 1$; then Equation 10.106 and Equation 10.108 become

$$B = \{h_{T,\infty}/L_{multi}\} = \{h_{T,\infty}/L_A\}. \tag{10.123}$$

From Equation 10.117,

$$Y_{Aw} = \left[1 - \frac{1}{1+B}\right] \tag{10.124}$$

When B << 1 (Stefan flow is neglected), $Y_{Aw} \approx B$.

10.8.2 Combustion of Multicomponent Drop

10.8.2.1 Nonvolatile (B) and Volatile (A) Components

In this case, the results are same as for isolated-drop combustion except for the phase equilibrium relation, which requires Raoult's law.

$$p_{Aw} = X_{A,\ell}\, p_A{}^{,sat}\,(T_w)$$

Thus, partial pressure at the surface is less than the saturation pressure of the pure component and, hence, the drop temperature will be less compared to the pure-component drop.

10.8.2.2 Combustible Volatile Components

10.8.2.2.1 Formulation and Interface Conservation Equations

From Chapter 8 we learned that the SZ formulation can be applied to two single-step reaction schemes in the gas phase for a binary-component drop. The vapors from fuel A and fuel B oxidize in the gas phase. Recall that $h_{c,O2}$, the heating values per unit mass of stoichiometric oxygen ($= h_c/\nu_{O2}$) remains almost constant at about 13800 kJ/kg for pure HCs. Thus, we will also assume that $h_{c,O2,A} = h_{c,A}/\nu_{O2,A} = h_{c,B}/\nu_{O2,B} = h_{c,O2}$ is almost constant for all components. Consider A as heptane and B as hexadecane.

$$C_7H_{16} + 11O_2 \rightarrow 7CO_2 + 8H_2O \tag{I}$$

$$C_{16}H_{34} + 24.5O_2 \rightarrow 16CO_2 + 17H_2O \tag{II}$$

$$\frac{\dot{w}'''_{F,A}}{(-1)} = \frac{\dot{w}'''_{O_2}}{-\nu_{O_2,A}} = \frac{\dot{w}'''_{CO_2}}{+\nu_{CO_2,A}} = \frac{\dot{w}'''_{H_2O}}{+\nu_{H_2O,A}} = \dot{w}'''_A \tag{10.125}$$

$$\frac{\dot{w}'''_{F,B}}{(-1)} = \frac{\dot{w}'''_{O_2}}{-\nu_{O_2,B}} = \frac{\dot{w}'''_{CO_2}}{+\nu_{CO_2,B}} = \frac{\dot{w}'''_{H_2O}}{+\nu_{H_2O,B}} = \dot{w}'''_B \tag{10.126}$$

Generalizing the two-component problem in Chapter 8, the SZ is defined as

$$\beta_{O_2-FA-FB-...} = \sum_k Y_{Fk}\, \nu_{O_2k} - Y_{O_2}, \quad k = FA, FB, ... \tag{10.127}$$

$$\beta_{ht-FA-FB-FC...} = \frac{h_t}{h_{c,O_2}} + \sum_k Y_{Fk}\, \nu_{O_2k} \tag{10.128}$$

$$\beta_{ht-O_2} = \frac{h_t}{h_{c,O_2}} + Y_{O_2}. \tag{10.129}$$

Note that $\beta_{O2-FA-FB} > 0$ closer to drop surface where $Y_{O2} = 0$ (fuel rich), and $\beta_{O2-FA-FB} < 0$ far from the drop surface where Y_{FA}, $Y_{FB} = 0$ (fuel lean). Thus $\beta_{O2-FA-FB} = 0$ represents the stoichiometric surface dividing fuel-rich and fuel-lean regimes.

Profiles:

$$\phi = \frac{\beta - \beta_\infty}{\beta_w - \beta_\infty} = \frac{1 - \exp\ (-\xi)}{1 - \exp\ (-\xi_w)} \tag{10.130}$$

where β could be defined by any of Equation 10.127 to Equation 10.129. The interface species and energy conservation equations remain the same as Equation 10.103 and Equation 10.105. The O_2 conservation equation is written

$$Y_{O_2w} + \left\{\frac{dY_{O_2}}{d\xi}\right\}_w = 0. \tag{10.131}$$

Divide Equation 10.105 by h_{cO2} and add to Equation 10.131. Then

$$\beta_w + \left\{\frac{d\beta}{d\xi}\right\}_w = F \quad \text{where} \quad F = -\frac{L_{multi}}{h_{c,O_2}}, \quad \beta = \beta_{ht\text{-}O_2}. \tag{10.132}$$

Similarly, with $\beta = \beta_{O_2\text{-FA-FB}}$, multiply Equation 10.103 for i = A by $\nu_{O_2,A}$ and for i = B by $\nu_{O_2,B}$ and subtract the O_2 interface conservation equation. Then

$$F = \varepsilon_{FA}\ \nu_{O_2A} + \varepsilon_{FB}\ \nu_{O_2B} = \nu_{O_2 multi},\ \beta = \beta_{O2\text{-FA-FB}}. \tag{10.133}$$

10.8.2.2.2 Solutions

(i) Burn Rate

Generally, for any SZ variable,

$$\xi_w = \ell n\left[1 + \left(\frac{\beta_w - \beta_\infty}{F - \beta_w}\right)\right] \tag{10.134}$$

With $\beta = \beta_{ht\text{-}O2}$, $F = -L_{multi}/h_{c,O_2}$

$$\xi_w = \ell n(1+B) \tag{10.135}$$

where

$$B = \frac{Y_{O_2,\infty} h_{c,O_2}}{L_{multi}} + \frac{h_{t\infty}}{L_{multi}} = \frac{Y_{O_2,\infty} h_{c,O_2}}{L_{multi}} + \frac{c_p(T_\infty - T_w)}{L_{multi}} \tag{10.136}$$

where L_{multi} and T_w are still unknowns. With the O_2-FA-FB combination

$$\begin{aligned}\xi_w &= n\left[1 + \left(\frac{(\beta_{h_t-O_2,w} - \beta_{h_t-O_2,\infty})}{([-L_{multi}/h_{c,O_2}] - \beta_{h_t-O_2,w})}\right)\right] \\ &= n\left[1 + \left(\frac{(\beta_{O_2-FA-FB,w} - \beta_{O_2-FA-FB,\infty})}{(\nu_{O_2,multi} - \beta_{O_2-FA-FB,w})}\right)\right].\end{aligned} \tag{10.137}$$

Equating the second and third terms in Equation 10.137, we find that the fuel mass fraction is governed by an additional relation:

$$B = \left(\frac{Y_{FA,w}\nu_{O_2A} + Y_{FB,w}\nu_{O_2B} - Y_{O_2,\infty}}{\nu_{O_2,multi} - \{Y_{FA,w}\nu_{O_2A} + Y_{FB,w}\nu_{O_2B}\}}\right) = \left(\frac{\nu_{O2Avg} - Y_{O_2,\infty}}{\nu_{O2,multi} - \nu_{O2Avg}}\right), \tag{10.138}$$

where

$$\nu_{O2Avg} = \{Y_{FA,w}\ \nu_{O_2A} + Y_{FB,w}\ \nu_{O_2B}\}. \tag{10.139}$$

If one assumes T_w, then Y_{Aw} and Y_{Bw} and hence $Y_{F,w}$ are known and ν_{O2avg} can then be obtained; however L_{multi} involves ε_A and ε_B are still unknown. It will be shown later that $\nu_{O_2,\ multi} = \nu_{O_2,\ avg}/Y_{F,W}$

(ii) Flame Location

We obtain an explicit relation for flame location by using $\beta_{O2\text{-}FA\text{-}FB} = Y_{FA}\nu_{O2,A} + Y_{FB}\nu_{O2,B} - Y_{O2}/\nu_{O2}$ and setting $\beta_{O2\text{-}FA\text{-}FB,f} = 0$ because at flame, $Y_{FA} = Y_{FB} = Y_{O2} = 0$. For flame temperature, we get the relation by using $\beta_{ht\text{-}O2}$ and evaluating at $r = r_f$. As in combustion of single component drop,

$$\Phi_f = \frac{-\beta_{O_2-FA-FB,\infty}}{\beta_{O_2-FA-FB,w} - \beta_{O_2-FA-FB,\infty}} = \frac{1-\exp\left(-\xi_w\left[\frac{a}{r_f}\right]\right)}{1-\exp\left(-\xi_w\right)}. \tag{10.140}$$

Because $\xi_w = \ln(1 + B)$, one can solve for ξ_f as

$$\xi_f = -\ln\left[1-\Phi_f\frac{B}{1+B}\right] \tag{10.141}$$

where

$$\phi_f = \left(\frac{Y_{O_2,\infty}}{\nu_{O_2Avg} + Y_{O_2,\infty}}\right). \tag{10.142}$$

The thin-flame approximation requires an explanation. It is possible that component A might react very fast, whereas component B may burn slowly. Thus, $Y_{F,A}$ will reach zero at $\xi_{f,A}$ (say, closer to the drop surface, at a lower temperature, or where Y_{O2}/Y_{FA} rather than Y_{O2}/Y_F is near stoichiometric) whereas $Y_{F,B}$ will reach zero at $\xi_{f,B}$, farther from the drop surface. But typically, once A starts burning, T raises and B will also be consumed rapidly unless A is of very low concentration.

(iii) Profiles for A and B between Flame and Drop Surface and $\varepsilon_k/Y_{k,w}$ Relation

If the gas-phase reactions are infinitely fast, then the gas phase is almost frozen except at the thin flame. We assume the flame location for $\xi_{f,A}$, where $Y_A = 0$, is the same as the flame location for $\xi_{f,B}$, where $Y_B = 0$. Then,

$$\left(\frac{Y_A - 0}{Y_{A,w} - 0}\right) = \left(\frac{Y_B - 0}{Y_{B,w} - 0}\right) = \left(\frac{\exp(-\xi_f) - \exp(-\xi)}{\exp(-\xi_f) - \exp(-\xi_w)}\right). \tag{10.143}$$

Using Equation 10.103,

$$Y_{Aw}\left[\frac{\exp(-\xi_f)}{\exp(-\xi_f)-\exp(-\xi_w)}\right]=\varepsilon_A \tag{10.144a}$$

$$Y_{Bw}\left[\frac{\exp(-\xi_f)}{\exp(-\xi_f)-\exp(-\xi_w)}\right]_w=\varepsilon_B \tag{10.144b}$$

then

$$\varepsilon_B/\varepsilon_A = Y_{B,w}/Y_{Aw}, \text{ for a binary-component drop,} \tag{10.145}$$

which is the same relation as that obtained for evaporation. More generally, $\varepsilon_k/\varepsilon_A = Y_{k,w}/Y_{Aw}$,

$$\varepsilon_A + \varepsilon_B + \varepsilon_C + \varepsilon_D + \ldots = \varepsilon_A\,[1 + Y_{B,w}/Y_{Aw} + Y_{B,w}/Y_{Aw}] = \{\varepsilon_A/Y_{AW}\}[1 - Y_{I,w}] = 1.$$

Thus,

$$\{\varepsilon_A/Y_{AW}\} = 1/[1 - Y_{I,w}] = 1/Y_{Fw} = C, \text{ for species A, B, C, …, K.} \tag{10.146}$$

Hence,

$$\nu_{O_2,multi} = \sum \varepsilon_{F,k}\nu_{O2k} = \sum Y_{FK,w}\,\nu_{O_2,k}/Y_{F,w} = \nu_{O_2,avg}/Y_{F,w}. \tag{10.147}$$

(iv) Fuel Mass Fraction at Drop Surface

Using Equation 10.147 in Equation 10.138 to eliminate $\nu_{O_2,avg}$ and solving for $Y_{F,w}$

$$Y_{Fw}=\left(1-\frac{1+s}{1+B}\right),\qquad s=\frac{Y_{O_2,\infty}}{\nu_{O_2,multi}} \tag{10.148}$$

where $\nu_{O_2,multi}$ is known from Equation 10.147.

10.8.2.3 Combustible and Noncombustible Components

For example consider ethanol-mwethanol-water slurry. Here the combustible components are ethanol (A), methanol (B) while noncombustible but still an evaporating component is H_2O (a). Then generalizing

$$B=\frac{Y_{O_2,\infty}h_{c,O_2}}{L_{multi,mod}}+\frac{C_p(T_\infty-T_w)}{L_{multi,mod}} \tag{10.149}$$

$$L_{multi,mod}=L_A\varepsilon_A+L_B\varepsilon_B+\ldots+L_a\varepsilon_a+L_b\varepsilon_b\ldots \tag{10.150}$$

where A, B, C … are combusting components; a,b,c … are evaporating components; and other definitions remain unchanged.

$$\nu_{O2,multi} = \Sigma_k \varepsilon_{F,k} \nu_{O2k}, \quad k = A,\ B\ \ldots\ etc$$

$$\nu_{O2,avg} = \Sigma_k Y_{k,w} \nu_{O2k}, \quad k = A,\ B\ \ldots\ etc$$

Example 11

Consider a multicomponent 100-μm drop containing heptane (component A, $X_{Hep,\ell}$ = 60% by mole) and hexadecane (component B, $X_{hexadec,\ell}$ = 40% by mole) combusting in an environment at $T_\infty = 1000$ K and $Y_{O2,\infty} = 0.23$. For heptane, M = 100.2, L = 316 kJ/kg K, h_c = 48156 kJ/kg, T_{BP} = 372 K and $\upsilon_{O2,A} = 3.50$; for hexadecane $C_{16}H_{34}$, M = 226.5, L = 232 kJ/kg K, h_c = 43990 kJ/kg, T_{BP} = 560 K, $\upsilon_{O2,B} = 3.46$, c_p = 1.17 kJ/kg K, $\rho D = 5 \times 10^{-5}$ kg/m K., and P = 1 bar.
(a) Estimate the temperature of the drop under combustion, flux ratios, and burn rate. (b) Repeat for $X_{Aliq} = 0.05$. See Example 10 for saturation relations.

Solution

(a) Species A = C_7H_{16}, Species B = $C_{16}H_{34}$, $X_{A,liq} = 0.6$, $h_{c,O2,A} = h_{CA}/\nu_{O2,A}$ = 48456/3.51 = 13805 kJ/kg of O_2, $h_{c,O2,B}$ = 43990/3.46 = 12714 kJ/kg of O_2, $h_{c,O2,avg}$ = (13805 + 12714)/2 = 13260 kJ/kg of O_2. From iterative solution T_w = 381.2 K, p_{AW} = 0.77 bar, X_{Aw} = 0.77, p_{Bw} = 0.002, X_{Bw} = 0.002, ε_{FA} = 0.995, M_{mix} = 84.2, Y_{FAw} = 0.92, Y_{FBw} = 0.005, $Y_{FW} = Y_{FAW} + Y_{FBw}$ = 0.92, $\nu_{O2\text{-}Multi} = \Sigma\varepsilon_{F,k}\nu_{O2,k} = \varepsilon_{FA}\nu_{O2,A} + \varepsilon_{FB,}\nu_{O2,B}$ = 3.51,

$$B = 12.34,\ \xi_w = \ln(1 + B) = \ln(1 + 12.34) = 2.59,\ \dot{m}_{t=0} = 8.14 \times 10^{-7}\ \text{kg/sec.}$$

(b) $X_{A,liq} = 0.05$, T_w = 486.7 K, ε_{FA} = 0.588, ε_B = 0.412, M_{mix} = 103, Y_{FAw} = 0.543, Y_{FBw} = 0.381, $Y_{FW} = Y_{FAW} + Y_{FBw} = Y_{Fw}$ = 0.924,

$$B = 13.04,\ \xi_w = \ln(1 + B) = \ln(1 + 13.04) = 2.64,\ \dot{m}_{t=0} = 8.30 \times 10^{-7}\ \text{kg/sec}$$

$$Y_{Iw} = 1 - Y_{FAw} - Y_{FBw} = 1 - 0.92 - 0.005 = 0.075.$$

Remarks

(i) Equal binary diffusion coefficients may not be justifiable because Ms′ of components differ by a large amount.
(ii) For case (b), even though the liquid mole fraction of heptane is very low, the contribution of heptane to total mass-loss rate (59%) is still considerable.
(iii) As the drop for case (a) evaporates X_A decreases, to say, X_{Aliq} = 5%, the saturation temperature of the drop is 487 K compared to initial values of 381 K (when X_{Aliq} = 60%). However, the flux ratio of A at X_{Aliq} = 5% is still about 60% and, hence, L_{multi} is still dominated by component A; therefore the B number does not change significantly. Thus, the d^2 law may still be followed for most of the period of combustion. The experimental

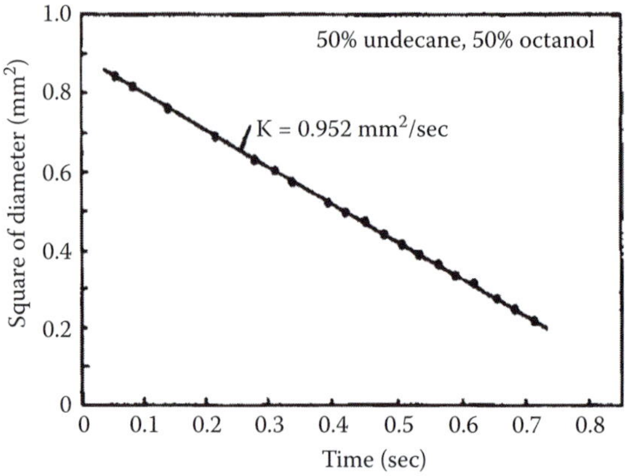

Figure 10.9 Existence of d² law for undecane:octanol drop during combustion [Law et al., 1982].

data of Law et al. [1982] seems to indicate the d^2 law for 50:50 undecane (T_{BP} = 469 K, L = 268.0 kJ/kg) – octanol (T_{BP} = 467 K, L = 545 kJ/kg) drop combustion (Figure 10.9). The saturation temperatures are similar but latent heats differ.

10.9 Summary

The droplet burn rates are given in terms of the transfer number B; given the environmental conditions (oxygen mass fraction and ambient temperature), fuel properties (latent heat, heating values, and chemical formulae), transport properties (thermal conductivity, diffusivity, or both), and the size of the drop, the burn rate of drop and burn time can be determined. The results are almost independent of finite kinetics of oxidation. The burn rate is found to be proportional to instantaneous particle diameter, burn rate per unit surface area is proportional to 1/d, and burn time is proportional to d_0^2. While the temperature of the flame located r_f away from the drop surface is equal to the adiabatic flame temperature of a stoichiometric mixture of liquid fuel and air, the drop remains at about the saturation temperature. The SZ method is modified to include multiple components, and results are presented for the burn rate and fixing components. Isolated-drop results can be simply applied provided the latent heat and stoichiometric coefficient are replaced by flux-ratio-weighted latent heat and individual species stoichiometric coefficient in estimating the B number. The formulae at the end of text summarizes the results of studies on drop combustion.

Chapter 11

Combustion in Boundary Layers

11.1 Introduction

In Chapter 9 and Chapter 10 we dealt with the combustion of carbon (solid) particles and liquid droplets, particularly in stagnant atmospheres. The droplets evaporate and form a flame enveloping the fuel source (i.e., droplets), and hence all the fuel is burned owing to this "enveloped flame." Thus the flame area (= 4 πr_f^2) times fuel burn rate per unit surface area of flame ($\dot{m}_f''$) equals drop area (=$4\pi a^2$) times fuel production rate per unit drop area ($\dot{m}''$). In this chapter, we will deal with chemically reacting systems involving combustion of carbon, plastics, and liquids in a flow situation, particularly under forced, free, and stagnation flow conditions. The applications include catalytic reactors, ablating reentry vehicles, burning of walls and trees, combustion of a pool of liquid, etc. The reactions can include both homogeneous (gas phase) and heterogeneous reactions (e.g., carbon oxidation and reduction). The main types of flows are represented in Figure 11.1 to Figure 11.4. Although numerical codes are currently available for laminar or turbulent configurations of any given shape, these problems serve as benchmark examples for checking several numerical programs under laminar conditions. Further, microcombustion systems involve flows in the laminar range. Important dimensionless groups can be obtained using a simplified approach. Unlike droplets' flame, the flame in this case does not envelope the source of fuel vapor and, as such, all the fuel is not burned within the same length of the fuel production regime. There are several methods of solving the governing equations for boundary-layer (BL) combustion: (1) exact, (2) integral, (3) nonintegral, and (4) numerical. Rarely can the exact solution be explicitly obtained.

Several problems involving convection will be illustrated: evaporation and combustion of liquid and solid polymer fuels from a horizontal plate under forced convection, vertically oriented plates (free convection), and stagnation flows, etc., including vapor deposition on the condensate phase.

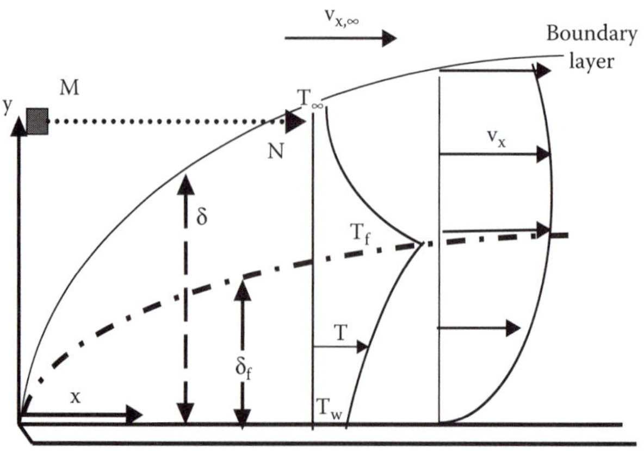

Figure 11.1 Schematic illustration of velocity boundary layer under forced convection.

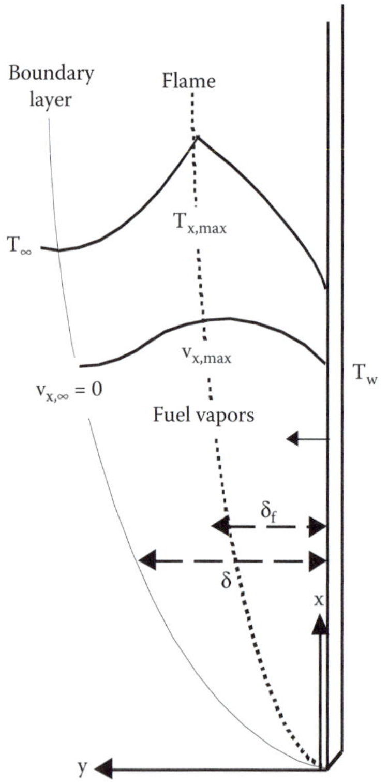

Figure 11.2 Free-convective burning.

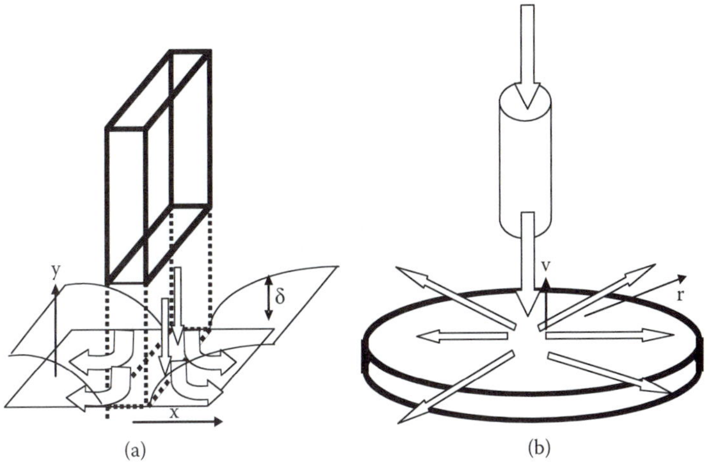

Figure 11.3 Stagnation flow: (a) 2-D, (b) axisymmetric.

11.2 Phenomenological Analyses

In Chapter 8 we briefly discussed BL equations. Here, we present simple relations for BL thicknesses and the physics behind transport phenomena within BLs.

11.2.1 Momentum

As the flow is past a horizontal plate, the velocity at the surface is zero whereas it tends to $v_{x,\infty}$ as $y \to \infty$ (v_x in Figure 11.1). For example, if $v_{x,\infty} = 10$ m/sec, the velocity v_x at a given x (say, 10 cm) is given by $v_x = 1, 2, 2.5, 3, \ldots, 9.9$ m/sec at $y = 1, 1.5, \ldots, 4$ mm, and at $y \to \infty$, $v_x = 10$ m/sec. If we define the BL thickness (δ) as the distance at which the velocity is 99% of $v_{x,\infty}$, then for the example, $\delta = 4$ mm at x = 10 cm, and its value increases with an increase in x. Just as heat diffuses from a hotter part to a colder portion via

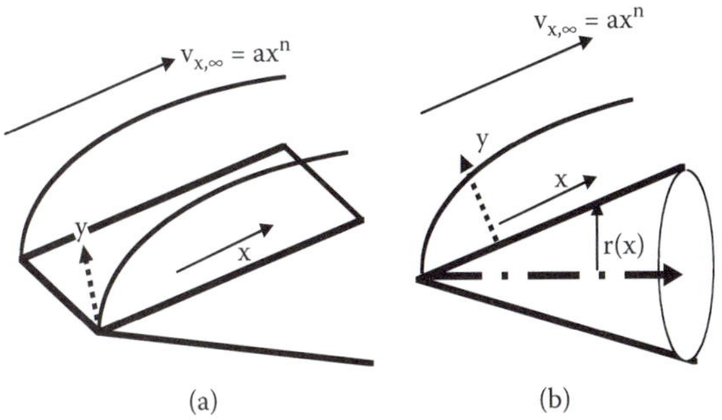

Figure 11.4 Forced convection over (a) wedge, (b) cone: axisymmetric.

thermal diffusion, species diffuse from higher concentrations to lower concentrations via mass diffusion, momentum/mass, i.e., $v_x(= mv_x/m)$, also diffuses from higher-momentum zones to lower-momentum zones via momentum diffusion. Thus, as fluid travels downstream, it keeps losing momentum to the plate owing to frictional effects, with lower and lower velocity at fixed y. Hence, as x increases, δ must increase and thus (velocity) BL thickness is a function of x. As the fluid mass travels from M to N downstream (Figure 11.1), it does not undergo momentum loss until it reaches point N; i.e., momentum has diffused from $\delta(x) = 0$ to $y = 0$ within a timescale of (x/v_x), and because momentum diffusivity (m^2/sec)* time (sec) = length^2, then $\nu^* (x/v_x) = \delta^2$, where ν is the dynamic viscosity; thus, BL thickness can be expressed as $\{\delta/x\} \approx 1/(Re_x)^{1/2}$ where $Re_x = v_x\, x/\nu$.

This momentum transfer rate to the surface is governed by Newton's law:

$$\tau = \mu \frac{dv_x}{dy} = \rho\nu \frac{dv_x}{dy}, \quad \textit{momentum transfer rate, } N/m^2. \tag{11.1}$$

Letting $\delta = \delta_v$ indicate velocity boundary layer, the momentum transfer rate is expressed as

$$\tau = \rho\nu \frac{dv_x}{dy} \approx \rho\nu \frac{v_{x\infty}}{\delta_V} = h_v\, v_{x\infty},\ h_v = \frac{\rho\nu}{\delta_V},\ \text{kg/m}^2\text{s}$$

where h_v may be called the *momentum transfer coefficient* just like the heat transfer coefficient. The momentum transfer is a consequence of drag on the plate. In analogy with Nu and Sh, h_v could have been presented in terms of the dimensionless momentum transfer number $Mm = h_v\, x/\mu$;

$$Mm = h_V\, x/\mu \approx x/\delta_V = (Re_x)^{1/2}. \tag{11.2}$$

However, the drag coefficient $C_f = \tau/[1/2\ \rho v_{x\infty}{}^2]$ is typically used to represent the stress directly. Thus,

$$C_f = \text{stress at the plate}/[\rho v_{x\infty}{}^2/2].$$

$$C_f/2 \approx [\mu\, v_{x\infty}/\delta_v]/[\rho v_{x\infty}{}^2] = 1/[Re_x\{\delta_v/x\}] = 1/(Re_x)^{1/2}. \tag{11.3}$$

Thus, $Mm = 2/C_f$. Figures 11.1 to 11.4 show the growth of the BL in several flow systems.

11.2.2 Energy

Similar analysis can be conducted for hot gas flowing over a cold plate in which the thermal energy (or "heat") diffuses and thermal BL, $\delta_T(x)$ increases with an increase in x. The heat transfer is governed by the Fourier law:

$$\dot{q}'' = \lambda \frac{dT}{dy} = \rho \alpha_T\, c_p \frac{dT}{dy}, \quad \textit{Heat transfer rate, } kW/m^2 . \tag{11.4}$$

Defining

$$\dot{q}'' = h_H \{T_w - T_\infty\} = \lambda \frac{dT}{dy} \approx \frac{\lambda\{T_w - T_\infty\}}{\delta_T} \tag{11.5}$$

where the heat transfer coefficient h_H is given as

$$h_H \approx \frac{\lambda}{\delta_T}, \ \frac{kW}{m^2 K} .$$

Using similar analysis as for momentum, thermal diffusivity (m²/sec) × time (sec) = length², then $\alpha_T \times (x/v_x) = \delta_T^2$; thus, $\{\delta_T/x\} \approx 1/\{Pr^{1/2}(Re_x)^{1/2}\}$, where $Pr = \nu/\alpha_T$. The δ_T, and hence h_H, is a strong function of velocity or the Reynolds number. Typically, h_H is presented in terms of a dimensionless Nusselt number, $Nu_x = h_H\, x/\lambda$.

$$h_H\, x/\lambda \approx Nu_x = x/\delta_T \approx Pr^{1/2}\, (Re_x)^{1/2} \tag{11.6}$$

There are several empirical relations for Nu as shown in Table F.8.12 under the Formulae section.

11.2.3 Mass

When air flows over a carbon plate, the oxygen diffuses to burn the carbon. Oxygen flows from a higher concentration (free stream) to a lower concentration (near the carbon plate) and, hence, the species BL, $\delta_k(x)$ also increases. The species diffusion is governed by Fick's law:

$$j_k'' = \rho D \frac{dY_k}{dy}, \quad \textit{species transfer rate, } kg/m^2 s. \tag{11.7}$$

Defining

$$\dot{m}_k'' = h_m \{Y_{kw} - Y_{k,\infty}\} = \rho D \frac{dY_k}{dy} \approx \frac{\rho D\{Y_{kw} - Y_{k,\infty}\}}{\delta_m} \tag{11.8}$$

where the mass transfer coefficient h_m is given as

$$h_m \approx \frac{\rho D}{\delta_m}, \ \frac{kg}{m^2 s}.$$

Similarly proceeding diffusivity (m^2/sec) × time (sec) = length^2, then D* $(x/v_x) = \delta_m^2$; thus $\{\delta_m/x\} \approx 1/[Sc^{1/2}(Re_x)^{1/2}]$, where $Sc = \nu/D$.
As before

$$h_m\ x/(\rho D) \approx Sh_x = x/\delta_m \approx \mathbf{Sc^{1/2}\ (Re_x)^{1/2}}. \tag{11.9}$$

11.2.4 Growths of BLs and Dimensionless Numbers

Recall that ν, α_T, and D are momentum, thermal and mass diffusivities with units of m^2/sec. Further

$$Pr = \nu/\alpha_T, \quad Sc = \nu/\mathbf{D}, \quad Le = Pr/Sc = \mathbf{D}/\alpha_T. \tag{11.10}$$

If Pr >> 1, $\nu >> \alpha_T$. Then, momentum diffuses more rapidly than heat. As one follows a fixed mass at fixed y as it flows along x, the momentum loss is higher than the energy transfer by diffusion, with the result that the velocity at a given y will decrease very rapidly with an increase in x. Hence, momentum loss would be felt at greater δ_v and the δ_v grows rapidly, i.e., at a given x, $\delta_v >> \delta_T$ if Pr >> 1. Similarly, $\delta_v >> \delta_m$ when Sc >> 1. When $Le = D/\alpha_T >> 1$, the mass diffuses very rapidly and hence $\delta_m >> \delta_T$. When Le = 1 (Sc and Pr are equal but it is not necessary that Pr = 1, Sc = 1), $\delta_m = \delta_T$. On the other hand, Pr = 1 implies that $\delta_T = \delta_V$.

11.2.5 Combustion

When flow occurs over a flat plate, a BL develops as shown in Figure 11.1. If the plate is made of naphthalene, it sublimates and burns in air. The vapor is transported away from plate via diffusion, mixes with O_2 diffusing from the free stream, and combustion occurs at a location δ_f within the BL. As outlined in Chapter 9 and Chapter 10 for carbon and drop combustion, a thin flame is formed at $y = \delta_f$ within the BL as shown in Figure 11.1 Other examples are combustion over a vertical wall (Figure 11.2) and stagnation point combustion (Figure 11.3) if an air jet is directed over a naphthalene plate, and combustion in opposed jets. Note that temperature gradients are steep within the thin flame and, as such, there may not be enough time for species to achieve instantaneous local chemical equilibrium and concentrations could be at super-equilibrium levels.

11.3 Generalized Conservation Equations and Boundary Conditions

Generalized conservation equations for 2-D jet, forced convection over flat and curved surfaces, free convection, and axisymmetric stagnation flow are presented. (Refer to Chapter 8 for additional details.)

11.3.1 Conservation Equations in Compressible Form

11.3.1.1 Mass

$$\frac{\partial}{\partial x}\left(\rho\, v_x\, x^k\right)+\frac{\partial}{\partial y}\left(\rho\, v_y\, x^k\right)=0 \tag{11.11}$$

where k = 0 for 2-D flows, e.g., forced flow over a flat plate (Figure 11.1), curved surfaces, free convection over a vertical plate (Figure 11.2), 2-D stagnation flow (Figure 11.3a), etc.

For axisymmetric flows (Figure 11.3b), flow over an axisymmetric body of revolution (Figure 11.4b), and a circular jet impinging on a plate (Figure 11.3a), etc., k = 1.

11.3.1.2 Momentum Conservation

For the x direction,

$$\rho\, v_x \frac{\partial v_x}{\partial x} + \rho\, v_y \frac{\partial v_x}{\partial y} = \frac{\partial}{\partial y}\left(\rho\, \nu \frac{\partial v_x}{\partial y}\right) + S_v(x, y), \quad \frac{N}{m^3}, \frac{kg}{m^2 s^2} \tag{11.12}$$

where $S_v(x, y)$ is the momentum source per unit volume for the x-momentum equation. Table 11.1 lists the source terms for various flow systems.

Table 11.1 Momentum Sources $S_v(x,y) = S_v(x)$ for Various Flows
$x'^* = x^{*(2k+1)}/(2k+1)$

Problem	Free-Stream Velocity	$S_\upsilon(x)$	$S_v^* = S_\upsilon(x,y) d_{ref}/\{\rho \upsilon_{ref}^2 x^{*2k}\}$
Forced convection, flat plate	$v_{x\infty}$	0	0
Forced convection, curved surface	$\upsilon_{x\infty} = a\, x^n$	$\rho_\infty\, v_{x\infty}\, (d\, v_{x\infty}/dx)$	$n\{\rho_\infty/\rho\} x^{*\{2n-1\}}$, $k = 0$, $v_{ref} = a\, d_{ref}{}^n$, $x'^* = x^*$
Free convection, vertical plate	0	$g(\rho_\infty - \rho)$	$g(\rho_\infty/\rho - 1)\{d_{ref}^3/\nu^2\}$, $k = 0$, $v_{ref} = \nu/d_{ref}$, $x'^* = x^*$; $(\rho_\infty/\rho - 1)$, $v_{ref} = (g d_{ref})^{1/2}$
Stagnation flow, 2-D	$\upsilon_{x\infty} = ax$	$\rho_\infty\, v_{x\infty}\, d\, v_{x\infty}/dx$	$\{\rho_\infty/\rho\} x^*$, $k = 0$, $v_{ref} = a\, d_{ref}$; $\{a\, d_{ref}/v_{ref}\}^2 (\rho_\infty/\rho)\, x'^*$, $\upsilon_{ref} \neq a\, d_{ref}$
Stagnation flow, axisymmetric	$v_{x\infty} = ax$	$\rho_\infty\, v_{x\infty}\, d\, v_{x\infty}/dx$	$= (1/3)^{(2/3)}\{\rho_\infty/\rho\}/x^*$, $k = 1$, $v_{ref} = a\, d_{ref}$, $x'^* = (x^{*3}/3)$ $(1/3)^{1/3}\{a\, d_{ref}/v_{ref}\}^2 (\rho_\infty/\rho)]$ $\{1/x'^{*1/3}\}$, $\upsilon_{ref} \neq a\, d_{ref}$

Note that (i) $v_{x\infty}$ is the velocity at edge of the BL for stagnation flows.

(ii) $v_{x\infty} = a\,x$, $a = v_j/d_i$, for a circular jet of diameter d_i impinging on a plate (11.13)

$a = 4\,v_j/d_{cyl}$ for stagnation flow over a cylinder of diameter d_{cyl} (cross flow) and of jet velocity v_j (11.14a)

$a = 3\,v_j/d_s$ for stagnation flow over a sphere of diameter d_s (11.14b)

$a = v_j/(2\,\ell)$, ℓ distance between jet exit and plate [Konishita, 1981] (11.14c)

The term $dv_{x\infty}/dx$ is also called the stretch rate (1/sec), which is a measure of how the stream is accelerated owing to reduction in density (increased velocity or stretching of molecules to a farther distance) or because of squeezing of flow over a narrower region (i.e., nozzle effect).

11.3.1.3 Species Conservation

For any species "k"

$$\rho\, v_x \frac{\partial Y_k}{\partial x} + \rho\, v_y \frac{\partial Y_k}{\partial y} = \frac{\partial}{\partial y}\left(\rho\, D \frac{\partial Y_k}{\partial y}\right) + S_k(x, y),\ \frac{kg}{m^3\ s} \tag{11.15}$$

where the species source

$S_k(x,y) = \dot{w}_k''(x,y)$, rate of production of species k in kg/m^3 sec. (11.16)

For an inert species, I

$$\rho\, v_x \frac{\partial Y_I}{\partial x} + \rho\, v_Y \frac{\partial Y_I}{\partial y} = \frac{\partial}{\partial y}\left(\rho\, D \frac{\partial Y_I}{\partial y}\right) \tag{11.17}$$

where the species source $S_k(x, y) = 0$ for $k = I$.

11.3.1.4 Energy Conservation (Thermal Enthalpy Form)

$$\rho\, v_x \frac{\partial h_t}{\partial x} + \rho\, v_y \frac{\partial h_t}{\partial y} = \frac{\partial}{\partial y}\left(\rho\, \alpha_T \frac{\partial h_t}{\partial y}\right) + S_{h_t}(x, y), \tag{11.18}$$

where α_T is thermal diffusivity; under pure mixing conditions $S_{ht}(x, y) = 0$. For reacting conditions and single-step reactions,

$$S_{ht}(x, y) = |\,\dot{w}_F'''\,|\,h_c. \tag{11.19}$$

11.3.1.5 Energy (Total Enthalpy Form)

$$\rho\, v_x \frac{\partial h}{\partial x} + \rho\, v_y \frac{\partial h}{\partial y} = \frac{\partial}{\partial y}\left(\rho\, \alpha_T \frac{\partial h}{\partial y}\right) \tag{11.20}$$

11.3.1.6 General Property

In general, for any property "b," per unit mass,

$$\rho\, v_x \frac{\partial b}{\partial x} + \rho\, v_y \frac{\partial b}{\partial y} = \frac{\partial}{\partial y}\left(\rho\, \chi \frac{\partial b}{\partial y}\right) + S_b(x, y), \quad \frac{kg * (property / mass)}{m^3\ s} \tag{11.21}$$

where $\chi = \nu$ for b = v_x, $\chi = \alpha_T$ for b = h_t and h, χ = D for b = Y_k, and χ = D for b = β (Shvab–Zeldovich [SZ] variable), with α_T = D, $S_b = 0$, and other source terms defined as before. Both mixing and reacting problems are also included.

11.3.1.7 SZ Formulation

Using the SZ formulation, and assuming α_T = D (Le = 1),

$$\rho\, v_x \frac{\partial \beta}{\partial x} + \rho\, v_y \frac{\partial \beta}{\partial y} = \frac{\partial}{\partial y}\left(\rho\, D \frac{\partial \beta}{\partial y}\right), \quad \frac{kg * (\beta\ units)}{m^3\ s} \tag{11.22}$$

where $\beta = \beta_{O_2\text{-}F}, \beta_{CO_2\text{-}F}, \beta_{H_2O\text{-}F}, \beta_{h_T\text{-}F}$.

11.3.2 Boundary Conditions

Equation 11.12, Equation 11.15, Equation 11.18, and Equation 11.22 are solved under the following conditions.

$$\text{(a)} \quad y = 0,\ x > 0,\ v_x = 0, Y_k = Y_{kw}, h_t = h_{tw}, \beta = \beta_w \tag{11.23}$$

$$\text{(b)} \quad y \to \infty,\ x > 0 \begin{cases} v_x = v_{x,\infty}, & or\ \dfrac{\partial v_x}{\partial y} = 0 \\ Y_k = Y_{k\infty}, & or\ \dfrac{\partial Y_k}{\partial y} = 0 \\ h_T = h_{T,\infty}, & or\ \dfrac{\partial h_T}{\partial y} = 0 \\ \beta = \beta_\infty, & or\ \dfrac{\partial \beta}{\partial y} = 0 \end{cases} \tag{11.24}$$

$$\text{(c)} \quad x = 0,\ v_x = v_{x,\infty},\ Y_k = Y_{k,\infty},\ \beta = \beta_\infty,\ y > 0 \tag{11.25}$$

(Note that for any property "b," the boundary conditions b → b_∞ as y → ∞ and ∂b/∂y → 0 are not independent of each other. Thus, we included "or" in Equation 11.24.)

11.3.3 Transformation Variables for Conversion to "Incompressible" Form

These conservation equations are in "compressible" form or with variable density. There are explicit solutions for velocity v_x as a function of x and y in fluid mechanics [Schlichting, 1955] but only with an assumption of constant density or for conservation equations in "incompressible" form. In order to make use of these solutions for our problem, we will convert the above compressible equations to incompressible form using the following transformations. The conservation equations given in equations in Subsection 11.3.1 can be transformed from compressible form into incompressible form using the following transformation [Annamalai and Sibulkin, 1979a].

Let

$$x' = x\,\{x^{*2k}/(2k+1)\} = x^{2k+1}/\{d_{ref}^{2k}\,(2k+1)\},\ m \tag{11.26a}$$

$$x^* = x/d_{ref} \tag{11.26b}$$

where d_{ref} is yet to be defined.

$$y' = \int_0^y (\rho x^{*k}/\rho_{ref})\, dy, m \tag{11.26c}$$

$$v'_y = (\rho/[\rho_{ref}\, x^{*k}])v_y + (v_x/x^{*2k})\left\{\int_0^{y^*} \partial/\partial x\,(\rho x^{*k}/\rho_{ref})dy\right\},\ m/sec \tag{11.26d}$$

$$\rho\, v_y = \rho_{ref}\, v'_y\, x^{*k} - (v_x/x^{*k})\left\{\int_0^y \{\partial(\rho\, x^{*k})/\partial x\}dy\right\},\ kg/(m^2\ sec) \tag{11.26e}$$

$$v'_x = v_x,\ m/sec \tag{11.26f}$$

For example, if k = 0 (2-D systems with x and y coordinates), then Equation 11.26 becomes

$$x' = x, m \tag{11.27a}$$

$$y' = \int_0^y \frac{\rho}{\rho_{ref}}\, dy, m \tag{11.27b}$$

$$v_y' = \frac{\rho}{\rho_{ref}} v_y + v_x \int_0^y \frac{\partial}{\partial x}\left(\frac{\rho}{\rho_{ref}}\right) dy,\ m/sec \tag{11.27c}$$

$$v_x' = v_x, m/sec \tag{11.27d}$$

The coordinates x′ and y′ (in units of "m") are stretched variables that account for the compressible effects. For example, if ρ in Equation 11.27b is assumed to be a constant and equal to ρ_{avg}, then $y' = (\rho_{avg}/\rho_{ref})\ y$. Typically for flames, $\rho_{avg}/\rho_{ref} \approx 1/4$, and hence y′ (stretched or compressed variable) = y/4 (real variable). Because y′ has a smaller dimension than y, it is more appropriate to call y′ a compressed y coordinate. Thus, our solutions in terms of x′ and y′ are in compressed coordinates. Using these newly defined variables, the differential equations listed in Equation 11.11 to Equation 11.25 can be rewritten in incompressible form (exercise problems 11.1 to 11.2).

11.3.4 Conservation Equations in Incompressible Form

11.3.4.1 Mass

$$\frac{\partial v'_x}{\partial x'} + \frac{\partial v'_y}{\partial y'} = 0 \tag{11.28}$$

11.3.4.2 Momentum

$$v_x{}' \frac{\partial v_x{}'}{\partial x'} + v_y{}' \frac{\partial v_x{}'}{\partial y'} = \frac{\partial}{\partial y'}\left(\frac{\rho^2\ \nu}{\rho_{ref}^{\ 2}} \frac{\partial v_x{}'}{\partial y'} \right) + \frac{S_v(x,y)}{\rho x^{*2k}}, \quad \frac{m}{sec^2} \tag{11.29}$$

where $\rho^2\nu$ is treated as constant. Typically $S_v(x,y) = S_v(x)$.

11.3.4.3 Species

$$v_x{}' \frac{\partial Y_k}{\partial x'} + v_y{}' \frac{\partial Y_k}{\partial y'} = \frac{\partial}{\partial y'}\left(\frac{\rho^2\ D}{Sc\ \rho_{ref}^{\ 2}} \frac{\partial Y_k}{\partial y'} \right) + \frac{S_k(x,y)}{\rho x^{*2k}}, \quad \frac{1}{sec} \tag{11.30}$$

11.3.4.4 Thermal Enthalpy

$$v_x{}' \frac{\partial h_t}{\partial x'} + v_y{}' \frac{\partial h_t}{\partial y'} = \frac{\partial}{\partial y'}\left(\frac{\rho^2 \alpha}{\rho_{ref}^{\ 2}} \frac{\partial h_t}{\partial y'} \right) + \frac{S_{ht}(x,y)}{\rho x^{*2k}}, \quad \frac{kJ}{kg\ sec} \tag{11.31}$$

where $S_{ht}(x, y) = |w_F'''(x, y)|\, h_c$.

11.3.4.5 Total Enthalpy

$$v_x{}' \frac{\partial h}{\partial x'} + v_y{}' \frac{\partial h}{\partial y'} = \frac{\partial}{\partial y'}\left(\frac{\rho^2 \alpha}{\rho_{ref}^{\ 2}} \frac{\partial h}{\partial y'} \right), \quad \frac{kJ}{kg\ sec} \tag{11.31a}$$

11.3.4.6 General Property "b"

In general, for any property "b,"

$$v_x' \frac{\partial b}{\partial x'} + v_y' \frac{\partial b}{\partial y'} = \frac{\partial}{\partial y}\left(\frac{\rho^2 \chi}{\rho_{ref}^2} \frac{\partial b}{\partial y'} \right) + \frac{S_b(x, y)}{\rho x^{*2k}}, \quad \frac{units\ of\ b}{sec} \tag{11.32}$$

where $\rho^2\chi$ is treated as constant.

11.3.4.7 SZ Variable

$$v_x' \frac{\partial \beta}{\partial x'} + v_y' \frac{\partial \beta}{\partial y'} = \frac{\partial}{\partial y'}\left(\frac{\rho^2 D}{\rho_{ref}^2} \frac{\partial \beta}{\partial y'} \right), \quad \frac{units\ of\ \beta}{sec} \tag{11.33}$$

11.3.4.8 Normalized SZ (NSZ) Variable

Defining $\phi = (\beta - \beta_\infty)/(\beta_{ref} - \beta_\infty)$,

$$v_x' \frac{\partial \phi}{\partial x'} + v_y' \frac{\partial \phi}{\partial y'} = \frac{\partial}{\partial y'}\left(\frac{\rho^2 D}{\rho_{ref}^2} \frac{\partial \phi}{\partial y'} \right), \quad \frac{1}{sec} \tag{11.34}$$

11.3.4.9 Boundary Conditions (BCs)

$$\text{(a)} \quad y' = 0,\ \ x' > 0,\ v_x' = 0,\ Y_k = Y_{kw},\ h_t = h_{tw},\ \beta = \beta_w, \phi = \phi_w \tag{11.35}$$

$$\text{(b)} \quad y' \to \infty,\ \ x' > 0 \ \begin{cases} v_x = v_{x,\infty} \ \ or \ \ \dfrac{\partial v_x}{\partial y'} = 0 \\ h_T = h_{T,\infty} \ \ or \ \ \dfrac{\partial h_T}{\partial y'} = 0 \\ Y_k = Y_{k,\infty} \ \ or \ \ \dfrac{\partial y_k}{\partial y'} = 0 \\ \beta = \beta_\infty \ \ or \ \ \dfrac{\partial \beta}{\partial y'} = 0 \\ \phi = 0 \ \ or \ \ \dfrac{\partial \phi}{\partial y'} = 0 \end{cases} \tag{11.36}$$

$$\text{(c)} \quad x = 0,\ v'_x = v_{x,\infty},\ Y_k = Y_\infty,\ \beta = \beta_\infty,\ \phi = 0,\ y' > 0 \tag{11.37}$$

Consider Equation 11.29, which is second order in y′ and first order in x′; hence, the solution for v_x' equation requires three boundary conditions (BCs), which are given in Equation 11.35 to Equation 11.37, provided v_y' (x′,y′) is known. v_y' (x′,y′) can be obtained by integrating Equation 11.28 and assuming $v_{y,w} = 0$ at $y = 0$. Unlike heat and fluid mechanics literature, $v_y \neq 0$ at $y = 0$ for evaporation and combustion problems, and it is related to the gradient of Y_k at the fuel surface through an interface conservation equation. Hence, it requires interface boundary conditions for the solution of v_y' at $y' = 0$ (Subsection 11.4.1). Similarly, Equation 11.30 to Equation 11.32 are second order in y′ and first order in x′, and hence the solution of each equation can be obtained using BCs given by Equation 11.35 to Equation 11.37, provided the wall boundary conditions $Y_{k,w}$, $h_{t,w}$, and β_w are known. Often, $Y_{k,w}$, $h_{t,w}$, and β_w (including $v_{y,w}$) are unknown, and they also require interface boundary conditions.

11.4 Interface Boundary Conditions

11.4.1 Species and Energy

The reader is referred to Section 8.6.4.1 and Chapter 9 for more details on interface conservation equations. They are briefly summarized here. The interface fuel species conservation yields the following:

$$\dot{m}'' = \rho_w v_{y,w} = -\{\rho D\ (\partial Y_F/\partial y)_{y=0}\}(Y_{F,c} - Y_{F,w}) \tag{11.38}$$

where $Y_{F,c}$ is fuel mass fraction in condensed phase. The oxygen or other species do not dissolve into liquid and hence,

$$\rho_w v_{y,w}\ Y_k - \{\rho D\ \partial Y_k/\partial y\}_{y=0} = \dot{m}_k'' = 0,\ k = O_2,\ CO_2,\ H_2O,\ N_2. \tag{11.39}$$

Following Chapter 8, Equation 11.38, Equation 11.39, etc., can be written in SZ form as

$$\rho v_y\ \beta - \rho D\ \partial\beta/\partial y = \dot{m}''\ F_\beta \tag{11.40}$$

where $F_\beta = -L/h_c$ for $\beta = \beta_{ht-O2}$ and $F_\beta = Y_{F,c}$ for $\beta = \beta_{O2-F}$. Solving for $\dot{m}''$ from Equation 11.40,

$$\dot{m}'' = \rho_w v_{yw} = -\rho D\ (\partial\beta/\partial y)_{y=0}/(F_\beta - \beta_w). \tag{11.41}$$

Rewriting in terms of NSZ,

$$\dot{m}'' = \rho_w v_{y,w.} = -\rho D\ B\ (\partial\phi/\partial y)_{y=0} = \frac{\rho^2 D \chi^{*k}\ B}{\rho_{ref}}\left(\frac{\partial\phi}{\partial y'}\right)_{y'=0} \tag{11.42}$$

where the transfer number

$$B = (\beta_{ref} - \beta_{\infty})/(F - \beta_{w}) \tag{11.43}$$

and typically for most problems in this chapter, $\beta_{ref} = \beta_{w}$. The F_β values for various cases are presented in Table 10.1 of Chapter 10 for liquid fuels and Chapter 9 for carbon combustion for various β definitions.

11.5 Generalized Numerical Solution Procedure for BL Equations in Partial Differential Form

(1) Solve for v_x and v_y using mass and momentum Equation 11.28 and Equation 11.29 for an assumed value of $v_y(0) = v_{yw}$. As $\dot{m}''' = \rho_w v_{y,w}$, we can find the burn rate for assumed v_{yw}.

(2) Then, one can solve the $\phi(x, y)$ or $\phi(x', y')$ equation with $\beta = \beta_{ht-O2}$ (Equation 11.34). For $\beta = \beta_{ht-O2}$, surface properties are known as $\beta_w = \beta_{ht-O2,w} = h_{t,w}/L + Y_{O2,w}/\nu_{O2} = 0$ for diffusion-controlled combustion and $F_\beta = -L/h_c$. Similarly, $\beta_{ht-O2,\infty} = h_{t,\infty}/L + Y_{O2,\infty}/\nu_{O2}$ is also known if one assumes that $T_w = T_{BP}$. Thus, gradient $(\partial\phi/\partial y)_{y=0}$ is evaluated. Using Equation 11.42 and knowing B, hence $\dot{m}''$ can be solved; however $\dot{m}'' = \rho_w v_{yw}$. If the assumed value v_{yw} is correct, then the values should match; if not, iterate.

(3) If $T_w \neq T_{BP}$, we follow procedures outlined in Chapter 8 and 10 to solve for T_w.

The following sections deal with normalization of conservation equations, similarity formulation, and derivation of simple solutions for stagnation flows and free and forced convections. Most of the derivations will be left as exercise problems.

11.6 Normalized Variables and Conservation Equations

11.6.1 Normalized Variables

The nondimensional variables used for the normalization are as follows. Using Equation 11.26,

$$x'^* = x'/d_{ref} = x^{*(2k+1)}/(2k+1),\ x'^* = x^* \text{ when } k = 0 \tag{11.44}$$

$$y'^* = y'/d_{ref} = \int_0^{y^*} (\rho/\rho_{ref}))\ x^{*\,k}\ dy^* \tag{11.45}$$

$$v_x^* = v'^*_x = v_x/v_{ref} = v_x/v_{ref} \tag{11.46}$$

$$v'^*_y = v'_y/v_{ref} = \{\rho/(\rho_{ref}\ x^{*k})\}v_y{}^* + \{v_x{}^*/x^{*2k}\}\left(\int_0^{y} \partial/\partial x^*(\rho x^{*K}/\rho_{ref})dy^*\right). \tag{11.47a}$$

Rewriting

$$(\rho/\rho_{ref})v_y^* = v'^*_y\ x^{*k} - \{v_x{}^*/x^{*k}\}\left\{\left(\int_0^{y^*} \partial/\partial x^*(\rho x^{*K}/\rho_{ref})dy^*\right)\right\} \quad (11.47b)$$

$$b^* = b/b_{ref} \quad (11.48)$$

where b any generic property; further, $\rho^2 D = \rho_{ref}^2\ D_{ref}$ = constant, $\rho^2 \nu = \rho_{ref}^2\ \nu_{ref}$ = constant, $\rho^2 \alpha_T = \rho_{ref}^2\ \alpha_{T,ref}$ = constant. The reference condition can either be selected at the wall (w) or at the free stream (∞); *properties at mean film conditions are preferable for more accurate evaluations.* Selecting $\beta_{ref} = \beta_w$, the normalized SZ variable can now be generalized for all other variables.

$$\phi = (\beta - \beta_\infty)/(\beta_w - \beta_\infty) = (f - f_\infty)/(f_w - f_\infty) = (Z_k - Z_{k,\infty})/(Z_{k,w} - Z_{k,\infty})$$
$$= (h - h_\infty)/(h_{,w} - h_\infty) \quad (11.49)$$

where "f" is the mixture fraction (chapter 8), Z_k is the element fraction and β_w is typically constant for diffusion-controlled combustion. Finally, equations, BCs, and interface BCs can be cast in nondimensional form (exercise problem 11.3).

11.6.2 Normalized Conservation Equations

11.6.2.1 Mass

$$\partial v'^*_x/\partial x'^* + \partial v_y'^*/\partial y'^* = 0 \quad (11.50)$$

11.6.2.2 Momentum (x direction)

$$v_x'^*\partial v_x'^*/\partial x'^* + v_y'^*(\partial v_x'^*/\partial y'^*) = [1/Re_{ref}]\ (\partial^2 v_x'^*/\partial y'^{*2}) + S_v(x'^*,y'^*) \quad (11.51)$$

where $Re_{ref} = v_{ref}\ d_{ref}/\nu$.

The dimensionless source S_v^* is analogous to the chemical source or Damkohler number in chemical reactions.

$$S_v(x'^*, y'^*) = \frac{S_v(x) d_{ref}}{\rho\ v_{ref}^2\ x^{*2k}} = \frac{\textit{Momentum Source}}{\textit{Convective flow of Momentum}} \quad (11.52)$$

See Table 11.1 for S_v (x, y) and S_v^* for typical flow problems.

If $v_{ref} = a\ d_{ref}{}^n$, $v_{x\,\infty}{}^* = \{a\ d_{ref}{}^n/v_{ref}\}\ x^{*n} = x^{*n} = [(2k + 1)\ x'^*]^{n/(2k+1)}$, flow over curved surface

$$S_v(x'^*,y'^*) = [\rho_\infty/\rho]\ v_{x\,\infty}'^*[dv_{x\,\infty}'^*/dx'^*] \quad (11.53)$$

$$S_v(x'^*,y'^*) = n(2k + 1)^{2n/(2k+1)-1}(\rho_\infty/\rho)\ x'^{*\{2n/(2k+1)\}-1}, n > 0 \quad (11.54)$$

If $v_{ref} \neq a\, d_{ref}$, $v_{x\,\infty}'^{*} = \{a\, d_{ref}{}^{n}/v_{ref}\}\, x^{*n}$, $S_v(x^*) = \{\rho_\infty\, v_{ref}^2/d_{ref}\}\, v^*_{x\infty}\, \mathbf{dv}^*_{x\infty}/dx^*$

$$S_v(x'^*) = n\{ad_{ref}{}^{n}/v_{ref}\}^2(\rho_\infty/\rho)\,(2k+1)^{\{2n/(2k+1)\}-1}\, x'^{*\{2n/(2k+1)\}-1}. \quad (11.55)$$

The nondimensional momentum sources are listed in Table 11.1.

11.6.2.3 Species

With $b = Y_k$ in Equation 11.32, $\chi = D$, the normalized form of the species equation is given as

$$v_x'^{*}\frac{\partial Y_k}{\partial x'^{*}} + v_y'^{*}\frac{\partial Y_k}{\partial y'^{*}} = \frac{\partial}{\partial y'^{*}}\left(\frac{1}{\mathrm{Re}_{ref}\, Sc_{ref}}\frac{\partial Y_k}{\partial y'^{*}}\right) + S_k(x^*, y^*) \quad (11.56)$$

$$S^*_k(x^*, y^*) = Da_I = \frac{S_k(x,y)d_{ref}}{\rho\, v_{ref}\, x^{*2k}} = \frac{\textit{Species chemical production rate}}{\textit{Convective flow rate}} = \frac{t_{res}}{t_{chem}} \quad (11.57)$$

where $S_k(x,y) = \dot{w}_k'''(x,y)$ and S^*_k = Damkohler number I (see Chapter 7). Finite chemistry problems have been handled by Sibulkin et al. [1981, 1982], particularly in determining extinction limits of a vertically burning surface.

11.6.2.4 Thermal Enthalpy

With $b = (h_t/h_c)$ in Equation (11.32), $\chi = \alpha_T$

$$v_x'^{*}\frac{\partial\left(\frac{h_t}{h_c}\right)}{\partial x'^{*}} + v_y'^{*}\frac{\partial\left(\frac{h_t}{h_c}\right)}{\partial y'^{*}} = \frac{\partial}{\partial y'^{*}}\left(\frac{1}{\mathrm{Re}_{ref}\,\mathrm{Pr}_{ref}}\frac{\partial\left(\frac{h_t}{h_c}\right)}{\partial y'^{*}}\right) + S_{ht}(x^*, y^*) \quad (11.58)$$

$$\text{where}\quad S_{ht}(x^*, y^*) = \frac{S_{ht}(x,y)\; d_{ref}}{\rho\, h_c v_{ref}\; x^{*2k}} \quad (11.59)$$

and for a single-step reaction,

$$S_{ht}(x'^*, y'^*) = Da_I = \frac{\left|\dot{w}_F''''\right|\, d_{ref}}{\rho\, v_{ref}\; x^{*2k}}. \quad (11.60)$$

11.6.2.5 Generic Property "b"

$v_x^*\,(\partial b^*/\partial x^*) + v_y^*(\partial b^*/\partial y^*) = \{1/(Re_{ref}\ \zeta_{ref})\}\partial^2 b^*/\partial y^{*2} + \{S_b(x,y)\ d_{ref}/(\rho x^{*2k}\ b_{ref} v_{ref}\},$ (11.61)

$$\zeta = \nu/\chi \quad (11.62)$$

$$b^* = b/b_{ref}$$

11.6.2.6 Normalized SZ Variable

$$v_x{}'^*\frac{\partial\phi}{\partial x'^*} + v_y{}'^*\frac{\partial\phi}{\partial y'^*} = \frac{\partial}{\partial y'^*}\left(\frac{1}{\mathrm{Re}_{ref}\,\zeta}\frac{\partial\phi}{\partial y'^*}\right) \quad (11.63)$$

$\zeta = Sc$ for $\phi = (Y_k - Y_{k,\infty})/(Y_{k,w} - Y_{k,\infty}),\ (Z_k - Z_{k,\infty})/(Z_{k,w} - Z_{k,\infty});$ (11.64a)

$\zeta = Pr$ for $\phi = (h - h_\infty)/(h_w - h_\infty)$ (11.64b)

If Sc = Pr, then $\phi = (\beta - \beta_\infty)/(\beta_w - \beta_\infty),\ (f_M - f_{M\infty})/(f_{Mw} - f_{M\infty})$; thus, for these cases,

$$v_x{}'^*\frac{\partial\phi}{\partial x'^*} + v_y{}'^*\frac{\partial\phi}{\partial y'^*} = \frac{\partial}{\partial y'^*}\left(\frac{1}{\mathrm{Re}_{ref}\,Sc_{ref}}\frac{\partial\phi}{\partial y'^*}\right). \quad (11.65)$$

11.6.2.7 Reference Conditions

The reference conditions v_{ref} and d_{ref} must be adjusted in such a way that the BCs for dependent variables become constants; for example, if the flat-plate problem is of interest, v_{ref} is selected as $v_{x\infty}$ so that as $y \to \infty$, $v_x^* \to 1$ and d_{ref} could be the total length of the plate. If a laminar jet problem is of interest (see Chapter 12), d_{ref} could be d_i, the injector diameter and v_{ref} can be set at v_{xi}, the injection velocity.

11.7 Similarity Solutions–BL Equations

11.7.1 Stream Functions and Similarity Variable

Equation 11.51 is a partial differential equation. One can solve for $v_x'^*(x'^*,y'^*)$ and plot v'^*_x vs. y'^* with x'^* as a parameter. All these parametric curves can be made to collapse to a single curve if we plot $v_x'^*$ vs. $y'^*/x'^{*\,\alpha} = \eta$, where α needs to be selected suitably in order to make the curves collapse. If such a curve can be found, then $v_x'^*(x'^*,y'^*) \to v_x'^*(\eta)$, then the governing differential equation for $v_x'^*$ should only be a function of η or a partial differential equation in terms of x'^*,y'^* should reduce to an ordinary differential equation in terms of the single independent variable η. In order to discover the suitable α, let us use the following similarity transformation.

From the mass conservation equation (Equation 11.11), define a stream function as

$$\partial\psi/\partial y = \rho v_x\, x^k/\rho_{ref}, \quad (11.66)$$

$$\partial\psi/\partial x = -\rho v_y\, x^k/\rho_{ref}. \quad (11.67)$$

Transforming to normalized variables,

$$\partial\psi/\partial y'^{*} = v_x'^{*},\ \partial\psi/\partial x'^{*} = -v_y'^{*}. \tag{11.68}$$

Defining a similarity variable as

$$\eta = C_1(y'^{*}/x'^{*\alpha}) \tag{11.69}$$

and converting the stream function $\psi(x^*, y'^*)$ in terms of a function $f(\eta)$

$$\psi\ (x'^{*}, y'^{*}) = f(\eta)\ x'^{*(1-\alpha)}/C_2 \tag{11.70}$$

where C_1, C_2, and α are arbitrary constants. The constant α is selected suitably to obtain an ordinary differential equation in terms of $f(\eta)$.

$$v_x^* = \partial\psi/\partial y'^{*} = f'\ (C_1/C_2)\ x'^{*1-2\alpha} = f'(C_1/C_2)\ \{1/(2k+1)\}^{(1-2\alpha)}\ x^{*(2k+1)(1-2\alpha)} \tag{11.71}$$

$$v_y'^{*} = -\partial\psi/\partial x'^{*} = \{1/(x'^{*\ \alpha}C_2)\}\ \{f'\alpha\eta - f(1-\alpha)\} \tag{11.72}$$

Derivation of similarity variable is left as an exercise in Problem 11.5 and adjustment of suitable constants in Problem 11.6 to Problem 11.10 for forced convection over a flat plate, curved surfaces, 2-D and axisymmetric stagnation flows, and free convection.

11.7.2 Conservation Equations in Terms of Similarity Variable

11.7.2.1 Momentum

Using similarity variables, the momentum equation (Equation 11.51) becomes

$$(f')^2\ (1-2\alpha)\ - ff''\ (1-\alpha) = (C_1\ C_2/Re_{ref})f''' + C_m \tag{11.73}$$

where

$$C_m = \{(C_2/C_1)^2\ S_v(x'^{*},y'^{*})/x'^{*1-4\alpha}\} = (C_2/C_1)^2\ (2k+1)^{1-4\alpha}\ S_v(x'^{*},y'^{*})/x^{*(2k+1)(1-4\alpha)}. \tag{11.74}$$

11.7.2.2 Species and SZ

Similarly proceeding for SZ and species

$$\{C_1\ C_2/Re_{ref}\}\ \phi'' + Sc\ f\ \phi'(1-\alpha) = 0. \tag{11.75}$$

Table 11.2 summarizes the constants C_1, C_2, C_m, and α for various flow problems of interest. Equation 11.75 can also be used for frozen chemistry. For example, with species k, Equation 11.75 is written as,

$\{C_1\ C_2/Re_{ref}\}\ \phi_k'' + Sc\ f\ \phi_k\ (1-\alpha) = 0$ where $\phi_k = \{Y_k - Y_{k\infty}\}/Y_{k,ref}$
where $Y_{k,ref}$ could be $Y_{kw} - Y_{k\infty}$.

Table 11.2 Similarity Variables and Governing Equations in Similarity Coordinates for Various Flows

Momentum: $(f')^2\,(1-2\alpha) - ff''\,(1-\alpha) = (C_1\,C_2/Re_{ref})f''' + C_m$ (a)

Energy and Species: $\{C_1\,C_2/Re_{ref}\}\,\phi'' + Sc\,f\,\phi'\,(1-\alpha) = 0$ (b)

Generic property with finite chemistry: $\phi_b'' + f\,Sc\,(1-\alpha)\,\phi_b' + Sc\,S_b^* = 0$ (c)

$\phi'(0)\,B/\{Sc\,(1-\alpha)\} = \{Re_{ref}/(C_2\,C_1)\}\,f(0)$ (d) (see Equation 11.75a and 11.75b for details)

$y' = \int_0^y \frac{\rho x^{*k}}{\rho_{ref}}\,dy$, $y' = \int_0^y \frac{\rho}{\rho_{ref}}\,dy$ for 2D, $y' = \int_0^y \frac{\rho x^*}{\rho_{ref}}\,dy$ for axisymmetric

$\eta = C_1(y'^*/x'^{*\alpha})$, $\psi(x'^*, y'^*) = f(\eta)\,x'^{*(1-\alpha)}/C_2$, $Gr_{ref} = g\,\{T_w - T_\infty\}\,d_{ref}^{\,3}/\{T_\infty\,\nu_{ref}^{\,2}\}$, $Re_{ref} = v_{ref}\,d_{ref}/\nu_{ref}$, $x'^* = x^{*(2k+1)}/(2k+1)$, k = 0 2D, k = 1 axisymmetric

$v_x'^* = f'\,(C_1/C_2)\,x'^{*1-2\alpha}$, $v_y'^* = -\partial\psi/\partial x'^* = \{1/(x'^{*\,\alpha}C_2)\}\,\{f'\alpha\eta - f(1-\alpha)\}$

$m^* = \dot{m}''/\{v_{ref}\,\rho_{ref}\} = v_y'^*(0)\,x^{*k} = K\,\{-f(0)\}/\{C_2\,x^{*\sigma}\}$, $K = (1-\alpha)\,(2k+1)^{\alpha}$, $\sigma = \alpha(2k+1) - k$; $-f(0) = m^*C_2x^{*\sigma}/K$

#	Problem	α	C_1	C_2	C_m	$\{C_1C_2\}/Re_{ref}$	K	σ
1.	**Forced Convection (k = 0)**							
1a.	(a) Curved $v_{x\infty} = ax^n$, $v_{ref} = a\,d^n_{ref}$	(1 – n)/2	$Re_{ref}^{\,1/2}$	$Re_{ref}^{\,1/2}$	$n\,(\rho_\infty/\rho)$, $n\,(T/T_\infty)$	1	**{(1 + n)/2}**	(1-n)/2
1b.	(b) Flat surface: $v_{x\infty}$ const or n =0, $v_{ref} = v_{x\infty}$	1/2	$Re_{ref}^{\,1/2}$	$Re_{ref}^{\,1/2}$	0	1	**(1/2)**	1/2
2.	**Stagnation 2-D** (k = 0),					1	1	0
	$v_{x\infty} = a\,x^n$, n = 1, $v_{ref} = a\,d_{ref}$	0	$Re_{ref}^{\,1/2}$	$Re_{ref}^{\,1/2}$	(ρ_∞/ρ), T/T_∞			
3.	**Stagnation Axisymmetric** (k = 1),							
	$v_{x\infty} = ax^n$, n = 1, $v_{ref} = a\,d_{ref}$, $x'^* = x^{*3}/3$, $C_1C_2 = Re$	1/3	$Re_{ref}^{\,1/2}\,3^{1/6}$	$\{Re_{ref}^{\,1/2}/3^{1/6}\}$	$(1/3)(\rho_\infty/\rho)$, $(1/3)\,T/T_\infty$	1	**$2/3^{2/3}$**	0
4.	**Free convection** (k = 0), $Re_{ref} = 1$, $v_{ref} = \{\nu_{ref}/d_{ref}\}$							
4a.	(a) Set $Re_{ref} = 1$	1/4	$\{4Gr_{ref}\}^{1/4}$	$\{1/(4Gr_{ref})^{1/4})\}$	$\{T - T_\infty\}/\{T_w - T_\infty\}$	1	3/4	1/4
4b.	(b) Set $Re_{ref} = 1$	1/4	$\{Gr_{ref}/4\}^{1/4}$	$(1/4)\,(4/Gr_{ref})^{1/4}$	$\{T - T_\infty\}/\{T_w - T_\infty\}$	1/4	3/4	1/4

11.7.2.3 Finite Chemistry

For finite chemistry Equation 11.62 with b replaced by ϕ_b can be reduced to

$$\phi_b'' + f\,Sc(1-\alpha)\phi_b' + Sc\,S_b^* = 0 \quad (11.75a)$$

where $\phi_b = \{b - b_\infty\}/b_{ref}$ where b_{ref} could be $b_w - b_\infty$

$$S_b^* = \{S_b(x, y)\}\,C_2 x'^{*2\alpha - 2k/(2k+1)} d_{ref}/\{\rho(2k+1)^{2k/(2k+1)} C_1 b_{ref} v_{ref}\}. \quad (11.75b)$$

Typically $b = Y_F$.

11.7.3 Boundary and Interface Conditions in Terms of Similarity Variable

It is seen from Equation 11.71 that $f'(0) = 0$ because $v_x^* = 0$ at $y^* = 0$. The mass burn rate is

$$\dot{m}'' = \rho_w v_y(0). \quad (11.76)$$

Using Equation 11.73 at $y = 0$ and

$$m^* = \dot{m}''/\{v_{ref}\,\rho_{ref}\} = \{\rho_w/\rho_{ref}\}\;v_y{}^*(0) = v_y'^*(0)x^{*k} \quad (11.77a)$$

$$\text{where } v_y'^*(0) = v_{y,w}'^* = -\{f(0)/C_2\}\,(1-\alpha)(1/x'^{*\alpha}). \quad (11.77b)$$

Using Equation 11.77 in Equation 11.76 and Equation 11.44,

$$m^* = \dot{m}''/\{v_{ref}\,\rho_{ref}\} = K\,\{-f(0)\}/\{C_2\,x^{*\sigma}\} \quad (11.78)$$

where

$$K = (1-\alpha)\,(2k+1)^{\alpha} \quad (11.79a)$$

$$\sigma = \alpha(2k+1) - k. \quad (11.79b)$$

The results for typical flow systems are summarized in Table 11.2.

Using Equation 11.42, converting to similarity coordinates, and simplifying,

$$\dot{m}^* = -B\,\phi'(0)\,x^{*k}\,C_1/\{Re_{ref}\,Sc_{ref}\,x'^{*\alpha}\} = -\{B/(1-\alpha)\}K\phi'(0)C_1/\{Re_{ref}Sc\,x^{*\sigma}\}. \quad (11.80)$$

Equating Equation 11.80 with Equation 11.78,

$$\phi'(0)\,B/\{Sc(1-\alpha)\} = \{Re_{ref}/(C_2\,C_1)\}\,f(0). \quad (11.81)$$

Using Equation 11.75 at $\eta = 0$ and f(0) from Equation 11.81,

$$\phi''(0) = \phi'^2(0)\,B. \quad (11.82)$$

The relation shown by Equation 11.82 is useful in integral analysis.

11.8 Applications of Generalized Similarity Equations to Various Flow Systems

11.8.1 Forced Convection over Flat Plate, Inclined Plate, and Curved Surfaces

Now, we illustrate the procedure for forced convection in the following example.

Example 1

Formulate the (a) momentum and (b) SZ conservation equations in terms of similarity coordinates for forced convection over a curved surface (Figure 11.4a).

Solution

(a) *Momentum:* For forced convection over a curved or inclined 2-D surface: $k = 0$ but $n \neq 0$, select $(C_1C_2) = Re_{ref}$. Then $x'^* = x^*$.

With Equation 11.54, the source term becomes

$$S_v(x'^*,y'^*) = n\ (\rho_\infty/\rho)\ x'^{*\{2n-1\}} = n\ (\rho_\infty/\rho)\ x^{*\{2n-1\}},\ k = 0,\ v_{ref} = a\ d_{ref}^{\ n},$$
$$Re_{ref} = v_{ref}\ d_{ref}/\nu_{ref} = a\ d_{ref}^{\ n+1}/\nu_{ref}.$$

From Equation 11.74,

$$C_m = (C_2/C_1)^2\ S_v(x'^*,y'^*)/x'^{*1-4\alpha} = (C_2/C_1)^2\ n\ (\rho_\infty/\rho)\ x^{*\{2n-1\}}/x^{*1-4\alpha}.$$

For C_m to be a constant in the above equation, we let

$$\alpha = (1 - n)/2, \text{ (see Schlichting, 1955, p. 156, and Table 11.2).}$$

Consider the x-velocity with $k = 0$ (Equation 11.71)

$$v_x^* = f(C_1/C_2)\ \{1/(2k + 1)\}^{(1-2\alpha)}\ x^{*(2k+1)(1-2\alpha)} = f'(C_1/C_2)\ x^{*(1-2\alpha)}$$
$$= f'(C_1/C_2)\ x^{*n}. \tag{A}$$

Since $v_{x\infty} = ax^n$, $v_{x\infty}^* = x^{*n}$ with $v_{ref} = ad_{ref}^{\ n}$. As $\eta \rightarrow \infty$, Equation A becomes

$$v_{x\infty}^* = f'(\infty)(C_1/C_2)\ x^{*n}. \tag{B}$$

Solving for $f'(\infty)$ and letting $C_2 = C_1$, $f'(\infty) = 1$. Because $(C_1C_2) = Re_{ref}$, then $C_1 = C_2 = Re_{ref}^{1/2}$. Thus,

$C_m = (C_2/C_1)^2\ n\ (\rho_\infty/\rho)\ x^{*\{2n-1\}}/x^{*1-4\alpha} = n\ (\rho_\infty/\rho)$ (see Table 11.2, 1a).

For an ideal gas fluid with constant molecular weight, Equation 11.73 becomes

$$f''' + ff''\ (1 + n)/2 + n\ \{(T/T_\infty) - f'^2\} = 0, \tag{C}$$

with $f'(0) = 0$, $f'(\infty) = 1$.

The similarity variables, $f(\eta)$, $v_x'^*$, $v_y'^*$ in Equation 11.69, Equation 11.70, Equation 11.71, and Equation 11.72 become

$$\eta = C_1(y'^*/x'^{*\alpha}) = Re^{1/2}\ y'^*/x^{*(1-n)/2} \tag{D}$$

$$f(\eta) = C_2\ \psi(x'^*, y'^*)/x'^{*1-\alpha} = Re^{1/2}\ \psi(x'^*, y'^*)/x'^{*(1+n)/2} \quad (E)$$

$$v_x^* = fx'^{*n} = f'x^{*n} \quad (F)$$

$$v_y'^* = [f'\{(1-n)/2\}\eta\}]/(Re^{1/2}\, x^{*(1-n)/2}) - f\,[(1+n)/2]/\{Re^{1/2}x'^{*(1-n)/2}\}.$$

$$\text{At } \eta = 0,\ v_y^*\ (0) = -f\ (0)[(1+n)/2]/\{Re^{1/2}x'^{*(1-n)/2}\}. \quad (G)$$

Further, Equation 11.79 becomes
$K = \{(1+n)/2\}$ and $\sigma = (1-n)/2$ (see Table 11.2).
Finally, the nondimensional burn rate (Equation 11.78)

$$m^* = \dot{m}''/\{v_{ref}\ \rho_{ref}\} = \{(1+n)/2\}\{-f(0)\}/\{Re_{ref}^{1/2}x^{*(1-n)/2}\}. \quad (H)$$

(b) *Normalized Variable* ϕ*:* with $\alpha = (1-n)/2$, Equation 11.75 becomes

$$\phi'' + Sc\ f\ \phi'\ \{(1+n)/2\} = 0. \quad (I)$$

With boundary conditions $\phi\ (0) = 1$, $\phi\ (\infty) = 0$ and Equation 11.81 becomes

$$\phi'(0)\ 2\ B/\{Sc\ (1+n)\} = f(0). \quad (J)$$

Remarks

(i) For a flat plate, $k = 0$, $n = 0$, $\alpha = 1/2$, $C_m = 0$. Equation C becomes

$$f''' + (1/2)\ ff'' = 0, \quad (K)$$

with $f'(0) = 0$, $f'(\infty) = 1$. Equation J becomes

$$\phi'' + Sc\ f\ \phi'\ \{1/2\} = 0$$

$$\phi'(0)\ 2\ B/Sc = f(0) \quad (L)$$

(ii) Equation 11.44 to Equation 11.48 become

$$x'^* = x^* \quad (N)$$

$$y'^* = \int_0^y (\rho/\rho_w)\ dy^*. \quad (N')$$

(Note that $y'^* = \int_0^y (\rho/\rho_w)\ dy^* \approx \int_0^y (T_w/T)\ dy^*$ for an ideal gas.)

(O)

$$\eta = R_e^{1/2}(y'^*/x'^{*\alpha}) = R_e^{1/2}(y'^*/x'^{*1/2}) \quad (P)$$

$$v_x'^* = v_x^*,\ v_y'^* = (\rho/\rho_w)v_y + (v_x^*)\ \partial/\partial x(\textstyle\int_0^y \partial/\partial x\ (\rho/\rho_w)\ dy^* \quad (Q)$$

From Equation 11.78 or Equation H,

$$m^* = \dot{m}''/\{v_{ref}\ \rho_{ref}\} = (1/2)\ \{-f(0)\}/\{Re_{ref}\ ^{1/2}x^{*1/2}\}. \quad (R)$$

11.8.2 2-D Stagnation Flow Systems (*k* = 0)

11.8.2.1 Infinite Chemistry

Example 1 can be used to derive similarity equations for this case. This is left as an exercise problem. The equations and BCs are given below. For 2-D flow, $v_{x\infty} = ax$. Thus, one may set n = 1 in Equation C and Equation I of Example 1 to obtain momentum and SZ equations. An important difference is that η is independent of x^*. (Checks with Schlichting, 1955, p. 97, BL thickness $\approx 2.4\ \{\nu/a\}^{1/2}$.)

11.8.2.2 Finite Chemistry

The 2-D stagnation flow problems lead to a similarity solution of species of equation with finite chemistry. The momentum equation remains the same. With finite chemistry, the source term in Equations 11.75a and 11.75b with $b = Y_f$, $S_b = |\dot{w}'''|$, $\alpha = 0$, $k = 0$ is given as

$$S_F^* = |\dot{w}'''|\ \ d_{ref}/(\rho\{Y_{F,w} - Y_{F,\infty}\}v_{ref}) \quad (11.83)$$

$$\phi_F'' + Sc + \phi_F' + Sc\ S_F^* = 0. \quad (11.84)$$

11.8.3 Axisymmetric Stagnation Flow Systems (*k* = 1)

Example 1 can be used to derive similarity equations for this case. This is left as an exercise (Problem 11.9). Results are tabulated in Table 11.2.

11.8.3.1 Species and Energy for Axisymmetric Jet with Finite Chemistry

Using Equation 11.75b, $k = 1$, $\alpha = 1/3$, $S_b(x,y) = |\dot{w}'''|$

$$S^*_F = \{3^{1/3}\ \dot{w}'''\ d_{ref}\}/\{\rho(Y_{F,w} - Y_{F,\alpha})v_{ref}\}. \quad (11.85)$$

Equation 11.75a becomes,

$$\phi'' + f\ Sc\ (2/3)\ \phi' + Sc\ S_F^* = 0. \quad (11.86)$$

11.8.3.2 Infinite Chemistry or SZ

$$\phi'' + (2/3)\ Sc\ f\ \phi' = 0 \quad (11.87)$$

11.8.4 Free Convection

Example 1 can be used as a methodology to derive similarity equations for this case. For this case, $v_{x\infty} = 0$. (Problem 11.10.)

11.8.4.1 Finite Chemistry

From Table 11.2, $C_1 = \{4\ Gr\}^{1/4}$, $C_2 = 1/\{4\ Gr\}^{1/4}$, $\alpha = 1/4$, $Re_{ref} = 1$, $k = 0$ in Equations 11.75a and 11.75b;

$$S_F{}^* = |\,\dot{w}'''|\ x^{*1/2} d_{ref}/((4\ Gr)^{1/2} \rho Y_{F,ref} v_{ref}\},\ Y_{F,\,ref} = Y_{F,w} - Y_{F,\infty} \tag{11.88}$$

$$\phi'' + f\ Sc\ (3/4)\ \phi'\} + Sc\ S_F{}^* = 0. \tag{11.89}$$

Note that $S_F{}^*$ is a function of $x^{*1/2}$ and there is no similarity solution.

11.8.4.2 Infinite Chemistry or SZ

See Table 11.2 for this case with $C_1C_2/Re_{ref} = 1/4$; $\alpha = 1/4$.

$$\phi'' + 3\ Sc\ f\ \phi' = 0 \tag{11.90}$$

11.9 Solutions for Boundary Layer Combustion of Totally Gasifying Fuels

11.9.1 Exact Numerical Solution Procedure for Various Flow Systems

The governing equations for momentum and species/energy/SZ are listed as Equations (a)–(c) in Table 11.2.

We need the solutions for f and ϕ as a function of η. The momentum equation is of third order and, hence, we need three BCs. The normalized SZ is of second order and, hence, we need only two BCs. The Boundary and Interface Conditions are listed below:
For a third-order equation in "f," we have three BCs as listed below

$$f'(0) = 0, \tag{11.91}$$

$$f'(\infty) = 1 \text{ for forced, stagnation flows and } f'(\infty) = 0 \text{ in free convection;} \tag{11.92}$$

and from Equation 11.81 (or Equation (d) in Table 11.2),

$$\phi'(0)\ B/\{Sc\ (1-\alpha)\} = \{Re_{ref}/(C_2\ C_1)\}\ f(0). \tag{11.93}$$

Note that f(0) is coupled to $\phi'(0)$. For the second-order NSZ (ϕ) equation

$$\phi\ (0) = 1 \tag{11.94a}$$

$$\phi(\infty) = 0. \tag{11.94b}$$

Thus, the problem can be solved, and all pertinent profiles and burn rates can be obtained, provided B is known. The procedure is as follows: Assume $\phi'(0)$; then with three BCs given by Equation 11.91 to Equation 11.93, solve third-order equation for f. Then, with two BCs given by Equation 11.94, solve the second-order equation in ϕ, compute $\phi'(0)$, and check with the assumed value. Iterate until convergence is obtained. Then, using Equation 11.80 the mass burn rate is evaluated. Once the profile is obtained in terms of η, use Equation 11.69, Equation 11.44, Equation 11.45, and Equation 11.26 to convert the solution into real coordinates. All the results for wall properties Y_{Fw}, T_w, and Y_{kw} derived in terms of B, etc., in Chapter 8 and Chapter 10 are valid for diffusion-controlled combustion. As presented in Chapter 10,

$$Y_{F,w} = 1-(1 + s)/(1 + B), \text{ for combustion} \tag{11.95}$$

$$Y_{F,w} = 1 - 1/(1 + B) \text{ for pure evaporation} \tag{11.96}$$

11.9.1.1 Flame Location

Suppose a flat liquid fuel of length L burns under convection. Using the definition of NSZ and setting $\phi = \phi_f$

$$\phi_f = \frac{\beta_{O2-Ff} - \beta_{O2-F,\infty}}{\beta_{O2-Fw} - \beta_{O2-F,\infty}} \tag{11.97}$$

$$\phi_f = \frac{s(1+B)}{B(1+s)}, \quad s = \frac{Y_{O2\infty}}{\nu_{O2}\, Y_{F,c}} \tag{11.98 a,b}$$

Then, from the known solution for $\phi(\eta)$, one can determine ϕ_F and hence can solve for η_f.

Example 2

In air, *n*-hexane fuel burns at T_∞ = 600 K, L = 335 kJ/kg, h_c = 45105 kJ/kg, c_p = 1.17 kJ/kg K, P = 100 kPa. The burning may take place under forced convection or free convection or stagnation flow. Assume that $T_w = T_{BP}$ at given P. Obtain wall properties Y_{Fw}, $Y_{CO2,w}$, Y_{H2Ow}, and Y_{N2w} and the flame properties T_f, $Y_{CO2,f}$, Y_{H2Of}, and Y_{N2f}. For estimation of B, etc., assume c_p for gas at $(T_w + T_f)/2$. How do the results change if one calculates T_w iterativly.

Solution

See Example 8, in Chapter 10. Similarly, the flame temperature and product mass fractions at the flame will be independent of the type of flow. The answers are exactly the same.

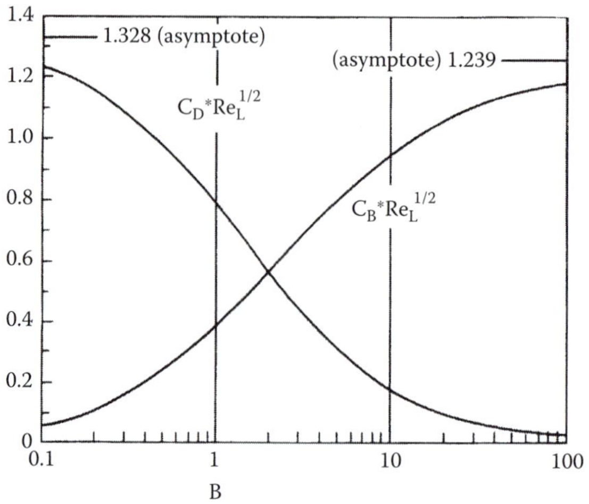

Figure 11.5 Nondimesional average burn rate and drag force under forced convection. The ordinate is also the same as –f(0). (Adapted from Williams, F.A., *Combustion Theory*, 2nd ed., Benjamin Cummins Publishing, Menlo Park, CA, 1985.)

11.9.1.2 Forced Convection

Using Table 11.2 for forced convection over a flat plate (see Equation R; Example 1), k = 0, $\alpha = 1/2$, $C_2 = Re_{ref}^{1/2}$

$$m^* = -(1/2)\ \{f(0)/Re_{ref}^{1/2}\}(1/x'^{*1/2}). \tag{11.99}$$

Rewriting $v_{ref} = v_{x\infty}$, $\rho_{ref} = \rho_\infty$, $x'^* = x^*$, $Re_{ref}\, x^* = Re_x = v_\infty x/\nu_\infty$

$$m^*(x^*) = \{\, \dot{m}''\ Re_x^{1/2}/(v_{x\infty}\rho_\infty)\} = -(1/2)\ \{f(0)\} \tag{11.100}$$

$$m^*_{avg} = \dot{m}''_{avg}\,(x)Re_x\,^{1/2}/(v_{x\infty}\rho_\infty)\} = 2\,\dot{m}''(x)\ \ Re_x\,^{1/2}/(v_{x\infty}\rho_\infty) = -\{f(0)\}. \tag{11.101}$$

The classical literature of Emmons contains a solution for f(0) forced convection when Pr = Sc = 1 [Emmons, 1956], which was recast in a different form by Williams. Adapting Williams' definition, $C_B = \dot{m}''_{avg}\,(x)/(v_{x\infty}\rho_\infty)$, $C_B\, Re_x^{1/2} = -f(0)$ and a plot for ($C_B\, Re_x^{1/2}$) or –f(0) vs. B is shown in Figure 11.5. In addition, defining $C_D = \{\int_0^L \mu(dv_x/dy)\, dx\}/\{L\, \rho_\infty v_{x\infty}^2\}$, ($C_D\, Re_L^{1/2}$) is also plotted in Figure 11.5.

For Example 1, at B=10, $C_B\, Re_L = 0.92$, $C_B = 0.0083 = \dot{m}''/(v_{x\infty}\,\rho_\infty)$

Hence $\dot{m}''_{avg} = 0.0083\ v_{x\infty}\,\rho_\infty = 0.00721 \times 10 \times 0.5807 = 0.00481$ kg/m² sec if $v_{x\infty} = 10$ m/sec and $\rho_\infty = 0.5807$ kg/m³.

11.9.1.3 Free Convection

Using Table 11.2,

$$\eta = C_1(y'^*/x'^{*\alpha}) = Gr_{ref}\ y'^*/\{4x'^{*\,(1/4)}\},\ \psi\,(x'^*,\ y'^*) = f(\eta)\ x'^{*(1-\alpha)}C_2$$
$$= 4^{3/4}f(\eta)\ x'^{*3/4}Gr_{ref}^{1/4}, \tag{11.102}$$

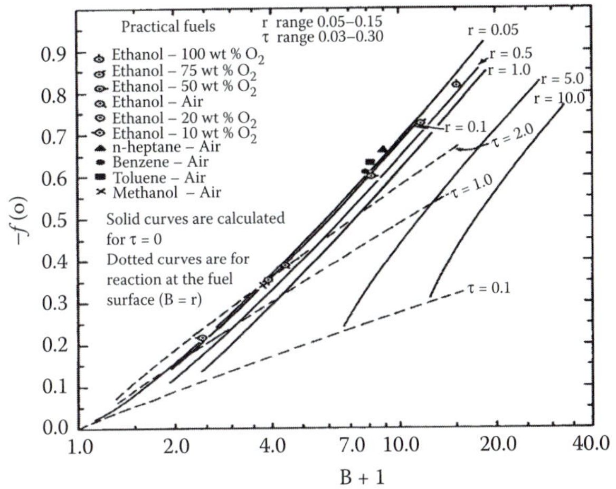

Dimensionless mass transfer, $f(o)$ vs. $B + 1$

Figure 11.6 The burn rate under free convection. (Adapted from Kim, J.S., J. de Ris, and W. Kroesser, Laminar Free Convective Burning of Fuel Surfaces, 13th Symposium on Combustion, pp. 949–961, 1971); ●, $r = s = Y_{O2\infty}/\{\nu_{O2}\, Y_{F,C}\}$; $\bullet/\{v_{ref}\, \sigma_{ref}\} = 3\, (Gr/4)^{1/4}\, \{f(0)\}/\{x^{*1/4}\}$.

11.9.1.3.1 Momentum

$$f''' - 2f'^2 + 3\, ff' + \theta = 0 \tag{11.103}$$

11.9.1.3.2 NSZ

$$\phi'' + 3f\, Sc\, \phi' = 0 \tag{11.104}$$

11.9.1.3.3 Burn Rate

$$m^* = \dot{m}''/\{v_{ref}\, \rho_{ref}\} = 3\, (Gr/4)^{1/4}\, \{-f(0)\}/\{x^{*1/4}\} \tag{11.105}$$

$$\phi'(0)\, B(1/3)/Sc = f(0) \tag{11.106}$$

Exact solutions for laminar free convection have been obtained by Kim et al. [1971]. Figure 11.6 shows the exact results. They have also obtained approximate integral solutions (see Subsection 11.9.2). In the figure, $\tau = \frac{cp(T_w - T_\infty)}{L}$, $v_{ref} = \nu_{ref}/d_{ref}$, $\rho_{ref} = \rho_w$, $\nu_{ref} = \nu_w$, $Gr_{ref} = g\, \{T_w - T_\infty\}\, d_{ref}^3/\{T_\infty\, \nu_w^2\}$, $T_w = T_g$ for gasification of plastics, and $T_w \approx T_{BP}$ for liquids. Hence, $\dot{m}''(x) = 3\, (\rho_w\, \nu_w/x)\, Gr_x^{1/4}\, \{-f(0)\}$, where $Gr_x\, \{g\, (T_w - T_\infty)\, x^3/(4T_\infty\, \nu_w^2)\}^{1/4}$, letting H be the height of plate, $\dot{m}''_{avg}(H) = (4/3)\, \dot{m}''(H)$ where f(0) is determined from Figure 11.6. Thus, plastics' burn rates under free convection can be evaluated using data in Table A.2D and with liquid burning using Table A.2A of Appendix A. The velocity contours in Figure 11.7 show that maximum velocity occurs between the surface and flame.

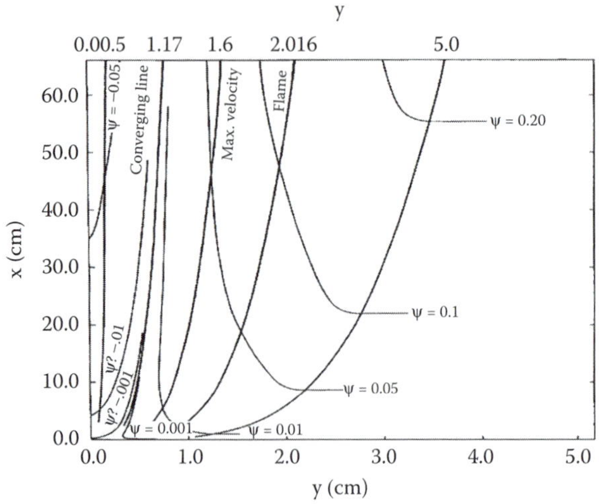

Figure 11.7 The velocity profiles and streamlines under free convection. (Adapted from Kim, J.S., J. de Ris, and W. Kroesser, Laminar Free Convective Burning of Fuel Surfaces, 13th Symposium on Combustion, pp. 949–961, 1971.)

11.9.1.4 Stagnation Flows

11.9.1.4.1 2-D System

$v_{x\infty}$ = ax, (for example, for flow over a large cylinder of diameter d, a = 4 * U/d, where U is the cross-flow jet velocity; d_{ref} could be selected as d).

$$(f')^2 - ff'' = f''' + (T/T_\infty) \tag{11.107}$$

$$\phi'' + Sc\ f\ \phi' = 0 \tag{11.108}$$

With K = 1, σ = 0,

$$m^* = \dot{m}''/\{v_{ref}\ \rho_{ref}\} = Re_{ref}^{1/2}\{-f(0)\} \tag{11.109}$$

$$\phi'(0)\ B/\{Sc\} = -f(0). \tag{11.110}$$

Kinoshita et al. [1981] related f(0) to the B number as

$$-f(0) = 0.543 s^{-0.02} B^{0.626} B^{-0.0678 \ln B}, \quad s = \frac{Y_{O2\infty}}{Y_{Fw} \nu_{O_2}}. \tag{11.111a, 11.111b}$$

For Plexiglas f(0) = 0.53 [Kinoshita et al., 1981]. See Equation 8.107 for $Y_{F,W}$.

11.9.1.3.2 Axisymmetric Flow

From Tables 11.1 and 11.2, k = 1,$v_{x\infty}$ = ax, $x'^* = x^{*3}/3$, $\eta = Re_{ref}^{\ 1/2}\ 3^{1/2}\ (y'^*/x^*)$, $\psi\ (x'^*, y'^*) = f(\eta)\{3^1/6/Re_{ref}^{\ 1/2}\}\ x'^{*2/3}$, v_{ref} = a d_{ref}, where a = $dv_{x\infty}/dx$.

Momentum

$$(f')^2\ (1/3)\ - ff''\ ((2/3) = f''' + (1/3)\ (T/T_\infty) \tag{11.112}$$

Energy and Species

$$\phi'' + Sc\ f\ \phi'\ (2/3) = 0 \tag{11.113}$$

$$m^* = -(2/3^{1/2})\ f(0)/Re_{ref}^{1/2} \tag{11.114}$$

$$\phi'(0)\ (3/2)\ B/Sc = f(0) \tag{11.115}$$

If (T/T_∞) is ignored for the momentum equation, whether a temperature peak exists or not, the burn rate depends only upon B.

11.9.2 Approximate Results

11.9.2.1 Simplified Solutions from Fluid Mechanics and Heat Transfer for Liquids, Solids, and Plastics

In this section, we use relations in the heat transfer and fluid mechanics literature to determine burn rates. If $f(0) \approx 0$ (i.e., the burn rate is very small and, as such, $v_y(0) \approx 0$; see Equation 11.77b), and we ignore the variation of ρ_∞/ρ in the momentum source term for forced convection over curved surfaces and stagnation flows under diffusion-controlled combustion. Then, the governing equations for momentum, species, normalized SZ, etc., and BCs for forced and stagnation flows with chemical reactions are similar to those in fluid mechanics and heat transfer without chemical reactions. Thus $\phi'(0) = \theta'(0)$, where $\theta = (T - T_\infty)/(T_w - T_\infty)$ for heat transfer problems. Note that free convection cannot be included here if there is a temperature peak in the gas phase because momentum equations with gas-phase combustion are different compared to those without combustion; however, if there are only heterogeneous reactions such as C burning in air with negligible gas-phase oxidation, we can use free-convection relations from the heat transfer literature.

Reader is referred to Section 8.6.4.3 for approximate results on burning rates for different flows.

A brief procedure is as follows:

1. Use the correlations for Nu from heat transfer literature. Since mass transfer is approximately zero, let $Nu = Nu_0$.
2. With $Le = 1$, use $Sh_0 = Nu_0$ and hence get $h_{m,0}$, mass transfer coefficient under low mass transfer conditions.
3. Obtain the burn rate using $\dot{m}''_0\ (x) = h_{m,0}(x)\ B$ (Chapter 8).
4. Use mass transfer correction and reestimate $\dot{m}''(x) = h_{m,0}(x)\ \ln(1+B)$

Example 3

Consider a flat hexane liquid pool burning in air with air velocity at about 0.1 m/sec. The air temperature is 600 K. Assume properties of air at the boiling point temperature of *n*-hexane. Other data are as follows: c_{pfuel} = 4.39 kJ/kg K, ρ_{liq} = 659 kg/m^3, T_{BP} = 342.15 K, L = 335 kJ/kg, h_c = 45105 kJ/kg, length of plate L = 0.1 m

Free-stream gas data : $Y_{O2\infty}$ = 0.23, T_∞ = 600 K, P = 1 bar

Transport property data : λ_{fuel} =0.0001525 kW/m K, λ_{gas} = 0.000083471 kW/m K, Pr = 1, c_{pgas} =1.17 kJ/kg K

Assume $T_w = T_{BP}$ and Sc = Pr; $Nu_o(x) = 0.332\ Re_x^{1/2}\ Pr^{1/3}$ and $Nu_o(x) = Sh_o(x)$. Determine the (a) mass burn rate using heat transfer analogy and (b) compare with the exact solution.

Solution

$$B = \{c_p\ (T_\infty - T_w)/L\} + (Y_{O2,\infty}\ h_c)/(\nu_{O2}\ L)$$

$$C_6H_{14} + 9.5\ O_2 \rightarrow 6\ CO_2 + 7\ H_2O$$

Molecular weight, M_F = 86.2

ν_{O2} = 9.5*32/M_F = 3.527 kg of O_2 per kg of fuel

Assuming $T_w = T_{BP}$, B = {1.17(600 − 342.2)/335} + (0.23 × 45,105)/(3.527 × 335) = 9.68.

As we selected reference conditions at the wall, we select all properties at the wall for air at about 400 K, $\rho_w = P/RT_w \approx 100 \times 28.97/(8.314 \times 342.2)$ = 1.018 kg m^{-3},

$\nu_w = 2.64 \times 10^{-5}$ m^2/sec,

ν/D = Sc = Pr = 0.71, D = 2.64 * 10^{-5}/1 = 2.64×10^{-5} m^2/sec,

$\rho_w D_w$ = 2.69 * 10^{-5} kg/m sec.

(a) *Approximate solution using heat transfer analogy*. For a flat plate, we find that $[h_{H,avg}\ L/\lambda] = Sh = h_{m,avg}\ L/\rho D = 0.664\ Re_\infty^{1/2}\ Pr^{0.333}$; because $h_{Havg} = 2 \times h_H(x)$, then

$Re_x = [v_{x,\infty}\ x]/\nu_w = 10 \times 0.1/2.64 \times 10^{-5} = 37{,}900$, Pr = 1; using Nu correlation for Sh,

$Sh = 0.664 \times 37{,}900^{1/2}\ 1^{\ 0.333} = 129$, $h_{m,avg} = Sh\ \rho D/L = 129 \times 2.69 \times 10^{-5}/0.1$ = 0.0348 kg/m^2 sec

$\dot{m}''_{avg} = h_{m,avg}$ B under low mass transfer = 0.0348 × 9.68 = 0.336 kg/m^2 sec

$\dot{m}''_{avg} = h_{m,avg}\ B \times \ln(1 + B)/B = h_{m,avg} \ln(1 + B)$ under high mass transfer = 0.0348 × ln (1 + 9.68) = 0.0823 kg/m^2 sec.

(b) *Williams exact solution*. Because $\rho^2\nu$ = constant, then $\rho_w^2\nu_w = \rho_\infty^2\nu_\infty$, so $\nu_\infty = \rho_w^2\nu_w/\rho_\infty^2 \approx T_\infty^2\nu_w/T_w^2$ = (600/342)2 * 2.64 * 10^{-5} = 8.12 × 10^{-5} m^2/sec

$$\rho_\infty = P/RT_\infty = 100*28.97/(8.314 * 600) = 0.5807\ \text{kg/m}^3.$$

$$C_B * Re_L^{1/2} = -f(0)$$

$$Re_L = (v_{x\infty} \times L/\nu_\infty) = 10 \times 0.1/8.12x10^{-5} = 12{,}300$$

From Figure, at B = 9.68, $C_B^* = C_B * Re_L^{1/2} = 0.92$, $C_B = 0.00829 = \dot{m}''/(v_{x\infty}\ \rho_\infty)$

Hence $\dot{m}''_{avg}$ = 0.00829 * $v_{x\infty}\ \rho_\infty$ = 0.00829 * 10 * 0.5807 = 0.0481 kg/m^2 sec.

Remarks

(i) This approximate solution under high mass transfer agrees better than low mass transfer conditions; the approximate solution has an error of 70.9%.
(ii) Regression velocity = $\dot{m}''/\rho_{liq}$ = {0.00481 kg/m^2 sec}/659 {kg/m^3} = 7.3 × 10^{-6} m/sec = 0.0073 mm/sec; thus, for complete combustion, time required for a film of 0.01-mm-thick hexane liquid layer (e.g., automobile engine), is 1 × 10^{-5} m/7.3 × 10^{-6} m/sec = 1.37 sec. Combustion stroke time for a 4-stroke 3000 RPM engine is only 10 msec. Thus, the HC will not be burned completely; cold liquid forms a film on the combustion chamber wall and will be emitted in exhaust.
(iii) The results are sensitive to the transport properties.
(iv) If B = 1, then from Figure, at B = 1, $C_B{}^* = C_{B\,\infty}\,Re_\infty^{1/2}$ = 0.4, Hence, $\dot{m}''_{avg}$ approx = 0.00241 kg/m^2 sec, whereas exact results $\dot{m}''_{avg,\ exact}$ = 0.00209 kg/m^2 sec with an error of +16%, i.e., the lower the B, the better the approximate result.

11.9.2.2 "Conventional" and "Nonconventional" Integral Technique

The conventional integral techniques employ polynomial profiles in dimensional forms of partial differential equations (Equation 11.28 to Equation 11.34), solve for boundary layer thicknesses and then determine the gradient at wall for use in mass burn rate relation. In "nonconventional" techniques we use the profiles directly in the differential equations involving similarity coordinates, i.e., Equation 11.73 and Equation 11.75, for determining the BL thickness and burn rate. The procedure is briefly outlined here (for details, see Annamalai et al., 1986, 1988.)

11.9.2.2.1 Species and NSZ Profiles

For free, forced, stagnation flows, etc., assume the following for use in Equation 11.75

$$\phi_s = a_0 + a_1\,\eta_s + a_2\,\eta_s^2 + a_3\,\eta_s^3 \qquad (11.116)$$

where $\eta_s = y'/\delta'_s$ where δ'_s is the species BL thickness in "compressed" coordinate. Note that η in exact method ≠ η_s in integral method. Using the BCs $\phi_s(0) = 1$, $\phi_s(1) = 0$, $\phi_s''(0) = -B\,\phi_s'(0)^2$ (Equation 11.84), $\phi_s'(1) = 0$, one can solve for four unknowns in Equation 11.116 (see exercise problems). Thus, for forced and free convection and stagnation flows,

$$a_0 = 1, \qquad (11.117)$$

$$a_1 = \frac{2}{B}\left[1-\left(\frac{3B}{2}+1\right)^{1/2}\right],\ \frac{2}{B}\left[1+\left(\frac{3B}{2}+1\right)^{1/2}\right] \qquad (11.118)$$

(a_1 = 3/2 as B → 0 or low mass transfer)

$$a_2 = -2a_1 - 3 \qquad (11.119)$$

$$a_3 = a_1 + 2 \qquad (11.120)$$

For problems with low mass transfer,
when B <<<< 1, $a_1 = 3/2$, $a_2 = 0$, a3 = 1/2, and

$$\phi_s = 1 + (-3/2)\ \eta_s + 1/2\ \eta_s^{\ 3}, \quad \text{or} \quad 1-\phi_s = 3/2\ \eta_s - 1/2\ \eta_s^3, \quad B <<< 1. \tag{11.121}$$

(Orloff et al. [1971] uses $\phi_s = \left(1 - \frac{y'}{\delta'_s}\right)^{3/2}$ particularly for free convection.)

This profile given in Equation 11.121 is similar to the temperature profiles adopted in nonreacting thermal BLs (Incropera and DeWitt, 2002).

Recall from Equation 11.42

$$\dot{m}'' = -\rho D\ B\ (\partial\phi/\partial y)_{y=0}.$$

$$\text{Thus,}\ \dot{m}'' = -\rho D\ (d\phi_s/d\ \eta'_s)_0\ (1/\delta'_s)\ B = (-a_1)\ \rho D\ (B/\delta'_s). \tag{11.122}$$

For evaporation and combustion problems, $B > 0$, $\dot{m}'' > 0$ and hence $a_1 < 0$. When $B < 0$ (e.g., condensation of vapors at the surface, particle deposition at the surface owing to chemical reaction), one can get two solutions for a_1, both being greater than zero and, hence, $\dot{m}'' < 0$.

11.9.2.2.2 Flame Location (η_f,s)

From the known polynomial profile $\phi_s(\eta_s)$ (Equation 11.116), where $\eta_s = y'/\delta_s'$ and with $\phi_s = \phi_f$, one can solve for $\eta_{f,s} = y_f/\delta_s'$ (Equation 11.112). Figure 11.8 shows the variation of $\eta_{f,s}$ with B, when $s = y_{O_2,\infty}/\nu_{O_2}$ as parameter. (Also see exercise problem.)

Remarks

Orloff has an approximate solution for $\eta_{s,f} = (1 - \Phi_f^{\ 0.6667}) = 0.834$, whereas the current profile yields 0.823, when B = 9.68 and S = 0.0653.

The conversion from y′ to y requires integration with knowledge of temperature variation.

11.9.2.2.3 Momentum or Velocity Profiles

(i) *Forced and stagnation flows:* Similarly, one may let $f_m' = v_x/v_{x\infty}$; and select the velocity profile for the momentum equation (exercise problems 11.11–11.13) for use in Equation 11.73. Define

$$F' = 1-f'_m = b_0 + b_1\ \eta + b_2\ \eta^2 + b_3\eta^3 \tag{11.123}$$

$$f'_m = 1-b_0-b_1\ \eta_m-b_2\ \eta_m^{\ 2} - b_3\ \eta_m^{\ 3}, \tag{11.124}$$

where $\eta_m = y'/\delta_m'$, $\eta_m = 0$, $f'_m = 0$, $F' = 1$, $\eta_m = 1$, $F' = 0$, $f'_m = 1$; $F'' = 0$, $f''_m = 0$. Thus $b_0 = 1$, $b_2 = -2b_1 - 3$, $b_3 = b_1 + 2$. We solve for b_1 using the fourth condition

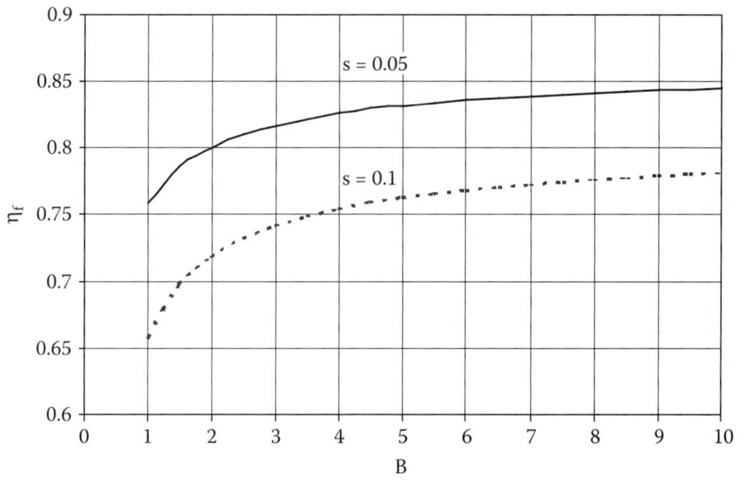

Figure 11.8 Variation of flame location for forced, free, and stagnation flows from polynomial profiles.

from the momentum Equation 11.73 at $\eta_m = 0$. Note that $\eta_m \neq \eta$ in exact method. Thus,

$$(f'(0))^2 (1-2\alpha) - f(0)f''(0)(1-\alpha) = (C_1 C_2/Re_{ref})f'''(0) + C_m \quad (11.125)$$

where $f' = df/d\eta$ but $df/d\eta \neq df_m/d\eta_m$. The procedure for determining b_1 is left as an exercise problem.

Letting $\delta_s/\delta_m = \tau$, a constant and, solving for b_1

$$b_1 = \{6 (C_1 C_2/Re_{ref}) + c^2 C_m/\tau^2\}/\{(a_1 B/\tau Sc) - 4 (C_1 C_2/Re_{ref})\}. \quad (11.126)$$

For example, if $C_m = 0$, $\alpha = 1/2$, $Sc = 1$, $b_1 = \{6(C_1 C_2/Re_{ref}) + d^2 C_m\}/(a_1 B/(\tau Sc) - 4 (C_1 C_2/Re_{ref})\}$, $\tau = 1$, $(C_1 C_2/Re_{ref}) = 1$ as in a forced convection and hence

$$b_1 = \{6/\{a_1 B - 4\}, \text{ forced convection.} \quad (11.127)$$

(ii) *Free Convection:* The velocity profile is different because of velocity being zero at $y' = 0$ and $y' = \delta'$ with a peak within $0 < y' < d'$. Further, the velocity is normalized to maximum velocity, which occurs within the BL.

With $\eta_m = y'/\delta'_m$, $f'_m = 0$ at $\eta_m = 0$, $\eta_m = 1$, $f'''_m = 0$ at $\eta_m = 1$, the velocity profile becomes,

$$f_m' = c_1 \eta_m (1-\eta_m)^2 = (27/4) \eta_m (1-\eta_m)^2. \quad (11.128)$$

$$\left(\text{Orloff et al. [1971] uses, } \frac{v_x}{v_{x\max}} = \frac{y}{\delta'_m}\left(1-\frac{y}{\delta'_m}\right)^{3/2}.\right)$$

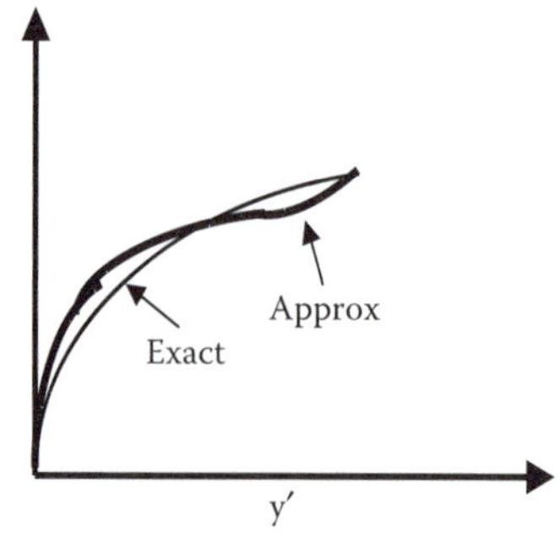

Figure 11.9 Exact and approximate profile in conventional integral techniques.

11.9.2.2.4 Solutions for δ_s' and δ_m''

When one plots f′ versus η for exact method, one may get f′ = 0.99 at η = d (Figure 11.10a) and ϕ = 0.01 at η = c, but for assumed profile f'_m and ϕ_s, $\phi_s = \phi = 0$ at $\eta_s = 1$ and f′ = 0.99 at η_m = 1 (Figure 11.10b). Then we set $\eta_s = \eta/c$ and $\eta_m = \eta/c$; c and d can be solved as follows.

The polynomial profiles satisfy the momentum and SZ equations but only at $\eta_m = 0$, and $\eta_m = 1$ in Equation 11.73 and $\eta_s = 0$ and $\eta_s = 1$ in Equation 11.75, but not within $0 < \eta_m < 1$, and $0 < \eta_s < 1$. Thus, the right-hand sides of Equation 11.73 and Equation 11.75 will have an error value. The error values for Equation 11.73 and Equation 11.75 depend on c and d and we determine the values c and d such that integrated error is zero (Figure 11.9). The solutions so obtained are the same as conventional integral solutions. As the error may have negative and positive values in the conventional method, the error may be squared and then c and d may be selected so

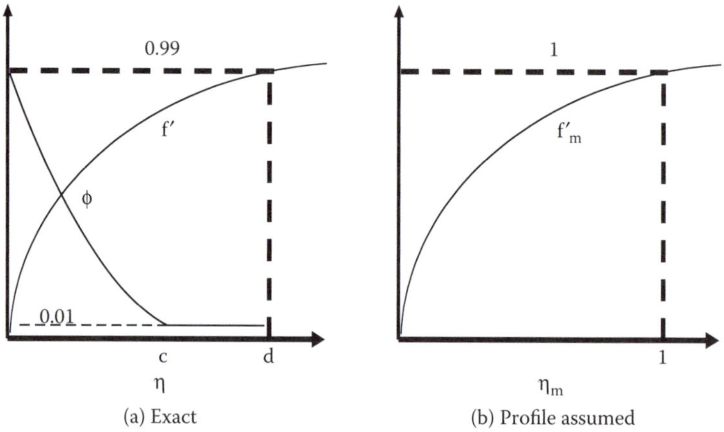

Figure 11.10 Variation of f with η as obtained from the exact solution. (a) Exact, (b) profile assumed $\eta_m = \eta$/d, "d" to be determined from error minimization.

that the integrated (error)^2 is minimized. However, solutions will be slightly different but may be more accurate [Annamalai et al., 1988].

11.10 Combustion Results for Fuels Burning under Convection

An example will be presented briefly to obtain explicit results for forced convection when Sc = Pr = 1 using the "nonconventional" integral technique.

Example 4

Consider,

$$\{C_1\ C_2/Re_{ref}\}\ \phi'' + Sc\ f\ \phi'\ (1 - \alpha) = 0 \qquad \text{(A)}$$

(a) Obtain expressions for BL thickness (δ_s) and burn rate ($\dot{m}''$) of a fuel burning under forced convection. Assume Sc = Pr = 1.
(b) Using same data as in Example 3, determine BL thickness and average burn rate if length (X) is 0.1 m.

Solution

$$C_1 = Re^{1/2},\ x' = x$$

Let $\gamma_1 = (1 - \alpha)$ Sc, $\{C_1\ C_2/Re_{ref}\} = \chi$; Equation A becomes,

$$\chi\Phi'' + f\ \phi'\ \gamma_1 = 0. \qquad \text{(B)}$$

If polynomial profiles are used, then there will be error "ε" on the right-hand side of Equation B. Integrating once and using integration by parts for the second term

$$0 - \chi\Phi'(0) + \gamma_1\ \{0 - f(0)\} - \gamma_1 \int_0^1 f'\ \phi\ d\eta = \int_0^1 \varepsilon(\eta)\ d\eta. \qquad \text{(C)}$$

Using Equation 11.81 and setting inegrated error to zero, $\phi'(0)\ B\chi/\gamma_1 = f(0)$ (D)

$$0 - \chi\Phi'(0) - \gamma_1\ \phi'\ (0)\chi B/\gamma_1 - \gamma_1 \int_0^1 f'\ \phi\ d\eta = 0. \qquad \text{(E)}$$

$$\text{Let } \eta_s = \eta/c \qquad \text{(F)}$$

$$f' = v_x/v_{ref} = v_x/v_{x\infty} = f'_m;\ \text{then,}$$

$$\Phi' = d\Phi/d\eta = d\Phi/d\eta_s\ d\eta_s/d\ \eta = \Phi_s'\ (1/c),\ \Phi'' = \Phi_s''\ (1/c^2) \qquad \text{(G), (H)}$$

$$0 - \Phi_s'(0)\chi\ \{1 + B\}/c - \gamma_1 \int_0^1 f_m'\ \phi_s\ c\ d\eta_s = 0 \qquad \text{(I)}$$

Solving for

$$c^2 = -\{\Phi_s'(0)\chi\{1 + B\}/\gamma_1\}/\left\{\int_0^1 f_m' \, \phi \, d\eta_s\right\} \tag{J}$$

(a) Forced Convection: Sc = Pr = 1, $(\delta_s/\delta_m) = \tau = 1$ or $\delta_s = \delta_m = \delta$ and hence $c = d$, $\eta_m = \eta_s$, $\gamma_1 = (1-\alpha) = 1/2$, $\chi = 1$, $1 - f_m' = \phi_s$.

Substitute the profiles in Equation J and get c^2 as

$$c^2 = 840\ a_1\ \{1 + B\}/[\{-54 + 9\ a_1 + 4a_1^2\}] \tag{K}$$

$$\eta_s = \eta/c = Re^{1/2}\ (y'^*/c\ x'^{*1/2}) = Re^{1/2}\ (y'^*/x'^{*1/2})\ [\{-54 + 9\ a_1 + 4a_1^2\}]^{1/2}/[840\ a_1 \times \{1 + B\}]^{1/2} \tag{L}$$

Using Equation L and $Re_x = v_{x\infty}\ x/\nu$

$$\frac{\delta_s'}{x} = \sqrt{\frac{840\ \ (-a_1)\,(B+1)}{\left(54-9a_1-4a_1^{\ 2}\right)Re_x}} \tag{M}$$

Using Equation 11.42

$$\dot{m}'' = -\rho D\ (a_1/\delta_s)B \tag{N}$$

$$\frac{\dot{m}''\,x}{\rho D\sqrt{Re_x}} = (-a_1)\ B\sqrt{\frac{\left(54-9a_1-4a_1^{\ 2}\right)}{840\ \ (-a_1)\,(B+1)}}\,. \tag{O}$$

(b)
$$B = \{c_p\ (T_\infty - T_w)/L\} + \{(Y_{O2,\infty}\ h_c)/(\nu_{O2}\ L)\}$$

$$C_6H_{14} + 9.5\ O_2 \rightarrow 6\ CO_2 + 7\ H_2O,$$

Molecular weight, $M_F = 86.2$
$\nu_{O2} = 9.5*32/M_F = 3.527$ kg of O_2 per kg of fuel

Assuming $T_w = T_{BP}$, B = {1.17(600 − 342.2)/335} + (0.23 * 45105)/(3.527 * 335) = 9.6
With $a_0 = 1$, $a_1 = -0.6073$, $a_2 = -1.78538$, $a_3 = 1.39$
$Re_x = v_{x,\infty}\ X/\nu_w = v_{x,\infty}\ X/D_w = 10 * 0.1/2.64 \times 10^{-5} = 3.79 \times 10^4$

Boundary layer thickness:

$$\frac{\delta_s'}{x} = \sqrt{\frac{840\ \ (-a_1)\,(B+1)}{\left(54-9a_1-4a_1^{\ 2}\right)Re_x}} = 0.04981,\ \delta_s' = 0.049 * 0.1 = 0.00482\ \text{m}$$

because $y' = \int_0^y \frac{\rho}{\rho_w} dy \approx \frac{\rho_{avg}}{\rho_w} y \approx \frac{T_w}{T_{avg}} y$, where we invoked the ideal gas law and constant M.

Thus, $y' = (352/1200)\, y = 0.285\, y$ and hence $\delta' = 0.285\, \delta$, thus, $\delta = 0.00482/0.285 = 0.0175$ m.

Burn rate:

$\dot{m}''(x) = \rho D\,(-a_1)\,(1/\delta'_s)\, B = 2.69 \times 10^{-5} * 0.6073 * (1/0.00482)\, 9.68 = 0.0317$ kg/m² sec.

Note that $\delta'_s \propto x^{1/2}$ and hence $\dot{m}''(x) \propto x^{-1/2}$

$$\dot{m}''_{avg} = y' = \frac{1}{X}\int_0^X \dot{m}''(x)\,dx = 2 * \dot{m}''(X) \text{ since } \dot{m}''(x) \propto x^{-1/2} = 0.0635 \text{ kg/m}^2$$

sec.

An exact solution yields $\dot{m}_{exact}'' = 0.0481$ kg/m² sec.

■ ■ ■

Combustion of Fully Pyrolyzing Plastics. The relations are similar to those for liquid fuels except that solid plastics gasify upon heating and the composition of gas is different from that of solid fuel. For example, a solid polymer such as Plexiglas decomposes into gaseous monomers $C_5H_8O_2$ (see Table A.2C of Appendix A for relevant polymer properties).

11.10.1 Chemical Reactions Involving Nonpyrolyzing Solids

11.10.1.1 Chemical Vapor Deposition

Example 5

A gaseous mixture of 98% H_2 and 2% $SiCl_4$ (mole %) at 0°C, 1 bar flows at a velocity of 0.15 m/sec over a flat plate maintained at T = 1473 K. The following heterogeneous reaction takes place at the surface

$$SiCl_4(g) + 2\,H_2(g) \Leftrightarrow Si\,(s) + 4\,HCl\,(g) \qquad \text{(A)}$$

Si(s) is deposited on the plate. The atomic K^0 is given as: log $K^0(T)$ for $SiCl_4(g) \Leftrightarrow Si\,(s) + 2Cl_2\,(g) = 16.5297$, for $HCl \Leftrightarrow 1/2\, H_2 + 1/2\, Cl_2 = 3.679$.

(a) Compute free-stream element mass fractions.
(b) Formulate the element conservation equations.
(c) Write down the interface mass balance equation.
(d) Calculate $K^0(T)$ for the given reaction.
(e) If equilibrium is assumed for heterogeneous Reaction A, then estimate the species mass fraction at wall and the average growth rate of Si(s) per unit surface area; $\rho_{si}(s) = 2420$ kg/m³.
(f) Estimate the rate of heat transfer at the interface. At 1473K, $h_{sicl_4,\bar{\bar{w}},}$ −3184 $h_{Hcl,\bar{\bar{w}},}$ −1528 $h_{H_2,\bar{\bar{w}},}$ 17544. At 273K $h_{sicl4,\bar{\bar{\infty}},}$ −3915 $h_{H_2,\bar{\bar{\alpha}},}$ −358 all in kJ/kg

Assume Sc = 0.7.

Solution

The $SiCl_4$ and H_2 are transported by convective and diffusive processes to the plate, where they undergo heterogeneous reactions. The reactions cause the $SiCl_4$ concentration to decrease at the surface. Thus, if $Y_{SiCl4,w}$ is known, then the transport rates from the bulk gas phase as well as the heterogeneous reaction rate can be estimated, which will lead to the determination of the deposition rate. Finally, the heterogeneous reaction rate can either be estimated if kinetics are known, or we can assume that heterogeneous reactions occur such that there is chemical equilibrium between Si, $SiCl_4$, H_2, and HCl at the given plate temperature.

(a) *The Ambient Element Si Mass Fractions*:

$$Z_{Si,\infty} = Y_{SiCl4,\infty}, M_{Si}/M_{SiCl4}\ Z_{H,\infty} = Y_{H2,\infty} + Y_{HCl\,\infty}\ M_H/M_{HCl},\ Z_{Cl,\infty}$$
$$= Y_{SiCl4,\infty}\ 4 * M_{Cl}/M_{SiCl4} + Y_{HCl\backslash\infty}\ M_{Cl}/M_{HCl}$$

$$M_{mix,\infty} = X_{H2,\infty} * M_{H2} + X_{SiCl4,\infty} * M_{SiCl4} = 0.98 * 2.01 + 0.02 * 169.92 = 5.374$$

$$Y_{SiCl4,\infty}, = X_{siCl4} * M_{SiCl4}/M_{mix,\infty} = 0.632,\ Y_{H2,\infty} = 0.368$$

Using the ideal gas law $\rho_\infty, = P\ M_{mix,\infty}/R_u\ T_\infty = 100 * 5.374/(8.314 * 273) = 0.237\ kg/m^3$.

For the gas mixture, using the linear law (Chapter 6), the viscosity is estimated as $\mu_{mix,\infty} = \mu_{H2,\infty} * X_{H2} + \mu_{siCl4,\infty} * X_{SiCl4} \approx \mu_{H2,\infty} = 8.41 * 10^{-6}\ kg/m^s$, $\nu_\infty = \mu_{mix,\infty}/\rho_\infty, = 0.351 \times 10^{-4}\ m^2/sec$.

(b) *Element Conservation and Si Deposition:*

So far, we have discussed the solutions with species conservation equations. Now we will present atom conservation equations, the important difference being the absence of chemical source terms, which become analogous to SZ variable equations. Consider the deposition of Si with any number of heterogeneous and homogeneous reactions involving Si, H, and Cl.

The momentum equations remain the same and normalized SZ, viz.

$$\phi'' + \gamma Sc\ f\ \phi' = 0$$

will remain the same except that

$$\phi = \{Z_k - Z_{k\,\infty}\}/(Z_{k\,w} - Z_{k\,w}),\ k = Si, H, Cl$$

(c) *Interface Conservation Conditions:*

The deposition rate combustion of solids in forced, free, and stagnation flows can be solved by using the BC at the surface: for Si, the mass in or deposition rate at the wall per unit surface area is given as,

$$\rho_w v_{y,w}\ Z_{Si,w} - \{\rho D\ \partial Z_{Si}/\partial y)_{y=0} = \dot{m}''_{si} = \dot{m}''$$

Note that $\rho_w v_{y,w} = \dot{m}''$, $v_{y,w} < 0$ and hence $\dot{m}'' < 0$ in this case

$$\dot{m}'' = \rho_w v_{y,w} = -\{\rho D\ \partial Z_{Si}/\partial y)_{y=0}/(1 - Z_{Si,w})$$

The hydrogen atom and Cl atoms do not get deposited; hence for H and Cl,

$$\rho_w v_{y,w} Z_k - \{\rho D\, \partial Z_k/\partial y\}_{y=0} = \dot{m}_k'' = 0, \; k = H, Cl$$

Generalizing for any element Z_k,

$$\rho v_y Z_k - \rho D\, \partial Z_k/\partial y = \dot{m}'' F$$

$$\dot{m}'' = \rho_w v_{y,w} = -\rho D\, (\partial Z_k/\partial y)_{y=0}/(F - Z_{kw})$$

where for Si, F = 1 and $Z_k = Z_{si}$, F = 0 for $Z_k = Z_H$ and Z_{Cl}.

These equations are similar to those derived for β.

$$\dot{m}'' = \rho_w v_{y,w} = -\rho D\, (\partial \phi/\partial y)_{y=0}\, B$$

$$\text{where, } B = \{Z_{k\,w} - Z_{k\,\infty}\}/(F - Z_{k\,w})$$

$$k = Si, H, Cl$$

$$2(d\phi/d\eta)_0\, B/\{Sc\} = f(0)$$

Thus, if $Z_{si,w}$ is known, then $Z_{H,w}$ and $Z_{Cl,w}$ are known from the definition of B number. Note that the atom mass fraction Z_k is related to species mass fractions. For example, if the species are HCl, H_2, Si(s), and $SiCl_4$ at the wall

At the wall, $Z_{Si,w} = Y_{SiCl4,w}\, M_{Si}/M_{SiCl4}$

$Z_{H,w} = Y_{H2,w} * 1 + Y_{HClw}\, M_H/M_{HCl}$,

$Z_{Cl,w} = Y_{SiCl4,w}\, 4 * M_{Cl}/M_{SiCl4} + Y_{HClw}\, M_{Cl}/M_{HCl}$.

Thus, if $Z_{Si,w}$ or $Y_{SiCl4,w}$ is known, B can be calculated and hence the deposition rate.

At the surface we assume chemical equilibrium (Equation A). Thus,

$$K^0(T) = (p_{HCl}/1)^4/\{(pH2/1)^2 (p_{SiCl4}/1)\} = (X_{HCl})^4 (P/1)/\{(X_{H2}/1)^2 X_{SiCl4}\}$$

where the X_k's are related to Y_k's. (1) Assume $Y_{SiCl4,w}$, (2) calculate B and hence Z_{si}, Z_H, Z_{Cl}, (3) solve for Y_{H2}, Y_{HCl}, (4) calculate molecular weight of mixture at the wall, (5) convert to mole fractions, and (6) check whether the equilibrium relation is satisfied; if not, assume a different $Y_{SiCl4,w}$. To get the mass deposition rate, calculate B; the procedure is similar to that for other problems.

For forced convection $\dot{m}'' = -(\rho D/d_{ref})(d\phi/d\eta)_{y=0}\, B\, Re^{1/2}/L^{*\,1/2} - (\rho D/L)(d\phi/d\eta)_{y=0}\, B\, Re_L^{1/2}$

$\dot{m}''/\{-(\rho D/L)(d\phi/d\eta)_{y=0}\, Re_L^{1/2}\} = B$

Note: $\dot{m}'' \propto L^{1/2}$;

(d) *Equilibrium Constant:*

ln $K^0(T)$ is estimated from atomic equilibrium constants as

ln $K^0(T) = \ln K^0_{Si(s)} + 4 * \ln(K^0_{HCl}) - \ln(K^0_{SiCl4}) - 2\ln(K^0_{H2}) = 0 + 4 * 3.679 - 16.5299 - 0 = -1.8039$ or

$K^0(T) = 0.016$. Iteratively solving

$$Y_{SiCl4},W = 0.0673,\ Y_{H2,w} = 0.392,\ Y_{HCl,w} = 0.541,$$

(e) *Average Deposition and Growth Rate:*

$$B = \{Y_{siw} - Y_{si,\infty}\}/\{1 - Y_{siw}\} = \{0.0111 - 0.104\}/\{1 - 0.0111\} = -0.0945$$

$Re_x = [0.15 * L/0.351 \times 10^{-4}]^{1/2} = 65.4\ L^{1/2}$, $D = 0.351 \times 10^{-4}/Sc = 0.501 \times 10^{-4}\ m^2/sec$

Using exact results, and assuming $\rho_w D_w = \rho_\infty D_\infty$, $(d\phi/d\eta)_{y=0} \approx 0.43$

$\dot{m}''(L)/\{-(\rho D/L)(d\phi/d\eta)_{y=0} Re_L^{1/2}\} = B = -0.0945$; using Nu correlations and Nu = Sh

$\dot{m}''(L) = 1.054 \times 10^{-4}/L^{1/2}$ kg/m² sec.

Using Si density, $v_w = 1.054 \times 10^{-4}/[L^{1/2} * 2420]$ m/sec.

Because B number is so small, an approximate solution will suffice.

(f) *Heat Transfer Rate:*

Similarly, we consider total enthalpy. The interface energy balance at the wall yields,

$$\{\dot{m}'' h_w - \{(\rho D)\ \partial h/\partial y\}_{y=0} = \dot{m}''\ h_{si(s)}\ F.$$

Where $F = [h_{si(s)} + \dot{q}''_{rad} / \dot{m}'']$

Solving,

$$-\{(\rho D)\partial\phi/\partial y\}_{y=0} = \dot{m}''\ B$$

where $B = \{h - h_w\}/\{\{F - h_w\} = \{Z_{k\ w} - Z_{k\ \infty})\}/(F - Z_{k\ w})$, $\phi = (h - h_\infty)/(h_w - h_\infty)$

$$h_w = h_{HCl}\ Y_{HCl,w} + h_{SiCl4}\ Y_{SiCl4,w} + h_{H2,w}\ Y_{H2,w},\ h_\infty$$
$$= h_{HCl\infty}\ Y_{HCl,\infty} + h_{SiCl4\infty}\ Y_{SiCl4,\infty} + h_{H2,\infty}\ Y_{H2,\infty}$$

$$h_w - h_\infty = 8660\ kJ/kg,\ hw - h_{si(s)} = 5830 - 1060 = 4770\ kJ/kg.$$

Emissivity of Si = 0.525 on wavelength [Toulokian, 1972]

$$q_r'' = 134.5\ kW/m^2.$$

Surface heat flux can be shown to be $\{1.873 \times 10^{-3}/X^{1/2}\} + 135$ kW/m².

11.10.1.2 Boundary Layer Combustion of Carbon

For example, consider combustion of charcoal in a barbeque grill. When we blow air, the large char particles seem to glow more because we enhance the mass transfer rate of O_2 by convection to the particle surface that promotes the burn rate. In this section we will consider combustion of carbon under convection: forced and free convection and stagnation flows. We will use two approaches: (1) equilibrium composition at the charcoal surface and (2) diffusion-controlled combustion.

11.10.1.2.1 Equilibrium Method

Carbon reactions were discussed in Chapter 9 and are summarized in the following text:

$$C(s) + \frac{1}{2}O_2 \rightarrow CO \quad \text{(I)}$$

$$C(s)O_2 \rightarrow CO_2 \quad \text{(II)}$$

$$C(s)CO_2 \rightarrow 2CO \quad \text{(III)}$$

$$C(s)H_2 \rightarrow CO + H_2 \quad \text{(IV)}$$

$$CO + \frac{1}{2}\,O_2 \rightarrow CO_2 \quad \text{(V)}$$

The reactions range from the Boudouard reaction (Reaction III) followed by the oxidation of carbon monoxide to carbon dioxide (Reaction V) in the gas phase, to the direct oxidation of carbon to carbon monoxide (Reaction I) and carbon dioxide (Reaction II). Particularly, the two dominant mechanisms of carbon combustion are direct oxidation to CO with frozen gas phase (single-film model or SFM; see Chapter 9) and the Boudouard reaction at the carbon wall followed by CO oxidation (double-film model or DFM; see Chapter 9). One may use similar procedures as adopted for Si deposition for carbon combustion. See the exercise for Problem 11.18 the case of chemical equilibrium at the surface of a carbon plate when subjected to forced convection. For SFM, one may assume that species at the surface are (a) CO, O_2, N_2, (b) CO_2, O_2, and N_2, and (c) CO_2, CO, N_2, and calculate the burn rates.

11.10.1.2.2 Finite Chemistry and Diffusion-Controlled Combustion

Any one of Reaction I to Reaction IV, taken alone, is the SFM, whereas Reaction III followed by Reaction V constitutes the DFM. In Chapter 9, it was shown that the burning rate, surface temperature (with or without radiation loss), and wall mass fraction of CO under DFM with Reaction I are exactly same as those of SFM models. It can be shown that similar results are valid for forced convection and stagnation flows (when temperature-dependent density terms are neglected in the momentum equation).

11.10.1.2.3 SFM — Oxidation and Reduction Reactions

The derivation of expressions for burn rate and plate temperature are similar to those outlined in Chapter 9. They will be briefly summarized for free and forced convection and stagnation flows.

Burn Rate:

Using Equations 11.42 and 11.69 the burn rate expression becomes,

$$\dot{m}^* = -[C_1/x'^{*\,\alpha}]\; B\; \Phi'(0) \quad (11.129)$$

where $\dot{m}^* = \dot{m}''\, d_{ref}/(\rho_w D_w)$, (11.130)

$$B = \left(\frac{Y_{O2,\infty}}{\nu_{O2,I}} + \frac{Y_{CO2,\infty}}{\nu_{CO2,III}} \right), \text{ SFM, diffusion control, oxidation and reduction.}$$

(11.131)

The burn rate results can be obtained once $\Phi'(0)$ is obtained from integration of the differential equation (Equation 11.75).

CO Mass Fraction:

Using the procedure outlined in the appendix of Chapter 9, we obtain (see also exercise Problem 11.19):

$$Y_{CO,W} = \frac{Y_{CO,\infty} + \frac{Y_{O2,\infty}[1+\nu_{O2,I}]}{\nu_{O2,I}} + \frac{Y_{CO2,\infty}[1+\nu_{CO2,III}]}{\nu_{CO2,III}}}{[1+B]} = 1 - \frac{Y_{I,\infty}}{1+B} \quad (11.132)$$

Surface or Wall Temperature:

From energy balance, one can show (derivation similar to the one shown in the appendix of Chapter 9; also see exercise Problem 11.20) that

$$-\frac{h_{T\infty}}{\tilde{h}_{C,I}} = \frac{c_p(T_w - T_\infty)}{\tilde{h}_{C,I}} = \left(\frac{Y_{O2,\infty}}{\nu_{O2,I}} + \frac{Y_{CO2,\infty}\tilde{h}_{C,III}}{\nu_{CO2,III}\tilde{h}_{C,I}} \right) \quad (11.133)$$

where

$$\tilde{h}_{c,I} = h_{c,I} - \frac{\dot{q}''_{rad}}{\dot{m}''}$$

$$\tilde{h}_{c,III} = h_{c,III} - \frac{\dot{q}''_{rad}}{\dot{m}''} \quad (11.134)$$

Note that $h_{c,III} < 0$, whereas $h_{c,I} > 0$. For char burning, the mass loss is due to chemical reaction and/or species concentration gradients, whereas for the usual case of liquid or volatile fuels, the mass loss is due to pyrolysis and/or evaporation driven by temperature gradients. Consequently, the char-burning B number is independent of surface temperature or temperature gradients in contrast to the case for pyrolysis and/or evaporation.

11.10.1.3 Double-Film Model

The previous section dealt with carbon combustion under SFM. Now consider the DFM with finite gas-phase chemistry. Following the procedure outlined in Chapter 9 for a carbon particle, one can show that the B number is the same as Equation 11.131.

As opposed to liquid drop combustion in which the drop temperature is too cold to support oxidation reaction and $Y_{O2w} = 0$, the flame could be thick here because the carbon surface temperature is high, as indicated in Figure 11.11.

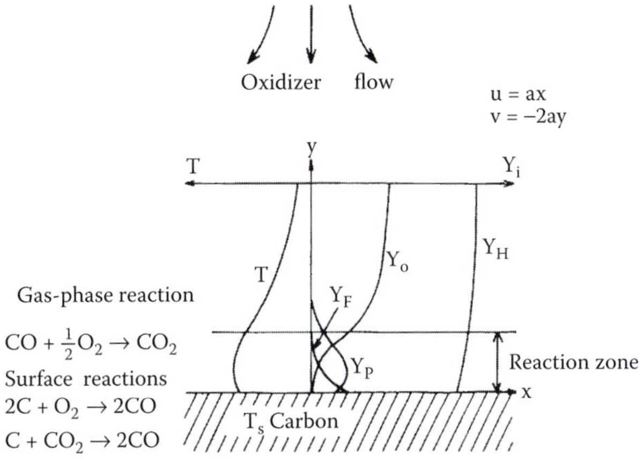

Figure 11.11 Flame structure with carbon combustion. (Adapted from Makino, 1990.)

For diffusion-controlled combustion, the carbon surface temperature is same as those in SFM with heterogeneous oxidation and reduction with CO_2 to CO.

$$B = Y_{CO2,\infty}/\nu_{CO2,III} + Y_{02,\infty}/\nu_{O2,I},\ \nu_{O2,I} = 1.3333,\ \nu_{CO2,III} = M_{CO2}/M_c = 3.666$$

As opposed to drops in which heat transfer drives the vapors away from the liquid surface, the mass transfer of CO_2 drives the combustion. For the case $Y_{CO2,\infty} = 0$, there is no difference between SFM with $\upsilon_1 = 1.333$ and the DFM combustion rate even though the latter produces CO_2, whereas the former produces CO. The diffusion rate of O_2 is slower for SFM because O_2 has to reach the surface instead of the flame as in DFM.

11.10.1.3.1 Flame Location

As derived in Chapter 9,

$$Y_{CO,W} = 1 - \frac{Y_{I,\infty}}{[1+B]}$$

$$\phi_f = 1 - \frac{1+B-Y_{I,\infty}}{1+B+s(1+B)-Y_{I,\infty}} = \frac{s(1+B)}{1+B+s(1+B)-Y_{I,\infty}}.$$

With $\phi = \phi_f$, the flame location can be solved using the solution for $\phi(\eta)$. Particularly with profiles for $\phi(\eta_s)$ used for the integral solution, solution $\eta_{s,f}$ can be obtained (Equation 11.116).

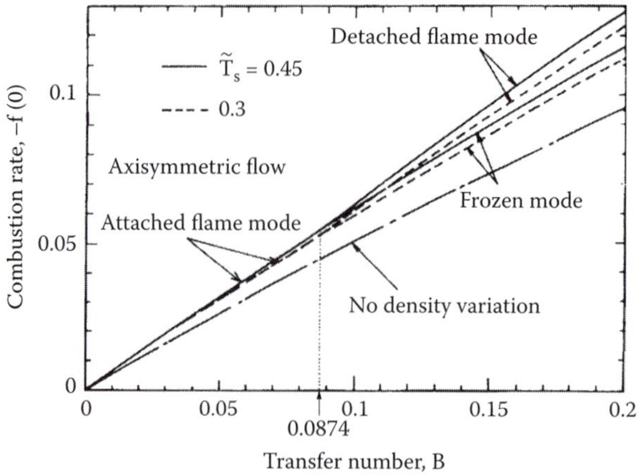

Figure 11.12 The f(0) for axisymmetric stagnation flows under frozen, reacting conditions for various B's. f_s = f(0). (Adapted from Makino.)

Similarly, the solution for flame temperature and CO_2 mass fraction at the flame are obtained (see Chapter 9) as:

$$\frac{c_p(T_f - T_\infty)}{h_{C,I}} = \frac{Y_{O2,\infty}}{\nu_{O2,V}} \frac{h_{C,V}}{h_{C,I}} + B \frac{s(1+B)}{1+B+s(1+B)-Y_{I,\infty}} \left[\frac{h_{C,III}}{h_{C,I}} - \frac{\dot{q}''_{rad}}{\dot{m}'' h_{C,I}} \right] \tag{11.134}$$

$$Y_{CO2,f} = B[1-\phi_f][1+\nu_{O2,I}][1+\nu_{O2,V}] = B[1-\phi_f]\nu_{CO_2,III}. \tag{11.135}$$

For the case of carbon combustion with oxidation and reduction surface reactions and homogeneous oxidation reactions (refer to Chapter 9; DFMs), a qualitative flame structure for carbon fuel is shown in Figure 11.11 whereas the f(0) is shown in Figure 11.12 versus B [Makino].

Example 5

Consider a flat carbon surface burning in vitiated gas with gas velocity at about 10 m/sec. The air temperature is 600 K. Assume properties of gas at the surface temperature of carbon. Other data are as follows: $c_{pcarbon}$ = 4.39 kJ/kg K, ρ_{carbon} = 1300 kg/m^3, $h_{c,I}$ = 9203.2 kJ/kg, $h_{c,III}$ = −14,360 kJ/kg, L, length of plate = 0.1 m.

Free-stream gas data: $Y_{O2\infty}$ = 0.23, $Y_{CO2\infty}$ = 0.5, and the rest is N_2; T_∞ = 600 K, P = 1 bar.

Transport property data for air at 600 K λ_{gas} = 0.0000469, Pr = 1, c_{pgas} = 1.051 kJ/kg K, $\mu = 3.06 \times 10^{-5}$ N sec/m^2, $\nu_w = 5.27 \times 10^{-5}$ m^2/sec.

Use DFM. Note that the burn rate per unit area and average burn rate will remain the same as in SFM except that the final products are CO_2. Assume diffusion-controlled combustion.

(a) Estimate the plate temperature including radiation heat loss with mass burn rate from heat transfer analogy.
(b) Determine burn rates using the integral technique.
(c) Estimate the plate and flame temperatures.
(d) Use integral solution to determine the (1) BL thickness in mm for carbon surface at L = 0.1 m if an average density is used in coordinate transformation and (2) the flame location.

Solution

(a) $\nu_{O2,\,I} = 1.332,\ \nu_{O2,\,III} = 3.664,\ v_{x,\infty} = 10$ m/sec

$$B = Y_{O2\infty}/\nu_{O2,I} + Y_{CO2\infty}/\nu_{CO2,III} = 0.23/1.332 + 0.5/3.664 = 0.3091$$

$$Re_L = \rho V\ L/\mu = V\ L/\nu = 10\text{ m/sec} * 0.1\text{ m}/5.27 \times 10^{-5} = 1.90E + 04$$

Approximate solution for burn rate from heat transfer analogy:

$$Nu_{average} = 0.664\ Re_L^{1/2}\ Pr^{1/3} = 0.664 * 19000^{1/2} = 9.15E{+}01$$

$$\text{Hence, } Sh_{average} = Nu_{average} = 9.15E{+}01$$

$$h_{m,\ average} = Sh_{average} * D_w/L = 91.5 * 5.27 \times 10^{-5}/0.10 = 0.0275\ \text{kg/m}^2\ \text{sec}$$

$$m''_{average,low\ B}, = h_{m,\ average}B = 0.0275 * 0.309 = 8.50E{-}03\ \text{kg/m}^2\ \text{sec}$$

$$m''_{average,high\ B} = h_{m,\ average}B\ [\ln(1 + B)/B] = 7.40E{-}03\ \text{kg/m}^2\ \text{sec}$$

(b) *Integral solution:*
Using Equations 11.116–11.120 with B = 0.3091

$$a_0 = 1;\ a_1 = -1.3576,\ a_2 = -0.2849,\ a_3 = 0.642$$

Compressed species BL thickness:
Using Equation M from Example 4

$$\delta_s'/x = 0.0366$$

$$\delta_s'/x = 0.0366,\ \text{with } L = 0.1\text{ m},\ \delta_s' = 0.00366\text{ m or } 3.66\text{ mm}$$

Compressed momentum BL thickness:

$$\delta_m' = \delta_s'\ s \text{ if } Pr = Sc = 1$$

Burn rate:
Using the profile given by Equation 11.116

$$\dot{m}'' = -\rho_w v_{y,w} = \rho D\ (d\ \phi/d\ \eta_s)_w\ (1/\delta_s')B = \rho D\ (-a_1)\ (1/\delta_s')B.$$

Local burn rate per unit area $\dot{m}''$ (x = L) = 0.00345 kg/m² sec
Average burn rate per unit area = $[\int_0^{0.1} \times \dot{m}''(x)\, dx]/x = 2 \times m''$ (x = L) = 0.00690 kg/m² sec

(c) *Plate or wall temperature:*

$$\frac{c_p(T_w - T_\infty)}{h_{C,I}} = \left(\frac{Y_{O2,\infty}}{\nu_{O2,I}} \left\{ 1 - \frac{\dot{q}''_{rad}}{\dot{m}'' h_{C,I}} \right\} + \frac{Y_{CO2,\infty}}{\nu_{CO2,III}} \left\{ \frac{h_{C,III}}{h_{C,I}} - \frac{\dot{q}''_{rad}}{\dot{m}'' h_{C,I}} \right\} \right)$$

As $\dot{m}'' \rightarrow \infty$ as x → 0, then radiation loss terms can be ignored. 1.051* $(T_w - 600)/9203 = 0.23/1.333 + (0.5/3.664) \times (-14360/9203) = -0.04025)$

T_w = 247.5 K at x ≈ 0. This temperature is lower than ambient temperature owing to extreme endothermicity due to reduction reaction of CO_2 at high concentrations of CO_2. Thus, diffusion-controlled approximation is invalid. At x = 0.1 m,

1.051 * $(T_w - 600)/9203 = 0.23/1.333\{1 - 5.67 \times 10^{-11}\ (T_w^4 - 600^4)/(0.00345 \times 9203)\} + (0.5/3.664)\{-14360/9203 - 5.67 \times 10^{-11}\ (T_w^4 - 600^4)/(0.00345 \times 9203)\}$; solving iteratively, T_w = 456.1 K; this temperature is higher than the temperature that would be obtained in the absence of radiation heat transfer because radiation heat input raises the temperature; but the assumption of infinite surface kinetics appears to be invalid.

Flame temperature:

$$\frac{c_p(T_f - T_\infty)}{h_{C,I}} = \frac{Y_{O2,\infty}}{\nu_{O2,V}} \frac{h_{C,V}}{h_{C,I}} - \frac{s(1+B)}{1+B+s(1+B)-Y_{I,\infty}} \left[\frac{h_{C,V}}{h_{C,I}} \frac{Y_{O2,\infty}}{\nu_{O2,V}} - \frac{c_p(T_w - T_\infty)}{h_{C,I}} \right]$$

12.056 * (Tf – 456.1)/9203 = {0.4026 * 10103.2/(9203)} + 0.4026 * (1 + 0.309) * (10103.2 * 04026/9203–1.0 * (456–600)/9203, T_f = 3120 K

(d) *BL thickness:*

From Part (b), δ_s'/x = 0.03656, with L = 0.1 m, the compressed (zipped) BL thickness is given as δ_s' = 0.003656 m or 3.656 mm.

$$dy' = \frac{\rho}{\rho_w} dy = \frac{T_w}{T} dy, dy = \frac{T}{T_w} dy'$$

Integrating $y_{k+1} - y_k = \{y'_{k+1} - y'_k\} * (T/T_w)$.

The BL thickness is estimated as 14 mm at x = 0.1 m. If we use an approximate procedure with an average density

$$y_f' = \int_0^{y_f} \frac{\rho}{\rho_w} dy \approx \frac{\rho_{avg}}{\rho_w} y_f \approx \frac{T_w}{T_{avg,w-f}} y_f, \ \delta'_s \approx \{456/\{0.5 * (456 + 3120)\}\delta_s,$$

$$\delta_s \approx \{0.5 * 3.656\ (456 + 3120)/456\},\ \delta_s = 14.3 \text{ mm}.$$

Flame location:

$\phi_f = s/\{s + Y_{COw}\}$,

$$Y_{CO,w} = 1 - \frac{Y_{I,\infty}}{[1+B]}$$

where $s = Y_{O2,\infty}/\upsilon_{O2,V} = 0.23/(16/28.01) = 0.5403$, $Y_{I,\infty} = 1 - 0.23 - 0.5 = 0.27$, $Y_{COw} = 1 - \{0.27/(1 + 0.3091)\} = 0.7938$, $\phi_f = 0.5403/(0.5403 + 0.7938) = 0.336$

If we adopt the ϕ profile given by Equation 11.116 and using the cubic solver (THERMOLAB Software with *Advanced Thermodynamics Engineering*; see CRC Web site), we have three solutions: $\eta_{s,f} = 1.419, -1.4707, 0.4945$; the only realistic solution is $\eta_{s,f} = y'_f/\delta_s' = 0.4945$.

$y_f' = 0.4945 * 0.003656 = 0.001808$ m

Figure 11.13 shows the flame structure. The exact solution requires "unzipping" the transformation.

$$dy' = \frac{\rho}{\rho_w} dy = \frac{T_w}{T} dy, dy = \frac{T}{T_w} dy', \text{ Integrating } y_{k+1} - y_k = \{y'_{k+1} - y'_k\} * (T/T_w)$$

$$y_f' = \int_0^{y_f} \frac{\rho}{\rho_w} dy \approx \frac{\rho_{avg}}{\rho_w} y_f \approx \frac{T_w}{T_{avg,w-f}} y_f \approx \{456/\{0.5 * (456 + 3120)\} * yf,$$

$$y_f = 0.00706 \text{ m} \quad \text{or} \quad 7.06 \text{ mm}$$

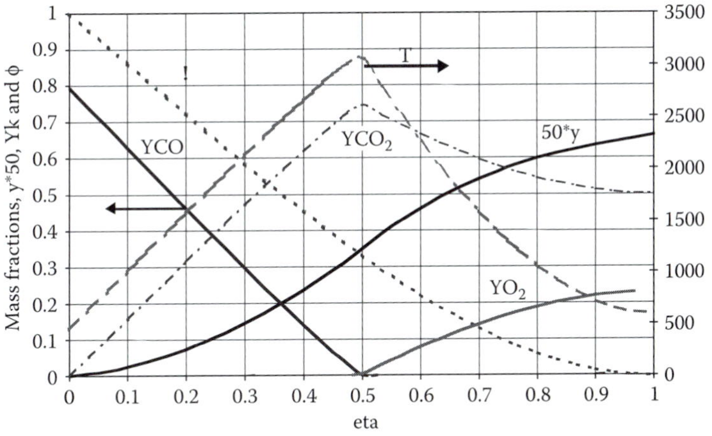

Figure 11.13 Flame structure under DFM for carbon combustion.

11.10.2 Free Convection

11.10.2.1 Physics of Free Convection

Consider an immersed object of volume V and density ρ in a bulk liquid of density ρ_b. It is well known in fluid mechanics that the upward force is equal to the weight of the displaced fluid (ρ_b g V) for an immersed object, whereas the downward force is ρ g V. Thus, the net upward force is ρ_b g V − ρ g V, which according to Newton's law should create a net upward acceleration of the mass:

$$\rho\ V\ d^2x/dt^2 = [\rho_b\ g\ V - \rho\ g\ V]$$

where x is the coordinate as indicated in Figure 11.2. If the bulk fluid is cold gas with density ρ_∞, whereas the object is hot gas of density ρ, then

$$\rho\ d^2x/dt^2 = g\ [\rho_\infty - \rho] \text{ or } d^2x/dt^2 = g\ [\rho_\infty/\rho - 1] = g\ [T/T_\infty - 1].$$

Multiply by 2 (dx/dt) dt and integrating

$$[v_x^2/2] = g\ [T/T_\infty - 1]x \tag{11.136}$$

where v_x is the free-convection velocity and $v_x \propto x^{1/2}$. Thus, hot fluid rises and the velocity of gases scale as $x^{1/2}$. Recall the simplified analyses presented in the introductory section.

$$h_H \approx \frac{\lambda}{\delta_T},\ \frac{kW}{m^2 K}$$

Using similar analysis as for momentum, time for thermal diffusion is given as (x/v_x) and because thermal diffusivity (m²/sec) * time (sec) = length², then $\alpha^* (x/v_x) = \delta_T^2$, where $v_x = \{[2g]\ [T/T_\infty - 1]x\}^{1/2}$; thus, $\{\delta_T/x\} \approx 1/[Pr^{1/2}\ Gr^{1/4}]$, where $Pr = \nu/\alpha$, $Gr = g\ \{T_w - T_\infty\}x^3/\{T_\infty\ \nu^2\}$. The thermal BL thickness δ_T and hence heat transfer coefficient are functions of velocity Gr. Typically, h_H is presented in terms of the dimensionless Nusselt number $Nu_x = h_H\ x/\lambda$. Thus, for free convection,

$$h_H\ x/\lambda \approx Nu_x = x/\boldsymbol{\delta_T} \approx [Pr^{1/2}\ Gr^{1/4}]. \tag{11.137}$$

Similarly, Sh may scale as

$$h_m\ x/(\rho D) \approx Sh_x = x/\boldsymbol{\delta_m} \approx [Sc^{1/2}\ Gr^{1/4}]. \tag{11.138}$$

Here, $\boldsymbol{\delta_T} = \boldsymbol{\delta_m}$.

This section presents some analytical results for burning rates and char surface temperature for a vertical flat plate in free-convection flow. In contrast to forced convection, the results for SFMs and DFMs under free convection yield different burning rates and surface temperatures. The governing equations are summarized in Table 11.2.

11.10.2.2 Simplified Solutions from Fluid Mechanics and Heat Transfer

If carbon combustion occurs under free convection, the burn rate is typically low and, as such, one can ignore the mass transfer effects; thus, one can adopt all the correlations used in the heat transfer literature for free convection (see Chapter 8 for correlation). If the mass transfer rate is high, then the same correlations may be used with mass transfer corrections.

11.10.2.3 Integral Solutions for Free-Convective Burning

Using the profiles given in Equations 11.116 and 11.128 for free convections, results are presented as follows [Annamalai and Colaluca, 1995].

$$v_{x,\max}{}^* = \frac{\dfrac{v_{x,\max} x}{\nu_w}}{Gr_x^{1/2}} = \frac{\left[1.3333 \; J \; (-a_1) \; (B+1)\right]^{0.5}}{\left[\dfrac{27}{4} I_{\phi,b} \, \mathrm{Pr}_w + 1.6667(-a_1) I_{m,b} \; (B+1)\right]^{0.5}}, \text{ SFM or DFM} \tag{11.139}$$

$$\delta'^* = \frac{\delta' \, Gr_x^{1/4}}{x} = \left[\frac{1.33 \; (B+1) \; (-a_1)}{\mathrm{Pr}_w \; v_{x,\max}{}^* \; (I_{\phi,b})}\right]^{1/2} \tag{11.140}$$

where

$$Gr_x = \frac{g x^3}{\nu_w^2} \frac{T_w - T_\infty}{T_\infty} \tag{11.141}$$

$$I_{m,b} = \int_0^1 f_m'^2 d\eta = \frac{c_1^2}{105} = \frac{27^2}{4^2 * 105} \tag{11.142}$$

$$I_{\phi,b} = \int_0^1 f_m \, \phi_s \; d\eta = \frac{11}{210} + \frac{a_1}{105}, \text{ SFM} \tag{11.143}$$

$$J = \int_0^1 \frac{T - T_\infty}{T_w - T_\infty} d\eta \tag{11.144}$$

$$J_{SFM} = \int_0^1 \Phi_s \, d\eta = \frac{a_1 + 6}{12}, \text{ SFM} \tag{11.145}$$

$$I_{\phi,0-\eta f} = \int_0^{\eta} \phi \, d\eta = \eta_f + (a_1/2)\eta_f^{\,2} + (-2a_1 - 3)\eta_f^{\,3}/3 + (2a_1 + 2)\eta_f^{\,4}/4 \tag{11.146}$$

$$I_{\phi,\eta f-1} = \int_{\eta_f}^{1} \phi \, d\eta = (1-\eta_f) + (a_1/2)(1-\eta_f^{\,2}) + (-2a_1 - 3)(1-\eta_f^{\,3})/3 + (2a_1 + 2)(1-\eta_f^{\,4})/4 \tag{11.147}$$

$$J_{DFM} = \left[1 + \frac{Y_{O2,\infty}}{\nu_{O2,V}} \frac{h_{c,V}}{c_p(T_w - T_\infty)}\right] I_{\phi 0-\eta f} + \frac{Y_{O2,\infty}}{\nu_{O2,V}} \eta_f + \left[1 + \frac{Y_{CO,w} h_{c,V}}{c_p(T_w - T_\infty)}\right] I_{\phi,\eta f-1} \tag{11.148}$$

The "J" values affect v_{xmax} and, hence, normalized BL thickness δ^*.

$$\dot{m}^* = \frac{\dot{m}'' x}{(\rho D)_w Gr_x^{\,1/4}} = \left[\frac{\Pr_w B^2 v_{x,\max}^{\;*} (-a_1)\,(I_{\phi,b})}{1.33\,(B+1)}\right]^{1/2} \tag{11.149}$$

Note that $v^*_{x,\,max}$ is affected by the type of model SFM or DFM and hence burn rates differ.

(i) SFM

Figure 11.14 plots v_{xmax}^*, δ^*, m* vs. B for SFM, while Figure 11.15 shows the profiles for ϕ and f'_m.

For char burning, the mass loss is due to chemical reaction, whereas for the usual case of liquid or volatile fuels, the mass loss is due to pyrolysis and/or evaporation. Consequently, the char-burning B number is independent of surface temperature T_w or temperature gradients, in contrast to the case for pyrolysis and/or evaporation.

(ii) Double Film model

Due to the presence of peak flame temperatures under DFM, free-convection velocity is increased and hence δ is reduced. The reduced δ increases the transport rates and, hence, the burn rate for DFM. The results show that the

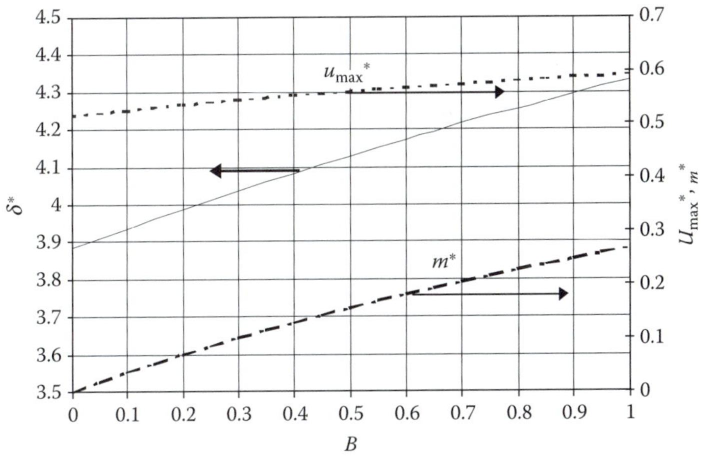

Figure 11.14 Variation of maximum velocity, boundary-layer thickness, and burn rate vs. B for free convection: SFM; $u^*_{max} = v^*_{x\ max}$.

mass burning rate is increased compared to the case of SFM; further, the wall temperature and the flame temperature calculated using the SFM differs from those calculated using DFM for free-convection flow over a vertical wall when radiation loss is accounted for.

The B numbers for the combustion of char/carbon (about 0.2) are of an order of magnitude lower than those for Plexiglas (1.8 to 8), methanol (2.5 to 8), and hydrocarbon fuels (4 to 8). The resulting thin free-convection BLs for char/carbon combustion make temporal measurements of species concentration and temperature difficult. Both SFM and DFM result in the same wall mass fraction of CO. Wall temperatures for DFM are about 40 K higher than

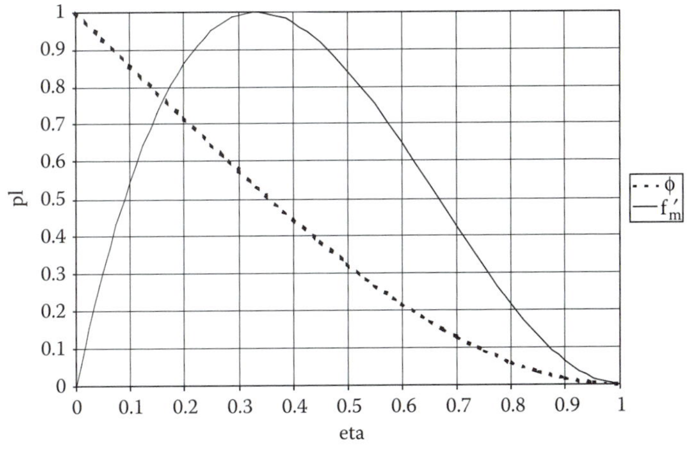

Figure 11.15 Variation of dimensionless velocity f'_m and ϕ_s for free convection: SFM or DFM.

for SFM when radiation loss is accounted for. The average burning rates are about four times greater for DFM than for SFM (Annamalai and Colaluca, 1995).

11.10.2.4 Combustion of Liquids and Pyrolyzing Solids over Vertical Walls

All the relations from Equation 11.139 to Equation 11.144 and 11.149 are valid except the following

$$Y_{F,w} = 1 - \frac{(1+s)}{(1+B)} \tag{11.150}$$

$$\phi_f = \frac{s}{(Y_{Fw}+s)} \tag{11.151}$$

$$J_{DFM} = \left[1 + \frac{Y_{O2,\infty}}{\nu_{O2}} \frac{h_c}{c_p(T_w - T_\infty)}\right] I_{\phi 0-\eta f} + \frac{Y_{O2,\infty}}{\nu_{O2}} \eta_{f,s} + \left[1 + \frac{Y_{F,w} h_c}{c_p(T_w - T_\infty)}\right] I_{\phi,\eta f-1} \tag{11.152}$$

where I's are evaluated using Equation 11.146 and Equation 11.147. For a vertical porous wall soaked with oil, the combustion causes strong free convection. An exact solution has been obtained by Kim et al. [1971] (Subsection 11.9.1.2). Integral solutions have been obtained by Annamalai and Sibulkin [1979].

11.10.3 Stagnation Flows

2-D Stagnation Flow:

$$x'^* = x'/d_{ref}, \quad x' = x, \ \alpha = 0 \tag{11.153}$$

$$f''' + ff'' + \{(T/T_\infty) - f'^2\} = 0, \text{ constant molecular weight,}$$

$$\phi'' + Sc\, f\, \phi' = 0 \tag{11.154}$$

The boundary conditions are listed below.

$$f'(0) = 0, \ \phi(0) = 1 \tag{11.155}$$

$$f'(\infty) = 1 \text{ or (or } f''(\infty) = 0), \ \phi(\infty) = 0 \tag{11.156}$$

$$\phi(0)B/Sc = f(0), \tag{11.157}$$

Thus, the problem can be solved for burn rate.

$$\dot{m}''\ x/\{Re_x^{1/2}\rho_w\ \nu_w\} = -f(0) \quad \text{where} \tag{11.158}$$

$$f(0) = -0.543\ \{Y_{O2\infty}/Y_{Fw}\ \nu_{O2}\}^{-0.02}\ B^{0.626 - 0.0678 \ln B} \tag{11.159}$$

[Konishita et al, 1981].

If one uses the heat and mass transfer analogy (see Chapter 7 and Chapter 8) for 2-D stagnation flow,

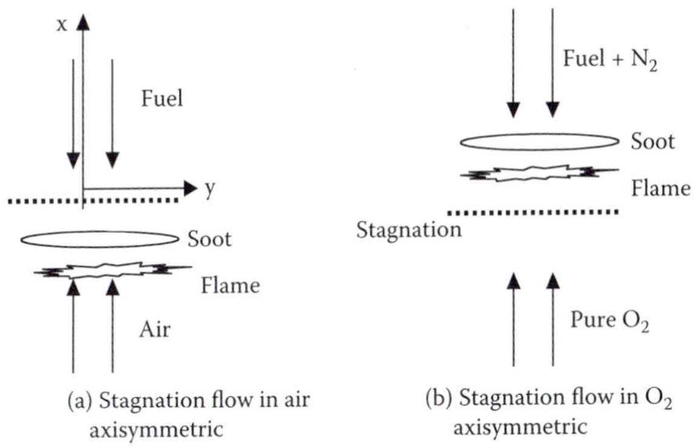

Figure 11.16 (a) Stagnation flow in air, (b) stagnation flow in pure O_2.

$$\frac{\dot{m}''x}{\rho v} = \left[0.57\,\mathrm{Pr}^{-3/5}\ln(1+B)\right]\mathrm{Re}_x^{\;1/2},\ \mathrm{Re}_x = \left(\frac{v_{x\infty}x}{\nu}\right)^{1/2}. \quad (11.160)$$

With B = 1.5, Pr = 1, Equations 11.158 and 11.160 yield almost the same results.

Axisymmetric:

$$f''' + (2/3)\,ff' + (1/3)\,\{(\rho_\infty/\rho)-(f')^2\} = 0 \quad (11.161)$$

$$\phi'' + f\,\mathrm{Sc}\,(2/3)\,\phi' = 0 \quad (11.162)$$

$$\dot{m}''\,x/\{\mathrm{Re}_x^{1/2}\rho_w\,\nu_w\} = -f(0) \quad (11.163)$$

where the solution for f(0) has been obtained by Konishita et al. [1981].

$$f(0) = -0.353\,\{Y_{O2\infty}/Y_{Fw}\,\nu_{O2}\}^{-0.02}\,B^{0.611\,-\,0.0651\,\ln B} \quad (11.164)$$

Thus, burn rate can be solved.

The flame location of η_f can be obtained by setting $\phi = \phi_f$; air flame is located closer to the air jet (Figure 11.16a). When O_2 jet is used, the flame moves closer to the fuel jet and soot is formed closer to fuel jet (Figure 11.16b).

11.11 Excess Fuel and Excess Air under Convection

11.11.1 Closed (Enveloped) and Open Flames

Droplet flames are typically envelope flames at low velocity and, as such, under steady state the total fuel burn rate at the flame (kg/sec) is the same as the fuel production rate. However, flames in convection are open flames. The fuel production per unit area within $L_{F,\,prod}$ exceeds fuel consumption within the same length (Figure 11.17). Then there will be the same fuel flow between y = 0 and y = y_f (Figure 11.17) called excess fuel, which continues to flow downstream and get burned in the wake region. Similarly, air entrained into the BL is more than the amount of air used for combustion and hence the flow between y = δ_f and Y = ∞ may be called excess air.

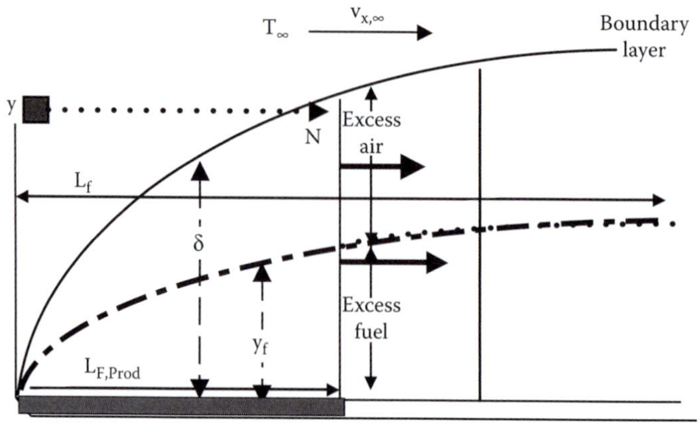

Figure 11.17 Excess fuel and excess air under convection.

11.11.2 Excess Fuel

The total fuel produced between x = 0 and x = $L_{F,Prod}$ is given as (Figure 11.17)

$$\int_0^L \dot{m}_F''(x)\,dx = \dot{m}_{F,\mathrm{Prod}}' \tag{11.165}$$

where

$$\dot{m}_F''(x) = -\rho D\ B\ (\partial\phi/\partial y)_0 = -\rho D\ B\ \phi'(0)\ (\partial\eta/\partial y) = -\rho D\ B\ \phi_s'(0)/\delta \tag{11.166}$$

At the flame,

$$\rho v_y\ Y_{F,f} - \rho D\ (\partial Y_F/\partial y)_{yf} = 0 - \rho D\ (\partial Y_F/\partial y)_{yf} = \dot{m}_{F,f}''(x) \tag{11.167}$$

where $\dot{m}_{F,f}''(x)$ is the fuel burn per unit area normal to y at the flame front (i.e., projected area being parallel to the plate). With $\beta = \beta_{02-F}$.

$$\phi = \{Y_F + s\}/\{Y_{F,w} + s\},\ 0 < y < yf, \tag{11.168}$$

where $Y_{F,w} = 1 - (1 + s)/(1 + B)$, then one can solve for fuel burned per unit area,

$$\dot{m}_{F,f}''(x) = -\rho D\ (1 + s)\{B/(1 + B)\}\phi_s'(\eta_f)/\delta \tag{11.169}$$

$$\int_0^L \dot{m}_{F,f}''(x)\,dx = \dot{m}_{F,f}' \tag{11.170}$$

It is apparent from Equation 11.166 and Equation 11.169 that fuel burned per unit area is less than the fuel produced per unit (because s <<< B). Hence, the integrated fuel burned between x = 0 and x = $L_{F,Prod}$ is less than the fuel produced, and the difference is called excess fuel [Pagni and Shih, 1974].

$$\dot{m}'_{F,prod} - \dot{m}_{F,f}(x) = \dot{m}'_{F,\,excess} \tag{11.171}$$

Hence at x = $L_{F,Prod}$, there is still fuel flow given by

$$\left[\int_0^{y_f} \rho u x\, Y_F\, dy\right]_{x=L} = \dot{m}'_{F,Excess} \tag{11.172}$$

The excess-fuel flow continues to burn downstream and is responsible for the "overhang" of flame ($L_f > L_{F,prod}$). This overhang exists for free, forced, and stagnation flows.

$F_{excess} = \{\dot{m}_{F,\,excess}/\dot{m}_F'\} = 1 - [(1+s)/(1+B)]\,[\phi'(\eta_f)/\phi'(0)] = 1 -$ (burn rate/prod rate) (11.173)

Using the integral profiles (Equation 11.116) for the derivatives,

$$\frac{\dot{m}'_{F,Excess}}{\dot{m}'_{F,Prod}} = 1 - \frac{1+s}{1+B}\sum_{i=1}^{I}\frac{a_i}{a_1}\left(i\,\eta_f^{(i-1)}\right) \approx 1 - 1.4\frac{s}{B} \approx 1 - 1.4\frac{L}{h_C} \tag{11.174}$$

[see Annamalai and Sibulkin, 1979b].

Figure 11.18 shows the plots of flame location and burned fuel fraction up to x = $L_{F,Prod}$. As "s" or $Y_{O2,\infty}$ increases, the increased O_2 increases the burned

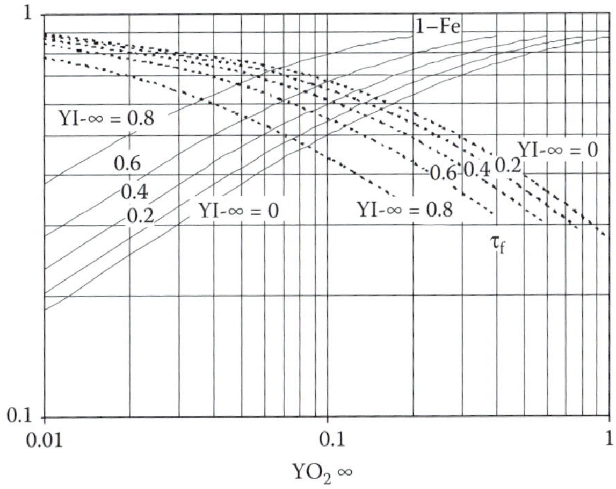

Figure 11.18 Variation of flame location and burned fuel fraction for carbon surfaces: free, forced, and stagnation flow; parameter: ambient inert mass fraction, $Y_{I,\infty}$. (Adapted from Annamalai and Sibulkin.)

fraction at a given B. Similarly, as B increases at a given s, the blowing rate increases, which increases the BL thickness and O_2 transport to the flame; hence, the burned fraction decreases, indicating less fuel burned within the same distance. Similarly, the flame moves away from the surface.

If there are only CO_2, O_2, and inerts (I) in the environment, then once $Y_{I,\infty}$ is specified, $Y_{CO2,\infty} + Y_{O2,\infty} = 1 - Y_{I,\infty}$. With $Y_{O2,\infty}$ as the independent variable and specified, $Y_{CO2,\infty}$ is estimated using the relation $Y_{CO2,\infty} = 1 - Y_{O2,\infty} - Y_{I,\infty}$. Using Equation 11.121 and Equation 11.173, η_f and $(1 - F_{excess})$ are calculated and plotted in Figure 11.18. It shows the excess-fuel fraction vs. $Y_{O2,\infty}$ at various free-stream concentrations of inerts. The more the excess fuel, the more the flame overhang; for example, when a vertically oriented plastic burns, pyrolysis occurs over a length x_p, whereas flame (x_f) extends owing to excess pyrolysate; the ratio x_f/x_p for vertically burning plastics was predicted by Annamalai and Sibulkin [1979b].

Example 6

(a) Compute the excess-fuel fraction at x = 0.1 m, for combustion under free convection. (b) If the burn rate follows the same trend as before, how much farther does the flame extend beyond 0.1 m? Fuel: Hexane. Other data is in Example 3.

Solution

(a) The excess-fuel fraction is given as

$$\frac{\dot{m}'_{F,Excess}}{\dot{m}'_{F,Prod}} = 1 - \frac{1+0.0652}{1+9.681}\left(1 + 2*\frac{-1.7854}{-0.61}0.823 + 3*\frac{1.393}{-0.61}0.823^2\right) = 0.88$$

If the burn rate expression is the same as before for overhang, $\int_0^{Lf} \dot{m}''_{F,f}(x)dx = \int_0^{L} \dot{m}''(x)dx$, then

$$\int_0^{Lf} \{\dot{m}''_{F,f}(x)/\dot{m}''(x)\}\ \dot{m}''(x)\,dx = (1-Fe)\int_0^{Lf} \dot{m}''(x)\,dx = \int_0^{L} \dot{m}''(x)dx$$

Because $\dot{m}''(x) \propto x^{-1/4}$ for free convection, then

$$(L_f/L) = 1/(1-Fe)^{4/3} = 1/0.12^{1.333} = 17.3,\ L_f = 1.73\ \text{m}.$$

Remarks

(i) For forced convection, $(L_f/L_{F,Prod}) = 1/(1-Fe)^2 = 1/0.12^2 = 60.4$, $L_f = 6.04$ m. This length is very long compared to free-convective burning because the burn rate decreases much faster with increase in length for forced convection.

(ii) Similar calculations could be performed for turbulent free convection. Instead of laminar solutions for burn rate, turbulent correlations could be used. However, the excess-fuel fraction is expected to be the same if the burn rate at the flame and the production rate at the wall are affected by similar corrections.

11.12 Summary

Whereas Chapter 9 and Chapter 10 present results on burning of liquid drops and solid-fuel particles in a stagnant atmosphere, this chapter presents burn rate results for the combustion of liquid fuels and plastics for laminar forced, free, and stagnation (2-D and axisymmetric) flows. In addition, a new methodology is given for obtaining conventional integral solutions directly by using similarity forms of differential equations. Results for excess fuel and excess air for open flames are presented.

Chapter 12

Combustion of Gas Jets

12.1 Introduction

In this chapter, combustion of gaseous jets will be discussed. For example, the following questions will be addressed: When natural gas is fired in a boiler, how long is the flame (turbulent jet)? When fuel is turned on in a gas stove, how tall is the flame (laminar jet)? What is the flame structure? As opposed to the combustion of solids and liquids, discussed in Chapter 9 to Chapter 11, the burning rate of a gas is known *a priori* because the gas generally flows at a specified rate through burners. The combustion of fuel and airflow within concentric tubes (Burke–Schumann [BS] flame) will be treated first as a diffusion-limited phenomenon and later extended to coflowing air. This will be followed by a discussion on jet combustion in stagnant air for any given Sc and coflowing air. The laminar jet combustion (Re < 2000) will be treated in detail, followed by turbulent combustion using a semiempirical approach.

12.2 Burke–Schumann (B–S) Flame

12.2.1 Overview

The earliest gas combustion studies were in a concentric enclosure (Figure 12.1). This consists of a cylindrical duct of radius "a." The fuel "F" flows through the duct at a velocity "v." It is surrounded by another concentric duct of radius "b" through which the oxygen-carrying gas flows. Our objective is to determine the flame height H and to obtain the flame structure or shape. If we light a flame at the exit of the duct, two kinds of flames may be established.

1. Overventilated flame: If methane flows in the inner duct at the rate of 0.1 kg/sec, the required oxygen flow is 0.4 kg/sec. If the oxidizer flow in the outer duct is greater than the stoichiometric amount of O_2(>0.4 kg/sec for example methane), the flame is said to be an *overventilated flame* (with more

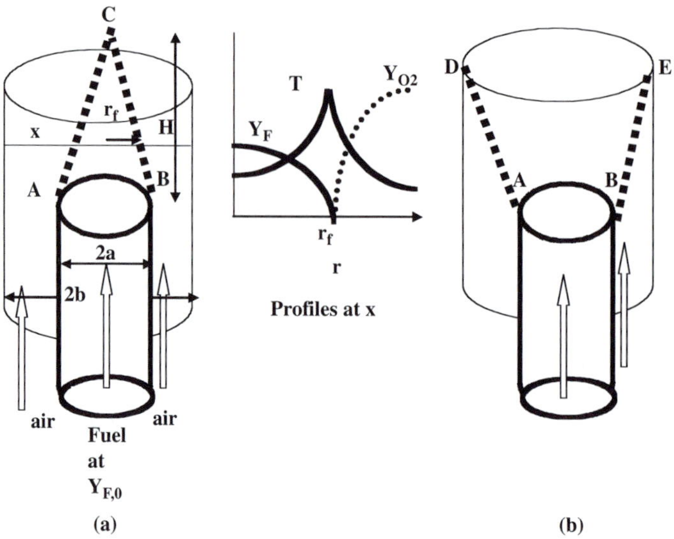

Figure 12.1 Burke–Schumann flame: (a) overventilated flame; (b) underventilated flame.

than the required O_2), which converges on the axis (shape ACB in Figure 12.1a) i.e., all the fuel is consumed and the flame "caps" the fuel. No fuel escapes ACB. Oxygen (O_2) diffuses from the outer tube toward the surface AC and BC, while the fuel diffuses from the axis toward AC and BC.

2. Underventilated flame: If the oxidizer flow is less than the stoichiometric amount, the flame extends to the wall of duct from A and B, i.e., the fuel diffuses toward oxygen. For the aforementioned example, an underventilated flame occurs if the O_2 flow is below 0.4 kg/sec. The flame is called *underventilated* (lesser oxidizer than the stoichiometric flow), which converges on the wall (Figure 12.1b, shape ADBE), i.e., all of the oxygen is consumed and the flame "caps" the oxygen flow. No O_2 escapes AD and BE. O_2 flows into the surface AD and BE, whereas fuel diffuses from the axis toward AD and BE.

Burke and Schumann analyzed this problem analytically and obtained solutions for the flame shape and flame height.

12.2.2 Assumptions

The following assumptions are made for the analysis of these flames:

(a) v_z is uniform ($a < r < b$) (Figure 12.1).
(b) ρD = constant.
(c) Axial diffusion rate $\rho D\ \partial Y_k/\partial x$ << radial diffusion rate $\rho D\ \partial Y_k/\partial r$.
(d) A thin flame is valid for a single-step global reaction

$$1 \text{ kg of fuel} + \upsilon_{O2} \text{ kg of oxygen} \rightarrow \upsilon_{CO2} \text{ kg of } CO_2 + \upsilon_{H2O} \text{ kg of } H_2O$$

For example,

1 kg of CH_4 + 4 kg of oxygen → 2.75 kg of CO_2 + 2.25 kg of H_2O

The following additional assumptions are also made:

(e) P = constant throughout the field of interest.
(f) Re = $\rho va/\mu >> a/H$ or $(\rho v H/\mu) >> 1$.

12.2.3 Governing Equations

12.2.3.1 Mass

For cylindrical ducts,

$$\frac{\partial(\rho v_x r)}{\partial x} + \frac{\partial(\rho v_r r)}{\partial r} = 0. \quad (12.1)$$

Because ρv_x is assumed to be a constant and $\rho v_x = \rho_0 v_{x,0}$,

$$\frac{\partial(\rho_0 v_{x,0} r)}{\partial x} + \frac{\partial(\rho v_r r)}{\partial r} = 0$$

$$\frac{\partial(\rho v_r r)}{\partial r} = 0. \quad (12.2)$$

Equation 12.2 yields

$$\rho v_r r = C. \quad (12.3)$$

At r = b there is no radial mass flow, therefore, $v_r = 0$. Hence, $\rho v_r r = \rho v_r b = C = 0$; thus, C = 0.

12.2.3.2 Species and Energy Conservation Equation

From Chapter 8, the species and energy equations can be written in terms of the SZ variable β as:

$$\underset{\text{I}}{\frac{\partial(\rho v_x r\beta)}{\partial x}} + \underset{\text{II}}{\frac{1}{r}\frac{\partial(\rho v_r r\beta)}{\partial r}} = \underset{\text{III}}{\frac{1}{r}\frac{\partial}{\partial r}\left(\rho D\frac{\partial\beta}{r\partial r}\right)} + \underset{\text{IV}}{\frac{\partial}{\partial x}\left(\rho D\frac{\partial\beta}{\partial x}\right)} \quad (12.4)$$

where term I to term IV will be compared for their relative significance. The term β may include β_{O2F}, β_{htF}, β_{htO2}, etc., as mentioned in Chapter 8. From the overall mass conservation equation, $v_r = 0$. Hence, term I is zero. Using order-of-magnitude analysis (r ~ a, x ~ H, β ~ 1) and comparing term III and term IV,

$$\frac{\rho D}{a^2}\left(a\frac{1}{a}\right) >> \rho D\frac{1}{H^2}. \quad (12.5)$$

If a << H, then term IV can be neglected in comparison with term III. Term II = 0 since $\rho v_r\, r = 0$. Comparing term I and term IV,

$$\frac{\rho v_x H}{\rho D} >> 1 \quad \text{or} \quad \frac{\rho v_x H}{\mu} >> 1 \quad \text{or} \quad \frac{\rho v_x a}{\mu} >> \left(\frac{a}{H}\right). \tag{12.6}$$

Term IV can be neglected if the Reynolds number based on H is greater than 1 (assumption f). Thus, we are left with term I and term III, i.e.,

$$\frac{\partial}{\partial x}(\rho v_x \beta) = \frac{\rho D}{r}\frac{\partial}{\partial r}\left(r\frac{\partial \beta}{\partial r}\right) \tag{12.7}$$

where β is any SZ variable. Now consider

$$\beta_{O_2-F} = Y_F - \frac{Y_{O_2}}{\nu_{O_2}} \tag{12.8}$$

where ν_{o2} is the stoichiometric oxygen in mass per unit mass of fuel.

12.2.3.3 Boundary Conditions

With assumption d

$$\beta_{O_2-F} = Y_F \qquad 0 < r < r_f, \quad \text{fuel rich regime}$$

$$\beta_{O_2-F} = -\frac{Y_{O_2}}{\nu_{O_2}} \qquad r_f < r < b, \quad \text{fuel lean regime} \tag{12.9}$$

In order to solve Equation 12.7, we require two boundary conditions w.r.t. r and one initial condition w.r.t. x. These conditions are:

(i) At x = 0, $\beta_{O_2-F} = Y_{F,o}$ or β_{0a}, $0 < r < a$ (12.10)

$$\beta_{O_2-F} = -\frac{Y_{O_2 0}}{\nu_{O_2}}, \quad \text{or} \quad \beta_{0,ab}, \quad a < r < b \tag{12.11}$$

(ii) At r = 0 for any x, because of symmetry,

$\partial\beta_{O2-F}/\partial r = \partial\beta/\partial r = 0$ or β_{O2-F} or β is a finite quantity at r = 0. (12.12)

(iii) At r = b, at any z, the radial oxygen flow must be zero because the walls are impermeable. Thus,

$$[\dot{m}_{O2,r}]_{r=b}'' = \rho\,(v_r + V_{O_2,r}) = 0. \tag{12.13}$$

Because $v_r = 0$, then $[V_{O2,r}]_{r=b} = 0$. Hence

$$\frac{\partial Y_{O_2}}{\partial r} = 0 \quad \text{at} \quad r = b. \tag{12.14}$$

Since $(\partial Y_F/\partial r)_{r=b} = 0$, then it follows that

$$\frac{\partial \beta_{O_2\text{-}F}}{\partial r} = 0 \quad \text{at } r = b. \tag{12.15}$$

Or more generally,

$$\partial\beta/\partial r = 0 \tag{12.16}$$

because nothing diffuses through an adiabatic impermeable surface.

Homogeneous boundary conditions in r imply that the solution is an Eigen-function.

12.2.4 Normalized Conservation Equations

Let us normalize Equation 12.7 and the boundary conditions:

$$\xi = x\ D/v_x a^2 = \{\rho D\ x/(\rho v_x a^2)\} = \pi\rho x\ D/\dot{m}_i \tag{12.17}$$

Note: $t \approx x/V_x$, residence time and v_x and D are not constants. Both V_x and D increase with x. Let

$$\eta = r/a \tag{12.18}$$

$$s = \frac{Y_{O_2,0}}{Y_{F,0}\nu_{O_2}} \tag{12.19}$$

$$\phi = \frac{\beta - \beta_\infty}{\beta_i - \beta_\infty} \tag{12.20}$$

where $\beta = \beta_{O2-F}$, β_{ht-F}, β_{ht-O2}, etc. For O_2–F coupling, $\beta_i = \beta_{O2-F,0,a}$ and $\beta_\infty = \beta_{O2-F,0,ab}$. For ht-F coupling, $\beta_i = \beta_{ht-F,0,a} = Y_{F,i}$ if T_{ref} for enthalpy is selected as T_i and $\beta_\infty = \beta_{ht-F,0,ab} = c_p\,(T_\infty - T_i)/h_c$. Using Equation 12.17 to Equation 12.20 in Equation 12.7,

$$\frac{\partial\phi}{\partial\xi} = \frac{1}{\eta}\frac{\partial}{\partial\eta}\left(\eta\frac{\partial\phi}{\partial\eta}\right) \tag{12.21}$$

The boundary conditions, Equation 12.10 to Equation 12.16, transform to

$$\text{at } \xi = 0, \phi = 1, \quad 0 < \eta < 1, \quad \text{and} \quad \phi = 0, \quad 1 < \eta < \eta_b \tag{12.22}$$

$$\text{at } \xi > 0, \frac{\partial\phi}{\partial\eta} = 0, \text{ at } \eta = \eta_b \quad \frac{\partial\phi}{\partial\eta} = 0, \quad \eta = 0 \quad \text{or} \quad \phi \text{ is finite} \tag{12.23}$$

12.2.5 Solution

12.2.5.1 Normalized SZ Variable

One can use the method of separation of variables for solving Equation 12.21 subject to the boundary conditions Equation 12.22 and Equation 12.23.

$$\phi = \frac{1}{\eta_b^2} + \frac{2}{\eta_b}\sum_{n=1}^{\infty} \frac{\exp\left(-k_n^{\,2}\xi\right)}{(k_n\,\eta_b)} \frac{J_1(k_n)J_0(k_n\,\eta)}{[J_0(k_n\,\eta_b)]^2} \tag{12.24}$$

where $\eta_b = (b/a)$, J_o, J_1, etc., are Bessel functions of first kind and of order 0 and 1, respectively, and k_n represents the roots of equation $J_1\{\kappa_n\ \eta_b\} = 0$. Then, the roots are (Table A.40)

$$\eta_b\, k_1,\ \eta_b\, k_2, \cdots\cdots \eta_b\, k_{10} = 3.8317, 7.0156, 10.135, 13.324, 16.471, 19.616, 22.760, 25.905, 29.047, 32.190. \tag{12.25}$$

This solution is valid irrespective of whether the flame is thick or thin. However, solutions for flame shape and height can be obtained if a thin-flame assumption is made. The same solution can be used to obtain a temperature profile.

12.2.5.2 Flame Structure

Using a Normalized SZ profile and with $\beta = \beta_{O2-F} = Y_F - Y_{O2}/\nu_{O2}$, $\beta_{ht-F} = h_t/hc + Y_F$, $\beta_{ht-O2} = h_t/hc - Y_{O2}/\nu_{O2}$ and thin-flame models, the flame structure, Y_F, T, and Y_{O2} vs. η can be obtained. At the flame location it can be shown that (e.g., thin-flame assumption; with $\beta_{htF} = h_t/hc + Y_F$, $\beta_i = \beta_{htF,0,a}$, and $\beta_\infty = \beta_{htF,0,ab}$) Equation 12.20 reduces to

$$\phi_f = s/(1 + s). \tag{12.26}$$

12.2.5.3 Flame Profile

Using the expression from Equation 12.26 in Equation 12.24, we obtain the flame profile η_f as a function of ξ with "s" as a parameter for a given ξ_b. One will find that a flame changes from underventilated (Figure 12.1b) to overventilated (Figure 12.1a) as $\eta_b = b/a$ is increased gradually.

12.2.5.4 Over- and Underventilated Flames — Criteria

Overventilated flames occur if the supplied oxygen flow in the outer duct exceeds the stoichiometrically required amount for the fuel flowing in the inner duct and vice versa.

Overventilated flames occur if

$$\rho\, v\, \pi (b^2 - a^2) Y_{O_{2,0}} > \rho\, v\, \pi\, a^2 Y_{F,o}\, \nu_{O2}. \tag{12.27}$$

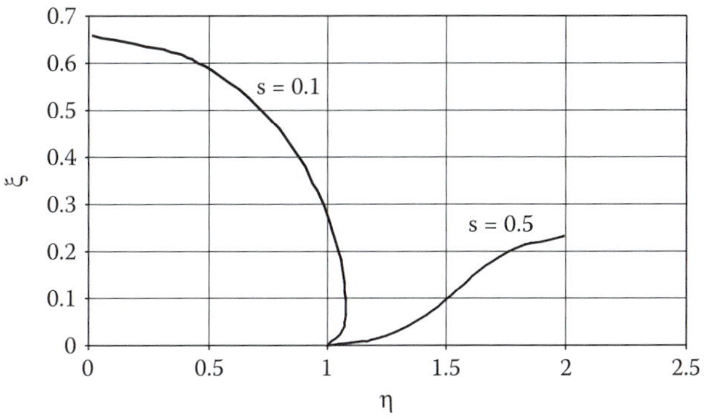

Figure 12.2 Burke–Schumann flame; exact solution for $\eta_b = 2$ with two different "s" values, $s = Y_{O2,0}/(Y_{F,0}\ \nu_{O2})$.

Underventilated flames occur if

$$\rho\ v\ \pi(b^2 - a^2)\ Y_{O_2,0} < \rho\ v\ \pi\ a^2\ Y_{F,0}\nu_{O_2}. \tag{12.28}$$

Normalized criteria for an overventilated flame to occur are given by

$$s = Y_{O2,0}/(Y_{F,0}\ \upsilon_{O2}) > 1/(\eta_b^2 - 1) \tag{12.29}$$

whereas for underventilated flames,

$$\eta_b^2 > [1 + \{(Y_{F,0}\ \upsilon_{O2})/Y_{O2,0}\}] = (s + 1)/s. \tag{12.30}$$

For $\eta_b = 2$, $s > 1/3$ for overventilated flames and $s < 1/3$ for underventilated flames (Figure 12.2). Note that as "s" is decreased (for example, by decreasing $Y_{O2,0}$ at the inlet), the flame shape suddenly changes to that of underventilated flames if "s" falls below 1/3 for b/a = 2.

12.2.5.5 Flame Heights

It is seen from Figure 12.2 that the flame converges to r = 0 at x = H for overventilated flames (e.g., $s = 0.1$). Thus, the flame height can be obtained from Equation 12.24 by setting $\phi = \phi_f = s/(1 + s)$ at $\eta = 0$. In order to get an explicit solution for H, we expand Equation 12.24 using only the first term in expansion and ignoring other terms. Thus,

$$\xi_{f,over} = \frac{1}{k_1^2} \ln \left[\frac{2(1+s)\ J_1(k_1)}{\left\{s\eta_b^2 - (1+s)\right\}\left\{J_0(k_1\eta_b)\right\}^2 k_1} \right] \tag{12.31}$$

where $k_1\ \eta_b = 3.8317$ for the first term.

For underventilated flames, we set $\phi = \phi_f = s/(1 + s)$ at $\eta = \eta_b = b/a$ in Equation 12.24 and then with only the one-term expansion, we get

$$\xi_{f,under} = \frac{1}{k_1^2} \ln \left[\frac{2(1+s)\, J_1(k_1)}{\{s\eta_b^2 - (1+s)\}\,\{J_0(k_1\eta_b)\}k_1} \right]. \tag{12.32}$$

Note that $\exp(k_1^2 * [\xi_{f,under} - \xi_{f,over}]) = J_0\,(k_1 * \eta_b)$ or $[\xi_{f,under} - \xi_{,over}]) = \ln [J_0\,(k_1 * \eta_b)]/k_1^2$.

Example 1

A tubular reactor of radius "a" is supplied with a mixture of methane at 57.5% and N_2 at 42.5%. It is surrounded by another tube of radius b = 2a through which air is fed. The air and fuel mixture flows in at T_0 and P_0.

(a) What is the percentage of excess or deficient air used in the experiment?
(b) State whether overventilated or underventilated flame occurs.
(c) Determine the nondimensional flame height using the exact procedure and using a one-term expansion.
(d) Obtain flame profiles.
(e) If $Y_{F0} = 0.115$, is the flame overventilated or underventilated? What is the flame structure?
(f) If the mass flow of the gas in the inner duct is 0.1 g/sec, $\rho D = 6 \times 10^{-05}$ kg/m sec, and a = 2 mm, what are the dimensional flame heights for the two cases?

Solution

(a) $\dot{m}_F$ = amount of fuel flow $\rho v_z\, \pi a^2\, Y_{F,0}$; airflow in = $\rho v_z\, \pi\, (b^2 - a^2)$; supplied $O_2 = \rho v_z\, \pi\, (b^2 - a^2) Y_{O2,0}$.
Required stoichiometric $\dot{m}_{O2,st} = \rho v_z\, \pi a^2\, Y_{F,0} \times \nu_{O2}$, $\dot{m}_{air,st} = \rho v_z\, \pi a^2\, Y_{F,0}\, \nu_{O2}/Y_{O2,0}$.
Excess- or deficient-air percent = $\{\rho v_z\, \pi\, (b^2 - a^2)\, Y_{O2,0} - \rho v_z\, \pi a^2\, Y_{F,0}\, \nu_{O2}\}/\{\rho v_z\, \pi a^2\, Y_{F,0}\, \nu_{O2}\} = \{(b^2 - a^2)\, Y_{O2,0} - Y_{F,0}\, \nu_{O2}\}/\{a^2\, Y_{F,0}\, \nu_{O2}\} = \{(b^2/a^2 - 1)\, [Y_{F,0}\, \nu_{O2}/Y_{O2,0}] - 1\} = [(\eta_b^2 - 1)\, s - 1]$,

$$s = Y_{O2,0}/(Y_{F,0}\, \upsilon_{O2}) = 0.23/(0.575 \times 4) = 0.1$$

Fraction of excess or deficient air = $[(\eta_b^2 - 1)\, \alpha s - 1] = (2^2 - 1) * (0.1) - 1 = -0.7$ or 70 % deficient. Thus, only 30% of the fuel will be burned.
(b) Because air is insufficient, an underventilated flame occurs.
(c) $k_1\, \xi_b = 3.832$; $k_1 = 3.832/2 = 1.916$; $J_1(k_1) = 0.581$, $J_0(k_1\, \xi_b) = -0.403$. Using Equation 12.32, underventilated flame height $\xi_{f,under} = 0.235$.
(d) With $\phi_f = s/(1 + s) = 0.1/1.1 = 0.0909$, using Equation 12.26, and given any ξ, one can solve for η iteratively. The flame structure is shown in Figure 12.2 for s = 0.1.
(e) $s = Y_{O2,0}/(Y_{F,0}\, \upsilon_{O2}) = 0.23/(0.115 \times 4) = 0.5$.

Using the criterion given by Equation 12.29, one can show that the flame is overventilated. Using Equation 12.29, $\xi_{f,over} = 0.658$.

(f) $\xi = x\ D/va^2 = \{\rho D\ x/(\rho va^2)\} = \pi\rho x\ D/\dot{m}_i = x\ \pi\ 6*10^{-05}/0.1$. With $\xi_{f,over} = 0.658$, we get the flame height H = 0.35 m and for the underventilated flame, H = 0.1244 m.

Remarks

(i) The answers in part d and part e use a one-term expansion. If an exact summation of series is conducted using Equation 12.24, the sum of the term in Equation 12.24 is 0.159, whereas the first term only yields 0.155 when $\xi_f = 0.235$.
(ii) Experimental data on flame heights is available in Chapter 7 of Bartok and Sarofim.

12.3 Modification to B–S Analyses

12.3.1 Flames in Infinite Surroundings with Equal Fuel and Air Velocity

By letting $b \rightarrow \infty$ or $\xi_b \rightarrow \infty$, one can obtain a solution for Equation 12.31 [Annamalai and Durbetaki, 1975]; with assumption that immediate surrounding air velocity is comparable to the fuel velocity.

The flame height is given as

$$\xi_f = \frac{1}{4\ \ln(1+s)} \tag{12.33}$$

$$H = \frac{\dot{m}_{F,o}}{4\pi\ \rho\ D \ln(1+s)} \tag{12.34}$$

$$H = \dot{m}_{F,0}/\{4\ \pi\ \rho\ D\ s),\ s << 1. \tag{12.35}$$

The experiments on methane flames reveal that at $\dot{Q}_{F,o} = 80$ cm³/min, H = 0.8 cm, whereas from theory we obtain H = 0.8 to 1.6 cm depending on the value used for ρ D in Equation 12.34. Note that even though the assumption ρv_x = constant is not justifiable, the theoritical results are in good agreement with experiments. It will be seen later that the expressions for jets in coflowing air are similar to the relations for flames in infinite surroundings.

12.3.2 Mass Flux as Function of Axial Distance

The BS analysis assumes that the mass flow per unit area, ρv_x, is equal to C for the flow through two concentric ducts. However, the buoyant flow will cause ρv_x to be a function of x (Chapter 11). Roper's analysis [1977a,b] extended by Sunderland et al. [1999] lets $\rho v_x = f(x)$ and presents a simplified solution. Roper solved the equations for rectangular and curved slot jets. Note that $\rho = \rho(r, x)$ because temperature changes with radius r at the given x and, hence, density will also change. Thus, the condition that $\rho v_x = f(x)$ requires v_x change with r. For example, if $\rho \propto 1/r$, then $v_x \propto f(x)$ r. Roper obtained exact solutions but with the assumption of constant density. One may get an approximate solution by extending the BS solution by treating $\rho v_x = f(x)$.

An approximate crude solution may be obtained by treating η_b' as locally constant and using the solution of the BS problem in Equation 12.24 with ξ and η replaced by

$$\xi' = \xi \{a/r_D(0)\}^2, \tag{12.36}$$

$$\eta' = \eta \{r_D(0)/r_D(X)\} \tag{12.37}$$

$$\{r_D(X)\}^2 = (A/V_x), \tag{12.38}$$

where A is constant. This is left as an exercise problem.

The height of the flame in Equation 12.31 and Equation 12.32 with η_b replaced by η_b' must be obtained by iteration.

12.4 Laminar Jets

12.4.1 Introduction

The BS analysis is concerned with combustion of gas within a constrained configuration. This section deals with the combustion of laminar jets, also called *free jets*, issued from a slit in the wall or hole in the wall into a unconstrained system. For example, consider the residential gas heater. Natural gas is issued from a small slit that draws in the surrounding (ambient) air and flows through the duct through momentum transfer and ejector action, as shown in Figure 12.3; the partially premixed fuel and air exits through the small holes and burns in the unconstrained surrounding stagnant air. The reader is referred to extensive research on partially premixed flames conducted by Qin et al. (2001, 2003, 2004), Xue et al. (2002, 2003), Choi et al. (2000, 2001, 2003), and Beta et al. (2005, 2006). Here, we are interested in estimating air entrained $\dot{m}_A$ for a given $\dot{m}_F$ so that the mixture at cross section XX is at the desired A:F ratio.

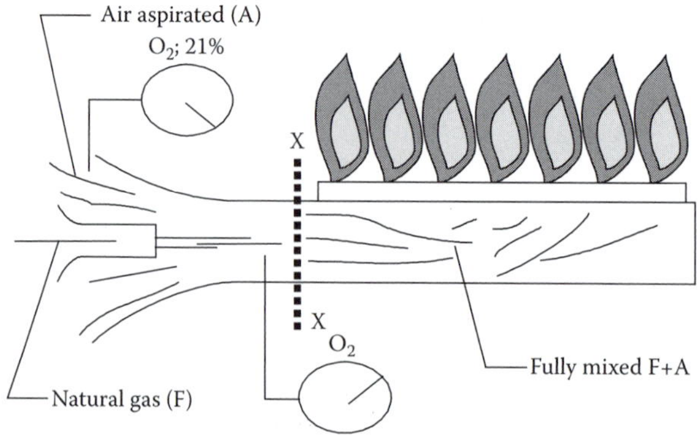

Figure 12.3 Schematic of a residential burner.

We have seen before that for liquid fuels, one must heat the liquid, evaporate the fuel, mix it with air, and then burn it. Thus, the combustion rate is controlled by both heat and mass transfer processes. In the case of combustion of gaseous fuels, the combustion intensity (or rate) of a gas burner is essentially controlled by the mixing rate of fuel with oxygen. We will first consider the jet issued from a small rectangular slit, followed by a circular jet.

12.4.2 Terminology of Jets

12.4.2.1 Potential Core

If one measures the velocity, v_x, along x as a function of x and y, it is possible to identify three regimes P, M, and A, which stand for potential core, mixing, and ambient. Considering momenta, the velocity profiles are almost flat within the potential core (P) (Figure 12.4) for $y < y_p(x)$ and $x < L_p$. Here, y_p is defined as that location at which $v_x = 0.99\ v_{x,max}$, where v_{xmax} is the maximum velocity, along y = 0 (i.e., along the axis of the flame). The value of v_{xmax} is unchanged up to $x = L_p$, which is called *potential core length*. Within the potential core, there are no significant gradients or transfer of heat and mass. Normally, the length of potential core, L_p, is approximately estimated as follows. Because there is momentum transfer from the high-velocity jet stream to the low-velocity ambient (just as heat is transferred from hotter to colder zones, Chapter 6), momentum diffusion occurs laterally across y, and hence the timescale of diffusion is on the order of

$$t_{diff,v} \approx (d_i^2/4\nu) \qquad (12.39)$$

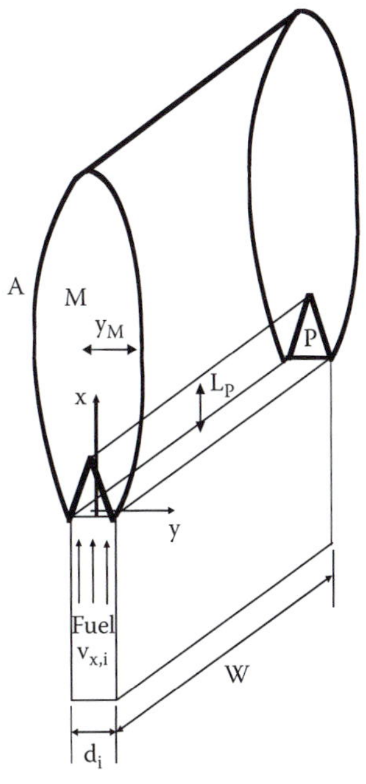

Figure 12.4 2-D jet.

Thus, a fixed mass traveling along the axis will lose all its momentum in a time $t_{diff,v}$. Within this timescale, the length traveled

$$L_p \approx v_{x,i}\, t_{diff,v} \tag{12.40}$$

where $v_{x,i}$ is the injection velocity.

$$L_p \approx v_{x,i}\, (d_i^2/4\nu). \tag{12.41}$$

Simplifying,

$$L_p/d \approx v_{x,i}\, d_i/(4\nu) = Re_i\, /4. \tag{12.42}$$

If $v_{x,i} = 0.2$ m/sec, $d_i = 0.001$ m, $\nu = 10^{-5}$ m²/sec, then $L_p/d = 5$ and $L_p = 0.005$ m, or 5 mm. Beyond L_p, the velocity at y = 0 will start decreasing from $v_{x,i}$. Note that total momentum at x = L_p is the same as the momentum at x = 0. We will assume that this length is much smaller compared with the flame length or mixing-layer length. Hence we presume that the velocity at y = 0 will start decreasing at x ≈ 0. In case the potential core length is not negligible, one may use (x – L_p) as the x coordinate with a virtual origin [Lin et al., 1999]. The value of L_p is determined using experimental velocity data for jets.

12.4.2.2 Mixing Layer

As the jet mixes with surrounding still air, it transfers momentum, which causes more air to be entrained into the jet. Thus, the jet velocity or momentum decreases, and hence the jet velocity for x > L_p decreases for $0 < y < y_M(x)$, where $y_M(x)$ is the mixing-layer thickness. The velocity profile appears as shown in Figure 12.5. The value of y_M at which $v_x = 0.01\, v_{x,max}$ is called the

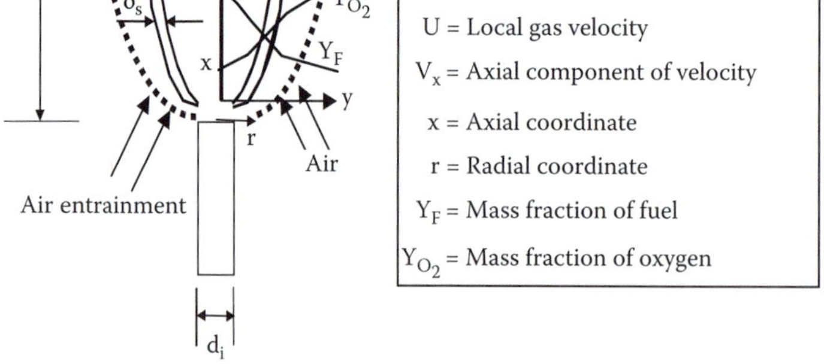

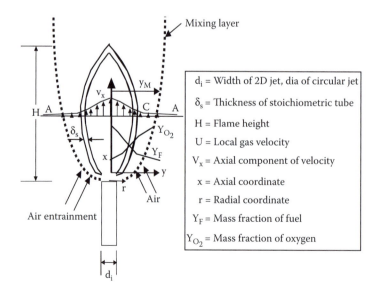

Figure 12.5 The axial velocity profile, and fuel and oxygen mass fraction profiles for a 2-D jet under mixing conditions.

mixing point, which is between the primary jet and the surrounding gases in the ambient. If one plots the contour of y_M values at which $v_x = 0.01\ v_{x,max}$, one obtains the mixing layer as shown in Figure 12.5. Outside the mixing layer, v_x is zero if the jet issues into still air. The mixing layer is a function of x, and begins at edge of the injector, and diverges out. If the gas injected is inert (e.g., N_2), then mixing it with air will not cause any chemical reaction. However, the O_2 concentration will change because of the mixing of air with N_2 (which sometimes can be used to estimate the air entrained into or mixed with the jet). In residential burners called *inshot burners*, the primary jet is a fuel jet, which entrains the surrounding air to about 30% of the stoichiometric air if the nozzle diameter is about 1 mm. However, caution must be exercised in using the current analysis for calculating the amount of air entrained in residential inshot burners because the pressure decreases below the atmospheric pressure in such a burner, which causes suction.

12.4.2.3 Global Chemical Reaction and Thin Flames

If a gaseous fuel is injected through the slit, then mixing of fuel with air occurs as the surrounding air is entrained. Such mixing will cause the fuel and oxygen mass fractions, Y_F (x, y) and Y_{O2} (x, y), to vary as functions of x and y. If not lighted, the flow is almost frozen (i.e., devoid of chemical reactions). For the case of reacting systems, we will assume a one-step reaction:

$$1 \text{ kg of fuel (F)} + \nu_{O2} \text{ kg of } O_2 \rightarrow \nu_P \ (= 1 + \nu_{O2}) \text{ kg of products} \tag{12.43}$$

Initially, the mixing of air and fuel begins at the edge of the slit, and as such combustion starts at the edge of the slit if lighted. Because the fuel is dispersed radially due to mixing, the flame moves radially outward. As the fuel is consumed the fuel concentration decreases along the axis and the flame then moves radially inward toward the axis and the flame merges with the axis at x = H (Figure 12.5). We are interested in determining the flame length or height H for a given fuel flow at the slit and the shape of the flame profiles. The flame is located inside the mixing layer.

Suppose the fuel flow is 2 g/sec and the total flow within the mixing layer is 12 g/sec at AA (Figure 12.5), then the air entrained within the distance x is 10 g/sec. Not all the air entrained through the mixing layer within x is necessarily used by the flame within x. If 1.4 g/sec of O_2 or 1.4/0.23 = 6 g/sec of air is used by the flame, then 10 − 6 = 4 g/sec is the excess air entrained by the jet at AA. 4 g/sec of air flows between the mixing layer and flame, i.e., AC.

12.5 Planar Laminar Jets

12.5.1 Overview

It has been shown that for Re < 2000, both 1g and 0g jets exhibit laminar-flow characteristics. The flame height increases with flow rates (or Re); transition to turbulent burning occurs at 2000 < Re < 3000; turbulent conditions prevail

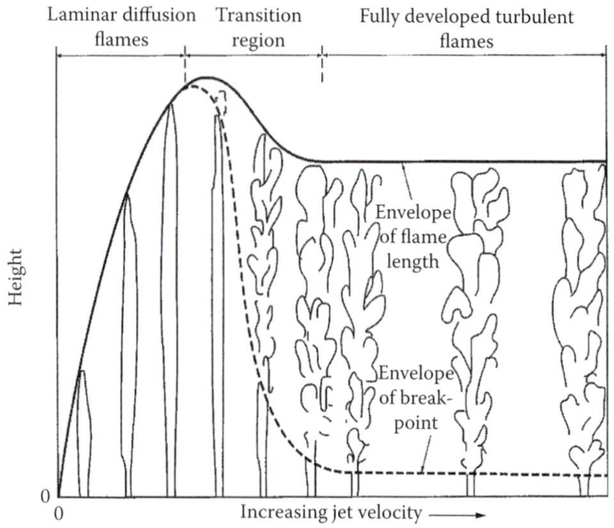

Figure 12.6 Effect of velocity on flame height. (Adapted from Hottel and Hawthorne, Proceedings of the Combustion Institute, Vol. 3, 254–266, 1949.)

at Re > 3000; the flame height remains constant with change in Re under turbulent regime, and finally the blow-off occurs (Figure 12.6) at high velocity.

12.5.2 Simplified Analysis of 2-D Laminar Jets

Because the fuel flows around the axis, it exists in the inner region, whereas oxygen is only in the outer region. Let us assume a thin-flame model.

Diffusion rate of fuel per unit flame area ≈ $\rho D\ Y_{F,0}(x)/r_f\ (x)$, where $Y_{F,0}$ is fuel concentration along axis.

Total diffusion rate of fuel over whole flame per unit depth ≈ $H\rho D\ Y_{F,0}(x)/r_f\ (x)$

The burn rate of fuel must be equal to flow rate per unit width $\dot{m}_F' = H\ \rho D\ Y_{F,0}(x)/r_f\ (x)$

If we assume that $Y_{F,0}\ (x)/r_f\ (x)$ scales with $Y_{F,i}/d_i$, then

$$\dot{m}_F' \approx H^* \rho D\ Y_{F,i}/d_i$$

Hence,

$$H \approx d_i\ \dot{m}_F'/\{\rho D\ Y_{F,i}\} \tag{12.44}$$

Thus, H should linearly increase with the flow rate of fuel for laminar jets, and for a given flow rate, it is proportional to the jet size. Figure 12.6 shows a qualitative variation of flame height versus flow rate. A similar derivation is valid for circular jets also.

12.5.3 Governing Differential Equations for Planar Jets

We will neglect the potential core in the subsequent analysis. For an isolated 2-D jet, axial diffusion terms are neglected compared to diffusion terms. The 2-D boundary layer type of conservation equations presented in Chapter 11 are made use of and then solved with the boundary conditions to give solutions for the axial and radial gas velocity (v_x and v_y), the species concentrations (Y_F, Y_{O2}, etc.), and the flame height (H). The liftoff height of the flame (L), and the blow-off velocity (v_{blow}) are discussed later in this chapter. Refer to Tillman's Ph.D. thesis [2000] for complete details of the derivations and the solution of the ordinary differential equations for velocity and species.

The governing equations in "stretched" coordinates for mass, momentum, generic property, and energy/species are the same as those presented in Equations 11.50, 11.51, 11.61, and 11.63 except with two differences:

(i) Variables b, β are changed to $b' = b - b_\infty$, $\beta' = \beta - \beta_\infty$
(ii) Reference quantities: $d_{ref} = d_i$, $v_{ref} = v_{xi}$, $\rho_{ref} = \rho_i$, $Re_{ref} = Re_i$

12.5.4 Normalized Conservation Equations

Equation 11.51 without source term, Equation 11.61 with b* replaced by ϕ_F and Equation 11.63 with $\zeta = Sc$.

Mass Conservation

$$\partial v_x'^*/\partial x'^* + \partial v_y'^*/\partial y'^* = 0 \tag{12.45}$$

Momentum Conservation (x direction)

$$v_x'^* \partial v_x'^*/\partial x'^* + v_y^*(\partial v_x'^*/\partial y'^*) = [1/Re_i]\,(\partial^2 v_x'^*/\partial y'^{*2}) \tag{12.46}$$

Energy, Species, and SZ Conservation

The normalized SZ equation is given as

$$v_x'^* \partial\phi/\partial x'^* + v_y'^*(\partial\phi/\partial y'^*) = [1/(Re_i\ Sc)]\,(\partial^2\phi/\partial y'^{*2}) \tag{12.47}$$

where $Sc = \nu/D$. It should be noted that Equation 12.45 to Equation 12.47 can be used either for mixing or combustion with the appropriate definition for ϕ. For fuel species with finite kinetics,

$$v_x'^*\frac{\partial\phi_F}{\partial x'^*} + v_y'^*\frac{\partial\phi_F}{\partial y'^*} = \frac{\partial}{\partial y'^*}\left(\{1/(Re_i\,Sc)\}\frac{\partial\phi_F}{\partial y'^*}\right) + \frac{|\dot{w}_F'''|\,d_{ref}}{\rho Y_{kref} v_{ref}},\quad \phi_F = \frac{Y_F}{Y_{F,ref}}$$

For 2-D systems, Equations 11.26 and 11.27 still define transformation from compressible to incompressible coordination with k = 0 for planar jet.

12.5.5 Boundary Conditions for Planar Jets

Because the momentum, energy, and species equations are second order with respect to y and first order with respect to x, two boundary conditions are needed in the y direction and one initial condition is needed in the x direction. The boundary conditions to be enforced for the isolated 2-D jet are as follows.

$$\text{As } y \to \infty,\ v_x \ \text{ or } \ \partial v_x/\partial y \to 0,\ b' \ \text{ or } \ \partial b'/\partial y \to 0 \qquad (12.49)$$

and using symmetry,
at y = 0,

$$\partial b'/\partial y \to 0,\ \partial v_x/\partial y \to 0. \qquad (12.50)$$

At x = 0,

$$v_x = v_{x,i},\ 0 < y < d_i/2,\ v_x = 0,\ d_i/2 < y < \infty \qquad (12.51)$$

$$b' = b_i',\ 0 < y < d_i/2,\ b' = 0,\ d_i/2 < y < \infty \qquad (12.52)$$

where $b' = Y_k - Y_{k\infty}$, $h_t - h_{t\infty}$ (mixing problem), $b' = \beta' = \beta - \beta_\infty$, $h' = h - h_\infty$, $f_M' = f_M - f_{M\infty}$, and $Z_k' = Z_k - Z_{k\infty}$ (combustion problem). β' can be β'_{O2-F}, β'_{hT-F}, β'_{CO2-F}, $\beta\ni_{H2O-F}$, β'_{P-F}, $Y_I - Y_{I,\infty}$, etc. Thus, we can solve the partial differential equations numerically. However, the explicit solution requires certain approximations. Satisfaction of the boundary conditions (Equation 12.51 and Equation 12.52) will require numerical solution to the problem. Instead, if one neglects the potential core or assumes that d_i is very small ($d_i \to 0$) but with a finite momentum M, an explicit solution can be obtained. The momentum M′ is defined as

$$2\int_0^\infty \rho v_x^2\, dy = 2\int_0^{d_i/2} \rho v_x^2\, dy + 2\int_{d_i/2}^\infty \rho v_x^2\, dy \approx 2\int_0^\infty \rho v_x^2\, dy = \dot{M}' \qquad (12.53)$$

where M′ is the momentum flux rate per unit depth (N/m or kg/sec²). Note that outside the injector at x = 0, $v_{x,i} = 0$. Thus, as we let $d_i \to 0$, $v_{x,i} \to \infty$ at finite M′. If a flat velocity profile is assumed,

$$\dot{M}' = \rho v_{x,i}^2\, d_i. \qquad (12.54)$$

Similarly, for generic property $b' = b - b_\infty$

$$2\int_0^\infty \rho v_x b' dy = 2\int_0^{d_i/2} \rho v_x b'\, dy + 2\int_{d_i/2}^\infty \rho v_x^2\, dy \approx 2\int_0^\infty \rho v_x b'\, dy = J_{bi}'. \qquad (12.55)$$

For flat velocity and property profiles

$$\dot{J}_i' = \rho v_{x,i} b'\, d_i. \qquad (12.56)$$

where $\dot{J}_i'$ is the flow per unit width in units of (kg/m³) × m/sec × m = kg/(sm).

12.5.6 Normalized Boundary Condition

After using "stretched" coordinates for Equations 12.49 to 12.56, the normalized conditions to be enforced are:

$$\text{At } y = 0\text{: } \partial v_x'^*/\partial y'^* = 0,\ \partial\phi/\partial y'^* = 0,\ \partial\phi_F/\partial y'^* = 0 \tag{12.57}$$

$$\text{At } y = \infty\text{: } v_x'^* = 0,\ \phi = 0,\ \phi_F = 0 \tag{12.58}$$

The conservation of momentum and SZ variables are rewritten as:

$$M^* = 2\int_0^\infty v_x'^{*2}\, dy'^* \tag{12.59}$$

$$J^* = 2\int_0^\infty v_x'^*\phi\, dy'^* \tag{12.60}$$

where

$$y^* = y'/d_i,\ \text{for a 2-D jet, } y' = \int_0^y \{\rho/\rho_i\}\, dy \approx \{\rho_{avg}/\rho_i\}\, y,\ x' = x,\ v_x^* = v_x/v_{xi}$$

$$M^* = \left\{2\int_0^\infty v_x(\rho v_x)dy\right\}/\dot{M}_i' = \text{momentum at any x/injected momentum} \tag{12.61}$$

$$J^* = \left\{2\int_0^\infty v_x(\rho b')dy\right\}/\dot{J}_i' = \text{property flux at any x/injected property flux.} \tag{12.62}$$

Typically, in the absence of buoyancy $M^* = 1$ and $J^* = 1$ for flat profiles. With a parabolic velocity profile at the burner exit, it can be shown that $M^* = 6/5$. Also, $J^* = 6/5$ for a parabolic velocity and species profile at the burner exit, and $J^* = 1$ for a parabolic velocity profile and a flat species profile at the burner exit.

12.5.7 Similarity Variables for Planar Jets

The governing differential equations, now transformed and normalized, can be reduced from partial differential equations to ordinary differential equations by selecting appropriate similarity variables. If Sc = 1, then the conservation equations are similar for Φ and v_x^*. (In fact, one can replace v_x^* in Equation 12.46 by Φ and obtain Equations 12.47). The boundary conditions are the same. Thus,

$$v_x^* = \Phi = (\beta - \beta_\infty)/(\beta_i - \beta_\infty),\ \text{Sc} = 1 \tag{12.63}$$

Using solutions from Schlichting [1955] for v_x^*, which are developed for jet mixing, one can get solutions for Φ. Knowing Φ profiles, one can obtain the flame structure or T, Y_F, and Y_{O2} profiles as in the case of droplet and boundary-layer combustion. Typically, solutions for v_x^* and Φ are given in terms of similarity variable η (x*, y*) instead of giving each directly in terms of x* and y*. The methodology of deriving partial differential equation in terms of similarity variables is briefly presented in Chapter 11. Table F.12.1 under "Chapter 12" in Formulae appendix summarizes the results for 2-D jets.

The appropriate similarity variable, η, modified from the definition given in Chapter 11 (also see Schlichting [1955]) is:

$$\eta = \{y'^*/x'^*\}^{2/3}\{Re_i/3\}^{2/3}M^{*1/3} \tag{12.64}$$

$$\eta = \{y'^*/x'^{*\,2/3}\}\,\{Re_i/3)^{2/3} \text{ with } M^* = 1 \tag{12.65}$$

where $Re_i = v_{x,i}\, d_i/\upsilon_i$,

$$f = (Rei/3)^{1/3}\,(1/M^*)^{1/3}\,\Psi(x'^*, y'^*)/x'^{*\,1/3} \tag{12.66}$$

$$f = \left(\frac{R_{e,i}}{3}\right)^{\frac{1}{3}} \frac{\psi\,(x'^*, y'^*)}{x'^{*\frac{1}{3}}},\ M^* = 1 \tag{12.67}$$

and

$$v_x'^* = v_x{}^* = \partial\psi/\partial y'^*,\ v_y'^* = -\partial\psi/\partial x'^* \tag{12.68}$$

12.5.8 Momentum Equation in Similarity Coordinates

The use of Equations 12.64 to Equation 12.68 allows the momentum equation given in Equation 12.46 to be rewritten as:

$$f'^2 + f\, f' + f''' = 0 \tag{12.69}$$

where

$$f' = df/d\eta = v_x^*(3/Rei)^{1/3}\, x^{*1/3}/M^{*1/3} \tag{12.70}$$

Transformed boundary conditions for the momentum equation are given as

$$f' \to 0 \text{ as } \eta \to \infty \tag{12.71}$$

$$f' \to 0 \text{ at } \eta = 0 \tag{12.72}$$

$$M^* = 2\int_0^\infty v_x'^{*2}\, dy'^* = 2\int_0^\infty f'^2\, d\eta \tag{12.73}$$

Typically M* = 1.0.

12.5.9 Momentum Solutions for Planar Jets

12.5.9.1 Velocities (Momentum Equation)

Equation 12.69 is then integrated with the boundary conditions (Equation 12.71 and Equation 12.73) to give the following solutions:

$$f=\left(\frac{3}{2}\right)^{\frac{1}{3}}\tanh\xi \tag{12.74}$$

Note: f is not mixture fraction (see Equation 12.66) and

$$\xi=\frac{1}{2}\left(\frac{3}{2}\right)^{\frac{1}{3}}\eta\,. \tag{12.75}$$

Substituting for η in terms of y'^* (Equation 12.65)

$$\xi=\frac{1}{2}\left(\frac{3}{2}\right)^{(1/3)}\frac{R_{e,i}^{\frac{2}{3}}y'^*M^{*1/3}}{x^{*\frac{2}{3}}}=0.2752\frac{\mathrm{Re}_i^{\frac{2}{3}}y'^*M^{*1/3}}{x^{*\frac{2}{3}}} \tag{12.76}$$

$$f'=\left(\frac{1}{2}\right)\left(\frac{3}{2}\right)^{2/3}\mathrm{Sec}h^2\,\xi \tag{12.77}$$

$$v_x{}^*=\frac{v_x}{v_{xi}}=\frac{\left(\frac{1}{2}\right)\left(\frac{3}{2}\right)^{2/3}\left(\frac{1}{3}\right)^{1/3}M^{*2/3}\,\mathrm{Re}_i{}^{1/3}\mathrm{Sec}h^2\,\xi}{x^{*1/3}}\,. \tag{12.78}$$

Setting $\xi = 0$, one can find the centerline velocity or maximum velocity as,

$$v_{x,\max}{}^*=\frac{v_{x,\max}}{v_{xi}}=\frac{\left(\frac{1}{2}\right)\left(\frac{3}{2}\right)^{2/3}\left(\frac{1}{3}\right)^{1/3}M^{*2/3}\,\mathrm{Re}_i{}^{1/3}}{x^{*1/3}}\,. \tag{12.79}$$

Hence Equation 12.78 can be written as

$$\frac{v_x}{v_{x\max}}=\mathrm{Sec}h^2\,\xi\,. \tag{12.80}$$

Velocity v_y is

$$v_y^*=\left(\frac{v_y'}{v_{xi}}\right)=\frac{0.5503M^{*1/3}}{x^{*2/3}R_{e,i}^{1/3}}\left[2\,\xi\,\mathrm{sec}h^2\,\xi-\tan\mathrm{h}\,\xi\right] \tag{12.81}$$

and as $y \to \infty$, one can show that flow is inward from the ambience and $y \to$ small values, which indicates that flow is outward (as we approach the axis).

12.5.9.2 Jet Half Width ($y^*_{1/2}$)

The distance at which the velocity v_x is half the centerline velocity v_{x0} is called the *jet half width*. Thus, using Equation 12.80, we determine,

$$\xi_{1/2} = 0.2752 \frac{R_{e,i}^{\frac{2}{3}} y^*_{1/2}}{x^{*\frac{2}{3}}} = \cosh^{-1}(\sqrt{2}), \tag{12.82}$$

$$y_{1/2} = \frac{3.634\, x^{*2/3}}{\mathrm{Re}_i^{2/3}} \cosh^{-1}(\sqrt{2}), \tag{12.83}$$

Thus, half width $y_{1/2}$ increases as x^m where m = 2/3 for a 2-D jet. It will be seen later that m = 1 for a circular jet.

12.5.10 Species, Temperature, and Φ Equations in Similarity Coordinates

The solution for ϕ will lead to the solution for T and Y_K for mixing problems and β for combustion problems. Recall that if Sc = 1, the momentum solution can be adopted. However, to solve for nonunity Schmidt number, a new variable $\Omega(\eta)$ needs to be introduced:

$$\Omega(\eta) = \phi\, x'^{*1/3} \tag{12.84}$$

Selection of this variable allows the species and energy equations to be reduced to the following:

$$\Omega'' + \Omega' \mathrm{Sc}\, f + \Omega\, \mathrm{Sc}\, f' = 0, \text{ for all species and SZ variables} \tag{12.85}$$

12.5.11 Solutions for Normalized SZ and Scalar Properties

Solutions are given as [Tillman, 2000]:

$$\phi = (\beta - \beta_\infty)/(\beta_i - \beta_\infty) = \{0.45428\, C\, (\mathrm{Re}_i/x^*)^{1/3}\, (J^*/M^{*1/3})\, \mathrm{sech}^{2\mathrm{Sc}}(\xi)\} \tag{12.86}$$

The constant C in Equation 12.86 is given as:

$$C = \left(\int_0^\infty \mathrm{sech}^4 \xi\, d\xi\right) \Big/ \left(\int_0^\infty \mathrm{sech}^{2+2\mathrm{Sc}} \xi\, d\xi\right) \tag{12.87}$$

As Sc → 0, C → 2/3. Solutions can be obtained for Sc = 0.5, 1.5, . . . and 1,2,3. . . and a curve fit can be made.

A curve fit yields

$C = 1 + 0.2906*(Sc - 1) - 0.0296\ (Sc - 1)^2$, $0 < Sc < 3.5$ (pass through C = 1 at Sc = 1), $R^2 = 0.9991$. (12.88)

Note that $C \to 1$ when $Sc \to 1$.

12.5.12 Mass Flow of Gas and Air and A:F at Any x

The mass flow of gases for the 2-D jet per unit width at any given x is given by:

$$\dot{m}'\ (y,\ x) = 2\int_0^y \rho v_x\, dy. \tag{12.89a}$$

If the jet burns in air, the total air entrained at x is given as:

$$\dot{m}'_A = \dot{m}'(\infty, x) - \dot{m}_i' \tag{12.89b}$$

and $\dot{m}'_{F,i} = \dot{m}'_i * Y_{F,i}$,
for both 2-D and circular jets. Because the entrained gas is normally air, using Equation 12.78 and Equation 12.76 the integrated A:F ratio = $\dot{m}_A / \dot{m}_{F,i}$ at given x (but not the local A:F ratio at given x and y) is

$$(A:F)_x = \frac{1}{Y_{F,i}}\left[\frac{\dot{m}'}{\dot{m}_{i'}} - 1\right] = \left[\frac{3.3019 M^{*1/3}}{R_{e,i}^{1/3}}\ x^{*\frac{1}{3}} - 1\right]\frac{1}{Y_{F,i}} \tag{12.89c}$$

where the fuel flow $\dot{m}_{F,i}$ is used for defining the A:F ratio. Equation 12.89c is valid for combustion also. In residential burners called inshot burners, primary air entrained is about 30% of the required air. Equation 12.89c is useful in determining the distance at which stoichiometric ratio (SR) = 0.3 due to pure entrainment. It is seen from Equation 12.89c that the integrated mass flow at any x* per unit depth is

$$\left(\frac{\dot{m}'}{\dot{m}_{i'}}\right) = 3.3016\, M^{*1/3}\frac{x^{*\frac{1}{3}}}{R_{e,i}^{1/3}},\quad x^* \geq (R_{e,i}/[3.3019M^*]^3)\,. \tag{12.89d}$$

Defining the entrainment coefficient as $C_{entrain} = d\ \dot{m}/\dot{m}_i\ /dx^*$, it is given as

$$C_{Entrain} = 1.1 M^{*^{1/3}}\frac{1}{x^{*\frac{2}{3}} R_{e,i}^{1/3}},\quad x^* \geq (R_{e,i}\,/\,[3.3016\,M^*]^3) \tag{12.90}$$

where $\dot{m}' = \dot{m}/W$, is the gas entrained per unit width of slit; and it should be a number greater than $\dot{m}_i/W$. Note that $\dot{m}'$ gradually increases with x due to entrainment.

12.5.13 Solutions for Pure Mixing Problems (Chemically Frozen Flow)

The expressions are summarized under "Chapter 12" in Formulae appendix; see also section 12.8 for profiles.

12.5.14 Solutions for Combustion

12.5.14.1 Normalized SZ Variable

With "f" from the momentum equation, one can solve for Ω from Equation 12.85. Thus,

$$\phi = \frac{0.4543 \quad J^* C \mathrm{Re}_i^{\frac{1}{3}}}{M^{*1/3} \; x^{*1/3}} \mathrm{Sech}^{2Sc} \; \xi \tag{12.91}$$

$$\phi_{\max} = \frac{0.4543 \quad J^* C \mathrm{Re}_i^{\frac{1}{3}}}{M^{*1/3} \; x^{*1/3}} \tag{12.92}$$

$$\phi/\phi_{\max} = \mathrm{sech}^{\,2Sc} \; \xi \tag{12.93}$$

Note that $\phi_{\max}$ is still a function of Sc. Figure 12.7 shows nondimensional plots for $v_{x\max}$, $\phi_{\max}$, and entrained flow $\{\dot{m}'$ (kg/m sec)/$\dot{m}_i'\}$ vs. x* for Sc = 0.5, 1, and 1.5. The effect of increase in Sc is to increase $\phi_{\max}$ at a given x* (Equation 12.92). For example, in case of a pure mixing problem, ϕ represents Y_F/Y_{Fi}. An increased Sc implies a reduced diffusion coefficient; thus, fuel is diffused away slowly to the ambient. Hence, $Y_{F\max}$ will be higher for higher Sc.

12.5.14.2 Flame Profile and Structure

Flame profile, i.e., y_f vs. x for 2-D jets can be obtained by setting $\beta_{O2-F} = 0$ or $\phi = \phi_f$ at the flame with the thin-flame assumption. With $\phi = \phi_f$, where

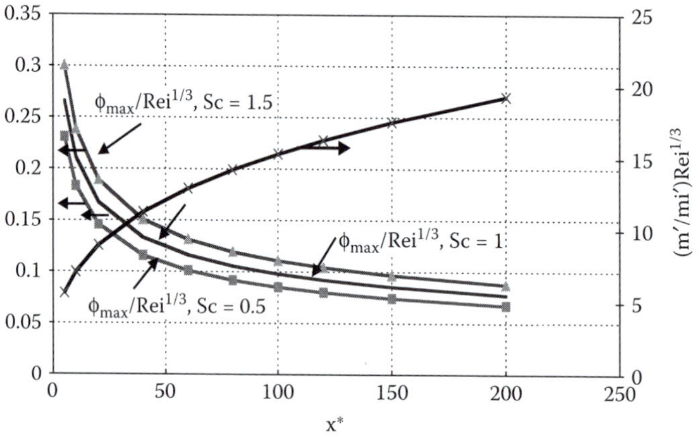

Figure 12.7 Nondimensional $v_{x\max}$, $\phi_{\max}$, entrained flow for a 2-D jet.

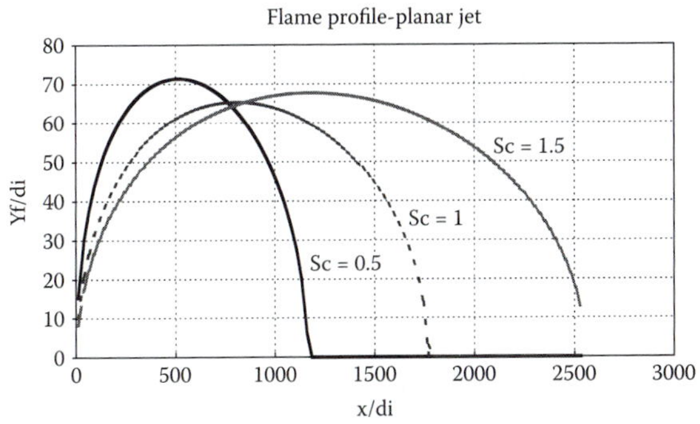

Figure 12.8 Flame profile for a 2-D jet for various Sc numbers, fuel C_3H_8, $Re_i = 4$.

$$\phi_f = \frac{\beta_{O2-F,f} - \beta_{O2-F,\infty}}{\beta_{O2-F,i} - \beta_{O2-F,\infty}} = \frac{0 - \beta_{O2-F,\infty}}{\beta_{O2-F,i} - \beta_{O2-F,\infty}} = \frac{s}{(1+s)} \quad (12.94)$$

and

$$s = \frac{Y_{O2,\infty}}{\nu_{O2} Y_{F,i}} \quad (12.95)$$

we obtain ξ_f at flame using Equation 12.91, and then using the definition of ξ (Equation 12.76),

$$\xi = 0.2752\ y_f'^* \ (Re_i/x^*)^{2/3}\ M^{*1/3} \quad (12.96)$$

we obtain $y_f'^*$ as a function of x. You will find that the flame is always overventilated because the jet is in infinite surroundings.

From Equation 11.45 with k = 0 and d_{reg} = di

$$y_f'^* = \int_0^{y_f^*} \left(\frac{\rho}{\rho_i}\right) dy^* \quad (12.97)$$

Thus, y_f^* vs. x_f^* can be plotted at any Sc. Figure 12.8 shows the results.

As Sc increases, i.e., as diffusion coefficient decreases, the transport rate of O_2 lateral to the fuel stream decreases, and hence fuel burn rate per unit length decreases, resulting in an increased flame height. The increased flame height can result in enhanced thermal NOx production.

Fuel, oxygen, mass fraction, product, and temperature are obtained using ϕ profiles. The procedure is similar to the one outlined in Chapter 10. Figure 12.9 shows the flame structure for C_3H_8 at $x^* = 40$ and $x^* = 200$.

12.5.14.3 Flame Heights

The flame length H_f is obtained by setting $\phi = \phi_f$ and $\xi_f = 0$ in Equation 12.93, we obtain $\phi_f = \phi_{max}$ and solving for $x_f^* = H_f^*$, from Equation 12.92

$$H_f^* = \left(\frac{H_f}{d_i}\right) = \frac{3}{32}\frac{C^3 J^{*3} R_{e,i}}{M^* \phi_f^3} = \frac{3}{32}\frac{C^3 J^{*3} R_{e,i}}{M^*}\left(\frac{1+s}{s}\right)^3 \quad (12.98)$$

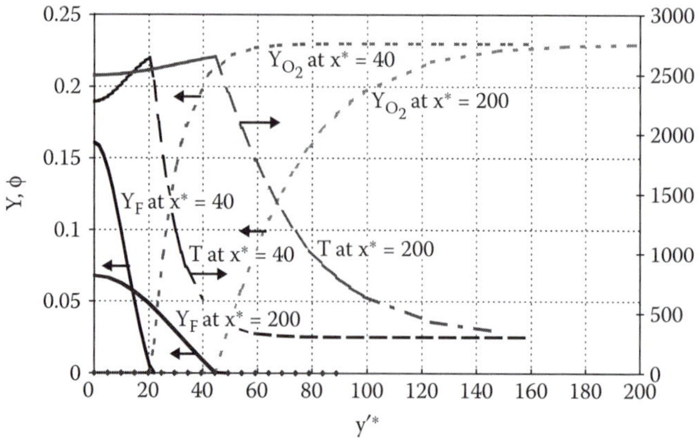

Figure 12.9 Flame structure for a 2-D jet. $Y_{O_2,\infty}$ = 0.23, C_3H_8, Re_i = 4.0, c_p = 1.175 kJ/kgk, Sc = 0.5.

where C = 1 for Sc = 1; typically, J* = 1, M* = 1;

$$H_f \;\alpha\; v_i d_i^2, \quad H_f \,\alpha\, \dot{Q}_i \quad or \quad H_f \;\alpha\; \dot{m}_i$$

(Later we will see that for turbulent jets, $H_{f,turb}$ α d_i and is independent of the flow rate for a given d_I.)

12.5.14.4 Mass Flow of Gas, Air and A:F at Any x

In boundary-layer combustion of solid and liquid fuels, there is an excess amount of fuel at the pyrolysis front, which results in overhang of the flame in the plume region. On the other hand, the laminar jets are overventilated flames, i.e., the flame closes at the axis with excess air flowing at flame tip because of the entrained airflow exceeding the amount required for combustion of the fuel. The excess air beyond the flame tip affects the gas temperature in that region, the oxidation rate of soot, and the production of thermal NOx. Further, the excess air affects the oxygen concentration in the interstitial space between the burners and hence the degree of interaction between the jets when multiple jets are used. It is, therefore, useful to have an expression for predicting the amount of air entrained by the burner at a given axial distance and to estimate the percentage of excess air at the flame tip. We will also show that excess-air percentage at the flame tip for 2-D and laminar jets is only a function of Sc. This can be accomplished by the following procedure:

By setting x* = H* in Equation 12.89c,

$$(A:F)_{Hf} = \left[\frac{3.3016 M^{*1/3}}{R_{e,i}^{1/3}} H_f^{*\frac{1}{3}} - 1\right]\frac{1}{Y_{F,i}} \tag{12.99}$$

where H_f^* is known in terms of fuel properties and Re of the jet. It is interesting to note that for 2-D jets, the C defined in Equation 12.87 approximately represents the ratio of air entrained at x* = H* for any Schmidt number to the air entrained at x* = H* for Sc = 1. Using Equation 12.98 in Equation 12.99,

$$(A:F)_{Hf} = \left[\frac{1.5CJ^*(1+s)}{s} - 1\right]\frac{1}{Y_{F,i}}. \quad (12.100)$$

Because stoichiometric airflow is given as $\dot{m}_{F,i}\, \nu_{O2}/Y_{O2,\infty} = (\dot{m}_{F,i}/\{sY_{F,i}\})$, $(A{:}F)_{st} = 1/\{sY_{F,i}\}$ and the excess-air fraction, Ex is given as

$$(Ex)_{Hf} = \frac{A:F}{(A:F)_{st}} - 1 = (1+s)[1.5\,CJ^* - 1]. \quad (12.101)$$

With J* = 1 and C = 1 for Sc = 1,

$$(Ex)_{Hf} = (1 + s)\,[1.5\ C\ J^* - 1] \approx [1.5\ C\ J^* - 1] \approx 0.5$$

and at Sc = 0, C = 2/3, J* = 1

$$(Ex)_{Hf} = (1 + s)\,[1.5\ C\ J^* - 1] = 0. \quad (12.102)$$

Figure 12.10 shows the excess-air fraction as a function of Sc. Restricting the discussion to a pure fuel jet issuing from a 2-D jet, it seen that as the Schmidt number becomes small, the air entrained approaches the stoichiometric amount at the tip of the flame. Then, for all fuels with Sc = 1 (C = 1), the ratio of air entrained at x* = H* to the fuel injected is 50% in excess of the stoichiometric amount. It should be noted that excess-air percentage at x* = H_f* will not change even if the fuel is diluted with inerts. If the fuel is diluted with inerts, the flame height will decrease, the amount of air entrained decreases because of the shorter height, but the amount of air needed for combustion will also decrease; therefore, the excess-air percentage at x* = H_f* will remain constant. Also a low Sc implies that the diffusion coefficient is

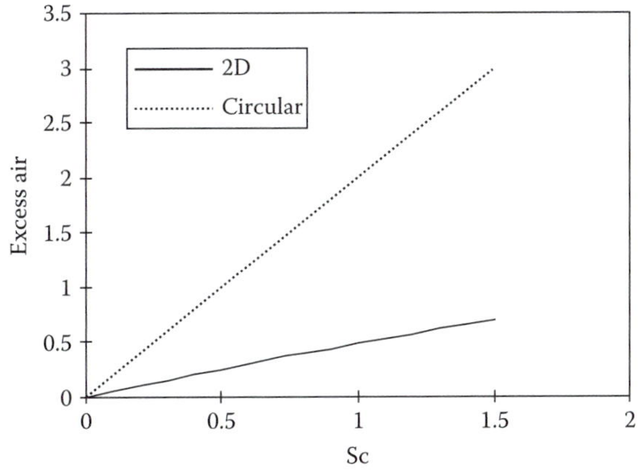

Figure 12.10 Excess-air fraction at flame tip vs. Sc.

higher compared to the momentum diffusivity and the stoichiometric contour (or flame contour) height is shorter than flame velocity contour.

12.6 Circular Jets

12.6.1 Simplified Relations for Circular Laminar Jets

The flame contour is similar to those shown in Figure 12.4 and Figure 12.5 except that a flame of radius $r_f(x)$ is symmetric around the axis. Because the fuel flows around the axis, the fuel exists in the inner region, whereas oxygen exists in the outer region. Let us assume a thin-flame model.

$$\text{Diffusion rate of fuel per unit flame area} \sim \rho D\ Y_{F,0}(x)/r_f\ (x) \quad (12.103)$$

where $Y_{F,0}$ is fuel concentration along the axis.

Total diffusion rate of fuel over whole flame = $H_f*2\ \pi\ r_f\ (x)*\rho D\ Y_F(x)/r_f\ (x)$.

If we assume that $Y_{F,0}$ (x) scales with $Y_{F,I}$, the diffusion rate = $H_f*2\ \pi*\rho D\ Y_{F,i}$.

(12.104)

The burn rate of fuel must be equal to flow rate $\dot{m}_{F,i} = \rho_i v_{xi}\ (\pi\ d_i^2/4)\ Y_{F,i}$.

(12.105)

Hence, equating Equation 12.104 and Equation 12.105,

$$H_f = \dot{m}_{F,i}/\{2\ \pi\ \rho D\ Y_{F,i}\} = \dot{m}_i/\{2\ \pi\ \rho D\}. \quad (12.106)$$

Thus, H_f is proportional to the flow rate of fuel for laminar jets, and for a given flow rate, it is independent of jet size.

For the isolated circular jet, the equations of mass, momentum, energy, and species are first presented in compressible form along with boundary conditions, which are very similar to 2-D jets. Unlike in a 2-D jet, conversion to incompressible form is not possible for a circular jet. The properties, including density, are assumed to be constant so that simple explicit solutions can be presented. Actual diffusive terms are ignored in comparision with radial diffusive terms.

12.6.2 Governing Differential Equations for Circular Jets

The conservation of mass, momentum, energy, and species equations for the circular jet into compressible form can be written as shown in the following subsections.

Mass Conservation

$$\partial/\partial x\ \{\rho\ v_x\ r) + \partial/\partial r\ (\rho\ v_r\ r) = 0 \quad (12.107)$$

Momentum Conservation (x direction)

$$\partial/\partial x \{\rho\ v_x\ r\ v_x) + \partial/\partial r\ (\rho\ v_r\ rv_x) = \partial/\partial r\ (\rho\nu\ r\ \partial v_x/\partial r) \qquad (12.108)$$

Energy, Species, and SZ Conservation

In general, for any generic property "b",

$$\partial/\partial x \{\rho\ v_x\ r\ b) + \partial/\partial r\ (\rho\ v_r\ r\ b) = \partial/\partial r\ (\rho D\ r\ \partial b/\partial r) \qquad (12.109)$$

where $b = Y_K$, h_T (pure mixing) or β (combustion).

As in a 2-D jet, Equation 12.109 can be converted as:

$$\partial/\partial x \{\rho\ v_x\ r\ b') + \partial/\partial r\ (\rho\ v_r\ r\ b') = \partial/\partial r\ (\rho D\ r\ \partial b'/\partial r) \qquad (12.110)$$

where

$b' = b - b_\infty = Y_k - Y_{k,\infty,}$ $h_T - h_{T,\infty}$ (mixing), $\beta - \beta_\infty$, $f_M - f_{M\infty}$ (combustion), etc.

12.6.3 Boundary Conditions for Circular Jets

The following three boundary conditions will be applied for the circular jet:

Far field and symmetric conditions

$$\text{As } r \to \infty,\ v_x \to 0,\ b' \to 0 \qquad (12.111)$$

$$\text{At } r = 0,\ \partial b'/\partial r \to 0,\ \partial v_x/\partial r \to 0 \qquad (12.112)$$

Total momentum flux

The momentum at any given x is simply given by:

$$\dot{M} = 2\pi \int_o^\infty \rho v_x^2\ r\ dr = \text{constant} \qquad (12.113)$$

Total property flux

The integral of the property b' is written as $\dot{J}$. Hence,

$$\dot{J} = 2\pi \int_o^\infty \rho v_x b'\ r\ dr = (d_i/2) \int_o^\infty \rho\, v_x b_i'\ r\ dr = \text{Constant} \qquad (12.114)$$

12.6.4 Normalization

To normalize the governing equations and appropriate boundary conditions, the following variables will be used: $r^* = r/d_i$, $x^* = x/d_i$, $v_x^* = v_x/v_{x,i}$, $v_r^* = v_r/v_{x,i}$.

12.6.5 Normalization of Governing Differential Equations

Using the governing normalized form:

Mass Conservation

$$\partial/\partial x^*(v_x^*\ r^*) + \partial/\partial r^*(v_r^*\ r^*) = 0 \tag{12.115}$$

Momentum Conservation (x direction)

$$v_x^*\partial v_x^*/\partial_x^* + v_x^*\partial v_r^*/\partial r^* = 1/(r^*Re_i)\ \partial/\partial r^*\ (r^*\partial v_x^*/\partial r^*) \tag{12.116}$$

Energy, Species, and SZ Conservation

$$v_x^*\ \partial\phi/\partial x^* + v_r^*\ \partial\phi/\partial r^* = 1/(r^*\ Sc\ Re_i)\ \partial/\partial r\ (r^*\ \partial\phi/dr^*) \tag{12.117}$$

If finite kinetics are involved for species k (e.g., fuel), then

$$v_x{}^*\frac{\partial\phi_k}{\partial x'^*} + v_r{}^*\frac{\partial\phi_k}{\partial r'^*} = \frac{\partial}{\partial r'^*}\left(\{1/(\mathrm{Re}_i\, Sc)\}\ \frac{\partial\phi_k}{\partial r'^*}\right) + S_k{}^*, k = F, \phi_k = \frac{Y_k}{Y_{k,ref}} \tag{12.118}$$

where

$$S_k{}^* = \frac{\dot{w}_{k'''} d_i}{\rho Y_{kref} v_{xi}},\quad \phi_k = Y_k/Y_{kref}. \tag{12.119}$$

Normalized Boundary Conditions

The boundary conditions must also be updated. The new conditions to be enforced are:

$$\text{As } r^* \to \infty,\ v_x^* \to 0,\ \phi \to 0 \tag{12.120a}$$

$$\text{At } r^* = 0,\ \partial\phi/\partial r^* \to 0,\ \partial v_x^*/\partial r^* \to 0 \tag{12.120b}$$

Following the procedure outlined for 2-D jets,

$$\dot{M}_i = \frac{\pi d_i^2}{4}\rho_i v_{x,i}{}^2,\ \mathrm{N = kg\ m/sec^2} \tag{12.121a}$$

$$\dot{J}_i = \frac{\pi d_i^2}{4}\rho_i v_{x,i}\ b'_i. \tag{12.121b}$$

For flat velocity profiles, and constant ρ,

$$M^* = \frac{\dot{M}}{\dot{M}_i} = \frac{2\pi\int_0^\infty \rho v_{xi}^2\, r\, dr}{\frac{\pi d_i^2}{4}\rho_i\ v_{x,i}{}^2} = 8\int_0^\infty v_x^{*2}\ r^*\ dr^* \quad \text{or} \tag{12.122}$$

$$\int_0^\infty v_x{}^{*2} r^* \, dr^* = \frac{1}{8} \frac{\dot{M}}{\dot{M}_i} = \frac{M^*}{8}. \tag{12.123}$$

Similarly, for flat velocity and property profiles,

$$\int_0^\infty v_x^* \phi r^* \, dr^* = \frac{1}{8} \frac{\dot{J}}{\dot{J}_i} = \frac{J^*}{8}. \tag{12.124}$$

Typically, $M^* = 1$ and $J^* = 1$. For variable density treatment in circular jet, see Fay [1954].

12.6.6 Similarity Variables for Circular Jets

Schlichting [1955] gives an appropriate similarity variable, η, as:

$$\eta = r^*/x^* \tag{12.125}$$

$$f(\eta) = \Psi(r^*, x^*)/\{C\, x^*\}, \quad \text{where} \quad C = 1/Re_i \tag{12.126}$$

$$v_x^* = (C/x^*)\, f'/\eta,\ v_r^* = \{C/x^*\}\,\{f' - f(\eta)/\eta)\},\ C = 1/Re_i \tag{12.127}$$

12.6.7 Momentum Equation in Similarity Coordinates

The use of this similarity variable allows the momentum equation given in Equation 12.116 to be rewritten as:

$$f''' - f''/\eta + ff''/\eta + f'/\eta^2 + f'^2/\eta - ff'/\eta^2 = 0 \tag{12.128}$$

The boundary conditions are similar to those for the 2-D jet except for M^* and J^*, which are given in Equation 12.123 and Equation 12.124, respectively.

12.6.8 Solution to Momentum Equation

12.6.8.1 Velocities

Using the appropriate boundary conditions allows the momentum equation in Equation 12.128 to be integrated to give the following solutions:

$$f = \xi^2/(1 + \xi^2/4\}, \tag{12.129}$$

where

$$\xi = \frac{\sqrt{3}}{8} M^{*1/2}\, Re_i \left(\frac{r^*}{x^*}\right) = \frac{\sqrt{3}}{8} M^{*1/2}\, Re_i\, \eta \tag{12.130}$$

$$f' = df/d\eta = (df/d\xi)\,(d\xi/d\eta) = 2\,(\sqrt{3}/8)\, M^{*1/2}\, Re_I\, \xi/(1 + \xi^2/4)^2 \tag{12.131}$$

Axial Velocity v_x

$$v_x^* = \frac{v_x}{v_{x,i}} = \frac{3}{32}\left(\frac{M^*\mathrm{Re}_i}{x^*}\right)\frac{1}{\left\{1+\frac{1}{4}\xi^2\right\}^2} \tag{12.132}$$

$$v_{x\max}^* = \frac{v_{x,\max}}{v_i} = \left(\frac{3}{32}\right)\left(\frac{M^*\mathrm{Re}_i}{x^*}\right)$$

$$\frac{v_x}{v_{x\max}} = \frac{1}{\left[1+\left(\xi^2/4\right)\right]^2} \tag{12.133}$$

Radial Velocity v_r^*

$$\frac{v_r}{v_{x,i}} = \frac{\sqrt{3}}{8}\left(\frac{M^{*1/2}}{x^*}\right)\frac{\left\{\xi - \frac{1}{4}\xi^3\right\}}{\left\{1+\frac{1}{4}\xi^2\right\}^2} \tag{12.134}$$

Half width $r^*_{1/2}$

Following the procedure for 2-D jets and using Equation 12.132, the half width $r^*_{1/2}$, at which $v_x = (1/2)\, v_{x\max}$ is given as $\xi_{1/2} = 2\,\{\sqrt{2}-1\}^{1/2}$ and expanding

$$r^*_{1/2} = \xi_{1/2} x^* \frac{8}{M^{*1/2}\sqrt{3}} = \frac{5.95\, x^*}{M^{*1/2}}\ . \tag{12.135}$$

It is seen that $r_{1/2} \propto x$ for circular jet, while $y_{1/2} \propto x^{2/3}$ for a 2-D jet.

12.6.9 NSZ Equation in Similarity Coordinates

The solution for ϕ will lead to the solution for T and Y_K for mixing problems and β for combustion problems. Define:

$$\Omega(\eta) = \phi\, x^* \tag{12.136}$$

Equation 12.117 reduces to

$$\Omega''\eta + \Omega' + \mathrm{Sc}\, f'\, \Omega + \mathrm{Sc}\, f\, \Omega' = 0. \tag{12.137}$$

Equation 12.137 is modified with inclusion of finite kinetics as,

$$\Omega_k''\, \eta + \Omega_k' + \mathrm{Sc}\, f\, \Omega_k' + \mathrm{Sc}\, f'\, \Omega_k + \mathrm{Sc}\, \mathrm{Re}_i\, S_k^*\{\eta\, x^{*3}\} \tag{12.138}$$

where

$$S_k^* = \frac{\dot{w}_k''' d_i}{\rho Y_{kref} v_{xi}},\ k = F. \tag{12.139}$$

12.6.10 Solutions for Species, Temperature, and Φ

Solutions for the energy, species, and SZ conservation equation are given by:

$$\phi = \{(2Sc + 1)/32\} \{Re_i J^*/x^*\}/(1 + \xi^2/4)^{2Sc} \quad (12.140)$$

$$\text{with } \xi = 0,\ \Phi_{max} = \Phi_0 = \{(2Sc + 1)/32\} \{Re_i J^*/x^*\} \quad (12.141)$$

12.6.11 Solutions for Combustion and Pure Mixing

Solutions for combustion will be presented first.

12.6.11.1 Solutions for Combustion

$$\phi = \left(\frac{\beta - \beta_\infty}{\beta_i - \beta_\infty}\right) = \frac{(2Sc+1)}{32}\frac{J^*\,Re_i}{x^*}\frac{1}{\left[1 + \frac{\xi^2}{4}\right]^{2Sc}} \quad (12.142)$$

$$\phi_0 = \phi_{max} = \frac{(2Sc+1)}{32}\frac{J^*\,Re_i}{x^*} \quad (12.143)$$

$$\frac{\phi}{\phi_{max}} = \frac{\beta - \beta_\infty}{\beta_i - \beta_\infty} = \frac{1}{\left[1 + \frac{\xi^2}{4}\right]^{2Sc}} \quad (12.144)$$

Rewriting,

$$\frac{\phi}{\phi_{Sc=1}} = \left(\frac{Y_I - Y_{I,\infty}}{Y_{I,Sc=1} - Y_{I,\infty}}\right) = \left(\frac{\beta - \beta_\infty}{\beta_{Sc=1} - \beta_\infty}\right) = \frac{(2Sc+1)}{3}\frac{\left[1 + \frac{\xi^2}{4}\right]^2}{\left[1 + \frac{\xi^2}{4}\right]^{2Sc}} \quad (12.145)$$

$$\frac{\phi_{max}}{\phi_{max,Sc=1}} = \left(\frac{Y_I - Y_{I,\infty}}{Y_{I,Sc=1} - Y_{I,\infty}}\right) = \left(\frac{\beta - \beta_\infty}{\beta_{Sc=1} - \beta_\infty}\right) = \frac{(2Sc+1)}{3}. \quad (12.146)$$

With $J^* = 1$ and $M^* = 1$,

$$\phi/\phi_v^{Sc} = \{(2Sc + 1)/3^{Sc}\}\{Re_i/(32x^*)\}^{(1-Sc)}. \quad (12.147)$$

Flame Profile: The flame profile, i.e., r_f vs. x, for circular jets can be obtained by setting $\beta_{O2-F} = 0$ or $\phi = \phi_f$ at the flame with a thin-flame assumption. With $\phi = \phi_f$, we obtain ξ_f at the flame, and then using the definition of ξ, we obtain r_f as a function of x.

Let

$$s = Y_{O2\infty}/(Y_{F,i}\upsilon_{O2}) \qquad (12.148)$$

Then,

$$\Phi_f = s/(Y_{F,i} + s) = \{Y_{O2-\infty}/\upsilon_{O2}\}/[Y_{F,i} + \{Y_{O2-\infty}/\upsilon_{O2}\}] = s/(s+1) \qquad (12.149)$$

$$\phi_f = \left(\frac{s}{s+1}\right) = \frac{(2Sc+1)}{32}\frac{J^*\,Re_i}{x_f^*}\frac{1}{\left[1+\frac{\xi_f^2}{4}\right]^{2Sc}} \qquad (12.150)$$

With Equation 12.130 and Equation 12.150, ξ_f vs. x* and hence r_f^* vs. x* can be obtained.

Flame Height: The flame height H_f is obtained by setting $\xi = 0$ in Equation 12.150.

$$H_f^* = \frac{(2Sc+1)}{32}\frac{J^*\,Re_i}{\phi_f} = \frac{(2Sc+1)}{32}\frac{J^*\,Re_i(s+1)}{s} \qquad (12.151)$$

$$H_f \;\alpha\; v_i d_i^2, \quad H_f \,\alpha\, \dot{Q}_i \quad or \quad \dot{m}_i \qquad (12.152)$$

For any circular jet, the flame height increases linearly with the injected flow rate. Experiments in zero g reveal a linear relation between flame height and the volume flow of gas (or Re) as shown in Figure 12.11. The 1g case reveals a lower height due to additional flow due to buoyancy and increased mass transfer.

12.6.11.2 Solutions for Mixing

Equations 12.142 to 12.147 can be used for pure mixing with $\phi = (Y_k - Y_{k,\infty})/(Y_{k,i} - Y_{k,\infty}) = (T - T_\infty)/(T_i - T_\infty)$. If $S_c < 1$, then $\phi_{k,o} < \phi_{K,o}(S_c = 1)$. Thus, the centerline fuel mass fraction with $S_c < 1$ is less than centerline fuel mass fraction with $S_c = 1$. Hence, stoichiometric contours move toward the axis, whereas they move away for $S_c > 1$. The flame height becomes the height of stoichiometric contour where $Y_F = (Y_{O_2}/v_{O_2})$.

Example 2

Pure gaseous C_3H_8 fuel is injected at a velocity of 20 cm/sec through a circular burner of diameter 0.2 cm into air. Assume $\rho D = 6.5\text{x}10^{-5}$ kg/m sec,

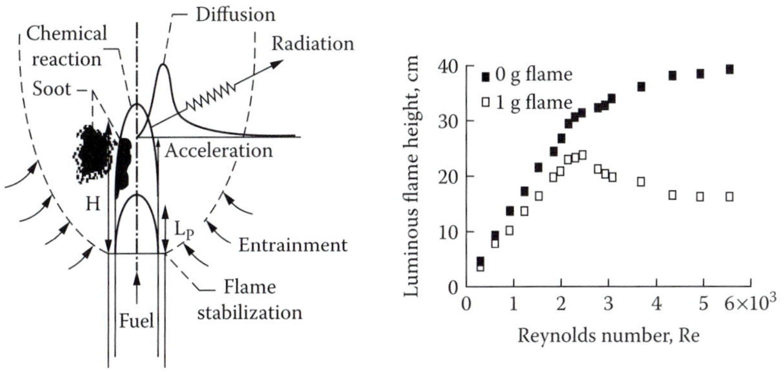

Figure 12.11 Circular jet in 0 and 1g. (Adapted from Bahadori, M.Y. et al., 1993; also see Ronney, P.D., 1998, www.GRC.NASA.GOV/www/RT1995/6000/67115.htm.)

c_p = 1.175 kJ/kg K, $\nu_i = 1.0 \times 10^{-5}$ m²/sec. Assume T_i = 300 K, T_∞ = 300 K, P = 100 kPa, h_c = 46357 kJ/kg, flat velocity, and species profiles at inlets. Plot

(a) y_f/d_i vs. x/di for Sc = 0.5, 1, and 1.5
(b) Y_F, Y_{O2}, Y_P, and T in K vs. y/d_i at x/d_i = 20 assuming Sc = 1

Solution

For propane,

$$C_3H_8 + 5\ O_2 = 3\ CO_2 + 4\ H_2O,\ \nu_{O2} = 5*32/44.11 = 3.63$$

$$Re_i = v_{xi}\ d_i/\nu = \{0.2\ m/sec*0.002\ m\}/1x10^{-5}\ m^2/sec = 40$$

Let x/di = x* = 20

$$y'^* = y^*\ \text{(incompressible approximation)}$$

$\phi_{max} = \phi_0 = \{(2Sc + 1)/32\}\ \{Re_i\ J^*/x^*\} = \{(2*1.5 + 1)/32\}*40/20 = 0.25$ at Sc = 1.5, = 0.18 at Sc = 1

$$s = Y_{O2\infty}/\nu_{O2} = 0.23/3.63 = 0.0634$$

(a) Flame Profile

Because air is the ambient, $Y_{O2\infty} = 0.23$ $\beta_{O2-F,i} = Y_{F,i} - Y_{O2,i}/\nu_{O2} = 1$, $\beta_{O2-F,\infty} = -0.0634$, $\phi_f = s/(Y_{F,i} + s) = 0.0635/(1 + 0.0635) = 0.0596$

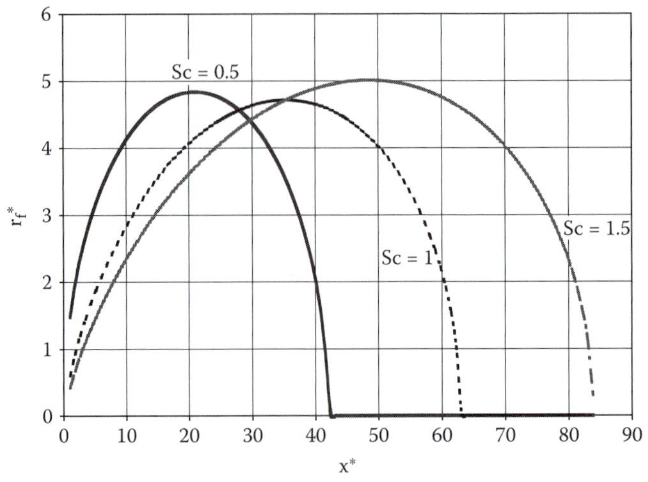

Figure 12.12 Flame profiles, r_f^* vs. x^* for circular jets.

Evaluating ϕ_{max} from Equation 12.143 at $x^* = 20$, $\phi_{max} = 0.25$ for Sc = 1.5; 0.18 at Sc = 1. Using Equation 12.144, $\phi_f/\phi_{max} = 0.0596/0.25 = 1/(1 + \xi_f{}^2/4\}^{2*1.5}$, $\xi_f = 1.565$ at Sc = 1.5 and 1.76 at Sc = 1; using Equation 12.130, $r_f^* = \xi_f/\{0.2165\ (Re_i/x^*)M^{*1/2}\} = 1.565/\{0.2165\ (40/20)\} = 3.62$ at Sc = 1.5 and 4.06 at Sc = 1. Thus r_f^* vs. x^* (Figure 12.12) can be obtained. Using Equation 12.151,

$$H_f{}^* = (2*1.5 + 1)*40/(32*0.0596) = 83.85 \text{ for Sc} = 1.5$$

Flame height, $H_f = 83.85 * 0.002 = 0.17$ m or 17 cm.
(b) Flame Structure

$$\frac{v_{x,max}}{v_i} = \frac{v_{x,max}}{v_i} = \left(\frac{3}{32}\right)\frac{R_{e,i}}{x^*} = \left(\frac{3}{32}\right)\frac{40}{20} = 0.18$$

$v_{x,max} = 0.18 * 0.2 = 0.036$ m/sec at $x^* = x_1{}^* = 20$; using Equation 12.133

$v_x/v_{x,max} = 0.69$ at $\xi = 0.9$, and hence $v_x = 0.69*0.036 = 0.03$ m/sec. Using Equation 12.130

$$r^* = \xi/\{0.2165\ M^{*1/2}Re_i/x^*\} = 0.9/\{0.2165*1*(40/20)\} = 2.08,$$

$$\frac{\phi}{\phi_{max}} = \frac{\beta - \beta_\infty}{\beta_i - \beta_\infty} = \frac{1}{\left[1+\frac{\xi^2}{4}\right]^{2Sc}} = \frac{1}{\left[1+\frac{0.9^2}{4}\right]^{2*1.5}} = 0.575 \ at \ Sc = 1.5, = 0.69 \ at \ Sc = 1$$

At $x^* = 20$, $\phi_{max} = 0.18$ at Sc = 1 with $\xi = 0.9$, $\phi/\phi_{max} = 0.692$, $\phi = 0.129$, $r^* = 2.08$. $\{\beta_{O2-F,i} - \beta_{O2-F,\infty}\} = 1.0635$, $\beta_{O2-F,\infty} = -0.0634$

$\{\beta_{O2-F} - \beta_{O2-F,\infty}\} = 1.0635 * 0.129 = 0.137$, $\beta_{O2-F} = 0.0634 + 0.137 = 0.0738$, $Y_F = 0.0738$ because $Y_{O2} = 0$ for $r^* < r_f = 3.62$

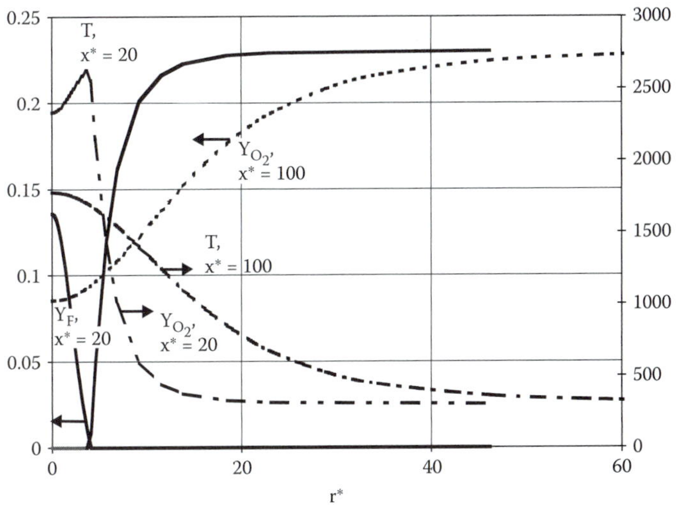

Figure 12.13a Flame structure at x* = 20, 100 for a circular jet.

$$\beta_{ht-F,i} = h_{t,i}/h_c + Y_{F,I} = 0 + 1 = 1$$

$$\beta_{ht-F,\infty} = c_p\,(T_\infty - T_i)/h_c + 0 = 0$$

$\beta_{ht-F} = 0.129$, $c_p(T - T_i)/h_c + Y_F = 1.175^*(T - 300)/46357 + 0.105 = 0.129$, T = 2477 K

Remarks

(i) At x* = 100, at a location greater than the flame height of 84, there is no fuel but only oxygen. Thus, axial oxygen concentration (at r* = 0) will increase from 0 beyond x* > 84. Figure 12.13a shows the resulting structure. The thermal NO will be produced in the flame and plume regions which depend upon the O_2 concentration and temperature profiles.

(ii) Figure 12.13b shows the concentrations and temperature distributions in H_2:air diffusion flame issued from a 5-cm-radius tube at a height of 2 cm [Fukutani et. al., 1991]. The flow rates are 30 and 143 cc/sec respectively for inner and outer cylinders. The flame is located at about 9 mm from the axis. See also section 12.9.

12.6.12 Mass Flow of Gas, Air, and A:F at Any x

Mass flow within a radius r = 0 to r = r:

$$\dot{m}\,(r,x) = \int_0^r 2\pi\, r\, v_x\, \rho\, dr$$

Normalizing,

$$\frac{\dot{m}}{\dot{m}_i} = \frac{32}{8}\left(\frac{x^*}{Re_i}\right)\left(\frac{(\xi^2/4)}{1 + (\xi^2/4)}\right) = \frac{32}{8}\left(\frac{x^*}{Re_i}\right)\frac{1}{\left\{1 + \dfrac{256\, x^{*2}}{3\, M^*\, Rei^2 r^{*2}}\right\}} \qquad (12.153)$$

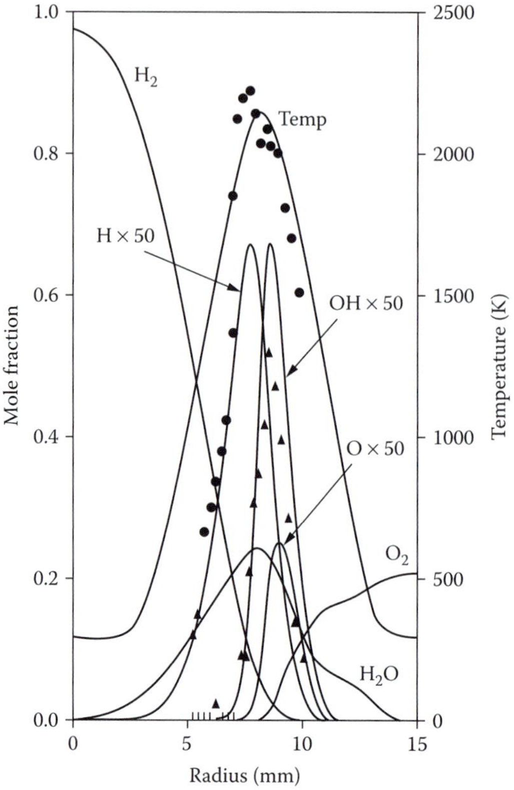

Figure 12.13b Temperature and concentrations from H_2: air diffusion flame at 2 cm for a tube of 5 mm inner radius and 20 mm outer radius; Re ≈ 150. Triangles: temperature; solid circles: concentrations. (Adapted from Fukutani, S., Kunioshi, N., and Jinno, H., 23rd International Symposium on Combustion, 567–573, 1991.)

As r* tends to infinity,

$$\left(\frac{\dot{m}}{\dot{m}_i}\right) = \frac{32}{\mathrm{Re}_i} x^*, \; x^* > \left(\frac{\mathrm{Re}_i}{32}\right) \tag{12.154}$$

Simplifying,

$$\dot{m} = 32 \; \upsilon \; (\mathrm{x}/\mathrm{d_i}) \; \rho \; \mathrm{v_i} \; \pi \; \mathrm{d_i}^{\;2}/\{4 \; \mathrm{v_i} \; \mathrm{d_i}\} = 8 \; \pi \; \mu \; \mathrm{x} \tag{12.155}$$

Amount of total gas within r* at x* is given as a fraction of total entrained at x* as

$$\frac{\dot{m}(r^*, x^*)}{\dot{m}(\infty, x^*)} = \frac{1}{\left\{1 + \dfrac{256\, x^{*2}}{3\, M^* \mathrm{Re}_i^2\, r^{*2}}\right\}} \tag{12.156}$$

Air required for combustion is given as

$$\dot{m}_{A,st} = \dot{m}_{Fi}\ \{A:F\}_{st} = \dot{m}_i\ Y_{F,i}\ \{A:F\}_{st}. \tag{12.157}$$

The total airflow at x* is

$$\left(\frac{\dot{m}_A}{\dot{m}_i}\right) = \frac{32}{Re_i}x^* - 1\ , x^* > \left(\frac{Re_i}{32}\right). \tag{12.158}$$

Knowing the fuel injected, as $\dot{m}_{F,i} = \left\{\frac{\pi d_i^2}{4}\rho\ v_{x,i}\right\} Y_{F,i}$

and the amount of total mass flow rate $\dot{m}$ at any x, we can determine A:F ratio.

$$\frac{\dot{m}}{\dot{m}_i} = \frac{32}{Re_i}x^*, (A:F)_x = \left[\frac{32}{Re_i}x^* - 1\right]\frac{1}{Y_{F,i}} \tag{12.159}$$

The air:fuel ratio at $x^* = H_f^*$ is

$$(A:F)_{Hf} = \left[\frac{32}{Re_i}H_f^* - 1\right]\frac{1}{Y_{F,i}},\quad x^* > \left(\frac{Re_i}{32}\right). \tag{12.160}$$

Using the result for flame height,

$$(A:F)_{Hf} = \left[\frac{(2Sc+1)(1+s)^{J^*}}{s} - 1\right]\frac{1}{Y_{F,i}}\quad J^* = 1. \tag{12.161}$$

Because stoichiometric airflow is given as $\dot{m}_{F,i}\ \nu_{O2}/Y_{O2,\infty} = (\dot{m}_{F,i}/\{sY_{F,i}\})$, then $(A:F)_{st} = 1/\{sY_{F,i}\}$, excess-air fraction, Ex, is given as

$$(Ex)_{Hf} = \frac{A:F}{(A:F)_{st}} - 1 = \left[\ (2Sc+1)(1+s)\ J^* - s\ \right] - 1 = (1+s)\left[(2Sc+1)\ J^* - 1\right]. \tag{12.162}$$

With J* = 1 and C = 1 for Sc = 1,

$$(Ex)_{Hf} = (1 + s)\ [3 - 1] \approx 2.0$$

and at Sc = 0, J* = 1

$$(Ex)_{Hf} = (1 + s)\ [1 - 1] = 0.$$

The variation of excess-air percentage at x* = H* (the tip of flame) with Schmidt number is shown in Figure 12.10 for 2-D and circular jets. For fuels with Sc = 1 at x* = H*, 2-D jets entrain 50% excess air, whereas circular jets entrain 200% excess air because a circular jet is completely surrounded by ambient. Thus, multiple 2-D jets may undergo more severe interactions compared to circular jets for the same interburner spacing.

12.7 Summary of Solutions for 2-D and Circular Jets

The solutions to the governing differential equations for both 2-D and circular jets are summarized in Table F.12.1 under "Chapter 12" Formulae appendix. Table F.12.2 presents a comparison of the results for laminar and the empirical results of turbulent jets. All solutions in this table are presented in the cases of absence of buoyancy forces. If buoyancy forces are included, M* and J* will be higher than the values listed.

12.8 Stoichiometric Contours for 2-D and Circular Jets, Liftoff, and Blow-Off

12.8.1 A:F Contours

For pure mixing problems Y_F and Y_{O_2}, and hence stoichiometric contour, where $Y_F = Y_{O_2}/\nu_{O_2}$ can be determined using solutions from Section 12.5.13. The stoichiometric contour (SR = 1) is extremely important to determine flame stability characteristics of 2-D and circular burners. If an ignition source is provided and a flame is established, the flame is most likely to be located along the stoichiometric contour. Therefore, if the stoichiometric contour is known, the flame geometry (height, maximum width, and axial location of maximum width) will also be known. Further, because the position of the stoichiometric contour in relation to the flame speed contour determines the flame stability characteristics of 2-D and circular burners, the stoichiometric contour location is needed to determine the liftoff height and blow-off velocity of the burners.

(a) *SR (stoichiometric ratio) = 1*

The following procedure is used to determine the stoichiometric contour particularly for a circular jet: First, by setting $\beta = Y_F$ and Y_{O2}, Y_F (r*, x*) and Y_{O2} (r*, x*) can be obtained using Equation 12.142 for fuel and oxygen species.

$$\phi = \left(\frac{Y_F - Y_{F,\infty}}{Y_{F,i} - Y_{F,\infty}}\right) = \left(\frac{Y_{O2} - Y_{O2,\infty}}{Y_{O2,i} - Y_{O2,\infty}}\right) = \left(\frac{\beta - \beta_\infty}{\beta_i - \beta_\infty}\right) = \frac{(2Sc+1)}{32}\frac{J^*\,\mathrm{Re}_i}{x^*}\frac{1}{\left[1+\frac{\xi^2}{4}\right]^{2Sc}} \tag{12.163}$$

The Y_F and Y_{O2} profiles can be obtained as

$$Y_F = Y_{F,\infty} + \frac{Y_{F,\max} - Y_{F,\infty}}{\left[1+\frac{\xi_{SR}^{\;2}}{4}\right]^{2Sc}} \tag{12.164}$$

$$Y_{O2} = Y_{O2,\infty} + \frac{Y_{O2max} - Y_{O2\infty}}{\left[1+\frac{\xi_{SR}^2}{4}\right]^{2Sc}}. \quad (12.165)$$

Hence, the stoichiometric contour y^*_{st} (x^*), where $Y_{F,st} = Y_{O2}/\nu_{O2}$ can be obtained.

(b) *SR ≠ 1*

Rewriting Equation 12.163

$$\frac{\phi_{SR}}{\phi_{max}} = \frac{Y_{k,SR} - Y_{k\infty}}{Y_{kmax} - Y_{k\infty}} = \frac{\beta_{SR} - \beta_{\infty}}{\beta_{max} - \beta_{\infty}} = \frac{1}{\left[1+\frac{\xi_{SR}^2}{4}\right]^{2Sc}} \quad (12.166)$$

where from Equation 12.163,

$$\phi_{max} = \{Y_{k,max} - Y_{k,\infty}\}/\{Y_{k,i} - Y_{k,\infty}\} = \{(2Sc + 1)/32\}\{Re_i J^*/x^*\} \quad (12.167)$$

$$Y_{O2,SR} = Y_{F,SR} * SR^* \nu_{O2} \quad (12.168)$$

where SR is arbitrary. Then, knowing SR,

$$\beta_{O2-f,SR} = \{Y_{F,SR} - Y_{O2,SR}/\nu_{O2}\} = Y_{F,SR}\{1 - SR\}. \quad (12.169)$$

And using Equation 12.166, one can plot the contours r_{SR}^* vs. x^* for a given SR in a mixing problem. The derivation is left as an exercise. Figure 12.14 shows the SR contours for C_3H_8 at Sc = 1.2. A method of estimating Sc for a fuel:air mixture is given in section 6.4.3.

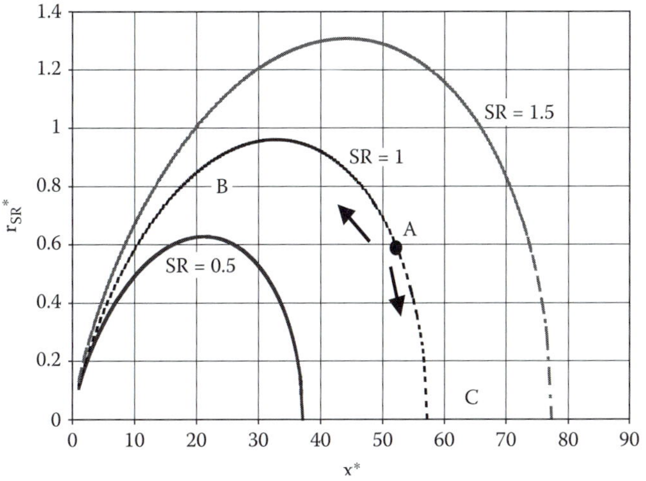

Figure 12.14 Stoichiometric ratio contours (circular jet). Fuel C_3H_8, $S_c = 1.2$, $Re_i = 160$.

These contours are useful in understanding the anchoring ability of flame to the burner. Let us assume that the mixture is flammable only for mixture with SR = 1 (Chapter 15), i.e., if one sparks the mixture along SR = 0.5 or 1.5, it will not be ignitable. Hence, if one ignites the mixture at A (SR = 1), then the flame will propagate only toward B and C, at a flame velocity of approximately 0.40 m/sec along SR = 1. However, the combustible mixture flows at v_x in a direction opposite to that of flame propagation from A to C. If we ignore the curvature effect and if v_x at A = 0.4 m/sec, then the flame will be stationary at A. The distance x_A is called liftoff distance. More details are given in later sections.

12.8.2 Stoichiometric Contours

If the SZ formulation is used for the combustion problem under finite kinetics at the stoichiometric surface, $Y_F \neq 0$ and $Y_{O2} \neq 0$, but $\beta_{F-O2} = 0$. At the thin-flame limit, $Y_F = 0$ and $Y_{O2} = 0$, and still $\beta_{F-O2} = 0$. Hence, $y^*_{st}(x^*)$ will be the same for these two cases and the stoichiometric contour (in real coordinates) under mixing conditions will also correspond to the flame location (in stretched coordinates) under combustion.

For the stoichiometric contour with SR = 1 for a 2-D jet, we obtain

$$y^*_{st}(x^*) = y_f(x^*) = \left[\frac{x^{*2/3}}{0.2752\,\mathrm{Re}_i^{2/3}\,M^{*1/3}}\right]\cosh^{-1}\left[\left(\frac{0.09375C^3\,\mathrm{Re}_i\,J^{*3}}{M^* x^* \phi_f^3}\right)^{\frac{1}{6Sc}}\right]. \tag{12.170}$$

Whereas following a similar procedure gives the stoichiometric contour for circular jets to be:

$$r^*_{st}(x^*) = r^*_f(x^*) = \left(\frac{16x^*}{\sqrt{3}\,\mathrm{Re}_i\,M^{*1/2}}\right)\left[\left(\frac{2(Sc+1)J^*\,\mathrm{Re}_i}{32x^*\phi_f}\right)^{1/(2Sc)} - 1\right]^{1/2} \tag{12.171}$$

where $\phi_f = \phi_{st}$ is defined in Equation 12.149. The flame profile in real coordinates will be exactly the same as the stoichiometric contour only if the gas properties remain constant for both mixing and combustion problems.

The maximum flame width and axial location of the maximum flame width can be found by manipulating Equation 12.170 and Equation 12.171. The details of the manipulation can be found in the Ph.D. thesis of Tillman [2000].

12.8.3 Liftoff Distance

When a gas jet is ignited, the flame is lifted off the base to a height of L for certain fuels. The liftoff distance can be predicted using the flame velocity and a knowledge of axial velocity v_x profile [Lee et al., 1994; Tillman et al.]. The balancing of the laminar flame speed and the axial gas velocity determines where the flame will stand along the stoichiometric contour. Therefore, finding

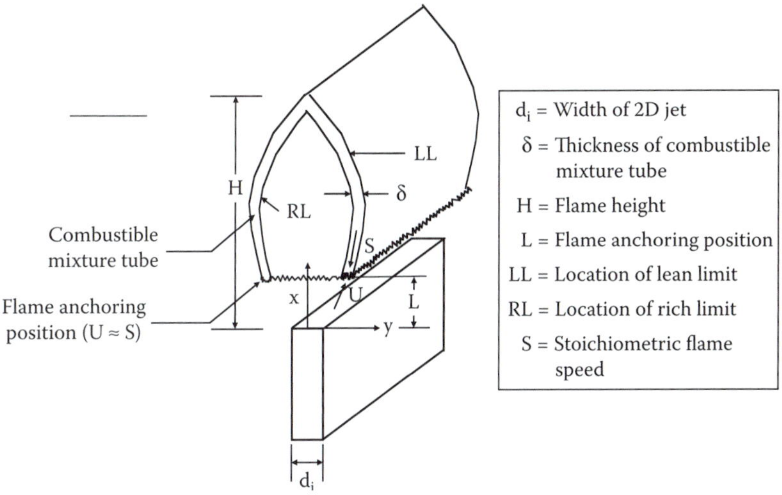

Figure 12.15 Qualitative illustration of a 3-D lifted flame in a 2-D jet. The flame is only stable for x > L. For x < L, v_x > S. For x = L, v_x = S and for x > L, v_x < S. (Adapted from Tillman, Ph.D. thesis, Texas A&M University, College Station, 2000.)

an expression for the flame speed contour (where $v_x = S$) is extremely important to predict the flame stability characteristics of the burner. The derivations for 2-D and circular jets are similar; therefore, only the 2-D procedure will be provided in the following text.

Using the solution for the axial velocity, v_x^*, velocity contours can be plotted by setting v_x equal to the specified flame velocity (S) (Equation 12.79). The stoichiometric contour ($\phi = \phi_{st} = \phi_f$) with SR = 1 can also be plotted. Hence, the intersection of this velocity contour $v_x = S$ with the stoichiometric contour at point "L" represents the anchoring position of the flame (Figure 12.16), or liftoff height L*. The velocity contour y^*_v (x*) = S can be plotted by solving for ξ and eventually y*. The resulting expression for 2-D jets is

$$y_v^* = \left(\frac{x^{*2/3}}{0.2752\,\mathrm{Re}_i^{2/3}\,M^{*1/3}}\right)\mathrm{sech}^{-1}\left[\left(\frac{S\,x^{*1/3}}{0.4543\,v_{x,i}\,\mathrm{Re}_i^{1/3}\,M^{*2/3}}\right)^{1/2}\right] \tag{12.172}$$

A similar procedure yields the following expression for the velocity contour for an isolated circular jet:

$$r_v^* = \frac{16x^*}{\sqrt{3}\,\mathrm{Re}_i\,M^{*1/2}}\left[\left(\frac{3\,v_{x,i}\,M^*\,\mathrm{Re}_i}{32x^*S}\right)^{1/2} - 1\right]^{1/2} \tag{12.173}$$

Equation 12.172 and Equation 12.173 give the velocity contour for 2-D and circular jets, where the axial gas velocity is equal to the laminar flame speed, S. Henceforth such a contour will be called the flame speed contour. By setting y_v^* and $r_v^* = 0$, the height of the flame speed contour, $H_{pl}^* = H_{f,pl}^*$ for planar

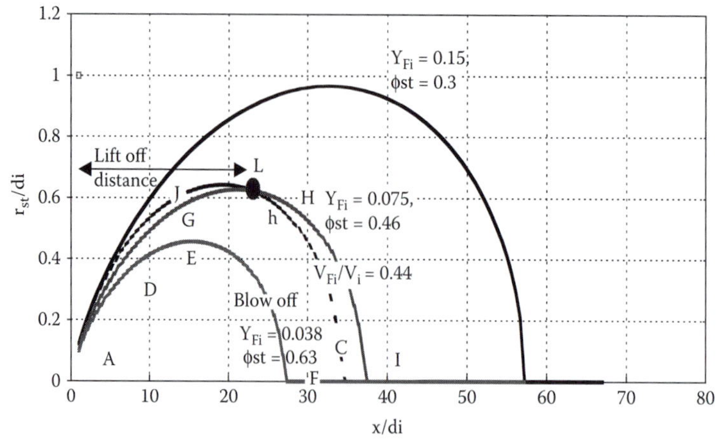

Figure 12.16 Stoichiometric and flame velocity contours and liftoff distance: circular jet.

jet can also be determined (See Table 12.1 under "Chapter 12" in Formulae appendix).Then, one may track ϕ along this contour and determine the distance L* (liftoff distance) at which $\phi = \phi_f$. The result for L* is as follows.

$$\{L^*/H_{f,pl}{}^*\} = [\{S/v_{x,i}\}\ \{C\ J^*/(\phi_f M^*)\}]^{3Sc/(1-Sc)},\ Sc > 1 \tag{12.174a}$$

where

$$\frac{H_{f,pl}{}^*}{d_i} = \frac{3}{32}\frac{C^3 J^3\ Re_i}{M^*}\left(\frac{1+s}{s}\right)^3 \tag{12.174b}$$

A similar derivation can be carried out for circular jets, and an expression for liftoff height can be obtained.

Example 3

Gaseous C_3H_8 fuel and N_2 mixture is injected at a velocity of 80 cm/sec through a circular burner of diameter 0.2 cm into air. Assume $\rho D = 6.5 \times 10^{-5}$ kg/m sec, $\nu_i = 1.0 \times 10^{-5}$ m²/sec, and Sc = 1.2. Assume T_i = 300 K, T_∞ = 300 K, P = 100 kPa, and flat velocity and species profiles at inlets. Plot

(a) Stochiometric contour y_{st}/d_i vs. x/di for Y_{Fi} = 0.15 (15% of fuel by mass), 0.075, and 0.038
(b) Velocity contour for v_x = 0.35 m/sec
(c) Liftoff distance at Y_{Fi} = 0.075

Solution

(a) The stoichiometric contour is given by $\beta_{O2-F,st} = Y_{F,st} - Y_{O2,st}/\nu_{O2} = 0$ and $Y_{O2,st}/Y_{F,st} = \nu_{O2}$ at the stoichiometric surface.

$$Re_i = v_{xi}\ d_i/\nu = \{0.8\ m/sec*0.002\ m\}/1 \times 10^{-5}\ m^2/sec = 160$$

$$s = Y_{O2\infty}/(Y_{Fi}\ \nu_{O2}) = 0.23/(0.15\ 3.63*) = 0.423$$

Following Example 2, $\beta_{O2\text{-}F,i} = 0.15$, $\beta_{O2\text{-}F,\infty} = 0.063$, $\phi_{st} = 0.3$.

At $x^* = 20$, $\phi_{max} = 0.85$ for Sc = 1.2, then $\phi_{st}/\phi_{max} = 0.3/0.85 = 1/(1 + \xi_{st}{}^2/4\}^{2*1.2}$; $\xi_{st} = 1.48$, $r_f^* = \xi_f/\{0.2165\ (Re_i/x^*)M^{*1/2}\} = 1.48/\{0.2165\ (160/20)\} = 0.86$ at Sc = 1.2

Thus, r_f^* vs. x^* can be obtained. Figure 12.16 shows the contours ADEF, GLI, etc., for specified $\phi_{st} = 0.63$, 0.46, and 0.3. The height of the stoichiometric contour is given by Equation 12.151:

$$H_{f,cir}{}^* = (2*1.2 + 1)*160/(32*0.297) = 57.2 \text{ for Sc} = 1.2$$

Stoichiometric height, $H_{f,cir} = 57.2*0.002 = 0.115$ m

(b) Velocity Contour

$$v_x^* = v_x/v_{x,i} = 0.35/0.8 = 0.44;$$

using $\phi_{v,cont} = v_x^* = 0.44$ in Equation 12.132,

$$\xi_v = 2[\{(3*M^*\ Re_i)/(32*\phi_{v,cont}\ x^*)\}^{1/2} - 1]^{1/2};$$

knowing x^* and $\phi_{v,cont}$ one can solve for ξ_v and hence plot r_v^* vs. x^* (Equation 12.130) as shown by contour JLC in Figure 12.16. The height of velocity contour, H_v^* can be obtained as $H_v^* = \{3/32\}\ (M^*\ Re_i/\phi_{v,cont})$. (i.e., setting $\xi_v = 0$).

$H_v^* = 34.3$ for the current case.

(c) Lift off distance: From Figure 12.16, $L^* = L/di \approx 23$.

Remarks

(i) Note that $H_v^* < H_{st}^*$ when Sc > 1, which implies that momentum diffuses faster than mass and hence the axial velocity.

(ii) From the plots, it is seen that if one ignites the mixture at a point on the stoichiometric contour, it will propagate with a laminar burn velocity at about 35 cm/sec (value will change depending on dilution) along GLI. If the local velocity (v_x) on the contour $v_x < 35$ cm/sec, then flame will travel toward the burner and stabilize at L; flame is lifted to $x^* = 22$ ($Y_{Fi} = 0.075$).

(iii) Flame cannot be sustained for $x^* < 22$, because v_x on the stoichiometric contour >35 cm/sec. If one increases the injection velocity to, say, 82 cm/sec, with the same $Y_{Fi} = 0.075$, the liftoff distance increases. At about 87 cm/sec, the liftoff distance equals the height of stoichiometric contour, and any further increase causes blow-off to occur. Thus, by setting $y_v^* = 0$ (2-D), $r_v^* = 0$ (circular), and $x^* = H_{st}^*$, we get the laminar blow-off velocity expressions.

12.8.4 Anchoring of Flames

Example 3 reveals the four possibilities of anchoring the flames (Figure 12.17). First, the stoichiometric contour (SC) can lie "inside" the flame speed or velocity

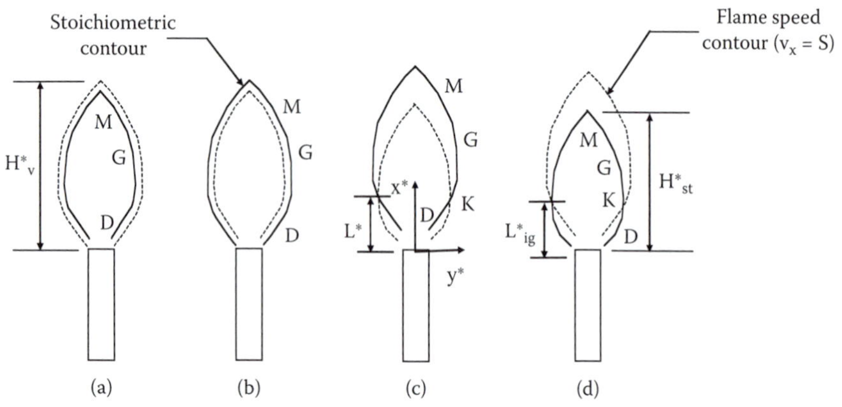

Figure 12.17 Qualitative illustrations of stoichiometric and flame speed contours: (a) unstable flame, (b) stable flame, (c) lifted flame, and (d) partial flame. (Tillman et al., 2000.)

contour (VC) (Figure 12.17a). Second, the SC can lie "outside" the VC (Figure 12.17b). And third and fourth, the contours can intersect at some distance away from the burner (Figure 12.17c and Figure 12.17d). Consider the first option when SC is parallel to VC since the Schmidt number = 1.0. If the SC lies "inside" the VC then at every point along the stoichiometric contour (line MGD in Figure 12.17a) the axial gas velocity must be higher than the flame speed. If an ignition source is provided at M, the flame attempts to propagate along the stoichiometric contour toward the burner at a velocity equal to the laminar flame speed, S, but the axial gas velocity is higher than the flame speed ($v_x > S$) along the stoichiometric contour, and thus the flame is moved downstream away from the burner where the axial gas velocity is equal to the flame speed. However, this must happen beyond the tip of the stoichiometric contour, where the mixture is no longer combustible. Thus, the first option will not result in stable flame. Consider the second option (Figure 12.17b). The SC lies "outside" the VC. Now, along the SC, the magnitude of the axial gas velocity is less than the flame speed ($v_x < S$). If an ignition source is provided at M, the flame will propagate toward the burner because the axial gas velocity along the SC is now less than the flame speed ($v_x < S$). This leads to a stable flame anchored at the burner base. The third option (Figure 12.17c), where the SC and the V_c intersects, gives the anchoring position for the flame ($v_x = S$ and $Y_{O2}/Y_F = \nu_{O2}$ at the intersection point). Further, downstream from the intersection point, the SC lies on the "outside" the flame speed contour, which is a stable profile as shown in Figure 12.17b. From the intersection point toward the burner base, the SC lies "inside" the VC which is an unstable profile as shown in Figure 12.17b. Therefore, the flame can only exist from the intersection point, K in Figure 12.17c, to the tip of the SC. Hence, L*, gives the liftoff height of the flame. Now, consider the fourth option. If an ignition source is provided at M, the flame attempts to propagate at S. However, the axial gas velocity at M is much higher than the flame velocity, and hence the mixture cannot be ignited at M. If an ignition source is provided at D, the axial gas velocity is less than S, and the flame propagates toward the burner base. The flame is therefore only ignitable up to position K, and hence, LI* (limited ignition) gives the ignitable height from the burner rim. To summarize, the fuel mixture is not ignitable for $LI^* < x < H_{st}^*$.

As will be explained in the following discussion, the Schmidt number, which physically represents the ratio of momentum to mass diffusion, plays the key role in defining the various options shown in Figure 12.17a to Figure 12.17b. If Sc > 1, lifted flames are predicted (Figure 12.17c). If Sc < 1, partial flame or ignitable heights are predicted (Figure 12.17d). If Sc = 1, no intersection between the contours is possible and the predicted flames are either (i) stable and anchored at the burner base (Figure 12.17b) or (ii) blown off (Figure 2.17a). Explicit expressions for plotting the stoichiometric contour and the flame speed contours shown in Figure 12.19a to Figure 12.19d can be determined by solving the governing differential equations for the single 2-D jet.

12.9 Jets in Coflowing Air: Jet Flame Structure in Strongly Coflowing Air for 2-D and Circular Jets

It is noted that there is a steep velocity gradient for a jet in stagnant air with a high strain rate. In coflowing air, the relative velocity is reduced. As an approximation, the fuel jet is treated as linear source for circular jets and as a plane source for 2-D jets.

12.9.1 Governing Equations

Mass:

$$\rho v_x = \rho_i v_{x,i} = \rho_{a0} v_{a0} = \text{constant} \tag{12.175a}$$

SZ variable:

$$(1/r^{(m-1)})\partial/\partial r[r^{(m-1)}\, \rho D\, \partial\beta'/\partial r] = \rho v_x \partial\beta'/\partial x \tag{12.175b}$$

where $\beta' = \beta - \beta_\infty$, $\beta_{O2-F} = Y_F - Y_{O2}/\nu_{O2}$, and m = 1, 2, 3 in Cartesian, cylindrical, and spherical coordinates.

Assume ρD = constant. The case of m = 3 will not be treated. Thus, even if there is flow at x = 0 outside the slot, the flow of β will be zero outside the slot at x = 0 because $\beta = 0$.

These equations are solved subject to

$$\partial\beta'/\partial r \rightarrow 0 \text{ as } r \rightarrow 0, \text{ and } \beta \rightarrow 0 \text{ as } r \rightarrow \infty \tag{12.176}$$

Further,

$$\int_0^\infty \rho\, v_x\, r^{(m-1)}\, (\beta - \beta_\infty)\, dr = C_1', \tag{12.177}$$

where $C_1 = \dot{m}_F/\{(2m-2)\pi\}$ in cylindrical and spherical coordinates, $C_1 = \dot{m}_F$ in Cartesian coordinates.

Normalizing

$$\phi = (\beta - \beta_\infty)/(\beta_i - \beta_\infty),\ x^* = x/d,\ r^* = r/d_i.$$

Then,

$$(1/r^{*(m-1)})(\partial/\partial r^*)[r^{*(m-1)}\ \partial\phi/\partial r^*] = \{Re_i\ Sc_i\}\ \partial\phi/\partial x^*, \text{ where } Re = \{d_i\ v_{x,i}/\nu\}. \quad (12.178)$$

Equation 12.177 becomes

$$\int_0^\infty r^{*(m-1)}\ \phi dr^* = C, \quad (12.179)$$

where $C = \dot{m}_F/\{(2m-2)\pi\ \rho_i\ v_{xi}\ (\beta_i - \beta_\infty)\ d_i^2\}$ in cylindrical and spherical coordinates, $C = \dot{m}_F/\{(\beta_i - \beta_\infty)\ \rho_i\ v_{xi}\ d_i\}$ in Cartesian coordinates.

These relations simplify to $\rho_F\ v_F/\{8\ \rho_{A0}\ v_{A,0}\ (\beta_i - \beta_\infty)\}$ (cylindrical) and $\rho_F\ v_F/\{\rho_{A0}\ v_{A,0}\ (\beta_i - \beta_\infty)\}$ (Cartesian).

12.9.2 Boundary Conditions

Equation 12.178 can be solved subject to the following boundary conditions:

$$\partial\phi/\partial r^* \to 0 \text{ as } r^* \to 0, \text{ and } \phi \to 0 \text{ as } r^* \to \infty$$

12.9.3 Solutions

Details of solutions are omitted.

12.9.3.1 Planar Jet in Coflowing Air

$$\eta = r^*/x^{*(1/2)},\ \phi = g/x^{*(1/2)} \quad (12.180)$$

where

$$g = A \exp[-(Re_i\ Sc_i/2)\ \eta^2/2]$$

$$A\int_0^\infty \exp[-y^2]\ dy = C\ [(Re_i\ Sc_i/4)^{1/2}]$$

$$Erf(y) = \{2/(\pi)^{1/2}\}\int_0^y \exp\{-y^2\}\ dy,\ erf(\infty) = 1.0$$

$$\phi = [\dot{m}_F/\{\rho_i\ v_{xi}\ d_i\ (\beta_i - \beta_\infty)\}]\ [(Re_i\ Sc_i/4\ x^*)^{1/2}]\ \{2/(\pi)^{1/2}\}]\ \exp[-(Re_i\ Sc_i/2)\ \eta^2/2], \quad (12.181)$$

Flame length is obtained by setting $\phi = \phi_f$, $\eta = 0$ in Equation 12.181.

$$H_f^* = [\dot{m}_F'/\{\rho_i\ \phi_f\ v_{xi}\ d_i\ (\beta_i - \beta_\infty)\}]^2\ [Re_i\ Sc_i/\pi] = [\rho_F\ v_{xi}/\{\rho_{a,0}\ \phi_f\ v_{a,0}\ (\beta_i - \beta_\infty)\}]^2\ [Re_i Sc_i/\pi] \tag{12.182}$$

12.9.3.2 *Cylindrical System*

$$\eta = r^*/x^{*(1/2)},\ \phi = g/x^*,\ g = A \exp[-(Re_i\ Sc_i/2)\ \eta^2/2] \tag{12.183}$$

$$A = 2\ C\ [(Re_i\ Sc_i/4)]$$

where

$$C = \dot{m}_F/\{2\pi\ \rho\ v_x\ (\beta_i - \beta_\infty)\ d^2\}$$

$$\phi = [\dot{m}_F/\{2\pi\ \rho_i\ v_{xi}\ (\beta_i - \beta_\infty)\ d_i^2\}]\ [(Re_i\ Sc_i/4)]\ (1/x^*)\exp[-(Re_i\ Sc_i/2)\ \eta^2/2]. \tag{12.184}$$

If $\eta' = (Re_i\ Sc_i/4)^{1/2}\ \{r^*/x^{*(1/2)}\} = (Re_i\ Sc_i/4)^{1/2}\ \{r^*/x^{*(1/2)}\}$

$$\phi = [\rho_{F/}\rho_{a,0}]\ [1/\{32\ (\beta_i - \beta_\infty)\}]\ [Re_i\ Sc_i][1/x^*]\ \exp[-\eta'^2] \tag{12.185}$$

[Mahalingam et al., 1990]

$$\phi_f = [\dot{m}_F/\{2\pi\ \rho\ v_x\ (\beta_i - \beta_\infty)\ d^2\}]\ [(Re_i\ Sc_i/4)]\ (1/x^*)\ \exp[-(Re_i\ Sc_i/2)\ \eta_f^2/2]$$

By setting $\phi = \phi_f$, $\eta' = 0$, the flame height is given as

$$H^*_f = [\dot{m}_F/\{2\pi\ \rho_i\ v_{xi}\ (\beta_i - \beta_\infty)\ d_i^2\}]\ [(Re_i\ Sc_i/4)]\ (1/\phi_f). \tag{12.186}$$

The result is the same as B–S solution when $r^* \to \infty$. Figure 12.13b shows experimental flame structure for H_2: air diffusion flame with coflowing air.

12.10 Turbulent Diffusion Flames

As the jet velocity is increased, the flame height (H) increases linearly with flow rate. At a velocity called *transition velocity*, the flame becomes brush-like at a particular height (y_B < H, break point), called *turbulent flame*, at Re ≈2000. The flame height then decreases due to increased diffusivity of fuel and oxidizer (Figure 12.6). With further increase in velocity, the flame height stabilizes to a constant value independent of gas velocity, whereas the break-point height decreases with increased velocity. The transition Re is measured to be 2000 to 10000 for gases ranging from H_2 (soot-free gas) to acetylene (sooty gas). While recent computational codes can provide more detailed turbulent combustion modeling and numerical results with an averaging procedure and k-ε models, simplified turbulent jet results and dimensionless groups governing jets can be obtained as a simple extension from laminar jet results with empirical turbulent diffusivity models.

12.10.1 Turbulent Planar Jets

12.10.1.1 Empirical Viscosity

In previous sections we dealt with momentum transfer through molecular diffusivity. In this section, results are presented for turbulent combustion through an empirical diffusivity relation. Let the turbulent momentum diffusivity or eddy momentum diffusivity be $\varepsilon_{T,M}$ (m^2/sec). Using phenomenological analysis on $\varepsilon_{T,M}$,

Let v′ characteristic fluctuating velocity, δ is mixing layer thickness if $\delta \propto d_{ref}\, x^{*m}$ and if $v'/v_{x,ref}$ is constant, then

$$\varepsilon_{T,M\ f} = v'\delta\, C_M\, v_{x,ref}\, d_{ref}\, x^{*\ m}$$

Defining $\varepsilon_{ref} = C_M\, v_{x,ref}\, d_{ref}$,

$$\varepsilon_{T,M}{}^{*} = \varepsilon_{T,M}/\varepsilon_{ref} = x^{*\ m} \tag{12.187}$$

where ε_{ref} is the turbulent diffusivity at $x = d_{ref}$, $m = \frac{1}{2}$ for planar and 0 for circular jets [Schlichting, 1955]; $v_{x,ref} = v_{xi}$, and $d_{ref} = d_i$. Hence,

$$\varepsilon_{ref} = C_M\, v_{x,i}\, d_i \tag{12.188}$$

$$\varepsilon_{ref}{}^{*} = \varepsilon_{ref}/\nu = C_M\, v_{x,i}\, d_i/\nu = C_M\, Re_i$$

12.10.1.2 Conservation Equations

Mass Conservation

$$\frac{\partial}{\partial x}(\rho v_x) + \frac{\partial}{\partial y}(\rho v_y) = 0 \tag{12.189}$$

where v_x and v_y are time-averaged velocities in x and y directions.

Momentum Conservation (x direction)

$$\rho v_x \frac{\partial v_x}{\partial x} + \rho v_y \frac{\partial v_y}{\partial y} = \frac{\partial}{\partial y}\left(\rho \varepsilon_{T,M} \frac{\partial v_x}{\partial y}\right) \tag{12.190}$$

Species Conservation

$$\rho v_x \frac{\partial Y_k}{\partial x} + \rho v_y \frac{\partial Y_k}{\partial y} = \frac{\partial}{\partial y}\left(\frac{\rho \varepsilon_{T,S}}{Sc_T} \frac{\partial Y_k}{\partial y}\right) + \dot{w}_{k'''} \tag{12.191}$$

Energy

$$\rho v_x \frac{\partial h_t}{\partial x} + \rho v_y \frac{\partial h_t}{\partial y} = \frac{\partial}{\partial y}\left(\frac{\rho \varepsilon_{T,M}}{Pr_T} \frac{\partial h_t}{\partial y}\right) + |\dot{w}_{F'''}|\, h_c \tag{12.192}$$

With transformations into incompressible coordinates, we have the following equations (Section 11.3.3):

Mass

$$\frac{\partial v_{x'}}{\partial x'}+\frac{\partial v_{y'}}{\partial y'}=0 \tag{12.193}$$

Momentum

$$v_{x'}\frac{\partial v_{x'}}{\partial x'}+v_{y'}\frac{\partial v_{x'}}{\partial y'}=\frac{\partial}{\partial y'}\left(\frac{\rho^2\varepsilon_{ref}}{\rho_w^2}\varepsilon^*_{T,M}\frac{\partial v_{x'}}{\partial y'}\right) \tag{12.194}$$

Species

$$v_x{}'\frac{\partial Y_k}{\partial x'}+v_y{}'\frac{\partial Y_k}{\partial y'}=\frac{\partial}{\partial y'}\left(\frac{\rho^2\varepsilon_{ref}\;\varepsilon^*_{T,M}}{Sc_T}\frac{\partial Y_k}{\partial y'}\right)+\frac{\dot{w}_k'''}{\rho} \tag{12.195}$$

Note that $\varepsilon^*_{T,s} = \varepsilon_{T,s}/\varepsilon_{ref} = x^{*\,m}$, and $Sc_T = \varepsilon_{T,m}/\varepsilon_{T,s}$ = momentum turbulent diffusivity/turbulent species or mass diffusivity = Pr_T.

Energy

$$v_{x'}\frac{\partial h_T}{\partial x'}+v_{y'}\frac{\partial h_T}{\partial y'}=\frac{\partial}{\partial y'}\left(\frac{\rho^2\varepsilon_{ref}}{Sc_T\rho_w^2}\varepsilon^*_{T,M}\frac{\partial h_T}{\partial y'}\right)+\frac{\dot{w}_F h_c}{\rho} \tag{12.196}$$

Note that $\rho^2\,\varepsilon_{ref}$ is not a constant as in the laminar jets because $\rho^2 \propto 1/T^2$, and $\varepsilon_{T,M}$ is not sensitive to temperature. Hence we have to assume constant property and select ρ at film temperatures.

Normalized SZ

$$v_{x'}\frac{\partial \phi}{\partial x'}+v_{y'}\frac{\partial \phi}{\partial y'}=\frac{\partial}{\partial y'}\left(\frac{\rho^2\varepsilon^*_{T,M}\;\varepsilon_{ref}}{Sc_T\rho_w^2}\frac{\partial \phi}{\partial y'}\right) \tag{12.197}$$

12.10.1.3 Normalized Conservation Equations

Mass Conservation

$$\partial v_x^*/\partial x^* + \partial v_y^*/\partial y^* = 0 \tag{12.198}$$

Momentum Conservation (x direction)

$$v_x^*\partial v_x^*/\partial x^* + v_y^*(\partial v_x^*/\partial y^*) = \{[1/Re_{i,T}]\,(\partial^2 v_x^*/\partial y^{*2})\,x^{*m}\} \tag{12.199}$$

where $Re_{i,T}$, turbulent Reynolds number, $= v_{xi}d_i/\varepsilon_{ref}$. Using Equation 12.188 for ε_{ref}, $Re_{i,T} = v_{xi}d_i/\varepsilon_{ref} = 1/C_M$.

Energy, Species, and SZ Conservation

The normalized SZ equation with $\rho_{ref} = \rho_i$, is

$$v_x'^{*}\partial\phi/\partial x^{*} + v_y'^{*}(\partial\phi/\partial y'^{*}) = \{[1/(Re_{i,T}\ Sc_T)]\ (\partial^2\phi/\partial y'^{*2})\ x^{*m}\} \quad (12.200)$$

with the transformation:

$$x''^{*} = x^{*\ (m+1)}/(m+1) \quad \text{or} \quad x^{*\ (m+1)}/(m+1). \quad (12.201)$$

$$v_y''^{*} = v_y'^{*}/x^{*m}. \quad (12.202)$$

Using Equation 12.201 and Equation 12.202, the conservation equations become:

Mass

$$\frac{\partial v_{x'}}{\partial x''} + \frac{\partial v_y''}{\partial y'} = 0 \quad (12.203)$$

Momentum

$$v_{x'}{}^{*}\frac{\partial v_{x'}{}^{*}}{\partial x''{}^{*}} + v_{y'}\frac{\partial v_{x'}{}^{*}}{\partial y''{}^{*}} = \frac{\partial}{\partial y''{}^{*}}\left(\frac{1}{Re_{i,T}}\frac{\partial v_{x'}{}^{*}}{\partial y''{}^{*}}\right) \quad (12.204)$$

Species

$$v_{x'}{}^{*}\frac{\partial Y_k}{\partial x''{}^{*}} + v_{y'}\frac{\partial Y_k}{\partial y''{}^{*}} = \frac{\partial}{\partial y''{}^{*}}\left(\frac{1}{Re_{i,T}\ Sc_T}\frac{\partial Y_k}{\partial y''{}^{*}}\right) + \frac{\dot{w}_k\ d_i}{\rho v_i x^{*m}} \quad (12.205)$$

Energy

$$v_{x'}{}^{*}\frac{\partial\left[\frac{h_t}{h_c}\right]}{\partial x''} + v_{y''}{}^{*}\frac{\partial\left[\frac{h_t}{h_c}\right]}{\partial y''{}^{*}} = \frac{1}{Re_{i,T}\ Sc_T}\frac{\partial^2\left[\frac{h_t}{h_c}\right]}{\partial^2 y''{}^{*}} + \frac{w_k'''\ d_i}{\rho v_i x^{*m}} \quad (12.206)$$

NSZ

$$v_{x'}{}^{*}\frac{\partial\phi}{\partial x'{}^{*}} + v_{y''}{}^{*}\frac{\partial\phi}{\partial y''{}^{*}} = \frac{1}{Re_{i,T}\ Sc_T}\frac{\partial^2\phi}{\partial y''{}^{*2}} \quad (12.207)$$

12.10.1.4 Solutions for Turbulent Planar Jets

These equations look similar to those used for planar laminar jets except that molecular kinematic viscosity ν is replaced by ε_{ref}. Thus, we replace Re in laminar jet solutions by $Re_{i,T}$ $(= v_{xi}\ d_i/\varepsilon_{ref} = 1/C_M)$ and Sc by Sc_T and use the same results. We take the solutions of planar laminar jet, replace ν by ε_{ref}, x by x''^{*} i.e., $x^{*(m+1)}/(m+1)$ and $m = 1/2$.

$$\eta = y^{*}\ (3/2)^{2/3}\ \{v_{x,i}\ d_i/3\varepsilon_{ref}\}^{2/3}/x^{*} \quad (12.208)$$

$$\xi = (1/2)\ (3/2)^{1/3}\ \eta = y^*\ (1/2)\ (3/2)(Re_{i,t}/3)^{(2/3)}/x^*$$
$$= (3/4)\ (1/3)^{2/3}\ Re_{I,t}\ ^{2/3}\ y^*/x^*\ \xi = [(3/4)\ (1/3)^{2/3}\ (Re_{i,T})\ ^{2/3}]\ y^*/x^* \quad (12.209)$$

The group within the square brackets is called the Richardson number (σ).

$$\sigma_{pl} = (3/4)\ (1/3)^{2/3}\ (Re_{i,T})\ ^{2/3} \quad (12.210)$$

Because σ_{pl} = 7.67 (Schlicting, experimental data), then Equation 12.210 yields,

$$Re_{i,T} = 98.11. \quad (12.211)$$

Now, C_M can be determined because

$$C_M = 1/Re_{iT} = 1/98.11 = 0.0102. \quad (12.212)$$

Then, we can rewrite Equation 12.209 as

$$\xi = \sigma_{pl}\ y^*/x^*. \quad (12.213)$$

Velocity

$$v_x^* = \left(\frac{v_x}{v_{x,i}}\right) = 0.866\sqrt{\frac{\sigma_{pl}}{x^*}}\ \mathrm{sech}^2\ \xi \quad (12.214)$$

$$v_y'^* = [0.433/\sigma_{pl}^{1/2}]\ \{1/x^{*1/2}\}\{2\ \xi\ \mathrm{sech}^2\xi - \tanh\ \xi\} \quad (12.215)$$

Mass flow at any x

Similarly, $m'/m_{inj}' = 2\int v_x^* dy^*$

$$\frac{\dot{m}'}{\dot{m}'_i} = 1.7322\frac{x^{*1/2}}{\sigma_{pl}^{1/2}} \quad (12.216)$$

$$\phi(Sc_T \neq 1) = \{0.866\ C\ (\sigma_{pl}/x^*)\ ^{1/2}\ (J^*/M^{*\ 1/3})\}\mathrm{Sech}^{2Sc_T}\ \xi \quad (12.217)$$

$$\phi_k/\phi_{k,ScT=1} = C\ \mathrm{Sech}^{2Sc_T}\ \xi/\{\mathrm{Sech}^2\ \xi\} \quad (12.218)$$

where C is a function of Sc_T. $C = 1 + 0.2906^*(Sc_T - 1) - 0.0296\ (Sc_T - 1)^2$

$$(A:F)_{x^*} = \frac{1}{Y_{F,i}}\left(1.7322\frac{x^{*1/2}}{\sigma_{pl}^{1/2}} - 1\right),\quad x^* > \sigma_{pl}/(1.7322)^2 \quad (12.219)$$

Flame heights

Setting $\phi = \phi_f\ s/(1 + s)$ in Equation 12.217, $s = Y_{O2,\infty}/(Y_{Fi}v_{O2})$, $Sc_T = 1.0$ and $\xi_f = 0$,

$$L_f^{*\ 1/2} = \frac{0.866}{\phi_f}\sqrt{\sigma_{pl}} \quad (12.220)$$

$$L_f = d_i\left[\frac{0.866}{\phi_f}\right]^2 \sigma_{pl} = \left[\frac{5.7522}{\phi_f^{\ 2}}\right] d_i \quad (12.221)$$

It is clear that flame length is independent of flow rate. Figure 12.6 qualitatively shows that the flame height is independent of flow velocity in the turbulent regime. Using Equation 12.221 in Equation 12.219, it can be shown that

$$(\mathrm{A{:}F}) = (1/Y_{F,i})[(1.5/\phi_f) - 1] \tag{12.222}$$

12.10.2 Turbulent Circular Jets

In this case, $\varepsilon_{T,M}$ = constant, i.e., independent of x. Assume incompressible flow with laminar diffusivity replaced by turbulent momentum diffusivity, $\varepsilon_{T,M} = \sigma_{cyl}\sqrt{(\pi/4)}\, d_i\, v_{x,i}$, a constant, or $Re_T = d_i\, v_{x,i}/\varepsilon_M = [1/\sigma_{cyl}]\sqrt{(4/\pi)}$. According to Reichhardt, $\sigma_{cyl} = 0.0161$ and hence $Re_{I,T} = 70.09$. Thus, these solutions are identical to those of laminar jets with ν_w and D_w replaced by the corresponding turbulent diffusivities that are treated as constants. Also we, replace Re by $Re_{I,T} = v_{ref} d_{ref}/\varepsilon = 70.09$, and Sc by Sc_T and use the same results. Note that ε = constant for circular jets. Again, constant properties are used. The solutions are same as those of laminar circular jets with υ replaced by $\varepsilon_{T,M}$.

$$x^* = \frac{x}{d_i}, \quad \sigma_{cyl} = 0.0161, \varepsilon_{T,M}/(d_i v_{xi}) = \sigma_{cyl}\sqrt{\pi/4}$$

12.10.2.1 Solutions

$$\eta = 0.2443\ [1/\sigma_{cyl}] M^{*1/2}\ r^*/x^* \tag{12.223}$$

Velocities

$$\frac{v_{x,\max}}{v_{x,i}} = \frac{0.1058}{\sigma_{cyl} x^*} \frac{6.57}{x^*} \tag{12.224}$$

$$\frac{v_x}{v_{x,\max}} = \frac{1}{\left(1+\xi^2/4\right)^2} \tag{12.225}$$

$$\left(\frac{v_r}{v_{x,i}}\right) = \frac{\sqrt{3}}{8} \frac{1}{x^*} \frac{\left\{\xi - \frac{1}{4}\ \xi^3\right\}}{\left\{1 + \frac{1}{4}\ \xi^2\right\}^2} \tag{12.226}$$

Jet half width ($r_{1/2}$)

$$r_{1/2}^* = \frac{\sigma_{cyl}\ x^*}{0.2443\ M^{*1/2}} 2\left[\sqrt{2}-1\right]^{1/2} = 0.0848 x^*, \ M^* = 1 \tag{12.227}$$

It is seen that the half width linearly increases with x*.

Air entrained and A:F

$$\dot{m} / \dot{m}_i = 28.36\, x^* \sigma_{cyl} \qquad (12.228)$$

where $\sigma_{cyl} = 0.0161$
air entrained = $m' - m'_i$
air entrained/gas injected = $\{m'/m'_i - 1\}$

$$\frac{\dot{m}_{air}}{\dot{m}_i} = \left[\frac{28.3593}{\sigma_{cyl}} x^* - 1\right], \quad x^* > \frac{\sigma_{cyl}}{28.36}, \quad \sigma_{cyl} = 0.0161$$

Defining entrainment coefficient as $C_{entrain} = d\,(\dot{m} / \dot{m}_i\}/d(x/di)$,

$$C_{entrain} = 28.36 \quad \sigma_{cyl} = 0.457$$

$$(A:F)_{x^*} = \frac{1}{Y_{F,i}}\left[\frac{28.36}{\sigma_{cyl}} \quad x^* - 1\right], \quad x^* > \frac{\sigma_{cyl}}{28.36}. \qquad (12.229)$$

Normalized average SZ profile

$$\phi = \frac{\frac{0.1058(2Sc_T+1)}{3}}{\sigma_{cyl} x} \frac{1}{\left(1 + \frac{\xi_f^2}{4}\right)^{2Sc_T}} \qquad (12.230)$$

Flame height
With $\xi_f = 0$ at the flame tip, $Sc_T = 1.0$

$$H^*_{f,turb} = \frac{0.1058}{(\phi_f \sigma_{cyl})}. \qquad (12.231)$$

Using Equation 12.231 in Equation 12.229

$$(A:F)_x = (1/Y_{F,i})[(3/\phi_f)-1]. \qquad (12.232)$$

12.11 Partially Premixed Flame

As opposed to pure diffusion flames, partially premixed flame occurs particularly for residential burners where oxygen for flame is provided by both the jet and free stream; oxygen needed from free stream is reduced thus moving jet flame toward the axis and reducing the volume of the fuel rich region. Partially premixed flames are hybrid flames that contain multiple reaction zones, some with a premixed-like structure and others with a nonpremixed structure. These flames can be established by varying a fuel-rich mixture or stream ($1 < \phi_r < \infty$, where ϕ is the equivalence ratio) in contact with a fuel-lean mixture or stream ($0 \le \phi_1 < 1$). In general, the global equivalence ratio is in the range fuel-lean to stoichiometric [Aggarwal and Puri, 1998].

Many partially premixed flame applications occur at higher pressures. Recent research has focused on their stabilization [Qin et al., 2001, 2003, 2004], stretch effects on them [Mukhopadhyay and Puri, 2003; Choi and Puri, 2000, 2001, 2003]; their propagation in mixing layers [Qin et al., 2003, 2004] and NOx emission [Xue et al., 2002, 2003]. Their structure has been probed using enhanced visualization [Mukhopadhyay et al., 2002, 2003] and nonintrusive diagnostic tools for gaseous [Xio et al. 2002; Qin et al., 2003] and prevaporized liquid fuels [Arana et al., 2004; and Berta et al., 2005, 2006].

12.12 Summary

Chapter 12 deals with Burke–Schumann flames and 2-D and circular jets. Governing equations and similarity solutions are presented for the flame structure of laminar 2-D and circular jets. Using the results for pure mixing problems, an explicit expression for liftoff distance was obtained. Further, the amount of excess air entrained by laminar jets is also presented, and excess-air percentage was shown to be a function of Sc only for 2-D and circular jets. An empirical approach was presented for turbulent jets for a given power law distribution of empirical turbulent diffusivity.

Chapter 13

Ignition and Extinction

13.1 Introduction

Ignition is a process during which combustible mixtures begin to burn. An explosion occurs during a short timescale (t), or also within a short distance, $\delta(\delta = v * t)$ as *a wave*. It could be in the form of a temperature explosion (a useful criterion in flow systems) or a pressure explosion (a useful criterion in rigid systems). Ignition is an explosive process in which T rises rapidly for an open system or T and P rise rapidly for a closed system. Autoignition is the process of initiating ignition using a hotter ambient without using an ignition source such as a spark or a flame from a matchstick. Ignition is of great importance for a number of different practical situations: spontaneous combustion in stocked coal, ignition of dust clouds leading to explosive propagation in coal grinding and conveying equipment, and ignition in a pulverized-coal flame. The basic essentials of ignition are: (1) fuel, (2) oxidizer, (3) heat, and (4) reaction kinetics called the *fire tetrahedron* in which sufficient heat must be generated or a minimum temperature achieved to sustain combustion.

13.2 Modes of Ignition

The different modes of ignition are:

Homogeneous ignition: Consider a liquid droplet at T_d heated in air at $T_\infty > T_d$. The droplet evaporates. The vapor diffuses and mixes with O_2. If the ambient temperature T_∞ is low, only evaporation occurs ($T < T_t$). A flame is formed if the ambient temperature T_∞ is high (Figure 13.1). As both the reactants and oxidants are in the gaseous phase during oxidation, the ignition is said to be homogeneous. Liquid fuels typically ignite homogeneously and require a thermal source. Further, if a solid fuel fully pyrolyzes (e.g., PMMA), the released combustible gases undergo homogeneous ignition.

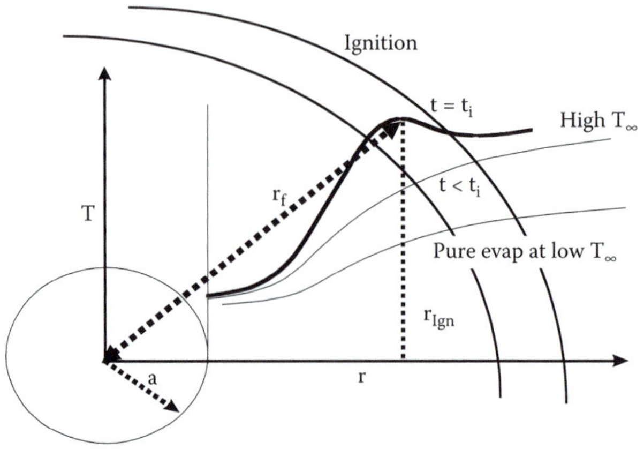

Figure 13.1 Ignition of drops — homogeneous.

Heterogeneous ignition: Consider a solid carbon particle at T_p that is heated in air at T_∞. The O_2 diffuses to the surface, reacts with carbon, and liberates CO or CO_2, for an exothermic reaction. Heat is liberated at the particle surface. The particle temperature T_p rises, and when $T_p > T_\infty$, heat loss occurs from the particles to air. Since reaction rates and, consequently, heat liberation rates are exponential with temperature, the heat loss increases almost linearly with T_p. Thus, above a certain T_∞, T_p jumps exponentially owing to the larger increase in heat liberation compared to the heat loss. Then heterogeneous ignition (i.e., oxidation involving two phases) is said to have occurred (Figure 13.2). For example, charcoal on a barbecue grill involves heterogeneous ignition. We can start the ignition by using a lighter fluid that on ignition creates hotter gases (T_∞) and subsequently ignites the char.

Ignition within rigid systems with an isothermal bath: Figure 13.3 illustrates nonadiabatic ignition processes involving a rigid system in which an isothermal

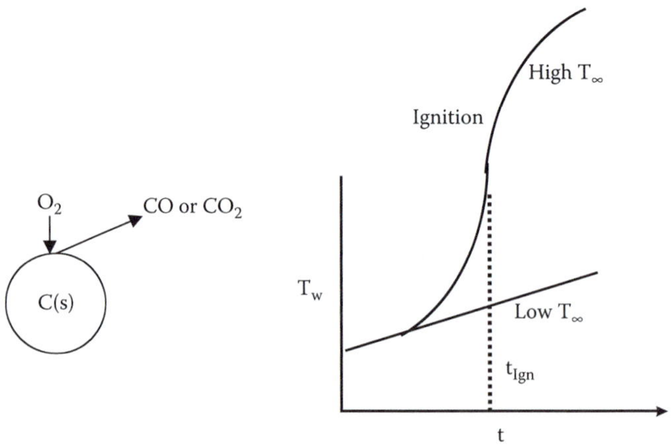

Figure 13.2 Ignition of carbon — heterogeneous.

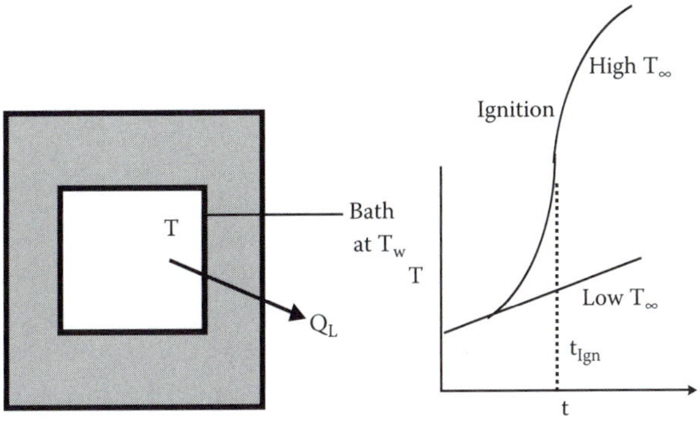

Figure 13.3 Homogeneous ignition of gaseous mixture.

bath is used to a transfer heat to a combustible mixture, after which chemical reactions occur, liberating heat and raising the temperature above the bath temperature. If the heat liberation rate is almost the same as the heat loss rate, then ignition may not occur. Figure 13.1 to Figure 13.3 show heat losses from reacting systems called *nonadiabatic ignition processes.*

Generally, the energy balance is

$$mc\frac{dT}{dt} = \dot{Q}_{gen} - \dot{Q}_{loss} \tag{13.1}$$

where m is the mass of the system and c is the specific heat (c_v if the system is rigid closed and c_p for a closed system with an isobaric process or an open system)

$$\dot{Q}_{gen} = \left|\dot{w}'''_F\right| h_c V, \quad \text{for a closed system of volume V (Figure 13.3)} \tag{13.2}$$

$$\dot{Q}_{loss} = h_H S(T - T_\infty), \text{ is the heat loss from a system of surface area S; } h_H, \text{ is the heat transfer coefficient.} \tag{13.3}$$

Note that $Q_{gen}\, \alpha \exp(-E/\bar{R}T)$, whereas $\dot{Q}_{\text{loss}}\, \alpha (T - T_\infty)$. Gases next to walls undergo high heat loss. Hence, a temperature gradient may exist inside the constant-volume or bomb reactor. We will first deal with problems of uniform temperature and later with those involving nonuniform temperature. A theory of ignition based on temperature rise is called *thermal theory of ignition.* *Hypergolic ignition*: When species are mixed at room T and P, they ignite (e.g., $H_2 + F_2 \rightarrow 2HF$).

13.3 Ignition of Gas Mixtures in Rigid Systems: Uniform System

We keep the reacting system in a bath at T_∞ and monitor the temperature rise of the system (Figure 13.3). At low T_∞, the temperature does not rise very much; as we increase T_∞, at some specified $T_{\infty,I}$, the temperature of the system rises rapidly resulting in the formation of a flame. We can determine the required $T_{\infty,I}$ experimentally. In order to determine the governing parameters for ignition, we can derive an expression for $T_{\infty,I}$ using the energy conservation equation.

13.3.1 Solution for Ignition

13.3.1.1 Numerical Solution

As outlined in Chapter 7, sensible energy rise of the mixture = heat generated by chemical reaction less the heat loss by convection and radiation. Hence

$$mc_v dT/dt = \dot{Q}_{gen} - \dot{Q}_{Loss} \tag{13.4}$$

where $\dot{Q}_{gen}$ and $\dot{Q}_{Loss}$ are given by Equation 13.2 and Equation 13.3. Assuming a first-order reaction with fuel and oxygen and a global one-step reaction,

$$\left|\dot{w}'''_F\right| = k\, Y_F Y_{O_2} \rho^2 \tag{13.5}$$

where

$$k = A_Y \exp(-E/\bar{R}T). \tag{13.6}$$

Rewriting Equation 13.2 and modifying Equation 13.3,

$$\dot{Q}_{gen} = A' \exp(-E/RT) \tag{13.7}$$

where $A' = A_Y\ V\ h_c\ Y_F\ Y_{O2}\ \rho^2$

$$\dot{Q}_{Loss} = h_H\ S\ (T - T_\infty) + \varepsilon\sigma\ S\ (T^4 - T_\infty^4) \tag{13.8}$$

where ε is emissivity, and σ is the Stefan–Boltzmann constant, $5.67 \times 10^{-11}\ \frac{kW}{m^2\ {}^\circ K^4}$. Equation 13.4 becomes

$$m\,c_v \frac{dT}{dt} = A' \exp\left(-\frac{E}{\bar{R}T}\right) - h_H S(T - T_\infty) - \varepsilon S\, \sigma (T^4 - T_\infty^{\ 4}) \tag{13.9}$$

Figure 13.4 shows qualitative plots of T and dT/dt vs. t for both inert and reacting systems. For inert heating, dT/dt (>0) is highest at t = 0, and as T increases, dT/dt decreases (Figure 14.3b). On the other hand for a reacting system, initially, the curve T vs. t and dT/dt vs. t are similar to those of an inert heating system; further, T can increase above T_∞ owing to chemical heat. Thus, when $T > T_\infty$, the system loses heat to the bath at T_∞, slowing down the rate of heating. When T is sufficiently high, the chemical heat liberation rate is so high that dT/dt will increase again, thus indicating a minimum in dT/dt vs. t as illustrated in Figure 13.4b. The point I denotes the inflection condition in the T vs. t curve.

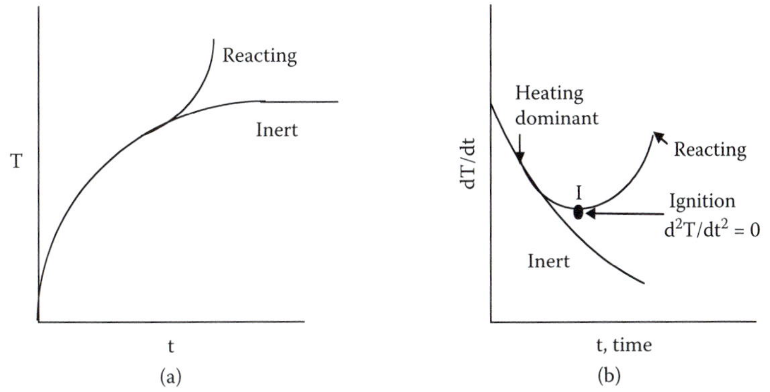

Figure 13.4 (a) Illustration of T vs. t for reacting and inert systems. (b) Illustration of dT/dt for reacting and inert systems.

We will assume that fuel and O_2 concentrations remain constant because there is negligible consumption prior to ignition. Hence, A′ = constant. One can numerically integrate Equation 13.9.

$$m c_v \{T_{new} - T_{old}\}/\Delta t = A' \exp(-E/RT_{avg}) - h_H S (T_{avg} - T_\infty) - \varepsilon\sigma S (T_{avg}^4 - T_\infty^4) \quad (13.10)$$

where

$$T_{avg} = (T_{new} + T_{old})/2$$

Select Δt, and input m, c_v, A, Y_{F0}, $Y_{O2,o}$, V, h_c, ρ, h_H (or Nu), ε, and σ; T_∞ = say 500, 550, and 600 K, obtain T vs. t for various values of T_∞ and plot T vs. t. Typical qualitative plots are shown in Figure 13.4.

Example 1

CH_4 along with 25% excess air is stored in a rigid vessel of diameter 5 cm at 298 K. It is quickly placed in a furnace at T_∞. The furnace temperature is altered stepwise from T_∞ = 600 K, to 760, 920, 1010, 1240, and 1400 K. Assume the following: dry air, LHV = 44500 kJ/kg; initial pressure and temperature of mixture: 1 bar, 298 K. Ignore radiation loss. Assume Nu = $h_H d/\lambda = 2$, $\lambda = 7.50E-05$ kW/m K, $c_v = 0.914$ kJ/kg K.

CH_4 kinetics:

$$d[CH_4]/dt \text{ (kmol/m}^3\text{ sec)} = k\,[\text{Fuel}]^a\,[O_2]^b \quad (A)$$

where $k = A_{[\,]} \exp(-E/\bar{R}T)$, E = 202,600 kJ/kmol, a = −0.3, b = 1.3, and $A_{[\,]} = 1.3 \times 10^8$ 1/sec.

(a) Calculate the equivalence ratio, fuel, and oxygen mass fractions
(b) Determine adiabatic flame temperature.
(c) Plot temperature vs. time.
(d) Plot dT/dt vs. time.

Solution

From Equation A,

$$d[CH_4]/dt\ (kmol/m^3\ sec) = k\ [Fuel]^a\ [O_2]^b \quad (B)$$

where $k = A \exp(-E/\bar{R}T)$, E = 202,600 kJ/kmol, a = −0.3, b = 1.3, (3.13) A = 1.3×10^8 (1/sec).
Converting to mass concentration basis and then to mass fraction basis (Chapter 5),

$$d(CH_4)/dt\ (kg/m^3\ sec) = k\ [M_F/\{(M_F)^a\ (M_{O2})^b\}]\ (Fuel)^a\ (O_2)^b \quad (C)$$

$$d\ (CH_4)/dt\ (kg/m^3\ sec) = A_Y\ \rho^{\ (a+b)}\ (Y_F)^a\ (Y_{O2})^b \exp(-E/RT) \quad (D)$$

where $A_Y = A_{[]} M_F^{\ (1-a)}/(M_{O2})^b = 1.3 \times 10^8 * 16.05^{(1.3)}/32^{1.3} = 2.068 \times 10^6$ 1/sec

(a) M_F = 16.05 kg/kmol, ν_{O2} = 3.99 kg of O per kg of fuel, equivalence ratio = 0.8, stoichiometric air = 17.34, actual air: 21.67; $Y_{N20} = 0.736$; $Y_{O20} = 0.22$, $Y_{F0} = 0.044$.

(b) $mc_v\ (T_{ad} - T_\infty) = m_{F,0}\ hc$, then $(T_{ad} - T_\infty) = \{Y_{F,0}\ h_c\}/c_v$

$$= 0.0441 \times 44500/0.914 = 1635\ K$$

$$T_{ad} = 1635 + 298 = 1933\ K$$

(c) $\rho_0 = 1.17$ kg/m³, surface area S = 0.0079 m², volume V = 6.55×10^{-5}m³, m = 7.65×10^{-5} kg, M_{mix} = 28.81 kg/kmol

With a time step dt = 0.13 sec, numerical integration was performed on Equation 13.10. Results for T and dT/dt are shown in Figure 13.5 and Figure 13.6.

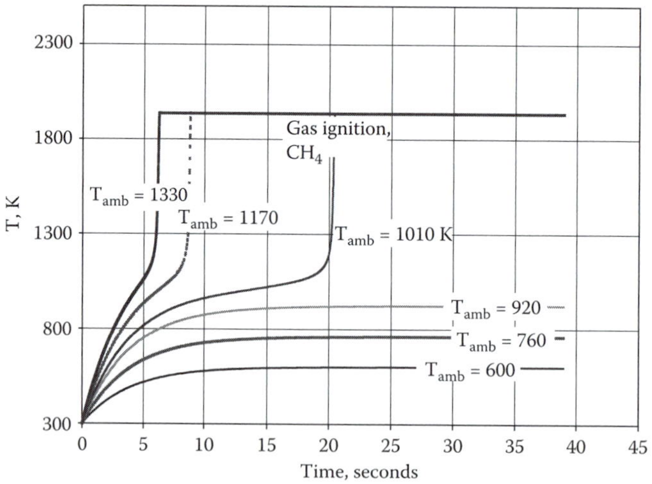

Figure 13.5 Illustration of T vs. t for methane — air system.

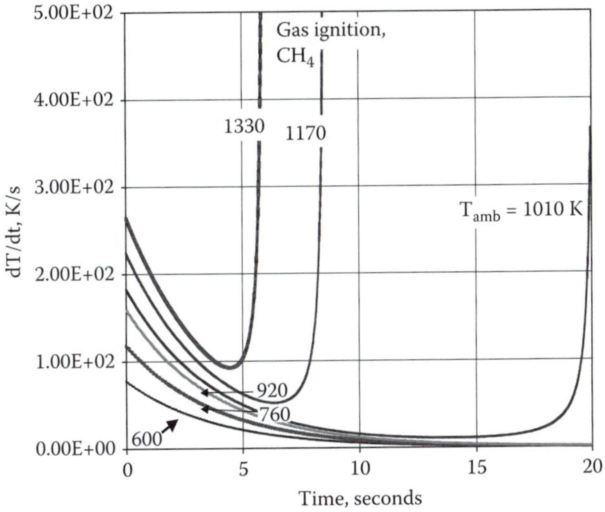

Figure 13.6 Illustration of T vs. t for methane — air system in a rigid vessel.

It is seen that at $T_\infty < 1010$ K, there is no runaway condition with the slope dT/dt tending to zero at large times. However, at $T_\infty = 1010$ K, dT/dt is initially positive; then it has a minimum at $t = t_I = 20$ sec, after which it increases again, resulting in rapid temperature rise or ignition. The condition where dT/dt reaches a minimum is called *inflection.* Using mathematics the reader can see that dT/dt reaches a minimum when $d/dt\ (dT/dt) = d^2T/dt^2 = 0$. For T_∞ = 1400, 1240, and 1010 K the minimum values of dT/dt are about 90, 40, and 0 K/sec, at which $d^2T/dt^2 = 0$. *Ignition temperature* is defined as the minimum ambient temperature that causes the inflection condition $d^2T/dt^2 = 0$, as well as $(dT/dt)_{min} = 0$. It is seen that the ignition temperature is determined as $T_{\infty,I}$ = 1010 K. In the next section we will derive an explicit expression for $T_{\infty,I}$. Note that if we keep the methane vessel in the furnace only for t = 10 sec with $T_{furnace}$ = 1010 K, ignition may not be achieved. Thus, the minimum ambient temperature condition presumes that there is infinite time available for ignition. If $T_\infty = T_{furnace} > 1010$ K, the time required to ignite decreases rapidly. Table A.39C tabulates T_{ign} of several fuels. Recall Equation 13.7.

A plot of heat liberation rate $\dot{Q}_{gen}$ vs. T is given by curve HGFIDECBA (Figure 13.7). The heat loss rate relation (Equation 13.7) under negligible radiation loss, is written as

$$\dot{Q}_{Loss} = h_H\ S\ (T - T_\infty) \tag{13.11}$$

where S is the surface area and h_H, the heat transfer coefficient. See lines FDC, IB, LA, GE, etc. (Figure 13.7), for heat loss $\dot{Q}_{Loss}$ vs. T. If T_∞ is very low, $\dot{Q}_{loss}$ is high at a given T and vice versa.

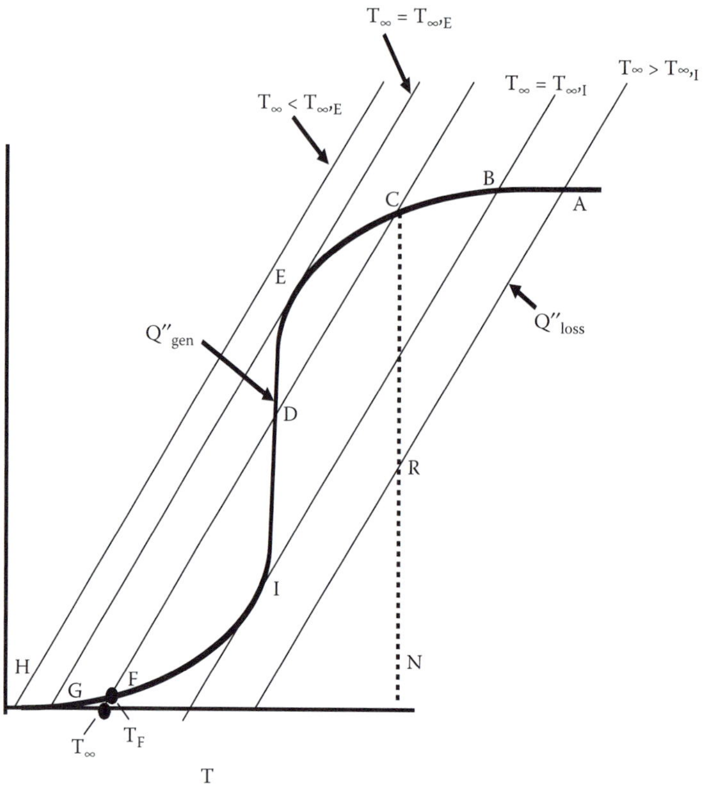

Figure 13.7 Effect of T on Q_{gen} and Q_{loss} with T_∞ as a parameter.

13.3.1.2 Steady-State Approximate Solution

When $\dot{Q}_{gen} >> \dot{Q}_{Loss}$, dT/dt is very high, e.g., at $T = T_N$ the heat generation is given by ordinate NC, which is greater than the heat loss given by NR. If $\dot{Q}_{gen} = \dot{Q}_{Loss}$, dT/dt = 0 and one has multiple solutions for T (points I, E, B, etc.).

If $T = T_\infty$, $\dot{Q}_{Loss} = 0$ but $\dot{Q}_{gen} > 0$ (e.g., point S). Hence the temperature rises to T_I. However, there is a difference between the solutions for point I and point F. First, consider point F. If the reactor temperature $T_F > T_\infty$ (say T_F = 600 K > T_∞ = 550 K) and suppose we drop a little cigarette ash into the reactor, temperature T_F may jump to 610 K. But $\dot{Q}_{loss} > \dot{Q}_{gen}$ at this T_F. Hence, the temperature returns to $T = T_F$ = 600 K. If we drop snow flakes, T may go to T_F = 590 K. But at 590 K, $\dot{Q}_{gen} > \dot{Q}_{loss}$; hence, the temperature will return to 600 K, i.e., with disturbance to the right or left of F, the T falls back to T_F. Solution T_F is called a *stable* solution. Similarly, you can show C, B, and A are stable or stationary states.

Now consider point I. If you disturb to the left (i.e., drop in snow flakes), T comes back to T_I. If you drop in cigarette ash, $\dot{Q}_{gen} > \dot{Q}_{loss}$; $\frac{dT}{dt} > 0$, T keeps increasing to point B, at which $\dot{Q}_{gen} = \dot{Q}_{loss}$ (you can show that B is a stable point). Thus, with perturbation, the point I is *semistable.* I is called the *ignition point.* T_∞ corresponding to point I is called the *ignition temperature,* whereas the system temperature is slightly higher. Thus, if one keeps the reactor at

T_∞, T will rise to T_I. As the disturbance is inherent, state I is not stable. Thus ignition occurs at I. Consider point D. You can repeat a similar analysis, and show that point D is always *unstable*. Consider point E. If you drop snow flakes into the flames, $\dot{Q}_{loss} > \dot{Q}_{gen}$, T decreases to T_G, and, extinction occurs. Hence E is also semistable. E is called the *extinction point*. Thus when $T_\infty = (T_\infty)_E$, extinction occurs.

13.3.1.2.1 Implicit Approximate Solutions for T_∞

Recall from Example 1 that the inflection condition exists at minimum T_∞ = 1010 K. At this point dT/dt = 0. Solutions so obtained (at which dT/dt = 0) are called *stationary solutions*. Equation 13.1 yields

$$\dot{Q}_{gen} = \dot{Q}_{loss} \tag{13.12}$$

The heat generation rate is still given by Equation 13.7, and if radiation loss is ignored, heat loss rate is given by Equation 13.11.

From Figure 13.7, when T_∞ is low, there is only one solution (point H) for T of combustible mixture that satisfies Equation 13.20. For $T_\infty = T_{\infty,L}$, there are two solutions for T: point I and point B. For $T_\infty > T_{\infty,I}$, there is only one solution (i.e., point A). For $T_{\infty,E} < T_\infty < T_{\infty,I}$, there are three solutions, points F, D, and C for T. If T vs. T_∞ is plotted, then the S curve somewhat similar to the one shown in Figure 7.11 of Chapter 7 is obtained, with the ordinate replaced by T and abscissa by T_∞. Also at I and E,

$$\frac{d\,\dot{Q}_{gen}}{dT} = \frac{d\,\dot{Q}_{loss}}{dT} \tag{13.13}$$

Using Equation 13.7 and Equation 13.11 in Equation 13.12, and then solving for T,

$$(V \quad A_Y \quad h_c \quad Y_F \quad Y_{O2} \quad \rho^2)\exp\left(-\frac{E}{\bar{R}T}\right) = h_H \quad S(T - T_\infty) \tag{13.14}$$

$$T = T_\infty + \left[\frac{V}{S}\frac{A_Y \; h_c \; Y_F Y_{O_2} \rho^2}{h_H}\exp\left(-\frac{E}{\bar{R}\,T}\right)\right] \tag{13.15}$$

where

$$V/S = d/6 \text{ for sphere of dia d} \tag{13.16a}$$

$$V/S = d/4 \text{ for a long cylinder of dia d} \tag{13.16b}$$

$$V/S = d/6 \text{ for a cube of side d} \tag{13.16c}$$

Equation 13.15 is interpreted as follows: reactor temperature at ignition = free-stream temperature + additional temperature rise due to chemical reaction. As ρ, Y_F, and Y_{O_2} are constants and letting ρ be constant (valid for a closed system because ρ = m/V), we can differentiate Equation 13.14 with respect to T; then using Equation 13.13

$$\left(\frac{V}{S} A\ h_c\ Y_F\ Y_{O_2}\ \rho^2\right)\left(\frac{E}{\bar{R}\ T^2}\right)\exp(-E/\bar{R}\ T) = h_H \quad (13.17)$$

Using Equation 13.14 in Equation 13.17 to eliminate $e^{-E/RT}$ and other terms,

$$\frac{E}{RT^2}(T - T_\infty) = 1.0 \quad (13.18)$$

Solving the quadratic Equation 13.18,

$$(\bar{R}T/E) = \frac{1}{2}\left\{1 \pm \sqrt{1-(4\bar{R}\ T_\infty/E)}\right\} \quad (13.19)$$

For the same $T_{\infty,I}$, there are two possible solutions for T. Using the minus sign for ignition with $T_\infty = T_{\infty I}$, and then using the result in Equation 13.17, $T_{\infty,I}$ at I can be evaluated.

Let

$$\theta = \bar{R}\ T/E \quad (13.20)$$

and let the fourth Damkohler number (Chapter 7) be defined as,

$$Da_{IV} = \frac{V}{S}\frac{A_Y\ h_c Y_F\ Y_{O_2}\rho^2}{h_H\ \frac{E}{\bar{R}}} = \frac{m}{S}\frac{A_Y\ h_c Y_F\ Y_{O_2}\rho}{h_H\ S\frac{E}{\bar{R}}} \quad (13.21)$$

where m, mass. Then Equation 13.18 can now be written as,

$$\theta^2 - (\theta - \theta_\infty) = 0 \quad (13.22)$$

Whereas Equation 13.19 is written as

$$\theta = \frac{1}{2}\left\{1 \pm \sqrt{1-4\ \theta_\infty}\right\},\ \theta_\infty < 1/4 \quad (13.23)$$

and using Equation 13.15, the normalized temperature of the system becomes

$$\theta = \theta_\infty + Da_{IV}\exp\left(-\frac{1}{\theta}\right) \quad (13.24)$$

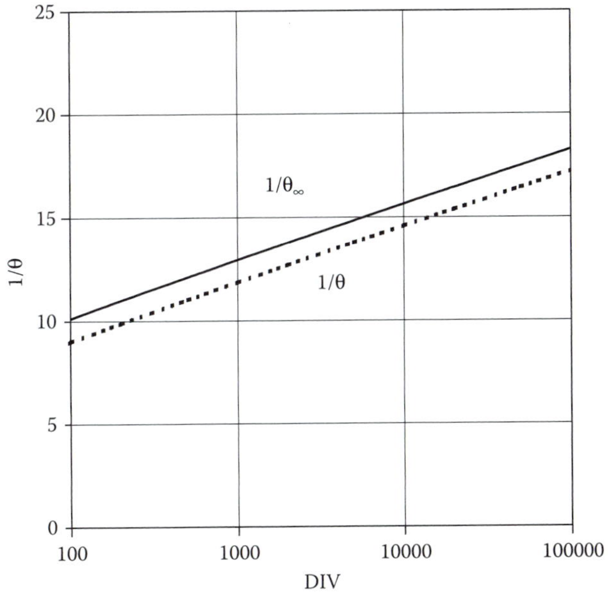

Figure 13.8 Nondimensional ignition temperature and particle temperature vs. Da_{IV}.

Assume θ_∞, find θ from Equation 13.23, and check whether Equation 13.24 is satisfied. If not, iterate. Figure 13.8 shows plots of $1/\theta_\infty$ and $1/\theta$ vs. ln (D_{IV}), and they are almost linear. The ignition temperature is a function of mixture composition because Y_F and Y_{O2} are changed with the equivalence ratio. If very lean or very rich, T is higher and a minimum exists near stoichiometry as shown qualitatively in Figure 13.9.

For a fixed-mass closed system, the higher the $T_{\infty,I}$, the higher the thermal energy required to initiate ignition. Figure 13.10 shows an experimental plot obtained for a coal–air cloud in a vessel as a function of coal concentration,

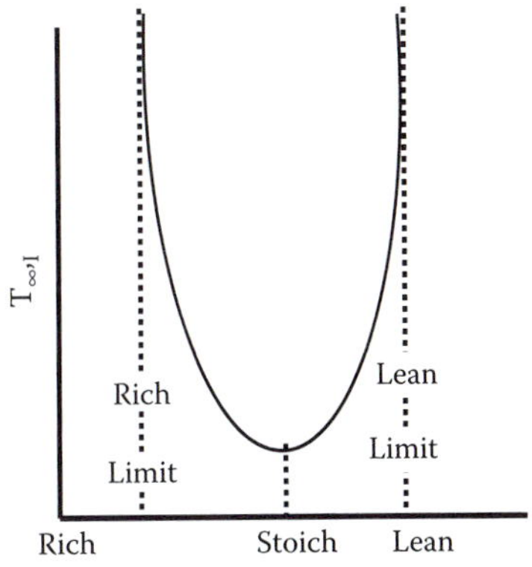

Figure 13.9 Ignition temperature vs. fuel composition.

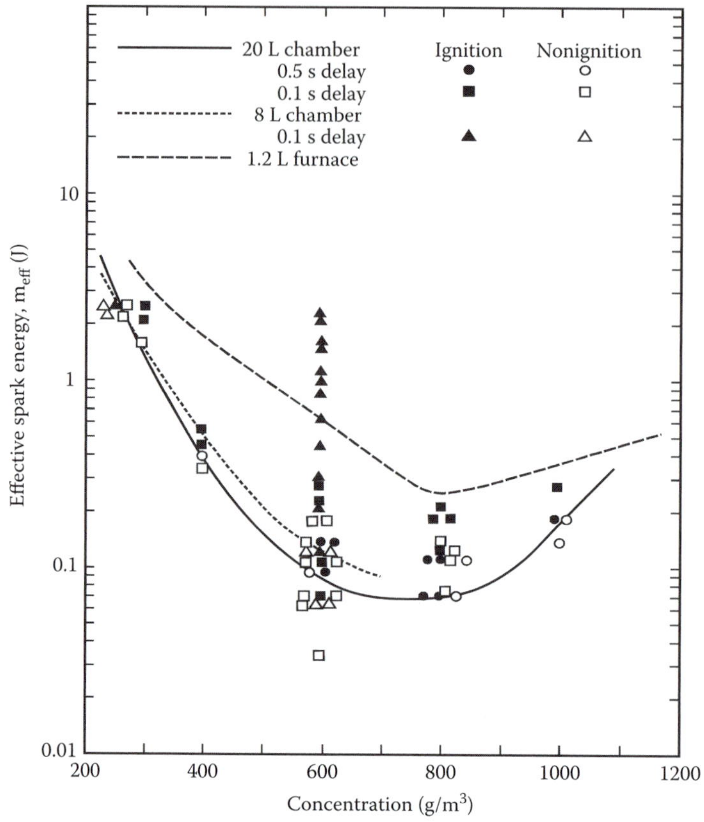

Figure 13.10 Spark ignition energy vs. coal dust concentration. (Adapted from Hertzberg, M., Conti, R.S., and Cashdoillar, K.L., Electrical Ignition Energies and Thermal Autoignition Temperatures for Evaluating Explosion Hazards of Dusts, Bureau of Mines Report RI 8988, Pittsburgh, PA, 1985.)

confirming the qualitative behavior depicted in Figure 13.9. The effect of size of the vessel on spark energy will be discussed later.

13.3.1.2.2 Explicit Approximate Solution

Typically, $(\theta - \theta_\infty) < \theta_\infty$ at ignition. Hence approximating θ^2 as

$$\theta^2 = (\theta - \theta_\infty + \theta_\infty)^2 \approx \theta_\infty^2\ \{1 + 2\ (\theta - \theta_\infty)/\theta_\infty\} \tag{13.25}$$

and using Equation 13.22, one solves for $(\theta - \theta_\infty)$

$$(\theta - \theta_\infty) \approx \theta_\infty^2/[1 - 2\theta_\infty] \approx \theta_\infty^2 \text{ as } \theta_\infty << 1. \tag{13.26}$$

Similarly, approximating exp $(-1/\theta)$ as, and using $\theta - \theta_\infty \approx \theta_\infty^2$

$$\exp(-1/\theta) \approx \exp\{(-1/\theta_\infty)\ [1 - (\theta - \theta_\infty)/\theta_\infty]\} = \exp\{-(1/\theta_\infty)\}\ \exp\{\theta - \theta_\infty)/\theta_\infty^2\ ]\}$$
$$\approx \exp\{(-1/\theta_\infty)\}\exp(1) \tag{13.27}$$

Using Equation 13.26 and Equation 13.27 in Equation 13.24, the inverse of ignition temperature is correlated as,

$$(1/\theta_\infty) \approx \ln\,[\mathrm{Da_{IV}}\,\exp(1)/\theta_\infty^{\,2}] = 1 + \ln\,[\mathrm{Da_{IV}}/\theta_\infty^{\,2}]. \qquad (13.28)$$

13.3.2 Ignition Energy

Ignition energy is the energy required to raise the temperature of a system of mass from initial temperature T_0 to ignition temperature T_I.

$$E_I = mc\,(T_I - T_0) \qquad (13.29)$$

where c is specific heat. Knowing m, T_I and T_0 one can estimate the required E_I.

Example 2

Consider Example 1 with similar data; using approximate solution,

(a) Determine the minimum furnace temperature at which the mixture can be ignited.
(b) Determine the gas mixture temperature at time of ignition.
(c) Determine the ignition energy in kJ.

Solution

(a) Consider an approximate relation first in (Equation 13.28)

$$(1/\theta_\infty) \approx \ln[\mathrm{Da_{IV}}\,\exp(1)/\theta_\infty^{\,2}] = 1 + \ln[\mathrm{Da_{IV}}/\theta_\infty^{\,2}].$$

Assume T_∞ = 1000 K; E/R = 24368 K; θ_∞ = 1000/24369 = 0.0410;

Initial density = 1.17 kg/m^3 (density at time of ignition = initial density if the system is closed).

$$Da_{IV} = \frac{V}{S}\,\frac{A\;hc\,Y_F^{\;nF}\,Y_{O_2}^{\;nO2}\,\rho^{\;nF+nO2}}{b^{E/\bar{R}}}$$

$$= \frac{6.55\times10^{-5}}{0.003}\;\frac{0.44^{-0.3}*0.2199^{1.3}*44500*1.17^{1.3-0.3}}{0.00785*24369} = 4.6\times10^{6}$$

$$(1/\theta_\infty) \approx \{1 + \ln[4.6 \times 10^6/0.0410^2]\} = 22.73,$$

θ_∞ = 0.04399, T_∞ = 0.0439 * 24369 = 1072 K

Recall that numerical analysis yielded 1010 K; approximate analysis yields results within 6% error margin.

(b) The gas temperature is estimated by using Equation 13.26

$$\theta \approx \theta_\infty + \theta_\infty^{\,2} = 0.04399 + 0.04399^2 = 0.04593,$$

$$T = 0.04539 * 24369 = 1119\ \mathrm{K}$$

(c) Mass = 6.5 * 10^{-5} * 1.17 = 7.65298E − 05 kg

Ignition energy = 7.65298E − 05 kg * 0.914 kJ/kg K * (1119 − 298) = 0.0755 kJ

As T is higher for lean and rich mixtures, the corresponding ignition energies are also higher (see Figure 13.10).

13.4 Constant-Pressure Systems

13.4.1 Analysis

The analysis can now be extended to a closed system with a constant pressure. All previous analyses apply if we replace c_v by c_p, but density does not remain constant for a closed system (see Chapter 7 for details).

$$m\ c_p\ dT/dt = \dot{Q}_{gen} - \dot{Q}_{Loss} \qquad (13.30)$$

Example 3

Consider Example 2. The vessel is assumed to be elastic, and expands upon a rise in temperature, keeping pressure constant. What are the minimum furnace temperature and ignition energy required for ignition? Initial pressure and temperature: 1 bar, 298 K. Ignore radiation loss. c_p = 1.2 kJ/kg K. Assume d = 5 cm at time of ignition.

Solution

Let us assume that T_I = 1000 K, then for P = 1 bar, density = 0.346 kg/m^3, and $Da_{IV} = 0.941 \times 10^6$ which is less compared to Example 2 as the density is lower.

$$(1/\theta_\infty) \approx \{1 + \ln [0.941 \times 10^6/0.0410^2]\} = 21.13,$$

$$\theta_\infty = 0.0473,\ T\infty = 0.0473 * 24369 = 1150\ K$$

(If the numerical analysis is repeated with c = c_p = 1.2 kJ/kg K, ignition occurs at about 1080 K; recall that constant-volume ignition occurs at about 1010 K.) The gas temperature at the time of ignition is given as,

$$\theta \approx \theta_\infty + \theta_\infty^2 = 0.0473 + 0.0473^2 = 0.0495,$$

$$T = 0.0495 * 24369 = 1210\ K$$

$$\text{Mass} = 6.5 * 10^{-5} * 1.17 = 7.65 \times 10^{-5}\ kg$$

$$\text{Ignition energy} = 7.65 \times 10^{-5}\ kg * 1.2\ kJ/kg\ K * (1210 - 298) = 0.0835\ kJ$$

Remarks

(i) The required temperature is higher because gas density decreases, resulting in lower chemical reaction rate; the required energy is higher owing to two effects: increased gas temperature for ignition and increased specific heat (i.e., the energy now includes work for deformation of the vessel to a larger volume).

(ii) In diesel engines, the pressure is almost constant and hence ignition occurs at constant pressure. Further, in flow systems the ignition occurs at constant pressure.

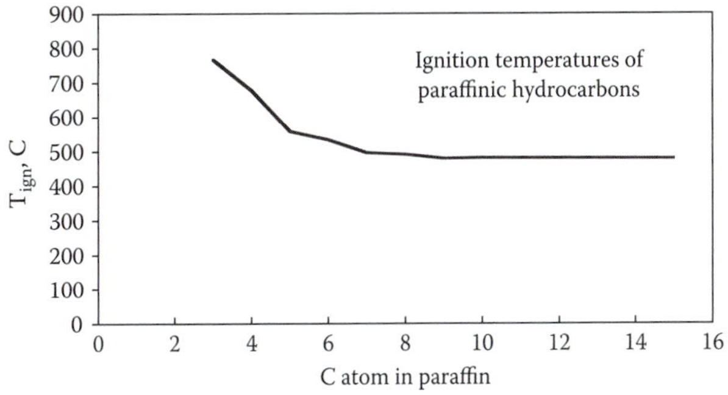

Figure 13.11 Experimental data on T_{ign} vs. carbon atoms in paraffins.

(iii) The ignition temperature is not a property of the given gas; it depends upon the equivalence ratio, local heat loss conditions, geomety and dimensions, transport and thermal properties, etc.

(iv) Figure 13.11, shows the experimental data on T_{ign} for HC for constant-P systems. In order to improve ignition characteristics, one can blend CH_4 with C_2H_6, ..., C_8H_{18} since higher HC fuels have lower ignition temperatures.

(v) Recall that the activation energy is related to bond energy. Methane has 4 C–H bonds per C atom, which are much stronger than C–C bonds. On the other hand, other hydrocarbons, say, C_2H_6, have 6 C–H bonds for 2 C atoms or 3 C–H bonds per C atom and C8H18 has 18 C–H bonds for 8 C atoms and (18/8) C–H bonds per C–C bond. Thus, methane is much stronger and hence, difficult to ignite. Thus, the ignition temperature decreases with an increase in carbon atoms for paraffinic HC (Figure 13.11).

(vi) For heterogeneous drop–air mixtures, the drops vaporize, mix with air, and ignite; ignition energy includes both sensible energy to raise the drop to boiling point, supply latent heat and raise the gas mixture to ignition temperature. If the reaction zone thickness is δ_R, the reaction zone volume around each drop $\approx 4\ \pi\ r_{st}^{\ 2}\ \delta_R$. For n drops per unit volume, the volume is $4n\pi\ (r_{st}/a)^2\ a^2\ \delta_R$; at fixed (r_{st}/a) (Chapter 10) and at fixed fuel mass per unit volume $n \propto \frac{1}{a^3}$, smaller drops provide large reaction volume per unit geometrical volume and, hence, ignition energy decreases with decreased size (Figure 13.12). If the drops are very large, ignition may occur individually around each drop. Typically, ignition refers to bulk ignition over a large volume when a large number of droplets or particles are involved. For more discussion see Chapter 16. Recall from Chapter 4 and Chapter 10 that the Rosin–Rammler distribution (Chapter 4) changes with time as evaporation, ignition, and combustion occur. For ignition of spray, a minimum amount of vapor needs to be generated before the whole spray ignites. Typically, ignition time is computed by assuming the time required for 20% of spray mass to evaporate, whereas combustion time is estimated assuming 90% of mass loss. Recall $R = \exp\{-(d_p/d_{p,Charac})^n\}$, where R is the fraction of volume having size greater than d_p and the higher the value of n, the greater the uniformity of spray. According to Chin et al. [1984], the higher the value of n, the higher the 20% evaporation time (longer flame stand-off distance) but the lower the 90% combustion time (better combustion efficiency). From the ignition point of view, uniform spray size distribution is better.

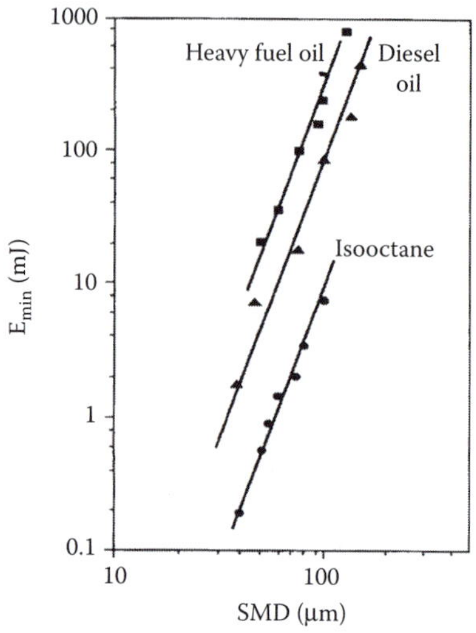

Figure 13.12 Ignition energy vs. drop size in a flow system, v = 15 m/sec, and P = 1 atm. (Adapted from Ballal, D.R. and Lefebvre, A.H., *Combust. Flame*, 35, 155–168, 1979.)

13.4.2 Ignition of Combustible Gas Mixture in Stagnation Point Flow

Recall from Chapter 12 that if a combustible mixture impinges on a heated flat plate at T_w and burns, the stagnation boundary-layer thickness (δ) is constant, and free-stream velocity increases with distance (Figure 13.13). Consider an inert mixture. As the mixture is heated, the temperature rises with x and reaches T_w asymptotically. Thus, dT/dx is very steep near x = 0 and gradually decreases to dT/dx = 0 as $x \rightarrow \infty$. Now consider a combustible mixture. If there is insignificant chemical reaction, a similar profile occurs except that the final T of the mixture is slightly above T_w. On the other hand at certain T_w, the slope dT/dx gradually decreases first, reaches dT/dx = 0 at certain x and then starts rising again because of significant chemical reaction. Thus, we have an inflection in T vs. x just as in Example 1. The following example illustrates ignition over a heated flat plate.

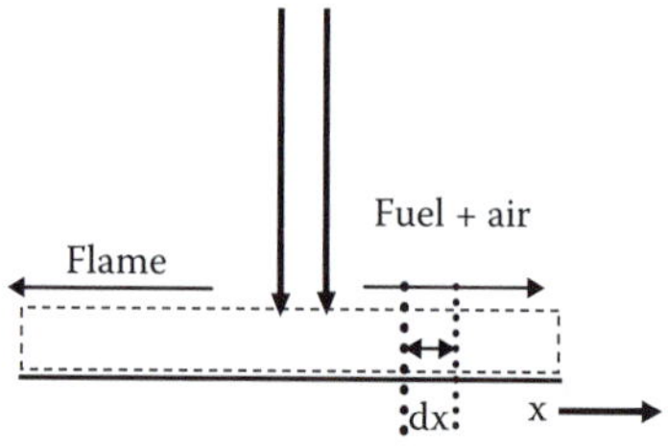

Figure 13.13 Stagnation flow of combustible mixture over heated plate.

Example 4

Consider impingement and ignition of a combustible mixture over a heated plate. Show that Equation 13.28 defines the ignition temperature with a slight modification in the Damkohler expression:

$$Da_{IV} = \frac{A_Y \; h_c \delta Y_{F,in} Y_{O_2,in} \rho^2}{h_H \; \frac{E}{R}}. \tag{A}$$

Solution

Consider an elemental distance dx. Assume unit width. Rewrite Equation 13.9 as

$$\dot{m}'' \delta c_p \, dT = \left|\dot{w}_F'''\right| h_c \, \delta dx - h_H (T - T_w) dx - \varepsilon\sigma\left(T_w^{\,4} - T_\infty^{\,4}\right) dx$$

where $\dot{m} \, \delta$cp dT is the thermal energy of gases within a short distance dx. Dividing by δdx

$$\dot{m}'' \; c_p \frac{dT}{dx} = \left|\dot{w}_F'''\right| h_c - h_H (T - T_w)/\delta - \varepsilon\sigma\left(T_w^{\,4} - T_\infty^{\,4}\right)/\delta \tag{B}$$

where T is the average temperature within the boundary layer, and δ is the boundary-layer thickness. Chemical heat generated within dx raises the temperature above T_w, which results in heat loss to the plate. Note that at the inflection point T of the gas must be closer to T_w. Thus, T_w may be used to estimate the heat generated. Again invoking the inflection condition, one can determine the ignition temperature numerically just as in Example 1. Let us simplify Equation B at the inflection point. Ignoring the radiation loss to the ambience and using an overall heat transfer coefficient,

$$\frac{dT}{dx} = 0; \quad \left|\dot{w}_F''\right| h_c \, \delta = h_H (T - T_w) \tag{C}$$

$$\frac{d^2T}{d^2x} = 0; \quad \frac{d}{dx}\left[\left|\dot{w}_F''\right| h_c \, \delta h_c \, \delta\right] = \frac{d}{dx}[h_H (T - T_w)] \tag{D}$$

$$\frac{d}{dT}\left[\left|\dot{w}_F'''\right| h_c \, \delta\right] \frac{dT}{dx} = \frac{d}{dT}[h_H (T - T_w)] \frac{dT}{dx} \tag{E}$$

$$\frac{d}{dT}\left[\left|\dot{w}_F'''\right| h_c \, \delta\right] = \frac{d}{dT}[h_H (T - T_w)] = h_H \tag{F}$$

where $|\dot{w}_F'''| = A_Y \; V \; Y_F \; Y_{O2} \; \rho^2 \exp(-E/RT)$. The consumption of fuel prior to the inflection is assumed to be negligible. These relations are similar to those of previous illustrations except that the characteristic vessel dimension (= V/S) is replaced by δ. Defining the fourth Damkohler number as

$$Da_{IV} = \frac{A_Y\ h_c \delta Y_{F,in} Y_{O_2,in}\ \rho^2}{h_H\ \frac{E}{R}} \tag{G}$$

where first-order reactions with fuel and O_2 are assumed. Results for ignition are given by Equation 13.26 and Equation 13.28.

■ ■ ■

The boundary-layer thickness at 99% of the free-stream value is [Schlichting, 1955] $\delta = 2.4\ \{\nu/a\}$, where $u_e = ax$, a is in 1/sec, and the assumption of unity for Sc has been made. $Nu_x = h_H x/\lambda = 0.57\ \{Re_x\}^{1/2}$, at Pr = Sc = 1 [Schlichting, 1955, p. 304]; $Re_x = u_e\ x/\nu$, and $u_e = ax$, where a is a stretch factor. Thus, $h_H = \lambda\ 0.57\ \{a/\nu\}^{1/2}$ is constant. Equation G of Example 4 becomes

$$Da_{IV} = \frac{2.4\ A_Y\ h_c\ \nu\ Y_{F,in} Y_{O_2,in}\ \rho^2}{0.57\ a\ \lambda\ \frac{E}{R}} = \frac{4.21 A_Y\ h_c\ \nu\ Y_{F,in} Y_{O_2,in}\ \rho^2}{a\ \lambda\ \frac{E}{R}}. \tag{13.31}$$

The higher the value of "a" (i.e., stretch), then lower the value of Da_{IV} and the higher the ignition temperature, because the heat loss occurs over a thinner boundary layer. Similar relations may be derived for an axisymmetric jet. The reader is referred to literature by Seshadri and Williams (1978); Seiser et al. (2000); and Fachini and Seshadri (2003).

13.5 Ignition of Solid Particle

13.5.1 Carbon/Char Particle

Recall the example of ignition of a gas mixture within a rigid vessel. We replace the gas mixture with a large solid particle of initial temperature T_{p0}, suspend it in a furnace at T_∞, and replace gas temperature with the particle temperature (T_p). The particle heats up; if it is inert, it reaches $T = T_\infty$ as $t \to \infty$ (Figure 13.4a). However if the particle is carbon, it is chemically active and T may rise above T_∞. Then the particle loses heat to the furnace. If heat generation is insignificant, the particle reaches a new equilibrium temperature $T > T_\infty$ but $T - T_\infty$ is still very small. At $T_{\infty,I}$, the particle temperature rises well above T_∞ resulting in "glowing" or *ignition.* We wish to determine the minimum T_∞ value at which ignition occurs. We assume the particle to be a sphere and consider unit surface area of the sphere for energy balance. The problem is very similar to ignition of the gas mixture at constant pressure except for the following differences:

1. The kinetic expression is for oxidation of carbon.
2. Unlike the gas mixture, it is not premixed. The oxygen has to diffuse from the surroundings for oxidation to occur at the surface of the particle.
3. Radiation loss may be significant.
4. The reaction occurs on the external surface of area S.

Figure 13.14 shows a sketch of the temperature profile in the gas phase.

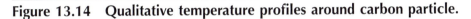

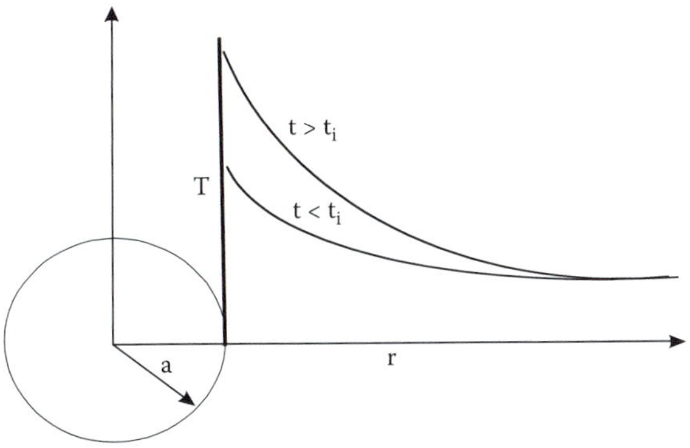

Figure 13.14 Qualitative temperature profiles around carbon particle.

13.5.1.1 Numerical Method

Sensible energy rise of the particle = heat generated by chemical reaction less the heat loss by convection and by radiation.

$$m_p\ c\ dT/dt = \dot{Q}_{gen} - \dot{Q}_{Loss} \tag{13.32}$$

where

$$m_p = \rho_p\ \pi\ d_p^{\ 3}/6.$$

$$\dot{Q}_{gen} = S\left|\dot{w}_c''\right| h_c \tag{13.33}$$

where $|\dot{w}_c''|$ is carbon consumption per unit surface area and S is the surface area = πd_p^2. Assuming an arbitrary order with oxygen, and from the law of stoichiometry and mass action

$$\frac{\dot{w}_C''}{(-1)} = \frac{\dot{w}_{O2}''}{-\nu_{O2}} = \dot{w}'' = kY_{O_{2,w}}{}^{nO2}\rho^{nO2},\ kg/m^2s \tag{13.34}$$

$$\left|\dot{w}_C''\right| = k\ Y_{O_{2,w}}{}^{nO2}\rho^{nO2},\ kg/m^2s \tag{13.35}$$

where

$$k = A_{het}\ \exp(-E/\bar{R}\ T_p), \tag{13.36}$$

$Y_{O2,\,w}$ is the oxygen mass fraction at the carbon surface. If n_{O2} = 1, then from Chapter 7 or Chapter 9

$$Y_{O_{2,w}} = \frac{h_m\ Y_{O_{2,\infty}}}{(k_{O_2}\ \nu_{O2}\rho_w + h_m)}. \tag{13.37}$$

The heat loss per unit area is given as,

$$\dot{Q}''_{loss} = h_H(T_p - T_\infty) + \varepsilon\sigma(T_p^4 - T_\infty^4). \qquad (13.38)$$

The energy equation becomes,

$$m_p c \frac{dT_p}{dt} = A_{het} S h_c Y_{O_2,w} \exp\left(-\frac{E}{\bar{R}T_p}\right) - h_H S(T_p - T_\infty) - \varepsilon S\sigma\left(T_p^{\,4} - T_\infty^{\,4}\right). \qquad (13.39)$$

The numerical methodology is similar to the one adopted for gas mixture ignition. One can determine the ignition temperature by plotting the particle temperature T_p vs. t at various free-stream temperatures, T_∞. The procedure will be later illustrated with an example.

13.5.1.2 Explicit Solutions When Y_{O2} at Char Surface is the Same as Free-Stream Mass Fraction

The problem is similar to that for a gas mixture except that we consider surface reactions (Equation 13.33, 13.35, and 13.36). The $\dot{Q}''_{gen}$ and $\dot{Q}''_{loss}$ with T_p are plotted in Figure 13.15.

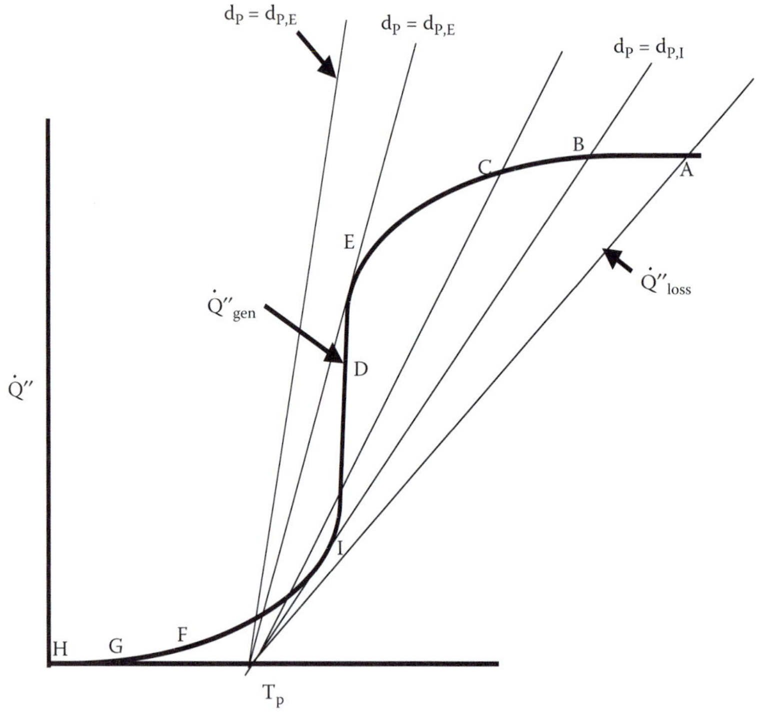

Figure 13.15 $\dot{Q}_{gen}$ **and** $\dot{Q}_{loss}$ **vs. T for convection-dominated, fixed** T_∞, **and varying** d_p.

Following Equation 13.19

$$(\bar{R}\ T_p/E) = \frac{1}{2}\left\{1 \pm \sqrt{1-(4\bar{R}\ T_\infty/E)}\right\}. \tag{13.40}$$

Modifying Equation 13.15 as

$$T_p = T_\infty + \left[\frac{A_{het}\ h_c\ Y_{O2,w}{}^{nO2}\ \rho_w{}^{nO2}}{h_H}\exp(-E/\bar{R}\ T_p)\right]. \tag{13.41}$$

Particle temperature at ignition = gas temperature + temperature rise due to chemical heat release.

Following Equation 13.21,

$$Da_{IV} = \frac{A_{Het}\ h_c\ Y_{O2,w}\,\rho}{h_H \frac{E}{\bar{R}}}. \tag{13.42}$$

Using Equations 13.40, 13.41, and 13.42, one can solve for two unknowns T_p and T_∞ for any given Da_{IV}. Note that $h_H \propto 1/d_p$, and hence smaller particles have higher heat loss per unit area of carbon particle surface (Figure 13.15); therefore, they may not ignite. There are some chars (resembling small sponges) that are highly porous and in which reaction can occur throughout the volume; we can modify Equation 13.42 as,

$$T = T_\infty + \left[\frac{A_{Het}\ S_v V h_c\ \eta\ Y_{O2,w}\,\rho_w}{h_H\ S}\exp(-E/\bar{R}\ T_p)\right] \tag{13.43}$$

where V = volume of particle, $\pi d_p^3/6$, S_v, internal surface area per unit volume of particle, $S = \pi d_p^2$, η is the effectiveness or Thiele modulus factor if only a part of the volume of the particle is used or if the oxygen fraction is not uniform within the particle. Figure B.3 in Appendix B presents the effectiveness factor.

Example 5

A carbon particle of diameter 50 μm initially at 298 K is placed in a furnace. Assume the following: dry air, pressure 1 bar. Ignore radiation loss. Nu = 2, λ = 7.50E − 05kW/m K, c = 2.2 kJ/kg K, and ρ_p = 750 kg/m³, $Y_{02,\infty}$ = 0.23, $c_{p,gas}$= 1.2kJ/kg K

$$\dot{w}''\ (\text{kg/m}^2\ \text{sec}) = k\ \rho Y_{O2,w} \tag{A}$$

where

k = A_{het} exp(−E/$\bar{R}$ T), E = 73700 kJ/kmol, and A_{het} = 75 m/sec.

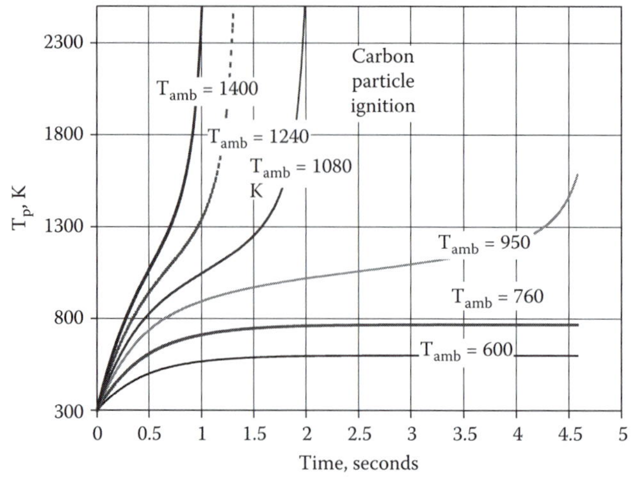

Figure 13.16 Temperature history of carbon particle at various ambient temperatures; no radiation loss.

Assume that $C + O_2 \rightarrow CO_2$, $h_c = 32760$ kJ/kg, $\upsilon_{02} = 2.66$ kg of O_2/kg of C

(a) Plot T_p and dT_p/dt vs. t.
(b) From the plots, determine the minimum furnace temperature at which the carbon can be ignited.
(c) Determine the minimum furnace temperature using the approximate ignition temperature relation.
(d) Estimate the ignition energy in kJ.

Solution

(a) The input data are as follows:
$h_c = 32761$ kJ/kg for $C + O_2 \rightarrow CO_2$; P = 1 bar; and $T_{p0} = 298$ K.

Area of particle, $S = 7.85 \times 10^{-7}$ m^2; volume $V = 6.54 \times 10^{-11}$ m^3; mass $m_p = 4.91 \times 10^{-8}$ kg; and $M_{gas} = 28.8$ (assumed).

$h_H = 2 * \lambda/d_p = 3.00E - 01$ kW/m^2 K and E/R = 8864.6 K

Figure 13.16 and Figure 13.17 show the T_p vs. t and dT_p/dt vs. t using numerical integration.

(b) It is seen from Figure 13.17 that at $T_\infty = T_{amb} = 950$ K ignition occurs.
(c) Let us assume for the initial iteration that $T_{\infty,I} = 1000$ K; then for P = 1 bar, density $\rho = 0.347$ kg/m^3. $\theta_\infty = 1000/8865 = 0.113$. Using Equation 13.42, Da_{IV} = 75.1 m/sec * 32761 kJ/kg * 0.23 * 0.347 kg/m^3/{0.003 kW/m K * 8865 K} = 73.8 which is considerably less compared to the gas mixture example since the size of the reacting system (diameter of particle) is very small. With $\theta_\infty = 0.113$ in Equation 13.28, $(1/\theta_\infty) \approx \{1 + \ln [737/0.113^2]\} = 9.67$, $\{\theta_\infty\}_{new} = 0.103$, $\{T_{\infty,I}\}_{new} = 0.103 * 8865 = 917$ K. Thus, from the approximate relation, $\{T_{\infty,I}\}_{approx} = 917$ K. Recall that numerical analysis yielded 950 K; approximate results yield an error of –3.5%

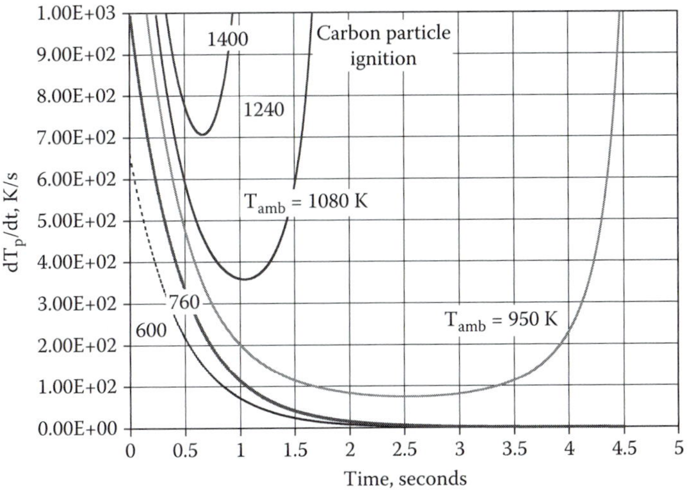

Figure 13.17 Rate of temperature rise, (dT_p/dt) history of carbon particle at various ambient temperatures; no radiation loss.

The particle temperature is given by Equation 13.26.

$$\theta_p \approx \theta_\infty + \theta_\infty^2 = 0.1035 + 0.1035^2 = 0.114,$$

$$\{T_p\}_{approx} = 0.114 * 8865 = 1010 \text{ K}$$

(d) Ignition energy = $m_p c\,(T_p - T_\infty) = 4.91 \times 10^{-8}$ kg * 2.2 kJ/(kg K) * (1010 − 298) K = 7.7×10^{-5} kJ

Remarks

(i) $Q_{loss} \propto h_H$ S and for a given Nu, $h_H \propto 1/d$; hence, h_H S $\propto$ d, whereas radiative loss $\propto d^2$; thus radiative loss is extremely large for larger particles. Figure 13.18 shows qualitative characteristics of heat generation and heat loss vs. T_p under radiation loss conditions. If radiation loss is included for 0.5-mm particles, then $T_{\infty,I}$ (i.e., the inflection condition) can be shown to be 1080 K, an increase of almost 130 K. In the next section an approximate relation will be given that includes radiation loss. Figure 13.19 and Figure 13.20 show T_p vs. time and dT_p/dt vs. time plots that include radiation loss.

(ii) We had assumed that the oxygen mass fraction at the surface is the same as free stream. Unlike the premixed gaseous combustible mixtures, oxygen is transported to the particle surface by a diffusion process. Particularly as the reaction approaches the runaway condition, the reaction rate is extremely high, resulting in rapid depletion of oxygen mass fraction at the surface. Further, near the extinction regime for a burning particle, T_p is extremely high and, hence, surface oxygen mass fraction is low. Hence, the assumption of $Y_{O2,w} = Y_{O2,\infty}$ is relaxed in the next section.

(iii) If time to react is very long, then the particle size may decrease with time; thus $dT_p/dt \propto \dot{m}/m$ and for a reaction-dominated case, $dT_p/dt \propto \dot{m}/m \propto d^2/d^3$ m $\propto 1/d$. Thus, dT_p/dt may increase with mass-induced inflection.

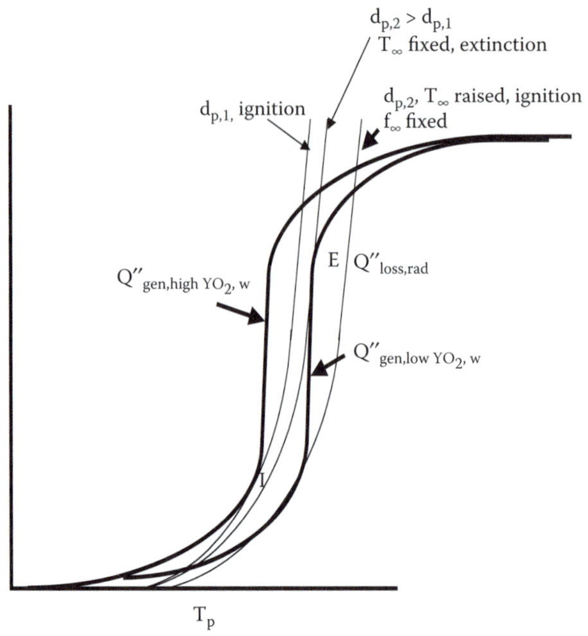

Figure 13.18 Radiation-dominated, fixed T_∞.

13.5.1.3 Implicit Steady-State Solutions When $Y_{O2,w} \neq Y_{O2,\infty}$

Using Equation 13.37 in Equation 13.35 and rearranging

$$\left|\dot{w}''_c\right| = \frac{kY_{O2,\infty}\,\rho_w}{\left[\dfrac{k\rho_w v_{O2}}{h_m} + 1\right]} \tag{13.44}$$

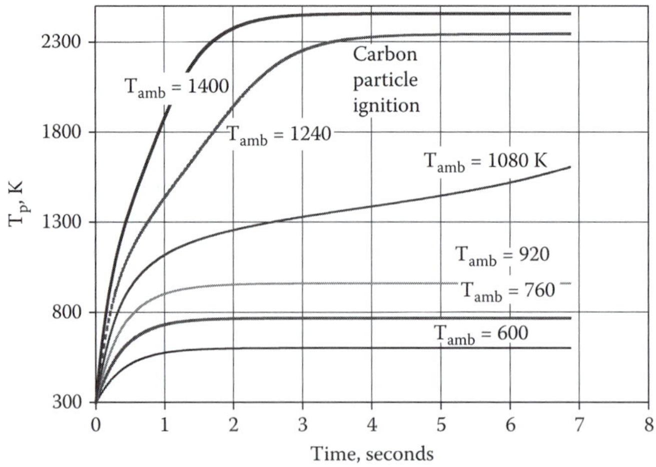

Figure 13.19 Temperature history of carbon particle at various ambient temperatures: radiation loss included.

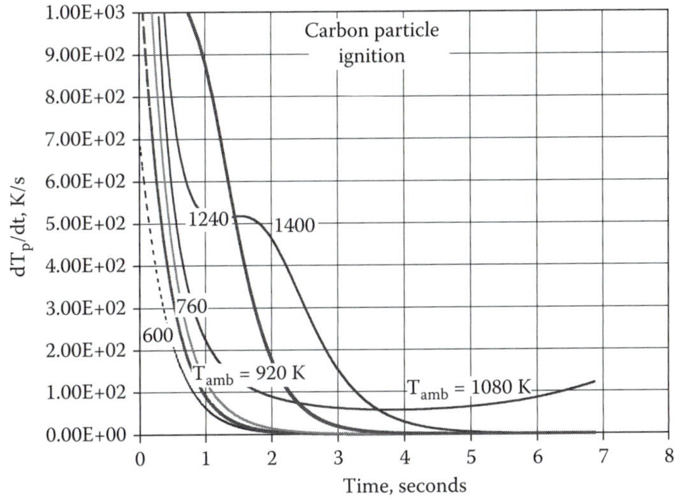

Figure 13.20 dT_p/dt history of carbon particle at various ambient temperatures: radiation loss included.

Using $(h_m d/\rho D) = Sh$, and defining Da_{III} as the third Damkohler number:

$$Da_{III} = \frac{A_{het}\, d_p}{Sh\, D_w} \tag{13.45}$$

Equation 13.37 and Equation 13.44 can be simplified as

$$Y_{O2,w} = \frac{Y_{O2,\infty}}{\left(D_{III}\, \nu_{O2}\, \exp\left(-\frac{E}{RT_p}\right) + 1\right)} \tag{13.46}$$

$$w_c'' = \frac{\dfrac{Da_{III}\, Sh\, Y_{O2,\infty} \rho_w D_w}{d_p}}{\left[Da_{III}\, \nu_{O2} + \exp\left(\frac{E}{RT_p}\right)\right]} \tag{13.47}$$

Note that as $dp \to 0$, $Da_{III} \to 0$, $h_m \to \infty$, and $Y_{O2,w} \to Y_{O2,\infty}$ (Equation 13.46). As $d_p \to$ large, $Da_{III} \to$ large, $Y_{O2,w} \to 0$, and the reaction is controlled by diffusion. The heat generated per unit surface area is given as

$$\dot{Q}''_{gen} = \dot{w}''_c\, h_c \tag{13.48}$$

Using Equation 13.47 in Equation 13.48

$$\dot{Q}''_{gen} = \frac{\left(\frac{Sh\, B \rho_w D_w}{d_p}\right) h_c}{\left\{1 + \left[\dfrac{\exp\left(\frac{E}{RT_p}\right)}{Da_{III,mod}}\right]\right\}} \tag{13.49}$$

where

$$B = Y_{O2,\infty}/\nu_{O2}, \text{ transfer number, and} \tag{13.50}$$

$$Da_{III,mod} = Da_{III}\ \nu_{O2} = [A_{het}\ \nu_{O2}\ d/Sh\ D_w]. \tag{13.51}$$

A plot of $\dot{q}''_{gen}$ vs. T_p is qualitatively similar to Figure 13.15

Let $G_{gen} = B\ \{Sh\ hc/(Nu\ c_p\ E/R)\} = B\ \{hc/(c_p\ E/R)\}$ = heat generated/convective heat loss, (13.52a)

$G_{rad} = \{\varepsilon\ \sigma\ d\ (E/R)^3\}/(Nu\ \lambda)$ = radiative heat loss/convective heat loss, (13.52b)

Equating heat generation with heat loss,

$$G_{gen}/\{1 + (1/Da_{III,mod})\exp(1/\theta_p)\} = (\theta_p - \theta_\infty) + G_{rad}\ (\theta_p^4 - \theta_\infty^4). \tag{13.53}$$

For a given θ_∞, three solutions are obtained iteratively for particle temperature θ_p for $\theta_{\infty,E} < \theta_\infty < \theta_{\infty,I}$, two solutions at $\theta_\infty = \theta_{\infty,I}$, two solutions at $\theta_\infty = \theta_{\infty,E}$, only one solution for $\theta_\infty > \theta_{\infty,I}$, and only one solution also for $\theta_\infty < \theta_{\infty,E}$. If θ_p vs. θ_∞ is plotted at the prescribed G_{gen} and G_{rad}, the S curve will be obtained. A representative S curve for carbon ignition is given in Figure 13.21.

13.5.1.4 Approximate Explicit Solution with Radiation Heat Loss

Rewrite Equation 13.38 as

$$\dot{Q}''_{loss} = \tilde{h}_H\ (T_p - T_\infty) \tag{13.54}$$

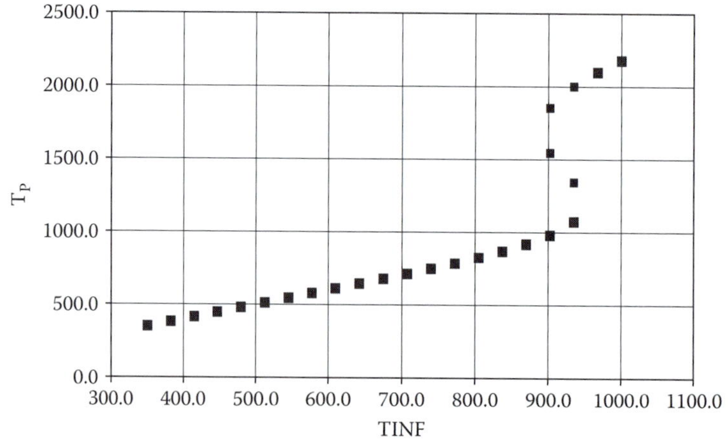

Figure 13.21 S curves for carbon ignition, T_p vs. T_∞.

where

$$\tilde{h} = h_H + h_{rad} \tag{13.55}$$

$$h_{rad} = \frac{\varepsilon\sigma\left(T_p^4 - T_\infty^4\right)}{\Delta T} = \frac{\varepsilon\sigma\left([T_\infty + \Delta T]^4 - T_\infty^4\right)}{\Delta T} \approx \frac{4\,\varepsilon\sigma\frac{\Delta T}{T_\infty}T_\infty^4}{\Delta T} = 4\varepsilon\sigma T_\infty^3 \tag{13.56}$$

Where $\Delta T = T_p - T_\infty$. We will treat h_{rad} as a constant in order to obtain explicit solutions. This is not a bad approximation as long as $T_p - T_\infty$ is very small. Thus, under steady state, setting Equation 13.49 equal to Equation 13.54

$$\dot{Q}_{g''en} = \frac{\left(\frac{Sh\,B\rho_w D_w}{d}\right)h_c}{\left\{1 + \left[\frac{\exp\left(\frac{E}{RT_p}\right)}{Da_{III,\text{mod}}}\right]\right\}} = \dot{q}_{l''oss} = \tilde{h}(T_p - T_\infty) = \left\{\frac{N\tilde{u}\lambda}{d}\right\}\{T_p - T_\infty\} \tag{13.57}$$

$$N\tilde{u} = \frac{\tilde{h}d}{\lambda} = Nu + Nu_{rad} \quad \text{where } Nu_{rad} = \frac{4d_p\varepsilon\sigma T_\infty^3}{\lambda}. \tag{13.58}$$

With $Le = \lambda/\rho Dc_p$ and defining

$$h_c^* = (Sh/N\tilde{u})\ B\ (\bar{R}\,h_c/[E\ Le\ c_p]) \tag{13.59}$$

We can transform Equation 13.57 as

$$\frac{h_c^*}{\left\{1 + \left[\frac{\exp\left(\frac{1}{\theta_p}\right)}{Da_{III,\text{mod}}}\right]\right\}} = \{\theta_p - \theta_\infty\}. \tag{13.60}$$

Differentiating with respect to θ_p (i.e., at ignition and extinction)

$$\frac{\frac{h_c^*\exp\left(\frac{1}{\theta_p}\right)}{Da_{III,\text{mod}}\,\theta_p^2}}{\left\{1 + \left[\frac{\exp\left(\frac{1}{\theta_p}\right)}{Da_{III.\text{mod}}}\right]\right\}^2} = 1. \tag{13.61}$$

Often,

$$\exp(1/\theta_p)/Da_{III,\text{mod}} >>> 1. \tag{13.62}$$

Simplifying Equations 13.60 and 13.61

$$Da_{IV} \exp\left(-\frac{1}{\theta_p}\right) \approx \{\theta_p - \theta_\infty\} \tag{13.63}$$

$$\frac{Da_{IV} \exp\left(-\frac{1}{\theta_p}\right)}{\theta_p^2} \approx 1 \tag{13.64}$$

where

$$Da_{IV} = Da_{III.mod}\ h_c^* = [A_{het}\ \bar{R}\ h_c\ Y_{O2,\infty} d]/[E\ \tilde{Nu}\ Le\ c_p\ D_w] \tag{13.65}$$

The fourth Damkohler number provides a dimensionless parameter to correlate all results, including numerical, approximate results with $Y_{O2,w} \neq Y_{O2,\infty}$, $Y_{O2,w} = Y_{O2,\infty}$, etc. Figure 13.22 correlates experimental char ignition data of several investigators.

Critical ignition conditions can be obtained by solving Equation 13.93 and Equation 13.61 for limiting values of $Da_{III\ mod}$.

The left-hand side of Equation 13.61 represents d (heat generation)/dθ_p, whereas the right-hand side represents d (heat loss)/dθ_p = 1. However, when $Da_{III,mod}$ falls below a certain value (e.g., particle size being extremely small, oxygen concentration very low), Equation 13.61 can no longer be satisfied, i.e., there is no rapid temperature jump, with the consequence that the reaction rate is slowed down and hence becomes kinetic limited. The solution procedure is left as an exercise problem. The maximum slope occurs when d^2 (heat generation)/dθ_p^2 = 0 and at this condition limiting $Da_{III,mod}$ is given by

$$\{Da_{III,mod}\}_{crit} = [\{1/(2\theta_p)\} - 1] \exp(1/\theta_p)/[1 + \{1/(2\theta_p)\}] \tag{13.66}$$

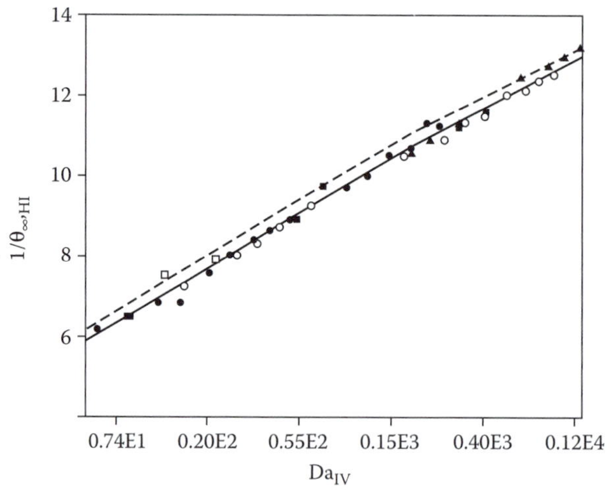

Figure 13.22 Correlation between the fourth Damkohler number for carbon oxidation and experimental data on carbon ignition temperatures. (Adapted from Du, X. and Annamalai, K., Transient ignition of isolated coal particle, *Combust. Flame*, 97, 339–354, 1994.)

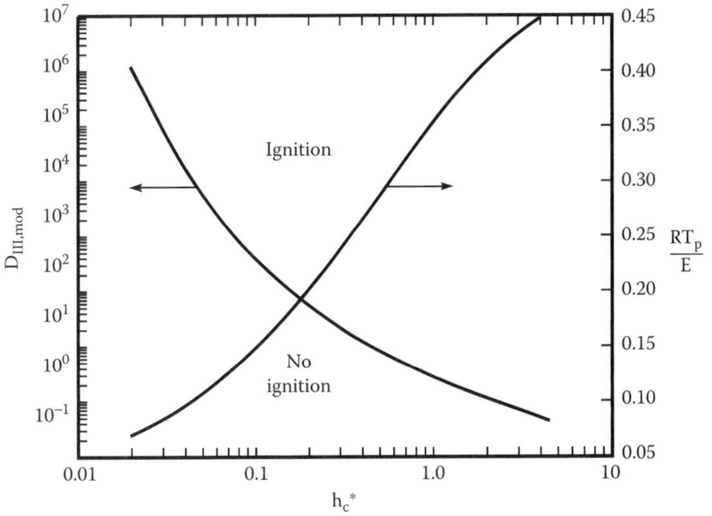

Figure 13.23 Critical condition of particle ignition. (From Annamalai, K. and Ryan, W., Interactive process during gasification and combustion; Part II: isolated coal and porous char particles, Prog. Energy Combust. Sci., 1993.)

where

$$\{1/\theta_p\} = 2\ \{(1/h_c^*) + 1\}^{1/2}. \tag{13.67}$$

Figure 13.23 shows the results. Given the preexponential factor and activation energy, the minimum size and concentration for nonignition can be determined.

In addition, particle size has the following effect on T_∞ at ignition. The heat generation rate per unit area is independent of particle size as long as we are in the kinetic regime (small particles), but heat loss per unit area is extremely high owing to convection because radiation heat loss is not significant. Thus, small particles will have higher ignition temperatures. On the other hand, as d becomes large, radiative heat loss becomes significant and, hence, heat generation per unit area may again not balance the radiation heat loss. Thus, large particles will also have higher ignition temperatures.

13.5.2 Coal Ignition

Coal is partially a pyrolyzing solid in that it contains approximately 40–95% as carbon (FC) (Table A.2B). Sometimes volatiles may be liberated earlier, before carbon oxidation reactions can proceed. If so, volatiles may ignite in the gas phase — this is called *homogeneous ignition* — if the right conditions (temperature and appropriate oxygen-to-fuel ratio) exist in the gas phase. On the other hand, the particle temperature required for pyrolysis may be high and, as such, carbon reactions may proceed first, resulting in *heterogeneous ignition*. The volatiles may be burned *in situ* along with carbon oxidation (base oxidation) or using a heterogeneous (surface oxidation) mechanism. It is now generally agreed that coal particles can ignite either homogeneously, by prior pyrolysis and subsequent ignition of the volatiles, or heterogeneously, by direct oxygen attack on the coal particle [Howard and Essenhigh, 1967a,b; Midkiff et al., 1986 a,b]. The amount of volatile matter, flammability of the volatiles, and transport from the particle determine whether ignition of an isolated coal particle occurs heterogeneously or homogeneously. If coal is ignited homogeneously, the volatiles burn in the gas phase (similar to drop flames) and screen oxygen from the particle surface.

Researchers have speculated for a long time on whether pulverized-coal particles ignite by a homogeneous or heterogeneous mechanism. Secondary ignition of char occurs, but only after pyrolysis is complete. Particularly when the diameter is extremely small, there are insufficient volatiles (amount of volatiles $\propto d^3$) to form a flammable mixture away from the particle surface and, as such, ignition occurs heterogeneously [Annamalai and Durbetaki, 1977]. For these particles, volatiles may be burned with carbon; thus, the heterogeneous reaction removes material that would otherwise be expelled as volatiles. Further, if volatile evolution occurs similar to jets from the pore, then oxygen can still reach the particle surface.

13.5.2.1 Heterogeneous Ignition of Coal

If volatiles burn *in situ* along with carbon, then the expressions deduced earlier can be used for heterogeneous ignition. If VM = 40% and FC = 60%, then for every kg of FC burned, the volatiles burned can be assumed to be 0.67 times the combustion of fixed carbon. Thus the burn rate of coal = the burn rate of carbon × {(VM + FC)/FC} = {1/FC} as VM + FC = 1 when the values are on DAF basis.

$$\dot{w}''_{coal}\ (\text{kg/m}^2\ \text{sec}) = \{k\ \rho Y_{O2,w}\}\{1/FC\} \qquad \text{(A)}$$

One can modify the heating value to include volatiles or h_c by $h_{c,coal}$ of coal and ν_{O2} by the stoichiometric coefficient for coal combustion.

13.5.3 Ignition of Plastics

Most plastics are fully pyrolyzing; thus, the monomers released from the polymer undergo homogeneous ignition. If data on ignition temperature of monomers vs. fuel-to-air ratio is available, one can formulate the governing conservation equations for a spherical particle or for flow situations (e.g., stagnation flows), obtain the monomer concentration vs. temperature in the gas phase as though the gas phase is frozen, and then, match with the data to determine the ignition condition and ignition time. Details of such a procedure are presented by Annamalai and Durbetaki [1976]. Flammability characteristics of plastics are characterized by required ignition temperature and ignition times; the lower the ignition time, the higher the flammability. If a hot ambience is available, the plastics ignite as soon they reach pyrolysis

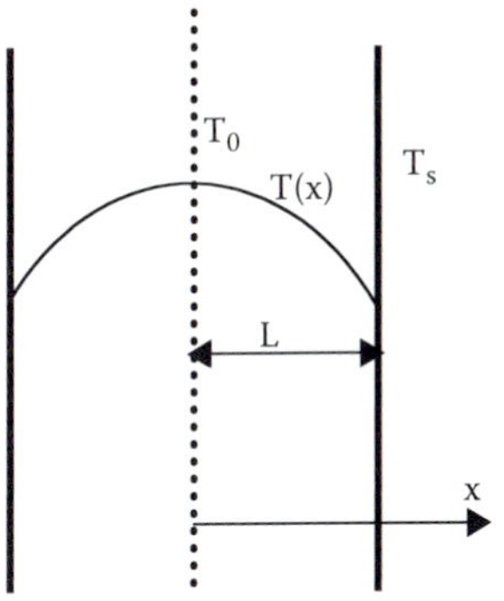

Figure 13.24 FK ignition for slab.

temperature. Hence, time required to ignite can be estimated using heat transfer literature by estimating the time required to reach pyrolysis temperature [Annamalai and Sibulkin, 1978].

13.6 Ignition of Nonuniform Temperature Systems— Steady-State Solutions

Recall that previous sections assumed T to be uniform. However, for large spherical particles, porous char, and large piles of coal particles, the temperature is nonuniform. Thus, chemical reactions occur through a region of varying temperature gradients. Frank-Kamenetski (FK) presented algebraic relations to estimate the required ignition conditions for a slab (thickness 2L; Figure 13.24), and gave numerical results for the ignition of a cylinder (radius R; Figure 13.25), and sphere (radius R; Figure 13.26) containing a combustible gas mixture with the surface temperature maintained at T_S.

13.6.1 Slab

Consider a slab of thickness 2L (Figure 13.24) with the surface at T_s. Let us assume that the slab contains a combustible mixture of gas and air. The heat

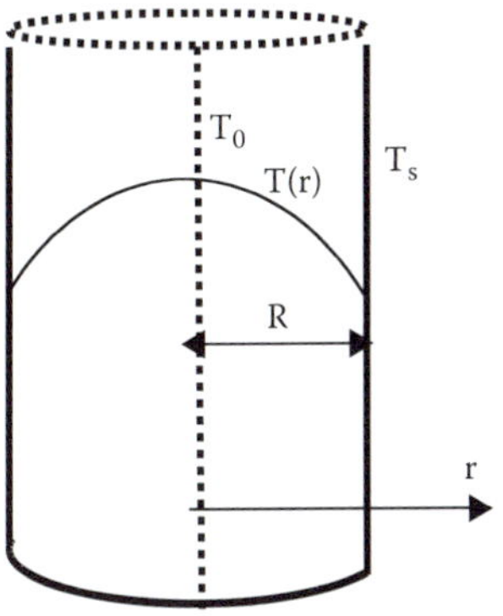

Figure 13.25 FK ignition for cylinder.

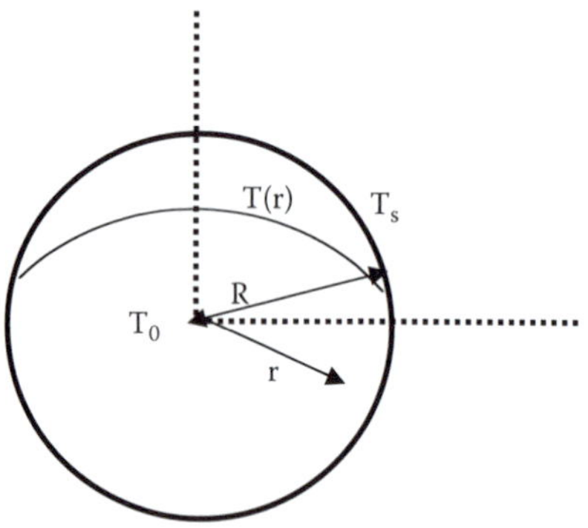

Figure 13.26 FK ignition for sphere.

generated is only a function of T, and there is no flow within the slab. The energy conservation equation is written as,

$$\lambda \frac{d^2T}{dx^2} = -\dot{q}''' \tag{13.68}$$

where $\dot{q}'''$ is heat liberated per unit volume due to chemical reaction. Assuming a global reaction,

$$\dot{q}''' = h_c A_Y Y_F^{nF} Y_{O_2}^{nO2} \rho^{nF+nO2} \exp(-E/\bar{R}\ T) = h_c A_1 \exp(-E/\bar{R}\ T) \tag{13.69}$$

where

$$A_1 = A_Y Y_F^{nF} Y_{O_2}^{nO2} \rho^{nF+nO2} \tag{13.70}$$

The boundary conditions (BCs) are as follows:

$$\frac{dT}{dx} = 0 \quad at\ x = 0 \tag{13.71}$$

$$T = T_s \quad \text{at } x = L \tag{13.72}$$

Thus, Equation 13.68 can be solved subject to two BCs given by Equation 13.71 and Equation 13.72.

13.6.1.1 Physical Processes

The solution behavior could be described as follows. Since dT/dx = 0 at x = 0, heat loss occurs only at x = L. Assume that the heat loss between the

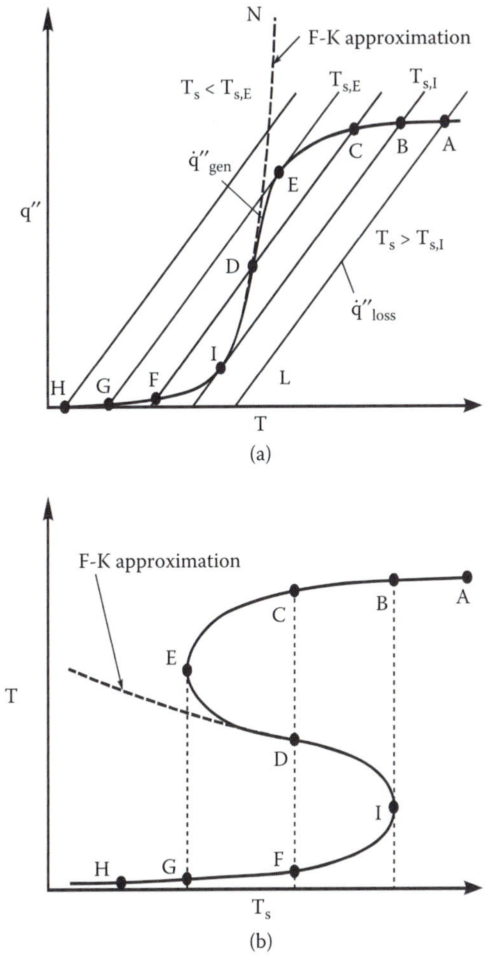

Figure 13.27 The FK approximation for Q_{gen}, Q_{loss} curves and S curves.

central axis at T_0 and the surface is proportional to $(T_0 - T_s)$. Let us assume the whole slab is at $T_{avg} = \frac{T_0+T_s}{2}$. The heat generation rate is proportional to $e^{-E/RT_{avg}} \approx e^{-E/RT_s}$, but still $T_0 \approx T_s + \Delta T$ where $\Delta T \ll T_s$. If $T = T_s$ in Figure 13.27a, $\dot{q}_{gen}$ is similar to HGFIDECBA. Similarly, assuming heat loss is proportional to $T_0 - T_s$, then, the heat loss curve has one stable solution (rapid reaction) point (A), two solutions (semistable and stable) at ignition (B, I), and three solutions (F, stable-slow reaction; D, unstable; and C, stable-rapid reaction) for high heat losses, two solutions at extinction (E, semistable extinction; G, stable-slow reaction), and one solution (stable, almost frozen) beyond extinction (H). In order to get an explicit solution, a few approximations are made for $\dot{q}_{gen}$ as given in the following subsections.

13.6.1.2 Normalized Governing Equations

Let

$$\left.\begin{aligned} \theta &= \frac{T}{(E/\bar{R})}, \\ \eta &= \frac{x}{L}, \end{aligned}\right\} \tag{13.73}$$

$$Da_{IV} = \frac{\bar{R}\ A_1\ h_c\ L^2}{\lambda\ E} = \frac{\text{heat generated}}{\text{heat loss by conduction}} \tag{13.74}$$

where A_1 is defined by Equation 13.70. Then Equation 13.68 becomes,

$$\frac{d^2\theta}{d\eta^2} = -Da_{IV}\exp\left(-\frac{1}{\theta}\right). \tag{13.75}$$

The BCs become

$$\frac{d\theta}{d\eta} = 0 \quad \text{at} \quad \eta = 0 \tag{13.76}$$

$$\theta = \theta_s \quad \text{at} \quad \eta = 1. \tag{13.77}$$

Equation 13.76 is difficult to solve explicitly with prescribed boundary conditions. FK assumed that the temperature T anywhere in the system is only slightly different from the surface temperature T_s. In situations when T (or θ) is not much different from T_s or θ_s, $(\theta - \theta_s)$ is very small. Rewrite

$$\frac{1}{\theta} = \frac{1}{\theta - \theta_s + \theta_s} = \frac{1}{\theta_s\left[1 + \frac{\theta - \theta_s}{\theta_s}\right]} = \frac{1}{\theta_s}\left[1 + \frac{\theta - \theta_s}{\theta_s}\right]^{-1} = \frac{1}{\theta_s} - \frac{\theta - \theta_s}{\theta_s^2} \tag{13.78}$$

$$Da_s = \left(\frac{Da_{IV}}{\theta_s^2}\right)\exp\left(-\frac{1}{\theta_s}\right). \tag{13.79}$$

Let

$$\psi = \frac{\theta - \theta_s}{\theta_s^2}. \tag{13.80}$$

Then Equations 13.75 to 13.77 become

$$\frac{d^2\psi}{d\eta^2} = -Da_s\, e^{\psi} \tag{13.81}$$

$$\left.\begin{array}{ll} \eta = 0 & \dfrac{d\psi}{d\eta} = 0 \\ \eta = 1, & \psi = 0 \end{array}\right\}. \tag{13.82}$$

13.6.1.3 Solution for Ignition

(a) If $Da_s = 0$, the reaction is frozen. There is no ignition.

(b) If Ψ is extremly small $e^{\psi} \approx 1$; then $\psi = Da_s/2\ (1 - n)$; $\psi_0 = Da_s/2$

(c) $e^{\psi} = 1 + \psi$; $\psi^2 = \{\cos\sqrt{Da_s}\ \eta\}/\{\sin\sqrt{Da_s} - 1\}$ [Strahle, 1993]; when $\sqrt{Da_s} = \pi/2$, $\psi \to \infty$

We will retain e^{ψ} and obtain the solution. Prior to presenting the solution, the consequence of the FK approximation needs to be discussed. Figure 13.27a shows the curve HGFIDECBA, which is the exact curve for $\dot{Q}_{gen}$ vs. T. As FK assumed that T is close to T_s (around the ignition point), the trend of curve HGFID is extended to N to get HGFIDN with the FK approximation. A consequence is the loss of the third stable high-temperature part of the solution (Figure 13.27b). Thus, for the heat loss curve LA there is no solution with the FK approximation. For IB there is one solution. For FDC there are two solutions only (points D and F). We are interested in finding the critical heat loss condition below which there is no steady solution. For curve IB, the slab can just ignite.

It is apparent that the use of the FK approximation gives only two QS solutions (stable and unstable) for $T_s < T_{s,I}$, one solution for $T_s = T_{s,I}$, and no QS solution at $T_s > T_{s,I}$. The literature mentions steady solution $\psi\ (\eta)$ profiles for various conditions (e.g., various T_S) and then looks for the minimum $T_s = T_{s,I}$ above which there is no steady solution. Thus, one is interested in $\psi\ (\eta)$ around the ignition point. Equation 13.81 could be integrated with two BCs (see Equation 13.82) (See exercise). Using the $d\psi/d\eta = 0$ at $\eta = 0$, the solution is given as

$$e^{\psi} = \frac{a}{\cos h^2 \left[\sqrt{\frac{a\ Da_s}{2}}\ \eta\right]} \tag{13.83a}$$

where, "a" is an arbitrary constant. Using the second BC in Equation 13.82, $\eta = 1$, $\psi = \psi_s = 0$, we get

$$\cos h^2 \sqrt{\frac{a\ Da_s}{2}} = a \tag{13.83b}$$

which is a transcendental equation for "a."

13.6.1.3.1 Solution for Da_{IV} at Prescribed θ_s

Let

$$\sigma = \sqrt{\frac{a\ Da_s}{2}}. \tag{13.84}$$

Using Equation 13.84, rewrite Equation 13.83b as

$$\frac{\cos h\ \sigma}{\sigma} = \sqrt{\frac{2}{Da_s}}. \tag{13.85}$$

A plot of {cos h σ/σ} vs. σ is given in Figure 13.28. It shows a minimum for σ such that

$$\sigma \tan h\ \sigma = 1 \tag{13.86}$$

and solving Equation 13.86, σ = 1.1997 and the corresponding {cos h σ/σ} = 1.5089; using this result in Equation 13.85

$$Da_s = 2\left(\frac{\sigma}{\cos\ h\ \sigma}\right)^2 = 0.8785 \tag{13.87}$$

$$a = 2\ \sigma^2/D_s = 2*1.22^2/0.878 = 3.2767. \tag{13.88}$$

Da_s = 0.8785 is called the *critical condition.* For Da_S < 0.8785, there appear to be two solutions (Figure 13.28). Thus, spontaneous ignition occurs for any $Da_s \geq 0.8785$. The corresponding centerline temperature is given by Equation 13.82 with $\eta = 0$.

$$e^{\psi_0} = \frac{a}{\cos h^2\left(\sqrt{\frac{a\ Da_s}{2}}\cdot 0\right)} = a = 2\ \frac{\sigma^2}{Da_s} = 3.2767 \tag{13.89}$$

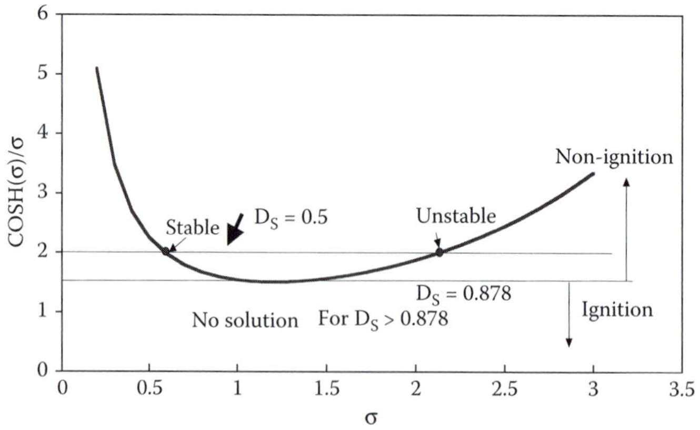

Figure 13.28 Combustion of slab: stable solution for combustion condition.

and hence solving for ψ_0

$$\psi_0 = \frac{\theta_0 - \theta_s}{\theta_s^2} = 1.1868. \tag{13.90}$$

13.6.1.3.2 Solution for $\theta_{s,I}$ at Prescribed Da_{IV}

For a prescribed pile geometry, Da_{IV} is fixed. Thus, one can specify a surface temperature θ_S and calculate Da_S using Equation 13.79, which may yield $Da_S <$ 0.8785. If so, the pile will not ignite or steady solution (corresponding to points H, G, F, etc.) for the nonigniting case will be obtained. When θ_S is gradually increased, Da_S increases and approaches the value of $Da_{S,Crit}$ = 0.8785 at which ignition occurs. The corresponding $\theta_{S,I}$ is called the *surface temperature* for ignition of the pile having slab geometry. It is evaluated using Equation 13.79 and rewriting

$$\ln[Da_{S,crit}] = \ln Da_{IV} - \frac{1}{\theta_S} - 2\ln\theta_s \quad \textit{where } Da_{s.crit} = 0.8785. \tag{13.91}$$

Approximating

$$(1/\theta_s) \approx \ln\,[Da_{IV}/Da_{S,crit}]. \tag{13.92}$$

Equation 13.92 and Equation 13.28 look very similar. Alternately, for a prescribed θ_s, one may get $Da_{IV\ crit}$ and L_{crit} from Equation 13.74, above which ignition always occurs.

Example 6

Consider a slab of thickness 2L = 3.4 m. It contains a combustible gas mixture of fuel F and oxidizer O_2.
E = 120000 kJ/kmol, h_c = 9000 kJ/kg, and λ = 5.00E 05 kW/m K. Assume A_1 = 10^5 {kg/(m^{3}s)}

(a) If T_s = 475 K, determine two values for σ and centerline temperature.
(b) What is the surface temperature T_s above which ignition always occurs?

Solution

(a) $E/\bar{R}$ = 120000/8.314 = 14433 K

$$(Da_{IV}) = \frac{\bar{R}\ A_1\ h_c\ L^2}{\lambda\ E} = \frac{3.6\times10^9\ \frac{kg}{m^3\ s}\ 9000\frac{kJ}{kg}\ 1.7m^2}{5\times10^{-5}\ \frac{kJ}{s\ m\ K}\ 14433\ K} = 3.6\times10^9$$

$$\theta_S = 475/14433 = 0.0345$$

With Equation 11.85, $Da_{IV} = 3.6\times10^9$, $\theta_s = 0.0346$, solutions are given in Table 13.1.

Table 13.1 Solutions for Given Da_{IV} (= 3.6 × 10^9), at Prescribed T_s

	σ	θ_s	ψ_0	θ_0	$T_{s,}$ K	T_0K
Solution 1	0.3450	0.03291	0.1178	0.03304	475	477
Solution 2	1.9495	0.03291	4.3518	0.03762	475	543

The solution 2 is unstable; as the pile heats up from 475 K, it reaches 477 K at the center; then it stops heating. Hence it cannot self-ignite at T_s = 475 K.

(b) For part (b) we need to determine the T_s above which the pile will always self-ignite. For this case, Da_{scrit} = 0.8785

$$Da_s = 0.8785 = Da_{IV} \cdot \frac{1}{\theta_s^2} \exp\left(-\frac{1}{\theta_s}\right)$$

Knowing Da_s and Da_{IV}, find θ_s as 0.0346 and T_s = 14433 × 0.0346 = 499 K. At that critical point $\psi_0 = 1.186 = \{\theta_0 - \theta_s\}/\theta_s^2$, T_0 = 520 K.

The *critical solutions* for ignition conditions are summarized in Table 13.2.

Remarks

For a given T_s, one can find a critical slab thickness 2L above which ignition always occurs.

13.6.1.3.3 Required Ambient $T_{\infty,I}$ at the Time of Ignition

Let us subject the pile to a free-stream ambient temperature of T_∞. The gas transfers heat to the pile, raises the surface temperature, and at a certain T_∞, the pile ignites. It is noted that $\theta_{S,I}$ for a given Da_{IV} will remain the same. Using the interface energy balance, the required $\theta_{\infty,I}$ is determined. At x = L

$$-[\lambda dT/dx]_{x=L} = h\,(T_s - T_\infty) \tag{13.93}$$

which transforms to

$$-\{d\,\psi/d\eta\}_{\eta=1} = Bi\,\{\theta_s - \theta_\infty)/\theta_s^2 \tag{13.94}$$

$$\{d\,\psi/d\eta\}_{\eta=1} = -Bi\,\psi_\infty \tag{13.95}$$

It can be shown that Bi $\psi_\infty = 2$.

13.6.2 Generalized Geometry

The general form of the QS energy conservation equation can be written as

$$d^2T/dx^2 + (s/x)\,dT/dx = -\dot{q}''' \tag{13.96}$$

Table 13.2 Solutions for Ignition Conditions

s	θ_s	ψ_0	θ_0	$T_{S,I}$K	T_0 K
1.1997	0.03457	1.1869	0.03599	499	520

Table 13.3 FK Ignition Conditions for Slab, Cylinder, and Sphere

Geometry	$Da_{s,crit}$	$\psi_{0,crit}$	$Bi\ \psi_\infty/2$	Da_{IV}
Slab	0.88	1.2	1	$\frac{\bar{R}\,A_1 b_c\,L^2}{\lambda\,E}$
Cylinder	2.0	1.39 = (ln 4)	1	$\frac{\bar{R}\,A_1\,b_C\,R^2}{\lambda\,E}$.
Sphere	3.32	1.61	1	$\frac{\bar{R}\,A_1\,b_C\,R^2}{\lambda\,E}$.

Note: $\left[\frac{1}{\theta_s}\right] = \ln\left[\frac{Da_{IV}}{Da_{s,crit}\,\theta_s^2}\right]$.

Source: Adapted from Annamalai, K., Ryan, W., and Dhanapalan, S., Interactive processes in gasification and combustion — Part III: coal/char particle arrays, streams, and clouds, *Prog. Energy Combust. Sci.*, Vol. 19, 383–446. 1995.

where $\dot{q}''' = \dot{w}'''\,h_c$, s = 0, 1, and 2 for a slab, cylinder, and sphere (Figure 13.24 to Figure 13.26), respectively, [Annamalai, et al., 1995]. As before, Equation 13.96 is solved subject to the condition that dT/dx = 0 at x = 0 and T = T_s at x = L. The energy conservation equation (Equation 13.134) can be written as

$$d^2\psi/d\eta^2 + (s/\eta)\,d\psi/d\eta = -Da_{IV}\exp(\psi) \qquad (13.97)$$

where Da_{IV} is the same as in Equation 13.74 with R, the radius of the cylinder and sphere replacing the term L for a slab, with the boundary conditions $d\psi/d\eta = 0$ at $\eta = 0$ and $\psi = \psi_0$ at $\eta = 0$. The exact solution for ψ is given in Carslaw and Jaeger, and Buckmaster and Ludford for the slab (s = 0), whereas Chambre presents the exact solution for a cylinder (s = 1) and tabulates values for spherical geometries. The solutions for each geometry at the critical point (at ignition) are given in Table 13.3. Figure B.6 in the Appendix presents the results for θ_s and θ_0 for slab, cylinder, and sphere.

For slab (s = 0)

$$\psi = \psi_0 - 2\ln\left[\cos h\left\{\eta\left(\frac{Da_s}{2}\exp(\psi_0)\right)^{1/2}\right\}\right]. \qquad (13.98)$$

For cylinder (s = 1) the solution is given as [Annamalai et al., 1995; also see exercise problem]

$$\psi = 2\ln\left[\frac{c_1+1}{c_1+\eta^2}\right];\quad c_1 = \left[\frac{4}{Da_s}-1\right] + \left[\left\{\frac{4}{Da_s}-1\right\}^2 - 1\right]^{1/2},\quad Da_s < 2 \qquad (13.99)$$

where η = r/R.

13.6.3 Some Applications

The previous results could be applied to coal piles, cylindrical streams, grain (biomass) storage, and ignition of combustible mixture around spark plugs.

Thus, large piles can self-ignite at low temperatures if heat is not removed; i.e., if a cloud of particles is stored in a larger vessel, either in a suspended state or as a pile. For example, grain elevators store biomass fuels, agricultural feedlots store animal waste, and power plants store coal; thus, ignition hazards can occur if not properly ventilated. The bureau of mines evaluated the self heating tendencies of bituminous coals collected from coal seams using adiabatic ovens and determined minimum initial temperature (called self heating temperature, SHT) that sustained the exothermic reactions [Smith, AC, 1989]. They are correlated with DAF oxygen percent in fuel:

SHT min in C = 139.6-6.6* percent of DAF oxygen, correlation coefficient: 0.934. HVCb: 35-70 C, HVAb: 65-100 C, higher rank, medium volatile, and low volatile: 100 C; they classified coals to be high combustion potential if SHT < 70 C, medium: 70-100 C and low: > 100 C

Consider a porous coal particle having a radius of R. The internal surface area per unit volume is given as $S_v = S/V$ and is called *BET area.* If oxygen diffuses throughout the particle with uniform concentration, then the present analysis can be used to determine the required surface temperature for ignition and the center temperature.

13.6.4 Biological Systems

We have seen that at extremely low T_∞ there are some steady (stable) solutions at low temperatures. Let us consider one such solution (e.g., point H in Figure 13.27). At this point

$$\dot{Q}_{gen} = \dot{Q}_{loss}.$$

As

$\dot{q}_{gen} = \dot{q}'''_{gen}$ V and $\dot{q}_{loss} = h\ S\ (T - T_\infty)$, then equality implies that

$$\dot{q}'''_{gen} = h\ S\ (T - T_\infty)/V. \tag{13.100}$$

For systems of characteristic size, V/S = d

$$\dot{q}'''_{gen} = h_H\ (T - T_\infty)/(V/S) = h_H\ (T - T_\infty)/d. \tag{13.101}$$

Thus, the smaller the size, the higher the required heat generation, if the temperature is to be maintained at T above T_∞. Many times $\dot{Q}_{gen}$ can be estimated from the knowledge of breathing rate and nasal exhaust gas composition [Annamali and Morris, 1997]. Typically, however, steady-state solutions are at the lower end. For biological systems, $\dot{q}'''_{gen}$ is called the *metabolic rate per unit volume.* Many times $\dot{Q}_{gen}$ can be estimated from the knowledge of breathing rate and nasal exhaust gas composition [Annamalai and Morris, 1997.] Typically, for many warm blooded animals, T = 310 K (98.4 F, 37°C). Hence, a dinosaur, having a large d, will have a lower metabolic rate per unit volume compared to an ant. So, each cell in the body of an ant has to achieve a higher rate of chemical reaction compared to the body of a dinosaur. The increased generation rate for an ant may result in a short life span for the cells and hence for the ant.

13.7 Summary

The thermal theory of ignition was presented for predicting ignition temperatures of gaseous and solid fuels. It was shown that the fourth Damkohler number plays a significant role in the steady-state ignition of uniform systems (with no temperature gradients within the system, e.g., solid-fuel particle) such as particles, combustible mixtures and nonuniform systems (with temperature gradients, e.g., solid-coal piles) such as slab, cylinder, and spherical geometries, containing combustibles. If heat loss and heat generation are well defined, the ignition temperature can be estimated using simple correlations.

Chapter 14

Deflagration and Detonation

14.1 Introduction

In Chapter 2, Chapter 3, and Chapter 7, we studied combustion in open systems. As the fuel is burned in a constant-diameter combustion chamber, the temperature increases, density decreases, and hence velocity increases. A change in density owing to combustion should cause a slight change in pressure in an open system and significant change in temperature, which in turn causes a velocity change. This will be discussed in this chapter.

Consider an automobile engine. If a gasoline–air mixture is ignited near the spark plug, a flame propagates into the unburned mixture. Here, the system is closed and the gas pressure changes with time.

Now consider a quartz tube that is closed at one end but open at the other (Figure 14.1a). We quickly charge the tube with premixed fuel and air and ignite it at the closed end. The flame propagates toward the unburned mixture. We use a digital camera to record the movement of the flame. Thus, we can measure the velocity of propagation into the unburned mixture as v_0. Suppose the density of burned gases is the same as that of unburned gases. Then, the velocity of gases in the burned zone with respect to a stationary observer will be zero. However, with respect to a point of reference on the flame front, the unburned mixture appears to travel toward the point with velocity v_0 (similar to a hypothetical situation involving an observer sitting on the wing of an aircraft feeling the air rushing toward him or her), and burned gases leave the point with velocity $v_\infty = v_0$ as $\rho_\infty = \rho_0$. Now, let the density of burned gases be lower than the density of unburned gases. The burned gases exit with velocity $v_\infty > v_0$. The problem can be analyzed if the flame is stationary and the unburned gases move into the flame at velocity v_0 and the products leave at velocity v_∞. There is a velocity variation of velocity within a flame thickness of δ (Figure 14.1).

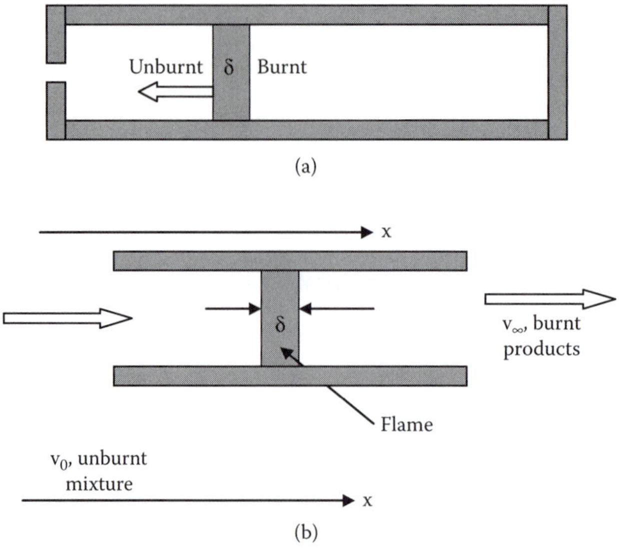

Figure 14.1 Schematic of flame propagation (a) propagating flame (flame moves to the left into unburned mixture), (b) stationary flame.

The molecules are closely spaced when they enter the flame front (where velocity is v_0), whereas they are spaced farther apart when they leave it (where the velocity is v_∞). Consider another case. If the cars are going on an expressway enter a town at a low speed, you will see that they are bunched together; if you find that the cars are spaced farther apart as they leave the town, then they must have speeded up. If the cars behind you move at faster rate (say, twice your car velocity), they will catch up with you. Such a situation may occur in gases also. The distance between the molecules is, say, ℓ; then the time required to space them apart by another ℓ will be ℓ/c, where c is the speed of the molecule (called *speed of sound*). Similarly, if matter at v_0 moves at a speed faster than sound ($v_0 > c$, i.e., the leading edge travels at supersonic velocity), then it covers distance ℓ within a time of ℓ/v_0 whereas molecules require a time of ℓ/c, with the consequence that the molecules get accumulated, i.e., the burned gases just behind the flame front get packed, with pressure and density increasing. Now time required for spacing apart is reduced and, finally, equilibrium sets in. So sometimes the density of burned gases could be higher; this case is called *detonation*. For example, consider a shuttle entering the atmosphere. If the speed of the shuttle $v > c$, the air frame moves at v; thus, air molecules close to the frame are pushed at v. However, the molecules disperse themselves by moving at maximum random (Chapter 1) velocity $c < v$. Thus, molecules get packed as they move at slower speed, resulting in a rapid pressure rise or shock front about 0.6 m ahead of the shuttle.

This chapter deals with the limiting results for v_0 without considering the chemistry in detail but using only momentum, mass, and energy balance equations.

14.2 Conservation Equations

For engineering applications, assume the following conditions for analysis: 1-D, adiabatic, no momentum loss, steady state, plane wave, and constant c_P. We use wave-fixed coordinates; i.e., we imagine ourselves sitting on top of the wave; we would then observe the unburned mixture enter ahead of us and burned products leave behind us (Figure 14.1b).

14.2.1 Mass

$$d(\rho v)/dx = 0 \tag{14.1}$$

After integration,

$$\rho_0 v_0 = \dot{m}''_0 = \rho_\infty v_\infty = \dot{m}''_\infty = \dot{m}'', \text{ a constant} \tag{14.2}$$

14.2.2 Momentum

The 1-D momentum balance equation (Chapter 8) reduces to

$$\rho v\ dv/dx = -dp/dx \tag{14.3}$$

Using Equation 14.2 for ρv and integrating Equation 14.3 for the regions extending from 0 to ∞, we get

$$v_\infty - v_0 = (p_0 - p_\infty)/\dot{m}'' \tag{14.4}$$

Because $\rho_0 v_0 = \rho_\infty v_\infty = m''$ Equation 14.4 becomes

$$\rho_0 v_0^2 + p_0 = \rho_\infty v_\infty^2 + p_\infty \tag{14.5}$$

Note that if we know v_0, p_0, and p_∞ we can determine v_∞.

14.2.3 Energy

Using the first law of thermodynamics for an adiabatic open system,

$$h_0 + v_0^2/2 = h_\infty + v_\infty^2/2 \tag{14.6}$$

$$h_0 - h_\infty = \sum (h_{f,k} + h_{T,k,0})\, Y_{k,0} - \sum (h_{f,k} + h_{T,k,\infty})\, Y_{k,\infty} \tag{14.7}$$

$$h_0 - h_\infty = \sum h_{f,k}\,(Y_{k,0} - Y_{k,\infty}) + (c_P\{T_0 - T_\infty\}) = q + (c_P\{T_0 - T_\infty\}) \tag{14.8}$$

where $q = \Sigma\, h_{f,k}\,(Y_{k,0} - Y_{k,4})$, the heat released per unit mass of mixture,

$$q = h_c\, Y_{F,0} \quad \text{(for lean to stoichiometric mixtures)} \tag{14.9}$$

$$q = h_{c,O2}(Y_{O2,0}/\nu_{O2}) \quad \text{(for stoichiometric to rich mixtures)} \tag{14.10}$$

Note that if chemical equilibrium is assumed, the composition of products and heat released, q, may not be same as that estimated for complete combustion [see Soloukhin, 1966]. Equation 14.6 becomes

$$c_P\,(T_\infty - T_0) = q - [v_\infty^2/2 - v_0^2/2] \qquad (14.11)$$

Note that if $v_\infty > v_0$, then the sensible or thermal enthalpy available for raising the temperature of gases is reduced because some energy is used in raising the kinetic energy. Once T_∞, q, and v_0 are known, then v_∞ can be solved from Equation 14.11 and, hence, p_∞ can be solved either from Equation 14.4.

Example 1

Consider a combustion-premixed stoichiometric mixture of C_2H_2 and air. The unburned gas mixture velocity is 1.410 m/sec; LHV = 48200 kJ/kg.

(a) What is the molecular weight of the input mixture and the products? Assume inlet temperature and pressure to be 298 K and 100 kPa.
(b) Ignoring velocity effects, calculate the adiabatic flame temperature and exit pressure. Assume c_p for N_2 at a mean temperature of 1500 K.
(c) If you include velocity effects, what will be the temperature and pressure on the burned side? Assume that the molecular weights of the unburned mixture are the same as those of the burned products.

Solution

(a) For the stoichiometric mixture, $C_2H_2 + 2.5\,(O_2 + 3.76\,N_2) \rightarrow 2\,CO_2 + 1\,H_2O + 9.4\,N_2$
Input mass per kmol of C_2H_2 = [26.05 + 2.5*32 + 9.4*28.02] = 369.4 kg
Input moles per kmol of C_2H_2 = 1 + 2.5 + 9.4 = 12.9
$M_{mix,in}$ = 369.4/12.9 = 28.64 kg/kmol of mix, Y_{C2H2} = 26.05/369.4 = 0.0705, Y_{O2} = 0.217, Y_{N2} = 0.713
Product moles per kmol of C_2H_2 = 2 + 1 + 9.4 = 12.4 kmol
$M_{mix,out}$ = 369.4/13.4 = 29.7 kg/kmol of prod
$\rho_0 = P_0/RT_0$ = 100/{(8.314/28.64)*298}= 1.16 kg/m^3, $\dot{m}''$ = 11.69 kg/m^2 sec
q = 48200 kJ/kg of fuel * 0.0705 kg of fuel/kg of mix = 3398 kJ/kg mix.
T_{mean} = 1500 K. Selecting c_p at 1500 K, c_p = 1.25 kJ/kg K (from Table A.8)
(b) From Equation 14.11 and ignoring kinetic energy

$$c_P(T_\infty - T_0) = q - [v_\infty^2/2 - v_0^2/2] \simeq q$$

1.25 * (T_∞ − 298) = 3398, T_∞ = 3016.4 K
If we ignore velocity effects, Equation 14.5 states that $p_\infty = p_0$
(c) If we include velocity effects Equation 14.11 yields
$c_P\,(T_\infty - T_0) = q - [v_0^2/2]\,[\rho_0^2/\rho_\infty^2 - 1] = q - [v_0^2/2]\,[T_\infty^2/T_0^2 - 1]$, which is a quadratic equation in T_∞; solving, T_∞ = 3016.3 K.
From the momentum equation (Equation 14.5)
$p_0 - p_\infty = \rho_\infty v_\infty^2 - \rho_0 v_0^2 = \rho_0^2 v_0^2\,\{1/\rho_\infty - 1/\rho_0\} = \{\rho_0^2 v_0^2/\rho_0\}\{T_\infty/T_0 - 1\}$ = {1.69^2/1.16}{3016/298 − 1} = 21.7 Pa, p_∞ = 100 − 21/1000 = 99.98 kPa. Thus, p is approximately constant.

Table 14.1 Detonation Limits—Mixture

Mixture	*Lean Percentage*	*Rich Percentage*
$CO–O_2$	38	90
$C_3H_8–O_2$	3	37
$H_2–O_2$	15	90
H_2–air	18	59
$NH_3–O_2$	25	75

Source: Lewis, B. and Von Elbe, G., *Combustion, Flames, and Explosion of Gases,* 2nd ed., Academic Press, New York, 1961.

14.2.4 The Equation of State

$$p = \rho RT \tag{14.12}$$

Typically ρ_0, p_0, $\gamma\,(=c_P/c_V)$, v_0, and q are known; the unknowns are v_∞, ρ_∞, p_∞, and T_∞. Using the four equations (Equation 14.2, Equation 14.4, Equation 14.11, and Equation 14.12) we can solve for all unknowns.

14.3 Solutions for Rayleigh and Hugoniot Curves

14.3.1 Rayleigh Lines

With $\rho_0 v_0 = \dot{m}'' = \rho_\infty v_\infty$, one can replace v_0 and v_∞ in Equation 14.4

$$p_\infty - p_0 = -\ \dot{m}''^2\ \{1/\rho_0 - 1/\rho_\infty\} \tag{14.13}$$

On the other hand, if we replace $\dot{m}''$ in terms of v_0 and v_∞, we get,

$$(v_\infty - v_0)^2 = (p_\infty - p_0)\ (1/\rho_0 - 1/\rho_\infty) \tag{14.14}$$

If $v_0 = 0$ (stationary mixture), then from Equation 14.14

$$v_\infty = \{(p_\infty - p_0)\ (1/\rho_0 - 1/\rho_\infty)\}^{1/2} \tag{14.15}$$

i.e., when the combustion wave spreads into the static combustible mixture, Equation 14.15 represents the velocity of the gases after combustion. Note that $v_\infty > 0$. Hence, if $\rho_\infty < \rho_0$ then $p_\infty < 0$ (deflagration). If $\rho_\infty > \rho_0$ then $p_\infty > p_0$ (detonation), so that $v_\infty > 0$. If $\rho_\infty > \rho_0$, $v_\infty < v_0$ and $p_\infty > p_0$ (detonation: molecules are piled up after combustion because waves move faster than the speed with which molecules can be dispersed, resulting in increased pressure and temperature). Suppose p_0 and ρ_0 are given, we can fix point I in Figure 14.2, which is a plot of p at a given $\dot{m}''$ for various assumed values of $1/\rho_\infty$. These lines are called constant-momentum or Rayleigh lines. These lines satisfy only the continuity and momentum balance equations; I: initial point (ρ_0, p_0), F_1:

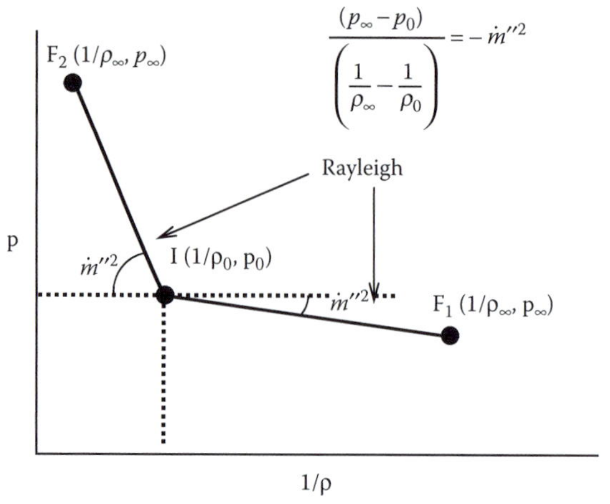

Figure 14.2 Rayleigh lines.

final point for deflagration ($\rho_\infty < \rho_0$ or $1/\rho_\infty > 1/\rho_0$), and F_2: final point for detonation ($\rho_\infty > \rho_0$ or $1/\rho_\infty < 1/\rho_0$).

14.3.2 Hugoniot Curves

Using the mass/continuity equation in Equation 14.11 to eliminate v's (i.e., $\rho_0 v_0 = \rho_\infty v_\infty = \dot{m}''$), we have,

$$c_p(T_\infty - T_0) = q - [\dot{m}''^2][1/\rho_\infty^2 - 1/\rho_0^2]. \tag{14.16}$$

Eliminating $\dot{m}''^2$ using Equation 14.4 and simplifying

$$c_p(T_\infty - T_0) = q + [1/2]\,[1/\rho_\infty + 1/\rho_0]\,[p_\infty - p_0]. \tag{14.17}$$

Using the state equation (Equation 14.12) to eliminate T in terms of ρ and P (i.e., $T_\infty = p_\infty/R\rho_\infty$ and $T_0 = p_0/R\rho_0$), and assuming R is constant we have after simplification

$$\{\gamma/(\gamma - 1)\}\,\{p_\infty/\rho_\infty - p_0/\rho_0\} = q + [1/2]\,[1/\rho_\infty + 1/\rho_0]\,[p_\infty - p_0] \tag{14.18}$$

where the identity $R = c_p - c_v$ has been used. If one plots p_∞ by assuming various values of ρ_∞ for a given q, the curves so obtained are called *Hugoniot curves.* Figure 14.3 shows the results for various q values.

Before we present the solutions, let us normalize the equations.

$$p_\infty^* = p_\infty/p_0, \tag{14.19a}$$

$$\rho_\infty^* = \rho_\infty/\rho_0, \tag{14.19b}$$

$$q^* = q\,(\rho_0/p_0), \tag{14.19c}$$

$$m^* = \dot{m}''^2/(p_0\rho_0), \tag{14.19d}$$

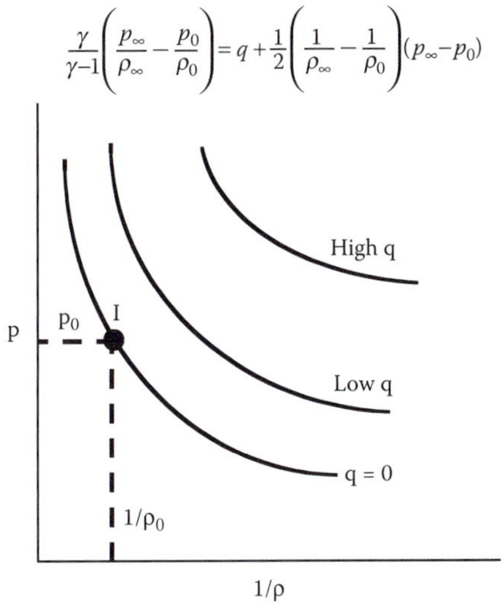

Figure 14.3 Qualitative Hugoniot curves.

$$M_0 = v_0/\{\gamma RT_0\}^{1/2} = \{\dot{m}''^2/(\gamma P_0\rho_0)\}^{1/2} = \{m^*/\gamma\}^{1/2} \qquad (14.20a)$$

$$M_\infty = v_\infty/\{\gamma RT_\infty\}^{1/2} = \{\dot{m}''^2/(\gamma P_\infty\rho_\infty)\}^{1/2} = \{m^*/(\gamma P_\infty{}^*\rho_\infty{}^*)\}^{1/2} \qquad (14.20b)$$

where c is the velocity of sound. The Hugoniot equation (Equation 14.18) reduces to

$$\{\gamma/(\gamma-1)\}\ \{(p_\infty^*/\rho_\infty^*) - 1\} = q^* + [1/2]\ [1/\rho_\infty^* + 1]\ [p_\infty^* - 1]. \qquad (14.21)$$

Figure 14.4 shows a dimensionless plot of Hugoniot curves with q^* as a parameter. Similarly one can show that $(s - s_0)/R = \{1/(\gamma-1)\}\ \ln\ \rho_\infty^* - \{\sigma/(\sigma-1)\}$ $\ln P_\infty$ and $(s - s_0)/R$ attains a local minimum at CJ point along the Hugoniot curve (Figure 14.4). Similarly, the Rayleigh equation (Equation 14.13) reduces to

$$(p_\infty^* - 1)/\{1/\rho_\infty^* - 1\} = -m^*. \qquad (14.22)$$

We can superpose Rayleigh lines and Hugoniot curves to find the intersecting points (which satisfy both the momentum and energy equations), as shown in Figure 14.5. We find two solutions on the deflagration branch (D and L, low $\dot{m}''$, mixture burns slower) and two solutions on the detonation branch (E and K, high $\dot{m}''$, mixture burns faster).

For some values of $\dot{m}''$, we cannot get steady-state solutions. At some values of $\dot{m}''$, the slope of the Rayleigh line is the same as the slope of the Hugoniot curve. These points are called CJ points (points J and M). One can obtain the solutions for CJ points as follows:

Using,

$$\left[\frac{dp^*}{d\left(\frac{1}{\rho^*}\right)}\right]_{Hugoniot} = \left[\frac{dp^*}{d\left(\frac{1}{\rho^*}\right)}\right]_{Rayleigh} \qquad (14.23)$$

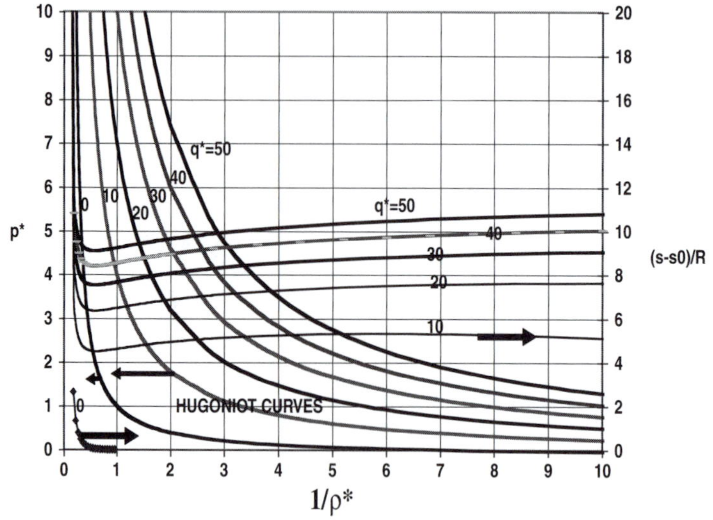

Figure 14.4 Dimensionless Hugoniot curves, $\gamma = 1.3$.

where the left-hand side is from Equation 14.21 and the right-hand side is from Equation 14.22 in order to solve $\dot{m}''$ corresponding to J and M.

14.3.3 Entropy

Recall from thermodynamics (chapter 1) that ds − δq/T = δσ as one follows a fixed mass traveling through a flame front. For an adiabatic system ds = δσ.

$$\text{where } \delta\sigma = d\,\{\Sigma Y_k\ s_k\} \text{ and } s = \{\Sigma Y_k\ s_k\}. \tag{14.24a}$$

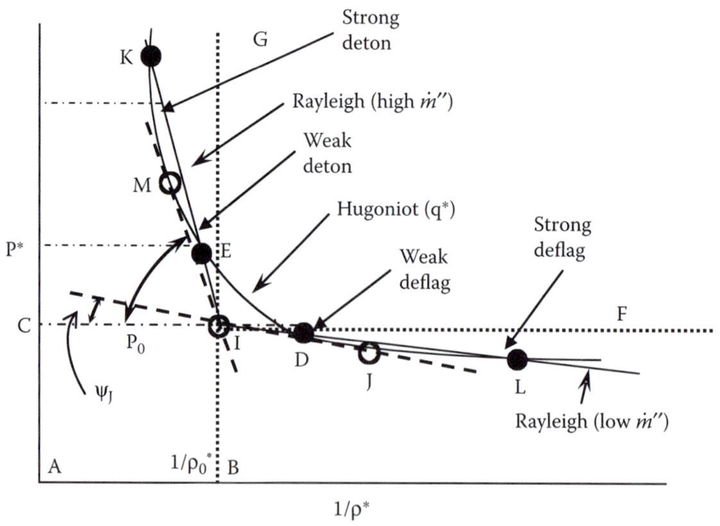

Figure 14.5 Superposition of Rayleigh and Hugoniot curves.

Thus for differential change

$$\delta\sigma = ds = d\left[\sum_{exit} Y_k s_k\right] > 0 . \qquad (14.24b)$$

One finds that entropy keeps increasing (i.e., $\delta\sigma > 0$) whether it is detonation or deflagration until the CJ points are reached owing to irreversible chemical reactions. However, the increase is smaller for detonation compared to deflagration due to increased pressure for detonation.

14.4 Flame Propagation into Unburned Mixture

Solving for p_∞^* from Equation 14.21, one obtains,

$$p_\infty^* = \frac{\left\{\left[2q^* + \left(\frac{\gamma+1}{\gamma-1}\right)\right] - \frac{1}{\rho_\infty^*}\right\}}{\left\{\left[\frac{(\gamma+1)}{\gamma-1}\right]\left[\frac{1}{\rho_\infty^*}\right] - 1\right\}} . \qquad (14.25)$$

Using Equation 14.25, the Hugoniot curve KED can be obtained at a given q^* as shown in Figure 14.5. Similarly, using Equation 14.22 the Rayleigh lines p^* vs. $(1/\rho^*)$ can be plotted for any given initial condition I, as shown by the lines IEK and IDL.

14.4.1 General Remarks

(1) If $\rho_\infty^* \to 1$, p_∞^* must tend to one, to keep m^* finite (see Equation 14.22).
(2) From Equation 14.25 it is seen that as $\rho_\infty^* \to 1$,

$$p_\infty^* = [q^*\{(\gamma - 1)\} + 1]. \qquad (14.26)$$

From Equation 14.25, $p_\infty^* \to 1$ as $(1/\rho_\infty^*) \to 1$ with $q^* = 0$ (see Figure 14.4 for Hugoniot curve with $q^* = 0$).

(3) The momentum equation also has to be satisfied. The Rayleigh curve from I must have a negative slope: p_∞^* vs. $(1/\rho_\infty^*)$ has a negative slope (see Equation 14.22).
(4) Observe the Rayleigh lines IEK (detonation side) and IDL (deflagration side). With $q^* > 0$, we find that at K, E, D, and L, we satisfy the mass momentum and energy conservation.
(5) Quadrants ABIC and FIG are meaningless because of the requirement of negative slope for p_∞^* vs. $1/\rho_\infty^*$ (the Rayleigh line).

14.4.2 Detonation Branch

On Figure 14.5, point I is the initial condition with $p^*_\infty = 1$ and $\rho^*_\infty = 1$. The detonation branch is that branch lying within the quadrant CIG where $\rho^* > 1$ or $(1/\rho_\infty) < 1$ and $p^* > 1$; there are two solutions at E and K. Point K is called *strong detonation* (observed) and point E is called *weak detonation.* Thus, by altering m^* (high range), we can obtain the corresponding values of p_∞^* for both strong and weak detonations, which can then be plotted

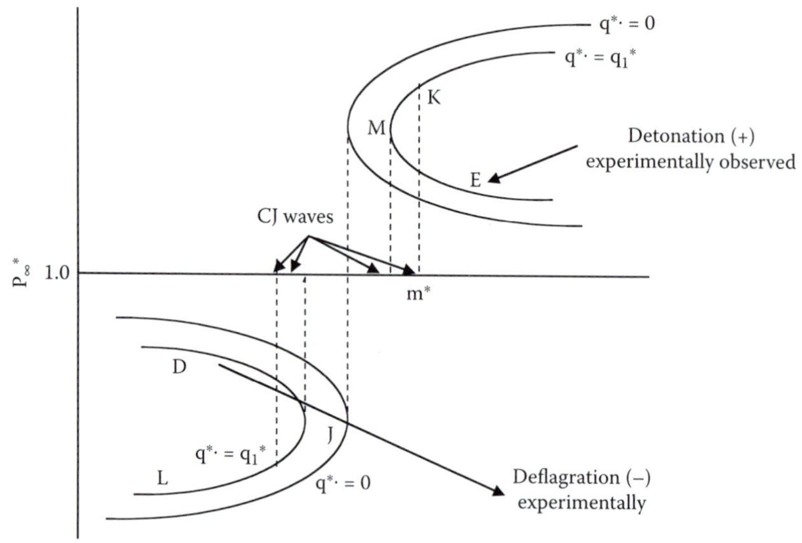

Figure 14.6 Solutions for P_∞* vs. m*.

against each other, as in the upper part of Figure 14.6. If m* is increased, then for the detonation branch, p_∞* at E will decrease and p_∞ at K will increase. Slope IEK in Figure 14.5, i.e., the slope of the Rayleigh line, is steeper (m* is larger or v_0 is larger). When m* is high, the only possible solution is on the detonation side. For detonation, $\rho_\infty > \rho_0$, therefore, $v_\infty < v_0$. But the value of v_0 is very large, i.e., the velocity of the flame propagating into static gas is very large. From Equation 14.25 it is clear that $(1/\rho_\infty^*)\ [(\gamma+1)/(\gamma - 1)] > 1.0$ and hence

$$[(\gamma-1)/(\gamma+1)] \leq (1/\rho_\infty^*) \leq 1.0. \tag{14.27}$$

When $(1/\rho_\infty^*) = 1.0$, Equation 14.21 yields a lower limit, $p_\infty^* = (q^*(\gamma - 1) + 1)$
When $(1/\rho_\infty^*) = (\gamma - 1)/(\gamma + 1)$, Equation 14.25 yields $p^*_\infty \rightarrow \infty$. Thus

$$[q^*(\gamma - 1) + 1] < p_\infty^* < \infty. \tag{14.28}$$

14.4.3 Physical Explanation for Detonation

Consider a tube closed at one end and open at the other (Figure 14.1). Let us say that the tube is quickly filled with combustible mixture. If we ignite at the closed end, a flame is initiated. The flame slowly starts with velocity v_1 at time t_1. The gases once burned expand owing to local increase of pressure. Then a pressure wave propagates, say, of 2 kN/m^2, at the velocity of sound, say, 300 m/sec. The flame accelerates to a velocity v_2 at time t_2 with increased heat liberation rate. Now the gases are traveling at faster velocity. The waves initiated at t_2 are traveling faster because pressure is a little higher. Let the pressure wave be of strength 3 kN/m^2. Because these travel faster, they catch up with the previous wave, and once they do so the strength of the pressure wave jumps to 3 + 2 kN/m^2. This process continues with pressure buildup. Sometimes, the pressure buildup continues along with increased chemical reaction rates, with increasingly faster flame propagation; thus, v_0 can approach an extremely high velocity with $M_0 > 1$. The pressure of burned gases will be extremely high, whereas the pressure of unburned gases will still be around 1 bar (for more discussion see Lewis and Von Elbe [1961]). If combustible mixture is very lean or very rich, there is an insufficient heat liberation rate to sustain detonation. Table 10.1 presents the lean and rich limits for detonation.

14.4.4 Deflagration Branch

Starting from I, there are two possible branches. We find that if $1/\rho_\infty^* > 1$ or $\rho_\infty < \rho_0$, the quadrant BIF is called the *deflagration branch*; when m^* is low, the possible solution is on the deflagration side. There are two solutions: at D and L. Point D is called *weak deflagration* (observed). Point L is called *strong deflagration.* For deflagration, $\rho_\infty < \rho_0$, therefore, $v_\infty > v_0$. Thus, the gas velocity increases after it passes the flame front.

Because $p_\infty < p_0$,

$$0 < p_\infty^* < 1.0 \tag{14.29}$$

From Equation 14.25

$$[1 + q^*(\gamma - 1)/\gamma] \le 1/\rho_\infty^* \le [2q^* + (\gamma + 1)/(\gamma - 1)] \tag{14.30}$$

Similarly, as m^* is increased, then for the deflagration branch, p_∞^* at D will decrease and p_∞ at L will increase, as observed in Figure 14.6. Thus, by altering m^* (low range) we can obtain the corresponding values of p_∞^* for both strong and weak deflagrations, which can then be plotted against each other as in the lower part of Figure 14.6. Figure 14.7 shows the deflagration results, whereas Figure 14.8 shows the detonation solutions for different values of q^*.

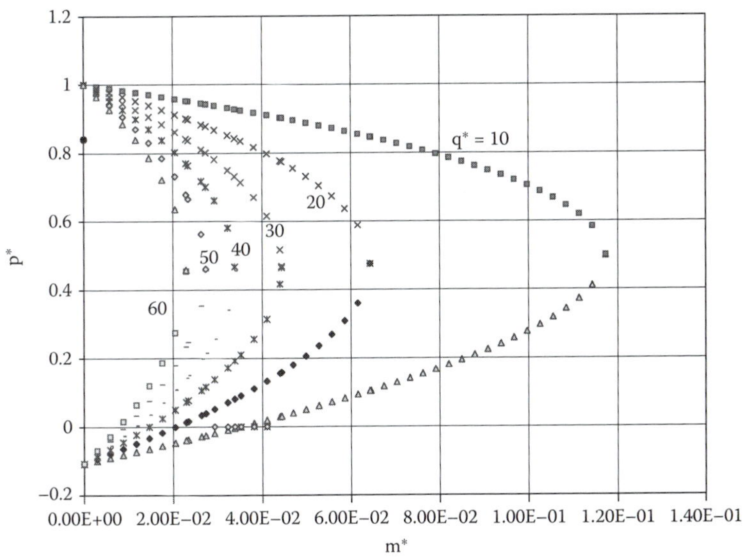

Figure 14.7 Deflagration solutions.

14.4.5 CJ Waves

We will now determine the maximum m* on the deflagration side and the minimum m* on the detonation side.

Draw tangents IJ (deflagration side) and IM (detonation side) from I to the Hugoniot curve KMEDJL as in Figure 14.5. Points J and M are called CJ points. One can show that $M_\infty = v_\infty/(\gamma R T_\infty) = 1$ at CJ points. M is the upper CJ point and J, the lower CJ point. Solutions for M will be denoted as (+), whereas those for J will be denoted as (–).

14.4.5.1 Explicit Results for CJ Waves

Differentiate Equation 14.22 with respect to $(1/\rho_\infty)$ and Equation 14.25 with respect to $(1/\rho_\infty)$. Equating the two, we get the solutions for CJ waves

$$\frac{1}{\rho_\infty^*}\left[\left(\frac{\gamma+1}{\gamma-1}\right)\frac{1}{\rho_\infty^*}-1\right]=\left[\frac{1}{\rho_\infty^*}(\gamma-1)-\gamma\right]\left[2q^*+\left(\frac{\gamma+1}{\gamma-1}\right)+\frac{1}{\rho_\infty^*}\right] \qquad (14.31)$$

At a given q*, solving for $(1/\rho_\infty^*)$ we get two solutions, $\rho_{\infty,-}^*$ (point J) and $\rho_{\infty,+}^*$, (point M) corresponding to J(–) and M(+). Once this obtained, then all the results for CJ waves can be obtained from Equation 14.19 to Equation 14.20. Figure B.4 in Appendix B and Figure 14.9 present the theoretical results on CJ waves; experimental data confirm the results for CJ detonation but not for deflagration.

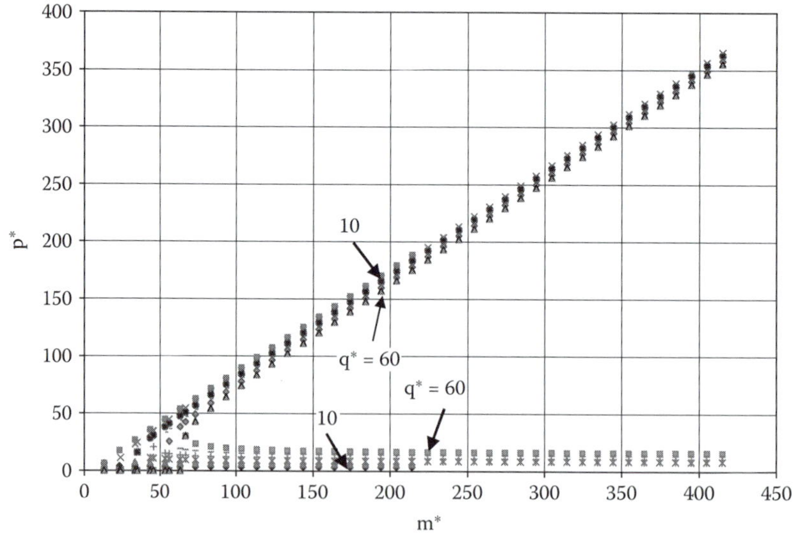

Figure 14.8 Detonation solutions.

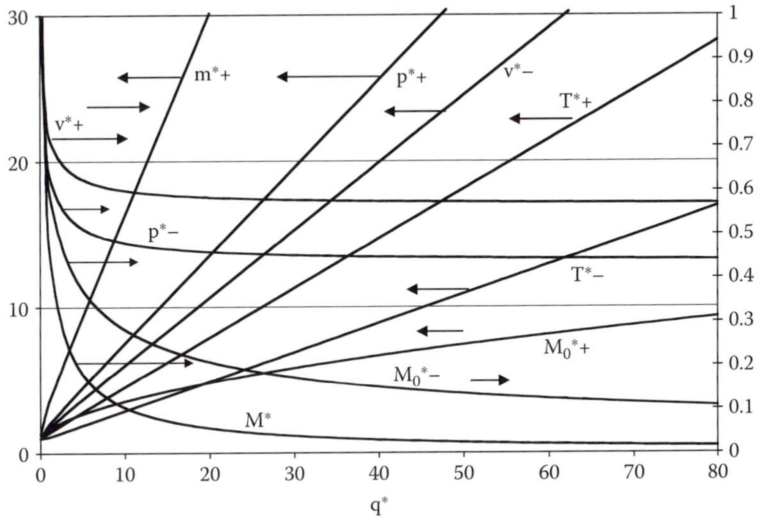

Figure 14.9 CJ charts.

Example 2

Consider a CH_4-air mixture with 20% excess air. (a) What is the maximum possible deflagration velocity if T_0 = 300 K, ρ_0 = 1.14 kg/m³, $p_0 = 10^5$ N/m², γ = 1.4, and LHV = 50020 kJ/kg? (b) What is the detonation velocity?

Solution

$$\text{(a) } q = h_C/(\text{air + fuel mass}) = \frac{50200}{20.8+1} = 2294 \text{ kJ/kg}$$

$$q^* = q\rho_0/p_0 = \frac{2294\times 10^3 \times 1.14}{10^5} = 26.24$$

From the charts in Figure 14.9, at q* = 26.24 we have,

$$(M_0)_- = 0.16 = (v_0)/c_0$$

$$c_0 = (\gamma p_0/\rho_0)^{1/2} = \sqrt{\frac{1.4\times 10^5 \frac{N}{m^2}}{1.14\frac{kg}{m^3}}} = 349.8 \text{ m/sec}$$

$$v_0 = 349.8 \times 0.16 = 56.8 \text{ m/sec}$$

This is the maximum possible velocity of propagation on the deflagration branch. The velocity of hot gases at this condition is such that it reaches the speed of sound after the gases pass through the wave, i.e., $M_{prod} \to 1$.

(b) From Figure 14.9, at q* = 26.2, detonation velocity: $(M_0)_+ = 6.16$, $(v_0)_+ = 6.16*349.8 = 2155$ m/sec, p* = 22.6

Table 14.2 Typical Variation of Properties for Deflagration and Detonation

Parameter	*Deflagration*	*Detonation*
T_∞/T_0	4–16	8–21
P_∞/P_0	0.976–0.98	13–55
ρ_∞/ρ_0	0.06–0.25	1.4–2.6
v_∞/c_0	0.0001–0.03	5–10
v_∞/v_0	4–16	0.4–0.7

Source: Friedman, American Rocket Society, v 23, 349, 1953.

Remarks

Note that for an H_2-air mixture, $q \approx 3455$ kJ/kg of mixture, $q^{\cdot} \approx 34.55$, $(M_0)_+ \approx 6.5$, $(v_0)_+ \approx 2431$ m/s, $(M_{0-}) = 0.15$, and $(v_{0-}) = 56.1$ m/sec.

Knowing q* for several HC fuels, one can estimate typical properties for deflagration and detonation (see Table 14.2).

■ ■ ■

Detonation velocities can be observed experimentally, but observing deflagration is unrealistic. Further, deflagration is observed for a short tube.

14.5 Summary

Combustion occurs at constant pressure for subsonic speeds with respect to unburned mixture; however, there is a pressure increase for supersonic speeds. This chapter presents possible solutions for pressure and density after combustion for any prescribed initial condition of unburned mixture, provided the flame propagation velocity (v_0) is specified. Solutions are presented for both deflagration (when the flame propagates at low velocities) and detonation (when the flame propagates at high velocities). Solutions for the maximum possible deflagration velocity and minimum detonation velocities (CJ waves) are also presented.

14.6 Appendix I: Spreadsheet Program for CJ Waves

We first consider CJ solutions (where Rayleigh lines are tangential to Hugoniot curves).

Using Equation 14.21 to Equation 14.23

$$m^* = \gamma + q^* (\gamma^2 - 1) \{1 \pm [1 + (2\ \gamma/(q^*(^*\ (\gamma^2 - 1))]^{1/2}\} \qquad (a)$$

where + denotes detonation and – denotes deflagration.

The Mach number corresponding to detonation or deflagration ($M_0 = v_0/c_0$), is given by

$$M_0 = [1 + \{q^* (\gamma^2 - 1)/2\gamma\}]^{1/2} \pm \{q^* (\gamma^2 - 1)/2\gamma\}^{1/2} \qquad (b)$$

$$p_\infty^* = 1 + \{q^* (\gamma^2 - 1)\}[1 \pm \{1 + (2\gamma/(q^* (\gamma^2 - 1)))\}^{1/2}] \quad \text{(c)}$$

$$1/\rho_\infty^* = 1 + \{q^* (\gamma - 1)/\gamma\}[1 \pm \{1 + (2\gamma/(q^* (\gamma^2 - 1)))\}^{1/2}], \text{ specific volume} \quad \text{(d)}$$

Because $\rho_\infty = p_\infty/RT_\infty$, $\rho_0 = p_0/R\ T_0$, then $P_\infty/\rho_\infty^* = T_\infty$
Alternately, the flow rate at the CJ points is given as

$$m^* = \dot{m}''^2/p_0\ \rho_o = \rho_0^2\ \{M_0^2\ c_0^2\}/p0\ \rho 0 = M_0^2\ \gamma \quad \text{(e)}$$

Note that $M_\infty = 1$ for CJ waves. Also for specified q* we know m* for deflagration and detonation.

14.7 Appendix II: The Solutions for v_∞ at a Given v_0 or $\dot{m}''$ or m*

Recall the energy equation (Equation 14.21) and momentum equation (Equation 14.22)

$$p^* = (2\ q^* + \delta - v^*)/(\delta v^* - 1) \quad \text{(A)}$$

where $\delta = (\gamma + 1)/(\gamma - 1)$

$$(p^* - 1)/(v^* - 1) = -m^* \quad \text{(B)}$$

Solving for p* from Equation B

$$p^* = -m^*(v^* - 1) + 1$$

and using in Equation A

$$v^*2 + v^* \{-\delta - m^* - 1 - \delta m^*\}/\{\delta\ m^*\} + \{1 + m^* + 2q^* + \delta\}/\{\delta\ m^*\} = 0$$

If $m^* < m^*_{CJ-}$, we get two solutions on the deflagration side. If $m^* > m^*_{CJ+}$, we get two solutions on the detonation side. There are no solutions for m^*_{CJ-} $- < m^* < m^*_{CJ-}$.

Deflagration $m^* < m^*_{CJ-}$: We obtain two solutions for v*; the first solution near the initial condition on the deflagration side (i.e., $M_0 < 1$) is called weak deflagration, whereas the farthest solution is called strong deflagration.

Detonation $m^* > m^*_{CJ+}$: Similarly, we obtain two solutions for v*; the first solution near the initial condition on the detonation side (i.e., $M_\infty > 1$ and around 1) is called weak detonation, whereas the farthest solution ($M_\infty << 1$) is called strong detonation. A spreadsheet program provides the results for any fuel: C_2H_2, CH_4, volatiles from coal, and volatiles from grains and biomass. This can be downloaded from the Taylor & Francis Web site.

Chapter 15

Flame Propagation and Flammability Limits

15.1 Introduction

Premixed fuel:air mixtures are used in residential, commercial, and industrial devices. The flow could be either laminar or turbulent. Further, the mixture could be either partially premixed (residential) or fully premixed (oxyacetylene welding torch). As the fuel is burned in a constant-diameter combustion chamber, the velocity increases owing to decrease in density. In Chapter 14, we formulated governing equations to relate the density and pressure changes due to change in velocity and temperature in an open system. Further, we saw that if a combustible gas:air mixture is ignited, the flame propagates either at deflagration velocity or detonation velocity into the unburned mixture. Limiting conditions for those velocities were obtained under a given heat input but without finite chemistry and detailed flame structure. We also determined that change in pressure is negligible under deflagration conditions but not so under detonation conditions. This chapter deals with combustion wave propagation in premixed fuel:air mixtures under deflagration conditions. We will also obtain relations for the flame speed (v_0 or S), lean (ϕ_L) and rich (ϕ_R) limits called the flammability limits, quench diameter, and minimum ignition energy. We obtain deflagration velocity or flame propagation velocity under subsonic conditions when kinetics limit the combustion rate. Even though typical combustion occurs under turbulent conditions, laminar theory is still required to model the turbulent flames and interpret the results.

A flame represents a spatial domain within which rapid exothermic reaction takes place accompanied by emission of light (Figure 15.1). Consider a quartz tube of constant diameter that is closed at one end but open with a small orifice at other end (to prevent the pressure rise and reduce reflected shock) (Figure 15.1a). We will provide premixed fuel and air and ignite the mixture

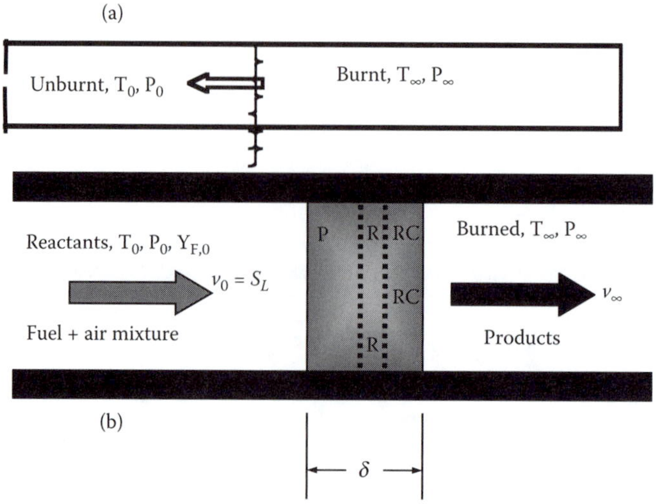

Figure 15.1 Flame propagation; three sections: (a) Burner schematic, (b) reactants, flame, and products. The flame consists of three zones: unburned preheat zone (P), reaction zone (R), and recombination zone (RC).

at the closed end. A flame is formed. Once the flame propagates toward the unburned mixture (Figure 15.1a), it leaves behind combustion products. The flame of thickness δ (which is about 1 mm thick for weak mixtures or 0.01 mm for fast burn [e.g., an H_2–O_2 system >> 10^{-4}mm, typical mean free path]) or a combustion wave separates the hot combustion products at T_∞ from the cold unburned mixture at T_0, $Y_{F,0}$ as seen in Figure 15.1 and Figure 15.2. Within the flame, there are steep gradients for T and Y_k. One finds that T(t), at fixed x_1 gradually increases with time as the flame approaches a location, indicating the preheat of the combustible mixture (Figure 15.2). Then suddenly

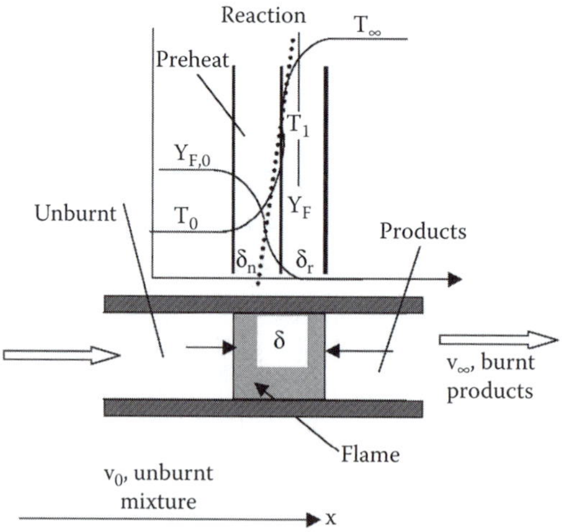

Figure 15.2 Schematic of plug flow reactor.

the rate of increase becomes almost exponentially with time, indicating an intense chemical reaction region (typically called the primary reaction zone, which is very thin). The rate of increase is slowed down as the local combustion approaches completion in the secondary reaction zone (e.g., CO oxidizing slowly), which is thicker. Once the combustion is complete, the local heat loss slightly decreases the temperature. Thus, there could be temperature peaks and similar events at other locations. If the time interval between successive peaks is τ and their spacing is Δx, the flame propagation velocity is $\Delta x/\tau$. (The propagation velocity is typically denoted as S or v_0 and sometimes S_L to indicate laminar burn velocity.)

Now consider a premixed uniform flow of premixed HC (hydrocarbons):air in a slightly diverging quartz tube. Let the flow take place from a smaller section to a larger section. If ignited at the larger section, in which velocity is low, the flame will propagate toward the smaller section and a position will be reached where the laminar flame speed S (order of 0.01 to 10 m/sec << sound speed) matches the incoming velocity of the reactant mixture and the flame will be stationary. With respect to a point of reference on the stationary flame front, the unburned mixture appears to travel toward the point with velocity v_0 (= S) and burned gases leave the point with velocity v_∞. The problem can now be analyzed for a constant-diameter tube with respect to a point of reference on the stationary flame (called the *reference coordinate system*). Unburned gases move into flame with velocity v_0, and products leave with velocity v_∞. Typically, $v_\infty > v_0$ because $\rho_\infty < \rho_0$. Thus, the flow is accelerated, which results in $p_\infty < p_0$ though $p_0 - p_\infty$ is very small for a deflagration wave. There is a variation of gas velocity within the flame thickness of δ (Figure 15.2). The flame velocity can be defined as the velocity of the unburned gas through the combustion wave but in a direction normal to the wave. During the flame propagation, there arise four distinct zones: unburned, preheat, reaction/recombination, and product sections [Glassman]. The preheat and reaction zones constitute the flame thickness. The reaction zone can be subdivided into a pyrolysis zone in which hydrocarbon fragments are formed, and a zone in which intermediates such as CO and H_2 are consumed.

15.2 Phemenological Analysis

15.2.1 Homogeneous Mixtures

In this subsection we wish to estimate the actual burning velocity, first by using phenomenological analysis and later by detailed analysis using thermo-kinetics.

The flame propagation mechanism is as follows: the reaction occurs within a narrow thickness of δ. Thus, the temperature gradient $(T_\infty - T_0)/\delta$ drives the heat flux $\lambda(T_\infty - T_0)/\delta$. The heat flux is used to raise the temperature of the unburned mix in the preheat zone to T_{ign} (Figure 15.2). As T rises, the velocity also increases; thus, for the same reaction time, the reaction-zone thickness increases due to a greater velocity. If a constant-diameter duct is made of highly conducting material (say, Cu), then the heat loss will be higher because of large reaction-zone thickness. The larger the temperature gradient, the greater the heat flux, the smaller the reaction-zone thickness, and the faster the speed of propagation; also, T_{ign} is reached within a shorter time. Large gradients are achieved when $T_\infty - T_0$ is large or when the mixture becomes stoichiometric. When a lean or rich mixture is used, the flame propagation rate slows because the difference $T_\infty - T_0$ is smaller and the reaction-zone thickness is larger. Similarly, when λ is higher, the flame will propagate faster (e.g., λ of He-O_2 is five times higher than N_2-O_2 system). On the other hand, c_p is also larger for He and, hence, T_∞ is reduced for same amount of heat released. When reaction kinetics are faster, T_{ign} is reduced and hence the flame propagates faster. Thus, complex processes occur during flame propagation. The flame propagation velocities of various fuels are tabulated in Table A.39D.

Consider a simplified model. A gaseous mixture of fuel, oxidizer, and inert enters a tube of cross-sectional area A at v_0 and leaves as products at v_∞. The combustion takes place within a distance of δ within which T increases from T_0 to T_∞, and Y_F decreases from $Y_{F,0}$ to 0. Using the following energy balance, T_∞ can be obtained from

$$q = Y_{F,0}\, h_c = c_p\,(T_\infty - T_0) \tag{15.1}$$

If a wave has to remain stationary at a fixed temperature, the heat transfer from flame must be balanced by heat generated by the chemical reaction within the space bounded by δ. Assuming a linear profile for T within δ and a uniform chemical reaction rate, $\dot{w}'''_{F,avg}$ at an average temperature within δ,

$$A\,\lambda\,(T_\infty - T_0)/\delta = \dot{w}'''_{F,avg}\, h_c\, A\, \delta \tag{15.2}$$

where T_∞ is the *adiabatic flame temperature*. Equation 15.2 presumes that all heat generated is used to raise the temperature of mass from T_0 to T_∞. If there is heat loss and if there is still flame propagation, then $T_b < T_\infty$, where T_b is the burned products' temperature with heat loss. Using Equation 15.1 in Equation 15.2 to eliminate T_∞, and solving for δ

$$\delta = \left[\lambda\, Y_{F,0} / \left\{c_p\, \dot{w}'''_{F,avg}\right\}\right]^{1/2} = \left[\alpha_T\, \rho_0\, Y_{F,0} / \left\{\dot{w}'''_{F,avg}\right\}\right]^{1/2} \tag{15.3}$$

Assuming the reaction to be uniform within δ, we have

$$\rho_0\, v_0\, A\, Y_{F,0}\, h_c = \rho_0\, A\, v_0\, q = \dot{w}'''_{F,avg}\, h_c\, A\, \delta \tag{15.4}$$

Using Equation 15.3 in Equation 15.4 to eliminate δ and solving for the laminar flame velocity $v_0 = S$,

$$S = v_0 = [1/\rho_0]\left[\lambda\, \dot{w}'''_{F,avg}\, \{Y_{F,0}\, c_p\}\right]^{1/2} = \left[\alpha_T\, \dot{w}'''_{F,avg} / \{\rho_0 Y_{F,0}\}\right]^{1/2} \tag{15.5}$$

and the total mass flow rate per unit area is given as,

$$\rho_0\ v_0 = \dot{m}'' = \left[\alpha_T \rho_0 \dot{w}'''_{F,avg} / Y_{F,0}\right]^{1/2}, \tag{15.6}$$

where $\dot{w}$,(kg/m^3 sec) = $A_Y{}^{\alpha 1}\ \rho^n \exp(-E/RT)\ [Y_F]^{nF}\ [Y_{O2}]^{nO2}$. The higher the flame temperature, the higher the $\dot{w}'''_{F,avg}$ and the higher the v_0.

If the characteristic reaction time t_{chem} is defined as

$$t_{chem} = \{\rho_0 Y_{F,0}\}/\dot{w}'''_{F,avg} = (\rho_0 Y_{F,0})/\dot{w}'''_{F,avg}$$

then Equation 15.5 can be written as,

$$S = v_0 = [\alpha_T\ \dot{w}'''_{F,avg}/\{\rho_0 Y_{F,0}\}]^{1/2} = [\alpha_T/t_{chem}]^{1/2} \tag{15.7a}$$

where $(F_0) = \rho_0 Y_{F,0}$ is the unburned fuel concentration in kg/m^3. Thus, if one moves with a unit mass of combustibles from an unburned zone into the flame, the characteristic time required to burn the fuel is given by t_{chem}. Rewriting Equation 15.7a,

$$\rho_0\ v_0 = \dot{m}'' = [\alpha_T\ \rho_0\ \dot{w}'''_{F,avg}/Y_{F,0}]^{1/2} = \rho_0[\alpha_T/t_{chem}]^{1/2}. \tag{15.7b}$$

The term $\rho_0 v_0$ represents the mixture flow rate in kg/m^2 sec, whereas $\rho_0 v_0$ Y_{F0} indicates fuel consumption (i.e, kg of fuel/m^2 sec). One may write Equation 15.3 in terms of velocity v_0 by eliminating $\dot{w}'''_{F,avg}$ between Equation 15.7 and Equation 15.3:

$$\delta \approx \alpha_T/v_0 \tag{15.8}$$

For example, if the flame velocity for propane is 0.35 m/sec, $\alpha_T = 10^{-4}$ m^2/sec, then $\delta \approx \alpha_T/v_0 = 3$ mm.

Zeldovich gives an approximate expression for flame velocity as [Kuo, 1986]

$$v_0 = [1/\rho_0]\ [\lambda\ \dot{w}'''_{F,avg}/\{\Lambda c_p\ Y_{F0}\}]^{1/2} = [\alpha_T\ \dot{w}'''_{F,avg}/\{\Lambda\ \rho_0\ Y_{F0}\}]^{1/2} \tag{15.9a}$$

where $\Lambda = \frac{1}{2}$, and $\dot{w}'''_{F,avg}$ is computed for $T_{Ign} < T < T_b$. Equation 15.9 yields twice the rate of Equation 15.7 because we assumed a linear temperature gradient over the whole thickness δ. Note that there are many reaction steps and intermediaries involved in chemical energy release (see Chapter 5); thus, in addition to diffusion of heat, the radicals can diffuse into the preheat zone. Borman and Ragland present the following expression (also known as Mallard and LeChatelier relation):

$$v_0 = \left\{\frac{\alpha_T}{t_{chem}}\right\}\left\{\frac{T_\infty - T_{ign}}{T_{ign} - T_0}\right\} = \left\{\frac{\alpha_T \dot{w}'''_{F,avg}}{\rho_0 Y_{F,0}}\right\}\left\{\frac{T_\infty - T_{ign}}{T_{ign} - T_0}\right\} \tag{15.9b}$$

where T_{ign} is the ignition temperature of the combustible mixture. Using Equation 15.3, $\rho_0\ v_0\ Y_{F0}/\ \dot{w}'''_{F,avg} = \delta$.

15.2.1.1 Space Heating Rate (SHR)

One may rewrite Equation 15.4 in terms of SHR as

$$\text{SHR} = \dot{w}'''_{F,avg}\ h_c\ (\text{kW/m}^3) = c_p\ h_c\ Y_{F0}\{\rho_0\ v_0\}^2/\lambda. \tag{15.9c}$$

15.2.1.2 Effect of Various Parameters on v_0 or S

From Equation 15.5, one can study the effects of the following on v_0 or flame velocity S:

(a) Pressure:

Because the chemical reaction rate $\dot{w}'''_{F,avg}$ is proportional to ρ^n, then,

$$v_0\ \alpha\ \rho^{(n/2-1)}. \tag{15.10a}$$

If n = 2 (i.e., HC fuels) then v_0 is independent of pressure. If n < 2, v_0 decreases with pressure.

(b) Thermal conductivity:

$$v_0\ \alpha\ (\lambda)^{1/2} \tag{15.10b}$$

The higher λ is, the higher the burning velocity is because heat feedback is very rapid. One can alter λ by mixing the fuel and oxidizer with the He and Ar (inert gas) mixture.

(c) Specific heat c_p:

The lower c_p is, the higher is T_∞, average temperature of reaction, reaction rate, α_T and v_0. c_p is changed by using a mixture of inerts, O_2, and fuel.

(d) Mixture stoichiometry:

Mixture stoichiometry has a significant effect on v_0 due to change in flame temperature. If the mixture is dilute, then $Y_{F,0}$ is very low. Further T_∞ is low, the reaction rate will be low and hence v_0 will be low. On the other hand, if the mixture is rich, then all the oxygen will be consumed whereas fuel will be left over. Then, for a rich mixture, we must modify Equation 15.1 as

$$[Y_{O2,0}/\upsilon_{O2}]\ h_{c,R} = c_p\ (T_\infty - T_0) \tag{15.11a}$$

where $h_{c,R}$ may be different from h_c because of incomplete combustion. The corresponding result is

$$\delta = [\lambda[Y_{O2,0}/\upsilon_{O2}]/\{c_p\ \dot{w}'''_F\ \}]^{1/2} \tag{15.11b}$$

$$v_0 = [1/\rho_0]\ [\lambda\ \dot{w}'''_F\ /\{[Y_{O2,0}/\upsilon_{O2}]c_p\}]^{1/2}. \tag{15.12}$$

For a very rich mixture $Y_{O2,0}$ is very low. Also T_∞ and average temperature of reaction are very low, and thus v_0 is low again. Hence, v_0 is nearly a maximum for a stoichiometric mixture.

Experimental studies on v_0 of CH_4–O_2–Ar mixture at T = 298 K, P = 1 atm yield the following curve fit [Rahim et al. 2002]: $v_0(\text{cm/sec}) = -313.9\ \phi^2 + 659\ \phi - 263.9$.

15.2.2 Heterogeneous Liquid Mixtures

Simplified analyses are presented here. For detailed treatment, the reader is referred to several numerical studies on ignition and propagation through fuel drop suspension systems by Aggarwal et al. [1983–1985].

15.2.2.1 Micronized Drops

If drops are extremely small and timescales for evaporation ($t_{evap} = d_0^2/\alpha_e$, α_e evaporation rate constant $= 4$ Sh ρ D $\ln(1 + B_e)/\rho_\ell$, $B_e = c_p\ (T_\infty - T_w)/L$; see Chapter 10) are considerabley smaller compared to kinetic timescales ($t_{chem} = \rho_0\ Y_{F,0}/\dot{w}''_{F,avg}$; see Chapter 7 and Chapter 5), then Equation 15.6 could be adopted for determining the deflagration velocity through sprays. This criterion is given as

$$\frac{\{d_0^{\ 2}\ \rho_\ell\}\dot{w}'''_{F,avg}}{\{4\ \mathrm{Sh}\,\rho\ \ D\ln(1+B_e)\}(\rho_0 Y_{F,0})} \lll 1.$$

15.2.2.2 Medium-Sized Drops

The evaporation/combustion timescale is much larger than the chemical timescale and, as such, the fuel consumption rate is controlled by drop combustion rates. If "n" is the number of drops per unit volume, then combustion rate per unit volume is given as {4 Sh n d ρ D $\ln(1 + B_e)$} with an average free-stream temperature for evaluation of B; hence, replacing the chemical source $\dot{w}'''_{F,avg}$ by $\dot{m}'''_c$ in Equation 15.6

$$\rho_0 v_0 = \dot{m}'' = [\alpha_T\ \{4\ \mathrm{Sh}\ n\ d\ \rho\ D\ \ln(1 + B_e)\}/(\rho_0 Y_{F,0})]^{1/2}, \qquad (15.12)$$

where $Y_{F,0} = Y_{\ell,0}$ is the initial liquid fuel mass fraction or fuel mass fraction. A more detailed treatment can be found in Williams. Equation 15.12 presumes that the drop number density is large enough to provide a uniform mass source. If drops are spaced farther apart (i.e., large interdrop spacing), the flame may jump from one drop to another. Propagation rate is slow and may approach that for lean mixures. Finally, if "ℓ" is very large, propagation ceases.

15.2.3 Heterogeneous Pulverized Coal: Air Mixtures

15.2.3.1 Micronized Particles

If solid particles are extremely small with high volatile matter (VM), for example, biomass that contains up to 80% VM, and timescales for pyrolysis (Chapter 5) are considerably smaller (i.e., extremely fast pyroysis rate $\dot{m}_p$) compared to kinetic oxidation timescales, then Equation 15.6 could be adopted for determining the deflagration velocity through pulverized fuel (pf) suspension.

15.2.3.2 Medium-Sized Particles

The pyrolysis timescale of solid fuel particles is high, and VM is much larger than the chemical timescale and, as such, the fuel consumption rate is controlled by pyrolysis rates. If "n" is the number of particles per unit volume, replacing the chemical source term in Equation 15.6,

$$\rho_0 \ v_0 = \dot{m}'' = [\{\alpha_T \ n \ \dot{m}_p\}/(\rho_0 \ Y_{F,0})]^{1/2},$$

where $Y_{F,0} = Y_{s,0}$ is the initial solid mass fraction or fuel mass fraction.

15.3 Rigorous Analysis

15.3.1 Conservation Equations

We will now write down the balance equations for unburned and burned regions involving flame propagation in a gaseous combustible mixture.

Assume:

- 1-D, constant area, steady
- Neglect KE, PE, viscous dissipation, and thermal radiation
- P = constant across flame (Mach # << 1)
- Diffusion of heat and mass follows the Fourier law and Fick's law, no thermal diffusion
- Le = 1
- constant c_P.

15.3.1.1 Mass

$$d(\rho v)/dx = 0 \tag{15.13}$$

After integration,

$$\rho_0 v_0 = \dot{m}_0'' = \rho_\infty v_\infty = \dot{m}_\infty'' = \dot{m}'', \text{ a constant.} \tag{15.14}$$

15.3.1.2 Momentum

From Chapter 14, the pressure change due to $\{\rho_\infty v_\infty^2 - \rho_0 v_0^2\} \approx 0$; thus $p_0 \approx p_\infty$ = constant.

15.3.1.3 Species

Using the species conservation equation for a 1-D steady-state system (Chapter 7),

$$\rho v \ (d\varepsilon_k/dx) = \dot{w}_k''' \tag{15.15}$$

where ε_k, flux ratio = $\rho Y_k \ (v + V_k)/(\rho v)$ = species k flux/average mass flux. With k = F and one-step reaction,

$$\rho v \ (d\varepsilon_F/dx) = \dot{w}_F''' = (-1) \ \dot{w}''', \text{ kg/m}^3 \text{ sec} \tag{15.16a}$$

$$\dot{w}''' = A_Y \ T^\alpha \rho^n \exp(-E/RT) \ [Y_F]^{nF} \ [YO_2]^{nO2}. \tag{15.16b}$$

15.3.1.4 Energy

We have shown before that for Le = 1 (Chapter 14) and constant c_p

$$\rho v\ c_p dT/dx = d/dx\ \{(\rho D dh_t/dx) + |\ \dot{w}_F'''\ | h_c. \qquad (15.17)$$

15.3.1.4.1 Boundary Conditions

The origin x = 0 can be selected within the flame, say, at $T = T_0 + 0.01(T_\infty - T_0)$.

$$\text{As } x \to -\infty,\ Y_k = Y_{k0} \text{ or } dY_k/dx = 0, T = T_0 \text{ or } dT/dx = 0 \qquad (15.18a)$$

$$\text{As } x \to \infty,\ Y_k = Y_{k\infty} \text{ or } dY_k/dx = 0,\ T = T_\infty \text{ or } dT/\ dx = 0 \qquad (15.18b)$$

15.3.2 General Solution

The differential equations are second order in Y_K and T and, hence, the two boundary conditions for each variable (Equations (15.18a) and Equation (15.18b)) can be used to obtain Y_K (x) and T(x). Integrating Eequation (15.16) with $\varepsilon_F = Y_{F0}$ as as $x \to -\infty$ and 0 as $x \to \infty$

$$\rho_0 v_0\ Y_{F0} = \int_{-\infty}^{\infty} A_Y\ T^\alpha \rho^n \exp\ (-E/RT)\ [Y_F]^{nF}\ [YO_2]^{nO2}\ dx. \qquad (15.19a)$$

Rewriting

$$\rho_0 v_0\ Y_{F0} = \int_{-\infty}^{\infty} A_Y\ T^\alpha \rho^n \exp\ (-E/RT)\ [Y_F]^{nF}\ [YO_2]^{nO2}\ dT/(dT/dx). \qquad (15.19b)$$

With knowledge of Y_K (x) and T(x) profiles, the right-hand side of Equation 15.19b can be integrated, and v_0 can be obtained.

15.3.3 Explicit Solutions

In order to develop explicit solutions, the following normalizations and a few relations between fuel mass fractions, oxygen mass fractions and temperature are presented, followed by relations between flux ratios and temperature. The following normalizations are adopted.

$$\theta = T/T_\infty \qquad (15.20)$$

$$h_c{}^* = h_c/(c_p\ T_\infty) \qquad (15.21)$$

$$x^* = (\rho v) \int dx/(\rho D) = (\rho v)x/(\rho D) \qquad (15.22)$$

$$|w_F{}^*| = |\ \dot{w}_F'''\ |\ \rho D/\ \dot{m}''^2 = A_Y\ T^{\alpha 1}\ \rho^{n1}\ Y_F{}^{nF}\ Y_{O2}{}^{nO2}$$

$$\exp(-E/RT)\ \{\rho D/\ \dot{m}''^2\} \qquad (15.23)$$

15.3.3.1 SZ Variable

Using the 1-D SZ formulation for all reacting species, with axial diffusive terms

$$\rho v d\beta/dx = d/dx \{(\rho D d\beta/dx) \quad (15.24)$$

As ρv = constant, then integrating

$$\beta = (\rho D/\rho v)(d\beta/dx) + C \quad (15.25)$$

As $x \rightarrow \infty$, $dY_k/dx \rightarrow 0$, $dT/dx \rightarrow 0$ and, hence, $(d\beta/dx) \rightarrow 0$ and $\beta \rightarrow \beta_\infty$. Hence $C = \beta_\infty$. The condition $dT/dx = 0$ is called the adiabatic boundary condition. Integrating further, simplifying, and as $x \rightarrow \infty, \beta$ is finite one obtains:

$$\beta = C = \beta_0 = \beta_\infty, \text{ a constant} \quad (15.26)$$

When

$$\beta = \beta_{ht-F} = h_t/h_c + Y_F \quad (15.27)$$

then using conditions at $x = -\infty$ and ∞

$$\beta_{ht-F} = h_{t,0}/h_c + Y_{F,0} = h_{t,\infty}/h_c + Y_{F,\infty} = h_{t,\infty}/h_c \quad (15.28)$$

which will yield the temperature at $x = \infty$.

Then Equation 15.26 becomes

$$\beta = \beta_{ht-F} = h_t/h_c + Y_F = \theta/h_c^* + Y_F = C \quad (15.29)$$

which shows that T and Y_F are linearly related with a single step kinetics approximation. This relation is useful in integrating Equation 15.19b.

15.3.3.2 Product Temperature for Lean Mixture

Applying Equation 15.28

$$T_\infty = T_0 + Y_{F,0}\, h_c/c_p \quad (15.30)$$

15.3.3.3 Relation between Y_F and T Profiles for Lean Mixtures

From Equation 15.29,

$$Y_F = (1/h_c^*)\,(1 - \theta) \quad (15.31a)$$

When $\theta = \theta_0$, $Y_F \rightarrow Y_{F,0}$

$$Y_{F,0}\, h_c^* = (1 - \theta_0) \quad (15.31b)$$

Using Equation 15.29 and Equation 15.30,

$$Y_F/Y_{F,0} = (1 - \theta)/(1 - \theta_0) \quad (15.32)$$

In the absence of heat loss, Equation 15.37 yields,

$$\{Y_F/Y_{F,0}\}(1 - \theta_0) + \theta = 1 \tag{15.33}$$

15.3.3.4 Product Temperature for Rich Mixture

Here, all the oxygen is consumed but fuel is left over. Using β_{ht-O2} at x = $-\infty$ and $x \to \infty$ for a rich mixture with a heating value of $h_{c,R}$.

$$\theta_0/h_{c,R}^* + \{Y_{O2,0}/\nu_{O2}\} = \theta_{\infty,R}/h_{c,R}^* + \{Y_{O2,\infty}/\nu_{O2}\} = 1/h_{c,R}^*$$

Thus,

$$Y_{O2,0}/\nu_{O2} = (1 - \theta_0)/h_{c,R}^*, \tag{15.34}$$

$$T_{\infty,R} = T_0 + \{Y_{O2,0}/\nu_{O2}\}\, h_{c,R}/c_p;\ \text{if } h_{c,R} \approx h_c, \text{ then } T_{\infty,R} = T_\infty$$

Note: $h_{c,R} \approx h_c$, if products contain CO_2, H_2O, and unburned fuel. Typically, $\theta_{\infty,R} < 1$.

15.3.3.5 Relation between Y_{O2} and T Profiles for Rich Mixtures

Similarly, Y_{O2} can also be related in terms of $(1 - \theta)/(1 - \theta_0)$. Using β_{ht-O2} = constant = $\beta_{ht-O2,\infty}$, then

$$Y_{O2}/\nu_{O2} = (1/h_c^*)\,(-\theta) \tag{15.35}$$

Using Equation 15.34,

$$Y_{O2}/Y_{O2,0} = (\theta_{\infty,R} - \theta)/(\theta_{\infty,R} - \theta_0) \tag{15.36}$$

If $\theta_{\infty,R} = 1$, then

$$Y_{O2}/Y_{O2,0} = (1 - \theta)/(1 - \theta_0) \tag{15.37}$$

15.3.4 Relation between Flux Ratio and Temperature

Normalizing Equation 15.16a and rewriting

$$(d\varepsilon_F/d\theta)\,(d\theta/dx^*) = -|w_F^*| \tag{15.38}$$

where the constancy of β yields a relation between temperature θ and the fuel mass fraction Y_F. Differentiating Equation 15.31a

$$dY_F/dx = -\{d\theta/dx\}/h_c^* \tag{15.39}$$

The flux ratio and mass fractions are related as follows:

$$\varepsilon_F = Y_F - (\rho D dY_F/dx)/\rho v = Y_F - dY_F/dx^* \tag{15.40}$$

Thus,

$$-dY_F/dx^* = \varepsilon_F - Y_F. \tag{15.41}$$

Using Equation 15.41 in Equation 15.39 and making use of the result for Y_F in terms of θ (Equation 15.31a)

$$\{d\theta/dx^*\} = h_c^* \{\varepsilon_F - Y_F\} = \varepsilon_F h_c^* - (1 - \theta). \tag{15.42}$$

Using the result in equation 15.38

$$(d\varepsilon_F/d\theta) = -|w_F^*|/\{(\varepsilon_F h_c^* - (1 - \theta)\}. \tag{15.43}$$

15.3.5 Solution for Flame Velocity for Lean Mixtures

The chemical source term contains Y_F and Y_{O2}. For a lean mixture, one may treat Y_{O2} as a constant because O_2 is in excess (similarly, for a rich mixture, Y_F can be treated as constant). Y_F is related to θ (Equation 15.31). Thus, $|\dot{w}_F^*|$ could be expressed in terms of θ. With $\rho = P/RT$, Equation 15.23 can be written as

$$|\dot{w}_F^*| = A_Y \{\rho D/ \dot{m}''^2\} Y_{O2}^{nO2} T_\infty^{\alpha 1} (p/RT_\infty)^{n1}$$
$$[1/hc^*]^{nF} \theta^{\alpha 1} (1/\theta)^{n1} (1 - \theta)^{nF} \exp(-E^*/\theta). \tag{15.44}$$

Using Equation 15.44 in Equation 15.43,

$$-(d\varepsilon_F/d\theta) \{\varepsilon_F h_c^* - (1 - \theta)\}$$
$$= \Lambda (1/\theta)^{n1} \theta^{\alpha 1} (1 - \theta)^{nF} \exp(-E^*/\theta) \tag{15.45}$$

where the normalized inverse flame velocity is given by

$$\Lambda = A_Y \{\rho D/ \dot{m}''^2\} Y_{O2}^{nO2} T_\infty^{\alpha 1} (p/RT_\infty)^{n1}[1/h_c^*]^{nF}. \tag{15.46}$$

The qualitative plot of ε_F vs. θ is shown as curve DCE in Figure 15.3. Now one can integrate Equation 15.45 within limits $\theta = \theta_0$ to 1.

$$-\int_{Y_{F0}}^{0} \{\varepsilon_F h_c^* - (1 - \theta)\} d\varepsilon_F = \Lambda \int (1/\theta)^{n1} (1 - \theta)^{nF} \exp(-E^*/\theta)\theta^{\alpha 1}\, d\theta \tag{15.47}$$

$\int_{Y_{F0}}^{0} \theta d\varepsilon_F$, area within ABECD ≈ area ABCD = $Y_{F,0}$

Hence, the left term of Equation 15.47 becomes,

$$-\int_{YF0}^{0} \{\varepsilon_F h_c^* - (1 - \theta)\} d\varepsilon_F = Y_{F,0}^2 h_c^*/2. \tag{15.48}$$

Using these results in Equation 15.47, and solving for Λ

$$\Lambda = \{Y_{F,0}^2 h_c^*/2\} \Big/ \left\{\int_{\theta 0}^{1} (1-\theta)^{nF} \exp(-E^*/\theta)\theta^{\alpha 1 - n1} d\theta\right\}. \tag{15.49}$$

The appendix of Chapter 5 lists a few approximations for integrating the denominator of Equation 15.49. Consider the integral in the denominator of

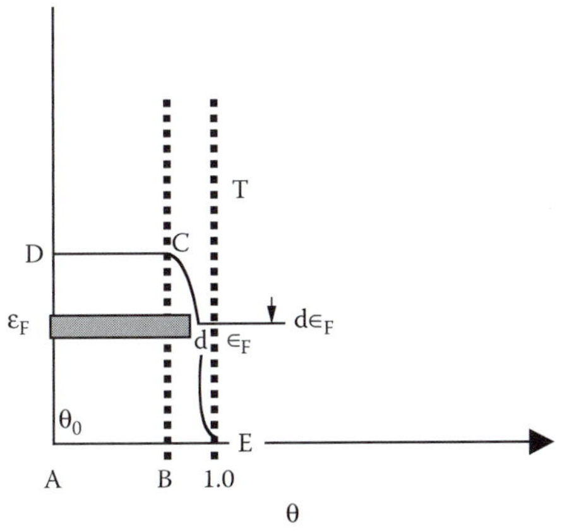

Figure 15.3 Variation of fuel flux ratio ε_F and schematic of integral.

Equation 15.49 and let $n_F = 1$, $\alpha_1 - n_1 = 0$. If one ignores values at lower limits, i.e., $\exp(-E^*) << \exp(-E^*/\theta_0)$ then

$$\int_{\theta 0}^{1} (1 - \theta) \exp(-E^*/\theta)\, d\theta \approx \{\exp(-E^*)/E^{*2}\}$$

$$[1 - 6/E^* + 36/E^{*2}], \text{ error } 3\% \text{ at } E^* = 20, 7\% \text{ at } E^* = 15.$$

Using this result in Equation 15.49 and solving for Λ

$$\Lambda = E^{*2}\, \{Y_{F,0}{}^2 h_c{}^*/2\}\, \exp(E^*)/\{1 - 6/E^* + 36/E^{*2}\}. \qquad (15.50a)$$

Eliminating $h_c{}^*$ with Equation 15.31b

$$\Lambda \approx [\{Y_{F,0}/2\}\{1 - \theta_0\}]\, E^{*2}\, \{\exp(E^*)\}. \qquad (15.50b)$$

Using the definition of Λ given in Equation 15.46,

$$\dot{m}''^2 = \{2A_Y/[Y_{F,0}{}^2 h_c{}^{*nF+1}]\}\, \{\exp(-E^*)/E^{*2}\}\, \rho D\, Y_{O2}{}^{nO2}\, (p/R)^{n1} \qquad (15.51)$$

where $\dot{m}'' = \rho_0\, v_0$. With $n_F = 1$, $\alpha_1 - n_1 = 0$

$$\dot{m}''^2 = \{2A_Y/[Y_{F,0}{}^2 h_c{}^{*nF+1}]\}\, \exp(-E^*)\, \rho D\, Y_{O2}{}^{nO2}\, (p/R)^{n1}/E^{*2}.$$

Solving for flame speed v_o,

$$S = v_0 = \frac{\exp(-E^*/2)}{\rho_0\, Y_{F0}\, h_c{}^*\, E^*} \left(\frac{P}{R}\right)^{n_1/2} \sqrt{2\, A_Y\, \rho D\, Y_{O2}{}^{nO2}}. \qquad (15.52)$$

15.3.6 Effects of Thermophysical and Chemical Properties of Mixture on Flame Velocity

15.3.6.1 Transport Properties

The two main properties that influence flame speed are λ and c_p. Increased conductivity for an adiabatic system causes rapid preheating of a combustible mixture ahead of the wave and hence increases the flame speed. Thus, as seen from Equation 15.52 with ρ,D=λ/c_p, S or $v_o \propto \{1/\rho_0\}\ \{\lambda/c_p\}^{1/2} = \{1/\rho_0^{1/2}\}\ \{\alpha_T\}^{1/2}$. It has been observed experimentally that when a mixture is changed from CH_4:O_2:He to CH_4:O_2:Ar, the flame velocity decreases. As Ar and He are monatomic gases, they have similar $\bar{c}_p$ value (= (5/2) $\bar{R}$) and, hence, similar flame temperatures. But $\{\alpha_{T,He}\}^{1/2} > \{\alpha_{T,Ar}\}^{1/2}$ essentially owing to the lower density of He:CH_4 mixtures. Hence, $S_{Ar} < S_{He}$.

If a mixture is changed from CH_4:O_2:Ar mixture to CH_4:O_2:N_2, $c_{p,Ar}$ (= (5/2) $\bar{R}$) < $c_{p,\ N2}$ (= (9/2) $\bar{R}$) and, hence, flame temperature is larger for the CH_4:O_2:Ar mixture. Therefore, the flame velocity is higher.

15.3.6.2 Order of Reaction

It is seen from Equation 15.52 that S or $v_o \propto \{1/\rho_0\}\ \{P\}^{n_1/2} \propto \{P\}^{n_1/2-1}$, and hence for the second-order reaction, n_1 = 2 and S is independent of pressure P. When n_1 = 1, then increasing P causes a decrease in flame velocity.

15.3.7 Numerical Simulation

Detailed chemical kinetics may be coupled with fluid mechanics. Chemistry complicates the problem because of nonlinear dependence of reaction rates on temperature (stiff equations) and because of steep gradients of temperature and species in the reaction zone; adaptive gridding is necessary.

Example 1

The gas mixture C_3H_8:air is supplied with 50% excess air, h_c (LHV) = 46,357 kJ/kg, c_p = 1.2 kJ/kg K, T_0 = 300 K, $A_{Y,g} = 4.8 \times 10^9$ $sec^{-0.1}$ kg $m^{2.25}$, n_F = 0.1, n_{O2} = 1.65, E = 125604 kJ/kmole, $\lambda_{1200\ K} = 7.63 \times 10^{-5}$ kW/m K for air.

$$|\dot{w}_F''| = A_{Y,g}\ \rho^{nF+nO2}\ Y_F^{\ nF}\ Y_{O2}^{\ nO2}\ \exp(-E/RT) kg/m^3s \qquad (A)$$

Determine the (a) dimensionless flame speed Λ assuming n_F = 1 and (b) laminar burning velocity.

Solution

$$C_3H_8 + 7.5(O_2 + 3.76N_2) \rightarrow 3\ CO_2 + 4\ H_2O + 2.5\ O_2 + 28.2\ N_2$$

$$Y_{F,0} = 44.11/(44.11 + 7 + 5 \times 32 + 28.2 \times 28.02)$$

$$Y_{O2,0} = 0.04106,\ Y_{O20} = 0.2234,\ Y_{N2,0} = 0.7355$$

$$\rho_0 = PM/(RT_0) = 1 \times 28.97/(0.08314 \times 300) = 1.16\ kg/m^3$$

$$Y_{O2,prod} = 2.5 \times 32/(3 \times 44.01 + 4 \times 18.02$$

$$+ 2.5 \times 32 + 28.2 \times 28.01) = 0.0744$$

$$1.2(T_\infty - 300) = 46357 \times 0.0411,\ T_\infty = 1886\ K$$

$$h_c^* = 46357/\{1.2 \times 1886\} = 20.48,\ \theta_0 = 300/1886 = 0.159,\ \theta_\infty = 1.0$$

$$E^* = E/(RT_\infty) = 125604/(8.314 \times 1886) = 8,$$

(a) From Equation 15.50b and assuming $\theta_0 << 1$

$\Lambda \approx Y_{F,0}\{E^{*2} \exp(E^*)\}/2 = 0.04106^* \ 8^2 \exp(8)/2 = 3964$ for $\alpha_1 = n_1$, $n_F = 1$.

(b) $n_F \neq 1$; so we will rewrite the reaction rate Equation A as

$$|\dot{w}_F''| = \{A_{Y,g}\ \rho^{nF+nO2-\alpha 1} Y_F^{(nF-1)}\ T^{\alpha 1}\}\ \rho^{\alpha 1}\ Y_F\ Y_{O2}^{nO2} \exp(-E/RT)$$

$$= A_Y\ \rho^{\alpha 1}\ Y_F\ YO2^{nO2} \exp(-E/RT)$$

where $A_Y = A_{Y,g}\ \rho^{nF+nO2-\alpha 1} Y_F^{\ (nF-1)}\ T^{\alpha 1}$

$T_{avg} = (1886 + 300)/2 = 1094$ K

$\rho_{avg} = PM/RT_{avg} = 1 \times 28.97/(0.08314 \times 1094)$

$= 0.32$ kg/m^3. $n_F = 0.1$, $n_{O2} = 1.65$, $\alpha_1 - n_1 = 0$

$Y_{Favg} = (Y_{Finlet} + 0)/2 = 0.04106/2 = 0.0205$

$Y_{O2avg} = (Y_{O20} + Y_{O2\infty})/2 = (0.2234 + 0.0744)/2 = 0.149$

$A_Y = A_{Y,g}\ \rho^{nF+nO2-\alpha 1} Y_F^{\ (nF-1)}\ T^{\alpha 1} = 4.8 \times 10^9 \times 0.32^{(0.1+1.65-1)} \times 0.0205^{(0.1-1)}$

$= 2.1429E + 09$

$\rho D = \lambda_{1200\ K}/c_p = 7.63 \times 10^{-5}/1.2 = 6.36 \times 10^{-5}$ kg/m sec.

Using Equation 15.51

$$\dot{m}''^2 = \{2\ A/[Y_{F,0}{}^2 h_c^{*\ 0.1+1}]\} \exp(-E^*)\ \rho D\ Y_{O2}{}^{nO2}\ (p/R)^{n1}/E^{*2}$$

$$= 2 \times \{2.1429E + 09/[0.0416^2 \times 20.48^{1.1}]\}$$

$$\exp(-8) \times 6.36 \times 10^{-5} \times 0.149^{1.65} \times (1 \times 28.97/0.08314)^0/8^2) = 0.0536$$

$$\dot{m}'' = 0.0536^{0.5} = 0.23\ \text{kg/m}^2\ \text{sec}$$

$$v_0 = S = \dot{m}''/\rho_0 = 0.23/1.16 = 0.20\ \text{m/sec.}$$

Remarks

(i) If the mixture is stoichiometric, v_0 is computed to be 0.49 m/sec or 50 cm/sec; a multistep reaction model yields similar values of 50 cm/sec — too close for comfort! The velocity estimation is sensitive to the flame temperature.
(ii) If the mixture is either too lean or too rich, the flame velocity decreases (Figure 15.4a and 15.4b).

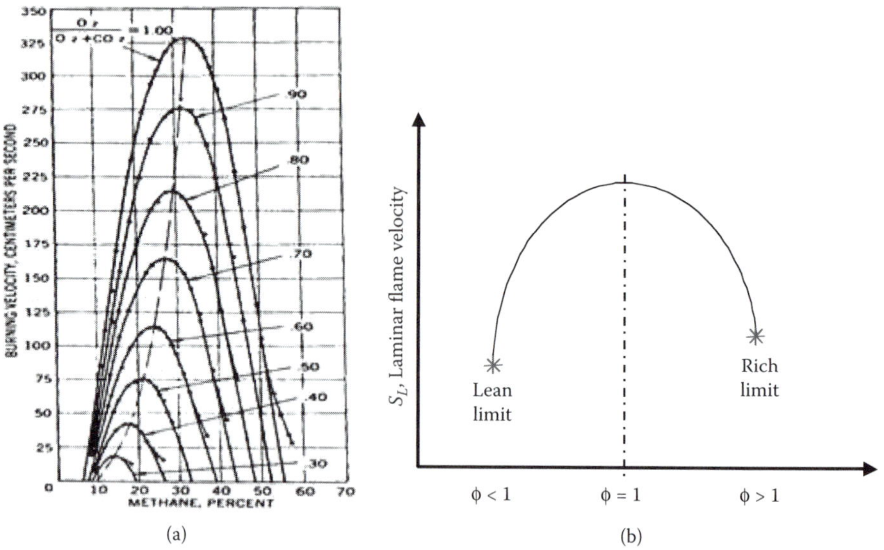

Figure 15.4 (a) Variation of flame velocity with methane percentage [V_{on}Elbe]; (b) Variation of laminar burn velocity with equivalence ratio (SR = 1/Φ).

15.4 Flame Stretching

Consider laminar flame propagation. The flame is of thickness δ. Within this zone, say, CH_4 burns to CO_2 and H_2O; let us assume that the reaction timescale is of the order of t_{chem}. With respect to the standing flame, the unburned gases move into the flame with a velocity of v_0. Thus, a hypothetical person traveling with a unit-mass of unburned gases with a velocity of v_0 would see that the unit mass rises in temperature to the flame temperature within t_{chem}. Consider the special case in which ρ is constant (cold gas assumption). Then $\delta \approx v_0\, t_{chem}$. Hence, $\delta_{cold} \approx v_0\, t_{chem}$. However, we know that the velocity is not constant as we ride on the mass. Hot gas velocity is higher compared to cold gas velocity. Now, $\delta_{hot} \approx t_{chem}\{v_0 + v_{hot}\}/2$, which is higher than δ_{cold}. The flame is said to be stretched owing to temperature rise because molecules are stretched apart, which decreases local concentrations of fuel and hence decreases the local collision rate or local reaction rate. Because the flame is spread over a larger distance at a higher gas temperature, it can increase heat loss to the walls. Comparing two fuels A and B, if $T_{adiab,A} < T_{adiab,B}$, then $\delta_A < \delta_B$ even though the reaction times may be the same.

Consider pure fuel issuing from a small stainless tube. It entrains air that mixes with fuel to form a premixed gas at location x. At x + dx there is more flow because more air is entrained. Suppose the temperature T_u at x differs from T_b at x + dx, the latter location has a higher velocity. The ratio of mass flows is

$$\{\rho v_{x+dx}\, A_{x+dx}\}/\{\rho\, v_x\, A_x\} \tag{15.53}$$

Using the Taylor series expansion,

$$[\rho\ [A\ v_x] + \rho\ [d\ (Av_x)/dx\ dx]/[\rho\ u_x\ A_x] = 1 + \{d\ (v_xA)/dx\}dx\ (1/\{Av_x\}).$$

If $dx \approx \delta_f$ then the second term is

$$\sigma' = \{d\ (v_xA)/dx\}\ [\delta_f/Av_x] = u\delta\ \{dA/dx\}/A + [dv_x/dx]\ [\delta_f/v_x]$$

$$\sigma' = \delta\ (d/dx)\ \ln\ \{A\} + \delta_f\ (d/dx)\ \ln\ \{v_x\}.$$

Ignoring changes in A, and defining strain rate $\sigma = (dv_x/d_x)$

$$\sigma' \approx [dv_x/dx]\,[\delta_f/v_x] = \sigma\ \delta_f/v_x$$

= change in velocity within flame thickness/ original velocity

Define

$$Ka = (\sigma\delta_f\ )/S_L.$$

Ka is called the Karlovitz number and it is a measure of the stretch rate compared to flame velocity. The greater the value of Ka, the greater the flame stretch and heat loss. As the mass flow increases locally (e.g., in jets with more entrainment), the temperature is further lowered from the adiabatic flame temperature based on conditions at x. If the mass flow increase is very high (i.e., high σ' or Ka), then the temperature is too low for flame propagation and extinction occurs. Such stretch occurs in stagnation flows also. For example, in a counterflow diffusion flame, the flame is stretched because of an increase in velocity; at high stretch, the heat loss is so high that extinction occurs [Seshadri and Williams, 1978; Seiser et al., 2000].

If one considers area effects only:

$$\sigma' = \{d(v_xA)/dx\}\ [\delta/Av_x] = u\delta\ \{dA/dx\}/A + [dv_x/dx]\ [\delta/v_x] = \delta\ (d/dx)\ \ln\{A\}$$

One example is expanding a spherical flame for which $\sigma' > 0$. (15.54)

Flame stretch is also defined as d{ln A}/dt. If the flame area becomes larger, the flame is stretched; then the flame velocity is reduced for the same mass flow. Particularly near a lean limit, a reduction in flame velocity causes extinction.

15.5 Determination of Flame Velocity

Figure 15.1 depicts a method of measuring flame velocity v_0 or S_L using a quartz tube of area A_T in which one can visually observe the flame moving at a velocity of v_f. Because one end is closed and other end has an orifice, burned gases expand, pushing the unburned gases out through the orifice. Thus, we can estimate v_u. If v_u is the velocity of unburned gases in the tube, then the velocity of the flame relative to the gases must be $v_f - v_u$. However, the flame front has a parabolic profile with a surface area of A_f and, hence, $S_L\ A_f$ is the rate at which the combustible mixture is consumed. Therefore,

$$S_L\ A_f = A_T\{v_f - v_u\} \tag{15.55}$$

For mixtures involving low flame velocities, one may choose the flat-flame burner technique in which a sintered porous bronze metal disk is used.

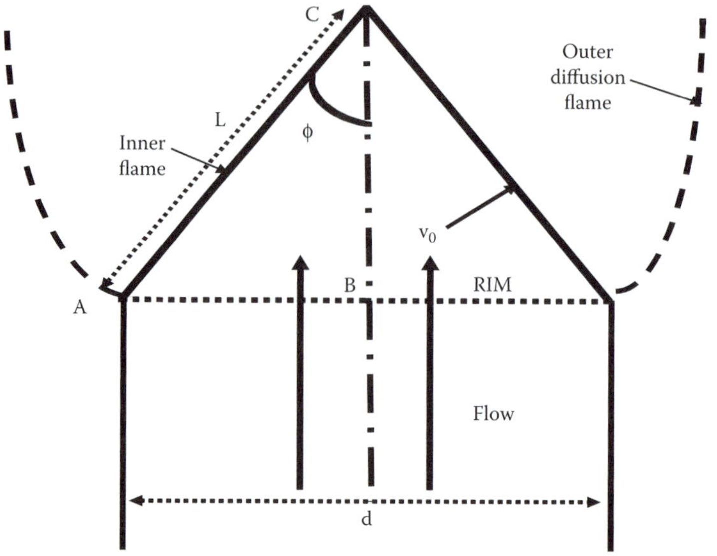

Figure 15.5 Flame velocity using inner flame of Bunsen burner.

Finally, a Bunsen burner can also be adapted for measuring flame velocity. In Bunsen burner, the premixed mixture flows at a velocity v through a tube of diameter d and area $\pi d^2/4$; it forms an inner flame or cone of surface area $A_f > \pi d^2/4$ (because of undisturbed core flow) and an outer diffusion flame because of air entrainment. Because the flame propagates normal to A_f, then

$$S_L * A_f = v\ \pi d^2/4 = \dot{V}, \text{ volumetric flow rate.} \tag{15.56}$$

Knowing the volumetric flow rate $\dot{V}$ and flame (inner cone) area A_f, one can determine S_L. As the conical lateral surface area with base diameter d and slanted length L is (1/2) πd L (Figure 15.5),

$$S_L * (1/2)\ \pi d\ L = v\ \pi d^2/4 \tag{15.57}$$

S_L * (1/2) L = v {d/(2L)} = v sin ϕ, where ϕ is half of the cone angle, (15.58)

where sin ϕ = AB/AC = {d/(2L)}. (15.59)

By enclosing the burner in a pressurized system and conducting experiments at various pressures one can determine the effect of P on S_L. Note: $S_L \propto p^{(n/2-1)}$. C_2H_4: air flames indicates S_L = 1 m/s at p = 0.1 bar, 0.5 m/s at p = 40 bars, while for CH_4, S_L drops from 0.6 m/s at p = 0.1 bar to 0.12 m/s at p = 20 bars.

15.6 Flammability Limits

It is apparent from previous sections that flame velocity decreases to a low value for both lean and rich mixtures due to a decreased average reaction temperature. Above a certain A:F ratio (lean limit) and below a certain A:F ratio (rich limit), there is no solution for flame velocity. These limits are called

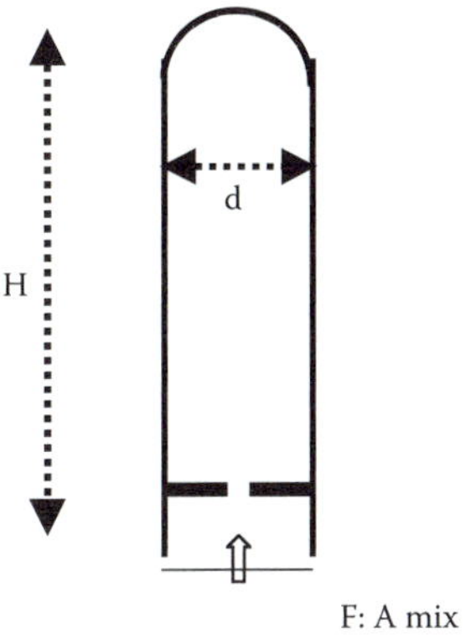

Figure 15.6 Flammability apparatus, H = 1.22 m, d = 5 cm.

flammability limits (Figure 15.4b). Figure 15.6 shows a glass tube apparatus for determining the flammability limits of a fuel: air mixture; a space is provided at the bottom of the tube to show whether a flame can propagate upward. This section presents simplified and detailed analyses for predicting the flammability limits.

15.6.1 Simplified Analyses

Assuming that extinction occurs at 1200 K for most gaseous fuels, one can determine the rich and lean flammability limits of fuel:air flames and express the limit in volume percentage or mass percentage of the fuel. Consider a gaseous mixture of fuel, oxidizer, and inert entering at v_0 and leaving as products at v_∞. The combustion takes place within a distance δ within which T increases from T_0 to T_∞ and Y_F decreases from $Y_{F,0}$ to $Y_{F,\infty}$. Using the following energy balance, one can obtain T_∞ as

$$Y_{F,0}\ h_c = c_p\ (T_\infty - T_0) \tag{15.60}$$

where one must use a weighted average of specific heats of a product mixture at a mean temperature of 900 K.

Note that

$$q = Y_{F,0}\ h_c = c_p\ (T_\infty - T_0). \tag{15.61}$$

If $Y_{F,0}$ is relatively small in a rich flame, then all of the fuel is burned and T_∞ is too low for the reaction to proceed. If $T_\infty < T_{F,L}$ (say 1200 K), then one may reach the lean limit. Thus

$$Y_{F,L} = c_p\ (T_{F,L} - T_0)/h_{c,L} = [c_p\ (T_{F,L} - T_0)/h_{c,L,O2}]/\nu_{O2} \tag{15.62}$$

where h_{cL} could correspond lean conditions, i.e., products CO_2 and H_2O and $h_{c,L,O2} = h_{c,L}/\nu_{O2}$, kJ/kg of O_2.

For a rich mixture, all the O_2 is used.

$$Y_{F,burnt} = Y_{O2,R}/\nu_{O2} = c_p\ (T_{F,R} - T_0)/h_{c,R} \tag{15.63}$$

$$Y_{O2,R} = c_p\ (T_{F,R} - T_0)\nu_{O2}/h_{c,R} = c_p\ (T_{F,R} - T_0)/h_{c,R,O2}. \tag{15.64}$$

If one assumes $T_{F,R} = T_{F,L}$, limiting temperature for flame propagation and heating values based on stoichiometric O_2 are the same either for rich and lean limits, then $Y_{F,L} = Y_{O2,R}\ \nu_{O2}$.

$$Y_{O2,R} + Y_{F,R} + Y_{I,R} = 1,$$

$$Y_{O2,R} + Y_{F,R} + Y_{O2,R}\ (77/23) = 1,$$

Solving

$$Y_{O2,R} = \{1 - Y_{F,R}\}^{*}\ 0.23$$

Hence

$$Y_{O2,R} = \{c_p\ (T_{F,R} - T_0)/h_{c,R,O2}\} = \{1 - Y_{F,R}\}^{*}\ 0.23 \tag{15.65}$$

Or

$$Y_{F,,R} = 1 - \{c_p\ (T_{F,R} - T_0)/(h_{c,R,O2}\ {}^{*}0.23)\} \tag{15.66}$$

The simplified analyses show the dependence of flammability limits on pressure through change in $T_{F,L}$ or $T_{F,R}$. Increased pressure typically causes a decrease in $T_{F,L}$ or $T_{F,R}$ and, hence, widening of limits. Also note that there is decreased heat input per unit mass for rich mixtures through production of CO.

Example 2

Let $T_{F,L} = T_{F,R} = 1515$ K. Consider a CH_4:air mixture, LHV with CO_2 and H_2O = 50,000 kJ/kg, LHV with CO and H_2O = 32,360 kJ/kg, $c_p = 1$, $\nu_{O2} = 4$.

Obtain lean and rich limits.

Solution

$$Y_{F,L} = c_p\ (T_{F,L} - T_0)/h_c = 1 \times (1515 - 300)/50000 = 0.024$$

For the rich limit, $CH_4 + 1.5O_2 = CO + 2\ H_2O$, $\nu_{O2} = 1.5 \times 32/16 = 3$

$$h_{c,R,O2} = h_{c,R}/3 = 32360/3 = 10{,}800 \text{ kJ/kg of } O_2$$

$$Y_{O2,R} = c_p\ (T_{F,R} - T_0)\ \nu_{O2}/h_{c,R} = 1(1515 - 300)\ 3/32360 = 0.11$$

$$Y_{F,R} = 1 - \{c_p\ (T_{F,R} - T_0)/(h_{c,R,O2}\ 0.23)\}$$

$$= 1 - (1(1515 - 300)/(10{,}800 \times 0.23)\} = 0.49$$

$$X_{F,L} = (0.024/16)/\{(0.24/16) + (0.976/29)\} = 0.043 \text{ or } 4.3\%$$

$$X_{F,R} = (0.49/16)/\{.49/16 + 0.51/29\} = 0.635 \text{ or } 64\%$$

For a stoichiometric mixture, $X_F = 1/\{1 + 2 + 2 * 3.76\} = 0.095$, $1 - X_F = 0.905$

$$\varphi_L = (0.0.043/0.957)/(0.095/0.905) = 0.43$$

$$\varphi_R = (0.64/0.36)/(0.095/0.905) = 16.94$$

Remarks

The lean fuel percentage is computed to be 4.3% with $c_p = 1$ kJ/kg K, whereas experimental data gives 5%. However, for the rich limit experimental data indicate that the methane percentage is 15 (by volume), whereas our analysis yields 64%. The reason is the inaccuracy in the c_p value for rich mixtures because fuel forms the dominant gas (c_p of CH_4 is 2.23 kJ/kg K). When c_p of a mixture is as 1.86 (= $0.7c_{pfuel} + 0.3c_{pair}$), the fuel volume percentage under the rich limit is 16%.

15.6.2 Rigorous Analyses

Suppose we have half the combustion wave with preheat from T_0 to T_{Ign} (Figure 15.7), a quarter of the flame zone with heat generation resulting in temperature rise from T_{Ign} to T_1 and a reaction rate proportional at an average temp of $(T_1 + T_{Ign})/2$, and the remaining quarter with heat loss proportional to the local temperature, then the peak temperature T_1 is less than T_∞ (the adiabatic flame temperature). Because the heat transfer coefficient is fixed by the geometry and flow conditions, T_1 is altered by the air:fuel ratio of the combustible mixture. The leaner or richer the mixture, the lower the peak temperature, heat generation, and flame velocity. At certain values of the A:F ratios, called flammability limits, the heat generation is insufficient to overcome heat loss and a flame will not propagate. This is the model used to obtain the relations for flammability limits. Note that the wall of a tube can extinguish the flame not only by heat loss but also by radical destruction; also flame stretching can cause increased heat loss.

Here, we will modify Spalding's analysis of flammability limits in order to be consistent with flame velocity formulations. We will use the SZ formulation and retain the chemical kinetics expression rather than using polynomial approximations. Consider A + B → C; let B be in excess. Thus, A will be completely consumed. For example, if a lean mixture of CH_4:air is used, then O_2 is in excess and, hence, all the fuel will be consumed. In the flame velocity formulation, we assumed adiabaticity. Here, we will formulate with heat loss, then assume an adiabatic condition for a part of the flame with a chemical reaction. Then we will assume frozen flow with heat loss for the remainder of the flame and determination of the peak temperature and flame velocity for a given combustible mixture.

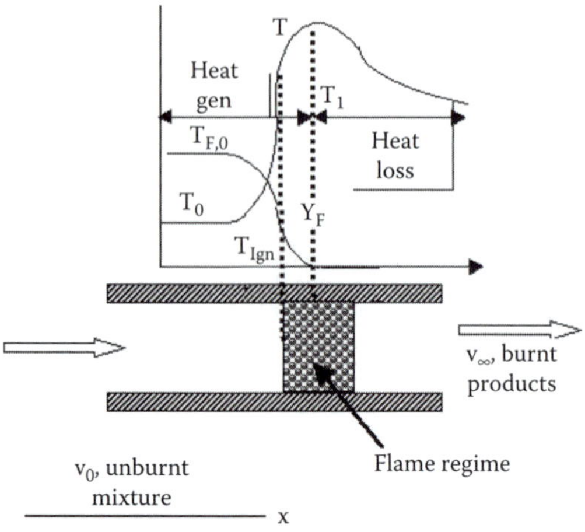

Figure 15.7 Flammability limit model.

15.6.2.1 Species B or O_2 in Excess for Lean Flammability Limit (LFL)

We modify the energy expression as

$$\rho v dh_t/dx = d/dx \{(\rho D dh_t/dx) + |\dot{w}_F|h_c - \dot{q}'''_{loss} \quad (15.67)$$

where $\dot{q}'''_{loss}$, heat loss rate per unit volume (kW/m^3)

$$\rho v d\beta/dx = d/dx \{(\rho D d\beta/dx) - \dot{q}'''_{loss}/h_c \quad (15.68)$$

where $\beta = h_t/h_c + Y_A = h_t^*/h_c^* + Y_A$ is in excess.

Let

$$\xi = \exp\{(\rho v/\rho D)\ x\} = \exp\{x^*\},\ d\xi/\xi = dx^* \quad (15.69)$$

Using Equation 15.68 and simplifying

$$(d^2\beta/d\xi^2) = \{\dot{q}'''_{loss}\ \rho D/(h_c\ \dot{m}''^2\xi^2)\} \quad (15.70)$$

Adiabatic Regime (O < ξ < 1, AC in Figure 15.8): Let us select $x^* = 0$ ($\xi = 1$) at the peak temperature; for the regime $0 < \xi < 1$, the heat loss is almost zero. Integrating Equation 15.70, using the boundary conditions, $\beta = \beta_0$ at $x = -\infty$ (unburned combustible mixture or $\xi = 0$), $\beta = \beta_1$ at $x = 0$ (burned combustible mixture or $\xi = 1$),

$$(\beta - \beta_0)/(\beta_1 - \beta_0) = \xi \quad (15.71)$$

With $\beta = \beta_{ht\text{-}A} = h_t/h_c + Y_A$, we obtain after normalization,

$$(\theta - \theta_0)/h_c^* + Y_A - Y_{A0} = \{\theta_1 - \theta_0)/h_c^* - Y_{A0}\}\xi \quad (15.72)$$

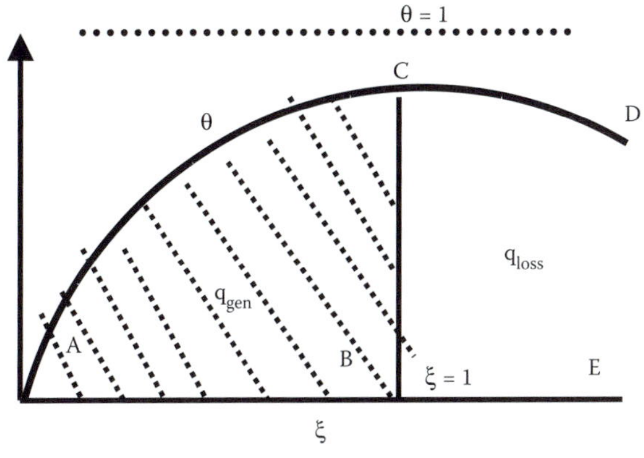

Figure 15.8 Qualitative temperature profile with distance.

The concentration Y_A is solved as

$$Y_A = Y_{A0} - (\theta - \theta_0)/h_c^* + \{(\theta_1 - \theta_0)/h_c^* - Y_{A0}\}\xi = \{1/h_c^*\}\{F(\xi) - \theta\} \quad (15.73)$$

where $F(\xi) = \theta_0 + Y_{A0}\, h_c^* + \xi\, \{(\theta_1 - \theta_0) - h_c^* Y_{A0}\}$.

Note that $Y_A \to Y_{A0}$, $F = F_0 = \theta_0 + Y_{A0}\, h_c^* = 1$ at $\xi = 0$ and $Y_A \to 0$, $F \to \theta$ as $\xi \to 1$. The variation of F is between 1 and θ_1. Differentiating Equation 15.73 and with $(dY_A/d\xi) = \{Y_A - \varepsilon_A\}/\xi$(Equations 15.41 and 15.69), one obtains

$$(d\theta/d\xi) = (\theta_1 - \theta_0) - h_c^*\, Y_{A0} + h_c^*\{\varepsilon_A - Y_A\}/\xi. \quad (15.74)$$

$$\text{Thus, at } \xi = 1,\ (d\theta/d\xi)_{\xi=1} = (\theta_1 - \theta_0) - h_c^*\, Y_{A0}. \quad (15.75)$$

The concentration Y_A in Equation 15.74 could be replaced in terms of θ using Equation 15.73. Thus,

$$\xi\, (d\theta/d\xi) = h_c^*\varepsilon_A + (\theta - \theta_0) - Y_{A0}h_c^*. \quad (15.76)$$

Qualitative temperature profile is shown in Figure 15.8.

From Equation 15.38 with A as species, and Equation 15.69

$$(d\varepsilon_A/d\xi) = -\dot{w}_A''\, \rho D/(\dot{m}''^2\xi) = -|w_A^*|/\xi \quad (15.78)$$

where $w_A^* = \dot{w}_A''\, \rho D/\, \dot{m}''^2$.

The procedure is similar to that adopted for the flame velocity derivation under zero heat loss. Details are left as an exercise problem. Using Equations 15.76, 15.27, and 15.73 in Equation 15.77 and integrating the resultant expression

$$-\int_{YA0}^{0} d\varepsilon_A\, \{\varepsilon_A\, hc^* - (1 - \theta)\} = \Lambda_A \int_{\theta 0}^{\theta 1} \theta^{\alpha 1 - n1}\, \{F - \theta\}^{nA} \exp(-E/RT)\, \theta^{\alpha 1}\, d\theta \quad (15.79)$$

where

$$\Lambda_A = A_Y \{\rho D/ \dot{m}''^2\} Y_B^{nB} T_\infty^{\alpha 1} (p/RT_\infty)^{n1}/h_c^{*nA} \quad (15.80)$$

$$\theta_0 + Y_{A0} h_c^* \approx 1, \text{ and}$$

F varies between 1 and θ_1 (in the flame velocity formulation in Section 15.3, F remains at 1). One may assume F to remain constant at an average value of $\{1 + \theta_1\}/2$. Equation 15.81 is similar to Equation 15.50 except that $F \neq 1$ and the upper integration limits on the right-hand side are not equal to 1.

$$-\int_{YA0}^{0} d\varepsilon_A \{\varepsilon_A h_c^* - (1 - \theta)\} = -\left\{\frac{\varepsilon_A^2 h_c^*}{2}\right\}_{Y_{A0}}^{0}$$

$$-\int_{YA0}^{0} \theta \, d\varepsilon_A \approx \{Y_{A0}{}^2 h_c^*/2 - Y_{A0} + \theta_1 Y_{A0}\}$$

$$= \Lambda_A \int_{\theta 0}^{\theta 1} (F - \theta)^{nF} \exp(-E^*/\theta) \, \theta^{\alpha 1 - n1} \, d\theta \quad (15.81)$$

Note that for a completely adiabatic system $\theta_1 = 1$, but with heat loss $\theta_1 < 1$. Let $n_A = 1$, $\alpha_1 - n_1 = 0$.

Integrating and approximating as before

$$\{Y_{A0}{}^2 h_c^*/2 - (1 - \theta_1)Y_{A0}\} = \Lambda_A \int_{\theta 0}^{\theta 1} (F - \theta) \exp(-E^*/\theta) \, d\theta. \quad (15.82)$$

Details of integration are left as an execise problem.

$$\int_{\theta 0}^{\theta 1} [F - \theta] \exp(-E^*/\theta) \, d\theta = \{\theta_1{}^2/E^{*2}\}\exp(-E^*/\theta_1)\}$$

$$\{F(E^* - 2\theta_1 + 6\theta_1{}^2/E^* - 24\theta_1{}^3/E^{*2} + 120\theta_1{}^4/E^{*3})$$

$$- \theta_1 E^* + 3\theta_1{}^2 - 12\theta_1{}^3/E^* + 60\theta_1{}^4/E^{*2}\} \quad (15.83)$$

where $F \approx (1 + \theta_1)/2$. Because $Y_{A,0} h_c^*/2 = (1 - \theta_0)/2$, the nondimensional flame velocity is given as

$$\Lambda_A = Y_{A,0} \, [(1 - \theta_0)/2 - (1 - \theta_1)]/[\{\exp(-E^*/\theta_1)\}$$

$$\{F(E^* - 2\theta_1 + 6\theta_1{}^2/E^* - 24\theta_1{}^3/E^{*2} + 120\theta_1{}^4/E^{*3})$$

$$- \theta_1 E^* + 3\theta_1{}^2 - 12\theta_1{}^3/E^* + 60\theta_1{}^4/E^{*2}\} \quad (15.84)$$

where θ_1 is yet to be determined. One may use a spreadsheet program assuming various values of θ_1 to get flame velocity solutions. The value of θ_1 that is less than 1 depends on the heat loss from the reactor. Figure 15.9 plots $\Lambda_A/Y_{A,0}$ vs. θ_1 for $E^* = 8$, $\theta_0 = 0.125$. It is seen that $\Lambda_A/Y_{A,0}$ has a peak at a certain θ_1, nonadiabatic regime ($1 < \xi < \infty$, CD in Figure 15.8). When heat loss is dominant (region CBED in Figure 15.8) and assuming a frozen condition (Equation 15.67)

$$d^2(h_t/h_c)/d\xi^2 = \{\dot{q}'''_{loss} \, \rho D/(h_c \; \dot{m}''^2 \xi^2)\}. \quad (15.85)$$

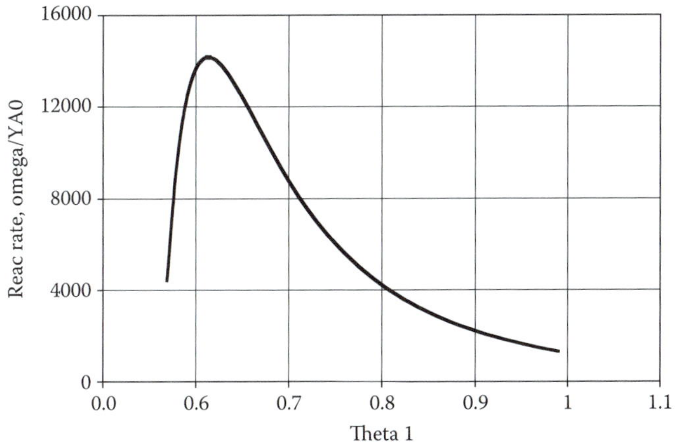

Figure 15.9 Nondimensional reaction rate.

Normalizing

$$d^2\theta/d\xi^2 = \{\dot{q}'''_{loss}\ \rho D/(c_p\ T_{\infty,ideal}\ \dot{m}''^2\xi^2)\}. \quad (15.86)$$

The BCs are: $\theta = \theta_1$ at $\xi = 1$ and $d\theta/d\xi = 0$ at $\xi = \infty$ (Figure 15.8). Assuming that heat loss is proportional to the polynomial in the temperature difference

$$\dot{q}'''_{loss} = C_L(T - T_0)^m = \{C_L T_\infty\}^m\ (\theta - \theta_0)^m,\ m = 1 \text{ to } 5. \quad (15.87)$$

For example, for convective heat loss, m = 1. (15.88)

For a cylindrical container of length dx and diameter D,

$$\dot{q}'''_{loss} = \{h_H\ (T - T_0)\pi d\ dx\}/\{\pi(d^2/4)\ dx\} = C_L\ (T - T_0)\} \quad (15.89)$$

where $C_L = 4\ h_H/d$. Similarly, for radiaton heat loss $\dot{q}_{loss} = \varepsilon\sigma\ (T^4 - T_0^4)\ \pi d\ (dx)$

$$\dot{q}'''_{loss} = 4\ \varepsilon\sigma\ (T^4 - T_0^4)/d \approx C_L\ (T - T_0)^m,$$

where $C_L = 4\varepsilon\sigma/D$ and m needs to be suitably determined.

Let us assume that the heat loss per unit volume is constant. Then

$$\dot{q}'''_{loss} \approx \{C_L T_\infty\}^m\ (\theta_1 - \theta_0)^m,\ \text{constant} \quad (15.90)$$

$$q_{loss}{}^* = \{C_L T_{\infty,ideal}\}^m\ \rho D/(c_p\ T_{\infty,ideal}\ \dot{m}''^2). \quad (15.91)$$

Using Equation 15.91 in Equation 15.86 and integrating twice

$$(d\theta/d\xi) = -\{q_{loss}{}^*\}(\theta_1 - \theta_0)^m/\xi \quad (15.92)$$

$$\theta - \theta_1 = \{q_{loss}{}^*\}(\theta_1 - \theta_0)^m\ \ln\ (\xi) \quad (15.93)$$

Further, at $\xi = 1$, $\{d\theta/d\xi\}_1 = (\theta_1 - \theta_0) - Y_{A0}\ h_c{}^*$.

Equating with Equation 15.92 and solving for θ_1,

$$\theta_1 = 1 - q_{loss}^*(\theta_1 - \theta_0)^m = 1 - C_L T_\infty^{m-1} (\theta_1 - \theta_0)^m \rho D/(c_p \dot{m}''^2). \quad (15.94)$$

For the convective and radiation cases, Equation 15.94 is simplified as

$$(\theta_1 - \theta_0) = K_{conv} (1 - \theta_1) \quad (15.95)$$

$$(\theta_1^4 - \theta_0^4) = K_{rad} \{1 - \theta_1\} \quad (15.96)$$

$$\text{where } K_{conv} = [A\ Y_B^{\ nB}\ T_\infty^{\alpha 1}\ (p/RT_\infty)^{\ n1}/h_c^{*nA}\}/\{(4\ h_H/dc_p)\Lambda_A\}] \quad (15.97)$$

$$K_{Rad} = \{A\ Y_B^{nB}\ T_\infty^{\alpha 1}\ (p/RT_\infty)^{n1}/h_c^{*nA}\}]/[(\Lambda_A\ T_\infty^{3}4\ \varepsilon\sigma/d\ c_p)]. \quad (15.98)$$

An iterative procedure is employed to solve for LFL.

Example 3

Determine the maximum excess-air percentage for a CH_4:air flame with A = 1.3×10^{10} (kmol, m, sec) and E = 202,600 kJ/kmol, $n_F = 0.3$, $n_{O2} = 1.3$. Assume d = 5 cm M = 28.97 for mixture. $\alpha_1 - n_1 = 0$, λ = 7.63E-05 kW/(m-K), h_c = 50,016 kJ/kg, and c_p = 1 kJ/kg K. Use the convective loss case.

Solution

Here, the heat loss is primarily convective, and we can assume θ_1. Use Equation 15.84 first to obtain the nondimensional flame velocity Λ_A, and then Equation 15.80 to obtain the dimensional velocity. Then check the assumed θ_1 with Equation 15.95; if the check fails, iterate and thus get the solution for flame velocity with heat loss. The excess air is then gradually increased. The problem yields no solution for excess air above 58.7% or for $\theta_1 < 0.8889$. The flame velocity at the LFL is estimated as 23 cm/sec.

15.6.2.2 Fuel (A) in Excess for Rich Flammability Limit (RFL)

The heat loss formulation is unaffected. Only the flame velocity formulation is affected. From exercise problem:

Let $n_A = 1$, $\alpha_1 - n_1 = 0$.

$$\dot{m}''^2 = \{2A\ n_{O2}^{\ 2}\ Y_F^{\ nF}\ (P/R)^{\alpha 1}\ \rho D/(h_{c,R}^{\ *2}(Y_{O2,0})\ E^{*2})\}\ \theta_1^{\ 2}\ \{\theta^{\alpha 1}\ \exp(-E^*/\theta_1)\}$$
$$\{F\ (E^* - 2\theta_1 + 6\theta_1^2/E^*) - \theta_1\ E^* + 3\theta_1^2 - \theta_1^3\ 12/E^*\}]/[-(Y_{O2,0}/\nu_{O2})$$
$$+ 2(\theta_1 - \theta_0)/h_c^*] \quad (15.99)$$

where $A = A_{given}\ (P/R)^{n1-\alpha 1} Y_{O2}^{(nO2-1)}$.

Care must be taken for the rich flammability limit, under which the fuel could form in significant amounts, affecting c_p of mixtures; further, the fuel may burn only to CO.

15.6.2.3 Spalding's Explicit Results

Spalding assumed that reaction rate is proportional to $[\{\theta_1 - \theta_0\}/\{1 - \theta_0\}]^n$, and heat loss rate is proportional to $[\{\theta_1 - \theta_0\}/\{1 - \theta_0\}]^m$.

$$\Lambda_{Spald} = A\ h_c\ \lambda_\infty \dot{w}_F'''\ (T_\infty, Y_{AO})/\{(T_\infty - T_0)(\dot{m}''\ c_p)^2\}$$

$$= \{\tfrac{1}{2}\}\ \{(n + 1)(n + 2)\}/\{(\theta_1 - \theta_0)/(1 - \theta_0)\}^{n+2} \tag{15.100}$$

$$\frac{\dot{Q}'''(T_\infty)}{\dot{w}F'''(T_\infty, Y_{A0})h_c} = K = \frac{2\left\{\frac{1-\theta_1}{1-\theta_0}\right\}\left\{\frac{\theta_1-\theta_0}{1-\theta_0}\right\}^{(n+2-m)}}{(n+1)(n+2)} \tag{15.101}$$

$$\Lambda_{Spald} = [\{\theta_1 - \theta_0\}/\{1 - \theta_0\}]^{-m}[\{1 - \theta_1\}/\{1 - \theta_0\}] \tag{15.102}$$

where there are two solutions for θ_1 for $K < K_c$, but there is no solution for $K > K_c$; $K_c = 4.95 \times 10^{-4}$ for $n = 11$, $m = 4$.

15.6.3 *Empirical Methods*

Consider fuel $C_c\ H_h\ O_o$. If z is stoichiometric and O_2 is in moles then the combustible mole percentage in the air:fuel mixture is expressed [Coward and Jones, 1952] as:

$$C_{st} = \frac{100}{1 + \left(\frac{z}{0.21}\right)} \tag{15.103}$$

LFL = 0.55 C_{st} and RFL = 3.50 C_{st}.

Shimy hypothesized that flammability limits of all hydrocarbons are related to the energy required to break the bond between two carbon atoms. Further, the lean limit of hydrocarbons depends on the number of C atoms in the chain but is independent of the number of hydrogen atoms. For paraffinic and Olefinic HC [Shimy 1970],

$$LFL = \frac{6}{c} + 0.2\ , \quad \text{mole percentage} \tag{15.104}$$

$$RFL = \frac{60}{h} + \frac{c}{20} + 2.2\,. \tag{15.105}$$

As C increases, the molecular weight increases with increased molar heating value and reduced mole percentage at the lean limit (c_p dominated by air). However, in the rich limit, unburned HC alters the c_p value and the relations are more complex. The following table shows a comparison of experimental, empirical, and text.

15.6.4 Temperature and Pressure Dependencies

If the premixed combustible mixture is heated, then the LFL is lowered and the RFL is increased as shown in the following text (all are volume fractions or mole fractions) [Zabetakis et al., 1959]

	Lean Flammability Limit, mole %					*Rich Flammability Limit, mole %*			
	Exp.	*Text*	*C_{st}*	*Bond*		*Exp.*	*Text*	*C_{st}*	*Bond*
$CH_{4\ 2}$	5.0–5.35	9.5	5.23	6.20	$CH_{4\ 2}$	14.2–15.0	9.6	33.26	17.25
C_3H_8	2.0–2.3	2.7	2.22	2.20	C_3H_8	7.3–9.5	10.2	14.11	9.85
C_4H_{10}	1.5–2.0	2.1	1.72	1.70	C_4H_{10}	6.5–9.0	8.2	10.95	8.40
C_6H_{14}	1.1–1.5	1.5	1.19	1.20	C_6H_{14}	6.9–7.7	5.6	7.57	6.79
C_7H_{16}	1.0–1.3	1.3	1.03	1.06	C_7H_{16}	6.7–7.0	4.9	6.56	6.30
$C_8H_{18\ 1}$	0.8–1.1	1.2	0.91	0.95	$C_8H_{18\ 1}$	3.2–6.5	4.2	5.78	5.93

$$LFL(T) = LFL(25C)\left(\frac{1-0.75*(T\,in C-25)}{\bar{h}_C}\right) \tag{15.106}$$

$$RFL(T) = RFL(25C)\left(\frac{1+0.75*(T\,in C-25)}{\bar{h}_C}\right) \tag{15.107}$$

Whereas the LFL is relatively unaffected with change in pressure except at very low pressures (<20 kPa), the RFL changes [Zabetakis et al., 1965a]

$$RFL(T) = RFL(1atm) + 20.6*(\log_{10} p + 1) \quad \text{where p in MPa.}$$

15.6.5 Flammability Limit of Multiple Fuel and Inert Mixtures

The Le Chatelier Rule (LCR) may be adopted to calculate the LFL and RFL for most mixtures with a few exceptions (e.g., ethylene and CO; ethylene and H_2) [Wierzba and Karim, 1998; Coward and Jones, 1952]. The LCR states that the LFL and RFL of mixtures of fuels follow the rules given below:

$$LFL = \frac{100}{\left(\sum_k \frac{X_k}{LFL_k}\right)} \tag{15.108}$$

$$RFL = \frac{100}{\left(\sum_k \frac{X_k}{LFL_k}\right)} \tag{15.109}$$

For example, LFL of HC mixture, which is 70% CH_4, 20% C_2H_6, and 10% C_3H_8 is given as 3.9% according to LCR.

15.7 Quenching Diameter

15.7.1 Definition

Inside a tube, a propagating flame transfers heat to the unburned mixture, raises the temperature to T_{ign}, and sustains the flame propagation rate. However, when the flame transfers heat to the colder walls, heat loss occurs to the ambient. As the heat loss increases, the average reaction temperature $T_R = (T_1 + T_0)/2$ decreases, reducing the chemical heat liberation rate. Heat loss to the walls is increased, particularly with a reduction in the diameter of the tube. As a result, when the diameter is reduced below d_{quench}, the heat loss becomes so significant that all the heat generated is just used to overcome heat loss rather than to preheat the unburned combustible mixture. Thus, the flame remains stationary without propagation. Any slight increase in heat loss or reduce in diameter results in extinction. For all practical purposes, at d_{quench}, the flame extinguishes rather than propagates. Further, the wall could act as the third body for radical recombination reactions and reduce radical concentrations. Hence, there exists a critical diameter, called the *quenching diameter*, below which the flame will not propagate.

For a tube of diameter d, the ratio of the flame surface to flame volume within a distance δ is $\pi d\ \delta/(\pi d^2\ \delta/4) = 4/d$. Thus, as d is reduced, the ratio of surface area to volume is too large for the flame to propagate. For a parallel plate separated by L_p, $W\ \delta/(W \delta Lp) = 1/L_p$, where W is the width of the plate.

Experimentally, one can determine d_{quench} using a Bunsen burner. If the combustible mixture is suddenly cut off, the flame flashes back at a certain mixture concentration. As we gradually decrease the burner diameter, then at a certain d_{quench}, the flame will not flash back. Or we could maintain the diameter constant and change the concentration until it is too lean or too rich for flashback to occur.

Various d_{quench} values are as follows: H_2, 0.6 mm for stoichiometric mixture of H_2 at 20 C, 1 bar; 1.9 mm for CH_4, 2.1 mm for propane (see Bartok and Sarofim [1991]).

Thus, if a premixed combustible mixture flows through a burner of diameter D and if a disk of a diameter slightly larger than D with many holes each of diameter $d < d_{quench}$ or filling a segment of tube within steel wool (called a *flash arrester*) is placed over the burner, the flame will not propagate back or flash back.

15.7.2 Simplified Analyses

Let us consider a duct of cross-sectional area A and circumference C. When heat generation over volume V equals the heat loss rate, then

$$|\dot{w}_F'''|V\ h_c = h_H\ A_{HL}\ (T_{avg} - T_w) \qquad (15.110)$$

where $T_{avg} = \{T_b + T_0\}/2$, $T_b < T_\infty$, T_b is the temperature of the burned products; V, volume of reaction zone at the time when the temperature is T_{avg}, h_H heat transfer coefficient, A_{HL} heat loss area, and T_w wall temperature. If $T_w = T_0$, then

$$| \dot{w}'''_F | h_c = h_H (A_{HL}/V) (T_b - T_0)/2$$

where the reaction rate per unit volume is related to the flame velocity v_0 as

$$v_0 = S = [\alpha_T \dot{w}'''_{F,avg}/\{\rho_0 Y_{F,0}\}]^{1/2}$$

$$\{\rho_0 Y_{F,0} v_0^2/\alpha_T\} h_c = h_H (A_{HL}/V) (T_b - T_0)/2. \quad (15.111)$$

(i) Parallel Plates: For flow between two plates of width W spaced apart by d, $V/A_{HL} = (d W\delta)/(2w\delta + 2d\delta) \simeq \delta/2$. With $Nu = h_H d/\lambda$, $d = [\lambda(T_b - T_0) Nu\alpha_T/\{\rho_0 Y_{F,0} v_0^2 h_c\}]/2$.

(ii) Cylindrical Geometry: Here, flow occurs through a cylindrical duct.

$$V = (\pi d^2/4)\ \delta,\ A_{HL} = \pi d\ \delta$$

Using Equation 15.110, and $\rho_0 Y_{F,0}\ v_0\ h_c = \rho_0 v_0\ c_p\ (T_\infty - T_0)$, then solving for d

$$d = \frac{\alpha_T}{v_0}\sqrt{2Nu\,\frac{T_b - T_0}{T_\infty - T_0}}. \quad (15.112)$$

If we assume $\sqrt{\frac{T_b - T_0}{T_\infty - T_0}} \approx 1$, then

$$d \approx \frac{\alpha_T}{v_0}\sqrt{2Nu} = \delta_f\sqrt{2Nu} \quad (15.113)$$

where δ_f is the flame thickness, or

$$\{\rho_0 Y_{F,0} v_0^2/\alpha_T\}\ A\ h_c = h\ C\ (T_b - T_0)/2. \quad (15.114)$$

(iii) Spherical Geometry: Consider an infinite source of combustible mixture, say, of spherical geometry. In this case, we ignite a small spherical region of diameter d at the center and then check whether the flame can propagate radially outward. If the radius is extremely small, the volume of the source is small and, hence, the amount of thermal energy released is small compared to the heat loss from the same region to the rest of the cold combustible mixture. On the other hand, if the diameter is large, it readily propagates outward. We must therefore determine the critical diameter below which the flame will not propagate. Using Equation 15.110 with $V = \pi d_{crit}^3/6$, $A_{HL} = \pi d_{crit}^2$ $Nu = h_H\, d/\lambda$,

$$d_{crit}^2 = 3\ Nu\ \lambda\ \alpha_T\ (T_b - T_0)/\{\rho_0 Y_{F,0} v_0^2\ h_c\}$$

which is similar to the expression for a cylindrical duct. Following the derivation for cylindrical geometry

$$d_{crit} = \frac{\alpha_T}{v_0}\sqrt{3\ Nu\ \frac{T_b - T_0}{T_\infty - T_0}} \tag{15.115}$$

$$d_{crit} \approx \frac{\alpha_T}{v_0}\sqrt{3\ Nu} = \delta_f\sqrt{3 Nu},\ (T_b - T_0)/(T_\infty - T_0) = 1 \tag{15.116}$$

where δ_f is the flame thickness. d_{crit} is somewhat analogous to d_{quench}. With $D = \alpha_T = 10^{-4}$ m²/sec, $v_0 = 0.4$ m/sec, Nu = 2, $d_{crit} = 0.6$ mm.

Also, it is seen that

$$\frac{d_{crit}}{\delta_f} \approx \sqrt{3 Nu} = \sqrt{6} = 2.45$$

where Nu = 2 for a quiescent atmosphere. As $d_{crit} = 0.6$ mm, then $\delta_f \approx 0.25$ mm or 250 μm.

15.7.3 Effect of Physical and Chemical Properties

Because $\rho_0 D_0$ is independent of P, it is apparent from Equation 15.113 and Equation 15.115 $d_{quench} \propto 1/\dot{w}_F'''(\theta_{1c})^{1/2} \propto 1/P^{n/2}$. Thus, the quench diameter is smaller at high pressures. Also, $d_{quench} \propto [\rho_0 D_0/\dot{w}_F'''(\theta_{1c})]^{1/2} \propto D_0^{1/2}/\rho_0^{(n-1)/2}$. If n = 1, it is independent of density and as the value of D_0 increases, the d_{quench} increases. Because $D_0 \propto 1/M^{1/2}$, the higher the molecular weight, the lower the D_0 and hence $d_{quench,CO2} < d_{quench,He}$, where CO_2 and He are inerts used with oxygen and fuel. The quench distance is sensitive to thermal diffusivity; the lower the diffusivity, the lower the d_{quench}. Kuo shows that $d_{quench,\ He} < d_{quench,\ Ar} < d_{quench,\ N2} < d_{quench,CO2}$ because $\alpha_{He} < \alpha_{Ar} < \alpha_{N2} < \alpha_{CO2}$.

15.8 Minimum Ignition Energy for Spark Ignition

Suppose we have a spherical vessel containing premixed gases, which is heated by a spark. The electrical power to the spark plug is obtained by discharging stored energy from capacitors. The energy discharged by a capacitor is

$$E_{capac} = (1/2)\ C\ (V_1^2 - V_2^2),\ \text{where} \tag{15.118}$$

V_1: voltage prior to spark, V_2: voltage after spark, and C: capacitance.

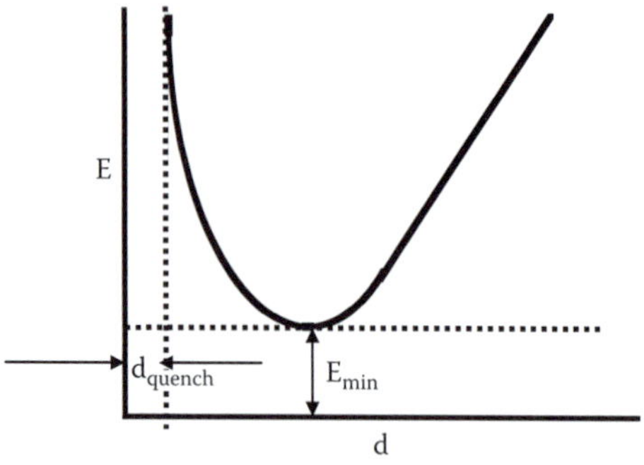

Figure 15.10 Ignition energy vs. spark gap.

As we supply spark energy, the combustible gas mixture is heated and ignites. However, the spark is applied over a small gap. If the gap becomes larger, we have to supply more energy. If it is too small, heat loss via the wire is so high that the spark energy increases. Thus, there exist a minimum (Figure 15.10).

$$E_{min} = \kappa d_{quench}^2 \tag{15.119}$$

where E_{min} is the energy required to raise the temperature of the combustible gases. $d_{quench} = [(4\ \rho_0\ D_0 Nu)(\theta_{1c} - \theta_0)Y_{F,0}/\{(1 - \theta_{1c})\ \dot{w}_F'''\ (\theta_{1c})\}]^{1/2}$, and κ is a constant. An empirical relation is given by Calcote et al. (1952): $E_{min}(J) = 2400\ d_{quench}$ in m. The direct correlation between E_{min} and d_{quench} is validated by extensive experimental data as shown in Figure 15.11, which shows the results for a stoichiometric mixture (for E_{min}) at 1 bar.

Recall that d_{quench} is as determined from the cylindrical duct correlation. The heat generation is $\pi\ d_{quench}^2\ \delta/4$. The flame thickness is roughly fixed at $\delta_f = \alpha_T/v_0$ for a given mixture. Energy within that volume is

$$E_{min} = \pi\rho_0 d_{quench}^2\ \delta\ c_p\ (T_I - T_0)/4 = \pi\ d_{quench}^2\ \rho_0\ \{\alpha_T/v_0\}\ c_p\ (T_I - T_0)/4$$

$$= \kappa\ d_{quench}^2; \tag{15.120}$$

where

$$\kappa = \pi\ \rho_0\{\alpha_T/v_0\}\ c_p\ (T_I - T_0)/4 = \pi\ (\lambda/v_0)\ (T_I - T_0)/4,\ kJ/m^2$$

$$= [1/\rho_0]\ [\lambda\ \dot{w}_F'''/\{Y_{F,0}\ c_p\}]^{1/2}. \tag{15.121}$$

Typically, $\kappa \approx 0.07$ kJ/m² [Kanury]. If d_{quench} equals 4 mm for the propane:air mixture, minimum ignition energy is estimated as $0.07^* (4 \times 10^{-3}\ m)^2 = 1.12 \times 10^{-6}$ J.

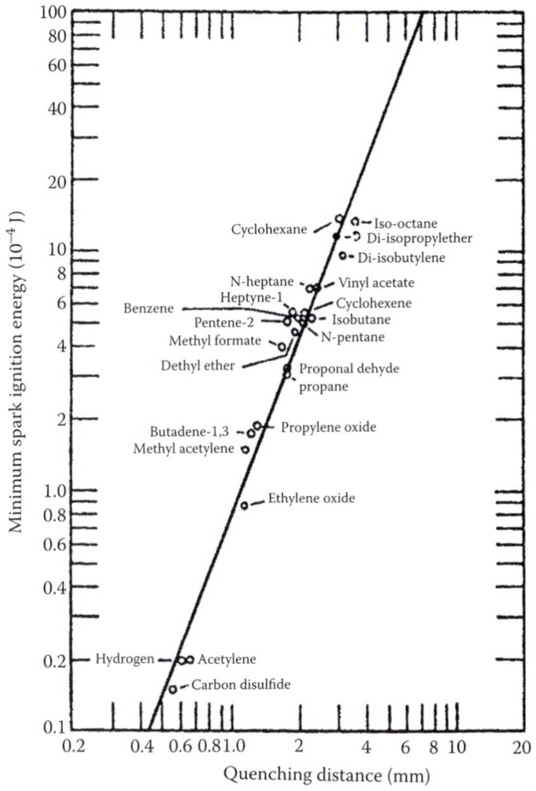

Figure 15.11 Minimum ignition energy vs. quenching distance for different fuels. (Adapted from Calcote, H.F., Gregory, C.A. Jr., Barnett, C.M., and Gilmer, R.B., *Ind. Eng. Chem.*, 44, 2656, 1952.)

Using the d_{quench} relation,

$$E_{min} = \pi\, d_{quench}{}^2\, \rho_0\, \{\alpha_T/v_0\}\, c_p\, (T_I - T_0)/4 = \pi\, \alpha_T\, [4\, Nu\, \rho_0{}^3 D_0 Y_{F,0}{}^{3/2}\{(\theta_{1c} - \theta_0)/(1 - \theta_{1c})\} c_p{}^{3/2}\, \dot{w}_F'''{}^{-3/2}(\theta_{1c}) \lambda^{-1/2}\, (T_I - T_0). \quad (15.122)$$

If n, order of reaction, $E_{min} \propto P^{3-2-3n/2} \propto P^{1-3n/2}$. With n = 1, $E_{min} \propto P^{3-2-3n/2} \propto P^{-1/2}$. As P is increased, as in automobile engines, the ignition energy decreases. Ignition energy decreases with reduced ignition temperature, reduced c_p, and reduced Nu number.

The amount of thermal energy within a sphere of diameter d_{crit} is the minimum ignition energy required to propagate the flame. Thus, it can be shown that

$$E_{min} = (\pi/6)\, \rho_0\, Y_{F,0}\, h_c\, (T_0/T_{avg})\, \delta_f^3\, \{3\, Nu\}^{3/2}. \quad (15.123)$$

The minimum ignition energy of HC gases is about 0.1 to 1 mJ, methane: 0.29 mJ and carpet sparking 40 mJ.

Example 4

A premixed gas mixture of C_3H_8:air is burned in a tube of diameter 1 cm., h_c (LHV) = 46,357 kJ/kg, c_p = 1.2 kJ/kg K, T_0 = 298 K, A = 4.8E + 08 (kg, m, s),

$n_F = 0.1$, $n_{O2} = 1.65$, E = 125,604, $\rho D = 6.36 \times 10^{-5}$ kW/m K for air, Nu = 3.66 (Increpora), $\alpha_1 = 0$. Nu for laminar flow through cylinder: 3.66.
$|\dot{w}''_F| = A_{given}\, \rho^{nf+nO2}\, Y_F^{nF}\, Y_{O2}^{nO2} \exp(-E/RT)$.

(a) Determine the LFL.
(b) Determine the RFL.
(c) Plot peak temperature and flame velocity vs. equivalence ratio.
(d) Study the effect of reducing the diameter on flame velocity and determine the quench diameter.

Solution

(a) We need to determine the flame velocity vs. various percentages of excess air and determine the LFL. Suppose 57% excess air is supplied, then

$$C_3H_8 + 5 \times 1.57(O_2 + 3.76N_2) \rightarrow 3CO_2 + 4\ H_2O + 2.85O_2 + 29.52N_2.$$

Here, fuel is the deficient species and is burned completely. Using a similar procedure as in Example 1,

$$Y_{A,0} = 0.0393,\ Y_{B0} = 0.224,\ Y_{N20} = 0.737,\ \rho_0 = 1.16\ \text{kg/m}^3.$$

Similarly, on the product side, $Y_{O2prod} = 0.0813$.

$$\rho_0 = PM/(\bar{R}T_0) = 1 \times 28.97/(0.08314 \times 300) = 1.16\ \text{kg/m}^3$$

$$T_\infty = Y_{A0}{}^*\ h_c/c_p + T_0 = 1816\ \text{K},\ \theta_0 = T_0/T_\infty = 0.164,\ h_c{}^* = 20.4,$$

$$E^* = E/R\ T_\infty = 8.318$$

As in Example 3,

$$T_{avg} = 1094\ \text{K},\ \rho_{avg} = 0.32\ \text{kg/m}^3.\ n_F = n_A = 0.1,\ n_{O2} = n_B$$
$$= 1.65,\ \alpha_1 = 0,\ Y_{Aavg} = 0.0393/2 = 0.0196,\ Y_{Bavg} = 0.153$$

$$A_Y = A_{given}\ \rho^{nF+nO2-\alpha 1} Y_A{}^{(nA-1)}\ T^{\alpha 1} = A_{given}\ \rho^{nA+nB-\alpha 1} Y_{F,avg}{}^{(nA-1)}\ T^{\alpha 1} = 2.3635E + 09$$

Assume $\theta_1 = 0.99$ ($T_1 = 1798$ K)

$$\dot{m}''^2 = \{2A_Y\ Y_B{}^{nB}\ (P/R)^{\alpha 1}\ \rho D/(h_c{}^{*2} Y_{A,0}\ E^{*2})\}\ \theta_1{}^2\ \{\theta^{\alpha 1} \exp(-E^*/\theta_1)\}$$
$$\{F\ (E^* - 2\theta_1 + 6\ \theta_1{}^2/E^*) - \theta_1\ E^* + 3\theta_1{}^2 - \theta_1{}^3\ 12/E^*\}]/$$
$$[-Y_{A,0} + 2(\theta_1 - \theta_0)/h_c{}^*].$$

Thus $\dot{m}''^2 = 0.050306966$, $\dot{m}'' = 0.224$, v = 0.224/1.16 = 0. 192 m/sec.

However, we have to check whether the assumed value of $\theta_1 = 0.99$ ($T_1 = 1798$ K) is correct by checking with heat loss calculations.

$$\theta_1 = 1 - \{4\ \text{Nu}\ [\rho D/(d\,\dot{m}'')]^2\}(\theta_1 - \theta_0)$$

$$= 1 - (4 \times 3.66 \times [6.36 \times 10^{-5}/0.01 \times 0.224]^2)(0.99 - 0.164)$$

$= 0.99025$, which is close to the assumed value.

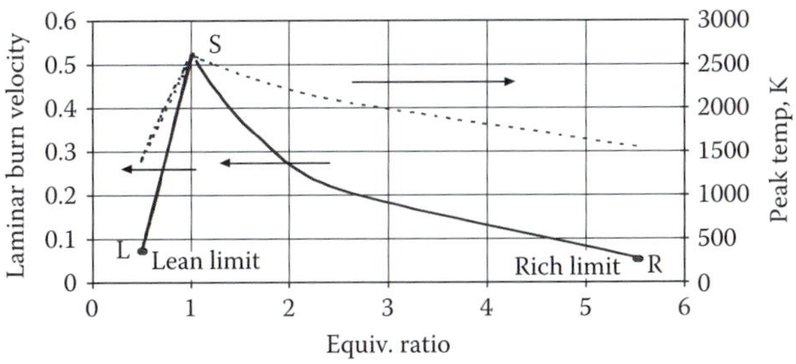

Figure 15.12 Variation of flame velocity and peak temperature with equivalence ratio.

(b) Similarly, we repeat with an excess-air percentage of –82% (i.e., rich mixture), we will have fuel left unburned. Thus, fuel-specific heat being much higher, (4.68 kJ/kg K at T = 1500) it was used in calculating mixture-specific heats. A:F = 2.05, T_∞ = 1652 K, θ_1 = 0.938, T_1 = 1551 K, v = 0.0543 m/sec, $c_{pmixture}$ = 1.62. However, there is no solution for excess air below –82%.

(c) If the mixture is stoichiometric, v_0 is computed to be 0.49 m/sec or 50 cm/sec; the multistep reaction model yields similar values 50 cm/sec. The velocity estimation is sensitive to the flame temperature.

(d) If we repeat with excess-air percentage at 40%, T_∞ = 1993 K, θ_1 = 0.995, T_1 = 1982, $\dot{m}''$ = 0.306 kg/m^2 sec, v = 0.262 m/sec. Thus, with reduced excess air, the flame velocity increases.

(e) If we repeat with the highest excess air at 105%, A:F = 32.33, T_∞ = 1472 K, θ_1 = 0.917, T_1 = 1350, $\dot{m}''$ = 0.0715 kg/m^2 sec, and v = 0.0611 m/sec. However, if excess-air percentage > 105%, there is no solution. Thus, the LFL corresponds to is A:F = 32.33. The corresponding LFL mole percentage of propane is 2%.

(f) Figure 15.12 plots T_1 and flame velocity v vs. equivalence ratio.

(g) The tube diameter for the current calculation is assumed to be 1 cm. If we fix the excess-air percentage to be 50%, then the propane percentage = 2.72 (mole), 4.11% (mass), A:F = 23.66, ϕ = 0.67, T_1 = 1870 K, v = 0.218 m/sec, and T_∞ = 1884 K. If the tube diameter is reduced below 3.9 mm at the same excess-air percentage, we find that there is no solution indicating quenching because of excessive heat losses. The diameter at which a flame will not propagate is called the quenching diameter. For 50% excess air, d_{quench} = 3.9 mm.

15.9 Stability of Flame in a Premixed Gas Burner

Consider a Bunsen burner with the gaseous fuel premixed with a controlled amount of air. Thus, A:F can be varied from a lean to a rich limit. Suppose we light the mixture as it comes out of the burner: assume that the flow is laminar, and the laminar velocity profile inside the tube is given as

$$v_x = [2\ \dot{Q}/\{\pi R^4\}]\ \{R^2 - r^2\} = 2\ v_{x,avg}\{1 - r^2/R^2\} \quad \text{where } v_{avg} = \dot{Q}/\{\pi R^2\} \qquad (15.124)$$

$$v_{x,max} = 2\,\dot{Q}/\{\pi R^2\} = 2\ v_{x,avg} \qquad (15.125)$$

Gradient "g" at the wall r = R is given by $|dv_x/dr|$ (15.126)

$$\text{where } |g| = 4\,\dot{Q}/\{\pi R^3\} = 4\,v_{x,avg}/R = 8\,v_{x,avg}/d \quad (15.127)$$

Let us denote the flame velocity as S instead of v_0. The flame can propagate toward the burner (called flashback) at S if the velocity of gases, v_{avg}, is very low. At high velocity, flame is blown off because the combustible concentration downstream is extremely low owing to dilution. Thus, by performing experiments in a Bunsen burner of diameter d with a premixed fuel:air mixture, one can gradually reduce the flow and determine the flow $\dot{Q} = \dot{Q}_F$ at which flashback occurs. At $\dot{Q}_F$, we can calculate a term called *velocity gradient* $|g| = |g_F| = 4\,\dot{Q}_F/\{\pi R^3\}$ at flashback. Similarly, one can keep increasing the flow and determine the flow $\dot{Q} = \dot{Q}_B$ at which blow-off occurs and calculate g_B at blow-off. Often the blow-off flow $\dot{Q}_B > \dot{Q}_F$ because the mixture flowing out of the tube mixes with the ambient air and can form a combustible mixture downstream of the burner.

15.9.1 Flash-Back Criteria

One can alter the mixture stoichiometry and plot $\dot{Q}_F$ vs. X_F as shown by solid line in Figure 15.13. If the burner diameter is altered, the plot changes. The points of $\dot{Q}_F = 0$ (i.e., no flame propagation as heat losses are high) are called quench diameter for the indicated fuel mole fraction X_F. For example, if the

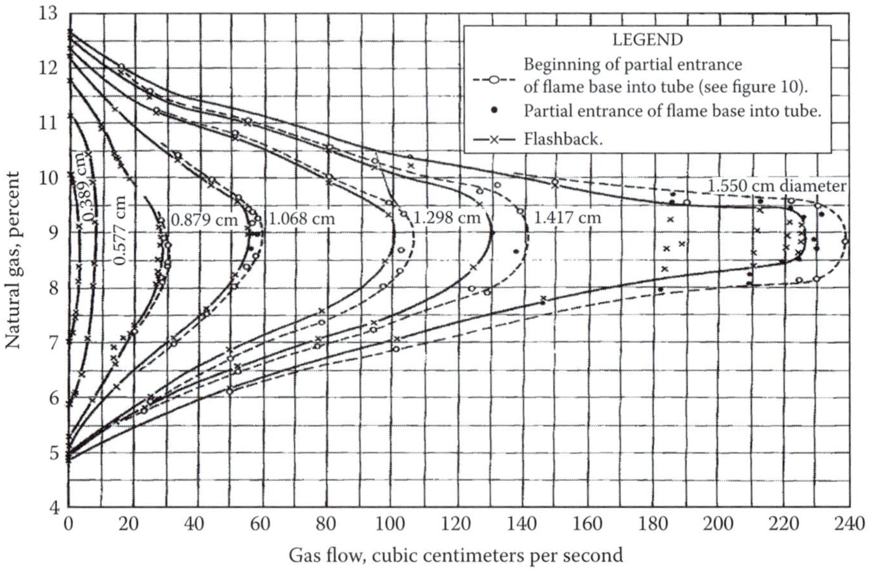

Figure 15.13 Flashback limits. (Adapted from Lewis, B. and Von Elbe, G., *Combustion, Flames, and Explosion of Gases*, 3rd ed., Academic Press, New York, 1987.)

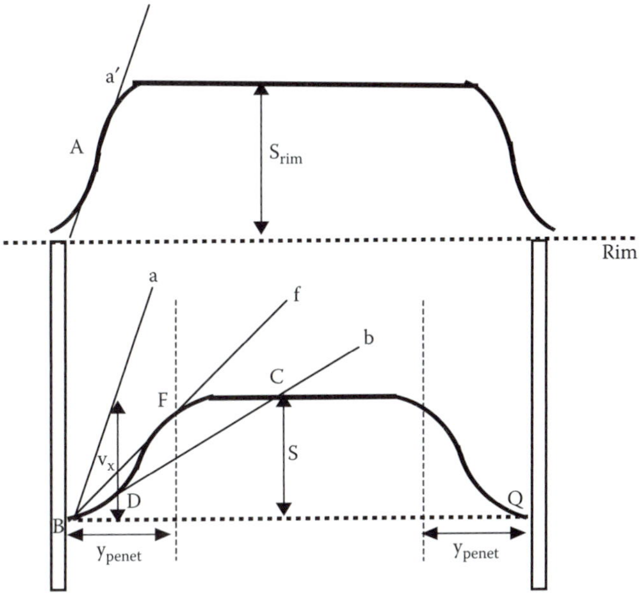

Figure 15.14 Variation of flame velocity and velocity within tube.

tube diameter is 0.577 cm, then for $X_F = 0.065$, flashback occurs at 4 cm³/sec; if X_F is reduced to 0.059, then flashback occurs at 0 cm³/sec. Thus, 0.577 cm is the quench diameter for a gas mixture with $X_F = 0.059$. The same results can be interpreted differently. If $X_F = 0.059$ but the burner diameter is 1.3 cm, flashback occurs at 25 cm³/sec.

The following analysis is conducted to obtain a theoretical relation for flash-back flow $\dot{Q}_F$. So far, we have used average velocity in order to interpret the flash-back process. However, the subsonic viscous flow velocity in the tube is parabolic, and variation needs to be accounted for in obtaining a theoretical $\dot{Q}_F$; near the wall, the velocity profile v_x vs. y is almost linear as represented by curves "a," "f," and "b" in Figure 15.14 for flows $\dot{Q}_a$, $\dot{Q}_F$, and $\dot{Q}_b$, where $\dot{Q}_a > \dot{Q}_F > \dot{Q}_b$. If the velocity is linear, then using Equation 15.127,

$$v_x(y) = g\, y = 4\,\dot{Q}\, y/\{\pi R^3\} = 4 v_{x,avg}\, y/R \qquad (15.128)$$

where $y = R - r$. Then, at flashback

$$v_x(y) = g_F y = 4\, v_{x,avg,F}\, y/R. \qquad (15.129)$$

The velocity profiles are given by BDC, BF, etc., in Figure 15.14. Typically, the flame is located within 1 mm of the rim lip. Figure 15.14 also plots the variation of flame velocity as BDFCQ with respect to distance y from the wall, just below the rim of the burner tube. The velocity is maximum of S in the central regime; however, it falls close to zero near the walls owing to the wall quench effect. For very low flow rates as indicated by curve "b," the flow velocity v_x < flame velocity for $y_D < y < y_C$. Thus, flashback will always occur.

However, for curve "f," the flow velocity v_F matches the flame velocity at point F (where the flame velocity is approximately S at a distance of approximately y_{penet}). Thus, to avoid flashback, flow rates must be kept above a value that yields the flow rate given by curve "f." (We will discuss the case of flame anchoring — when the velocity exceeds v_F — later.) Thus, the critical velocity gradient (g_F) corresponding to F and the corresponding flow is called the flash-back flow $\dot{Q}_F$. At flashback,

$$\text{flow velocity } v_x = S = g_F\, y_{penet} \quad \text{or} \quad g_F = S/y_{penet}. \tag{15.130}$$

Assuming $y_{penet} \approx d_{quench}/3$ for cylindrical tubes [Berlad and Potter],

$$g_F = 3\, S/d_{quench} \tag{15.131}$$

15.9.2 Blow-Off

At high flows (curve (a), Figure 15.14), $v_x > S$ at any point within the tube and, hence, the flame cannot be stabilized within the tube. Therefore, the flame will move toward the rim of the burner and then move away from it. However, once the flame moves away from the rim, quench effects will be reduced above the rim with the result that the flame velocity profile with "y" is altered owing to reduced heat loss. Hence, the flame velocity may become higher because of reduced heat loss. Thus, for the same flow as at "a," the flame could be stabilized just outside the rim at location A (Figure 15.14). This is illustrated by curve "a" for S_{rim} vs. y with $S_{rim} > S$. Hence, as the flow velocity increases, the flame seat (the point A) is farther away from the rim. However, as air is entrained, dilution effects occur that alter the equivalence ratio, particularly at the periphery, and lower the flame velocity. As more air is entrained, even the core of the combustible mixture gets diluted. Beyond the potential core, the mixture is diluted with reduced flame velocity $S \ll S_{rim}$. Eventually, a condition is reached beyond which velocity $v_x > S$ at all points, resulting in blow-off. Whereas flashback can be predicted well by the velocity gradient and the concept of quenching, blow-off prediction is rather complicated due to dilution effects. The blow-off condition for lean flames is approximately

$$(dv_x/dy)_{rim} \times \delta_{preheat}/S = 1 \tag{15.132}$$

where $\delta_{preheat}$ is the preheat zone thickness.

Whereas the flame is almost flat inside the burner (outside y_{penet}), the flame profile changes if it is anchored above the rim of the tube. As mentioned before, the flame is anchored at A and oriented obliquely to the flow along AB as shown in Figure 15.15. Suppose the flow rate is higher than the flash-back flow. If one ignites the mixture at A where $v_x = S$, then a flame will be established at A and will attempt to propagate along a curve at which the combustible mixture exists. The combustible mixture exists within ABG, which

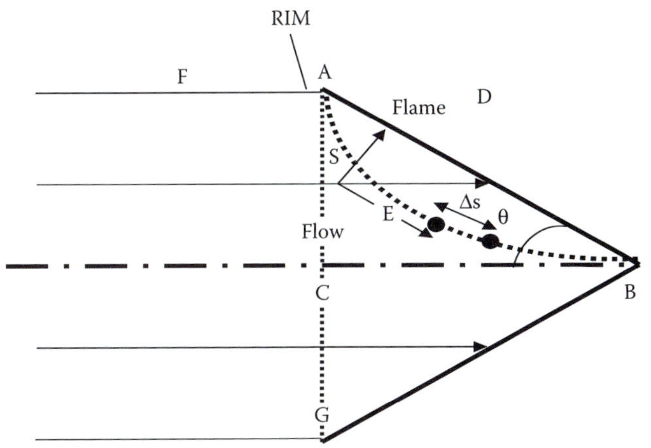

Figure 15.15 Flame shape at exit of tube.

is the potential core for the jet. Outside the core the mixture is diluted with air. However, the flame cannot exist at D in Figure 15.15 because the mixture is not flammable. The flame cannot propagate back to F, because there is no flashback. The flame cannot propagate along AC, because $v_x > S$. The flame orients itself in such a way that $v_x \times$ area AC = S × AB. Thus, area AC/area AB = $\sin \theta = S/v_x$, $v_x > S$, $v_t = v_x \cos \theta$, $S = v_x \sin \theta$.

Example 5

The premixed gas mixture C_3H_8 and 50% excess air flows through a Bunsen burner of diameter 1 cm. The flow is assumed to be laminar. Here, fuel is the deficient species and is burned completely. Using the results of the previous example, the propane percentage = 2.72% (mole), 4.11% (mass), A:F = 23.66, $\phi = 0.67$, $T_\infty = 1884$ K, and v = 22.5 m/sec. The burning velocity is smaller due to heat loss. For 50% excess air, assume $d_{quench} = 3.5$ mm. Thus, the burning velocity profile is known from the data: S = 0.225 m/sec for $0 < r < 0.3$ cm, 0.15 m/sec at r = 0.31 cm, 0 for $0.325 < r < 0.5$ cm.

(a) Plot S vs. r.
(b) Plot v_x vs. r near the wall, assumimg that the gradient of velocity is constant for $v_{xavg} = 0.25$ m/sec, 0.14 m/sec, and 0.1 m/sec.
(c) Determine the flash-back and blow-off limits.

Solution

(a) Line AFHC represents the burning velocity S profile vs. r (Figure 15.16).
(b) The mixture velocity near the wall is $v_x = 4\, v_{x,avg}\, (R - r)/R$. For $v_{avg} = 0.1$ m/sec, $4 \times 0.1\,(1 - r/R)$. Line GHD (Figure 15.16) is for $v_{xavg} = 0.1$ m/sec. Similarly, one obtains BFD and ED for $v_{x,avg} = 0.14$ m/sec, 0.25 m/sec.
(c) Consider $v_{xavg} = 0.1$ m/sec. $v_x > S_L$, $0 < r < 0.22$ cm, vx $< S_L$ for $0.22 < r < 0.31$ cm and $v_x > S$ for $0.31 < r < 0.5$ cm. The flame could be anchored at points

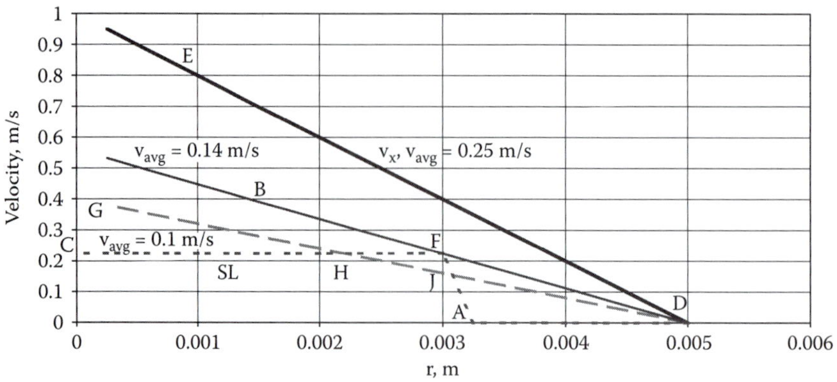

Figure 15.16 Flame-anchoring mechanism.

J and H. However, flashback will occur for the regime $0.22 < r < 0.31$ cm. On the other hand, blow-off will always occur at $v_{xavg} = 0.25$ m/sec (line DE), as $v_x > S_L$ for $r > 0$. Increasing the velocity gradually from 0.1 m/sec to 0.14 m/sec, we find that we can avoid flashback for $v_{x,avg} = 0.14$ m/sec. At this velocity the flame could be anchored at point F. Thus, for velocities above 0.14 m/sec flashback will not occur.

(d) At a first glance it seems that blow-off will occur even for $v_{x,avg} = 0.14$ m/sec for the regime $0 < r < 0.3$ cm. Once a flame is anchored at F, it tries to propagate in all directions. Suppose the flame is no longer flat. The flame front attempts to move toward the axis (i.e., toward a combustible mixture) in the normal direction with velocity S. Thus, the mass consumption rate of fuel must be equal to the mass of fuel into the flame front or S × elemental area normal to flame velocity × area AB = vx × area AG normal to v_x (Figure 15.15). For the elemental flame in area dA_1 within AB and dA_2 within FG, $S \times dA_1 = v_x \times dA_2$ or $\sin\theta = S/v_x$. Suppose v_x is constant. This angle will remain constant and the flame front is linear with r as shown by line AB.

The previous example illustrates that the flame could be anchored at a specified average velocity. But owing to air entrainment, the velocity v_x is reduced outside the tube. For example, at $v_{xavg} = 0.14$ m/sec, $v_{xmax} = 0.28$ m/sec. If air entrainement is included, v_{xmax} is reduced from 0.28 m/sec, say, to 0.25 m/sec; thus, the profile becomes flatter (see Chapter 12). Further, the equivalence ratio is also changed, particularly near the periphery of the burner, which decreases the S. Blow-off occurs when the velocity $v_{x,n}$ (normal component) > $S(\phi)$ everywhere. Note that S is a function of the equivalence ratio.

Consider a combustible mixture flowing through the tube. When the jet leaves the burner rim, there is a transfer of momentum and outward transfer of fuel that alters the fuel:air mixture ratio as well as the flow velocity. The stream boundary layer moves outward with a gradual reduction of velocity in the center of the jet owing to momentum transfer. The flame velocity vs. distance from the center is plotted in Figure 15.17. Quenching effects are dominant near the burner. Thus, the flame is located slightly above the rim. As we increase the velocity further, the anchor point moves toward C away

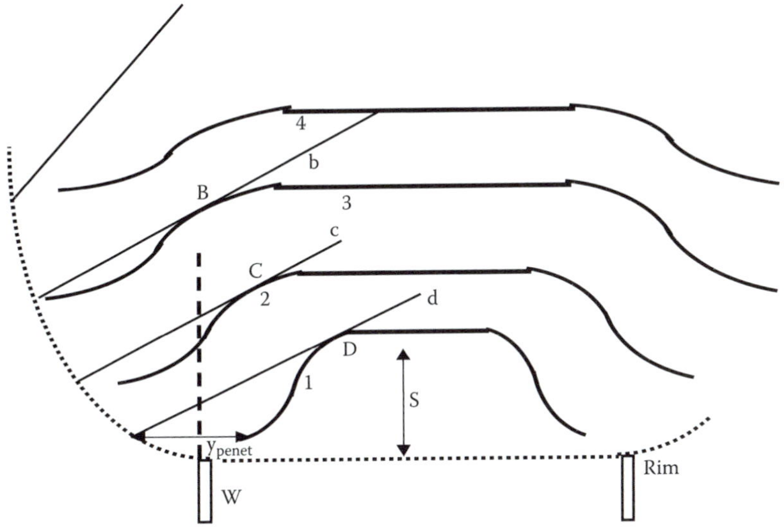

Figure 15.17 Variation of flame velocity and axial gas velocity at burner rim and away from burner rim — stoichiometric and lean mixtures. (From Lewis, B. and Von Elbe, G., *Combustion, Flames, and Explosion of Gases*, 3rd ed., Academic Press, New York, 1987.)

from the rim. Quenching effects are minimal away from the rim, and a lean mixture forms near the mixing layer. The flame moves away from the central axis; however, farther from the rim, the dilution is so strong that S is decreased and the flow velocity > flame velocity at any r. Now, one can draw v_x vs. r outside the rim (including jet effects) as shown by curves "b," "c," and "d" for increasing gas flows. The critical condition for blow-off is reached for curve "b," which is tangential to curve "3" of S vs. r at B. Thus, g_B is given approximately by $S/d_{jet\ stream}$, where d is the distance from the the jet stream boundary. The value of S will correspond to S of the stoichiometric mixture.

If the jet diameter is very small (see Chapter 12),

$$v_x^*/v_{xmax}^* = 1/(1 + \xi^2/4)^2 \approx 16/\xi^4 \text{ for large } \xi \qquad (15.133)$$

$$v_{x\ ,max}^* = (3/32)Re_i\ M^*/x^* \qquad (15.134)$$

where M* is typically assumed to be 1, and

$$\xi = (3/8)\ Re_i M^{*1/2}\ r^*/x^*. \qquad (15.135)$$

The mixing-layer radius is

$$r_m^* = 48\ x^*/\{3\ Re_i M^{*1/2}\} \qquad (15.136)$$

$$\partial v_x^*/\partial r^* = v_{x\ ,max}^*\ \{-\xi/(1 + \xi^2/4)^3\}\ \{(3/8)\ Re_i M^{*1/2}/x^*\} \qquad (15.137)$$

When ξ is large (i.e., at a larger radius),

$$dv_x^*/dr^* \approx v_{x,max}^*$$

$$\{-64\ x^{*\ 4}/r^{*5}\}\ /\{(3/8)\ Re_i\}^4 \quad (15.138)$$

If $r^* = r_m^* - z^*$, and $z^* << r_m^*$ then

$$dv_x^*/dz^* \approx \{(3/32)\}\{3^5 64\ Rei^2/(x^{*2} 48^5\}/\{(3/8)\}^4 \quad (15.139)$$

$$v_x^* = z^*\ \{(3/32)\}\ \{3^5 64\ Rei^2/(x^{*2} 48^5)\}/\{(3/8)\}^4 \quad (15.140)$$

Thus, v_x^* vs. z^* is linear; the slope decreases with an increase in x^* at the given flow rates. At any given x^*, as the flow rate increases, the slope increases. Suppose the initial anchor point is at D (Figure 15.17). As v_{xi} is increased, v_x increases at a given x^* and hence $v_x > S$; the anchor point moves downstream to C. Thus, $x_C > x_D$. However, as the dilution effect takes place, S decreases and $v_x > S$ within the flammable range of the mixtures.

Consider a rich mixture flowing through the tube. There is no flashback due to richness. The stream boundary layer moves outward with a gradual reduction of velocity in the center of the jet due to momentum transfer. The flame velocity vs. distance from the center is plotted in Figure 15.18. As the rich fuel mixes with air, the stoichiometric contour is located close to the outer boundary; it first moves away from the burner axis with decreased flow velocity. The flame velocity for a rich mixture near the axis is less than the flame velocity at the stoichiometric surface (a peak in profile). However, near the burner rim, quenching effects are dominant and, hence, there is no peak (curve 1 is located at x_1 from rim), whereas away from the rim, the quenching

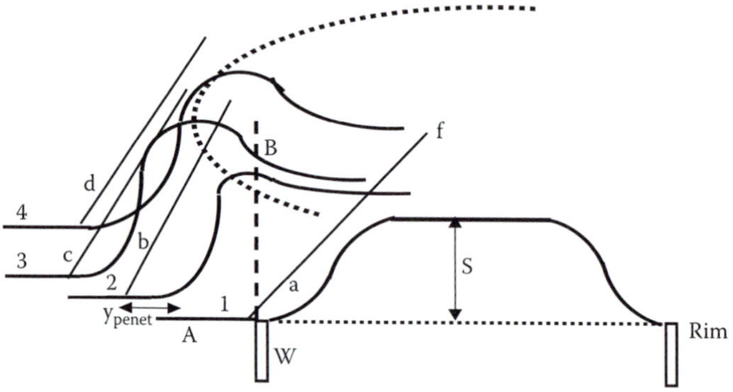

Figure 15.18 Variation of flame velocity and axial gas velocity at burner rim and away from burner rim — rich mixture in tube. (From Lewis, B. and Von Elbe, G., *Combustion, Flames, and Explosion of Gases*, 3rd ed., Academic Press, New York, 1987.)

effects are minimal with a peak in S vs. r (curves 2, 3, and 4 b at $x_2 < x_3 < x_4$, respectively). Near the rim, the stoichiometric contour moves outward with a peak in velocity shifting outward (curves 2 and 3). However, farther from the rim, it moves inward with a peak in S vs. r shifting inward (curve 4) and closing at the stoichiometric contour tip H due to dilution effects. Now, one can draw v_x vs. r outside the rim (including jet effects), as shown in Figure 15.18 by curves "a," "b," "c," and "d," for increasing gas flows. The critical condition for blow-off is reached for curve "c," which is a tangent to curve "3" of S vs. r at B. Thus, g_B is given approximately by S/z, where z is the distance from the jet stream boundary. The value of S will correspond to that of the stoichiometric mixture.

15.10 Turbulent Flame Propagation

Turbulence involves an average velocity and a fluctuating velocity. An eddy is characterized by a rotating flow. If a combustible mixture is trapped into the eddy, combustion requires a timescale t_{chem}. If $t_{eddy} > t_{chem}$, then combustion will be almost complete before being replaced by another eddy. The ratio t_{eddy}/t_{chem} is called the turbulent Damkohler number, Da_I. Thus, at high Da_I, equilibrium chemistry may be adopted. Defining the turbulent $Re_{t,I}$ based on the integral length scale L_I as

$$\mathrm{Re}_{t,I} = \sqrt{\frac{u'_{RMS} L_I}{\nu}}$$

the higher the Re_t value, the greater the inertial forces compared to viscous forces. The Re_t value can range from 10 to 1000. If $Re_t < 1$, the turbulence is quickly damped out. As $t_{chem} u_{RMS} = \delta$, it can be shown that

$$\sqrt{\mathrm{Re}_{t,I}\, Da_t} = \frac{L_I}{\delta}\sqrt{\frac{u'^2_{RMS} t_{chem}}{\nu}}$$

Large values of $Re_{t,I}$ and Da_t are associated with the reaction sheet concept because most reactions are completed within a short distance.

For 2300< Re< 6000, where Re = v d/ν, transport of heat and species occur by "macroscopic" movement of mass within the flame thickness. The kinematic viscosity ν is replaced by eddy diffusivity ε_T for smaller eddies (compared to the flame thickness). Thus,

$$v_o \text{ or } S \propto D^{1/2},$$

$$S_T \propto \varepsilon_T^{1/2}$$

$$S_T/v_0 = \{\varepsilon_T/D\}^{1/2} = \{\varepsilon_T/\nu\}^{1/2}, \tag{15.141}$$

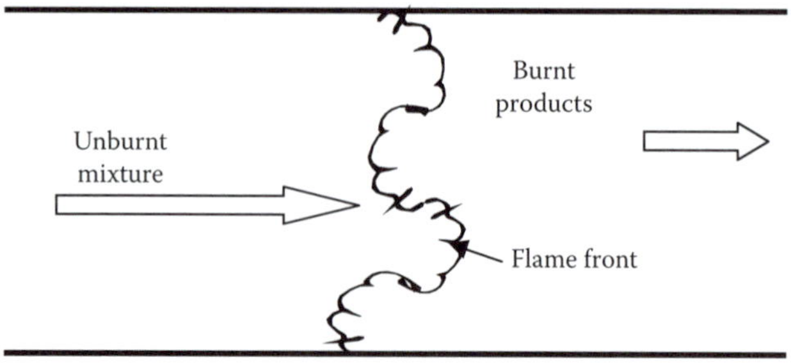

Figure 15.19 Tubulent flame front.

From measuremnets, $\{\varepsilon_T/\nu\} = 0.01$ Re. Hence,

$$S_T/v_0 = 0.1\{Re\}^{1/2},\ 2300 < Re < 6000 \tag{15.142}$$

As the eddy size changes, and for smallest eddy $Re_{t,k}\ Da_t > 1$, then all eddies are larger than δ, and the reaction sheet becomes wrinkled.

When Re > 6000, eddies of size ℓ are large (large compared to flame thickness, δ_f), and the flame front area is distorted (Figure 15.19). The surface area of the flame front is increased, as is the volume of the reaction zone (area × δ_f). Thus, flame velocity × area of duct = volume of reaction zone × reaction rate = area of distorted flame front × flame thickness × reaction rate

$$S_T/S_L = A_T/A_L = u'\ \ell \propto \varepsilon_T \propto Re \tag{15.143}$$

Other empirical models include

$$S_T/S_L = 1 + 0.63(u'/S_L)^{1/2} Re \tag{15.144}$$

15.11 Summary

When a premixed combustible mixture is ignited, the flame propagates with a certain speed. The burned gases provide heat feedback to the unburned mixture, thus continuously igniting it, resulting in flame propagation. Using the conservation equations for species and energy, laminar flame propagation velocities and flammability limits (rich and lean) are predicted. Empirical formulae are also presented. The laminar flame relation is extended to turbulent flame propagation using a suitable scaling law for turbulent diffusivity. A flame will not propagate if the tube diameter falls below d_{quench}. A methodology is presented for predicting flashback and blow-off limits.

Chapter 16

Interactive Evaporation and Combustion

16.1 Introduction

Spray ignition and combustion studies are extremely important to determine flame stability behavior at widely varying loads, to ensure efficient utilization of fossil energy, and to better understand the mechanisms of formation and destruction of pollutants. Extensive fundamental and applied research has been conducted on the combustion of oil sprays and pulverized-fuel particle sprays. Both heterogeneous (e.g., drop combustion either in groups, arrays, or individually) and homogeneous (evaporation of small drops less than 10 μm) combustion can occur in a spray.

Isolated-particle or isolated-drop models described in Chapter 9 and Chapter 10 yield the parameters controlling the ignition and combustion characteristics of individual drops or particles. Sprays normally involve a large number of drops. The simplest analysis is to treat the fuel spray as an aggregate of the behavior of individual fuel droplets. Thus, the spray combustion rate can be estimated to be the sum of the combustion rates of each of the drops (as if each drop burned individually).

However, a drop in the spray does not behave like an isolated drop, because its behavior will be modified owing to the presence of other drops. Interactions between the droplets alter the drag coefficient, change the flow field, increase or decrease the ignition delay depending on the denseness of the spray, compete for heat and oxygen, and alter the fuel vapor distribution around the drops. The behavior of sprays is vastly different from that of isolated droplets [McCreath and Chigier, 1973; Yule et. al., 1982].

Apart from the experimentally observed behavior of cloud combustion processes, spray combustion modeling requires a knowledge of interactive transport processes that occur during evaporation and combustion. Because the dimensions of a typical combustor are on the order of 10 to 100 cm, whereas the drops are on the order of a few micrometers, computational resolution on the order of the drop size is impossible because of computational limitations. A cluster of drops containing thousands of representative drops is considered in spray modeling depending on the location of interest in the combustor. Particularly in the vicinity of the burner in which ignition occurs, the spray can be dense.

How does one define ignition for an oil spray? Is the ignition of one drop in a spray or one particle in the coal suspension (Chapter 13) sufficient to define the ignition of the whole spray or suspension undergoing transport processes in a nonuniform field? In other words, does the isolated-particle or isolated-drop ignition model, which involves a large ignition source (infinite environment) and a small test source (single drop or particle) reveal the ignition characteristics of sprays igniting in a finite environment? Cloud models (large test source) seem to be more appropriate [Annamalai et. al., 1984, 1992, 1995; Laster et. al., 1991; Ryan, 1991]. Hertzberg et al. (1985) argues for models in which the test source volume is comparable to the ignition source volume for realistic interpretation of the experimental data. Further, the isolated-drop ignition model results reveal increasing ignition temperature as size is reduced, which is contrary to industrial experience. Thus, the usefulness of isolated-droplet modeling or single-drop experimental results for ignition is limited in the modeling or the interpretation of experimental results for sprays, especially in the vicinity of the injector where the spray is very dense. Such a dense mode of combustion in the oxygen-deficient zones leads to increased combustion time, reduced flame stability, incomplete combustion, and increased emission of HC, CO, and other pollutants. Thus, interactions must be accounted for.

Array studies consider interactions among a finite number of drops ranging from two to nine and are an improvement over isolated-drop studies [Labowsky, 1976a, 1976b, 1978a, 1980; Raju and Sirignano,1990; Twardus and Brzustowski, 1977; Xiong et al., 1984]. Although these studies are rather academic in their application to sprays, their results provide a fundamental knowledge of the processes involved in the interactions. When a cluster is considered, interactions are among a large number of drops. The term *cloud/ group* or *cluster* in this chapter refers to a large number of drops, usually much higher than ten.

16.2 Simplified Analyses

16.2.1 Interactive Processes

Consider a spray in which drops evaporate and the resulting vapors mix with surrounding air. There are areas with densely packed drops, creating fuel-rich regimes, and locations at which drops are sparsely distributed, creating an overall fuel-lean regime. In a fuel-rich regime, the fuel vapor mass fraction is dominant compared to the oxygen concentration. Thus, these drops may not burn individually and may simply evaporate under interactions. Recall from Chapter 8 that for any gaseous fuel injection, say, CH_4 and air, the Shvab–Zeldovich formulation, $\beta_{O2\text{-}F} = Y_F - Y_{O2}/\nu_{O2}$ = {fuel mass fraction used for combustion when all local O_2 is consumed}, yields a fuel-rich regime for $\beta_{O2-F} > 0$ and fuel-lean regime for $\beta_{O2-F} < 0$, with $\beta_{O2-F} = 0$ representing stoichiometric or flame-sheet regime. Typically, a flame sheet is established at the stoichiometric surface. When liquid is injected, the locations near the burner will have larger drops, and those farther will have smaller drops owing to evaporation and combustion, which starts near the burner. Thus, steady state exists at any given location in a burner operating steadily and drops are statistically always present, establishing a steady profile.

16.2.2 Combustion

An envelope flame (Figure 16.1a) surrounds a single drop burning in a quiescent atmosphere. The flame acts as a sink for fuel and oxygen and as a source of thermal energy. If another burning drop is brought near the

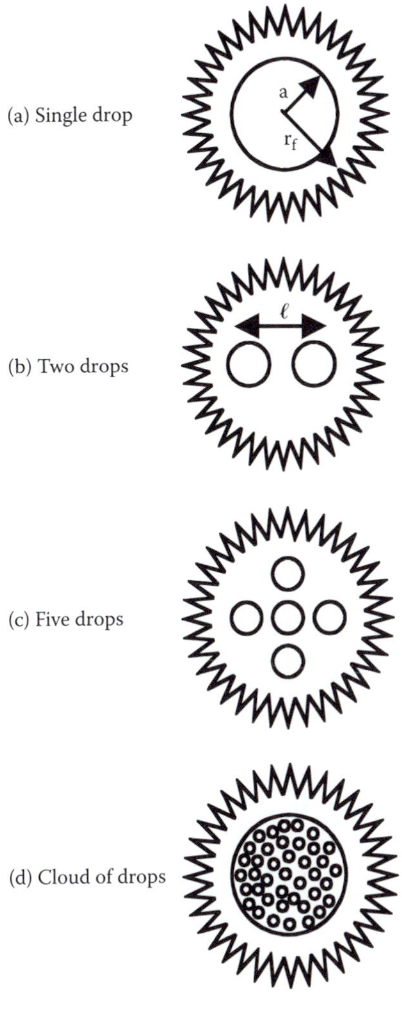

Figure 16.1 Combustion for isolated drop and multiple drops.

existing one, the vapors released by the drops compete for oxygen molecules. If the interdrop spacing (ℓ) is small, a common flame is formed around the drops (Figure 16.1b, somewhat similar to a common flame around two candles). Similarly, a common or group flame may be formed around an array of drops (Figure 16.1c) or a dense cluster of drops (Figure 16.1d).

Following Williams, a simple criterion for interactive combustion (or combustion with a common flame) is given as follows. If "r_f" is the flame radius for a drop of radius "a" and the interdrop spacing is "ℓ," then the flames will merge if

$$l < 2r_f. \tag{16.1}$$

Relations for flame radius for an isolated drop are given in Chapter 10. The number of drops per unit volume (n), fuel volume fraction (σ), mass loading κ, air-to-fuel ratio (A:F), and equivalence ratio (ϕ) are given as

$$n = 1/\ell^3 \tag{16.2}$$

$$\sigma = (\text{fuel volume fraction}) = (4/3)\ \pi\ a^3\ n \tag{16.3}$$

$$\kappa = \sigma\ \rho_\ell/\rho_g,\ \text{fuel-to-gas ratio or mass loading, kg fuel per kg gas} \tag{16.4}$$

$$A{:}F = 1/\kappa \tag{16.5}$$

$$\phi = (A{:}F)_{st}/(A{:}F) = (\nu_{O2}/Y_{O2})/(A{:}F). \tag{16.6}$$

The criteria for interactive combustion in terms of these parameters are presented in Table 16.1.

From Chapter 10, the fuel mass fraction at the drop surface is given as

$$Y_{F,w} = 1 - \{1 + s\}/\{1 + B\} \tag{16.7}$$

where

$$s = Y_{O2}/\upsilon_{O2} \tag{16.8}$$

$$B = (h_c s/L) + c_p(T_\delta - T_w)/L = (h_c/L)\ (Y_{O2,\delta}/\upsilon_{O2}) + c_p(T_\delta - T_w)/L \tag{16.9}$$

$$r_f/a = \ln(1 + B)/\ln(1 + s) \approx B/s \text{ if B and s are small.} \tag{16.10}$$

Table 16.1 Approximate Criteria for Interactive Combustion

Parameter	*Relational Operator* >,<	*Criterion*	*Estimate for Liquid Clouds*
ℓ/a	<	$2r_f/a$	75.0
n	>	$(1/2r_f)^3$	2.0×10^7 drops m^{-3}
σ	>	$(\pi/6)\ (a/r_f)^3$	10^{-5} m^3 of fuel/m^3
κ	>	$(\pi/6)\ [\rho_\ell/\rho_g]\ (a/r_f)^3$	0.01
A:F	<	$(6/\pi)(\rho_g/\rho_\ell)\ (r_f/a)^3$	100.0
ϕ	>	$(\pi/6)(\rho_\ell/\rho_g)(\nu_{O2}/Y_{O2,\infty})\ (a/r_f)^3$	0.015

Note: Estimates are based on: $Y_{O2} = 0.23$, $\nu_{O2} = 3.5$, $B = 10$, $\rho_\ell/\rho_g = 1000$, and $a = 50 \times 10^{-6}$ m. The flame-to-drop radius ratio is estimated to be 37.7 under QS conditions (Chapter 10).

The flame radius is inversely proportional to the oxygen mass fraction. Because most of the criteria are dependent on the cubic power of (r_f/a), it is apparent that interactive processes are very sensitive to oxygen concentration. It should be noted that the flame-to-drop radius ratio is based on QS theory. Buoyancy and unsteady effects are expected to reduce the flame radius significantly. If the ratio (r_f/a) predicted by QS theory is reduced to half the original value, then the estimates in Table 16.1 differ by an order of magnitude. For fuel sprays, the core is fuel rich and usually the interactive criteria are satisfied. Table 16.1 indicates that interactions start occurring at equivalence ratio as low as $\varphi \approx 0.02$ with $r_f/a = 35$ or 0.2 with $r_f/a = 18$.

16.2.3 Evaporation

In the case of combustion, the drops compete for oxygen and the flame acts as a sink for oxygen and as a source of heat, whereas the drops act as heat sinks and as sources of mass. For evaporation, the drops simply compete for heat in the interstitial space. At the same time, the decrease in temperature coupled with saturated conditions for the gas vapor mixture in the interstitial space between the drops reduces the evaporation rate. The temperature profile for an isolated drop under QS evaporation (Chapter 10) is as follows:

$$\phi = (T_\infty - T)/(T_\infty - T_w) = (1 - \exp(-\xi))/(1 - \exp(-\xi_w)) \qquad (16.11a)$$

where

$$\xi = m/(2\ \mathrm{Sh}\ p\ r\ D\ r), \qquad (16.11b)$$

$$\xi_w = m/(2\ \mathrm{Sh}\ p\ r\ D\ a). \qquad (16.12)$$

If the mass transfer rate is very small, then Equation 16.11a can be approximated as,

$$\phi = (T_\infty - T)/(T_\infty - T_w) = a/r. \qquad (16.13)$$

Equation 16.13 is the exact solution when radial convective (Stefan) flow is neglected in the diffusive and convective transport equations. Incidentally, Equation 16.13 is also the solution for a point heat sink in heat transfer problems governed by the Laplace equation.

If one plots ϕ vs. r, then at r = 10a, the temperature difference between the drop wet bulb temperature (T_w) and T(r) is 90% of the difference between the ambient and the wet bulb temperature of the drop. If another drop is located at r = 20a, then the temperature at the midpoint between the drops is about 90% of the far-stream temperature. As the mass transfer rate is proportional to the difference $T_\infty - T_w$ for an isolated drop undergoing evaporation, such a reduction in the interstitial space temperature will reduce the evaporation rate of drops in interactive fields to about 95% of the isolated-drop rate. As the interdrop spacing is reduced further, the evaporation rate per drop continues to decrease. Thus, interactive evaporation occurs if the interdrop spacing is such that l/a < 20. Then, with r_f/a replaced by 10 in third column of Table 16.1, one can obtain the corresponding criteria for interactive evaporation processes, except for φ, which is meaningless for evaporation problems. It should be noted that the criteria for interactive processes in combustion involved merging the flames, whereas the criteria for interactive evaporation was based on a 10% decrease in the drop temperatures.

16.2.4 Correction Factor

Once interactive processes occur, the mass-loss rate of each drop is altered from the isolated-drop mass-loss rate. The effect of these interactions is to either increase or decrease the mass-loss rate of the drops. Let the mass-loss rate of an interacting drop be denoted as $\dot{m}_s$(single) and the drop mass-loss rate under isolated conditions, as $\dot{m}_{iso}$(isolated). Then, the correction factor η is

$$\eta = \dot{m}_s / \dot{m}_{iso} \tag{16.14}$$

The focus of many previous investigators has been to evaluate the correction factor under interactive conditions as the interdrop spacing and number of drops are varied. Interactive effects usually retard the drop or particle mass-loss rate ($\eta < 1$); under pure diffusive transport, the single-component array analysis always yields $\eta < 1$. However, occasionally they can enhance mass loss ($\eta > 1$) under restrictive conditions such as those involving buoyant convective effects and when drops of dissimilar composition are considered.

16.3 Arrays and Point Source Method

16.3.1 Evaporation of Arrays

16.3.1.1 Non-Stefan Flow (NSF) Problems

We will use PSM for solving array problems. The fuel vapor species F from a drop with surface fuel mass fraction $Y_{F,w}$ evaporates at the rate of $\dot{m}_F$. The strength of the source $\dot{m}_F$ and surface mass fraction ($Y_{F,w}$) are unknown, which are to be solved from the Laplace equation for each species and the boundary conditions. Let us consider a non-Stefan flow (NSF) problem. Then,

$$\frac{d}{dr}\left(\rho D r^2 \frac{dY_F}{dr}\right) = 0 \tag{16.15a}$$

subject to two boundary conditions:

$$\text{At r, } \dot{m}_F = -4\pi r^2 \rho D \frac{dY_F}{dr} \tag{16.15b}$$

$$\text{and} \quad \text{as r} \to \infty,\ Y_F = Y_{F\infty}. \tag{16.15c}$$

The following solution is obtained [Annamalai and Ryan, 1991, 1992; Annamalai et. al., 1993]:

$$Y_F - Y_{F,\infty} = \frac{\dot{m}_F}{4\pi\rho D r}. \tag{16.16}$$

The term $(Y_F - Y_{F,\infty})$ may be called *evaporation potential.* This is the same solution obtained when one considers a point mass source of strength $\dot{m}$ placed at any arbitrary location in space. Equation 16.15a is a linear equation. For the linear equations, one can superimpose solutions. Let us drop subscript F from Y and $\dot{m}$. If one considers many drops (Figure 16.2), then Y at any point r is considered as a sum of Y from 1, Y from 2, etc. Thus

$$Y(\vec{r}) - Y_\infty = \frac{\dot{m}_i}{4\pi\rho D|\vec{r} - \vec{r}_i|} + \sum_{j=2, j\neq i} \frac{\dot{m}_j}{4\pi\rho D\left|\vec{r}_i - \vec{r}_j\right|}. \tag{16.17}$$

We can obtain the value of Y_w at point P on the surface of drop "i," i.e.,

$$Y_w(\vec{r}_i + \vec{a}_i) - Y_\infty = \frac{\dot{m}_i}{4\pi\rho D a_i} + \sum_{j=2, j\neq i} \frac{\dot{m}_j}{4\pi\rho D\left|\vec{r}_i + \vec{a}_i - \vec{r}_j\right|}$$

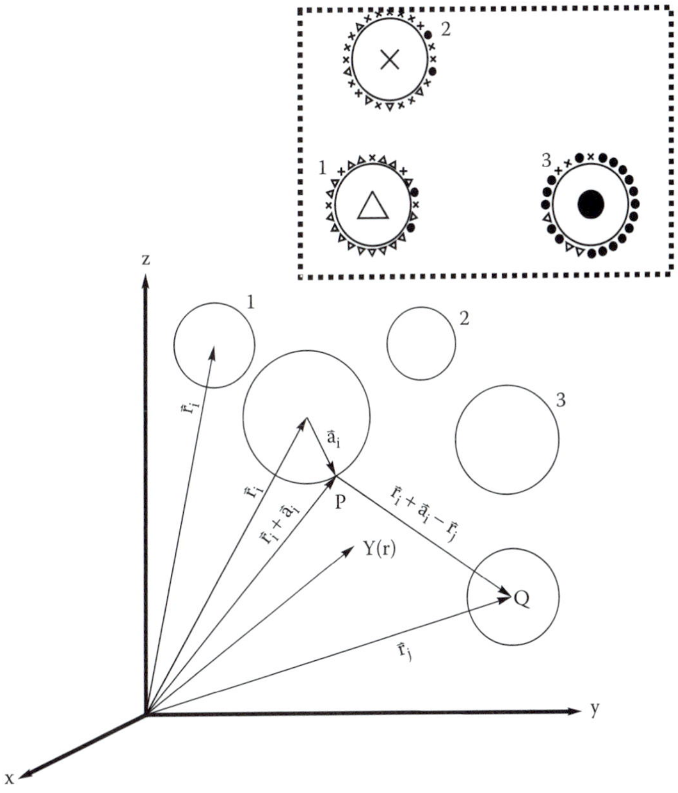

Figure 16.2 Illustration of point source method (PSM).

For a point mass, $|a| \ll |\vec{r}_i - \vec{r}_j| = l_{ij}$, interdrop spacing; using the binomial expansion for $a_i + r_i - r_j$ and considering the first term only

$$Y_w(\vec{r}_i + \vec{a}_i) \approx \frac{\dot{m}_1}{4\pi \rho D a_i} + \sum_{j=1, j\neq i} \frac{\dot{m}_j}{4\pi \rho D \left| \vec{r}_i - \vec{r}_j \right|}. \tag{16.18}$$

The physical interpretation is as follows. The mass fraction at the surface of the drop i = mass fraction due to vapor produced by drop i of strength $\dot{m}_i$ + mass fraction increase at drop i owing to mass source of strength $\dot{m}_i$ located at r_1 + mass fraction increase at drop i owing to mass source of strength $\dot{m}_2$ located at r_2 + ….

Thus, one can visualize the surface of drop 1 filled with vapor molecules from drop 3 (denoted as "•") and molecules from drop 2 (denoted as "x"; see insert of Figure 16.2).

(The problem is analogous to a kitchen range in which there are multiple electrical coils. The heat flux felt just above coil say 2 is due to direct heat flux from the coil 2 + heat fluxes arriving at 2 from other coils.) Solving for $\dot{m}_1$ and as drop sizes are so small, Y_w is uniform around any given drop

$$\frac{\dot{m}_1}{4\pi \rho D a_1} = Y_w - \sum_{j=2,} \frac{\dot{m}_j}{4\pi \rho D |\vec{r}_1 - \vec{r}_2|}. \tag{16.19}$$

It can be shown that all the drops in arrays are at the same temperature T_w and hence phrase equilibrium requires that Y_w's are same for all drops under steady state. Further, Y_w or T_w is exactly the same as for isolated drops. Generalizing Equation 16.19,

$$\frac{\dot{m}_i}{4\pi \rho D a_i} = Y_w - \sum_{j=1,\, j\neq i}^{N} \frac{\dot{m}_j}{4\pi \rho D |\vec{r}_i - \vec{r}_j|}_{j=1,}. \tag{16.20}$$

Using the definition given in Equation 16.14

$$\eta_i = \frac{\dot{m}_i}{\dot{m}_{iso,i}}$$

where

$$\dot{m}_{iso,i} = 4\pi \rho D a_i B \tag{16.21}$$

$$B = Y_{Fw} \quad \text{or} \quad \frac{b_{T,\infty}}{L} \text{ for evaporation.} \tag{16.22}$$

Dividing Equation 16.20 by $\dot{m}_{iso,i}$ and rearranging

$$\eta_i = 1 - \sum_{j=1, j\neq i}^{N} \frac{n_j a_j}{|\vec{r}_i - \vec{r}_j|} = 1 - \sum_{j=1, j\neq i}^{N} \frac{n_j a_j}{|\ell_{ij}|}, \quad i = 1, 2, \ldots, N \tag{16.23}$$

Equation 16.23 provides N linear relations for an N-drop array in terms of dimensionless interdrop spacing = ℓ_{ij}/a_j, a geometrical parameter. The correction factors seem to be independent of fuel properties.

16.3.1.2 Stefan Flow (SF) Problems

$$\nabla \cdot \left[\rho \vec{v} Y_k - \rho D \nabla Y_k\right] = 0 \tag{16.24}$$

Define evaporation potential for SF as

$$\psi = \frac{\ell n\left[\frac{F - Y_k}{F - Y_{k\infty}}\right]}{\ell n\left[\frac{F - Y_{kw}}{F - Y_{k\infty}}\right]} = -\frac{\ell n\left[\frac{F - Y_k}{F - Y_{k\infty}}\right]}{\ell n\ [1 + B]}, \quad F = Y_{k\ell}, \quad B = \frac{Y_{kw} - Y_{k\infty}}{F - Y_{k\infty}} \tag{16.25}$$

where F's and B are defined for drops in Chapter 10 (Table 10.1 of Chapter 10). Typically F = 1 as $Y_{k\ell}$ mass fraction of fuel species in liquid (ℓ) = 1 for pure liquid drop evaporation. Using Equation 16.25 in Equation 16.24, then

$$\nabla^2 \psi = 0 \tag{16.26}$$

which is similar to the conservation equation for the NSF problem (Equation 16.15). With k = F (fuel species), Equation 16.25 for evaporation becomes

$$\psi = \frac{\ln\left[\frac{1 - Y_F}{1 - Y_{F\infty}}\right]}{\ln\left[\frac{1 - Y_{Fw}}{1 - Y_{F\infty}}\right]} \quad \text{or} \quad \frac{\ln\left[\frac{1 - Y_F}{1 - Y_{F\infty}}\right]}{-\ln(1 + B)} \quad \text{where} \quad B = \ln\left[\frac{Y_{Fw} - Y_{F\infty}}{1 - Y_{Fw}}\right], \tag{16.27}$$

where $\psi = 1$ at the drop surface, 0 far from drops.

Also, the mass flow at any radius is given as

$$\dot{m}_F = 4\pi r^2 \left[\rho v\ Y_F - \rho D \frac{dY_F}{dr}\right] = \rho D 4\ \pi\, r^2 \frac{d\ln(1 - Y_F)}{dr}$$

Using Equation 16.27

$$\dot{m}_F = -\ln(1 + B) \rho D 4 \pi\, r^2 \frac{d\psi}{dr}$$

Proceeding as in an NSF problem, the solution for ψ is given as

$$\psi(\vec{r}) = \frac{\dot{m}_1}{4\pi\rho D \ln(1+B) \,|\vec{r}-\vec{r}_1|} + \frac{\dot{m}_2}{4\pi\rho D \ln(1+B)\,|\vec{r}-\vec{r}_2|} + \frac{\dot{m}_3}{4\pi\rho D \ln(1+B)\,|\vec{r}-\vec{r}_3|} + \ldots \tag{16.28}$$

Then, at any drop surface, say, drop 1, $\psi = 1$, as before

$$1 = \frac{\dot{m}_1}{4\pi\rho D \ln(1+B)\, a_1} + \frac{\dot{m}_2}{4\pi\rho D \ln(1+B)\,|\vec{r}_1-\vec{r}_2|} + \frac{\dot{m}_3}{4\pi\rho D \ln(1+B)\,|\vec{r}_1-\vec{r}_3|} + \ldots \tag{16.29}$$

Now we can convert Equation 16.29 in terms of the correction factor using Equation 16.14 but with $\dot{m}_{iso} = 4\pi\rho D a \ln(1+B)$ and obtain the final equation in the same form as Equation 16.23, which is valid whether we consider Stefan flow or not. Table 16.2 summarizes the correction factor relations for selected array configurations. Again, the correction factors seem to be independent of fuel properties, whereas the experimental data show that in binary drop arrays for propanol, decane and hexadecane drops show dependency on fuel type, presumably owing to transient effects that are not considered in the steady-state theoretical model (Figure 16.3).

Example 1

Consider a linear array of three *n*-hexane drops of radii 100 (= 2 a_1), 50(= 2 a_2), and 100(= 2 a_3) μm with drops of different radii all being at the same temperature separated by $\ell_{12} = \ell_{23} = \ell = 300$ μm kept in an ambient of 600 K.

$\rho D = 3.73 \times 10^{-05}$ kg/(msec), $c_p = 1.021$ kJ/kg K $h_c = 45105$ kJ/kg, L = 335 kJ/kg. Assume that the gas phase has only two components: fuel vapor and N_2.

(a) Compute the isolated-drop evaporation rate in kg/sec.
(b) Compute correction factors.
(c) Compute the drop evaporation rate in kg/sec for a drop in the array.

Solution

(a) Following the same procedure as in Chapter 10 for evaporation of an isolated drop, $T_w = 297.9$ K, $B = h_{t,\infty}/L = c_p(T_\infty - T_w)/L = 0.921$, where B is not less than 1. Hence, SF must be accounted for.

Table 16.2 Correction Factors for Combustion Arrays of Particles/Drops, Figure 16.4

Array	Peripheral, η_{peri}	Center, η_{cent}
Binary	$\frac{1}{1+\frac{a}{l}}$	—
Ternary (triangular)	$\frac{1}{1+\frac{2a}{l}}$	—
Tetrahedral	$\frac{1}{1+\frac{3a}{l}}$	—
Symmetric N-particle array	$\frac{1}{1+\frac{(N-1)a}{l}}$	—
Linear three particle (Figure 16.4 a)	$\frac{1-\frac{a}{l}}{1+\frac{a}{2l}-\frac{2a^2}{l^2}}$	$1-\eta_{peri}\frac{2a}{l}$
Five particles on x-y plane (Figure 16.4 b)	$\frac{1-\frac{a}{l}}{1+\frac{2^{1/2}a}{l}+\frac{a}{2l}-\frac{4a^2}{l^2}}$	$1-\eta_{peri}\frac{4a}{l}$
Four linear particles	$\frac{1-\frac{1}{2}\frac{a}{l}}{1+\frac{4a}{3}\frac{a}{l}-\frac{a^2}{l^2}\left(\frac{9}{4}-\frac{1}{3}\right)}$	$\frac{1-\eta_{peri}\frac{3}{2}\frac{a}{l}}{1+\frac{a}{l}}$
Seven-particle symmetric array (Figure 16.4 c)	$\frac{1-\frac{a}{l}}{1+\left(2^{3/2}+\frac{1}{2}\right)\frac{a}{l}-6\frac{a^2}{l^2}}$	$1-\eta_{peri}\frac{6a}{l}$
Cubical with particles at center (nine particles) (Figure 16.4 d)	$\frac{1-\frac{2}{3^{1/2}}\frac{a}{l}}{1+\left(3+\frac{3}{2^{1/2}}+\frac{1}{3^{1/2}}\right)\frac{a}{l}-\frac{32}{3}\frac{a^2}{l^2}}$	$1-\eta_{\mathrm{peri}}\frac{16a}{3^{1/2}l}$
Polygon of N particles with no center particle	$\left(1+\frac{aS(N)}{\ell}\right)$, $S(N)=\sin\left(\frac{\pi}{N}\right)\sum_{j=1}^{N-1}\left[\sin\left(\frac{j\pi}{N}\right)\right]^{-1}$	—
Polygon of N particles with center particle	$\left(\frac{1-\left(a/\ell\right)\mathrm{Sin}\left(\frac{\pi}{N_p}\right)}{1+\left(a/\ell\right)S(N_p)\mathrm{Sin}\left(\frac{\pi}{N_p}\right)-4\left(a/\ell\right)^2 N_p\mathrm{Sin}^2\left(\frac{\pi}{N_p}\right)}\right)$	$1-2N_n\eta_p\left(a/\ell\right)\mathrm{Sin}\left(\frac{\pi}{N_p}\right)$

Source: Adapted from Annamalai, Ryan, and Dhanapalan 1995.

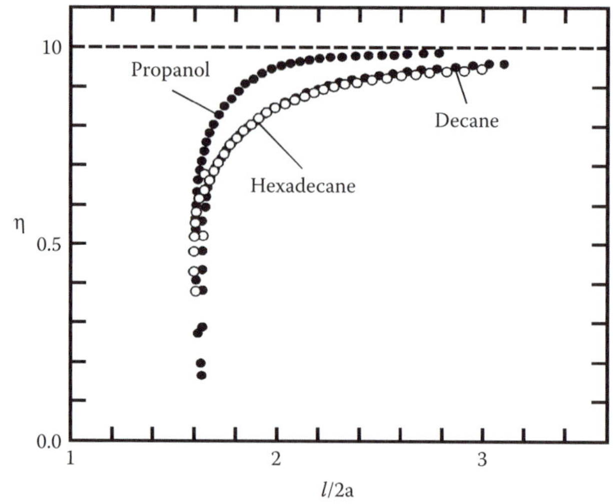

Figure 16.3 Effect of fuel type on correction factor for binary array (from Xiong, T.Y., Law, C.K., and Miyasaka, K., Interactive Vaporization and Combustion of Binary Droplet Systems, 20th Symposium (International) on Combustion, 1781–1787, 1984) at low pressures (90 m of Hg) and $Y_{O2,\infty} = 0.40$.

$Y_{F,w} = 1 - 1/(1 + B) = 0.48$, $\xi_{w,iso} = \ln(1 + B) = 0.653$, $\dot{m}_{iso,1} = \dot{m}_{iso,3} = 4\ \pi\rho D\ a_1$, $\xi_w = 1.50*10^{-8}$ kg/sec, $\dot{m}_{iso,2} = 0.749 * 10^{-8}$ kg/sec

(b) Taking the origin at the drop-2 center (see Figure 16.4a) and using Equation 16.23 for i = 1, 2, 3

$$\eta_1 = 1 - (a_2\ \eta_2/\ell_{12}) - a_3\eta_3/\ell_{13} = 1 - (a_2\ \eta_2/\ell) - a_3\eta_3/2\ell \quad \text{(A)}$$

$$\eta_2 = 1 - (a_1\ \eta_1/\ell_{21}) - a_3\eta_3/\ell_{23} = 1 - (a_1\ \eta_1/\ell) - a_3\eta_3/\ell \quad \text{(B)}$$

$$\eta_3 = 1 - (a_2\ \eta_2/\ell_{32}) - a_1\eta_1/\ell_{31} = 1 - (a_2\eta_2/\ell) - a_1\eta_1/2\ell \quad \text{(C)}$$

As $a_1 = a_3$, and due to symmetry, $\eta_1 = \eta_3$. Solving Equation A and B

$$\eta_1 = \{1 - a_2/\ell\}/[1 - 2a_1\ a_2/\ell^2 + a_1/2\ell] = \eta_3 \quad \text{(D)}$$

$$\eta_2 = 1 - 2(a_1\ \eta_1/\ell). \quad \text{(E)}$$

With $a_2 = 25$, $\ell = 300$, $a_1 = a_3 = 2\ a_2$, $a_1/\ell = 1/6$,

$$\eta_3 = \eta_1 = 0.868,\ \eta_2 = 0.711.$$

(c) The evaporation rates of drops 1, 2, and 3 are $\dot{m}_1 = \dot{m}_3 = \dot{m}_{iso,1}\ \eta_1 = 1.5 \times 10^{-8} \times 0.868 = 1.30 \times 10^{-8}$ kg/sec, $\dot{m}_2 = 0.75 \times 10^{-8} \times 0.714 = 0.53 \times 10^{-8}$ {kg/sec}.

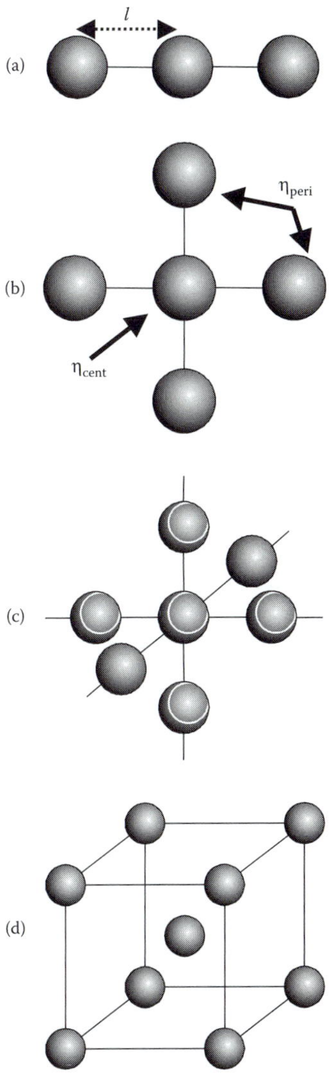

Figure 16.4 Combustion for isolated drop and multiple drops. (a) linear three drop, (b) five drop, (c) seven drop symmetrical, (d) cubicle

16.3.1.3 Diameter Law and Evaporation Time

Using the isolated-drop evaporation rate relations, for $\dot{m}_{iso}$ and Equation 16.24

$$\dot{m}_s = -\{d/dt\}\,\{\rho_\ell\ \pi\ d^3/6\} = \eta(t)\ \dot{m}_{iso} \tag{16.30}$$

with $\dot{m}_{iso} = 2\ \pi\rho D\ d\ \ln\ (1 + B)$ in Equation 16.30 and integrating

$$d_0^{\,2} - d^2 = \alpha_{iso} \int_0^t \eta(t)\,dt \tag{16.31}$$

where $\alpha_{iso} = \{8\ \pi\rho D/\rho_\ell\}\ d\ \ln\ (1 + B) \approx 4\ \mathrm{Sh}\ \rho D\ \ln(1 + B)/\rho_\ell$.

Letting d → 0 as t → t_b and using Equation 16.31, the burning time t_b is given by the following implicit expression:

$$d_0^2 = \alpha_{iso} \int_0^{t_b} \eta(t)\, dt. \tag{16.32}$$

As drop radius decreases with time and hence η(t) increases with time (i.e., less severe interaction), the d^2 law is no longer linear with time. However, the maximum evaporation time can be determined by using the initial interaction factor

$$d_0^2 = \alpha_{iso} \int_0^{te} \eta_0 \, dt = \alpha_{iso}\, \eta_0\, t_{e,max} \tag{16.33}$$

$$t_{e,max} = d_0^2/(\eta_0 \alpha_{iso}). \tag{16.34}$$

Using the following result from Table 16.2

$$\eta(t) = 1/(1 + d/(2\ell)) \tag{16.35}$$

in Equation 16.32, the d^2 law for binary drop systems can be written as

$$d^2 = d_0^2 - \alpha_{iso}\, t + (d_0^3 - d^3)/3\, \ell. \tag{16.36}$$

Using this equation, the evaporation time for a binary array can be expressed as

$$t_e = d_0^2/\alpha_{iso} + d_0^3/(3\, \ell\, \alpha_{iso}). \tag{16.37}$$

Thus, it is seen from Equation 16.37 that the evaporation time is increased by a factor that is proportional to the volume of the drop, as a result of drop interactions. Writing Equation 16.37 as a ratio of time under interaction to isolated-drop evaporation time

$$t_e/t_{e,iso} = 1 + d_0/3\, \ell. \tag{16.38}$$

With $B_e = 2$, $\alpha_{iso} = 4.4 \times 10^{-7}$ m²/sec, t = 23 msec for an isolated 100-μm diameter drop, t = 24 msec with $\ell/d_0 = 10$, 25 msec with $\ell/d_0 = 5$, and 27 msec at $\ell/d_0 = 1$ (the point of contact).

16.3.2 Combustion of Arrays

The governing SZ equation for the NSF problem is given as

$$\nabla^2 \beta = 0 \tag{16.39}$$

where the SZ variables are defined in Chapters 8, 10, 11, and 12.

On the other hand, for an SF problem involving combustion, the SZ equation is given as

$$\nabla \cdot \rho \vec{v} \beta - \nabla \cdot (\rho D \, \nabla \beta) = 0 \tag{16.40}$$

The solution procedure is similar to that for evaporation for SF and NSF problems.

16.3.2.1 Combustion under NSF

Adopting the spherical coordinate system,

$$\phi(\vec{r}) - \phi_\infty = \frac{a_i \, \eta_i}{|\vec{r} - \vec{r}_i|} + \sum_{j=2, j\neq i} \frac{a_j \, \eta_j}{4\pi \rho D |\vec{r}_i - \vec{r}_j|} \tag{16.41}$$

where the combustion potential is given as

$$\phi = \{\beta - \beta_\infty\}/\{\beta_W - \beta_\infty\} \tag{16.42}$$

and η is still given by Equation 16.14 but with the mass-loss rate corresponds to combustion conditions. The solution for the combustion potential (ϕ) can now be used to obtain the flame contours for a binary array by setting $\phi = \phi_f$. Thus, the flame location is only dependent on the interdrop spacing and the stoichiometry. It has been shown in the literature that the relation given by Equation 16.23 is equally valid for the correction factor of combusting drops; but the B for drop combustion

$$B = \frac{h_{T,\infty}}{L} + \frac{Y_{O_2,\infty}}{\nu_{O_2}} \frac{h_c}{L} \tag{16.43}$$

For combustion or gasification of char with O_2 and CO_2 producing CO,

$$B = \frac{Y_{O_2,\infty}}{\nu_{O_2}} + \frac{Y_{CO_2,\infty}}{\nu_{CO_2}} \tag{16.44}$$

16.3.2.2 SF in Combustion

Following the brief procedure outlined by Annamalai and Ryan [1992], one can define the combustion potential ψ for the SF problem as

$$\psi = \frac{\ell n\left[\frac{F-\beta}{F-\beta_\infty}\right]}{\ell n\left[\frac{F-\beta_w}{F-\beta_\infty}\right]} = -\frac{\ell n\left[\frac{F-\beta}{F-\beta_\infty}\right]}{\ell n\,(1+B)}, \quad B = \frac{\beta_w - \beta_\infty}{F - \beta_w} \tag{16.45}$$

where β is a scalar variable and F is selected so that the transfer number defined for combustion drops in Table 10.1 of Chapter 10 can be used. The governing SZ equation is

$$\nabla^2 \psi = 0. \tag{16.46}$$

Adopting the spherical coordinate system,

$$\psi(\vec{r}) - \psi_\infty = \frac{a_i\, \eta_i}{|\vec{r} - \vec{r}_i|} + \sum_{j=2, j \neq i} \frac{a_j\, \eta_j}{4\pi\, \rho\, D |\vec{r}_i - \vec{r}_j|}\,. \tag{16.47}$$

The correction factor relation remains the same as Equation 16.23 except that B is defined for combustion and

$$\dot{m}_{iso} = 4\, \pi \rho\, D\, a \ln(1 + B)\,. \tag{16.48}$$

Example 2

Using the same data as in Example 1 and h_c = 45105 kJ/kg.

(a) Compute the isolated-drop combustion rate in kg/sec.
(b) Compute correction factors.
(c) Compute the drop combustion rates in kg/sec in the array.
(d) Is the flame an individual flame or group flame at $\ell = 300\ \mu$m? at $\ell = 6800\ \mu$m.

Solution

(a) Following Chapter 10, one can obtain isolated-drop results as: T_w = 333 K, $B = c_p\,(T_\infty - T_w)/L + h_c\, Y_{O2\infty}/(L\nu_{O2})$ = 1.021*(600 – 333)/335 + 0.23*45105/(335*3.52) = 9.59, B is not less than 1; hence, SF must be accounted for.

$Y_{F,w} = 1 - (1 + s)/(1 + B) = 0.905$, $X_{Fw} = 0.761$, $\xi_{w,iso} = \ln(1 + B) = 2.36$

$\dot{m}_{iso,1} = \dot{m}_{iso,3} = 4\ \pi\rho D\ a_1\ \xi_w = 5.42 \times 10^{-8}$ kg/sec, $\dot{m}_{iso,2} = 2.71 \times 10^{-8}$ kg/sec

(b) The correction factors depend only on geometry and are independent of whether the process is evaporation or combustion, SF or NSF. Thus, from Example 1,

$$\eta_1 = 0.868,\ \eta_2 = 0.711.$$

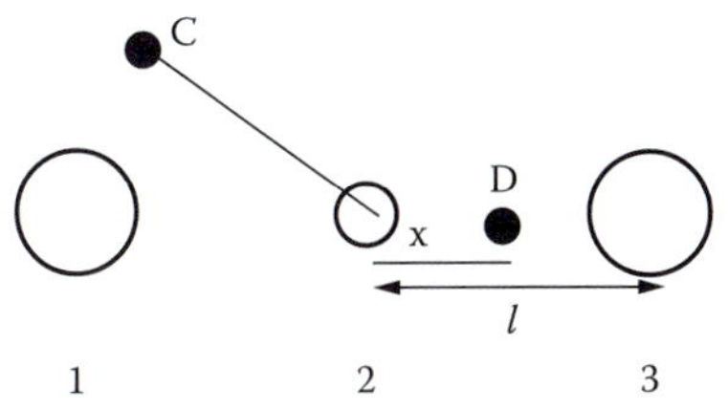

Figure 16.5 Three drop-Linear array.

(c) The combustion rates of drops 1, 2, and 3 are given by $\dot{m}_1 = \dot{m}_3 = \dot{m}_{iso,1}$ η_1 = 5.42 × 10^{-8} * 0.868 = 4.71 × 10^{-8} kg/sec, $\dot{m}_2$ = 2.71 × 10^{-8}*0.714 = 1.92 × 10^{-8} kg/sec

(d) With $\beta = \beta_{O2\text{-}F}$ and $\beta_{O2\text{-}F,f} = 0$ at flame

$$\psi_f = \frac{\ell n\left[\frac{F-\beta_f}{F-\beta_\infty}\right]}{\ell n\left[\frac{F-\beta_w}{F-\beta_\infty}\right]} = -\frac{\ell n\left[\frac{1-\beta_{O2-F,f}}{1-\beta_\infty}\right]}{\ell n\ (1+B)} = -\frac{\ell n\left[\frac{1-0}{1+s}\right]}{\ell n\ (1+B)}.$$

where F = 1 if $\beta = \beta_{O2\text{-}F}$ (Table 10.1, Chapter 10), s = $Y_{O2\infty}/\nu_{O2}$ = 0.065, at flame ψ_f = –ln{1/(1 + 0.065)}/2.36 = 0.0267. The fuel-rich region is along the axis. Let us calculate ψ at D (Figure 16.5). Using Equation 16.47

$$\psi_D = \frac{a_1\ \eta_1}{|\ell + x|} + \frac{a_2\ \eta_2}{x} + \frac{a_3\ \eta_3}{|\ell - x|}.$$

With a_1 = 50 × 10^{-6} m, a_2 = 25 × 10^{-6} m, and a_3 = 50 × 10^{-6} m, all being at the same temperature separated by ℓ = 300 × 10^{-6} m, one can plot ψ and $\beta_{O2\text{-}F}$ vs. x/a_2 (Figure 16.6). It is seen that ψ = 1 and $\beta_{O2\text{-}F}$ = 0.9, as expected,

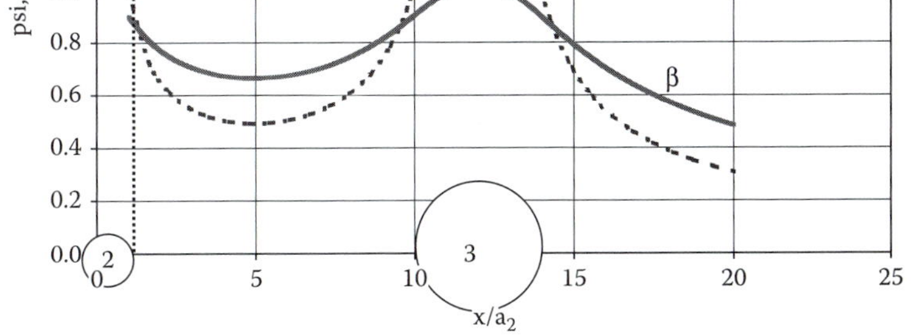

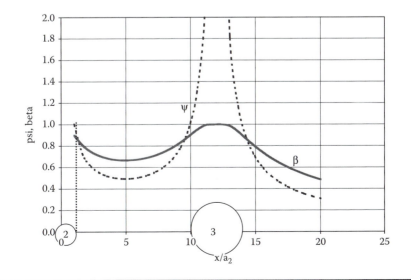

Figure 16.6 The ψ and β profiles along the axis with center drop at 2.

at drop surfaces 2 and 3. However, the values of $\psi > 1$ are not valid within the drop (a drop is supposed to be a point mass). With $x/a_2 = 6.37$, $\psi = 0.514$, and $\beta_{O2\text{-}F} = 0.684$; within $0 < x < \ell$ along the axis, $0.9 < \beta_{O2\text{-}F} < 1$ indicating rich mixture. For an individual flame to occur, $\beta_{O2\text{-}F} < 0$ or, at least, $\beta_{O2\text{-}F} = 0$. Thus, only group combustion occurs around the array.

However, when ℓ is increased to 6800×10^{-6} m or 6.8 mm, then at $x/a_2 = 10.9$ or $x/\ell = 0.91$, $\beta_{O2\text{-}F} \approx 0$, indicating that the flame location lies within $0 < x < \ell$. The combustion under this condition is called *incipient group combustion.* If ℓ is increased further, individual flames will be formed.

■ ■ ■

As an array moves through the combustion chamber, the interdrop spacing may not remain the same. Xin and Megaridis [1994] investigated the interaction of an array of droplets injected quickly into quiescent air and determined the number density and fuel vapor distribution with time. The leading drops slowed down rapidly owing to the high drag coefficient, whereas trailing drops continued to travel with a higher velocity, resulting in an increase of number density (or gradual compaction) with time. Further, the leading array evaporates rapidly compared to the trailing array, changing the size distribution, and the ambient gas properties of the trailing drops are affected by the behavior of the leading drops.

16.4 Combustion of Clouds of Drops and Carbon Particles

As opposed to an array in which discrete point sources contribute to local properties, the sources in a cloud are assumed to be so numerous that source strengths are continuous in a cloud problem.

16.4.1 Conservation Equations

16.4.1.1 Overall Mass

$$\nabla \cdot \rho \vec{v} = \dot{m}''' \tag{16.49a}$$

$$\text{where} \quad \dot{m}''' = n\,\dot{m}_s,\ \ell \approx 1/n^{1/3} \tag{16.49b}$$

Equation 16.49 presumes that the mass source is continuous, which implies closer spacing of drops. If drops are spaced at, say, $\ell/a = 100$, then every drop of radius "a" must distribute its vapor uniformly over a radius of 50a. The QS governing equations for fuel and oxygen species mass fractions become

$$\nabla \cdot \rho \vec{v} Y_F = \nabla \cdot (\rho D\, \nabla Y_F) + \dot{m}''' Y_{F,\ell} - \left|\dot{m}'''_{F,cb}\right| \tag{16.50}$$

where $Y_{F,\ell} = 1$ for pure fuel drops. Using Equation 16.49a in Equation 16.50 to eliminate $\dot{m}'''$ and defining $Y_F' = Y_F - Y_{F,\ell}$,

$$\nabla \cdot \rho \vec{v} Y_F' = \nabla \cdot (\rho D\ \nabla Y_F') - \left|\dot{m}'''_{F,cb}\right| \tag{16.51}$$

$$\nabla \cdot \rho \vec{v} Y_{O2} - \nabla \cdot (\rho D\ \nabla Y_{O2}) = -\left|\dot{m}'''_{F,cb}\right| \nu_{O2} \tag{16.52}$$

and the energy equation becomes

$$\nabla \cdot (\rho \vec{v} h_t) - \nabla \cdot \left(\frac{\lambda}{c_p} \nabla h_t \right) = -\dot{m}'''\, L + \left|\dot{m}'''_{F,cb}\right| h_c \tag{16.53}$$

where L is the latent heat. Using Equation 16.49a in Equation 16.53 and defining $h_t' = h_t + L$,

$$\nabla \cdot (\rho \vec{v} h_t') - \nabla \cdot \left(\frac{\lambda}{c_p} \nabla h_t' \right) = \left|\dot{m}'''_{F,cb}\right| h_c. \tag{16.54}$$

Using Shvab–Zeldovich transformation with Le = 1, Equation 16.51, Equation 16.52, and Equation 16.54 become

$$\nabla \cdot \rho \vec{v} \beta' - \nabla \cdot (\rho D\ \nabla \beta') = 0 \tag{16.55}$$

where $\beta'_{O2\text{-}F} = Y_F' - Y_{O2}/\nu_{O2} = Y_F - Y_{O2}/\nu_{O2} - Y_{F,\ell} = \beta_{O2\text{-}F} - F_{O2\text{-}F}$ (16.56a)

$\beta'_{ht\text{-}F} = \{h_t'/h_c + Y_F'\} = (h_t + L)/h_c + Y_F - Y_{F,\ell} = \beta_{ht\text{-}F} - F_{ht\text{-}F}$ (16.56b)

$\beta'_{ht\text{-}O2} = (h_T + L)/h_c + Y_{O2}/\nu_{O2} = h_t/h_c + Y_{O2}/\nu_{O2} - \{-L/h_c\}$

$= \beta_{ht\text{-}O2} - \{F_{ht\text{-}O2}\}$ (16.56c)

where $F_{O2\text{-}F} = Y_{F,\ell}$, $\{F_{ht\text{-}O2}\} = \{-L/h_c\}$, $F_{ht\text{-}F} = Y_{F,\ell}$. (16.56d)

(See also Table 10.1 of Chapter 10 for F values.) If the mass source would have been zero, then $Y_F' = Y_F$, $h_t' = h_t$, $Y_{O2}' = Y_{O2}$, $\beta' = \beta$ and Equations 16.51, 16.54, and 16.55 become conventional conservation equations.

Consider a 2-D laminar stream of droplets injected with a small amount of air as a jet (Figure 16.7). The jet entrains air and drops get distributed with higher number density along the axis and lower density farther from the axis. One may use symmetry conditions at y = 0 with $\partial\beta'/\partial y = 0$ and $\beta' = \beta_\infty' \to 1 - Y_{O2,\infty}/\nu_{O2}$, as $y \to \infty$. Then $\beta_{O2\text{-}F} \to Y_F - Y_{O2}/\nu_{O2}$ profile could be obtained. The values of $\beta_{O2\text{-}F}$ could be positive (fuel-rich regime), zero (stoichiometric regime, ABC), or negative (fuel-lean regime); $\beta_{O2\text{-}F} = 0$, when $Y_F = Y_{O2}/\nu_{O2}$ (stoichiometric surface, ABC). Typically, the stoichiometric surface can be easily ignited [Laster and Annamalai, 1991] and it yields the highest temperature

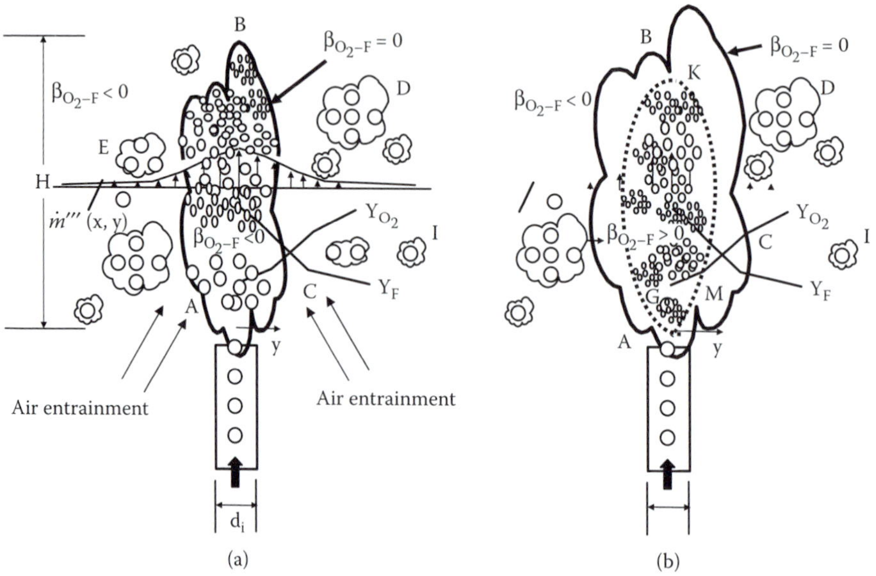

Figure 16.7 Stream of drops burning in air: (a) group flame located just around drop cloud called sheath combustion, (b) group flame located away from drops.

with the highest possible reaction rate. As discussed in Chapter 10, a thin flame is established along the surface ABC, where $\beta_{O2\text{-}F} = 0$; note that it is also possible to get $\beta_{O2\text{-}F} = 0$ at D, E, I, etc. When a flame surrounds a cluster having a large number of drops, the flame is called a *group flame.* If the drops are far apart, there may be individual flames and the mass source is no longer continuous and, as such, they exist at discrete point sources. For fuel-rich regimes, drops are closely spaced and only evaporation occurs around each drop. Within ABC in Figure 16.7a, there may be only evaporation. The mass source for each drop is given as

$$\dot{m}''' = n\,\dot{m}_s,\ \ \dot{m}_s = 4\pi\rho\ D\ a\ \ln\,(1 + B_e),\ \text{(from Chapter 10)}, \tag{16.57}$$

$$B_e = h_T/L. \tag{16.58}$$

This approximation is called the *dense cloud approximation.* If a flame surrounds the drop (as in fuel-lean regimes, e.g., I), then

$$\dot{m}_{iso} = 4\pi\rho\ D\ a\ \ln(1 + B),\ B \approx \{c_p\ (T - T_w)/L\} + (Y_{O2}\ h_c)/(\nu_{O2}\ L) \tag{16.59}$$

where T and Y_{O2} are local temperature and concentration within the cloud. This approximation is called the *dilute cloud approximation.* However, the mass sources in Equation 16.49 may no longer be continuous. O_2 is not available to fuel-rich regimes; thus, a group flame exists along ABC and at

D, E, etc. Note that fuel drops need not be present for a fuel-rich regime as shown in Figure 16.7b, in which the drops within GKM supply a large amount of fuel vapor that diffuses or gets convected into the regime between ABC and GKM with a group flame around ABC.

One can define the NSZ combustion potential as

$$\phi = (\beta' - \beta'_\infty)/(\beta'_{ref} - \beta_\infty') \tag{16.60}$$

where it is presumed that β'_{ref}'s are values at any prescribed reference locations. Labowsky (1976 a, b) has shown that the thermodynamic equilibrium of evaporation requires that all drops be at the same temperature if they evaporate or burn under steady state when an array is placed in an infinite environment. Hence, $\beta_{ref} = \beta_w$ and is not a function of the space variable within the array or cloud.

16.4.1.2 Group Combustion for Simple Geometries

Figure 16.7 shows the rationale behind the occurrence of group combustion for a flow configuration. A closed-form solution is difficult for such a system owing to the random distribution of drops and complexities of flow fields. It is useful to obtain explicit results for the burn rate or correction factor for simple geometries in simplified flow systems so that the parameters governing the group behavior can be identified. For example, if there are drops only within ABC but none outside, then there is a mass source term only for drops within ABC but for none outside; i.e., there is discontinuity in the mass sources. For example, if we let the shape of ABC be spherical with no convective flows in and out, we can obtain a closed-form solution to the burn rate of cloud. Let us consider a spherical cloud burning quiescently in an infinite environment. Prior to presenting the governing equations, let us introduce the terminology (Figure 16.8). The burn rate of the cloud is given by $\dot{m}_c$, whereas that of a single drop is given by $\dot{m}_s$.

16.5 Terminology

16.5.1 Isolated-Drop Combustion (ISOC)

If a drop is far away from other drops, then the drop burns in the isolated-combustion mode and the combustion intensity of each drop is unaffected by the presence of the other drops. The burn rate is $\dot{m}_{s,iso}$.

16.5.2 Individual Flame Combustion (IFC)

When the number of drops per unit volume (n) is increased, the burning characteristics of the constituent drops change owing to decreased Y_{02}. However, if the ℓ is large enough, each drop will still have its own flame, the flame radius being larger than the isolated-flame radius. This is termed IFC. The burn rate is $\dot{m}_s$ (Figure 16.8a).

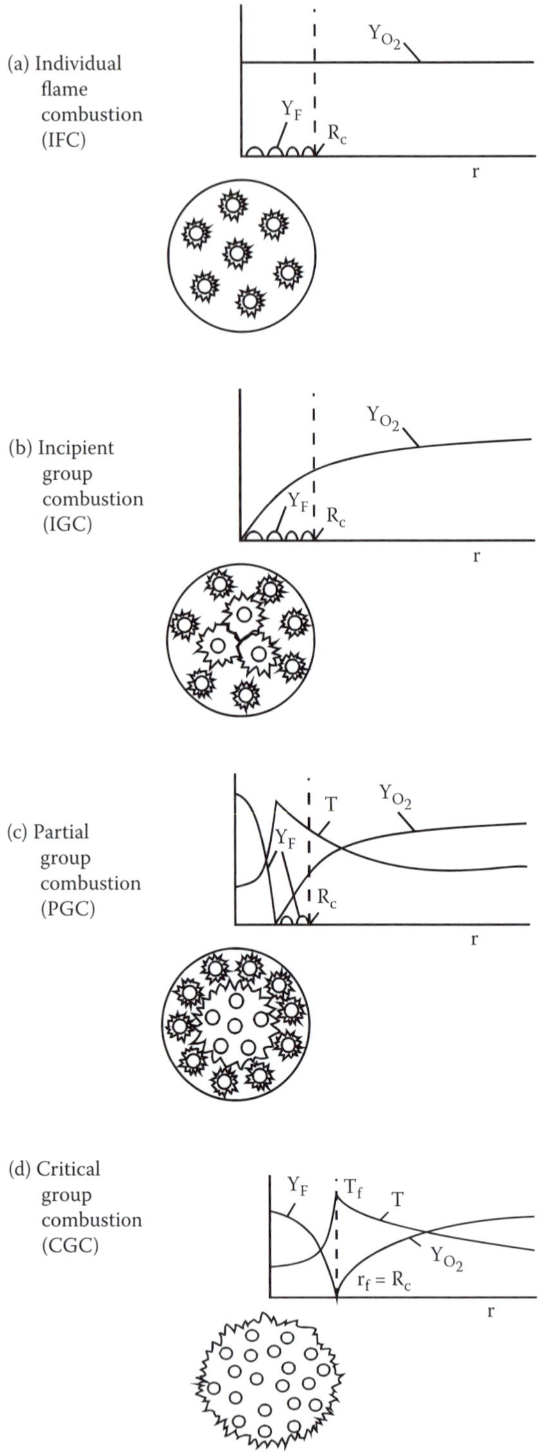

Figure 16.8 Terminology in group combustion. (Adapted from Annamalai, K., and Ryan, W., 1992.)

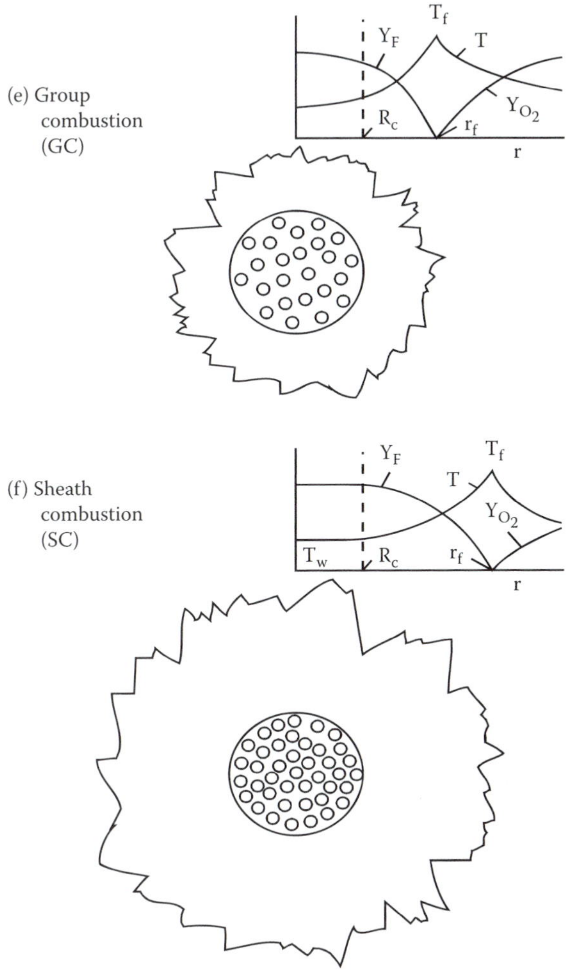

Figure 16.8 (Continued).

16.5.3 Incipient Group Combustion (IGC)

As the n is increased still further, the outer drops will continue to burn in IFC. These drops consume oxygen, lowering the Y_{02} within the cloud, resulting in increased flame radii around the drops. A condition will be reached when the flame radii of the center drops of the cloud just touch each other. This is termed IGC (Figure 16.8b).

16.5.4 Partial Group Combustion (PGC)

As n is increased further, the inner drops, starved of oxygen, simply evaporate. The vapors from the inner drops diffuse outward and establish a group flame inside the cloud. However, the outer drops burn in the IFC mode. This is termed PGC (Figure 16.8c).

16.5.5 Critical Group Combustion (CGC)

As n is further increased, the evaporation rate or the blowing rate of vapors is such that the oxygen cannot penetrate into the cloud and the flame just stands at the cloud surface. This is termed CGC. The temperature at the edge of the cloud T_c is equal to the flame temperature (Figure 16.8d).

16.5.6 Total Group Combustion or Group Combustion (GC)

If n is further increased, the blowing rate of vapors prevent the oxygen from penetrating into the cloud and a flame is established outside the cloud. This is termed GC. The temperature of gases in the cloud is such that $T_w < T < T_f$ (Figure 16.8e).

16.5.7 Sheath Combustion (SC)

If n is increased to a large value (very dense cloud), eventually a condition is reached in which the temperature at the cloud surface is at the wet bulb temperature of the drops ($T_c = T_w$). The cloud now behaves as an equivalent single drop of radius R_c, with the equivalent drop density being equal to the cloud mass density ($\rho_{cl} = \rho_\ell \sigma$). This type of combustion is called sheath combustion. Thus, we can use the single-drop results presented in Chapter 10 for a very dense spherical cloud. The burn rate of the cloud is given by $\dot{m}_{c,SC}$ (Figure 16.8f).

$$\dot{m}_{c,SC} = 4\,\pi\,\mathrm{Sh}\,\rho\,D\,\ln(1+B) \tag{16.61}$$

In contrast to the combustion of a single drop, in which the d^2 law is followed as the gas density is assumed to be small compared to the liquid density, the cloud mass density is comparable to gas density; however, as an approximation, the d^2 law may be used:

$$D_{cl}^2 = D_{cl,0}^2 - \alpha_{cl} t \tag{16.62a}$$

where the combustion-rate constant is given as

$$\alpha_{cl} = 4\,\mathrm{Sh}\,\rho\,D\,\ln(1+B_c)/\rho_{cl}. \tag{16.62b}$$

16.6 Governing Equations for Spherical Cloud

Consider a collection of a large number of drops contained in an imaginary spherical enclosure of radius R_c. If this cloud of drops is introduced into a hot furnace at T_∞, a heat wave propagates into the cloud raising the temperature of the drops inside the cloud and producing vapors which diffuse outward. The oxygen diffuses toward the cloud in which the oxygen concentration is less.

As the cloud is a fuel-rich, low-temperature region, the vapors ignite in the hot ambience, typically outside the cloud, and hence a single flame is formed around the entire cloud at radius $r_f > R_c$, instead of a flame being established around every drop inside the cloud. If it is hypothesized that the drops are maintained at constant size throughout the vaporization process, with each drop being replenished with fuel, a steady state condition is reached in the cloud. Thus, fundamental information can be obtained through a simplified QS analysis.

The earliest work on QS group combustion was conducted by Suzuki and Chiu (1971), which was then followed by Labowsky and Rosner (1978b) and later by Bellan et al. (1983, 1987). Suzuki and Chiu obtained group combustion results for a spherical cloud of uniformly distributed monosized liquid propellant drops composed of *n*-butylbenzene $C_{10}H_{14}$.

They divided the regime into inner (cloud) and outer (single phase outside the cloud) regions. Simplified results for mass flux, the Shvab–Zeldovich variable, cloud center properties, integral expressions for the cloud burning rate and flame locations, and the condition for critical group combustion of spherical, cylindrical, and planar monosized clouds are presented.

Consider a spherical cloud of drops (Figure 16.9). The gas phase can be divided into three zones: (1) a thin-film region surrounding the drops in which mass and thermal gradients are high (Zone I), (2) the remaining interstitial

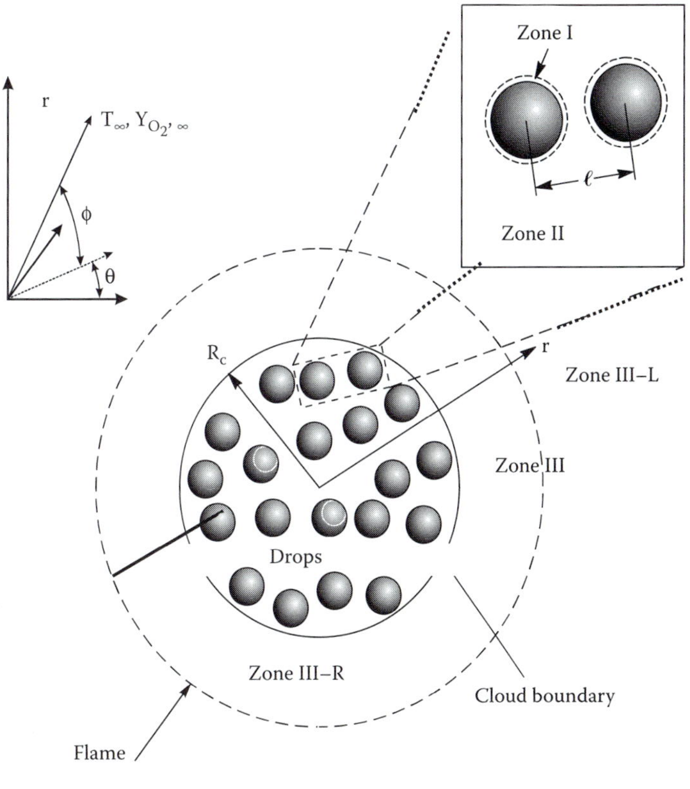

Figure 16.9 Spherical cloud model.

gas in which the gradients of mass and heat are not as significant (Zone II, $\beta_{O2\text{-}F} = Y_F - Y_{O2}/\nu_{O2} >> 0$), and (3) and a single-phase zone surrounding the cloud (Zone III, where $\beta_{O2\text{-}F} > 0$ in Zone III-R and $\beta_{O2\text{-}F} < 0$ in Zone III-L). The gas film thickness may be affected by the presence of other drops. Group theory presumes that the relation between Nu and the local film thickness is given in order to determine the mass source strength in terms of the local bulk gas properties. The cloud is treated as a homogeneous phase with point mass sources for fuel and point sinks for oxygen, i.e., negligible film thickness.

We will expand the governing equations given in Equation 16.49 to Equation 16.60 for a spherical coordinate system (see Exercise Problem 16.10 for derivation from generic equations) in the following subsections.

16.6.1 Mass

$$\frac{1}{r^2}\frac{d}{dr}(r^2 \rho v) = \dot{m}''', \quad r < R_c \tag{16.63}$$

$$\frac{1}{r^2}\frac{d}{dr}(r^2 \rho v) = 0, \quad r > R_c \tag{16.64}$$

where $\dot{m}''' = n\,\dot{m}_s = \Sigma n_i\,\dot{m}''_{s,i}\,4\,\pi\,a_i^2 = \Sigma S_{v,i}\,\dot{m}''_{s,i}$ where $S_{v,i}$ is the surface area of drops of size i per unit volume and $\dot{m}''_{s,i}$, the mass-loss rate of drop i per unit surface area of drop size i. For monosized drops, $\dot{m}''' = S_v\,\dot{m}''_s$.

16.6.2 Fuel Species

$$\frac{1}{r^2}\frac{d}{dr}\left(r^2 \rho v\, Y'_F\right) = \frac{1}{r^2}\frac{d}{dr}\left(r^2 \rho v\,\frac{dY'_F}{dr}\right) - \left|\dot{m}'''_{F,ch}\right| \tag{16.65}$$

16.6.3 Modified SZ Variable

$$\frac{1}{r^2}\frac{d}{dr}(\dot{m}\beta') - \frac{1}{r^2}\frac{d}{dr}\left(4\,\pi\rho D\, r^2\,\frac{d\beta'}{dr}\right) = 0, \ r < R_c \tag{16.66}$$

Because Zone II is at the average temperature and mass fraction within the interstitial space, it is consistent to assume that chemical reactions take place uniformly in Zone II, and the point mass source strength for each drop is given as

$$\dot{m}_s = 4\,\pi\rho\ D\ a\ \ln(1 + B(r)) \tag{16.67}$$

where

$$\frac{\beta_w - \beta}{F - \beta_w} = \frac{\beta'_w - \beta'}{-\beta'_w} = \mathbf{B}(r), \text{ transfer number as function of r.} \quad (16.68)$$

For a group flame surrounding the cloud, Equation 16.68 reduces to

$$B(r) = h_t(r)/L \text{ (dense cloud approximation or absence of flame around each drop), } r < R_c \quad (16.69)$$

and h_t is the averaged thermal enthalpy in the gas phase adjacent to the drops. For the case of an individual flame around a drop inside the cloud, Equation 16.68 reduces to

$$B(r) = h_t\,(r)/L + Y_{O2}\,(r)\,h_c/(L\,\nu_{O2}), \text{ (dilute cloud approximation), } r < R_c. \quad (16.70)$$

Define

$$\phi = \beta'/\beta_\infty'. \quad (16.71)$$

Using Equation 16.71 in Equation 16.68,

$$\frac{\beta'_w - \beta'}{-\beta'_w} = \frac{\beta'}{\beta'_w} - 1 = \phi\frac{\beta'_\infty}{\beta'_w} - 1 = \mathbf{B}(r) \quad (16.72)$$

and simplifying

$$\phi = (1 + B(r))\gamma,\ r < R_c \quad (16.73)$$

where

$$\gamma = \frac{\beta'_w}{\beta'_\infty} = \frac{\beta_w - F}{\beta_\infty - F} \quad (16.74)$$

where B(r) could be either defined by Equation 16.69 or Equation 16.70.

With $\beta' = \beta_{O2\text{-}F} - F_{O2\text{-}F} = \beta_{O2\text{-}F} - Y_{F\ell}$, Equation 16.74 reduces to

$$\gamma = \frac{Y_{Fw} - Y_{F\ell}}{\frac{Y_{O2,\infty}}{\nu_{O2}} - Y_{F\ell}} = \frac{1}{1+B} \quad (16.75)$$

$$Y_{Fw} = Y_{F\ell} - \frac{Y_{F\ell} + s}{1+B} \quad (16.76)$$

where $Y_{F\ell} = 1$ and the relation is valid for an individual drop burning or evaporating inside the cloud.

With Equation 16.71 in Equation 16.66,

$$\frac{d}{dr}(\dot{m}\phi) - \frac{d}{dr}(4\pi\rho D\, r^2\phi) = 0, \tag{16.77a}$$

$$\dot{m}_c \frac{d\phi}{dr} - \frac{d}{dr}(4\pi\rho D\, r^2\phi) = 0, \quad r > R_c. \tag{16.77b}$$

The boundary conditions are

$$d\phi/dr = 0 \text{ at } r = 0 \tag{16.78a}$$

ϕ is a continuous function at $r = R_c$

$$\text{(i.e., } \{d\phi/dr\}_{r=Rc-} = \{d\phi/dr\}_{r=Rc+}, \{\phi\}_{r=Rc-} = \{\phi\}_{r=Rc+}) \tag{16.78b}$$

$$\phi \rightarrow 1,\ r \rightarrow \infty. \tag{16.78c}$$

Note that $\dot{m}$ (r) continuously increases from 0 (at r = 0) to $\dot{m}_c$ (at $r = R_c$) with r, and then $\dot{m} = \dot{m}_c$ remains constant for $r > R_c$. With boundary conditions given by Equations 16.78a–c, Equation 16.77 can be solved for ϕ. ϕ is related to B(r) for $r < R_c$ (Equation 16.73) and then the B(r) profile is used in the mass conservation equation (Equation 16.63) to determine the mass-loss rate from the cloud. Thus

$$\frac{d\dot{m}}{dr} = 4\,\pi\, r^2 \dot{m}''' = 4\,\pi\, r^2 n\, \dot{m}_s = 4\pi\, r^2 n\{4\,\pi\,\rho\, D a \ln\{1 + B(r)\}.$$

Net flow of mass at any radius $r < R_c$ is

$$\dot{m}(r) = \int_0^r 4\,\pi\, r^2 n(r)\,\{4\,\pi\,\rho\; D a(r) \ln\{1 + B(r)\}, \quad r < R_c \tag{16.79}$$

The drop sizes may not be uniform within the cloud. For example, the drops near the outer boundary that are burning rapidly may have smaller size compared to the core drops. For the right-hand side of Equation 16.79, the product of the particle number density and the particle radius can be assumed to follow a power law.

$$na/(na)_{Rc} = (r/R_c)^{-q},\ -1 < q < 1 \tag{16.80}$$

where $(na)_{Rc}$ is the value of "na" at $r = R_c$. As r increases, na decreases. Three possible situations arise: (1) a monosized uniform cloud, q = 0, (2) a cloud with the product of size and number density as a power law function of (R_c/r), and

(3) a power law with a fixed cloud mass. The solutions are left as exercise problems. Normalized equations are presented below, followed by solutions.

$$\text{Let } \xi = R_c/r \tag{16.81}$$

$$M = \dot{m}_c/(4\pi\ \rho\ D\ R_c) \tag{16.82}$$

Nondimensional sheath combustion rate:

$$M_{SC} = \dot{m}_{c,SC}/4\pi\ \rho\ D\ R_c = \ln(1 + B) \tag{16.83}$$

Nondimensional cloud burning rate if each drop burns in the isolated-combustion mode:

$$M_{ISOC} = \dot{m}_{c,iso}/4\pi\ \rho\ D\ R_c \tag{16.84}$$

$$G_1 = 4\ \pi\ (na)_{Rc}\ R_c^{\,2} \tag{16.85}$$

$\alpha(r)$ = {mass flow at any radius/mass flow at r = R_c} = $\dot{m}(r)/\dot{m}_c$
Using Equation 16.79, the cloud mass loss rate is expressed as

$$\frac{M}{M_{SC}} = G_1 \int_0^{Rc} \left[\frac{na}{\{na\}_{Rc}}\right] \frac{\ln\{1+B(\xi)\}}{\{\ln\{1+B\}} \frac{d\xi}{\xi^4}. \tag{16.86}$$

The mass conservation equation becomes

$$-M\xi^4 \frac{d\alpha}{d\xi} = G_1\,\xi^q\ \ln(1+B),\ \ \xi \ge 1$$
$$= 0,\ \ \xi \le 1 \tag{16.87}$$

whereas the nondimensional SZ equation becomes

$$M\xi^4 \frac{d(\alpha\phi)}{d\xi} + \xi^4 \frac{d^2\phi}{d\xi^2} = 0,$$
$$\alpha = \alpha(\xi),\ \ \xi > 1,\ \alpha = 1,\ \xi \le 1. \tag{16.88}$$

Derivations of solutions are left as an exercise (see Problem 16.12). The rest of the sections deal only with q = 0 or G_1 = G (i.e., uniformly sized and spaced drop cloud) [Annamalai and Ramalingam].

16.7 Results

16.7.1 G Number

The physical meaning is as follows:

$$G_1 = G = 4\ \pi\ na\ R_c^{\,2} = 3\ N\ a/R_c = 3\ \sigma\ R_c^{\,2}/a^2$$
$$= 3\ \{m_{cloud}/m_{drop}\}\ \{a/R_c) = S_v\ R_c^{\,2}/a. \tag{16.89}$$

For pure evaporation,

G = characteristic heat consumption rate/heat transport rate to cloud.

For combustion of a given fuel,

G = characteristic O_2 consumption rate/O_2 transport rate to cloud.

The G number can also be expressed in terms of the total number of drops and interdrop spacing of ℓ. Assuming the cloud to consist of packed spheres of radius $\ell/2$ with a packing factor of F_p, the cloud radius is estimated as

$$R_c = (N/F_p)^{(1/3)}(\ell/2). \tag{16.90}$$

The interdrop spacing can also be related to number of drops per unit volume as

$$\ell = (6\ F_p/\pi\ n)^{1/3}. \tag{16.91}$$

The group combustion number for a monosized uniform cloud is rewritten as

$$G = 5.43\ N^{2/3}/(\ell/a). \tag{16.92}$$

16.7.2 Nondimensional Mass Flow Rate

$$\alpha(x) = \frac{\dot{m}}{\dot{m}_c} = \xi^{\frac{-3}{2}} \frac{I_{\left[\frac{3}{2}\right]}\{G^{1/2}\xi^{-3/2}\}}{I_{\left[\frac{3}{2}\right]}\{G^{1/2}\}}$$

$$= \xi^{\frac{-3}{4}} \frac{\left\{\mathrm{Cosh}(G^{1/2}\xi^{-3/2}) - \frac{\mathrm{Sinh}(G^{1/2}\xi^{-3/2})}{G^{1/2}\xi^{-3/2}}\right\}}{\left\{\mathrm{Cosh}(G^{1/2}) - \frac{\mathrm{Sinh}(G^{1/2})}{G^{1/2}}\right\}}, \quad \xi \geq 1, \quad q = 0 \tag{16.93}$$

16.7.3 NSZ Variable

$$\phi = \frac{\beta'}{\beta'_\infty} = \exp[-M\{1 + f(\xi)\}], \ \xi \geq 1, \ r < R_c \tag{16.94a}$$

$$= \exp(-M\xi), \ \xi \leq 1, \ r > R_c \tag{16.94b}$$

$$f(\xi) = \frac{\left[\sinh\sqrt{G} - \xi \sinh\left(\frac{\sqrt{G}}{\xi}\right)\right]}{\sqrt{G}\left(\cosh\sqrt{G} - \frac{1}{\sqrt{G}} Sinh(\sqrt{G})\right)} \tag{16.95}$$

16.7.4 Cloud Mass-Loss Rate and Correction Factor

The solution for the cloud mass-loss rate is given in two forms:

$$M/M_{SC} = \dot{m}_c/\dot{m}_{c,SC} \tag{16.96}$$

$$M/M_{ISOC} = \eta = \dot{m}_c/\dot{m}_{c,ISOC} \tag{16.97}$$

where η is known as the correction factor. As a ratio of sheath form,

$$\dot{m}_c/\dot{m}_{c,SC} = M/M_{SC} = 1 - [\tanh G^{(1/2)}/G^{(1/2)}] \tag{16.98}$$

As G becomes large, the ratio approaches unity. For the purpose of discussion, clouds having G < 0.1 will be referred to as dilute, whereas clouds having G > 100 will be referred to as dense. Limiting cases can be given as

$$M/M_{SC} = G/(3 + G), \quad G >> 100 \text{ or } G << 1 \tag{16.99}$$

As a ratio of isolated burn rate

$$\eta = \frac{M}{M_{ISOC}} = \frac{M}{M_{SC}}\frac{3}{G} = \frac{3}{G}\left[1 - \frac{\tanh G^{(1/2)}}{G^{(1/2)}}\right] \tag{16.100}$$

or

$$\frac{M}{M_{SC}} = \frac{G\eta}{3} \tag{16.101}$$

with limiting cases given as

$$\eta = 3/(3 + G) \quad \text{for} \quad G >> 100 \quad \text{or} \quad G << 1 \tag{16.102}$$

Figure 16.10 and Figure 16.11 show the results for M/M_{SC} and the correction factor. For monosized and uniform clouds q = 0, G_1 = G. For a uniform cloud

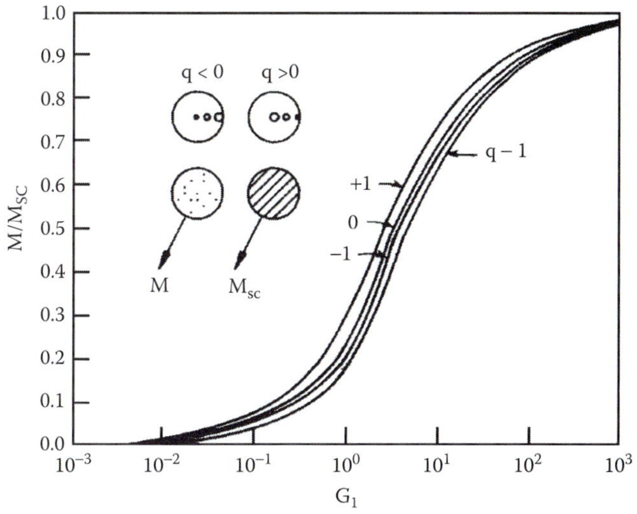

Figure 16.10 The variation of M/M_{SC} vs. G for a spherical cloud of particles or drops. (Adapted from Annamalai, K. and Ramalingam, S., Group combustion of char particles, *Combustion and Flame*, 70: 307–332, 1987.)

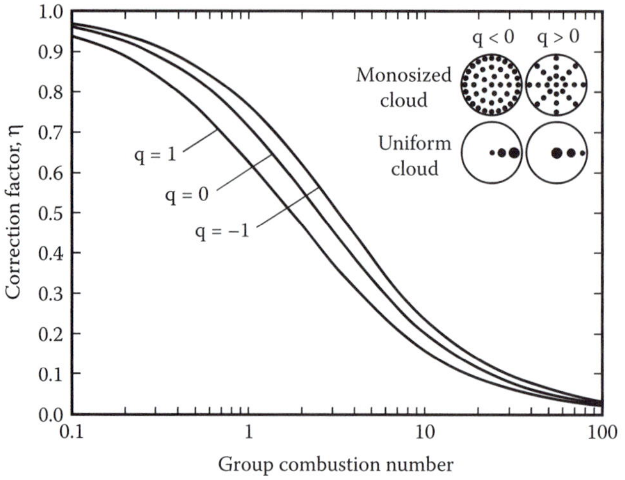

Figure 16.11 Correction factor vs. G1 for a spherical cloud.

(number density "n" invariant), q = 0 for a monosized uniform cloud, q < 0 for decreasing drop size or decreasing number density toward center, and q > 1 for increasing drop size or increasing number density toward the center. As G approaches zero, every drop in the cloud is surrounded by gas at T_∞, and $\eta \rightarrow 1$. As G becomes large, every drop in the interior of the cloud is at the same temperature as the drop temperature (T_w), the vapor mass fraction corresponds to the saturated condition, and only the drops near the cloud surface evaporate. Thus, dense clouds behave similar to a large isolated drop placed in an ambience of T_∞ and $Y_{O2,\infty}$ and $M/M_{SC} \rightarrow 1$. At large value of G, only a thin layer at the periphery of the cloud participates in vaporization [Correa and Sichel 1982a, b; Deutch et al., 1976].

For the evaporation process, Correa and Sichel found that the cloud vaporizes following the d^2 law. Thus

$$D_c^2 = D_{c,0}^2 - \alpha_e\, t \tag{16.103}$$

where $\alpha_e = 4$ Sh D $\ln(1 + B_e)/\rho_{cl}$ and $D_c = 2\, R_c$.

16.7.5 NSZ Variable at Cloud Center

In order to get the properties at the center, we need to determine f(r = 0) (see Equation 16.95).

$$f(\xi = \infty) = f(r = 0) = \frac{\left[\frac{\sinh\sqrt{G}}{\sqrt{G}} - 1\right]}{\left(\cosh\sqrt{G} - \frac{1}{\sqrt{G}}\,\mathrm{Sinh}\left(\sqrt{G}\right)\right)} \tag{16.104}$$

Using Equation 16.104 and Equation 16.94,

$$-\frac{\ln\phi(r=0)}{\ln(1+B)} = -\frac{\ln\left[\frac{\beta_0-F}{\beta_\infty-F}\right]}{\ln(1+B)} = \frac{\ln[\{(1+B(r=0)\}\gamma]}{\ln(1+B)} = \frac{M}{MSc}(1+f(r=0)). \quad (16.105)$$

Using Equation 16.104,

$$-\frac{\ln\phi(r=0)}{\ln(1+B)} = 1.5\frac{G}{3+G} = 0.5G \quad \text{as} \quad G \to 0 \ or\ \phi(r=0) \to 1 \quad (16.106)$$

$$-\frac{\ln\phi(r=0)}{\ln(1+B)} = \frac{1}{\left(1-1/\sqrt{G}\right)}\frac{G}{3+G} = \frac{1}{\left(1-1/\sqrt{G}\right)} \quad as\ G \to \infty. \quad (16.107)$$

16.7.6 Flame Radius

The flame radius in diffusion-limited combustion is obtained by determining the loci of points with $\phi = \phi_f$, using $Y_F = Y_{O2} = 0$ at the flame location. Using Equation 16.94b

$$\phi_f = \frac{\beta_{O2-F,f} - Y_{F,\ell}}{\beta_{O2-F,\infty} - Y_{F,\ell}} = \frac{1}{1+s} = \exp(-M\xi_f) = \exp(-\frac{M}{M_{SC}}M_{SC}\,\xi_f),\ \xi_f < 1 \quad (16.108)$$

$$\frac{M}{M_{SC}}\frac{R_c}{r_f} = -\frac{\ln\phi_f}{\ln(1+B)} = \frac{\ln\,(1+s)}{\ln(1+B)} = \left(\frac{a}{r_f}\right)_{iso}. \quad (16.109)$$

(a) Flame at Cloud Surface: It is seen from Equation 16.109 that $r_f \to R_c$ at a particular M/M_{SC}. The value of the G number at which $M/M_{SC} = (a/r_f)_{iso}$ is called the *critical G number*. The G_c at which critical group combustion sets in can be obtained by setting $M/M_{SC} = (a/r_f)_{iso}$ in Equation 16.109. As $M/M_{SC} \approx G/(3 + G)$ for a dilute cloud, then

$$G_C \approx 3(a/r_f)_{iso}/\{1 - (a/r_f)_{iso}\} \approx 3(a/r_f)_{iso}. \quad (16.110)$$

With $(Y_{O2,\infty}/\nu_{O2}) = 0.066$, $B = 10.7$ (dodecane), and $(r_f/a)_{iso} = 38.5$, M/M_{SC} is estimated to be 0.026 and G_c is estimated to be 0.0801, whereas according to Labowsky and Rosner [1978] $G_c = 0.11$. This critical G_c is around 0.1 for most fuels.

(b) Sheath Flame: As G increases beyond G_c, the flame moves away from the cloud surface and beyond a certain value; the flame standoff distance approaches an asymptotic value given by isolated-drop combustion theory. The location of the flame is farthest from the cloud when $M/M_{SC} \to 1$, which corresponds to the case of an isolated drop having same radius as the cloud radius, i.e., a cloud under sheath combustion mode behaves similar to a large liquid drop but with mass density corresponding to cloud bulk density.

(c) Partial Group Combustion (PGC) Flame: The flame can penetrate inside the cloud and PGC ensues when $G < G_c$. Equation 16.94a equal to ϕ_f and simplifying

$$\phi_f = \frac{\beta_{O2-F,f} - Y_{F,\ell}}{\beta_{O2-F,\infty} - Y_{F,\ell}} = \frac{1}{1+s} = (1+B)^{\left(-\frac{M}{M_{SC}}1+f(\xi_f)\right)}. \qquad (16.111)$$

For any assigned $G < G_c$, M/M_{sc} is known from Equation 16.98; then $f(\xi_f)$ is obtained from Equation 16.111; the corresponding ξ_f is solved from Equation 16.95.

(d) Flame at Center of Cloud: As G is reduced to extremely low values called G_{IGC}, the flame moves toward the center of cloud, and we approach the limits of isolated-drop combustion. When the flame is located inside the cloud at the center, then one may use (1) flame radius = $\ell/2$ and, hence, flame-to-cloud-radius ratio is $\ell/2R_c$ or (2) use $\phi_f = 1/(1 + Y_{O2,\infty}/\nu_{O2})$ to get ϕ_f; let $\xi \to \infty$, and use Equation 16.104 to get $f(\infty)$, and the corresponding G_{IGC} is evaluated from Equation 16.105. For small values of G

$$\ln(\phi_f)/\ln(1 + B) = -(3/2)\ \{G/(G + 3)\}. \qquad (16.112)$$

The value of $G = G_{IGC}$ for dodecane is evaluated as 0.053, whereas Labowsky and Rosner (1978b) G_{IGC} is determined to be 0.068 (or more exactly, 0.055).

16.7.7 Spray Classification

The previous subsections dealt with the effects of the G number on the different regimes of cloud combustion. Rather than classifying sprays simply into dense and dilute clouds, one may use the following classification:

Regime	*Class*	*Range of G*	*Remarks*
I	Ultradilute, very low G	$G < 0.01$	Isolated drop
II	Dilute, low G	$0.01 < G < 0.06$	Isolated to incipient
III	Marginally dense, moderate G	$0.06 < G < 0.1$	Partial group combustion
IV	Dense	$0.1 < G < 10$	Group combustion
V	Very dense, high G	$10 < G < 100$	Group to sheath combustion
VI	Ultradense, very high G	$G > 100$	Sheath combustion

Example 3

Consider a monosized uniform spherical cloud of radius $R_c = 1$ cm consisting of *n*-hexane drops of equal diameter 100 (= 2 a) μm kept in an ambience of 600 K and $Y_{O2,\infty} = 0.23$. The number density is such that $n = 3.7 * 10^9$ drops/m^3. See Example 1 for properties.

(a) Compute isolated-drop combustion rate in kg/sec and flame radius to drop radius ratio for an isolated drop.
(b) Compute M/M_{ISOC}, M/M_{SC} for the cloud and cloud burn rate in kg/sec.
(c) If ℓ is increased to ℓ_{crit}, at which $r_f/R_c = 1$, what is the value of ℓ_{crit}? Estimate the number density.
(d) What is the interdrop spacing ℓ_{IGC} when flame moves to the center (IGC) or as it approaches individual drop combustion?

Solution

(a) From previous examples, $T_w = 333.2$ K, $B = 9.59$, $s = 0.065$, $r_f/a = \ln(1 + B)/\ln(1 + s) = 37.3$, $\xi_w = \ln(1 + B) = 2.36$, $\dot{m}_{iso,1} = 4\pi\rho D\, a\xi_w = 3.1*10^{-8}$ kg/sec.
(b) $G = 4\pi na\, R_c^2 = 4*3.1427*3.7 \times 10^9*50.0 \times 10^{-6}*0.01^2 = 232$, from Equations 16.100 and 16.98 $M/M_{ISOC} = \eta = 0.0121$,
$M/M_{SC} = 0.93$; Sheath burn rate $\xi w = \ln(1 + B) = 2.36$, $\dot{m}_{SC} = 4\pi\rho D\, R_c\, \xi_w = 1.08 \times 10^{-5}$ kg/sec.
N = Total number of drops within $R_c = 0.1$ m $= (4/3)\, \pi\, R_c^3\, n = 15{,}500$ drops.
(c) $r_f/R_c = (r_f/a)_{iso}\, (M/M_{SC}) = 1$, $(r_f/a)_{iso} = \ln(1 + B)/\ln(1 + s) = 37.3$, $M/M_{SC} = (a/r_f)_{iso} = 1/37.3 = 0.0267$; r_f of cloud = 37.3 cm. The G value corresponding to $M/M_{SC} = 0.0267$ is given as 0.083; $G_{crit} = 4\pi n\, a\, R_c^2 = 0.083$, $n_{crit} = 1.32 \times 10^6$ drops/m^3, $\ell_{crit} = 1/n^{1/3} = 0.00404$ m, $\ell_{crit}/a = 183$, $\eta = 0.968$; N = 5.5.
(d) Using iterative procedure $G_{IPC} = 0.0456$, $M/M_{ISOC} = 0.982$, $M/M_{SC} = 0.0149$, $\dot{m}_c = 1.62 \times 10^{-7}$ kg/sec, n = 725390 drops/m^3, $\ell_{IGC} = 0.0112$ m, $\ell_{IGC}/a = 224$, $r_f/R_c = 0.0149$, $\eta = 0.98$; N = 3.

Remarks

For $0.073 < G < 0.267$, the group flame is inside the cloud.

16.8 Relation between Group Combustion and Drop Array Studies

We have seen that array studies use the point source method (PSM) to determine the effect of interactions on combustion.

Section 16.7 used correction factors based on group combustion theory, which uses the continuum approach, to determine the correction factor. As drop array studies use a finite number of drops and correlate the results in terms of the interdrop spacing, it is necessary to derive the G number in terms of ℓ/a and N. Such a relation is presented by assuming that spheres of radii $\ell/2$ are densely packed. The following equation

$$G = 6F_p^{1/3}N^{2/3}/(\ell/a) \tag{16.113}$$

where F_p is the packing factor for spheres of radii $\ell/2$ and Equation 16.90 is a suitable definition for relating the results of group combustion theory to drop array studies.

Consider drops arranged in a body-centered cubic-type array (N = 9) (Figure 16.4d). Using Equation 16.113, G can be found as a function of ℓ/a. Referring to Figure 16.11 with q = 0, η can be found as a function of G and hence ℓ/a.

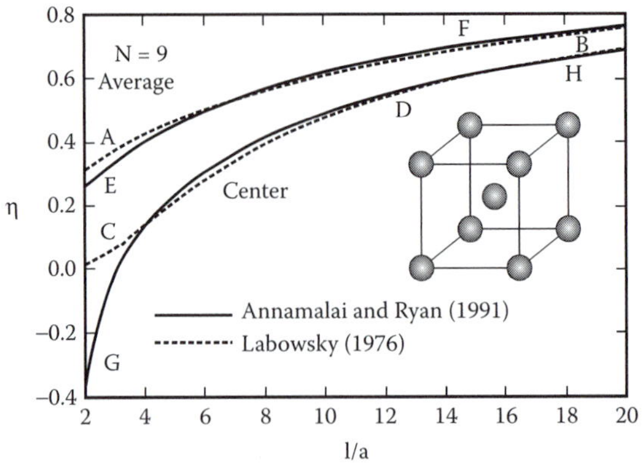

Figure 16.12 Combustion of nine-drop array: extension from group theory.

Figure 16.12 shows the results for the correction factor vs. ℓ/a for a nine-drop array evaluated from the group combustion model and the comparison of these results to the rigorous results of Labowsky (1976), who used the MOI. The curve AB represents the average correction factor for the array of nine drops whereas the solid curve CD represents the correction factor for the primary drop of the array (the center drop of the array). Curve EF is based on group combustion theory.

With the assumption of a uniform source distribution within the entire cloud, a simple empirical expression for the average correction factor is given as

$$\eta = [1 - (2a/\ell) + 1.81N^{2/3}/(\ell/a)]^{-1},\ 3 < N < 9 \qquad (16.114)$$

The agreement with exact results for average correction factor is very good.

The error for the correction factor of the primary drop (drop at center) with closer interdrop spacing increases as more drops are included. Thus, the PSM is limited in predicting the correction factor of the primary drop (the center drop), especially for arrays containing more than seven drops.

16.9 Interactive Char/Carbon Combustion

Here, we consider combustion of a char cloud. The char undergoes oxidation to CO, reduction with CO_2 to CO, and the CO produced oxidizes in the gas phase forming a gas flame. We discussed SFM and DFM in Chapter 9 for isolated carbon or char particles. Here, we discuss combustion of a group of char or carbon particles.

16.9.1 Terminology

The terminology for group combustion with the DFM is analogous to cloud combustion of fuel drops except that fuel here is CO produced by reduction reaction with CO_2 [Annamalai et. al., 1995].

16.9.1.1 SFM

This is somewhat similar to the interactive drop evaporation problem. In drops, there is interaction through decreased interstitial temperature. Here, the interaction is due to decreased Y_{O2}. For the purpose of illustration, it is assumed that each char particle oxidizes to CO and that the Boudouard reaction ($CO_2 + C \rightarrow 2CO$) is negligible. By definition, the gas phase is frozen.

16.9.1.2 ISOC

If the cloud is very dilute, then the bulk gas-phase oxygen mass fraction far from each individual char particle is almost the same as the ambient-oxygen mass fraction. Each particle burns with $Y_{O2,w} = 0$, and far from the particles, $Y_{O2} = Y_{O2,\infty}$. Burning under this condition is called isolated particle combustion (ISOC), and the burning rate of each particle is the same as that of an isolated particle.

16.9.1.3 IFC

For the SFM, this terminology is a misnomer as there is no gas flame. It should be interpreted as individual particle combustion but not as an isolated-particle combustion, i.e., interactive behavior still exists because the particle temperature and oxygen mass fraction fields are affected by the presence of the other particles. The particle number density, however, is not sufficient to completely deplete the oxygen mass fraction at the center of the cloud and $0 < Y_{O2} < Y_{O2,\infty}$.

16.9.1.4 IGC

If the number of particles per unit volume is increased, the O_2 mass fraction at the cloud center decreases until a condition is reached at which $Y_{O2}\{r = 0\} = 0$. This is called incipient group combustion.

16.9.1.5 PGC/CGC/GC

It is difficult to define group combustion for the SFM as there are no gas-phase reactions. Thus, PGC, CGC, and GC are lumped into the same group. Increasing the particle number density further decreases the O_2 concentration within the cloud, and the location at which $Y_{O2} = 0$ (implied to be approximately zero) moves outward from the cloud center. In addition, the oxygen mass fraction at the cloud periphery decreases as the particle number density is increased.

16.9.1.6 SC

For very large particle number densities, the oxygen mass fraction at the cloud surface approaches zero. Under this condition, the reaction rates of the particles are independent of the temperature of the cloud or the diffusion rates of the individual particles, but depend on the diffusion rate into the cloud as a single entity. This type of combustion is called SC. Once again, for SC the entire cloud can be treated as a single large particle having the same density as the cloud mass density, in which porosity is due to the number density of particles within the cloud.

16.9.2 Model

Combustion with a frozen gas phase has a strong analogy to porous char combustion. For the frozen gas phase, cloud combustion in ISOC mode is equivalent to Zone-I combustion of porous char because the reaction occurs with the oxygen concentration being the same as the ambience throughout the cloud, PGC is equivalent to Zone II because $Y_{O2} > 0$ for only a portion of the cloud, and SC of the cloud is equivalent to Zone-III reactions of porous char because the oxygen concentration is nearly zero throughout the cloud. In order to differentiate the zones of reactions in the cloud from zones of reactions for porous char, a suffix "c" is added to the zones of clouds (Zone I-c, Zone II-c, and Zone III-c). To gain a basic understanding of group behavior, Annamalai and Ramalingam (1987) formulated a quasi-steady GC theory for a spherical cloud of char or carbon particles burning in a quiescent environment. For cylindrical stream, see Annamalai et. al. (1988). A spherical cloud was selected because the behavior of clouds under dilute and dense limits can be determined using the isolated-particle model. The results are exactly the same as before for drops except for the change in B number definition, and β definition.

16.9.3 Results

The potential for the cloud is given as

$$\phi = \frac{\beta'}{\beta'_\infty} = \frac{\beta - F}{\beta_\infty - F} = \frac{\{(1+B(\xi)\}}{1+B}, \quad B(\xi) = \frac{Y_{CO2}(\xi)}{\nu_{CO2,III}} + \frac{Y_{O2}(\xi)}{\nu_{O2,I}}, \quad \xi \geq 1$$

$$B = \frac{Y_{CO2,\infty}}{\nu_{CO2,III}} + \frac{Y_{O2,\infty}}{\nu_{O2,I}} \tag{16.115}$$

where F = −1, and Y_{CO2} and Y_{O2} are functions of ξ.

Let M_{SC} be the nondimensional sheath combustion rate given by

$$M_{SC} = \dot{m}_{c,SC}/4\pi\,\rho\,D\,R_c = \ln(1+B). \tag{16.116}$$

Also, the expression for ϕ is the same as Equation 16.94, and correction factor expressions are the same as for drops. Further, the flame radius to cloud radius ratio (for DFM) is still given by Equation 16.109.

16.9.4 Analogy between Porous Char Particle Combustion and Cloud Combustion of Char Particles

Recall the source terms for cloud combustion:

$$\dot{m}''' = n\,\dot{m}_s = n\,\dot{m}''_s 4\pi\,a^2$$

for monosized particles, $S_v\,\dot{m}''_s$, where S_v is the surface area of particles per unit volume.

$$G_1 = G = 4\pi\,na\,Rc^2 = S_v\,Rc^2/a$$

In the porous char combustion literature, the Thiele modulus is used to describe combustion of porous char instead of the G number.

$$\psi_T^2 = G = S_v\, a_{porous}^2/a_{ch},$$

S_v, internal surface area of porous char per unit volume of the particle,
a_{ch}, characteristic radius of porous char particle = $D_{eff}/k_{O2}(T_p)$
a_{porous}, radius of porous char particle,

For kinetic-controlled combustion within porous char (Zone-I reaction, $\psi_T^2 < 2$, $G < 2^{1/2}$; see Chapter 9), the k_{O2} specific reaction rate constant for O_2 consumption = $A_{O2}\exp(-E/RT_p)$, m/sec and D_{eff} is the effective diffusion coefficient of O_2 within porous char.

16.10 Multicomponent Array Evaporation

In Chapter 10, we presented isolated multicomponent drop evaporation and combustion models. Here, we consider evaporation and combustion of multicomponent drop arrays. Labowsky (1976) formulated a MOI model and obtained results for a three-drop array consisting of volatile and nonvolatile components; Annamalai et. al., presented the PSM results are for (1) correction factor for each component in the drop of interest, (2) the net correction factor for the drop, and (3) average correction factor for the array. For an array containing J number of multicomponent drops, the correction factor of the i-th component in the j-th drop is given as

$$\eta_{ij} = (\dot{m}_{s,i})_j/(\dot{m}_{iso,i})_j$$

where the isolated drop has the same composition as the j-th drop in the array. Under pure diffusive transport, the single-component array analysis always yields $\eta < 1$, whereas the interactive effects in a multicomponent array can occasionally enhance the mass-loss rate ($\eta > 1$). For simplicity, the procedure is illustrated only for two components, A and B, although the method could be extended to a large number of components.

16.10.1 Array of Arbitrary Composition

Derivations are very similar to those for the single-component drop array. The species i from drop j with a surface fuel mass fraction of $(Y_{i,w})_j$ evaporates at the rate of $\dot{m}_{i,j}$. The strength of the source $\dot{m}_{i,j}$ and $(Y_{i,w})_j$ are unknown, which are to be solved from the Laplace equation for each species and the boundary conditions [Annamalai et. al., 1993]. For NSF conditions and for a binary-component droplet,

$$\nabla^2 Y_i = 0, \quad i = A, B \tag{16.117}$$

where we had selected two components A and B as an illustration. The solution procedure is similar to that for an array of drops, which was discussed in earlier sections. The solution is obtained subject to the following conditions:

$$\text{At r = r, } \dot{m}_{F,i} = -4\pi r^2 \rho D \frac{dY_{F,i}}{dr} \tag{16.118}$$

$$\text{and as } r \to \infty, Y_i = Y_{i\infty} \tag{16.119}$$

and at the drop surface

$$Y_i = \{Y_{i,w}\}_k, \; k = 1....K \text{ at } \vec{r}_{kw} = \vec{r}_k + \vec{a}_k. \tag{16.120}$$

Using the superposition technique, the mass fraction of component i at P (Figure 16.2) because of all drops j = 1 … (J 1) is given as

$$Y_i = \sum_{j=1}^{N} \frac{\dot{m}_{i,j}}{4\pi\rho D|\vec{r} - \vec{r}_j|}. \tag{16.121}$$

(If $Y_{i,\infty} \neq 0$, Equation 16.118 can still be used with Y_i replaced by $Y_i - Y_{i,\infty}$.) At the wall (or surface) of the k-th droplet,

$$(Y_{i,w})_k = [\dot{m}_{k,i}/4\pi\rho D a_k] + \sum_{\substack{j=1 \\ j\neq k}}^{J} \dot{m}_{i,j} \Big/ \left(4\pi\rho D \left| \underset{\rightarrow}{r_{k,w}} - \underset{\rightarrow}{r_j} \right|\right) \tag{16.122}$$

With $\vec{r}_{kw} = \vec{r}_k + \vec{a}_k$ and assuming $\vec{r}_k - \vec{r}_j >>>> \vec{a}_k$,

$$(Y_{iw})_k = \frac{\dot{m}_{k,i}}{4\pi\rho D a_k} + \sum_{\substack{j=1 \\ j\neq k}}^{N} \frac{\dot{m}_{i,j}}{4\pi\rho D a_k} \quad \text{where} \quad i = A, B \tag{16.123}$$

It can be shown that the drop temperature and mass fractions of species i at the surface of the k-th drop remain exactly the same as those obtained for an isolated drop of similar composition [Labowsky, 1980; Chapter 10]. From Chapter 10, for an isolated drop

$$(\dot{m}_{i,iso})_k = 4\pi\rho D a_k (Y_{i,w})_k \; i = A, B \tag{16.124}$$

Using the definition of correction factor, the expression for the correction factor of the i-th component in the k-th drop is given as

$$\eta_{ik} = 1 - \sum_{\substack{j=1 \\ j\neq k}}^{N} \frac{\eta_{ij} a_j (Y_{iw})_j}{|\ell_{kj}|(Y_{iw})_k}, \quad i = A, B \tag{16.125}$$

where ℓ_{kj} is the interdrop spacing $=|\vec{r}_k - \vec{r}_j|$. Multiplying Equation 16.124 by ε_i, the flux ratio summing over all i's and using the identity (see Chapter 10)

$$\frac{Y_{A,w}}{\varepsilon_A} = \frac{Y_{B,w}}{\varepsilon_B} = C \tag{16.126}$$

where

$$Y_{i,w} = \frac{b_{t,\infty}\varepsilon_i}{\sum Li\varepsilon_i + b_{t,\infty}} \tag{16.127}$$

the following result is obtained with $Y_{I,\infty} = 1.0$ (I denotes inerts)

$$\eta_k = 1 - \sum_{\substack{j=1 \\ j \neq k}}^{N} \frac{\eta_j a_j (1 - Y_{Iw})_j}{|\ell_{kj}|(1 - Y_{Iw})_k} \tag{16.128}$$

where η_k is overall correction factor for kth in drop, including all components.

Equation 16.128 provides the net correction factor for the evaporation rate of drop k. For arrays of multicomponent droplets of similar composition, the inclusion of SF will not alter the result for the correction factor as long as SF is included for both the array and isolated-drop analyses [Annamalai and Ryan, 1992]; however, if the compositions are dissimilar, SF will affect the correction factor.

16.10.2 Array of Drops of Volatile (A) and Nonvolatile (B) Components

Consider a three-drop array of radii a_1, a_2, and a_3 mounted at the apexes of a scalene triangle (sides ℓ_{12}, ℓ_{13}, and ℓ_{23} with $|\vec{r}_2 - \vec{r}_1| = \ell_{12}$ and $|\vec{r}_3 - \vec{r}_1| = \ell_{13}$. Using Equation 16.125 for component A

$$\eta_{A1} = 1 - \frac{a_2 (Y_{Aw})_2 \eta_{A2}}{\ell_{12} (Y_{Aw})_1} - \frac{a_3 (Y_{Aw})_3 \eta_{A3}}{\ell_{13} (Y_{Aw})_1} \tag{16.129}$$

$$\eta_{A2} = 1 - \frac{a_1 (Y_{Aw})_1 \eta_{A1}}{\ell_{12} (Y_{Aw})_2} - \frac{a_3 (Y_{Aw})_3 \eta_{A3}}{\ell_{23} (Y_{Aw})_2} \tag{16.130}$$

$$\eta_{A3} = 1 - \frac{a_1 (Y_{Aw})_1 \eta_{A1}}{\ell_{13} (Y_{Aw})_3} - \frac{a_2 (Y_{Aw})_2 \eta_{A2}}{\ell_{23} (Y_{Aw})_3}. \tag{16.131}$$

From isolated-drop results, $(Y_{A,w})_1$, $(Y_{A,w})_2$, and $(Y_{A,w})_3$ can be determined (see Chapter 10) and the three unknowns η_{A1}, η_{A2}, and η_{A3} are solved from the three linear equations (Equation 16.129 to Equation 16.131).

Table 16.3 Fuel Properties

Property	Fuel		
	Heptane (C_7H_{16})	*Octane (C_8H_{18})*	*Hexadecane ($C_{16}H_{34}$)*
Density (kg/m^3)	684	703	773
Boiling point (K)	372	399	560
Heat of vaporization (kJ/kg)	316	300	225
E	15.89	15.93	24.66
F	2,911.32	3,120.0	10,660.2
G	56.4	63.5	+54.1

Note: $\ell_{12}/a1 = 5.0$, $\ell_{13}/a1 = 3.6$, = 33.7E, $T_\infty = 300$ K.
Source: Adapted from Annamalai et al.

The average for the array is given as

$$\eta_{avg,A} = \frac{\sum_{k}^{K} \eta_{A,k}\, \dot{m}_{iso,A,k}}{\dot{m}_{iso,A}} \tag{16.132}$$

where

$$\dot{m}_{iso,A} = \sum_{k} \dot{m}_{iso,A,k}. \tag{16.133}$$

Example 4

Consider a linear array of three drops of diameter 100 (= 2 a_1 with 20% octane), 150 (= 2 a_2) with 45% octane, and 50 (= 2 a_3) μm of 5% octane. The other component is nonvolatile (B).

See Table 16.3 for properties of fuels.

$$\ln p^{sat} = E + \frac{F}{T+G},\ p^{sat},\ \text{mm of Hg; T, K}$$

Solution

As in Chapter 10 for an isolated multicomponent drop, $T_\infty = 300$ K, $T_{w,1} = 296.1$ K, $(Y_{A,w})_1 = 0.01279$, $T_{w,2} = 292.8$ K, $(Y_{A,w})_2 = 0.0235$, $T_{w,3} = 298.9$ K, and $(Y_{A,w})_3 = 0.003767$. Equation 16.129 to Equation 16.131 are then used to solve for η_{A1}, η_{A2}, and η_{A3}. Table 16.4 shows the comparison of the results between PSM and the more accurate MOI (does not assume point mass). The maximum error is 4%.

Remarks

(i) The negative value for drop 3 implies condensation of vapors. The temperature and mass fraction of vapor at the location r_3 would have been

Table 16.4 Comparison of Correction Factor Obtained by MOI and PSM

Drop No.	*Octane Mole Fraction (Liq)*	*Drop Size* a_i/a_1	*Correction Factor* (η) *MOI*	*PSM*
1	0.20	1.0	0.54	0.56
2	0.45	1.5	1.02	1.02
3	0.05	0.5	2.93	2.90

298 K and 0.006 if only drop 2 was present as an isolated drop. Thus, the presence of drop 3 increased the local temperature to 298.9 K and reduced $Y_{A,w}$ to 0.004, which implies the removal of vapor from location r_3. Drop 3 acts as a mass sink and a heat source owing to higher drop temperature. Drop 3 had the lowest octane content and the lowest equilibrium vapor mass fraction. Hence, the vapors originating from drop 1 and drop 2 diffuse outward and condense on drop 3.

(ii) Further, drop 2 has a correction factor that is greater than unity because it has a high octane content and is in close proximity to a mass sink and heat source (i.e., drop 3), resulting in an evaporation rate greater than the evaporation rate of a similar isolated drop.

16.10.3 Binary Array of Drops of Volatile Components

16.10.3.1 Binary Array of Drops of Volatile and Nonvolatile Components

Two drops of the same size but differing composition with octane as the volatile component (drop 1, octane: 80%; drop 2, octane: 20%) were considered. Using Equation 16.125 and solving

$$\eta_{A1} = \left[I - \frac{(Y_{A,w})_2}{(Y_{A,w})_1} \times \frac{a}{\ell} \right] \Big/ \left(\left[I - \frac{a^2}{\ell^2} \right] \right) \tag{16.134}$$

$$\eta_{A2} = \left[I - \frac{(Y_{A,w})_1}{(Y_{A,w})_2} \times \frac{a}{\ell} \right] \Big/ \left(\left[I - \frac{a^2}{\ell^2} \right] \right). \tag{16.135}$$

16.10.3.2 Experimental Data Binary Array

Xiong et al. (1984) conducted evaporation experiments on a two-drop array of (1) high-volatile heptane drop (drop 1) and (2) low-volatile hexadecane drop (drop 2) at an ambient temperature of 24°C. The results show that the hexadecane drop had a negative vaporization constant in that the size of hexadecane drop increases as a result of the condensation of vapors from the heptane drop

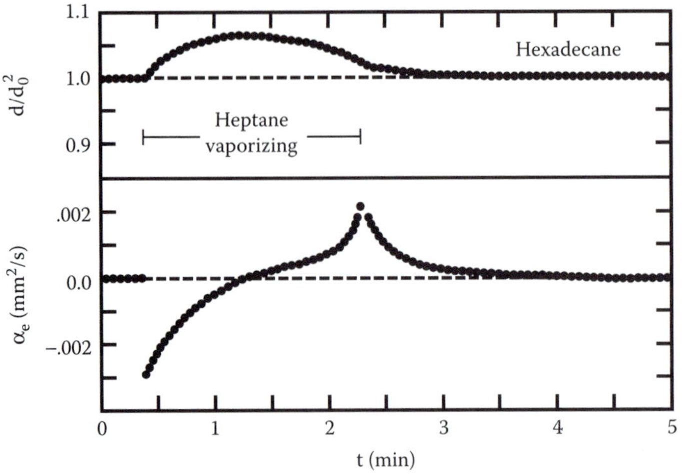

Figure 16.13 Evaporation of binary array of multicomponent drops — transient experimental data. (Adapted from Xiong, T.Y., Law, C.K., and Miyasaka, K., Interactive Vaporization and Combustion of Binary Droplet Systems, 20th Symposium (International) on Combustion, 1781–1787, 1984.)

(Figure 16.13). Hence, drop 2 acts as the mass sink and heat source for drop 1, which is at a lower temperature. It is not possible to emulate Xiong's experimental conditions in the current analysis because one cannot achieve the phase equilibrium in the array under steady state unless each drop in the array consists of both the components, heptane (A) and hexadecane (B). In order to simulate Xiong's data, a binary array of drops consisting of 95% heptane and 5% hexadecane (drop 1) and 5% heptane and 95% hexadecane (drop 2) is selected. Using the fuel properties listed in Table A.2A of Appendix A, the isolated-drop results for T_w and Y_w for the binary-component system are then obtained. The drop temperatures and surface mass fractions are as follows: $T_{w,1} = 278$, $(Y_{A,w})_1 = 0.065$, $T_{w,2} = 297$ K, and $(Y_{A,w})_2 = 0.010$ at $T_\infty = 300$ K. Drop 2, being hotter, supplies heat to drop 1 and, at the same time, the requirement of phase equilibrium at the surface of drop 2 results in condensation of heptane vapors on drop 2. Thus, at closer interdrop spacing, the correction factor for heptane at drop 1 is higher than unity, whereas for drop 2 it is negative, indicating condensation accompanied by an increase in diameter. These results qualitatively confirm the results shown in Figure 16.14. The present analysis yields steady-state results, whereas the experimental data are for the transient case in which the composition and size change with time.

It is noted that the nonisothermal conditions selected for the present problem involve weak temperature differences. However, most practical applications involve higher temperature differences during the evaporation. Thus, SF must be included. Annamalai et al. [1993] present the transformation for the evaporation of a multicomponent array of drops, which could be used to convert the SF problem into an NSF problem.

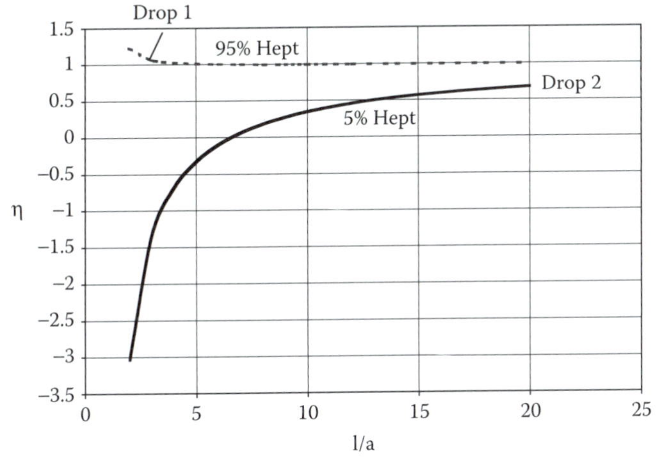

Figure 16.14 Evaporation of binary array of multicomponent drops — steady-state theoretical results.

16.11 Summary

Interactive processes occur for dense sprays and suspensions. The correction factor is defined as a ratio of burn rate of drop under interaction to the burn rate of the same drop under isolated conditions. PSM is presented for estimating the correction factor for arrays. The theory is then extended using the continuum method in order to obtain correction factors for a spherical cloud of drops. Burn rate relations are also presented for very dense clouds in which the whole cloud remains at about saturation temperature, with the flame standing away from the cloud. The regimes of interactive combustion are classified as follows: isolated, incipient group combustion, partial group combustion, group combustion, and sheath combustion. QS evaporation of a multicomponent drop array has also been studied using PSM.

Chapter 17

Pollutants Formation and Destruction

17.1 Introduction

A pollutant is a substance that affects the health of a biological system directly or indirectly through the environment. With billions of automobiles on the roads, and numerous power plants producing electricity for the energy-starved and increasing population of the world, the levels of combustion pollutants emitted are at ever increasing levels. Combustion from fuels has numerous by-products, many of which are pollutants (e.g., from car exhaust and power and chemical plants). In small quantities, these pollutants are harmless and pose no threat to the environment. The leading pollutants are carbon oxides (CO_x: CO, CO_2, etc., called greenhouse gases), nitrogen oxides (NO_x: NO, NO_2 [nitrogen dioxide], etc.), sulfur oxides (SO_x: SO_2, SO_3, etc.), particulate matter having size less than 10 μm (called PM 10) released from coal-fired plants and soot from engines (e.g., diesel), formaldehyde and acetaldehyde, polycyclic aromatic hydrocarbons (PAHs such as primarily naphthalene that are carcinogenic), Hg and unburned hydrocarbons (UHCs; not completely and partially burned/oxidized hydrocarbons [PHCs such as partially combusted $C_xH_y \cdot$ COOH carboxylic acids, ketones $C_xH_y \cdot$ CO, aldehydes, and $C_xH_y \cdot$ CHO] due to low temperature reactions as in lean eddies) mostly due to limited residence time and quenching by walls as in automobile engines. The high NO_x and SO_x emissions from fossil fuels are caused by the high fuel-bound N and S content in coal. Sources of SO_2 are fossil fuel power plants (66%), industrial processes (16%), nonutility stationary fuel combustion (14%), and transportation services (4%). Vehicles release large amounts of thermal NO_x (because of high temperature) and volatile organic compounds (VOCs; typically, HCs composed of carbon compounds with H/C 2:1 to 0:1). The organic

Table 17.1 Pollutants Emitted from a 1000 MW Power Plant

Particulates	4,490
Sulfur oxides	139,000
Nitrogen oxides	20,880
Carbon monoxide	210
Hydrocarbons	520

Note: Tons (SI) per year, or tpy.

compounds originate primarily from unburned fuel and lubricating oil. The organic species include sulfur, oxygen, carbon, and nitrogen-containing compounds. There are numerous types of pollutants found in combustion products. Table 17.1 shows typical emission of various pollutants from a 1000 MW plant, whereas the percentages of NO_x from various sources are presented in Table 17.2. Total emissions from U.S. power plants alone are 10.6 Mt of SO_2, 4 Mt of NO_x, 2.5 Gt of CO_2, and 45.6 t of mercury [http://www.0101010.org/2004/05/what-does-2-billion-tons-of-carbon.html]. The SO_2 and NO_x from other sources are 5.1 and 16.4 Mt respectively.

17.2 Emission-Level Expressions and Reporting

The emission levels are reported in various forms depending upon combustion systems.

17.2.1 Reporting as ppm

Many analyzers yield gas composition in mole percentage (or volume percentage) on dry basis. Because pollutants are in trace amounts, they are also reported in parts per million (ppm). For the pollutant species k,

$$\text{Species k in ppm} = X_k \times 10^6,$$
$$\text{(interpreted as molecules per million dry molecules).} \tag{17.1}$$

where X_k is in dry mole fraction. For Hg content in solid fuel such as coal, the ppm indicates the mass of species in g per million g of solid fuel.

Table 17.2 Sources of NO

Sector	*Uses*	*Fraction*
Electricity	Utility plants; oil/coal	40%
Transportation	Automobiles, ships, aircraft, oil production	25%
Commercial	Gas, oil production	15%
Residential		20%

17.2.2 O_2 Normalization or Corrected ppm Concentrations

Oxygen is measured at the same point as NO_x (= NO + NO_2) and quoted along with the NO_x percentage. If the mass of the emitted NO_x during a combustion process is held constant, then by simply increasing the amount of air supplied one can artificially reduce the NO mole fraction or ppm. Therefore, often the amount of NO emitted is normalized by the amount of O_2 provided to a combustor. For a generic fuel participating in the chemical reaction,

$$C_cH_hO_oN_n\ S_s + aO_2 + 3.76a\ N_2 + b\ H_2O(g) \rightarrow c\ CO_2 + (h/2 + b)H_2O + (a - a_{st})\ O_2 + 3.76a\ N_2 + y\ NO + s\ SO_2.$$

The N in NO can originate either from fuel or atmospheric nitrogen and

$$a_{st} = (c + s + h/4 - o/2). \tag{17.2}$$

Because NO_x and SO_x occur in trace amounts, the O_2 requirements for N to NO_x and S to SO_x are excluded in product moles.

$$\text{Total dry moles, } N_{P,\ dry} = (a - a_{st}) + c + 3.76a = 4.76a + c - a_{st} \tag{17.3}$$

$$X_{NO} = y/N_{P,\ dry} \tag{17.4}$$

where $N_{p,dry}$ is the number of product dry moles per kmole of fuel.

$$X_{NO,std} = y/N_{P,dry,std} \tag{17.5}$$

The O_2 mole fraction is given as

$$X_{O2} = (a - a_{st})/N_{p,dry} \tag{17.6}$$

$$X_{O2,std} = (a_{std} - a_{st})/N_{p,dry,std} \tag{17.7}$$

where

$$N_{P,dry,std} = 4.76\ a_{std} + c - a_{st}. \tag{17.8}$$

We have seven unknowns: a_{st}, a, a_{std}, $N_{p,dry}$, $N_{p,dry,\ std}$ $X_{NO,std}$, y and the seven equations are Equation 17.2 to Equation 17.8. Using Equation 17.3 in Equation 17.6 to eliminate "a,"

$$X_{O2} = (N_{p,dry} - c - 3.76a_{st})/\{4.76 \times N_{p,dry}\}. \tag{17.9}$$

Solving for $N_{p,dry}$,

$$N_{p,dry} = (c + 3.76\ a_{st})/(1 - X_{O2} \times 4.76). \tag{17.10}$$

Similarly,

$$N_{p,dry,std} = (c + 3.76a_{st})/(1 - X_{O2,std} \times 4.76). \tag{17.11}$$

Taking ratios of Equation 17.10 and Equation 17.11,

$$N_{p,dry}/N_{p,dry,std} = (1 - X_{O2,std} \times 4.76)/(1 - X_{O2} \times 4.76). \quad (17.12)$$

Taking ratios of Equation 17.5 and Equation 17.4 and using Equation 17.12,

$$\begin{aligned} X_{NO,std}/X_{NO} &= N_{P'dry}/N_{P'dry,std} = (1-4.76\,X_{O2,std})/(1-4.76\,X_{O2}) \\ &= (0.21 - X_{O2,\,std})/(0.21 - X_{O2}). \end{aligned} \quad (17.13)$$

Generalizing Equation 17.13 for an ambient oxidizer concentration of $X_{O2,a}$

$$X_{NO,std}/X_{NO} = (X_{O2,\,a} - X_{O2,std})/(X_{O2,a} - X_{O2}) \quad (17.14)$$

where X_{NO} is the uncorrected mole fraction at oxygen mole fraction X_{O2} and $X_{NO,std}$ is the corrected mole fraction at standard oxygen mole fraction $X_{O2,std}$. The standard O_2 percentage is 3% for utilities (low excess air) and 6% for gas turbines (high excess air).

17.2.3 Emission Index (g/kg of Fuel)

Alternately, the emission index EINO expressed as g/kg of fuel can be used in place of O_2 normalization:

$$\begin{aligned} \text{EI of species k} &= \text{mass of pollutant species k/mass of fuel} \\ &= \{m_k/m_F\} = X_k N_{P,dry} \times M_k/M_F. \end{aligned} \quad (17.15)$$

Consider

$$\begin{aligned} &C_cH_hO_oN_n\,S_s + aO_2 + 3.76a\ N_2 + b\ H_2O(g) \rightarrow x\ CO_2 + (c - x)CO \\ &\quad + (h/2 + b)H_2O + (a - a_{st})O_2 + 3.76a\ N_2 + c\ NO + s\ SO_2. \end{aligned}$$

If X_{CO} and X_{CO2} are measured, then

$$X_{CO2} = x/N_{P,dry},\ X_{CO} = (c - x)/N_{P,dry} \quad (17.16a,b)$$

Adding Equation 17.16a and Equation 17.16b, product dry moles is solved as

$$N_{P,dry} = c/(X_{CO2} + X_{CO2}) \quad (17.17)$$

and using the result in Equation 17.15 with k = NO,

$$\begin{aligned} \text{EINO (g/kg of fuel)} &= \{c \times X_{NO} \times M_{NO_2}/(X_{CO2} + X_{CO})\} \\ &\quad (\text{kg/kmol})\ 1000/M_F\ (\text{kg/kmol}) \end{aligned} \quad (17.18)$$

where

$$\begin{aligned} M_F &= c \times 12.01 + h \times 1.01 + n \times 14.01 \\ &\quad + 0 \times 16 + s \times 32 \text{ kg/kmol of fuel} \end{aligned} \quad (17.19)$$

For reporting NO emission, the U.S. Environmental Protection Agency (EPA) stipulates that M_k for NO should be that of NO_2 (M_{NO2} = 46.01) instead of 30 because NO is eventually converted into NO_2 in the atmosphere, which plays a major role in the destruction of O_3. For any other species k,

$$\text{EI of any pollutant species k in g/kg of fuel} \\ = \{c \times M_K X_k\} \times (1000 \text{ g/kg})/[\{X_{CO} + X_{CO2}\}M_F]. \quad (17.20)$$

If empirical or chemical formulae are not available (e.g., coal, biomass), and if the mass fraction of C element for fuel is available, then multiply the numerator and denominator of Equation 17.20 by M_C (=12.01)

$$\text{EI of species k in g/kg of DAF fuel} \\ = \{Y_{C,DAF} \times M_k X_k\} \times (1000 \text{ g/kg})/[12.01 \{X_{CO} + X_{CO2}\}] \quad (17.21)$$

where $Y_{C,DAF} = c \times M_C/M_F$, mass fraction of carbon k on dry-ash-free (DAF) basis.

Coal contains ash (A) and water or moisture (M). Thus, the dry-ash-free combustible fraction is $Y_{DAF} = 1 - Y_A - Y_M$, kg of DAF combustibles per kg of fuel as received. Multiplying both sides by $(1 - Y_A - Y_M)$, Equation 17.21 can be written as

$$\text{EI of species k in g/kg of fuel as received} \\ = \{Y_C \times M_k X_k\} \times (1000 \text{ g/kg})/[12.01 \{X_{CO} + X_{CO2}\}] \quad (17.22)$$

$$\text{"k" in kmole per kmole of fuel} = \frac{c\, X_k}{(X_{CO2} + X_{CO})}. \quad (17.23)$$

17.2.4 Emissions in Mass Units per Unit Heat Value (g/GJ)

Pollutant species k can be reported as,

$$\text{"}k\text{" in } \frac{g}{GJ} = \frac{c\, X_k}{(X_{CO2} + X_{CO})} \frac{M_k \times 1000}{M_F HHV_F \text{ (GJ/kg)}} = \text{fuel on atom basis} \quad (17.24)$$

$$\text{"}k\text{" in } \frac{g}{GJ} = \frac{Y_C\, X_k}{(X_{CO2} + X_{CO})} \frac{M_k \times 1000}{12.01 \times HHV_F \text{ (GJ/kg)}}. \quad (17.25)$$

For gaseous fuels, heating values are available in GJ/m^3. Thus, standard temperature and pressure are used to calculate the volume in m^3; 1 kmol occupies $\bar{v}_{STP}$ = 22.4 m^3 at STP of 0°C, 101 kPa and 24.5 m^3/kmol at STP of 25°C and 101 kPa. As HHV (GJ/kg) × M_F = HHV (GJ/m^3) $\bar{v}_{STP}$, then for gaseous fuels Equation 17.24 is written as

$$\text{"}k\text{" in } \frac{g}{GJ} = \frac{c\, X_k}{(X_{CO2} + X_{CO})} \frac{M_k * 1000}{HHV'_F \;(GJ/m^3) * \bar{v}_{STP}} \quad (17.26)$$

For k = NO, we use M_{NO_2} for expressing NO on mass basis.

$$NO \text{ in } \frac{g}{GJ} = \frac{c\, X_{NO}}{(X_{CO2} + X_{CO})} \frac{M_{NO2} \times 1000}{M_F HHV_F \text{ (GJ/kg)}} \quad (17.27)$$

$$NO \text{ in } \frac{g}{GJ} = \frac{Y_C\, X_{NO}}{(X_{CO2} + X_{CO})} \frac{M_{NO2} \times 1000}{12.01^* \, HHV_F \text{ (GJ/kg)}} \quad (17.28)$$

For automobiles, instead of GJ, kW h or HP units are used. Thus,

NO in g per kWh = NO in g per GJ × 0.0036 (17.29)

NO in g per HP = NO in g per GJ × 0.00268. (17.30)

17.2.5 Reporting as kg per Million m³ of Gas

Using Equation 17.23,

EI of species k in g/m³ of gaseous fuel

$= \{Y_{C,DAF} * M_k\ X_k\} * (1000 \text{ g/kg}) * M_F/[12.01 * \bar{v}_{STP}\ \{X_{CO} + X_{CO2}\}].$ (17.31)

17.2.6 Conversion of NO to mg of NO₂/m³

If NO_x in ppm is known, then

mg of NO_2/m^3 = [NO_x in ppm] × 10^{-6} × N_{prod} × 46.0kg/kmol × (10^6 mg/kg)/ $\{N_{p,dry} \times \bar{v}_{STP}\}$ = [NO_x in ppm] 46.01kg/kmol/{24.5} = 1.88 * [NO_x in ppm]

where STP is defined at 25°C, 1 bar called SATP, Table A.1). If STP at 0°C, 1 bar is used, mg of NO_2/m^3 = 1.88 × {NO_x in ppm}; if CST is adopted, mg of NO_2/m^3 = 2.03{NO_x in ppm}. For flue gas volume, see Equation 4.16b.

Example 1

The NO emission from a coal-fired utility boiler is reported as 400 ppm (dry) at 4% O_2, 0.5% CO, and 8% CO_2. The dry-ash-free empirical formula is given as $CH_{0.8}O_{0.5}$. The ash and moisture percentages are known to be 5 and 25. HHV is 20,000 kJ/kg.

(a) Determine NO at 3% O_2.
(b) Calculate the emission index (g/kg of fuel).
(c) Determine NO in g per GJ and in g per kW h.

Solution

(a) $X_{NO,std}/X_{NO} = NO_{ppm,\,std}/NO$ in ppm $= (X_{O_2,a} - X_{O_2,std})/(X_{O_2,a} - X_{O_2}) = (0.21 - 0.03)/(0.21 - 0.04) = 1.059$

$$X_{NO,std} = X_{NO} \times 1.059 = 400 \text{ ppm} * 1.059 = 424 \text{ ppm}$$

(b) NO in g/kg = $X_K \times (1000\ g/kg) \times c \times M_K/\{M_F \times (X_{CO} + X_{CO2})\}$ = 400 × 10^{-6} × 1000 × 1 × 46.01/{15.2 × (0.08 + 0.005)} = 14.27 g/kg DAF
1 kg of as-received fuel contains 1 − 0.05 − 0.25 = 0.7 kg of DAF fuel
Thus, NO in g/kg as received = 424 g/kg DAF × 0.70 kg DAF/kg as received = 295 g/kg as received

(c) NO in g/GJ = (295 g/kg as received) × kg (20 GJ) = 14.75 g/GJ
NO in g/kW h = 475.6 × 0.0036 GJ/{kW h × GJ} = 1.71 g/kW h

17.2.7 Fuel N Conversion Efficiency

For a fuel $C_cH_hN_nO_oS_s$, the fuel N conversion efficiency is the fraction of fuel nitrogen that is converted to NO_x.

$$N_{conv} = \frac{(c/n)x_{NO}}{(x_{CO2} + x_{CO})} = \frac{Y_c * 14 * x_{NO}}{Y_n * 12 * (x_{CO2} + X_{CO})} \tag{17.32}$$

where:

N_{conv} = overall equivalent fuel N conversion efficiency
x_{NO} = mole fraction of NO

The majority of NO in coal combustion originates from fuel N, whereas the remainder comes from N in air. The N conversion efficiency is defined as the fraction of fuel nitrogen that produces total NO_x.

Another relation for emission of NO in terms of mass per unit heat value is presented in Chapter 2 (when CO is absent). If NO and CO_2 are known in mole fractions, then for a fuel $C_cH_hN_nO_oS_s$

$$\textit{Equivalent fuel N converted NO} = \frac{c\ X_{NO}}{n\ X_{CO2}} = \frac{\%\ \text{carbon}\ X_{NO}}{\%\ \text{nitrogen}\ X_{CO2}} \tag{17.33}$$

In order to make use of Equation 17.33, x_{CO} in Equation 17.32 must be set to zero. For sulfur conversion replace x_{NO} with x_{SO_2}, and c/n and Y_c/Y_n with c/s and Y_c/Y_S ratios.

17.3 Effects of Pollutants on Environment and Biological Systems

Because NO_x and SO_x are the most dominant components of emission of pollutants, their effects on health and environment will be briefly discussed.

17.3.1 Health Effects

Humans first recognize NO_2 by odor when it is present at approximately 0.4 ppm. Its worst effect is to promote the deterioration of the human respiratory system in the presence of NO_x. If NO > 3 ppm, it can measurably decrease lung function. NO at 0.1 ppm causes lung irritation and measurable decreases in lung function in asthmatics. NO_2 has a more serious effect on the respiratory system than NO. NO_2 affects the blood hemoglobin, thus depriving body tissue of oxygen.

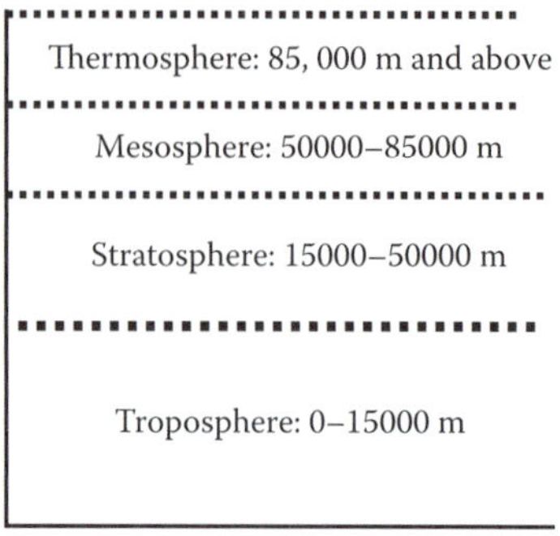

Figure 17.1 Earth's atmosphere.

17.3.2 NO and Ozone Destruction

Ozone (O_3; a Greek word meaning "to smell") is regulated by ozone precursors such as NO_x and VOCs. Environmental damage from NO does not appear to occur directly, but rather through the subsequent conversion to NO_2. Ozone exists at 15,000 to 30,000 m above the Earth (Figure 17.1), acting as an ultraviolet filter. It is typically formed around the equator and transported to polar regions by wind. Ozone could also be formed at lower altitudes owing to reactions between NO and VOC. Figure 17.2 shows the daytime formation of O_3, NO, and NO_2 vs. time. It is seen that the O_3 level is highest around noon because of energy input through photons from the sun. The ultraviolet (0.01 to 0.40 μm) radiation from the sun is not absorbed by atmospheric O_2 and is damaging to organisms. However, O_3 can absorb such radiation as illustrated in

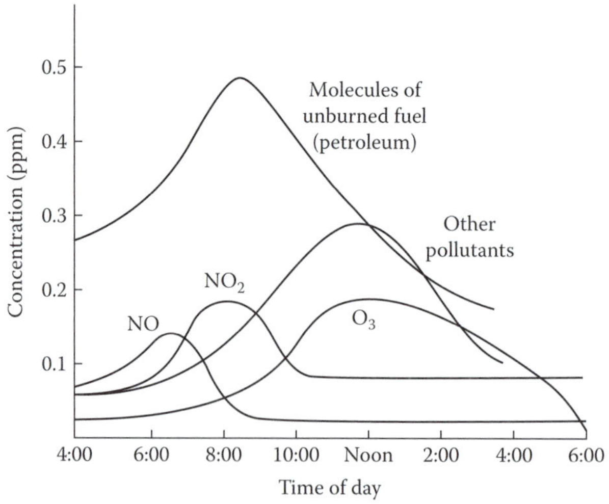

Figure 17.2 Daytime concentrations of pollutants. (Adapted from Zumdahl, C.S., *Chemistry*, DC Heath and Co., Lexington, MA, 1986.)

the following text. The thermosphere (85,000 m and above, Figure 17.1) absorbs the ultraviolet radiation and produces O atoms, or

$$O_2 \text{ (bond energy 496 MJ/kmol)} + h\nu \text{ from the sun} \rightarrow 2O \quad (17.34)$$

$$O + O_2 \Leftrightarrow O_3 \quad (17.35)$$

Thus, O_3 formation reduces the amount of ultraviolet (200 to 300nm) photons reaching the Earth. In the stratosphere (15,000 to 50,000 m), O atoms undergo the following reaction:

$$O + O_2 \Leftrightarrow O_3^* \text{ (excited state about 106 MJ kmol more than stable } O_3) \quad (17.36)$$

The O_3^* can produce stable O_3 after reaction with a third body M.

$$O_3{}^* + M \rightarrow O_3 + M \quad (17.37)$$

where M = O_2, N_2, etc., or the $O_3{}^*$ can also dissociate back to O and O_2. Further

$$O_3 \rightarrow (3/2)\ O_2,\ \Delta G = -163 \text{ MJ/kmol} \quad (17.38)$$

which is a slow reaction without light emission. On the other hand

$$O_3 + h_P\nu \text{ (ultraviolet band)} \rightarrow O_2 + O \quad (17.39)$$

is a fast reaction.

The O_3 concentration is about 10^{18} molecules per m^3 in the stratosphere. O_3 levels are also measured in Dobson units (DU). One Dobson unit is defined for 0.01 mm thickness at STP (standard temperature and pressure), i.e., its thickness when compressed to STP (the Earth's atmosphere). A normal level is 300 to 500 DU (corresponding to 1/8 in). Around the equator, O_3 is about 250 DU; when an ozone hole is formed in the polar regions, this concentration could be as low as 100 DU.

Example 2

Determine the wavelength of a photon required to dissociate an N_2 molecule if its bond energy N≡N is 945,000 kJ/kmol.

Solution

Bond per molecule = 945,000 kJ/kmol ÷ 6.023 × 10^{26} molecules per kmol = 1.57 × 10^{-21} kJ/molecule.

$$h_P\nu = 1.57 \times 10^{-21} \text{ kJ/molecule},$$
$$\nu = 1.57 \times 10^{-21} \text{ (kJ/molecule)}/6.626 \times 10^{-37} \text{ (kJ sec/molecule)}$$
$$= 2.37 \times 10^{15} \text{ Hz},$$
$$\nu = c/\lambda,\quad \lambda = 2.998 \times 10^8 \text{ m/sec}/(2.37 \times 10^{15}\text{/sec})$$
$$= 1.27 \times 10^{-7} \text{ m or 127 nm}$$

Thus, wavelengths should be shorter than 127 nm to dissociate N_2.

■ ■ ■

Air pollution due to chlorofluorocarbons (CCl_2F_2, which are nonreactive at sea level) catalyzes the reaction of $2O_3 \rightarrow 3O_2$ in the stratosphere owing to photodissociation. The reaction scheme is as follows: $O_3 + O \rightarrow 2O_2$

This reaction occurs in the presence of sunlight, but is slow. With Cl as a catalyst, it gets accelerated due to the following reaction.

$$\text{CFC} + \text{uv light} \rightarrow R + Cl\cdot$$
$$Cl\cdot + O_3 \rightarrow ClO\cdot + O_2$$
$$ClO\cdot + O \rightarrow Cl\cdot + O_2$$

N_2O (nitrous oxide) in the stratosphere reacts with O atoms to form NO (nitric oxide), which then reacts to deplete the level of stratospheric ozone. Also, N_2O rises to the stratosphere and breaks down to NO (nitrogen monoxide or nitric oxide) through photochemical reactions via sunlight. In addition to Cl, the NO in the stratosphere acts as catalysts in converting O_3 to O_2:

$$NO + O_3 \rightarrow NO_2 + O_2 \tag{17.40}$$

then

$$NO_2 + O \rightarrow NO + O_2. \tag{17.41}$$

Thus, ozone is depleted by the first reaction, whereas NO is formed by the second reaction. Once O_3 is depleted, ozone holes can be formed in the stratosphere. Recent studies reveal that sulfates also play a role in destruction via ClO, BrO, OH, etc. In addition to CH_4 (CFCs), N_2O also plays a role in the destruction of O_3. N_2O is still not regulated at this time; however, its regulation is inevitable owing to a fivefold increase of N_2O over the last 10 yr [Bowman, 1992a]. NO_x emissions from supersonic aircraft can greatly affect the deterioration of the ozone layer because of the emission location.

17.3.3 Photochemical Smog

Photochemical smog (PCS) is a brownish gray haze caused by interaction between solar radiation, HC, and NO_x. PCS may contain O_3, acrolein (CH_2=CH-CHO) and formaldehyde (HCHO, which causes irritation of eyes, nose and throat, coughing, breathing difficulties, chest tightness, headaches, and damages to agricultural crops as well as reduced driving visibility), peroxyacetyl nitrate (PAN), and peroxybenzyl nitrate (PBzN, benzene chemicals are carcinogenic). Nitrogen oxides, emitted from power stations, stationary sources, and vehicles, were first discovered to produce certain types of smog in the late 1940s by A.J. Haagen-Smit [1952]. He discovered that nitrogen oxides combined with hydrocarbons (released from automobiles) to form ozone (O_3) in the presence of sunlight in the troposphere. Therefore, owing to high NO_x and hydrocarbon emissions, smog is a serious problem in urban areas [Glarborg et al., 1994]. NO_x accumulates to superequilibrium concentrations at 25°C because the decomposition to N_2 and O_2 occurs slowly. The NO formed is converted to NO_2, and O_3 appears.

$$NO + 1/2\ O_2 \rightarrow NO_2 \tag{17.42}$$

$$NO_2 + h_p\nu \rightarrow NO + O \tag{17.43}$$

$$O + O_2 + M \rightarrow O_3 + M,\ (M = N_2,\ O_2) \tag{17.44}$$

$$O + 1/2\ O_2 \rightarrow O_3 \tag{17.45}$$

Note that NO_2 is just used as an intermediary in forming O_3.

$$O_3 + HC \rightarrow \text{photochemical smog.}$$

In conclusion, NO_x disrupts the ozone balance in two adverse ways: it produces ground-level ozone (photochemical smog), which is harmful, and destroys naturally occurring ozone high in the atmosphere, where it is beneficial. Thus, in the trophosphere the problem is production of O_3, which is destructive to organisms. O_3 at 10 to 15 ppmv (parts per million volume) can kill mammals. Smog (approximately 0.08 ppm) can damage cells in the lung's airways, causing inflammation; 3 ppmv can create an O_3 alert. Reaction with gasoline vapor and NO_x creates CH_3NO_3 (methyl nitrate), which is harmful to the eyes (causing them to water) and the respiratory system. Ozone can be formed using SO_2, which produces the O atom through complex reactions.

$$SO_2 \text{ from coal} + OH \rightarrow SO_2OH$$
$$SO_2OH + O_2 \rightarrow SO_3 + OOH$$
$$OOH \rightarrow OH + O$$
$$O + O_2 \rightarrow O_3$$

17.3.4 Acid Rain

Rain can become acidic because of carbonic acid formed from CO_2. Nitrogen and sulfur oxides are also known to contribute to acid rain and snow. Acid rain and snow are caused by the reaction of water with SO_x and NO_x to form sulfuric and nitric acid, H_2SO_4 and HNO_3, respectively, with a pH value less than 5:

$$NO_2 + OH + H_2O \rightarrow HNO_3$$
$$SO_3 + OH + H_2O \rightarrow H_2SO_4$$

Acid rain and snow can damage vegetation upon contact, make soil less fertile, reduce growth rate, cause leaves to fall, and corrode metals (because of nitrate salts formed from nitrogen oxides). Furthermore, acid rain and snow can dissolve aluminum and cadmium out of soil minerals, and thus allow them to enter roots and kill trees. Acid rain can contaminate drinking water supplies as well as make lakes and rivers lethal to fish owing to low pH levels (pH < 7 acidic). NO_x can form nitric acid in the lungs and is therefore more toxic than similar CO concentrations.

17.3.5 CO_2 Greenhouse Effect

Atmospheric CO_2 is essential for the survival of biological species, including humans, because photosynthesis converts CO_2 into plant materials. On the other hand, CO_2 absorbs infrared radiation (0.76 to 100 μm) in addition to CH_4. In addition, H_2O clouds blanket Earth (especially during the winter months) and absorb radiation > 20,000 nm (infrared).

"Greenhouses" can occur naturally in the atmosphere (as water vapor, carbon dioxide, methane from coal mines, nitrous oxide, and ozone) or in man-made form, that is, as (a) carbon dioxide released from fossil fuel combustion (solid waste and fossil fuels [oil, natural gas, and coal]), and wood and wood products); (b) methane formed during the production and transport of coal, natural gas, and oil, and from the decomposition of organic wastes in municipal solid-waste landfills as well as from the raising of livestock; (c) nitrous oxide emitted during agricultural and industrial activities, as well as during combustion of solid waste and fossil fuels; and (d) hydrofluorocarbons (HFCs), perfluorocarbons (PFCs), and sulfur hexafluoride (SF_6), which are generated in a variety of industrial processes. The ability to trap heat by the various gases differs. For example, for equal concentrations of CH_4 and CO_2, CH_4 is 21 times more potent in global warming potential (GWP) compared to CO_2. N_2O is considered a greenhouse gas because it absorbs infrared radiation. N_2O is measured as 286 ppbv in the atmosphere (e.g., sampling in Antarctic ice). Table 17.3 illustrates the GWP of greenhouse gases (CO_2, CH_4, N_2O, chlorofluorocarbons [CFCs], and ozone), which are measured in terms of CO_2 potentials. The CO_2 contribution is 50%; CH_4, 20%; CFCs, 15%; O_3, 10%; and N_2O, 5%.

17.3.6 Particulate Matter

The EPA defines exhaust particulate matter (PM) as the mass of material collected by filtration at a temperature less than 52°C [Abdul-Khalek et al., 1998]. The sources of PM include stationary combustion sources (67%), vehicles (20%), solid-waste disposal (5%), and natural sources (5%). The tunable infrared laser differential absorption spectrometer and the mass spectrometer are used to measure the diameter of particulates 1.5 to 40 nm. In addition, the spectrometer gives the chemical composition of PM.

Table 17.4 shows the National Ambient Air Quality Standards (NAAQS) is the U.S. at PM10, ozone and CO. Almost 80% is fly ash (with 40% of fly ash < 10 μm), whereas 20% falls out as bottom ash in a coal-fired plant. The particles with a size less than 10 μm are suspended in air and can cause health problems. The ash begins to sinter (a process in which particles stick together without melting) at T < 650°C, whereas melting occurs at T > 1000°C with formation of liquid slag at a temperature of about 1400°C. The alkali compounds Na_2O,

Table 17.3 Global Warming Potentials, GWP (100-Year Time Horizon)

Gas	*GWP*
Carbon dioxide (CO_2)	1
Methane (CH_4)	21
Nitrous oxide (N_2O)	310
HFC-23	11,700
HFC-32	550
HFC-125	2,800
HFC-134a	1,300
HFC-143a	3,800
HFC-152a	140
HFC-227ea	2,900
HFC-236fa	6,300
HFC-4310mee	1,300
CF_4	6,500
C_2F_6	9,200
C_4F_{10}	7,000
C_6F_{14}	7,400
SF_6	23,900

Source: EPA.

K_2O, etc., volatilize at T > 1100°C, forming sulfates Na_2SO_4, and K_2SO_4, and condense on water tubes that are at about 500°C [Borman and Ragland, 1998].

A stack dust sampling monitor for Cu smelter is shown in Figure 17.3 [Haglund et al., 2000]. EPA standards require quarterly measurements of stack

Table 17.4 Clean Air Act (CAA)

Pollutant	*Classification*	*Threshold*
Ozone	Serious	50 tpy or more of NO_x or VOC
Ozone	Severe	25 tpy or more of NO_x or VOC
Ozone	Extreme	10 tpy or more of NO_x or VOC
Ozone transport regions		50 tpy or more of NO_x or VOC
CO	Serious	50 tpy or more of CO if stationary sources contribute to CO
PM	Serious	70 tpy or more of PM 10

Note: tpy = tons per year; VOC = volatile organic compounds; PM 10 = PM having d < 10 μm.

Source: James Holkamp.

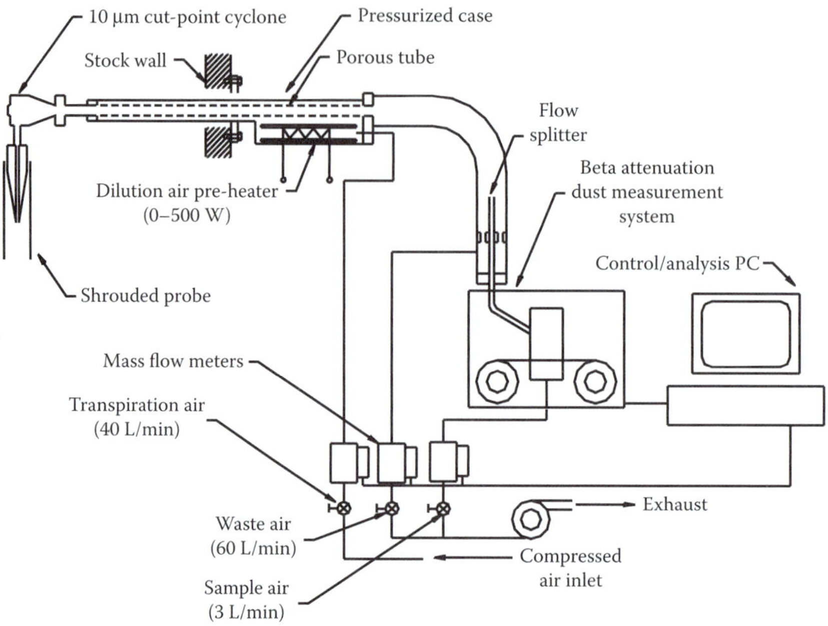

Figure 17.3 Stack dust sampling monitor. (From Haglund, J.S., McFarland, A.R., Wooldridge, M.S., and Vardiman, S., A Continuous Emission Monitor for Quantitative Measurement of PM10 Emissions from Stationary Sources, Air and Waste Management Association's 93rd Annual Conference and Exhibition, June 18–22, 2000, Salt Lake City, Utah.)

dust emissions. The measurements are performed manually by sampling aerosol from the stack using a probe and an in-stack filter. The design and use of the probes are given by the EPA in either Method 5 or Method 201A. The former calls for sampling of particulate in all particle sizes (total suspended particulate), and the latter calls for the sampling of only "respirable" particulate, or particles of 10-μm aerodynamic diameter and smaller (PM 10). Excellent correlation between the time-averaged dust concentration measurement of the system shown in Figure 17.3 and EPA Method 201A was observed when the concentrations ranged from 0 to 30 mg/m^3.

17.4 Pollution Regulations

The Clean Air Ambient Standards (CAA) requires the EPA to maintain a list of air pollutants and to establish primary and secondary NAAQS. The EPA standard for particulate matter is as follows: The original PM 10 standard is in effect and implements a PM 2.5 standard of an annual average level of 15 μg/cm^3 and a 24-h standard of 65 μg/cm^3 [EPA, 1995].

Table 17.5 lists the NAAQS for the U.S. Table 17.6 summarizes EPA regulations on selected power sectors. NO and NO_2 have a similar chemical makeup, yet they are formed in separate ways and have significantly different

Table 17.5 National Ambient Air Quality Standards (NAAQS) for the U.S.

Pollutant	*Averaging Time*	*Primary Standard*
Carbon monoxide	8 h	10,000 μg/m³ (9 ppm)
	1 h	40,000 μg/m³ (35 ppm)
Hydrocarbons[a]	3 h	160 μg/m³ (0.24 ppm)
Lead	Monthly average	1.5 μg/mil
Nitrogen dioxide	Annual average	100 μg/m³ (0.05 ppm)
	1 h	500 μg/m³(0.25 ppm)
Photochemical oxidants[b]	1 h	240 μg/m³ (0.12 ppm)
Sulfur dioxide	Annual average	80 μg/m³ (0.03 ppm)
	24 h	365 μg/m³ (0.14 ppm)
Ozone	1 h	23.5 μg/m³ (0.12 ppm)
Total suspended particulate	Annual geometric mean	75 μg/m³
	24 h	260 μg/m³
PM	24 h	150 μg/m³ for PM 10 and 65 for PM 2.5
	Long term	150 g/m³ for PM 10 and 15 for PM 2.5

[a]Corrected for CH_4.

[b]Corrected for SO_2 and NO_2.

Source: Davis, M.L. and Cornwell, D.A., *Introduction to Environmental Engineering*, Chemical Engineering Series, 2nd ed., McGraw-Hill, Singapore, 1991.

properties. Typically, NO, NO_2, and N_2O_4 are grouped as NO_x. In most stationary combustion sources 95% of all NO_x is NO. In most stationary combustion sources (e.g., coal combustion) 95% of all NO_x is NO. N_2O is formed from fossil fuel combustion at T of 800 to 950°C, but disappears at T > 1300°C owing to the reaction $N_2O + H = N_2 + OH$. Emission of N_2O is higher for fluidized beds (T = 800 – 950°C but low for coal-fired boiler burners (T > 1300°C). N_2O is lower for wood chips but higher for bituminous coals. N_2O can also react with char.

Table 17.6A EPA Regulations for Gas-Fired Units Built after 1973 (g/GJ)

	NO_x	*Particulates*
Units < 250 GJ/h	—	68
Units > 250 GJ/h	90	14

Note: Multiply by 0.002324 to get lb/mm Btu.

Table 17.6B EPA Regulations for Oil-Fired Units (g/GJ)

	NO_x	*Particulates*	SO_x
Utility	130	13	344
Industrial	172	43	344

Table 17.6C EPA Regulations for Coal-Fired Units (g/GJ)

	NO_x	*Particulates*	SO_x
Utility	260	13	70% at 560 g; if more than 90% removal

Table 17.6D EPA Regulations for Automobiles, 2004–2006 (g/mi)

NO_x	*HC*	*CO*
1.7	0.125	1.7

Table 17.6E EPA Regulations for Heavy-Duty Trucks, 1998 (g/hp h)

NO_x	*HC*	*CO*	*Particulates*
4	4	15.5	0.05

Table 17.6F Particulate Emissions Standards for Heavy-Duty Diesel Engines

Model Year	*Emissions Standard*
1970	Smoke opacity
1988	0.60 g/bhp-h
1991	0.25 g/hp-h
1994	0.10 g/bhp-h
1996	0.05 g/bhp-h
1998	0.10 g/bhp-h

17.5 NO_x Sources and Production Mechanisms

17.5.1 Nitrogen Oxide Compounds

There are seven oxides of nitrogen generated by the chemical bonding of nitrogen and oxygen: nitric oxide (NO), nitrogen dioxide (NO_2), nitrogen trioxide (NO_3), nitrous oxide (N_2O), dinitrogen trioxide (N_2O_3), dinitrogen tetroxide (N_2O_4), and dinitrogen pentoxide (N_2O_5) [Yaverbaum, 1979]. Of these oxides, NO and NO_2 are the most important because they are emitted in large quantities. NO_2 causes brownish haze. In 1975, N_2O, nitrous oxide, was not even considered to be an air contaminant and was thought to have no environmental effect at low concentrations. It can be argued that most N_2O formation is by natural means because combustion is a major source of NO_x and a minor direct source of N_2O. However, combustion could be contributing to N_2O increase by indirect means. N_2O has a long lifetime (100 to 200 yr) in the troposphere, compared to the lifetime of NO_x (minutes to days). Typically, NO is produced in thousands of ppm, whereas NO_2 is produced in tens of ppm [Borman and Ragland, 1998]. In past years, evidence indicated that automobile emissions were the major producer of NO_x, but as automobile standards were raised and enforced, electricity-generating plants and other stationary sources contributed a higher fraction of total NO_x emitted into the atmosphere.

The total worldwide NO_x production rate is on the order of 520 Mt per yr. Of this 520 Mt, natural sources far outweigh the man-made sources, contributing approximately 500 Mt. Concentrations of NO and NO_2 have been reported to be approximately 20 to 40 ppb in urban areas of the U.S. Typically, the percentage of NO sources are: 50% from vehicles, 35% from industrial and commercial boilers, and 14% from area sources. Tobacco smoke NO_x levels are on the order of 250 ppm. NO is measured using a chemiluminescent analyzer that is based on photon emission from the reaction NO + O_3; the more the NO, the more the photons emitted on reaction with O_3.

17.5.2 Sources of NO_x

Fuel combustion devices account for 99% of technology-related NO_x. For most combustion devices, NO levels are much higher than those of NO_2, with NO representing approximately 95% of the total NO_x [Yaverbaum, 1979]. The NO_2 is believed to be formed by HO_2 in the low temperature regime and with reaction $NO + HO_2 \rightarrow NO_2 + OH$. Once NO has been released into the atmosphere, it has a short lifetime of approximately 4 days, primarily because NO readily oxidizes to NO_2 in the presence of oxygen. This oxidation process is greatly accelerated by photochemical effects.

17.5.2.1 Mechanisms of Production of NO_x

There are three mechanisms by which nitrogen and oxygen can combine and form NO_x: (1) when nitrogen in the fuel combines with oxygen in the combustion air (fuel NO_x), (2) when fuel hydrocarbons break down and recombine with atmospheric nitrogen (prompt NO_x or Fenimore mechanism), and (3) when the intense heat of combustion causes atmospheric nitrogen to combine with atmospheric oxygen (thermal NO_x). The importance of each of the three mechanisms varies, depending on numerous conditions including fuel, combustion environment, and temperature. In order to minimize the NO_x level, a detailed knowledge of each mechanism is required.

17.5.2.1.1 Thermal NO_x

Thermal NO_x is formed by the oxidation of atmospheric (molecular) nitrogen. This mechanism dominates when clean fuels (those with little or no nitrogen content) are combusted. As air is typically 21% oxygen and 79% nitrogen (mole basis), a large amount of superfluous nitrogen is inevitably supplied to the combustion system along with the needed oxygen. The bonds between the atmospheric nitrogen atoms can be broken at elevated temperatures and they can be subsequently oxidized to thermal NO_x. Thermal NO's strong temperature dependence is due to the fact that the N_2 molecules have triple bond (bond energy of 945 kJ/k mol, Table A.31A and Table A.31B of Appendix A). The computed equilbrium NO in ppm vs. T for CH_4 burning with 20% excess air and 0% excess air is shown in Figure 17.4. It is seen here that thermal NO_x levels are not significant below 1500 K and increase dramatically with temperature above 1500 K. With diesel, about 40% of thermal NO_x is produced after peak pressure. Further, the presence of soot may reduce NO_x owing to the reduction reaction of NO + C to N_2 and CO, as shown in Figure 17.5. Numerous techniques used to reduce thermal NO_x simply reduce the combustion temperature. Based on kinetics, thermal NO_x is very temperature sensitive and is formed at combustion temperatures of approximately 1900 K to 2200 K (3000 F to 3600 F) [Cooper and Alley, 1990]. From kinetics, it was found that the NO concentration reaches equilibrium very quickly (less than 0.00001 sec) at high temperatures for T = 2800 K to 3000 K, but not so quickly at lower temperatures.

17.5.2.1.1.1 Zeldovich Model — The Zeldovich model, which is time dependent, better predicts thermal NO levels from combustion processes than the equilibrium model. It was first postulated by Zeldovich in 1946 and is

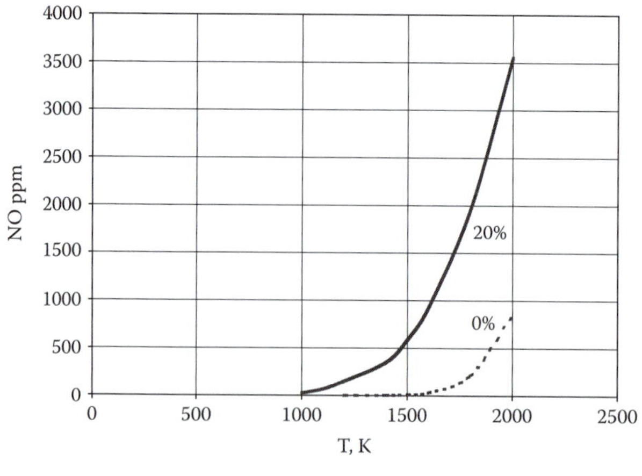

Figure 17.4 Thermal NO_x: equilibrium calculations for CH_4 combustion.

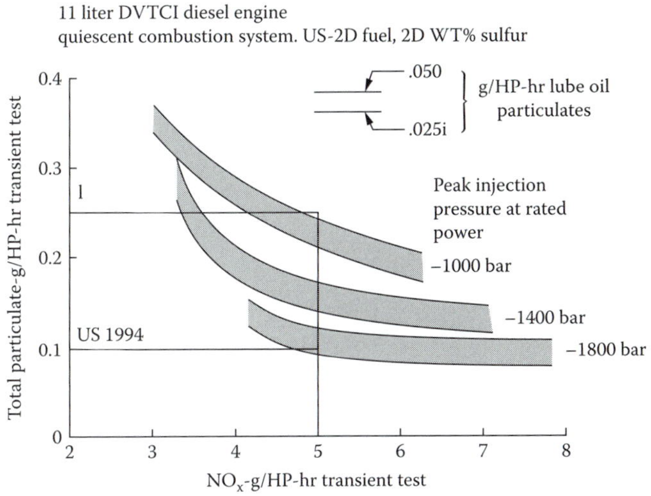

Figure 17.5 Effect of particulate matter on NO_x from diesel engine. (Adapted from Zelenka, P., Kriegler, W., Herzog, P.L., and Cartellieri, W.P., Ways Toward the Clean Heavy-Duty Diesel, SAE Paper 900602, 1990.)

referred to as the Zeldovich or thermal mechanism. This mechanism is given by the chain of reactions shown in the following text.

The Zeldovich mechanism for thermal NO_x consist of three reactions as follows:

$$O + N_2 \rightarrow NO + N \tag{I}$$

$$N + O_2 \rightarrow NO + O \tag{II}$$

$$N + OH \rightarrow NO + H \tag{III}$$

where this last reaction is small for a lean mixture as OH is small. The O atom could be formed from Reaction IV (cited later) and OH by $H + O_2 \rightleftarrows OH + O$. The H is produced by dissociation of HC fuel. The individual reactions in the Zeldovich reaction have high activation energy and do not become significant until the temperature reaches 1800 K.

$$\begin{aligned}\frac{d[NO]}{dt} &= k_{I_f}[N_2]\,[O] - k_{I_b}[N]\,[NO] + k_{II_f}[N]\,[O_2] \\ &\quad - k_{II_b}[O]\,[NO] + k_{III_f}[N\ ]\,[OH] - k_{III_b}[NO]\,[H]\end{aligned} \tag{17.46}$$

Units: K, kmole, m^3, sec

$$\begin{aligned}k_{If} &= 1.8 \times 10^{11} \exp(-38370/T) \\ k_{Ib} &= 3.8 \times 10^{10} \exp(-425/T) \\ k_{IIf} &= 1.8 \times 10^{7}\ T \exp(-4680/T) \\ k_{IIb} &= 3.8 \times 10^{6}\ T \exp(-20820/T) \\ k_{IIIf} &= 7.1 \times 10^{10} \exp(-450/T) \\ k_{IIIb} &= 1.7 \times 10^{11} \exp(-24560/T)\end{aligned} \tag{17.47}$$

The O and H atom concentrations can be determined using the following equilibrium relations.

$$O_2 \Leftrightarrow 2O \quad \text{(IV)}$$

$$H2 \Leftrightarrow 2H \quad \text{(V)}$$

The OH, O, and H concentrations are assumed to remain at equilibrium level and N concentration is assumed to be at steady state. Typically, the first reaction is rate limiting. As most combustors (e.g., gas turbines) operate in a lean regime, [OH] is small Reaction III can sometimes be ignored. Thus, a simplified model assumes that there are two chemical reactions which are the primary thermal NO_x formation mechanisms (see Chapter 5 for a simplified analysis).

$$\frac{d[\mathrm{NO}]}{dt} = k_{I_f}[\mathrm{N_2}][\mathrm{O}] - k_{I_b}[\mathrm{N}][\mathrm{NO}] + k_{II_f}[\mathrm{O_2}][\mathrm{N}] - k_{IIb}[\mathrm{NO}][\mathrm{O}] \quad (17.48)$$

$$\frac{d[\mathrm{N}]}{dt} = k_{I_f}[\mathrm{N_2}][\mathrm{O}] - k_{I_b}[\mathrm{N}][\mathrm{NO}] - k_{II_f}[\mathrm{N}][\mathrm{O_2}] + k_{II_b}[\mathrm{O}][\mathrm{NO}] \quad (17.49)$$

Using the equilibrium relation for $O_2 \Leftrightarrow 2O$ (Reaction IV) and solving for [N] by setting d[N]/dt = 0 in Equation 17.49,

$$\frac{d[\mathrm{NO}]}{dt} = 2k_{If}\mathrm{K}_{IV}{}^{0}(\mathrm{T})\left(\frac{1}{RT}\right)^{\frac{1}{2}}[\mathrm{N_2}]\sqrt{[\mathrm{O_2}]} - 2K_{IV}{}^{0}(\mathrm{T}) \left(\frac{1}{RT}\right)^{\frac{1}{2}} \frac{k_{II_b}k_{I_b}}{k_{II_f}} \frac{[\mathrm{NO}]^2}{\sqrt{[\mathrm{O_2}]}} \quad (17.50)$$

and $K_{IV}{}^0$ is an equilibrium constant. Note that there are four variables in Equation 17.50. $K_{IV}{}^0$, which is only dependent on temperature and $[O_2]$, both of which are dependent on time and excess-air percentage; $[N_2]$, which we can assume to be constant during the combustion process; and [NO], which we wish to calculate. Now,

$$\frac{d[\mathrm{NO}]}{dt} = \frac{2\,k_{If}\;\{K_{IV}\left(\frac{\mathrm{p}^0}{\bar{R}\mathrm{T}}\right)[\mathrm{O2}]\}^{1/2}[\mathrm{N2}]}{\{1 + \frac{k_{Ib}[\mathrm{NO}]}{k_{IIf}\,[\mathrm{O2}]}\}}\left[1 - \frac{k_{Ib}[\mathrm{NO}]}{k_{IIf}\,[\mathrm{O2}]}\frac{k_{IIb}[\mathrm{NO}]}{k_{If}[\mathrm{N2}]}\right] \quad (17.51)$$

where p^0 = 1 bar. Because NO is in trace amounts, $[k_{Ib}/k_{IIf}]\,[NO]/[O_2] \lll 1$, $[k_{IIb}/k_{IIf}]\,[NO]/[N_2] \ll 1$

$$\frac{d[\mathrm{NO}]}{dt} = 2k_{1f}[\mathrm{N2}]\left[\mathrm{K}^0{}_{IV}\,[\frac{1}{\bar{R}\mathrm{T}}]\right]^{1/2}[\mathrm{O2}]^{1/2} \quad (17.52)$$

The Zeldovich mechanism predicts that thermal NO_x formation has a strong dependence on T through an activation temperature of 38,370 K, moderate dependence upon oxygen concentration, and strong dependence on N_2 concentration. Thus, when pure O_2 is used, thermal NO is zero. If air is used as the oxygen supply to the flame, the air:fuel ratio greatly affects the thermal NO_x levels.

Equation 17.52 is based upon the presumption that "O" atoms exist at equilibrium concentration; if kinetics are involved, then O atom concentration may exist at superequilibrium concentration in low-T zones (by an order of 100 to 1000 and subequilibrium in high-T zones owing to limited residence times) [Miller and Bowman, 1989]. Many NO_x reduction techniques use this principle to minimize NO_x formation. The overall or global order of reaction rate is 1.5 (Equation 17.52). Increased pressure will cause increased NO and vice versa, whereas equilibrium considerations reveal zero pressure dependence. When an infinite time is assumed, the Zeldovich mechanism predicts NO levels very close to those predicted by chemical equilibrium. The Zeldovich model accurately predicts that the shorter the residence time, the lower the level of thermal NO_x. However, this simplified trend can be complicated by the steep temperature gradients present in most combustion systems (e.g., gradient of T around a burning droplet).

The literature on the subject suggests that NO concentration should reach its final equilibrium value fairly quickly and that NO concentration should peak slightly to the lean side of stoichiometric.

17.5.2.1.2 Global Reactions

Instead of a theroretically derived equation (Equation 17.52), one can present an empirical experimental correlation as

$$\frac{d(\mathrm{NO})}{dt} = \frac{1.9\times10^{17}}{T^{0.5}}\exp\left(\frac{-69{,}096}{T}\right)(\mathrm{O_2})^{0.5}(\mathrm{N_2}),\ \mathrm{kmol/m^3s} \tag{17.53}$$

17.5.2.2 Fuel NO_x

Fuel NO_x is formed by the oxidation of fuel-bound nitrogen in the fuel. This type of fuel would be coal, heavy oil, biomass, etc. Table 17.7 shows the N percentage in fossil fuels.

Table 17.7 Nitrogen (Percentage by Weight) in Fossil Fuels

1.	Natural gas	0
2.	Crude oil	0.15–0.5
3.	# 6 fuel oil	0.1–0.5
4.	# 2 fuel oil	0.01
5.	Heavy distillate	0.6–2.15
6.	Coal	1–1.5
7.	Feedlot biomass (cattle manure)	2–4
8.	Litter biomass (chicken waste)	2–5

The reaction of fuel N to NO occurs by the production of nitrogen compounds HCN, amines (e.g., NH_3), and cyanides (CN) from the pyrolysis of fuels. It should also be noted that all fuel N is bound as either carbon–nitrogen (HCN) or nitrogen–hydrogen bonding (NH_3). The compounds then either form NO or N_2 depending on availability of oxygen, as detailed in the prompt NO mechanism. The fuel N evolution from coal has been studied in detail by Pohl and Sarofim [1977] (Figure 17.6a). The fuel N retained in char and dry-weight loss are indicated in Figure 17.6a. Studies on pyrolysis of coals suggest significant conversion of fuel N to HCN and lesser amounts to NH_3. Figure 17.6b shows the pathways for fuel NO formation when fuel N is converted into HCN. When both HCN and NH_3 are present, the following simplified reaction schemes are used.

Fuel N is released as HCN, NH_3, N_2, during pyrolysis.

$$HCN + O_2 \rightarrow NO + \ldots \tag{I}$$

$$HCN + NO \rightarrow N_2 + \ldots \tag{II}$$

$$NO + \text{char} \rightarrow N_2 + \ldots \tag{III}$$

$$NH_3 + O_2 \rightarrow NO + \ldots \tag{IV}$$

$$NH_3 + NO \rightarrow N_2 + \ldots \tag{V}$$

The reaction rate constants are described by an Arrhenius-type expression, in units of k, kmol, m^3, sec

$$\begin{aligned} k_I &= 1\times10^{11}\ \exp(-33800/T) \\ k_{II} &= 3.0\times10^{12}\ \exp(-30190/T) \\ k_{III} &= 4.1\times10^{-4}\ \exp(-17440/T) \\ k_{IV} &= 4.0\times10^{6}\ \exp(-16120/T) \\ k_{V} &= 1.8\times10^{8}\ \exp(-13590/T) \end{aligned} \tag{17.54}$$

The timescales for homogeneous and heterogeneous NO reactions using the kinetics data are shown in Figure 17.7 as a function of temperature. The NO reduction reaction with char has the largest timescale, indicating that it is a slow process. On the other hand, the gas-phase reactions are fast. The NH_3 reactions are faster than the HCN reactions at low temperatures. At around 1800 K, the ammonia reactions have a slightly larger timescale compared to HCN, but the difference is not appreciable. These timescales will be very helpful in discussing the NO results obtained during the parametric study. As the timescale for ammonia oxidation is less compared to that for HCN oxidation, the NO level is higher when ammonia and HCN are used in the NO reaction scheme.

All of the nitrogen in fuels is not converted to fuel NO_x (Figure 17.8). Conversion efficiencies of fuel N in coal and fuel oils are between 10% and 60% [Cooper and Alley, 1990]. When 0.5% pyridine (C_5H_5N) is added to oil, NO

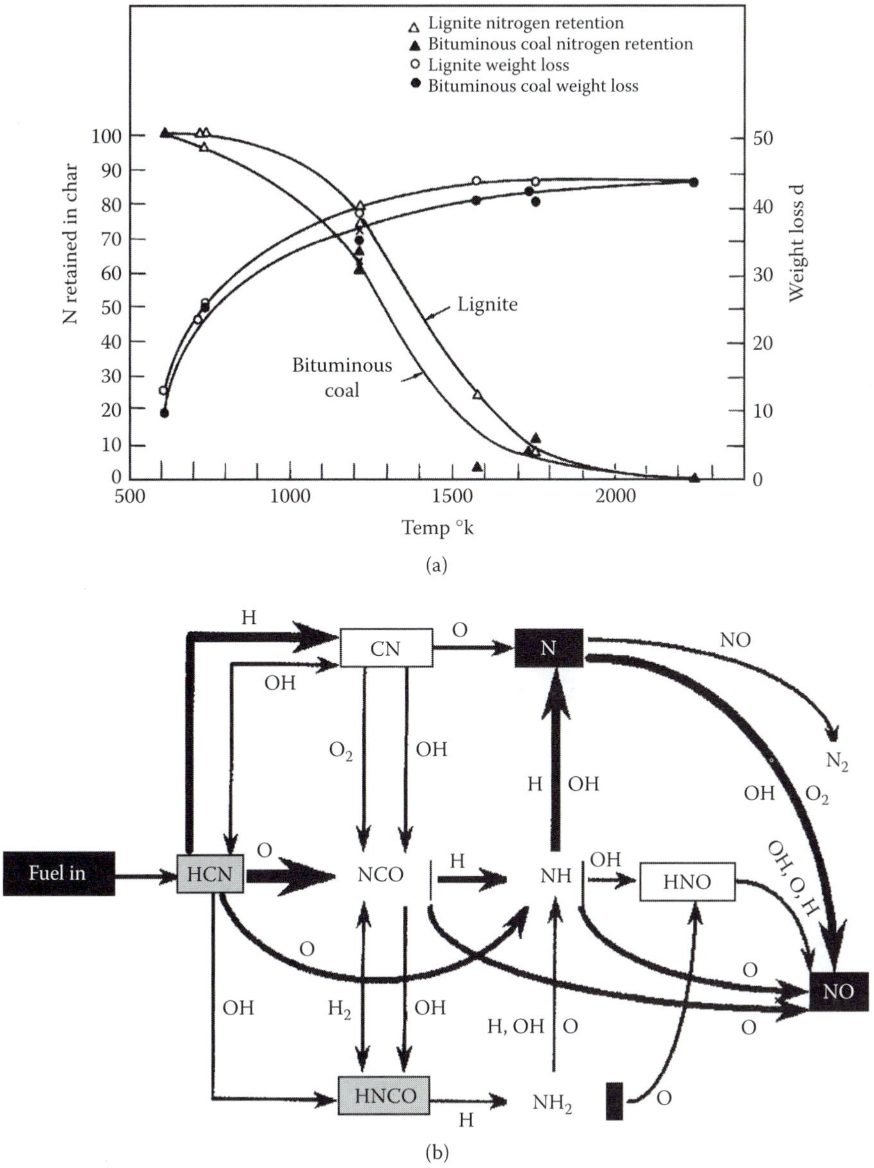

Figure 17.6 (a) Nitrogen retained in charcoal, 38 to 45 μm. (Adapted from Pohl, J.H. and Sarofim, A.F., Devolatalization and Oxidation of Coal Nitrogen, 16th International Symposium on Combustion, The Combustion Institute, Pittsburgh, PA, 491–501, 1977.) (b) Fuel NO formation pathways. (Adapted from Bowman, C.T., *Chemistry of Gaseous Pollutants Formation and Destruction; Fossil Fuel Combustion*, Bartok and Sarofim, Eds., John Wiley & Sons, New York, 1991, chap. 4.)

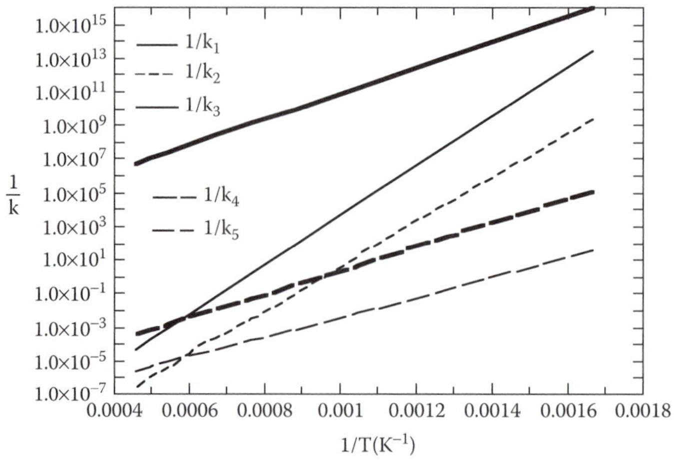

Figure 17.7 Comparison of timescales based on kinetic data for homogeneous and heterogeneous NO reactions; k_1:HCN oxidation; k_2: NO reduction on HCN; k_3: NO reduction on charcoal; k_4: NH_3 oxidation; and k_5:NO reduction on NH_3.

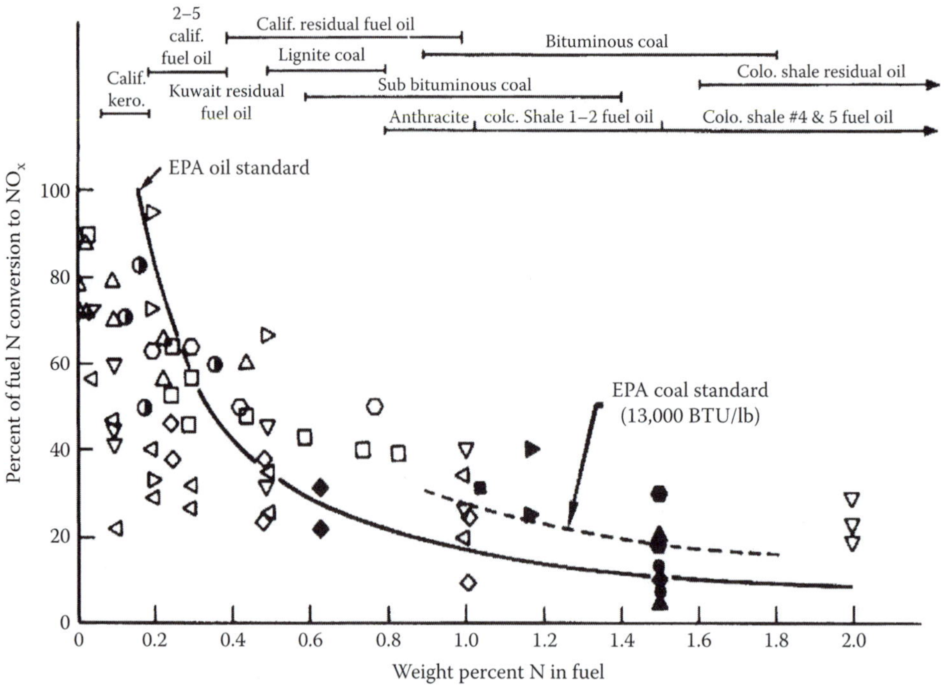

Figure 17.8 Fractional fuel nitrogen conversion to NO. Open: oil; solid: coal; semisolid: laminar premixed. (Adapted from Pohl, J.H. and Sarofim, A.F., Proceedings of the Stationary Source Combustion Symposium, EPA report 600-12-76-15a, 1976; Fenimore, C.P., *Combust. Flame*, 19, 289, 1972.)

increases by 10 times [Martin and Berkau, 1971]. Fuel NO_x conversion efficiencies are dependent on the air:fuel ratio and degree of fuel–air mixing, and are not affected by small temperature changes unlike thermal NO_x. The competing reactions of N oxidation to NO oxidation and NO reduction to N_2 depend highly on stoichiometric conditions; a rich flame promotes N_2 formation, whereas a lean flame promotes NO formation. The fraction of fuel N that oxidizes to fuel NO_x is also inversely proportional to the nitrogen content of the fuel. This trend is seen in Figure 17.8. This figure shows that as fuel-bound nitrogen increases, the conversion rate to fuel NO_x decreases because most of the NO could be converted into N_2. No matter how much of the fuel-bound nitrogen is converted, it is clear that the oxidation to fuel NO_x is rapid and occurs at a speed comparable to that of the heat-release reactions that occur during combustion.

Recall that thermal NO_x is very sensitive to temperature because the N-N bond in molecular nitrogen is very strong. This is not the case with the fuel N mechanism, in which the temperature dependence is weak as the fuel nitrogen bonds, N-C and N-H, are relatively weak (Table A.31A and Table A.31B of Appendix A) having activation energies approximately half that of the N-N bond. Therefore, fuel NO_x is formed at relatively low temperatures when compared to thermal NO_x. Yet, at any temperature that can sustain combustion, significant amounts of fuel NO_x are formed. In coal-fired boilers, 80% burn-out and significant NO_x formation occurs within 2 quarl diameter. High-volatile coal of 1.6% N produces 450 to 550 ppm of NO_x (long jet flame with no recirculation). With increased swirl, NO_x is almost doubled, owing to more availability of O_2 particles swirled to the edge of the recirculation zone (RCZ). On the other hand, the mixing of fresh air with increased RCZ gases will reduce NO_x output.

17.5.2.3 Prompt NO_x

Experiments show that NO_x levels often exceed those predicted by the Zeldovich mechanism, especially when hydrocarbon fuels are used at low temperatures and richer mixtures. It is clear that there is a significant amount of NO_x that is formed by neither thermal nor fuel NO_x mechanisms. This NO_x has been found to form rapidly (promptly) in the flame front and has been accordingly termed prompt NO_x [Fenimore, 1961, 1972]. Levels of prompt NO_x are usually quite lower than fuel and thermal NO_x levels. In the prompt NO mechanism, hydrocarbons (40 to 60 ppm) react with molecular N_2 in the air to form HCN. The steps are: (1) The O and OH radicals promote thermal NO formation; (2) the HC radicals react with N_2 to form CN compounds and N atoms (Reactions I and III, see below); (3) The N atoms yield NO by Reactions II and III of the thermal NO_x scheme and CN can form NO with O atoms or O_2.

$$\begin{aligned} &CH + N_2 \Leftrightarrow HCN + N, \; k\ (cm^3/(gmol\ sec)) \\ &k = 2.86 \times 10^8\ T^{1.1} \exp(-10267/T),\ \text{rate} \end{aligned} \tag{I}$$

$$CH_2 + N_2 \Leftrightarrow HCN + NH \tag{II}$$

$$C + N_2 \Leftrightarrow CN + N \tag{III}$$

For prompt NO_x, Reaction I is the rate-limiting step. Then N, NH, and HCN react with the O atom to form NO where O and N are present at equilibrium concentration. The atomic N may react with O_2 and OH to produce NO just as in thermal NO_x (Chapter 5); further HCN, CN, NH, etc., are thought to readily oxidize to NO [Glassman, 1977]. The HCN that is formed through reaction with hydrocarbons is then either converted to N_2, or it is combined with O radicals to form NO. The process is detailed in the following steps:

$$HCN + O \Leftrightarrow NCO + H \quad \text{(IV)}$$

$$NCO + H \Leftrightarrow NH + CO \quad \text{(V)}$$

$$NH + H \Leftrightarrow H + H_2 \quad \text{(VI)}$$

$$N + OH \Leftrightarrow NO + H \quad \text{(VII)}$$

$$O + N_2 + M \Leftrightarrow N_2O + M \quad \text{(VIII)}$$

$$N_2O + O \Leftrightarrow NO + NO \quad \text{(IX)}$$

It is seen that the prompt NO_x mechanism is very similar to the fuel NO_x mechanism. This is because both mechanisms involve an HCN intermediate. The amount of HCN is extremely small due to limited HC fragments compared to HCN in N-containing coal. Several parameters involving the formation of prompt NO_x are as follows: The highest levels of prompt NO are produced by fuel with low H/C ratios. Further, prompt NO_x may contribute significant amount at high pressure (almost 95% of the total NO for CH:air fuel-rich flames) as compared to low pressure [Miller and Bowman, 1989]. It is also known that prompt NO_x formation is relatively independent of temperature, mixture ratio, and residence time. When the fuel mixture is rich, fuel NO_x and thermal NO_x may be reduced, but prompt NO_x could become significant, although still in low amounts.

If nitrogen-containing fuels are used, the prompt NO levels are usually negligible compared to fuel NO. Typically, prompt NO accounts for 5% of the total NO from a utility boiler burning coal with 1% nitrogen and the percentage could be as high as 5 to 10% for N absent fuels [Drake et. al., 1980]. However, if thermal and fuel NO control techniques improve, the prompt NO mechanism may contribute a significant percentage of the total NO.

17.6 NO_x Formation Parameters

There are numerous parameters that affect the levels of NO_x produced by a combustion device. They are listed and discussed in the following subsections.

17.6.1 Type of Facility

Cyclone furnaces, which are characterized by extremely high heat-release rates, have NO_x levels of 900 to 1500 ppm owing to the high T required for slag formation and are often prohibited. Conversely, stoker furnaces have low

NO_x emissions, ranging from 250 to 400 ppm, yet are much less combustion efficient than cyclone furnaces. It is clear that the type of facility takes into account numerous parameters such as burner and combustion chamber design as well as burner configuration and operating conditions. All of these variables affect T and mixture stoichiometry, which directly affect NO_x levels.

17.6.2 Operational Conditions

Combustion Temperature: Because the formation of thermal NO_x is kinetically controlled, the temperature — time history of the combustion gases is an important parameter. Furthermore, prompt NO_x is formed where the maximum temperature levels may be as low as 1600 K. Beyond this threshold, however, prompt NO_x is relatively independent of temperature. Therefore, variables that affect combustion temperature, such as fuel selection, air preheat, and load, should be considered in NO_x control.

To reduce NO_x levels, the combustion gas temperature–time profile should be made so that the residence time around the maximum-temperature and oxygen-rich zone is kept to a minimum. To do this, the combustion chamber should be designed to rapidly dissipate the heat released during combustion. There are many ways of achieving this heat dissipation, including tangential firing and insertion of ceramic rods that reduce the temperature and reduce average O_2 concentration, staged combustion, modified burners, and off-stoichiometric combustion.

Load: Reducing the load causes lower heat-release rates and corresponding cooler temperatures. Therefore, load reduction reduces NO_x and should be kept at levels just high enough to meet energy demands.

Air Preheat: Increased air preheat increases the combustion temperature. As a result, the NO_x levels also increase.

Air:Fuel Ratio/Oxygen Concentration: If nitrogen-rich fuels are used because clean fuels are unavailable or economically unfeasible, then the most important factor in reducing fuel NO_x is the combustion oxygen concentration. Increase in O_2 concentration promotes the formation of both thermal and fuel NO_x.

17.6.3 Fuel

Whether or not the fuel contains nitrogen directly determines the levels of fuel NO_x. Fuels with little or no fuel-bound nitrogen, such as methane and natural gas, do not produce significant levels of fuel NO_x. Conversely, in the combustion of fuels such as coal, oil, and manure, which have a high nitrogen content, the fuel NO_x mechanism could be dominant. Therefore, using fuels without fuel-bound nitrogen is the easiest way to eliminate fuel NO_x.

The fuel-bound nitrogen is either tied up in the volatile matter or the coal char. The fuel NO produced by solid-fuel combustion (e.g., coal, biomass, etc.) is formed by two separate routes: oxidation of nitrogen compounds in the volatile matter (volatile NO) and oxidation of char nitrogen (char NO). Volatile NO is formed early in the combustion process as the solid fuel is heated and emits volatile matter. Typically, the nitrogen-bearing volatile matter is in the form of N_2, NH_3, and HCN. The volatile NO usually contributes

60 to 80% of the total fuel NO from coal combustion. The rest is accounted for by char NO. The highest potential for NO reduction by burner design lies in the minimization of volatile NO.

As the nitrogen content of the coal increases, the fractional conversion to fuel NO_x decreases. Typically, this fractional conversion ranges from 15 to 30%. Fuels that are characterized by high flame temperatures produce more NO_x than fuels that burn at low temperatures. Usually, fuels with higher carbon/hydrogen ratios produce higher flame temperatures. The tangentially fired units produce almost 50% of the NO_x in wall-fired units. Cofiring of coal with agricultural biomass fuels can reduce NO_x due to smaller fuel N.

17.7 Stationary Source NO_x Control

Following are four ways NO_x emissions can be reduced from a stationary source in order to meet emission standards: (1) reduction of N from fuels, (2) creation of a combustion environment that inhibits NO_x formation (combustion controls or modifications, e.g., reduction of O_2 at locations by reduction of excess-air and rich mixtures, exhaust gas recirculation [EGR], staged combustion, or low-NO_x burners [LNB]) in which fuel N reacts with O_2, (3) temperature reduction to lower thermal NO (water–steam injection, reduced air preheat), and (4) postcombustion removal of NO_x prior to release into the atmosphere. The postcombustion controls require no combustor modification for use with several processes: (1) selective noncatalytic reduction (SNCR), injecting NH_3, thermal $DeNO_x$ process by Exxon, $T < 1100$ to 1200 K, and low O_2%, or other N-containing additives (urea or cyanuric acid) to combustion gases without presence of catalysts [Bowman, 1992]; (2) selective catalytic reduction (SCR; in reaction with NH_3 but in the presence of a catalyst); (3) reburn systems by injection of natural gases and coal; and (4) combinations of the preceding techniques. It should be noted that not all NO_x reduction techniques work for all three of the NO_x formation mechanisms, and NO_x reduction can affect thermodynamic efficiency as well as increase emissions of other pollutants. This is due to the fact that high combustion temperatures, for which thermodynamic efficiencies are at a maximum, readily promote thermal NO_x formation. The highest attainable flame temperatures, and therefore the highest (thermal) NO_x levels, are found in stoichiometric flame conditions. These near-stoichiometric mixtures promote complete combustion and minimize emissions of CO, hydrocarbons, and soot particles. Figure 17.9 shows the operating conditions for low NO_x for gas turbines.

17.7.1 Combustion Modifications

The following are techniques that alter the combustion environment so that the chemical and thermal conditions inhibit NO_x formation.

Flue Gas Recirculation: Flue gas recirculation (FGR) is a process in which a portion of the low-temperature flue gases is recirculated back to the combustion zone. These combustion gases are relatively inert and therefore dilute the combustion reactants. As a result, the combustion oxygen partial pressure and flame temperature decrease with recirculation. Thus, both thermal NO_x

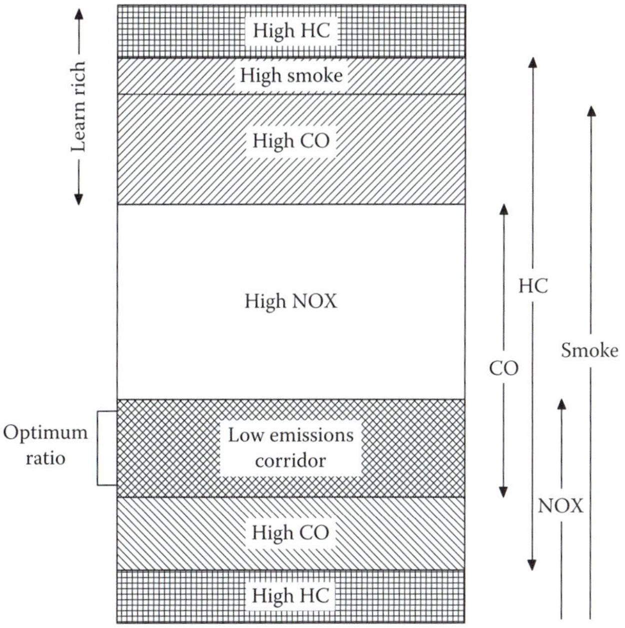

Figure 17.9 Operating conditions for low NO_x, low NO, and low HC. (Ryan, Siemens Westinghouse.)

due to low T and fuel NO_x decrease. In automobiles, this is achieved by mixing exhaust gases with fresh air. The exhaust gas acts as a diluent and increases the thermal heat capacity.

One major problem with FGR is that, at high recirculation rates, flame instability occurs due to the reduced O_2%. This condition results in increased CO and hydrocarbon levels. Figure 17.10 shows a stratified burner for low

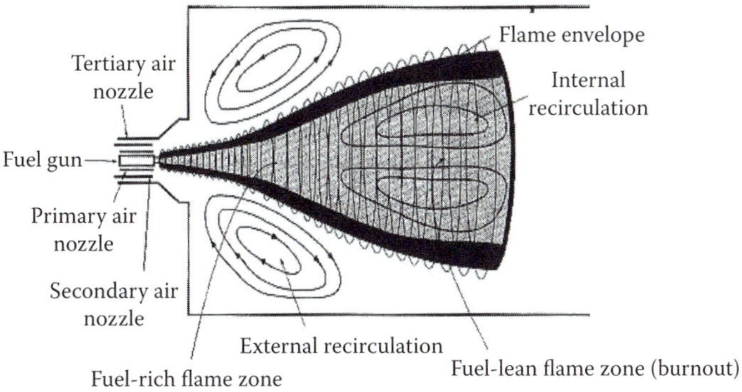

Figure 17.10 Stratified burner for NO_x reduction for gas-fired combustors. Fuel jet in fuel gun surrounded by primary, secondary, and tertiary air. (Adapted from Toqan, M.A., Beer, J.M., Johnson, P., Sun, N., Tsta, A., Shihadeh, A., and Teare, D., 24th International Symposium on Combustion, The Combustion Institute, Pittsburgh, PA, 1391–1397, 1992.)

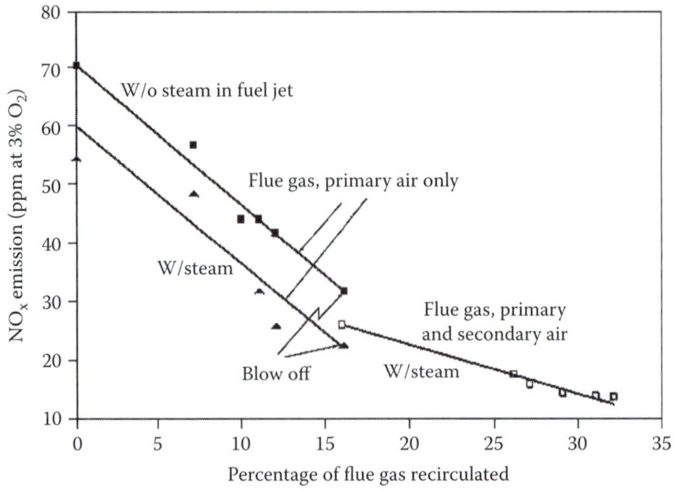

Figure 17.11 Effect of flue gas recirculation; steam:fuel ratio, 0.12:1. (Adapted from Toqan, M.A., Beer, J.M., Johnson, P., Sun, N., Tsta, A., Shihadeh, A., and Teare, D., 24th International Symposium on Combustion, The Combustion Institute, 1391–1397, 1992.)

NO_x production when gas is used as fuel. The reduction is achieved through reduction in T. Figure 17.11 shows the effect of diluting the secondary and primary air in the burner quarl on NO_x emission. As the percentage of O_2 is reduced, NO_x emission decreases. However, at a large percentage, blow-off occurs.

Water and Steam Injection: Water and steam injection (WSI) at 0.5 to 1.5 × fuel flow into gas turbines to reduce NO_x emissions to about 100 ppm at 15% O_2 is a common practice. The injected steam acts as additional mass loading, increasing the power by 15%, lowering the flame temperature, and reducing the partial pressure of O_2, which reduces the amount of atmospheric nitrogen that can bind with oxygen. Major side effects of WSI include decreases in thermodynamic efficiency, increased corrosion rates, and flame instability.

Low Excess Air: Combusting at low excess air (LEA) levels decreases oxygen concentrations. However, flame instabilities may occur, and CO levels may increase as a result of LEA.

Off-Stoichiometric Combustion: Off-stoichiometric combustion (OSC) is also called biased firing. OSC involves firing numerous burners, some operating under rich conditions and others under lean conditions. OSC burners allow the overall combustion stoichiometry to match that in conventional burners, resulting in high thermodynamic efficiency. However, substoichiometric conditions prevail locally in the primary combustion zone. Complete combustion takes place downstream of this substoichiometric region.

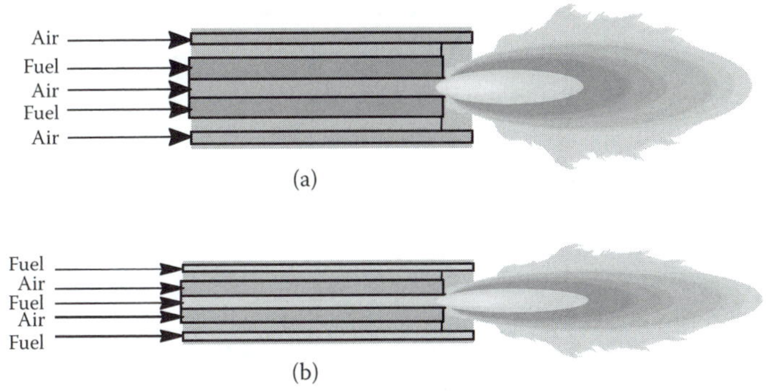

Figure 17.12 Two-stage combustion. (a) staged air, (b) staged fuel. (Adapted from Baukal, 2000.)

The effect of OSC is reduced oxygen levels in the primary zone which reduces NO_x by 50%.

Two-Stage Combustion: In two-stage combustion (TSC), the first stage is fuel rich because only a portion of the secondary air is supplied (Figure 17.12). Incomplete combustion occurs in the first stage but with reduced NO_x and the remaining N in the fuel converted into N_2. The remaining secondary air is then added, so that complete combustion occurs in the second stage (Figure 17.12a). The second stage is therefore operating under fuel-lean conditions.

There are numerous benefits to TSC. By intentionally delaying the stoichiometric air injection, the first stage operates under conditions that minimize NO_x levels. That is, oxygen concentrations are low in the primary combustion zone, and temperatures are cooler because of partly combusted fuel. Therefore, much as in OSC, both fuel and thermal NO_x levels are minimized in the first stage. In the lean second stage, CO and hydrocarbons levels are reduced, yet some nitrogen is oxidized to NO_x. Regardless, 58% NO_x reduction can be achieved when 10% of the secondary air is added in the second stage.

A large percentage of coal-induced NO_x is associated with volatile matter. In TSC, the volatile matter, which is released early in the combustion process, will be burned in the fuel-rich first stage. Therefore, the volatile bound nitrogen compounds tend to form molecular nitrogen molecules instead of NO_x owing to reduced O_2 availability.

Increased Flame Luminosity: Increased flame luminosity (IFL) is a technique that lowers flame temperature by increasing radiative heat losses. This is done by selecting materials that have high luminosity and are to be located near the flame front. Ceramic rods can be inserted in a hot zone of gas forced flames that radiate heat at high temperatures. IFL can also be achived because of increased luminosity from soot (see Figure 17.13). For example, the sooting tendency of $C_2H_4 > C_3H_8 > CH_4 > CO/H_2$ in nonpremixed turbulent jet diffusion

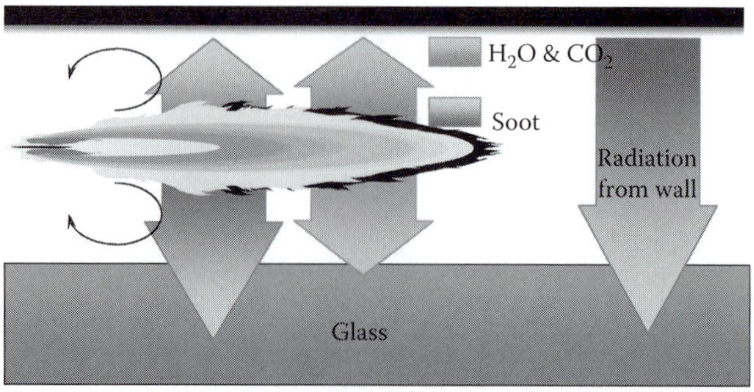

Figure 17.13 Flame luminosity and radiation loss for NO_x reduction in glass furnace. (Adapted from Baukal, 2000.)

flames and, hence, radiation loss fraction of $C_2H_4 > C_3H_8 > CH_4 > CO/H_2$ at the same N_2 percentage. When the N_2 percentage is increased, there are two effects: (1) the temperature decreases, and (2) the sooting tendency decreases, which will induce the heat loss fraction via radiation to decrease. For a CO/H_2 mixture there is no sooting tendency and, hence, the first effect is dominant and NO decreases (note that there is no fuel NO). However, for C_2H_4, the second effect is dominant, and hence NO increases. The interpretation is complicated by the presence of prompt NO for HC fuels (CH_4 to C_2H_4) and the mixing between air and fuel in nonpremixed turbulent jet flames.

Lean Premixing for NO_x Reduction for N-free Gaseous Fuels: Lean premixing with a large amount of excess air reduces flame temperature and hence results in low NO_x. However, there is serious system pressure oscillation: humming, hooting, screeching, buzzing, and flashback. The 7 kPa (1 psi) oscillation requires only 0.01% of chemical energy. Oscillations can originate from fuel:air oscillation. Oscillations can be up to 70 bar (1000 psi) in rockets, but only 7 kPa in gas turbines.

Mode of Firing: Another method that lowers combustion temperatures by increased radiative heat loss is tangential firing. This is a setup in which the burners are located at the four corners of the combustion unit and then fired tangentially. By firing in this fashion, the heat of combustion is dissipated very rapidly to the chamber walls, thereby lowering the temperature.

Low-NO_x Burners: Low-NO_x burners (LNBs) are those that incorporate all or a part of all the preceding controls: TSC, OSC, and FGR. LNBs essentially reduce O_2% and use overfire air to complete combustion. It is most widely used due to low-cost retrofitting; however, it is not suitable for cyclone burners. Regardless of the operating principle an LNB uses, there must be a compromise between minimum NO_x levels and flame stability, as well as the formation of other pollutants (CO and HC; see Figure 17.11).

Fuel Blending: Another cost-effective solution is cofiring, i.e., blending coal with renewable low-N biomass fuels and firing in existing coal-fired burners.

17.7.2 Postcombustion Exhaust Gas Treatment or Flue Gas Denitrification

This technique is used to eliminate NO_x in flue gas after the primary combustion zone. These techniques can be used independently or in conjunction with the combustion modifications if extremely low NO_x emissions are required. It should be noted that flue gas scrubbing, which is very effective in removing oxides of sulfur from the flue gases, is not an effective means of denitrification. This is because NO is insoluble in water; a similar problem exists with Hg removal (see Section 17.12).

17.7.2.1 Selective Non-Catalytic Reduction (SNCR)

The SNCR of NO process is based on the injection of additives into the combustion gases after the primary combustion zone. The most common additives are ammonia, urea [$(NH_2)_2$ CO], and cyanuric acid. In all three cases, the process requires the generation of OH, O, or H radicals, which control the reactions [Bowman, 1992]. If the temperature at which the ammonia or additive is injected is too high, then the ammonia may oxidize and produce additional NO instead of reducing it. Consequently, if the temperature is too low, then the ammonia may not completely react and may be emitted (called *ammonia slip*) in the exhaust gas. This temperature window (1100 to 1200 K) is critical to the correct operation of the process [Lyon, 1986].

In the case of SNCR, there are three ways to achieve low NO emission: (1) the $DeNO_x$ process using ammonia injection is more effective at low oxygen concentrations [reaction schemes in Jodal et. al., 1992]; (2) the rapid reduction of the NO_x ($RAPRENO_x$) process, also called the cyanuric acid process, is more effective at higher oxygen concentrations; (3) the NO_xOUT process with urea as an agent results in the production of both NH_3 and isocyanuric and, hence, results are in between the previously mentioned results. Figure 17.14 illustrates the reaction schemes for the SNCR process.

Thermal $DeNO_x$: Experimental studies at atmospheric pressure have shown that the thermal $DeNO_x$ process (patented by Exxon; Lyon, 1975, 1976) performs best at low levels of O_2 even though some O_2 (approximately 0.5%) is necessary for any significant NO_x reduction (Figure 17.15). Reductions of NO of at least 95% are possible for temperatures from 1100 to 1200 K with NH_3/NO molar ratios ranging from 0.5 to 3 for O_2 levels up to 5%. "Ammonia slip" can be a problem if oxygen levels are below 5%; consequently, the overall performance of the process seems to be best at around 5% O_2. However, the optimum level of O_2 content will be a function of temperature, additives present, ratio of NH_3 to NO, residence time at the ideal temperature range, and the initial amount of NO. Oxygen levels above 5% possibly inhibit the NO_x reduction because an increase in O causes the oxidation of NH_2 to NO, thereby shifting the selectivity of the process.

Controlling a final temperature of 1100 to 1200 K (i.e., temperature window of only about 100 K) self-sustains selective NO_x reduction below 50 ppm with

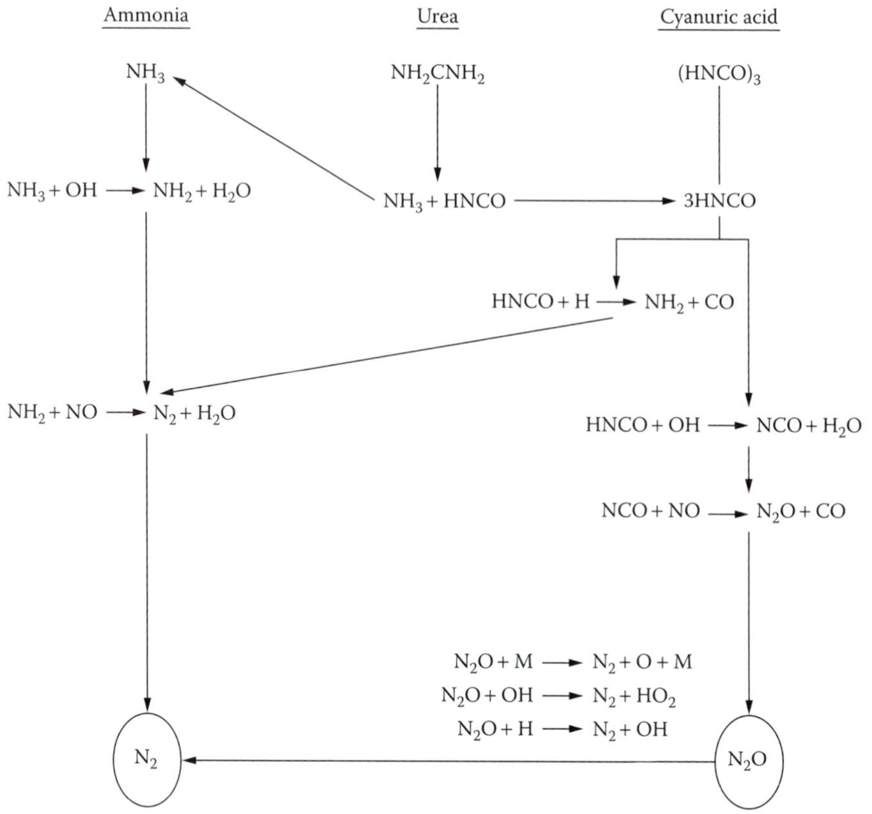

Figure 17.14 Reaction schemes for SNCR.

less than 2 ppm NH_3 slip, 50 ppm CO, and no residual hydrocarbons [Lyon and Hardy, 1986].

RAPRENO$_x$ and NO$_x$OUT methods tend to produce much more N_2O than the thermal DeNO$_x$ method.

To facilitate quick and inexpensive predictions with the thermal DeNO$_x$ method, two competitive reaction formulations have been used for modeling purposes [Rosenburg, 1980; Lyon and Benn, 1978]. One may also use an empirically based model developed by Duo [1992], which includes the following forward-direction-only competitive reactions:

$$4NH_3 + 4NO + O_2 \rightarrow 4N_2 + 6H_2O \text{ (fast)} \tag{A}$$

$$4NH_3 + 5O_2 \rightarrow 4NO + 6H_2O \text{ (slow)} \tag{B}$$

Reaction A depicts the second-order oxidation of NO to N_2. The first-order oxidation of NH_3 is modeled in Reaction B. The rate constants were expressed in Arrhenius form as follows:

$$k_A = 2.45 \times 10^{17} \exp(-244{,}400/(\bar{R}T)) \text{ m}^3\text{, kmole, sec, kJ}$$
$$k_B = 2.21 \times 10^{14} \exp(-317{,}000/(\bar{R}T),) \text{ m}^3\text{, kmole, sec, kJ}$$

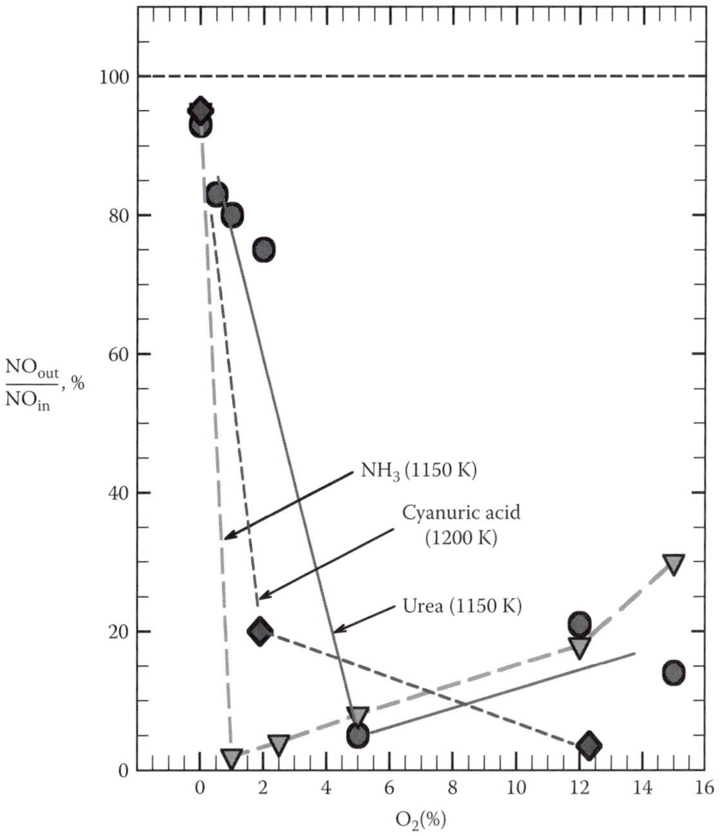

Figure 17.15 Nitric oxide reduction as a function of oxygen concentration with SNCR: urea, NH_3, cyanuric acid. (Adapted from Caton, J.A., Narney, J.K., II, Cariappa, C., and Laster, W.R., The selective non-catalytic reduction of nitric oxide using ammonia at up to 15% oxygen, *Can. J. Chem. Eng.*, Vol. 73, No. 3, pp. 345–350, 1995.)

It can be seen from the values of the two activation energies and the overall rate constants how the simple model was able to predict the temperature window for NO reduction. It is apparent that required $\{NH_3/NO\} = 1$ for fast and $NH_3/NO = 0.667$ for slow reactions.

Example 3

Show that for NH_3 concentration, 1 ppm = 0.75 mg/m^3. Let STP conditions be represented by 0°C, 1 bar.

Solution

STP volume: 0°C, 1bar: 22.71 m^3/kmol or 24.5 m^3/kmol at 25°C, 1 bar.

Conversion: 1 ppm NH_3 = volume fraction = 10^{-6} m^3 of NH_3/m^3 = 10^{-6} kmol of NH_3/kmol of flue gas = $17 \times 10^{-6} \times 10^6$ mg/22.71 m^3 of NH_3/m^3 = 0.75 mg NH_3/m^3.

■ ■ ■

Cyanuric Acid Process (RAPRENO$_x$ Process): In this process, known as the RAPRENO$_x$ process, solid cyanuric acid is injected into the hot exhaust gases. The solid cyanuric acid sublimates at about 650 K to gaseous isocyanic acid (HNCO). HNCO converts to NCO and NH_2 by the following reactions [Caton et. al.]

The higher oxygen levels enhance the cyanuric acid process. The nitric oxide removal increases rapidly as the oxygen concentration increases from 0 to 2%, and then increases more gradually as the oxygen concentration continues to rise from 2 to 12% (Figure 17.15).

The Urea Process (The NO$_x$OUT Process): NO$_x$OUT is the name of the process that uses urea. In this process, typically either solid urea or a urea solution (with water) is injected into the exhaust gases. The percentage of removal increases from 0 to about 90% as the oxygen concentration increases from 0 to 5%.

17.7.2.2 Selective Catalytic Reduction (SCR)

This technique is almost identical to ammonia injection. The major difference is that a catalyst is used along with NH_3, and NO_x reduction takes place, in a large part, on the surface of a catalyst (heterogeneous reactions), whereas ammonia reactions are entirely gas-phase reactions (homogeneous reactions).

The catalysts include metal oxide, iron, vanadium, chrome, manganese, cobalt, copper, and nickel and/or barium, which are supported on a carrier metal, typically aluminum. These catalytic systems mostly operate at low temperatures between about 500 and 750 K. The advantages of this form of reduction are the low temperatures, the easy control of the reaction, and the possible small size of the catalyst (which is required in a small-scale application such as a car). The obvious disadvantages are the high price, aggressive gases endangering the catalyst, and poisoning of sulfur (e.g., by coal products). The presence of S reduces the effectiveness of the SNCR process owing to the poisoning effect of SO_3 and the plugging effect. Further, NH_3 can combine with SO_2 and form ammonium bisulfate (NH_4HSO_4).

17.7.2.3 Reburn Methodology

In reburning, additional fuel (typically, natural or gas, about 15% of the total fuel) is injected downstream from the lean primary combustion zone to create a fuel-rich reburn zone in which NO_x is reduced through (reverse prompt NO_x) mechanism (reaction of NO_x with excess fuel to reduce NO_x to N_2, NH_3, and HCN). After the reburn zone, additional air is injected in the burnout zone to complete the combustion process. Typically, 60% reduction is achieved with natural gas as the reburn fuel. A diagram of the entire process with the different combustion zones is shown in Figure 17.16. This process is somewhat similar to air staging, in which the fuel is first burned in a rich primary zone to minimize the production of NO_x and, later, overfire air is injected to complete the combustion process [EERC, 1988]. When overfire air is used, NH_3 and HCN may produce additional NO and N_2; further, any unmixed NO may react with NO in

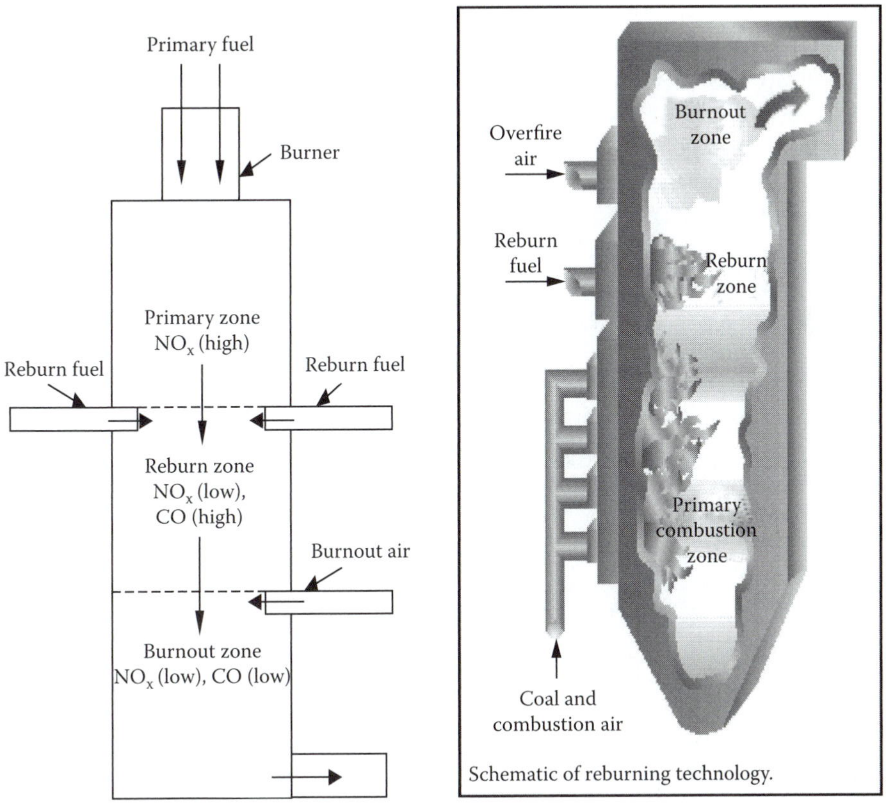

Figure 17.16 Reburn schematic: laboratory-style downward-fired furnace; industrial-type upward-fired furnace.

the overfire zone and form N_2. The primary zone (PZ) is the zone in which main fuel is burned with slightly excess air and NO_x is produced. The reburn zone (RZ) is the zone into which reburn fuel is supplied with deficient air so that the reburn air after mixing with main product gases produce the ϕ values of RZ from 1.05 to 1.2. The overfire zone (OZ) is the zone into which air is injected to complete combustion.

Typical reburn conditions: Main burner coal with N of 1% and natural gas as reburn fuel. Coal: 95% (by mass); 60 to 80% through 200 mesh or 60 μm. Primary air as coal carrier gas: 15 to 25% of combustion air; A:F 10; approximate flue gas flow: 13 to 15 kg per kg fuel or approximately 14 m^3 per kg fuel; temperature of PZ: 1670 K; overall excess air: 10 to 35%.

Reburn zone T: 1370 to 1590 K; 5% by mass or 0.05 kg per kg of main fuel; reburn air flow: 0.5 per kg of main fuel [Breen et. al., 2002]; A:F: 10, so that the flue gas mass flow is 0.55 kg per kg of main fuel; flue gas volume from reburn system: 0.55 m^3 per kg main fuel. So the reburn gas volume/main flue gas volume = 0.55/(14 + 0.55) = 0.03 or 3% total out of which the natural gas volume is 0.3% only; out of 1000 m^3 of flue gas, we have only 3 m^3 of natural gas. Hence, mixing is extemely important because the natural gas has to penetrate to a large zone to reduce NO_x.

Other reburn fuels include coal [Young et al., 1997], oil, agricultural biomass, slurry fuels, and animal manure [Ben and Annamalai, 2001; Rudiger et. al., 1986]. Typical reburn zone stoichiometry SR = 0.9 (ϕ = 1.1) or RZ equivalent ratio: 1.1 is shown as optimum. The more the primary zone excess air, the more the fuel that needs to be injected to create SR = 0.9. Typical reburn fuel: 15 to 25% of total fuel. The fuel-rich RZ requires a high temperature for reactions to proceed.(>1700 K or 2600°F). Sometimes, NH_3 is added with natural gas to amplify NO reduction.

The review of the literature has shown that the effectiveness of the reburn process depends on a number of variables:

Temperature: Under reducing conditions of the reburn zone, a higher temperature will result in better NO reduction.
Turbulence or reburn jet mixing: A high degree of mixing in the reburn zone is necessary for NO reduction.
Time (Residence): The longer the reburn zone residence time, the lower the NO_x emission. The residence times required for gaseous reburn fuels are shorter than those required for solid fuels. The first three variables are informally known as the three "T's" required for effective reburn.
Fuel type: The fuel type does have an effect on the NO levels.
Oxygen concentration: A lower oxygen concentration in the reburn zone will result in lower NO.
Reburn zone stoichiometry: This is the most important parameter. Experiments show that there is an optimum reburn SR, usually between 0.7 and 0.9. This does not seem to be a constant across all experiments, but changes with experimental conditions.

Table 17.8 lists the reduction percentage achieved with reburn technology.

The amounts of NO reduction for all of the fuels tested are presented in Figure 17.17 for coal and feedlot biomass (FB), also known as cattle manure. The results show that the level of NO reduction for feedlot biomass is greater

Table 17.8 Percentage of Reduction in NO_x: Demonstration pf or Operating Reburn Installations at Coal-Fired Boilers in the U.S.

#	*Type of Burner*	*% Reburn Heat in*	*% Reduction*	*NO_x with Reburn Lb/mm Btu*
1	**Gas Reburning**			
	Tangential	18	50–67	0.25
	Cyclone	20–23	58–60	0.39–0.56
	Wall without LNB	18	63	0.27
2	**Coal Reburn**			
	Cyclone (micronized)	30 (17)	52 (57)	0.39 (0.59)
	Tangential (micron) with LNB	14	28	0.25

Source: U.S. Department of Energy, 1999.

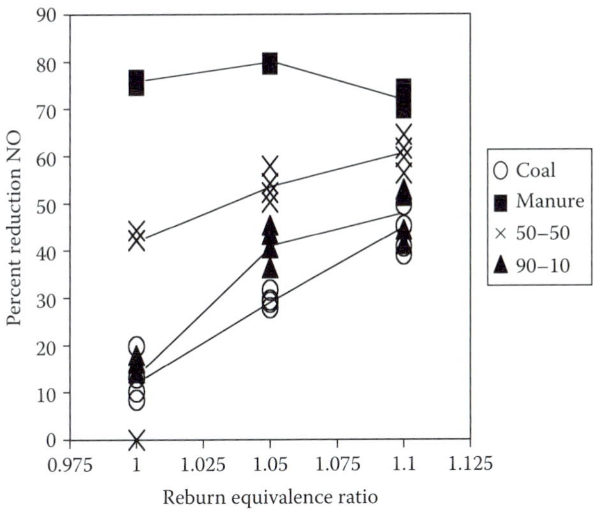

Figure 17.17 NO reduction with coal, biomass, coal:biomass blends. (Adapted from Thien, B. and Annamalai, K., National Combustion Conference, Oakland, CA, March 25–27, 2001.)

than coal at all reburn equivalence ratios. A reduction of approximately 80% was achieved for pure biomass, whereas coal experienced a reduction of between 10 to 40% depending on the equivalence ratio. It is believed that the greater effectiveness of the feedlot biomass is due to its greater volatile content on a dry-ash-free basis, and its release of fuel nitrogen in the form of NH_3 instead of HCN.

17.8 CO_2 Sequestration

Measurements from air bubbles in glaciers show that CO_2 was at 280 ppm before the Industrial Revolution, but by 2003 the reading showed $CO_2 > 360$ ppm. Thus, the CO_2 is expected to reach 500 ppm by 2050 and 700 ppm by 2100 [Winters, 2003]. CO_2 emission reduction can be achieved by (1) a reduction in energy consumption, (2) use of C-free fuels (e.g., H_2), (3) adopting alternate renewable energy sources, and (4) CO_2 sequestration. Out of these options, the renewable solar (wavelength 0.3 to 3 μm, visible range 0.40 to 0.76 μm) and wind energy options are insufficient, because the maximum power that could be generated per m^2 by wind and solar energy is on the order of 700 W, which is far less than the about 50 kW needed per m^2 (estimated from current CO_2 level in ppm, tracking it back to fuel burned and amount of heat generated). Hence, we need "carbo-nated" fuel for current energy needs. Carbon sequestration then becomes a necessity.

It is estimated that the carbon removal cost is \$100 per t of carbon or 7.5 c per kW h of power from coal plants or \$1 per gal of gasoline used. The target is \$20 per ton or \$0.20 per gal of gasoline. The Kyoto Treaty required a specified reduction of CO_2 from 1990 levels: 8% for the EU, 7% for the U.S., and 6% for Japan.

Another approach is to capture CO_2, bury it at the bottom of the ocean at a high pressure and keep it pinned in liquid form. The ocean absorbs 2 Gt of CO_2 per yr. Seams of CO_2 and CH_4 are naturally found in geological formations, trapped by clay or shale. Hence, the CO_2 can also be used to push out oil from reservoirs and flush out CH_4 from coal streams.

17.9 Carbon Monoxide: CO

About 84% of CO is from vehicles and 2% from stationary sources. The CO emission is governed by formation and destruction reactions. CO is the principal mode of HC combustion. CO formation from fuel appears to be rapid.

$$C_nH_m + (n/2)\ O_2 \rightarrow n\ CO + m/2\ H_2, \tag{17.55}$$

or

$$C_nH_m + (m/4 + n/2)\ O_2 \rightarrow n\ CO + m/2\ H_2O, \tag{17.56}$$

where CO oxidizes by

$$CO + OH \rightarrow CO_2 + H. \tag{17.57}$$

Neglecting backward reaction (valid in the initial period of CO oxidation),

$$d[CO]/dt = -k\ [CO][OH],\ \text{kmol/m}^3\text{sec}, \tag{17.58}$$

where [Bausch and Drysdale, 1974]

$$k = A\ T^B \exp(-C/T),\ A = 1.5 \times 10^4,\ B = 1.3,\ C = -385.0,\ \text{kmol, m}^3\text{, K} \tag{17.59}$$

The [OH] is calculated from the equilibrium assumption:

$$1/2H_2 + 1/2O_2 \Leftrightarrow OH,\ H_2O \Leftrightarrow H_2 + 1/2O_2, \tag{17.60}$$

(or use $O + H_2O \rightarrow OH + OH$, $O + O + M \rightarrow O_2 + M$, [Bartok and Sarofim, 1991]). Using Equation 17.60, one can solve [OH] and use in Equation 17.58. Simplifying

$$[OH] = K_{OH}{}^0\ [H_2]^{1/2}\ [O_2]^{1/2}, \tag{17.61a}$$

$$[H2] = K_{H2O}{}^0\ [1/RT]^{1/2}\ [H_2O]/[O_2]^{1/2} \tag{17.61b}$$

$$-d[CO]/dt = k_{glob}[CO][H_2O]\ [O_2]^{1/4}, \tag{17.62}$$

where

$$k_{glob} = k\ K_{OH}{}^{0} K_{H2O}{}^{0\ 1/2}. \tag{17.63}$$

The value of k_{glob} from this formulation is about 100 times less than the empirical rate expression because of equilibrium assumption.

Global expressions are valid only for the conditions under which they are obtained, e.g.,

$$-d[CO]/dt = 2.24 \times 10^{12} \exp(-4800/T)\ K_{OH}{}^{0}\ [CO]\ [H2]^{0.5}\ [O_2]^{0.25} + 5 \times 10^{8} \exp(-4800/T)\ [CO]^2. \tag{17.64}$$

17.10 SO_x Formation and Destruction

SO_x represents both SO_2 and SO_3; these react with water to produce sulfuric acid, which prevents the use of the SCR process for NO_x control. The single largest SO_2 emissions sector is utilities, which in 1985 released some 15.6 Mt of SO_2 into the atmosphere. This is approximately 70% of the 22.4 Mt total SO_2 emission by anthropogenic sources. In contrast, the transportation sector released 8.8 Mt of NO_x, or about 43% of the 20.5 Mt of total NO_x released in 1985 [Clement and Kagel, 1990]. Because coal is responsible for the largest quantity of SO_x formation, this section will concentrate on coal combustion. The production of SO_2 from coal combustion now reaches upward of 30 Mt a year, yet owing to increasingly effective control techniques the amount of SO_2 emissions is stabilizing but still at unacceptable levels.

Sulfur is released during devolatilization (under reducing conditions) in the form of H_2S, which is subsequently oxidized to SO_2. SO_2 is also formed during char burnout (under oxidizing conditions). Virtually all of the sulfur initially present in the fuel is oxidized during the combustion process. Most is emitted as gaseous SO_2 and SO_3, and some as aerosols, H_2SO_4, and inorganic sulfates in ash [Bowman, 1991]. Some is retained in the ash as inorganic sulfates. The proportion retained in the ash may be increased by adding limestone, dolomite, or other additives during the combustion process. Typical sulfur content of various fuels is given in Table 17.9.

17.10.1 Elements of SO_x Formation from Coal

Reducing conditions dominate the fuel-rich devolatilization pyrolysis process, and oxidizing conditions dominate the oxygen-rich char combustion process. The major point of sulfur release occurs during the devolatilization step. After devolatilization and during char combustion, very little sulfur is released until near the end of the char burnout, when a secondary point of sulfur release occurs. The sulfur so released is mostly in the form of H_2S under reducing

Table 17.9 Typical Sulfur Content of Various Fuels

Gasoline	220 ppm
Jet A	0.3
JP4	0.4
Kerosene #1	0.12
Premium #2	0.3
#2 heating oil	0.4
Railroad diesel	0.5
Marine diesel	1.2
Residual oil	2.2
Heavy residual oil	0.5–4
Coals	1.1–7.1

Note: Dry weight percentage except where noted.

conditions, or it may be oxidized under favorable conditions. Devolatilization reactions include [Moffat, 1982]:

$$FeS_2 \text{ \{red\}} \rightarrow FeS_x + H_2S \tag{17.65}$$

$$\text{R-S-H \{red\}} \rightarrow H_2S + \ldots \tag{17.66}$$

$$\text{Cyclic sulfur or organic sulfur, \{red\}} \rightarrow H_2S + \ldots \tag{17.67}$$

Under reducing {red} conditions, the following additional reactions occur:

$$Fe_3O_4 + H_2S \rightarrow FeS \tag{17.68}$$

$$Fe + H_2S \rightarrow FeS \tag{17.69}$$

The H_2S also reacts with some of the mineral components of the ash:

$$CaCO_3 + H_2S \rightarrow CaS + H_2O + CO_2 \tag{17.70}$$

$$CaO + H_2S \rightarrow CaS + H_2O \tag{17.71}$$

Oxidizing [oxy] reactions include [Moffat, 1982]:

$$H_2S \text{ \{Oxy\}} \rightarrow SO_2 + H_2O \tag{17.72}$$

$$CS_2 \text{ \{Oxy\}} \rightarrow 2SO_2 + CO_2 \tag{17.73}$$

$$C \text{ \{Oxy\}} \rightarrow CO + [Oxy] \rightarrow CO_2 \tag{17.74}$$

$$CaS \text{ \{Oxy\}} \rightarrow 1/4\ CaS + 3/4\ CaSO_4 + O_2 \rightarrow CaSO_4 \tag{17.75}$$

$$CaS \text{ \{Oxy\}} \rightarrow 1/4\ CaS + 3/4\ CaSO_4 \rightarrow SO_2 + CaO \tag{17.76}$$

$$\text{M S \{Oxy\}} \rightarrow \text{M O} + SO_2 \tag{17.77}$$

$$\text{FeSx \{Oxy\}} \rightarrow xSO_2 + \text{iron oxides} \tag{17.78}$$

$$\text{FeS \{Oxy\}} \rightarrow SO_2 + SO_3 + \text{iron oxides} \tag{17.79}$$

where M is a metal species.

SO_2 formation in the flame is rapid, occurring on a timescale comparable to that of fuel oxidation reactions. The rapid progress of the SO_2 formation process is due principally to rapid biomolecular reactions between sulfur species and OH radicals [Bowman, 1991], including:

$$SO + OH \rightarrow SO_2 + H \tag{17.80}$$

$$H_2O + H \rightarrow OH + H_2 \tag{17.81}$$

$$SO + H_2O \rightarrow SO_2 + H_2 \tag{17.82}$$

The principal reaction responsible for SO_3 formation in flames is:

$$SO_2 + O + M \rightarrow SO_3 + M \tag{17.83}$$

This reaction proceeds rapidly near the flame zone (e.g., volatile flame), where oxygen atom concentration is high because of the high temperature. Some of the SO_3 is subsequently converted to SO_2 by the following reactions:

$$SO_3 + O \rightarrow SO_2 + O_2 \tag{17.84}$$

$$SO_3 + H \rightarrow SO_2 + OH \tag{17.85}$$

Whereas the formation of SO_3 is rapid near the flame zone, the conversion of SO_3 to SO_2 is relatively slow. The levels of SO_3 in the quenched combustion gases are determined by the quenching process. Because the rate constant for the SO_3 formation reaction is weakly dependent on temperature, the quenching process is controlled by reduction of the oxygen atom concentration (which is temperature dependent).

17.10.1.1 Simplified Schemes

The overall chemical reaction is,

$$S + O_2 \rightarrow SO_2. \tag{17.86}$$

At room temperature, SO_2 is a stable, colorless gas. In the presence of water, SO_2 forms H_2SO_4 (sulfuric acid),

$$SO_2 + H_2O + 1/2\ O_2 \rightarrow H_2SO_4. \tag{17.87}$$

The water needed for this reaction is readily available in the atmosphere in the form of rain. As a result, the rainwater becomes more acid and results in the familiar acid rain. If water is not present, SO_2 will form SO_3 (sulfur trioxide) in the following way:

$$SO_2 + 1/2\ O_2 \rightarrow SO_3 \tag{17.88}$$

In the presence of water, SO_3 will also form sulfuric acid and additional acids, which can create problems in the scrubber:

$$SO_3\ (g) + H_2O \rightarrow H_2SO_4\ (aq) \tag{17.89}$$

17.10.2 SO_x Reduction Methods

Coals with sulfur less than 0.6% do not require scrubbers (e.g., Wyoming low-sulfur coal). Reduction of sulfur emissions can be achieved through the use of one or more of the following methods:

1. Flue gas desulfurization systems
2. Switching to low-sulfur coals
3. Fuel desulfurization methods

Flue Gas Desulfurization in Fluidized Beds: Several researchers have studied the parameters that influence the behavior of SO_2 abatement by chemical reactions. An often-studied means to accomplish this with the combustion of coal in a fluidized bed combustor (FBC) is with the addition of a calcium source, usually limestone or dolomite particles to sulfonate SO_2 to $CaSO_4$. The most important variables affecting the removal of SO_2 are the following: (1) Ca:S molar ratio fed to the fluidized bed, (2) bed temperature, (3) sorbent type, (4) sorbent particle size, and (5) O_2 concentration in the bed [Henttonen, et al. 1992].

The chemical adsorbtion of SO_2 by limestone occurs in two steps during the FBC of coal–limestone mixtures. The first step is the calcination reaction, which occurs around 900 C:

$$CaCO_3 \rightarrow CaO(s) + CO_2,\ \Delta H = 183050\ kJ \tag{17.90}$$

$$CaO(s) + SO_2 + 1/2\ O_2 \rightarrow CaSO_4 \tag{17.91}$$

Calcium sulfate, $CaSO_4$, is the main product of SO_2 sorbency in FBC. Suppose one injects 1 kmol of $CaSO_4$ (Ca = 1, S = 1, O = 4). The products are found to be CaO(s), SO_2, 1/2 O_2, and $CaSO_4$(s). We have three atom balance equations. We need the following equilibrium relation to determine the composition:

$$K^0 = 1/(p_{SO2}/1)\ (p_{O2}/1)^{1/2} \tag{17.92}$$

One can determine the extent of decomposition of $CaSO_4$ as a function of temperature. $CaSO_4$ does not decompose at room temperature, but decomposes completely after about 1500 K; thus, although sulfur is tied as inorganic, it can still decompose and produce SO_2.

The ideal temperature window for the most favorable sulphanation rates has been identified as 1073 K to 1473 K [Lisauskas et al., 1985].

A common parameter for investigation is how the mol ratio of Ca added in the limestone to the sulfur in fuel, which is referred to as the Ca:S mol ratio, affects SO_2 capture. Wang et al. [1991] investigated the combustion characteristics of coal with mol ratios of from 1 to 6. A linear relationship with the Ca:S mol ratio and the SO_2 reduction was noticed. Nearly linear trends in SO_2 reduction with increasing Ca:S mol ratio was seen in work by Lisauskas et al. [1985].

Following experiments with 23 and 99 μm size particles, it was concluded that the finer particles reduced SO_2 emissions to 10% of the original value of total sulfur vs. 40 to 65% for the larger size [Hippinen et al., 1992]. Particle sizes of 2 to 16 μm are stated to be best for sulphanation by another team [Badin et al., 1985]. The effect is attributed to the larger surface area per unit mass of sorbent for reactions with smaller particles.

Flue Gas Desulfurization (FGD) in Boilers: When powdered limestone ($CaCO_3$) is fired into the combustion chamber, the limestone $CaCO_3$ (s) decomposes into lime CaO (s), which then reacts with SO_2 to form sulfite:

$$CaCO_3\ (s) \rightarrow CaO(s)\ (lime) + CO_2 \tag{17.93}$$

$$CaO(s) + SO_2(g) \rightarrow CaSO_3(g),\ \text{calcium sulfite} \tag{17.94}$$

The process could be either wet as in boilers or dry as in fluidized beds. The flue gas is drawn into the spray tower by an induced draft fan, where it mixes with the limestone slurry. The flue gas enters the stack and is emitted to the atmosphere. The sprayed limestone slurry collects in the bottom of the tower and is recirculated back into the spray tower with the appropriate addition of fresh limestone from the slurry tank. The sulfite is used as landfill. The flue gas is scrubbed with a slurry that contains lime (CaO) and limestone ($CaCO_3$), as well as the salts of calcium sulfite ($CaSO_3$) and calcium sulfate ($CaSO_4$). The SO_2 in the flue gas reacts with the slurry to form additional salts, which are later recycled with the addition of fresh lime or limestone. The chemical reactions are thought to be:

$$CaO + H_2O \rightarrow Ca(OH)_2 \tag{17.95}$$

$$Ca(OH)_2 + CO_2 \rightarrow CaCO_3 + H_2O \tag{17.96}$$

$$CaCO_3 + CO_2 + H_2O \rightarrow Ca(HCO_3)_2 \tag{17.97}$$

$$Ca(HCO_3)_2 + SO_2 + H_2O \rightarrow CaSO_3 \cdot 2H_2O + 2CO_2 \tag{17.98}$$

$$CaSO_3 \cdot 2H_2O + 1/2\ O_2 \rightarrow CaSO_4 \cdot 2H_2O \tag{17.99}$$

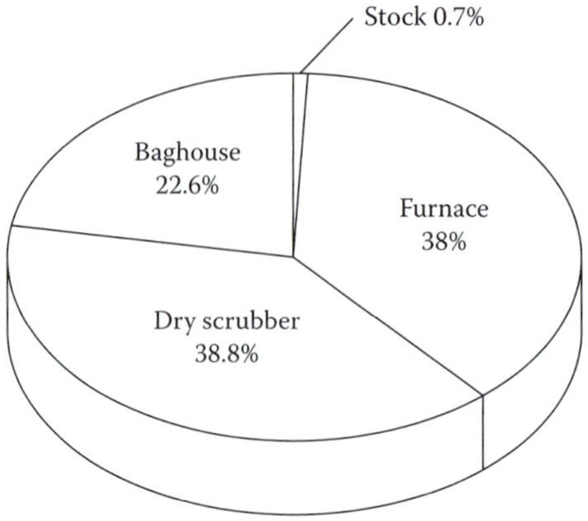

Figure 17.18 SO_x% capture and emission in power plants. (Adapted from EPA.)

where $CaSO_3 \cdot 2H_2O$, called calcium sulfite dehydrate, is precipiated from a retention tank and sold as landfill.

The wet scrubber has the advantage of high SO_2 removal efficiencies, good reliability, and low flue-gas energy requirements. One of the main disadvantages is the waterlogged sludge waste material, which poses difficult and expensive disposal problems.

Figure 17.18 shows the percentage of removal of SO_2 when coal is combusted. The in-furnace removal is due to capture of SO_2 by coal ash.

Fuel Desulfurization (FDS): FDS methods include conventional coal washing. Coal washing or cleaning is a relatively simple and cost-effective method of removing 30 to 50% of the sulfur contained in coal. Coal cleaning is only effective for removing the pyritic sulfur in coal because it exists as distinct particles, contrary to organic sulfur, which is part of the coal matrix. Pyrite is approximately three times as dense as the organic fraction of coal. This density difference forms the basis for coal cleaning. Typically, coal cleaning begins by crushing the coal to a fine size; the finer the size, the greater the amount of sulfur that is removed. Essentially, the pulverized coal is immersed in liquid baths, typically water, with the clean coal rising to the top owing to the density variation. The more dense residue sinks to the bottom and is discarded. Generally, physical coal-cleaning methods are not sufficient to reduce SO_2 emissions to within the required guidelines; therefore, it is also necessary to employ other techniques.

17.11 Soot

Soot mainly consists of carbon and PAHs (polycylic aromatic hydrocarbons) with approximate C:H = 8 (C: 95 to 99.5%, H: 0.2 to 1.3%, O: 0.2 to 0.5%, N: 0 to 0.7%, S: 0.1 to 1%, ash < 1%). It appears as finely divided carbon particles in flames and cause yellow-to-white luminous flames. Soot is believed to cause cancer. If local O:C < 1 in a flame, soot is formed at T > 1000 K. According to overall stoichiometry, if the O:C atom ratio > 1 (e.g., $C + 1/2\ O_2 \rightarrow CO$), then no solid carbon is present. Experiments indicate that the soot can be formed at O:C ≈ 1.7 to 2.5. For C_2H_2, soot is readily formed at O/C = 1.2. Thus, rich flames and diffusion flames produce soot, which consists of fine particles primarily formed in diffusion type of flames (see Chapter 12). Premixed flames are less likely to form soot owing to the O_2 availability. Higher pressure favors soot formation. The experimental data on sooting equivalence ratios (i.e., φ at the time of sooting) of premixed flames involving nonalcohol fuels such as toluene (lowest φ), *n*-octane, isodecane, benzene, *n*-hexane, isoioctane, propane, ethane, propylene, ethylene, and acetylene (highest φ) indicate almost a direct correlation with gaseous fuel mole fractions, as shown in Figure 17.19. Soot strongly absorbs radiation and is also a good radiator (e.g., candle flames). The initial particle size range is from 1 to 2 nm; collision of rings causes coagulation and clustering together similar to a chain, making the soot grow to agglomerates with sizes ranging from 100 to 1000 nm. The agglomerates can also be formed from 10 to 40 nm "primary spheres" with a density of less than 2000 kg/m³ [Haynes, 1991]. The specific surface area is about 40 to 140 m²/g with ρ = 1800 kg/m³ and d = 20 to 50 nm diameter. As HC reacts with the particles, the growth rate of soot could be of the order of 5×10^3 kg/m² sec [Borman and Ragland, 1998]. Soot also catalytically oxidizes NO and SO_2 in the atmosphere to NO_2, SO_4^2. Soot is characterized by soot volume fraction, f (soot volume/geometrical volume, $\approx 10^{-7}$); diameter, d; and particle number density, n. The d increases due to surface growth (at constant n) or by coagulation (changing n) or decreases due to soot oxidation.

The amount of soot depends on the rate of formation and destruction. At inception, the heavier HC is transformed into lighter HC. The process is as

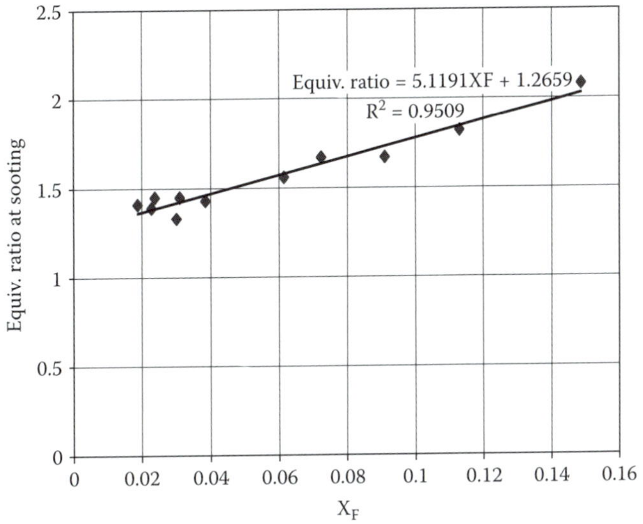

Figure 17.19 Sooting equivalence ratio replotted with fuel mole fraction in mixture:premixed flames. (From Street, J.C. and Thomas A., Carbon formation in premixed flames, *Fuel*, 34, 4–36, 1955; replot by authors.)

follows: production of the precursor species C_2H_2 to larger PAH (polycyclic aromatic hydrocarbon); $C_6H_6 + H \rightarrow C_6H_5 + H_2$ (abstraction of H, making carbon heavy in C_6H_5); addition of C_2H_2 to AR (aromatic ring); and use of C_2H_2 to form connected aromatic rings (similar to benzene rings), dimer (two rings), trimer (three rings), etc.

The higher the particulate content (which includes ash and soot), the lower is NO_x because soot reacts with NO_x and produces CO. In diesel engines, about 1/10th to 1/20th of soot produced escapes with exhaust.

Soot oxidizes with OH, O, O_2 Nagle–Strickland 1273 to 2273 K [Borman and Ragland, 1998]. The Nagle-Strickland [Neoh et al., 1981] kinetics obtained for pyrolytic graphite in O_2, T:1273 to 2273 K, have also been validated at T as high as 4000 K and are given as follows:

$$\dot{w}'' = \frac{12 * k_A\ p_{O2}}{(1+k_B\ p_{O2})\{1+\frac{k_D}{k_C\ p_{O2}}\}} + \frac{12 * k_C\ p_{O2}\ \frac{k_D}{k_C\ p_{O2}}}{\{1+\frac{k_D}{k_C\ p_{O2}}\}}$$

w'', carbon oxidation rate in g/cm² sec
$k_A = 20\ \exp(-15{,}100/T)$, T in K, g atm/cm² sec atm
$k_B = 21.3\ \exp(2{,}065/T)$, T in K, 1/atm
$k_C = 4.46 \times 10^{-3}\ \exp(-7{,}650/T)$, T in K, g/cm² sec
$k_D = 1.51 \times 10^{-5}\ \exp(48{,}820/T)$, T in K, g atm/cm² sec
p_{O2}, partial pressure in atm

In diffusion flames such as jet flames, carbon is formed only if the height exceeds a certain value. Higher HC decomposes to lower HC; soot is formed in CH_4 owing to decomposition at high temperature. Hence, it may also be quickly consumed because of oxidation. On the other hand, C_2H_2 decomposes at a lower temperature, but soot will be consumed slowly as a result. Thus, soot will accumulate for C_2H_2, and soot concentration will be higher. Soot is oxidized at the tip of diffusion flames.

17.12 Mercury Emissions

17.12.1 Mercury Sources

Fossil fuels (e.g., coal, oil, and natural gas), vegetation, and waste products contain Hg in trace amounts. Most coals typically contain 0.1 μg of Hg per g of coal, whereas bituminous coals contain 160 μg of Cl per g of coal. The existence of Hg in elemental or oxidized form depends on the Cl content of coal. During combustion, the Hg in coal is vaporized as elemental mercury, yielding vapor concentrations in the range of 1 to 20 ng/m³ (1 to 20 ppbw, parts per billion weight or mass basis) in the combustion zone. It is carried by air currents to many parts of the globe. While worldwide emission of Hg is about 4900 tons, the total annual emissions of mercury (Hg) in the U.S. were estimated to be 159 t [Keating et al., 1997] per yr including combustion sources, which account for about 85% of the total emissions. Because coal-fired utilities (320 GW capacity) emit about 50 t of mercury per year [Senior et al., 2000], the ambient levels have climbed up to 1.6 ng/m^3 of air as compared to preindustrial levels of 0.5 to 0.8 ng/m^3. The increased emissions result in Hg contamination of aqueous systems, along with increased levels of Hg in fish tissues (above the safe limit of 1 ppbw). Hence, there is an option on placing an Hg emission cap of 0.03 t per Mt of coal. The EPA plans to reduce U.S. emissions to 15 tons per year beginning in 2018. Thus the Hg emission compliance requires 90% of plants in the U.S. to install some emission controls to reduce Hg emissions.

17.12.2 Mercury Forms and Effect of Cl

Mercury in the environment can exist in three forms: elemental mercury (Hg^0), inorganic mercury compounds (Hg+) (e.g., mercuric chloride, $HgCl_2$), and organic mercury (Hg^2+) compounds (e.g., methyl-mercury, when it is produced from inorganic Hg through bacterial action in water). *Speciation* is a term used to describe the amounts of elemental mercury, oxidized mercury, and mercury attached to particulate in flue gas. Elemental Hg is unreactive and only slightly soluble in water; wet scrubbing cannot remove elemental Hg, whereas the oxidized forms of mercury, such as mercuric chloride, are soluble in water and can be captured. Hg vapor is believed to undergo oxidation reaction to $HgCl_2$ under kinetics control through reaction with atomic chlorine [Senior et al., 2000a, b; Helble et al., 2000; Sliger et al., 2000; Widmer et al., 2000]. The chlorine chemistry can be both heterogeneous and homogeneous. Further, $HgCl_2$ is formed when the stack gases with Hg (g) cool down in the presence of chlorine. Some fly ash has been observed to oxidize elemental mercury in both laboratory-scale apparatus [Norton et al., 2000; Ghorishi et al., 2000] and pilot-scale baghouses. Hg speciation in gases is affected by coal chlorine content and ash composition, gas temperature and, finally, the residence time; further, a fraction of the mercury is also adsorbed on the particulate matter. Activated carbon injection in the baghouse has been shown to remove Hg by almost 80 to 95%. But disposal cost may increase, because the ash contains higher S and more Hg, thus reducing its potential for secondary products such as cement and road and wallboard materials, etc. Coal cleaning for S removal has also resulted in about 80% of Hg removal for Eastern and Midwestern bituminous coal because Hg exists as $HgCl_2$, but in only 10 to 15% removal for Powder River Basin coal because of low Cl content. The Hg from sub-bituminous coal is more difficult to control because it exists mostly in elemental form. About 70% of utility boilers are equipped with ESPs, 70% for capturing fly ash emissions; however, they have limited ability to remove Hg (only 24%). This is because Hg exists in vapor form in flue gas and so can easily pass through ESPs.

The Cl content in bituminous coals range from 200 to 2000 ppm (dry-mass basis), whereas for low-rank coals (subbituminous and lignite), it ranges from 20 to 200 ppm (an order of magnitude lower). The presence of acid gases (HCl, SO_2, NO, and NO_2) in the flue gas has also been shown to cause Hg oxidation in the presence of fly ash [Carey et al., 1998; Miller et al., 1998]. Hg removal plotted against coal chlorine content [Afonso, 2001] reveals increasing capture with Cl because of HgCl formation. Further, the lesser the Cl (less than about 150 ppm on a dry basis), the more the elemental Hg. When Cl exceeded 200 ppm, Hg is captured by the particulate phase.

The literature review shows that (1) the speciation of Hg at stack depends on the chlorine content of the coal and chlorine chemistry via heterogeneous or homogeneous mechanisms; (2) coal cleaning has an average mercury removal efficiency of only 21%; further, whereas 77% of Eastern and Mid-western U.S. bituminous coal is cleaned, only 10 to 15% of Powder River Basin coal is cleaned and lignite is not cleaned; and (3) Hg reduction methods include injection of hypochlorite solution or direct injection of Cl gas into the scrubber (or in the gaseous feed) to promote Hg capture.

Typically, Hg is concentration is 1 to 20 ng/m^3 in the gas phase. According to equilibrium calculations, Hg exists as vapor at T > 700 C and as $HgCl_2$ for T < 450 C; HCl is 1 to 130 ppm in flue gas depending upon Cl content. The bituminous coals have high Cl, 500 to 1300 ppm, whereas lignite and Wyoming subbituminous have low Cl (20 to 60 ppm).

It is believed that apart from Cl, low-NO_x devices result in more unburned C, which is believed to adsorb mercury for both bituminous and subbituminous coals [DeVito et al., 1998; Butz et al., 1999]. Operation from stoichiometric to substoichiometric may result in more capture of Hg with increasing SR.

17.12.3 Determination of Hg

Mercury in flue gas generally exists as Hg^0 and Hg^2+. Currently, impingers are available to selectively dissolve Hg^0 and Hg^2+ [Linak et al., 2001]. The relative solubilities of Hg^0 and $HgCl_2$ are 1 and 1,400,000, respectively [Wilhelm, 1999]. Because $HgCl_2$ is more soluble, it is dissolved by first passing the flue gas through a 1-*M* NaOH solution. Hg^0 is then captured by passing it through a strong oxidizing solution of acidified $KMnO_4$ (10% H_2SO_4:4% $KMnO_4$). Once the mercury gases are dissolved, mercury is digested, and then its level is measured in the elemental form by the cold-vapor atomic absorption (CVAA) method. The Hg reported from each of the impingers is reported as Hg^0 and Hg^2.

17.12.4 Reactions with Hg

The N release from solid fuel during pyrolysis is presumed to occur as HCN; similarly the Cl evolution is assumed to occur as HCl. For simulating the oxidation of Hg$^\circ$ to $HgCl_2$ in the reburn zone, a multistep reaction scheme could be used, and Cl, Cl_2, HCl, Hg$^\circ$, and $HgCl_2$ can be treated as trace species. It has been shown in the literature [Kramlich et al., 1998] that the Cl atom concentration is higher at higher temperatures, which provides Cl atoms at superequilibrium concentration (owing to slow recombination reaction) at lower temperature for Hg oxidation. The reaction mechanism of Hg^0 oxidation is given in Table 17.10 [Niksa et al., 2001]. In the reaction mechanism shown in Table 17.2, the rate constants for these reactions are of the form $k = AT^n \exp(-E/RT)$.Characteristic reaction timescales (≈1/k) indicate that Reaction 5

Table 17.10 Hg^0 Reaction Mechanism

	Reaction	A	n	E kcal/mol
1	$Hg^0 + Cl_2 \leftrightarrow HgCl + Cl$	1.39×10^{14} cm^3/mol sec	0	34.0
2	$Hg^0 + Cl + M \leftrightarrow HgCl + M$	9.00×10^{15} cm^6/mol^2 sec	0.5	0
3	$HgCl + Cl_2 \leftrightarrow HgCl_2 + Cl$	1.39×10^{14} cm^3/mol sec	0	1
4	$HgCl + Cl + M \leftrightarrow HgCl_2 + M$	1.16×10^{15} cm^6/mol^2 sec	0.5	0
5	$Hg^0 + HCl \leftrightarrow HgCl + H$	4.94×10^{14} cm^3/mol sec	0	79.3
6	$HgCl + HCl \leftrightarrow HgCl_2 + H$	4.64×10^{3} cm^3/mol sec	2.5	19.1

can be neglected because of a very high-energy barrier and, hence, large timescale [Sliger et al., 2000], and Reaction 6 can be neglected because of a very low preexponential factor. Therefore, only Reaction 1 to Reaction 4, typically, are included in the formulation.

17.13 Summary

An overview of the types of pollutants, including NO, SO_x, particulate matter, and Hg, and their formation and control methods are briefly presented. Various emission-reporting methods were summarized. The fuel, thermal, and prompt NO_x mechanisms and the governing relations were given.

Chapter 18

An Introduction to Turbulent Combustion

18.1 Introduction

The earlier chapters dealt with combustion in a quiescent atmosphere and laminar-flow systems. There the transport properties provide a basic understanding of combustion and essential parameters governing combustion behavior, particularly at the molecular level. In the examination of laminar flows, several assumptions and approximations are made that greatly reduce the complexity of the solution and yield very simple general solutions. However, most practical systems deal with combustion under turbulent conditions, in which an aggregate mass consisting of a large number of molecules is randomly transported just as molecules are randomly transported. For example, the aggregate transfer of momentum (rather than molecules themselves being transported across) results in rapid transfer of momentum compared to molecular transport. If "turbulent" viscosity, which is not a property, has its value defined, it is much larger than molecular viscosity. Hence, transport of momentum species and energy are affected by turbulence. Turbulence plays a major role in pollutant formation and destruction. The combustion process in most practical devices, such as jet engines and industrial power plants, involves turbulence. However, turbulent flows do not lend themselves easily to analysis.

18.2 Turbulence Characteristics

A few characteristics of turbulence are summarized below [Chigier, 1981; Kuo, 1986]:

- Properties, including flame profiles, fluctuate with time at any given location x and, hence, flame thickness is increased.
- Typically, fluctuations occur at large Re.

- Fluctuations are random and irregular.
- Turbulence is rotational, accompanied by fluctuating vorticity.
- The fluctuations possess turbulent kinetic energy, k.
- The turbulent kinetic energy (k) is dissipated by viscous stresses and, hence, k is converted into thermal energy; thus, sustaining k requires a supply of energy from the main flows.
- Because transport occurs through bulk mass migration, the turbulent diffusivity (ε_T) is much greater (almost 100 times and increasing with Re) than the molecular diffusivity in laminar flows. This results in increased mixing and, consequently, increased momentum, heat, and mass transfer rates.
- Turbulent diffusivity is not a property of the fluid but is dependent on the type of flow.
- Isotropic turbulence occurs if fluctuating velocities are of similar magnitudes in all three directions.

For nonreacting flows, these characteristics impose a sizable obstacle to the solution of the problem. When a combusting flow is considered, several more effects of turbulence are observed. The fluctuations result in a reaction zone that is rather thick in comparison with the thin flame of the laminar case. Changes are also noticed in the flame speed; the increased transport properties for turbulent flow cause much greater flame speed as well as shorter flame heights in jets.

To solve the problem by starting from first principles, some simplifications must be made. Because turbulent flow involves almost instantaneous random variation in properties, techniques are needed to create an average value that is much easier to deal with. Two such averaging techniques are commonly used: time averaging or Reynolds averaging and mass averaging or Favre averaging.

18.3 Averaging Techniques

The first averaging technique is conventional time averaging or Reynolds averaging. This involves separating the quantities into a mean and a fluctuating component. For a property q, the mean value $\overline{q(x_i)}$ at location x_i in a turbulent flow field is given by (Figure 18.1):

$$\overline{q(x_i)} = \lim_{\Delta t \to \infty} \frac{1}{\Delta t} \int_{t_0}^{t_0+\Delta t} q(x_i, t) dt \tag{18.1}$$

where q could be u, h, T, h_{tot}, etc., ($h_{tot} = h + v^2/2$), Δt is interpreted as a time interval much longer than fluctuation time interval. If we let $\Delta t \to \infty$, then instantaneous turbulent transport cannot be derived. That may cause inconsistency in certain equations. Because turbulence causes fluctuations in various physical and chemical properties, any property q can be expressed as the sum of a mean component and a fluctuating component q using Reynolds averaging.

$$q(x_i, t) = \overline{q(x_i)} + q'(x_i, t) \tag{18.2}$$

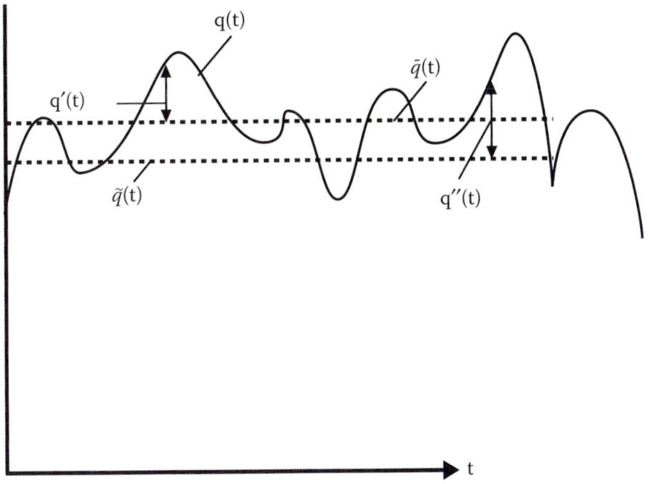

Figure 18.1 Definitions of simple time- and mass-weighted averages.

Equation 18.2 is called the *Reynolds decomposition rule* (Figure 18.1). Note that because the mean value is the average over time, the average of its fluctuating component $\overline{q'(x_i)}$ must be zero. The governing equations can be rewritten by replacing the laminar properties (e.g., v, Y_k, h_t, etc.) by the instantaneous value q_I. At a first glance, time averaging appears to be the most desirable way to handle property variations; however, a lump of mass of density $\rho(x_i,t)$ and property, say, $q(x_i,t) = v(x_i,t)$, will impart a momentum of $\rho(x_i,t)\ v(x_i,t)$ on a pitot tube measuring the pressure. Similarly, an extremely sensitive thermocouple measuring temperature will realize an energy gain of $\rho(x_i,t)\ c_p\ T(x_i,t)$. Thus, for a pitot tube, the average momentum over a period is $\overline{\rho(x_i,t)v(x_i,t)}$, i.e., most sampling probes yield values based more on a mass-weighted basis than on a time basis. Therefore, converting to a mass-weighted set of equations is beneficial.

For compressible flows or flows with variable density, the second technique called Favre averaging (or mass-weighted averaging) is usually employed for an arbitrary property. Mass-weighted averaging or Favre averaging is similar to Reynolds averaging in that the properties are represented by a mean and a fluctuating component. The Favre mean for a quantity q is given by

$$\tilde{q}(x_i) = \frac{\overline{\rho(x_i,t)q(x_i,t)}}{\bar{\rho}(x_i)} \tag{18.3}$$

where

$$\overline{\rho(x_i,t)q(x_i,t)} = \lim_{\Delta t\to\infty}\frac{1}{\Delta t}\int_{t_0}^{t_0+\Delta t}\rho(x_i,t)q(x_i,t)dt.$$

Adding the fluctuation term q (x_I,t) with respect to mass average yields the instantaneous property

$$q(x_i,t) \equiv \tilde{q}(x_i) + q''(x_i,t) \tag{18.4a}$$

where q = u, h, T, h_{tot} (= h + ke), etc. (Figure 18.1).

Multiply $q(x_i,t)$ by $\bar{\rho}$, then expand in terms of Reynolds- and Favre-averaged quantities

$$\bar{\rho}(x_i)\, q(x_i,t) = \bar{\rho}(x_i)\{\bar{q}(x_i) + q'(x_i,t)\} = \bar{\rho}(x_i)\{\tilde{q}(x_i) + q''(x_i,t)\}$$

and conduct time averaging

$$\overline{\bar{\rho}(x_i)\, q(x_i,t)} = \bar{\rho}(x_i)\, \bar{q}(x_i) = \overline{\bar{\rho}(x_i)}\, \tilde{q}(x_i) + \bar{\rho}(x_i)\overline{q''(x_i,t)}.$$

Due to the definition of mass-weighted average,

$$\overline{\rho(x_i,t)\, q''(x_i,t)} = \left[\lim_{\Delta t\to\infty}\frac{1}{\Delta t}\int_{t_0}^{t_0+\Delta t}\rho(x_i,t)\, q''(x_i,t)dt\right] = 0. \tag{18.4b}$$

So

$$\overline{\rho(x_i,t)q''(x_i,t)} = 0, \text{ Favre averaging} \tag{18.5a}$$

$$\overline{q'(x_i,t)} = 0, \text{ Reynolds averaging.} \tag{18.5b}$$

Because $\rho(x_i,t) = \bar{\rho}(x_i) + \rho'(x_i,t)$, rewriting Equation 18.4b,

$$\overline{\rho(x_i,t)\, q''(x_i,t)} = \left[\lim_{\Delta t\to\infty}\frac{1}{\Delta t}\int_{t_0}^{t_0+\Delta t}\{\bar{\rho}(x_i) + \rho'(x_i,t)\}\, q''(x_i,t)dt\right]$$

$$= \lim_{\Delta t\to\infty}\frac{1}{\Delta t}\left\{\bar{\rho}(x_i)\int_{t_0}^{t_0+\Delta t} q''(x_i,t)dt + \int_{t_0}^{t_0+\Delta t}\rho'(x_i,t)\, q''(x_i,t)dt\right\}.$$

Simplifying and recognizing that $\bar{\rho}(x_i)\int_{t_0}^{t_0+\Delta t} q''(x_i,t)dt = \bar{\rho}(x_i)\overline{q''(x_i,t)}$

$$\overline{q''(x_i,t)} = \frac{\overline{\rho'(x_i,t)\, q''(x_i,t)}}{\overline{\rho(x_i)}} \tag{18.6}$$

18.3.1 *Relation between Favre Averaging and Reynolds Averaging*

If the density is constant, both averaging techniques yield similar results. Reynolds averaging is typically adopted for incompressible flows, whereas Favre averaging is suitable for compressible flows.

Because

$$q(x_i,t) = \tilde{q}(x_i) + q''(x_i,t) \tag{18.7a}$$

taking the time average

$$\bar{q}(x_i) = \tilde{q}(x_i) + \overline{q''}(x_i,t) \tag{18.7b}$$

and using Equation 18.6 one finds that

$$\tilde{q}(x_i) - \bar{q}(x_i) = -\overline{q''}(x_i) = \frac{\overline{\rho'(x_i,t)q''(x_i,t)}}{\overline{\rho(x_i)}}. \tag{18.8}$$

The difference between Favre- and Reynolds-averaged quantities is given in terms of density fluctuation and Favre-based property fluctuations.

18.3.2 *A Few Rules of Averaging*

Multiply $q(x_i,t)$ by $\bar{\rho}$ and average

$$\bar{\rho}q(x_i,t) = \bar{\rho}\overline{q(x_i)} + \bar{\rho}\overline{q'}(x_i,t) = \bar{\rho}\overline{q(x_i)} + 0 = \bar{\rho}\tilde{q}(x_i) + \bar{\rho}\overline{q''}(x_i,t) \tag{18.9a}$$

$$\overline{(\rho(x_i,t)q(x_i,t))} = \bar{\rho}(x_i)\,\tilde{q}''(x_i) \tag{18.9b}$$

$$\overline{(\rho(x_i,t)q_i(x_i,t)q_j(x_i,t))} = \bar{\rho}(x_i,t)\,[\tilde{q}_i(x_i)\tilde{q}_j(x_i) + \overline{q_i''(x_i,t)q_j''(x_i,t)}] \tag{18.9c}$$

$$\frac{\overline{\partial[q(x_i,t)]}}{\partial x} = \frac{\overline{\partial[\bar{q}(x_i) + q'(x_i,t)]}}{\partial x} = \frac{\partial[\bar{q}(x_i)]}{\partial x} \tag{18.9d}$$

$$\frac{\overline{\partial[q(x_i,t)]}}{\partial x} = \frac{\overline{\partial[\bar{q}(x_i) + q'(x_i,t)]}}{\partial x} = \frac{\partial[\bar{q}(x_i)]}{\partial x}, \text{ e.g., } q = P. \tag{18.9e}$$

18.4 Instantaneous and Average Governing Equations

Recall that the forces between the particles and gas are the drag force, particle weight, and buoyant force. The weight and buoyant force can be combined as the $v^*(\rho_P - \rho)$. The settling velocity (terminal velocity) is given by $v_{settle} = \{(\rho_P - \rho)d^2g/18\ m_p\}$. For a 10-$\mu$m particle, v_{settle} = 3.02 mm/sec. In a room of

1-m height, a 100-μm particle takes 3.33 sec to hit the floor. For Re ~ 1.7, the assumption of Stokes drag is reasonable.

18.4.1 Mass

Rewriting Equation 8.26 [Smith et al., 1982; Smoot and Smith, 1985]

$$\frac{\partial \rho}{\partial t} + \nabla \cdot (\rho \vec{v}) = S^m \tag{18.10}$$

where S^m is the mass generation on a volumetric basis (e.g., particle gasifying and droplets evaporating).

Using the Favre averaging procedure and ignoring the mass source,

$$\frac{\partial \bar{\rho}}{\partial t} + \frac{\partial}{\partial x_j}(\bar{\rho}\tilde{u}_j) = 0 \tag{18.11}$$

where

$$\frac{\partial}{\partial x_j}(\bar{\rho}\tilde{v}_j) = \frac{\partial}{\partial x_1}(\bar{\rho}\tilde{v}_1) + \frac{\partial}{\partial x_2}(\bar{\rho}\tilde{v}_2) + \frac{\partial}{\partial x_3}(\bar{\rho}\tilde{v}_3)\,.$$

18.4.2 Momentum

Recall Equation 8.150; rewriting

$$\frac{\partial \rho \vec{v}}{\partial t} + \nabla \cdot (\rho \vec{v}\vec{v}) = -\nabla p + \nabla \cdot \bar{\bar{\tau}} + \rho \vec{g} + \vec{S}_p^v \tag{18.12}$$

where ρ is the gas density; $\vec{v}$, velocity vector; t, time; p, local pressure; $\vec{g}$, body force acting on unit mass of fluid; $\vec{S}_p^v$, net momentum imparted by particles to the gas per unit volume of gas (= 0 for flows with no particles); ∇, operator for spatial gradient; and $\bar{\bar{\tau}}$, shear stress tensor. For a Newtonian fluid with variable density [Hirschfelder et al., 1964], redefining Equation 8.139,

$$\bar{\bar{\tau}} = \mu[\nabla \vec{v} + (\nabla \vec{v})^T] + \left(\kappa - \frac{2}{3}\mu\right)(\nabla \cdot \vec{v})\bar{\bar{U}} \tag{18.13}$$

where μ is the mean dynamic viscosity of the gas mixture; $\bar{\bar{U}}$, a unit tensor; and κ, the volume viscosity that describes viscous dissipation due to normal shear stress. It is usually assumed that the volume viscosity is negligible such that

$$\bar{\bar{\tau}} = \mu[\nabla \vec{v} + (\nabla \vec{v})^T] - \frac{2}{3}\mu(\nabla \cdot \vec{v})\bar{\bar{U}}\,. \tag{18.14}$$

Shear stress on the face normal to direction 2 (= y) but in direction 1 (= x)

$$\tau_{21} = -2\mu\sigma_{21} = -2\mu\left[\frac{1}{2}\left(\frac{\partial vy}{\partial x} + \frac{\partial vx}{\partial y}\right)\right] = -\mu\left(\frac{\partial vy}{\partial x} + \frac{\partial vx}{\partial y}\right). \quad (18.15)$$

Ignoring momentum sources due to particles, etc., the averaged momentum conservation in the "i" direction is given as

$$\frac{\partial}{\partial t}(\bar{\rho}\tilde{v}_i) + \frac{\partial}{\partial x_k}(\bar{\rho}\tilde{v}_i\tilde{v}_k) = -\frac{\partial \bar{p}}{\partial x_i} + \frac{\partial}{\partial x_k}(\bar{\tau}_{ik} - \overline{\rho v_i'' \, v_k}), + \bar{\rho}g, \text{ I = 1, 2, 3.} \quad (18.16)$$

Note that repeated "k" within a term implies summation in tensor notation. For the "j" direction, the averaged momentum equation is

$$\frac{\partial}{\partial t}(\bar{\rho}\tilde{v}_j) + \frac{\partial}{\partial x_k}(\bar{\rho}\tilde{v}_j\tilde{v}_k) = -\frac{\partial \bar{p}}{\partial x_j} + \frac{\partial}{\partial x_k}(\bar{\tau}_{jk} - \overline{\rho\tilde{v}_j'' \, \tilde{v}_k}) + \bar{\rho}g, \quad \text{j = 1, 2, 3.} \quad (18.17)$$

18.4.3 Enthalpy, Kinetic Energy, and Stagnation Enthalpy

Adopting Equation 8.160 for enthalpy,

$$\rho\frac{\partial h}{\partial t} + \rho\vec{v}\cdot\nabla h = \frac{\partial p}{\partial t} - \bar{\bar{\tau}}:\nabla\vec{v} - \nabla\cdot\vec{q} + \rho\sum Y_k f_k \cdot V_k. \quad (18.18)$$

Ignoring body force we can rewrite the conservation equation for specific enthalpy h [Bird et al., 1960; Smoot and Smith, 1985], including radiant energy and source terms for heat transfer from particles to gas, as

$$\frac{\partial(\rho h)}{\partial t} + \nabla\cdot(\rho h\vec{v}) = q_{rg} - \nabla\cdot\vec{q} + \frac{\partial p}{\partial t} - \bar{\bar{\tau}}:\nabla\vec{v} + S_p^h \quad (18.19)$$

where q_{rg} is the radiation energy, $\bar{\bar{\tau}}:\nabla\vec{v}$ enthalpy production by viscous dissipation S_p^h, net heat given by particles (= 0 if no particles); and $\vec{q}$, heat flux (Equation 8.33);

$$\vec{q} = -\frac{\lambda}{c_p}\nabla h + \sum_{k=1}^{N}\rho_k\vec{V}_k h_k. \quad (18.20)$$

The first and second terms on the right-hand side of Equation 18.20 are heat flux by conduction based on the Fourier law and heat flux by species diffusion; λ and c_p, heat conductivity and specific heat of gas, respectively; N, the total number of species in the fluid; h_i, the specific enthalpy of species i; and $\rho_k \vec{V}_k$, mass diffusion flux of species described by Fick's law.

18.4.3.1 Kinetic Energy

If there is a source of momentum from particles to gas, the generalized kinetic energy equation is given as Equation 8.156 with expansion of $\vec{\vec{P}}:\nabla\vec{v}$

$$\frac{\partial\left[\frac{\rho v^2}{2}\right]}{\partial t}+\nabla\cdot\left\{\vec{v}\left(\frac{\rho v^2}{2}\right)\right\}=-\vec{v}\cdot\nabla\cdot\left[p^0\vec{\vec{U}}\right]-\vec{v}\cdot\nabla\cdot\vec{\vec{\tau}}+\sum\rho Y_k\vec{f}_k\cdot\vec{v}+\vec{v}\cdot\vec{S}_p^v. \tag{18.21}$$

Ignoring the particle momentum source,

$$\partial\,\{\rho v^2/2\}/\partial t+\partial\,[v_j\{\rho v_k^2/2\}]/\partial x_j=\rho(D\ v^2/2)/dt$$
$$=-v_j\,\partial p/\partial x_j-v_j\,\partial\tau_{ij}/\partial x_j+\sum\rho_k Y_k\ v_j f_{kj} \tag{18.22}$$

$$\tau_{21}=-2\mu\,\sigma_{21}=-2\mu\left[\frac{1}{2}\left\{\frac{\partial v_2}{\partial x_1}+\frac{\partial v_1}{\partial x_2}\right\}\right]=-\mu\left\{\frac{\partial v_2}{\partial x_1}+\frac{\partial v_1}{\partial x_2}\right\} \tag{18.23}$$

where

$$\sigma_{21}=\left[\frac{1}{2}\left\{\frac{\partial v_2}{\partial x_1}+\frac{\partial v_1}{\partial x_2}\right\}\right] \tag{18.24}$$

and ignoring body forces, writing in terms of total derivative

$$\rho(D\ v^2/2)/dt=(D\{v\rho^2/2\})/dt=-v_j\,\partial p/\partial x_j-v_j\,\partial\tau_{ij}/\partial x_j. \tag{18.25}$$

18.4.3.1.1 Average Kinetic Energy

Multiply average momentum Equation 18.17 by $\tilde{v}_j$

$$\tilde{v}_j\frac{\partial}{\partial t}(\bar{\rho}\tilde{v}_i)+\tilde{v}_j\frac{\partial}{\partial x_k}(\bar{\rho}\tilde{v}_i\tilde{v}_k)=-\tilde{v}_j\frac{\partial\bar{p}}{\partial x_i}+\tilde{v}_j\frac{\partial}{\partial x_k}(\bar{\tau}_{ik}-\overline{\rho\tilde{v}_i{}''\tilde{v}_k}). \tag{18.26}$$

If j = i

$$\frac{\partial}{\partial t}(\bar{\rho}\tilde{v}_i\tilde{v}_i/2)+\frac{\partial}{\partial x_k}(\bar{\rho}\tilde{v}_i\tilde{v}_i\tilde{v}_k/2)=-\tilde{v}_I\frac{\partial\bar{p}}{\partial x_i}+\tilde{v}_I\frac{\partial}{\partial x_k}(\bar{\tau}_{ik}-\overline{\rho\tilde{v}_I{}''\tilde{v}_k}) \tag{18.27}$$

Adding for all components

$$\frac{\partial}{\partial t}(\bar{\rho}\tilde{v}\,\tilde{v}/2)+\frac{\partial}{\partial x_k}(\bar{\rho}v\tilde{v}\tilde{v}_k/2)=-\tilde{v}_i\frac{\partial\bar{p}}{\partial x_i}+\tilde{v}_i\frac{\partial}{\partial x_k}(\bar{\tau}_{ik}-\overline{\rho\tilde{v}_i{}''\tilde{v}_k}) . \tag{18.28}$$

Multiply Equation 18.17 by $\tilde{v}_i$,

$$\tilde{v}_i\frac{\partial}{\partial t}(\bar{\rho}\tilde{v}_i)+\tilde{v}_i\frac{\partial}{\partial x_k}(\bar{\rho}\tilde{v}_j\tilde{v}_k)=-\tilde{v}_i\frac{\partial\bar{p}}{\partial x_j}+\tilde{v}_i\frac{\partial}{\partial x_k}(\bar{\tau}_{jk}-\overline{\rho\tilde{v}_j{}''\tilde{v}_k}) . \tag{18.29}$$

Adding Equation 18.28 with Equation 18.29,

$$\begin{aligned}&\tilde{v}_j\frac{\partial}{\partial t}(\bar{\rho}\tilde{v}_i)+\tilde{v}_j\frac{\partial}{\partial x_k}(\bar{\rho}\tilde{v}_i\tilde{v}_k)+\tilde{v}_i\frac{\partial}{\partial t}(\bar{\rho}\tilde{v}_i)+\tilde{v}_i\frac{\partial}{\partial x_k}(\bar{\rho}\tilde{v}_j{}''\tilde{v}_k)\\&=-\tilde{v}_j\frac{\partial\bar{p}}{\partial x_i}+\tilde{v}_j\frac{\partial}{\partial x_k}(\bar{\tau}_{ik}-\overline{\rho\tilde{v}_i{}''\tilde{v}_k})-\tilde{v}_i\frac{\partial\bar{p}}{\partial x_j}+\tilde{v}_i\frac{\partial}{\partial x_k}(\bar{\tau}_{jk}-\overline{\rho\tilde{v}_j{}''\tilde{v}_k}).\end{aligned} \tag{18.30}$$

Simplifying,

$$\begin{aligned}&\frac{\partial}{\partial t}(\bar{\rho}\tilde{v}_i\tilde{v}_j)-\bar{\rho}\tilde{v}_i\frac{\partial\tilde{v}_j}{\partial t}+\frac{\partial}{\partial x_k}(\bar{\rho}\tilde{v}_i\tilde{v}_k\tilde{v}_j)-\bar{\rho}\tilde{v}_i\tilde{v}_k\frac{\partial\tilde{v}_j}{\partial x_k}+\tilde{v}_i\tilde{v}_j\frac{\partial\bar{\rho}}{\partial t}+\bar{\rho}\tilde{v}_i\frac{\partial\tilde{v}_j}{\partial t}+\tilde{v}_i\tilde{v}_j\frac{\partial\bar{\rho}\tilde{v}_k}{\partial x_k}\\&+\bar{\rho}\tilde{v}_k\tilde{v}_i\frac{\partial\tilde{v}_j}{\partial x_k}=-\tilde{v}_j\frac{\partial\bar{p}}{\partial x_i}+\tilde{v}_j\frac{\partial}{\partial x_k}(\bar{\tau}_{ik}-\overline{\rho\tilde{v}_i{}''\tilde{v}_k})-\tilde{v}_i\frac{\partial\bar{p}}{\partial x_j}+\tilde{v}_i\frac{\partial}{\partial x_k}(\bar{\tau}_{jk}-\overline{\rho\tilde{v}_j{}''\tilde{v}_k}).\end{aligned}$$

(18.31)

Using continuity,

$$\begin{aligned}\frac{\partial}{\partial t}(\bar{\rho}\tilde{v}_i\tilde{v}_j)+\frac{\partial}{\partial x_k}(\bar{\rho}\tilde{v}_i\tilde{v}_k\tilde{v}_j)-\bar{\rho}\tilde{v}_i\tilde{v}_k\frac{\partial\tilde{v}_j}{\partial x_k}&=-\tilde{v}_j\frac{\partial\bar{p}}{\partial x_i}+\tilde{v}_j\frac{\partial}{\partial x_k}(\bar{\tau}_{ik}-\overline{\rho\tilde{v}_i{}''\tilde{v}_k})\\&-\tilde{v}_i\frac{\partial\bar{p}}{\partial x_j}+\tilde{v}_i\frac{\partial}{\partial x_k}(\bar{\tau}_{jk}-\overline{\rho\tilde{v}_j{}''\tilde{v}_k}).\end{aligned} \tag{18.32}$$

Note

$$p=p^0-\kappa\left(\frac{\partial v_1}{\partial x_1}+\frac{\partial v_2}{\partial x_2}+\frac{\partial v_3}{\partial x_3}\right). \tag{11.33}$$

Often, bulk viscosity $\kappa \approx 0$ for gases (Chapter 8). Thus, $p \approx p^0$, hydrostatic pressure. Ignoring momentum sources due to drops or particles

$$\frac{\partial}{\partial t}\left(\frac{\rho\tilde{v}_k\tilde{v}_k}{2}\right)+\frac{\partial}{\partial_{xj}}\left[\tilde{v}_j\left(\frac{\rho\tilde{v}_k\tilde{v}_k}{2}\right)\right]=-\tilde{v}_j\frac{\partial\bar{p}}{\partial_{xj}}-\tilde{v}_i\frac{\partial\bar{\tau}_{ij}}{\partial_{xj}}-\tilde{v}_i\frac{\partial(\rho\tilde{v}_i{}''\tilde{v}_j{}'')}{\partial_{xj}}. \quad (11.34)$$

18.4.3.2 Stagnation Enthalpy

From Chapter 8 (Equation 8.161)

$$\rho\frac{\partial\left[h+\frac{v^2}{2}\right]}{\partial t}+\rho\vec{v}\cdot\nabla\left\{\left(h+\frac{v^2}{2}\right)\right\}=\frac{\partial p}{\partial t}-\sum_{j=1}^{3}\sum_{i=1}^{3}\frac{\partial(\tau_{ji}v_i)}{\partial x_j}-\nabla\cdot\vec{q}+\sum_k\rho Y_k\vec{f}_k\cdot(\vec{v}+\vec{V}_k)$$

(18.35)

where τ denotes stress components due to shear alone (i.e., τ_{ij})v.

Let h_{stg}, stagnation enthalpy = h + ke = h + $v^2/2$ = h + (1/2) v_i v_I; also, $p = p^0$. Including the enthalpy source and neglecting body forces,

$$\frac{\partial(\rho h_{stg})}{\partial t}+\nabla\cdot(\rho h_{stg}\vec{v})=\frac{\partial p}{\partial t}-\sum_{j=1}^{3}\sum_{i=1}^{3}\frac{\partial(\tau_{ji}v_i)}{\partial x_j}-\nabla\cdot\vec{q}+\sum_k\rho Y_k\vec{f}_k\cdot(\vec{v}+\vec{V}_k)+S_p^h.$$

(18.36)

Averaging,

$$\frac{\partial}{\partial t}(\bar{\rho}\tilde{h}_{stg})+\frac{\partial}{\partial x_j}(\bar{\rho}\tilde{h}_{stg}\tilde{u}_j)=\frac{\partial\bar{p}}{\partial t}+\tilde{v}_j\frac{\partial\bar{p}}{\partial x_j}+\overline{\tilde{v}_{j''}\frac{\partial p}{\partial x_j}}$$
$$+\frac{\partial}{\partial x_j}(-\bar{q}_j-\overline{\rho h''_{stg}\tilde{v}_{j''}})-\bar{\tau}_{ij}\frac{\partial\tilde{v}_i}{\partial x_j}-\overline{\tau_{ij}\frac{\partial\tilde{v}_{i''}}{\partial x_j}}. \quad (18.37)$$

Change of enthalpy = flow work performed due to adverse pressure (mean and fluctuation) + conductive energy transport + enthalpy transport due to turbulent diffusion + frictional heat due to viscous friction.

18.4.4 Reynolds Stress Transport

The momentum transport in direction "1" is

$$\frac{\partial(\rho v_1)}{\partial t}+\frac{\partial(\rho v_k v_1)}{\partial x_k}=-\left(\frac{\partial(P_{k1})}{\partial x_k}\right)+\rho Y_\ell f_{\ell 1}. \quad (18.38)$$

Multiply by v_2

$$v_2 \frac{\partial(\rho v_1)}{\partial t} + v_2 \frac{\partial(\rho v_k v_1)}{\partial x_k} = -v_2\left(\frac{\partial(p_{k1})}{\partial x_k}\right) + v_2 \rho Y_\ell f_{\ell 1}.$$

Similarly, multiply x_2 momentum by v_1

$$v_1 \frac{\partial(\rho v_2)}{\partial t} + v_1 \frac{\partial(\rho v_k v_2)}{\partial x_k} = -v_1\left(\frac{\partial(p_{k2})}{\partial x_k}\right) + v_1 \rho Y_\ell f_{\ell 2}.$$

Adding and simplifying

$$\frac{\partial(\rho v_1 v_2)}{\partial t} + \frac{\partial(\rho v_k v_1 v_2)}{\partial x_k} + v_2 v_1\left(\frac{\partial \rho}{\partial t} + \frac{\partial(\rho v_k)}{\partial x_k}\right) = \frac{\partial(\rho v_1 v_2)}{\partial t} + \frac{\partial(\rho v_k v_1 v_2)}{\partial x_k}$$
$$= -v_2\left(\frac{\partial(p_{k1})}{\partial x_k}\right) + v_2 \sum_\ell \rho Y_\ell f_{\ell 1} - v_1\left(\frac{\partial(p_{k2})}{\partial x_k}\right) + v_1 \rho Y_\ell f_{\ell 2}. \tag{18.39}$$

Let subscript 1 be i and 2 be j (e.g., $v_1 = v_i$ and $v_2 = v_j$) ($p_{ki} = p + \tau_{ki}$).

$$\frac{\partial(\rho v_i v_j)}{\partial t} + \frac{\partial(\rho v_k v_i v_j)}{\partial x_k} = -v_j\left(\frac{\partial p}{\partial x_i}\right) - v_j\left(\frac{\partial \tau_{ki}}{\partial x_k}\right)$$
$$+ v_j \rho Y_\ell f_{\ell i} - v_i\left(\frac{\partial p}{\partial x_j}\right) - v_i\left(\frac{\partial \tau_{kj}}{\partial x_k}\right) + v_i \rho Y_\ell f_{\ell j}.$$

Again, using an averaging procedure ($v_j = \tilde{v}_j + v_j''$, $\tau_{ki} = \bar{\tau}_{ki} + \tau'_{ki}$) [Jones and Whitelaw, 1982],

$$\frac{\partial}{\partial t}(\rho \tilde{v}_i \tilde{v}_j + \overline{\rho v_i'' v_j''}) + \frac{\partial}{\partial x_k}(\rho \tilde{v}_k \tilde{v}_i \tilde{v}_j + \tilde{v}_k \overline{\rho v_i'' v_j''} + \tilde{v}_j \overline{\rho v_k'' v_i''} + \tilde{v}_i \overline{\rho v_j'' v_k''}$$
$$+ \tilde{v}_k \overline{\rho v_i'' v_j'' v_k''}) = -\tilde{v}_j\left(\frac{\partial p}{\partial x_i}\right) - \overline{v_j''\left(\frac{\partial p}{\partial x_i}\right)} - \tilde{v}_i\left(\frac{\partial p}{\partial x_j}\right) - \overline{v_i''\left(\frac{\partial p}{\partial x_j}\right)} - \tilde{v}_j\left(\frac{\partial \bar{\tau}_{ki}}{\partial x_k}\right)$$
$$- \overline{v_j''\left(\frac{\partial \bar{\tau}_{ki}}{\partial x_k}\right)} - \overline{v_j''\left(\frac{\partial \tau'_{ki}}{\partial x_k}\right)} - \overline{v_i''\left(\frac{\partial \tau_{kj}}{\partial x_k}\right)} - \tilde{v}_i\left(\frac{\partial \bar{\tau}_{kj}}{\partial x_k}\right) - \overline{v_i''\left(\frac{\partial \bar{\tau}_{kj}}{\partial x_k}\right)}$$
$$+ \tilde{v}_j \sum_\ell \rho Y_\ell f_{\ell i} + \tilde{v}_i \rho Y_\ell f_{\ell j}. \tag{18.40}$$

Using Equation 18.33 to eliminate the Favre-averaged velocities,

$$\frac{D}{Dt}\overline{(\rho v_i'' v_j'')} + \frac{\partial}{\partial x_k}\overline{(\rho v_i'' v_j'' v_k'')}$$

$$= -\overline{v_j''\left(\frac{\partial p}{\partial x_i}\right)} - \overline{v_i''\left(\frac{\partial p}{\partial x_j}\right)} - \left\{\overline{v_j''\left(\frac{\partial \tau'_{ik}}{\partial x_k}\right)} + \overline{v_i''\left(\frac{\partial \tau'_{jk}}{\partial x_k}\right)}\right\} \tag{18.41}$$

$$-\overline{(\rho v_k'' v_i'')}\frac{\partial \tilde{v}_j}{\partial x_k} - \overline{(\rho v_k'' v_j'')}\frac{\partial \tilde{v}_i}{\partial x_k} + \tilde{v}_i \bar{\rho} Y_\ell f_{\ell i}.$$

Because $p = \bar{p} + p'$, the term $\overline{v_i''\left(\frac{\partial p}{\partial x_j}\right)}$ is simplified as

$$\overline{v_j''\left(\frac{\partial p}{\partial x_i}\right)} = \int v_j''(x_i,t)\left(\frac{\partial(\bar{p}(x_i) + p'(x_i,t))}{\partial x_i}\right)dt = \overline{v_j''(x_i)}\left(\frac{\partial \bar{p}(x_i)}{\partial x_i}\right) + \overline{v_j''\left(\frac{\partial p'}{\partial x_i}\right)}.$$

Using this result in Equation 18.41,

$$\frac{D}{Dt}\overline{(\rho v_i'' v_j'')} = \bar{\rho}\frac{\partial}{\partial t}\overline{(v_i'' v_j'')} + \bar{\rho}\tilde{v}_k\frac{\partial}{\partial x_k}\overline{(v_i'' v_j'')} = -\frac{\partial}{\partial x_k}\overline{(\rho v_i'' v_j'' v_k'')}$$

$$-\overline{v_j''(x_i)}\left(\frac{\partial \bar{p}(x_i)}{\partial x_i}\right) - \overline{v_j''\left(\frac{\partial p'}{\partial x_i}\right)} - \overline{v_i''(x_i)}\left(\frac{\partial \bar{p}(x_i)}{\partial x_j}\right) - \overline{v_i''\left(\frac{\partial p'}{\partial x_j}\right)}$$

$$-\left[\overline{v_j''\left(\frac{\partial \tau'_{ik}}{\partial x_k}\right)} + \overline{v_i''\left(\frac{\partial \tau'_{jk}}{\partial x_k}\right)}\right] + \bar{\rho}\overline{(v_k'' v_i'')}\frac{\partial \tilde{v}_j}{\partial x_k} - \bar{\rho}\overline{(v_k'' v_j'')}\frac{\partial \tilde{v}_i}{\partial x_k} + \tilde{v}_i\sum_\ell \bar{\rho} Y_\ell f_{\ell i}.$$

$$\tag{18.42}$$

The rate of turbulent kinetic energy dissipation per unit fluid mass is given as

$$\varepsilon = \left[\overline{\frac{\mu}{\rho}\left(\frac{\partial v_i'}{\partial x_j}\right)\left(\frac{\partial v_i'}{\partial x_j}\right)}\right] = \left[\overline{\frac{\mu}{\rho}\left(\frac{\partial v_1'}{\partial x_1}\right)^2}\right] + \left[\overline{\frac{\mu}{\rho}\left(\frac{\partial v_2'}{\partial x_2}\right)^2}\right] + \left[\overline{\frac{\mu}{\rho}\left(\frac{\partial v_3'}{\partial x_3}\right)^2}\right].$$

The term { } in Equation 18.41 is expanded as

$$\left\{\overline{v_j''\left(\frac{\partial \tau'_{ik}}{\partial x_k}\right)} + \overline{v_i''\left(\frac{\partial \tau'_{jk}}{\partial x_k}\right)}\right\} = -\bar{\mu}\left[\overline{v_j''\frac{\partial}{\partial x_k}\left(\frac{\partial v_k''}{\partial x_i} + \frac{\partial v_i''}{\partial x_k}\right) + v_i''\frac{\partial}{\partial x_k}\left(\frac{\partial v_k''}{\partial x_j} + \frac{\partial v_j''}{\partial x_k}\right)}\right]$$

$$\frac{\partial}{\partial t}(\rho\tilde{v}_i\tilde{v}_j+\overline{\rho v_i'' v_j''})+\frac{\partial}{\partial x_k}(\rho\tilde{v}_k\tilde{v}_i\tilde{v}_i+\tilde{v}_k\overline{\rho v_i'' v_i''})=-\tilde{v}_i\left(\frac{\partial p}{\partial x_i}\right)-\overline{v_i''\left(\frac{\partial p}{\partial x_j}\right)}$$
$$-\tilde{v}_j\left(\frac{\partial\overline{\tau}_{ki}}{\partial x_k}\right)-\overline{v_j''\left(\frac{\partial\overline{\tau}_{ki}}{\partial x_k}\right)}-(\overline{\rho v_k'' v_i''})\frac{\partial\tilde{v}_i}{\partial x_k}+\tilde{v}_i\sum_{\ell}\rho Y_\ell f_{\ell i}. \quad (18.43)$$

18.4.5 Turbulent Kinetic Energy ($k = (1/2)\,\overline{\bar{\rho} v_i'' v_i''}$)

With i = j in one of the equations (e.g., v_j^* momentum equation in direction I with $v_j = v_i$)

$$\frac{D}{Dt}(\overline{(1/2)\rho v_i'' v_i''})+\frac{\partial}{\partial x_k}(\overline{(1/2)\rho v_i'' v_i''\tilde{v}_k})$$
$$=-\overline{v_i''\left(\frac{\partial p}{\partial x_I}\right)}-\overline{v_i''\left(\frac{\partial \tau'_{ik}}{\partial x_k}\right)}-\overline{(\rho v_k'' v_i'')}\frac{\partial\tilde{v}_j}{\partial x_k} \quad (18.44)$$

$$\frac{D\{(1/2)\overline{\rho_k}\}}{Dt}+\frac{\partial\left\{\overline{(1/2)\rho_k v_k''}\right\}}{\partial x_k}$$
$$=-\overline{v_i''\left(\frac{\partial p}{\partial x_I}\right)}-\overline{v_i''\left(\frac{\partial \tau'_{ik}}{\partial x_k}\right)}-\overline{(\rho v_k'' v_i'')}\frac{\partial\tilde{v}_j}{\partial x_k}. \quad (18.45)$$

Material derivative of turbulent ke + turbulent ke convected = flow work by perturbed velocity against pressure gradients, viscous stresses, and production of turbulent ke through product of turbulent stress and shear strain. Averaging the mechanical energy of flows (see Chapter 8),

$$\frac{\partial\left[\frac{\rho v^2}{2}\right]}{\partial t}+\vec{v}\cdot\left\{\nabla\left(\frac{\rho v^2}{2}\right)\right\}=\frac{D\left[\frac{\rho v^2}{2}\right]}{Dt}=-\frac{\rho v^2}{2}\nabla\cdot\vec{v}-\vec{v}\cdot\nabla\cdot\left[p^0\vec{\vec{U}}\right]$$
$$-\vec{v}\cdot\nabla\cdot\vec{\vec{\tau}}+\rho Y_k\vec{f}_k\cdot\vec{v}$$

$$\frac{D}{Dt}\left(\frac{1}{2}\bar{\rho}\tilde{u}_i\,u_i+\overline{u_i''\,u_i''}\right)=-\tilde{u}_i\frac{\partial\bar{p}}{\partial x_i}+\overline{u_i''\frac{\partial\tau'_{ik}}{\partial x_k}}+\frac{1}{2}\overline{\rho u_i'' u_k''}\frac{\partial\tilde{u}_i}{\partial x_k}. \quad (18.46)$$

Rate of change in ke = ke change due to pressure (as in diffuser) + ke change due to viscous frictions + net ke gain due to fluctuating motions or (momentum transfer between adjacent layers).

18.4.6 Species

$$\frac{\partial}{\partial t}(\bar{\rho}\tilde{Y}_k)+\frac{\partial}{\partial x_i}(\bar{\rho}\tilde{Y}_k\tilde{u}_i)=\frac{\partial}{\partial x_i}[D\bar{\rho}\frac{\partial \tilde{Y}_k}{\partial x_i}-\overline{\rho Y_k{}'' u_i{}''}]+\frac{\partial}{\partial x_i}\overline{D\rho\frac{\partial Y_k{}''}{\partial x_i}}+\overline{\dot{\omega}_k} \quad (18.47a)$$

where

$$\dot{w}_k=-A_k\rho^2 Y_i Y_j \exp(-E/\bar{R}T). \quad (18.47b)$$

Note that this term is difficult to model and, consequently, the laminar expression is shown here only for completeness. An appropriate chemical source term has to be determined for each case as mentioned in the text. Examination of the averaged equations shows that because average pressure can be measured directly, it is best left as a time-averaged value. The stress tensor (τ_{ij}) and the heat flux vector (q_j) are also left in the time-averaged form. As both methods are shown here, it may be interesting to examine how the values are related. By setting the two equal and simplifying, this relation between Reynolds and Favre averaging is found for a property q.

$$\tilde{q}-\bar{q}=\frac{\overline{\rho' q''}}{\bar{\rho}}=\frac{\overline{\rho' q'}}{\bar{\rho}} \quad (18.48)$$

18.4.7 Turbulence Models

When the expressions for turbulent flow are examined, it is obvious that more information is needed to solve the Navier–Stokes equations than in the laminar case. This is due to the presence of Reynolds stress or secondary correlation terms (i.e., v_iv_j or v_iv_j), which result from the cross products of the velocity formed by time averaging or mass averaging. Turbulence models are used for closure of the Navier–Stokes equations and to allow for their solution. These models are empirical in nature and provide an accurate description of the turbulence for a specified type of problem. They can be separated into three different categories: (1) algebraic models, (2) higher-order models, and (3) (k-ε) or two-partial-differential-equation models.

18.4.7.1 Algebraic Models

In algebraic models of turbulence, the Reynolds stress terms appearing in the time-averaged equations are taken to be simple functions of the time-mean-flow variables themselves and/or of the spatial coordinate. Using Boussinesq's assumption,

$$-\overline{\rho v_i{}' v_j{}'}\equiv(\tau_{yx})_{Reynolds}\equiv\rho v_t\left(\frac{\partial v_i}{\partial x_j}+\frac{\partial v_j}{\partial x_i}\right) \quad (18.49)$$

where v_t denotes turbulent eddy diffusivity. Note that with these algebraic models of turbulence, no additional partial differential equations need to be solved to predict the time-averaged flow field. But turbulent diffusivity is unknown. If turbulence is assumed to be isotropic, the number of stress terms would be reduced from six to one.

18.4.7.2 Higher-Order Models

Higher-order methods treat Reynolds stresses (and other double correlations) as new dependent variables governed by their own partial differential equations and boundary conditions. The triple correlations present in the partial differential equations (i.e., $-v_iv_jv_k$) are then modeled by the local dependent variables. This approach attempts to bring additional generality to turbulence models and hence increase the universality of the solution. However, this procedure is computationally impractical for most 3-D flows, particularly those with several reacting species or phases.

18.4.7.3 The (k-ε) Model

The k-ε model of turbulence is intermediate in complexity and generality. It involves the solution of two additional convective–diffusive partial differential equations which describe the transport of two scalar quantities used to characterize local turbulence. These two scalar quantities are:

1. The mass-averaged turbulent kinetic energy per unit mass (i.e., $\overline{v''v''}/2 = k$)
2. The mass-averaged turbulent energy dissipation rate per unit mass, written as ε.

18.4.7.3.1 Two-Equation Model

No density fluctuations are accounted for, but isothermal flows are assumed. Given a linear relationship between Reynolds stress and rate of strain, Reynolds stress is given as

$$\bar{\rho}\widetilde{v_i''v_j''} = \delta_{ij}(\bar{\rho}k) + (\mu_t\frac{\partial \tilde{v}_k}{\partial x_k}) - \mu_t(\frac{\partial \tilde{v}_i}{\partial x_j} + \frac{\partial \tilde{v}_j}{\partial x_i}) \quad (18.50)$$

where the turbulent kinetic energy $k = \overline{v_i''v_i''}/2$. The turbulent viscosity μ_t is given as

$$\mu_t = C_\mu \bar{\rho}\frac{k^2}{\varepsilon} \quad (18.51)$$

where $C_\mu = 0.09$, and k is obtained from the turbulent kinetic energy equation. Just as heat flux is described by the Fourier law, similar relations exist for scalar quantities,

$$\bar{\rho}\widetilde{u_j''\phi_\alpha''} = -\frac{\mu_t}{\sigma_t}\frac{\partial \tilde{\phi}_\alpha}{\partial x_j}, \quad \text{turbulent transport of property,} \quad (18.52)$$

where σ_t = Pr_t for energy flux, ϕ_α = T; σ_t = Sc_t, Schmidt number for species transport; and ϕ_α = Y_k.

If φ = k, ε then

$$\bar{\rho}\widetilde{v_j''k} = -\frac{\mu_t}{\sigma_k}\frac{\partial k}{\partial x_j}$$

$$\bar{\rho}\widetilde{v_j''\varepsilon} = -\frac{\mu_t}{\sigma_\varepsilon}\frac{\partial \varepsilon}{\partial x_j}.$$

The k required for estimation of Reynolds stress can be obtained by using the revised form of Equation 18.45:

$$\bar{\rho}\tilde{u}_j\frac{\partial k}{\partial x_j} = \frac{\partial}{\partial x_j}\left[\left(\frac{\mu_t}{\sigma_k}+\mu\right)\frac{\partial k}{\partial x_j}\right] - \bar{\rho}v_i''v_j''\frac{\partial \tilde{v}_i}{\partial x_j} + \frac{\mu_t}{\bar{\rho}^2}\frac{\partial \bar{\rho}}{\partial x_i}\frac{\partial \bar{p}}{\partial x_i} - \bar{\rho}\varepsilon. \quad (18.53)$$

ε is obtained from turbulent dissipation equation:

$$\bar{\rho}\tilde{u}_j\frac{\partial \varepsilon}{\partial x_j} = \frac{\partial}{\partial x_j}\left[\left(\frac{\mu_t}{\sigma_\varepsilon}+\mu\right)\frac{\partial \varepsilon}{\partial x_j}\right] - C_{\varepsilon_1}\frac{\varepsilon}{k}(\bar{\rho}u_i''u_j''\frac{\partial \tilde{u}_i}{\partial x_j} + \frac{\mu_t}{\bar{\rho}^{-2}}\frac{\partial \bar{\rho}}{\partial x_i}\frac{\partial \bar{p}}{\partial x_i}) - C_{\varepsilon_2}\bar{\rho}\frac{\bar{\varepsilon}^2}{k}.$$

(18.54)

The following set of constants is commonly used with the k-ε model:

C_μ = 0.09, $C_{\varepsilon 1}$ = 1.44, $C_{\varepsilon 2}$ = 1.92,

σ_k = 1.00, σ_ε = 1.30, and Sc_t = Pr_t = 0.70.

Although this approach is considerably more flexible than the algebraic models and computationally more feasible than the higher-order models, it still requires several empirical constants that cannot be considered universal (applicable to all problems).

18.5 Governing Differential Equations: Axisymmetric Case and Mixture-Fraction PDF Combustion Model

The equations used in the combustion model can be described by a general partial differential equation (PDE), Equation 18.55 (Table 18.1). The variable φ determines the nature of the equation. For example, the general equation becomes a mass conservation equation when φ = 1. Momentum, mixture fraction, and enthalpy conservation equations can be derived by an appropriate choice of the variable φ. The general PDE for the axisymmetric case is now

Table 18.1 Generalized Partial Differential Equations

φ	Γ_φ	$S\varphi$	
1	0	S_p^m	Mass
$\tilde{u}$	μ_e	$-\frac{\partial p}{\partial x}+\frac{\partial}{\partial x}\left(\mu_e\frac{\partial \tilde{u}}{\partial x}\right)+\frac{1}{r}\frac{\partial}{\partial r}\left(r\mu_e\frac{\partial \tilde{v}}{\partial x}\right)+S_p^u+\tilde{u}S_p^m$	Momentum, x
$\tilde{v}$	μ_e	$-\frac{\partial p}{\partial r}+\frac{\partial}{\partial x}\left(\mu_e\frac{\partial \tilde{u}}{\partial r}\right)+\frac{1}{r}\frac{\partial}{\partial r}\left(r\mu_e\frac{\partial \tilde{v}}{\partial r}\right)-2\mu_e\frac{\tilde{v}}{r^2}+\frac{\bar{\rho}}{r}\tilde{w}\tilde{w}+S_p^v+\tilde{v}S_p^m$	Momentum, y
$\tilde{w}$	μ_e	$-\frac{\bar{\rho}\tilde{v}\tilde{w}}{r}-\frac{\tilde{w}}{r^2}\frac{\partial}{\partial r}(r\mu_e)$	
$\tilde{\eta}_i$	$\frac{\mu_e}{\sigma_{\eta_i}}$	$S_p^{\eta_i}$ i = 1, 2, 3 for different mixture fractions	Mix fractions
g_{η_i}	$\frac{\mu_e}{\sigma_g}$	$\frac{C_{g1}\mu_e}{\sigma_g}\left[\left(\frac{\partial \tilde{\eta}_i}{\partial x}\right)^2+\left(\frac{\partial \tilde{\eta}_i}{\partial r}\right)^2\right]-\frac{C_{g2}\rho\varepsilon g_{\eta_i}}{k}$ i = 1, 2, 3 for different mixture fractions	
$\tilde{h}$	$\frac{\mu_e}{\sigma_h}$	$q'_{rg}+\tilde{u}\frac{\partial \bar{p}}{\partial x}+\tilde{v}\frac{\partial \bar{p}}{\partial r}+S_p^h+\tilde{h}S_p^m$	Energy
k	$\frac{\mu_e}{\sigma_k}$	$\psi-\bar{\rho}\varepsilon$	Turbulent kinetic energy
ε	$\frac{\mu_e}{\sigma_\varepsilon}$	$\frac{\varepsilon}{k}(C_1\psi-C_2\bar{\rho}\varepsilon)$	Eddy diffusivity

given, followed by a table defining all conservation equations and k-ε turbulence model equations.

$$\frac{\partial}{\partial x}(\bar{\rho}\tilde{u}\varphi)+\frac{1}{r}\frac{\partial}{\partial r}(r\bar{\rho}\tilde{v}\varphi)-\frac{\partial}{\partial x}\left(\Gamma_\varphi\frac{\partial \varphi}{\partial x}\right)-\frac{1}{r}\frac{\partial}{\partial r}\left(r\Gamma_\varphi\frac{\partial \varphi}{\partial r}\right)=S_\varphi \qquad (18.55)$$

k-ε model constants:

$C_\mu = 0.09$, $C_1 = C_{\varepsilon 1} = 1.44$, $C_2 = C_{\varepsilon 2} = 1.92$, $\sigma_k = 0.9$, $\sigma_\varepsilon = 1.22$

Combustion model constants:

$C_{g1} = 2.8$, $C_{g2} = 1.92$, $\sigma_h = 0.8$, $E = 9.793$

where

$$\psi=\mu_e\left\{2\left[\left(\frac{\partial \tilde{u}}{\partial x}\right)^2+\left(\frac{\partial \tilde{v}}{\partial r}\right)^2+\left(\frac{\tilde{v}}{r}\right)^2\right]+\left[\left(\frac{\partial \tilde{u}}{\partial r}\right)^2+\left(\frac{\partial \tilde{v}}{\partial x}\right)^2\right]+\left[r\frac{\partial}{\partial r}\left(\frac{\tilde{w}}{r}\right)\right]^2+\left(\frac{\partial \tilde{w}}{\partial x}\right)^2\right\} \qquad (18.56)$$

$$\mu_e = \mu^t + \mu^l \quad (18.57)$$

$$\mu^t = \frac{C_\mu \bar{\rho} k^2}{\varepsilon} \quad (18.58)$$

$$S_p^m = \sum_{i=1}^{3} S_p^{\eta_i}. \quad (18.59)$$

The symbol ~ represents Favre-averaged variables. For example, Favre-averaged u-velocity is defined as

$$\tilde{u} = \frac{\overline{\rho(x_i, t)\, u(x_i, t)}}{\bar{\rho}}. \quad (18.60)$$

The Favre weighted average is commonly used in variable-density flows such as combustion.

18.5.1 Chemical Kinetics in Turbulent Flames

The chemical kinetics of the turbulent gas phase can be classified into slow chemistry, fast chemistry, and intermediate chemistry according to the reaction timescale and the turbulent mixing timescale. In order to incorporate the effects of turbulence into the chemical kinetics of reactions, the species continuity equations are time averaged. The average or mean reaction rates are calculated based on the relative timescales of kinetics and turbulent fluctuations (details on averaging are given in Section 18.3).

18.5.2 Kinetics in Low Turbulence

If turbulent fluctuations are small ($u''(x_i,t) << u(x_i)$), they will not affect the kinetic reaction rate. In this case, the local time-mean reaction rate is equivalent to the local instantaneous reaction rate. This means that time-mean values for temperature and species composition can be used to calculate the reaction rate. This may be applicable to regions in which the system is particle laden, with very low turbulence. Because most coal combustion systems have high turbulence levels, this technique is not applicable there.

18.5.2.1 Slow Kinetics

When chemical reaction timescales are very large compared to the characteristic timescale of the local turbulence, turbulent fluctuations do not affect the chemical reaction rates. This is due to the fact that the response times of the reactions are much longer than the timescale of turbulence. For such reactions, time-mean properties are used to calculate the mean reaction rates. Char oxidation, for example, is very slow compared to the turbulence timescale and, thus, the turbulence effect can be neglected.

18.5.2.2 Fast Kinetics

When chemical reaction timescales are very small compared to the timescale of local turbulence (i.e., mixing), it can be assumed that mixing is the rate-limiting process, i.e., the reaction is complete as soon as the reactants are mixed. Some gas-phase combustion reactions come under this category. The assumption that the reactions are mixing limited allows the mean gas-phase species composition to be calculated from statistical considerations [Bilger, 1974] rather than from kinetics. There is no need for the estimation of reaction rates. If local mass fractions of elements and enthalpy are known, then equilibrium chemistry (Chapter 3) can be used to determine T and species mass fractions. The element mass fractions are known if the local mixture fraction f is known (Chapter 8). However f itself varies locally with time and, hence, we must determine its mean and variance.

In boiler burners, fuel and air are fired separately and, hence, they have to mix prior to combustion. The fuel could be solid, liquid, or gaseous. Liquid fuels must be heated and vaporized first, then mixed with air. Often, the combustion rate is limited by the rate of mixing between fuel and air rather than by the chemical kinetics of oxidation between the fuel and O_2. Thus, the problem reduces to a simple physical phenomenon of mixing at the molecular level before any reaction can take place. In burners fired by natural gas ($\approx CH_4$) (primary-fuel flow), the gaseous fuel mixes with secondary air (oxidizer). If fast kinetics is assumed for the gas-phase reactions, then the reaction rates may be only functions of the local extent of primary-fuel–gas mixing with secondary air. Once the extent of primary-fuel mixing with local gas is (i.e., mixture fraction f) known locally, the local elemental composition and enthalpy level can be calculated. These quantities can be used with a Gibbs free energy scheme to calculate equilibrium species composition and gas temperatures.

18.5.2.3 Intermediate Kinetics

When the timescale of reactions is of the same order as the turbulence timescale, neither the mixing nor the reaction kinetics may be considered as rate limiting, and the turbulence must be accounted for directly. This type of kinetics is called *intermediate kinetics.* All homogeneous NO formation and reduction reactions fall under this category. For intermediate kinetics, appropriate mean rates for NO gas-phase reactions must be used in calculation of the time-mean mass fraction of NO. The question here is how to find these mean reaction rates.

One technique used with intermediate kinetics to determine the mean reaction rates is the calculation of deviation of actual species concentration from its equilibrium or fully reacted value $\bar{Y}_i^f$ [Bilger, 1979]. This is accomplished by considering the time-mean mass fraction of species $\bar{Y}_i$ to consist of a time-mean fully reacted value $\bar{Y}_i^f$ and a perturbation from the fully reacted value $\bar{y}_i^p$, i.e.,

$$\bar{Y}_i = \bar{Y}_i^f + \bar{y}_i^p$$

A transport equation similar to the species transport equation is used to calculate the perturbation $\overline{y}_i^{\,p}$.

18.6 Turbulent Combustion Modeling (Diffusion Flames)

For intermediate kinetics, the major difficulty lies in the fact that the knowledge of the local averaged species densities and averaged temperature is not sufficient to define an appropriate chemical source term, $\dot{w}$, even when the instantaneous reaction rate laws are completely known. The most troublesome of these is the temperature. Even relatively small fluctuations in the local temperature about its averaged value can result in large changes in the source term owing to the presence of the Arrhenius factor exp(E/RT). According to Williams [1985], nearly isothermal conditions must exist, i.e., $(T - \tilde{T})/\tilde{T}$ must be no greater than 0.01 to 0.1, for regular chemical source term modeling to work.

Several researchers have tried to remedy this problem by modifying the chemical source term. This procedure conceals some of the details of the interaction between the turbulence and the chemical kinetics. In addition, according to Williams [1985], this procedure does not account for the possibility of a fluid element being transported by turbulence to a different location (say, high T), reacting there, and then being transported back to the current location (say, low T). This so-called *counter-gradient diffusion*, which causes the turbulent diffusivities to change sign inside the flow, has been found to occur both experimentally and theoretically throughout large regions of turbulent flames. Thus, it is of interest to determine time spent by the combustible mixture at various locations.

To partly remedy these problems, researchers have turned to statistical methods. These methods employ a probability density function (PDF) to provide a statistical description of either the entire turbulent field or just the quantities involved in the chemical source term (i.e., temperature, enthalpy, and species mass fractions). Several PDFs have been developed by researchers, including ones based on rectangular waves, the Gaussian distribution, and the Student's t-distribution. Each of these PDFs is used in conjunction with the governing equations.

Example 1

(a) Recall the mixture fraction definition from Chapter 8 and Chapter 12. If the measured local "f" follows f(t) = D + A sin B t, Plot f vs. time t if A = 0.04, B = 15 sec^{-1}, and D = 0.1, and plot $f(\tau)$ if $\tau = t/t_{period}$.
(b) For propane, determine f_{stoich}.
(c) Determine the probability of f being less than 0.07 or Prob(f < 0.07).
(d) Determine probability of f being less than f + df but more than f – df, i.e., $\text{Prob}(f_{stoich} - df < f < f_{stoich} + df)$, where df = 0.05.
(e) If P(f,x), PDF, is defined as d(Prob)/df, then estimate the PDF at $f = f_{stoich}$.
(f) Determine $\sqrt{\{f'^2\}} = [\{f - f_{avg}\}^2]^{1/2}$.

Solution

(a) Because f(t) = 0.1 + 0.04 sin 15 t, the period of oscillation, $t_{period} = 2 \times \pi/15 = 0.419$ sec, and

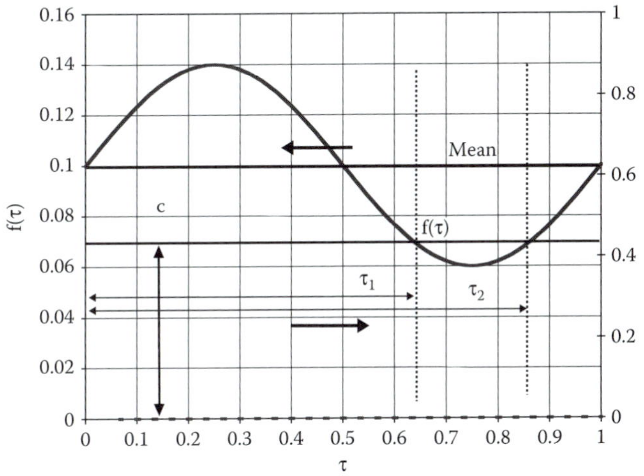

Figure 18.2 Illustration of mixture fraction variation.

$f(\tau) = 0.1 + 0.04 \sin [\{15 \times t_{period}\}\{t/t_{period}\}] = 0.1 + 0.04 \sin [\{15 \times t_{period}\ \tau]$, $0 < \tau < 1$.

The plot is shown in Figure 18.2. It is seen that $f(\tau)$ fluctuates between 0.06 and 0.14 with a mean of 0.1.

(b) $C_3H_8 + 5\ O_2 \rightarrow 3\ CO_2 + 4\ H_2O$

$$\nu_{O2} = 5 \times 32/(3 \times 12.01 + 8 \times 1.01) = 3.627.$$

For a pure mixing problem with pure fuel supplied with stoichiometric air,

$$(A{:}F)_{stoich} = 3.627/0.23 = 15.77.$$

From Chapter 2 and Chapter 10, $f_{stoich} = 1/\{(A{:}F)_{stoich} + 1\} = 0.06$.

(c) From Figure 18.2 it is seen that $f < 0.07$ for $\tau_1 = 0.63 < \tau < 0.86$.

Thus, the fraction of time spent by a mixture having $f < 0.07$ is $0.86 - 0.63 = 0.23$. More generally, let us assign $y(\tau) = 0$ if $f < 0.07$ and 0 if $f > 0.07$

$$\Pr\text{ob}\ (f < 0.07) = \int_0^1 y(\tau)d\tau = \int_0^{0.63} y(\tau)d\tau + \int_{0.63}^{0.86} y(\tau)d\tau + \int_{0.86}^{1} y(\tau)d\tau$$

$$= 0 + 1 \times \{0.86 - 0.63\} + 0 = 0.16$$

where the value of $y(\tau) = 1$ for $\tau_1 = 0.63 < \tau < 0.86$, $y(\tau) = 0$ for $\tau < \tau_1 = 0.63$ and $\tau > \tau_2 = 0.86$. Because of the way the function $y(\tau)$ is selected, the time interval $\tau_2 - \tau_1$ defines a probability.

(d) $\text{Prob}(f - df < f < f + df) = \text{Prob}(0.065 < f < 0.075) = \text{Prob}(f < 0.075) - \text{Prob}(f > 0.065) = \{\tau_{H2} - \tau_{H2}\}\ \{\tau_{L2} - \tau_{L2}\} = \{0.28\} - \{0.16\} = 0.12$

(e) PDF = P (f,x) at $f = 0.07$ = d(Prob)/df = 0.12/(2*0.005) = 12

(f) $$\sqrt{f'^2} = \left[\int_0^1 \left(f - f_{avg}\right)^2 d\tau\right]$$

where f_{avg} = D = 0.1. We can integrate for the given function. In order to generalize for any function, we can numerically integrate and get $\sqrt{f'^2} = 0.028$.

$$\text{Prob}\,(c - dc < f < c + dc) = \int_{\tau H1}^{\tau H2} y(\tau)d\tau - \int_{\tau L1}^{\tau L2} y(\tau)d\tau$$

$$= \{\tau_{H2} - \tau_{H1}\} - \{\tau_{L2} - \tau_{L1}\} = (\tau_{L1} - \tau_{H1}) + (\tau_{H2} - \tau_{L2})$$

Remarks

(i) The mixture fraction f(t) varies with location x. Near the fuel injector in a boiler burner, f remains around 1, whereas around air injection points, f is 0. Thus f = f(t,x), and the functional form for f(t) can vary with x.
(ii) If f = 0.07, then P(f, x) at 0.7 = 8.1. Thus PDF varies with f.
(iii) The PDF P = P (f, x).
(iv) Prob(a < f < b) = $\int_a^b \text{P}(f,x)df$.

18.7 Probability Density Function

We saw from Example 1 that the PDF is defined as

PDF = P(f,x) = d{Prob(u)}/df at any given location x,

where Prob(u) = probability that fluctuating variable u < f = {number of times u is less than f}/{number of times u < f + number of times u > f}. One can also get the rms value of fluctuations. The clipped Gaussian distribution function is a normal Gaussian distribution with clippings at f = 0 and f = 1 for the mixture fraction because 0 ≤ f ≤ 1. The PDF at any x is normally defined as

$$P(f,x) = \frac{1}{\sigma\sqrt{2\pi}} \exp\left[-\frac{1}{2}\left(\frac{f - f_{mp}}{\sigma}\right)^2\right], \tag{18.61}$$

where f_{mp} (x) is the most probable value of f (i.e., P = P (f,x) is the maximum) and σ (x) is the variance of f at a given location x. The values of f_{mp} (x) and σ will be determined later from $\tilde{f}$ *and* $\overline{f''^2}$.

The cumulative probability

$$F(c,x) = P(f < c, x) = \int_{-\infty}^{c} \frac{1}{\sigma\sqrt{2\pi}} \exp\left[-\frac{1}{2}\left(\frac{f - f_{mp}}{\sigma}\right)^2 df. \tag{18.62}\right.$$

However, f ranges from 0 to 1 only. Thus,

$$P(f,\ x) = \int_{-\infty}^{0} \frac{1}{\sigma\sqrt{2\pi}} \exp\left[-\frac{1}{2}\left(\frac{f - f_{mp}}{\sigma}\right)^2\right] df, \ \ if \ \ f = 0 \tag{18.63}$$

$$P(f,x)=\frac{1}{\sigma\sqrt{2\pi}}\exp\left[-\frac{1}{2}\left(\frac{f-f_{mp}}{\sigma}\right)^2\right],\quad if \quad 0<f<1. \tag{18.64}$$

$$P(f,x)=\int_1^{\infty}\frac{1}{\sigma\sqrt{2\pi}}\exp\left[-\frac{1}{2}\left(\frac{f-f_{mp}}{\sigma}\right)^2\right]df,\quad if \quad 1<f<\infty. \tag{18.65}$$

Because

$$erf(y)=\frac{2}{\sqrt{\pi}}\int_0^y \exp[-y^2]\,dy,\quad erf(\infty)=1 \tag{18.66}$$

with $z = (f - f_{mp})/\sigma$, the cumulative probability is given as

$$F(z_c,x) = P(f<z_c,x)=\frac{1}{\sqrt{2\pi}}\int_{-\infty}^{z_c}\exp\left[-\frac{z^2}{2}\right]dz$$

$$=\frac{1}{\sqrt{\pi}}\int_{-\infty}^{y_c}\exp[-y^2]\,dy \quad where \quad y=\frac{z}{\sqrt{2}}=\frac{f-f_{mp}}{\sigma\sqrt{2}}$$

$$F(y_c,x)=P(f<y_c,x)=0.5\left[1\pm erf\left|y_c\right|\right],\quad +sign \quad if \quad y_c>0, -sign$$

$$if \quad y_c<0,\ y_c=\frac{z_c}{\sqrt{2}}=\frac{c-f_{mp}}{\sigma\sqrt{2}}. \tag{18.67}$$

The area within z (z = ± 1, i.e., one standard deviation) is about 68.3% of the total area.

18.7.1 Property q and Average q

The convolution process determines the mean value of a variable q. Let f be a random variable (primary mixture fraction) and P(f, x), its PDF. If variable q is a function of the random variable f, then the mean value of q can be determined using the convolution operation, i.e.,

$$\bar{q}=\int_{-\infty}^{+\infty}q(x,f)p(f)df$$

The limits of integration are mathematically correct but unrealistic for the present problem because the mixture fractions can vary only between 0 and 1. Therefore, the integral can be split into three parts, i.e.,

$$\bar{q} = \int_{-\infty}^{0} q(0,x)p(f,x)df + \int_{0}^{1} q(f,x)p(f,x)df + \int_{1}^{+\infty} q(1,x)p(f,x)df \quad (18.68)$$

$$= \alpha_0 q_0(x,0) + \int_{0}^{1} q(f,x)p(f,x)df + \alpha_1 q_1(x,1)\,. \quad (18.69)$$

In these equations, f = 0 means a secondary stream, and f = 1 means a primary stream. The α's refer to the intermittency of a particular stream, in other words, the probability of occurrence of a stream (say, primary fuel stream with f = 1, secondary stream with f = 0, etc.) at a certain location. Thus, if f = 1 at some location, it represents pure fuel, whereas f = 0 represents pure air. It should be mentioned that the PDF and the intermittencies vary throughout the reactor. If q = f, then

$$\tilde{f} = \alpha_0 f(x,0) + \int_{0}^{1} f(x)p(f,x)df + \alpha_1 f(x,1) = \int_{0}^{1} f(x)p(f,x)df + \alpha_1$$

(18.70)

$$\tilde{f}''^2 = \alpha_0 (f-\tilde{f})^2(x,0) + \int_{0}^{1} (f-\tilde{f})^2 p(f,x)df + \alpha_1 (f-\tilde{f})^2(x,1)$$
$$= \alpha_0 \tilde{f}^2(x,0) \int_{0}^{1} (f-\tilde{f})^2(x)p(f,x)df + \alpha_1\alpha_1(1-\tilde{f})^2. \quad (18.71)$$

The values of f can also be obtained from (for steady state problems)

$$\bar{\rho}\tilde{v}_k \frac{\partial \tilde{f}}{\partial x_k} = \frac{\partial}{\partial x_k}\left(\frac{\mu_t}{\sigma_t}\frac{\partial \tilde{f}}{\partial x_k}\right) \quad (18.72)$$

$$\bar{\rho}\tilde{v}_k \frac{\partial \tilde{f}''^2}{\partial x_k} = \frac{\partial}{\partial x_k}\left(\frac{\mu_t}{\sigma_t}\frac{\partial \tilde{f}''^2}{\partial x_k}\right) + \left(2\frac{\mu_t}{\sigma_t}\frac{\partial \tilde{f}}{\partial x_k}\right)^2 - C_D\left(2\frac{\bar{\rho}\varepsilon}{k}\tilde{f}''^2\right). \quad (18.73)$$

Recall that

$$\tilde{f}(x) = \int_0^1 f(x)\ P(f,x)\ df = \int_0^1 f(x)\ \frac{1}{\sigma\sqrt{2\pi}} \exp\left[-\frac{1}{2}\left(\frac{f-\mu}{\sigma}\right)^2 df\right.$$

$$\tilde{f}''^2(x) = \int_0^1 \left(f(x) - \tilde{f}(x)\right)^2 P(f,x)\ df$$

$$= \int_0^1 \left(f(x) - \tilde{f}(x)\right)^2 \frac{1}{\sigma\sqrt{2\pi}} \exp\left[-\frac{1}{2}\left(\frac{f-\mu}{\sigma}\right)^2 df\right. \qquad (18.74)$$

where σ is the turbulent Pr or Sc number, and the constant $C_D = 2$.

18.7.2 Reaction Rate Expression

The time-mean reaction rate is obtained by convolution over the PDF of mixture fractions. To illustrate this procedure, consider the reaction

$$O + H \rightarrow OH.$$

The instantaneous forward rate is

$$\dot{w}_{OH} = \rho^2 Y_O Y_H\ A\ \exp(-E/RT).$$

If all the variables in the rate expression are functions of mixture fractions, then the mean reaction rate is given as

$$\bar{\dot{w}}_{OH}(x) = \int_{\eta_1}\int_{\eta_2}\int_{\eta_3} \dot{w}_{OH}(f,x) P(x,f) df. \qquad (18.75)$$

The PDF is the statistical distribution of the variable at any local point in space because of fluctuations of the variable in time. Once the time-mean reaction rate for a species is found, the species balance equation can be solved to find the local time-mean concentration of that species.

18.7.3 Qualitative PDFs for a Few Problems

For the case of planar jet mixing, the f value is close to 1 on the fuel side and f = 0 on the air side with f between 0 and 1 after partial mixing. For a pure fuel jet injected into stagnant or coflowing air, f = 0 near the mixing layer and 0 < f < 1 near the axis. For fuel flow with multiple air jets, the PDF is close to the Gaussian distribution curve. The joint probability density function (JPDF) for two-mixture fraction is

$$P = P\ (c,x) = d\{\text{Prob}(f < c)\}/dc. \qquad (18.76)$$

Consider the velocity = v_1. If $v_1 < c_1$, then

$$P = P\ (c_1) = d\{Prob(v_1 < c_1)\}/dc_1. \tag{18.77}$$

Consider the velocity = v_2. If $v_2 < c_2$, then

$$P = P\ (c_2) = d\{Prob(v_2 < c_2)\}/dc_2. \tag{18.78}$$

What is the probability for both $v_1 < c_1$ and $v_2 < c_2$?

$$P = P_j\ (c_1,\ c_2,\ x) = d^2\{Prob(v_1 < c_1,\ v_2 < c_2,\ x),\}/dc_1c_2 \tag{18.79}$$

Consider any property q that depends upon c_1 and c_2 at a given location x. The average based on the JPDF is

$$\tilde{q}(x) = \int_0^1 q(c_1,c_2,x)\ \ P_j(c_1,c_2,x)\ \ dc_1c_2. \tag{18.80}$$

Suppose we write $Prob(v_1 < c_1,\ v_2 < c_2) = Prob(v_1 < c_1)*prob(v_2 < c_2)$ as though these two events are independent of each other.

$$P(c_1,c_2,x) = \frac{d^2}{dc_1dc_2}\left\{Prob(c_1,x)*Prob(c_2,x)\right\}$$

$$= \frac{d}{dc_2}*\left[Prob(c_2,x)\frac{d}{dc_1}\left\{Prob(c_1,x)\right\}\right]$$

$$P(c_1,c_2,x) = \left[\frac{d}{dc_2}Prob(c_2,x)\frac{d}{dc_1}\left\{Prob(c_1,x)\right\}\right] \tag{18.81}$$

$$P(c_1,c_2,x) = \left[\frac{d}{dc_2}Prob(c_2,x)\frac{d}{dc_1}\left\{Prob(c_1,x)\right\}\right] = P(c_1,x)\ P(c_2,x) \tag{18.82}$$

$$\tilde{q}(x) = \int_0^1 q(c_1,c_2,x)\,P_j(c_1,c_2,x)\ \ dc_1c_2 = \int_0^1 q(c_1,c_2,x)P(c_1,x)P(c_2,x)\,dc_1c_2 \tag{18.83}$$

18.7.4 Mixture Fraction Governing Equations

18.7.4.1 Single Mixture Fraction

Consider the steady element conservation equation (Chapter 8)

$$\rho v_x\frac{\partial Z_j}{\partial x} + \rho v_y\frac{\partial Z_j}{\partial y} = \frac{\partial}{\partial x}\left(\rho D\frac{\partial Z_j}{\partial x}\right) + \frac{\partial}{\partial y}\left(\rho D\frac{\partial Z_j}{\partial y}\right) \tag{18.84}$$

where Z_j is the mass fraction of element j in the mixture. Generalizing in tensor form including nonsteady terms,

$$\frac{\partial(\rho Z_j)}{\partial t}+\frac{\partial(\rho v_k Z_j)}{\partial x_k}=-\frac{\partial}{\partial x_k}\left(\rho D\,\frac{\partial Z_j}{\partial x_k}\right). \tag{18.85}$$

The Favre-averaged element transport equation is given as

$$\frac{\partial(\bar{\rho}\tilde{v}_k\tilde{Z}_j)}{\partial x_k}=-\frac{\partial}{\partial x_k}\left(\overline{\rho D\,\frac{\partial Z_j}{\partial x_k}}-\overline{\rho v_k{}'' Z_j{}''}\right),\qquad j = 1,\ 2,\ \ldots\ L-1$$

where L denotes the number of elements. Apart from Z, other conserved scalars are SZ (but with a single step reaction) and the mixture fraction f. For a multistep reaction the element mass fraction equation is more useful, where, D is assumed to be the same for all species. If D's are not the same, there may be source terms and hence Z_j x could no longer be a conserved scalar. In turbulent flows, the turbulent diffusion is more significant than the molecular diffusion D. If a conserved scalar is represented by the symbol q, then ignoring molecular diffusion terms

$$\frac{\partial(\bar{\rho}\tilde{v}_k\tilde{q})}{\partial x_k}=\frac{\partial}{\partial x_k}\left(\overline{\rho v_k''\, q''}\right). \tag{18.85a}$$

The mixture fraction f is also a conserved scalar. Its Favre variance is described by

$$\bar{\rho}\tilde{v}_k\,\frac{\partial(q''^2)}{\partial x_k}=-2\overline{(\rho v_k''\,\xi'')}\left(\frac{\partial\tilde{q}}{\partial x_i}\right)-\left(\frac{\partial\overline{(\rho v_k''\,q''^2)}}{\partial x_k}\right)-2\overline{\left(\rho D\frac{\partial q''}{\partial x_k}\frac{\partial q''}{\partial x_k}\right)}. \tag{18.85b}$$

Solutions for $\tilde{q}$ and $\sqrt{q''^2}$ can be obtained from the above two equations.

18.7.4.2 Mixture Fraction with Source Terms

The Favre-averaged transport equation for mixture fraction $\tilde{f}$ is

$$\frac{\partial\bar{\rho}\tilde{f}}{\partial t}+\nabla\cdot(\bar{\rho}\tilde{f}\tilde{\vec{V}})=\nabla\cdot\left(\frac{\mu_e}{\sigma_\eta}\nabla\tilde{f}\right)+\overline{S^f},\qquad i = 1,\ 2,\ldots,N \tag{18.86}$$

where $\tilde{f}$ is the Favre mean of f, $S^{\bar{f}}$ is the source term in Reynolds mean, and

$$S^f=\rho\dot{f} \tag{18.87}$$

$\dot{f}$ denotes the generation rate of f.

For example, for two streams of pure gas fuel and air, one can obtain f and the stoichiometric contour f_{st} can be determined. Therefore, for $f < f_{st}$, it is a fuel-lean regime and hence $Y_F = 0$; for $f > f_{st}$, $Y_{O2} = 0$. In Chapter 8, we have shown that Y_k is a unique function of $Y_k(f)$. Similarly, we can get T = T(f). If f fluctuates, so do Y_k's and T's. In Example 1 we saw the probability that f lies between f – df and f + df is a function of f. Further, the waveform may change whether it is near the fuel or near the wall; thus, the PDF at any location is given by

$$\bar{\rho}\tilde{Y}_k(x) = \int_0^1 Y_k(f)\rho\, P(f,x)df \tag{18.88}$$

$$\overline{\rho Y_k^{\,2}(x)} = \int_0^1 \{Y_k(f) - \bar{Y}_k(f)\}^2 \rho\, P(f,x)\, df. \tag{18.89}$$

One can replace Y_k by h, Z_k, etc., so that we have a relation for Y_k in terms of f:

$$f/f_i = f/Y_{F,i,0} = (\beta - \beta_\infty)/(\beta_I - \beta_\infty) = \{h - h_\infty\}/\{h_i - h_\infty\}$$

$$1 \text{ kg of fuel (F)} + \nu_{O2} \text{ kg of } O_2 \rightarrow \nu_P\ (= 1 + \nu_{O2}) \text{ kg of products}$$

$$f/Y_{F,i,0} = [Y_F + (Y_P/\nu_P)]/[Y_{Fi,0}] = [Y_F - (Y_{O2}/\nu_{O2}) + Y_{O2,\infty}/\nu_{O2}]/[Y_{Fi,0} + Y_{O2,\infty}/\nu_{O2}]$$
$$f_{stoich}/Y_{F,i,0} = [Y_{O2,\infty}/\nu_{O2}]/[Y_{Fi,0} + Y_{O2,\infty}/\nu_{O2}] = s/(1 + s)$$

$$s = Y_{O2,\infty}/(\nu_{O2}\, Y_{Fi,0})$$

Note $s = f_{stoich}/\{Y_{F,i,0} - f_{stoich}\}$, $1 + s = Y_{F,i,0}/\{Y_{F,i,0} - f_{stoich}\}$, when $f > f_{stoich}$ (e.g., fuel-rich regime),

$$Y_{O2} = 0$$

$$[Y_F + Y_{O2,\infty}/\nu_{O2}]/[Y_{Fi,0} + Y_{O2,\infty}/\nu_{O2}] = f/Y_{F,i,0}$$

$$[Y_F/Y_{Fi,0}] = \{f/Y_{F,i,0}\}\{1 + s\} - s$$

$$Y_F = f\{1 + s\} - s\, Y_{F,i,0} = (1 + s)\, \{f - f_{stoich}\} = [Y_{F,i,0}/\{Y_{F,i,0} - f_{stoich}\}]\{f - f_{stoich}\},$$

$$\bar{\rho}\tilde{Y}_F(x) = \int_0^{fstoich} f\rho P(f:x)\, df + \int_{fstoich}^1 \frac{Y_{F,i,0}(f - f_{stoich})}{(Y_{F,i,0} - f_{stoich})}\rho\ \ P(f:x)\, df$$

(18.90)

$$f/Y_{F,i,0} = [Y_F + (Y_P/(\nu_{O2+1})]/[Y_{Fi,0}]$$

The relation f vs. Y_F is linear

$$Y_P/(\nu_{O2+1}) = (f - Y_F) = f - [Y_{F,i,0}/\{Y_{F,i,0} - f_{stoich}\}]\{f - f_{stoich}\}$$
$$= f_{stoich}\ \{Y_{F,i,0} - f\}/\{Y_{F,i,0} - f_{stoich}\}$$

$$Y_P = f_{stoich}\ (\nu_{O2+1})\{Y_{F,i,0} - f\}/\{Y_{F,i,0} - f_{stoich}\}$$

When $f < f_{stoich}$ (e.g., fuel-rich regime),

$$Y_F = 0$$

$$[-Y_{O2}/\nu_{O2} + Y_{O2\infty}/\nu_{O2}]/[Y_{Fi,0} + Y_{O2\infty}/\nu_{O2}] = f/Y_{F,i,0}$$

$$[-Y_{O2}/(\nu_{O2} Y_{Fi,0})] = \{f/Y_{F,i,0}\}\{1 + s\} - s$$

$$Y_{O2} = -\{f\nu_{O2}\}\{1 + s\} + s(\nu_{O2} Y_{Fi,0}) = (s + 1)\ \nu_{O2}\ \{f_{stoich} - f\}$$

$$= [Y_{F,i,0}/\{Y_{F,i,0} - f_{stoich}\}]\nu_{O2}\ \{f_{stoich} - f\}$$

$$Y_P = f\ (\nu_{O2+1}).$$

18.7.4.3 Favre Averaging

Because Y_k expressions are known in terms of f, one can get average value of Y_k. For example, using expressions for Y_P when $f < f_{stoich}$ and $f > f_{stoich}$, we have

$$\bar{\rho}\tilde{Y}_P(x) = (\nu_{O2}+1)\int_0^{f_{stoich}} f\,\rho P\,(f:x)\,df + (\nu_{O2}+1) \times \int_{f_{stoich}}^{1} \frac{f_{stoich}\ (Y_{F,i,0} - f)}{(Y_{F,i,0} - f_{stoich})}\rho P\,(f:x)\,df \qquad (18.91)$$

Thus, at a location x, the mixture fraction f varies with time and hence if fuel rich, $Y_F > 0$, and if fuel lean, $Y_F = 0$; thus, the mean may not be $Y_F = 0$. In particular, the stoichiometric location where $Y_F = 0$ and $Y_{O2} = 0$ will oscillate with time and, hence, the flame may oscillate accompanied by pressure oscillations.

18.7.5 Equilibrium Chemistry

Instead of a one-step reaction, if equilibrium chemistry is used, then one can replace Y_k by h, Z_k, etc.

$$f/f_i = f/Y_{F,i,0} = \{h - h_\infty\}/\{h_i - h_\infty\} = \{Z_k\ Z_{k\infty}\}/\{Z_{ki} - Z_{k\infty}\}$$

These relations express Z_k and h in terms of f. With inputs of Z_k and h, one may determine Y_k, T, etc., using equilibrium chemistry (Chapter 3).

$$\bar{\rho}\tilde{Y}_k^{\,e}(x) = \int_0^1 Y_k^{\,e}(f)\,\rho\,P\,(f:x)df \qquad (18.92)$$

Example 2

(a) A combustor is fired with propane and air in an unmixed state. Plot the measured local "f" at a location x follows f(t) = D + A sin B t vs. time t, where A = 0.04, B = 15 sec^{-1}, D = 0.1; plot f(τ) if $\tau = t/t_{period}$. If P(f, x), the PDF, is defined as d(Prob)/df, then estimate its value at f = 0.06, 0.07, …, 0.14 (fuel-rich regime) and plot P (f, x) vs. f.

(b) Calculate $\tilde{Y}_F$ assuming density is constant and the same as the mean value.

Solution

(a) Because f(t) = 0.1 + 0.04 sin 15 t the period of oscillation $t_{period} = 2\pi/15$ = 0.419 sec, and
f(τ) = 0.1 + 0.04 sin [{15 × t_{period}}{t/t_{period}}] = 0.1 + 0.04 sin [{15 × t_{period} τ], 0 < τ < 1
(see Figure 18.2).

(b) $C_3H_8 + 5\ O_2 \rightarrow 3\ CO_2 + 4\ H_2O$

$$\nu_{O2} = 5{*}32/(3{*}12.01 + 8{*}1.01) = 3.627$$

$(A{:}F)_{stoich} = 3.627/0.23 = 15.77$, $f_{stoich} = s/(s + 1) = 1/\{(A{:}F)_{stoich} + 1\} = 0.06$

$$\bar{\rho}\tilde{Y}_F(x) = \int_0^{fstoich} f\,\rho P\,(f:x)\,df + \int_{fstoich}^{1} \frac{Y_{F,i,0}(f - f_{stoich})}{(Y_{F,i,0} - f_{stoich})}\rho P\,(f:x)\,df$$

where dProb = Prob(f − df < f < f + df) = $\{dProb/df\}_f{*}df$ = pdf at f*df (see Example 1)

$$\bar{\rho}\tilde{Y}_F(x) = 0 + \int_{0.06}^{0.14} \frac{Y_{F,i,0}(f - f_{stoich})}{(Y_{F,i,0} - f_{stoich})}\rho P\,(f:x)\,df$$

$$+ \int_{0.14}^{1} \frac{Y_{F,i,0}(f - f_{stoich})}{(Y_{F,i,0} - f_{stoich})}\rho P\,(f:x)\,df$$

Note: *Prob(a < f < b)* = $\int_a^b P(f,x)df$. Thus, *Prob (0.14 < f < 1) = 0,*

$$\bar{\rho}\tilde{Y}_F(x) = 0 + \int_{0.06}^{0.14} \frac{Y_{F,i,0}(f - f_{stoich})}{(Y_{F,i,0} - f_{stoich})}\rho P\,(f:x)\,df + 0$$

where the intermittency at f = 0 to 0.04 and f = 0.14 to 1 are assumed to be zero.

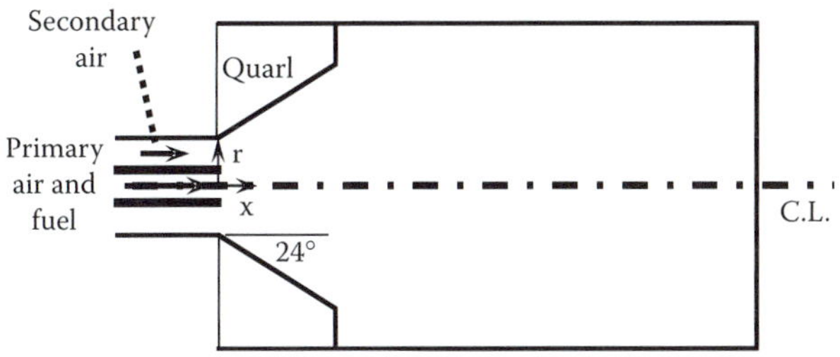

Figure 18.3 Schematic for combustion in wall-fired boilers.

Remarks

(i) Note that $f/f_i = \phi = (h - h_\infty)/(h_I - h_\infty)$. Thus, the total enthalpy h vs. f is also linear. This relation is useful for equilibrium chemistry.

(ii) If we use $\beta = \beta_{hT\text{-}O2}$, $f/f_i = \phi = \{c_p\,(T - T_i)/h_c\ \ [c_p(T_\infty - T_i)/h_c + (Y_{O2,\infty}/\nu_{O2})]\}/\{\,-[c_p(T_\infty - T_i)/h_c + (Y_{O2,\infty}/\nu_{O2})]\}$. Again, T vs. f is linear in the fuel-rich regime.

(iii) Using $\beta = \beta_{hT\text{-}F}$, $f/f_i = \phi = \{c_p\,(T - T_\infty)/h_c\}/\{[c_p(T_i - T_\infty)/(h_c + Y_{F,i})]\}$. Again, T vs. f is linear in the fuel-lean regime.

For example, one can obtain the mean mixture fraction and variance for two pure streams, fuel and air. Once the means $\tilde{f}, \sqrt{f''^2}$ are known, the stoichiometric contour $\tilde{f}_{st}$ can be determined.

18.7.6 *Two-Mixture Fraction Model*

In coal-fired burners, the primary air is used to transport coal under suspension mode. Coal first devolatilizes and mixes with the oxidizer. A schematic of a wall-fired burner is shown in Figure 18.3. Typically, coal is transported by primary air, and secondary air is given a swirl motion. The fuel mixture leaves through a quarl (a diffuser) made of refractory material, which provides radiant heat input. If fast kinetics is assumed for the gas-phase reactions, the reaction rates may be only functions of the local extent of primary gas mixing, local extent of coal off-gas mixing, or both. Once the extent of primary mixing with local gas and/or coal gas mixing with local gas is known locally, the local elemental composition and enthalpy level can be calculated. These quantities can be used with a Gibbs free energy scheme to calculate equilibrium species composition and gas temperatures. The effects of turbulence are incorporated by decomposing the primary and coal off-gas mixture fractions (i.e., f and η) into mean and fluctuating quantities (i.e., mean and variance). In essence, the mean and variance of primary and coal off-gas mixture fractions are used to obtain local PDFs for each mixture fraction. Convolution is then invoked on each dependent variable, such as temperature, density, equivalence ratio, and species composition, to calculate mean values of these dependent variables. This technique has been used by many researchers to obtain time-mean equilibrium concentrations of major species.

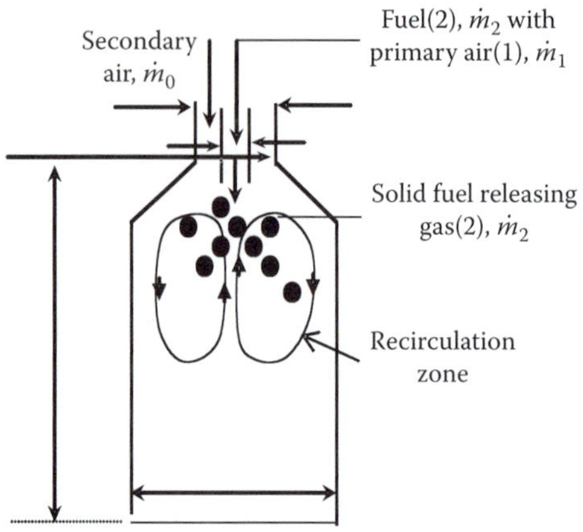

Figure 18.4 Illustration of two-mixture fraction.

In a two-mixture fraction model, another mixture fraction for coal off-gas is added. In a boiler burner in which coal is carried by primary air, the mixture fractions are defined as follows:

$$f_1 = \frac{\dot{m}_1}{\dot{m}_0 + \dot{m}_1} \tag{18.93}$$

$$f_2 = \frac{\dot{m}_2}{\dot{m}_0 + \dot{m}_1 + \dot{m}_2} \tag{18.94}$$

where $\dot{m}_0$, $\dot{m}_1$, and $\dot{m}_2$ refer, respectively, to secondary air, primary air, and coal off-gas with the same elemental gas composition as the basic elemental distribution of coal (Figure 18.4). Note that $(1 - f_1)$ refers to the fraction of gas originating from secondary stream, whereas $(1 - f_2)$ refers to the fraction originating from primary and secondary gas streams. Thus, $f_2 = 1$ implies pure coal off-gas without any air from the primary or secondary stream, whereas $f_2 = 0$ implies the presence of only air, from the primary stream, secondary stream, or both.

18.7.7 Three-Mixture Fraction: Calculation of Time-Mean Reaction Rates

Mean reaction rates for NO gas-phase reactions are calculated using the PDF approach, in which instantaneous reaction rates are convoluted over the PDFs of the three-mixture fractions. Once mean rates are available, overall mean reaction rates for each species are calculated. These overall mean reaction rates for nitrogenous species are used in species equations, which are solved to determine nitrogenous species concentration. Reaction V describes the reduction reaction of NO on the char surface. Because the timescale of the char reduction reaction is much larger than the turbulence timescale, slow kinetics is assumed. Hence, no convolution process is needed and the mean reaction rate for the NO-char reduction process is calculated from knowledge of mean gas properties.

When a coal and biomass blend is burned, we use three-mixture fractions defined as

$$\eta_1 = \frac{m_1}{m_0 + m_1} \tag{18.95}$$

$$\eta_2 = \frac{m_2}{m_0 + m_1 + m_2} \tag{18.96}$$

$$\eta_3 = \frac{m_3}{m_0 + m_1 + m_2 + m_3} \tag{18.97}$$

where m_0, m_1, m_2, and m_3 denote local masses of primary air, off-gas from dry coal, water vapor from wet coal, and off-gas from dry biomass in coal, respectively, assumed to be invariant with respective chemical compositions [Dhanapalan et al., 1997a,b]. Alternatively, one may use coal only but include moisture evaporation from wet coal, i.e., in this case, m_3 denotes local mass of water vapor that originates from wet coal.

With the fast chemistry assumption, local instantaneous temperature and chemical properties of the gas phase can be evaluated using Gibbs free energy minimization method once the mass-mixing extent, total enthalpy, and pressure are known locally, i.e.,

$$\phi = \phi(\eta_1, \eta_2, \eta_3, h, p). \tag{18.98}$$

The local enthalpy h can be partitioned into two parts: the adiabatic part of local enthalpy h_η and the enthalpy residual h_r

$$h = h_\eta + h_r\,. \tag{18.99}$$

h_η is a function of η_1, η_2, and η_3 only, according to Crocco's similarity as shown in Equation 18 of Chapter 4; h_r is calculated with the turbulence-mean enthalpy and mean mixture fractions as in Equation 18.19 of Chapter 4. Thus,

$$\phi = \phi(\eta_1, \eta_2, \eta_3, h_r, p). \tag{18.100}$$

Local turbulence-mean chemical properties, gas density, and temperature are determined using a convolution over the JPDF of the mixture fractions as follows:

$$\tilde{\phi} = \int_{-\infty}^{+\infty}\int_{-\infty}^{+\infty}\int_{-\infty}^{+\infty} \phi(\eta_1, \eta_2, \eta_3, h_r, \bar{p})\, P(\eta_1, \eta_2, \eta_3)\, d\eta_1 d\eta_2 d\eta_3 \tag{18.101}$$

where $P(\eta_1,\eta_2,\eta_3)$ is the local JPDF. If η_1, η_2, and η_3 are statistically independent, the JPDF can be separated as

$$P(\eta_1,\eta_2,\eta_3) = P(\eta_1)P(\eta_2)P(\eta_3) \tag{18.102}$$

where $P(\eta_1)$, $P(\eta_2)$, and $P(\eta_3)$ are the independent PDFs for $\eta_1,\eta_2,\ldots,\eta_N$, respectively.

Thus,

$$\tilde{\phi} = \int_{-\infty}^{+\infty}\int_{-\infty}^{+\infty}\int_{-\infty}^{+\infty} \phi(\eta_1,\eta_2,\eta_3,h_r,\bar{p})P(\eta_1)P(\eta_2)P(\eta_3)d\eta_1 d\eta_2 d\eta_3 \,. \tag{18.103}$$

Owing to turbulent eddy transportation (of smallest eddy on the order of 1 mm > ℓ_{mean}), intermittency for η_1, η_2, and η_3 may occur. Intermittencies for η_i are defined as the fraction of time when η_i equals 0 or 1 at a local point, mathematically expressed by the equations

$$a_i = \int_{-\infty}^{0} P(\eta_i)d\eta_i,\ \ b_i = \int_{1}^{+\infty} P(\eta_i)d\eta_i \,,\ (i = 1,2,3). \tag{18.104}$$

Because the physical upper and lower limits of mixture fractions are 0^+, 1^-, and 1. By letting N = 3, the expansion expression for N mixture fractions in reduced to

$$\tilde{\phi} = \int_{-\infty}^{+\infty}\int_{-\infty}^{+\infty}\int_{-\infty}^{+\infty} \phi(\eta_1,\eta_2,\eta_3,h_r,\bar{p})P(\eta_1)P(\eta_2)P(\eta_3)d\eta_1 d\eta_2 d\eta_3 = \sum_{i=1}^{15} I_i \tag{18.105}$$

where

$$I_1 = a_1 a_2 a_3 \phi(0,0,0,h_r,\bar{p})$$

$$I_2 = b_1 a_2 a_3 \phi(1,0,0,h_r,\bar{p})$$

$$I_3 = b_2 a_1 \phi(0,1,0,h_r,\bar{p})$$

$$I_4 = b_3 \phi(0,0,1,h_r,\bar{p})$$

$$I_5 = \int_{0^+}^{1^-}\int_{0^+}^{1^-}\int_{0^+}^{1^-} \phi(\eta_1,\eta_2,\eta_3,h_r,\bar{p})p(\eta_1)p(\eta_2)p(\eta_3)d\eta_1 d\eta_2 d\eta_3$$

$$I_6 = a_1\int_{0^+}^{1^-}\int_{0^+}^{1^-} \phi(0,\eta_2,\eta_3,h_r,\bar{p})p(\eta_2)p(\eta_3)\, d\eta_2 d\eta_3$$

$$I_7 = a_2 \int_{0^+}^{1^-} \int_{0^+}^{1^-} \phi(\eta_1, 0, \eta_3, h_r, \bar{p}) p\,(\eta_1) p\,(\eta_3)\, d\eta_1 d\eta_3$$

$$I_8 = a_3 \int_{0^+}^{1^-} \int_{0^+}^{1^-} \phi(\eta_1, \eta_2, 0, h_r, \bar{p}) p\,(\eta_1) p\,(\eta_2)\, d\eta_1 d\eta_2$$

$$I_9 = a_1 a_2 \int_{0^+}^{1^-} \phi(0, 0, \eta_3, h_r, \bar{p}) p\,(\eta_3)\, d\eta_3$$

$$I_{10} = a_1 a_3 \int_{0^+}^{1^-} \phi\,(0, \eta_2, 0, h_r, \bar{p}) p\,(\eta_2)\, d\eta_2$$

$$I_{11} = a_2 a_3 \int_{0^+}^{1^-} \phi\,(\eta_1, 0, 0, h_r, \bar{p}) p\,(\eta_1)\, d\eta_1$$

$$I_{12} = b_1 \int_{0^+}^{1^-} \int_{0^+}^{1^-} \phi\,(1, \eta_2, \eta_3, h_r, \bar{p}) p(\eta_2) p(\eta_3)\, d\eta_2 d\eta_3$$

$$I_{13} = b_1 a_2 \int_{0^+}^{1^-} \int_{0^+}^{1^-} \phi\,(1, 0, \eta_3, h_r, \bar{p}) p\,(\eta_3) d\eta_3$$

$$I_{14} = b_1 a_3 \int_{0^+}^{1^-} \int_{0^+}^{1^-} \phi\,(1, \eta_2, 0, h_r, \bar{p}) p\,(\eta_2) d\eta_2$$

$$I_{15} = b_2 \int_{0^+}^{1^-} \int_{0^+}^{1^-} \phi\,(0, 1, \eta_3, h_r, \bar{p}) p\,(\eta_3) d\eta_3$$

It is found that they are exactly the same as those proposed by Brewster et al. [1988] for the three-mixture-fraction PDF method.

The PDFs for three-mixture fractions are assumed to be Gaussian:

$$P(f_i) = (2\pi\sigma_{\eta_i})^{-1/2} \exp\left[-\frac{(f_i - \mu_{\eta_i})^2}{2\sigma_{\eta_i}} \right], \quad (\text{i} = 1,2,3)$$

The parameters σ_{η_i} and μ_{η_i} vary spatially and are determined at each local point using values of the turbulent-mean mixture fractions $\tilde{\eta}_i$ and their variances $\tilde{g}_{\eta_i}$. The $\tilde{\eta}_i$ and $\tilde{g}_{\eta_i}$ values (i = 1, 2, 3) must be calculated by solving their respective transport equations, which are reduced in PCGC-2 as

$$\frac{\partial}{\partial x}(\bar{\rho}\tilde{u}\tilde{\eta}_i) + \frac{1}{r}\frac{\partial}{\partial r}(r\bar{\rho}\tilde{v}\tilde{\eta}_i) = \frac{\partial}{\partial x}\left(\frac{\mu_e}{\sigma_\eta}\frac{\partial \tilde{\eta}_i}{\partial x} \right) + \frac{1}{r}\frac{\partial}{\partial r}\left(r\frac{\mu_e}{\sigma_\eta}\frac{\partial \tilde{\eta}_i}{\partial r} \right) + \overline{S^{\eta_i}} \quad (18.106)$$

$$\frac{\partial}{\partial x}(\bar{\rho}\tilde{u}\tilde{g}_{\eta_i}) + \frac{1}{r}\frac{\partial}{\partial r}(r\bar{\rho}\tilde{v}\tilde{g}_{\eta_i}) = \frac{\partial}{\partial x}\left(\frac{\mu_e}{\sigma_g}\frac{\partial \tilde{g}_{\eta_i}}{\partial x} \right) + \frac{1}{r}\frac{\partial}{\partial r}\left(r\frac{\mu_e}{\sigma_g}\frac{\partial \tilde{g}_{\eta_i}}{\partial r} \right) \quad (18.107)$$

where σ_η and σ_g are constants equal to 0.9, $\overline{S^{\eta_i}}$ is the source term approximately equal to the Reynolds mean production rate of mass m_i in the gas phase.

Therefore, to calculate turbulence-mean chemical properties, temperature, and density, one needs first to solve the transport equations for Favre-mean mixture fractions $\tilde{\eta}_i$ and its variance $\tilde{g}_{\eta_i}$. The Gaussian PDF can then be obtained at any location with local values of $\tilde{\eta}_i$ and $\tilde{g}_\eta$. The instantaneous chemical properties, temperature, and density can be obtained using Gibbs free energy minimization method.

It is noted that not all chemical reactions of the gas phase can be assumed with fast chemical kinetics. The timescales of NO production and reduction reactions are of the same order as that of turbulent mixing; thus, both kinetics and mixing must be considered when calculating turbulence-mean concentration of NO.

18.8 Premixed and Partially Premixed Turbulent Flames: Modeling Approaches

In the previous section, PDF methods based on the equilibrium chemistry approach were discussed. These are suitable for only one type of flame, the diffusion flame, in which fuel and oxidizer enter the combustion chamber through separate inlets. In premixed combustion, fuel and oxidizer are homogeneously mixed before entering the combustion chamber. Partially premixed combustion is a mix of both premixed and diffusion flames. It occurs when an additional fuel or oxidizer stream enters a premixed system. A diffusion flame, if lifted off, is also classified as a partially premixed flame.

What follows is a very brief description of some models that have been used in modeling premixed and partially premixed models. Basic governing equations are provided with appropriate references to the literature. Non-diffusion flames can be categorized based on the chemistry model.

18.8.1 Fast Kinetics

Magnussen [1976] presented a model that is applicable to all types of flames. This model is known as *eddy breakup* (EBU) model, in which the rates for each reaction are computed based on local turbulence levels (k and epsilon), stoichiometric coefficients, and species concentration. These reaction rates are used as source terms in the solution of standard species transport equations described earlier. Some empirical constants may need fine-tuning depending on the problem being solved.

Another model used specifically for premixed flames is based on tracking the flame front. This model solves a transport equation for a variable c, called the *progress variable.*

$$\partial(\bar{\rho}\,\tilde{c}) + \nabla\cdot(\bar{\rho}\,\tilde{c}\,\tilde{u}) = \nabla\cdot(\bar{\rho}\,D_t\nabla\tilde{c}) + \rho_u U_t\,|\nabla\tilde{c}| \tag{18.108}$$

In this equation, c takes the value 1 if the mixture is burned and 0 if it is not burned. To obtain properties such as mixture temperature, the following simple relation is used:

$$T = c\,T_b + (1 - c)\,T_u \tag{18.109}$$

where T_b is the temperature of burned products and T_u is the temperature of unburned reactants.

The most important parameter in this model is the turbulent flame speed U_t, which depends on the laminar flame speed of the mixture as well as the flame sheet wrinkling due to turbulence. Laminar flame speed is a function of fuel composition, temperature, molecular diffusion, and chemical kinetics.

Partially premixed flames are computed using an approach that combines the features of both PDF and progress variable methods. Mean field variables are computed using convolution over the mixture fraction and progress variable space. For example, mean temperature will be calculated as

$$T = c\int T(f)p(f)df + (1 - c)\int T(f,c)\;p(f,c)\;df \tag{18.110}$$

where the limits of integration are 0 and 1.

Figure 18.5 shows a representative plot of a progress variable and associated temperature contours in a premixed combustor. The two-equation standard k-ε model was used with axisymmetric geometry.

18.8.2 Finite-Rate Kinetics

If the assumption of fast kinetics cannot be made, finite-rate models are employed. These models can be used for all three types of flames discussed in the preceding text.

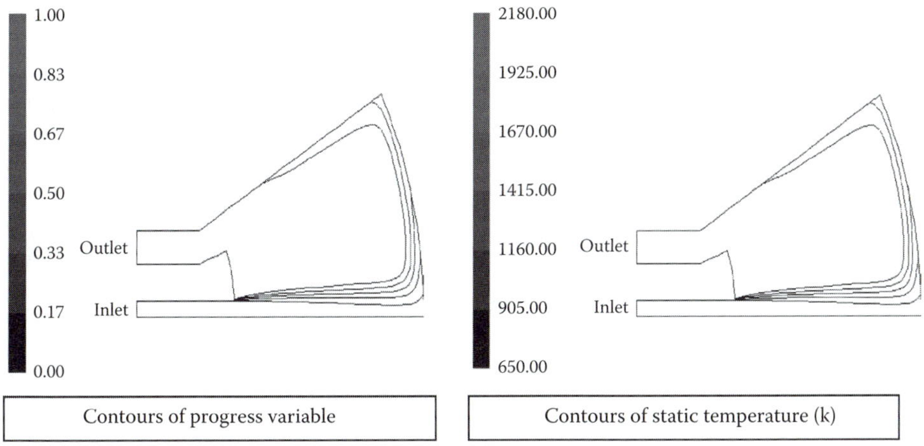

Figure 18.5 Contours of progress variable and static temperature in a premixed combustor. (Courtesy of Fluent Inc., Lebanon, New Hampshire.)

Magnussen [1981] presented an *eddy dissipation concept* model that includes the effect of turbulence on reaction rate source terms in the species transport equations. This model assumes that reaction occurs in small turbulent structures. Computationally, this model is quite expensive and is used only when the fast chemistry assumption is not valid, such as in CO and NO reaction mechanisms or for modeling ignition and extinction. This model is an extension of the EBU model.

Another method of solving finite-rate turbulent chemistry problems uses a transport equation for a JPDF of species, temperature, and pressure [Pope, 1985]. In this approach, Favre averaging is used to derive the composition PDF transport equation. Modeling the small scales of mixing is a major challenge in this approach because the single-point PDF does not provide information on gradients such as molecular diffusion.

18.9 Summary

The characteristics of turbulence are reviewed in this chapter. Turbulence acts as a promoter of mixing and as a dissipator of energy. As turbulent flows are continually flowing fluids, they are subject to the same governing equations as laminar flows. Two types of averaging are available, time averaging and mass averaging, to allow the laminar-flow-governing equations to be used with turbulent flows. Of these, mass (or Favre) averaging is preferred in most cases as it simplifies the governing equations, is more representative of the data obtained by probes and provides a way to circumvent variable-density complications.

Acknowledgments

A part of this chapter was contributed by Dr. Muhammad Sami of Fluent, Inc., and Dr. Gengsheng Wei of Flowscience, Inc.

18.10 Appendix I: Cylindrical Coordinate System with Particle-Laden Flow

Assuming particle volume is negligible, the continuity, momentum, and thermal energy equations are

$$\frac{\partial}{\partial x}(\rho u)+\frac{1}{r}\frac{\partial}{\partial r}(r\rho v)=S_p^m \tag{18.111}$$

$$\frac{\partial}{\partial x}(\rho uu)+\frac{1}{r}\frac{\partial}{\partial r}(r\rho vu)=-\frac{\partial p}{\partial x}+\frac{\partial}{\partial x}\left(2\mu\frac{\partial u}{\partial x}\right)+\frac{1}{r}\frac{\partial}{\partial r}\left(\mu r\frac{\partial u}{\partial r}\right)+\frac{1}{r}\frac{\partial}{\partial r}\left(\mu r\frac{\partial v}{\partial x}\right)$$
$$-\frac{\partial}{\partial x}\left(\frac{2}{3}\mu\nabla\cdot\vec{v}\right)+S_p^u+u_{pg}S_p^m \tag{18.112}$$

$$\frac{\partial}{\partial x}(\rho uv)+\frac{1}{r}\frac{\partial}{\partial r}(r\rho vv)=-\frac{\partial p}{\partial r}+\frac{\partial}{\partial x}\left(\mu\frac{\partial v}{\partial x}\right)+\frac{\partial}{\partial x}\left(\mu\frac{\partial u}{\partial r}\right)+\frac{1}{r}\frac{\partial}{\partial r}\left(2\mu r\frac{\partial v}{\partial r}\right)-2\mu\frac{v}{r^2}$$
$$+\rho\frac{w^2}{r}-\frac{1}{r}\frac{\partial}{\partial r}\left(\frac{2}{3}\mu r\nabla\cdot\vec{v}\right)+\frac{2\mu}{3r}\nabla\cdot\vec{v}+S_p^v+v_{pg}S_p^m \tag{18.113}$$

$$\frac{\partial}{\partial x}(\rho uw)+\frac{1}{r}\frac{\partial}{\partial r}(r\rho vw)=\frac{\partial}{\partial x}\left(u\frac{\partial w}{\partial x}\right)+\frac{1}{r}\frac{\partial}{\partial r}\left(\mu r\frac{\partial w}{\partial r}\right)-\rho\frac{vw}{r}-\frac{w}{r^2}\frac{\partial}{\partial r}(r\mu) \tag{18.114}$$

$$\frac{\partial}{\partial x}(\rho uh)+\frac{1}{r}\frac{\partial}{\partial r}(r\rho vh)=q_r+\frac{\partial}{\partial x}\left(\frac{\lambda}{c_p}\frac{\partial h}{\partial x}\right)+\frac{1}{r}\frac{\partial}{\partial r}\left(r\frac{\lambda}{c_p}\frac{\partial h}{\partial r}\right)+u\frac{\partial p}{\partial x}+v\frac{\partial p}{\partial r}$$
$$+S_p^{h,c}+h_{pg}S_p^m \tag{18.115}$$

where (u, v, w) are velocity components in (x, r, θ), S_p^u and S_p^v represent the aerodynamic-drag effects of the particles on the axial and radial components of momentum; $u_{pg}S_p^m$ and $v_{pg}S_p^m$, the momentum exchange due to particle mass efflux; $S_p^{h,c}$, represents the enthalpy source due to particle heat convection; $h_{pg}S_p^m$ and, the enthalpy due to particle mass efflux.

18.10.1 Favre-Averaged Governing Equations

Averaging the continuity, momentum, and thermal energy equations, we obtain the Favre-averaged equations as

$$\frac{\partial}{\partial x}(\bar{\rho}\tilde{u})+\frac{1}{r}\frac{\partial}{\partial r}(r\bar{\rho}\tilde{v})=\overline{S_p^m} \tag{18.116}$$

$$\frac{\partial}{\partial x}(\bar{\rho}\tilde{u}\tilde{u})+\frac{1}{r}\frac{\partial}{\partial r}(r\bar{\rho}\tilde{v}\tilde{u})=-\frac{\partial\bar{p}}{\partial x}+\frac{\partial}{\partial x}\left(2\mu_e\frac{\partial\tilde{u}}{\partial x}\right)+\frac{1}{r}\frac{\partial}{\partial r}\left(\mu_e r\frac{\partial\tilde{u}}{\partial r}\right)+\frac{1}{r}\frac{\partial}{\partial r}\left(\mu_e r\frac{\partial\tilde{v}}{\partial x}\right)$$
$$+\overline{S_p^u}+\tilde{u}_{pg}\overline{S_p^m} \tag{18.117}$$

$$\frac{\partial}{\partial x}(\bar{\rho}\tilde{u}\tilde{v})+\frac{1}{r}\frac{\partial}{\partial r}(r\bar{\rho}\tilde{v}\tilde{v})=-\frac{\partial\bar{p}}{\partial r}+\frac{\partial}{\partial x}\left(\mu_e\frac{\partial\tilde{v}}{\partial x}\right)+\frac{\partial}{\partial x}\left(\mu_e\frac{\partial\tilde{u}}{\partial r}\right)+\frac{1}{r}\frac{\partial}{\partial r}\left(2\mu_e r\frac{\partial\tilde{v}}{\partial r}\right)$$
$$-2\mu_e\frac{\tilde{v}}{r^2}+\rho\frac{\tilde{w}^2}{r}+\overline{S_p^v}+\tilde{v}_{pg}\overline{S_p^m} \tag{18.118}$$

$$\frac{\partial}{\partial x}(\bar{\rho}\tilde{u}\tilde{w})+\frac{1}{r}\frac{\partial}{\partial r}(r\bar{\rho}\tilde{v}\tilde{w})=\frac{\partial}{\partial x}\left(\mu_e\frac{\partial\tilde{w}}{\partial x}\right)+\frac{1}{r}\frac{\partial}{\partial r}\left(\mu_e r\frac{\partial\tilde{w}}{\partial r}\right)-\rho\frac{\tilde{v}\tilde{w}}{r}-\frac{\tilde{w}}{r^2}\frac{\partial}{\partial r}(r\mu_e) \tag{18.119}$$

$$\frac{\partial}{\partial x}(\rho\tilde{u}\tilde{h})+\frac{1}{r}\frac{\partial}{\partial r}(r\bar{\rho}\tilde{v}\tilde{h})=q_{rg}+\frac{\partial}{\partial x}\left(\frac{\lambda_e}{c_p}\frac{\partial\tilde{h}}{\partial x}\right)+\frac{1}{r}\frac{\partial}{\partial r}\left(r\frac{\lambda_e}{c_p}\frac{\partial\tilde{h}}{\partial r}\right)+\tilde{u}\frac{\partial\bar{p}}{\partial x}+\tilde{v}\frac{\partial\bar{p}}{\partial r}$$
$$+\bar{S}_p^{h,c}+\tilde{h}_{pg}\bar{S}_p^m \tag{18.120}$$

where μ_e, the effective viscosity, is obtained as

$$\mu_e=\mu+\mu_t \tag{18.121}$$

μ_t being turbulence viscosity.

λ_e is the effective conductivity

$$\lambda_e=\lambda+\frac{c_p\mu_t}{\sigma_h} \tag{18.122}$$

where σ_h is the Prandtl number for turbulence, generally equal to 0.9 ~ 1.0. The Boussinesq assumption

$$\overline{\rho u''v''}=-\mu_t\left(\frac{\partial\tilde{u}}{\partial r}+\frac{\partial\tilde{v}}{\partial x}\right) \tag{18.123}$$

and

$$\overline{\rho u''h''}=-\frac{\mu_t}{\sigma_h}\frac{\partial\tilde{h}}{\partial x},\quad \overline{\rho v''h''}=-\frac{\mu_t}{\sigma_h}\frac{\partial\tilde{h}}{\partial y} \tag{18.124}$$

are used to solve the closure problem in obtaining the Favre-averaged equations.

18.10.2 k-ε Turbulence Model

To evaluate turbulence viscosity μ_t, the classic k-ε model of Launder and Spalding [1972] is used. μ_t is related to turbulent kinetic energy k and turbulent kinetic energy dissipation rate ε by the Prandtl–Kolmogorov relationship

$$\mu_t=\frac{\rho C_\mu k^2}{\varepsilon} \tag{18.125}$$

The transport equations for k and ε in the cylindrical coordinate system with considerations of axis-symmetric flow are:

$$\frac{\partial}{\partial x}(\rho\tilde{u}k)+\frac{1}{r}\frac{\partial}{\partial r}(r\bar{\rho}\tilde{v}k)=\frac{\partial}{\partial x}\left(\frac{\mu_e}{\sigma_k}\frac{\partial k}{\partial x}\right)+\frac{1}{r}\frac{\partial}{\partial r}\left(r\frac{\mu_e}{\sigma_k}\frac{\partial k}{\partial r}\right)+G-\bar{\rho}\varepsilon \quad (18.126)$$

$$\frac{\partial}{\partial x}(\rho\tilde{u}\varepsilon)+\frac{1}{r}\frac{\partial}{\partial r}(r\bar{\rho}\tilde{v}\varepsilon)=\frac{\partial}{\partial x}\left(\frac{\mu_e}{\sigma_\varepsilon}\frac{\partial \varepsilon}{\partial x}\right)+\frac{1}{r}\frac{\partial}{\partial r}\left(r\frac{\mu_e}{\sigma_\varepsilon}\frac{\partial \varepsilon}{\partial r}\right)+\frac{\varepsilon}{k}(C_{\varepsilon 1}G-C_{\varepsilon 2}\bar{\rho}\varepsilon) \quad (18.127)$$

where G, the production term of k, is given by

$$G=2\mu_e\left[\left(\frac{\partial\tilde{u}}{\partial x}\right)^2+\left(\frac{\partial\tilde{v}}{\partial r}\right)^2+\left(\frac{\tilde{v}}{r}\right)^2\right]+\mu_e\left[\frac{\partial\tilde{u}}{\partial r}+\frac{\partial\tilde{v}}{\partial x}\right]^2+\mu_e\left[\frac{1}{r}\frac{\partial}{\partial r}\left(\frac{\tilde{w}}{r}\right)\right]^2+\mu_e\left(\frac{\partial\tilde{w}}{\partial x}\right)^2 \quad (18.128)$$

The coefficients C_μ, $C_{\varepsilon 1}$, $C_{\varepsilon 2}$, σ_k, and σ_ε are constants with values equal to 0.09, 1.44, 1.92, 0.9, and 1.22, respectively.

To account for particle effect on gas-phase turbulence, Melville and Bray's correlation [1979] is used:

$$\mu_{t,with\ particles}=\mu_{t,no\ particles}(1+\frac{\bar{\rho}_b}{\bar{\rho}})^{-0.5} \quad (18.129)$$

where $\bar{\rho}_b$ is the mean particle bulk density. The presence of particles reduces gas-phase turbulence slightly according to this relationship.

Chapter 1 — Problems

Problem 1.1:

(a) How much energy is required to change the velocity of a 2.2 liter, 140-hp GM Saturn car from v = 0 to 100 km/h (assume the car mass to be 1200 kg)? If the heating value of gasoline is 45,000 kJ/kg, how much fuel is required for acceleration if the thermal efficiency is 0.4? Ignore friction.

(b) How much energy is required to change the velocity of a Boeing 747 from 0 to 950 km/h and to lift it through a height of 11,000 km (assume aircraft mass to be 436,000 kg with passengers)? If the higher heating value of kerosine is 46,200 kJ/kg, how much fuel in kg is required if the thermal efficiency is 0.4?

Problem 1.2: An oven in a kitchen is operated with 4-kW heaters. The size of the oven is 60 × 50 × 50 cm. For simplicity, we will assume that the wire trays are of negligible mass. You must determine the rate of heating of the air and as well as the time required to reach 150°C inside the oven. Note that the oven is not tightly sealed.

(a) Write down the mass and energy conservation equations and the state equation.

(b) State the justifiable assumptions and present the simplified conservation equations.

(c) Explain the terms, including those crossing the boundaries.

(d) Derive a relation for the oven temperature as a function of time.

(e) Based on your results, suggest two modifications in design to the manufacturer of the oven for reducing the heat-up time without changing the capacity of the heaters.

Problem 1.3: Consider an automobile engine with a gas tank volume of 20 gal (1 gal = 3.785 l). Instead of gasoline, we wish to use natural gas (methane) as a fuel for the automobile. The natural gas enters the adiabatic tank from a pressurized gas line at 300 K and the charging process stops once the pressure reaches 50 bar. Determine the mass of gas that can be accommodated. Treat the gas as ideal. Make reasonable assumptions. c_{p0} = 36 kJ/kmol K, $\bar{R}$ = 8.314 kJ/kmol.

Problem 1.4: Consider the human body, which is an open system. The body loses heat at the rate of 0.110 kW at constant temperature. The mass flow rates in and out of the nose are assumed to be equal and is given as 6 g/min.

(a) Determine the entropy generation rate in kW/K assuming the control volume is just inside the skin surface.

(b) Determine the entropy generation rate in kW/K assuming the control volume is just outside the skin surface.

Assume the following: the body temperature is 37°C (same as the nasal exhaust temperature), outside ambient temperature is 25°C (same as the nasal intake temperature), pressure of air intake and exhaust are equal at 1 bar. The properties of intake and exhaust are same as those of air. c_p of air = 1 kJ/kg K.

Problem 1.5: Natural gas has the following composition based on molal percentage: CH_4 = 91.27, ethane = 3.78, N_2 = 2.81, propane = 0.74, CO_2 = 0.68, *n*-Butane = 0.15, *i*-Butane = 0.1, He = 0.08, *i*-pentane = 0.05, *n*-pentane = 0.04, H_2 = 0.02, C_6 and heavier (assume the species molecular weight to be 72 kg $kmol^{-1}$) = 0.26, and Ar = 0.02. Determine (a) molecular weight of mixture, (b) the composition percentage on mass basis, (c) density of mixture in kg/m^3 at STP, and (d) the specific gravity of the gas at 25°C and 1 bar.

Problem 1.6: If a = u − Ts and g = h − Ts, (a) determine g and a for O_2 in kJ/kg assuming P = 1 bar at (i) T = 1200 K and (ii) 2000 K. (b) What is the difference between "g" and "a"? Is it the same as RT? Why? Explain the difference in the answers for (i) and (ii).

Problem 1.7: Consider a car fired with gasoline and air:fuel ratio of 18 with a heating value of 45,000 kJ/kg (or 120,000 kJ/gal). Under normal steady driving conditions for every kg of fuel fired, approximately 1/3 of the heat value of gasoline is lost as heat loss via the radiator. If the measured exhaust temperature is 600 K and inlet air temperature is 300 K, how much work is developed per kg of gasoline? Assume complete combustion and that the properties of products are same as those of air. Take the control volume to include air intake, fuel intake, engine, and exhaust pipe.

Problem 1.8: A car radiator is maintained at steady state when driven over a long distance. Water enters the radiator at T_H (90°C), and cold water at T_c (40°C) due to heat loss to the ambient.

(a) Start from generalized mass and energy conservation and entropy balance equations and simplify them for the current problem.
(b) Determine the heat loss in kJ/kg of radiator water.
(c) If the ambient temperature is 25°C (T_0), what is the entropy generated per kg of radiator water circulated? Select the c.v. to be that around radiator such that the boundary temperature is the ambient temperature.

Assume c_p of H_2O = 4.184 kJ/kg K.

Chapter 2 — Problems

Problem 2.1: Are the following reactions exothermic or endothermic at 298 K?

(a) The hydrogen–fluorine reaction (used in chemical lasers)

(b) Nitrous oxide reaction with methane

$$4\ N_2O + CH_4 = CO_2 + 2\ H_2O + 4\ N_2$$

Problem 2.2: Consider combustion of H_2 with 25% excess air. Then, the ratio of volume of standard air supplied to standard volume of fuel is (select from the following choices):

(a) 0.5

(b) 3.0

(c) 4.76

(d) 43

Problem 2.3: Natural gas with HHV of 38,650 kJ per SCM (standard cubic meter) is assumed to have a molecular weight of 16.05 kg. The HHV on a unit-mass basis is:

(a) 0.5

(b) 3.0

(c) 4.76

(d) 43

Problem 2.4: Suppose the natural gas consumption for a typical house is 2.5 m^3/h in the winter months. What is the kmol flow per hour at STP? If the natural gas is a mixture of 95% of CH_4 and 5% C_2H_6 (mole basis), what is the theoretical air in m^3/h? How much CO_2 (in m^3/h and kg/h) is released into the atmosphere from the house? Assume STP conditions: T = 15.5°C, P = 101 kPa.

Problem 2.5: Suppose carbon is gasified using steam to synthetic gas, which is a mixture of 1 CO + 1 H_2 (used in methanol production). What is its higher heating value?

Problem 2.6: Determine (a) the stoichiometric air:fuel ratio, both on a mole and mass basis for C_2H_5OH (ethanol) and (b) the actual A:F on mole and mass basis if 50% excess air is supplied.

Problem 2.7: California gasoline is a mixture of 15% isobutane (C_4H_{10}), 20% hexane (C_6H_{14}), 45% isooctane (C_8H_{18}), and 20% decane ($C_{10}H_{22}$), all assumed to be in mass percentage. Calculate the $(A{:}F)_{mole}$ with 50% excess air. How does it differ from that of pure octane?

Problem 2.8: One kmol of acetylene (C_2H_2, H–C≡C–H) is dissociated into 2 C (s) and 1 H_2 in an SSSF reactor at 298 K, 100 kPa.

(a) Determine the enthalpy of formation using the bond energy method.
(b) What is the amount of heat transfer across the reactor if h_f from bond energy is used?

Problem 2.9: Determine the amount of cellulose ($C_6H_{10}O_5$) formed from CO_2 and H_2O during photosynthesis. If the higher heating value of cellulose is 17,500 kJ/kg, then how much energy in kJ is required from the sun to form 1 kg of cellulose at 298 K?

Problem 2.10: Calculate the h_f^0 of coal ($CH_{0.87}N_{0.018}O_{0.12}$) if the LHV is given as 34,576 kJ/kg.

Problem 2.11: The composition of natural gas from Texas is as follows: CH_4 = 76.33% (molar), C_2H_6 = 13.57%, C_3H_8 = 5.31%, C_4H_{10} = 1.46%, C_5H_{12} = 0.31%, C_6H_{14} = 0.10%, with the remainder being N_2 and CO_2.

(a) Determine the stoichiometric air:fuel ratio.
(b) What is the heating value of fuel in kJ/m³ at 1 bar and 15.6°C?
(c) Determine the adiabatic flame temperature in K.

Problem 2.12: Fuel CH_4 was supplied to a reactor along with air. Dry gas analysis yields the following composition: CO_2 = 4% and O_2 = 7%.

Determine (a) the CO content (%) in the products, (b) A:F, (c) the equivalence ratio, and (d) air required in m³/h for combustion of 15 m³/h of fuel at SATP conditions.

Problem 2.13: For Problem 2.12, obtain a relation for liquid and vapor fractions formed by the condensation of H_2O as the exhaust is cooled. Let $p^{sat}{}_{H2O(g)}(T)$ = $\ln P = A - B/T$, where $A = \ln P^{sat}_{ref} + h_{fg}/RT_{ref}$, $B = h_{fg}/R$, and $P = P^{sat}$. For H_2O, A = 13.082 and B = 4962 K.

(a) If the initial pressure is 2 bar, what is the dew point temperature?
(b) If T is 280 K, how many moles of liquid and vapor will be present in the products?

Problem 2.14: Entropy is generated because of irreversible oxidation of glucose within the cells of the human body, if it is assumed that the reaction occurs at 37°C (310 K). Assume for a person at steady state, the entropy of glucose at 37°C is 288.96 kJ/kmol K, T_{in} = 300 K, P_{in} = 1 bar, with 400% excess air.

(a) Determine the heat loss per kmol of glucose burned, assuming no work.
(b) What is the entropy generation per kmol of glucose?

Verify the calculations using the THERMOLAB software.

Problem 2.15: The No. 6 fuel oil has the following composition: S = 2.3% (by mass), H = 9.7%, and C = 88%. It burns in air. The products of combustion are SO_2, CO_2, N_2, and H_2O.

(a) If the fuel has HHV of 41,840 kJ/kg and the power plant is rated at 500 MW operating at a thermal efficiency of 38%, determine the SO_2 emission in tons per year.
(b) Calculate the adiabatic flame temperature in K and compare the result using THERMOLAB. What will be the result if you assume constant specific heat?
(c) How much SO_2 and CO_2 are produced in kg per GJ of heat released? How many million tons per year?

Problem 2.16: A fuel burns to yield the following products on dry volume percentage: CO: 3.4, CO_2: 12.1, H_2: 1.5, N_2: 82.5, and O_2: 0.5.

(a) If the fuel is hydrocarbon, determine the empirical formulae and A:F ratio in kmol per kmol of empirical fuel.
(b) Can the fuel be $C_{12}H_{26}$? Or C_6H_{13}?

Problem 2.17: Octane (C_8H_{18}) is burned at constant volume in an Otto cycle. The compression ratio, r_v = 10; equivalence ratio, ϕ = 0.8; inlet air temperature, T_1 = 350 K; and P_1 = 1.0 bar. What is the temperature at the end of the combustion process? Assume that c_p data for compression process 1 to 2 corresponds to air and that $c_{p,\ vapor}$

$$= 2.51 \frac{kJ}{kgK}, \quad \text{and} \quad h_f^{\circ}, C_8H_{18}\,(g) = -208{,}450 \frac{kJ}{kmole}$$

Problem 2.18:

(a) Books on combustion mention that a plot of the O_2% (on a dry basis) vs. the H-to-C ratio for many fuels is a flat curve (i.e., O_2% in the exhaust does not vary with fuel composition at fixed excess air %). If so, plot the O_2% vs. H:C ratio of fuel (H:C = 1, 1.5, 2, ..., 4) with excess air as a parameter (0%, 10%, ..., 50%). Verify if this statement is true. Rationalize the results.

(b) Suppose that the measured O_2% for an H:C fuel is 5%. What is your estimate of the excess air used during combustion?
(c) Plot the CO_2% (dry) vs. H:C for 0, 10, ..., 50% excess air. The CO_2% values are called the maximum possible values because we assume that for given excess air %, all the fuel is burned.
(d) In an experiment on the combustion of coal (H:C = 0.1), 20% excess air is supplied. What are the expected O_2% and CO_2%? If the measured CO_2% and O_2% for coal combustion (H:C = 0.1) are 12% and 8%, what is your conclusion?

Problem 2.19: A 1:1 CO and H_2O mixture is sent into an adiabatic reactor. $CO + H_2O \rightarrow CO_2 + H_2$. Plot T, $\bar{g}_{CO} + \bar{g}_{H2O}$, $\bar{g}_{CO2} + \bar{g}_{H2}$, σ, S vs. the fraction of CO reacted (for an adiabatic reactor).
(a) What is σ when 1 kmol of CO is combusted?
(b) What is the degree of reaction when σ is a maximum?
(c) What should be the criteria for a reaction or a process to take place in a particular direction? Do you believe that whenever $\sigma > 0$, the particular process has taken place? If not, what about the criterion $d\sigma > 0$? Which is better?

Problem 2.20: White solid ammonium perchlorate, NH_4HClO_4 (mol wt = 117.5, density = 1950 kg/m³, $hf^0 = -295{,}770$ kJ/kmol, $s^0 = 184.2$ kJ/kmol K) is a solid rocket fuel (used in space shuttles) and is produced by the following reaction:

$$NH_3\ (g) + HCl\ (g) + 4\ H_2O \rightarrow NH_4HClO_4\ (s) + 4H_2\ (g)$$

(a) What are the standard densities of HCl and H_2O (g) in kg/m³?
(b) How much HCl and H_2O in m³ are required to produce 5000 kg of NH_4HClO_4?
(c) Determine the heat to be added or removed if all reactants enter at 25°C in gaseous state.
(d) If NH_4HClO_4 (s) decomposes at T > 250°C to the gases N_2, Cl_2, O_2, and H_2O (g), determine the heat added or removed. Assume that the products leave at 827°C and NH_4HClO_4 (s) is initially at 25°C.

Problem 2.21: Hydrazine (N_2H_4) is a colorless liquid and is a rocket fuel. It freezes at 2°C and boils at 113.5°C.
(a) Determine the enthalpy of formation if the lower heating value is 534,000 kJ/kmol of hydrazine. Assume

$$N_2H_4\ (liq) + O_2 \rightarrow N_2\ (g) + 2\ H_2O\ (g)$$

(b) What is its higher heating value?

Problem 2.22: Ethanol is made by the fermentation of sugars. The raw material can be starch from potatoes, corn, etc. The yeast enzyme, zymase, converts the glucose into ethanol and carbon dioxide. The fermentation reaction is:

$$C_6H_{12}O_6\ (s) \rightarrow 2C_2\ H_5OH\ (l) + 2CO_2\ (g)$$

(a) What is the heat liberated or absorbed for this reaction? Make reasonable assumptions.
(b) If each bushel of corn (25 kg) can produce up to 2.5 gal of ethanol fuel, how much heat is liberated or absorbed per bushel of corn? Assume that corn has same formula as glucose.

Problem 2.23: Liquid ethanol (C_2H_5OH) is burned with a stoichiometric ratio (SR) of 0.9 (or equivalence ratio of 1.11). Because of the O_2 deficiency, the products are CO_2, CO, H_2O, and N_2. Assume that fuel enters at 298 K and 1 bar, whereas air enters at 500 K and 1 bar. If the temperature of the products is 1800 K, determine the heat loss or gain if the fuel flow is 23 kg per sec. What percentage of carbon in the fuel is oxidized to CO_2?

Problem 2.24: A car runs with natural gas and 20% excess air. Assume the gas composition to be CH_4. Under normal steady driving conditions, for every kg of fuel burned, approximately 15,000 kJ is supplied as work, whereas 15,000 kJ is lost as heat in the radiator. Determine the exhaust temperature. Assume complete combustion. Assume the control volume includes the air intake, fuel intake, engine, and exhaust pipe.

Problem 2.25: Consider the chemical reaction of a generic fuel, namely,

$$\begin{aligned} &CH_hO_oN_n\,S_s + a(O_2 + BN_2 + FH_2O(g) + G\,Ar) \rightarrow CO_2 \\ &\quad + (h/2 + Ba)\,H_2O + (a - (1 + s + h/4 - o/2))\,O_2 \\ &\quad + (Ba + n/2)\,N_2 + Ga\,Ar + s\,SO_2 \end{aligned} \tag{A}$$

where the ratios B, F, and G in air are defined as (molar or volumetric) amounts of N_2, O_2, H_2O, and Ar. All fuel N converts to N_2. Obtain an expression for (a) $X_{CO2,\ max}$ (maximum mole fraction with stoichiometric air) in terms of $X_{O2,a}$ (ambient mole fraction) and fuel atom composition, (b) X_{CO2} with excess air in terms of $X_{O2,a}$ and fuel atom composition, and (c) $X_{CO2}/X_{CO2,max}$ in terms of X_{O2} and $X_{O2,a}$.

Problem 2.26: Consider the reactions

$$\begin{aligned} C(s) + 2\,H_2 &\rightarrow CH_4 + Q_4, \\ C(s) + O_2 &\rightarrow CO_2 + Q_2, \quad \text{and} \\ H_2 + 1/2\,O_2 &\rightarrow H_2O + Q_5. \end{aligned}$$

Show that

$$CH_4 + 2O_2 \rightarrow CO_2 + 2H_2O + Q_6,$$

where

$$Q_6 = Q_2 + 2\,Q_5 - Q_4 = h^{\circ}_{f,CO_2} + 2h^{\circ}_{f,H_2O} - h^{\circ}_{f,CH_4}.$$

Problem 2.27: Consider a six-cylinder engine with stroke = 130 mm, bore = 108 mm, and gasoline consumption = 232 g/kWh.

(a) Determine the CO_2 produced in g per kWh.
(b) What is the heat release rate in kW per m^3, if combustion occurs within the clearance volume (= 10% of the stroke volume)?

Problem 2.28: Determine the adiabatic flame temperatures at T_∞ = 298, 500, 1000, and 1250 K of the following polymers assuming complete combustion. Use THERMOLAB for fuels (ii) to (iv).

(i) Polyvinyl chloride (PVC), (C_2H_3Cl, fuel temperature 613 K, HV = 19,370 kJ/kg)
(ii) Polyethylene (PE), (C_2H_4, 753 K, 47,320)
(iii) Polystyrene (PS), (C_8H_8, 661K, 141,580)
(iv) Polymethyl methacrylate (PMMA) or Plexiglas, ($C_5H_8O_2$, 660 K 26,560)

Explain the differences in flame temperatures.

Problem 2.29: Formulate the appropriate equations for developing a computer program for combustion of $C_cH_hN_nO_oS_s$ with assumed ϕ as (a) $\phi < 1$ (lean mixture), (b) $\phi > 1$ (very rich mixture) and $\phi = \phi_{CO}$ is such that the reaction produces only CO, and (c) $1 < \phi < \phi_{CO}$ (moderately rich mixture) such that a mixture of CO and CO_2 is produced. Input data: c, h, n, o, s, ϕ; output data: dry gas analysis % and dew point temperature.

Chapter 3 — Problems

Problem 3.1: Consider the equilibrium reaction: $CO_2 + H_2 \Leftrightarrow CO + H_2O$.
At a specified temperature an increase in pressure will
(a) increase the CO_2 concentration, or
(b) decrease the CO_2 concentration, or
(c) not cause any change in CO_2 concentration?

Problem 3.2: The equilibrium constant for the reaction $1/2\ N_2 + 1/2\ O_2 \Leftrightarrow NO$ increases with temperature. If the Earth warms up from 300 K to 310 K, then the equilibrium amount of NO (trace amount)
(a) increases, or
(b) decreases?

Problem 3.3: A gas mixture b CO + c H_2 is sometimes produced from reaction between CH_4 and H_2O. Determine the equilibrium values at 100 and 1000°C and state whether a high or low pressure is desired for more production of synthesis gas.

Problem 3.4: A gas mixture b CO_2 + c H_2 is produced from a reaction between CH_4 and H_2O. H_2 production occurs from the following reactions.

$$CxHy + x\ H_2O \rightarrow x\ CO + \{y/2 + x\}\ H_2 \tag{A}$$

$$C + H_2O \rightarrow CO + H_2 \tag{B}$$

$$CO + H_2O \rightarrow CO_2 + H_2 \tag{C}$$

Determine the heats of reactions if x = 1, y = 4 for Reaction A. Which reactions are favored at low temperature?

Problem 3.5: Initially 1 kmol of CO_2 is supplied to a constant-pressure reactor with P = 100 kPa.
(a) Using the definition of Gibbs function, determine ΔG^0 for reaction $CO_2 \Leftrightarrow CO + 1/2\ O_2$ at 3000 K and hence the equilibrium constant K^0 (3000 K).

(b) Determine the transition temperature in K for the equilibrium reaction $CO_2 \Leftrightarrow CO + 1/2\ O_2$.
(c) Determine the mole fraction for each species CO, CO_2, and O_2 at equilibrium and the degree of dissociation $\{= (CO_{2in} - CO_{2exit})/CO_{2in}\}$.
(d) What will the results be if 1 kmol of C (s) and 1 kmol of O_2 are supplied to the reactor? The products at the exit are still the same as in part (c) and T and P are the same as before.

Problem 3.6: With equilibrium constants for reactions at 1800 K for

$$C(s) + 2H_2 \Leftrightarrow CH_4,$$
$$C(s) + 1/2\ O_2 \Leftrightarrow CO,$$
$$H_2O \Leftrightarrow 1/2\ O_2 + H_2.$$

(a) Evaluate the equilibrium constant at 1800 K for the reaction

$$CH_4 + H_2O \Leftrightarrow 3H_2 + CO.$$

(b) Evaluate the equilibrium constants at 1800 K for the reaction:

$$C\ (s) + H_2O \Leftrightarrow CO + H_2.$$

Problem 3.7: In order to evaluate the pollution potential of various fuels, methanol (l) (CH_3OH), gasoline (l) ($CH_{2.5}$ or alternate formulae C_8H_{20}, HHV = 48,300 kJ/kg), diesel (l) ($CH_{1.9}$, HHV = 45,714 kJ/kg), kerosene (l) ($CH_{1.86}$, HHV = 46,200 kJ/kg), and coal (s) ($CH_{0.9}N_{0.02}O_{0.1}$, HHV = 35,847 kJ/kg), determine the equilibrium NO for a stoichiometric mixture under adiabatic combustion. Run the THERMOLAB program and rank the fuels in order of (a) increasing ppm and (b) increasing kg/GJ. Note that the code accepts one of enthalpy of formation, HHV, and LHV.

Problem 3.8: Determine ΔG^0 vs. T for the reaction $CO_2 = CO + (1/2)\ O_2$. Plot $\log K^0$ vs. (1/T). Use T = 500, 700, 1000, 2000, and 3000 K.

Problem 3.9: Methanol is produced through the reaction

$$CO + 3H_2 \rightarrow CH_3OH. \qquad (A)$$

The required H_2 and CO are produced by producing synthesis gas (H_2/CO = 3) using steam reforming reaction

$$CH_4 + H_2O\ (g) \rightarrow CO + 3H_2. \qquad (B)$$

(a) Are Reaction A and Reaction B exothermic or endothermic at 298 K? Justify.

(b) Reaction B is carried out at 750 to 1000°C and at low pressure to increase yield. If equilibrium is assumed at 1000°C, what is the conversion at 1 bar and at 0.1 bar?

Problem 3.10: The feedstock for producing a gaseous mixture of CO and H_2 is ethane, and

$$C_2H_6 + H_2O\ (g) \rightarrow 3CO + 7H_2. \quad (A)$$

Additional hydrogen can be produced by the water–gas shift reaction

$$CO + H_2O \rightarrow CO_2 + H_2. \quad (B)$$

Another source of H_2 is

$$C\ (s) + H_2O \rightarrow CO + H_2. \quad (C)$$

(a) Are Reaction A to Reaction C exothermic or endothermic at 298 K? Justify.
(b) Reaction B is carried out at 350°C (in order to have a finite reaction rate) and if equilibrium is assumed at 350°C, determine the conversions to H_2 at P = 1 bar and P = 0.1 bar. Comment on the effect of pressure.

Problem 3.11: Determine the equilibrium composition of CaO (s), O_2, SO_2, and $CaSO_4$ (s) vs. T from 300 K to 1500 K for Wyoming coal (formula $C_1\ H_{0.755}\ N_{0.0128}\ O_{0.182}\ S_{0.00267}$) and burning under 20% excess air in a bed containing limestone ($CaCO_3$). Assume that SO_2 exists as a trace species. $CaCO_3$ decomposes completely to CaO (s) and CO_2, and CaO (s) is used to capture SO_2 from gases under equilibrium.

$$CaCO_3\ (s) \rightarrow CaO\ (s) + CO_2 \quad (A)$$

$$CaO\ (s) + SO_2 + 1/2\ O_2 \Leftrightarrow CaSO_4\ (s) \quad (B)$$

Problem 3.12: Consider carbon and oxygen entering a reactor. The reactor is maintained at the temperatures 1000, 1500, 2000, 2500, and 3000 K. The products are C (s), CO, CO_2, and O_2. Determine the equilibrium composition vs. temperature.

Problem 3.13: If the equilibrium constant K^0 follows the curve

$$\ln K^0 = A + B/T,$$

then using the tabulated values, determine the constants A and B for the reactions:

(a) $1/2\ N_2 + 1/2\ O_2 \Leftrightarrow NO$
(b) $CO_2 \Leftrightarrow CO + 1/2\ O_2$
(c) $CO_2 + H_2 \Leftrightarrow CO + H_2O$
(Choose T between 1500 to 3000 K.)

Problem 3.14: Compute the equilibrium adiabatic flame temperature for $C_{12}H_{24}$ at 90 and 35 atm (conditions for gas turbines) as a function of ϕ using the THERMOLAB spreadsheet program for the products containing Ar, CO, CO_2, H_2, H_2O, NO, OH, O_2, and SO_2. Air composition: Ar = 1%, O_2 = 21%, and N_2 = 78%. Let ϕ range from 0.5 to 1.2.

Problem 3.15: Determine the trace amount of NO, NO_2, O, and N in air if air is assumed to be in chemical equilibrium at 300 and 3000 K. Justify treating NO as a trace species.

Problem 3.16:

(a) Determine the values of A (Helmholtz function) and G (Gibbs function) at 3000 K and 10 bar if CO = 5, O_2 = 3, CO_2 = 4 kmol. Assume a fixed mass and that G changes.

(b) What is the value of G after 0.2 kmol of CO oxidizes at the same specified P and T to form the equilibrium products CO, O_2, and CO_2?

(c) What is the value of A after 0.2 kmol of CO oxidizes at the same specified T and V? (Hint: Calculate P when V is held constant.)

(d) Comment on the change in G and A for parts (b) and (c).

Problem 3.17: Gaseous propane is burned with 60% of theoretical air in a steady-flow process at 1 atm. Both the fuel and air are supplied at 25°C. The products, which consist of CO_2, CO, H_2O, H_2, and N_2, are in equilibrium.

(a) For an adiabatic process, determine the composition of the products.

(b) If products leave the combustion chamber at 1500 K, determine the amount of heat transfer per kg of propane.

Problem 3.18: For every kmol of coal fired, 0.234 kmol of moisture enters with the fuel. Twenty percent excess air is used and combustion is complete except for trace amounts of NO. Assume that the fuel S is burned to SO_2. The products leave at 2800 K. Determine the mole fraction of trace amounts of NO assuming the following equilibrium reaction; pressure is unknown.

$$1/2\ N_2 + 1/2\ O_2 \Leftrightarrow NO$$

Assume that coal has the following atomic formulae:

Carbon	1
Hydrogen	0.755
Nitrogen	0.0128
Oxygen	0.182
Sulfur	0.00267

Problem 3.19: Consider 1 kmol of H_2 and 1/2 mol of O_2 entering a reactor maintained at 3000 K and 1 bar. As the H_2 reaction proceeds, the number of O_2 moles decreases, whereas that of H_2O increases.

(a) Calculate $\hat{g}_{H2}(T,\ p_{H2}), \hat{g}_{O_2}(T,\ p_{O_2})$ and $\hat{g}_{H_2O}(T, p_{H_2O})$ vs. N_{CO_2}.
(b) Plot $\bar{g}_{H2} + \frac{1}{2}\bar{g}_{O2}, \bar{g}_{H2O}$ and $G(= N_{H2}\ \bar{g}_{H2} + N_{O_2}\bar{g}_{O2} + \bar{g}_{H2O})$ vs. N_{H2O}.
(c) Comment on the results in parts (a) and (b).

Problem 3.20: Methanol is produced from natural gas (90–93% CH_4) using steam reforming; methane is converted into CO and H_2 (called synthesis gas) at 500°C and 20 to 30 bar pressure as follows:

$$CH_4 + H_2O\ (g) \rightarrow CO + 3H_2. \qquad (A)$$

The syngas reacts on the surface of a Cu–Zn catalyst:

$$CO + 2H_2 \Leftrightarrow CH_3OH. \qquad (B)$$

The overall reaction is:

$$CH_4 + H_2O\ (g) \rightarrow CH_3OH + H_2. \qquad (C)$$

(a) Determine the equilibrium amounts of CO and H_2 when 1 kmol each of CH_4 and H_2O are admitted into a reactor at 298 K and 700 K respectively.
(b) Determine the equilibrium amounts of CH_3OH and H_2 when 1 kmol of CH_4 and H_2O are admitted to a reactor at 298 K and 700 K, respectively.

Problem 3.21: Plot the degree of dissociation when 1 kmol of H_2O dissociate at various temperatures (T = 400, 600, …, 3000 K).

Problem 3.22: Determine the maximum possible flame temperatures when certain plastics are burned. Assume that fuel and air are initially at 298 K. Consider the following plastics:

(i) Polyvinyl chloride (C_2H_3Cl, HV = 19,370 kJ/kg)
(ii) Polyethylene (C_2H_4, 47,320 kJ/kg)
(iii) Polystyrene (C_8H_8, HV = 41,580 kJ/kg, e.g., coffee cups)
(iv) PMMA or Plexiglas ($C_5H_8O_2$, HV 26,560 kJ/kg; e.g., aircraft windows).

Assume:

(a) complete combustion
(b) chemical equilibrium

Use the THERMOLAB program.

Problem 3.23: Consider the stoichiometric combustion of 1 kmol of CH_4 with air that are initially at 298 K (state 1). The products leaving the adiabatic combustor are determined to be CO_2, CO, H_2O, H_2, O_2, and N_2. Assume chemical equilibrium.

(a) Determine the product temperature, the moles of products formed, and the percentage of each species for an open system.
(b) What will be the equilibrium temperature, composition, and pressure if 1 kmol of CH_4 is initially mixed with stoichiometric air at P = 1 bar and then ignited in a closed system? Assume the same final species as before.

Problem 3.24: Mercury emissions from coal-fired plants can be captured in the form of $HgCl_2$ using a water spray because $HgCl_2$ is soluble in water. Consider the following reactions:

(1) $HgCl_2\,(g) \Leftrightarrow Hg + Cl_2$
(2) $HgCl_2\,(g) \Leftrightarrow Hg + 2Cl$
(3) $HgCl \Leftrightarrow Hg + 1/2\ Cl_2$
(4) $HgCl \Leftrightarrow Hg + Cl$
(5) $HCl \Leftrightarrow 1/2\ H_2 + Cl_2$

Determine:

(a) The equilibrium constants at 298 K for all of the reactions
(b) A and B in the relation $\ln K = A + B/T$
(c) Transition temperature in°C

(Hint: Use the vant Hoff equation.)

Problem 3.25: Consider the following reactions:

$$CO_2 \Leftrightarrow CO + 1/2\ O_2 \tag{A}$$

$$H_2O \Leftrightarrow H_2 + 1/2\ O_2 \tag{B}$$

$$CO_2 + H_2 \Leftrightarrow CO + H_2O \tag{C}$$

(a) Show that $K^0_C = \{K^0_A/K^0_B\}$.
(b) Evaluate K^0_C at 1000 K using tables.
(c) Verify relation in (a) by using the values for K^0_A and K^0_B.

Problem 3.26: Consider a fuel blend of 20% H_2, 40% CH_4, and 40% propane (by volume). Using the THERMOLAB software, determine the adiabatic equilibrium flame temperature and NOx fraction at 15% excess air.

Chapter 4 — Problems

Problem 4.1: If the empirical formula for glucose is CH_2O, but its molar mass is 180, what is its formula?

Problem 4.2: Two gaseous fuels A and B have molecular weights and molar heating values such that $M_A/M_B = 2.25$ and $\overline{HV}_A/\overline{HV}_B = 1.5$. If the heat release rate of A is $\dot{H}_A$ and B is $\dot{H}_B$. Then $\dot{H}_A/\dot{H}_B$ for the same mass burn rates are

(a) > 1
(b) < 1
(c) $= 1$
(d) None of the answers listed from (a) to (c) because the information given is insufficient.

Problem 4.3: The ultimate analysis of a fuel provides C = 37.5%, H = 12.5%, O = 50%. Then, the empirical formula of the fuel is

(a) $CH_{0.33}O_{1.4}$
(b) CH_4O
(c) C_3HO

Problem 4.4: Coal with the chemical formula CH releases volatile matter (VM) that has an empirical formula of CH_2. The volatile matter percentage on a mass basis is

(a) 40
(b) 60
(c) 50

Problem 4.5: To develop a new fuel (NF) for power production, the cost of NF in comparison to an existing fuel (say, natural gas [NG]) must be considered. What are the best units to express the cost of NF and NG? Cost per kg? Cost per GJ?

Problem 4.6: Consider the cost of ethanol production from agricultural biomass. If ethanol fuel costs \$2.25 per gal and is cheaper compared to gasoline (\$2.50 per gal), what should be the cost of gasoline before ethanol becomes competitive?

Problem 4.7: The elemental analysis of a black liquor is C = 36.40, Na = 18.60, S = 4.80, H = 3.50, K = 2.02, Cl = 0.24, N = 0.14, and O (by difference) = 34.30.

(a) Develop an empirical chemical formula and determine the HHV using the Boie equation.

(b) If the Boie equation is correct, develop an expression for the enthalpy of formation, h_f^0, in kJ for the empirical fuel and determine the value.

Problem 4.8: The cost of biomass fuel including the payment to the farmer, collection, transportation, storage, processing (e.g., grinding) is \$20 per dt (dt: dry tons) to \$60 per dt, with an average of \$40 per dt. Express this in \$ per GJ if the ash content is 10%, assuming wood to be a typical biomass fuel. Assume that ultimate analysis of DAF wood yields the following: C = 50%, H = 6%, O = 43%, N = 0.5%, and S = 0.5%. Use Boie equation for estimation of HHV. How does this compare with the cost of Wyoming subbituminous coal fuel with 3.6% ash, 16.9% moisture (subbituminous A, dry analysis: C = 60.4%, H = 6.0%, N = 1.2%, O = 27.4%, S = 1.4%) if the delivered cost per ton is \$25, including moisture and ash?

Problem 4.9: Air is used to gasify DAF Wyoming coal given by the empirical formula $C_1\ H_{.755}\ N_{0.0128}\ O_{0.182}\ S_{0.00267}$ and having a heating value (HHV) of 30,829 kJ/kg. Sufficient air is supplied to just convert all C into CO. H is assumed to be converted into H_2, whereas N and S are converted into N_2 and H_2S. Ignoring the S oxidation, determine the higher heating value of gases produced (in units of MJ per SCM). Typically, low-Btu gas is defined as 3 to 7 MJ/SCM, medium-Btu gas as 14 to 21 MJ/SCM, and high-Btu gas as 30 to 40 MJ/SCM.

(a) How do you classify the gas produced by the coal gasifier?

(b) What is the HHV if pure O_2 is used instead of air for gasification? How do you classify this gas?

Problem 4.10: Tulsa natural gas contains (by volume %) CH_4: 93.4; C_2H_4: 2.70; C_3H_8: 0.60; C_4H_{10}: 0.2; CO_2: 0.7; N_2: 2.40.

(a) Determine the mixture molecular weight.

(b) Determine the enthalpy of formation of mixture per kmol and per SCM.

(c) Determine LHV and HHV per SCM of fuel and per SCM of stoichiometric air.

(d) Determine the Wobbe number.

(e) Compute the adiabatic flame temperature for complete combustion and equilibrium using the THERMOLAB software.

(f) Determine an empirical chemical formula in the form $C_cH_hN_nO_o$ and in the normalized form $CH_mN_nO_p$.

Problem 4.11: Recycled paper sludge (RPS) has the following composition: C: 23%, H: 3.8%, O: 50.4%, Al: 6.9%, Ca: 6.9%, Si: 7%, Ti: 1.5 [Agblevor, F.A.

and Mahafkey K., 2003, 26th Symposium on Biotechnology for Fuels and Chemicals, Chattanooga, TN, May 9–12]. (a) Determine the HHV in kJ/kg using the Boie equation and per kg of stoichiometric oxygen. (b) How do these compare with n-octane fuels? (c) Do you believe that the CO_2 emission per GJ is smaller compared with the use of n-octane as fuel?

Problem 4.12: A high-volatile coking coal (mass %, C = 71.6, H = 4.3, N = 1.6, S = 1.7, O = 3.8, moisture = 9, ash = 8, HHV = 28,400 kJ/kg) is fired into a boiler.

(a) What is the maximum amount of fuel-based NO and SO_2 in ppm? How much of each is emitted in kg per MJ of heat released?
(b) If the power plant has a rating of 600 MW (thermal) and a boiler efficiency of 80%, determine the tons of NO and SO_2 emitted per year.

Assume that all the fuel nitrogen and sulfur are converted into NO (also called fuel NO) and SO_2 and there is no thermal NO (i.e., NO produced solely from reactions involving molecular nitrogen contained in the atmosphere). Normally, most of the NO released from coal-fired power plants originates from fuel N, whereas the NO released from automobiles and aircraft is from thermal NO.

Problem 4.13: For the above problem, assume that the coal is fired into a boiler with such an A:F ratio that there is 6% O_2 in the exhaust.

(a) Determine the SCM of dry flue gas per gross GJ. Is it 350–370 SCM/GJ, as assumed in practice?
(b) Obtain the NO and SO_2 emissions in mg/SCM of flue gas.

Problem 4.14: There is growing concern regarding the emission of NO (nitric oxide). Assume that NO exists at equilibrium concentration in trace amounts. The EPA recommends reporting of NO values at 3% oxygen (dry mole) levels in the products for all fuels. In order to evaluate the pollution potential of methanol (CH_3OH), gasoline ($CH_{2.5}$, HHV = 48,300 kJ/kg), diesel ($CH_{1.9}$, HHV = 45,714 kJ/kg), kerosene ($CH_{1.86}$, HHV = 46,200 kJ/kg), and coal ($CH_{0.9}N_{0.02}O_{0.1}$, HHV = 35,847 kJ/kg), estimate the maximum NO emitted for the following cases. Express NO in ppm.

(a) Complete combustion at an equivalence ratio such that O_2 in the products is 3% (dry mole) but with trace amounts of NO in the products at 2000 K.
(b) Adiabatic combustion at an equivalence ratio such that O_2 is 3% (dry mole) but with trace amounts of NO. Assume the inlet temperature to be 298 K.
(c) Compare the results with those obtained from the THERMOLAB program. Verify the program with sample calculations.
(d) Rank the fuels in the order of increasing NO.

Problem 4.15: Consider the reaction of cattle manure with water, which produces methane at 40°C in the digestion process.

$$CH_{1.4}O_{0.5}N_{0.07} + a\,H_2O \rightarrow b\,CH_4 + c\,CO_2 + d\,NH_3$$

Ignore NH_3 in products. How much heat is absorbed or released by such a reaction? If the conversion efficiency is 70% and if all the heat required for such a process is supplied by the chemical reactions, how would you compare this process with pure pyrolysis process? This particular process is called anaerobic digestion process. Assume HHV = kJ/kg.

Problem 4.16: Caffeine has the following composition (on a unit mass basis): C, 49; H, 5; O, 17; N, 29.

(a) Write the chemical formula if the molecular weight is 194 kg/kmol.
(b) Determine the heating value in kJ/kg using the Boie equation.

Problem 4.17:

(a) Show that for any solid or liquid fuel with a formula CH_hO_o, the burned fraction $\varepsilon_b \approx [1/\phi]\{-(1\{X_{O2}/X_{O2,a}\})\}$, where $X_{O2,a}$ is the ambient mole fraction of O_2; ϕ, the equivalence ratio; and X_{O2}, the flue gas mole fraction.
(b) How is the formula modified for a blend containing gaseous fuel CH_sO_t and the above solid/liquid fuel if "f" is the mole fraction of the gaseous fuel and (1 − f) is the fraction of solid or liquid fuel?
(c) If f = 0.8 and combustion efficiency η_{blend} = 0.97 (for part (b)), what is η_{solid}? Assume that the unburned fuel in the combustor has the same chemical composition as the fuel supplied.
(d) What is the relation for ε_b if the unburned fuel in the combustor has the same chemical composition of the solid in the gas and solid-fuel blend?

Problem 4.18: A low-Btu gas has the following composition: CH_4: 3 (mole % or volume %); CO: 13; CO_2: 10; H_2: 5; N_2: 69; O_2: 0.

(a) Determine (i) υ_{O2} (for combustible part of fuel), (ii) $Y_{F,i}$ (combustible mass fraction in fuel), and (iii) $\upsilon_{O2}Y_{F,i}$
(b) Determine the empirical chemical formula and repeat part a for the fuel, including inert components.

Problem 4.19: The Wobbe index is a measure of the interchangeability of gaseous fuels. Assume a mixture of propane and hydrogen. We wish to change from propane to a mixture of propane and H_2. Determine the HHV in MJ/m^3 and the Wobbe number vs. % H_2.

Problem 4.20: Oxygen-blown coal gasification yields: 25% H_2, 40% CO, 20% H_2O, and 7050 kJ/m^3 lower heating value. If the Boie equation is used, will it yield the same heating value as that given here? What will be the Wobbe number for the mixture?

Problem 4.21: Synthesis gas is a gas produced from coal. Its composition is CH_4: 3.3%, CO: 21.9%, CO_2: 29.3%, H_2: 44%, N_2: 1.5%. It is burned with air.

(a) Determine the molecular weight of synthesis gas, M_F.
(b) If the specific gravity (sg) of synthesis gas is defined as a ratio of density at (T, P) to the air density at the same T and P, show that sg = M_F/M_{air}. It is reported as 0.72. Do you agree?

(c) Determine stoichiometric A:F in standard m³/standard m³ of fuel (note that gases are ideal). Start with 1 kmol of synthesis gas for the reaction.
(d) Determine the higher (HHV) or gross (GHV) and lower heating value (LHV) of fuel in kJ/m³ of fuel and kJ per m³ of stoichiometric air. GHV or HHV is reported as 247 Btu/ft³. Do you agree?

Problem 4.22: Bio-oil is an oil produced from fast pyrolysis of biomass. Bio-oil has an HHV of 15–20 MJ/kg and diesel has an HHV value of 43 MJ/kg. If the heating value per unit stoichiometric O_2 remains constant, how much stoichiometric air may be required for bio-oil? If bio-oil is represented as CH_hO_o, then determine the "h" and "o" values.

Problem 4.23: What is the maximum amount of water that can be tolerated in a fuel having a heat value of 20,000 kJ/DAF kg basis if the minimum temperature for the reaction is about 1200 K? Assume that fuel has a chemical formula of $CH_{1.78}\ O_{0.64}\ N_{0.0825}S_{0.0141}$.

Problem 4.24: In gasification processes, the following empirical methodology is adopted. The fuel, say, $CH_{1.4}O_{0.6}$ (a typical biomass fuel), releases 80% of its carbon matter as gases, retaining 20% of C as solid. Thus, when the fuel is reacted ($CH_{1.4}O_{0.6}$) with air, it produces CO, H_2, H_2O, CO_2, C (s), and N_2.
(a) Write the reaction equation assuming the number of moles of CO and CO_2 to be equal and the ratio of the number of moles of H_2O and H_2 = 1/6.
(b) Determine the volume of gases produced at STP per kg of fuel.
(c) Determine the heating value of the gases in kJ per SCM.

Chapter 5 — Problems

Problem 5.1: The term $\exp\left(-\frac{E}{\bar{R}T}\right)$ represents the fraction of all collisions having energy

(a) higher than E
(b) lower than E
(c) exactly equal to E

Problem 5.2: Sucrose ($C_{12}H_{22}O_{11}$) reacts with excess H_2O (l) (called the hydrolysis of sucrose) to produce glucose ($C_6H_{12}O_6$). What is the approximate order of the reaction if you assume the reaction to be elementary? What is the heat of reaction? (In case the heat of formation is not available, use the Boie equation.)

$$C_{12}H_{22}O_{11} \text{ (sucrose)} + H_2O \rightarrow 2\ C_6H_{12}O_6$$

Problem 5.3: Consider a high-temperature reactor that is supplied with O_2 molecules and O atoms. The following reactions can occur:

$$O_2 \xrightarrow{k_f} 2O \qquad \text{(A)}$$

$$O_2 \xrightarrow{k_b} O_2 \qquad \text{(B)}$$

Write an expression for the net production of O_2.

Problem 5.4:

(a) Determine the number of collisions of CO molecules with O_2 molecules per unit volume and per unit time at 300 K and 3000 K. Assume P = 1000 kPa, CO: 10% and O_2: 20% (mole basis).
(b) If each collision results in the reaction

$$CO + O_2 \rightarrow CO_2 + O$$

how many kmol of CO will be consumed per unit volume and per unit time at 300 K and 3000 K?

(c) If E = 120,000 kJ/kmol and the steric factor = 1, compute the reaction rates.

Problem 5.5:

(a) Determine the RMS velocity (also called peculiar velocity) of CH_4 at 300 K and 1500 K in a stagnant medium.

(b) If a medium flows with a velocity of 100 m/sec, what are the absolute RMS velocities with respect to a stationary observer at 300 K and 1500 K?

(c) Calculate the average translational (random) kinetic energy (ATE) per molecule and per kmol.

(d) Compare the ATE per kmol with the activation energy 120,000 kJ/kmol.

Problem 5.6: Consider reactions for H_2–O_2

$$H_2 + O_2 \Leftrightarrow HO_2 + H \quad \text{(R1)}$$

$$H + O_2 \Leftrightarrow OH + H \quad \text{(R2)}$$

$$OH + H_2 \Leftrightarrow H_2O + H \quad \text{(R3)}$$

$$H + O_2 + M \Leftrightarrow HO_2 + M \quad \text{(R4)}$$

Write the overall rate expression for d[H]/dt.

Problem 5.7: Chain initiation reactions are those that involve the initiation of radicals R from a stable species S. S = A_2 and R = A, B. Consider the following reactions:

$A_2 + M \rightarrow A + A + M$	specific reaction rate, k_1	(R1)
$A + B_2 \rightarrow AB + B$	specific reaction rate, k_2	(R2)
$B + A_2 \rightarrow AB + A$	specific reaction rate, k_3	(R3)
$A + B_2 + M \rightarrow A + B + M$,	specific reaction rate, k_4	(R4)
$A + B + M \rightarrow AB + M$	specific reaction rate, k_5	(R6)

Consider only Reactions R1, R2, R3, and R6. Solve for [A] and [B] using the steady-state radical hypothesis. Show that:

(a) Radical concentration under steady state is

$$[A] = \left(\frac{k_1 [A_2][M]}{2k_2 [B_2]} + \left(\frac{k_1 k_3}{k_6 k_2} \right)^{1/2} \frac{[A_2]}{[B_2]^{1/2}} \left\{ 1 + \frac{k_6 k_1}{4 k_2 k_3} \left\langle \frac{[M]^2}{[B_2]} \right\rangle \right\}^{1/2} \right)$$

(b) Rate of production of $[B_2]$ is

$$\frac{[dB_2]}{dt} = -\frac{k_1[A_2][M]}{2} - \left\{\frac{k_2 k_1 k_3}{k_6}\right\}^{1/2} [A_2][B_2]^{1/2} \left\{1 + \frac{1}{4}\left\langle\frac{k_1 k_6 [M]^2}{k_2 [B_2] k_3}\right\rangle\right\}^{1/2}$$

Problem 5.8:

(a) CO gas burns with O_2 to produce CO_2. If the reaction is a single-step reaction, what is the overall order of reaction of the following?

(i) $CO + (1/2)\ O_2 \rightarrow CO_2$ (ii) $2\ CO + O_2 \rightarrow 2CO_2$

(b) The global reaction scheme for CO yields the following expression:

$$d[CO]/dt = k\ [CO]\ [O_2]^{(1/2)}\ [H_2O]^{(1/2)}$$

where $k = A \exp(-E/\bar{R}T)$; $A = 1.3 \times 10^{11}$ m³/kmol sec; and E = apparent or global activation energy = 125,500 kJ/kmol. Assume that the gases in an automobile have the following composition at TDC (top dead center): CO = 10%, O_2 = 5%, and H_2O = 2% (all mole %), T = 1200 K, and P = 10 bar. If the O_2 concentration and H_2O concentration remain unaffected during CO oxidation, what is the half-life time?

Problem 5.9: Curve fit $\dot{w}''' = A \exp(-E/\bar{R}T)$ the reaction rate per unit volume, with $\dot{w}''' = B\ T^m$. How will you determine B and m for the best fit of the reaction rate expression?

Problem 5.10: A mixture of propane and air with an equivalence ratio of 1.05 and an initial pressure of 1 atm is burned in a steady-flow reactor. The equilibrium products of combustion leave the reactor at a pressure of 1 atm and a temperature of 2000 K. Calculate the rate of destruction of H atoms in the equilibrium products of combustion (with O and OH at equilibrium) due to the elementary reaction $H + O_2 \Leftrightarrow OH + O$. The forward reaction rate is

$$k_f = AT^n \exp(-E_a/R_u T) \quad \text{where} \quad A = 1.59 \times 10^{17} \frac{\text{cm}^3}{\text{gmol}^{-s}},$$

$$n = -0.927, \quad \text{and} \quad E = 70.6 \frac{\text{kJ}}{\text{gmol}}$$

Problem 5.11:

(a) Calculate the equilibrium constant for the reaction

$$O + H_2O \Leftrightarrow 2OH$$

using the tabulated values for elements found in the literature.

(b) If the forward reaction rate constants are known, evaluate the backward reaction rate constants A and E. Make reasonable approximations.

Problem 5.12: Nitric oxide (NO) is a regulated emission species. The Zeldovich mechanism for NO formation is:

$$O + N_2 \underset{k_{1b}}{\overset{k_{1f}}{\Leftrightarrow}} N + NO \qquad (1)$$

$$N + O_2 \underset{k_{2b}}{\overset{k_{2f}}{\Leftrightarrow}} O + NO \qquad (2)$$

(a) Write an expression for d [NO]/dt.

(b) If $k_{1b} << k_{1f}$ and $k_{2b} << k_{2f}$, simplify the expression in part a.

(c) Using the partial equilibrium assumption

$$(1/2)\ O_2 \Leftrightarrow O$$

determine [O] as a function of $[O_2]$ and rewrite the expression in part b.

(d) If $k_{1f} = 7 \times 10^{10} \exp(-38{,}000/T)(m^3/kmol\ sec)$, $k_{2f} = 1.3 \times 10^7 \exp(8.4 \times 10^{12}/T)(m^3/kmol\ sec)$, calculate the amount of NO formed in ppm within 1 msec if O_2 = 4% and N_2 = 73% (mole basis), 2400 K and 1 bar. Assume steady state for [N] radicals.

Problem 5.13: Nitrogen is released as HCN from coal. The HCN undergoes oxidation to NO

$$HCN + O_2 \xrightarrow[k_A]{} NO + HCO \qquad (A)$$

At the same time the HCN combines with NO to produce N_2

$$HCN + NO \xrightarrow[k_B]{} N_2 + HCO \qquad (B)$$

These reactions are known as De Soete's kinetics. Reaction A is favored if oxygen is abundant, and Reaction B is favored if O_2 is deficient. Calculate the production rate of NO if HCN = 1%, O_2 = 5%, NO = 0.2% (mole %), and T = 1500 K. $k_A = 1.0 \times 10^{11} \exp(-280{,}328/\bar{R}T)$, $k_B = 3.0 \times 10^{12} \exp(-251{,}040/\bar{R}T)$, and P = 1 bar. For Reaction A, order with respect to HCN = 1, order with O_2 = 1. For Reaction B, order with HCN = 0, order with NO = 1. What is the NO production rate if T = 2,000 K?

Problem 5.14: Consider the following reaction:

$$H + H + M \Leftrightarrow H_2 + M.$$

The value of the forward rate coefficient is given by

$$k_f = \frac{6.40\times10^5}{T} \frac{m^6}{g\text{-}mol^2\ sec}, \quad T \text{ is } K.$$

The standard free energy of formation for the hydrogen atom is given by

$$g^0_{f,H}(T) = 227{,}000 - 60\,T, \frac{J}{g\text{-}mol}, \quad where\ T\ is\ K, and\ P^0 = 1\,atm.$$

For a mixture of 20 mole % H_2 and 80 mole % N_2 at a total pressure of 5 atm (5.066×10^5 Pa) and a temperature of 1000 K, find the values of the rate coefficient for the reverse reaction and the rate of change of H atom concentration. (Hint: [H] << $[H_2]$.)

Problem 5.15: A reversible reaction set is:

$$2\ A + B \xrightarrow{k_f} 2C,\ -83{,}600\ kJ\ (exothermic)$$

$$2C \xrightarrow{k_b} B + 2A.$$

The forward reaction rate is:

$$d[A]/dt = -k_f\,[A]^2\,[B]$$

where

$$\ln k_f = 55.26 - 208090/(\bar{R}T)\}.$$

$\bar{R}$ is in kJ/kmol K, T in K, and k_f in $m^6/\{kmol^2\ sec\}$.

(a) Write the overall expression for d[C]/dt including forward and backward reaction rates.

(b) Write the relation between K_0, k_f, k_b, and any other parameter (if necessary).

(c) What is the activation energy for the backward reaction? State the assumptions made.

(d) At 250°C, the equilibrium constant $K^0 = 6200$. Calculate the value of k_b at 250°C. State the assumptions made.

(e) What is the reaction rate of species C in kmol/m³ sec at 350°C if [A] = 0.002, [B] = 0.001, and [C] = 0.008 kmol/m³?

Problem 5.16: Consider the following reaction mechanism:

$$O_3 + M \xrightarrow{k_1} O_2 + O + M, \quad k_1 = 10^5 \exp(-10{,}000/T)\, m^3/g\text{-}mol\ sec$$

$$O + O_3 \xrightarrow{k_2} O_2 + O_2, \quad k_2 = 10^6 \exp(-2000/T)\, m^3/g\text{-}mol\ sec$$

$$O + N_2 \xrightarrow{k_3} N + NO, \quad k_3 = 2\times10^8 \exp(-40{,}000/T)\, m^3/g\text{-}mol\ sec$$

$$N + O_2 \xrightarrow{k_4} O + NO, \quad k_4 = 4\times10^7 \exp(-400/T)\, m^3/g\text{-}mol\ sec$$

(a) Assuming that the O and N atom concentrations are in steady state, derive expressions for [O], [N], and d[NO]/dt in terms of the rate coefficients k_1, k_2, k_3, k_4, and the concentrations of the major species $[O_2]$, $[N_2]$, and $[O_3]$. Neglect reverse reactions.

(b) Consider a mixture of 70 mole % N_2, 20 mole % O_2, and 10 mole % O at a temperature of 2000 K and a pressure of 1 bar. Calculate the initial rate of increase of nitric oxide, d[NO]/dt, and the steady-state values of [O] and [N]. Assume that the radical concentrations [O] and [N] have reached their steady-state values, that the concentrations $[O_2]$, $[N_2]$, $[O_3]$ have not changed significantly from their initial values in the time it takes the radical concentrations to reach steady state, and that [O], [N] << $[O_2]$, $[N_2]$, $[O_3]$.

Problem 5.17: The hydrogen–oxygen reaction involves many steps before it can produce water. Consider the three elementary steps:

$$H + O_2 \rightarrow OH + O, \quad k_1 = 1.86 \times 10^{11} \exp(-70{,}250/\bar{R}T) \qquad (R1)$$

$$O + H_2 \rightarrow OH + H, \quad k_2 = 1.82 \times 10^{7} \exp(-37{,}240/\bar{R}T) \qquad (R2)$$

$$OH + H_2 \rightarrow H_2O + H, \quad k_3 = 2.19 \times 10^{10} \exp(-21{,}550/\bar{R}T) \qquad (R3)$$

where k_1, k_2, and k_3 are specific reaction rate constants and the units are in m^3/(kmol sec) and activation energies are in kJ/kmol, R in kJ/kmol K, and T in K. There is no backward reaction. Assume the ideal gas relation to be valid for the mixture.

(a) Using the steady-state radical hypothesis for [OH], derive an expression for OH radical concentration in terms of H, O_2, and H_2 concentrations and k_1 and k_3. Neglect k_2 as compared to k_1 and k_3.

(b) If the H atom concentration is determined by the equilibrium condition

$$H_2 \Leftrightarrow 2H \qquad (E1)$$

what is the relation for OH concentration derived in part (a) in terms of $[H_2]$ and $[O_2]$ and k_1, k_3, R, and K^0_{E1} (equilibrium constant)?

(c) Using the earlier results, derive an expression for $d[H_2O]/dt$ in terms of k_1, K_{E1}^0, and the concentrations of H_2 and O_2.

(d) Assume that

$$\frac{d[H_2O]}{dt} = k_3[OH][H_2] = k_1[O_2][H_2]^{1/2} K^{0\ 1/2}_{4} \left(\frac{P^0}{\bar{R}T}\right)^{1/2}$$

where P^0 is the reference pressure assumed to be 1 bar. If the temperature of the reaction is 1500 K, $[H_2]$ and $[O_2]$ correspond to the initial stoichiometric reactant mixture using pure O_2 as oxidant source at a pressure of 10 bar, determine the initial production rate of H_2O in kmol/m^3 sec.

R = 0.08314 bar m^3/(kmol sec) = 8.314 kJ/(kmol K)units.

Problem 5.18: Determine the half-life time of the following reaction if the reaction occurs in air at 1200 K, and if the mixture is lean and $X_{CH4,0} = 0.1$:

$$d[CH_4]/dt = k[CH_4]^a\ [O_2]^b$$

where a = 0.3, b = 1.3, k = A exp $(-E/\bar{R}T)$, A = 1.3×10^8 (1/s) (g-mol/cm^3)$^{(1ab)}$, and E = 48.4 kcal/g-mol.

Problem 5.19: Consider caffeine $C_8H_{10}N_4O_2(s)$ with a density of 1200 kg/m^3. Brewed coffee of one serving (240 cm^3) contains 135 mg of caffeine while a tea bag contains only 50 mg. Within 45 minutes of consumption, caffeine is absorbed in the stomach and intestines. Caffeine slows the metabolism and when transported to the brain it stays there for 9 to 15 hours. Caffeine A is distributed to all tissues the body and is "eliminated by first-order kinetics," by reaction with catalysts B: $d[A]/dt = -k(T)\ [A]^{nA}\ [B]^{nB}$. Assume that catalyst concentration is large. Letting $k'(T) = k(T)\ [B]^{nB}$, assuming first order and 12 hours to reduce the caffeine, estimate k' (T). (Also see Problem 4.16.)

Problem 5.20: You might have observed trees and plants coming back to life (with leaves) during spring while some other plants and trees seem to stay "ever green." Explain why. Formulate a justifiable hypothesis.

Chapter 6 — Problems

Problem 6.1: In a multicomponent mixture, one component F diffuses into all other components that are stagnant (e.g., N_2 in an evaporation problem). Let Y_F be the mass fraction of F and v_F the absolute velocity of F. Then the average velocity of mixture, v, at any point in space is

(a) $v_F/(1 - Y_F)$
(b) $v_F\ Y_F$
(c) v_F/Y_F
(d) None of the above

Problem 6.2: Prove that if species k is the only one to diffuse through a stagnant mixtures consisting of species A, B, C, etc., with binary diffusion coefficients of D_{AS}, D_{BS}, etc., then

$$D_{kS} = (1 - Y_k)/\left\{\sum_j \left(Y_j/D_{jS}\right)\right\}$$

where S is the stagnant medium.

Problem 6.3:

(a) Calculate the mole fractions for reactants of a stoichiometric H_2–air mixture.
(b) Determine μ_{H2}, μ_{air}, λ_{H2}, λ_{air}, D_{H2air}, μ_{H2+air}, λ_{H2+air} and SC_{H_2} air for a stoichiometric H_2–air (reactant) mixture at 1200 K and 5 bar. Select c_p values for each species at 1200 K.
(c) If air is assumed to be stagnant, determine D_{H2-O2}, and then D_{H2-N2} and D_{H2}–air using Wick's formulae. Compare with the answers in part b.

Problem 6.4: Calculate λ for H_2 at 1500 K using (a) Euken's equation and (b) the Hirschfelder equation. (See N.C. Blair and J.B. Mann, J. Chem. Phys., 32, 1459 (1960).)

Problem 6.5: Determine the binary diffusion coefficient for a CH_4–C_2H_6 system at 40°C and 1 bar.

Problem 6.6: Often, natural gas is assumed to be simply methane. In order to calculate the burning rate it is necessary to know the binary diffusion coefficient D_{CH4-O2} for methane in oxygen at 300 and 1200 K. Determine the value of D_{CH4-O2}. The experiments done by Walker et al., *J. Chem. Phys.*, 32, p. 436, 1960 yield $(D_{CH4-O2})_{300} = 0.22 \times 10^4$ m²/sec and (D_{CH4-O2}) 1200 K = 2.7 $\times 10^4$ m²/sec. Compare your result with experiments. Assume P = 100 kPa.

Problem 6.7: A liquid n-heptane droplet evaporates in a gas turbine combustor operating at 100 kPa. The droplet radius is 50 μm and it is at a temperature of 353.5 K. Assume Sh = 2. If D = 1×10^4 m²/sec and $\rho_l = 684$ kg/m³, the vapor pressure follows the relation,

$$\text{Ln } P^{sat} \text{ (kPa)} = A - B/(T + C)$$

where T in K for A, B, and C are 13.587, 2911.32, and 56.51, respectively.

(a) Calculate the mass fraction of vapor and the gas density of the mixture at the drop surface. Assume that the only other component at the drop surface is air (M_{air} = 28.97).
(b) What is the initial rate of evaporation in kg/sec?
(c) If the mass-loss rate during the droplet lifetime is the same as the initial mass-loss rate, calculate the lifetime of the droplet.
(d) What is the answer for the lifetime if the d² law is used?

Problem 6.8: A mixture of gases A, A_2, and N_2 flows over a catalytic surface (e.g., platinum surface in automobile engines in which carbon monoxide is catalytically oxidized). At the catalytic surface, A combines with itself to form A_2 ($A + A \rightarrow A_2$, e.g., $O + O \rightarrow O_2$). There is no net mass transfer across the catalytic interface (i.e., Pt does not lose its mass). The diffusional velocities are

$$V_A = -\frac{D_{AN_2}}{Y_A}\frac{\partial Y_A}{\partial y}, \quad V_{A2} = -\frac{D_{A_2N_2}}{Y_{A_L}}\frac{\partial Y_{A_L}}{\partial y}, \quad \frac{D_{A_2N_2}}{D_{AN_2}} = 1/8$$

(a) What is the average velocity at the interface (use the definition of average velocity)?
(b) What is the ratio of flow of species A to A_2 at the catalytic surface?
(c) If $Y_{A,W} = 0.5\ Y_{A2,W}$, what is the ratio of mass fraction gradients of A to A_2?

Problem 6.9: Assume that you work for Genetics, Inc. It is interested in the transport processes in lungs. The lung contains over 600 million air sacs with a total surface area of 60 to 100 m² (for a nonsmoker). You are asked to calculate the mass transfer rate of O_2 from the air sacs in the lungs to blood in the capillaries. Assume that the following data are provided:

Gas: air
Temperature of air: 37°C
Pressure: 100 kPa
Velocity of air: 5 m/sec
Length of air sac: 10 cm
O_2 mass fraction near surface of air sac: 0
O_2 mass fraction in air: 0.23
Geometry: Flat plate
Assume properties of air at 300 K; Le = 1.0

Calculate the mass transfer rate to the blood per unit width of the air sac. State how you will determine the increase in the oxygen content of the blood when it passes through the air sac.

Problem 6.10: A fuel cell uses H_2 and air for producing electricity directly (without a steam power plant). The H_2 and O_2 molecules diffuse toward the electrodes. When they draw a large current from the fuel cell, the voltage falls because the diffusion to the electrodes limits the chemical reaction at the electrodes. Hence, it becomes necessary to increase the mass transfer rate of oxygen to the electrodes, $\dot{m}_{O2}$. Assume that $\dot{m}_{O2}$ is proportional to the product of density and the binary diffusion coefficient.

(a) First, air is used as the source for oxygen but is pressurized by five times and sent to the fuel cell. Comment regarding whether $\dot{m}_{O2}$ can be increased. Justify.
(b) Second, O_2 (21%) and He (79%) are used instead of air (O_2 = 21% and N_2 = 79%) and pressure is kept at 100 kPa. Will this increase $\dot{m}_{O2}$? Comment and justify.

$M_{He} = 4$, $M_{N2} = 28$, $M_{O2} = 32$, and $M_{H2} = 2$. Assume that all molecules have the same diameter. The ideal gas law is applicable. Assume the collision integral to be unity.

Problem 6.11: For a spherical *n*-hexane (C_6H_{12}, M = 84.2 kg/kmol) drop of 100-μm diameter at 319.5 K, determine the following for pure vaporization in air. Let P = 2 bar, $D_{C6H12-air} = 2.13 \times 10^5$ m²/sec, Sh = 1.8, $M_{air} = 28.97$ kg/kmol, and $Y_{C6H12,\infty} = 0$. For *n*-hexane, the saturation vapor pressure relation is

$$\ln\, p^{sat}(bar) = A - \frac{B}{T + C}$$

where A = 10.15, B = 3573.3 K, C = 0, and T is the drop temperature in K.

(a) Calculate $Y_{C6H12,w}$, the mass fraction of fuel vapor at drop surface.
(b) Calculate the initial vaporization rate of *n*-hexane in kg/sec.

Problem 6.12: See Figure P.6.12. Typically, the carrier gas for an anesthesia agent (e.g., desflurane) is 100% O_2. By controlling the bypass valve, one can control flows through A and B. Properties of desflurane are as follows: $C_3OF_6H_2$

and M = 168. The liquid desflurane is typically at 1 atm and 20°C, and heated to 25°C to 30°C in section A. Thus, vapor pressure varies and concentration of vapor in the carrier gas is altered. Typically, total flow rate is 60 l/min and liquid volume = 100 cc. When T_{liq} decreases, the bypass valve allows more through A but less through B. Analyze the heat and mass transfer processes so that the concentration of anesthetic can be predicted at exit.

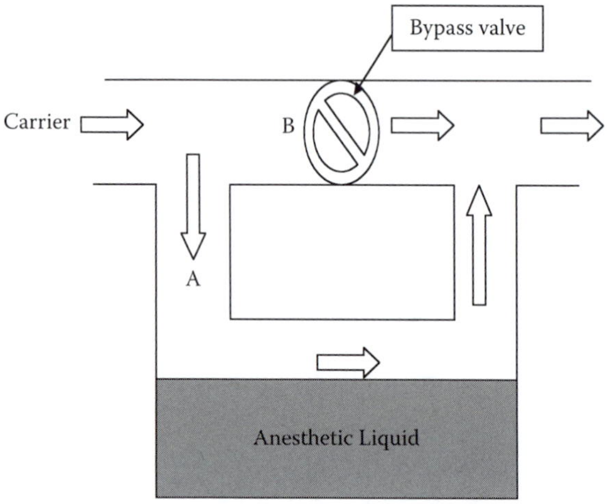

Problem P.6.12 Flow of anesthetic.

Problem 6.13: Compute the variation of Sc for CH_4: air combustible mixture at 298 K and 1200 K with stoichiometric ratio (SR) varying from 1 to 4. Treat air as one species. Comment on the results.

Problem 6.14: Natural gas is a mixture of many components. Assume it consists of $CH_4(1)$: 0.654, $C_2H_2(2)$: 0.224, and CO(3): 0.118. In order to determine flame velocity and Re number, thermal conductivity and viscosity data of the fuel are required. Determine λ_{mix} and μ_{mix} at 300 K and 1 atm.

Chapter 7 — Problems

Problem 7.1: Describe the relation between dP/dt in a closed system and the mixture equivalence ratio (ϕ). If the ϕ fluctuates from cycle to cycle, what are the fluctuations in dP/dt from cycle to cycle?

Problem 7.2: The first Damkohler number is a ratio of residence time to the chemical reaction time. True or false? Discuss the implication to combustion.

Problem 7.3: Under diffusion-controlled combustion, a spherical carbon particle producing CO will burn faster (i.e., carbon burn rate) compared to a carbon sphere of similar diameter burning to CO_2. Free-stream oxygen remains the same for both cases. True or false? (Hint: Use oxygen transport and resistance concept.)

Problem 7.4: Consider the rate of reaction

$$d[F]/dt = -A \exp(-E/\bar{R}T)\,[F]^{nF}\,[O_2]^{nO2}$$

(a) Convert the relation to mass form.
(b) Simplify for a closed system with a fixed volume. What are the results if all molecular weights are the same?
(c) For a fixed mass, the volume may change (e.g., fuel and air in a piston–cylinder assembly). Simplify the expression.

Problem 7.5: Consider the following Shvab–Zeldovich equation under steady state conditions:

$$\rho\vec{v} \cdot \nabla\beta = \nabla \cdot (\rho D \nabla\beta)$$

Apply this to a 1-D plug flow reactor (PFR), where axial (x direction) diffusion is ignored. At the inlet, $\beta = \beta_i$.

(a) Obtain the solution for β.

(b) Show that in a plug flow reactor with chemical reaction,

$$h_0 - h_\infty = q + (c_P\{T_0 - T_\infty\})$$

where q = heat released per kg mixture and T_0 and T_∞ are temperatures before and after complete combustion.

Problem 7.6: Global oxidation of CO occurs with the following reactions in a perfectly stirred reactor:

$$CO + (1/2)\, O_2 \rightleftarrows CO_2$$

$$\frac{d[CO]}{dt} = -k_f[CO][O_2]^{0.25}[H_2O]^{0.5}$$

$$\frac{d[CO_2]}{dt} = -k_b[CO_2]$$

$$k_f = 2.24 \times 10^{12}\ \exp\left\{-\frac{167400}{\bar{R}T}\right\},\ in\ kJ, kmol, m^3, \text{sec}\ units$$

$$k_B = 5 \times 10^{8}\ \exp\left\{-\frac{450400}{\bar{R}T}\right\},\ in\ kJ, kmol, m^3, \text{sec}\ units$$

where the H_2O concentration remains invariant because it virtually acts as catalyst.

(a) Ignore the backward reaction. If the Damkohler number $Da_I = t_{res}/t_{chem}$ and if the reactor temperature is fixed, then Da_I is larger for an inlet mixture mixed with steam than for a dry air. True or false?

(b) If CO exists in trace amounts, obtain an expression and plot variation of [CO] with time at fixed temperature, T, for products of propane: air mixture at $\phi = 0.9$.

Problem 7.7: $C_{12}H_{26}$ is fired into a cylindrical gas turbine combustion chamber. Plot ε_b vs. P given the following data:

$\phi = 1.0$	$E = 0.15 \times 10^6\ \frac{kJ}{kmol}$	$h_c = 44360\ \frac{kJ}{kg}$
$T_{in} = 600$ K	$c_P = 1.3\ \frac{kJ}{kg\ K}$	$A = 10^9\ \frac{m^3}{kg\ K}$

P = 10 bar order of reactions: 2

Diameter of combustor: 12 cm

Length: 20 cm

Determine the maximum possible mixture flow without extinction.

Problem 7.8: Consider a constant-volume reactor. If the reactor is charged with propane and 20% excess air and instaneously heated to 900 K, plot the variation of T(t), $Y_F(t)$, and $Y_{O2}(t)$ as oxidation reaction proceeds.

$$\left|w_F'''\right| = k_Y \rho^{nCH4 + nO2} [Y_F]^{nF} [Y_{O2}]^{nO2}, \quad k_Y = A_Y \exp\left(-\frac{E}{\bar{R}T}\right), \quad A_Y = A\left\{\frac{M_F}{M_F^{nF}} \frac{1}{M_{O2}^{nO2}}\right\}$$

$$\text{where} \quad \left|w_F'''\right| \frac{\text{kmol}}{\text{m}^3\text{sec}} = k[F]^{nF} \; [O_2]^{nO2}, \quad k = A\exp\left(-\frac{E}{\bar{R}T}\right)$$

Plot NO vs. time if NO is treated as a trace species with d(NO)/dt (kg/m^3 sec) = $7.2 \times 10^{14} {}^*\rho^{1.5}\exp(-573{,}000/RT)^* (1/T)^{0.5}\, Y_{N2}\, Y_{O2}^{0.5}$, where R is in units of kJ/kmol K.

Problem 7.9: Consider the oxidation of propane with stoichiometric air in a plug flow adiabatic reactor. Assume P = 1 bar, and

$$\dot{w}''' \text{ (kmol/m}^3\text{ sec)} = k\ [\text{Fuel}]^a\ [O_2]^b \qquad \text{(A)}$$

where k = A $\exp(-E/\bar{R}T)$, a = 0.1, b = 1.65, A = 4.836×10^9, E = 125,604 kJ/kmol, and c_{pmix} = 1.2 kJ/kg K = constant. The lower heating value is 46,357 kJ/kg and M_{mix} = 29 kg/kmol.

(a) Convert Equation A in terms of the mass fractions and express the reaction rate in kg/m^3 sec.
(b) Obtain a relation for the change in fuel mass fraction over time (i.e., dt = dx/v in PFR).
(c) Present the energy conservation for the adiabatic reactor. Reduce the equation to obtain an expression for dT/dt.
(d) Let T_{in} = 600, and 900 K. Obtain numerical answers for dT/dt, T, Y_F, and Y_{CO2} vs. t. Continue integration until T < 2000 K. Use a spreadsheet program. Note t = $\{dx/v_p\}$. Plot T, $Y_F/Y_{F,0}$ and $Y_{CO2}/Y_{CO2,final}$ vs. time for an adiabatic reactor using single-step kinetics. If ignition occurs when dT/dt = 1000 K/sec, what are the ignition times and the temperatures of mixture at which it occurs?

Problem 7.10: Starting from the relation

$$dE_{cv}/dt = \dot{Q}_{cv}\ \dot{W}_{cv} + \sum_{k,i} \dot{m}_k h_{,k} - \sum_{k,i} \dot{m}_k h_{,k}$$

show that

$$\dot{w}\ V\ M_{CO}\ HV_{CO} \approx [T - T_{in}][\ \dot{m}\ ]\ [(Y_{CO,in}\ c_{pcO} + Y_{O2,in}\ \bar{c}_{p,O2}$$

$$+ Y_{CO2,in}\ \bar{c}_{p,CO2} + Y_{N2,in}] - \dot{Q}_{c.v.}$$

Problem 7.11: Use the relation $\rho v dh/dx = d/dx\ \{(\lambda\ dT/dx - \rho \Sigma Y_k V_k\ h_k\}$, where $h = (\Sigma Y_k\ h_k)$, and $h_0 - h_\infty = q + (c_p\{T_0 - T_\infty\})$ to show that $\{T_0 - T_\infty\} = q/cp$.

Problem 7.12: Formulate a differential equation relating T to Y_F and time for a constant-volume nonadiabatic reactor if the heat transfer is

$$\dot{Q} = h_H (T_w - T)\ A_s\ ,$$

where T is the temperature of the reactor and A_s is the surface area.

Problem 7.13: Ethane burns under very lean conditions (excess air: 100%) at a constant pressure of 10 bar. The oxygen concentration is almost unaffected. The ethane reaction rate is

$$d[F]/dt\ (kmol/m^3\ sec) = A_{[\ ]}\ exp\ \{-15098/T\}\ [F]^{0.1}\ [O2]^{1.65}$$

where $A_{[\ ]} = 6.19 \times 10^9$, T in K, and [F] is expressed in kmol/m^3. Convert the rate expression into the following mass form:

$$d(F)/dt\ (kg/m^3\ sec) = A_{(\ )}\ exp\{-15098/T\}\ \rho^n\ Y_F^{0.1}\ Y_{O2}^{1.65}$$

where ρ is the density of mixture, (F) is the concentration in kg of fuel per m^3, and Y_F is the mass fraction of the fuel. Assume that the molecular weight of mixture is almost unaffected and remains at its initial value. $M_{C2H6} = 30.08$ kg/kmol, $M_{O2} = 32$, and $M_{N2} = 28.02$.

(a) Determine the activation energy E in kJ/kmol, initial mixture molecular weight, M_{mix}, and the values of $A_{(\ )}$ (in units of kg, m^3, sec) and n.
(b) Write an expression for $\rho dY_F/dt$.
(c) Evaluate $\rho dY_F/dt$ at t = 0 in units of kg/m^3 sec if T = 1200 K and the rate of temperature change at time t = 0 is 24,000 K/sec. Assume that $A_{(\ )} = 2.2 \times 10^{\ 7}$ (in units of kg, m^3, sec) and n = 1.75.

Problem 7.14:

(a) Starting from the relation $|\dot{w}_F^{m'''}|\,V\,h_c + \dot{Q} = mc_p dT/dt$, show that for constant-pressure adiabatic combustion systems, the variation of fuel mass fraction concentration is

$$Y_{F0} + \frac{c_p T_0}{h_c} = \left\{ Y_F + \frac{c_p T}{h_c} \right\}$$

Treat the molecular weight and c_p as constants.
(b) What is the expression for flame temperature?

(c) Calculate the maximum temperature attained for a diesel–air mixture with 30% excess air. Use c_p of products at 1200 K with $c_p = \Sigma Y_k\ c_{pk}$ and T_0 = 600 K. Assume complete combustion. Diesel fuel can be approximated as $C_{14.4}H_{24.9}$ and the lower heating value for gaseous diesel, h_c is 46,074 kJ/kg (assume negligible latent heat).

Problem 7.15: Determine the temperature at which maximum reaction rate occurs for a rigid closed system assuming the following reaction rate expression:

$$dY_F/dt = -A\ \exp(-E/RT)\ \rho^{(nF+nO21)}\,[Y_F]^{nF}\ [Y_{O2}]^{nO2}.$$

Assume a very lean mixture so that variations with O_2 can be ignored.

Problem 7.16: The Sherwood number relation is $Sh = (2 + 0.6\,\mathrm{Re}^{1/2})$ where

Re = v_∞ d/ν; d, the diameter of particle; v_∞, the free-stream velocity; and ν, the kinematic viscosity.

Also, $Y_{O_2,\infty} = 0.10$, v_∞ = 0 m/sec, d_0 = 100 μm, $\rho_p = 1550$ kg/m^3, ν = 1 × 10^{-4} m^2/sec, B_{O_2} = 5×10^7 m/sec, T_w = 1200 K, and E = 150,000 kJ/kmol.

(a) Evaluate the burning rate per unit area $\dot{m}''$ of a spherical particle for the following cases:
   (i) diffusion control
   (ii) kinetics control
   (iii) mixed control

(b) Obtain a relation between the diameter of particle and the time taken for kinetic-controlled combustion. What is the total burning time for diffusion- and kinetic-controlled combustion?

(c) What are the results if U_∞ = 20 m/sec? (Note: Re can change during combustion; use average Re value.)

(d) If carbon particles burn under diffusion control, what is the length of the flame in a boiler burner? Assume that the relative velocity between the gases and particles is negligible and the gases travel with a velocity of 10 m/sec.

Chapter 8 — Problems

Problem 8.1: Methane has a heating value of 50,000 kJ/Kg. The Shvab–Zeldovich variable β_{ht-O_2} is

(a) $h_t/50{,}000 - Y_{O2}/4$
(b) $h_t/Y_{O2} + 50{,}000/4$
(c) $h_t/50{,}000 + Y_{O2}/4$

Problem 8.2: Consider the 2-D continuity equation

$$\frac{\partial \rho}{\partial t} + \frac{\partial(\rho v_x)}{\partial x} + \frac{\partial(\rho v_y)}{\partial y} = 0.$$

If one applies it to a stagnant closed system such as an automobile engine, where $v_x = 0$ and $v_y = 0$, but volume V(t) varies the results yield ρ = constant. Do you agree with this result for an automobile engine? Why?

Problem 8.3: A heterogeneous reaction occurs at the surface of a carbon plate with the reaction C (s) + 1/2 $O_2 \rightarrow CO$. There is no gas-phase reaction. If y is the coordinate normal to the burning surface in a flow system and $\dot{m}$ is the burning rate of carbon under finite kinetics, write the interface boundary condition for oxygen species.

Problem 8.4: Consider the nonsteady 3-D (r, θ, and Φ) conservation equations in spherical coordinates for mass, species k (k = fuel F, O_2, CO_2, H_2O, N_2), elements C and H, mixture fraction, total enthalpy (h), and thermal enthalpy in boundary-layer-type equations.

(a) Simplify the equation for a spherical drop of radius "a" with the following assumptions:
  (i) Variations in θ and Φ directions are negligible compared to variations in "r" direction (i.e., spherically symmetric problem).
  (ii) ρD = constant.

(b) Assuming that the chemical reactions occur with a single-step reaction for the fuel *n*-hexane, reduce the species and thermal enthalpy conservation equations for part (a) in terms of the Shvab–Zeldovich variable β, i.e., define β for O_2-F, CO_2-F, H_2O-F, P-F, and h_t-F. The variable P is a sum of CO_2 and H_2O.

(c) Let $\phi = \dfrac{\beta - \beta_\infty}{\beta_w - \beta_\infty}$, where β is any SZ variable and ϕ is the normalized SZ (NSZ) variable. Reduce the equations in part b with ϕ as dependent variable. Note β_w denotes property at the surface (wall) of droplet.

(d) Simplify the equation obtained in part b assuming steady-state combustion.

Problem 8.5: Liquid *n*-hexane fuel burns over a flat plate in air. Use the 2-D, steady-state boundary-layer-type formulation.

(a) Write the generalized form of SZ equations for β.

(b) Expand $\beta_{O2\text{-}F}$, $\beta_{CO2\text{-}F}$, $\beta_{H2O\text{-}F}$, $\beta_{P\text{-}F}$, $\beta_{ht\text{-}F}$, $\beta_{ht\text{-}O2}$, and $\beta_{P\text{-}F}$ using ν_k, ν_P, and h_c.

(c) Reduce part a using the normalized SZ variable: $\phi = \dfrac{\beta - \beta_\infty}{\beta_w - \beta_\infty}$, and assume $\beta_w - \beta_\infty$ is independent of "x."

(d) If the interface boundary condition for SZ is

$$\dot{m}''\,\beta_w - \rho D\,(\partial\beta/\partial y)_0 = \dot{m}''\,F_\beta$$

and $B = \{\beta_w - \beta\infty\}/\{F_\beta - \beta_w\}$, write down F_β for $\beta = \beta_{O2\text{-}F,w}$ and $\beta_{ht\text{-}O2}$.

Problem 8.6: Starting from

$$\frac{\partial h}{\partial t} + \left(\rho v_x\,\frac{\partial h}{\partial x}\right) + \left(\rho v_y\,\frac{\partial h}{\partial y}\right) = \frac{\partial}{\partial x}\left(\frac{\lambda}{c_p}\,\frac{\partial h}{\partial x}\right) + \frac{\partial}{\partial y}\left(\frac{\lambda}{c_p}\,\frac{\partial h}{\partial y}\right)$$

prove that

$$\rho\frac{\partial h_T}{\partial t} + \left(\rho v_x\,\frac{\partial h_T}{\partial x}\right) + \left(\rho v_y\,\frac{\partial h_T}{\partial y}\right) = \frac{\partial}{\partial x}\left(\frac{\lambda}{c_p}\,\frac{\partial h_T}{\partial x}\right) + \frac{\partial}{\partial y}\left(\frac{\lambda}{c_p}\,\frac{\partial h_T}{\partial y}\right) + |\,\dot{w}_{F'''}\,|\,h_c$$

assuming a one-step reaction. h: total enthalpy; h_t: thermal enthalpy.

Problem 8.7: Given the following total energy conservation equation

$$\frac{\partial}{\partial t}\left(\frac{1}{2}\,\rho v^2 + \rho\,u\right) + \vec{\nabla}\cdot\left(\frac{1}{2}\,\rho v^2\,\vec{v} + \rho\,u\,\vec{v}\right)$$
$$= -\vec{\nabla}\cdot(\vec{v}\cdot\vec{\vec{P}} + \vec{q}'') + \sum \rho_k\,\vec{f}_k\cdot(\vec{v} + \vec{V}_k) \qquad \text{(A)}$$

where f_k is the body force per unit mass, prove the following.

$$\rho \frac{Du}{Dt} = \rho \frac{\partial u}{\partial t} + \rho \, \vec{v} \, . \, \vec{\nabla} u = - \vec{\nabla} \, . \, \vec{q}'' - \vec{\vec{P}} : \vec{\nabla} \vec{v} + \rho \sum Y_k \, \vec{f}_k . \vec{V}_k \tag{B}$$

Hint: Expand Equation A as

$$\frac{\partial}{\partial t}\left(\frac{1}{2} \rho v^2 + \rho u\right) = \frac{v^2}{2}\left(\frac{\partial \rho}{\partial t}\right) + \rho \frac{\partial}{\partial t}\left(\frac{v^2}{2}\right) + \rho \frac{\partial u}{\partial t} + u \frac{\partial \rho}{\partial t}. \tag{C}$$

Use the continuity equation. Then multiply the momentum equation with $\vec{v}$.

Problem 8.8: From thermodynamics, the relation

$$T ds = du + p \, d\left(\frac{1}{\rho}\right) - \sum_k \mu_k dY_k$$

is valid, where μ_k is the chemical potential per unit mass for the species k. Show that entropy changes are in general zero only if (a) $\lambda = \mu = \kappa = D_{kl} = \dot{w} = 0$ (i.e., frozen reaction) or (b) $\lambda = \mu = \kappa = D_{kl}$ and the fluid is in chemical equilibrium (criteria of chemical equilibrium can be in terms of chemical potential).

(Hint: Obtain ds/dt from $\frac{du}{dt}, \frac{d\left(\frac{1}{\rho}\right)}{dt}, \frac{dY_k}{dt}$.)

Problem 8.9: Propane is issued from a circular nozzle while air (in excess of stoichiometric) is injected in the annular space surrounding the fuel jet. See Figure P.8.9. CO_2, H_2O, and C_3H_8 were measured at various locations. At locations (x_{stoich}, y_{stoich}), there were no C_3H_8 or O_2, but measured mass fractions of CO_2, H_2O, and N_2 were found to be exactly the same as those found at the exit of a combustor supplied with C_3H_8 and stoichiometric air, i.e., $C_3H_8 + 5(O_2 + 3.76N_2) \rightarrow 3CO_2 + 4H_2O + 18.80N_2$. Calculate f, Z_C, and Z_H along (x_{stoich}, y_{stoich}). This stoichiomteric contour is called the flame contour for jet flames.

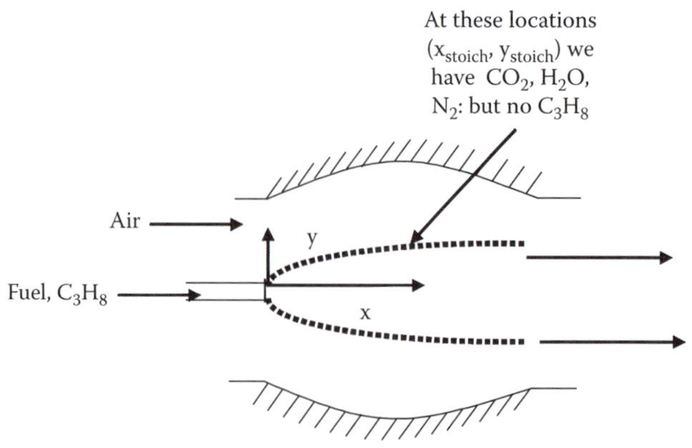

Figure P.8.9 Mixture fraction for Problem 8.9.

Problem 8.10: Consider a plug flow isothermal reactor in which methane with excess air is admitted at an arbitrary T and P. The overall global oxidation reaction is [Annamalai and Ryan, *Prog. Energy Combust. Sci.*, Part II, 383–446, 1993].

$$d[CH_4]/dt\ (kmol/m^3\ sec) = k\ [Fuel]^a\ [O_2]^b \qquad (A)$$

where $k = A \exp(-E/\bar{R}T)$, E = 202,600 kJ/kmol, a = 0.3, b = 1.3, and A = 1.3 × 10^8 1/sec. Assume $Y_{CO2} = 1.0 \times 10^6$, $Y_{H2O} = 1.0 \times 10^6$, and a single-step global reaction.

(a) Convert Equation A to a differential equation in terms of mass fractions. Assume that all molecular weights are equal.
(b) Present Shvab–Zeldovich equations for $\beta_{O2\text{-}F}$, $\beta_{CO2\text{-}F}$, $\beta_{hT\text{-}F}$, and $\beta_{H2O\text{-}F}$ and obtain a simple solution.
(c) Obtain numerical answers with 10% excess air, T = 1000 K, and P = 2 bar. Use a spreadsheet to determine Y_{CH4}, Y_{O2}, Y_{CO2}, and Y_{H2O} as a function of time t, t = $\{dx/v_p\}$.
(d) Plot Y_F, Y_{O2}, T vs. x or time.
(e) Mention briefly what will happen if CH_4 oxidation occurs through multiple reactions.

Problem 8.11: When coal is introduced into hot air, it releases combustible gases such as CH_4, C_2H_6, etc. We will assume the whole pyrolysate to be represented by CH_2. Once the pyrolysate is released, the residual material is called char. Char then oxidizes to CO. The gases CH_2 and CO then oxidize in the gas phase:

$$CH_2 + 1.5O_2 \rightarrow CO_2 + H_2O \qquad (I)$$

$$CO + 0.5O_2 \rightarrow CO_2. \qquad (II)$$

We will assume that a coal plate is subjected to flowing air and treat the problem as a 2-D boundary-layer problem.

(a) Write the gas-phase conservation equations for the species CO, CH_2, CO_2, O_2, H_2O, N_2, and the thermal enthalpy h_T.
(b) For these two reactions in the gas phase, define a suitable Shvab–Zeldovich variable β_{ijk} for CO_2 and O_2 and β_{ij} for H_2O and reduce the conservation equations in part a to Shvab–Zeldovich form.

$$\rho v_x \frac{\partial \beta}{\partial x} + \rho v_y \frac{\partial \beta}{\partial y} = \frac{\partial}{\partial y}\left(\rho D \frac{\partial \beta}{\partial y}\right) \quad (1)$$

Le = 1 is typically invoked for $\beta_{ht\text{-}F}$ or when one couples the species reaction with thermal enthalpy.

Problem 8.12: Refuse-derived fuels (RDFs) are made from waste materials containing plastics, paper, rubber, etc. The ultimate analysis is as follows: C = 45%, N = 1%, H = 7%, O = 15%, and S = 0.5%. What is the empirical chemical formula? Determine the heating value using the Boie equation (Chapter 4). Compare with the HV as received (with ash of 18%, moisture of 4%) is given as 22,000 kJ/kg. If the heat of pyrolysis is assumed as 500 kJ/kg and the RDF is shaped as a cylinder of 2 cm (diameter) × 4 cm (length), what is the estimated burning rate if the pyrolysis temperature is 600 K? Assume a gas velocity of 5 m/sec.

Problem 8.13: Obtain the Shvab–Zeldovich formulation for hydrazine N_2H_4, a rocket fuel, if H burns to H_2O and N to N_2 with a lower heating value of 534,000 kJ per kmol. Assume H_2O is in the gaseous form.

Problem 8.14: Consider the combustion of a spherical cloud of drops burning in air. The drops act as point sources, supplying species at a rate $\dot{m}_{k,p}$. The chemical species production rate per unit volume is $\dot{w}_{k,chem}$. The required latent heat for the evaporation of each drop $\dot{q}_s$ is supplied by the gas phase. Consider a spherical shell of radii r and r + dr. Select the c.v. for the space between r and r + dr but exclude drops of radii "a." Use the absolute velocity $v_k = V_k + v$. See Figure P.8.14. Starting from the global form of the first law of thermodynamics for energy and species k, present the conservation equations for Y_k and h_t, and prove that

$$\frac{\partial}{\partial t}[\rho\, 4\pi r^2 Y_k] + \frac{\partial}{\partial r}[\dot{m} Y_k] = \frac{\partial}{\partial r}\left(\rho D\, 4\pi r^2 \frac{\partial Y_k}{\partial r}\right) + \dot{m}_{k,p}'''\, 4\pi r^2$$

$$+\ \dot{w}_k'''\, 4\pi r^2, \quad k = F, P, O_2$$

where $\dot{m}'''_{k,p} = n\dot{m}_k = 0$ for k = O_2, CO_2, and H_2O
$\dot{w}_k = -|\dot{w}_F|$ if k = F, $\dot{w}_k = -|\dot{w}_{O2}|$ if k = O_2

$$\frac{\partial}{\partial t}\left[\rho\, 4\pi r^2 h_t\right] + \frac{\partial}{\partial r}[\dot{m}h_t] = \frac{\partial}{\partial r}\left(\lambda\, 4\pi r^2 \frac{\partial h_t}{\partial r}\right)$$

$$-\dot{q}'''4\pi r^2 - \left|\dot{w}_F'''\right| h_c 4\pi r^2, \quad k = F, P, O_2$$

where $\dot{q}''' = \dot{q}_s\, n = \dot{m}'''_{F,p}\, L = \dot{q}''_s\, 4\pi a^2\, n$, and n is the number of drops per unit volume.

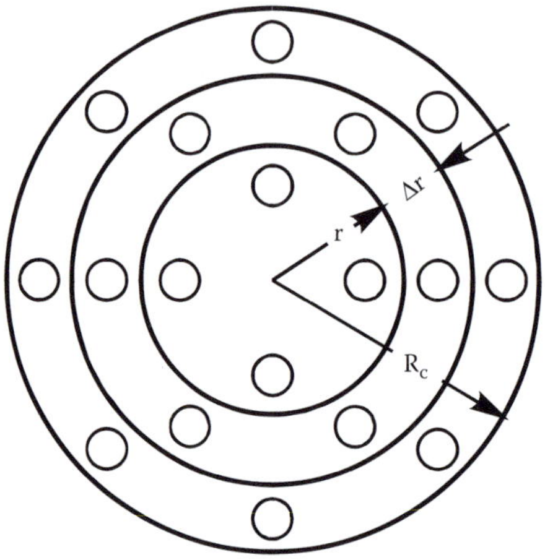

Figure P.8.14 Problem 8.14.

Problem 8.15:

(a) Calculate the Gr number for a coal particle of diameter 1 mm burning in air if the flame temperature is 1800 K and the ambient is at 1100 K.

$$Gr = g\, d_p^3\, (T/T)/U^2$$

Assume ν at 1500 K for air.

(b) If the effect of free convection on Nu is $Nu/Nu^0 = (1 + 0.533\, Gr^{0.52})$ $Nu^0 = 2$. Assume Pr at 1500 K for air. Determine Nu.

(c) For forced convection, the Nu correlation is $Nu/Nu^0 = (1 + 0.276\, Re^{1/2}\, Pr^{1/3})$. Calculate the equivalent Re for the free convection, which will have the Nu value as in part b.

(d) If an experiment on a 1-mm diameter particle is performed for Re = 1, do you believe that free-convection effects are important?

Problem 8.16:

(a) Consider the burning of a solid combustible wall under free convection in a 2-D configuration. Start from the generalized NSZ equation with coordinates x_1, x_2, and x_3. Formulate the boundary-layer-type momentum equation in the x_1 direction (coordinate along the vertical surface).

(b) Estimate the average burn rate of Plexiglas of 20 cm height burning in air. Assume diffusion-controlled combustion. Use heat transfer correlations along with mass transfer corrections.

Chapter 9 — Problems

Problem 9.1: Under diffusion-controlled combustion, a spherical carbon particle producing CO will burn faster compared with a similar carbon sphere burning to CO_2. True or false? Explain.

Problem 9.2: A charcoal particle of diameter 100 μm burns within 200 m/sec. What is the combustion time if a 5-cm diameter sphere is used in barbeque grills? Assume diffusion-controlled combustion. Express the answer in hours.

Problem 9.3: For zone-III reactions, a porous carbon sphere of diameter d has the same reaction rate as a nonporous carbon sphere of the same diameter. True or false? Explain.

Problem 9.4: Consider the combustion of Kentucky coal. Assume that the chemical formula is $CH_{0.769}N_{0.0216}O_{0.0692}$. What is the stoichiometric amount of oxygen in kg/kg of coal (include the N going to NO) assuming that C burns to CO_2 and H burns to H_2O? If the coal burns heterogeneously under diffusion control and if its density is 1300 kg/m^3, what is the burning time for 75-μm particles? If the velocity of coal and gas is 15 m/sec, what is the combustor length required for complete combustion? (Note that boilers fire pulverized fuel (pf) suspensions with 75% of mass passing through a 75-μm sieve.)

Problem 9.5: Activated carbon is manufactured from corn fiber. First, corn fiber is densified and then fed into rotary kilns for carbonization at 700 to 1000°C. The resulting corn char is activated in $H_2O/CO_2/O_2$ at 700 to 1000°C to get the desired pore structure and (large) internal surface area.

(a) Can CO_2, H_2O, and O_2 molecules enter submicro-, micro-, meso-, and macropores? Compare the dimensions of these pores to the molecular mean free path and collision diameter of molecules.

(b) How is the pore structure altered from 700 to 1000°C? Discuss qualitatively.

Problem 9.6: Coal water slurry is fired into a diesel engine used for locomotive trucks. Once water evaporates, it can be assumed that the burning time of coal is controlled by the time to turn it into char. Suppose $T_p = 1056$ K, $d_p = 6 \times 10^{-6}$ m, $Y_{O_2,\infty} = 0.22$, $T_\infty = 1030$ K, $\rho D = 0.5 \times 10^{-4}$ kg/m sec, $c_p = 1.18$ kJ/kg K, $h_c = 9242$ kJ/kg ($C + 1/2\ O_2 \rightarrow CO$), Sh = 2, $A_{O2} = 600$ m/sec, $M_{mix} = 30$, $\rho_p = 650$ kg/m^3, E = 66,400 kJ/kmol, P = 59.9 bar. Calculate the following:

(a) R_{diff}
(b) R_{chem}
(c) The burning rate of carbon (kg/sec)
(d) If the burning rate is kinetically controlled, what is the burning time?
(e) If the burning becomes diffusion controlled, what is the particle temperature T_p? Include radiation loss with $\varepsilon = 1$. Assume Nu = 2.
(f) What is the heat loss rate from particle to gas, if Le = 1.0, Nu = 2.0 for case (e)?
(g) What is the heat liberation or generation rate due to reaction for case (e)?

Problem 9.7: Black liquor is sprayed at 120°C into a boiler at 900°C; after drying and pyrolysis, it is converted into swollen char with 25% carbon and 75% inorganics (sodium salts such as Na_2CO_3, Na_2S, Na_2SO_4, and potassium salts such as K_2CO_3, K_2S, K_2SO_4, etc.). The main reactions that convert the carbon in the char particle of a black liquor are:

$$C + 1/2\ O_2 \rightarrow CO \quad \text{(I)}$$

$$C + O_2 \rightarrow CO_2 \quad \text{(II)}$$

$$C + CO_2 \rightarrow 2CO \quad \text{(III)}$$

$$C + H_2O \rightarrow CO + H_2 \quad \text{(IV)}$$

$$C + (1/2)Na_2SO_4 \rightarrow CO_2 + (1/2)Na_2S \quad \text{(A)}$$

$$C + (1/4)Na_2SO_4 \rightarrow CO + (1/4)Na_2S \quad \text{(B)}$$

It is mentioned in the literature that Reaction III and Reaction IV dominate, particularly when the O_2 concentration is low.

(a) Formulate the interface boundary conditions for Reactions I to IV and A to B.
(b) How will you solve for the combustion rate of char and the net heat release rate (or absorption) if the gas film is frozen and assuming that Reaction I to Reaction IV are significant?

Problem 9.8:

(a) Write the reaction equation for an arbitrary amount of CO and CO_2 mixture, if both are produced by heterogeneous reactions under finite kinetics.
(b) Write the interface balance equations and describe a procedure for determining the burning rate. Assume the single-film model.

Problem 9.9:

(a) Write the interface energy and species conservation equation for the diffusion-controlled burning of carbon for the following two schemes: single-film model and double-film model. The carbon is at a uniform temperature.

(b) The free-stream CO_2 mass fraction is negligible, whereas $Y_{O2,\infty} = 0.23$. Determine (i) CO mass fraction at the wall, (ii) T_w, and (iii) $\dot{m}$ for both models.

Problem 9.10:

(a) Use the interface energy and species conservation for CO_2 and the SZ variable for $\beta_{ht\text{-}CO2}$ to obtain a relation for burning rate ξ_w under DFM. Neglect radiation.

(b) Equate the result with $\xi_w = \ln(1 + B)$ and show that

$$\frac{c_p (T_w - T_\infty)}{h_{c,V}} = B\left[\frac{h_{C,I}}{h_{C,V}} - \frac{\dot{q}''_{rad}}{\dot{m}'' h_{C,V}}\right] - \frac{Y_{CO_2,\infty}(1+\nu_{O2,I})}{\nu_{CO_2,III}}$$

Problem 9.11: Consider the Reactions I to V mentioned in the text. In addition consider the following reaction:

$$H2 + 1/2\, O_2 \rightarrow H_2O \qquad (V)$$

Obtain an expression for the burning rate under diffusion-limited combustion. Use the SZ formulation for the two gas-phase reactions. (Hint: Chapter 8.)

Problem 9.12:

(a) Begin from the governing equations for the combustion of carbon C + $1/2\, O_2$ = CO in the frozen gas phase (SFM). Write the interface condition and solve for the burning rate. Show that

$$\frac{\dot{m}}{4\pi \rho D a} = \xi_w = \ell n[1 + B_{kin}]$$

where

$$B_{kin} = \left(\frac{\left(\dfrac{Y_{O_2,\infty}}{\nu_{O_2,V}} - Y_{CO,\infty}\right) - \left(\dfrac{Y_{O_2,w}}{\nu_{O_2,V}} - Y_{CO,w}\right)}{(1+\nu_{O_2,I}) + \dfrac{\nu_{O_2,I}}{\nu_{O_2,V}} - \left(\dfrac{Y_{O_2,w}}{\nu_{O_2,V}} - Y_{CO,w}\right)}\right)$$

(Hint: Use β_{O2-CO}.)

(b) Reduce the expression for the case when the oxygen mass fraction reaches extremely low values at the particle surface. Combustion under this condition is called diffusion-controlled combustion under SFM.

(c) Obtain an expression for the burning rate of carbon with the frozen chemical reaction in terms of CO mass fractions at the surface and at the ambient. Compare with the results in part a and obtain an expression for $Y_{CO,w}$. (Hint: Use Y_{CO}.)

(d) Obtain an expression for the temperature of carbon with the frozen chemical reaction for part (b).

(e) Determine the burning rate and CO mass fraction at the particle surface when the particle diameter = 100 μm and $Y_{O2,\infty} = 0.23$, when Reaction I dominates, there is diffusion-controlled combustion, and $\rho_w D_w$ (not a function of pressure) = 0.15×10^4 kg/msec. Neglecting radiation loss and assuming $T_\infty = 1000$ K, determine the value of T_w.

Problem 9.13: Obtain expressions for (a) burning rate, (b) wall mass fraction of CO for carbon (s) in Reaction III and Reaction IV, and CO_2 for Reaction II, and (c) surface or wall temperatures. Use the single-film model.

Problem 9.14: Show that the particle/surface/wall temperature of burning carbon particles for the single-film model (frozen gas phase) with the Boudouard reaction is

$$\frac{c_p(T_w - T_\infty)}{h_{C,I}} = \frac{Y_{O_2,\infty}}{v_{O_2,I}} + \frac{Y_{CO_2,\infty}}{v_{CO_2,III}} \frac{h_{C,III}}{h_{C,I}} - \frac{B\dot{q}''_{rad}}{\dot{m}'' h_{C,I}}$$

Figure P.9.14 T_W of gas leaving the vector is the same as the adiabatic flame temperature. Figure for Problem 9.14.

Problem 9.15: Carbon burns with Reaction I and Reaction III proceeding at a finite rate. Gas-phase oxidation (Reaction V) is frozen. Derive an expression for $Y_{CO,w}$; reduce it to the following equation, under diffusion control.

$$Y_{CO,w} = 1 - \frac{Y_{I,\infty}}{[1+B]}$$

Problem 9.16: Show that for a double-film model, the NSZ at the flame location is provided by

$$\phi_f = \frac{1+B}{2+B}, \text{ if } \quad Y_{CO_2\infty} = 0.$$

Problem 9.17:

(a) Show that flame temperature for diffusion-controlled carbon burning under DFM and $Y_{CO_2\infty} = 0$ is obtained by using

$$\frac{c_p(T_f - T_\infty)}{h_{C,I}} = \frac{Y_{O_2,\infty}}{\nu_{O_2,V}}\frac{h_{C,V}}{h_{C,I}} + B\frac{s(1+B)}{1+B+s(1+B)-Y_{I,\infty}}\left[\frac{h_{C,III}}{h_{C,I}} - \frac{\dot{q}''_{rad}}{\dot{m}''h_{C,I}}\right]$$

(b) Show that the expression for CO_2 mass fraction at the flame is $Y_{CO_2,f} = B[1-\phi_f]\,\nu_{O_2,III}$

Chapter 10 — Problems

Problem 10.1: An octane (C_8H_{18}) drop burns in pure oxygen. Under diffusion-controlled combustion, the mass fraction of products (i.e., CO_2 + H_2O) at the flame location is equal to unity. True or false? Discuss.

Problem 10.2: *n*-pentane and *n*-hexane have approximately the same latent heats (L or h_{fg}). However, their normal boiling points (NBPs) are 36°C and 68°C, respectively. Which one vaporizes more rapidly in air at 300 K? Explain.

Problem 10.3: Undecane (U) and octanol (O) have approximately the same boiling points (T_{BP}). However, the latent heats are different (U: 265, O: 548 kJ/kg). Which one vaporizes more rapidly in air at 300 K? Explain.

Problem 10.4: Consider a liquid fuel stored in a cylindrical tank of diameter d. The height of the rim above the liquid surface is H. Evaporation occurs and at the same time the vapor diffuses out to the open atmosphere. However, if diffusion is slow, then the vapor accumulates and forms a flammable mixture. Suppose that the tank contains a mixture of *n*-pentane, benzene, and methanol, calculate the ratio of timescales of diffusion for $t_{pentane}/t_{methanol}$ and $t_{benzene}/t_{methanol}$. Compare the evaporation timescales. Which of these fuels is highly flammable?

Problem 10.5: A spherical drop of water evaporates in air. The mass fraction of vapor follows the following law:

$$\frac{Y_v - Y_{v,\infty}}{Y_{v,w} - Y_{v,\infty}} = \frac{1 - \exp(-\xi)}{1 - \exp(-\xi_w)}$$

where $\xi = \dot{m}/4\pi\rho Dr$; $\xi_w = \dot{m}/4\pi\rho Da$; $\xi_w = \ln(1 + B_\infty)$; $\dot{m}$, evaporation rate; a, drop radius; D, diffusivity; Y_v, vapor mass fraction at r; $Y_{v,w}$, vapor mass fraction at the wall or particle surface; and r = a.

(a) Obtain an expression for the mass-average velocity v at r = a.
(b) Obtain an expression for the diffusion velocity of water vapor (V_v) at r = a.
(c) What is the expression for the absolute velocity of water vapor (V_v) at r = a.
(d) Obtain the numerical values for part a, part b, and part c given the following:

$B_\infty = C_p\,(T_\infty - T_w)/L$, $T_\infty = 1000\,K$, $T_w = 360\,K$, $L = 2303\ kJ/kg$,

$c_p = 1\ kJ/kg\ K$, $Y_{v,w} = 0.80$, $Y_{v,\infty} = 0$, $a = 50\ \mu m$, $\rho = 1\ kg/m^3$,

and $D = 5\times10^{-4}\,m^2/sec$.

(e) How much time is needed to completely vaporize the drop?

[Antoine Relation for H_2O: $\ln P^{sat} = 11.9302 - 3985.5/(234.15 + T \text{ in } °C)$.]

Problem 10.6: During the spray cooling of gases, atomization is used to break water into fine droplets. The droplet diameter and velocity are 75 μm and 40 m/sec. The water droplet is injected into a hot gas at 1500 K. Assume the gas to have the same composition as air and $\rho D = \lambda/c_p$ at 800 K for air.

(a) Calculate the lengths of cooling chamber for the following two cases: (i) $T_w = T^{sat}$ at P = 1 bar. (ii) T_w as determined from the thermodynamic phase equilibrium at the drop surface.
(b) Find the water vapor mass fraction at the surface.
(c) What is the average velocity (v, called the Stefan flow velocity), diffusion velocity (V_v), and absolute velocity of water vapor (v_v) at the surface?
(d) Which velocity is higher? v or V_v?
(e) Plot the fuel mass fraction, oxygen mass fraction, and temperature profiles vs. nondimensional radius (preferable to select nondimensional radius as a/r).

Problem 10.7:

(a) Determine the mole fraction of vapor at the drop surface if the drop temperature is 350 K for *n*-octane, R = 0.0729, and drop is burning at P = 2 bar. Assume all molecular weights are equal and determine $Y_{F,w}$.
(b) Calculate B number if L = 362 kJ/kg K, h_c = 44,350 kJ/kg, $\upsilon_{O2} = 3.51$, and c_p = 1.1 kJ/kg K.

Problem 10.8: Consider the combustion of a drop.

Fuel: methanol, CH_3OH; T_w = 338 K; h_c = 21160 kJ/kg; L = 1260 kJ/kg; c_p = 1.2 kJ/kg K; ρD = 0.00003 kg/msec; Y_{O2} = 0.23; d_0 = 100.0 μm; and T_4 = 300 K.

Determine the following:

(a) Initial burning rate (kg/sec).
(b) Burning rate constant, α_c (m²/sec).
(c) Burning time (msec).
(d) Flame radius in m and flame temperature in K.

(e) Mass fractions of CO_2, H_2O, and N_2 at the flame.
(f) Mass fraction of fuel vapor at surface.
(g) Mole fraction of vapor at surface, assuming the other species to have a molecular mass of 29 kg/kmol and the partial pressure of fuel vapor.
(h) The droplet temperature (T_w), assuming the Clayperon equation to be valid (do not iterate); how does this compare with the data provided?
(i) Calculate the flame length, assuming that the droplet travels with a velocity of 10 m/sec. If carbon of similar diameter burns over a period of 400 msec, what is the flame length?

Problem 10.9: For *n*-octane, the vapor pressure relation is

$$p_F\,(bar) = 8.07 \times 10^5 \exp(-5265/T), 0.01 < p_F < 0.5\,bar$$

For a spherical drop of 100-μm diameter determine the following for pure vaporization in air at 900 K. Let c_p = 1 kJ/kg K, P = 1 bar, and $\rho D = 5.1 \times 10^5$ kg/msec. Determine:
(a) T_w and $Y_{F,W}$ using iterative procedures.
(b) B, transfer number for evaporation. Assume that the remaining inert gas has a molecular weight of 28.
(c) Vaporization rate constant.
(d) Vaporization time.

Problem 10.10: Consider the Shvab–Zeldovich variable $\beta_{P\text{-}F} = Y_F + \{Y_P/(1 + \nu_{O2})\}$ for $C_{12}H_{26}$. Calculate the mass fraction of products at the thin flame. (Hint: Consider the mixture fraction at the flame surface.)

Problem 10.11: Methanol burns in air by evaporating at 338 K (T_w).
(a) If the burning configuration is a droplet of diameter 100 μm, what is the burning rate? Assume ρ1 at 1200 K, and use the value of D for O_2 in N_2 at 1200 K.
(b) Calculate the mass fractions of products CO_2, H_2O, and the inert N_2 in the thin flame [Hint: Use the definition $\phi = \frac{\beta_{kp} - \beta_{kp,\infty}}{\beta_{kp,ue} - \beta_{kp,\infty}}$ and apply the results for k = CO_2, H_2O and p = fuel or oxidizer]. Show that the mass fractions thus calculated are the same as the mass fractions calculated from the following stoichiometric equation:

$$CH_3OH + 1.5\,O_2 + 1.5\left(\frac{79}{21}\right)N_2 \rightarrow CO_2 + 2H_2O + 1.5\left(\frac{79}{21}\right)N_2$$

(c) Calculate the flame temperature using these mass fractions and the corresponding thermal enthalpies.
(d) What is the mass fraction of fuel vapor at the surface?

Problem 10.12: Write interface energy and species conservation equations for a mixture of 90% dodecane ($C_{12}H_{24}$) and 10% water (H_2O) (oil–water slurry). Assume the slurry to be at uniform temperature. Consider only evaporation. Species to be considered are C_{12} H_{24}, H_2O, and N_2. Note that there can be a mass fraction gradient inside the slurry for oil and H_2O.

Problem 10.13: Starting from the differential equation for mixture fraction (f):

$$\frac{df_M}{d\xi} + \frac{d^2 f_M}{d\xi^2} = 0,$$

show that a plot between h, Z_k vs. f_M for the gas phase around a spherical drop is linear.

Problem 10.14: Suppose $f^* = \{f_M - f_{M,\infty}\}/\{f_{M,w} - f_{M,\infty}\} = \phi$. Show that $f^* = 1 - (1 + B)^{(a/r)}$ where a is the drop radius. Obtain an expression for f^*_f at the flame location.

Problem 10.15: Consider

$$\xi_w = \ell n\left[1 + \left(\frac{(\beta_w - \beta_\infty)}{(F - \beta_w)}\right)\right] = \ln(1 + B)$$

where

$$\beta_w + (d\beta/d\xi)_w = F.$$

Starting with $\beta = \beta_{CO2\text{-}O2}$, obtain an expression for Y_{CO2w} in terms of B and the free-stream oxygen mass fraction, and stoichiometric coefficients for the reaction $F + O_2 \rightarrow$ products.

Problem 10.16: Consider a 100-μm hexane drop burning in air. Use a spreadsheet to solve the problem. Group all inputs and shade them in yellow (fuel, stoichiometric coefficient, h_{fg}, c_p, etc.); shade all outputs (flame temperature in K and flame position in r_f/a) in pink. The user should be able to change the fuel and get the output. Obtain:

(a) graph 1 for T vs. r/a.
(b) graph 2 for Y_F, Y_{O2} vs. r/a.

Assume that $\lambda_{mix} = 0.4\,\lambda_{fuel} + 0.6\lambda_{gas}$ at the film temperature; film temperature for properties to be $(T_f + T_w)/2$; and that $\lambda_{fuel} \approx \lambda_{fuel,T1}$, $(T/T_1)^{1/2}$ if data are not available at T. Let $\lambda_{gas} = \lambda_{air}$. Let $\rho D = \lambda_{mix}/c_{pfuel}$. Use $c_{p\ fuel}$ to estimate ρD, burn rate, and burn-rate constant; for all other properties, including B, use $c_{p\ air}$.

Problem 10.17: Two identical diesel engines of identical stroke lengths are fired with (1) alcohol, and (2) *n*-heptane.

(a) What is the expected ratio of engine speed (if it is assumed that combustion must be completed within stroke length)?

(b) If you want the same speed for both engines, what is the simplest solution?

Problem 10.18: A diesel engine was running at a certain speed with an average drop radius of a_1 in its spray. Suddenly, the drop radius increased to a_2 due to an injector problem with the same total mass of fuel. Derive an expression for the ratio of engine speed, N_2 to N_1. If $a_1 = 25\ \mu m$ and $a_2 = 50\ \mu m$, compute N_1/N_2.

Problem 10.19: Drop combustion in diesel engines may occur at a high P of 60 bar. Assume that the drops are of *n*-heptane. Determine the combustion rate at 60 bar. Compare this rate with the rate at low pressure. Assume diffusion control. Note that you must account for variations in the saturation temperature and the heat of vaporization.

Problem 10.20: Consider a droplet burning in air. The droplet is initially composed of a 50% mixture of (1) high vapor pressure (p_w) fuel with low heating value (h_c) and high latent heat (L), and (2) low vapor pressure with a high heating value and low latent heat. Assume that both components have the same stoichiometric coefficients (ν_{O2}).

(a) Sketch the behavior of the mass burning rate per unit area of drop.

(b) Sketch the wet bulb (T_w) temperature of the drop vs. time.

(c) Sketch the flame location vs. time.

(d) Discuss your answer using formulae and by applying the physics of the problem.

Problem 10.21:

(a) Starting from the equation

$$\frac{1}{r^2}\frac{d}{dr}(\rho v r^2 Y_F) = \frac{1}{r^2}\frac{d}{dr}\left(\rho D r^2 \frac{dY_F}{dr}\right) + \dot{w}_F'''$$

and with

$$\xi = \frac{\dot{m}}{4\pi\rho D r}$$

show that

$$\frac{dY_F}{d\xi} + \frac{d^2Y_F}{d\xi^2} = -\frac{\dot{w}_F'''(4\pi)^2\rho D r^4}{\dot{m}^2}$$

and

$$\frac{d\varepsilon_F}{d\xi} = -\frac{\dot{w}_F'''(4\pi)^2\rho D r^4}{\dot{m}^2}$$

where

$$\varepsilon_F = \dot{m}_F/\dot{m}\,.$$

(b) If a finite reaction occurs at a single surface $r = r_f$ or $\xi = \xi_f$ (Dirac delta approximation), obtain a relation for the change in flux fraction of fuel ε_F after the reaction.

Problem 10.22: Consider a multicomponent drop having components A, B, C, and D. Show that if $\varepsilon_A/Y_A = \varepsilon_B/Y_B = \varepsilon_C/Y_C = \varepsilon_D/Y_D$, then $\{\varepsilon_A/Y_{AW}\} = 1/Y_{F,w} = C$ without using the interface equation for inert and fuel species.

Problem 10.23: Consider a multicomponent drop containing heptane (component A, $X_{Hep,\ell}$) and hexadecane (component B, $X_{Hexadec,\ell}$) combusting in an environment at T_∞ and $Y_{O2,\infty}$. Obtain a solution for the flux ratio of A in terms of the heating values, latent heats, stoichiometric O_2 coefficients, T_∞, $Y_{O2,\infty}$, Y_{Aw}, and Y_{Bw}.

Problem 10.24: Consider a 50% (mole %) ethanol and 50% gasoline mixture. Assume that gasoline is isooctane.

(a) How will you determine the mole fractions of isooctane vapor (X_2) at 25°C and 2 bar? It is known that $\ln P_1^{sat} = A_1 - B_1/T$ and $\ln P_2^{sat} = A_2 - B_2/T$.

(b) Do you believe that vapor bubbles will have equal mole fractions of species 1 and 2? If not, justify your answer.

Problem 10.25: A pure fuel drop undergoes pure evaporation in an ambient in which the fuel mass fraction is $Y_{F,\infty}$. Starting from fundamental equations, show that the expression for fuel vapor mass fraction at surface of drop is

$$Y_{F,w} = \{Y_{F,\infty} + B\}/\{1 + B\}, \quad \text{where } B = h_t/L.$$

Problem 10.26: Consider a kerosene fuel droplet ($CH_{2.2}$, LHV = 42,800 kJ/kg) that evaporates, diffuses, and burns in dry air. Assume that it is a pure fuel with a single component. The element mass fraction profile is

$$\phi = \frac{\beta - \beta_\infty}{\beta_w - \beta_\infty} = \frac{Z_k - Z_{k,\infty}}{Z_{k,w} - Z_{k,\infty}} = \frac{1 - \exp(-\xi)}{1 - \exp(-\xi_w)}$$

where $\xi_w = \ln(1 + B)$, Z_k, k-th element mass fraction. For carbon

(a) What is $Z_{C,\infty}$?

(b) How will you determine $Z_{k,w}$ in terms of $Y_{C,l}$ (carbon mass fraction in liquid), and B (hint: write the interface boundary condition at the drop surface for the carbon mass)?

Problem 10.27:

(a) Consider the relation $\nabla \cdot (\rho \vec{v} Y_N - \rho D \nabla Y_N) = 0$. If $v_N = 0$, then show that

$$\nabla^2 \phi = 0 \quad \text{where} \quad \phi = \frac{\ln \frac{Y_N}{Y_{N\infty}}}{\ln \frac{Y_{N,w}}{Y_{N\infty}}}$$

where Y_N = inert mass fraction (for example, in pure evaporation $Y_F = 1 - Y_N$, and Y_F is the fuel mass fraction).

(b) If the governing equation is $\nabla \cdot (\rho \vec{v} \beta - \rho D \nabla \beta) = 0$, what transformation will yield $\nabla^2 \phi = 0$?

Problem 10.28: Find the burning time of a diesel drop of diameter 30 μm for an engine with CR = 15, γ = 1.4, and $T_{ambient}$ = 300 K. The properties of diesel are assumed to be the same as those of *n*-heptane. Select $\rho D = \lambda/cp$ at 1200 K for air.

Problem 10.29: Find the evaporation rate of a water drop of 100-μm diameter when it is sprayed into a fire at T = 1200 K. Assume that c_p = 1.2 kJ/kg K, L = 2450 kJ/kg, and select $\rho D = 2.5 \times 10^5$ kg/(m sec).

Chapter 11 — Problems

Problem 11.1: Consider the following equation in compressible form of mass conservation:

$$\partial\,(\rho v_x\, x^k)/\partial x + \partial\left(\rho v_y\, x^k\right)/\partial y = 0 \qquad \text{(A)}$$

k = 0, 1 for 2-D stagnation, forced-convection, free-convection, 2-D jets, and k = 1 for axisymmetric stagnation flows.

Using Equations 11.26a to 11.26f (in the text), show that Equation A becomes

$$\partial v_x'/\partial x' + \partial v_y'/\partial y' = 0, \quad \text{where } x' \text{ is in m; } y', \text{ in m; } v', \text{ in m/sec.}$$

Problem 11.2: Consider energy, species and Shvab–Zeldovich represented by property "b" $\rho v_x \frac{\partial b}{\partial x} + \rho v_y \frac{\partial b}{\partial y} = \frac{\partial}{\partial y}(\rho \chi \frac{\partial b}{\partial y}) + S_b(x, y)$, 2-D stagnation, forced, free convection, and 2-D jet, where $S_b(x, y)$ is the source term for the conserved scalar variable b. For example, if $b = Y_F$, then the source term, S_Y = kg/m³ sec, and $b = v_x$, S_v = (kg/m³)m/sec² = kg/(m²sec²) or N/m³, momentum source or force in N per unit volume. Using the above equation and Equations 11.26a to 11.26f transform b(x, y) to the incompressible coordinate form b(x′, y′), i.e.,

$$v_x'(\partial b/\partial x') + v_y'(\partial b/\partial y') = \partial/\partial y'\{(\rho/\rho_{ref})^2\,(\chi\,\partial b/\partial y')\} + S_b(x, y')/\rho\, x^{*2k}$$

Problem 11.3:

(a) Given (i) the incompressible forms of conservation equations for mass as

$$\partial v_x'/\partial x' + \partial v_y'/\partial y' = 0, \text{ where } x' \text{ is in m; } y', \text{ in m; and } v', \text{ in m/sec,} \qquad \text{(A)}$$

and property b as

$$v_x'(\partial b/\partial x') + v_y'(\partial b/\partial y') = \partial/\partial y'\{(\rho/\rho_{ref})^2\chi(\partial b/\partial y')\} + S_b(x, y')/(\rho\, x^{*2k}); \qquad \text{(B)}$$

(ii) the normalizations given in Equations 11.44 to 11.49 (in the text);

and (iii) $\rho^2 D = \rho_{ref}^2 D_{ref}$ = constant, $\rho^2 \nu = \rho_{ref}^2 \nu_{ref} = \rho^2 \chi$ = constant, and $\rho^2 \alpha = \rho_{ref}^2 \alpha_{ref} = \rho^2 \nu = \rho_{ref}^2 \nu_{ref}$ = constant, $\zeta = \nu/\chi$

show that for any generic property b, Equation B becomes

$$v_x'^* (\partial b^*/\partial x^*) + v_y'^* (\partial b^*/\partial y'^*) = \{1/(\mathrm{Re}\ \zeta)\} \partial^2 b^*/\partial y'^{*2} + \{S_b(x, y)\ d_{ref}/(\rho x^{*2k}\ b_{ref} v_{ref}\}. \tag{C}$$

(b) Simplify Equation C for $b = Y_k$, $\varsigma_k = \frac{\nu}{D_{k1}} = Sc_{k1} S_k(x, y) = \dot{w}'''_k$, and $\dot{w}'''_k$ = 0 (pure mixing) and nonequal binary diffusion coefficient
(c) Simplify Equation C if $b = \beta$, $b_{ref} = \beta_{ref}$
(d) Simplify Equation C for $b = v_x$, momentum source $S_b(x, y) = \rho v_{x\infty} dv_{x\infty}/dx$, kg/m² sec² or N/m³, $v_{x\,\infty} = a\ x^n$, $v_{ref} = a\ d_{ref}^n$ for flow over curved surfaces, and n = 0 for flat surfaces such that

$$v_x'^*(\partial v_x^*/\partial x^*) + v_y'^*(\partial v_x'^*/\partial y') = (1/\mathrm{Re}\zeta)\ \partial^2 v_x'^*/\partial y'^{2}+ S_v(x'^*, y'^*)$$

where

$$S_v(x'^*, y'^*) = S_v(x, y) * d_{ref}/\{v_{ref}^{\ 2*}\ \rho x^{*2k}\} = [\rho_\infty/\rho]\ v_{x\,\infty}{}^*[d\ v_{x\,\infty}{}^*/dx^*]$$

Problem 11.4: Consider the generalized momentum equations presented in Problem 11.3. Deduce the momentum source term for (a) free convection, with $S_v(x, y) = g\ (\rho_\infty - \rho)$; (b) 2-D stagnation flow, k = 0, n = 1; and (c) axisymmetric stagnation flows, k = 1, n = 1.

Problem 11.5: Convert the boundary-layer type of equations derived in Problem 11.3 (free, forced, and stagnation flows) into ordinary differential equations using the generalized similarity transformation given by Equations 11.60 to 11.70 (in the text).

The constants α and γ are selected suitably so that an ordinary differential equation for "f" in terms of η can be obtained.

(a) Deduce the mass conservation equation in terms of ψ.
(b) Show that the generalized momentum equation given by

$$v_x'^* \partial v_x'^*/\partial x'^* + v_y^*(\partial v_x'^*/\partial y'^*) = \{[1/\mathrm{Re}_{ref}]\ (\partial^2 v_x'^*/\partial y'^{*2})\} + S_v(x^*, y^*)$$

becomes

$$(f')^2\ (\gamma - \alpha) - ff''\ \gamma = (1/\mathrm{Re})(f'''(C_1\ C_2)(x'^{*\gamma-3\alpha-2\gamma+2\alpha+1}) + (C_2/C_1)^2\ S_v(x'^*, y'^*)/x'^{*2\gamma-2\alpha-1} + F_1(\xi,\eta),\ \text{where } F_1(\xi,\eta) = -(\partial f/\partial\eta) \times x'^*\ (\partial^2 f/\partial\eta\partial\xi) + \{(\partial f/\partial\xi)(x'^*\}\ (\partial^2 f/\partial\eta^2).$$

(c) Set F, $(\xi,\eta) = 0$ and select the constants such that the preceding equation becomes an ordinary differential equation in η.

(d) Show that the generic property b given by Equation 11.61 (in the text) becomes

$$\{C_1C_2/(\text{Re Sc})\}\,(\partial^2\phi/\partial\eta^2) + f(1-\alpha)\,(\partial\phi/\partial\eta) + S_b^* = F_2(\xi,\eta) \text{ where,}$$

$$F_2(\xi,\eta) = (\partial\phi/\partial\xi)\{\partial f/\partial\eta\}C_1^2\,(x'^*)^{1-2\alpha} - (\partial f/\partial\xi)(x'^{*1-2\alpha})\,(\partial\phi/\partial\eta)C_1^2$$

where $\phi = (b - b_\infty)/(b_w - b_\infty)$, $S_b = \{S_b(x, y)\}\, C_2\, x'^{*2\alpha-2k/(2k+1)}\, d_{ref}/(\rho\, C_1\, b_{,ref} v_{ref})$

(e) Show that the constants in the preceding equation are selected such that $\gamma + \alpha = 1$ and the partial differential equations becomes ordinary differential equations in η.

(f) If $C_1\, C_2 = \text{Re}$, how do the generic property b and momentum equations simplify?

Problem 11.6: Using the solution of Problem 11.4, (a) get expressions for η, ψ, and $v_y'^*$, and verify with Table 11.2; (b) derive the similarity differential equations for forced convection over a flat plate for momentum and NSZ, and show that they are

$$ff'' + 2f''' = 0, \quad \text{for forced convection}$$

$$\Phi'' + \text{Sc}\, f\, \phi'\,(1/2) = 0, \quad \text{for forced convection}$$

where $\eta = \text{Re}^{1/2}\, y'^*/x'^{*1/2}$

$\psi\,(x'^*, y'^*) = f(\eta)\, x'^{*1/2}/\text{Re}^{1/2}$

$v_y'^* = (f'\eta/2)(x'^{*-1/2}/\text{Re}^{1/2}) - (f/\text{Re}^{1/2})\{(1/2)\, x'^{*-1/2}\}$

$v_y'^* = \{\rho/\rho_{ref}\}v_y^* + \{v_x^*\}(\int_0^{y^*} \partial/\partial x^*\,(\rho/\rho_{ref})\,dy^*)$

$\rho\, v_y^* = \rho_{ref}\, v_y'^* - \{v_x^*\}\{(\int_0^{y^*}(\partial\rho/\partial x^*)dy^*)\}$

At $y = 0$, $v_y'^*(0) = -\{(1/2)\, f(0)/(\text{Re}^{1/2})\}\,(x'^{*-1/2})$

$\rho\, v_y^*(0) = \rho_{ref} v_y'^*(0) = -\{(1/2)\, f(0)/(\text{Re}^{1/2})\}(x'^{*-1/2})$, $x'^* = x^*$, $y'^* = y'/d_{ref}$

Problem 11.7: Using the solution of Problem 11.5, derive the similarity differential equations for forced convection over curved surface for both (a) momentum and (b) NSZ and show that they are

(a) $C_m = n^*\,(\rho_\infty/\rho)$, as $C_2 = C_1 = \text{Re}^{1/2}$, $\text{Re} = v_{ref}\, d_{ref}/\nu = \{a\, d_{ref}^n\, d_{ref}\}/\nu$

$n(f')^2 - ff''\,\{(1+n)/2\} = f''' + C_m$; for any molecular weight

$f''' + ff''\,(1+n)/2 + n^*\,\{(T/T_\infty) - f'^2\} = 0$, for constant molecular weight, ideal gas law

(b) $[(n+1)/2]\,\phi'' + \text{Sc}\, f\, \phi'\,(1+n)/2 = 0$

where $C_m = n^*\,\{a\, d_{ref}{}^n/v_{ref}\}^2\,(\rho_\infty/\rho)$,

$\eta = \text{Re}^{1/2}\,(y^*/x^{*(1-n)/2})$

$f(\eta) = \text{Re}^{1/2}\,\psi(x'^*, y'^*)/x'^{*(1+n)/2}$

$$v_x^* = f' = x^{*n},\ \{v_x/v_{x\infty}(x^*)\} = f',\ v_{ref} = a\ d_{ref}^{\ n}$$
$$v_y'^* = f'[(1-n)/2\}\eta\}/(Re^{1/2}\,x'^{*(1-n)/2}) - f[(1+n)/2]/\{Re^{1/2}x'^{*(1-n)/2}\}$$

Problem 11.8: Derive the similarity variable from generic formulation for the 2-D stagnation flows (k = 0). In stagnation flow, a jet impinges on a surface (or two opposed jets impinge on each other). Show that momentum equation reduces to

$$f''' + ff'' + \{(T/T_\infty) - f'^2\} = 0,\ \text{ideal gas, constant molecular weight}$$
$$\phi'' + Sc\ f\ \phi' = 0$$

where

$$\eta = Re^{1/2}\ y^*$$
$$f(\eta) = Re^{1/2}\ \psi(x'^*,\ y'^*)/x'^*$$
$$v_x^* = f'\{1/(2k+1)\}^{(1-2\alpha)}\ x^{*(2k+1)(1-2\alpha)} = f'\ x'^*$$
$$v_x^* = f'\ x^*,\ \{v_x/v_{x\infty}(x^*)\} = f',\ v_{ref} = a\ d_{ref}$$
$$v_y'^* = -f/\{Re^{1/2}\}$$
$$\dot{m}''/\{v_{ref}\ \rho_{ref}\} = -f(0)/Re^{1/2}$$
$$Re = \{a\ d_{ref}\ d_{ref}/\nu\}$$

Problem 11.9: For the axisymmetric stagnation flow (k = 1), show that the momentum equation and NSZ equation reduce to

$$f''' + (2/3)\ ff' + (1/3)\ \{(\rho_\infty/\rho) - (f')^2\} = 0,$$
$$\phi'' + f\ Sc\ (2/3)\ \phi' = 0$$

where

$$k = 1,\ \eta = 3^{1/2}\ Re^{1/2}\ (y'^*/x^*)$$
$$v_x^* = f'\ x^*,\ C_2 = \{Re_{ref}^{\ 1/2}/3^{1/6}\}$$
$$f(\eta) = 3^{1/2}\ Re^{1/2}\psi(x^*,\ y^*)/x^{*2}$$
$$v_y'^* = f'\{1/(Re^{1/2}x^*3^{1/2})\}\eta - 2f/(Re^{1/2}\ 3^{1/2}x^*)$$

Problem 11.10: Derive the following momentum and NSZ equations in terms of similarity coordinates from the generic formulation for free convection.

$$f''' - 2f'^2 + 3\ ff' + \theta' = 0$$
$$\phi'' + 3\ Sc\ f\ \phi' = 0$$

where

$$\eta = \{Gr/4\}^{1/4}\ (y^*/x^{*1/4})$$
$$f(\eta) = 1/\{4\ (Gr/4)^{1/4}\}\ (\psi/x^{*3/4})$$
$$v_x^* = f'\ 4\ \{Gr/4\}^{1/2}\ x^{*1/2}$$
$$v_y^* = -f'\ (x^{*\alpha}/C_2)(\partial\eta/\partial x^*) - (f/C_2)((1-\alpha)\ x^{*-\alpha})$$
$$C_1 = \{Gr/4\}^{1/4}$$
$$C_2 = (1/4)\ (4/Gr)^{1/4}\}$$

Problem 11.11: In order to solve boundary-layer types of problems, integral techniques are often adopted. Such a technique requires adaptation of the NSZ profile ($\Phi_s''^{(\eta)}$), which satisfies the boundary conditions but not necessarily the governing differential equation at every point. Consider a polynomial profile for forced, 2-D stagnation, axisymmetric, and free convection

$$\Phi_s = a_0 + a_1\, \eta_s + a_2\, \eta_s^2 + a_3\, \eta_s^3 \tag{A}$$

(a) Show from governing "Φ" equation that

$$\Phi_s''(0) = -B\, \Phi_s'^2(0)$$

where $\Phi_s' = d\, \Phi_s/d\eta_s$, $\eta_s = y'/\delta_s$ (note that $\eta_s \neq \eta$; $d\Phi/d\eta \neq d\Phi_s/d\eta_s$ although $\Phi = \Phi_s$)

(b) Other boundary conditions are $\Phi = 1$ at $\eta_s = 0$; $\Phi_s = 0$, at $\eta_s = 1$; $\Phi_s' = 0$, at $\eta_s = 1$

Show that the coefficients a_0, a_1, a_2, and a_3 are given by Equations 11.117–11.120 in Chapter 11.

(c) Explain why $a_1 < 0$.

(d) Obtain a solution for $\eta_{s,f}$ in terms of Φ_f that would be valid for free, forced, 2-D stagnation and axisymmetric convection.

(e) Use cubic solver (THERMOLAB software, available from the CRC Web site) to obtain three solutions for $\eta_{s,f}$ at B = 10.

Problem 11.12: As in Problem 11.11 for the NSZ profile, we need velocity profile for f′ that satisfies the boundary conditions but not necessarily the governing differential equation at every point.

Let

$$\eta_v = y'/\delta_m$$

and

$$f_M' = v_x/v_{x,\infty}$$

(a) Using momentum conservation equation show that

$$\Phi_s'(0)\, B\, f_m''(0) = Sc\, \tau\, f_m'''(0), \quad \text{where } (\delta_s/\delta_m) = \{\eta_m/\eta_s\} \text{ at } = \tau \tag{A}$$

where $\Phi_s' = d\, \Phi_s/d\eta_s$, $f_m' = d\, f_m/d\eta_m$, $\eta_m = y'/\delta_m$ (note that $\eta_m \neq \eta$; note that $f' = df/d\eta = f'_m = df_m/d'_m = v_x/v_{x\infty}$).

(b) If $F_m' = 1 - f_v'$ and a polynomial profile for F_M' is

$$F'_m = \{1 - f_m'\} = b_0 + b_1\, \eta_m + b_2\, \eta_m^2 + b_3\, \eta_m^3 \tag{B'}$$

$$f_m' = 0,\ F_m' = 1 \quad \text{at } \eta_m = 0 \tag{B}$$

$$f_m' = 1,\ F_m' = 0 \quad \text{at } \eta_m = 1 \tag{C}$$

$$f_m'' = 0,\ F_m' = 0 \quad \text{at } \eta_m \tag{D}$$

show that

$$b_2 = 3 - 2b_1;\ b_3 = -2 + b_1;\ b_1/a_1 = 6/[a_1\{B'\ a_1 - 4\}] = \{a_1B - 4\}/[\{B'a_1 - 4\}] \quad \text{(E)}$$

where $B' = B/(Sc\ \tau)$, and from Problem 11.11, $a_1 = -(2/B)\ \{[1 + 3\ B/2]^{1/2} - 1\}$

(c) Show that if Sc = 1, $1 - f_m' = F_m' = \phi_s$

(d) For case (c) what is f_m' at the flame location (see previous problem)?

(e) If the second-order polynomial is used and if we ignore the boundary condition given by Equation A, show that

$$f_m' = 2\eta_m - \eta_m^2$$

(f) If you select the result for velocity profile from part e, will that profile depend on the flow system, i.e., forced, free, or stagnation flows?

Problem 11.13: For forced convection with Sc = 1, and with integral techniques, show that the species or SZ or velocity boundary-layer thickness is given by

$$\delta_s'/x = [840\ a_1\ \{1 + B\}]^{1/2}/[Re_x\{-54 + 9\ a_1 + 4a_1^2\}]^{1/2} \quad \text{where } \delta_s' = \int_0^{\delta_S} \frac{\rho}{\rho_w}\, dy$$

and

$$\{\dot{m}''\ x\}/\{\rho D\ Rex^{1/2}\} = (-a_1)\ B\ [\gamma_1\{-54 + 9\ a_1 + 4a_1^2\}]^{1/2}/\{x[420\ a_1(1 + B)]^{1/2}\}$$

Problem 11.14: Consider a flat carbon surface burning in vitiated gas with a gas velocity of about 10 m/sec. The air temperature is 600 K. Assume the properties of the gas at the surface temperature of carbon.

Data are as follows: $c_{pcarbon} = 4.39$ kJ/kg K, $\rho_{carbon} = 1300$ kg/m^3, $T_w = 611$ K, $h_{c,I}$ 9203.2 kJ/kg, $h_{C,III} = -14{,}360$ kJ/kg X, and length of plate = 0.1 m.

Free-stream gas data: $Y_{O2\infty} = 0.23$, $Y_{CO2\infty}$: = 0.5, rest is N2, $T_\infty = 600$ K, and P = 1 bar.

Transport property data for air at 600 K: $\lambda_{gas} = 0.0000469$, Pr = Sc = 1, c_{pgas} = 1.051 kJ/kg K, $\mu = 3.06 \times 10^{-5}$ N sec/m^2, and $v_w = 5.27 \times 10^{-5}$ m^2/sec.

Obtain an integral solution for the average burn rate. Use the profiles given in Problem 11.11.

Problem 11.15: The following equations are well known for free convection:

$$\Phi'' + (3/4)Sc\ f\ \phi' = 0 \quad \text{(A)}$$

$$4f''' + ff''\ 3 + \theta' - 2f'^2 = 0 \quad \text{(B)}$$

where $\theta' = \{T - T_\infty\}/\{T_w - T_\infty\}$, $Gr = g\,\{T_w - T_\infty\}\, d_{ref}{}^{3}/\{T_\infty\, \nu_{,ref}{}^{2}\}$

$$\eta = \{4\ Gr\}^{1/4}(y^*/x^{*1/4}) \tag{C}$$

$$\phi'(0)B/\{\gamma Sc\} = f(0) \tag{D}$$

Use the profile Φ_s given in Problem 11.11 and velocity profile $f'_m = (27/4)\ \eta_M\ (1 - \eta_m)^2$.

Prove

(a) $\delta_s/x = c\ x^{*1/4}/\{4\ Gr\}^{1/4}$, and
(b) $\dot{m}''\ x/(\rho D\ Rex^{1/2}) = -a_1\ B/c$

$$c^4 = \frac{[4(9+a_1)\gamma\, Sc + 10(-a_1)\ \{1+B\}]210\{1+B\}(-a_1)}{\left[\gamma\, Sc(9+a_1)\right]^2 J}, \quad \gamma = 3/4$$

where $J = \int_0^1 \theta'\ d\eta_m = \int_0^1\ \{(T - T_\infty)/(T_w - T_\infty)\}\ d\eta_m$

Problem 11.16: Polymethyl methacrylate (PMMA) is kept horizontally in a wind tunnel and allowed to burn. The following data are given.

PMMA formula (a polymer): $C_5H_8O_2$ (monomers become as fuel vapor before they burn), $\rho_s = 1180\ kg/m^3$.

All properties evaluated at 1500 K.

$T_w = 600$ K.
$L = 1590$ kJ/kg.
$h_c = 29560$ kJ/kg.

Free-stream conditions: velocity, $v_{x\infty} = 30$ m/sec, $Y_{O2,\infty} = 0.23$, $T_\infty = 300$ K, and $Pr = Sc = 1.0$.

(a) Plot the burning rates per unit area ($\dot{m}''$) vs. x starting from x = 1 to 10 cm in intervals of 1 cm. If $2\,\dot{m}''x/(\rho_w\ D_w) = -2^{1/2}\ f(0)$, using Figure 11.5 determine the burn rate.
(b) Plot the profile of the top surface vs. minutes after burning started. What is the flame temperature?
(c) What is the fuel mass fraction at the surface?
(d) Determine the flame location η_f and flame distance y_f at x = 5 cm.

Problem 11.17: Consider air at 300 K at a velocity of $V_{x\infty} = 10$ m/sec flowing past a flat plate.

(a) Estimate Re.
(b) The velocity profile is approximately given as $v_x^* = [v_{x,\infty} - v_x]/\{v_{x,\infty} - v_{x,w}] = [(3/2)\ (y/\delta) - 1/2\ (y/\delta)^3]$, plot the velocity profile as $f(y/\delta)$.
(c) If the boundary-layer thickness is given by $\delta = 4.64\ x/Re_x{}^{1/2}$, where $Re_x = vx\rho/\chi$, compute the boundary-layer thickness at x = 10 and 100 cm.

(d) Calculate the friction coefficient if $c_f = \tau_s/[1/2\rho\ V^2] = 0.646/Re_x^{1/2}$.
(e) If the plate surface is maintained at T_w, the temperature profile $T^* = [T_\infty - T]/\{T_{,\infty} - T_{,w}] = [(3/2)\ (y/\delta_t) - 1/2\ (y/\delta_t)^3]$ and $\delta_t/\delta = Pr^{-1/3}/1.026$, $Pr \geq 1$ [Incropera and DeWitt, 2002]. Plot T^* as f (y/δ_t).
(f) $Nu_x = hx/\lambda = 0.332\ Re_x - Pr^{1/3}$. Obtain h at x = 10 cm, d = 100 cm. Determine Nu_x at x = 10 and 100 cm.
(g) Give an approximate ϕ profile if Pr = Sc = 1.

Problem 11.18.
(a) Following Example 11.5 in text on Si deposition rate, formulate the carbon combustion rate assuming chemical equilibrium at the surface of a carbon plate subjected to forced convection. Assume that the species at the surface are CO, O_2, and N_2, where N_2 is inert.
(b) If the carbon plate is at 2000 K, what is the expected value of $Y_{O2,w}$ when carbon burns in air? Do not calculate, i.e., state whether Y_{O2} is the same as free-stream concentration or very low.
(c) Reformulate part a, assuming that the species at the surface are CO_2, O_2, and N_2, where N_2 is inert. What happens to the answer of part b?

Problem 11.19: Prove that for boundary-layer type of problems of carbon burning under frozen flow in the gas phase,

$$Y_{CO,w} = \frac{Y_{CO,\infty} + \dfrac{Y_{O2,\infty}[1+\nu_{O2,I}]}{\nu_{O2,I}} + \dfrac{Y_{CO2,\infty}[1+\nu_{CO2,III}]}{\nu_{CO2,III}}}{[1+B]} = 1 - \frac{Y_{I,\infty}}{1+B}$$

Both oxidation with O_2 and reduction with CO_2 occur at the carbon surface. (Hint: Use CO balance first and then use the O_2 and CO_2 balance equations.)

Problem 11.20: Prove that for boundary-layer type of problems of carbon burning under frozen flow in gas phase, the surface temperature including radiation loss is given by rewriting

$$-\frac{h_{T\infty}}{\tilde{h}_{c,I}} = \frac{c_p(T_w - T_\infty)}{\tilde{h}_{C,I}} = \left(\frac{Y_{O_2,\infty}}{\nu_{O_2,I}} + \frac{Y_{CO_2,\infty}\tilde{h}_{C,III}}{\nu_{CO_2,III}\tilde{h}_{C,I}}\right) \qquad \text{(A)}$$

where

$$\tilde{h}_{c,I} = h_{c,I} - \frac{\dot{q}''_{rad}}{\dot{m}''}$$
$$\tilde{h}_{c,III} = h_{c,III} - \frac{\dot{q}''_{rad}}{\dot{m}''} \qquad \text{(B)}$$

Note that $h_{C,III} < 0$, whereas $h_{C,I} > 0$.
(Hint: Use interface energy balance first and then use Y_{O2} and h_t coupling.)

Problem 11.21: Consider the following heterogeneous and homogeneous reactions involving carbon and CO. CO is produced by oxidation and Boudouard reactions. CO is subsequently oxidized in the gas phase to CO_2.

$$C(s) + \frac{1}{2}O_2 \rightarrow CO \qquad \text{(I)}$$

$$C(s) + CO_2 \rightarrow 2CO \qquad \text{(III)}$$

$$CO + \frac{1}{2}O_2 \rightarrow CO_2 \qquad \text{(V)}$$

Show that if $\beta_{CO2\text{-}O2} = Y_{CO2}/\nu_{CO2,V} + Y_{O2}/\nu_{O2,V}$, $\phi = \{\beta_{CO2\text{-}O2} - \beta_{CO2\text{-}O2,\infty}\}/\{\beta_{CO2\text{-}O2,w} - \beta_{CO2\text{-}O2,\infty}\}$, then

$$\phi = \left(\frac{\left(\frac{Y_{O2}}{\nu_{O2,I}} + \frac{Y_{CO2}}{\nu_{CO2,III}}\right) - \left(\frac{Y_{O2,\infty}}{\nu_{O2,I}} + \frac{Y_{CO2,\infty}}{\nu_{CO2,III}}\right)}{\left(\frac{Y_{O2,w}}{\nu_{k,I}} + \frac{Y_{CO2,w}}{\nu_{CO2,III}}\right) - \left(\frac{Y_{O2,\infty}}{\nu_{O2,I}} + \frac{Y_{CO2,\infty}}{\nu_{CO2,III}}\right)} \right) = \frac{\frac{h_T}{\tilde{h}_{C,I}} - \frac{h_{T,\infty}}{\tilde{h}_{C,I}}}{\frac{h_{T,w}}{\tilde{h}_{C,I}} - \frac{h_{T,\infty}}{\tilde{h}_{C,I}}}$$

Problem 11.22: Consider the boundary-layer growth. Consider the c.v. within ABC.

(a) If the velocity profile is linear at BC, what is the mass flow at BC? What is the momentum flow at BC?

(b) What is the mass flow at AB? What is the momentum flow into the c.v. at AB?

(c) What is the momentum loss at AC (assume Newton's law)? Using momentum balance, show that

$$\delta(x) = \left\{\frac{4x\mu}{\rho v_x}\right\}^{1/2} \Big/ (\rho v_{x\infty})^{1/2}.$$

(d) What will be the relation for Sh if $\delta(x)$ is the same for species transport.

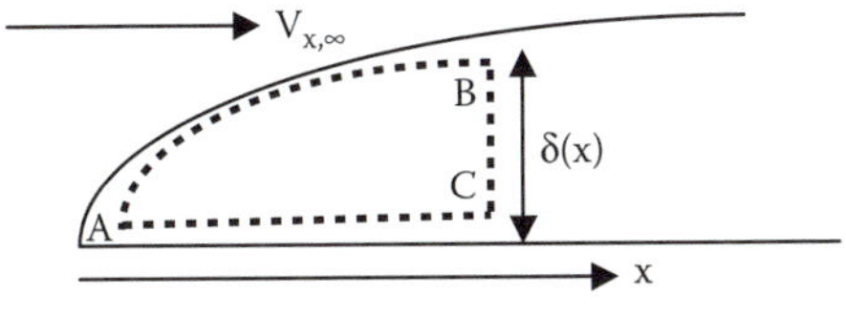

Problem 11.23: A tank of methanol ruptures and spills downward over a vertical wall. You wish to estimate the burn rate in air. If air is at 25°C, calculate the total burn rate over the vertical wall of 20-cm height based on the exact solution and integral solution.

Problem 11.24: Carbon burns over a vertical wall of 20-cm height. Assume air as the ambience, use an SFM, and CO_2 as the product of oxidation.

(a) What is the expected wall temperature and total burn rate? $T_\infty = 300$ K.

(b) If it burns under DFM, estimate the excess fuel, and assuming the burn rate continues under the same rate before x_g (gasification front), estimate the flame length x_f.

Chapter 12 — Problems

Problem 12.1: Underventilated flames will normally occur for Burke–Schumann solutions if the ratio of the outer diameter (b) and inner diameter (a) approaches unity. True or false? Explain.

Problem 12.2: Find the height of a methane–air flame in infinite surroundings. The diameter of the burner is 0.373 cm. The flow rate of CH_4 through the burner is 60 cm^3/min. The inlet temperature is 298 K, and the pressure is 100 kPa. (Experimental values are found to be around 1 cm. Assume $\rho D = 6.5 \times 10^5$ kg/m sec = constant.)

Problem 12.3: A mixture of N_2 (10% by mass) and C_2H_6 flows through an inner duct of 2-cm diameter while air flows through an outer duct of diameter 4 cm. Assume that the Burke–Schumann analysis is applicable (i.e., v = constant throughout the cross section of the duct).

(a) Is the flame overventilated or underventilated?

(b) What is the overall equivalence ratio?

Problem 12.4:

(a) Show that with $\beta_{ht\text{-}F} = h_t/hc + Y_F$, $\beta_I = \beta_{ht\text{-}F,0,a}$ and $\beta_\infty = \beta_{htF,0,ab}$ in Burke–Schumann flames,

$$\phi_f = s/(1 + s)$$

(b) Show that the flame temperature is same as the adiabatic flame temperature when the control volume is taken between the flame and entry section of the duct.

(c) For Problem 12.3, draw a temperature profile of T vs. η at given $\xi = (1/2)^* \xi_f$.

Problem 12.5: CH_4 is fired in a boiler burner using a circular jet. The utility has decided to change to hydrogen fuels. However, they are concerned whether the flame with H_2 fuel will touch the opposite wall of the boiler. Let the flame length with CH_4 be L1 and with H_2 be L2. Assume that the Re and free oxygen mass fraction remain the same for both fuels. What will L_2 be compared to L_1? Discuss.

Problem 12.6: Consider the similarity solution for pure mixing in a 2-D jet of width d. Draw the stoichiometric profile y^*_{st} vs. x^*, $x^* = x/d$ for propane:air mixture if the velocity of the jet is 10 cm/sec. What is the recommended distance for locating a spark plug so that ignition can be easily initiated?

Problem 12.7: Gaseous H_2 is injected with a velocity of 2 cm/sec into pure oxygen. Assuming incompressible flow, calculate:

(a) The Y_F, Y_{O2}, Y_F, and T profiles for a 2-cm-wide slit jet at 6 cm from the wall. Plot the profiles. Assume $\rho D = \lambda/c_p$ for air at 1200 K.

(b) The flame length for the jet in part a (use properties at 300 K). Let T_∞ = 300 K, P_∞ = 104 kPa.

Problem 12.8: If the fuel in Problem 12.7 is to be propane (C_3H_8) with all other conditions remaining the same, calculate the flame length for the slot and the circular jets and compare these with the values in Problem 12.5.

Problem 12.9: CH_4 is injected through 30-mm-diameter jet at 0.01 m/sec. Determine the radial location of the stoichiometric contour at x = 3 cm. The atmosphere is pure O_2.

Problem 12.10: Gaseous CH_4 is injected with a velocity of 5 cm/sec into air. Assume $\rho D = 6.5 \times 10^5$ kg/m sec and HHV = 50,000 kJ/kg. Assuming incompressible flow, calculate:

(a) The Y_F, Y_{O2}, Y_F, and T profiles for a 2-cm-wide slot jet at 50 cm from the wall. Plot the profiles. Assume $\rho D = \lambda/c_p$ for air at 1200 K.

(b) The flame length for the jet in part a (use properties at 300 K). Let T_∞ = 300 K, P_∞ = 101 kPa.

Problem 12.11: Gaseous CH_4 is injected at 60 cm³/min through a burner of diameter 0.373 cm into air. Assume $\rho D = 6.5 \times 10^5$ kg/m sec. Let T_∞ = 300 K, P_∞ = 104 kPa. HHV = 50,000 kJ/kg. Assuming incompressible flow, calculate:

(a) The Φ, Y_{O2}, Y_F, and T profiles at 5 cm from the injector.

(b) The flame length for the jet in part a.

Problem 12.12: Gaseous CH_4 is injected with a velocity of 5 cm/sec through a 2-cm-wide slot into air. Assume $\rho D = 6.5 \times 10^{-5}$ kg/m sec. Assume that $T_\infty = 300$ K, $P_\infty = 104$ kPa. Assuming compressible flow, calculate:

(a) The Φ/Φ_0 (or $= v_x/v_{x,max}$) Y_F, Y_{O2}, Y_F, and T profiles at 50 cm from the wall. Plot the profiles.

(b) The flame length for the jet in part a.

(*Note:* dy $= \rho/\rho_i$ dy or dy $=$ dy (ρ_i/ρ) $=$ dy (T/T_i)). Knowing the T or Φ profiles vs. y, one can integrate and obtain T vs. y.)

Problem 12.13: If $f/f_i = \phi = (h - h_\infty//(h_I - h_\infty) = \phi = \{\beta - \beta_\infty\}/\{\beta_I - \beta_\infty\}$, where $f_i = 1.0$, $Y_{Fi} = 1$.

(a) Show that $f_{stoich} = \phi_f = 1/\{(A{:}F)_{stoich} + 1\}$, where $(A{:}F)_{stoich} = \nu_{O2}/Y_{O2,\infty}$. (Hint: use $\beta_{O2\text{-}F}$.)

(b) Consider a propane jet flame burning in air. Evaluate f_{stoich}.

(c) Plot Y_F vs. f, Y_{O2} vs. f, and T vs. f; assume c_{pair} at 1200 K; $h_c = 46{,}357$ kJ/kg. (Hint: use $\beta_{O2\text{-}F}$, $\beta_{hT\text{-}O2}$, and $\beta_{hT\text{-}F}$.) Note that f ranges from 0 to 1. $T_\infty = 300$ K, P = 100 kPa.)

Problem 12.14: Pure gaseous C_3H_8 is injected at a velocity of 2 cm/sec through a burner of slit width 0.2 cm into air. Assume $\rho D = 6.5 \times 10^{-5}$ kg/m sec, $c_p = 1.175$ kJ/kg K, $\nu = 1.0 \times 10^5$ m²/sec, and Sc = 1. Assume $T_i = 300$ K, $T_\infty = 300$ K, P = 100 kPa, $h_c = 46{,}357$ kJ/kg, and flat velocity and species profiles at the inlets.

(a) Plot Y_F, Y_{O2}, Y_F vs. y/d_i at $x/d_i = 40$ and 200 assuming Sc = 1.

(b) Plot the stoichiometric contours (where $Y_{O2}/Y_F = \nu_{O2}$) y_{st}/d_i vs. x/d_i for Sc = 0.5, 1, 1.5. Estimate the distance x_{st} at which the contours meet the axis.

Problem 12.15:

(a) Consider a circular burner fired with a mixture of 90% CH_4 and 10% N_2 (mass %) (i.e., $Y_{Fi} \neq 1$). The fuel is then switched to an H_2 and N_2 mixture. Assume laminar combustion. The free-stream oxygen mass fraction is the same for both cases. What percentage of H_2 must be used so that the flame length remains the same? Assume that the kinematic viscosities are the same for both fuels.

(b) Present a sketch of the flame profile.

Problem 12.16: It has been predicted that the nondimensional flame heights, $\eta = \pi z \rho D/\dot{m}_i$, of overventilated and underventilated Burke–Schumann flames are 0.68 and 0.52, respectively, at $\alpha = 0.5$ and 0.1 [Williams, 1985]. Assume $\xi_b = 2.0$. Verify whether the approximate expressions using one-term expansions satisfy these results. Also, obtain the flame profiles.

Problem 12.17: Because the amount of soot formation is proportional to the volume of the fuel-rich zone, we must estimate the volume of the fuel-rich zone. Assume Sc = 1. Derive an expression for the fuel-rich volume (with appropriate equations and a sketch of the flame profile) in integral form with appropriate limits of integration. Assume laminar combustion.

Problem 12.18: Consider a situation in which CH_4 along with an inert is injected through a rectangular jet of width D_i at $x = 0$ and air is injected through an outer duct of width D_o at $x = 0$. They mix downstream at $x > 0$.

(a) Formulate the pure mixing problem and specify the boundary conditions. Discuss the relation between mixture fraction and fuel mass fraction in the mixing problem.
(b) Formulate a combustion problem using SZ variables, and specify the boundary conditions. Discuss the relation between the mixture fraction and fuel mass fraction in the combustion problem.
(c) Discuss the relation between products, fuel, and mixture fraction for the combustion problem.
(d) Consider a thin flame. Discuss the relation between products, fuel, oxygen, and mixture fraction for the combustion problem.

Problem 12.19: Consider a cylindrical turbulent jet of CH_4, with $Y_{O2,\infty} = 0.15$, jet properties: $d_i = 0.003$ m, $v_{xi} = 50$ m/sec, and $\sigma_{cyl} = 0.0161$. Obtain the flame structure. Use the empirical expressions.

Problem 12.20: Consider an He jet in stagnant air. Treat air as one species. Obtain Y_a, ρ_a, Y_{He}, and ρ_{He} profiles at various x values. Note that the steep density gradients typically result in jet stability problems.

Problem 12.21.

(a) It is desired to calculate the total NO_x production in a 2-D laminar jet. Using the thin-flame solution of species and T profiles for N_2, O_2, and T, and treating NO as a trace species, obtain an expression for N production rate using single-step kinetics. Mention how will you solve for NO production in (kg/m sec) and hence in g/GJ. Consider an elemental volume.
(b) How will the answer in part a change if equilibrium is assumed for $CO_2 \Leftrightarrow CO + 1/2O_2$ and $H_2O \Leftrightarrow H_2 + 1/2O_2$, assuming that fuel concentration is unaffected on the rich side and O_2 concentration on the lean side? Assume any extra production of O_2 to be in trace amounts on the rich and lean side. Treat CO and H_2 as trace species. Assume the mixture molecular weight to be same as M_{N2}.

Chapter 13 — Problems

Problem 13.1: Let the ignition temperature of a carbon particle of diameter d_1 be T_1. The particle diameter is increased to $2d_1$. Let the ignition temperature be T_2. Compare T_1 and T_2. Discuss.

Problem 13.2: It is known that carbon ignites at an ambient temperature $T_{\infty,I}$. To extinguish the burning carbon, one must lower this to an ambient temperature $T_{\infty,E}$. Then

(a) $T_{\infty,I} < T_{\infty,E}$
(b) $T_{\infty,I} > T_{\infty,E}$
(c) $T_{\infty,I} = T_{\infty,E}$

Discuss.

Problem 13.3: Consider a porous slab of char of thickness 2L. The slab is so thin that the temperature is uniform. However, as O_2 penetrates, the mass fraction decreases toward the center of the slab. The carbon consumption rate per unit volume is

$$|w'''| = S_v\ k\, Y_{O_2}\ \rho = k'\ Y_{O_2}, k' = S_v\ k$$

where S_v is the internal surface area per unit volume of slab; k = B exp (E/RT); B, intrinsic preexponential factor; E, intrinsic activation energy; and Y_{O_2}, oxygen mass fraction at any given x from the center of slab.

The governing equation is

$$\rho D\, d^2\, Y_{O_2}/dx^2 = w'''$$

(a) Obtain a solution for Y_{O2} as a function of x. Assume that the O_2 mass fraction at x = L is $Y_{O2,L}$.
(b) What is the mass consumption rate of porous char ($\dot{m}$) for a slab of height H and width W? If $\dot{m}$ is set proportional to ($H\ W\ Y_{O_2,L}\ A_{app}\ \exp(-E_{app}/RT)$), what is the relation between E_{app} and E?

Problem 13.4: Determine the ignition temperature of a cylindrical stream of coal particles of diameter 50 μm. Assume that coal particles burn heterogeneously.

$$\text{Coal} + O_2 \rightarrow x\, CO_2 + Y\ H_2O$$

The coal composition is: ash 8.0%, carbon 76.3%, hydrogen 5.0%, and oxygen 8.2% (dry). Neglect N and S. Use the Boie equation for the heating value. Assume A_{het} = 450 m/sec and E = 66,000 kJ/kmol.

(a) Let R of the stream be 5 mm. Determine $T_{\infty,I}$ for A:F in the stream of 1:1, 2:1, and 4:1.

(b) Let R be 10 mm. Determine $T_{\infty,I}$ for A:F of 1:1, 2:1, and 4:1 (ρ_{coal} = 1300 kg/m^3 and air density = 1.1 kg/m^3).

Problem 13.5: Most explosive conditions are presumed to exist in the pulverizer mills at 0.5 g/l. What is the probable ignition temperature for a cylindrical cloud of diameter 20 cm? What is the temperature for a 20-cm-diameter spherical cloud?

Problem 13.6: Calculate $T_{\infty,I}$ for carbon spheres of d_0 = 0.1 mm and 1 mm if $C + O_2 \rightarrow CO_2$. Assume the following data: A_{hetO2} = 450 m/sec, E = 66,000 kJ/kmol, Y_{O2} = 0.23, D = 1 × 10^4 m^2/sec, c_p = 1.1 kJ/kg k, ignition in quiescent atmosphere, and negligible radiation. What is the ignition temperatures if the reaction proceeds as $C + 1/2\ O_2 \rightarrow CO$?

Problem 13.7: If $Q_{Gen} = h_c^*/[1 + \{\exp(1/\theta)/\ D_{III,mod}]$. Ignoring change in concentrations of fuel and oxidizer on Q_G,

(a) Show that maximum slope of Q_G vs. θ occurs such that

$$\{1/\theta\} = 2\ (1 + x)/(x - 1) \quad \text{where } x = \exp(1/\theta)/D_{III,mod}.$$

(b) If heat loss $Q_L = \theta - \theta_\infty$, show that at the limiting $D_{III,mod}$,

$$\{1/\theta\} = 2\ \{(1/hc^*) + 1\}^{1/2},$$

and the limiting D_{III} is given by

$$D_{III,mod} = [\{1/(2\theta)\} - 1]\ \exp\ (1/\theta)/[1 + \{1/(2\theta)\}].$$

Chapter 14 — Problems

Problem 14.1:

(a) Calculate the CJ detonation velocity for a stoichiometric mixture of propane and O_2 with $T_0 = 298$ K and $P_0 = 1$ atm.
(b) For part a, calculate p_∞ and T_∞.
(c) Compare the results for part a and part b with those for H_2–O_2 mixture.

Problem 14.2: Consider an H_2–O_2–N_2 mixture with $N_2/O_2 = 5$ and stoichiometric combustion at 298 K and 1 atm.

(a) Neglect dissociation. Calculate the adiabatic flame temperature and the magnitude of heat of reaction per kg of the mixture.
(b) Using the answer of part a, calculate the detonation velocity as though the mixture has properties of air. How does this compare with the experimental value of 1822 m/sec at $T_2 = 2685$ K and $P_2 = 14.4$ atm?
(c) Calculate the equilibrium flame temperature, the composition assuming products to be H_2, O_2, OH, and H and the magnitude of heat of reaction per kg of mixture.
(d) Using the answer of part c, calculate the detonation velocity as though the mixture has properties of air. How does this compare with the experimental value given in part b?

Problem 14.3: Consider the combustion of a premixed stoichiometric mixture of C_2H_2 and air. Let the unburned gas mixture velocity be 10 m/sec.

(a) What are the molecular weights of the input mixture and products? Assume the inlet temperature and pressure to be 298 K and 100 kPa.
(b) Including velocity effects, calculate the adiabatic flame temperature and exit pressure. Assume c_p for N_2 at a mean temperature of 1500 K.

Chapter 15 — Problems

Problem 15.1: Determine θ_0 and the flame temperature T_∞ if a CH_4:air mixture is supplied with 10% excess air. Given h_c (LHV) = 50,016 kJ/kg, c_p = 1.17 kJ/kg K, and T_0 = 300 K. How much heat is liberated per unit mass of mixture?

Problem 15.2: In a simplified analysis, we assume that chemical reaction occurs over δ, which includes the preheat region. If the reaction is confined to $\chi\delta$, where $\chi < 1$, but the temperature profile remains linear within δ, provide a simplified expression for v_0.

Problem 15.3: When a premixed fuel:air mixture is ignited, the flame is established at a distance L from the burner. The theory behind such a phenomenon is that as the mixture is lit, the flame propagates toward the burner with a velocity S; however, the flame is stabalized at L because the gas velocity is equal and opposite to S. The flame velocity is a function of stoichiometry. Assuming that the flame is anchored at x and y such that the fuel is locally in stoichiometric proportion to oxygen, and gas velocity at that point is equal to S, derive an expression for that location. (Hint: Chapter 12.)

Problem 15.4: Show that $\int_{\theta 0}^{1} (1 - \theta) \exp(-E^*/\theta)\, d\theta \approx \{\exp(-E^*)/E^{*2}\} [1 - 6/E^* + 36/E^{*2}]$; ignore values due to lower limits, i.e., $\exp(-E^*) >> \exp(-E^*/\theta_0)$.

Problem 15.5. Show that:

$$\int_{\theta 0}^{\theta 1} [F - \theta] \exp(-E^*/\theta)\, d\theta = \{\theta_1^2/E^{*2}\} \exp(-E^*/\theta_1)\}\{F(E^* - 2\,\theta_1 + 6\,\theta_1^2/E^* - 24\theta_1^3/E^{*2} + 120\theta_1^4/E^{*3}) - \theta_1 E^* + 3\,\theta_1^2 - 12\theta_1^3/E^* + 60\,\theta_1^4/E^{*2}\},$$

where $F = \theta_0 + Y_{A0}\, h_c^* + \xi\, \{(\theta_1 - \theta_0) - h_c^* Y_{A0}\}$. We may treat $F = F_{avg} \approx \{1 + \theta_1\}/2$ as constant.

Problem 15.6: Using the derivation for lean flammability limit (LFL), obtain an expression for the rich flammability limit (RFL) assuming $n_B = 1$ and $\alpha_1 - n_1 = 0$.

Problem 15.7: Consider the solution for nondimensional speed for any order of reaction with fuel

$$\Lambda = \left\{ Y_{F,0}{}^2 h_c^* /2 \right\} \Big/ \left\{ \int_{\theta 0}^{1} (1-\theta)^{nF} \exp(-E^*/\theta)\theta^{\alpha 1 - n1} d\theta \right\} \quad (1)$$

How does this equation simplify if the order of reaction is zero. (Hint: See Chapter 5 for exponential integral.) Approximate it.

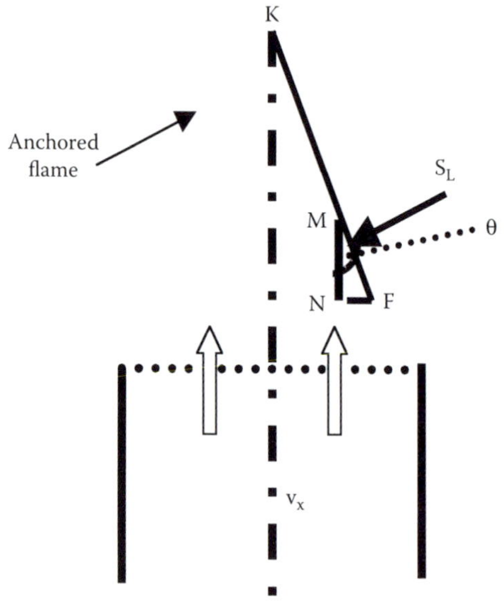

Figure P.15.7 Flammability limits modeling.

Problem 15.8: A premixed gas mixture containing C_3H_8 and 50% excess air flows at an average velocity of 25 cm/sec through a Bunsen burner of diameter 1 cm. The flow is assumed to be laminar. Determine the flashback and blow-off limits.

Problem 15.9: Calculate the minimum required ignition energy for propane. Determine θ_0 and the flame temperature T_∞ if a C_3H_8:air mixture is supplied with 50% excess air. Given h_c (LHV) = 46,357 kJ/kg, c_p = 1.2 kJ/kg K, and T_0 = 300 K.

Problem 15.10: Consider a CH_4:air mixture at stoichiometric conditions for which LHV = 44,500 kJ/kg. Calculate the flame speed using

(a) Simplified theory and assuming $T_{avg} = \{T_{ad} + T_0\}/2$.
(b) Rigorous theory, assuming that the reaction rate is first order with fuel.
(c) Compare these results with experimental data.

Problem 15.11: Air is used to gasify DAF Wyoming coal that can be characterized by the empirical formula $C_1H_{0.755}N_{0.0128}O_{0.182}S_{0.00267}$ and a heating value (HHV) of 30,829 kJ/kg. Oxygen-blown coal gasification yields: 25% H_2, 40% CO, 20% H_2O, and a 200 Btu/ft^3 or 7050 kJ/m^3 lower heating value. What are the approximate flammability limits of these mixtures (a) based on empirical formula of the fuel and (b) based on heating values and corresponding LFL and RFL if extinction occurs at 1400 K? Discuss the results.

Chapter 16 — Problems

Problem 16.1: Determine the correction factor for a binary array of unequally sized drops. Plot the results as a function of ℓ/a_1 with parameter a_2/a_1.

Problem 16.2: Consider drops mounted on the apexes of a hexagon. Determine the correction factor for the array using point source method. How will the results change if there is a drop at the center?

Problem 16.3: Derive an expression for the maximum possible G number in a monosized uniform cloud of radius R_c and drop size of radius a.

Problem 16.4: As the drop burns, the drop dia decreases, ℓ/a increases due to an increase in "a" and hence η changes. Consider central drop in a three drop array. Get an expression for η and predict d vs. t.

Problem 16.5: Derive an expression for correction factor for equisized drops at the apexes of an isosceles triangle.

Problem 16.6: Derive an expression for average correction factor for a five-drop array and four peripheral drops located at a cube of side ℓ from central drop.

Problem 16.7: Consider CV enclosed by AB, CD, exterior of EF, GH, etc., of many drops (Figure P.16.7). Use a 1-D, unsteady spherical system. The vapor flows out from boundaries EF and GH at T_w, whereas heat transfer occurs across the interfaces. The gas flows in and out at r and r + dr. Formulate the species and energy conservation equations, and prove the following:

$$\frac{\partial\left(4\pi r^2\,\rho Y_k\right)}{\partial t}+\frac{\partial}{\partial r}(\dot{m}Y_k)-\frac{\partial}{\partial r}\left(4\pi r^2\,\rho D\,\frac{\partial Y_k}{\partial r}\right)=\{\dot{w}'''_{k,p}+\dot{w}'''_{k,c}\}4\pi r^2 \qquad \text{(a)}$$

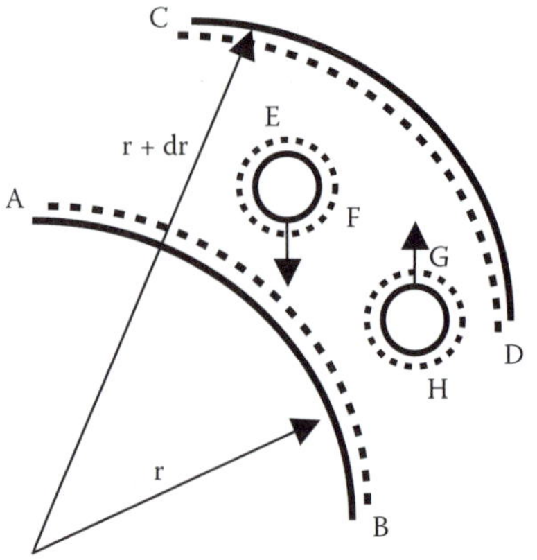

Figure P.16.7 Drops enclosed within control volume.

where $\dot{w}_{k,p}$ species k produced by evaporating drops and $\dot{w}_{k,c}$, species k produced by chemical reaction in the gas phase

$$\frac{\partial\left(4\pi r^2 \rho c_p T\right)}{\partial t} + \frac{\partial}{\partial r}(\dot{m} c_p T) - \frac{\partial}{\partial r}\left(4\pi r^2 \lambda \frac{\partial T}{\partial r}\right) \tag{b}$$

$$= -\{\dot{q}_w'' \; 4\pi a^2\} n 4\pi r^2 + |\dot{w}_F'''| h_c 4\pi r^2$$

where $\dot{q}_w$ is the heat flux entering from the gas phase to drop/particle per unit area of particle; n, the number of particles per unit volume; and "a," the radius of particle.

Problem 16.8: Starting from

$$\phi = \frac{\beta'}{\beta'_\infty} = \frac{\beta - F}{\beta_\infty - F} \tag{A}$$

and with the definition of F_β from Table 10.1 and

$$\frac{\beta_w - \beta}{F - \beta_w} = B(r), \tag{B}$$

show that:

(a) For evaporation with $B(r) = h(r)/L$ or combustion with $B(r) = h_t(r)/L + Y_{O2}(r)\, h_c/(L\, \nu_{O2})$ and individual flames around a drop, $r < R_c$,

$$\phi = (1 + B(r))\gamma, \quad r < R_c,$$

where $\gamma = 1/[B + 1]$

(b) $Y_{F,w} = 1 - (1 + s)/(1 + B)$ for an individual drop burning or evaporating inside the cloud, where $s = Y_{O2,\infty}/\nu_{O2}$

(Hint: Use the definition of B(r) given in Equation B for pure evaporation and individual flames with $\beta = \beta_{ht\text{-}O2}$, and $\beta = \beta_{O2\text{-}F}$, for flames.)

Problem 16.9: Starting from

$$M\xi^4 \frac{d(\alpha\phi)}{d\xi} + \xi^4 \frac{d^2\phi}{d\xi} = 0,$$

$$\alpha = \alpha(\xi), \xi > 1, \quad \alpha = 1, \xi \le 1.$$

(a) Prove that

$$\phi = \exp\{-M\xi\}, \quad 0 < \xi < 1$$

$$\phi = \exp\,[-M\{1 + F(\xi)\}],\ F(\xi) = \int_1^{\xi} \alpha(\xi)d\xi,\ 1 < \xi < \infty,\ \phi = \gamma \ln\{1 + B(\xi)\},$$

where γ is a constant $= 1/(1 + B)$ (Equation 16.75) for both evaporation in a dense cloud and combustion of drops with individual flames around drops in a dilute cloud.

(b) Using the result from part a for $1 < \xi < \infty$ and using the following equation:

$$-M\xi^4 \frac{d\alpha}{d\xi} = G_1\, \xi^q \ln(1 + B(\xi)),\ \xi \ge 1,$$

show that

$$-\frac{d}{d\xi}\left[\xi^{4-q} \frac{d\alpha}{d\xi}\right] = -MG_1\alpha.$$

(c) Derive the solutions for $\alpha(\xi)$, M/M_{SC}, and M/M_{ISOC}.

Problem 16.10:

(a) Consider a spherical cloud of radius R_c consisting of uniform monosized droplets with a number density of n_0 and radius of a_0. The G number is given as $G = 4\pi\, a_0\, n_0\, R_c^2$. Obtain an expression for the mass of liquid fuel in the cloud.

(b) To study the effect of size distribution for the same cloud mass, we fix $n = n_0$, but let the drop size vary according to $n_0 a = A(1/r)^q$. If $G_1 = 4\pi\,(na)_{Rc}\, R_c^2$, then show that $G_1 = G\,(1 - q)^{1/3}$.

(c) If we have a cloud in which the number density varies but the drop size remains constant, $na_0 = A(1/r)^q$, and total mass must remain as in part a. If $G_1 = 4\pi\,(na)_{Rc}\, R_c^2$, then show that $G_1 = G\,(3 - q)/3$, where $G = 4\pi\, a_0\, n_0\, R_c^2$.

Problem 16.11: Consider a monosized uniform spherical cloud of radius R_c = 1 cm, consisting of char/carbon particles of equal diameter 100 (= 2 a) μm. The number density is such that G = 10.

(a) Estimate the number density.
(b) Assume ρ_C = 700 kg/m³, λ = 76.3 × 10⁶ kW/m K for air at 1200 K, c_p = 1.18 kJ/kg K, ρ_D = 3 × 10⁵ kg/msec, $Y_{O2,\infty}$ = 0.23, T_∞ = 1000 K; ignore radiation loss, $h_{c,III}$ = −14,360 kJ/kg, and $h_{c,I}$ = 9,203 kJ/kg. Use the double-film model. Obtain T vs. R_c/r.
(c) Repeat part a and part b for G = 100.

Problem 16.12:

(a) Consider a spherical cloud of monosize (a = constant) and uniformly distributed (n = constant). Let the mass be "m." Write the expression correction factor in terms of G.
(b) Now let {n(r) a(r)}/$(na)_{RC}$ = $(Rc/r)^q$ such that mass is same as in part a. (i) Express a_{Rc}/a in terms of q for uniform number density, and (ii) express n_{Rc}/n in terms of q for a monosized cloud.

Problem 16.13: Obtain an expression for the correction factor of an n-sided polygon.

Problem 16.14: Consider four drops (1, 2, 3, 4) in a linear array of equal size, and spaced equally apart.

(a) Plot the average correction factor central drops' (2, 3) in terms of drop spacing, l.
(b) If the isolated drop rate is 1.50 × 10⁸ kg/sec, determine the average burn rate of the whole array.
(c) Plot d^2/d^2_0 vs. time for the central (2, 3) and peripheral drops (1, 4).

Problem 16.15: Consider a 3-D array of seven *n*-hexane drops of equal diameter 100 (= 2a) μm, all being at the same temperature, separated by ℓ = 300 μm, and kept in an ambience of 600 K (Figure 16.4 of Chapter 16).

ρD = 3.73 E–05 kg/(m sec), c_p = 1.021 kJ/kg K, and h_c = 45,105 kJ/kg. Assume that the gas phase has only two components: fuel vapor and N_2.

(a) Compute the isolated drop evaporation rate in kg/sec.
(b) Compute the correction factors for center and peripheral drops.
(c) Compute the average correction factor and average mass source per unit volume if all seven drops are located within a sphere of radius ℓ = 300 μm.

Problem 16.16: Derive expressions for correction factors for (a) a three-drop array of dissimilar sizes with uniform composition and (b) the same three-drop array of dissimilar sizes with dissimilar composition. Compare the results with all drops being at the same ambient temperatures, separated by $\ell_{12} = \ell_{23} = \ell = 300\ \mu m$, and kept in an ambience of 300 K.

$\rho D = 3.73$ E-05 kg/(m sec), $c_p = 1.021$ kJ/kg K, and $h_c = 45{,}105$ kJ/kg. Assume that the gas phase has only two components: fuel vapor and N_2.

Problem 16.17: Consider a 50% (mole %) ethanol (species – 1): 50% gasoline (species – 2) mixture. Assume that gasoline is isooctane.

(a) Describe the procedure for determining the mole fractions of isooctane vapor (X_2) at 25°C and 2 bar? It is known that $\ln P_1^{sat} = A_1 - B_1/T$ and $\ln P_2^{sat} = A_2 - B_2/T$.

(b) Do you believe that vapor bubble will have equal mole fractions of species 1 and 2? If not, justify your answer. Look at Table A-2 for normal boiling points.

Problem 16.18: Incipient group combustion (IPGC) occurs when the flame is located just around the center drop in a cloud of radius R_c. Assume pure fuel. Using the solution for ϕ (r) for $r < R_c$ and with $\phi = \beta_{O2\text{-}F}/\beta_{O2\text{-}F,\infty}$, obtain a solution for G at IPGC. Assume *n*-hexane drops, 600 K, $T_\infty = 600$ K, $c_p = 1.021$ kJ/kg K, $T_w = 333$ K, and $Y_{O2,\infty} = 0.23$.

Chapter 17 — Problems

Problem 17.1: Starting from the Zeldovich mechanism,

$$N_2 + O \rightarrow N + NO$$

$$O_2 + N \rightarrow O + NO$$

and assuming

$$\frac{1}{2} O_2 \Leftrightarrow O$$

and $[O_2] >> [NO]$, show that

$$\frac{d[NO]}{dt} = 2k_{1f}K_c[N_2]\sqrt{[O_2]_\infty} - 2K_c\frac{k_{2b}k_{1b}}{k_{2f}}\frac{[NO]^2}{\sqrt{[O_2]_\infty}}.$$

Problem 17.2: Methane and air enter a reactor (at 1F, 1A, Figure P.17.2) with an equivalence ratio of 0.8. A fraction of the exhaust gas is withdrawn at station 4 and mixed with the incoming mixture at station 3. If the exhaust gas recirculation fraction is (FGR) of the product gases, derive an expression for O_2% in the exhaust (station 5). What will equilibrium NO be for such a scheme if the inlet temperature is 298 K and 10% excess air are used?

Assume that CH_4 is burned with vitiated gas so that the O_2% in the vitiated gas is 19.3% (station 3). If flue gas leaving the boiler has 3% O_2, what percent of flue gas must be recirculated?

Problem 17.3: NH_3 is added to waste gas and passed over a metal catalyst to capture NO_x (selective catalytic reduction, SCR). Consider the posttreatment of NO_x with NH_3.

$$4NO + 4NH_3 + O_2 = 4N_2 + 6H_2O \text{ (liquid)} \quad \text{(a)}$$

$$6NO_2 + 8NH_3 = 7N_2 + 12H_2O \text{ (liquid)} \quad \text{(b)}$$

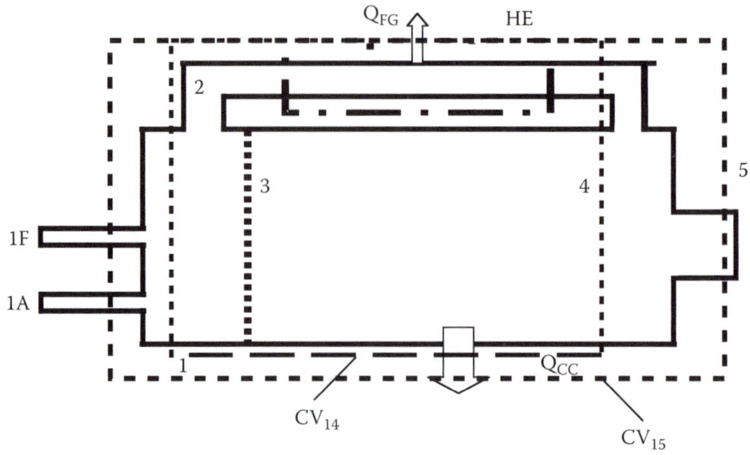

Figure P.17.2 Schematic of flue gas recirculation.

Are these reactions exothermic or endothermic? How much heat must be supplied or absorbed if the reaction occurs at 298 K?

Problem 17.4: Limestone is pulverized and added to water and sprayed. Calcium sulfite $CaSO_3$ (1/2H_2O) is produced, which is reacted with O_2 to remove SO_2. This process is called the wet limestone–gypsum method. Consider the posttreatment of SO_x with limestone:

$$SO_2 + CaCO_3 + 1/2\ O_2 = CaSO_3.1/2\ H_2O + CO_2 \quad \text{(a)}$$

$$CaSO_3.1/2\ H_2O + 1/2\ O_2 + 3/2\ H_2O = CaSO_4.2H_2O\ \text{(gypsum)} \quad \text{(b)}$$

Are these reactions exothermic or endothermic? If so, how much heat must be supplied or absorbed if the reaction occurs at 298 K?

Problem 17.5: Calculate the NO and NO_2 formed when 0.123 kmol of O_2 and 0.829 kmol of N_2 are heated at constant P to 1600 K.

Problem 17.6:

(a) NO_x is measured to be 200 ppm at 6% concentration. Determine the NO_x formed at 3% O_2.

(b) Provide units for conversion of NO in ppm to g/GJ if 350 m^3 (STP at 0°C, 1.013×10^5 Pa dry gas, and 6% oxygen content, 1 kmol = 22.42 m^3) of flue gas is produced per GJ.

Problem 17.7: NH_3 is used as a selective noncatalytic agent for the reduction of NO_x in coal-fired plants. If the emission from coal-fired plants is 260 g/GJ (0.6 lb per mmBtu), how much NH_3 is required to reduce NO_x by 75%?

Coal HHV = 26,535 kJ/kg; FB HHV = 8,300 kJ/kg

N in manure = 0.0186 kg per kg, N in manure as NH_3 = 0.67 kg of N as NH_3 per kg of N.

Assume the following reaction:

$$NH_3 + aNO \rightarrow bN_2 + cH_2O.$$

Problem 17.8:

(a) Calculate the density of CO_2 at STP.

(b) If yearly U.S. emissions are 2.5 Gt, how many m^3 of STP gas can be spread over the U.S.?

(c) Determine the CO_2 concentration increase in the atmosphere per year. If the current concentration is 0.035%, what will be concentration in 10 yr?

Problem 17.9: Show that the O_2 normalized emission is given as

$$X_{NO,std}/X_{NO} = (X_{O_2,a} - X_{O_2,std})/(X_{O_2,a} - X_{O_2}).$$

Problem 17.10: The overall selective catalytic and noncatalytic reduction reactions (SCR and SNCR) are as follows:

$$\text{SCR: } 4\ NO + 4\ NH_3 + O_2 \rightarrow 4\ N_2 + 6\ H_2O,\ 600 < T < 800\ K$$

$$\text{SNCR: } 4\ NO + 4\ CH(NH_2)_{2aq} + O_2 \rightarrow 4\ N_2 + 2\ CO_2 + 2\ H_2O,\quad T \approx 1270\ K$$

Estimate (a) the heat added/removed, and (b) the amount of NH_3 and $CH(NH_2)_2$ required for a 5000 MW power plant if NO_x had to be reduced from 0.19 kg/GJ to 0.1 kg/GJ. The power plant efficiency is 33%.

Problem 17.11: Many states encourage the use of renewable energy (e.g., wind energy, solar energy, agricultural crops) so that the use of fossil fuels (coal, oil, and gas) can be reduced. Incentives are offered in terms of an REC (renewable energy credit or certificate). Each REC equals 1 MWhr of electricity.

(a) What is the energy equivalent of 1 REC in BTU?

(b) If a power plant operates with an efficiency of 33% using coal as a fuel with a heat value of 18610 kJ/kg, how much coal consumption in kg will be reduced for each REC?

(c) If coal produces 0.36 kg of NO per GJ, how much reduction in NO_x will be achieved for each REC?

Chapter 18 — Problems

Problem 18.1: Plot $u(t) = D + A \sin B t$ vs. time t if $A = 5$ m/sec, $B = 1.5$ sec^{-1}, $D = 10$ m/sec. Determine Prob ($u > 14$ m/sec), Prob ($u > 13.5$ m/sec), Prob ($u > 14.5$ m/sec), and Prob ($13.5 < u < 14.5$).

Problem 18.2: Obtain the convolution integrals for the variable q using the three-mixture-fraction model:

$$\bar{q} = \int_{-\infty}^{+\infty}\int\int q(f_1, f_2, f_3)\, p(f_1)\, p(f_2)\, p(f_3)\, df_1\, df_2 df_3 = \sum_{i=1}^{15} I_i$$

Problem 18.3: Derive the Reynolds averaged and Favre averaged conservation equations for momentum in x direction for a boundary-layer type of problem over a flat plate.

Formulae

This section summarizes the standard relations of various concepts covered in the text.

Chapter 1: Introduction

Specific Gravity

$$\rho_{SG} = \frac{\rho_k}{\rho_{air}} = \frac{M_k}{M_{air}} \text{ for gases } (\rho \text{ air at STP})$$

$$\rho_{SG} = \frac{\rho_{fuel}}{\rho_{water}} \text{ for liquids and solids}$$

$$M_{MIX} = \sum X_i M_i,\ X_k = p_k/P$$

$$[X_k] = p_k/\bar{R}\,T, \quad Y_k = X_k\ M_k/M_{Mix}$$

Mass Conservation

$$\frac{dm_{CV}}{dt} = \dot{m}_i - \dot{m}_e$$

First Law or Energy Conservation Equation

$$\frac{dE_{CV}}{dt} = \dot{Q}_{CV} - \dot{W}_{CV} + \dot{m}_i[h + ke + pe]_i - \dot{m}_e[h + ke + pe]_e, \delta W_{rev} = -\text{vdp, open}$$

$$dE = \delta Q - \delta W, \delta W_{rev} = \text{pdv, closed}$$

Second Law

$$\frac{dS_{CV}}{dt} = \frac{\dot{q}}{T_b} + \dot{m}_i s_i - \dot{m}_e s_e + \dot{\sigma}, \text{ open}$$

$ds = \dfrac{dq}{T_b} + \delta\sigma$, for a fixed-mass (or Lagrangian) system or closed system.

$$ds = \frac{du}{T} - \frac{Pdv}{T}, \; ds = \frac{dh}{T} - \frac{vdP}{T}$$

$s_k^o(T) = \int_0^T c_{pko}(T)\dfrac{dT}{T}$, $\bar{s}_K(T, p_K) = \bar{s}_K^o(T) - \bar{R}\ln\left(\dfrac{p_K}{1}\right)$, variable $c_p(T)$, $p_k = X_k P$

$$s_{\text{Mixture}} = \sum N_K \bar{s}_K(T, p_K) = \sum m_K s_K(T, p_K)$$

Chapter 2: Combustion

Thermochemistry

Air Composition mole%: N_2 = 79%, O_2 = 21%

mass %: N_2 = 77%, O_2 = 23%

Molecular weight of air: 28.97 kg/kmol or 28.97 lb/lb-mol

Mole Balance

$$dN_k/dt = \dot{N}_{k,i} + \dot{N}_{k,gen} - \dot{N}_{k,e}$$

Energy Conservation in Molar Forms

$$\frac{dE}{dt} = \dot{Q} - \dot{W} + \sum \dot{N}_i(\bar{h} + \overline{ke} + \overline{pe})_i - \sum \dot{N}_e(\bar{h} + \overline{ke} + \overline{pe})_e$$

$$\overline{ke}\left[\frac{J}{kmol}\right] = M\left[\frac{kg}{kmol}\right] ke\left[\frac{J}{kg}\right], \; ke\left[\frac{J}{kg}\right] = \frac{1}{2}V^2,$$

$$\overline{pe} = M\left[\frac{kg}{kmol}\right] pe\left[\frac{J}{kg}\right], \; pe\left[\frac{J}{kg}\right] = gZ$$

$$\frac{dE}{dt} = \dot{Q} - \dot{W} + \sum \dot{N}_i \bar{h} - \sum \dot{N}_e \bar{h} \text{ if ke and pe are neglected}$$

$$\bar{h}_o(T) = \bar{h}_f(298K) + \int_{298}^{T} \bar{c}_{po}(T)dT = \bar{h}_f(298) + \left\{\bar{h}_{t,T}(T) - \bar{h}_t(298K)\right\}$$

E = U + KE + PE; use $\bar{u}_k = \bar{h}_k - P\bar{v}_k$; $\bar{u}_k = \bar{h}_k - \bar{R}T$ if the species k is an ideal gas

Table F.2.1 Combustion Formulae

Process or Variable	*Formulae*
Stoichiometric combustion	Complete combustion and no O_2 in products
Excess-air percentage	$\left\{\frac{A{:}F-(A{:}F)_{stoich}}{(A{:}F)_{stoich}}\right\}\times 100=(SR-1)\times 100$
Stoichiometric ratio (SR)	$\frac{\text{air supplied}}{\text{stoichiometric air}}=SR$
Equivalent ratio, ϕ	$1/(SR)$
Total enthalpy $(\bar{h})$	$(\bar{h}_f{}^0)+(\bar{h}_{t,298-T})$
Enthalpy of reaction (Δh_R^0)	$H_{P,T}-H_{R,T}$
Enthalpy of combustion $(\Delta h^0{}_c)$	$\Delta H_C=H_P-H_R$, also $\Delta H_C^0\equiv\Delta H_R^0$
Heating value (HV)	$HV=-\Delta H^\circ{}_{R,298}=H_{R,298}-H_{P,298}$
Higher heating value (HHV)	$HHV=H_R-H_P$, with H_2O in products in liquid state
Lower heating value (LHV)	$LHV=H_R-H_P$, with H_2O in products in gaseous state
Combustion efficiency	$\eta_{comb}=\frac{\text{actual heat release}}{\text{theoretical heat release}}$
A:F from gas analysis	$(A{:}F)_{dry}=(1/X_{O_2,a})[(-h/4+o/2+n/2)X_{O_2}+(1+h/4-o/2)]/(1-X_{O_2}/X_{O_2,a})$, Fuel: $CH_hN_nO_o$
$X_{co}/X_{co_2,max}$	$(1-X_{O_2}/X_{O_2,a})=(X_{O_2,a}-X_{O_2})/X_{O_2,a}$
$X_{NO,std}/X_{NO}$	$(X_{O_2,a}-X_{O_2,std})/(X_{O_2,a}-X_{O_2})$.

$Xo_{2,a}$: ambient mole or fraction of O_2

Entropy Balance in Molar Forms

$$dS_{cv}/dt=\dot{Q}_{cv}/T_b+\sum_{k,i}\dot{N}_k\hat{s}_k-\sum_{k,e}\dot{N}_k\hat{s}_k+\dot{\sigma}$$

Chapter 3: Chemical Equilibrium

Methods of Computing dU, dH, dA, and dG for Reacting Systems

$$dU=TdS-PdV+\sum\mu_j dN_j$$

$$dH=TdS+VdP+\sum\mu_j dN_j$$

$$dA=-SdT-PdV+\sum\mu_j dN_j$$

Table F.3.1 Chemical Equilibria

Process, Variable, or Condition	*Formulae*
Direction of reaction	$\delta\sigma \geq 0$, $dG_{T,P} \leq 0$ or $(\Sigma\mu_k\, dN_k) \leq 0$ $dG_{T,P} = (\Sigma\mu_k\, dN_k) = (\Sigma\mu_k\, v_k) \leq 0$,
Equilibrium condition	$\delta\sigma = 0$ $\Sigma\hat{g}_k\, dN_k = dG_{T,P} = 0$, where $\Sigma\hat{g}_k\, dN_k = dA_{T,V} = 0$
Chemical force potential (e.g., $H_2O \rightarrow H_2 + 1/2\ O_2$)	$F_{chem,Prod} = \hat{g}_{H_2} + \frac{1}{2}\hat{g}_{O2}$, $F_{chem,react} = \hat{g}_{H_2O}$ $F_{chem,\ react} > F_{chem,\ Prod}$ or $\frac{p_{H2}^{1/2}\ p_{O2}^{1/2}}{p_{H2O}} \leq K^0(T)$
Gibbs function useful for chemical reactions (P in bar)	$\hat{g}_i = \bar{g}_i^0(T) + \bar{R}T\ln(PX_i/1)$, see Table A.9 to A.19 for $\bar{g}_i^0$ $\hat{g}_i = \bar{g}_i(T,P) + \bar{R}T\ln X_i$
Equilibrium constant $K^0(T)$	$K^0(T) = \exp(-\Delta G^0/\bar{R}T)$, $\Delta G^0 = G^0_{RHS} - G^0_{LHS}$
Equilibrium constant, $K^0(T)$, from "elementary" reactions	$K^0(T) = \exp(\Delta A - (\Delta B/T))$, where T is in K, $\Delta A = \Sigma v_k A_k$, and $\Delta B = \Sigma v_k B_k$, (e.g., $CO_2 \Leftrightarrow CO + 1/O_2$), $\Delta A = 1\times A_{CO} + (1/2)\times A_{O_2} - 1\times A_{CO_2}$), use Table A.35
vant Hoff relation $\frac{d\ln K^0(T)}{dT} = \frac{\Delta\bar{h}_R^{-0}}{\bar{R}T^2}$	$\ln\left(\frac{K^0}{K^0_{ref}}\right) \approx -\left(\frac{\Delta H_R^0(T)}{R}\right)\left(\frac{1}{T} - \frac{1}{T_{ref}}\right)$ $\ln K^0(T) \approx A - B/T$, $A = \ln K^0_{ref}(T_{ref}) + \{\Delta H_R^0/(\bar{R}T_{ref})\}$, $B = (\Delta H_R^0/\bar{R})$.

RHS: right hand side; LHS: left hand side of reaction

$$dG = -SdT - VdP + \sum \mu_k dN_k$$

$$\mu_k = -T\,(\partial S/\partial N_k)_{H,P} = -T\,(\partial S/\partial N_k)_{U,V} = (\partial U/\partial N_k)_{S,V} = (\partial A/\partial N_k)_{T,V} = (\partial G/\partial N_k)_{T,P} = \hat{g}_k$$

$$a\ A + bB \Leftrightarrow c\ C + dD$$

$$\Delta G^0(T) = c\bar{g}_C^0(T) + d\bar{g}_D^0(T) - a\bar{g}_A^0(T) - b\bar{g}_B^0(T) \text{ where, } \bar{g}_k^0(T) = \bar{h}_k^0(T) - T\bar{s}_k^0(T)$$

$$G = \Sigma N_k\hat{g}_k,\ \hat{g}_k = \bar{g}(T,p_k) = \bar{h}_k(T) - T\hat{s}_k(T,p_k),\ \hat{s}_k(T,p_k) = \bar{s}_k^{\ 0}(T) - \bar{R}\ln(p_k/p^0)$$

Equilibrium constant for $aA + bB \Leftrightarrow cC + dD$ is

$$K^0 = \exp\left(\frac{-\Delta G^0(T)}{\bar{R}T}\right) = \frac{\left(\frac{p_C}{p^0}\right)^c\left(\frac{p_D}{p^0}\right)^d}{\left(\frac{p_A}{p^0}\right)^a\left(\frac{p_B}{p^0}\right)^b} = \frac{X_C^cX_D^d}{X_A^aX_B^b}\left(\frac{P}{p^0}\right)^{c+d-a-b} = \frac{N_C^cN_D^d}{N_A^aN_B^b}\left(\frac{P}{p^0\ N}\right)^{c+d-a-b}$$

Transition temperature occurs $K^0 = 1$ or $\log K^0 = 0$ or $\Delta G^0(T) = 0$.

Degree of Dissociation $= \frac{D_{in} - D_{out}}{D_{in}}$, where D is the species being dissociated.

Chapter 4: Fuels

Wobbe Number (for Gas Fuel)

$$Wo\# \frac{HHV'\left(\frac{MJ}{m^3}\right)}{\sqrt{\rho_{SG}}}$$

Dry Analysis

Dry basis: exclude moisture or H_2O
Dry-ash-free (DAF): exclude moisture and ash
HV_{DAF} = HV as received/[1 − moisture fraction-ash fraction]
$HV_{vdA} = [HV_{Fuel} - HV_{Fc}(1 - VM)/VM]$

Boie Equation

$$HHV\ (kJ/kg\ of\ fuel) = 35160\ C_m + 116225H_m - 11090O_m + 6280N_m + 10465S_m$$

where C_m, H_m, etc., are the carbon mass fraction, hydrogen mass fraction, etc.

Ash Tracer Technique

$$\text{Burned fraction (AF)} = \frac{A - A_0}{A(1 - A_0)}$$

where A is the dry-ash fraction after burning and A_0 is the initial dry-ash fraction.

Size Distribution

Rosin–Rammler relation

$$d(Y_p)/d(d_p) = nbd_p^{n-1}\exp(-bd_p^n)$$

where n and b are constants and $d(Y_p)$ is mass fraction of the particles within size d_p and $d_p + d(d_p)$.

Cumulative volume or weight fraction of mass of particles having size $0 < d_p < d_p$

$$CVF = 1 - \exp(-bd_p^n) = 1-\exp\left(-(d_p/d_{p,Charac})^n\right)$$

where $d_{p,Charac}$ denotes the characteristic drop or particle size, for which CVF = 63.2%.

$$\text{Sauter mean diameter (SMD)} = \text{total volume/total surface area} = \frac{\sum N_i d_{pi}^3}{\sum N_i d_{pi}^2}$$

$$= \frac{1}{\sum Y_i / d_{pi}}$$

Chapter 5: Chemical Kinetics

Law of Stoichiometry $aA + bB \rightarrow cC + dD$

$$\frac{\dot{w}'''_A}{-a} = \frac{\dot{w}'''_B}{-b} = \frac{\dot{w}'''_C}{c} = \frac{\dot{w}'''_D}{d} = \dot{w}''' > 0$$

Global Law of Mass Action

$$\dot{w}''' = k[CH_4]^{\alpha}[O_2]^{\alpha} > 0,\ CH_4 + 2O_2 \rightarrow CO_2 + 2H_2O$$

where $k = AT^n \exp\left(-\frac{E}{\bar{R}T}\right)$, Arrhenius law; $\alpha + \beta$: global order of reaction.

Unimolecular Reaction

$$\frac{\dot{w}'''_{Br2}}{(-1)} = \frac{\dot{w}'''_{Br}}{2} = \dot{w}''',\ Br_2 \rightarrow 2Bv$$

Elementary Law of Mass Action

$$\dot{w}'''_{Br} = 2k\,[Br_2]^1$$

Bimolecular Reaction

$$\frac{\dot{w}'''_{H2}}{(-1)} = \frac{\dot{w}'''_{I2}}{(-1)} = \frac{\dot{w}'''_{HI}}{(+2)} = \dot{w}''',\ H_2 + I_2 \rightarrow 2HI$$

$$\dot{w}''' = k\,[H_2]^1\,[I_2]^1$$

Consecutive or Series Reactions

$$A + B \underset{k_1}{\rightarrow} AB \underset{k_2}{\rightarrow} E + F$$

$$d[AB]/dt = k_1\,[A]\,[B] - k_2\,[AB]$$

Competitive or Parallel Reactions

$$A + B \overset{k_1}{\rightarrow} AB$$

$$A + B \overset{k_2}{\rightarrow} E + F$$

$$d[A]/dt = -k_1\,[A]\,[B] - k_2\,[A]\,[B]$$

Opposing or Backward Reactions

$O_2 \rightarrow 2O$, forward reaction, k_f
$O_2 \leftarrow 2O$, backward reaction, k_b
$O_2 \rightleftarrows 2O$
$d[O_2]/dt = -kf\,[O_2] + kb\,[O]^2$
If $d[O_2]/dt = 0$, then equilibrium

Chain Initiation

$$A_2 + M \rightarrow A + A + M, \text{ radical production}$$

Chain Propagation

$$A + B_2 \rightarrow AB + B, \text{ sustains radicals}$$

Chain Branching

$$A + B_2 + M \rightarrow A + B + M, \text{ radicals multiply}$$

Chain Terminating

$$A + B + M \rightarrow AB + M, \text{ termination of radicals}$$

Reaction Rate Theory: Homogeneous Reactions

$$A + B \rightarrow \text{products}$$

$$Z_{AB} = \pi\sigma_{AB}^2\ V_r\ [A]\ N_{avog}\ [B]\ N_{Avog}$$

$$N_{Avog} = 6.023 \times 10^{26}[\text{molecules/kmol}],\ \sigma_{AB} = \frac{\sigma_A + \sigma_B}{2},$$

$$V_r = \left(\frac{8k_B T}{\pi\mu}\right)^{\frac{1}{2}}, \text{ and } \mu = \frac{m_A m_B}{m_A + m_B}$$

$$\left|\frac{d[A]}{dt}\right| = \frac{Z_{AB}\left(e^{-E/\bar{R}T}\right) S}{N_{Avog}} = \pi\sigma_{AB}{}^2 V_r [A][B] N_{avog} S e^{-E/\bar{R}T} = A_{homog} e^{-E/\bar{R}T}[A][B]$$

where S is the *steric factor*

$$\left|\frac{d[A]}{dt}\right| = k(T)[A][B], \text{ where } k(T) = A_{homog} e^{-E/\bar{R}T}, \text{ Arrhenius Law}$$

and $A_{homog} = \pi \sigma_{AB}{}^2 V_r N_{Avog} S$

$$\dot{w}_A''' = (-1) A_{Homog} \exp\left\{-\frac{E}{\bar{R}T}\right\}[A][B]$$

How to Find A and E

$k = A_{Homog} \exp(-E/\bar{R}T)$, plot ln k vs. 1/T, Slope = $-E/\bar{R}$

Relation between k_f and k_b

$$\frac{d\,[\ln K^o(T)]}{dT} \approx \frac{E_f}{\bar{R}T^2} - \frac{E_b}{\bar{R}T^2}$$

$$K^o(T) = \exp\left(-\frac{\Delta G^o}{\bar{R}T}\right) \Rightarrow \frac{d \ln K^o(T)}{dT} \approx \frac{\Delta h_R^o}{\bar{R}T^2}$$

Consider $CO_2 \Leftrightarrow CO + 1/2\ O_2$.

$\{k_f/k_b\} = K^0(T)\,\{P^0/(\bar{R}T)\}^{(1+1/2-1)}$, where $\bar{R} = 0.08314$ bar m^3/kmol K and $P^0 = 1$ bar

Relation between E_b and E_f

$$\frac{d \ln K^0}{dT} = \frac{\Delta \bar{h}_R^{-0}}{\bar{R}T^2},$$

$$E_f - E_b = \Delta\bar{h}_R^o;\ \text{for exothermic } \Delta\bar{h}_R^o < 0,\ E_b > E_f, \text{endothermic } \Delta\bar{h}_R^o \geq 0,\ E_b < E_f$$

Four-Step Paraffin Reactions (Table A.39B)

$$C_nH_{2n+2} \rightarrow \frac{n}{2} C_2H_4 + H_2$$

$$C_2H_4 + O_2 \rightarrow 2CO + 2H_2$$

$$CO + \frac{1}{2} O_2 \rightarrow CO_2$$

$$H_2 + \frac{1}{2} O_2 \rightarrow H_2O$$

$$\frac{d[\text{fuel}]}{dt} = -k[B]^\alpha [C]^\beta [D]^\gamma,\ k = A_{homog} \exp\left(-\frac{E}{\bar{R}T}\right)$$

Timescales for Reactions

1. Half-life time ($t_{1/2}$)

$$[A] = \frac{1}{2}[A]_{initial},\ t_{1/2} \simeq \{\ln k(T)\}/2 \text{ for first order}$$

2. Physical delay (t_{phy}) = time required to heat a cold combustible mixture reaction rates close to zero to T_{ign}.
3. Induction delay (t_{ind}) = time to build up radicals to a concentration required for sustaining reaction rates.

Heterogeneous Reactions (Solid + Gas)

Pyrolysis

Thermal decomposition process (change in chemical structure) that might release gases from solid/liquid

Naphthalene sublimation: $C_{10}H_{18}(s) \rightarrow C_{10}H_{18}(g)$ on heating.

Plexiglas pyrolysis: $C_5H_8O_2$.... (polymer, s) $\rightarrow$ $C_5H_8O_2$(monomer, g)

$$\dot{m}\,(g) = A\ m_v^{nv} \exp(-E/RT)$$

Coal Pyrolysis:

$$\dot{m}_v = -dm_V/dt = A\ m_v^{nv}\ \exp(-E/\bar{R}T)$$

where m_v is the volatile mass left in dry coal ($m_v = m - m_{char} - m_{ash}$), $n_v = 1$ for first-order pyrolysis; for Plexiglas, $n_v = 1$, $m_{char} = m_{ash} = 0$.

Chapter 6

Fourier Law:

$q'' = -\lambda\ dT/dy$, $q'' = h_H\ (T_w - T_\infty)$, where h_H is the heat transfer coefficient

Ranz–Marshall Correlation for flow past spheres:

$$Sh = 2 + 0.6\ Re^{1/2}Sc^{1/3}$$

Table F.6.1 Mass Transfer

$$Sc = \frac{v}{D}, \quad Pr = \frac{v}{\alpha_T} \quad Le = (Sc/Pr) = \frac{\lambda}{\rho D c_p} \quad Re = \frac{Vd}{v} \quad Sh = \frac{h_m d_p}{\rho D}$$

Generalized Form of Fick's Law for a Binary Mixture

$$V_{kl} = -D_{kl} \nabla Y_k / Y_k$$

$$t_{diff,T} = \frac{x^2}{\alpha_T} \quad \alpha_T = \frac{\lambda}{\rho c} \quad V_{diff,T} = \frac{L}{t_{diff,T}} = \frac{\alpha_T}{L}$$

$$\text{Thermal energy penetrated } U_x = x \cdot \rho c = \rho c (t\alpha_T)^{1/2} = \left(\frac{t\lambda}{\rho c}\right)^{1/2},$$

Fick's Law

$$j_k''(\text{kg/m}^2\text{sec}) = -\rho\, D_{k1} \frac{dY_k}{dy}$$

$$Y_k = \frac{X_k M_k}{M_{mix}}, \quad M_{mix} = \sum M_k X_x, \quad c_k = \frac{p_k}{\bar{R}T}, \quad R_k = \frac{\bar{R}}{M_k}, \quad \rho_k = \frac{p_k}{R_k T} \quad \bar{v} = \sum X_k \bar{v}_k$$

$$v = \sum Y_k v_k, \; \rho_k = \rho Y_k$$

Diffusion

$$V_k = v_k - v \qquad j_k'' = \rho_k V_k = \rho_k (v_k - v) \qquad \dot{m}_k'' = \rho_k v_k = \rho_k (v + V_k)$$

$$m_k''\ (\text{kg/m}^2\text{s}) = h_m\ \{\text{kg/m}^2\ \text{s}\}\{Y_{k,w} - Y_{k,\infty}\}, \qquad m_k'' = \rho_k \left[v - D_{kl} \left\{ \frac{d(\ln Y_k)}{dy} \right\} \right]$$

Collision Number and Mean Free Path

$$Z_{coll} = \pi\sigma^2 n' V_{avg} \qquad \ell_{mean} = \frac{1}{\sqrt{2}\pi n' \sigma^2} \simeq \frac{1}{\pi n' \sigma^2}$$

$$\frac{N_v'}{N} = \left(\frac{2}{\pi^{\frac{1}{2}}}\right) \left\{ \left[\frac{E}{\bar{R}T}\right]^{\frac{1}{2}} \exp\left[-\left(\frac{E}{\bar{R}T}\right)\right] \right\}$$

$$v_{avg} = \left[\frac{8}{3\pi}\right]^{\frac{1}{2}} v_{rms}, \quad v_{mps} = \sqrt{\frac{2}{3}} V_{rms}, \quad v_{rms} = \left[\frac{3\bar{R}T}{M}\right]^{\frac{1}{2}}$$

$$\mu\left(\frac{\text{kg}}{\text{m sec}}\right) = \frac{2.6693 \times 10^{-6} M^{1/2} T^{1/2} (\text{in K})}{\left\{\sigma^2 (\text{in angstrom})\, \Omega_\mu\right\}}$$

$$\Omega_u = -0.104 \left[\frac{\varepsilon}{k_B T}\right]^2 + 0.9663 \left[\frac{\varepsilon}{k_B T}\right] + 0.7273, \; 0 < \frac{\varepsilon}{k_B T} < 3.5$$

Table F.6.1 Mass Transfer (Continued)

Monatomic Thermal Conductivity

$$\lambda'\left(\frac{\text{kJ}}{\text{m sec K}}\right)=\frac{15\,\text{R}\left(\frac{\text{kJ}}{\text{kmole K}}\right)\mu}{4\,\text{M}}$$

Polyatomic Gases

$\lambda = \lambda' + \lambda''$

$\lambda'' = 0.88\,\lambda'\,\{0.4\,c_p/R - 1\}$

Euken Formulae

$\lambda/\lambda' = \{(4/15)\,c_p/R + 5/15\}$ or $\{(4/15)\,c_v/R + 3/5\}$

$$D_{kl}\left(\frac{\text{m}}{\text{sec}^2}\right)=\frac{2.628\times10^{-7}\text{m}^2\left[\frac{\text{bar angstrom}^2\text{kg}^{1/2}}{\text{sec kmol}^{1/2}\text{K}^{3/2}}\right]T^{3/2}}{P\sigma_{kl}^2\Omega_D M_{kl}^{1/2}}\text{, T in K, P in bar}$$

$$\sigma_{kl}(angstrom)=\frac{\sigma_k+\sigma_l}{2}\;;\;\frac{1}{M_{kl}}=\left(\frac{1}{2}\right)\left(\frac{1}{M_k}+\frac{1}{M_l}\right)\;;\;\left(\frac{\varepsilon}{k_B}\right)_{kl}=\left\{\left(\frac{\varepsilon}{k_B}\right)_k\left(\frac{\varepsilon}{k_B}\right)_l\right\}^{1/2}$$

$$\Omega_D=-0.0759\left[\frac{\varepsilon}{k_BT}\right]^2+0.8505\left[\frac{\varepsilon}{k_BT}\right]+0.6682,\;0<\frac{\varepsilon}{k_BT}<3.5$$

Mixtures

$$\mu_{mix}=\sum_i\frac{X_i\mu_i}{\sum_j X_j\phi_{ij}},\quad\phi_{ij}=\frac{1}{\sqrt{8}}\left[1+\frac{M_i}{M_J}\right]^{-1/2}\left[1+\left(\frac{K_i}{K_j}\right)^{1/2}\left(\frac{M_i}{M_j}\right)^{1/4}\right]^2,\;\phi_{ji}=\phi_{ij}\left(\frac{\mu_j}{\mu_i}\right)\left(\frac{M_i}{M_j}\right)$$

Binary System

$\lambda_{1,mix} = X_A\lambda_A + X_B\lambda_B$, $1/\lambda_{2,mix} = X_A/\lambda_A + X_B/\lambda_B$, $\lambda_{mix} = a\lambda_{1,mix} + (1 - a)\,\lambda_{2,mix}$

where

$a = 0.4056\,X_A^2 + 0.0417\,X_A + 0.3327$, $R^2 = 0.9945$

Multiple Components

$$\lambda_{mix}=\sum_i\frac{X_i\lambda_i}{\sum_j X_j\phi_{ij}}$$

$\lambda'_{mix} = \lambda'_{mix} + \lambda''_{mix}$

$\lambda'_{mix} = \Sigma_i\,Xi\,\lambda_i'/\{\Sigma_j\,X_j\,\psi_{ij}\}$

$\psi_{ij} = \varphi_{ij}\,[1 + \{2.41\,(M_i - M_j)\,(M_i - 0.162\,M_j)/(M_i + M_j)^2\}]$

$\lambda''_{mix} = \Sigma_i\,Xi\,\lambda_i''/\{\Sigma_j\,X_j\,\varphi_{ij}\}$

(Note: $\varphi_{ji} = \{\varphi_{ij}\,\lambda_j'/\lambda_i'\}$)

$\lambda_{mix}/\lambda'_{mix} = (4/15)\,c_v/R + 0.6$

$$\lambda_{mix}=0.5\left(\sum_k\lambda_k+\frac{1}{\sum_k\frac{X_k}{\lambda_k}}\right)$$

(*continued*)

Table F.6.1 Mass Transfer(Continued)

Diffusion in Multicomponent System

$$\nabla Y_k = -\sum_{l=1}^{n} \frac{Y_k Y_l \left[\vec{V}_k - \vec{V}_l\right]}{D_{kl}}$$

Diffusion of Species k into a Mixture of Species, Say, "b,c,d ..."
Wick's Law

$$D_{k,\text{stag mix}} = \frac{\{1 - Y_k\}}{\left\{\sum_{j \neq k} \frac{Y_j}{D_{kj}}\right\}}$$

where k is the moving species.

Chapter 7: Global Applications of Combustion

Conservation And Balance Equations

Mass conservation (species) $\frac{dm_k}{dt} = \dot{m}_{k,i} - \dot{m}_{k,e} + \dot{m}_{k,\text{chem}}, \dot{m}''_{k,\text{gen}} = \dot{\omega}'''_{k,m} V$,

Molar balance equation $\frac{dN_K}{dt} = \dot{N}_{k,\text{inlet}} - \dot{N}_{k,\text{exit}} + \dot{N}_{k,\text{gen}}, \dot{N}_{k,\text{gen}} = \dot{\omega}'''_k V,$

Closed System

Simplified Method

$$\frac{dT}{dt} \approx \frac{V \,|\, \dot{\omega}'''_F \,|\, \bar{H}\bar{V} + \dot{Q} - \frac{PdV}{dt}}{mc_v}, \text{constant V}$$

$$\frac{dT}{dt} \simeq \frac{V \,|\, \dot{\omega}'''_f \,|\, \bar{H}\bar{V} + \dot{Q}}{mc_p}, \text{constant p}$$

Constant V System Profiles

$$\{T - T_0\}/\{Y_F - Y_{F0}\} = \{T - T_0\}/\{(Y_{O2}/\nu_{O2}) - (Y_{O_2,0}/\nu_{O_2,0})\} = -h/c_v,$$

Constant P System Profiles

$$\{T - T_0\}/\{Y_F - Y_{F0}\} = \{T - T_0\}/\{(Y_{O_2}/\nu_{O_2}) - (Y_{O_2,0}/\nu_{O_2,0})\} = -h/c_p,$$

Rigorous Method (Constant V)

$$\frac{dT}{dt} = \frac{1}{N\bar{c}_v}\left[\dot{Q} - P\frac{dV}{dt} + \bar{R}T\frac{dN}{dt} + |\dot{\omega}_{F,M}|\,HV\right]$$
$$= \frac{1}{mc_v}\left[\dot{Q} - P\frac{dV}{dt} + \bar{R}T\frac{dN}{dt} + |\dot{\omega}_{F,M}|\,HV\right]$$

$$\bar{c}_v = \sum_k X_k \bar{c}_{v,k},\ c_v = \sum_k Y_k c_{v,k},\ X_k = \frac{MY_k}{M_k}, \bar{c}_p = \sum_k X_k \bar{c}_{p,k},\ c_p = \sum_k Y_k c_{p,k},\ \bar{R} = \bar{c}_p - \bar{c}_v$$

$$\frac{dN}{dt} = \sum_k \frac{dN_k}{dt}$$

Plug Flow Reactor (PFR)

$$\frac{d\dot{N}''_k}{dx} = \dot{\omega}'''_k,\ \frac{d\dot{m}''_k}{dx} = \dot{\omega}'''_k\ M_k = \dot{\omega}'''_{k,m},$$

$$\dot{m}''\frac{d\varepsilon_k}{dx} = \dot{\omega}'''_k,\ \varepsilon_k\ \text{flux ratio} = \frac{\dot{m}_k}{\dot{m}} = Y_k - \frac{\rho D}{\rho v}\frac{dY_k}{dx}$$

$$\dot{C}''_p\frac{dT}{dx} = \dot{\omega}'''(-\Delta h^0_R(T)) + \dot{Q}''\frac{C}{A}$$

where $C = \pi D,\ A = \frac{\pi D^2}{4},\ \dot{C}''_p = \sum_k \dot{N}''_k\,\bar{c}_{p,k} = \sum_k \dot{m}''_k\,c_{p,k}$

$$|\dot{\omega}'''_F| = A_0\exp\left(-\frac{E}{\bar{R}T}\right)\frac{\rho_{mix}^{(\alpha+\beta)}}{M_F^\alpha M_{O_2}^\beta}Y_F^\alpha Y_{O_2}^\beta\ ,\ \text{kmole/m}^3\ \text{sec}$$

$$|\dot{\omega}'''_{F,m}| = A_0\exp\left(-\frac{E}{\bar{R}T}\right)M_F\,\frac{\rho_{mix}^{(\alpha+\beta)}}{M_F^\alpha M_{O_2}^\beta}Y_F^\alpha Y_{O_2}^\beta\ ,\ \text{kg/m}^3\ \text{sec}$$

$$\frac{dY_F}{dt} \approx \frac{\dot{\omega}'''_{F,m}}{\rho} = -\frac{\dot{\omega}'''_m}{\rho},\ \text{if diffusion neglected}$$

Perfectly Stirred Reactor (PSR)

$$\text{Burned fraction},\ \varepsilon_b = \frac{\dot{m}_{F,b}}{\dot{m}_{F,in}} = \frac{\{\dot{m}_{F,in} - \dot{m}_{F,e}\}}{\dot{m}_{F,in}} = 1 - \frac{Y_F}{Y_{F,in}} = \frac{|\dot{w}'''_m|\,t_{res}}{\rho Y_{F,in}},\ Y_F = Y_{F,e}$$

First Damkohler number

$$D_I = \frac{t_{res,\ conv}}{t_{chem}} = A_m t_{res} \nu_{O_2}^{\beta} Y_{F,in}^{\alpha+\beta-1} \rho^{\alpha+\beta-1}$$

where $t_{res,\ conv} = t_{res} = \frac{m}{\dot{m}} = \frac{\rho V}{\dot{m}}$, ν_{O_2}: stoichiometric O_2

$$\varepsilon_b = D_I \exp\left(-\frac{E}{\bar{R}T}\right)\left(\frac{\rho}{\rho_{in}}\right)^{(\alpha+\beta)} (1-\varepsilon_b)^{\alpha} \left(\frac{1}{\phi} - \varepsilon_b\right)^{\beta}, \phi = \frac{(A{:}F)_{stoich}}{(A{:}F)}$$

Carbon or Coal Combustion: Resistance Concept

$$\dot{m}''_{O_2} = h_m (Y_{O_2,\infty} - Y_{O_2,w}) = \frac{Y_{O_2,\infty}}{R_{diff} + R_{chem}}$$

$$R_{diff} = 1/h_m,\ R_{chem} = 1/(\rho_w k_{O2}),\ k_{O2} = k\nu_{O2}, k = A_{het}\exp(-E/\bar{R}T)$$

h_m is mass Transfer coefficient, $Sh = \frac{h_m d}{\rho D}, |\dot{w}''| = k\rho Y_{O_2 w}$

$$Y_{O_2,w} = \frac{h_m Y_{O_2,\infty}}{k_{O_2}\rho_w + h_m}, B = \frac{Y_{O_2,\infty}}{\nu_{O_2}}$$

$$\left|\dot{\omega}''_{O_2}\right| = \dot{m}''_{O_2} = \frac{Y_{O_2,\infty}}{R_{diff} + R_{chem}}, \left|\dot{\omega}''_c\right| = \frac{\left|\dot{\omega}''_{O_2}\right|}{\nu_{O_2}} = \dot{m}''_C = \frac{B}{R_{diff} + R_{chem}}$$

Burning Rate Relations

$$\dot{m}_c = \dot{m}''_c \pi d^2 = -\frac{dm_c}{dt} = Sh\pi\rho D(2a)B$$

d² Law

$$d^2 - d_0^2 = \alpha_c t, \quad \text{where} \quad \alpha_c = \frac{4Sh\rho_w D_w B}{\rho_p} \quad \text{and burning time } t = \frac{d_0^2}{\alpha_c}$$

Heat Generation and Heat Loss

$$\dot{Q}_{gen} = \dot{m}h_c,\ \dot{Q}_{loss} = h_H A(T_w - T_\infty) + \varepsilon\sigma A\left(T_w^4 - T_\infty^4\right)$$

Particle Temperature

$$(T_w - T_\infty) \approx \left\{\frac{Y_{O_2\infty} - Y_{O_2w}}{v_{O_2}}\right\}\left\{\frac{h_c}{c_p}\right\}, h_H = \left\{\frac{Nu\lambda}{d}\right\}$$

Chapter 8

Vectors and Tensors

Velocity and Gradient

$$\vec{V} = \sum_{j=1}^{3} \delta_i v_j$$

$$\vec{\Delta} = \sum_{i=1}^{3} \vec{\delta}_i \frac{\partial}{\partial x_i}.$$

Gradient of Vector and Its Transpose

$$\vec{\Delta}\vec{V} = \sum_{i=1}^{3} \vec{\delta}_i \frac{\partial}{\partial x_i} \sum_{j=1}^{3} \vec{\delta}_j v_j = \sum_{i=1}^{3}\sum_{j=}^{3} \vec{\delta}_i \vec{\delta}_j \frac{\partial v_j}{\partial x_i}, \text{ i.e.,}$$

$$\vec{\Delta}\vec{v} = \begin{pmatrix} \frac{\partial v_1}{\partial X_1} & \frac{\partial v_2}{\partial X_1} & \frac{\partial v_3}{\partial X_1} \\ \frac{\partial v_1}{\partial X_2} & \frac{\partial v_2}{\partial X_2} & \frac{\partial v_3}{\partial X_3} \\ \frac{\partial v_1}{\partial X_3} & \frac{\partial v_2}{\partial X_3} & \frac{\partial v_3}{\partial X_3} \end{pmatrix}$$

where δ_{11}, δ_{21}, δ_{31}, etc., are associated with $\partial v_1/\partial x_1$, $\partial v_2/\partial x_1$, $\partial v_1/\partial x_3$, etc.

$$\vec{\nabla}\vec{V}^T = \begin{pmatrix} \frac{\partial v_1}{\partial x_1} & \frac{\partial v_1}{\partial x_2} & \frac{\partial v_1}{\partial x_3} \\ \frac{\partial v_2}{\partial x_1} & \frac{\partial v_2}{\partial x_2} & \frac{\partial v_2}{\partial x_3} \\ \frac{\partial v_3}{\partial x_1} & \frac{\partial v_3}{\partial x_2} & \frac{\partial v_3}{\partial x_3} \end{pmatrix}.$$

Dot Products Involving Tensor and Vectors

Dot Products

$$\delta_{ij} = \begin{cases} 1 \text{ when } i = j \\ 0 \text{ when } i \neq j \end{cases}, \text{ Kronecker delta}$$

$$\vec{\delta}_\ell \cdot \vec{\delta}_i = 1 \quad \text{if} \quad i = \ell, \vec{\delta}_\ell \cdot \vec{\delta}_i = 0 \quad \text{if} \quad i \neq \ell$$

$$\nabla \cdot \vec{V} = \sum_{\ell=1}^{3} \vec{\delta}_\ell \frac{\partial}{\partial x_\ell} \cdot \sum_{j=1}^{3} \vec{\delta}_j v_j = \sum_{\ell=1}^{3} \frac{\partial v_\ell}{\partial x_\ell}$$

Stress Tensor

$$\vec{\vec{p}}, \text{stress tensor} = \begin{pmatrix} p_{11} & p_{12} & p_{13} \\ p_{21} & p_{22} & p_{23} \\ p_{31} & p_{32} & p_{33} \end{pmatrix}$$

$$\bar{U}, \text{unit tensor} = \begin{pmatrix} 1 & 0 & 0 \\ 0 & 1 & 0 \\ 0 & 0 & 1 \end{pmatrix}$$

$$\vec{\vec{U}} \cdot \vec{V} = \sum \sum \vec{\delta}_i \vec{\delta}_j \vec{\delta}_{ij} \cdot \sum \vec{\delta}_k v_k = \sum \vec{\delta}_k v_k = \vec{V}$$

$$\nabla \cdot \vec{\vec{P}} = \sum \sum \delta_k \frac{\partial}{\partial x_k} \cdot \sum_j \sum_i \vec{\delta}_i \vec{\delta}_j p_{ij} = \sum_\ell \sum_j \vec{\delta}_j \frac{\partial p_{\ell j}}{\partial x_\ell}, p_{ii} = p^\circ \delta_{ij} + \tau_{ij}$$

$$\vec{V} \cdot \nabla \vec{V} = \sum_{l=1}^{3} \vec{\delta}_\ell v_e \cdot \sum_{i=1}^{3} \sum_{j=1}^{3} \vec{\delta}_i \vec{\delta}_j \frac{\partial v_j}{\partial x_i} = \sum_{\ell=1}^{3} \sum_{j=1}^{3} \vec{\delta}_j \frac{\partial v_j}{\partial x_\ell}$$

$$\vec{V} \cdot \nabla \cdot \vec{\vec{P}} = \sum_{j=1}^{3} \sum_{i=1}^{3} v_i \frac{\partial p_{ij}}{\partial x_j}$$

$$\nabla \cdot (\vec{\vec{P}} \cdot \vec{V}) = \sum_{j=1}^{3} \sum_{i=1}^{3} \frac{\partial (p_{ji} v_i)}{\partial x_j}$$

$$\nabla \cdot (p^{\circ}\vec{\vec{U}} \cdot \vec{V}) = \nabla \cdot (p^{\circ}\vec{V}) = \sum_{i=1}^{3} \frac{\partial (p^{\circ} v_i)}{\partial x_j} = \vec{V} \cdot \nabla p^{\circ} + p^{\circ} \nabla \cdot \vec{V}$$

$$\vec{v} \cdot \nabla \cdot \left[p^{0} \vec{\vec{U}} \right] = \vec{v} \cdot \nabla p^{0}$$

Double Dot Products

$$\vec{\vec{P}} : \vec{\nabla} \vec{V} = \sum_i \sum_j p_{ij} \frac{\partial v_j}{\partial x_i}$$

Material Derivatives

$$\frac{D}{dt} = \frac{\partial}{\partial t} + (\vec{v} \cdot \nabla),$$

For kinetic energy per unit mass,

$$\rho \frac{D\left[\frac{v^2}{2}\right]}{Dt} = \rho \left[\frac{D\left[\frac{v^2}{2}\right]}{Dt} \right] = \rho \frac{\partial \left[\frac{v^2}{2}\right]}{\partial t} + \rho (\vec{v} \cdot \nabla) \left[\frac{v^2}{2} \right]$$

Simple Constitutive Equations

Diffusive Momentum-Transfer Equation — Newton's Law

$$\tau = -\mu \frac{dv}{dy} = -\rho \nu \frac{dv}{dy}, \ \vec{\vec{\tau}} = -\rho \nu \vec{\nabla} \vec{V}$$

Diffusive Heat-Transfer Equation — Fourier's Law

$$\dot{q}'' = -\lambda \frac{dT}{dy} = -\rho \alpha_T c_p \frac{dT}{dy}, \ \vec{\dot{q}}'' = -\rho \alpha_T c_p \vec{\nabla} T$$

Diffusive Species-Transfer Equation — Fick's Law

$$j_k'' = -\rho D \frac{dY_k}{dy}, \ \vec{j}_k'' = -\rho D \vec{\nabla} Y_k$$

Table F.8.1 Equation of Continuity in Several Coordinate Systems[a] (V_x, V_y, V_z or V_1,V_2,V_3, replaced by u_x, u_y, u_z in Cartesian Systems)

Overall mass

Rectangular Coordinates (*x*, *y*, *z*) (A)

$$\frac{\partial \rho}{\partial t}+\frac{\partial}{\partial x}(\rho u_x)+\frac{\partial}{\partial y}(\rho u_y)+\frac{\partial}{\partial z}(\rho u_z)=0$$

Cylindrical Coordinates (*r*, *θ*, *z*)

$$\frac{\partial \rho}{\partial t}+\frac{1}{r}\frac{\partial}{\partial r}(\rho u_r)+\frac{1}{r}\frac{\partial}{\partial \theta}(\rho u_\theta)+\frac{\partial}{\partial z}(\rho u_z)=0 \quad \text{(B)}$$

Spherical Coordinates (*r*, *θ*, *ϕ*)

$$\frac{\partial \rho}{\partial t}+\frac{1}{r^2}\frac{\partial}{\partial r}\left(\rho r^2 u_r\right)+\frac{1}{r\sin\theta}\frac{\partial}{\partial \theta}(\rho u_\theta \sin\theta)+\frac{1}{r\sin\theta}\frac{\partial}{\partial \phi}(\rho u_\phi)=0 \quad \text{(C)}$$

Species Equations

Rectangular Coordinates (*x*, *y*, *z*) (A)

$$\rho\left(\frac{\partial Y_i}{\partial t}+u_x\frac{\partial Y_i}{\partial x}+u_y\frac{\partial Y_i}{\partial y}+u_z\frac{\partial Y_i}{\partial z}\right)+\frac{\partial}{\partial x}(\rho Y_i V_{ix})+\frac{\partial}{\partial y}(\rho Y_i V_{iy})+\frac{\partial}{\partial z}(\rho Y_i V_{iz})=\omega_i$$

(where $V_{ir}=-\frac{D}{Y_i}\frac{\partial Y_i}{\partial r}$, $V_{i\theta}=-\frac{D}{Y_i r}\frac{\partial Y_i}{\partial \theta}$, and $V_{iz}=-\frac{D}{Y_i}\frac{\partial Y_i}{\partial z}$

according to Fick's Law of mass diffusion)

Cylindrical Coordinates (*r*, *θ*, *z*)

$$\rho\left(\frac{\partial Y_i}{\partial t}+u_r\frac{\partial Y_i}{\partial r}+\frac{u_\theta}{r}\frac{\partial Y_i}{\partial \theta}+u_z\frac{\partial Y_i}{\partial z}\right)+\frac{1}{r}\frac{\partial}{\partial r}(r\rho Y_i V_{ir})+\frac{1}{r}\frac{\partial}{\partial \theta}(\rho Y_i V_{i\theta})+\frac{\partial}{\partial z}(\rho Y_i V_{iz})=\omega_i \quad \text{(B)}$$

(where $V_{ir}=-\frac{D}{Y_i}\frac{\partial Y_i}{\partial r}$, $V_{i\theta}=-\frac{D}{Y_i r}\frac{\partial Y_i}{\partial \theta}$, and $V_{iz}=-\frac{D}{Y_i}\frac{\partial Y_i}{\partial z}$, according to Fick's law

of mass diffusion)

Spherical Coordinates (*r*, *θ*, *ϕ*)

$$\rho\left(\frac{\partial Y_i}{\partial t}+u_r\frac{\partial Y_i}{\partial r}+\frac{u_\theta}{r}\frac{\partial Y_i}{\partial \theta}+\frac{u_\phi}{r\sin\theta}\frac{\partial Y_i}{\partial}\right)+\frac{1}{r^2}\frac{\partial}{\partial r}\left(r^2\rho Y_i V_{ir}\right)$$

$$+\frac{1}{r\sin\theta}\frac{\partial}{\partial \theta}(\sin\theta\rho Y_i V_{i\theta})+\frac{1}{r\sin\theta}\frac{\partial}{\partial}(\rho Y_i V_i)=\omega_i$$

(where $V_{ir}=-\frac{D}{Y_i}\frac{\partial Y_i}{\partial r}$, $V_{i\theta}=-\frac{D}{Y_i r}\frac{\partial Y_i}{\partial \theta}$, and $V_{i\phi}=-\frac{D}{Y_i r\sin\theta}\frac{\partial Y_i}{\partial \phi}$, according to Fick's

Law of mass diffusion)

$r \ge 0$, $2\pi \ge \theta \ge 0$.

$r \ge 0$, $2\pi > \phi \ge 0$, $\pi \ge \theta \ge 0$.

Table F.8.2 The Momentum Equation in Rectangular Coordinates (x, y, z)[a] The Equation in Tables F.8.2 to F.8.4 are Called the Navier-Stokes Equations

In terms of τ

x component

$$\rho\left(\frac{\partial u_x}{\partial t}+u_x\frac{\partial u_x}{\partial x}+u_y\frac{\partial u_x}{\partial y}+u_z\frac{\partial u_x}{\partial z}\right)=-\frac{\partial p}{\partial x}+\left(\frac{\partial \tau_{xx}}{\partial x}+\frac{\partial \tau_{yx}}{\partial y}+\frac{\partial \tau_{zx}}{\partial z}\right)+\rho f_x \quad \text{(A)}$$

y component

$$\rho\left(\frac{\partial u_x}{\partial t}+u_x\frac{\partial u_x}{\partial x}+u_y\frac{\partial u_y}{\partial y}+u_z\frac{\partial u_y}{\partial z}\right)=-\frac{\partial p}{\partial y}+\left(\frac{\partial \tau_{xy}}{\partial x}+\frac{\partial \tau_{yx}}{\partial y}+\frac{\partial \tau_{zy}}{\partial z}\right)+\rho f_y \quad \text{(B)}$$

z component

$$\rho\left(\frac{\partial u_z}{\partial t}+u_x\frac{\partial u_z}{\partial x}+u_y\frac{\partial u_z}{\partial y}+u_z\frac{\partial u_z}{\partial z}\right)=-\frac{\partial p}{\partial z}+\left(\frac{\partial \tau_{xz}}{\partial x}+\frac{\partial \tau_{yz}}{\partial y}+\frac{\partial \tau_{zz}}{\partial z}\right)+\rho f_z \quad \text{(C)}$$

In terms of velocity gradients for a Newtonian fluid with constant ρ and μ

$$\rho\left(\frac{\partial u_x}{\partial t}+u_x\frac{\partial u_x}{\partial x}+u_y\frac{\partial u_x}{\partial y}+u_z\frac{\partial u_x}{\partial z}\right) \quad \text{(D)}$$

x component

$$=-\frac{\partial p}{\partial y}+\mu\left(\frac{\partial^2 u_x}{\partial x^2}+\frac{\partial^2 u_x}{\partial y^2}+\frac{\partial^2 u_x}{\partial z^2}\right)+\rho f_x$$

y component (E)

$$\rho\left(\frac{\partial u_y}{\partial t}+u_x\frac{\partial u_y}{\partial x}+u_y\frac{\partial u_y}{\partial y}+u_z\frac{\partial u_y}{\partial z}\right)$$

$$=-\frac{\partial p}{\partial y}+\mu\left(\frac{\partial^2 u_y}{\partial x^2}+\frac{\partial^2 u_y}{\partial y^2}+\frac{\partial^2 u_y}{\partial z^2}\right)+\rho f_y$$

z component (F)

$$\rho\left(\frac{\partial u_z}{\partial t}+u_x\frac{\partial u_z}{\partial x}+u_y\frac{\partial u_z}{\partial y}+u_z\frac{\partial u_z}{\partial z}\right)$$

$$=-\frac{\partial p}{\partial y}+\mu\left(\frac{\partial^2 u_z}{\partial x^2}+\frac{\partial^2 u_z}{\partial y^2}+\frac{\partial^2 u_z}{\partial z^2}\right)+\rho f_z$$

f, body force per unit mass

Table F.8.3 The Momentum Equation in Cylindrical Coordinates (r, θ, z)[a]

r component

$$\rho\left(\frac{\partial u_r}{\partial t}+u_r\frac{\partial u_r}{\partial r}+\frac{u_\theta}{r}\frac{\partial u_r}{\partial \theta}+u_z\frac{\partial u_r}{\partial z}\right) \quad \text{(A)}$$

$$=-\frac{\partial p}{\partial r}+\left(\frac{1}{r}\frac{\partial}{\partial r}(r\tau_{rr})+\frac{1}{r}\frac{\partial \tau_{r\theta}}{\partial \theta}-\frac{\tau_{\theta\theta}}{r}+\frac{\partial \tau_{rz}}{\partial y}+\frac{\partial \tau_{rz}}{\partial z}\right)+\rho f_r$$

θ component

$$\rho\left(\frac{\partial u_\theta}{\partial t}+u_r\frac{\partial u_\theta}{\partial r}+\frac{u_\theta}{r}\frac{\partial u_\theta}{\partial \theta}+u_z\frac{\partial u_\theta}{\partial z}\right) \quad \text{(B)}$$

$$=-\frac{1}{r}\frac{\partial p}{\partial \theta}+\left(\frac{1}{r^2}\frac{\partial}{\partial r}\left(r^2\tau_{r\theta}\right)+\frac{1}{r}\frac{\partial \tau_{\theta\theta}}{\partial \theta}+\frac{\partial \tau_{\theta z}}{\partial z}\right)+\rho f_\theta$$

z component

$$\rho\left(\frac{\partial u_z}{\partial t}+u_r\frac{\partial u_z}{\partial r}+\frac{u_\theta}{r}\frac{\partial u_z}{\partial \theta}+u_z\frac{\partial u_z}{\partial z}\right) \quad \text{(C)}$$

$$=-\frac{\partial p}{\partial z}+\left(\frac{1}{r}\frac{\partial}{\partial r}\left(r\tau_{rz}\right)+\frac{1}{r}\frac{\partial \tau_{\theta z}}{\partial \theta}+\frac{\partial \tau_{zz}}{\partial z}\right)+\rho f_z$$

In Terms of Velocity Gradients for a Newtonian Fluid with Constant ρ and μ:

r component

$$\rho\left(\frac{\partial u_r}{\partial t}+u_r\frac{\partial u_r}{\partial r}+\frac{u_\theta}{r}\frac{\partial u_r}{\partial \theta}-\frac{u_\theta^2}{r}+u_z\frac{\partial u_r}{\partial z}\right) \quad \text{(D)}$$

$$=-\frac{\partial p}{\partial r}+\mu\left[\frac{\partial}{\partial r}\left(\frac{1}{r}\frac{\partial}{\partial r}(ru_r)\right)+\frac{1}{r^2}\frac{\partial^2 u_r}{\partial \theta^2}-\frac{2}{r^2}\frac{\partial u_\theta}{\partial \theta}+\frac{\partial^2 u_r}{\partial z^2}\right]+\rho f_r$$

$\rho u_\theta^2/r$, *centrifugal force* in the r direction

θ component

$$\rho\left(\frac{\partial u_\theta}{\partial t}+u_r\frac{\partial u_\theta}{\partial r}+\frac{u_\theta}{r}\frac{\partial u_\theta}{\partial \theta}+\frac{u_r u_\theta}{r}+u_z\frac{\partial u_\theta}{\partial z}\right) \quad \text{(E)}$$

$$=-\frac{1}{r}\frac{\partial p}{\partial \theta}+\mu\left[\frac{\partial}{\partial r}\left(\frac{1}{r}\frac{\partial}{\partial r}(ru_\theta)\right)+\frac{1}{r^2}\frac{\partial^2 u_\theta}{\partial \theta^2}+\frac{2}{r^2}\frac{\partial u_r}{\partial \theta}+\frac{\partial^2 u_\theta}{\partial z^2}\right]+\rho f_\theta$$

$\rho u_r u_\theta/r$, *Coriolis force* in the θ direction (F)

Table F.8.4 The Momentum Equation in Spherical Coordinates (r, θ, ϕ)[a]

In Terms of τ

r component

$$\rho\left(\frac{\partial u_r}{\partial t}+u_r\frac{\partial u_r}{\partial r}+\frac{u_\theta}{r}\frac{\partial u_r}{\partial \theta}+\frac{u_\phi}{r\sin\theta}\frac{\partial u_r}{\partial \phi}-\frac{u_\theta^2+u_\phi^2}{r}\right) \tag{A}$$

$$=-\frac{\partial p}{\partial r}+\mu\left(\frac{1}{r^2}\frac{\partial}{\partial r}\left(r^2\tau_{rr}\right)+\frac{1}{r\sin\theta}\frac{\partial}{\partial \theta}\left(\tau_{r\theta}\sin\theta\right)+\frac{1}{r\sin\theta}\frac{2\tau_{r\phi}}{\partial \phi}-\frac{\tau_{\theta\theta}+\tau_{\phi\phi}}{r}\right)+\rho f_r$$

θ component

$$\rho\left(\frac{\partial u_\theta}{\partial t}+u_r\frac{\partial u_\theta}{\partial r}+\frac{u_\theta}{r}\frac{\partial u_\theta}{\partial \theta}+\frac{u_\phi}{r\sin\theta}\frac{\partial u_\theta}{\partial \phi}+\frac{u_r u_\theta}{r}-\frac{u_\phi^2\cot\theta}{r}\right) \tag{B}$$

$$=-\frac{1}{r}\frac{\partial p}{\partial \theta}+\mu\left(\frac{1}{r^2}\frac{\partial}{\partial r}\left(r^2\tau_{r\theta}\right)+\frac{1}{r\sin\theta}\frac{\partial}{\partial \theta}(\tau_{\theta\theta}\sin\theta)+\frac{1}{r\sin\theta}\frac{2\tau_{\theta\phi}}{\partial \phi}+\frac{\tau_{r\theta}}{r}-\frac{\cot\theta}{r}\tau_{\phi\phi}\right)+\rho f_\theta$$

ϕ component

$$\rho\left(\frac{\partial u_\phi}{\partial t}+u_r\frac{\partial u_\phi}{\partial r}+\frac{u_\theta}{r}\frac{\partial u_\phi}{\partial \theta}+\frac{u_\phi}{r\sin\theta}\frac{\partial u_\phi}{\partial \phi}+\frac{u_\phi u_r}{r}+\frac{u_\theta u_\phi}{r}\cot\theta\right) \tag{C}$$

$$=-\frac{1}{r\sin\theta}\frac{\partial p}{\partial \phi}+\left(\frac{1}{r^2}\frac{\partial}{\partial r}\left(r^2\tau_{r\phi}\right)+\frac{1}{r}\frac{\partial \tau_{\theta\phi}}{\partial \theta}+\frac{1}{r\sin\theta}\frac{\partial \tau_{\phi\phi}}{\partial \phi}+\frac{\tau_{r\phi}}{r}+\frac{2\cot\theta}{r}\tau_{\theta\phi}\right)+\rho f_\phi$$

In Terms of Velocity Gradients for a Newtonian Fluid with Constant ρ and μ:

r component

$$\rho\left(\frac{\partial u_r}{\partial t}+u_r\frac{\partial u_{r\varpi}}{\partial r}+\frac{u_\theta}{r}\frac{\partial u_r}{\partial \theta}+\frac{u_\phi}{r\sin\theta}\frac{\partial u_r}{\partial \phi}-\frac{u_\theta^2+u_\phi^2}{r}\right) \tag{D}$$

$$=-\frac{\partial p}{\partial \phi}+\mu\left(\nabla^2 u_r-\frac{2}{r^2}u_r-\frac{2}{r^2}\frac{\partial u_\theta}{\partial \theta}-\frac{2}{r^2}u_\theta\cot\theta-\frac{2}{r^2\sin\theta}\frac{\partial u_\phi}{\partial \phi}\right)+\rho f_r$$

θ component

$$\rho\left(\frac{\partial u_\phi}{\partial t}+u_r\frac{\partial u_\phi}{\partial r}+\frac{u_\theta}{r}\frac{\partial u_\phi}{\partial \theta}+\frac{u_\phi}{r\sin\theta}\frac{\partial u_\phi}{\partial \phi}+\frac{u_\phi u_r}{r}+\frac{u_\theta u_\phi}{r}\cot\theta\right) \tag{E}$$

$$=-\frac{1}{r\sin\theta}\frac{\partial p}{\partial \phi}+\left(\frac{1}{r^2}\frac{\partial}{\partial r}\left(r^2\tau_{r\phi}\right)+\frac{1}{r}\frac{\partial \tau_{\theta\phi}}{\partial \theta}+\frac{1}{r\sin\theta}\frac{\partial \tau_{\phi\phi}}{\partial \phi}+\frac{\tau_{r\phi}}{r}+\frac{2\cot\theta}{r}\tau_{\theta\phi}\right)+\rho f_\phi$$

ϕ component[b]

$$\rho\left(\frac{\partial u_\phi}{\partial t}+u_r\frac{\partial u_\phi}{\partial r}+\frac{u_\theta}{r}\frac{\partial u_\phi}{\partial \theta}+\frac{u_\phi}{r\sin\theta}\frac{\partial u_\phi}{\partial \phi}+\frac{u_\phi u_r}{r}+\frac{u_\theta u_\phi}{r}\cot\theta\right) \tag{F}$$

$$=-\frac{1}{r\sin\theta}\frac{\partial p}{\partial \phi}+\mu\left(\nabla^2 u_\phi-\frac{u_\phi}{r^2\sin^2\theta}+\frac{2}{r^2\sin\theta}\frac{\partial u_r}{\partial \phi}+\frac{2\cos\theta}{r^2\sin^2\theta}\frac{\partial u_\theta}{\partial \phi}\right)+\rho f_\phi$$

Table F.8.5 Components of the Stress Tensor for Newtonian Fluids in Rectangular Coordinates (x, y, z)[a]

$$\tau_{xx} = \mu\left[2\frac{\partial u_x}{\partial x} - \tfrac{2}{3}(\nabla \cdot V)\right] \qquad \text{(A)}$$

$$\tau_{VV} = \mu\left[2\frac{\partial u_Y}{\partial y} - \tfrac{2}{3}(\nabla \cdot V)\right] \qquad \text{(B)}$$

$$\tau_{ZZ} = \mu\left[2\frac{\partial u_z}{\partial z} - \tfrac{2}{3}(\nabla \cdot V)\right] \qquad \text{(C)}$$

$$\tau_{xy} = \tau_{yx} = \mu\left[\frac{\partial u_x}{\partial y} + \frac{\partial u_y}{\partial x}\right] \qquad \text{(D)}$$

$$\tau_{yz} = \tau_{xz} = \mu\left[\frac{\partial u_z}{\partial x} + \frac{\partial u_x}{\partial z}\right] \qquad \text{(E)}$$

$$\nabla \cdot V = \frac{\partial u_x}{\partial x} + \frac{\partial u_Y}{\partial y} + \frac{\partial u_z}{\partial z} \qquad \text{(F)}$$

Table F.8.6 Components of the Stress Tensor for Newtonian Fluids in Cylindrical Coordinates (r, θ, z)[a]

$$\tau_{rr} = \mu\left[2\frac{\partial u_r}{\partial r} - \tfrac{2}{3}(\nabla \cdot V)\right] \qquad \text{(A)}$$

$$\tau_{\theta\theta} = \mu\left[2\left(\frac{1}{r}\frac{\partial u_\theta}{\partial \theta} + \frac{u_r}{r}\right) - \tfrac{2}{3}(\nabla \cdot V)\right] \qquad \text{(B)}$$

$$\tau_{zz} = \mu\left[2\frac{\partial u_z}{\partial z} - \tfrac{2}{3}(\nabla \cdot V)\right] \qquad \text{(C)}$$

$$\tau_{r\theta} = \tau_{\theta r} = \mu\left[r\frac{\partial}{\partial r}\left(\frac{u_\theta}{r}\right) + \frac{1}{r}\frac{\partial u_r}{\partial \theta}\right] \qquad \text{(D)}$$

$$\tau_{\theta z} = \tau_{z\theta} = \mu\left[\frac{\partial u_\theta}{\partial z} + \frac{1}{r}\frac{\partial u_z}{\partial \theta}\right] \qquad \text{(E)}$$

$$\tau_{zr} = \tau_{zr} = \mu\left[\frac{\partial u_z}{\partial r} + \frac{\partial u_r}{\partial z}\right] \qquad \text{(F)}$$

$$\nabla \cdot V = \frac{1}{r}\frac{\partial}{\partial r}(ru_r) + \frac{1}{r}\frac{\partial u_\theta}{\partial \theta} + \frac{\partial u_z}{\partial z} \qquad \text{(G)}$$

Table F.8.7 Components of the Stress Tensor for Newtonian Fluids in Spherical Coordinates (r, θ, ϕ)[a]

$$\tau_{rr} = \mu\left[2\frac{\partial u_r}{\partial r} - \tfrac{2}{3}(\nabla \cdot V)\right] \quad \text{(A)}$$

$$\tau_{\theta\theta} = \mu\left[2\left(\frac{1}{r}\frac{\partial u_\theta}{\partial \theta} + \frac{u_r}{r}\right) - \tfrac{2}{3}(\nabla \cdot V)\right] \quad \text{(B)}$$

$$\tau_{\phi\phi} = \mu\left[2\left(\frac{1}{r\sin\theta}\frac{\partial u_\phi}{\partial \phi} + \frac{u_r}{r} + \frac{u_\theta \cot\theta}{r}\right) - \tfrac{2}{3}(\nabla \cdot V)\right] \quad \text{(C)}$$

$$\tau_{r\theta} = \tau_{\theta r} = \mu\left[r\frac{\partial}{\partial r}\left(\frac{u_\theta}{r}\right) + \frac{1}{r}\frac{\partial u_r}{\partial \theta}\right] \quad \text{(D)}$$

$$\tau_{\theta\phi} = \tau_{\phi\theta} = \mu\left[\frac{\sin\theta}{r}\frac{\partial}{\partial \theta}\left(\frac{u_\phi}{\sin\theta}\right) + \frac{1}{r\sin\theta}\frac{\partial u_\theta}{\partial \phi}\right] \quad \text{(E)}$$

$$\tau_{\phi r} = \tau_{r\phi} = \mu\left[\frac{1}{r\sin\theta}\frac{\partial u_r}{\partial \phi} + r\frac{\partial}{\partial r}\left(\frac{u_\phi}{r}\right)\right] \quad \text{(F)}$$

$$\nabla \cdot V = \frac{1}{r^2}\frac{\partial}{\partial r}\left(r^2 u_r\right) + \frac{1}{r\sin\theta}\frac{\partial}{\partial \theta}(u_\theta \sin\theta) + \frac{1}{r\sin\theta}\frac{\partial u_\phi}{\partial \phi} \quad \text{(G)}$$

Table F.8.8 Summary of Differential Operations Involving the ∇ Operator in Rectangular Coordinates (x, y, z)[a]

$$\nabla \cdot V = \frac{\partial u_x}{\partial x} + \frac{\partial u_r}{\partial y} + \frac{\partial u_z}{\partial z} \quad \text{(A)}$$

$$\nabla^2 s = \frac{\partial^2 s}{\partial x^2} + \frac{\partial^2 s}{\partial y^2} + \frac{\partial^2 s}{\partial z^2} \quad \text{(B)}$$

$$\tau : \nabla V = \tau_{xx}\left(\frac{\partial u_x}{\partial x}\right) + \tau_{yy}\left(\frac{\partial u_V}{\partial y}\right) + \tau_{zz}\left(\frac{\partial u_z}{\partial z}\right) + \tau_{xV}\left(\frac{\partial u_x}{\partial y} + \frac{\partial u_V}{\partial x}\right)$$
$$+ \tau_{yz}\left(\frac{\partial u_y}{\partial z} + \frac{\partial u_z}{\partial y}\right) + \tau_{zx}\left(\frac{\partial u_z}{\partial x} + \frac{\partial u_x}{\partial z}\right) \quad \text{(C)}$$

$$[\nabla s]_x = \frac{\partial s}{\partial x} \quad \text{(D)}$$

$$[\nabla s]_y = \frac{\partial s}{\partial y} \quad \text{(E)}$$

(continued)

Table F.8.8 Summary of Differential Operations Involving the ∇ Operator in Rectangular Coordinates (*x*, *y*, *z*) (Continued)

$$[\nabla s]_z = \frac{\partial s}{\partial z} \quad \text{(F)}$$

$$[\nabla \times V]_x = \frac{\partial u_z}{\partial y} - \frac{\partial u_y}{\partial z} \quad \text{(G)}$$

$$[\nabla \times V]_y = \frac{\partial u_x}{\partial z} - \frac{\partial u_z}{\partial x} \quad \text{(H)}$$

$$[\nabla \times V]_z = \frac{\partial u_V}{\partial x} - \frac{\partial u_x}{\partial y} \quad \text{(I)}$$

$$[\nabla \cdot \tau]_x = \frac{\partial \tau_{xx}}{\partial x} + \frac{\partial \tau_{xV}}{\partial y} + \frac{\partial \tau_{xz}}{\partial z} \quad \text{(J)}$$

$$[\nabla \cdot \tau]_V = \frac{\partial \tau_{yx}}{\partial x} + \frac{\partial \tau_{xy}}{\partial y} + \frac{\partial \tau_{yz}}{\partial z} \quad \text{(K)}$$

$$[\nabla \cdot \tau]_z = \frac{\partial \tau_{zy}}{\partial x} + \frac{\partial \tau_{zy}}{\partial y} + \frac{\partial \tau_{zz}}{\partial z} \quad \text{(L)}$$

$$[\nabla^2 V]_x = \frac{\partial^2 u_x}{\partial x^2} + \frac{\partial^2 u_x}{\partial y^2} + \frac{\partial^2 u_z}{\partial z^2} \quad \text{(M)}$$

$$[\nabla^2 V]_V = \frac{\partial^2 u_V}{\partial x^2} + \frac{\partial^2 u_V}{\partial y^2} + \frac{\partial^2 u_V}{\partial z^2} \quad \text{(N)}$$

$$[\nabla^2 V]_z = \frac{\partial^2 u_z}{\partial x^2} + \frac{\partial^2 u_z}{\partial y^2} + \frac{\partial^2 u_z}{\partial z^2} \quad \text{(O)}$$

$$[V \cdot \nabla V]_x = u_x \frac{\partial u_x}{\partial x} + u_V \frac{\partial u_x}{\partial y} + u_z \frac{\partial u_z}{\partial z} \quad \text{(P)}$$

$$[V \cdot \nabla V]_V = u_x \frac{\partial u_V}{\partial x} + u_V \frac{\partial u_V}{\partial y} + u_z \frac{\partial u_V}{\partial z} \quad \text{(Q)}$$

$$[V \cdot \nabla V]_z = u_x \frac{\partial u_z}{\partial x} + u_V \frac{\partial u_z}{\partial y} + u_z \frac{\partial u_z}{\partial z} \quad \text{(R)}$$

For Newtonian Fluids

$$\frac{\tau : \nabla V}{\mu} = 2\left[\left(\frac{\partial u_x}{\partial x}\right)^2 + \left(\frac{\partial u_V}{\partial y}\right)^2 + \left(\frac{\partial u_z}{\partial z}\right)^2\right] + \left[\frac{\partial u_V}{\partial x} + \frac{\partial u_x}{\partial y}\right]^2 + \left[\frac{\partial u_z}{\partial y} + \frac{\partial u_r}{\partial z}\right]^2 + \left[\frac{\partial u_x}{\partial z} + \frac{\partial u_z}{\partial x}\right]^2 - \frac{2}{3}\left[\frac{\partial u_x}{\partial x} + \frac{\partial u_y}{\partial y} + \frac{\partial u_z}{\partial z}\right]^2 \quad \text{(S)}$$

Table F.8.9 Summary of Differential Operations Involving the ∇ Operator in Cylindrical Coordinates (r, θ, z)[a]

$$\nabla \cdot V = \frac{1}{r}\frac{\partial}{\partial r}(r u_r) + \frac{1}{r}\frac{\partial u_\theta}{\partial \theta} + \frac{\partial u_z}{\partial z} \qquad \text{(A)}$$

$$\nabla^2 s = \frac{1}{r}\frac{\partial}{\partial r}\left(r\frac{\partial s}{\partial r}\right) + \frac{1}{r^2}\frac{\partial^2 s}{\partial \theta^2} + \frac{\partial^2 s}{\partial z^2} \qquad \text{(B)}$$

$$\begin{aligned} \tau : \nabla V = {} & \tau_{rr}\left(\frac{\partial u_r}{\partial r}\right) + \tau_{\theta\theta}\left(\frac{1}{r}\frac{\partial u_\theta}{\partial \theta} + \frac{u_r}{r}\right) \\ & + \tau_{zz}\left(\frac{\partial u_z}{\partial z}\right) + \tau_{r\theta}\left[r\frac{\partial}{\partial r}\left(\frac{u_\theta}{r}\right) + \frac{1}{r}\frac{\partial u_r}{\partial \theta}\right] \\ & + \tau_{\theta z}\left(\frac{1}{r}\frac{\partial u_z}{\partial \theta} + \frac{\partial u_\theta}{\partial z}\right) + \tau_{rz}\left(\frac{\partial u_z}{\partial r} + \frac{\partial u_r}{\partial z}\right) \end{aligned} \qquad \text{(C)}$$

$$[\nabla s]_r = \frac{\partial s}{\partial r} \qquad \text{(D)}$$

$$[\nabla s]_\theta = \frac{1}{r}\frac{\partial s}{\partial \theta} \qquad \text{(E)}$$

$$[\nabla s]_z = \frac{\partial s}{\partial z} \qquad \text{(F)}$$

$$[\nabla \times V]_r = \frac{1}{r}\frac{\partial u_z}{\partial \theta} - \frac{\partial u_\theta}{\partial z} \qquad \text{(G)}$$

$$[\nabla \times V]_\theta = \frac{\partial u_r}{\partial z} - \frac{\partial u_z}{\partial r} \qquad \text{(H)}$$

$$[\nabla \times V]_z = \frac{1}{r}\frac{\partial}{\partial r}(r u_\theta) - \frac{1}{r}\frac{\partial u_r}{\partial \theta} \qquad \text{(I)}$$

$$[\nabla \cdot \tau]_r = \frac{1}{r}\frac{\partial}{\partial r}(r\tau_{rr}) + \frac{1}{r}\frac{\partial}{\partial \theta}\tau_{\theta\theta} + \frac{\partial \tau_{rz}}{\partial z} \qquad \text{(J)}$$

$$[\nabla \cdot \tau]_\theta = \frac{1}{r}\frac{\partial \tau_{\theta\theta}}{\partial \theta} + \frac{\partial \tau_{r\theta}}{\partial r} + \frac{2}{r}\tau_{r\theta} + \frac{\partial \tau_{\theta z}}{\partial z} \qquad \text{(K)}$$

$$[\nabla \cdot \tau]_z = \frac{1}{r}\frac{\partial}{\partial r}(r\tau_{rz}) + \frac{1}{r}\frac{\partial \tau_{\theta z}}{\partial \theta} + \frac{\partial \tau_{zz}}{\partial z} \qquad \text{(L)}$$

$$[\nabla^2 V]_r = \frac{\partial}{\partial r}\left(\frac{1}{r}\frac{\partial}{\partial r}(r u_r)\right) + \frac{1}{r^2}\frac{\partial^2 u_r}{\partial \theta^2} - \frac{2}{r^2}\frac{\partial u_\theta}{\partial \theta} + \frac{\partial^2 u_r}{\partial z^2} \qquad \text{(M)}$$

$$[\nabla^2 V]_\theta = \frac{\partial}{\partial r}\left(\frac{1}{r}\frac{\partial}{\partial r}(r u_\theta)\right) + \frac{1}{r^2}\frac{\partial^2 u_\theta}{\partial \theta^2} + \frac{2}{r^2}\frac{\partial u_r}{\partial \theta} + \frac{\partial^2 u_\theta}{\partial z^2} \qquad \text{(N)}$$

$$[\nabla^2 V]_z = \frac{1}{r}\frac{\partial}{\partial r}\left(r\frac{\partial u_z}{\partial r}\right) + \frac{1}{r^2}\frac{\partial u_z}{\partial \theta^2} + \frac{\partial^2 u_z}{\partial z^2} \qquad \text{(O)}$$

(continued)

Table F.8.9 Summary of Differential Operations Involving the ∇ Operator in Cylindrical Coordinates (r, θ, z) (Continued)

$$[V\cdot\nabla V]_r = u_r\frac{\partial u_r}{\partial r} + \frac{u_\theta}{r}\frac{\partial u_r}{\partial \theta} - \frac{u_\theta^2}{r} + u_z\frac{\partial u_r}{\partial z} \quad \text{(P)}$$

$$[V\cdot\nabla V]_\theta = u_r\frac{\partial u_\theta}{\partial r} + \frac{u_\theta}{r}\frac{\partial u_\theta}{\partial \theta} + \frac{u_r u_\theta}{r} + u_z\frac{\partial u_\theta}{\partial z} \quad \text{(Q)}$$

$$[V\cdot\nabla V]_z = u_r\frac{\partial u_z}{\partial r} + \frac{u_\theta}{r}\frac{\partial u_z}{\partial \theta} + u_z\frac{\partial u_z}{\partial z} \quad \text{(R)}$$

For Newtonian Fluids

$$\frac{\tau:\nabla V}{\mu} = 2\left[\left(\frac{\partial u_r}{\partial r}\right)^2 + \left(\frac{1}{r}\frac{\partial u_\theta}{\partial \theta} + \frac{u_r}{r}\right)^2 + \left(\frac{\partial u_z}{\partial z}\right)^2\right]$$
$$+\left[r\frac{\partial}{\partial r}\left(\frac{u_\theta}{r}\right) + \frac{1}{r}\frac{\partial u_r}{\partial \theta}\right]^2 + \left[\frac{1}{r}\frac{\partial u_z}{\partial \theta} + \frac{\partial u_\theta}{\partial z}\right]^2 \quad \text{(S)}$$
$$+\left[\frac{\partial u_r}{\partial z} + \frac{\partial u_z}{\partial r}\right]^2 - \frac{2}{3}\left[\frac{1}{r}\frac{\partial}{\partial r}(r u_r) + \frac{1}{r}\frac{\partial u_\theta}{\partial \theta} + \frac{\partial u_z}{\partial z}\right]^2$$

Table F.8.10 Summary of Differential Operations Involving the ∇ Operator in Spherical Coordinates (r, θ, φ)[a]

$$\nabla\cdot V = \frac{1}{r^2}\frac{\partial}{\partial r}\left(r^2 u_r\right) + \frac{1}{r\sin\theta}\frac{\partial}{\partial \theta}(u_\theta \sin\theta) + \frac{1}{r\sin\theta}\frac{\partial u_\phi}{\partial \phi} \quad \text{(A)}$$

$$\nabla^2 s = \frac{1}{r^2}\frac{\partial}{\partial r}\left(r^2\frac{\partial s}{\partial r}\right) + \frac{1}{r^2\sin\theta}\frac{\partial}{\partial \theta}\left(\sin\theta\frac{\partial s}{\partial \theta}\right) + \frac{1}{r^2\sin^2\theta}\frac{\partial^2 s}{\partial \phi^2} \quad \text{(B)}$$

$$\tau:\nabla V = \tau_{rr}\left(\frac{\partial u_r}{\partial r}\right) + \tau_{\theta\theta}\left(\frac{1}{r}\frac{\partial u_\theta}{\partial \theta} + \frac{u_r}{r}\right) + \tau_{\phi\phi}\left(\frac{1}{r\sin\theta}\frac{\partial u_\phi}{\partial \phi} + \frac{u_r}{r} + \frac{u_\theta\cot\theta}{r}\right)$$
$$+\tau_{r\theta}\left(\frac{\partial u_\theta}{\partial r} + \frac{1}{r}\frac{\partial u_r}{\partial \theta} - \frac{u_\theta}{r}\right) + \tau_{r\phi}\left(\frac{\partial u_\phi}{\partial r} + \frac{1}{r\sin\theta}\frac{\partial u_r}{\partial \phi} - \frac{u_\phi}{r}\right) \quad \text{(C)}$$
$$+\tau_{\theta\phi}\left(\frac{1}{r}\frac{\partial u_\phi}{\partial \theta} + \frac{1}{r\sin\theta}\frac{\partial u_\theta}{\partial \phi} - \frac{\cot\theta}{r}u_\phi\right)$$

$$[\nabla s]_r = \frac{\partial s}{\partial r} \quad \text{(D)}$$

$$[\nabla s]_\theta = \frac{1}{r}\frac{\partial s}{\partial \theta} \quad \text{(E)}$$

Table F.8.10 Summary of Differential Operations Involving the ∇ Operator in Spherical Coordinates (r, θ, ϕ) (Continued)

$$[\nabla s]_\phi = \frac{1}{r\sin\theta}\frac{\partial s}{\partial \phi} \tag{F}$$

$$[\nabla \times V]_r = \frac{1}{r\sin\theta}\frac{\partial}{\partial \theta}(u_\phi \sin\theta) - \frac{1}{r\sin\theta}\frac{\partial u_\theta}{\partial \phi} \tag{G}$$

$$[\nabla \times V]_\theta = \frac{1}{r\sin\theta}\frac{\partial u_r}{\partial \phi} - \frac{1}{r}\frac{\partial}{\partial r}(r u_\phi) \tag{H}$$

$$[\nabla \times V]_\phi = \frac{1}{r}\frac{\partial}{\partial r}(r u_\phi) - \frac{1}{r}\frac{\partial u_r}{\partial \theta} \tag{I}$$

$$[\nabla \cdot \tau]_r = \frac{1}{r^2}\frac{\partial}{\partial r}\left(r^2 \tau_{rr}\right) + \frac{1}{r\sin\theta}\frac{\partial}{\partial \theta}(\tau_{r\theta}\sin\theta) + \frac{1}{r\sin\theta}\frac{\partial \tau_{r\phi}}{\partial \phi} - \frac{\tau_{\theta\theta} + \tau_{\phi\phi}}{r} \tag{J}$$

$$[\nabla \cdot \tau]_\theta = \frac{1}{r^2}\frac{\partial}{\partial r}\left(r^2 \tau_{r\theta}\right) + \frac{1}{r\sin\theta}\frac{\partial}{\partial \theta}(\tau_{\theta\theta}\sin\theta) + \frac{1}{r\sin\theta}\frac{\partial \tau_{\theta\phi}}{\partial \phi} + \frac{\tau_{r\theta}}{r} - \frac{\cot\theta}{r}\tau_{\phi\phi} \tag{K}$$

$$[\nabla \cdot \tau]_\phi = \frac{1}{r^2}\frac{\partial}{\partial r}\left(r^2 \tau_{r\phi}\right) + \frac{1}{r}\frac{\partial \tau_{\theta\phi}}{\partial \theta} + \frac{1}{r\sin\theta}\frac{\partial \tau_{\phi\phi}}{\partial \phi} + \frac{\tau_{r\phi}}{r} + \frac{2\cot\theta}{r}\tau_{\theta\phi} \tag{L}$$

$$[\nabla^2 V]_r = \nabla^2 u_r - \frac{2u_r}{r^2} - \frac{2}{r^2}\frac{\partial u_\theta}{\partial \theta} - \frac{2u_\theta \cot\theta}{r^2} - \frac{2}{r^2 \sin^2\theta}\frac{\partial u_\phi}{\partial \phi} \tag{M}$$

$$[\nabla^2 V]_\theta = \nabla^2 u_\theta + \frac{2}{r^2}\frac{\partial u_r}{\partial \theta} - \frac{u_\theta}{r^2 \sin^2\theta} - \frac{2\cos\theta}{r^2 \sin^2\theta}\frac{\partial u_\phi}{\partial \phi} \tag{N}$$

$$[\nabla^2 V]_\phi = \nabla^2 u_\phi - \frac{u_\phi}{r^2 \sin^2\theta} + \frac{2}{r^2 \sin\theta}\frac{\partial u_r}{\partial \phi} + \frac{2\cos\theta}{r^2 \sin^2\theta}\frac{\partial u_\theta}{\partial \phi} \tag{O}$$

$$[V \cdot \nabla V]_r = u_r \frac{\partial u_r}{\partial r} + \frac{u_\theta}{r}\frac{\partial u_r}{\partial \theta}\frac{\partial u_r}{\partial \phi} - \frac{u_\theta^2 + u_\phi^2}{r} \tag{P}$$

$$[V \cdot \nabla V]_\theta = u_r \frac{\partial u_\theta}{\partial r} + \frac{u_\theta}{r}\frac{\partial u_\theta}{\partial \theta} + \frac{u_\phi}{r\sin\theta}\frac{\partial u_\theta}{\partial \phi} + \frac{u_r u_\theta}{r} - \frac{u_\phi^2 \cot\theta}{r} \tag{Q}$$

$$[V \cdot \nabla V]_\phi = u_r \frac{\partial u_\phi}{\partial r} + \frac{u_\theta}{r}\frac{\partial u_\phi}{\partial \theta} + \frac{u_\phi}{r\sin\theta}\frac{\partial u_\phi}{\partial \phi} + \frac{u_\phi u_r}{r} + \frac{u_\theta u_\phi \cot\theta}{r} \tag{R}$$

For Newtonian Fluids

$$\begin{aligned} \frac{\tau : \nabla V}{\mu} = {} & 2\left[\left(\frac{\partial u_r}{\partial r}\right)^2 + \left(\frac{1}{r}\frac{\partial u_\theta}{\partial \theta} + \frac{u_r}{r}\right)^2 + \left(\frac{1}{r\sin\theta}\frac{\partial u_\phi}{\partial \phi} + \frac{u_r}{r} + \frac{u_\theta \cot\theta}{r}\right)^2\right] \\ & + \left[r\frac{\partial}{\partial r}\left(\frac{u_\theta}{r}\right) + \frac{1}{r}\frac{\partial u_r}{\partial \theta}\right]^2 + \left[\frac{\sin\theta}{r}\frac{\partial}{\partial \theta}\left(\frac{u_\phi}{\sin\theta}\right) + \frac{1}{r\sin\theta}\frac{\partial u_\theta}{\partial \phi}\right]^2 \\ & + \left[\frac{1}{r\sin\theta}\frac{\partial u_r}{\partial \phi} + r\frac{\partial}{\partial r}\left(\frac{u_\phi}{r}\right)\right]^2 \\ & - \frac{2}{3}\left[\frac{1}{r^2}\frac{\partial}{\partial r}\left(r^2 u_r\right) + \frac{1}{r\sin\theta}\frac{\partial}{\partial \theta}(u_\theta \sin\theta) + \frac{1}{r\sin\theta}\frac{\partial u_\phi}{\partial \phi}\right]^2 \end{aligned} \tag{S}$$

Conservation of Energy

Table F.8.11 Energy Equation in Several Coordinate Systems for Multicomponent Mixture[a,c]

Rectangular Coordinates (x, y, z)

$$\rho c_p\left(\frac{\partial T}{\partial t}+u_x\frac{\partial T}{\partial x}+u_y\frac{\partial T}{\partial y}+u_z\frac{\partial T}{\partial z}\right)-\left(\frac{\partial p}{\partial t}+u_x\frac{\partial p}{\partial x}+u_y\frac{\partial p}{\partial y}+u_z\frac{\partial p}{\partial z}\right)$$
$$=\lambda\left(\frac{\partial^2 T}{\partial x^2}+\frac{\partial^2 T}{\partial y^2}+\frac{\partial^2 T}{\partial z^2}\right)-\sum_{i=1}^{N}\omega_i\Delta h^\circ_{f,i}$$
$$-\left[\frac{\partial}{\partial x}\left(\rho T\sum_{i=1}^{N}c_{pi}Y_iV_{ix}\right)+\frac{\partial}{\partial y}\left(\rho T\sum_{i=1}^{N}c_{pi}Y_iV_{iy}\right)+\frac{\partial}{\partial z}\left(\rho T\sum_{i=1}^{N}c_{pi}Y_iV_{iz}\right)\right]$$
$$+\mu\left\{2\left[\left(\frac{\partial u_x}{\partial x}\right)^2+\left(\frac{\partial u_y}{\partial y}\right)^2+\left(\frac{\partial u_z}{\partial z}\right)^2\right]+\left[\frac{\partial u_y}{\partial x}+\frac{\partial u_x}{\partial y}\right]^2+\left[\frac{\partial u_z}{\partial y}+\frac{\partial u_y}{\partial z}\right]^2\right.$$
$$\left.+\left[\frac{\partial u_x}{\partial z}+\frac{\partial u_z}{\partial x}\right]^2-\frac{2}{3}\left[\frac{\partial u_x}{\partial x}+\frac{\partial u_y}{\partial y}+\frac{\partial u_z}{\partial z}\right]^2\right\}+\rho\sum_{i=1}^{N}Y_i(f_{ix}V_{ix}+f_{iy}V_{iy}+f_{iz}V_{iz}) \quad \text{(A)}$$

Cylindrical Coordinates (r, θ, z)

$$\rho c_p\left(\frac{\partial T}{\partial t}+u_r\frac{\partial T}{\partial r}+\frac{u_\theta}{r}\frac{\partial T}{\partial \theta}+u_z\frac{\partial T}{\partial z}\right)-\left(\frac{\partial p}{\partial t}+u_r\frac{\partial p}{\partial r}+\frac{u_\theta}{r}\frac{\partial p}{\partial \theta}+u_z\frac{\partial p}{\partial z}\right)$$
$$=\lambda\left[\frac{1}{r}\frac{\partial}{\partial r}\left(r\frac{\partial T}{\partial r}\right)+\frac{1}{r^2}\frac{\partial^2 T}{\partial \theta^2}+\frac{\partial^2 T}{\partial z^2}\right]-\sum_{i=1}^{N}\omega_i\Delta h^\circ_{f,i}$$
$$-\left[\frac{1}{r}\frac{\partial}{\partial r}\left(r\rho T\sum_{i=1}^{N}c_{pi}Y_iV_{ir}\right)+\frac{1}{r}\frac{\partial}{\partial \theta}\left(\rho T\sum_{i=1}^{N}c_{pi}Y_iV_{i\theta}\right)+\frac{\partial}{\partial z}\left(\rho T\sum_{i=1}^{N}c_{pi}Y_iV_{iz}\right)\right]$$
$$+\mu\left\{2\left[\left(\frac{\partial u_r}{\partial r}\right)^2+\left(\frac{1}{r}\frac{\partial u_\theta}{\partial \theta}+\frac{u_r}{r}\right)^2+\left(\frac{\partial u_z}{\partial z}\right)^2\right]+\left[r\frac{\partial}{\partial r}\left(\frac{u_\theta}{r}\right)+\frac{1}{r}\frac{\partial u_r}{\partial \theta}\right]^2\right.$$
$$\left.+\left[\frac{1}{r}\frac{\partial u_z}{\partial \theta}+\frac{\partial u_\theta}{\partial z}\right]^2+\left[\frac{\partial u_r}{\partial z}+\frac{\partial u_z}{\partial r}\right]^2-\frac{2}{3}\left[\frac{1}{r}\frac{\partial}{\partial r}(ru_r)+\frac{1}{r}\frac{\partial u_\theta}{\partial \theta}+\frac{\partial u_z}{\partial z}\right]^2\right\} \quad \text{(B)}$$
$$+\rho\sum_{i=1}^{N}Y_i\left(f_{ir}V_{ir}+f_{i\theta}V_{i\theta}+f_{iz}V_{iz}\right)$$

(continued)

Table F.8.11 Energy Equation in Several Coordinate Systems for Multicomponent Mixture (Continued)

Spherical Coordinates (r, θ, ϕ)

$$\rho c_p\left(\frac{\partial T}{\partial t}+u_r\frac{\partial T}{\partial r}+\frac{u_\theta}{r}\frac{\partial T}{\partial \theta}+\frac{u_\phi}{r\sin\theta}\frac{\partial T}{\partial \phi}\right)-\left(\frac{\partial p}{\partial t}+u_r\frac{\partial p}{\partial r}+\frac{u_\theta\partial p}{r\partial \theta}+\frac{u_\phi}{r\sin\theta}\frac{\partial p}{\partial \phi}\right)$$

$$=\lambda\left[\frac{1}{r^2}\frac{\partial}{\partial r}\left(r^2\frac{\partial T}{\partial r}\right)+\frac{1}{r^2\sin\theta}\frac{\partial}{\partial \theta}\left(\sin\theta\frac{\partial T}{\partial \theta}\right)+\frac{1}{r^2\sin^2\theta}\frac{\partial^2 T}{\partial \phi^2}\right]-\sum_{i=1}^{N}\omega_i\Delta h^\circ_{f,i}$$

$$-\left[\frac{1}{r^2}\frac{\partial}{\partial r}\left(r^2\rho T\sum_{i=1}^{N}c_{pi}Y_iV_{ir}\right)+\frac{1}{r\sin\theta}\frac{\partial}{\partial \theta}\left(\rho T\sin\theta\sum_{i=1}^{N}c_{pi}Y_iV_{i\theta}\right)\right.$$

$$\left.+\frac{1}{r\sin\theta}\frac{\partial}{\partial \phi}\left(\rho T\sum_{i=1}^{N}c_{pi}Y_iV_{i\phi}\right)\right]$$

$$+\mu\left\{2\left[\left(\frac{\partial u_r}{\partial r}\right)^2+\left(\frac{1}{r}\frac{\partial u_\theta}{\partial \theta}+\frac{u_r}{r}\right)^2+\left(\frac{1}{r\sin\theta}\frac{\partial u_\phi}{\partial \phi}+\frac{u_r}{r}+\frac{u_\theta\cot\theta}{r}\right)^2\right]\right.$$

$$\left.+\left[\frac{1}{r\sin\theta}\frac{\partial u_r}{\partial \phi}+r\frac{\partial}{\partial r}\left(\frac{u_\phi}{r}\right)\right]^2-\frac{2}{3}\left[\frac{1}{r^2}\frac{\partial}{\partial r}(ru_r)+\frac{1}{r\sin\theta}\frac{\partial}{\partial \theta}(u_\theta\sin\theta)+\frac{1}{r\sin\theta}\frac{\partial u_\phi}{\partial \phi}\right]^2\right\} \qquad \text{(C)}$$

$$+\rho\sum_{i=1}^{N}Y_i(f_{ir}V_{ir}+f_{i\theta}V_{i\theta}+f_{i\phi}V_{i\phi})$$

[a] Adapted from Bird et al., 1960.

[b] $\nabla^2=\frac{1}{r^2}\left(r^2\frac{\partial}{\partial r}\right)+\frac{1}{r^2\sin\theta}\frac{\partial}{\partial\theta}\left(\sin\theta\frac{\partial}{\partial\theta}\right)+\frac{1}{r^2\sin^2\theta}\left(\frac{\partial^2}{\partial\phi^2}\right)$.

[c] (1) Ideal gas law, (2) constant c_p and λ, (3) Dufour effect negligible, (4) no radiation effect [Kuo, 1986].

Simplified Conservation Equations

Overall Mass Conservation Equation (Continuity Equation)

$$\frac{\partial\rho}{\partial t}+\frac{\partial(\rho v_x)}{\partial x}+\frac{\partial(\rho v_y)}{\partial y}+\frac{\partial(\rho v_z)}{\partial z}=0$$

$$\frac{\partial\rho}{\partial t}+\vec{\nabla}\cdot\rho\vec{V}=0$$

Species Conservation Equation

Species Density Form

$$\frac{\partial\rho_k}{\partial t}+\frac{\partial(\rho_k v_{kx})}{\partial x}+\frac{\partial(\rho_k v_{ky})}{\partial y}+\frac{\partial(\rho_k v_{kz})}{\partial z}=\dot{\omega}_k'''$$

$$\frac{\partial\rho_k}{\partial t}+\vec{\nabla}\cdot\rho_k\vec{v}_k=\dot{\omega}_k'''$$

Mass Fraction Form

$$\rho\frac{\partial Y_k}{\partial t}+\rho v_x\frac{\partial Y_k}{\partial x}+\rho v_y\frac{\partial Y_k}{\partial y}=\frac{\partial}{\partial x}\left\{\rho D\frac{\partial Y_k}{\partial x}\right\}+\frac{\partial}{\partial y}\left\{\rho D\frac{\partial Y_k}{\partial y}\right\}+\dot{\omega}_k'''$$

$$\rho\frac{\partial Y_k}{\partial t}+\rho\vec{v}\cdot\vec{\nabla}Y_k=-\vec{\nabla}\cdot\vec{j}_k''+\dot{\omega}_k'''$$

Energy Conservation Equation

Total Enthalpy Form

$$\rho\frac{\partial h}{\partial t}+\rho v_x\frac{\partial h}{\partial x}+\rho v_y\frac{\partial h}{\partial y}=\frac{\partial}{\partial x}\left\{\frac{\lambda}{c_P}\frac{\partial h}{\partial x}\right\}+\frac{\partial}{\partial y}\left\{\frac{\lambda}{c_P}\frac{\partial h}{\partial y}\right\}$$

$$\rho\frac{\partial h}{\partial t}+\rho\vec{v}\cdot\vec{\nabla}h=-\vec{\nabla}\cdot\left\{\frac{\lambda}{c_P}\vec{\nabla}h\right\}$$

Thermal Enthalpy Form

$$\rho\frac{\partial h_T}{\partial t}+\rho v_x\frac{\partial h_T}{\partial x}+\rho v_y\frac{\partial h_T}{\partial y}=\frac{\partial}{\partial x}\left\{\frac{\lambda}{c_P}\frac{\partial h_T}{\partial x}\right\}+\frac{\partial}{\partial y}\left\{\frac{\lambda}{c_P}\frac{\partial h_T}{\partial y}\right\}+\left|\dot{\omega}_F'''\right|h_c$$

$$\rho\frac{\partial h_T}{\partial t}+\rho\vec{v}\cdot\vec{\nabla}h_T=-\vec{\nabla}\cdot\left\{\frac{\lambda}{c_P}\vec{\nabla}h_T\right\}+\left|\dot{\omega}_F'''\right|h_c$$

Element Conservation Equation

$$\rho\frac{\partial Z_j}{\partial t}+\rho v_x\frac{\partial Z_j}{\partial x}+\rho v_y\frac{\partial Z_j}{\partial y}=\frac{\partial}{\partial x}\left\{\rho D\frac{\partial Z_j}{\partial x}\right\}+\frac{\partial}{\partial y}\left\{\rho D\frac{\partial Z_j}{\partial y}\right\},$$

$$\rho\frac{\partial Z_j}{\partial t}+\rho\vec{v}\cdot\vec{\nabla}Z_j=\vec{\nabla}\cdot\rho D\vec{\nabla}Z_j$$

where $Z_j=\sum_k\alpha_{j,k}Y_k,\alpha_{j,k}$: element mass j in unit mass of species k

Boundary-Layer Conservation Equations

Momentum equation in the X direction

$$\frac{\partial\rho}{\partial t}+\rho v_x\frac{\partial v_x}{\partial x}+\rho v_y\frac{\partial v_y}{\partial y}=\frac{\partial}{\partial y}\left\{\mu\frac{\partial v_x}{\partial y}\right\}-\frac{\partial P}{\partial x}$$

Species Conservation Equation

$$\rho\frac{\partial Y_k}{\partial t}+\rho v_x\frac{\partial Y_k}{\partial x}+\rho v_y\frac{\partial Y_k}{\partial y}=\frac{\partial}{\partial y}\left\{\rho D\frac{\partial Y_k}{\partial y}\right\}+\dot{\omega}_k'''$$

Energy Conservation Equation

Thermal Enthalpy Form

$$\rho \frac{\partial h_T}{\partial t} + \rho v_x \frac{\partial h_T}{\partial x} + \rho v_y \frac{\partial h_T}{\partial y} = \frac{\partial}{\partial y}\left\{ \frac{\lambda}{c_P} \frac{\partial h_T}{\partial y} \right\} + \left| \dot{\omega}_F''' \right| h_c$$

Element Conservation Equation

$$\rho \frac{\partial Z_j}{\partial t} + \rho v_x \frac{\partial Z_j}{\partial x} + \rho v_y \frac{\partial Z_j}{\partial y} = \frac{\partial}{\partial y}\left\{ \rho D \frac{\partial Z_j}{\partial y} \right\}$$

Mixture Fraction

$$f = \sum_{fuel\ element\ j} Z_j, \quad j = C, H \ for\ HC.$$

Air Mass Fraction

$$= 1 - f$$

Local Fuel–Air Ratio for Mixing Problem

$$\left.\frac{F}{A}\right)_{local} = \frac{f}{1-f}$$

Local Equivalence Ratio

$$\Phi_{\mathrm{local}} = \left\{ \frac{f}{1-f} \right\} \frac{1}{(F/A)_{\mathrm{stoichiometric}}}$$

Mixture Fraction Conservation Equation

$$\frac{\partial \rho}{\partial t} + \rho v_x \frac{\partial f}{\partial x} + \rho v_y \frac{\partial f}{\partial y} = \frac{\partial}{\partial x}\left\{ \rho D \frac{\partial f}{\partial x} \right\} + \frac{\partial}{\partial y}\left\{ \rho D \frac{\partial f}{\partial y} \right\}$$

Relationship between Enthalpy, Mixture Fraction, and Z_j within the Boundary Layer

$$\frac{h - h_\infty}{h_w - h_\infty} = \frac{f - f_\infty}{f_w - f_\infty} = \frac{Z_j - Z_{j,\infty}}{Z_{j,w} - Z_{j,\infty}}$$

Shvab–Zeldovich Formulations

$$\rho v_x \frac{\partial \beta}{\partial x} + \rho v_y \frac{\partial \beta}{\partial y} = \frac{\partial}{\partial y}\left\{ \rho D \frac{\partial \beta}{\partial y} \right\}$$

Shvab–Zeldovich Variable

$$\beta_{ik} = \frac{Y_i}{\nu_i} - \frac{Y_k}{\nu_k} \quad (\nu = \text{negative number if reactant, positive if product.})$$

$$\beta_{h_t-k} = \frac{h_t}{h_c} - \frac{Y_k}{\nu_k}$$

Normalized Shvab–Zeldovich Variable

$$\phi = \frac{\beta - \beta_\infty}{\beta_w - \beta_\infty}$$

$$\dot{m}_k'' = \rho_k v_k = \rho_k v - \rho D\,(\partial Y_k/\partial y)$$

$$\dot{m}'' = -\rho DB\,(\partial\phi/\partial y)_w \approx h_m B = (B/R_{diff}),\ R_{diff} = (1/hm)$$

$B = (\beta_w - \beta_\infty)/(F - \beta_w)$, F from interface balance, see Chapter 10 formulae

$\beta_w - \rho D\,(\partial\beta/\partial y)_0 = F$, where F changes depending upon β; see Chapter 10 for list of F

$Sh/Sh_0 = h_m/h_{mo} = \{\ln(1 + B)\}/B$

$Y_{F,w} = 1 - \{(1 + s)/(1 + B)\}$, $s = Y_{O2,\infty}/\nu_{O2}$

Table F.8.12 Some Recommended Heat Transfer Relations for Fluid Flows

These properties are recommended to be evaluated at some mean temperatures; adapted from Kanury and modified.

$$\mathrm{Re} \equiv \frac{ux}{\nu},\ Nu_x \equiv \frac{h(x)x}{k},\ \bar{N}u_x \equiv \frac{\bar{h}(x)\,x}{k} = \frac{Nu(x)}{n+1} \text{ if } Nu(x) \propto x^n,$$

$$Gr \equiv \frac{gx^3\beta\Delta T}{\upsilon^2},\ \mathrm{Pr} \equiv \frac{\nu}{x}$$

$$Sh_x \equiv \frac{h_m(x)x}{\rho D},\ Sh_{0,x} \equiv \frac{h_{m,0}(x)x}{\rho D},\ \frac{Sh_x}{Sh_{0,x}} = \frac{\ln(1+B)}{B}$$

Modified Reynolds Analogy, external

$$f_F = \frac{P_1 - P_2}{(\rho V^2/2)}\frac{D_H}{L}, \text{where } D_H \text{ is the hydraulic mean depth},\ f_F\text{: friction factor}$$

For internal Flows

$$St = \frac{f_F}{2},\ St = \frac{h_H}{V\rho c_p},\ f_F = \frac{\tau}{\rho V^2/2}$$

Modified Chilton-Colburn Analogy [Spalding, D.B., 1955.]

$$\frac{f_F}{2} = St\,\mathrm{Pr}^{2/3} = \frac{\dot{m}''Sc^{2/3}}{\rho V\ln(1+B)}$$

For noncircular geometry, use hydraulic diameter D = 4 × A/C, where C is the perimeter and A is the sectional area.

Table F.8.12 Some Recommended Heat Transfer Relations for Fluid Flows (Continued)

Nature of Flow	*Diagram*	*Result*
Forced Convection Critical Reynolds Numbers: Flat Plate: $3.2-10^5$, Cylinder: $5-10^5$, Sphere: $3-10^5$		
1. Laminar flow parallel to a flat plate heated over $\mathrm{Re} \equiv u_{xo}x/v < 3 \cdot 10^5$	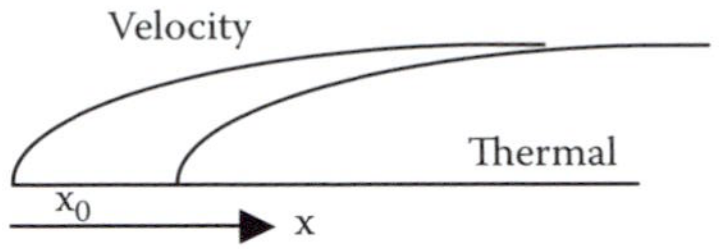	$Nu_x = \dfrac{0.332\mathrm{Re}^{1.2}\mathrm{Pr}^{1/3}}{\sqrt[3]{\left[1-\left(\frac{x_0}{x}\right)^{3/4}\right]}}$
2. Same as item 1 but $x_0 = 0$		$Nu_x = 0.332\mathrm{Re}^{1/2}\mathrm{Pr}^{1/3}$, $\mathrm{Re}_x < 5 \times 10^5$ $\overline{Nu}_x = 2 * Nu_x$
3. Same as item 1 but flow is turbulent		Tubulent Regime only, $\bar{N}u_x = 0.037\mathrm{Re}^{4/5}\mathrm{Pr}^{1/3}$ $0.6 < \mathrm{Pr} < 60$, $5 \times 10^5 < \mathrm{Re}_x < 10^7$
4. Laminar plane (2-D) stagnation point flow: $\mathrm{Re} \equiv ux/v$	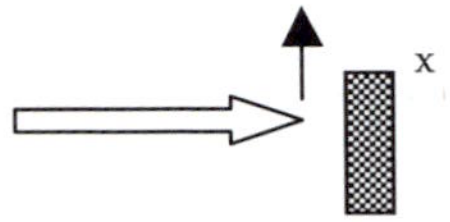	$Nu_x = 0.57\mathrm{Re}^{1/2}\mathrm{Pr}^{2/5}$ around stagnation line $(dv_x/dx)_{x=0} = a = 4\ U/d$
5. Axisymmetric stagnation point flow $\mathrm{Re} \equiv uR/v$ R = Radius of curvature over sphere $V = ax$, $a = 3u/d$	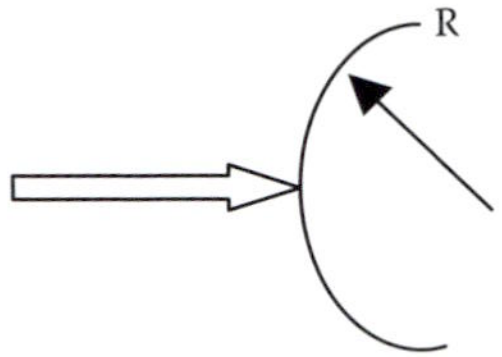	$Nu_x = 0.93\mathrm{Re}^{1/2}\mathrm{Pr}^{2/5}$ $Nu_{stagn} \cong d\sqrt{\dfrac{a}{\alpha_T}}, Sh = d\sqrt{\dfrac{a}{D}}$

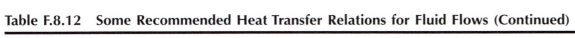

(continued)

Table F.8.12 Some Recommended Heat Transfer Relations for Fluid Flows (Continued)

Nature of Flow	*Diagram*	*Result*
6. Flow around a sphere $Re \equiv u\, d/v$ $17 < Re < 7 \cdot 10^4$	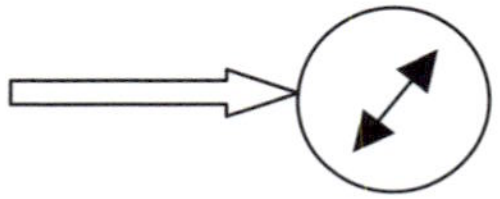	$\overline{Nu}_D = 2 + [0.4Re_D^{1/2} + 0.06Re^{2/3}]Pr^{0.4}\left(\frac{\mu_\infty}{\mu_w}\right)^{1/4}$ 3.5 < Re < 80000, 0.7 < Pr < 380
7. Flow around a cylinder $Re \equiv u_x d/v$	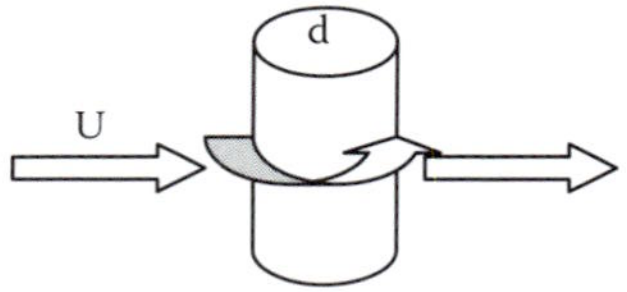	$\overline{Nu}_D = 0.3 + \frac{0.62Re_D^{1/2}Pr^{1/3}}{[1+(0.4/Pr)^{2/3}]^{1/4}}\left[1+\left(\frac{Re_D}{282{,}000}\right)^{5/8}\right]^{4/5}$ Re Pr > 0.2, [Churchill and Bernstein]
8. Internal flows through pipe	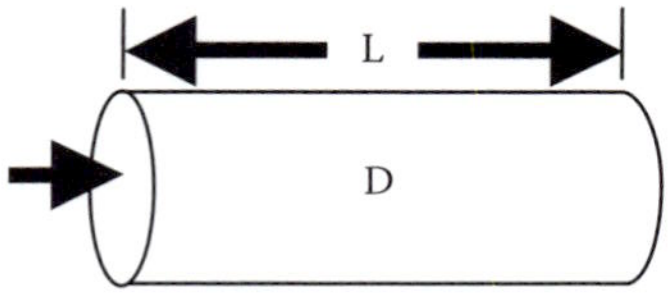	Nu = 3.66 laminar constant surface Ts Nu = 4.36 laminar constant heat flux $Nu_D = 0.023Re_D^{0.8}Pr^n$, Turbulent
9. Free convection on a flat plate $Gr \equiv \frac{gx^3\beta\Delta T}{v^2}$, vertical		$\bar{Nu}_x = 0.59\,(Gr\,Pr)^{1/4}$, $10^4 < Gr\,Pr < 10^9$ $\bar{Nu}_x = 0.13(Gr\,Pr)^{1/3}$, turbulent
$Gr \equiv \frac{gx^3\beta\Delta T}{v^2}\cos\alpha$; inclined, α with vertical	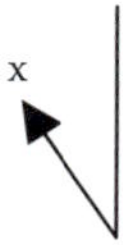	

10. Laminar hot horizontal plate heated face up. $Gr \equiv \frac{gL^3 \beta \Delta T}{\nu^2}$	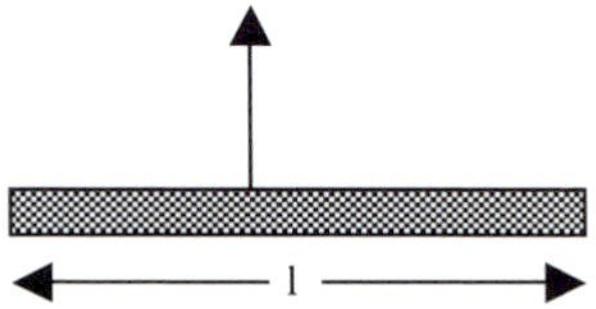	$\overline{Nu}_L = 0.54\,(Gr\,\mathrm{Pr})^{1/4}$, $10^5 < Gr\,\mathrm{Pr} < 2 \cdot 10^7$ $\overline{Nu}_L = 0.14\,(Gr\,\mathrm{Pr})^{1/3}$, turbulent, $2 \cdot 10^7 < Gr\,\mathrm{Pr} < 3 \cdot 10^{10}$
11. Horizontal plate heated face down. $Gr \equiv \frac{gL^3 \beta \Delta T}{\nu^2}$	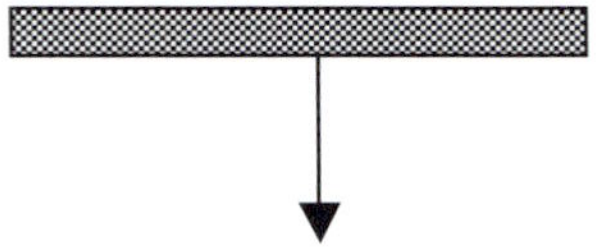	$\overline{Nu}_L = 0.27\,(Gr\,\mathrm{Pr})^{1/4}$, $3 \cdot 10^5 < Gr\,\mathrm{Pr} < 3 \cdot 10^{10}$
12. Laminar free convection around a heated horizontal cylinder $Gr \equiv gd^3 \beta \Delta T / \nu^2$	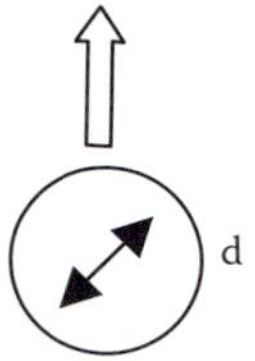	$\overline{Nu}_d = 0.525\,(Gr\,\mathrm{Pr})^{1/4}$, $10^3 < Gr\,\mathrm{Pr} < 10^9$
13. Free convection around a sphere $Gr \equiv gd^3 \beta \Delta T / \nu^2$	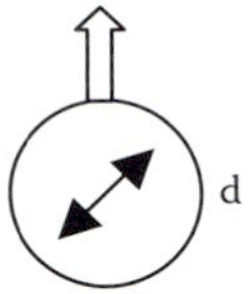	$\overline{Nu}_d = 2\{1 + 0.3 Gr^{1/4}\,\mathrm{Pr}^{1/3}\}$, $Ra < 10^9$ $Gr = 0, Nu_d = 2$ (stagnant case) $\overline{Nu}_d = 0.098\,(\mathrm{Gr\,Pr})^{0.345}$, $3 \cdot 10^8 < Gr\,\mathrm{Pr} < 5 \cdot 10^{11}$
14. Horizontal parallel plates; lower plate hot $Gr \equiv gL^3 \beta \Delta T / \upsilon^2$		$Nu_L = 1$, $Gr \equiv gL^3 \beta \Delta T / \upsilon^2$

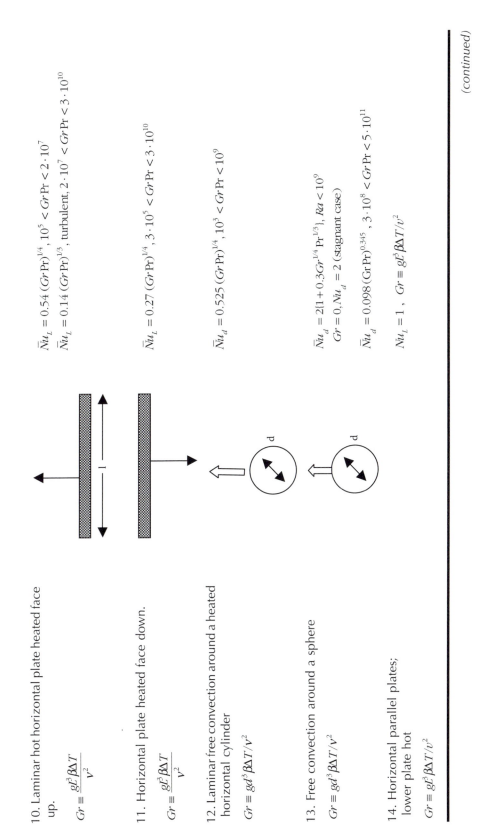

(continued)

Table F.8.12 Some Recommended Heat Transfer Relations for Fluid Flows (Continued)

Nature of Flow	*Diagram*	*Result*
	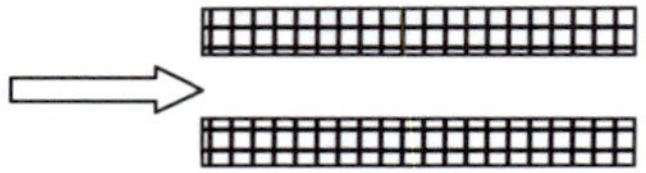	$\bar{Nu}_\Delta = 0.21(\mathrm{Gr\,Pr})^{1/4}$, $10^4 < Gr < 3.2 \cdot 10^5$
		$\bar{Nu}_d = 0.075(\mathrm{Gr\,Pr})^{1/3}$, $3.2 \cdot 10^5 < Gr < 10^7$
15. Vertical parallel plates		$Nu_L = 1$, $Gr < 2 \cdot 10^3$
		$Nu_L = 0.2(Gr\,\mathrm{Pr})^{1/4}$, $2 \cdot 10^3 < Gr < 2.1 \cdot 10^5$
		$Nu_L = 0.0071(Gr\,\mathrm{Pr})^{1/3}$, $2.1 \cdot 10^5 < Gr < 1.1 \cdot 10^7$

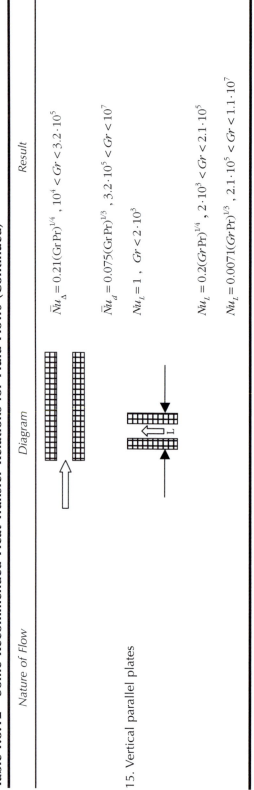

Note: Burning problems for various geometries solved using heat transfer analogy and the above relations.

Chapter 9: Carbon Combustion

SZ Variable and "F" Functions for Carbon Combustion

$$\beta_w + \left(\frac{d\beta}{d\xi}\right)_w = F$$

$$\xi_w = \frac{\dot{m}}{4\pi\rho D a} = \ln(1 + B_F),\ B_F = (\beta_W - \beta_\infty)/(F - \beta_W);\ B_F = B \text{ for thin flames}$$

For DFM or SFM

$$B = (Y_{O_2,\infty}/\nu_{O_2,I}) + (Y_{CO_2,\infty}/\nu_{CO2,III})$$

$$-h_{t,\infty} = (h_{c,I} Y_{O_2,\infty}/\nu_{O2,I}) + (Y_{CO_2,\infty} h_{c,III}/\nu_{CO2,III}) - (\dot{q}''_{rad} B/\dot{m}''),\ h_{c,III} < 0,\ h_{t,\infty} = c_p(T_\infty - T_w)$$

$$Y_{COw} = 1 - \frac{Y_{I,\infty}}{1+B}$$

$$\phi = \frac{\beta - \beta_\infty}{\beta_w - \beta_\infty} = \frac{1 - \exp(-\xi)}{1 - \exp(-\xi_w)}$$

Flame Radius for DFM

$$\frac{a}{r_f} = 1 - \frac{\ln\left[1 + B\frac{B(1+\nu_{CO2,III}) - \frac{Y_{O2,\infty}}{\nu_{O2,V}}}{s + B(1+\nu_{CO2,III}+s) - \frac{Y_{O2,\infty}}{\nu_{O2,V}}}\right]}{\ln[1+B]},\ \text{with } Y_{CO_2,\infty} = 0,\ \frac{r_f}{a} = \frac{\ln[1+B]}{\ln\left[1+\frac{B}{2}\right]},\ \frac{Y_{O_2,\infty}}{S = \nu_{O_2,V}}$$

$$\frac{c_p(T_f - T_\infty)}{h_{C,I}} = \frac{Y_{O2,\infty}}{\nu_{O2,V}}\frac{h_{C,V}}{h_{C,I}} + B\frac{s(1+B)}{1 + B + s(1+B) - Y_{I,\infty}}\left[\frac{h_{C,III}}{h_{C,I}} - \frac{\dot{q}''_{rad}}{\dot{m}'' h_{C,I}}\right]$$

CO_2 Mass Fraction at Flame for DFM

$$Y_{CO_2,f} = \beta[1 - \phi_f]\nu_{Co_2,III}$$

d^2 Law

$$d^2 = d_0^2 - \alpha_c t,\ \alpha_c = 8\rho D \ln(1 + B)/\rho_C = 4 Sh_0 \rho D \ln(1 + B)/\rho_C$$

Chapter 10: Drop Combustion

$$\frac{d\beta}{d\xi} + \frac{d^2\beta}{d\xi^2} = 0$$

where

$$\xi = \frac{\dot{m}}{4\pi\rho D r} \quad \text{and} \quad \beta_{ij} = \frac{Y_i}{\nu_i} - \frac{Y_j}{\nu_j} \text{ for species, } \beta_{hT-i} = \frac{h_T}{h_c} - \frac{Y_i}{\nu_i}$$

Normalized Shvab–Zeldovich Variable

$$\phi = \frac{\beta - \beta_\infty}{\beta_w - \beta_\infty} = \frac{1-\exp(-\xi)}{1-\exp(-\xi_w)}$$

$$\xi_w = \frac{\dot{m}}{4\pi\rho D a} = \ln(1+B)$$

Diffusion-Controlled Combustion

$$a < r \le r_f \quad Y_{O_2} = 0$$
$$r_f \le r < r_\infty \quad Y_F = 0$$

$$\beta_w + \left(\frac{d\beta}{d\xi}\right)_w = F$$

$$\xi_w = \frac{\dot{m}}{4\pi\rho D a} = \ln(1+\mathrm{B}), \quad \text{where } \beta = \left\{\frac{\beta_w - \beta_\infty}{F - \beta_w}\right\}$$

Drop Combustion Rate

(*Diffusion-Controlled Combustion*)

$$\xi_w = \frac{\dot{m}}{4\pi\rho D a} = \frac{\dot{m}}{2Sh\pi\rho D a} = \ln(1+B)$$

Table F.10.1 SZ Variable and "F" Functions for Liquids and Polymer Flames

#	*Shvab—Zeldovich β*	F	Remarks	Thin Flames
1	β_{ht-O2}	$-L/h_c$	$\beta_{ht-O2} = h_t/h_c + Y_{O2}/\nu_{O2}$	$\beta_w = 0$
2	β_{ht-F}	$Y_{F,\ell} - L/h_c$	$\beta_{ht-F} = h_t/h_c + Y_F$	
3	β_{O2-F}	$Y_{F,\ell}$	$\beta_{O2-F} = Y_F - Y_{O2}/\nu_{O2}$	$\beta_w = Y_{Fw}$
4	β_{CO2-F}	$Y_{F,\ell}$	$\beta_{CO2-F} = Y_F + Y_{CO2}/\nu_{CO2}$	
5	β_{H2O-F}	$Y_{F,\ell}$	$\beta_{H2O-F} = Y_F + Y_{H2O}/\nu_{H2O}$	
6	β_{ht-CO2}	$-L/h_c$	$\beta_{ht-CO2} = h_t/h_c - Y_{CO2}/\nu_{CO2}$	
7	β_{ht-H2O}	$-L/h_c$	$\beta_{ht-H2O} = h_t/h_c - Y_{H2O}/\nu_{H2O}$	
8	β_{CO2-O2}	0	$\beta_{CO2-O2} = Y_{CO2}/\nu_{CO2} + Y_{O2}/\nu_{O2}$	$\beta_w = Y_{CO2,w}/\nu_{CO2}$
9	β_{H2O-O2}	0	$\beta_{ht-H2O} = Y_{H2O}/\nu_{H2O} + Y_{O2}/\nu_{O2}$	$\beta_w = Y_{H2O,w}/\nu_{H2O}$

Note: The F's for thin or thick flames, $B = (\beta_w - \beta_\infty)/(F - \beta_w)$; $B_\beta = B$ for thin flames; $Y_{F\ell}$ is typically unity for pure fuels.

Transfer Number

$$B=\frac{h_{T,\infty}}{L}+\frac{Y_{O_2,\infty}}{\nu_{O_2}}\frac{h_c}{L}=\frac{c_p(T_\infty-T_w)}{L}+\frac{Y_{O_2,\infty}}{\nu_{O_2}}\frac{h_c}{L}, \text{ combustion}$$

For evaporation rate: set $h_c = 0$ in B number

d^2 Law

$$d^2=d_0^2-\alpha_k t$$

α_k = rate constant

k = e for evaporation; and k = c for combustion

$$\alpha_k = 8\,\rho D \ln(1+B)/\rho_\ell = 4\,Sh_0\,\rho D \ln(1+B)/\rho_\ell$$

Flame Location (Radius)

(See Table F.10.2)

$$\xi_f=\frac{\dot{m}}{4\pi\rho D r_f}=\ell n(1+s)$$

$$\phi_f=\frac{Y_{O_2,\infty}/\gamma_{O_2}}{\left(\frac{Y_{O_2,\infty}}{\gamma_{O_2}}+Y_{F,w}\right)}=\frac{(\beta_{h_T-F})_f-(\beta_{h_T-F})_\infty}{(\beta_{h_T-F})_w-(\beta_{h_T-F})_\infty},\quad s=Y_{O_2,\infty}/\nu_{O_2}>0,$$

Table F.10.2 Drop Combustion Results

Drop	*Evaporation*	*Combustion*
Transfer number, B	$B=\frac{c_p(T_\infty-T_w)}{L}$	$B=\frac{c_p(T_\infty-T_w)}{L}+\frac{h_c s}{L}$
Mass fraction at drop surface, $Y_{F,\,wall}$	$1-\frac{1}{1+B}$	$1-\frac{1+s}{1+B}$,
Nondimensional temperature $\frac{c_p(T_f-T_w)}{L}$, flame; $\frac{c_p(T_{st}-T_w)}{L}$, evap	$-\left[\frac{s}{1+s}\right]\left[1+\frac{c_p(T_\infty-T_w)}{L}\right]$	$\left[\frac{s}{1+s}\right]\left[\frac{h_c}{L}-1-\frac{c_p(T_\infty-T_w)}{L}\right]$
Nondimensional flame location, r_f/a, flame; r_{st}/a, evap	$\frac{\ln(1+B)}{\ln(1+s)}$	$\frac{\ln(1+B)}{\ln(1+s)}$
Mass loss rate constant, α(m²/sec), $d^2-d_0^2=\alpha t$	$\alpha_e=\frac{4Sh\rho D}{\rho_\ell}\ln(1+B)$	$\alpha_c=\frac{4Sh\rho D}{\rho_\ell}\ln(1+B)$
Life time (sec)	$\frac{d_0^2}{\alpha_e}$	$\frac{d_0^2}{\alpha_c}$

Note: The Sherwood number, for flow-past spheres $Sh=2+0.6\,Re^{1/2}\,Sc^{1/3}$

Chapter 11: Boundary Layer Combustion

Table F.11.1 Similarity Variables and Governing Equations in Similarity Coordinates for Various Flows

Momentum: $(f')^2 (1-2\alpha) - ff'' (1-\alpha) = (C_1 C_2 / Re_{ref}) f''' + C_m$

Energy and Species: $\{C_1 C_2/Re_{ref}\} \phi'' + Sc\, f\, \phi' (1-\alpha) = 0$

Generic Property, b: $\phi_{b\_} + f\, Sc \{(1-\alpha)\, \phi_b'\} + Sc\, S_b^* = 0,\ \phi_b = \{b - b_\infty\}/b_{ref}$

$C_m = \{(C_2/C_1)^2 S_v(x^*,y^*)/x'^{*1-4\alpha}\} = (C_2/C_1)^2 (2k+1)^{1-4\alpha} S_v(x^*,y^*)/x^{*(2k+1)(1-4\alpha)},\ y' = \int_0^y \frac{\rho x^{*k}}{\rho_{ref}} dy,\ y' = y' = \int_0^y \frac{\rho}{\rho_{ref}} dy$ for 2D, $y' = \int_0^y \frac{\rho x^*}{\rho_{ref}} dy$ for axisymm

$\eta = C_1 (y'^*/x'^{*\alpha}),\ \psi(x'^*, y'^*) = f(\eta)\, x'^{*(1-\alpha)}/C_2,\ Gr_{ref} = g\{T_w - T_\infty\} d_{ref}^3/\{T_\infty \nu_{ref}^2\},\ Re_{ref} = v_{ref} d_{ref}/\nu_{ref},\ x'^* = x^{*(2k+1)}/(2k+1),\ k = 0$ 2D, $k = 1$ axisymm

$m^* = \dot{m}''/\{v_{ref} \rho_{ref}\} = v_y'^*(0)\, x^{*k} = K\{-f(0)\}/\{C_2 x^{*\sigma}\},\ K = (1-\alpha)(2k+1)^\alpha,\ \sigma = \alpha(2k+1) - k;\ -f(0) = \{C_2 x^{*\sigma}\} \dot{m}''/\{v_{ref} r_{ref} K\}$

$\phi'(0) B/\{Sc(1-\alpha)\} = \{Re_{ref}/(C_2 C_1)\} f(0),\ \theta_1 = \{T - T_\infty\}/T_\infty,\ \theta = \{T - T_\infty\}/\{T_w - T_\infty\}$

#	Problem	α	C_1	C_2	C_m	$\{C_1C_2\}/Re_{ref}$	K	σ
1	**Forced convection (k = 0)**							
1a	a) Curved $v_{x\infty} = ax^n$, $v_{ref} = a\, d_{ref}$	$(1-n)/2$	$Re_{ref}^{1/2}$	$Re_{ref}^{1/2}$	$n(\rho_\infty/\rho)$, $n(T/T_\infty)$	1	$\{(1+n)/2\}$	$(1-n)/2$
1b	b) Flat surface : $v_{x\infty}$ const or $n = 0$, $v_{ref} = v_{x\infty}$	1/2	$Re_{ref}^{1/2}$	$Re_{ref}^{1/2}$	0	1	(1/2)	
2.	**Stagnation, 2-D** (k = 0), $v_{x\infty} = a x^n$, n = 1, $v_{ref} = a\, d_{ref}$	0	$Re_{ref}^{1/2}$,	$Re_{ref}^{1/2}$	(ρ_∞/ρ), T/T_∞	1	1	0
3.	**Stagnation axisymmetric** (k = 1), $v_{x\infty} = ax^n$, n = 1, $v_{ref} = a\, d_{ref}$, $x'^* = x^{*3}/3$,	1/3	$Re_{ref}^{1/2}\, 3^{1/6}$	$Re_{ref}^{1/2}/\{3^{1/6}\}$	$(1/3)(\rho_\infty/\rho)$, $(1/3)\, T/T_\infty$	1	$2/3^{2/3}$	0
4.	**Free Conv.** (k = 0), $Re_{ref} = 1$, $v_{ref} = \{n_{ref}/d_{ref}\}$							
4a	a) $Re_{ref} = 1$	1/4	$\{4Gr_{ref}\}^{1/4}$	$\{1/(4Gr_{ref})^{1/4})\}$	$\{T - T_\infty\}/\{T_w - T_\infty\}$	1	3/4	1/4
4b	b) $Re_{ref} = 1$	1/4	$\{Gr_{ref}/4\}^{1/4}$	$(1/4)\,(4/Gr_{ref})^{1/4}$	$\{T - T_\infty\}/\{T_w - T_\infty\}$	1/4	3/4	1/4

$S_b^* = \{S_b(x,y)\}\, C_2\, x'^{*2\alpha - 2k/(2k+1)}\, d_{ref}/(\rho(2k+1)^{2k/(2k+1)} C1\, b_{ref}\, v_{ref})$

Chapter 12: Combustion of Jets

Table F.12.1 Comparison of Single Laminar Jet Results for Liftoff and Blow-off, and Flame Structure for Circular and 2-D Planar Jets under Nonbuoyant Conditions and Nonunity Sc

Parameter	*Circular Jet*	*2-D Jet*	*Remarks*
Properties	Constant	Variable	For 2-D, ρ^2 D = const.
M*, parabolic inlet profile	4/3	6/5	
J_b^*, parabolic velocity, parabolic buoyancy profile	4/3	6/5	
J_b^*, parabolic velocity, flat. b profile	1	1	
Similarity coordinate, η	$\eta = r^*/x^*$	$\eta = \frac{M^{*1/3} y^*}{x^{*2/3}}\left[\frac{Re_i}{3}\right]^{(2/3)}$, $y^* = y'/d$, $y' = \int_0^y (\rho/\rho_i)\, dy$ (3c) $v_y' = (\rho/\rho_i)v_y + v_x\, \partial/\partial x(\int_0^y \partial/\partial x\, (\rho/\rho_i)\, dy)$	
Modified similarity variable, ξ	$\xi = (-3/8)\, Re_i M^{*1/2}\, r^*/x^*$ $\xi = (-3/8)\, Re_i M^{*1/2} \eta$	$\xi = \frac{1}{2}\left(\frac{3}{2}\right)^{(1/3)} \frac{Re_{e,i}^{\frac{2}{3}}\, y^* M^{*1/3}}{x^{*\frac{2}{3}}} = 0.2752 \frac{Re_{e,i}^{\frac{2}{3}}\, y^* M^{*1/3}}{x^{*\frac{2}{3}}}$	
Momentum equation in η	$f''' - f''/\eta + ff''/\eta + f'/\eta^2 + f'^2/\eta - ff'/\eta^2 = 0$ $f' = 2(\sqrt{3/8})\, Re_i\, \xi\, M^{*2}/(1 + \xi^2/4\}^2$ $v_x^* = f'(\xi)\, Rei/r^*$ $v_r^* = \{1/(Rei\ x^*)\}[f'(x) - f'(\xi)/\eta]$	$f'^2 + ff'' + f''' = 0$ $f = (2)\, (3/2)^{(2/3)}\, sech^2\, \xi$ $f = (3/2)^{1/3} \tanh \xi$ $v_x^* = f = (x)\, M^{*2/3}\, \{Re/(3x^*)\}^{1/3}$ $v_y^* = \frac{M^{*1/3}}{(3x^*)^{2/3} Re_i^{*1/3}}\left[2f'(\xi)\,\eta - f(\xi)\right]$	

(continued)

Table F.12.1 Comparison of Single Laminar Jet Results for Liftoff and Blow-off, and Flame Structure for Circular and 2-D Planar Jets under Nonbuoyant Conditions and Nonunity Sc (Continued)

Parameter	*Circular Jet*	*2-D Jet*	*Remarks*
Species equation in η	$\Psi'' \eta + \Psi' + Sc\, f'\Psi + Sc\, f\, \Psi' = 0$, where $\Psi = \phi\, x^*$	$\Psi'' + \Psi' Sc\, f + \Psi\, Sc\, f = 0$ where $\Psi = \phi x^{*1/3}$	
Axial velocity and maximum velocity $v_x^* = v_x/v_{x,i}$, v_{xmax}^*	$v_x^* = \frac{3}{32}\frac{M^*}{x^*}\frac{Re_i}{\left[1+\frac{\xi^2}{4}\right]^2}$ $v_{x,\,max}^* = (3/32)(Re_i M^*/x^*)$	$v_x^* = 0.4543\, M^{*2/3}\, \{Re_i/x^*\}^{1/3}\, sech^2\, \xi$ $v_{x,\,max}^* = 0.4543 M^{*2/3}\, \{Re_i/x^*\}^{1/3}$	
Velocity contour, with $v_x^* = S/v_{xi}$	$r_{vel}^* = x^*\, [2^{1/2}/(-3/8)\, Re_i M^{*1/2}]\, [(3/32)(Re_i\, M^*/x^*)^{1/2} - 1]^{1/2}$	$y_{vel}^* = 3.6337\, [x^{*2/3}//(Re^{2/3}\, M^{*1/3})] sech^{-1}\, \{1.4836\, x^{*1/6} (S/v_{x,i})^{1/2}/[Re^{1/6}\, M^{*1/3}]\}$	S, special axial velocity
Lateral velocity, v_y or v_r $v_r^* = v_r/v_{xi}$	$v_r{}^* = \frac{\sqrt{3}}{8}\frac{M^{*1/2}}{x^*}\frac{\xi\left\{1-\frac{\xi^2}{4}\right\}}{\left[1+\frac{\xi^2}{4}\right]^2}$ $v_r = 0$ at $\xi = 0, 2, \infty - v_r > 0$ $(0 < \xi < 2)$, <0, $\xi > 2$ $v_{r,max}^* = \{9/96)\, (1/x^*)\, M^{*\,2}$ at $\xi = 2/3$	$v_y{}^* = 0.5503\, \{(M^*/Rei)^{1/3}/x^{*2/3}\}\, \{2\xi\, sech^2\xi - \tanh\xi\}$, $v_y^* = 0$ at $\xi = 0$, 1.0887 and $v_\infty^* = -0.5503\, \{(M^*/Rei)^{1/3}/x^{*2/3}\}$, $v_{y,max}^* = 0.1788 (M^*/Re)^{1/3}/x^{*\,2/3}$ at $\xi = 0.5218$	
Mixing layer thickness, r_m^* or y_m^*, $v_x/v_{x,max} = 0.01$	$48\, x^*/\{-3\, Re_i M^{*2}\}$	$10.8765\, x^{*\,2/3}/\{Re_i^{2/3}\, M^{*1/3}\}$	Note mix layer αx, cir; but $x^{2/3}$ for 2D
Mass flow within r or y,x; $\dot{m}(r,x)$ or $\dot{m}(y,x)$	$\dot{m}\, (r,x)/\, \dot{m}_i = 32\, (x^*/Re_i)[\xi^{2/4}]\, [1/(1 + \xi^{2/4})]$	$\dot{m}\, (y,x)/\, \dot{m}_i = \{3.3016\, x^{*1/3}\, M^{*1/3}/Re_i^{1/3}\}\, \tanh\xi$	Estimated mass flow within l/2 for multiple burners
Total mass flow; $\dot{m}(x)$	$\dot{m}(\infty,x)/\dot{m}_i = 32\, (x^*/Re_i)$	$m^* = \{3.3016\, x^{*1/3}\, M^{*1/3}/Re_i^{1/3}\}$	

Height, H^*	$H_{st}^* = H_f = \left[\frac{2Sc+1}{32}\right]\frac{Re_i\, J^*}{\Phi^* M^*}$, where ϕ_f is known as mixture fraction.	$H_{st}^* = H_f^* = \left[0.09375 C^3\right]\frac{Re_i\, J^{*3}}{\Phi_f^{*3} M^*}$ H_{flame} $(Sc = 1) = 0.09375\ Re_i(J^{*3}/M^*)/\phi_f^3$ $H_v^* = 0.09375\ \{v_{x,i}/S\}^3\ Re_i\, M^{*2}$	$\phi_f = s/(1 + s)$, $s = Y_{O2,\infty}/(v_{O2}\ Y_{F,i})$ $= 1/(A:F)_{st}$, $\phi_f = 1/[1 + (A:F)_{st}]$
Air entrained at $x^* = H_{stoich}^*$	$\dot{m}_A(H^*_{stoich,}\ \infty)/\ \dot{m}_i = \{J^*\ (2\ Sc + 1)/\phi_f - 1\}$	$\dot{m}'_A\ (H^*_{stoich},\ \infty)/\dot{m}'_i - = 1.5\ C\ J_F^*/\phi_f - 1$,	
Flame width, y_f^* or r_f^*	$r_f^*/H^* = \{16/(3^{1/2}\ Rei\ M^{1/2})\}\{x^*/H^*\}$ $\{(H^*/x^*)^{1/2} - 1\}$	$y_f^* = [x^{*2/3}/\{0.2752\ Re_i^{2/3}\ M^{*2/3}\}]\ \cosh^{-1}\ [H_{pl}^*/x^*]^{1/(6Sc)}$	
Maximum width, $y_{f,max}^*$ or $r_{f,max}^*$	$r_{f,max}^* = \{(2Sc + 1)J^*/(2\sqrt{3}\ M^{*2}\ \phi_f)\}$ $\{4Sc - 1\}^{2Sc - 1/2}/\{4Sc\}^{2Sc}$, $Sc > 0.25$ $r_{f,max}^*(Sc = 1) = 9/\{32\ M^{*2}\ \phi_f\}$	$y_{f,max}^* = \{0.7499\ \gamma\,(\text{sech}^{4Sc}\ \gamma)(J^{*2}/M^{*4/3})(C/\phi_f)^2\}$ (valid when ρ is constant $0 < y < y_f$) where $\gamma \tanh \gamma = 1/\{4Sc\}$, $\gamma = 0.235/Sc + 0.2738$, $0.2 < Sc < 2$, $\gamma = 0.5218$ at	Note the max width/ height ratio $1/Re_i$ for circular jet
Lift off height, $\{L^*/H_f^*\}$	$[\{S/v_i\}\{(2Sc + 1)J^*/(3\phi_f\ M^*)\}]^{Sc/(1-Sc)}$ Note : $L^*\%\ v_l\ v_l^{-Sc/(1-Sc)}\%\ v_l^{(1-2Sc)/(1-Sc)}$ For $Sc = 1$, lift off $\rightarrow \infty$	$L^*/H_f^* = [\{S/v_i\}\{CJ^*/(\phi_f M^*)\}]^{3Sc/(1-Sc)}$ $L^*\%\ v_l\ v_l^{-3Sc/(1-Sc)}\%\ v_l^{(1-4Sc)/(1-Sc)}$ For $Sc = 1$, lift off $\rightarrow \infty$	Set $L_f = H_f$ for blow-off
Blow-off velocity	$S\{(2Sc + 1)/3\}J^*/(\phi_f\ M^*)$ $Sc = 1$, $SJ^*/(\phi_f\ M^*)$	$SCJ^*/(\phi_f M^*)$ $SJ^*/(\phi_f M^*)$	

Note: $v_x^* = v_x/v_i$; $x^* = x/d_i$; $H^* = H/d_i$; x, axial distance from burner exit; S, laminar burning velocity; $\phi = (\beta - \beta_\infty)/(\beta_i - \beta_\infty)$ for combustion; $\phi = \frac{Y_k - Y_{k,\infty}}{Y_{ki} - Y_{k,\infty}} = \frac{T - T_\infty}{T_i - T_\infty}$ for pure mixing; $\phi_f = \phi_{st}$; $H_f = H_{st}$; M^* = momentum at any x/injected momentum; J^* = species flux at any x/injected species flux; Set $J^* = M^* = 1$ in absence of buoyancy forces and inlet profiles; J^* and $M^* > 1$ under buoyancy and parabolic profiles; $Fr = v^2/\{gH_{stoich}\}$, $Fr \rightarrow 0$, $M^* > 1$;

$Re_i = v_{xi}\ d_i/\nu$; $\phi = (\beta - \beta_\infty)/(\beta_i - \beta_\infty) = (f - f_\infty)/(f_i - f_\infty) = (Z_k - Z_{k\infty})/(Z_{k,i} - Z_{k-\infty}) = (h - h_\infty)/(h_{,i} - h_\infty)$ for combustion; $\phi = Y_F/Y_{F,i}$, $\{Y_{O2,\infty} - Y_{O2}\}/Y_{O2,\infty}$ $(f - f_\infty)/(f_i - f_\infty)$, $(Z_k - Z_\infty)/(Z_{k,i} - Z_\infty)$, for mixing).

Table F.12.2 Comparison of Laminar and Empirical Turbulent Jet

Problem	ξ	$v_x{}^*{}_{,max} = \Phi_0$	$v_x{}^*/v_x{}^*{}_{,max} = \Phi/\Phi_0$	$\dfrac{\dot{m}_{ent}}{\dot{m}_{air}}$	H_f^*
Laminar planar	$0.2752\ Re_i^{2/3}\ y^*/x^{*(2/3)}$	$0.4543\ (Re_i/x^*)^{1/3}$	$\mathrm{sech}^2\xi$	$3.3019\ x^{*1/3}/Re_i^{1/3}$	$0.09376 Re_i/\phi_f^3$, $Sc = 1$
Planar turbulent jet	$\sigma y^*/x^*, \sigma = 7.67$	$0.866\ \{\sigma/x^*\}^{1/2}$	$\mathrm{sech}^2\xi$	$1.7322\ (x^*/\sigma)^{1/2}$	$0.750\ \sigma/\phi_f^2$
Laminar circular	$\sqrt{/8}\ Re_i\ r^*/x^*$	$(3/32) Re_i/x^*$	$1/[1 + (\xi^2/4)]^2$	$32\ x^*/Re_i$	$(3/32)\ Rei/\phi_f$
Circular turbulent jet	$0.2443\ r^*/(\sigma_c x^*),\ \sigma_c - 0.0161$	$(0.1058/[\sigma_c\ x^*])$	$1/\{1 + (\xi^2/4)\}^2$	$28.3593\ x^*\sigma_c$	$(0.1058/(\sigma_c),/\phi_f)$

Chapter 13: Ignition

$$mc_p \frac{dT}{dT} = \dot{q}_{gen} - \dot{q}_{loss}$$

$$\dot{q}_{loss} = h\ S\ (T - T_\infty) + \varepsilon\sigma S(T^4 - T_\infty^4)$$

System Temperature in Terms of Ambient Temperature at Ignition

$$(\bar{R}T/E) = \frac{1}{2}\left\{1 \pm \sqrt{1-(4\bar{R}\,T_\infty/E)}\right\},\ D_{IV} = \frac{V\ A\,hc\,Y_F\,Y_{O_2}\,\rho^2}{h\,S\,\frac{E}{R}}$$

$$= \frac{m\ A\,hc\,Y_F\,Y_{O_2}\,\rho}{h\,S\,\frac{E}{R}},\ \theta_\infty = \theta - D_{IV}\exp\left(-\frac{1}{\theta}\right),\ \theta = (\bar{R}T/E)$$

Approximate Ignition Temperature

$$(1/\theta_\infty) \approx \ln[D_{IV}\ \exp(1)/\theta_\infty^2] = 1 + \ln[D_{IV}/\theta_\infty^2].$$

Particle Ignition

$$D_{III} = \frac{B_{O_2}\,d}{Sh\,D_w},\ D_{III.mod} = D_{III}\ \nu_{O2} = [A_{het}\ \nu_{O2}\ d/Sh\ D_w]$$

$$D_{IV} = \frac{A_{Het}\,h_C\,Y_{O_2,w}\,\rho}{h_H\,\frac{E}{R}},$$

$$Y_{O_2w} = \frac{Y_{O_2,\infty}}{\left(D_{III,mod}\exp\left(-\frac{E}{RT_p}\right)+1\right)}$$

$$= \varepsilon\sigma\left(T_w^4 - T_\infty^4\right)/h_{rad}\left(T_w - T_\infty\right),$$

$$\tilde{h} = h + h_{rad},$$

$$N\tilde{u} = \frac{\tilde{h}d}{\lambda} = Nu + Nu_{rad} \quad \text{where} \quad Nu_{rad} = \frac{4d\varepsilon\sigma T_\infty^2}{\lambda}$$

At ignition,

$$\frac{Da_{IV}\exp\left(\frac{1}{\theta_P}\right)}{\theta_P^2} \approx 1.0$$

Table F.13.1 Critical Ignition Conditions for Piles

Geometry	$Da_{s,crit}$	$\psi_{0,crit}$	$(Bi\psi_\infty/2)$	Da_{IV}
Slab	0.88	1.19	1	$\frac{\bar{R} A_1 b_c L^2}{\lambda E}$
Cylinder	2.0	1.39(= ln 4)	1	$\frac{\bar{R} A_1 b_c R^2}{\lambda E}$
Sphere	3.32	1.61	1	$\frac{\bar{R} A_1 b_c R^2}{\lambda E}$

$Bi = h_H 2L/\lambda$, $h_H\ 2R/\lambda$, slab, cylinder, and sphere; $L = R$ for cyl, sphere

FK Ignition (Figure B.6 in Appendix)

$$\Psi = \frac{\theta - \theta_s}{\theta_s^2}, \frac{\cos h\,\sigma}{\sigma} = \sqrt{\frac{2}{Da_{IV}}}, Da_s = \left(\frac{Da_{IV}}{\theta_s^2}\right)\exp\left(-\frac{1}{\theta_s}\right)$$

$$\text{Critical: } (Da_{IV})_{crit} = \frac{\bar{R} A_1 b_c L^2}{\lambda E}\frac{1}{\theta_\infty^2}\exp\left(-\frac{1}{\theta_\infty}\right) \text{ for slab}$$

Ambient T at Ignition

$$-\psi_\infty = \frac{\theta_s - \theta_\infty}{\theta_s^2} = \frac{2}{Bi}\left(\frac{Da_s}{2}\exp(\psi_0)\right)^{1/2} \tanh\left\{\left(\frac{Da_s}{2}\exp(\psi_0)\right)^{1/2}\right\},$$

Chapter 14: Deflagration and Detonation

Rayleigh Lines

$$p_\infty - p_0 = -\dot{m}''^2\{1/\rho_0 - 1/\rho_\infty\}, (p_\infty^* - 1)/\{1/\rho_\infty^* - 1\} = -m^*, \quad m^* = \dot{m}''^2 / P_0\rho_0$$

$$v_\infty = \{(p_\infty - p_0)(1/\rho_0 - 1/\rho_\infty)\}^{1/2} \quad \text{if} \quad v_0 = 0$$

Hugoniot Curves

$$\gamma/(\gamma - 1)\}\{(p_\infty^*/\rho_\infty^* - 1\} = q^* + [1/2][1/\rho_\infty^* + 1][p_\infty^* - 1]$$

$$M_0 = v_0/c_0 = \{\dot{m}''^2/(\gamma P_0 \rho_0)\}^{1/2} = \{m^*/\gamma\}^{1/2}, \quad q^* = q\rho_0/P_0$$

$$M_\infty = v_\infty/c_\infty = \{\{\dot{m}''^2/(\gamma P_\infty \rho_\infty)\}^{1/2} = \{m^*/(\gamma P_\infty^* \rho_\infty^*)\}^{1/2}$$

CJ Points

$$p_\infty^* = \frac{\left\{\left[2q^* + \left(\frac{\gamma+1}{\gamma-1}\right)\right] - \frac{1}{\rho_\infty^*}\right\}}{\left\{\left[\frac{(\gamma+1)}{\gamma-1}\right]\left[\frac{1}{\rho_\infty^*}\right] - 1\right\}}$$

Chapter 15: Flame Speed and Flammability

Flame Thickness

$$\delta_f \approx [\lambda\, Y_{F,0}/\{c_p\, \dot{w}'''_{F,avg}\}]^{1/2} = \{\alpha_T\, \rho_0 Y_{F,0}/\{\dot{w}'''_{F,avg}\}]^{1/2} = \alpha_T/v_0$$

$$S = v_0 \approx [1/\rho_0][\lambda \dot{w}'''_F/\{Y_{F,0}\, c_p\}]^{1/2} = [\alpha_T\, \dot{w}'''_F/\{\rho_0 Y_{F,0}\}]^{1/2}$$

$$v_0 \approx [\alpha_T\, \dot{w}'''_{F,avg}/\{\rho_0 Y_{F,0}\}]^{1/2} \approx [\alpha_T/t_{chem}]^{1/2}$$

$$v_0 = \left\{\frac{\alpha_T \dot{w}'''_{F,avg}}{\rho_0 Y_{F,0}} \frac{(T_\infty - T_{ign})}{(T_{ign} - T_0)}\right\}$$

$$\dot{m}''^2 = \{2A/[Y_{F,0}{}^2 h_c{}^{*\, nF+1}]\}\, \exp(-E^*)\rho D Y_{O2}{}^{nO2}(p/R)^{n1}/E^{*2}$$

$$\frac{\dot{Q}'''(T_\infty)}{\dot{w}'''_F(T_\infty, Y_{A0}) h_c} = K = \frac{2\left\{\frac{1-\theta_1}{1-\theta_0}\right\}\left\{\frac{\theta_1-\theta_0}{1-\theta_0}\right\}^{(n+2-m)}}{(n+1)(n+2)},$$

$Kc = 4.95 \times 10^{-4}$ for $n = 11$, $m = 4$,
$\theta = T/T_\infty$, peak temperature $\theta_1 = T_1/T_\infty$ with heat loss

Empirical or Approximate Methods

$$C_m H_x O_y + z\, O_2 \rightarrow m\, CO_2 + \frac{x}{2} H_2O, \quad z = m + x/4 - y/2$$

$$C_{st}(\text{mole }\%) = \frac{100}{1+\left(\frac{z}{0.21}\right)}$$

Lean: LFL = 0.55 C_{st}

Rich: RFL = 3.50 C_{st}

$$\text{LFL} = \frac{6}{n_C} + 0.2, \quad \text{RFL} = \frac{60}{n_H} + \frac{n_C}{20} + 2.2, \; n_C\text{: carbon atom in fuel}$$

$$Y_{F,L} \approx [c_p(T_{F,L} - T_0)/h_{c,L,O2}]/\nu_{O2} \text{ in units of mole \%}$$

$$Y_{F,,R} \approx 1 - \{c_p(T_{F,R} - T_0)/(h_{c,R,O2}*0.23)\}$$

Quench Diameter or Distance

$$d = \frac{\alpha_T}{v_0}\sqrt{2Nu\frac{T_b - T_0}{T_\infty - T_0}}, \quad d_{quench} \approx 4.6\ \delta_f$$

Spheres

$$d_{crit} \approx \frac{\alpha_T}{v_0}\sqrt{3Nu} = \alpha_f\sqrt{3Nu}$$

$$d_{quench} = [(4\ \rho_0\ D_0 Nu)(\theta_{1c} - \theta_0)Y_{F,0}/\{(1 - \theta_{1c})\ \dot{w}_F'''\ (\theta_{1c})\}]^{1/2}$$

$$E_{min} = \kappa\ d_{quench} \text{ where } \kappa = \pi\ \rho_0\ \{\alpha_T/v_0\}\ c_p(T_I - T_0)/4$$

$$= \pi(\lambda/v_0\}(T_I - T_0)/4,\ kJ/m^2,\ \kappa \approx 0.07\ kJ/m^2$$

Laminar Flow

$$v_{x,max} = 2\dot{Q}/\{\pi R^2\} = 2\ v_{x,avg},\ |g| = 4\ \dot{Q}/\{\pi R^3\} = 4\ v_{x,avg}/R = 8\ v_{x,avg}/d.$$

Chapter 16: Interactive Processes

$$\eta = \frac{\dot{m}_S}{\dot{m}_{ISO}}, \quad \dot{m}_{iso} = 4\ \pi\ \rho D a\ \ln(1+B), \quad B = \frac{h_{T,\infty}}{L} + \frac{Y_{O_2,\infty}}{v_{O_2}}\frac{h_C}{L}$$

Arrays

$$\eta_1 = 1 - \sum_{j=2} \eta_j \frac{a_j}{|\gamma_j - \gamma_1|}$$

$$\Psi = \frac{\ln\left[\frac{F - Y_k}{F - Y_{k\infty}}\right]}{\ln\left[\frac{F - Y_{kw}}{F - Y_{k\infty}}\right]} = -\frac{\ln\left[\frac{F - Y_k}{F - Y_{k\infty}}\right]}{\ln[1 + B]}, \quad F = Y_{k\ell}, \quad B = \frac{Y_{kw} - Y_k}{F - Y_k}, \text{ evaporation}$$

$$\Psi = \frac{\ln\left[\frac{F - \beta}{F - \beta_\infty}\right]}{\ln\left[\frac{F - \beta_w}{F - \beta_\infty}\right]} = -\frac{\ln\left[\frac{F - \beta}{F - \beta_\infty}\right]}{\ln[1 + B]}, \quad B = \frac{\beta_w - \beta_\infty}{F - \beta_\infty}, \text{ combustion}$$

(See Chapter 10 for "F" definitions.)

Average Correction Factor

$$\eta = [1 - (2a/l) + 1.81N^{2/3}/(l/a)]^{-1},\ 3 < N < 9$$

Multicomponent Evaporation

$$\eta_{ik} = 1 - \sum_{\substack{j=1 \\ j \neq k}}^{N} \eta_{ij} \frac{a_j (Y_{i,w})_j}{|\ell_{kj}|(Y_{i,w})_k},\ i = A, B;\ \eta_k = 1 - \sum_{\substack{j=1 \\ j \neq k}}^{N} \frac{\eta_j a_j (1 - Y_{I,w})_j}{|\ell_{kj}|(1 - Y_{I,w})_k}$$

$$\frac{Y_{A,W}}{\varepsilon_A} = \frac{Y_{B,W}}{\varepsilon_B} = C$$

Clouds

n $\simeq 1/\ell^3$, σ = (fuel volume fraction) = $(4/3)\ \pi\ a^3\ n$, $\kappa = \sigma\ \rho_\ell/\rho_g$, fuel-to-gas mass ratio
$R_c = (N/F_p)^{(1/3)}(\ell/2)$, F_p is the packing factor for a hypothetical materials of radii $\ell/2$.

$$\ell = (6\ F_p/\pi\ n)^{1/3}$$
$$G = 4\pi n a R_c^2 = 5.43\ N^{2/3}/(\ell/a)$$

$$\dot{m}_{c,sc} = 4\ \pi\ Sh\ \rho\ D\ \ln(1 + B),$$
$$\alpha_{cl} = 4\ Sh\ \rho\ D\ \ln(1 + B_c)/\rho_{cl}\ ,\ \rho_{cl} = \rho_\ell\ \sigma,$$
$$M = \dot{m}_c/(4\pi\ \rho\ D\ R_c),\ M_{SC} = \dot{m}_{c,SC}/4\pi\ \rho\ D\ R_c = \ln(1 + B),$$
$$M_{ISOC} = \dot{m}_{c,iso}/4\pi\ \rho\ D\ R_c$$

$$G = 4\ \pi\ na\ R_c^2 = 3\ N\ a/R_c = 3\ \sigma\ R_c^2/a^2$$
$$= 3\{m_{cloud}/m_{drop}\}\{a/Rc\} = S_v\ R_c^2/a,$$

$$\dot{m}_c/\dot{m}_{c,SC} = M/M_{SC} = 1 - [\tanh G^{(1/2)}/G^{(1/2)}],$$

$$\eta = \frac{M}{M_{ISOC}} = \frac{M}{M_{SC}}\frac{3}{G} = \frac{3}{G}\left[I - \frac{\tanh G^{(1/2)}}{G^{(1/2)}}\right],\ \eta = 3/(3 + G)$$

$$\frac{M}{M_{SC}} = \frac{G\eta}{3},\ M/M_{SC} = G/(3 + G)\ \text{for}\ G >> 100\ \text{or}\ G << 1$$

Flame Location

$$r_f/R_c = (r_f/a)_{iso}\ (M/M_{SC}),\ \xi_f \le 1$$

Effectiveness Factor

$$\eta_{eff} = \dot{m}_{s,avg}/[\dot{m}_{c,iso} \text{ with properties at cloud surface}]$$

$$\eta_{eff} = \frac{M}{M_{ISOC,C}} = \frac{M}{M_{ISOC}}\frac{M_{ISOC}}{M_{ISOC,C}} = \eta\frac{\ln\{1+B\}}{\ln\{1+B(r=R_c)\}} = \frac{\eta}{1-\frac{M}{M_{SC}}}$$

$$\eta_{eff} = \frac{\eta}{1-\frac{G\eta}{3}} = \frac{3\eta}{3-G\eta} = \frac{3}{G}\left[\frac{G^{(1/2)}}{\tanh G^{(1/2)}} - 1\right]$$

Chapter 17: Pollutants Emission

X_{CO_2}/X_{CO_2}, max $= (1 - X_{O_2}/X_{O_2,a}) = (X_{O_2,a} - X_{O_2})/X_{O_2,a}$; $X_{O_2,a}$: ambient O_2 mole fraction

$$X_{NO,std}/X_{NO} = (X_{O_2,a} - X_{O_2,std})/(X_{O_2,a} - X_{O_2}).$$

Emission Reporting

EI of species k in g/kg of fuel as-received = $\{Y_C{*}M_k\ X_k\}{*}$ (1000 g/kg)/[12.01 $\{X_{CO} + X_{CO2}\}$] in units of g (kg of fuel)$^{-1}$.

$$\text{“k” in kMole per kmole of fuel} = \frac{c\,X_k}{(X_{CO_2} + X_{CO})}, \text{ C: fuel C atoms}$$

$$\text{“k” in } \frac{g}{GJ} = \frac{Y_C\, X_k}{(X_{CO_2} + X_{CO})}\,\frac{M_k \times 1000}{12.01 \times HHV_F (GJ/kg)}$$

NO in g per kWh = NO in g per GJ*0.0036

NO in g per HP = NO in g per GJ*0.00268

EI of species k in g/m^3 of gaseous fuel = $\{Y_{C,DAF}{*}M_kX_k\}{*}$(1000 g/kg)$*M_F/$[12.01$*\ \bar{v}_{STP}$ $\{X_{CO} + X_{CO2}\}$]

$$N_{conv} = \frac{c \times x_{NO}}{n(x_{CO_2} + x_{CO})}$$

Chapter 18: Turbulent Combustion Modeling

$$\overline{q(x_i)} = \lim_{\Delta t \to \infty}\frac{1}{\Delta t}\int_{t_0}^{t_0+\Delta t} q(x_i,t)dt,\quad q(x_i,t) = \overline{q(x_i)} + q'(x_i,t)$$

Favre Averaging

$$\tilde{q}(x_i) = \frac{\overline{\rho(x_i,t)q(x_i,t)}}{\bar{\rho}(x_i)},\ \overline{\rho(x_i,t)q(x_i,t)} = \lim_{\Delta t \to \infty} \frac{1}{\Delta t} \int_{t_0}^{t_0+\Delta t} \rho(X_i,t)q(X_i,t)dt$$

$$q(x_i,t) \equiv \tilde{q}(x_i) + q''(x_i,t)$$

$$\overline{\rho(x_i,t)\ q''(x_i,t)} = \left[\lim_{\Delta t \to \infty} \frac{1}{\Delta t} \int_{t_0}^{t_0+\Delta t} \rho(x_i,t)\ q''(x_i,t)dt \right] = 0$$

$$\overline{q''(x_i,\ t)} = -\frac{\overline{\rho'(x_i,\ t)q''(x_i,\ t)}}{\overline{\rho(x_i)}}$$

$$\tilde{q}(x_i) - \bar{q}''(x_i) = -q''(x_i)\frac{\overline{\rho'(x_i,\ t)q''(x_i,\ t)}}{\overline{\rho(x_i)}}$$

Algebraic Model

$$-\rho\overline{v_{i'}v_{j'}} \equiv (\bar{\tau}_{yx})_{\text{Reynolds}} \equiv \rho v_t\left(\frac{\partial v_i}{\partial x_j} + \frac{\partial v_j}{\partial x_i}\right)$$

$$P = P\ (c,x) = d\{\text{Prob}(f<c)\}/dc$$

$$\tilde{q}(x) = \int_0^1 q(c_1,c_2,x)\ P_j(c_1,c_2,x)\ dc_1c_2 = \int_0^1 q(c_1,c_2,x)\ P(c_1,x)P(c_2,x)\,dc_1c_2$$

Appendix A

Table A.1A Basic Units, Conversions, and Molecular Properties

Multiplier	*Prefix*	*Symbol*
10^{18}	exa	E
10^{15}	peta	P
10^{12}	tera	T
10^{9}	giga	G
10^{6}	mega	M
10^{3}	kilo	k
10^{2}	hecta	h
10	deca	da
10^{-1}	deci	d
10^{-2}	centi	c
10^{-3}	milli	m
10^{-6}	micro	m
10^{-9}	nano	n
10^{-12}	pico	p
10^{-15}	femto	f
10^{-18}	atto	a

Area

1 acre = 4046.9 m^2

1 m^2 = 10^{-6} km^2 = 10^4 cm^2 = 10^6 mm^2

1 m^2 = 10.764 ft^2 = 1550 $in.^2$

1 ft^2 = 144 $in.^2$ = 0.0929 m^2

1 hectare = 10,000 m^2 = 2.5 acres = 108,000 ft^2

Density

1 g/cm^3 = 1 kg/L = 1000 kg/m^3 = 62.43 lb_m/ft^3 = 0.03613 lb_m/in^3

1kg/m^3 = 0.06243 lb_m/ft^3, 1lb_m/ft^3 = 16.018 kg/m^3

Specific gravity = density/reference density
For liquids, reference density of water at 15.74°C (60°F) = 999 kg/m^3, 62.4 lb/ft^3
For gases, reference density of air at 15.74°C (60°F) = 1.206 kg/m^3

Energy
1 eV ≈ 1.602 × 10^{-19} joules.
1 mBtu = 1 kiloBtu = 1000 Btu, 1 mmBtu = 1000 kiloBtu = 10^6 Btu
1 trillion Btu = 10^9 Btu or 1 giga Btu
1 quad = 10^{15} Btu or 1.05 × 10^{15} kJ or 2.93 × 10^{11} kW h = 172.4 million barrels of crude oil
1 kW h = 0.0036 GJ = 3.6 MJ = 3,412BTU, 1 hp h = 0.00268 GJ = 2.68 MJ = Btu
1 Btu = 778.14 ft lb$_f$ = 1.0551 kJ, 1 kJ = 0.94782 Btu = 25,037 lb$_m$ ft/sec^2
1 calorie = 4.1868 J, one (food) calorie = 1000 calories or 1 kcal
1 kJ/kg = 0.43 Btu/lb, 1 Btu/lb = 2.326 kJ/kg, 1kg/GJ = 1g/MJ = 2.326 lb$_m$/mmBtu

$1\frac{Btu}{SCF} = 37\frac{kJ}{m^3}$, 1 m^3/GJ = 37.3 ft^3/mmBTU, 1 lb$_m$/mmBTU = 0.430 kg/GJ = 0.430 g/MJ
1 Therm = 10^5 Btu = 1.055 × 10^5 kJ
1 hp = 0.7064 Btu/sec = 0.7457 kW = 745.7 W = 550 lbf · ft/sec = 42.41 Btu/min
1 boiler HP = 33475 Btu/h, 1 Btu/h = 1.0551 kJ/h
1 barrel (42 gallons) of crude oil = 5,800,000 Btu = 6120 MJ
1 gallon of gasoline = 124,000 Btu = 131 MJ
1 gallon of heating oil = 139,000 Btu = 146.7 MJ
1 gallon of diesel fuel = 139,000 Btu = 146.7 MJ
1 barrel of residual fuel oil = 6,287,000 Btu = 6633 MJ
1 cubic foot of natural gas = 1,026 Btu = 1.082 MJ
1 gallon of propane = 91,000 Btu = 96 MJ
1 short ton of coal = 20,681,000 Btu = 21821 MJ

Forces
1 lbf = 4.4482 N = 32.174 $\frac{\text{lb}_m \cdot \text{ft}}{\text{s}^2}$ or g_c = 32.174 lb$_m$ ft/sec^2 lb$_f$

Ideal Gas Law
PV = RT; PV = m R T; $PV = n\bar{R}T$, $P\bar{v} = \bar{R}T$,

$$\bar{R} = 8.314\frac{\text{kPa m}^3}{\text{k mole K}} = 0.08314\frac{\text{bar m}^3}{\text{k mole K}} = 1.986\frac{\text{Btu}}{\text{lbmole°R}} = 1545\frac{\text{ftlbf}}{\text{lbmole°R}}$$

= 0.7299 atm ft^3/lb-mol R

Length/Velocity
1 in. = 0.0254 m
1 ft = 12 in = 0.3048 m
1 mile = 5280 ft = 1609.3 m
1 statute mile = 0.87 nmi = 1.609 km
1 nautical mile = 1.15 smi = 1.85 km
1 mi/h = 1.46667 ft/sec = 0.447 m/sec = 1.609 km/h
1 m/sec = 3.2808 ft/sec = 2.237 mi/h = 1.96 kt = 1.15 smi/h = 3.63 km/h
Speed of light in vacuum, c = 2.998 × 10^8 m/sec
Sound speed = $\sqrt{\gamma RT}$

Mass
1 teragram (Tg) = 1 million metric tonnes

Mass of an electron = 0.5 MeV (1 Mev = 10^6 eV; for mass, use $E = mc^2$)
= 9.109×10^{-31} kg
Mass of proton = 940 MeV = 1.67×10^{-27} kg, Mass of neutron = 1.675×10^{-27} kg,
1 lb_m = 0.45359 kg = 7000 grains
1 short ton = 2000 lb = 907.2 kg
1 long ton = 2240 lb or 1016.1 kg
1 metric ton = 1000 kg
1 ounce = 28.3495 g
1 kg = 2.2046 lb

Molecular Properties
1 Angstrom = 1.0×10^{-10} m
$N_{Avog} = 6.023 \times 10^{26}$ molecules/kmol for a molecular substance (e.g., oxygen)
= 6.023×10^{26} atoms/atom mole for an atomic substance (e.g., He)
Boltzmann constant, $k_B = 1.38 \times 10^{-26}$ kJ/molecule K
Planck's constant, $h_P = 6.626 \times 10^{-37}$ kJ-sec/molecule
Stefan–Boltzmann constant, $\sigma = 5.66961 \times 10^{-11}$ kW/m^2 K^4
Charge of an electron = 1.602×10^{-19} coulombs, orbit radius (nm) = $0.0529n^2$, n: orbit number
Energy level of an orbit (ev) = $13.56/n^2$

Numbers
$\ln x = 2.303 \log_{10} x$
$\log_{10} x = 0.4343 \ln x$
e = 2.718
$\pi = 3.142$
1 deg = 0.0175 radians

Pressures
1 bar = 10^5 Pa, 1 mm Hg = 133.3 Pa
1 in Hg = 3.387 k Pa = 0.491 psi
1 in water (4°C) = 0.03613 psi

$$1 \text{ atm} = 14.696 \frac{\text{lbf}}{\text{in.}^2} = 1.0133 \text{ bar} = 10.3323 \text{ mm of } H_2O \text{ (4°C)} = 760 \text{ mm of Hg(0°C)}$$

$$1 \text{ psi} = 1\frac{\text{lbf}}{\text{in.}^2} = 144\frac{\text{lbf}}{\text{ft}^2} = 6.894 \text{ kPa} = 6894 \text{ Pa} = 27.653 \text{ in water (4°C)}$$

Specific Heats
1 Btu/lb °F = 4.1868 kJ/kg°C
1 kJ/kg °C = 0.23885 Btu/lb°F

Temperature
T(K) = T(°C) + 273.15
T(°R) = T(°F) + 459.67
1°R = 0.556 K, 1 K = 1.8°R
To convert electron volts into the corresponding temperature in kelvin, multiply by 11,604.

Volume
1 m^3 = 1000 liters
1 fluid ounce = 29.5735 cm^3 = 0.0295735 l

1 m^3/kg = 1000 l/kg = 16.02 ft^3/lb, 1 m^3/GJ = 37.26 ft^3/mmBTU
1 ft^3/lb_m = 0.062428 m^3/kg
1 U.S. gallon = 128 fluid ounce = 3.786 l
1 barrel = 42 U.S. gallons = 35 imperial gallons = 158.98 l = 5.615 ft^3 = 231 in.^{3} = 0.1337 ft^3

Volume of 1 kmol (SI) and 1 lb-mol (English) of an ideal gas at STP conditions as defined below:

Scientific or SATP	*U.S. Standard (1976) or ISA*	*Chemists' Standard or CSA*	*NTP (gas industry)*
25°C (77°F), 101.3 kPa (14.7 psi, 29.92 in. of Hg)	15°C (60°F), 101.33 kPa (1 atm, 14.696 psi, 29.92 in. of Hg)	0°C (32°F), 101.33 kPa (1 atm, 14.7 psi, 29.92 in. of Hg)	20°C (65°F), 101.33 kPa (1 atm)
24.5 m^3/kmol (392 ft^3/lb-mol)	23.7 m^3/kmol (375.6 ft^3/lb-mol)	22.4 m^3/kmol (359.2 ft^3/lb-mol)	23.89 m^3/kmole (382.7 ft^3/lb-mole)

Note: SATP, Standard Ambient Temperature and Pressure; ISA, International Standard Atmosphere; NTP, Normal Temperature and Pressure

Air Composition

Species	*Mole %*	*Mass %*	*Molecular Weight*
Ar	0.934	1.288287	39.948
CO_2	0.0314	0.047715	44.01
N_2	78.084	75.51721	28.01
O_2	20.9476	23.14489	32
Ne	0.001818	0.001267	20.18
He	0.000524	7.24E-05	4.0026
Krypton	0.000114	0.00033	83.8
Xe	8.70E–06	3.94E–05	131.3
H_2	0.00005	3.48E-06	2.016
CH_4	0.0002	0.000111	16.043
N_2O	0.00005	7.6E–05	44.013
SO_2, NO_2, CO, I_2	0.000235	—	—

Note: Molecular weight (mass) of air = 28.96 kg/kmol.

Table A.1B Atomic Weights for Common Elements

Name	*Symbol*	*Atomic Number*	*Atomic Weight*
Aluminum	A1	13	26.98
Antimony	Sb	51	121.76
Argon	Ar	18	39.95
Arsenic	As	33	74.92
Barium	Ba	56	137.32
Beryllium	Be	4	9.01
Bismuth	Bi	83	208.98
Boron	B	5	10.811
Bromine	Br	35	79.90
Cadmium	Cd	48	112.41
Calcium	Ca	20	40.08
Carbon	C	6	12.01
Cesium	Cs	55	132.91
Chlorine	Cl	17	35.45
Chromium	Cr	24	52.00
Cobalt	Co	27	58.93
Copper	Cu	29	63.55
Fluorine	F	9	19.00
Germanium	Ge	32	72.61
Gold	Au	79	196.97
Helium	He	2	4.00
Hydrogen	H	1	1.01
Indium	In	49	114.82
Iodine	I	53	126.90
Iridium	If	77	192.22
Iron	Fe	26	55.85
Krypton	Kr	36	83.80
Lead	Pb	82	207.20
Lithium	Li	3	6.94
Magnesium	Mg	12	24.31
Manganese	Mn	25	54.94
Mercury	Hg	80	200.59
Molybdenum	Mo	42	95.94
Neon	Ne	10	20.18

(continued)

Table A.1B Atomic Weights for Common Elements (Continued)

Name	*Symbol*	*Atomic Number*	*Atomic Weight*
Nickel	Ni	28	58.69
Nitrogen	N	7	14.01
Oxygen	O	8	16.00
Palladium	Pd	46	106.42
Phosphorus	P	15	30.97
Platinum	Pt	78	195.08
Plutonium	Pu	94	244.00
Potassium	K	19	39.10
Radium	Ra	88	226.00
Radon	Rn	86	222.00
Rhodium	Rh	45	102.91
Selenium	Se	34	78.96
Silicon	Si	14	28.09
Silver	Ag	47	107.87
Sodium	Na	11	22.99
Strontium	Sr	38	87.62
Sulfur	S	16	32.07
Tantalum	Ta	73	180.95
Thallium	Tl	81	204.38
Tin	Sn	50	118.71
Titanium	Ti	22	47.87
Tungsten	W	74	183.84
Uranium	U	92	238.03
Vanadium	V	23	50.94
Xenon	Xe	54	131.29
Zinc	Zn	30	65.39
Zirconium	Zr	40	92.22

Table A.2A Properties of Selected Fuels

Substance	Formulae	M	T_c K	P_c bar	$\overline{v}_c$ m³/kmol	TBP K	L kJ/kg	ν_{O2} kJ/kg	h_f kJ/kmol	s^0_{298} (kJ/kmol K)	$h_{c,high}$ kJ/kg	$h_{c,low}$ kJ/kg	c_{pc} kJ/kgK	$c_{p(g)}$ kJ/kgK	ρ_c kg/m³	B #
Acetic acid	$C_2H_4O_2$	60.06	594.4	57.90	0.1710	391.1	394.6	1.066	–484300.0	159.8(l)		15098	2.05			8.54
Acetone (g)	CH_3COCH_3	58.08	508.1	47.00	0.2090	329.4	501.7	2.204	–217100	295.3(g)		38083	2.17	1.28	790	
Acetylene (g)	C_2H_2	26.04	308.8	61.40	0.1130	189.2	653.6	3.072	226748.0	200.82	49908	48218		1.69		5.32
Ammonia	NH_3	17.03	405.6	112.80	0.0725	239.7	1371.8	1.409	–45940.0	192.8	3394			2.06		
Anilene	C_6H_7N	93.14	699	540.00	0.2911	457.6	455.7	2.663	87500	317.9		6283		1.16		
Benzene (g)[c]	C_6H_6	78.11	562.1	48.90	0.2590	353.3	433	3.073	82927.0	269.2	42267	40577	1.72	1.05	879	7.88
Benzoic acid (s)	$C_7H_6O_2$ (s)	122.13	751	44.70	0.3440	522.4	312.9	1.965	–385200	167.73	26423	26392	1.82			
n-Butane (g)	C_4H_{10}	58.12	425.2	38.00	0.2550	272.6	387.1	3.579	–126140	310.03	49503	45717	2.42		579	7.51
Butanol (l)	C_4H_9OH	74.14	562.95	44.18	0.2746	390.9	561.0	2.590	–274900	225.8(l)			2.39		595	
Butene	C_4H_8	56.12	419.15	40.20	0.2410	210.2	393.3	3.421	1,172	307.44	48453	45316	2.10			
Butyl Benzene	$C_{10}H_{14}$	134.24	660.55	28.87	0.4970	456.5	292.4	3.218	–14000	321.2(l)	44123	41828	1.81			
Camphor	$C_{10}H_{16}O$ (s)	152.26						2.837								
Carbon[a]	C (s)	12.01				5100.2	59700	2.664	716130	5.74	32766	32766	0.71		2000	
Carbon monoxide	CO	28.01	132.9	35.00	0.0931	81.7	215.8	0.571	–110480.0	197.7	10103	10103		1.05		
Cetane-n	$C_{16}H_{34}$	226.5	720.6	14.19	0.9300	560.0	228.9	3.461	–373760		47658	44358		1.60	773	
Cyanogen	C_2N_2	52.04	400	59.80		252.0	448.3	1.230	306700	241.9				1.09		
Cyclopentane	C_5H_{10}	70.13	511.8	45.02	0.2583	322.4	388.7	3.422	–76400.0	204.5(l)		44019	1.84			7.69

(continued)

Table A.2A Properties of Selected Fuels (Continued)

Substance	*Formulae*	*M*	T_c *K*	P_c *bar*	$\bar{v}_c$ *m³/kmol*	*TBP K*	*L kJ/kg*	ν_{O2} *kJ/kg*	h_f *kJ/kmol*	s^0_{298} *(kJ/kmol K)*	$h_{c,high}$ *kJ/kg*	$h_{c,low}$ *kJ/kg*	c_{pc} *kJ/kgK*	$c_{p(g)}$ *kJ/kgK*	ρ_c *kg/m³*	*B #*
Cyclohep-tane	C_7H_{14}	98.21	604.2	38.20	0.3530	391.6		3.421	−118100.0							
Cyclohe-xane (g)[d]	C_6H_{12}	84.18	553.8	40.80	0.3080	192.4	356.0	3.421	−123400				1.84		NA	
Cyclohe-xene	C_6H_{10}	82.16	560.4			356.2	407.4	3.311	−5000	214.6(l)			1.81			
n-Decane(l)	$C_{10}H_{22}$	142.3	619	21.23	0.6031	447.0	281; 361	3.486	−249,659	540.53	48000	44598	2.21		730	11.10
Decene	$C_{10}H_{20}$	140.3	617	22.20	0.5840	443.8	NA	3.307	−124,139	539.65	47547	44410				
Diesel	$C_{14.4}H_{24.9}$	198.09						3.332							850	
Diesel	$C_{10.8}H_{18.7}$	148.6						3.332								
Diesel-light (l)	$C_nH_{1.8n}$ (l)	170					270.0	3.356			44800	42500	2.20	1.70	810	
Diesel-heavy (l)	$C_nH_{1.7n}$ (l)	200					230	3.322				41400	1.9		850	
n-Dodecane (g)	$C_{12}H_{26}$	172.72	659.15	18.14	0.7540	489.5	255.3	3.428	−292,162	622.8	47828	44467	2.18			
1-Dodecene	$C_{12}H_{24}$	168.32	658	19.30		486.6	NA	3.422	−165,352	138.41			2.14			
Ethane (g)[e]	C_2H_6	30.54	305.4	48.80	0.1483	184.5	489.4	3.667	−84,667	229.49	51077	46754	NA	1.72	370	5.71
Ethanol(l)	C_2H_5OH	46.07	516.2	63.80	0.1671	351.5	920	2.084	−277,690	160.7	29690	26800	2.50	1.93	785	3.71
Ethene	C_2H_4	28.06	282.34	50.41	0.1311	169.5	482.2	3.421	52283	219.83	50296	47158	NA	1.53		
Ether (diethyl) (g)	$C_4H_{10}O$	74.14	466.7	36.40	0.2800	307.6	365.4	2.590	−252100.0	342.7			NA	1.61		
Ethyl acetate	$C_4H_8O_2$	88.12	523.3	38.82	0.2860	350.3	362.5	1.271	−443600	257.7(l)			1.94			
Ethyl benzene	C_8H_{10}	106.2	617.2	36.09	0.3738	409.4	338.1	3.164	29900.0		43395	41322	1.73			9.28

Ethyl chloride	C_2H_5Cl	64.51	460.4	52.69	0.2000	285.4	383.7	1.612	−112100.0	276				0.97		
Ethylene (g)	C_2H_4	28.05	282.4	50.40	0.1304	169.4	483.1	3.422	52400.0	219.3	50302	47164		1.53		6.24
Ethylene glycol	$C_2H_6O_2$	62.07	645.1	75.30	0.1910	469.1	845.7	1.289	−392200.0	303.8		15793		1.33		
Freon 134a	CF_3CH_2F	102.04	374.2	40.60	0.1980			0.784								
Freon 152a	CHF_2CH_3	66.06	386.4	45.20	0.1795			1.332								
Fructose (s)	$C_6H_{12}O_6$	180.16							−1266000		48380					
HFC-32	CH_2F_2	52.03	351.6	58.30		222.2		0.923			9350		0.843			
HFC-125	CHF_2CF_3	120.03	339.2	35.95		225.1		0.667			4020		0.796			
Gasoline	C_7H_{17}	101.24	543.8	25.70	0.4680	372.4[a]	350	3.556	−224000		48582	44886	2.36		740	
Gasoline	$C_{8.26}H_{15.5}$	114.8						3.383							720	
Gasoline (l)	$C_nH_{1.87n}$	110					305.0	0.427				44000	2.40	1.70	750	
Glucose (s)	$C_6H_{12}O_6$	180.1							−1268000	212	48347					
Glycerine	$C_3H_8O_3$	92.11	765.35	43.00		563.2	662.4	1.216	−577900.0	206.3(l)		38498	2.38			
Glycol-ethylene	$C_2H_6O_2$	62.07	645.1	75.30	0.1910	470.5	845.7	0.773	−392200.0	303.8				1.33		
Glycol-1,2 Propylene	$C_3H_8O_2$	76.1	626	61.00	0.2390	460.8	715.9	1.051	−429800.0				2.51			
n-Heptane	C_7H_{16}	100.2	540.2	27.40	0.4320	371.7	316.7	3.513	−187,820	425.26	48438	44924	2.22	1.66	684	9.57
Heptene	C_7H_{14}	98.21	537.3	29.20	0.4090	366.8	NA	3.421	−62,132	327.6(l)	47800	44662	2.16			
Hexadeane (cetane)	$C_{16}H_{34}$	226.4	720.6	14.19	0.9300	559.8	232.0	3.463	−374800.0			44221	2.22	1.64	777	14.02
n-Hexane	C_6H_{14}	86.18	507.4	29.70	0.3700	341.0	338; 366	3.528	−167,193	386.81	48675	45100	2.27		659	8.86

(continued)

Table A.2A Properties of Selected Fuels (Continued)

Substance	Formulae	M	T_c K	P_c bar	$\overline{v}_c$ m³/kmol	TBP K	L kJ/kg	ν_{O2} kJ/kg	h_f kJ/kmol	s^0_{298} (kJ/kmol K)	$h_{c,high}$ kJ/kg	$h_{c,low}$ kJ/kg	c_{pc} kJ/kgK	$c_{p(g)}$ kJ/kgK	ρ_c kg/m³	B #
Hexene	C_6H_{12}	84.18	504	32.10	0.3551	336.6	392	3.421	–41,673	385.97	47937	44800	1.84		673	
Hydrazine (g)	N_2H_4	32.06					1384		95000	238						
Hydrogen	H_2	2.02	33.2	13.00	0.0651	20.4	448.6	7.921	0	130.7	141505	119718		1.44		7.01
Hydrogen sulfide	H_2S	34.08	373.2	89.40	0.0985	213.7	554.3	0.469	–20600.0	205.8				1.00		
Isobutane	C_4H_{10}	58.12	408.1	36.50	0.2627	260.9	368.2	3.579	–134200.0							
Isodecane (2-methyl pentane)	$C_{10}H_{22}$	142.3	623	25.10		433.7	268.0	3.486	–258600	423.8(l)			2.21			
Isoheptene	C_7H_{14}	98.2				366.5		0.000		424.38						
Isohexane	C_6H_{14}	86.2	497	30.40	0.3610	330.9	322.5	3.528	–174600.0	290.6(l)			2.25			
Isooctane	C_8H_{18}	114	544	25.60	0.4680	372.4	308.0	3.503	–224000			44608	2.10	1.63	692	9.80
Isopentane	C_5H_{12}	72.2	460	33.80	0.3060	300.9	342.2	3.548	–153600.0	260.4(l)			2.32		626	
Isopropanol	C_3H_7OH	60.1	508	47.64	0.2201	355.4	663.4	2.396	–272600	309.2				1.49		
Methane (g)	CH_4	16.04	190.6	46.00	0.0992	109.2	510.2	3.990	–74600.0	186.19	55509	50021		2.23	300	5.21
Methanol (g)[f]	CH_3OH	32.04	512.6	81.00	0.1180	337.8	1101.0	1.498	–201000.0	239.9	23845	19800	2.60	1.72	792	2.99
Methyl chloride	CH_3Cl	50.49	416.3	66.80	0.1389	249.3	426.8	0.634	–81900	98.19				0.81		
Methylene chloride	CH_2C_{l2}	84.93	416.3	66.80	0.1930	313.0	329.8	0.377	–95400	86.18	497.45			0.60		
Natural gas	$C_nH_{3.8n}N_{0.1n}$	18								114.2	543.9	45000		2.0		
Naphthlene	$C_{10}H_8$ (s)	128.2	748.4	40.50	0.4100	491.1	337.8	2.995	78500	72.15	460.4	40221	1.29			9.83

Neopentane	C_5H_{12}	72.17	433.8	32.00	0.3030	282.6	315.1	3.567	−168000	60.1	508.3					
Nitromethane	CH_4NO_2	62.06	588	58.70	0.1730	374.3	547.7	0.516	−74300	275	12247	11165		0.92		
Nonane	C_9H_{20}	128.3	595	22.73	0.5477	423.7	293.8	3.492	−229,032	502.08	48134	44686	2.22		718	10.53
Nonene	C_9H_{18}	126.27	594		0.5260	420.1		3.421	−103,512	501.24	47612	44475				
n-Octane (g)[t]	C_8H_{18}	114.2	568.8	24.80	0.4920	398.8	302, 33	3.503	−208,447	463.67	48268	44800	2.23		703	10.16
Octene	C_8H_{16}	112.24	578.15	25.84	0.4680	394.2	303.7	3.421	−82,927	462.79	47693	44556	2.15			
Octanol	$C_8H_{18}O$	130.23				467	545									
Pentadecane	$C_{15}H_{32}$	212.47	708	14.80	0.9660	543.7	235.7	3.464								
n-Pentane (g)	C_5H_{12}	72.15	469.6	33.70	0.3040	309.0	360.2	3.548	−146,440	348.4	49032	45355	2.32		626	8.20
1-Pentene	C_5H_{10}	70.15	464.8	35.60	0.2934	303.2	359.3	3.421	−20,920	347.61	48152	45000	2.20		641	
Phenol	C_6H_5OH	94.12	694.2	61.30		455.0	485.4	2.380	−96400	144 (s)	33176	31774				
Propane (g)	C_3H_8	44.1	369.8	42.50	0.2030	231.1	426.0	3.628	−103,847	269.91	50340	46348	2.77	1.67	500	6.71
Propanol (ℓ)	C_3H_7OH	60.1	536.7	51.70	0.2185	370.4	688.5	2.396	303000	322.6		30690		1.42	800	
Propyne	C_3H_4	40.065	402.4	56.30	0.1640	250.0		3.195	184900			46200				
Propyl benzene	C_9H_{12}	120.21	638.35	32.00	0.4400	432.4		3.194	7900.0	287.8 (l)	43800	41603	1.79			
Propylene (g)	C_3H_6	42.08	365	46.20	0.1810	225.4	437.8	3.422	20414	266.94	48919	45781			514	6.83
Styrene (l)	C_8H_8	104.2	647.6	39.99	0.3518	418.3	351.4	3.071	103800.0			38736	1.75			8.67
Sucrose (s)	$C_{12}H_{22}O_{11}$	342.34						1.122	−222000	360	16494	16109				
Sulphur	S	32.06	1313.2			717.6	1403.6	0.000								
Tetradecane	$C_{14}H_{30}$	198.44	693	15.70	0.8940	526.5	242.7	3.467								
Toluene (l)	C_7H_8	92.14	591.7	41.10	0.3158	383.8	365; 412	3.126	12400.0	221	42847	40936	1.68	1.10	867	8.54

(continued)

Table A.2A Properties of Selected Fuels (Continued)

Substance	Formulae	M	T_c K	P_c bar	$\bar{\upsilon}_c$ m^3/kmol	TBP K	L kJ/kg	ν_{O2} kJ/kg	h_f kJ/kmol	s^0_{298} (kJ/kmol K)	$h_{c,high}$ kJ/kg	$h_{c,low}$ kJ/kg	c_{pc} kJ/kgK	$c_{p(g)}$ kJ/kgK	ρ_c kg/m³	B #
Tridecane	$C_{13}H_{28}$	184.41	675	16.80	0.8230	508.7	250.5	3.471					2.21			
Triptane (2,2,3 trimethyl butane)	C_7H_{16}	100.23	531.1	29.50	0.3980	354.3	288.4	3.512	−204400	292.2(l)			2.13			
Undecane	$C_{11}H_{24}$	156.35	639	19.80	0.6890	469.1	268.0	3.479	−270,286	578.94	47926	44532	2.21		740	
1-Undecene	$C_{11}H_{22}$	154.3	637	19.70				3.422	−144,766	578.06	47512	44360				
Urea	$CO(NH_2)_2$	60.06							−333510	104.6	10523					
Urethane	$C_3H_7NO_2$ (s)	89.11						1.347								
p-Xylene	C_8H_{10}	106.2	616.3	35.11	0.3791	411.5	337.3	3.164	18000.0			41276	1.71			8.48

Note: Unless otherwise mentioned, most properties are at 298 K; B number for combustion is at $T_w = T_{BP}$ with $h_c = h_{clow}$; L is at T_{BP} with $c_p = 1.2$ kJ/kg K and $Y_{O2\infty} = 0.23$; density of condensates is at 15 to 20°C with $h_{c,low}$ (kJ/kg) = $h_{c,high}$ (kJ/kg) − (H atoms*22005/M_F); h_c (liquid fuel in kJ/kg) = h_c (gaseous fuel in kJ/kg) − L (kJ/kg). Also, latent heat at any T is L(T) = L (T_{ref}) − ($c_{pc} - c_{pg}$) (T − T_{ref}), where c_{pc} is the specific heat of condensate phase [Abram et al]; also, $L_{BP}/RT_{BP} \approx 8$ to 10.

a: carbon vapor pressure 0.13 kPa at 3860 K.

b: gasoline boiling range 320–490 K.

c: s^0 for C_6H_6(l) = 173.3, $h_{c,\ high}$ = 41833 kJ/kg

d: s^0 for C_6H_6(l) = 204.4, $h_{c,\ high}$ = 46580 kJ/kg

e: s^0 for ethanol(l) = 2827, $h_{c,\ high}$ = 30585 kJ/kg

f: s^0 for methanol(l) = 126.8, $h_{c,\ high}$ = 34932 kJ/kg

t: s^0 for octane(l) = 358, $h_{c,\ high}$ = 48268 kJ/kg

Source: Some data has been taken from Lide, D.R., Ed., *CRC Handbook of Chemistry and Physics*, 80th ed., CRC Press, Boca Raton, FL, 1999.

Table A.2B Analysis of Typical U.S. Solid Fuels

		Proximate analysis, %					Ultimate analysis, %			HHV (or gross)	
Classification	*State, County*	*Moisture*	VM	*FC*	*Ash*	*S*	*H*	*C*	*N*	*O*	*MJ/kg*
Meta-antracite	Rhode Island,	13.2	2.6	65.3	18.9	0.3	1.9	64.2	0.2	14.5	21.64
	Newport		3.8	96.2		0.4	0.6	94.7	0.3	4.0	31.89
Anthracite	Pennsylvania,	4.3	5.1	81.0	9.6	0.8	2.9	79.7	0.9	6.1	29.93
	Lackawanna		5.9	94.1		0.9	2.8	92.5	1.0	2.8	34.81
Semianthracite	Arkansas, Johnson	2.6	10.6	79.3	7.5	1.7	3.8	81.4	1.6	4.0	32.26
			11.7	88.3		1.9	3.9	90.6	1.8	1.8	35.86
Low-volatile	West Virginia,	2.9	17.7	74.0	5.4	0.8	4.6	83.2	1.3	4.7	33.47
bituminous	Wyoming		19.3	80.7		0.8	4.6	90.7	1.4	2.5	36.46
Medium-volatile	Pennsylvania,	2.1	24.4	67.4	6.1	1.0	5.0	81.6	1.4	4.9	33.26
bituminous	Clearfield		26.5	73.5		1.1	5.2	88.9	1.6	3.2	36.23
High-volatile A	West Virginia,	2.3	36.5	56.0	5.2	0.8	5.5	78.4	1.6	8.5	32.63
bituminous	Marion		39.5	60.5		0.8	5.7	84.8	1.7	7.0	35.28
High-volatile B	Kentucky,	8.5	36.4	44.3	10.8	2.8	5.4	65.1	1.3	14.6	27.14
bituminous	Muhlenbur		45.0	55.0		3.4	5.5	80.6	1.7	8.8	33.61
High-volatile C	Illinois, Sangamon	14.4	35.4	40.6	9.6	3.8	5.8	59.7	1.0	20.1	25.12
bituminous			46.6	53.4		5.0	5.6	78.6	1.3	9.5	33.07
Subbituminous	Wyoming,	16.9	34.8	44.7	3.6	1.4	6.0	60.4	1.2	27.4	24.75
A	Sweetwater		43.7	56.3		1.8	5.2	76.0	1.5	15.5	31.12
Subbituminous B	Wyoming,	22.2	33.2	40.3	4.3	0.5	6.9	53.9	1.0	33.4	22.33
	Sheridan		45.2	54.8		0.6	6.0	73.4	1.3	18.7	30.40
Subbituminous C	Colorado, El	25.1	30.4	37.7	6.8	0.3	6.2	50.5	0.7	35.5	19.89
	Paso		44.6	55.4		0.5	5.0	74.1	1.1	19.3	29.19
Lignite	North Dakota,	36.8	27.8	29.5	5.9	0.9	6.9	40.6	0.6	45.1	16.27
	McLean		48.4	51.6		1.6	5.0	70.9	1.1	21.4	28.42

Note: Approximate heating values from the Boie equation are HHV_F, $kJ/kg_{fuel} = 35160Y_C + 116225Y_H - 11090Y_O + 6280Y_N + 10465Y_S$, LHV_F, $kJ/kg_{fuel} = 35160Y_C + 94438Y_H - 11090Y_O + 6280Y_N + 10465Y_S$ Y_c, carbon mass fraction = {C%/100}.

[a]Organic sulfur
[b]By difference

Table A.2C Thermophysical and Thermochemical Properties of Some Plastics

Fuel Materials	Formula	T_g K	q_g kJ/Kg	h_c kJ/Kg	ν_{O2}	h_c/ν_{O2} kJ/Kg	$Y_{O2\infty}/\nu_{O2}$	B	s/B	q_g/h_c	ρ kg/m[a]	c_p kJ/kg K	λ W/m K
Solids/Plastics													
Polypropylene	C_3H_6	703	2030	45480	3.42	13280	0.0672	1.29	0.0523	0.0446	902	1.926	0.117
Polyethylene	C_2H_4	753	2330	47320	3.42	13820	0.0672	1.16	0.058	0.0492	910-965	2.301	0.335–0.519
Polystyrene	C_8H_8	661	1760	41580	3.07	13540	0.0749	1.55	0.0484	0.0423	1065	1.340	0.119
Nylon 6/6	$C_{12}H_{22}N_2O_3$		2350	33160	2.33	14210	0.099	1.27	0.0780	0.0709	1130-1420	1.674	0.243
Polycarbonate	$C_6H_{11}O_3$		2070	31820	2.27	14050	0.102	1.41	0.0729	0.0651	1200	1.21	0.193
Polyacrylonitrile	C_3H_3N	513–713		30990	2.24	13870	0.103						
Phenyl formaldehyde	$C_{15}H_{12}O_2$			26730	2.43	11020	0.095						
PMMA	$C_5H_8O_2$	660	1590	26560	1.92	13850	0.120	1.78	0.0674	0.06	1180	2.41	0.159
PVC	C_2H_3Cl	553–673	2470	19370	1.40	13800	0.163	1.15	0.141	0.128	1380		
Polyvinylidene fluoride	$C_2H_2F_2$	723		8480	1.25	6800	0.184						
Fir wood	$C_{4.8}H_8O_4$		2430	17970	1.18	15170	0.194	1.75					
α-cellulose	$C_6H_{10}O_5$	573	370	14820	1.19	12510	0.194	6.96	0.0279	0.0279			
Polyoxymethylene	CH_20		2430	17900	1.07	16790	0.216	1.47	0.147	0.136			

Source: Annamalai and Sibulkin, *Combust. Sci. Technol.*, 1978; q_g: heat of garification

Table A.2D Ultimate Analyses and Heating Values of Biomass Fuels

	Ultimate Analysis, % Dry Weight						HHV MJ/kg	
	C	H	O	N	S	Residue	Measured HV_M	Estimated* HV_{EB}
Field crops								
Alfalfa seed straw	46.76	5.40	40.72	1.00	0.02	6.07	18.45	18.27
Bean straw	42.97	5.59	44.93	0.83	0.01	5.54	17.46	16.68
Corn cobs	46.58	5.87	45.46	0.47	0.01	1.40	18.77	18.19
Corn stover	43.65	5.56	43.31	0.61	0.01	6.26	17.65	17.05
Cotton stalks	39.47	5.07	39.14	1.20	0.02	15.10	15.83	15.51
Rice straw (fall)	41.78	4.63	36.57	0.70	0.08	15.90	16.28	16.07
Rice straw (weathered)	34.60	3.93	35.38	0.93	0.16	25.00	14.56	12.89
Wheat straw	43.20	5.00	39.40	0.61	0.11	11.40	17.51	16.68
Orchard prunings								
Almond prunings	51.30	5.29	40.90	0.66	0.01	1.80	20.01	19.69
Black walnut	49.80	5.82	43.25	0.22	0.01	0.85	19.83	19.50
English walnut	49.72	5.63	43.14	0.37	0.01	1.07	19.63	19.27
Vineyard prunings								
Cabernet sauvignon	46.59	5.85	43.90	0.83	0.04	2.71	19.03	18.37
Chenin blanc	48.02	5.89	41.93	0.86	0.07	3.13	19.13	19.14
Pinot noir	47.14	5.82	43.03	0.86	0.01	3.01	19.05	18.62
Thompson seedless	47.35	5.77	43.32	0.77	0.01	2.71	19.35	18.60
Tokay	47.77	5.82	42.63	0.75	0.03	2.93	19.31	18.88
Energy crops								
Eucalyptus								
Camaldulensis	49.00	5.87	43.97	0.30	0.01	0.72	19.42	19.19
Globulus	48.18	5.92	44.18	0.39	0.01	1.12	19.23	18.95
Grandis	48.33	5.89	45.13	0.15	0.01	0.41	19.35	18.84
Casuarina	48.61	5.83	43.36	0.59	0.02	1.43	19.44	19.10
Cattails	42.99	5.25	42.47	0.74	0.04	8.13	17.81	16.56
Popular	48.45	5.85	43.69	0.47	0.01	1.43	19.38	19.02
Sudan grass	44.58	5.35	39.18	1.21	0.08	9.47	17.39	17.63
Forest residue								
Black locust	50.73	5.71	41.93	0.57	0.01	0.97	19.71	19.86
Chaparral	46.9	5.08	40.17	0.54	0.03	7.26	18.61	17.98
Madrone	48	5.96	44.95	0.06	0.02	1	19.41	18.82
Manzanita	48.18	5.94	44.68	0.17	0.02	1	19.3	18.9
Ponderosa pine	49.25	5.99	44.36	0.06	0.03	0.3	20.02	1937
Ten oak	47.81	5.93	44.12	0.12	0.01	2	18.93	18.82
Redwood	50.64	5.98	42.88	0.05	0.03	0.4	20.72	20.01
White fir	49	5.98	44.75	0.05	0.01	0.2	19.95	19.22

(continued)

Table A.2D Ultimate Analyses and Heating Values of Biomass Fuels (Continued)

	Ultimate Analysis, % Dry Weight						HHV MJ/kg	
	C	H	O	N	S	Residue	Measured HV_M	Estimated* HV_{EB}
Food- and fiber-processing waste								
Almond hulls	45.79	5.36	40.6	0.96	0.01	7.2	18.22	17.89
Almond shells	44.98	5.97	42.27	1.16	0.02	5.6	19.38	18.14
Babassu husks	50.31	5.37	42.29	0.26	0.04	1.73	19.92	19.26
Sugarcane bagasse	44.8	5.35	39.55	0.38	0.01	9.79	17.33	17.61
Coconut fiber dust	50.29	5.05	39.63	0.45	0.16	4.14	20.05	19.2
Cocoa hulls	48.23	5.23	33.09	2.98	0.12	10.25	19.04	19.56
Cotton gin trash	39.59	5.26	36.33	2.09		16.68	16.42	16.13
Macadamia shells	54.41	4.99	39.69	0.36	0.01	0.56	21.01	20.55
Olive pits	48.81	6.23	43.48	0.36	0.02	1.1	21.39	19.61
Peach pits	53	5.9	39.14	0.32	0.05	1.59	20.82	21.18
Peanut hulls	45.77	5.46	39.56	1.63	0.12	7.46	18.64	18.82
Pistachio shells	48.79	5.91	43.41	0.56	0.01	1.28	19.26	19.25
Rice hulls	40.96	4.3	35.86	0.4	0.02	18.34	16.14	15.45
Walnut shells	49.98	5.71	43.35	0.21	0.01	0.71	20.18	19.45
Wheat dust	41.38	5.1	35.19	3.04	0.19	15.1	16.2	16.78

* Estimated from Boie equation, see Table 2B footnote; M: measured, EB: estimated Boie

Source: Adapted from Eberling and Jenkins, 1985; Annamalai, K., Sweeten, J., and Ramalingam, S.C., Estimation of the gross heating values of biomass fuels, *Trans. Soc. Agric. Engineers*, 30, 1205–1208, 1987.

Table A.2E Mineral Matter Properties

	FeO	SiO_2	Al_2O_3	MgO
Fusion temperature, K	1644	1996	2310	3098
Specific heat of solid, kJ/kg K	0.95	1.43	1.882	1.66
Specific heat of liquid, kJ/kg K	1.506	1.506	1.381	1.506
Density, kg/m³	5240	4000	4000	3000
Heat of fusion, kJ/kg K	435	1464	1068	1921
Heat of sublimation, Q_S, solid to gas, kJ/kg	3637	9791	5870	13489
Heat of vaporization, Q_L, liquid to gas, kJ/kg	3203	9644	4803	11569
P_S[a]	6.44E+11	3.41E+10	2.56E+08	5.26E+08
P_L[b]	6.44E+11	3.408E+11	2.56E+08	5.26E+08
QS, kJ/kmol	610027	505009	523000	551451
QL, kJ/kmol	610027	505009	523000	551451
Molecular weight, kg/kmol	71.85	60.06	101.9	40.32
Normal boiling point, K	2700	2250	3253	3538

[a] p^{sat} (sublimation) = $P_s \exp(-Q_S/RT)$.

[b] p^{sat} (evaporation) = $P_L \exp(-Q_L/RT)$.

Table A.2F Melting and Boiling Points of Ash Components

Inorganic	T_{MP}, *K (°F)*	T_{BP}, *K (°F)*
Alumina Al_2O_3	2303.15 K (3686°F) 2318.15 K (3713°F)	3253.15 K (5396°F)
Calcium CaO(lime)	2843.15 K (4658°F) 2853.15 K (4676°F)	3123.15 K (5162°F)
Fe	1808 K (2794°F)	2236 K (3565°F)
Siderite $FeCO_3$ (ferrous carbonate)	1873 K (2911°F)	
Wuestite, FeO	1688 K (2560°F)	
Haematite Fe_2O_3 (ferric oxide)	1811.15 K (2800.4°F) (1838.15 K (2849°F))	
Magnetite, Fe_3O_4		
Magnesium MgO	3073.15 K (502794 72°F)	3873.15 K (6512°F)
Manganese MnO	1353.15 K (1976°F)	
Phosphorus P_2O_5	613.15 K (644°F) Sublimation temperature = 633.15 K (680°F)	
Potassium K_2O (volatile oxide)	Decomposes at 623.15 K (662°F)	< 1373.15 K (2012°F)
Potassium carbonate K_2CO_3	1164.15 K (1635.8°F)	
Silica SiO_2	1983.15 K (3110°F)	2503.15 K (4046°F)
Sodium Na_2O (volatile oxide)	Decompose at temperature > 673.15 K (752°F) Sublimated at 1548.15 K (2327°F)	< 1373.15 K (2012°F)
Sodium carbonate, Na_2CO_3	1124.15 K (1563.8°F)	
Sulfur SO_3	289.95 K (62.24°F)	317.95 K (112.64°F)
Titanium oxide	1973.15 K (3092°F)	> 3273.15 K (5432°F)

Note: Other minerals in coal: Albite: $NaAlSi_3O$; Anhydrite: $CaSO_4$; Anorthite: $CaAl_2Si_2O_3$; Calcite: $CaCO_3$; Corundum solution: $Al_2O_3—Cr_2O_3—Fe_2O_3$; Dolomite: CaO + MgO; Fayalite: (Fe_2SiO_4, 2 units FeO, 1 unit SiO_2), T_{MP} = 1205°C; FeO + Al_2O_3 + SiO_2 contamination can result in low T_{MP} i.e., 1080°C; Ferrosilite: ($Fe_2Si_2O_6$, 2 units FeO, 2 units SiO_2); Gehlinite: $Ca_2Al_2SiO_7$; Gillespite: $CaCrSi_4O_{10}$; Gypsum: $CaSO_4$. $2H_2O$; Kaolinite: $Al_2Si_2O_{2.2}H_2O$; Leucite: $KAlSi_2O_6$; Liquid slag: $SiO_2—Al_2O_3—CaO—FeO—Fe_2O_3$-CrO—$Cr_2O_3—Na_2O—K_2O$; Metakaolnite: $Al_2Si_2O_2$; Mullite: $Al_2Si_2O_{13}$; Nephline: $NaAlSiO_4$; Pyrite: FeS_2; Triolite: FeS.

Source: www.chemfinder.com, www.abcr.de, Merck Index

Table A.3 Lennard–Jones Parameters

Species	Molecular Weight	ε/k_B, K	σ, nm	ε/k_B, K	σ, nm	b_o, $m^3/kmol \times 10^3$
Al	26.98	2750	0.2655			
Air	28.96	78.6	0.3711	99.2	0.3522	55.11
Ar	39.95	93.3	0.3542	119.8	0.3405	49.80
B	10.81	93.3	0.2265			
BO	26.81	3331	0.2944			
B_2O_3	69.62	596	0.4158			
Br	79.92	236.6	0.3672			
Br_2	159.83	507.9	0.4296	520	0.427	
C	12.01	2092	0.3385			
CCl_4	153.84	327	0.5881			
CF_4	88.01	137	0.3882			
CH	13.009	68.6	0.3370			
CH-CH	26.018	185	0.4221			
CH_2-CH_2	28.06	205	0.4232			
CH_2Cl_2	84.94	406	0.4759			
CH_3Cl	50.49	855	0.3375			
CH_3OH	32.05	481.8	0.3626			
CH_4	16.043	148.6	0.3758	148.2	0.3817	70.16
$CHCl_3$	119.39	327	0.543			
CN	26.02	75.0	0.3856			
CO	28.01	91.7	0.3690	100.2	0.3763	67.22
CO_2	44.01	195.2	0.3941	189.0	0.4486	113.9
COS	60.08	335	0.413			
CS_2	76.14	488	0.4438			
i-C_4H_{10}	58.14	313	0.5341			
n-C_4H_{10}	58.14	410	0.4997			
C_2H_2	26.038	231.8	0.4033			
C_2H_4	28.054	224.7	0.4163	199.2	0.4523	116.7
C_2H_6	30.07	230	0.4418		0.422	
C_2N_2	52.04	339	0.438			
n-C_3H_7OH	60.1	576.7	0.455			
C_3H_8	44.09	254	0.5061			
i-C_4H_{10}	58.12	330.1	0.528			
C_4H_{10}-n	58.12	531.4	0.469			
C_5H_{12}	72.15	345	0.5769			
C_6H_6	78.11	412.3	0.5349			
C_6H_6	78.11	440	0.5270			
C_6H_{12}	84.16	297.1	0.618			
c-C_6H_{12}	84.16	324	0.6093			
C_6H_{14}	86.17	413	0.5909			
C_8H_{18}	114.22	320	0.7451			
Cl	35.45	130.8	0.3613			

Table A.3 Lennard–Jones Parameters (Continued)

Species	*Molecular Weight*	*ε/k_B, K*	*σ, nm*	*ε/k_B, K*	*σ, nm*	*b_o, m³/kmol × 10³*
Cl_2	70.91	316.0	0.4217			
F	18.99	112.6	0.2986			
F_2	37.99	112.6	0.3357			
H	1.008	37.0	0.2708			
HBr	80.92	449.0	0.3353			
H_2	2.016	59.7	0.2827	29.2	0.287	29.76
H_2O	18.016	809.1	0.2641			
H_2O_2	34.016	389.3	0.420			
HCl	36.47	344.7	0.3339			
HCN	27.03	569.1	0.3630			
He	4.00	10.22	0.2551	10.22	0.2556	21.07
HF	20.01	330.0	0.3148			
I_2	253.82	550	0.4982			
Kr	83.80	190	0.361			
N	14.007	71.4	0.3298			
NH	15.015	65.3	0.3312			
NH_3	17.031	558.3	0.2900			
NO	30.008	116.7	0.3492	131.0	0.317	40.0
N_2	28.013	71.4	0.3798	95.05	0.3698	63.78
N_2O	44.02	232.4	0.3828	189.0	0.459	122.0
Ne	20.18	32.8	0.2820	35.60	0.2749	26.21
O	16.000	106.7	0.3050			
OH	17.008	79.8	0.3147			
O_2	31.999	106.7	0.3467	117.5	0.358	57.75
S	32.07	847	0.384			
SF_6	146.7	200.9	0.551			
SO	48.07	301	0.399			
SO_2	64.07	335.4	0.411			
SO_2	64.07	252	0.4290			
Xe	131.3	229	0.4055			

Note: If data are not found, use the approximate formulae: $\varepsilon/k_B = 0.77\ T_c$, $1.15\ T_{BP}$, $1.9\ T_{MP}$ where T_{BP} is the boiling point in K; T_{MP} melting point in K. σ (nm) $= 0.841\ \bar{v}_c^{1/3}$ (m³/kmol), $1.166\ \bar{v}_b^{1/3}$, $1.122\ \bar{v}_m^{1/3}$. $\bar{v}_c$, $\bar{v}_b$, $\bar{v}_m$ are specific volumes—critical, boiling, and melting, respectively. The first set (columns 3 and 4) is from viscosity data and the second (columns 5 and 6) from second virial coefficient data. $b_0 = (2/3)\ \pi\sigma^3\ N_{avog} \times 10^{-27}$ m³/kmol, where σ is in nm. It is suggested that the viscosity data be used exclusively for transport coefficient calculations and that the second virial coefficient be used exclusively for equation of state calculations.

Source: Data taken from Svehla, R.A., NASA Technical Report R–132, Lewis Research Center, Cleveland, OH, 1962. Part of data and tables taken from Hirshfelder, J.O., Curtis, C.F., and Bird, R.B., *Molecular Theory of Gases and Liquids*, John Wiley & Sons, New York, 1954. (With permission); also see Reid, R.C., Prausnitz, J.M., and Poling, B.E., *The Properties of Gases and Liquids*, 4th ed., McGraw-Hill, New York, 1987.

Table A.4A Properties (Density and c_p) of Some Solid Fuels and Solids at 25 °C

Substance	Solids ρ kg/m³	c_p kJ/kg K
Carbon, diamond	3250	0.51
Carbon, graphite	2000–2500	0.61
Coal	1200–1500	1.26
Coal ash	2100–2700	0.96–1.13
Ice (0°C)	917	2.04
Paper	700	1.2
Plexiglas (PMMA)	1180	1.44
Polystyrene	920	2.3
Polyvinyl chloride	1380	0.96
Rubber, soft	1100	1.67
Sand, dry	1500	0.8
Silicon	2330	0.70
Wood, hard (oak)	720	1.26
Wood, soft (pine)	510	1.38
Wool	100	1.72
Aluminum	2700	0.90

Table A.4B Properties (Density and c_p) of Some Liquid Fuels and Liquids at 25°C

Substance	*Liquids* ρ *kg/m*^{3}	c_P *kJ/kg K*
Ammonia	604	4.84
Benzene	879	1.72
Butane	556	2.47
CCl_4	1584	0.83
Ethanol	783	2.46
Gasoline	750	2.08
Glycerin	1260	2.42
Kerosene	815	2.0
1-Methanol	787	2.55
n-octane	692	2.23
Oil, engine	885	1.9
Oil, light	910	1.8
Propane	510	2.54
R-12	1310	0.97
R-22	1190	1.26
R-134a	1206	1.43
Water	997	4.18
Liquid metals		
Lead, Pb	10660	0.16
Mercury, Hg	13580	0.14
Potassium, K	828	0.81
Sodium, Na	929	1.38
Tin, Sn	6950	0.24

Note: For petroleum oils, at 15°C, API = {141.5/SG}-131.5; SG: Specific gravity, c (kJ/kgK) = {1.685 + (0.039 * T in °C)}/{SG}$^{1/2}$, 0 < T < 205; °C = 0.75 < SG < 0.96 [Baukal et al.].

Source: Adapted from Sonntag, R., Borgnakke, C., and Wylen, G.J., Fundamentals of Classical Thermodynamics, 5th ed., John Wiley & Sons, New York, 1998.

Table A.5 Properties of Various Ideal Gases at 25°C, 100 kPa

Gas	Chemical Formula	Molecular Mass	R kJ/kg K	ρ kg/m^3	c_{p0} kJ/kg K	c_{v0} kJ/kg K	k
Acetylene	C_2H_2	26.038	0.3193	1.05	1.699	1.380	1.231
Air		28.97	0.287	1.169	1.004	0.717	1.400
Ammonia	NH_3	17.031	0.4882	0.694	2,130	1.642	1.297
Argon	Ar	39.948	0.2081	1.613	0.520	0.312	1.667
Benzene	C_6H_6	78.11	0.1064	3.151	0775	0.67	1.157
Butane	C_4H_{10}	58.124	0.1430	2.407	1.716	1.573	1.091
Carbon monoxide	CO	28.01	0.2968	1.13	1.041	0.744	1.399
Carbon dioxide	CO	44.01	0.1889	1.775	0.842	0.653	1.289
Ethane	C_2H_6	30.07	0.2765	1.222	1.766	1.490	1.186
Ethanol	C_2H_5OH	46.069	0.1805	1.883	1.427	1.246	1.145
Ethylene	C_2H_4	28.054	0.2964	1.138	1,548	1.252	1.237
Helium	He	4.003	2.0771	0.1615	5.193	3.116	1.667
Hydrogen	H_2	2.016	4.1243	0.0813	14.209	1.008	1.409
Methane	CH_4	16.043	0.5183	0.648	2.254	1.736	1.299
Methanol	CH_3OH	32.042	0.2595	1.31	1.405	1.146	1.227
Neon	Ne	20.183	0.4120	0.814	1.03	0.618	1.667
Nitric oxide	NO	30.006	0.2771	1.21	0.993	0,716	1.387
Nitrogen	N_2	28.013	0.2968	1.13	1.042	0.745	1.400
Nitrous oxide	N_2O	44.013	0.1889	1.775	0.879	0.690	1.274
n-Octane	C_8H_{18}	114.23	0.07279	0.092	1.711	1.638	1.044
Oxygen	O_2	31.999	0.2598	1.292	0.922	0.662	1.393
Propane	C_3H_8	44.094	0.1886	1.808	1.679	1.490	1.126
R-12	CCl_2F_2	120.914	0.06876	4.98	0.616	0.547	1.126
R-22	$CHClF_2$	86.469	0.09616	3.54	0.658	0.562	1.171
R-134a	$C_2F_4H_2$	102.03	0.08149	4.20	0.852	0.771	1.106
Sulfur dioxide	SO_2	64.059	0.1298	2.618	0.624	0.494	1.263
Sulfur trioxide	SO_3	80.053	0.10386	3.272	0.635	0.531	1.196
Water/steam	H_2O	18.02	0.4614	0.727	1.86	1.40	1.329

Table A.6A Curve Fit for Thermodynamic Properties of Gases[a]

Species	*T(K)*	a_1	a_2	a_3	a_4	a_5	a_6	a_7
CO	1,000–5,000	0.03025078E+02	0.14426885E–02	–0.05630827E–05	0.10185813E–09	–0.06910951E–13	–0.14268350E+05	0.06108217E+02
	300–1,000	0.03262451E+02	0.15119409E–02	–0.03881755E–04	0.05581944E–07	–0.02474951E–10	–0.14310539E+05	0.04848897E+02
CO_2	1,000–5,000	0.04453623E+02	0.03140168E–01	–0.12784105E–05	0.02393996E–08	–0.16690333E–13	–0.04896696E+06	–0.09553959E+01
	300–1,000	0.02275724E+02	0.09922072E–01	–0.10409113E–04	0.06866686E–07	–0.02117280E–10	–0.04837314E+06	0.10188488E+02
H_2	1,000–5,000	0.02991423E+02	0.07000644E–02	–0.05633828E–06	–0.09231578E–10	0.15827519E–14	–0.08350340E+04	–0.13551101E+01
	300–1,000	0.03298124E+02	0.08249441E–02	–0.08143015E–05	0.09475434E–09	0.04134872E–11	–0.10125209E+04	–0.03294094E+02
H	1,000–5,000	0.02500000E+02	0.00000000E+00	0.00000000E+00	0.00000000E+00	0.00000000E+00	0.02547162E+06	–0.04601176E+01
	300–1,000	0.02500000E+02	0.00000000E+00	0.00000000E+00	0.00000000E+00	0.00000000E+00	0.02547162E+06	–0.04601176E+01
OH	1,000–5,000	0.02882730E+02	0.10139743E–02	–0.02276877E–05	0.02174683E–09	–0.05126305E–14	0.03886888E+05	0.05595712E+02
	300–1,000	0.03637266E+02	0.01850910E–02	–0.16761646E–05	0.02387202E–07	–0.08431442E–11	0.03606781E+05	0.13588605E+01
H_2O	1,000–5,000	0.02672145E+02	0.03056293E–01	–0.08730260E–05	0.12009964E–09	–0.06391618E–13	–0.02989921E+06	0.06862817E+02
	300–1,000	0,03386842E+02	0.03474982E–01	–0.06354696E–04	0.06968581E–07	–0.02506588E–10	–0.03020811E+06	0.02590232E+02
N_2	1,000–5,000	0.02926640E+02	0.14879768E–02	–0.05684760E–05	0.10097038E–09	–0.06753351E–13	–0.09227977E+04	0.05980528E+02
	300–1,000	0.03298677E+02	0.14082404E–02	–0.03963222E–04	0.05641515E–07	–0.02444854E–10	–0.10209999E+04	0.03950372E+02
N	1,000–5,000	0.02450268E+02	0.10661458E–03	–0.07465337E–06	0.01879652E–09	–0.10259839E–14	0.05611604E+06	0.04448758E+02
	300–1,000	0.02503071E+02	–0.02180018E–03	0.05420529E–06	–0.05647560E–09	0.02099904E–12	0.05609890E+06	0.04167566E+02
NO	1,000–5,000	0.03245435E+02	0.12691383E–02	–0.05015890E–05	0.09169283E–09	–0.06275419E–13	0.09800840E+05	0.06417293E+02
	300–1,000	0.03376541E+02	0.12530634E–02	–0.033027SOE–04	0.05217810E–07	–0.02446262E–10	0.09817961E+05	0.05829590E+02
NO_2	1,000–5,000	0.04682859E+02	0.02462429E–01	–0.10422585E–05	0.01976902E–08	–0.13917168E–13	0.02261292E+05	0.09885995E+01
	300–1,000	0.02670600E+02	0.07838500E–01	–0.08063864E–04	0.06161714E–07	–0.02320150E–10	0.02896290E+05	0.11612071E+02
O_2	1,000–5,000	0.03697578E+02	0.06135197E–02	–0.12588420E–06	0.01775281E–09	–0.11364354E–14	–0.12339301E+04	0.03189165E+02
	300–1,000	0.03212936E+02	0.11274864E–02	–0.05756150E–05	0.13138773E–08	–0.08768554E–11	–0.10052490E+04	0.06034737E+02
O	1,000–5,000	0.02542059E+02	–0.02755061E–03	–0.03102803E–07	0.04551067E–10	–0.04368051E–14	0.02923080E+06	0.04920308E+02
	300–1,000	0.02946428E+02	–0.16381665E–02	0.02421031E–04	–0.16028431E–08	0.03890696E–11	0.02914764E+06	0.02963995E+02

Note: Enthalpies of elements in natural form at 298 K, 1 atm = 0; h^0 includes enthalpy of formation.

Source: Adapted from Turns, S.R., *An Introduction to Combustion*, 2nd ed., McGraw-Hill, New York, 2000. Originally from Kee, R.J., Rupley, F.K., and Miller, J.A., The Chemkin Thermodynamic Data Base, Sandia Report, SAND87–8215B, reprinted March 1991.

a: $\bar{c}_{p0}/\bar{R} = c_p/R\, a_1 + a_2T + a_3T^2 + a_4T^3 + a_5T^4$, $\frac{\bar{s}_0}{\bar{R}} = (s_j/R) = a_1 \ln(T) + a_2T + \frac{a_3}{2}T^2 + \frac{a_4}{3}T^3 + \frac{a_5}{4}T^4 + a_7$ total entropy, includes s_0^0 at 298K $\frac{\bar{h}_0}{\bar{R}T} = (h_0/RT) = a_1 + \frac{a_2}{2}T + \frac{a_3}{3}T^2 + \frac{a_4}{4}T^3 + \frac{a_5}{4}T^4 + \frac{a_6}{T}$, total enthalpy, includes $\bar{h}_f^0$

Table A.6B Curve Fit for Thermodynamic Properties of Gaseous Fuels and Gases

Chemical Species		T_{max}	*A*	10^3B	10^6C	$10^{-5}D$
Paraffins						
Methane	CH_4	1,500	1.702	9.081	–2.164	
Ethane	C_2H_6	1,500	1.131	19.225	–5.561	
Propane	C_3H_8	1,500	1.213	28.785	–8.824	
n-Butane	C_4H_{10}	1,500	1.935	36.915	–11.402	
iso-Butane	C_4H_{10}	1,500	1.677	37.853	–11.945	
n-Pentane	C_5H_{12}	1,500	2.464	45.351	–14.111	
n-Hexane	C_6H_{14}	1,500	3.025	53.722	–16.791	
n-Heptane	C_7H_{16}	1,500	3.570	62.127	–19.486	
n-Octane	C_8H_{18}	1,500	8.163	70.567	–22.208	
1-Alkenes						
Ethylene	C_2H_4	1,500	1.424	14.394	–4.392	
Propylene	C_3H_6	1,500	1.637	22.706	–6.915	
1–Butene	C_4H_8	1,500	1.967	31.630	–9.873	
1–Pentene	C_5H_{10}	1,500	2.691	39.753	–12.447	
1–Hexene	C_6H_{12}	1,500	3.220	48.189	–15.157	
1–Heptene	C_7H_{14}	1,500	3.768	56.588	–17.847	
1–Octene	C_8H_{16}	1,500	4.324	64.960	–20.521	
Miscellaneous organics						
Acetylene	C_2H_2	1,500	6.132	1.952		–1.299
Benzene	C_6H_6	1,500	–0.206	39.064	–13.301	
Cyclohexane	C_6H_{12}	1,500	–3.876	63.249	–20.928	
Ethanol	C_2H_6O	1,500	3.518	20.001	–6.002	
Ethylbenzene	C_8H_{10}	1,500	1.124	55.380	–18.476	
Methanol	CH_4O	1,500	2.211	12.216	–3.450	
Toluene	C_7H_8	1,500	b.290	47.052	–15.716	
Styrene	C_8H_8	1,500	2.050	50.192	–16.662	
Miscellaneous inorganics						
Air		2,000	3.355	0.575		–0.016
Ammonia	NH_3	1,800	3.578	3.020		–0.186
Dinitrogen tetroxide	N_2O_4	2,000	11.660	2.257		–2.787
Sulfur dioxide	SO_2	2,000	5.699	0.801		–1.015
Sulfur trioxide	SO_3	2,000	8.060	1.056		–2.028

Note:

$c_{P,0}/R = A + BT + CT^2 + D\,T^{-2}$ from 298 K to T_{max}.

$(h_0 - h_{0,ref})/R = A(T - T_{ref}) + (B/2)(T^2 - T_{ref}^2) + (C/3)\,(T^3 - T_{ref}^3) - D(1/T - 1/T_{ref})$

$(s_0 - s_{0,ref})/R = A\ln(T/T_{ref}) + B(T - T_{ref}) + (C/2)(T^2 - T_{ref}^2) - (D/2)\,(1/T^2 - 1/T_{ref}^2)$, where h_{ref} and s_{ref} are typically set to zero at T_{ref} or h^0_f and s^0_f and $T_{ref} = 298$ K.

Source: Adapted from Smith and Van Ness, *Introduction to Chemical Engineering Thermodynamics*, 4th ed., McGraw-Hill, New York, 1987. (Selected from Spencer, H.M., *Ind. Eng. Chem.*, 40, 2152, 1948; Kelley, K.K., *U.S. Bur. Mines Bull.*, 584, 1960; Pankratz, L.B., *U.S. Bur. Mines Bull.*, 672, 1982).

Table A.7 Saturation Properties of H_2O (T and P Tables)

T as Independent Variable				*P as Independent Variable*			
T, °C	*P^{sat}, MPa*	*T, °C*	*P^{sat}, MPa*	*P, MPa*	*T^{sat}, °C*	*P, MPa*	*T^{sat}, °C*
0.01	0.000611	190	1.2544	0.00061	0.01	0.8	170.43
5	0.000872	195	1.3978	0.001	6.98	0.85	172.96
10	0.001228	200	1.5538	0.0015	13.03	0.9	175.38
15	0.001705	205	1.723	0.002	17.5	0.95	177.69
20	0.002339	210	1.9062	0.0025	21.08	1	179.91
25	0.003169	215	2.104	0.003	24.08	1.1	184.09
30	0.004246	220	2.318	0.004	28.96	1.2	187.99
35	0.005628	225	2.548	0.005	32.88	1.3	191.64
40	0.007384	230	2.795	0.0075	40.29	1.4	195.07
45	0.009593	235	3.06	0.01	45.84	1.5	198.32
50	0.012349	240	3.344	0.015	53.97	1.75	205.76
55	0.015758	245	3.648	0.02	60.06	2	212.42
60	0.01994	250	3.973	0.025	64.97	2.25	218.45
65	0.02503	255	4.319	0.03	69.1	2.5	223.99
70	0.03119	260	4.688	0.04	75.87	3	233.9
75	0.03858	265	5.081	0.05	81.33	3.5	242.6
80	0.04739	270	5.499	0.075	91.78	4	250.4
85	0.05783	275	5.942	0.1	99.63	5	263.99
90	0.07014	280	6.412	0.125	105.99	6	275.64
95	0.08455	285	6.909	0.15	111.37	7	285.88
100	0.10135	290	7.436	0.175	116.06	8	295.06
105	0.12082	295	7.993	0.2	120.23	9	303.4
110	0.14327	300	8.581	0.225	124	10	311.06
115	0.16906	305	9.202	0.25	127.44	11	318.15
120	0.19853	310	9.856	0.275	130.6	12	324.75
125	0.2321	315	10.547	0.3	133.55	13	330.93
130	0.2701	320	11.274	0.325	136.3	14	336.75
135	0.313	330	12.845	0.35	138.88	15	342.24
140	0.3613	340	14.586	0.375	141.32	16	347.44
145	0.4154	350	16.513	0.4	143.63	17	352.37
150	0.4758	360	18.651	0.45	147.93	18	357.06
155	0.5431	370	21.03	0.5	151.86	19	361.54
160	0.6178	374.14	22.09	0.55	155.48	20	365.81
165	0.7005			0.6	158.85	21	369.89
170	0.7917			0.65	162.01	22	373.8
175	0.892			0.7	164.97	22.09	374.14
180	1.0021			0.75	167.78		
185	1.1227						

Source: Adapted from Wylen, G.J. and Sonntag, R., *Fundamentals of Classical Thermodynamics*, 3rd ed., John Wiley & Sons, New York, 1986.

Table A.8 Transport and Thermal Properties of Air, Nitrogen, and Oxygen, 1 atm

T (*K*)	ρ (*kg/m*^{3})	c_p (*kJ/kg K*)	$\mu \times 10^7$ (*N sec/m*^{2})	$\nu \times 10^6$ (*m*^{2}*/sec*)	$\lambda \times 10^3$ (*W/m K*)	$\alpha_T \times 10^6$ (*m*^{2}*/sec*)	*Pr*
Air							
100	3.5562	1.032	71.1	2	9.34	2.54	0.786
150	2.3364	1.012	103.4	4.426	13.8	5.84	0.758
200	1.7458	1.007	132.5	7.59	18.1	10.3	0.737
250	1.3947	1.006	159.6	11.44	22.3	15.9	0.72
300	1.1614	1.007	184.6	15.89	26.3	22.5	0.707
350	0.995	1.009	208.2	20.92	30	29.9	0.7
400	0.8711	1.014	230.1	26.41	33.8	38.3	0.69
450	0.774	1.021	250.7	32.39	37.3	47.2	0.686
500	0.6964	1.03	270.1	38.79	40.7	56.7	0.684
550	0.6329	1.04	288.4	45.57	43.9	66.7	0.683
600	0.5804	1.051	305.8	52.69	46.9	76.9	0.685
650	0.5356	1.063	322.5	60.21	49.7	87.3	0.69
700	0.4975	1.075	338.8	68.1	52.4	98	0.695
750	0.4643	1.087	354.6	76.37	54.9	109	0.702
800	0.4354	1.099	369.8	84.93	57.3	120	0.709
850	0.4097	1.11	384.3	93.8	59.6	131	0,716
900	0.3868	1.121	398.1	102.9	62	143	0.72
950	0.3666	1.131	411.3	112.2	64.3	155	0.723
1,000	0.3482	1.141	424.4	121.9	66.7	168	0.726
1,100	0.3166	1.159	449	141.8	71.5	195	0.728
1,200	0.2902	1.175	473	162.9	76.3	224	0.728
1,300	0.2679	1.189	496	185.1	82	238	0.719
1,400	0.2488	1.207	530	213	91	303	0.703
1,500	0.2322	1.23	557	240	100	350	0.685
1,600	0.2177	1.248	584	268	106	390	0.688
1,700	0.2049	1.267	611	298	113	435	0.685
1,800	0.1935	1.286	637	329	120	482	0.683
1,900	0.1833	1.307	663	362	128	534	0.677
2,000	0.1741	1.337	689	396	137	589	0.672
2,100	0.1658	1.372	115	431	147	646	0.667
2,200	0.1582	1.417	740	468	160	714	0.655
2,300	0.1513	1.478	766	506	175	783	0.647
2,400	0.1448	1.558	792	547	196	869	0.630
2,500	0.1389	1.665	818	589	222	960	0.613
3,000	0.1135	2.726	955	841	486	1,570	0.536
Nitrogen (N_2)							
100	3.4388	1.07	68.8	2	9.58	2.6	0.768
150	2.2594	1.05	100.6	4.45	13.9	5.86	0.759
200	1.6883	1.043	129.2	7.65	18.3	10.4	0.736

Table A.8 Transport and Thermal Properties of Air, Nitrogen, and Oxygen, 1 atm (Continued)

T (K)	ρ (kg/m^3)	c_p (kJ/kg K)	$\mu \times 10^7$ (N sec/m^2)	$v \times 10^6$ (m^2/sec)	$\lambda \times 10^3$ (W/m K)	$\alpha_T \times 10^6$ (m^2/sec)	Pr
250	1.3488	1.042	154.9	11.48	22.2	15.8	0.727
300	1.1233	1.041	178.2	15.86	25.9	22.1	0.716
350	0.9625	1.042	200	20.78	29.3	29.2	0.711
400	0.8425	1.045	220.4	26.16	32.7	37.1	0.704
450	0.7485	1.05	239.6	32.01	35.8	45.6	0.703
500	0.6739	1.056	257.7	38.24	38.9	54.7	0.7
550	0.6124	1.065	274.7	44.86	41.7	63.9	0.702
600	0.5615	1.075	290.8	51.79	44.6	73.9	0.701
700	0.4812	1.098	321	66.71	49.9	94.4	0.706
800	0.4211	1.22	349.1	82.9	54.8	116	0.715
900	0.3743	1.146	375.3	100.3	59.7	139	0.721
1,000	0.3368	1.167	399.9	118.7	64.7	165	0.721
1,100	0.3062	1.187	423.2	138.2	70	193	0.718
1,200	0.2807	1.204	445.3	158.6	75.8	224	0.707
1,300	0.2591	1.219	466.2	179.90	81.0	256	0.701
Oxygen (O_2)							
100	3.945	0.962	76.4	1.94	9.25	2.44	0.796
150	2.585	0.921	114.8	4.44	13.8	5.8	0.766
200	1.93	0.915	147.5	7.64	18.3	10.4	0.737
250	1.542	0.915	178.6	11.58	22.6	16.0	0.723
300	1.284	0.92	207.2	16.14	26.8	22.7	0.711
350	1.100	0.929	233.5	21.23	29.6	29.0	0.733
400	0.962	0.942	258.2	26.84	33	36.4	0.737
450	0.8554	0.956	281.4	32.9	36.3	44.4	0.741
500	0.7698	0.972	303.3	39.4	41.2	55.1	0.716
550	0.6998	0.988	324	46.3	44.1	63.8	0.726
600	0.6414	1.003	343.7	53.59	47.3	73.5	0.729
700	0.5498	1.031	380.8	69.26	52.8	93.1	0.744
800	0.481	1.054	415.2	86.32	58.9	116	0.743
900	0.4275	1.074	447.2	104.6	64.9	141	0.74
1,000	0.3848	1.09	477	124	71	169	0.733
1,100	0.3498	1.103	505.5	144.5	75.8	196	0.736
1,200	0.3206	1.115	532.5	166.1	81.9	229	0.725
1,300	0.296	1.125	588.4	188.6	87.1	262	0.721

Note: Thermal conductivity of oils λ (kW/m K) = 1.172×10^{-9} (1 – 0.00054 × T in°C) × ρ in kg/m^3 at 15.6°C.

Source: Incropera, F.P. and DeWitt, D.P., *Fundamentals of Heat and Mass Transfer*, 5th ed., Reprinted by permission, ©2002, John Wiley & Sons.

Table A.9 Ideal Gas Properties of Air

T, K	*h, kJ/kg*	P_r	*u, kJ/kg*	v_r	s^0, *kJ/(kg·K)*	g^0, *kJ/(kg·K)*
200	199.97	0.3363	142.56	1707.0	1.29559	–59.148
210	209.97	0.3987	149.69	1512.0	1.34444	–72.3624
220	219.97	0.4690	156.82	1346.0	1.39105	–86.061
230	230.02	0.5477	164.00	1205.0	1.43557	–100.1611
240	240.02	0.6355	171.13	1084.0	1.47824	–114.7576
250	250.05	0.7329	178.28	979.0	1.51917	–129.7425
260	260.09	0.8405	185.45	887.8	1.55848	–145.1148
270	270.11	0.9590	192.60	808.0	1.59634	–160.9018
280	280.13	1.0889	199.75	738.0	1.63279	–177.0512
285	285.14	1.1584	203.33	706.1	1.65055	–185.26675
290	290.16	1.2311	206.91	676.1	1.66802	–193.5658
295	295.17	1.3068	210.49	647.9	1.68515	–201.94925
300	300.19	1.3860	214.07	621.2	1.70203	–210.419
305	305.22	1.4686	217.67	596.0	1.71865	–218.96825
310	310.24	1.5546	221.25	572.3	1.73498	–227.6038
315	315.27	1.6442	224.85	549.8	1.75106	–236.3139
320	320.29	1.7375	228.42	528.6	1.76690	–245.118
325	325.31	1.8345	232.02	508.4	1.78249	–253.99925
330	330.34	1.9352	235.61	489.4	1.79783	–262.9439
340	340.42	2.149	242.82	454.1	1.82790	–281.066
350	350.49	2.379	250.02	422.2	1.85708	–299.488
360	360.58	2.626	257.24	393.4	1.88543	–318.1748
370	370.67	2.892	264.46	367.2	1.91313	–337.1881
380	380.77	3.176	271.69	343.4	1.94001	–356.4338
390	390.88	3.481	278.93	321.5	1.96633	–375.9887
400	400.98	3.806	286.16	301.6	1.99194	–395.796
410	411.12	4.153	293.43	283.3	2.01699	–415.8459
420	421.26	4.522	300.69	266.6	2.04142	–436.1364
430	431.43	4.915	307.99	251.1	2.06533	–456.6619
440	441.61	5.332	315.30	236.8	2.08870	–477.418
450	451.80	5.775	322.62	223.6	2.11161	–498.4245
460	462.02	6.245	329.97	211.4	2.13407	–519.6522
470	472.24	6.742	337.32	200.1	2.15604	–541.0988
480	482.49	7.268	344.70	189.5	2.17760	–562.758
490	492.74	7.824	352.08	179.7	2.19876	–584.6524
500	503.02	8.411	359.49	170.6	2.21952	–606.74
510	513.32	9.031	366.92	162.1	2.23993	–629.0443
520	523.63	9.684	374.36	154.1	2.25997	–651.5544
530	533.98	10.37	381.84	146.7	2.27967	–674.2451
540	544.35	11.10	389.34	139.7	2.29906	–697.1424
550	555.74	11.86	396.86	133.1	2.31809	–719.2095
560	565.17	12.66	404.42	127.0	2.33685	–743.466
570	575.59	13.50	411.97	121.2	2.35531	–766.9367

Table A.9 Ideal Gas Properties of Air (Continued)

T, K	*h*, kJ/kg	P_r	*u*, kJ/kg	v_r	s^0, kJ/(kg·K)	g^0, kJ/(kg·K)
580	586.04	14.38	419.55	115.7	2.37348	–790.6
590	596.52	15.31	427.15	110.6	2.39140	–814.4
600	607.02	16.28	434.78	105.8	2.40902	–838.4
610	617.53	17.30	442.42	101.2	2.42644	–862.6
620	628.07	18.36	450.09	96.92	2.44356	–886.9
630	683.63	19.84	457.78	92.84	2.46048	–866.5
640	649.22	20.64	465.50	88.99	2.47716	–936.2
650	659.84	21.86	473.25	85.34	2.49364	–961.0
660	670.47	23.13	481.01	81.89	2.50985	–986.0
670	681.14	24.46	488.81	78.61	2.52589	–1011.2
680	691.82	25.85	496.62	75.50	2.54175	–1036.6
690	702.52	27.29	504.45	72.56	2.55731	–1062.0
700	713.27	28.80	512.33	69.76	2.57277	–1087.7
710	724.04	30.38	520.23	67.07	2.588 0	–1113.4
720	734.82	32.02	528.14	64.53	2.60319	–1139.5
730	745.62	33.72	536.07	62.13	2.61803	–1165.5
740	756.44	35.50	544.02	59.82	2.63280	–1191.8
750	767.29	37.35	551.99	57.63	2.64737	–1218.2
760	778.18	39.27	560.01	55.54	2.66176	–1244.8
780	800.03	43.35	576.12	51.64	2.69013	–1298.3
800	821.95	47.75	592.30	48.08	2.71787	–1352.3
820	843.98	52.59	608.59	44.84	2.74504	–1406.9
840	866.08	57.60	624.95	41.85	2.77170	–1462.1
860	888.27	63.09	641.40	39.12	2.79783	–1517.9
880	910.56	68.98	657.95	36.61	2.82344	–1574.1
900	932.93	75.29	674.58	34.31	2.84856	–1630.8
920	955.38	82.05	691.28	32.18	2.87324	–1688.0
940	977.92	89.28	708.08	30.22	2.89748	–1745.7
960	1000.55	97.00	725.02	28.40	2.92128	–1803.9
980	1023.25	105.2	741.98	26.73	2.94468	–1862.5
1000	1046.04	114.0	758.94	25.17	2.96770	–1921.7
1020	1068.89	123.4	776.10	23.72	2.99034	–1981.3
1040	1091.85	133.3	793.36	23.29	3.01260	–2041.3
1060	1114.86	143.9	810.62	21.14	3.03449	–2101.7
1080	1137.89	155.2	827.88	19.98	3.05608	–2162.7
1100	1161.07	167.1	845.33	18.896	3.07732	–2224.0
1120	1184.28	179.7	862.79	17.886	3.09825	–2285.8
1140	1207.57	193.1	880.35	16.946	3.11883	–2347.9
1160	1230.92	207.2	897.91	16.064	3.13916	–2410.5
1180	1254.34	222.2	915.57	15.241	3.15916	–2473.5
1200	1277.79	238.0	933.33	14.470	3.17888	–2536.9

(continued)

Table A.9 Ideal Gas Properties of Air (Continued)

T, K	*h, kJ/kg*	P_r	*u, kJ/kg*	v_r	s^0, *kJ/(kg·K)*	g^0, *kJ/(kg·K)*
1220	1301.31	254.7	951.09	13.747	3.19834	−2600.7
1240	1324.93	272.3	968.95	13.069	3.21751	−2664.8
1260	1348.55	290.8	986.90	12.435	3.23638	−2729.3
1280	1372.24	310.4	1004.76	11.835	3.25510	−2794.3
1300	1395.97	330.9	1022.82	11.275	3.27345	−2859.5
1320	1419.76	352.5	1040.88	10.747	3.29160	−2925.2
1340	1443.60	375.3	1058.94	10.247	3.30959	−2991.3
1360	1467.49	399.1	1077.10	9.780	3.32724	−3057.6
1380	1491.44	424.2	1095.26	9.337	3.34474	−3124.3
1400	1515.42	450.5	1113.52	8.919	3.36200	−3191.4
1420	1539.44	478.0	1131.77	8.526	3.37901	−3258.8
1440	1563.51	506.9	1150.13	8.153	3.39586	−3326.5
1460	1587.63	537.1	1168.49	7.801	3.41247	−3394.6
1480	1611.79	568.8	1186.95	7.468	3.42892	−3463.0
1500	1635.97	601.9	1205.41	7.152	3.44516	−3531.8
1520	1660.23	636.5	1223.87	6.854	3.46120	−3600.8
1540	1684.51	672.8	1242.43	6.569	3.47712	−3670.3
1560	1708.82	710.5	1260.99	6.301	3.49276	−3739.9
1580	1733.17	750.0	1279.65	6.046	3.50829	−3809.9
1600	1757.57	791.2	1298.30	5.804	3.52364	−3880.3
1620	1782.00	834.1	1316.96	5.574	3.53879	−3950.8
1640	1806.46	878.9	1335.72	5.355	3.55381	−4021.8
1660	1830.96	925.6	1354.48	5.147	3.56867	−4093.0
1680	1855.50	974.2	1373.24	4.949	3.58335	−4164.5
1700	1880.1	1025	1392.7	4.761	3.5979	−4236.3
1750	1941.6	1161	1439.8	4.328	3.6336	−4417.2
1800	2003.3	1310	1487.2	3.994	3.6684	−4599.8
1850	2065.3	1475	1534.9	3.601	3.7023	−4783.9
1900	2127.4	1655	1582.6	3.295	3.7354	−4969.9
1950	2189.7	1852	1630.6	3.022	3.7677	−5157.3
2000	2252.1	2068	1678.7	2.776	3.7994	−5346.7
2050	2314.6	2303	1726.8	2.555	3.8303	−5537.5
2100	2377.7	2559	1775.3	2.356	3.8605	−5729.4
2150	2440.3	2837	1823.8	2.175	3.8901	−5923.4
2200	2503.2	3138	1872.4	2.012	3.9191	−6118.8
2250	2566.4	3464	1921.3	1.864	'3.9474	−6315.3

Note: For an ideal gas, $P_2/P_1 = (T_2/T_1)^{\{k/k-1)\}} = (v_1/v_2)^k$ for isentropic process, $v_2/v_1 = (T_1/T_2)^{\{1/(k-1)\}} = (P_1/P_2)^{1/k}$ for constant c_p and c_v, and $P_2/P_1 = Pr_2/Pr_1$, $v_2/v_1 = (v_{r1}/v_{r2})$ for variable $c_p(T)$ and $_v(T)$; $g^0 = h - Ts^0$; P_r: relative pressure; v_r: relative volume; $P_r = 0.00368 \exp(s^0/R)$, $v_r = 2.87\ T/p_r$. See *Advanced Thermodynamics Engineering* by Annamalai and Puri, 2002.

Source: Adapted from Wark, K., *Thermodynamics*, 4th ed., McGraw-Hill, New York, 1983, pp. 785–786; originally from Keenan, J.H. and Keye, J., *Gas Tables*, John Wiley & Sons, New York, 1948.

Table A.10 Properties of Carbon (C)

T, K (K)	$\bar{c}_{p0}$ (kJ/kmol K)	$\bar{h}_{t,T} - \bar{h}_{t,298}$ (kJ/kmol)	$\bar{h}_f^0(T)$ (kJ/kmol)	$\bar{s}^0(T)$ (kJ/kmol K)	$\bar{g}^0(T)$ (kJ/kmol)
0	0	–1,054	0	0	–1054
100	1.653	–996	0	0.879	–1084
200	5.029	–669	0	3.012	–1271
298	8.527	0	0	5.686	–1694
300	8.594	17	0	5.740	–1,705
400	11.929	1,046	0	8.682	–2,427
500	14.627	2,381	0	11.648	–3,443
600	16.895	3,962	0	14.523	–4,751
700	18.577	5,740	0	17.263	–6,344
800	19.832	7,661	0	19.828	–8,201
900	20.794	9,699	0	22.221	–10,301
1000	21.543	11,816	0	24.451	–12,636
1100	22.192	14,004	0	26.535	–15,185
1200	22.719	16,246	0	28.489	–17,940
1300	23.125	18,543	0	30.326	–20,880
1400	23.451	20,870	0	32.054	–24,005
1500	23.719	23,230	0	33.681	–27,292
1600	23.937	25,614	0	35.217	–30,732
1700	24.121	28,016	0	36.673	–34,328
1800	24.280	30,439	0	38.058	–38,065
1900	24.418	32,874	0	39.376	–41,940
2000	24.539	35,321	0	40.631	–45,940
2100	24.648	37,777	0	41.832	–50,069
2200	24.744	40,250	0	42.978	–54,302
2300	24.836	42,727	0	44.083	–58,663
2400	24.920	45,216	0	45.141	–63,122
2500	24.995	47,710	0	46.158	–67,685
2600	25.071	50,216	0	47.141	–72,351
2700	25.142	52,726	0	48.087	–77,108
2800	25.213	55,241	0	49.003	–81,967
2900	25.280	57,768	0	49.890	–86,913
3000	25.342	60,300	0	50.748	–91,943
3100	25.409	62,835	0	51.580	–97,064
3200	25.472	65,379	0	52.388	–102,262
3300	25.535	67,923	0	53.170	–107,539
3400	25.602	70,488	0	53.936	–112,894
3500	25.665	73,053	0	54.677	–118,315
3600	25.732	75,622	0	55.400	–123,820
3700	25.794	78,199	0	56.107	–129,399
3800	25.861	80,780	0	56.798	–135,051
3900	25.928	83,362	0	57.471	–140,777
4000	25.995	85,964	0	58.128	–146,549

(continued)

Table A.10 Properties of Carbon (C) (Continued)

T, K (K)	$\bar{c}_{p0}$ (kJ/kmol K)	$\bar{h}_{t,T} - \bar{h}_{t,298}$ (kJ/kmol)	$\bar{h}_f^0(T)$ (kJ/kmol)	$\bar{s}^0(T)$ (kJ/kmol K)	$\bar{g}^0(T)$ (kJ/kmol)
4100	26.066	88,567	0	58.768	–152,384
4200	26.137	91,178	0	59.400	–158,303
4300	26.209	93,797	0	60.015	–164,269
4400	26.280	96,420	0	60.618	–170,298
4500	26.355	99,052	0	61.208	–176,383
4600	26.430	101,692	0	61.789	–182,539
4700	26.506	104,315	0	62.358	–188,769
4800	26.585	106,993	0	62.919	–195,018
4900	26.665	109,654	0	63.467	–201,334
5000	26.744	112,324	0	64.007	–207,710

Note: Carbon (C) (reference state is graphite) M = 12.011, $\bar{h}_{f,298} = 0$.

Table A.11 Ideal Gas Properties of Methane (CH_4)

T (K)	$\bar{c}_{p0}$ (kJ/kmol K)	$\bar{h}_{t,T} - \bar{h}_{t,298}$ (kJ/kmol)	$\bar{h}_f^0(T)$ (kJ/kmol)	$\bar{s}^0(T)$ (kJ/kmol K)	$\bar{g}^0(T)$ (kJ/kmol)
0	0	–10,025	66,906	0	–84875
100	33.259	–6,699	69,990	149.394	–96488
200	33.476	–3,368	72,032	172.473	–112713
298	35.639	0	74,873	186.146	–130322
300	35.710	59	74,931	186.368	–130701
400	40.501	3,862	77,973	197.250	–149888
500	46.342	8,201	80,818	206.911	–170105
600	52.229	13,129	83,329	215.882	–191250
700	57.794	18,636	85,475	224.354	–213262
800	62.932	24,673	87,266	232.413	–236107
900	67.601	31,204	88,730	240.099	–259735
1000	71.797	38,179	89,881	247.446	–284117
1100	75.530	45,551	89,939	254.467	–309213
1200	78.835	53,271	91,437	261.182	–334997
1300	81.747	61,304	91,927	267.609	–361438
1400	84.308	69,609	92,257	273.763	–388509
1500	86.559	78,153	92,483	279.659	–416186
1600	88.538	86,910	92,621	285.311	–444438
1700	90.287	95,855	92,667	290.729	–473234

Table A.11 Ideal Gas Properties of Methane (CH_4) (Continued)

T (K)	$\bar{c}_{p0}$ (kJ/kmol K)	$\bar{h}_{t,T} - \bar{h}_{t,298}$ (kJ/kmol)	$\bar{h}_f^0(T)$ (kJ/kmol)	$\bar{s}^0(T)$ (kJ/kmol K)	$\bar{g}^0(T)$ (kJ/kmol)
1800	91.826	104,960	92,650	295.934	–502571
1900	93.190	114,215	92,579	300.938	–532417
2000	94.399	123,595	92,462	305.750	–562755
2100	95.479	133,089	92,320	310.382	–593563
2200	96.441	142,687	92,157	314.846	–624824
2300	97.303	152,373	91,969	319.151	–656524
2400	98.077	162,143	91,776	323.310	–688651
2500	98.776	171,979	91,579	327.327	–721189
2600	99.403	181,895	91,374	331.214	–754111
2700	99.972	191,866	91,169	334.979	–787427
2800	100.491	201,891	90,964	338.624	–821106
2900	101.215	211,961	90,768	342.159	–855150
3000	101.391	222,083	90,579	345.586	–889525
3100	101.784	232,241	90,383	348.920	–924261
3200	102.144	242,438	90,211	352.155	–959308
3300	102.479	252,668	90,056	355.305	–994689
3400	102.780	262,931	89,906	358.368	–1030370
3500	103.064	273,224	89,784	361.351	–1066355
3600	103.282	283,541	89,676	364.175	–1102339
3700	103.562	293,888	89,596	367.092	–1139202
3800	103.788	304,256	89,525	369.857	–1176051
3900	103.993	314,645	89,492	372.556	–1213173
4000	104.186	325,055	89,479	375.192	–1250563
4100	104.366	335,481	89,483	377.765	–1288206
4200	104.533	345,925	89,525	380.284	–1,3v26,118
4300	104.692	356,389	89,588	382.744	–1,364,260
4400	104.838	366,866	89,680	385.154	–1,402,662
4500	104.977	377,355	89,801	387.510	–1,441,290
4600	105.106	387,861	89,948	389.819	–1,480,156
4700	105.228	398,375	90,123	392.083	–1,519,265
4800	105.341	408,902	90,328	394.300	–1,558,588
4900	105.449	419,442	90,558	396.472	–1,598,121
5000	105.550	429,994	90,818	398.601	–1,637,861

Note: M = 16.04 and $\bar{h}^0_{f,298}$(kJ/kmol) = –74 850. Total enthalpy $\bar{h}$(T) must be estimated using $\bar{h}_{f,298}{}^0 + \bar{h}_{t,T} - \bar{h}_{t,298}$. Use $\bar{h}_f^0$ (T) tabulated in 4th column only for estimating heat of reaction$\Delta\bar{h}_R^0$ or neating value at anv given T. For most calculations, the reader can ignore column 4 tabulating $\bar{h}_f^0$(T) at given T and use only the above formulae in estimating total enthalpy $\bar{h}$

Table A.12 Ideal Gas Properties of Carbon Monoxide (CO)

T (K)	$\bar{c}_{p0}$ (kJ/kmol K)	$\bar{h}_{t,T} - \bar{h}_{t,298}$ (kJ/kmol)	$\bar{h}_f^0(T)$ (kJ/kmol)	$\bar{s}^0(T)$ (kJ/kmol K)	$\bar{g}^0(T)$ (kJ/kmol)
200	28.687	–2,835	–111,308	186.018	–150,580
298	29.072	0	–110,541	197.548	–169,410
300	29.078	54	–110,530	197.728	–169,805
400	29.433	2,979	–110,121	206.141	–190,018
500	29.857	5,943	–110,017	212.752	–210,974
600	30.407	8,955	–110,156	218.242	–232,531
700	31.089	12,029	–110,477	222.979	–254,597
800	31.860	15,176	–110,924	227.180	–277,109
900	32.629	18,401	–111,450	230.978	–300,020
1,000	33.255	21,697	–112,022	234.450	–323,294
1,100	33.725	25,046	–112,619	237.642	–346,901
1,200	34.148	28,440	–113,240	240.595	–370,815
1,300	34.530	31,874	–113,881	243.344	–395,014
1,400	34.872	35,345	–114,543	245.915	–419,477
1,500	35.178	38,847	–115,225	248.332	–444,192
1,600	35.451	42,379	–115,925	250.611	–469,140
1,700	35.694	45,937	–116,644	252.768	–494,310
1,800	35.910	49,517	–117,380	254.814	–519,689
1,900	36.101	53,118	–118,132	256.761	–545,269
2,000	36.271	56,737	–118,902	258.617	–571,038
2,100	36.421	60,371	–119,687	260.391	–596,991
2,200	36.553	64,020	–120,488	262.088	–623,115
2,300	36.670	67,682	–121,305	263.715	–649,404
2,400	36.774	71,354	–122,137	265.278	–675,854
2,500	36.867	75,036	–122,984	266.781	–702,458
2,600	36.950	78,727	–123,847	268.229	–729,209
2,700	37.025	82,426	–124,724	269.625	–756,103
2,800	37.093	86,132	–125,616	270.973	–783,133
2,900	37.155	89,844	–126,523	272.275	–810,295
3,000	37.213	93,562	–127,446	273.536	–837,587
3,100	37.268	97,287	–128,383	274.757	–865,001
3,200	37.321	101,016	–129,335	275.941	–892,536
3,300	37.372	104,751	–130,303	277.090	–920,187
3,400	37.422	108,490	–131,285	278.207	–947,955
3,500	37.471	112,235	–132,283	279.292	–975,828
3,600	37.521	115,985	–133,295	280.349	–1,003,812
3,700	37.570	119,739	–134,323	281.377	–1,031,897
3,800	37.619	123,499	–135,366	282.380	–1,060,086
3,900	37.667	127,263	–136,424	283.358	–1,088,374

Table A.12 Ideal Gas Properties of Carbon Monoxide (CO) (Continued)

T (K)	$\bar{c}_{p0}$ (kJ/kmol K)	$\bar{h}_{t,T} - \bar{h}_{t,298}$ (kJ/kmol)	$\bar{h}_f^0(T)$ (kJ/kmol)	$\bar{s}^0(T)$ (kJ/kmol K)	$\bar{g}^0(T)$ (kJ/kmol)
4,000	37.716	131,032	–137,497	284.312	–1,116,757
4,100	37.764	134,806	–138,585	285.244	–1,145,235
4,200	37.810	138,585	–139,687	286.154	–1,173,803
4,300	37.855	142,368	–140,804	287.045	–1,202,467
4,400	37.897	146,156	–141,935	287.915	–1,231,211
4,500	37.936	149,948	–143,079	288.768	–1,260,049
4,600	37.970	153,743	–144,236	289.602	–1,288,967
4,700	37.998	157,541	–145,407	290.419	–1,317,969
4,800	38.019	161,342	–146,589	291.219	–1,347,050
4,900	38.031	165,145	–147,783	292.003	–1,376,211
5,000	38.033	168,948	–148,987	292.771	–1,405,448

Note: MW = 28.010 and $\bar{h}^0_{f,298}$ (kJ/kmol) = –110,541; $\bar{g}^0 = \bar{h}_{f,298}{}^0 + (\bar{h}_{t,T} - \bar{h}_{t,298}) - T\,\bar{s}^0$; $\bar{g}^0 \neq \bar{g}_f{}^0$, $\bar{s}_k(T,p_k) = \bar{s}^0 - \bar{R}\ \ln(p_k/1)$; $\bar{g}_k(T.p_k) = \bar{g}^0 + \bar{R}\,T\,\ln(p_k/1)$;

Source: A few tables for CO, CO_2, H_2O, N_2, and O_2 adapted from Turns, S.R., *An Introduction to Combustion*, 2nd ed., McGraw-Hill, New York, 2000.

Table A.13 Ideal Gas Properties of Carbon Dioxide (CO_2)

T (K)	$\bar{c}_{p0}$ (kJ/kmol K)	$\bar{h}_{t,T} - \bar{h}_{t,298}$ (kJ/kmol)	$\bar{h}_f^0(T)$ (kJ/kmol)	$\bar{s}^0(T)$ (kJ/kmol K)	$\bar{g}^0(T)$ (kJ/kmol)
200	32.387	–3,423	–393,483	199.876	–436,944
298	37.198	0	–393,546	213.736	–457,239
300	37.280	69	–393,547	213.966	–457,667
400	41.276	4,003	–393,617	225.257	–479,646
500	44.569	8,301	–393,712	234.833	–502,662
600	47.313	12,899	–393,844	243.209	–526,572
700	49.617	17,749	–394,013	250.680	–551,273
800	51.550	22,810	–394,213	257.436	–576,685
900	53.136	28,047	–394,433	263.603	–602,742
1,000	54.360	33,425	–394,659	269.268	–629,389
1,100	55.333	38,911	–394,875	274.495	–656,580
1,200	56.205	44,488	–395,083	279.348	–684,276
1,300	56.984	50,149	–395,287	283.878	–712,438
1,400	57.677	55,882	–395,488	288.127	–741,042
1,500	58.292	61,681	–395,691	292.128	–770,057
1,600	58.836	67,538	–395,897	295.908	–799,461

(*continued*)

Table A.13 Ideal Gas Properties of Carbon Dioxide (CO_2) (Continued)

T (K)	$\bar{c}_{p0}$ (kJ/kmol K)	$\bar{h}_{t,T} - \bar{h}_{t,298}$ (kJ/kmol)	$\bar{h}_f^0(T)$ (kJ/kmol)	$\bar{s}^0(T)$ (kJ/kmol K)	$\bar{g}^0(T)$ (kJ/kmol)
1,700	59.316	73,446	–396,110	299.489	–829,231
1,800	59.738	79,399	–396,332	302.892	–859,353
1,900	60.108	85,392	–396,564	306.132	–889,805
2,000	60.433	91,420	–396,808	309.223	–920,572
2,100	60.717	97,477	–397,065	312.179	–951,645
2,200	60.966	103,562	–397,338	315.009	–983,004
2,300	61.185	109,670	–397,626	317.724	–1,014,641
2,400	61.378	115,798	–397,931	320.333	–1,046,547
2,500	61.548	121,944	–398,253	322.842	–1,078,707
2,600	61.701	128,107	–398,594	325.259	–1,111,112
2,700	61.839	134,284	–398,952	327.590	–1,143,755
2,800	61.965	140,474	–399,329	329.841	–1,176,627
2,900	62.083	146,677	–399,725	332.018	–1,209,721
3,000	62.194	152,891	–400,140	334.124	–1,243,027
3,100	62.301	159,116	–400,573	336.165	–1,276,542
3,200	62.406	165,351	–401,025	338.145	–1,310,259
3,300	62.510	171,597	–401,495	340.067	–1,344,170
3,400	62.614	177,853	–401,983	341.935	–1,378,272
3,500	62.718	184,120	–402,489	343.751	–1,412,555
3,600	62.825	190,397	–403,013	345.519	–1,447,017
3,700	62.932	196,685	–403,553	347.242	–1,481,656
3,800	63.041	202,983	–404,110	348.922	–1,516,467
3,900	63.151	209,293	–404,684	350.561	–1,551,441
4,000	63.261	215,613	–405,273	353.161	–1,590,577
4,100	63.369	221,945	–405,878	353.725	–1,621,874
4,200	63.474	228,287	–406,499	355.253	–1,657,322
4,300	63.575	234,640	–407,135	356.748	–1,692,922
4,400	63.669	241,002	–407,785	358.210	–1,728,668
4,500	63.753	247,373	–408,451	359.642	–1,764,562
4,600	63.825	253,752	–409,132	361.044	–1,800,596
4,700	63.881	260,138	–409,828	362.417	–1,836,768
4,800	63.918	266,528	–410,539	363.763	–1,873,080
4,900	63.932	272,920	–411,267	365.081	–1,909,523
5,000	63.919	279,313	–412,010	366.372	–1,946,093

Note: Molecular weight = 44.011 and $\bar{h}^0_{f,298}$ (kJ/kmol) = –393,546.

Table A.14 Ideal Gas Properties of Acetylene (C_2H_2)

T (K)	$\bar{c}_{p0}$ (kJ/kmol K)	$\bar{h}_{t,T} - \bar{h}_{t,298}$ (kJ/kmol)	$\bar{h}_f^0(T)$ (kJ/kmol)	$\bar{s}^0(T)$ (kJ/kmol K)	$\bar{g}^0(T)$ (kJ/kmol)
0	0.000	–10,012	227,296	0.000	216,724
100	29.347	–7,096	226,911	163.184	203,322
200	35.585	–3,925	226,915	184.987	185,814
298	44.095	0	226,731	200.849	166,883
300	44.229	84	226,727	201.121	166,484
400	50.480	4,833	226,513	214.748	145,670
500	54.869	10,117	226,204	226.518	123,594
600	58.287	15,778	225,773	236.831	100,415
700	61.149	21,753	225,254	246.040	76,261
800	63.760	28,004	224,710	254.379	51,237
900	66.111	34,497	224,149	262.023	25,412
1000	68.275	41,221	223,635	269.102	–1,145
1100	70.245	48,145	223,145	275.705	–28,395
1200	72.053	55,262	222,706	281.897	–56,278
1300	73.693	62,551	222,288	287.729	–84,761
1400	75.178	69,994	221,924	293.244	–113,812
1500	76.530	77,584	221,589	298.478	–143,397
1600	77.747	85,299	220,442	303.457	–173,496
1700	78.847	93,127	221,011	308.206	–204,087
1800	79.852	101,065	220,769	312.741	–235,133
1900	80.760	109,094	220,555	317.084	–266,630
2000	81.605	117,215	220,371	321.248	–298,545
2100	82.358	125,411	220,208	325.247	–330,872
2200	83.065	133,683	220,053	329.097	–363,594
2300	83.697	142,022	219,928	332.800	–396,682
2400	84.312	150,423	219,806	336.377	–430,146
2500	84.860	158,883	219,702	339.829	–463,954
2600	85.370	167,393	219,601	343.167	–498,105
2700	85.851	175,958	219,518	346.398	–532,581
2800	86.295	184,565	219,442	349.531	–567,386
2900	86.718	193,213	219,363	352.565	–602,490
3000	87.111	201,907	219,296	355.594	–638,139
3100	87.487	210,635	219,237	358.376	–673,595
3200	87.847	219,405	219,179	361.159	–709,568
3300	88.190	228,204	219,112	363.866	–745,818
3400	88.508	237,040	219,058	366.502	–782,331
3500	88.805	245,906	218,991	369.075	–819,121
3600	89.107	254,801	218,928	371.581	–856,155
3700	89.400	263,726	218,852	374.024	–893,427
3800	89.667	272,680	218,786	376.414	–930,957
3900	89.931	281,663	218,706	378.744	–968,703
4000	90.194	290,667	218,622	381.024	–1,006,693

(continued)

Table A.14 Ideal Gas Properties of Acetylene (C_2H_2) (Continued)

T (K)	$\bar{c}_{p0}$ (kJ/kmol K)	$\bar{h}_{t,T} - \bar{h}_{t,298}$ (kJ/kmol)	$\bar{h}_f^0(T)$ (kJ/kmol)	$\bar{s}^0(T)$ (kJ/kmol K)	$\bar{g}^0(T)$ (kJ/kmol)
4100	90.437	299,700	218,535	383.254	−1,044,905
4200	90.667	308,754	218,430	385.438	−1,083,350
4300	90.910	317,833	218,317	387.572	−1,121,991
4400	91.136	326,934	218,200	389.668	−1,160,869
4500	91.358	336,059	218,070	391.719	−1,199,941
4600	91.558	345,205	217,932	393.727	−1,239,203
4700	91.776	354,372	217,786	395.698	−1,278,673
4800	91.985	363,560	217,622	397.635	−1,318,352
4900	92.199	372,769	217,455	399.530	−1,358,192
5000	92.370	381,999	217,275	401.396	−1,398,245

Note: Molecular weight = 26.04 and $\bar{h}^0_{f,298}$ (kJ/kmol) = 226736.

Table A.15 Ideal Gas Properties of Ethylene (C_2H_4)

T (K)	$\bar{c}_{p0}$ (kJ/kmol K)	$\bar{h}_{t,T} - \bar{h}_{t,298}$ (kJ/kmol)	$\bar{h}_f^0(T)$ (kJ/kmol)	$\bar{s}^0(T)$ (kJ/kmol K)	$\bar{g}^0(T)$ (kJ/kmol)
0	0.000	−10,519	60,994	0.000	41,761
100	33.271	−7,192	57,852	180.435	27,045
200	35.359	−3,803	55,543	203.849	7,707
298	42.886	0	52,467	219.225	−13,049
300	43.062	79	52,405	219.488	−13,487
400	53.049	4,883	49,342	233.237	−36,132
500	62.480	10,669	46,610	246.107	−60,105
600	70.664	17,334	44,254	258.241	−85,331
700	77.714	24,761	42,250	269.676	−111,732
800	83.843	32,849	40,589	280.466	−139,244
900	89.203	41,505	39,212	290.654	−167,804
1000	93.902	50,664	38,129	300.302	−197,358
1100	98.019	60,266	37,279	309.449	−227,848
1200	101.629	70,254	36,639	318.135	−259,228
1300	104.784	80,575	36,141	326.398	−291,462
1400	107.554	91,199	35,802	334.289	−324,526

Table A.15 Ideal Gas Properties of Ethylene (C_2H_4) (Continued)

T (K)	$\bar{c}_{p0}$ (kJ/kmol K)	$\bar{h}_{t,T} - \bar{h}_{t,298}$ (kJ/kmol)	$\bar{h}_f^0(T)$ (kJ/kmol)	$\bar{s}^0(T)$ (kJ/kmol K)	$\bar{g}^0(T)$ (kJ/kmol)
1500	109.976	102,077	35,551	341.774	−358,304
1600	112.106	113,181	35,376	348.941	−392,845
1700	113.981	124,487	35,292	355.795	−428,085
1800	115.629	135,972	35,263	362.355	−463,987
1900	117.093	147,607	35,284	368.648	−500,544
2000	118.390	159,385	35,346	374.686	−537,707
2100	119.541	171,280	35,434	380.493	−575,475
2200	120.575	183,288	35,535	386.079	−613,806
2300	121.495	195,393	35,664	391.459	−652,683
2400	122.323	207,585	35,790	396.647	−692,088
2500	123.068	219,852	35,915	401.656	−732,008
2600	123.742	232,195	36,049	406.497	−772,417
2700	124.353	244,601	36,179	411.178	−813,300
2800	124.905	257,065	36,309	415.710	−854,643
2900	125.407	269,579	36,422	420.103	−896,440
3000	125.867	282,144	36,526	424.362	−938,662
3100	126.290	294,750	36,631	428.496	−981,308
3200	126.675	307,398	36,710	432.513	−1,024,364
3300	127.026	320,084	36,769	436.412	−1,067,796
3400	127.353	332,804	36,819	440.211	−1,111,633
3500	127.654	345,557	36,836	443.906	−1,155,834
3600	127.934	358,334	36,836	447.508	−1,200,415
3700	128.189	371,142	36,798	451.014	−1,245,330
3800	128.428	383,974	36,752	454.437	−1,290,607
3900	128.650	396,827	36,660	457.776	−1,336,219
4000	128.855	409,701	36,547	461.035	−1,382,159
4100	129.047	422,597	36,405	464.219	−1,428,421
4200	129.227	435,513	36,225	467.332	−1,475,001
4300	129.394	448,441	36,008	470.374	−1,521,887
4400	129.553	461,391	35,769	473.353	−1,569,082
4500	129.700	474,353	35,485	476.265	−1,616,560
4600	129.838	487,331	35,175	479.118	−1,664,332
4700	129.968	500,319	34,823	481.913	−1,712,392
4800	130.093	513,322	34,439	484.649	−1,760,713
4900	130.206	526,339	34,024	487.331	−1,809,303
5000	130.315	539,364	33,568	489.963	−1,858,171

Note: Molecular weight = 28.05 and $\bar{h}_{f,298}$ (kJ/kmol) = 52280.

Table A.16 Ideal Gas Properties of Hydrogen Atom (H)

T (K)	$\bar{c}_{p0}$ (kJ/kmol K)	$\bar{h}_{t,T} - \bar{h}_{t,298}$ (kJ/kmol)	$\bar{h}_f^0(T)$ (kJ/kmol)	$\bar{s}^0(T)$ (kJ/kmol K)	$\bar{g}^0(T)$ (kJ/kmol)
200	20.786	–2,040	217,346	106.305	194,676
298	20.786	0	217,977	114.605	183,825
300	20.786	38	217,989	114.733	183,595
400	20.786	2,117	218,617	120.713	171,809
500	20.786	4,196	219,236	125.351	159,498
600	20.786	6,274	219,848	129.351	146,640
700	20.786	8,353	220,456	132.345	133,689
800	20.786	10,431	221,059	135.121	120,311
900	20.786	12,510	221,653	137.569	106,675
1,000	20.786	14,589	222,234	139.759	92,807
1,100	20.786	16,667	222,793	141.740	78,730
1,200	20.786	18,746	223,329	143.549	64,464
1,300	20.786	20,824	223,843	145.213	50,024
1,400	20.786	22,903	224,335	146.753	35,426
1,500	20.786	24,982	224,806	148.187	20,679
1,600	20.786	27,060	225,256	149.528	5,792
1,700	20.786	29,139	225,687	150.789	–9,225
1,800	20.786	31,217	226,099	151.977	–24,365
1,900	20.786	33,296	226,493	153.101	–39,619
2,000	20.786	35,375	226,868	154.167	–54,982
2,100	20.786	37,453	227,226	155.181	–70,450
2,200	20.786	39,532	227,568	156.148	–86,017
2,300	20.786	41,610	227,894	157.072	–101,679
2,400	20.786	43,689	228,204	157.956	–117,428
2,500	20.786	45,768	228,499	158.805	–133,268
2,600	20.786	47,846	228,780	159.620	–149,189
2,700	20.786	49,925	229,047	160.405	–165,192
2,800	20.786	52,003	229,301	161.161	–181,271
2,900	20.786	54,082	229,543	161.890	–197,422
3,000	20.786	56,161	229,772	162.595	–213,647
3,100	20.786	58,239	229,989	163.276	–229,940
3,200	20.786	60,318	230,195	163.936	–246,300
3,300	20.786	62,396	230,390	164.576	–262,728
3,400	20.786	64,475	230,574	165.196	–279,214
3,500	20.786	66,554	230,748	165.799	–295,766
3,600	20.786	68,632	230,912	166.954	–314,425
3,700	20.786	70,711	231,067	166.954	–329,042
3,800	20.786	72,789	231,212	167.508	–345,764
3,900	20.786	74,868	231,348	168.048	–362,542
4,000	20.786	76,947	231,475	168.575	–379,376
4,100	20.786	79,025	231,594	169.088	–396,259
4,200	20.786	81,104	231,704	169.589	–413,193

Table A.16 Ideal Gas Properties of Hydrogen Atom (H) (Continued)

T (*K*)	$\bar{c}_{p0}$ (*kJ/kmol K*)	$\bar{h}_{t,T} - \bar{h}_{t,298}$ (*kJ/kmol*)	$\bar{h}_f^0(T)$ (*kJ/kmol*)	$\bar{s}^0(T)$ (*kJ/kmol K*)	$\bar{g}^0(T)$ (*kJ/kmol*)
4,300		83,182	231,805	170.078	–430,176
4,400	20.786	85,261	231,897	170.556	–447,208
4,500	20.786	87,340	231,981	171.023	–464,287
4,600	20.786	89,418	232,056	171.480	–481,413
4,700	20.786	91,497	232,123	171.927	–498,583
4,800	20.786	93,575	232,180	172.364	–515,795
4,900	20.786	95,654	232,228	172.793	–533,055
5,000	20.786	97,733	232,267	173.213	–550,355

Note: Molecular weight = 1.01 and $\bar{h}^0_{f,298}$ (kJ/kmol) = 217,977.

Table A.17 Ideal Gas Properties of Hydrogen (H_2)

T (*K*)	$\bar{c}_{p0}$ (*kJ/kmol K*)	$\bar{h}_{t,T} - \bar{h}_{t,298}$ (*kJ/kmol*)	$\bar{h}_f^0(T)$ (*kJ/kmol*)	$\bar{s}^0(T)$ (*kJ/kmol K*)	$\bar{g}^0(T)$ (*kJ/kmol*)
200	28.522	–2,818	0	119.137	–26,645
298	28.871	0	0	130.595	–38,917
300	28.877	53	0	130.773	–39,179
400	29.120	2,954	0	139.116	–52,692
500	29.275	5,874	0	145.632	–66,942
600	29.375	8,807	0	150.979	–81,780
700	29.461	11,749	0	155.514	–97,111
800	29.581	14,701	0	159.455	–112,863
906	29.792	17,668	0	162.950	–129,965
1,000	30.160	20,664	0	166.106	–145,442
1,100	30.625	23,704	0	169.003	–162,199
1,200	31.077	26,789	0	171.687	–179,235
1,300	31.516	29,919	0	174.192	–196,531
1,400	31.943	33,092	0	176.543	–214,068
1,500	32.356	36,307	0	178.761	–231,835
1,600	32.758	39,562	0	180.862	–249,817
1,700	33.146	42,858	0	182.860	–268,004
1,800	33.522	46,191	0	184.765	–286,386
1,900	33.885	49,562	0	186.587	–304,953
2,000	34.236	52,968	0	188.334	–323,700
2,100	34.575	56,408	0	190.013	–342,619
2,200	34.901	59,882	0	191.629	–361,702

(*continued*)

Table A.17 Ideal Gas Properties of Hydrogen (H_2) (Continued)

T (K)	$\bar{c}_{p0}$ (kJ/kmol K)	$\bar{h}_{t,T} - \bar{h}_{t,298}$ (kJ/kmol)	$\bar{h}_f^{\,0}(T)$ (kJ/kmol)	$\bar{s}^{\,0}(T)$ (kJ/kmol K)	$\bar{g}^0(T)$ (kJ/kmol)
2,300	35.216	63,388	0	193.187	–380,942
2,400	35.50	66,925	0	194.692	–400,336
2,500	35.811	70,492	0	196.148	–419,878
2,600	36.091	74,087	0	197.558	–439,564
2,700	36.361	77,710	0	198.926	–459,390
2,800	36.621	81,359	0	200.253	–479,349
2,900	36.871	85,033	0	201.542	–499,439
3,000	37.112	88,733	0	202.796	–519,655
3,100	37.343	92,455	0	204.017	–539,998
3,200	37.566	96,201	0	205.206	–560,458
3,300	37.781	99,968	0	206.365	–581,037
3,400	37.989	103,757	0	207.496	–601,729
3,500	38.190	107,566	0	208.600	–622,534
3,600	38.385	111,395	0	209.679	–643,449
3,700	38.574	115,243	0	210.733	–664,469
3,800	38.759	119,109	0	211.764	–685,594
3,900	38.939	122,994	0	212.774	–706,825
4,000	39.116	126,897	0	213.762	–728,151
4,100	39.291	130,817	0	214.730	–749,576
4,200	39.464	134,755	0	215.679	–771,097
4,300	39.636	138,710	0	216.609	–792,709
4,400	39.808	142,682	0	217.522	–815,720
4,500	39.981	146,672	0	218.419	–836,214
4,600	40.156	150,679	0	219,300	–858,101
4,700	40.334	154,703	0	220.165	–880,073
4,800	40.516	158,746	0	221.016	–902,131
4,900	40.702	162,806	0	221.853	–924,274
5,000	40.895	166,886	0	222.678	–946,504

Note: Molecular weight = 2.02 and $\bar{h}_{f,298}^{\,0}$ (kJ/kmol) = 0.

Table A.18 Ideal Gas Properties of Water (H_2O)

T (K)	$\bar{c}_{p0}$ (kJ/kmol K)	$\bar{h}_{t,T} - \bar{h}_{t,298}$ (kJ/kmol)	$\bar{h}_f^0(T)$ (kJ/kmol)	$\bar{s}^0(T)$ (kJ/kmol K)	$\bar{g}^0(T)$ (kJ/kmol)
200	32.255	–3,227	–240,838	175.602	–280,192
298	33.448	0	–241,845	188.715	–298,082
300	33.468	62	–241,865	188.922	–298,460
400	34.437	3,458	–242,858	198.686	–317,861
500	35.337	6,947	–243,822	206.467	–338,132
600	36.288	10,528	–244,753	212.992	–359,112
700	37.364	14,209	–245,638	218.665	–380,702
800	38.587	18,005	–246,461	223.733	–402,826
900	39.930	21,930	–247,209	228.354	–425,434
1,000	41.315	25,993	–247,879	232.633	–448,485
1,100	42.638	30,191	–248,475	236.634	–471,951
1,200	43.874	34,518	–249,005	240.397	–495,803
1,300	45.027	38,963	–249,477	243.955	–520,024
1,400	46.102	43,520	–249,895	247.332	–544,590
1,500	47.103	48,181	–250,267	250.547	–569,485
1,600	48.035	52,939	–250,597	253.617	–594,693
1,700	48.901	57,786	–250,890	256.556	–620,204
1,800	49.705	62,717	–251,151	259.374	–646,001
1,900	50.451	67,725	–251,384	262.081	–672,074
2,000	51.143	72,805	–251,594	264.687	–698,414
2,100	51.784	77,952	–251,783	267.198	–725,009
2,200	52.378	83,160	–251,955	269.621	–751,851
2,300	52.927	88,426	–252,113	271.961	–778,929
2,400	53.435	93,744	–252,261	274.225	–806,241
2,500	53.905	99,112	–252,399	276.416	–833,773
2,600	54.340	104,524	–252,532	278.539	–861,522
2,700	54.742	109,979	–252,659	280.597	–889,478
2,800	55.115	115,472	–252,785	282.595	–917,639
2,900	55.459	121,001	–252,909	284.535	–945,996
3,000	55.779	126,563	–253,034	286.420	–974,542
3,100	56.076	132,156	–253,161	288.254	–1,003,276
3,200	56.353	137,777	–253,290	290.039	–1,032,193
3,300	56.610	143,426	–253,423	291.777	–1,061,283
3,400	56.851	149,099	–253,561	293.471	–1,090,547
3,500	57.076	154,795	–253,704	295.122	–1,119,977
3,600	57.288	160,514	–253,852	296.733	–1,149,570
3,700	57.488	166,252	–254,007	298.305	–1,179,322
3,800	57.676	172,011	–254,169	299.841	–1,209,230
3,900	57.856	177,787	–254,338	301.341	–1,239,288
4,000	58.026	183,582	–254,515	302.808	–1,269,495
4,100	58.190	189,392	–254,699	304.243	–1,299,849

(continued)

Table A.18 Ideal Gas Properties of Water (H_2O)(Continued)

T (K)	$\bar{c}_{p0}$ (kJ/kmol K)	$\bar{h}_{t,T}-\bar{h}_{t,298}$ (kJ/kmol)	$\bar{h}_f^0(T)$ (kJ/kmol)	$\bar{s}^0(T)$ (kJ/kmol K)	$\bar{g}^0(T)$ (kJ/kmol)
4,200	58.346	195,219	–254,892	305.647	–1,330,343
4,300	58.496	201,061	–255,093	307.022	–1,360,979
4,400	58.641	206,918	–255,303	308.368	–1,391,746
4,500	58.781	212,790	–255,522	309.688	–1,422,651
4,600	58.916	218,674	–255,751	310.981	–1,453,684
4,700	59.047	224,573	–255,990	312.250	–1,484,847
4,800	59.173	230,484	–256,239	313.494	–1,516,132
4,900	59.295	236,407	–256,501	314.716	–1,547,546
5,000	59.412	242,343	–256,774	315.915	–1,579,077

Note: Molecular weight = 18.02 and $\bar{h}_{f,298}^0$ (kJ/kmol) = –241,845, enthalpy of vaporization (kJ/kmol) = 44,010.

Table A.19 Ideal Gas Properties of Nitrogen Atom (N)

T (K)	$\bar{c}_{p0}$ (kJ/kmol K)	$\bar{h}_{t,T}-\bar{h}_{t,298}$ (kJ/kmol)	$\bar{h}_f^0(T)$ (kJ/kmol)	$\bar{s}^0(T)$ (kJ/kmol K)	$\bar{g}^0(T)$ (kJ/kmol)
200	20.790	–2,040	472,008	144.889	441,611
298	20.786	0	472,629	153.189	426,979
300	20.786	38	472,640	153.317	426,672
400	20.786	2,117	473,259	159.297	411,027
500	20.786	4,196	473,864	163.935	394,858
600	20.786	6,274	474,450	167.725	378,268
700	20.786	8,353	475,010	170.929	361,332
800	20.786	10,431	475,537	173.705	344,096
900	20.786	12,510	476,027	176.153	326,601
1,000	20.786	14,589	476,483	178.343	308,875
1,100	20.792	16,668	476,911	180.325	290,940
1,200	20.795	18,747	477,316	182.134	272,815
1,300	20.795	20,826	477,700	183.798	254,518
1,400	20.793	22,906	478,064	185.339	236,060
1,500	20.790	24,985	478,411	186.774	217,453
1,600	20.786	27,064	478,742	188.115	198,709
1,700	20.782	29,142	479,059	189.375	179,834

Table A.19 Ideal Gas Properties of Nitrogen Atom (N) (Continued)

T (*K*)	$\bar{c}_{p0}$ (*kJ/kmol K*)	$\bar{h}_{t,T} - \bar{h}_{t,298}$ (*kJ/kmol*)	$\bar{h}_f^0(T)$ (*kJ/kmol*)	$\bar{s}^0(T)$ (*kJ/kmol K*)	$\bar{g}^0(T)$ (*kJ/kmol*)
1,800	20.779	31,220	479,363	190.563	160,836
1,900	20.777	33,298	479,656	191.687	141,722
2,000	20.776	35,376	479,939	192.752	122,501
2,100	20.778	37,453	480,213	193.766	103,173
2,200	20.783	39,531	480,479	194.733	83,747
2,300	20.791	41,610	480,740	195.657	64,228
2,400	20.802	43,690	480,995	196.542	44,618
2,500	20.818	45,771	481,246	197.391	24,923
2,600	20.838	47,853	481,494	198.208	5,141
2,700	20.864	49,938	481,740	198.995	–14,720
2,800	20.895	52,026	481,985	199.754	–34,656
2,900	20.931	54,118	482,230	200.488	–54,668
3,000	20.974	56,213	482,476	201.199	–74,755
3,100	21.024	58,313	482,723	201.887	–94,908
3,200	21.080	60,418	482,972	202.555	–115,129
3,300	21.143	62,529	483,224	203.205	–135,419
3,400	21.214	64,647	483,481	203.837	–155,770
3,500	21.292	66,772	483,742	204.453	–176,185
3,600	21.378	68,905	484,009	205.054	–196,660
3,700	21.472	71,048	484,283	205.641	–217,195
3,800	21.575	73,200	484,564	206.215	–237,788
3,900	21.686	75,363	484,853	206.777	–258,438
4,000	21.905	77,537	485,151	207.328	–279,146
4,100	21.934	79,724	485,459	207.868	–299,906
4,200	22.071	81,924	485,779	208.398	–320,719
4,300	22.217	84,139	486,110	208.919	–341,584
4,400	22.372	86,368	486,453	209.431	–362,499
4,500	22.536	88,613	486,811	209.936	–383,470
4,600	22.709	90,875	487,184	210.433	–404,488
4,700	22.891	93,155	487,573	210.923	–425,554
4,800	23.082	95,454	487,979	211.407	–446,671
4,900	23.282	97,772	488,405	211.885	–467,836
5,000	23.491	100,111	488,850	212.358	–489,050

Note: Molecular weight = 14.01 and $\bar{h}_{f,298}{}^0$ (kJ/kmol) = 472,629.

Table A.20 Ideal Gas Properties of NH_3

T (K)	$\bar{c}_{p0}$ (kJ/kmol K)	$\bar{h}_{t,T} - \bar{h}_{t,298}$ (kJ/kmol)	$\bar{h}_f^0(T)$ (kJ/kmol)	$\bar{s}^0(T)$ (kJ/kmol K)	$\bar{g}^0(T)$ (kJ/kmol)
0	0	–10,058	38,920	0	–56,248
100	33.263	–6,732	41,790	155.687	–68,491
200	33.740	–3,393	43,706	178.824	–85,348
298	35.627	0	45,898	192.602	–103,585
300	35.673	67	45,940	192.820	–103,969
400	38.664	3,778	48,041	203.480	–123,804
500	41.991	7,812	49,869	212.464	–144,610
600	45.221	12,171	51,388	220.409	–166,264
700	48.275	16,849	52,643	227.610	–188,668
800	51.149	21,820	53,656	234.245	–211,766
900	53.840	27,070	54,459	240.429	–235,506
1000	56.346	32,581	55,074	246.233	–259,842
1100	58.702	38,338	55,526	251.714	–284,737
1200	60.877	44,317	55,827	256.914	–310,170
1300	62.886	50,505	56,003	261.868	–336,113
1400	64.685	56,886	56,074	266.596	–362,538
1500	66.316	63,438	56,061	271.115	–389,425
1600	67.802	70,145	55,986	275.445	–416,757
1700	69.120	76,994	55,844	279.596	–444,509
1800	70.132	83,956	55,660	283.575	–472,669
1900	71.107	91,019	55,438	287.395	–501,222
2000	72.048	56,338	55,191	291.064	–571,980
2100	72.923	105,424	54,915	294.600	–559,426
2200	73.764	112,759	54,614	298.014	–589,062
2300	74.580	120,177	54,279	301.311	–619,028
2400	75.371	127,675	53,919	304.499	–649,313
2500	76.128	135,252	53,543	307.591	–679,916
2600	76.860	142,900	53,124	310.595	–710,837
2700	77.559	150,620	52,685	313.507	–742,039
2800	78.232	158,410	52,216	316.340	–773,532
2900	78.881	166,268	51,714	319.097	–805,303
3000	79.496	174,184	51,195	321.783	–837,355
3100	79.977	182,159	50,647	324.394	–869,652
3200	80.450	190,180	50,095	326.942	–902,224
3300	80.923	198,250	49,530	329.423	–935,036
3400	81.387	206,363	48,944	331.850	–968,117
3500	81.852	214,275	48,354	334.193	–1,001,591
3600	82.308	222,735	47,748	336.527	–1,034,952
3700	82.764	230,568	47,133	338.787	–1,069,134
3800	83.211	239,287	46,493	341.000	–1,102,703
3900	83.655	247,630	45,848	343.167	–1,136,911
4000	84.098	256,019	45,187	345.293	–1,171,343

Table A.20 Ideal Gas Properties of NH_3 (Continued)

T (K)	$\bar{c}_{p0}$ (kJ/kmol K)	$\bar{h}_{t,T} - \bar{h}_{t,298}$ (kJ/kmol)	$\bar{h}_f^0(T)$ (kJ/kmol)	$\bar{s}^0(T)$ (kJ/kmol K)	$\bar{g}^0(T)$ (kJ/kmol)
4100	84.538	264,450	44,514	347.372	–1,205,965
4200	84.969	272,927	43,823	349.414	–1,240,802
4300	85.400	281,445	43,129	351.418	–1,275,842
4400	85.822	290,006	42,401	353.389	–1,311,096
4500	86.245	298,608	41,677	355.322	–1,346,531
4600	86.659	307,252	40,924	357.222	–1,382,159
4700	87.073	315,938	40,166	359.088	–1,417,966
4800	87.479	324,666	39,392	360.929	–1,453,983
4900	87.885	333,436	38,602	362.736	–1,490,160
5000	88.282	342,243	37,794	364.514	–1,526,517

Note: Molecular weight = 17.03 and $\bar{h}_{f,298}{}^0$ (kJ/kmol) = –46,190.

Table A.21 Ideal Gas Properties of Nitric Oxide (NO)

T (K)	$\bar{c}_{p0}$ (kJ/kmol K)	$\bar{h}_{t,T} - \bar{h}_{t,298}$ (kJ/kmol)	$\bar{h}_f^0(T)$ (kJ/kmol)	$\bar{s}^0(T)$ (kJ/kmol K)	$\bar{g}^0(T)$ (kJ/kmol)
200	29.374	–2,901	90,234	198.856	47,625
298	29.728	0	90,297	210.652	27,523
300	29.735	55	90,298	210.836	27,101
400	30.103	3,046	90,341	219.439	5,567
500	30.570	6,079	90,367	226.204	–16,726
600	31.174	9,165	90,382	231.829	–39,635
700	31.908	12,318	90,393	236.688	–63,067
800	32.715	15,549	90,405	241.001	–86,955
900	33.489	18,860	90,421	244.900	–111,253
1,000	34.076	22,241	90,443	248.462	–135,924
1,100	34.483	25,669	90,465	251.729	–160,936
1,200	34.850	29,136	90,486	254.745	–186,261
1,300	35.180	32,638	90,505	257.548	–211,877
1,400	35.474	36,171	90,520	260.166	–237,764
1,500	35.737	39,732	90,532	262.623	–263,906
1,600	35.972	43,317	90,538	264.937	–290,285
1,700	36.180	46,925	90,539	267.124	–316,889
1,800	36.364	50,552	90,534	269.197	–343,706
1,900	36.527	54,197	90,523	271.168	–370,725
2,000	36.671	57,857	90,505	273.045	–397,936

(continued)

Table A.21 Ideal Gas Properties of Nitric Oxide (NO) (Continued)

T (K)	$\bar{c}_{p0}$ (kJ/kmol K)	$\bar{h}_{t,T} - \bar{h}_{t,298}$ (kJ/kmol)	$\bar{h}_f^{\,0}(T)$ (kJ/kmol)	$\bar{s}^{\,0}(T)$ (kJ/kmol K)	$\bar{g}^{0}(T)$ (kJ/kmol)
2,100	36.797	61,531	90,479	274.838	–425,332
2,200	36.909	65,216	90,447	276.552	–452,901
2,300	37.008	68,912	90,406	278.195	–480,640
2,400	37.095	72,617	90,358	279.772	–508,539
2,500	37.173	76,331	90,303	281.288	–536,592
2,600	37.242	80,052	90,239	282.747	–564,793
2,700	37.305	83,779	90,168	284.154	–593,140
2,800	37.362	87,513	90,089	285.512	–621,624
2,900	37.415	91,251	90,003	286.824	–650,242
3,000	37.464	94,995	89,909	288.093	–678,987
3,100	37.511	98,744	89,809	289.322	–707,857
3,200	37.556	102,498	89,701	290.514	–736,850
3,300	37.600	106,255	89,586	291.670	–765,959
3,400	37.643	110,018	89,465	292.793	–795,181
3,500	37.686	113,784	89,337	293.885	–824,517
3,600	37.729	117,555	89,203	294.947	–853,957
3,700	37.771	121,330	89,063	295.981	–883,503
3,800	37.815	125,109	88,918	296.989	–913,152
3,900	37.858	128,893	88,767	297.972	–942,901
4,000	37.900	132,680	88,611	298.931	–972,747
4,100	37.943	136,473	88,449	299.867	–1,002,685
4,200	37.984	140,269	88,283	300.782	–1,032,718
4,300	38.023	144,069	88,112	301.677	–1,062,845
4,400	38.060	147,873	87,936	302.551	–1,093,054
4,500	38.093	151,681	87,755	303.407	–1,123,354
4,600	38.122	155,492	87,569	304.244	–1,153,733
4,700	38.146	159,305	87,379	305.064	–1,184,199
4,800	38.162	163,121	87,184	305.868	–1,214,748
4,900	38.171	166,938	86,984	306.655	–1,245,375
5,000	38.170	170,755	86,779	307.426	–1,276,078

Note: Molecular weight = 30.01 and $\bar{h}_{f,298}^{\,0}$ (kJ/kmol) = 90,297.

Table A.22 Ideal Gas Properties of Nitrogen Dioxide (NO_2)

T (K)	$\bar{c}_{p0}$ (kJ/kmol K)	$\bar{h}_{t,T} - \bar{h}_{t,298}$ (kJ/kmol)	$\bar{h}_f^0(T)$ (kJ/kmol)	$\bar{s}^0(T)$ (kJ/kmol K)	$\bar{g}^0(T)$ (kJ/kmol)
200	32.936	–3,432	33,961	226.016	–15,537
298	36.881	0	33,098	239.925	–38,400
300	36.949	68	33,085	240.153	–38,880
400	40.331	3,937	32,521	251.259	–63,469
500	43.227	8,118	32,173	260.578	–89,073
600	45.737	12,569	31,974	268.686	–115,545
700	47.913	17,255	31,885	275.904	–142,780
800	49,762	22,141	31,880	282.427	–170,703
900	51.243	27,195	31,938	288.377	–199,246
1,000	52.271	32,375	32,035	293.834	–228,361
1,100	52.989	37,638	32,146	298.850	–257,999
1,200	53.625	42,970	32,267	303.489	–288,119
1,300	54.186	48,361	32,392	307.804	–318,686
1,400	54.679	53,805	32,519	311.838	–349,670
1,500	55.109	59,295	32,643	315.625	–381,045
1,600	55.483	64,825	32,762	319.194	–412,787
1,700	55.805	70,390	32,873	322.568	–444,878
1,800	56.082	75,984	32,973	325.765	–477,295
1,900	56.318	81,605	33,061	328.804	–510,025
2,000	56.517	87,247	33,134	331.698	–543,051
2,100	56.685	92,907	33,192	334.460	–576,361
2,200	56.826	98,583	32,233	337.100	–609,939
2,300	56.943	104,271	33,256	339.629	–643,778
2,400	57.040	109,971	33,262	342.054	–677,861
2,560	57.121	115,679	33,248	344.384	–732,846
2,600	57.188	121,394	33,216	346.626	–746,736
2,700	57.244	127,116	33,165	348.785	–781,506
2,800	57.291	132,843	33,095	350.868	–816,489
2,900	57.333	138,574	33,007	352.879	–851,677
3,000	57.371	144,309	32,900	354.824	–887,065
3,100	57.406	150,049	32,776	356.705	–922,639
3,200	57.440	155,791	32,634	358.529	–958,404
3,300	57.474	161,536	32,476	360.297	–994,346
3,400	57.509	167,285	32,302	362.013	–1,030,461
3,500	57.546	173,038	32,113	363.680	–1,066,744
3,600	57.584	178,795	31,908	365.302	–1,103,194
3,700	57.624	184,555	31,689	366.880	–1,139,803
3,800	57.665	190,319	31,456	368.418	–1,176,571
3,900	57.708	196,088	31,210	369.916	–1,213,486
4,000	57.750	201,861	30,951	371.378	–1,250,553
4,100	57.792	207,638	30,678	372.804	–1,287,760
4,200	57.831	213,419	30,393	374.197	–1,325,110

(continued)

Table A.22 Ideal Gas Properties of Nitrogen Dioxide (NO_2) (Continued)

T (K)	$\bar{c}_{p0}$ (kJ/kmol K)	$\bar{h}_{t,T} - \bar{h}_{t,298}$ (kJ/kmol)	$\bar{h}_f^0(T)$ (kJ/kmol)	$\bar{s}^0(T)$ (kJ/kmol K)	$\bar{g}^0(T)$ (kJ/kmol)
4,300	57.866	20,204	30,095	375.559	–1,561,602
4,400	57.895	224,992	29,783	376.889	–1,400,222
4,500	57.915	230,783	29,457	378.190	–1,437,974
4,600	57.925	236,575	29,117	379.464	–1,475,861
4,700	57.922	242,367	28,761	380.709	–1,513,867
4,800	57.902	48,159	28,389	381.929	–1,552,002
4,900	57.862	253,947	27,998	383.122	–1,590,253
5,000	57,798	259,730	27,586	384.290	–1,628,622

Note: Molecular weight = 46.01 and $\bar{h}_{f,298}{}^0$(kJ/kmol) = 33,098.

Table A.23 Ideal Gas Properties of Nitrogen (N_2)

T (K)	$\bar{c}_{p0}$ (kJ/kmol K)	$\bar{h}_{t,T} - \bar{h}_{t,298}$ (kJ/kmol)	$\bar{h}_f^0(T)$ (kJ/kmol)	$\bar{s}^0(T)$ (kJ/kmol K)	$\bar{g}^0(T)$ (kJ/kmol)
200	28.793	–2,841	0	179.959	–38,833
298	29.071	0	0	191.511	–57,070
300	29.075	54	0	191.691	–57,453
400	29.319	2,973	0	200.088	–77,062
500	29.636	5,920	0	206.662	–97,411
600	30.086	8,905	0	212.103	–118,357
700	30.684	11,942	0	216.784	–139,807
800	31.394	15,046	0	220.927	–161,696
900	32.131	18,222	0	224.667	–183,978
1,000	32.762	21,468	0	228.087	–206,619
1,100	33.258	24,770	0	231.233	–229,586
1,200	33.707	28,118	0	234.146	–252,857
1,300	34.113	31,510	0	236.861	–276,409
1,400	34.477	34,939	0	239.402	–300,224
1,500	34.805	38,404	0	241.792	–324,284
1,600	35.099	41,899	0	244.048	–348,578
1,700	35.361	45,423	0	246.184	–373,090
1,800	35.595	48,971	0	248.212	–397,811
1,900	35.803	52,541	0	250.142	–422,729
2,000	35.988	56,130	0	251.983	–447,836
2,100	36.152	59,738	0	253.743	–473,122
2,200	36.298	63,360	0	255.429	–498,584
2,300	36.428	66,997	0	257.045	–524,207
2,400	36.543	70,645	0	259.598	–552,390
2,500	36.645	74,305	0	260.092	–575,925

Table A.23 Ideal Gas Properties of Nitrogen (N_2) (Continued)

T (K)	$\bar{c}_{p0}$ (kJ/kmol K)	$\bar{h}_{t,T} - \bar{h}_{t,298}$ (kJ/kmol)	$\bar{h}_f^0(T)$ (kJ/kmol)	$\bar{s}^0(T)$ (kJ/kmol K)	$\bar{g}^0(T)$ (kJ/kmol)
2,600	36.737	77,974	0	261.531	–602,007
2,700	36.820	81,652	0	262.919	–628,229
2,800	36.895	85,338	0	264.259	–654,587
2,900	36.964	89,031	0	265.555	–681,079
3,000	37.028	92,730	0	266.810	–707,700
3,100	37.088	96,436	0	268.025	–734,442
3,200	37.144	100,148	0	269.203	–761,302
3,300	37.198	103,865	0	270.347	–788,280
3,400	37.251	107,587	0	271.458	–815,370
3,506	37.302	111,315	0	272.539	–844,207
3,600	37.352	115,048	0	273.590	–869,876
3,700	37.402	118,786	0	274.614	–897,286
3,800	37.452	122,528	0	275.612	–924,798
3,900	37.501	126,276	0	276.586	–952,409
4,000	37.549	130,028	0	277.536	–980,116
4,100	37.597	133,786	0	278.464	–1,007,916
4,200	37.643	137,548	0	279.370	–1,035,806
4,300	37.688	141,314	0	280.257	–1,063,791
4,400	37.730	145,085	0	281.123	–1,091,856
4,500	37.768	148,860	0	281.972	–1,120,014
4,600	37.803	152,639	0	282.802	–1,148,250
4,700	37.832	156,420	0	283.616	–1,176,575
4,800	37.854	160,205	0	284.412	–1,204,973
4,900	37.868	163,991	0	285.193	–1,233,455
5,000	37.873	167,778	0	285.958	–1,262,012

Note: Molecular weight = 28.01, $\bar{h}_{f,298}{}^0$ (kJ/kmol) = 0.

Table A.24 Ideal Gas Properties of Oxygen Atom (O)

T (K)	$\bar{c}_{p0}$ (kJ/kmol K)	$\bar{h}_{t,T} - \bar{h}_{t,298}$ (kJ/kmol)	$\bar{h}_f^0(T)$ (kJ/kmol)	$\bar{s}^0(T)$ (kJ/kmol–K)	$\bar{g}^0(T)$ (kJ/kmol)
200	22.477	–2,176	248,439	152.085	216,604
298	21.899	0	249,197	160.945	201,235
300	21.890	41	249,211	161.080	200,914
400	21.500	2,209	249,890	167.320	184,478
500	21.256	4,345	250,494	172.089	167,498
600	21.113	6,463	251,033	175.951	150,089
700	21.033	8,570	251,516	179.199	132,328
800	20.986	10,671	251,949	182.004	114,265

(continued)

Table A.24 Ideal Gas Properties of Oxygen Atom (O) (Continued)

T (K)	$\bar{c}_{p0}$ (kJ/kmol K)	$\bar{h}_{t,T} - \bar{h}_{t,298}$ (kJ/kmol)	$\bar{h}_f^0(T)$ (kJ/kmol)	$\bar{s}^0(T)$ (kJ/kmol–K)	$\bar{g}^0(T)$ (kJ/kmol)
900	20.952	12,768	252,340	184.474	95,938
1,000	20.915	14,861	252,698	186.679	77,379
1,100	20.898	16,952	253,033	188.672	58,610
1,200	20.882	19,041	253,350	190.490	39,650
1,300	20.867	21,128	253,650	192.160	20,517
1,400	20.854	23,214	253,934	193.706	1,223
1,500	20.843	25,299	254,201	195.145	−18,222
1,600	20.834	27,383	254,454	196.490	−37,804
1,700	20.827	29,466	254,692	197.753	−57,517
1,800	20.822	31,548	254,916	198.943	−77,352
1,900	20.820	33,630	255,127	200.069	−97,304
2,000	20.819	35,712	255,325	201.136	−117,363
2,100	20.821	37,794	255,512	202.152	−137,528
2,200	20.925	39,877	255,687	203.121	−157,792
2,300	20.831	41,959	255,852	204.047	−178,152
2,400	20.840	44,043	256,007	204.933	−198,599
2,500	20.851	46,127	256,152	205.784	−219,136
2,600	20.865	48,213	256,288	206.602	−239,755
2,700	20.881	50,300	256,416	207.390	−260,456
2,800	20.899	52,389	256,535	208.150	−281,234
2,900	20.920	54,480	256,648	208.884	−302,087
3,000	20.944	56,574	256,753	209.593	−323,008
3,100	20.970	58,669	256,852	210.280	−344,002
3,200	20.998	60,768	256,945	210.947	−365,065
3,300	21.028	62,869	257,032	211.593	−386,191
3,400	21.061	64,973	257,114	212.221	−407,381
3,500	21.095	67,081	257,192	212.832	−428,634
3,600	21.132	69,192	257,265	213.427	−449,948
3,700	21.171	71,308	257,334	214.007	−471,321
3,800	21.212	73,427	257,400	214.572	−492,750
3,900	21.254	75,550	257,462	215.123	−514,233
4,000	21.299	77,678	257,522	215.662	−535,773
4,100	21.345	79,810	257,579	216.189	−557,368
4,200	21.392	81,947	257,635	216.703	−579,009
4,300	21.441	84,088	257,688	217.207	−600,705
4,400	21.490	86,235	257,740	217.701	−622,452
4,500	21.541	88,386	257,790	218.184	−644,245
4,600	21.593	90,543	257,840	218.658	−666,087
4,700	21.646	92,705	257,889	219.123	−687,976
4,800	21.699	94,872	257,938	219.580	−709,915
4,900	21.752	97,045	257,987	220.028	−731,895
5,000	21.805	99,223	258,036	220.468	−753,920

Note: Molecular weight = 16.00 and $\bar{h}_{f,298}^0$ (kJ/kmol) = 249,197.

Table A.25 Ideal Gas Properties of Hydroxyl (OH)

T (K)	$\bar{c}_{p0}$ (kJ/kmol K)	$\bar{h}_{t,T} - \bar{h}_{t,298}$ (kJ/kmol)	$\bar{h}_f^0(T)$ (kJ/kmol)	$\bar{s}^0(T)$ (kJ/kmol –K)	$\bar{g}^0(T)$ (kJ/kmol)
200	30.140	–2,948	38,864	171.607	1,716
298	29.932	0	38,985	183.604	–15,729
300	29.928	55	38,987	183.789	–16,097
400	29.718	3,037	39,030	192.369	–34,926
500	29.570	6,001	39,000	198.983	–54,506
600	29.527	8,955	38,909	204.369	–74,681
700	29.615	11,911	38,770	208.925	–95,352
800	29.844	14,883	38,599	212.893	–116,446
900	30.208	17,884	38,410	216.428	–137,916
1,000	30.682	20,928	38,220	219.635	–159,722
1,100	31.186	24,022	38,039	222.583	–181,834
1,200	31.662	27,164	37,867	225.317	–204,231
1,300	32.114	30,353	37,704	227.869	–226,892
1,400	32.540	33,586	37,548	230.265	–249,800
1,500	32.943	36,860	37,397	232.524	–272,941
1,600	33.323	40,174	37,252	234.662	–296,300
1,700	33.682	43,524	37,109	236.693	–319,869
1,800	34.019	46,910	36,969	238.628	–343,635
1,900	34.337	50,328	36,831	240.476	–367,591
2,000	34.635	53,776	36,693	242.245	–391,729
2,100	34.915	57,254	36,555	243.942	–416,039
2,200	35.178	60,759	36,416	245.572	–440,514
2,300	35.425	64,289	36,276	247.141	–465,150
2,400	35.656	67,843	36,133	248.654	–489,942
2,500	35.872	71,420	35,986	250.114	–514,880
2,600	36.074	75,017	35,836	251.525	–539,963
2,700	36.263	78,634	35,682	252.890	–565,184
2,800	36.439	82,269	35,524	254.212	–590,540
2,900	36.604	85,922	35,360	255.493	–616,023
3,000	36.759	89,590	35,191	256.737	–641,636
3,100	36.903	93,273	35,016	257.945	–667,372
3,200	37.039	96,970	34,835	259.118	–693,223
3,300	37.166	100,681	34,648	260.260	–719,192
3,400	37.285	104,403	34,454	261.371	–745,273
3,500	37.398	108,137	34,253	262.454	–771,467
3,600	37.504	111,882	34,046	263.509	–797,765
3,700	37.605	115,638	33,831	264.538	–824,168
3,800	37.701	119,403	33,610	265.542	–850,672
3,900	37.793	123,178	33,381	266.522	–877,273
4,000	37.882	126,962	33,146	267.480	–903,973
4,100	37.968	130,754	32,903	268.417	–930,771
4,200	38.052	134,555	32,654	269.333	–957,659

(*continued*)

Table A.25 Ideal Gas Properties of Hydroxyl (OH) (Continued)

T (K)	$\bar{c}_{p0}$ (kJ/kmol K)	$\bar{h}_{t,T} - \bar{h}_{t,298}$ (kJ/kmol)	$\bar{h}_f^0(T)$ (kJ/kmol)	$\bar{s}^0(T)$ (kJ/kmol K)	$\bar{g}^0(T)$ (kJ/kmol)
4,300	38.135	138,365	32,397	270.229	–984,635
4,400	38.217	142,182	32,134	271.107	–1,011,704
4,500	38.300	146,008	31,864	271.967	–1,038,859
4,600	38.382	149,842	31,588	272.809	–1,066,094
4,700	38.466	153,685	31,305	273.636	–1,093,419
4,800	38.552	157,536	31,017	274.446	–1,120,820
4,900	38.640	161,395	30,722	275.242	–1,148,306
5,000	38.732	165,264	30,422	276.024	–1,175,871

Note: Molecular weight = 17.01 and $\bar{h}^0_{f,298}$ (kJ/kmol) = 38,985.

Table A.26 Ideal Gas Properties of Oxygen (O_2)

T (K)	$\bar{c}_{p0}$ (kJ/kmol K)	$\bar{h}_{t,T} - \bar{h}_{t,298}$ (kJ/kmol)	$\bar{h}_f^0(T)$ (kJ/kmol)	$\bar{s}^0(T)$ (kJ/kmol K)	$\bar{g}^0(T)$ (kJ/kmol)
200	28.473	–2,836	0	193.518	–41,540
298	29.315	0	0	205.043	–61,103
300	29.331	54	0	205.224	–61,513
400	30.210	3,031	0	213.782	–82,482
500	31.114	6,097	0	220.620	–104,213
600	32.030	9,254	0	226.374	–126,570
700	32.927	12,503	0	231.379	–149,462
800	33.757	15,838	0	235.831	–172,827
900	34.454	19,250	0	239.849	–196,614
1,000	34.936	22,721	0	243.507	–220,786
1,100	35.270	26,232	0	246.852	–245,305
1,200	35.593	29,775	0	249.935	–270,147
1,300	35.903	33,350	0	252.796	–295,285
1,400	36.202	36,955	0	255.468	–320,700
1,500	36.490	40,590	0	257.976	–346,374
1,600	36.768	44,253	0	260.339	–372,289
1,700	37.036	47,943	0	262.577	–398,438
1,800	37.296	51,660	0	264.701	–424,802
1,900	37.546	55,402	0	266.724	–451,374
2,000	37.788	59,169	0	268.656	–478,143
2,100	38.023	62,959	0	270.506	–505,104

Table A.26 Ideal Gas Properties of Oxygen (O_2) (Continued)

T (K)	$\bar{c}_{p0}$ (kJ/kmol K)	$\bar{h}_{t,T} - \bar{h}_{t,298}$ (kJ/kmol)	$\bar{h}_f^0(T)$ (kJ/kmol)	$\bar{s}^0(T)$ (kJ/kmol–K)	$\bar{g}^0(T)$ (kJ/kmol)
2,200	38.250	66,773	0	272.280	–532,243
2,300	38.470	70,609	0	273.985	–559,557
2,400	38.684	74,467	0	275.627	–587,038
2,500	38.891	78,346	0	277.210	–614,679
2,600	39.093	82,245	0	278.739	–642,476
2,700	39.289	86,164	0	280.218	–670,425
2,800	39.480	90,103	0	281.651	–698,520
2,900	39.665	94,060	0	283.039	–726,753
3,000	39.846	98,036	0	284.387	–755,125
3,100	40.023	102,029	0	285.697	–783,632
3,200	40.195	106,040	0	286.970	–812,264
3,300	40.362	110,068	0	288.209	–841,022
3,400	40.526	114,112	0	289.417	–869,906
3,500	40.686	118,173	0	290.594	–898,906
3,600	40.842	122,249	0	291.742	–928,022
3,700	40.1994	126,341	0	292.863	–957,252
3,800	41.143	130,448	0	293.959	–986,596
3,900	41.287	134,570	0	295.029	–1,016,043
4,000	41.429	138,705	0	296.076	–1,045,599
4,100	41.566	142,855	0	297.101	–1,075,259
4,200	41.700	147,019	0	298.104	–1,105,018
4,300	41.830	151,195	0	299.087	–1,134,879
4,400	41.957	155,384	0	300.050	–1,164,836
4,500	42.079	159,586	0	300.994	–1,194,887
4,600	42.197	163,800	0	301.921	–1,225,037
4,700	42.312	168,026	0	302.829	–1,255,270
4,800	42.421	172,262	0	303.721	–1,285,599
4,900	42.527	176,510	0	304.597	–1,316,015
5,000	42.627	180,767	0	305.457	–1,346,518

Note: Molecular weight = 32.0 and $\bar{h}_{f,298}^0$ (kJ/kmol) = 0.

Table A.27 Ideal Gas Properties of Ozone (O_3)

T (K)	$\bar{c}_{p0}$ (kJ/kmol K)	$\bar{h}_{t,T} - \bar{h}_{t,298}$ (kJ/kmol)	$\bar{h}_f^0(T)$ (kJ/kmol)	$\bar{s}^0(T)$ (kJ/kmol K)	$\bar{g}^0(T)$ (kJ/kmol)
0.000	0.000	–10,351	145,348	0.000	132,742
100	33.292	–6,990	151,214	200.681	116,035
200	35.058	–3,636	143,319	224.112	94,635
298	39.238	0	142,674	238.823	71,924
300	39.330	71	142,666	239.065	71,445
400	43.744	4234	142,365	251.007	46,924
500	47.262	8,791	142,335	261.161	21,304
600	49.857	13,652	142,457	270.019	–5,266
700	51.752	18,740	142,662	277.855	–32,666
800	53.154	23,987	142,900	284.859	–60,807
900	54.208	29,359	143,164	291.185	–89,615
1000	55.024	34,823	143,436	296.943	–119,027
1100	55.660	40,359	143,708	302.219	–148,989
1200	56.174	45,949	143,976	307.085	–179,460
1300	56.593	51,589	144,239	311.595	–210,392
1400	56.948	57,266	144,494	315.804	–241,767
1500	57.245	62,978	144,737	319.741	–273,541
1600	57.501	68,714	144,971	323.444	–305,703
1700	57.722	74,475	145,197	326.938	–338,227
1800	57.919	80,257	145,398	330.243	–371,087
1900	58.095	86,061	145,582	333.381	–404,270
2000	58.250	91,876	145,750	336.364	–437,759
2100	58.396	97,709	145,904	339.209	–471,537
2200	58.526	103,554	146,026	341.929	–505,597
2300	58.647	109,412	146,135	344.531	–539,916
2400	58.764	115,286	146,222	347.034	–574,503
2500	58.869	121,164	146,277	349.431	–609,321
2600	58.969	127,060	146,323	351.745	–644,384
2700	59.066	132,959	146,335	353.971	–679,670
2800	59.158	138,871	146,327	356.121	–715,175
2900	59.245	144,792	146,298	358.196	–750,883
3000	59.329	150,720	146,248	360.209	–786,814
3100	59.409	156,657	146,176	362.154	–822,927
3200	59.480	162,603	146,084	364.041	–859,235
3300	59.563	168,552	145,976	365.874	–895,739
3400	59.639	174,515	145,842	367.652	–932,409
3500	59.714	180,481	145,695	369.384	–969,270
3600	59.781	186,456	145,532	371.066	–1,006,289
3700	59.852	192,439	145,352	372.707	–1,043,484
3800	59.919	198,426	145,160	374.301	–1,080,825
3900	59.986	204,422	144,955	375.861	–1,118,343
4000	60.063	210,426	144,729	377.380	–1,156,001

(continued)

Table A.27 Ideal Gas Properties of Ozone (O_3) (Continued)

T (K)	$\bar{c}_{p0}$ (kJ/kmol K)	$\bar{h}_{t,T} - \bar{h}_{t,298}$ (kJ/kmol)	$\bar{h}_f^0(T)$ (kJ/kmol)	$\bar{s}^0(T)$ (kJ/kmol K)	$\bar{g}^0(T)$ (kJ/kmol)
4100	60.120	216,434	144,494	378.861	–1,193,803
4200	60.183	222,447	144,252	380.313	–1,231,775
4300	60.245	228,467	143,997	381.727	–1,269,866
4400	60.308	234,496	143,729	383.116	–1,308,121
4500	60.371	240,530	143,457	384.472	–1,346,501
4600	60.434	246,571	143,176	385.798	–1,385,007
4700	60.492	252,617	142,884	387.099	–1,423,655
4800	60.555	258,672	142,591	388.371	–1,462,416
4900	60.614	264,730	142,294	389.622	–1,501,325
5000	60.672	270,793	141,984	390.848	–1,540,354

Note: Molecular weight = 48.0 and $\bar{h}^0_{f,298}$ (kJ/kmol) = 143,093.

Table A.28 Properties of Sulfur (Crystal)

T (K)	$\bar{c}_{p0}$ (kJ/kmol K)	$\bar{h}_{t,T} - \bar{h}_{t,298}$ (kJ/kmol)	$\bar{h}_f^0(T)$ (kJ/kmol)	$\bar{s}^0(T)$ (kJ/kmol K)	$\bar{g}^0(T)$ (kJ/kmol)
0	0.000	–4,406	0	0.000	–4,406
100	12.803	–3,720	0	12.406	–4,961
200	19.410	–1,962	0	23.522	–6,666
298	22.598	0	0	31.928	–9,515
300	22.644	42	0	32.070	–9,579
400	25.660	2,845	1,795	40.045	–13,173
500	28.531	5,552	3,012	46.078	–17,487
600	31.397	8,552	3,598	51.534	–22,368
700	34.267	11,832	3,665	56.589	–27,780
800	37.137	15,405	57,936	61.354	–33,678
900	40.003	19,259	55,915	65.894	–40,046
1000	42.873	23,405	53,614	70.258	–46,853
1100	45.744	27,836	51,036	74.475	–54,087
1200	48.614	32,552	48,179	78.580	–61,744
1300	51.480	37,560	45,032	82.584	–69,799
1400	54.350	42,848	41,614	86.504	–78,258
1500	57.220	48,430	37,907	90.353	–87,100
1600	60.086	54,292	33,920	94.136	–96,326
1700	62.957	60,446	29,644	97.864	–105,923
1800	65.994	66,885	25,087	101.546	–115,898
1900	68.693	73,609	20,246	105.182	–126,237
2000	71.563	80,626	15,117	108.776	–136,926

Note: Molecular weight = 32.06 and $\bar{h}^0_{f,298}$ (kJ/kmol) = 0.

Table A.29 Ideal Gas Properties of SO_2

T K)	$\bar{c}_{p0}$ (kJ/kmol K)	$\bar{h}_{t,T} - \bar{h}_{t,298}$ (kJ/kmol)	$\bar{h}_f^0(T)$ (kJ/kmol)	$\bar{s}^0(T)$ (kJ/kmol K)	$\bar{g}^0(T)$ (kJ/kmol)
0.0	0.0	–10,552	294,307	0.000	–307,394
100	33.52639	–7,217	294,641	208.915	–324,951
200	36.37151	–3,736	295,482	232.923	–347,163
298	39.874	0	296,842	248.103	–370,777
300	39.94465	75	296,863	248.350	–371,272
400	43.49268	4,251	300,261	260.337	–396,726
500	46.60139	8,757	302,738	270.383	–423,277
600	49.04903	13,544	304,696	279.102	–450,759
700	50.961	18,548	306,294	286.813	–479,063
800	52.43389	23,719	362,305	293.717	–508,097
900	53.5803	29,024	362,238	299.964	–537,786
1000	54.48405	34,430	362,138	305.658	–568,070
1100	55.2037	39,915	362,016	310.884	–598,899
1200	55.79364	45,463	361,874	315.716	–630,238
1300	56.27898	51,070	361,715	320.202	–662,035
1400	56.68902	56,718	361,552	324.386	–694,264
1500	57.03629	62,404	361,385	328.310	–726,903
1600	57.33754	68,124	361,209	332.000	–759,918
1700	57.60113	73,873	361,029	335.486	–793,295
1800	62.01525	79,642	360,862	338.783	–827,009
1900	58.04045	85,437	360,694	341.916	–861,045
2000	58.22873	91,249	360,535	344.900	–895,393
2100	58.40027	97,081	360,380	347.745	–930,026
2200	58.55508	102,931	360,238	350.464	–964,932
2300	58.702	108,792	360,104	353.071	–1,000,113
2400	58.83959	114,671	359,979	355.573	–1,035,546
2500	58.96511	120,562	401,706	357.975	–1,071,218
2600	59.08645	126,461	359,765	360.293	–1,107,143
2700	59.19942	132,378	359,673	362.523	–1,143,276
2800	59.3082	138,302	359,598	364.715	–1,179,742
2900	59.413	144,239	359,531	366.761	–1,216,210
3000	59.51322	150,185	359,481	368.778	–1,252,991
3100	59.60945	156,143	359,431	370.732	–1,289,968
3200	59.706	162,109	359,401	372.623	–1,327,127
3300	59.79354	168,084	359,376	374.464	–1,364,489
3400	59.88141	174,067	359,368	376.250	–1,402,025
3500	59.96927	180,058	359,368	377.987	–1,439,739
3600	60.05295	186,058	359,372	379.677	–1,477,621
3700	60.13663	192,071	359,389	381.321	–1,515,659
3800	60.21613	198,087	359,410	382.928	–1,553,881
3900	60.29562	204,112	359,439	384.493	–1,592,253
4000	60.375	210,146	359,477	386.020	–1,630,776

(*continued*)

Table A.29 Ideal Gas Properties of SO_2 (Continued)

T (K)	$\bar{c}_{p0}$ (kJ/kmol K)	$\bar{h}_{t,T} - \bar{h}_{t,298}$ (kJ/kmol)	$\bar{h}_f^0(T)$ (kJ/kmol)	$\bar{s}^0(T)$ (kJ/kmol K)	$\bar{g}^0(T)$ (kJ/kmol)
4100	60.45043	216,187	359,519	387.514	−1,669,462
4200	60.52993	222,237	359,565	388.970	−1,708,279
4300	62.27884	228,292	359,619	390.396	−1,747,253
4400	60.67637	234,358	359,678	391.790	−1,786,360
4500	60.752	240,429	359,740	393.154	−1,825,606
4600	60.82281	246,509	359,803	394.488	−1,864,978
4700	60.89394	252,592	359,874	395.798	−1,904,501
4800	60.96925	258,688	359,924	397.083	−1,944,152
4900	61.03619	264,789	360,012	398.338	−1,983,909
5000	61.10732	270,893	360,092	399.572	−2,023,809

Note: Molecular weight = 64.07 and $\bar{h}^0_{f,298}$ (kJ/kmol) = −296,842.

Table A.30 Ideal Gas Properties of SO_3

T (K)	$\bar{c}_{p0}$ (kJ/kmol K)	$\bar{h}_{t,T} - \bar{h}_{t,298}$ (kJ/kmol)	$\bar{h}_f^0(T)$ (kJ/kmol)	$\bar{s}^0(T)$ (kJ/kmol K)	$\bar{g}^0(T)$ (kJ/kmol)
0	0	−11,698	390,032	0	−407,463
100	34.07868	−8,360	391,857	212.267	−425,352
200	42.3379	−4,577	393,986	238.153	−447,973
298	50.65987	0	395,765	256.663	−472,251
300	50.80213	92	395,794	256.977	−472,766
400	57.67226	5,531	399,417	272.571	−499,262
500	63.10309	11,581	401,882	286.048	−527,208
600	67.2578	18,108	403,677	297.938	−556,420
700	70.39162	24,999	405,015	308.553	−586,753
800	72.76394	32,162	460,704	318.114	−618,094
900	74.57143	39,530	460,278	326.791	−650,347
1000	75.96889	47,062	459,784	334.724	−683,427
1100	77.0651	54,714	459,248	342.021	−717,274
1200	77.93955	62,467	458,675	348.766	−751,817
1300	78.64246	70,300	458,085	355.033	−787,008
1400	79.21567	78,191	457,483	360.883	−822,810
1500	79.68846	86,136	456,880	366.364	−859,175
1600	80.08176	94,127	456,269	371.518	−896,067
1700	80.4123	102,152	455,659	376.384	−933,466
1800	80.69262	110,207	455,064	380.987	−971,335

(*continued*)

Table A.30 Ideal Gas Properties of SO_3 (Continued)

T (K)	$\bar{c}_{p0}$ (kJ/kmol K)	$\bar{h}_{t,T} - \bar{h}_{t,298}$ (kJ/kmol)	$\bar{h}_f^0(T)$ (kJ/kmol)	$\bar{s}^0(T)$ (kJ/kmol K)	$\bar{g}^0(T)$ (kJ/kmol)
1900	80.9353	118,290	454,483	385.359	−1,009,657
2000	81.14031	126,394	453,914	389.514	−1,048,399
2100	81.32022	134,516	453,362	393.48	−1,087,557
2200	81.47922	142,658	452,834	397.267	−1,127,094
2300	81.61729	150,812	452,324	400.89	−1,167,000
2400	81.73862	158,979	451,839	404.367	−1,207,267
2500	81.84741	167,159	451,378	407.706	−1,247,871
2600	81.94364	175,347	450,939	410.919	−1,288,807
2700	82.02732	183,548	450,525	414.011	−1,330,047
2800	82.10263	191,753	450,140	416.994	−1,371,595
2900	82.17376	199,966	449,780	419.877	−1,413,442
3000	82.23652	208,187	449,449	422.663	−1,455,567
3100	82.2951	216,413	449,131	425.362	−1,497,974
3200	82.3453	224,647	448,851	427.977	−1,540,644
3300	82.39133	232,965	448,583	430.513	−1,583,493
3400	82.43317	241,124	448,349	432.973	−1,626,749
3500	82.47501	249,371	448,132	435.362	−1,670,161
3600	82.51266	257,621	447,931	437.684	−1,713,806
3700	82.54614	265,872	447,763	439.948	−1,757,701
3800	82.57542	274,127	447,604	442.148	−1,801,800
3900	82.60471	282,387	447,466	444.295	−1,846,129
4000	82.62982	290,650	447,353	446.387	−1,890,663
4100	82.65492	298,913	447,253	448.424	−1,935,390
4200	82.68002	307,181	447,169	450.416	−1,980,331
4300	82.70094	315,448	447,098	452.362	−2,025,474
4400	82.71768	323,720	447,052	454.265	−2,070,811
4500	82.7386	331,992	447,014	456.123	−2,116,327
4600	82.75534	340,268	446,989	457.943	−2,162,035
4700	82.77207	348,544	446,981	459.721	−2,207,910
4800	82.78881	356,824	446,981	461.466	−2,253,978
4900	82.80136	365,100	446,993	463.173	−2,300,213
5000	82.81391	373,384	447,019	464.847	−2,346,616

Note: Molecular weight = 80.06 and $\bar{h}^0_{f,298}$ (kJ/kmol) = −395,765.

Table A.31A Mean Bond Energies — Simple Bonds, MJ/kmol

	Br–	*Cl–*	*C–*	*C=*	*C≡*	*F–*	*H–*	*I–*	*N–*	*N=*	*N≡*	*O–*	*O=*	*P–*	*S–*	*Si–*
H			413				436	299	391			463		318	339	323
C			348	615	812	448	413	240	292	615	891	351	804		267	301
N	243	200	339	615	891		391		251	418	945					
O			360	804	1072	190	463	234	628	631		139	498			368
S		251	259	477			339								266	226
F	237	253	441			151	563		270			185			327	
Cl	218	242	328			253	432	208	200			203		326	251	
Br	193	218	276	400		237	366	175	243					267	218	
I			240				299	151				234				
P	268	327					318							200		

Note: Chemical bond strength between atoms decreases as atomic number increases. For diatomic gases, N_2 (atomic mass: 14; 945 MJ per kmol) > O_2 (16; 498 MJ) > F_2 (19; 158 MJ).

Similarly, bond strength of HF > HCl > HBr > HI; note that F = 19, Cl = 35.5, Br = 79.9, I = 126.9.

Thus, the heavier the atom, the weaker the bond.

Weak chemical bond: < 200 MJ per kmol; Average: around 500 MJ; Strong: > 800 MJ.

Sublimation: C (s) → C (g) + 7184000 kJ.

Evaporation: H_2O (l) → H_2O (g) + 44010 kJ.

Source: Pauling, L., *The Nature of Chemical Bond*, 3rd ed., Cornell University Press, Ithaca, 1960.

Table A.31B Mean Bond Energies and Resonance Energies for Complex Bonds

Bond	MJ/kmol	Bond	MJ/kmol	Resonance Energy	MJ/kmol
H–CH	422	H–O_2	196	Benzene[a]	205
H–CH_2	465	H–O_2H	369	COOH group	117
H–CH_3	438	HCCH	965	CO_2	138
H–$CHCH_2$	465	$H_2C{=}CH_2$	733	Naphthlene $C_{10}H_8$	368
H–C_2H_5	420	H_3C-CH_3	376	Andibe $C_6H_5NH_2$	291
H–CHO	364	O=CO	532	Furfural C_4H_2O CHO	125
H–NH_2	449	O–NO	305		
		O–O_2	167		
H–OH	498				

Note: Formaldehyde: 699 MJ, higher aldehydes: 720 MJ; ketone: 766 MJ.

[a] 150 [Winterbone], OH: 429 [Winterbone]

Source: CRC Handbook of Chemistry and Physics, 74th ed., CRC Press, Boca Raton, FL, 1993, pp. 6-203–6-204.

Table A.32 Enthalpy of Formation, Gibbs Energy of Formation, Entropy, and Enthalpy of Vaporization

Substance	Formula	h_f^0, kJ/kmol	g_f^0, kJ/kmol	$\bar{s}^0$, kJ/kmolK	$\bar{h}_{fg}^0$ kJ/kmol	[a]T_{Flame}, K
Acetylene (ethyne)	C_2H_2 (g)	226,736	209,170	200.85		2559.4
Ammonia	NH_3 (g)	–46,190	–16,590	192.33		
Ammonium Nitrate(s)	NH_4NO_3	–1531000		151	26000	
Ammonium Perchlorate(s)	NH_4HClO_4	–295770		–184.2		
Benzene	C_6H_6 (g)	82,930	129,660	269.20	33,830	2354.2
Benzene	C_6H_6(l)					
n-Butane	C_4H_{10} (g)	–126,150	–15,170	310.03	21,060	2279.6
Carbon	C(s)	0	0	5.74		2315.9
Carbon (diamond) (CS)		1890				
Carbon dioxide	CO_2 (g)	–393,520	–394,380	213.67		971.4
Carbon monoxide	CO (g)	–110,530	–137,156	197.56		2404.2
Chlorine(atomic)	Cl (g)	121,010		165.1		
Chlorine	Cl_2 (g)	0		222.9		
n-Decane	$C_{10}H_{22}$	–249530	32970	545.7	40020	2287.0
Diesel(light)	$CH_{1.8}$ (g)	–21,506			3706	2286.6
Dinitrogen pentoxide	N_2O_5	–43,000				
DME	C_2H_6 O(g)	–184,100	–112,600	226.4		2287.6
Dodecane	$C_{12}H_{26}$ (g)	–292,162	50,200	622.8		2279.5
Ethane	C_2H_6 (g)	–84,680	–32,890	229.49		2269.0

(*continued*)

Table A.32 Enthalpy of Formation, Gibbs Energy of Formation, Entropy, and Enthalpy of Vaporization (Continued)

Substance	*Formula*	h_f^0, *kJ/kmol*	g_f^0, *kJ/kmol*	$\bar{s}^0$, *kJ/kmol K*	$\bar{h}_{fg}^0$ *kJ/kmol*	[a]T_{Flame}, *K*
Ethyl alcohol	C_2H_5OH (g)	–235,310	–168,570	282.59	42,340	2246.2
Ethyl alcohol	C_2H_5OH (l)	–277,690	–174,890	160.70		2203.2
Ethylene (ethene)	C_2H_4 (g)	52,280	68,120	219.83		2383.2
Ethylene glycol	$(CH_2OH)_2$	–389,320	–304470	323.55	52,490	2214.9
Gasoline	$CH_{1.87}$ (l)	–3363			4239	2325.3
Glucose	$C_6H_{12}O_6$	–1,260,268		212		2130.3
Heptane	C_7H_{16}	–187,820	8000	425.3		2276.8
Hydrogen	H_2 (g)	0	0	130.57		2060
Hydrogen fluoride	HF	–273,300	221,508	173.8		
Hydrogen peroxide	H_2O_2 (g)	–136,310	–105,600	232.63	61,090	
Hydrogen—monatomic	H (g)	218,000	203,290	114.61		
Hydrogen chloride	HCl (g)	–92,310		186.8		
Hydroxyl	OH (g)	39,040	34,280	183.75		
Lead	Pb (c)	0	0	64.81		
Lead oxide	PbO_2 (c)	–277,400	–217,360	68.6		
Lead sulfate	$PbSO_4$ (c)	–919,940	–813,200	148.57		
Manganese	Mn (c)	0	0	31.8		
Manganese dioxide	MnO_2 (c)	–520,030	–465,180	53.14		
Manganese trioxide	Mn_2O_3 (c)	–958,970	–881,150	110.15		
Mercuric chloride (l)	$HgCl_2$ (l)	–215,075		170.4		
Mercuric chloride (g)	$HgCl_2$ (g)	–146,295		294.7		
Mercuric oxide	HgO (c)	–90,210	–58,400	70.45		
Mercury (l)	Hg (l)	0	0	77.24		
Mercury (g)	Hg (g)	61,295		174.9		
Methane	CH_4 (g)	–74,850	–50,790	186.16		2229
Methyl alcohol	CH_3OH (g)	–200,890	–162,140	239.70	37,900	2231.3
Methyl alcohol	CH_3OH (1)	–238,810	–166,290	126.80		2157.7
Naphthalene	$C_{10}H_8$ (s)	–77,100		167	71500	2291.2
Nitric oxide	NO	90,290	86,595	210.65		
Nitrogen	N_2(9)	0	0	191.50		
Nitrogen dioxide	NO_2	33,100	51,240	239.91		
di-Nitrogen oxide	N_2O(g)	82,100	104,200	220		
di-Nitrogen tetroxide	N_2O_4(g)	9160	97,900	306		

(continued)

Table A.32 Enthalpy of Formation, Gibbs Energy of Formation, Entropy, and Enthalpy of Vaporization (Continued)

Substance	Formula	h_f^0, kJ/kmol	g_f^0, kJ/kmol	$\bar{s}^0$, kJ/kmol K	$\bar{h}_{fg}^0$ kJ/kmol	[a]T_{Flame}, K
Nitrogen—monatomic	N (g)	472,680	455,510	153.19		
n-Nonane	C_9H_{20}	–228,870	24730	506.4	37,690	2286.4
Isooctane(g)	C_8H_{18} (g)	–232,191				2279.8
Isooctane(l)	C_8H_{18} (l)	–263,111	–25948	425.2	31,095	2272.1
n-Octane(g)	C_8H_{18} (g)	–208,450	17,320	463.67	41,460	2285.7
n-Octane(l)	C_8H_{18} (l)	–249,910	6,610	360.79		2275.4
Oxygen	O_2 (g)	0	0	205.04		
Oxygen—monatomic	O (g)	249,170	231,770	160.95		
Ozone	O_3 (g)	143093		238.8		
Palmitic acid(fat)	$C_{16}H_{32}O_2$	–834,694		452.37		2259.3
n-Pentane	C_5H_{12} (g)	–146,440	–8,200	348.40	31,410	2282.1
Propane	C_3H_8 (g)	–103,850	–23,490	269.91	15,060	2276.9
Propylene (propene)	C_3H_6 (g)	20,410	62,720	266.94	18,490	2346.4
Silver oxide	Ag_2O(c)	–31,050	–11,200	121.7		
Styrene	C_8H_8 (l)	-58600		238	43930	2313.2
Sucrose	$C_{12}H_{22}O_{11}$ (s)	-2,221,000		392.40		2159.6
Sulfur	S	0	0	32.06		
Sulfur dioxide	SO_2 (g)	–296,842	–300,194	248.12		
Sulfur trioxide	SO_3 (g)	–395,765	–371,060	256.77		
Sulfuric acid	H_2SO_4 (l)	–813,990	–690,100	156.90		
Sulfuric acid	(aq, m = 1)	–909,270	–744,630	20.1		
Toluene	C_7H_6(l)	12400		221		2315.0
Urethane	$C_3H_7NO_2$ (s)	–517,000				2114.3
Water(g)	H_2O (g)	–241,820	–228,590	188.72		
Water(l)	H_2O (l)	–285,830	–237,180	69.95	44,010	
Zinc	Zn (c)	0	0	41.63		
Zinc oxide	ZnO (c)	–343,280	–318,320	43.64		

Note: T = 25° C and P = 1 atm; multiply kJ/kmol by 0.43 to get BTU/lbml. See also Table A.2A for many other substances.

[a]T-Flame: Adiabatic flame temperature in air under chemical equilibrium: NO, OH, CO, CO_2, H_2O, H_2, N_2, O_2; computed using THERMOLAB spreadsheet software.

Table A.33 Values of Enthalpy of Combustion, Gibbs Free Energy Change, Entropy Change, and Chemical Availability

Fuel Formula	*M* kg/ kmol	ΔH_c^0 (a) MJ/kg	ΔG^0 MJ/kg	Δs^0 kJ/kgK	$(\Delta H^0 - \Delta G^0)/\Delta G^0$ %	$Avail_F$ MJ/ kmol
Acetylene (C_2H_2)	26.038	–48.3	–47.1	–3.7	+2.4	1265.6
Benzene (C_6H_6)	78.114	–40.6	–40.8	0.5	–0.4	3298.5
Carbon (graphite) (C)	12.011	–32.8	–32.9	0.2	–0.2	410.26
Carbon monoxide (CO)	28.01	–10.1	–9.2	–3.1	+10.1	275.10
Ethane (C_2H_6)	30.07	–47.5	–48.0	1.5	–1.0	1493.9
Ethanol (C_2H_5OH)	46.069	–27.8	–28.4	2.1	–2.2	1359.6
Ethylene (C_2H_4)	28.054	–47.2	–46.9	–1.1	+0.7	1359.6
Ethylene glycol $(CH_2OH)_2$	62.07	–17.1	–18.6	5.1	–8.1	1226.4
Glucose ($C_6H_{12}O_6$)(s)	180.2	–15.6	–15.9	0.001		
Hydrogen (H_2)	2.016	–120.0	–113.5	–22.0	+5.8	235.2
Isooctane (C_8H_{18})	114.23	–44.7	–45.8	3.7	–2.4	5375.8
Methane (CH_4)	16.043	–50.0	–49.9	–0.3	+0.2	830.2
Methanol (CH_3OH)	32.042	–21.1	–21.5	1.4	–1.9	722.3
n-Butane (C_4H_{10})	58.12	–45.8	–46.6	2.7	–1.7	2802.5
n-Decane ($C_{10}H_{22}$)	142.29	–44.6	–45.7	3.5	–2.3	6726.4
n-Heptane (C_7H_{16})	100.21	–45.0	–45.9	3.2	–2.1	4764.3
n-Hexane (C_6H_{14})	86.18	–45.1	–46.1	3.1	–2.0	4110
n-Nonane (C_9H_{20})	128.26	–44.7	–45.7	3.4	–2.2	6072.3
n-Octane (C_8H_{18})	114.23	–44.8	–45.8	3.3	–2.2	5418.6
n-Pentane (C_5H_{12})	72.15	–45.4	–46.3	2.9	–1.9	3455.8
Palmitic acid ($C_{16}H_{32}O_2$)	256.5	–39.1	–39.1			
Propane (C_3H_8)	44.097	–46.4	–47.1	2.3	–1.5	2149
Propylene (C_3H_6)	42.081	–45.8	–45.9	0.4	–0.3	1999.9
Sucrose ($C_{12}H_{22}O_{11}$) (s)	342.3	–16.5				
Sulfur (S)	32.064	–9.2	–9.3	0.3	–0.9	609.6
Sulfur monoxide (SO)	48.063	–6.3	–5.8	–1.6	+8.5	—

Note: Dry air during combustion of fuels at standard temperature, $T_o = 25°C$, and pressure, $p_0 = 1$ atm*. Heating value $= -\Delta H_c^0$, e.g., $CO + (1/2)\,O_2 \rightarrow CO_2$, $\Delta G° = \{\bar{g}^{\circ}_{CO_2} - \bar{g}^{\circ}_{CO} - (1/2)\bar{g}^{\circ}_{O_2}\}/28.01 = -9.2$ MJ/kg of CO, $\Delta s° = (\Delta H° - \Delta G°)/T_0$.

Assumed ambient mole fractions: $CO_2 = 0.0003$, $H_2O = 0.0303$, $N_2 = 0.7659$; $O_2 = 0.2035$; $Avail_F$ denotes the maximum possible work that could be developed by the fuel as in an ideal fuel cell.

[a] Multiply by 430 to get BTU/lb.

Source: Gyftopoulos, E.P. and Beretta, G.P., *Thermodynamics, Foundations and Application*, Macmillan Publishing, New York, 1991. Originally from the data presented by Weast, R.C., Ed., *CRC Handbook of Chemistry and Physics*, 66th ed., CRC Press, Boca Raton, FL, 1985; Chemical Availability were calculated by KA's using spreadsheet software. Other data from Bejan, A., *Advanced Engineering Thermodynamics*, John Wiley & Sons, New York, 1988.

Table A.34 Values of Adiabatic Flame Temperature, Entropy Generation at 298 K and 1 atm; Composition of Some of the Product Gases for the Combustion of Various Hydrocarbons in a Perfectly Insulated Steady-State Burne

Fuel	*Formula*	T_{adiab} K	$(To\sigma)/\Delta g^o$ %	CO_2 kmol/MJ	CO mol/MJ	H mol/MJ	NO mol/MJ	NO_2 mmol/MJ	N_2O mmol/MJ
Acetylene	C_2H_2	2598.0	22.6	1.17	457	51.4	87.3	19.8	6
Benzene	C_6H_6	2382.6	27.5	1.61	270	30.1	49.6	10.4	2.7
Carbon	C	2326.0	26.0	2.25	280	0	44.6	9.5	2.5
Ethane	C_2H_6	2300.5	29.0	1.23	155	53.2	35.6	6.6	1.9
Ethylene	C_2H_4	2416.6	26.2	1.27	254	56.6	52.9	10.8	2.8
Hydrogen	H_2	2448.5	20.9	0	0	231	46.3	7.4	2.3
Isooctane	C_8H_{18}	2312.1	30.2	1.35	175	45.0	37.2	7.1	2.0
Methane	CH_4	2266.0	28.3	1.12	124	57.5	31.3	5.6	1.7
n-Butane	C_4H_{10}	2311.8	29.6	1.31	170	48.6	37.1	7.01	2.0
n-Decane	$C_{10}H_{22}$	2316.4	30.1	1.36	179	44.9	37.8	7.3	2.0
n-Heptane	C_7H_{16}	2316.7	29.9	1.34	178	46.2	37.8	7.2	2.0
n-Hexane	C_6H_{14}	2313.2	29.8	1.34	175	46.4	37.4	7.1	2.0
n-Nonane	C_9H_{19}	30.0	1.36	179	45.2	37.8	7.3	2.0	2.0
n-Octane	C_8H_{18}	2316.1	30.0	1.35	178	45.6	37.8	7.2	2.0
n-Pentane	C_5H_{12}	2313.2	29.7	1.32	173	47.4	37.4	7.1	2.0
Propane	C_8H_8	2307.9	29.4	1.28	165	50.3	36.6	6.9	2.0
Propylene	C_3H_6	2378.5	27.6	1.33	227	50.8	47.0	9.4	2.5

Note: Combustion for a mixture of each hydrocarbon is with the stoichiometric amount of dry air at 298 K and 1 atm. Values are determined assuming chemical equilibrium for the gases in the outlet stream with respect to the reaction mechanisms $CO_2 \rightarrow CO + (1/2)O_2$, $H_2O \rightarrow H_2 + (1/2)O_2$, $N_2 + O_2 \rightarrow 2NO$, $(1/2)N_2 + O_2 \rightarrow NO_2$ and $N_2 + (1/2)O_2 \rightarrow N_2O$. The entropy generation σ and change in Gibbs free energy during combustion Δg^o are per unit amount of fuel.

Source: From Gyftopoulos, E.P. and Beretta, G.P., *Thermodynamics, Foundations and Application*, Macmillan Publishing, New York, 1991.

Table A.35 Values of the Constants A_k and B_k for the Equilibrium Constants of Various Substances in Ideal Gas States when Formed from Elements

Substance	*Formula*	*A_k (K)*	*B_k (K) $\times 10^{-3}$*
Acetylene	C_2H_2	6.325	26.818
Ammonia	NH_3	–13.951	–6.462
Carbon	C	18.871	86.173
Carbon (diatomic)	C_2	22.870	100.582
Carbon dioxide	CO_2	–0.010	–47.575
Carbon monoxide	CO	10.098	–13.808
Carbon tetrafluoride	CF_4	–18.143	–112.213
Chlorine (atomic)	Cl	7.244	14.965
Chloroform	$CHCl_3$	–13.284	–12.327
Ethylene	C_2H_4	–9.827	4.635
Fluorine (atomic)	F	7.690	9.906
Freon 12	$CC1_2F_2$	–14.830	–58.585
Freon 21	$CHC1_2F$	–12.731	–34.190
Hydrogen (atomic)	H	7.104	26.885
Hydrogen chloride	HCl	0.713	11.30
Hydronium ion	H_3O^+	–8.312	71.295
Hydroxyl	OH	1.666	4.585
Hydroxyl ion	OH^-	–6.753	–20.168
Mercuric oxide	HgO	–4.935	2.602
Mercurous chloride	HgCl	9.59	–7.506
Mercury dichloride	$HgCl_2$	–11.483	24.053
Methane	CH_4	–13.213	–10.732
Nitric oxide	NO	1.504	10.863
Nitrogen (atomic)	N	7.966	57.442
Nitrogen dioxide	NO_2	–7.630	3.870
Nitrogen oxide	N_2O	–8.438	10.249
Oxygen (atomic)	O	7.963	30.471
Oxygen ion	O^-	0.528	10.048
Ozone	O_3	–8.107	17.307
Proton	H^+	13.437	188.141
Water	H_2O	–6.866	–29.911

Note: Standard pressure, $p^0 = 1$ atm, is used in the approximate expression $K^0(T) = \exp(\Delta A - (\Delta B/T))$, T in K, $298 < T < 5000$; $\Delta A = \Sigma \nu_k A_k$, and $\Delta B = \Sigma \nu_k B_k$. A_k, and B_k constants are computed from reactions involving elements in their natural form and product species; for example, elemental equilibrium C(s) + 1/2 O_2 CO ⇔ with $K^0_{CO}(T) = \exp\{A_{CO} - (B_{CO}/T)\}$, $A_{CO} = 10.098$, and $B_{CO} = -13808$. Then for reaction $CO_2 \Leftrightarrow CO + 1/O_2$, $\Delta A = 1 \times A_{CO} + (1/2) \times A_{O_2} - 1 \times A_{CO_2}$, $A_{O2} = 0$, $\Delta A = 10.108$, and $\Delta B = 33767$. For any elemental species in natural form, $A_k = 0$ and $B_k = 0$.

Mercury Reactions: $Hg + 1/2O_2 \Leftrightarrow HgO$, $Hg + Cl_2 \Leftrightarrow HgCl_2$
$Hg + 1/2Cl_2 \Leftrightarrow HgCl$

Source: Regression of data from the *JANAF Thermochemical Tables*, 2nd ed., Stull, D.R. and Prophet, H., Project Directors, NSRDS–NBS37. U.S. Department of Commerce National Bureau of Standards, Washington, D.C., 1971.

Also see [http://EN.wikipedia.org/wiki]

Table A.36A Logarithms to the Base 10 of the Equilibrium Constant $K^0(T)$

T, K	(1)	(2)	(3)	(4)	(5)	(6)	(7)	(8)	(9)
298	–71.224	–81.208	–159.600	–15.171	–40.048	–46.054	–45.066	–5.018	–41.355
500	–40.316	–45.880	–92.672	–8.783	–22.886	–26.130	–25.025	–2.139	–30.725
1000	–17.292	–19.614	-43.0516	–4.062	–10.062	–11.280	–10.221	–0.159	–23.039
1200	–13.414	–15.208	–34.754	–3.275	–7.899	–8.789	–7.764	+0.135	–21.752
1400	–10.630	–12.054	–28.812	–2.712	–6.347	–7.003	–6.014	+0.333	–20.826
1600	–8.532	–9.684	–24.350	–2.290	–5.180	–5.662	–4.706	+0.474	–20.126
1700	–7.666	–8.706	–22.512	–2.116	–4.699	–5.109	–4.169	+0.530	–19.835
1800	–6.896	–7.836	–20.874	–1.962	–4.276	–4.617	–3.693	+0.577	–19.577
1900	–6.204	–7.058	–19.410	–1.823	–3.886	–4.177	–3.267	+0.619	–19.345
2000	–5.580	–6.356	–18.092	–1.699	–3.540	–3.780	–2.884	+0.656	–19.136
2100	–5.016	–5.720	–16.898	-1.586	–3.227	–3.422	–2.539	+0.688	–18.946
2200	–4.502	–5.142	–15.810	–1.484	–2.942	–3.095	–2.226	+0.716	–18.773
2300	–4.032	–4.614	–14.818	–1.391	–2.682	–2.798	–1.940	+0.742	–18.614
2400	–3.600	–4.130	–13.908	–1.305	–2.443	–2.525	–1.679	+0.764	–18.47
2500	–3.202	–3.684	–13.070	–1.227	–2.224	–2.274	–1.440	+0.784	–18.337
2600	–2.836	–3.272	–12.298	–1.154	–2.021	–2.042	–1.219	+0.802	–18.214
2700	–2.494	–2.892	–11.580	–1.087	–1.833	–1.828	–1.015	+0.818	–18.1
2800	–2.178	–2.536	–10.914	–1.025	–1.658	–1.628	–0.825	+0.833	–17.994
2900	–1.882	–2.206	–10.294	–0.967	–1.495	–1.442	–0.649	+0.846	–17.896
3000	–1.606	–1.898	–9.716	–0.913	–1.343	–1.269	–0.485	+0.858	–17.805
3100	–1.348	–1.610	–9.174	–0.863	–1.201	–1.107	–0.332	+0.869	–17.72
3200	–1.106	–1.340	–8.664	–0.815	–1.067	–0.955	–0.189	+0.878	–17.64
3300	–0.878	–1.086	–8.186	–0.771	–0.942	–0.813	–0.054	+0.888	–17.566
3400	–0.664	–0.846	–7.736	–0.729	–0.824	–0.679	+0.071	+0.895	–17.496
3500	–0.462	–0.620	–7.312	–0.690	–0.712	–0.552	+0.190	+0.902	–17.431

Note: (1) $H_2 \Leftrightarrow 2H$; (2) $O_2 \Leftrightarrow 2O$; (3) $N_2 \Leftrightarrow 2N$; (4) $1/2O_2 + 1/2N_2 \Leftrightarrow NO$; (5) $H_2O \Leftrightarrow H_2 + 1/2O_2$; (6) $H_2O \Leftrightarrow OH + 1/2H_2$; (7) $CO_2 \Leftrightarrow CO + 1/2O_2$; (8) $CO_2 + H_2 \Leftrightarrow CO + H_2O$; (9) $N_2 + 2O_2 \Leftrightarrow 2\ NO_2$; (9a) $OH \Leftrightarrow 1/2H_2 + 1/2O_2$ $\log_{10} K^0$ for Reaction 9a = {$\log_{10} K^0$ for Reaction 5 – $\log_{10} K^0$ for Reaction 6}; $K^0(T) = 10$ (values from tables). If these reactions are reversed (e.g., from $H_2 \Leftrightarrow 2H$ to $2H \Leftrightarrow H_2$), then reverse the sign for the numbers in table (e.g., at 2000 K, change from –5.580 for $H_2 \Leftrightarrow 2H$ to + 5.580 for $2H \Leftrightarrow H_2$).

Source: Based on data from the JANAF Tables, NSRDS–NBS–37, 1971, and its revisions published in *Journal of Physical and Chemical Reference Data* through 1982.

Table A.36B Logarithms to the Base 10 of the Equilibrium Constant $K^0(T)$ (Reactions Involving Solid Carbon)

T, K	(10)	(11)	(12)	(13)	(14)
298	24.0479	69.0915	11	–20.52	–16.02
300	23.9285	68.668			
400	19.1267	51.5365	6.65	–13.02	–10.12
500	16.2528	41.2582	4.08	–8.64	–6.62
600	14.3362	34.4011	2.36	–5.69	–4.26
700	12.9648	29.5031	1.12	–3.59	–2.59
800	11.9319	25.8266	0.2	–1.98	–1.31
900	11.1256	22.9665	–0.53	–0.74	–0.33

(*continued*)

Table A.36B Logarithms to the Base 10 of the Equilibrium Constant $K^0(T)$ (Reactions Involving Solid Carbon) (Continued)

T, K	*(10)*	*(11)*	*(12)*	*(13)*	*(14)*
1000	10.4777	20.6768	–1.05	0.26	0.48
1100	9.445	18.8026	–1.49	1.08	1.15
1200	9.4983	17.24	–1.91	1.74	1.68
1300	9.1176	15.9165	–2.24	2.3	2.13
1400	8.79	14.7816	–2.54	2.77	2.5
1500	8.5045	13.7986	–2.79	3.18	2.83
1600			–3.01	3.56	3.14
1700			–3.2	3.89	3.41
1750	7.9182	12.0392			
1800			–3.36	4.18	3.64
1900			–3.51	4.45	6.86
2000	7.4623	10.3258	–3.64	4.69	4.05
2100			–3.75	4.91	4.22
2200			–3.86	5.1	4.37
2300			–3.96	5.27	4.51
2400			–4.06	5.43	4.64
2500	6.8008	8.2212	–4.15	5.58	4.76
2600			–4.23	5.72	4.87
2700			–4.3	5.84	4.97
2800			–4.37	5.95	5.06
2900			–4.43	6.05	5.14
3000	6.3372	6.8437	–4.49	6.16	5.23
3100			–4.55	6.25	5.3
3200			–4.61	6.33	5.37
3300			–4.66	6.41	5.44
3400			–4.71	6.49	5.5
3500	5.9968	5.7954	–4.75	6.56	5.56

Note: (10) $C + 1/2\ O_2 \Leftrightarrow CO$; (11) $C + O_2 \Leftrightarrow CO_2$; (12) $C + 2H_2 \Leftrightarrow CH_4$; (13) $C + CO_2 \Leftrightarrow 2CO$; (14) $C + H_2O \Leftrightarrow CO + H_2$.

Table A.37 Collision Integrals for Calculation of Transport Properties

kBT/ε	Ω_μ	Ω_D	k_bT/ε	Ω_μ	Ω_D
0.3	2.785	2.662	3.6	0.9932	0.9058
0.4	2.492	2.318	3.8	0.9811	0.8942
0.5	2.257	2.066	4	0.97	0.8836
0.6	2.065	1.877	4.2	0.96	0.874
0.7	1.908	1.729	4.4	0.9507	0.8652
0.8	1.78	1.612	4.6	0.9422	0.8568
0.9	1.675	1.517	4.8	0.9343	0.8492
1	1.587	1.439	5	0.9269	0.8422
1.1	1.514	1.375	6	0.8963	0.8124
1.2	1.452	1.32	7	0.8727	0.7896
1.3	1.399	1.273	8	0.8538	0.7712
1.4	1.353	1.233	9	0.8379	0.7556
1.5	1.314	1.198	10	0.8242	0.7424
1.6	1.279	1.167	20	0.7432	0.664
1.7	1.248	1.14	30	0.7005	0.6232
1.8	1.221	1.116	40	0.6718	0.596
1.9	1.197	1.094	50	0.6504	0.5756
2	1.175	1.075	60	0.6335	0.5596
2.2	1.138	1.041	70	0.6194	0.5464
2.4	1.107	1.012	80	0.6076	0.5352
2.6	1.081	0.9878	90	0.5973	0.5256
2.8	1.058	0.9672	100	0.5882	0.513
3	1.039	0.949	200	0.532	0.4644
3.2	1.022	0.9328	300	0.5016	0.436
3.4	1.007	0.9186	400	0.4811	0.417

Note: Curve fit to the table: $\varepsilon/(k_BT) = x$, $\Omega_D = -0.0759x^2 + 0.8505x + 0.6682$, $R^2 = 0.9996$, $0.1 < x < 3.3$, $\Omega_\mu = -0.104x^2 + 0.9663x + 0.7273$, $R^2 = 0.9996$.

Source: Adapted from Hirschfelder, J.O., Curtis, C.F., and Bird, R.B., *Molecular Theory of Gases and Liquids*, John Wiley & Sons, New York, 1954. With the permission of Wiley.

Table A.38 Binary Diffusion Coefficients at 1 atm

Substance A	*Substance B*	*T(K)*	$D_{AB}\cdot 10^5$ (m^2/sec)
Ammonia	Air	273	2.14
Ar	O_2	293	7.26
Benzene	Air	273	0.77
Carbon dioxide	Air	273	1.38
Carbon dioxide	Nitrogen	273	1.44
		293	1.63
	CO	273	1.37
	CH_4	273	1.53
	N_2O	273	0.96
	O_2	273	1.39
		293	1.60
CO	O_2	273	1.85
Cyclohexane	Air	318	0.86
n-Decane	Nitrogen	363	0.84
n-Dodeane	Nitrogen		
Ethanol	Air	273	1.02
n-Hexane	Nitrogen	288	0.757
Hydrogen	Air	273	6.11
Hydrogen	CH_4	273	6.25
Hydrogen		298	7.26
	C_2H_4	273	4.86
		298	6.02
Hydrogen	C_2H_6	273	4.59
Hydrogen		298	5.37
	cis-2-C_4H_8	298	3.78
Hydrogen	CO	273	6.51
Hydrogen	CO_2	273	5.5
		293	6.0
		298	0.646
Hydrogen	O_2	273	6.97
	SF_6	298	4.20
Nitrogen	H_2	273	6.74
	H_2	293	7.6
	O_2	273	1.81
		293	2.2
	CO	273	1.92
	CO_2	273	1.44
		293	1.6
	C_2H_6	298	1.48

(*continued*)

Table A.38 Binary Diffusion Coefficients at 1 atm (Continued)

Substance A	*Substance B*	*T(K)*	$D_{AB}\cdot 10^5$ *(m²/sec)*
	C_2H_4	298	1.63
	n-C_4H_{10}	298	0.96
	i-C_4H_{10}	298	0.908
	cis-2-C_4H_8	298	0.95
Methanol	Air	273	1.32
	Air	273	0.611
Water vapor	Air	288	2.82
		281	2.07
Water	Air	273	2.2
n-Octane	Air	273	1.32
n-Octane	N_2	303	0.505
Toluene	Air	303	0.88
		303	0.71
2,2,4-Trimethylpentane (Isooctane)	Nitrogen	303	0.705
2,2,3-Trimethyl heptane	Nitrogen	363	0.684

Note: $D = D_{ref}\{T/T_{ref}\}^{1.75}\{P_{ref}/P\}$, T in K.

$$D_{AB} = \frac{0.02628\, T^{3/2}}{PM_{AB}^{1/2}\sigma_{AB}^2\Omega_D}, \quad D_{AB}[=]\, m^2/s, \quad T\,[=]K, \quad P[=]Pa,$$

D_{AB} is in units of m²/sec, T in K, P in Pa.

For A diffusing into multicomponent mixture: $D_{Ak} = 1/\Sigma\{X_k/D_{Ak}\}$, A is in low concentration, $(1/M_{AB}) = (1/2)\ \{(1/M_A) + (1/M_B)\}$, $\sigma_{AB} = \{\sigma_A + \sigma_B)/2$, $(\varepsilon/k_B)_{AB} = [(\varepsilon/k_B)_A\ (\varepsilon/k_B)_B]^{1/2}$.

Source: Handbook of Chemistry and Physics, 38th ed., Chemical Rubber Publishing, Cleveland, OH, 1956–1957; Perry, R.H., Green, D.W., and Maloney, J.O. *Perry's Chemical Engineers' Handbook*, 6th ed., McGraw-Hill, New York, 1984.

Table A.39A Two-Step Oxidation for Best Fit between Computed and Experimental Flammability Limits: Step 1 Kinetics; kmole, m^3, s

Fuel	*Preexponential Factor*	*E, MJ/kmol*	*a*	*b*
CH_4	2.8×10^9	202.6	−0.3	1.3
CH_4	1.5×10^7	125.5	−0.3	1.3
C_2H_6	7.3×10^{10}	125.5	0.1	1.65
C_3H_8	5.6×10^{10}	125.5	0.1	1.65
C_4H_{10}	4.9×10^9	125.5	0.15	1.6
C_5H_{12}	4.3×10^9	125.5	0.25	1.5
C_6H_{14}	3.9×10^9	125.5	0.25	1.5
C_7H_{16}	3.5×10^9	125.5	0.25	1.5
C_8H_{18}	3.2×10^9	125.5	0.25	1.5
C_8H_{18}	5.4×10^{10}	167.4	0.25	1.5
C_9H_{20}	2.9×10^9	125.5	0.25	1.5
$C_{10}H_{22}$	2.6×10^9	125.5	0.25	1.5
CH_3OH	2.0×10^{10}	30.0	125.5	1.5
C_2H_5OH	1.0×10^{10}	125.5	0.15	1.6
C_6H_6	1.3×10^9	125.5	−0.1	1.85
C_7H_8	1.0×10^9	125.5	−0.1	1.85

Step 1: Fuel + $\nu_{O2}O_2 \rightarrow \nu_{CO}CO + \nu_{H2O}H_2O$

$$\bar{R} = 0.008314 \frac{MJ}{\text{kmole } K}, \quad \text{kmole}, m^3, s$$

Note: $\dot{w}''' = A_{[\,]}\, T^{\,n} \exp\{-E/(\bar{R}\,T)\}\, [A]^a [B]^b$, kmol/$m^3$ sec, [ ] kmol/m^3,

n-Octane: $A_{[\,]} = 4.6 \times 10^{11}$, E = 125,500 kJ/kmol, A = fuel, B = O_2, a = 0.25, b = 1.5

n-Decane: $A_{[\,]} = 2.6 \times 10^9$, E = 125,500 kJ/kmol, A = fuel, B = O_2, a = 0.25, b = 1.5

Source: Westbrook, CK and Dryer, FL, 1981.

Step 2: CO + (1/2) $O_2 \rightarrow CO_2$

$d[CO]/dt = -A_{[\,]}\, [CO]^{\alpha} [O_2]^{\beta} [H_2O]^{\gamma} \operatorname{Exp}\{-E/(\bar{R}\,T)\}$, kmol, m^3, sec; α, β, γ: see Table A39.B

Source: Adopted from Bartok and Sarofim, Chapter 3

Table A.39B Four-Step Paraffin Oxidation Schemes and Kinetic Constants

Fuel	*B*	*C*	*D*	α	β	γ	$A_{[\,]}$	*E, kJ*
C_nH_{2n+2}	C_nH_{2n+2}	O_2	C_2H_4	0.5	1.07	0.4	2.5704E+14	207,665
C_2H_4	C_2H_4	O_2	C_nH_{2n+2}	0.5	1.07	0.4	3.7154E+12	209,340
CO	CO	O_2	H_2O	0.5	1.07	0.4	2.2387E+12	167,472
H_2	H_2	O_2	C_2H_2	0.85	1.42	−0.56	2.4547E+11	171,659

Four Step Kinetics

$$C_nH_{2n+2} \rightarrow (n/2)\, C_2H_4 + H_2 \qquad \text{(I)}$$

$$C_2H_4 + O_2 \rightarrow 2\, CO + 2\, H_2 \qquad \text{(II)}$$

$$CO + (1/2)\, O_2 \rightarrow CO_2 \qquad \text{(III)}$$

$$H_2 + (1/2)\, O_2 \rightarrow H_2O \qquad \text{(IV)}$$

$$\frac{d[\text{Fuel}]}{\text{dt}}, \frac{\text{kmole}}{m^3 s} = -A_{[]}[B]^{\alpha}[C]^{\beta}[D]^{\gamma} \exp\left[-\frac{E}{\bar{R}T}\right], \quad \bar{R} = 8.314 \frac{\text{kJ}}{\text{kmole } K}, \quad \text{kmole}, m^3, s$$

Table A.39C Ignition Temperatures In Air, Deg K

Substance	*Formulae*	*K*	
Diesel #2		527	
Residual fuel oil, #6 (C: 85.7%, H: 16.5%)		681	
Gasoline		553	NACB (North American Comb Handbook)
Gasoline		723	
Gasoline		618	Steam Book, B&W co (1963)
Gasoline 73 ON, M=72		572	NACA 1300
Gasoline 100 ON		741	NACA 1300
Kerosene		548	Steam Book, B&W co (1963)
JP1, M=150		522	NACA 1300
JP-3		511	Chigier
JP4, M=126		534	NACA 1300
JP5, M=170		515	NACA 1300
Kerosene or JP-8 or Jet A		501	Chigier
PARAFFINS			
methane	CH4	810,880	Bartok and Sarofim; Conti and Hertzberg

Table A.39C Ignition Temperatures In Air, Deg K (Continued)

Substance	*Formulae*	*K*	
ethane	C2H6	745	Bartok and Sarofim
propane	C3H8	743	Bartok and Sarofim
Butane	C4H10	638	Bartok and Sarofim
Pentane	C5H12	558	Bartok and Sarofim
Hexane	C6H14	534	Bartok and Sarofim
Heptane	C7H16	496	Bartok and Sarofim
Octane	C8H18	479	Bartok and Sarofim
Isooctane	C8H18	691	Chigier
Nonane	C9H20	479	Bartok and Sarofim
Decane	C10H22	481	Bartok and Sarofim
Hexadecane (n-Cetane)	C16H34	478	Bartok and Sarofim
ALCOHOLS			
methanol	CH3OH	658	Bartok and Sarofim
ethanol	C2H5OH	638	
propanol	C3H7OH	713	Bartok and Sarofim
Benzyl alcohol		701	NACA 1300
OLEFINS			
Acetylene	C2H2	578	Bartok and Sarofim
1 Butene	C4H8	658	Bartok and Sarofim
ethylene	C2H4	763	Bartok and Sarofim
Propylene	C3H6	728	Bartok and Sarofim
Pentene	C5H10	548	Bartok and Sarofim
Hexene	C6H12	526	Bartok and Sarofim
CYCLO-Compounds			
Cyclopropane		771	NACA 1300
Cyclo pentane	C5H10	658	Bartok and Sarofim
Aniline	C6H5NH2	888	Bartok and Sarofim
Cyclo hexane	C6H12	543	Bartok and Sarofim
MethylCyclo pentane	C6H12	596	Bartok and Sarofim
MethylCyclohexane	C7H14	538	Bartok and Sarofim
Benzene	C6H6	865	
Toluene	C7H8	841	Bartok and Sarofim
Styrene	C8H8	763	Bartok and Sarofim
Ethyl Benzene	C8H10	733	Bartok and Sarofim
O-Xylene	C8H10 or C6H4 (CH3)2	774	Bartok and Sarofim
m xylene	C8H10	836	Bartok and Sarofim
p-Xylene	C8H10	836	Bartok and Sarofim
Naphthalene	C10H8	799	Bartok and Sarofim
Tetralin	C10H22	658	Bartok and Sarofim

(*continued*)

Table A.39C Ignition Temperatures In Air, Deg K (Continued)

Substance	*Formulae*	*K*	
OTHERS			
Acetone	(CH3)2CO	738	Bartok and Sarofim
Acetic acid	CH3COOH	738	Bartok and Sarofim
Ammonia	NH3	924	Bartok and Sarofim
camphor		739	NACA1300
Charcoal		618	Steam Book, B&W co (1963)
CO	CO	882,	Bartok and Sarofim
Diethyl ether (DEE)	C4100	433	B&W Handbook
FC (Bit Coal)		678	Steam Book, B&W co (1963)
FC (Semi Bit)		738	Steam Book, B&W co (1963)
Hydrogen	H2	673, 773	Bartok and Sarofim; Conti and Hertzberg
is-octane	C8H18	478	Bartok and Sarofim
LPG		723	
Mapp gas		728	NACB (North American Comb Handbook)
S		518	Steam Book, B&W co (1963)
Town gas		643	NACB (North American Comb Handbook)

Table A.39D Burning Velocities of Various Fuels

	$\phi = 0.7$	$\phi = 0.8$	$\phi = 0.9$	$\phi = 1.0$	$\phi = 1.1$	$\phi = 1.2$	$\phi = 1.3$	$\phi = 1.4$	S_{max}	ϕ at S_{max}
Saturated hydrocarbons										
Ethane	30.6	36.0	40.6	44.5	47.3	47.3	44.4	37.4	47.6	1.14
Propane			42.3	45.6	46.2	42.4	34.3		46.4	1.06
n-Butane		38.0	42.6	44.8	44.2	41.2	34.4	25.0	44.9	1.03
Methane		30.0	38.3	43.4	44.7	39.8	31.2		44.8	1.08
n-Pentane		35.0	40.5	42.7	42.7	39.3	33.9		43.0	1.05
n-Heptane		37.0	39.8	42.2	42.0	35.5	29.4		42.8	1.05
2, 2, 4-Trimethylpentane		37.5	40.2	41.0	37.2	31.0	23.5		41.0	0.98
2, 2, 3-Trimethylpentane		37.8	39.5	40.1	39.5	36.2			40.1	1.00
2, 2-Dimethylbutane		33.5	38.3	39.9	37.0	33.5			40.0	0.98
Isopentane		33.0	37.6	39.8	38.4	33.4	24.8		39.9	1.01
2, 2-Dimethylpropane			31.0	34.8	36.0	35.2	33.5	31.2	36.0	1.10
Unsaturated hydrocarbons										
Acetylene		107	130	144	151	154	154	152	155	1.25
Ethylene	37.0	50.0	60.0	68.0	73.0	72.0	66.5	60.0	73.5	1.13
Propyne		62.0	66.6	70.2	72.2	71.2	61.0		72.5	1.14
1, 3-Butadiene			42.6	49.6	55.0	57.0	56.9	55.4	57.2	1.23
n-1-Heptyne		46.8	50.7	52.3	50.9	47.4	41.6		52.3	1.00
Propylene			48.4	51.2	49.9	46.4	40.8		51.2	1.00
n-2-Pentene		35.1	42.6	47.8	46.9	42.6	34.9		48.0	1.03
2, 2, 4-Trimethyl-3-pentene		34.6	41.3	42.2	37.4	33.0			42.5	0.98

Substituted alkyls										
Methanol		34.5	42.0	48.0	50.2	47.5	44.4	42.2	50.4	1.08
Isopropyl alcohol		34.4	39.2	41.3	40.6	38.2	36.0	34.2	41.4	1.04
Triethylamine		32.5	36.7	38.5	38.7	36.2	28.6		38.8	1 06
n-Butyl chloride	24.0	30.7	33.8	34.5	32.5	26.9	20.0		34.5	1.00
Allyl chloride	30.6	33.0	33.7	32.4	29.6				33.8	0.89
Isopropyl mercaptan		30.0	33.5	33.0	26.6				33.8	0.44
Ethylamine		28.7	31.4	32.4	31.8	29.4	25.3		32.4	1.00
Isopropylamine		27.0	29.5	30.6	29.8	27.7			30.6	1 01
n-Propyl chloride		24.7	28.3	27.5	24.1				28.5	0.93
Isopropyl chloride		24.8	27.0	27.4	25.3				27.6	0.97
n-Propyl bromide	No ignition									
Silanes										
Tetramethylsilane	39.5	49.5	57.3	58.2	57.7	54.5	47.5		58.2	1.01
Trimethylethoxysilane	34.7	41.0	47.4	50.3	46.5	41.0	35.0		50.3	1.00
Aldehydes										
Acrolein	47.0	58.0	66.6	65.9	56.5				67.2	0.95
Propionaldehyde		37.5	44.3	49.0	49.5	46.0	41.6	37.2	50.0	1.06
Acetaldehyde		26.6	35.0	41.4	41.4	36.0	30.0		42.2	1.05
Ketones										
Acetone		40.4	44.2	42.6	38.2				44.4	0.93

(continued)

Table A.39D Burning Velocities of Various Fuels (Continued)

	$\phi = 0.7$	$\phi = 0.8$	$\phi = 0.9$	$\phi = 1.0$	$\phi = 1.1$	$\phi = 1.2$	$\phi = 1.3$	$\phi = 1.4$	S_{max}	ϕ at S_{max}
Methyl ethyl ketone		36.0	42.0	43.3	41.5	37.7	33.2		43.4	0.99
Esters										
Vinyl acetate	29.0	36.6	39.8	41.4	42.1	41.6	35.2		42.2	1.13
Ethyl acetate		30.7	35.2	37.0	35.6	30.0			37.0	1.00
Ethers										
Dimethyl ether		44.8	47.6	48.4	47.5	45.4	42.6		48.6	0.99
Diethyl ether	30.6	37.0	43.4	48.0	47.6	40.4	32.0		48.2	1.05
Dimethoxymethane	32.5	38.2	43.2	46.6	48.0	46.6	43.3		48.0	1.10
Diisopropyl ether		30.7	35.5	38.3	38.6	36.0	31.2		38.9	1.06
Thio ethers										
Dimethyl sulfide		29.9	31.9	33.0	30.1	24.8			33.0	1.00
Peroxides										
Di-t-butyl peroxide		41.0	46.8	50.0	49.6	46.5	42.0	35.5	50.4	1.04
Aromatic compounds										
Furan	48.0	55.0	60.0	62.5	62.4	60.0			62.9	1.05
Benzene		39.4	45.6	47.6	44.8	40.2	35.6		47.6	1.00
Thiophane	33.8	37.4	40.6	43.0	42.2	37.2	24.6		43.2	1.03
Cyclic compounds										
Ethylene oxide	57.2	70.7	83.0	88.8	89.5	87.2	81.0	73.0	89.5	1.07
Butadiene monoxide		36.6	47.4	57.8	64.0	66.9	66.8	64.5	67.1	1.24

Propylene oxide	41.6	53.3	62.6	66.5	66.4	62.5	53.8		67.0	1.05
Dihydropyran	39.0	45.7	51.0	54.5	55.6	52.6	44.3	32.0	55.7	1.08
Cyclopropane		40.6	49.0	54.2	55.6	53.5	44.0		55.6	1.10
Tetrahydropyran	44.8	51.0	53.6	51.5	42.3				53.7	0.93
Cyclic compounds										
Tetrahydrofuran			43.2	48.0	50.8	51.6	49.2	44.0	51.6	1.19
Cyclopentadiene	36.0	41.8	45.7	47.2	45.5	40.6	32.0		47.2	1.00
Ethylenimine		37.6	43.4	46.0	45.8	43.4	38.9		46.4	1.04
Cyclopentane	31.0	38.4	43.2	45.3	44.6	41.0	34.0		45.4	1.03
Cyclohexane			41.3	43.5	43.9	38.0			44.0	1.08
Inorganic compounds										
Hydrogen	102	120	145	170	204	245	213	290	325	1.80
Carbon disulfide	50.6	58.0	59.4	58.8	57.0	55.0	52.8	51.6	59.4	0.91
Carbon monoxide					28.5	32.0	34.8	38.0	52.0	2.05
Hydrogen sulfide	34.8	39.2	40.9	39.1	32.3				40.9	0.90
Propylene oxide	74.0	86.2	93.0	96.6	97.8	94.0	84.0	71.5	97.9	1.09
Hydrazine	87.3	90.5	93.2	94.3	93.0	90.7	87.4	83.7	94.4	0.98
Furfural	62.0	73.0	83.3	87.0	87.0	84.0	77.0	65.5	87.3	1.05
Ethyl nitrate	70.2	77.3	84.0	86.4	83.0	72.3			86.4	1.00
Butadiene monoxide	51.4	57.0	64.5	73.0	79.3	81.0	80.4	76.7	81.1	1.23
Carbon disulfide	64.0	72.5	76.8	78.4	75.5	71.0	66.0	62.2	78.4	1.00
n-Butyl ether		67.0	72.6	70.3	65.0				72.7	0.91

(*continued*)

Table A.39D Burning Velocities of Various Fuels (Continued)

	$\phi = 0.7$	$\phi = 0.8$	$\phi = 0.9$	$\phi = 1.0$	$\phi = 1.1$	$\phi = 1.2$	$\phi = 1.3$	$\phi = 1.4$	S_{max}	ϕ at S_{max}
Methanol	50.0	58.5	66.9	71.2	72.0	66.4	58.0	48.8	72.2	1.08
Diethyl cellosolve	49.5	56.0	63.0	69.0	69.7	65.2			70.4	1.05
Cyclohexene										
Monoxide	54.5	59.0	63.5	67.7	70.0	64.0			70.0	1.10
Epichlorohydrin	53.0	59.5	65.0	68.6	70.0	66.0	58.2		70.0	1.10
n-Pentane		50.0	55.0	61.0	62.0	57.0	49.3	42.4	62.9	1.05
n-Propyl alcohol	49.0	56.6	62.0	64.6	63.0	50.0	37.4		64.8	1.03
n-Heptane	41.5	50.0	58.5	63.8	59.5	53.8	46.2	38.8	63.8	1.00
Ethyl nitrite	54.0	58.8	62.6	63.5	59.0	49.5	42.0	36.7	63.5	1.00
Pinene	48.5	58.3	62.5	62.1	56.6	50.0			63.0	0.95
Nitroethane	51.5	57.8	61.4	57.2	46.0	28.0			61.4	0.92
Isooctane		50.2	56.8	57.8	53.3	50.5			58.2	0.98
Pyrrole		52.0	55.6	56.6	56.1	52.8	48.0	43.1	56.7	1.00
Aniline		41.5	45.4	46.6	42.9	37.7	32.0		46.8	0.98
Dimethyl formamide		40.0	43.6	45.8	45.5	40.7	36.7		46.1	1.04

Note: T = 25°C (air–fuel temperature); P = 1 atm (0.31 mole % H_2O in air); burning velocity S as a function of equivalence ratio ϕ in cm sec^{-1}. The data are for premixed fuel–air mixtures at 100°C and 1-atm pressure; 0.31 mole % H_2O in air; burning velocity S as a function of Ø in cm sec^{-1}.

Source: The compilation of laminar flame speed data given in tables is from Gibbs and Calcote, *J. Chem. Eng. Data*, 4, 2226, 1959.

Table A.40 The Z Values for First Ten Roots of Bessel Function $J_N(Z) = 0$; N = 0, 1, 2, 3, 4, and 5

	J_0	J_1	J_2	J_3	J_4	J_5
1	2.4048	3.8317	5.1356	6.3802	7.5883	8.7715
2	5.5201	7.0156	8.4172	9.7610	11.0647	12.3386
3	8.6537	10.1735	11.6198	13.0152	14.3725	15.7002
4	11.7915	13.3237	14.7960	16.2235	17.6160	18.9801
5	14.9309	16.4706	17.9598	19.4094	20.8269	22.2178
6	18.0711	19.6159	21.1170	22.5827	24.0190	25.4303
7	21.2116	22.7601	24.2701	25.7482	27.1991	28.6266
8	24.3525	25.9037	27.4206	28.9084	30.3710	31.8117
9	27.4935	29.0468	30.5692	32.0649	33.5371	34.9888
10	30.6346	32.1897	33.7165	35.2187	36.6990	38.1599

Note:

$$I_{1/2}(z) = \left[\frac{2}{\pi z}\right]^{1/2} \text{Sinh}(z), \quad I_{-1/2}(z) = \left[\frac{2}{\pi z}\right]^{1/2} \text{Cosh}(z)$$

$$I_{v-1}(z) - I_{v+1}(z) = \left[\frac{2v}{z}\right] I_v(z)$$

Appendix B

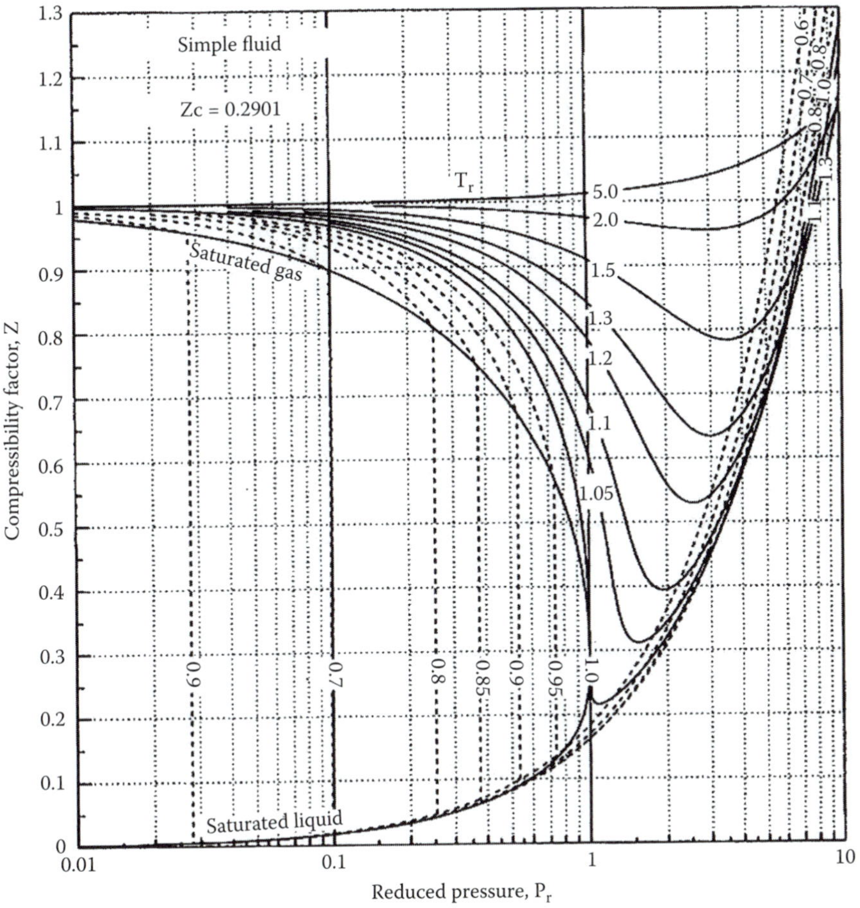

Figure B.1 Simple fluid compressibility factor. [Adapted from R. Sonntag, C. Borgnakke and G.J. Wiley, Fundamentals of Classical Thermodynamics, 5th Ed. John Wiley & Sons, 1998, pp 772, 763–765.

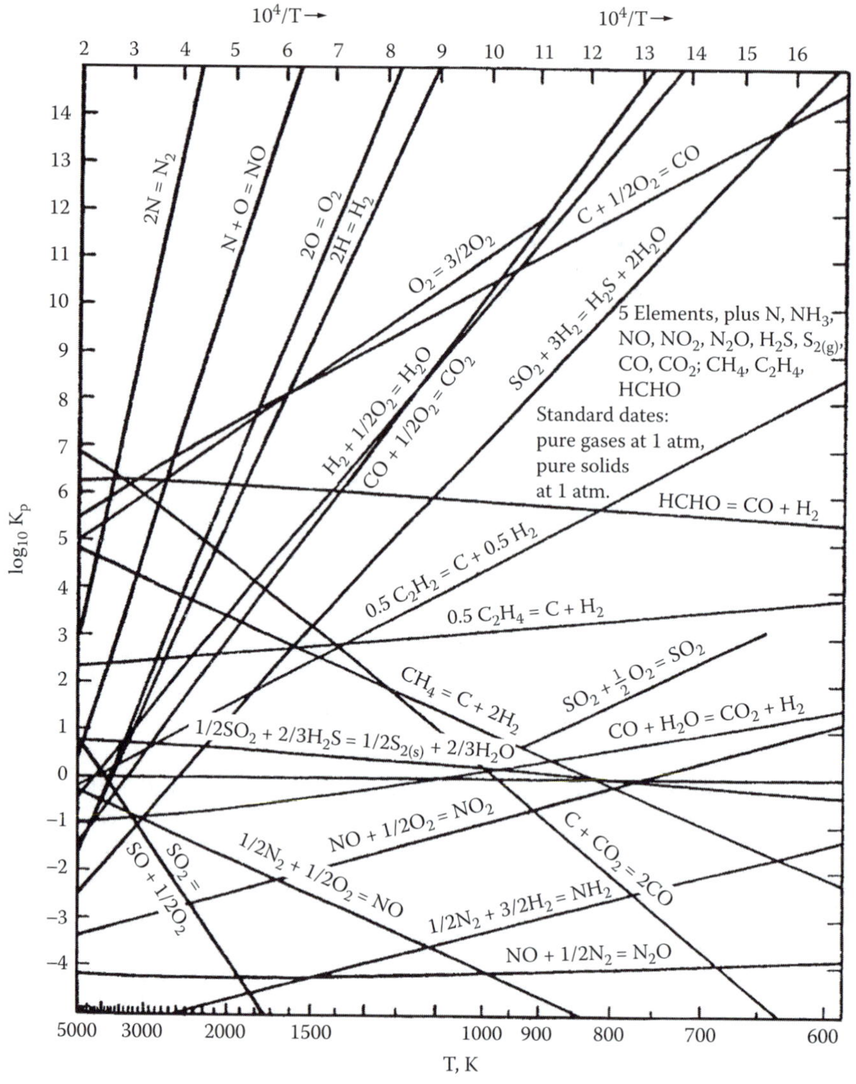

Figure B.2 Plots of ln K°(T) vs. 1/T for various reactions. [Adapted from M. Modell and R.C. Reid, Thermodynamics and its Applications, Second Edition, Prentice Hall, 1983.]

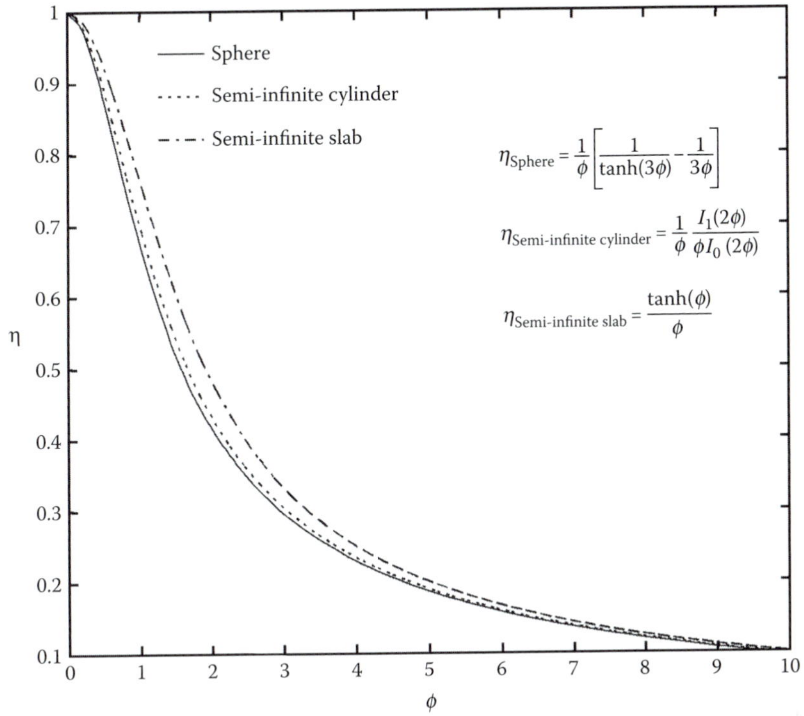

Figure B.3 Effectiveness factors for various geometries: G = ϕ^2; ϕ = characteristic length {rate of reaction/(rate of diffusion × bulk concentration)}$^{1/2}$; characteristic length = semithickness (slab); r/2 (cylinder), r/3 (sphere). See Chapter 9.

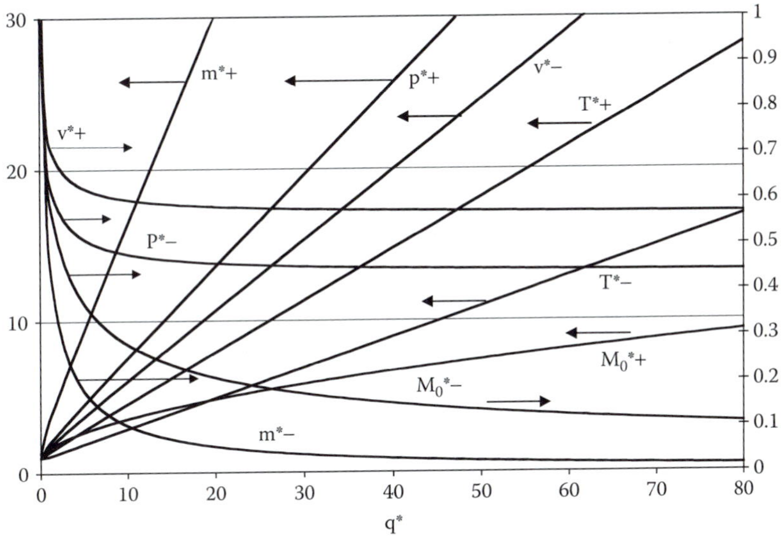

Figure B.4 Deflagration and detonation charts: CJ Charts.

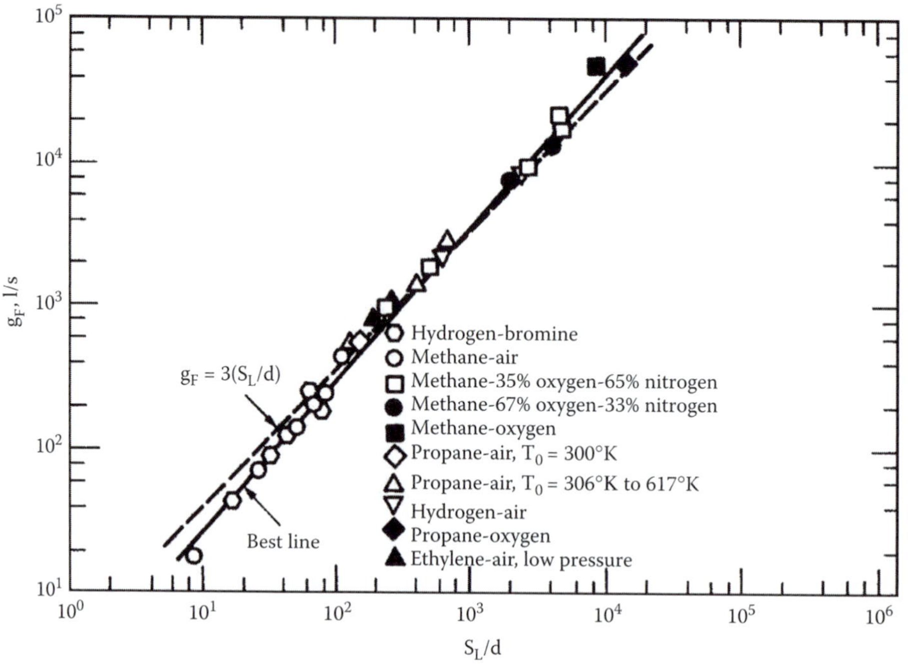

Figure B.5 Determination of quench distance from correlations. [Adapted from Berlad, A.L. and Potter, A. E. *Combust. Flame*, 1, 1, 127–128, 1957.]

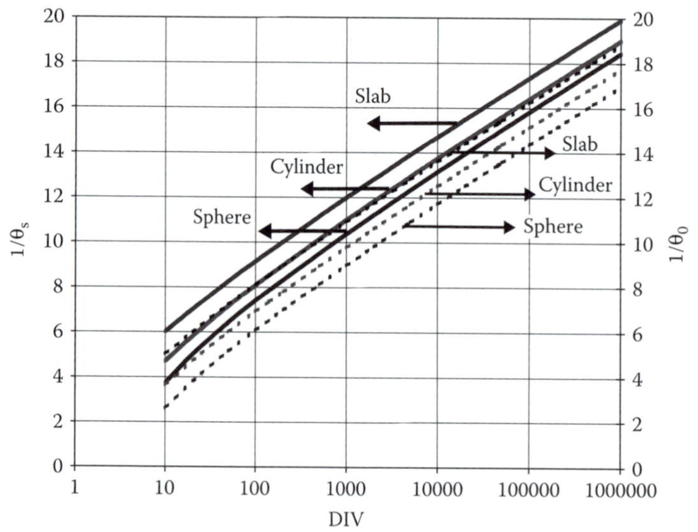

Figure B.6 Surface and center temperatures of stored combustibles vs. Damkohler number at ignition. [Adapted from Annamalai et al.]

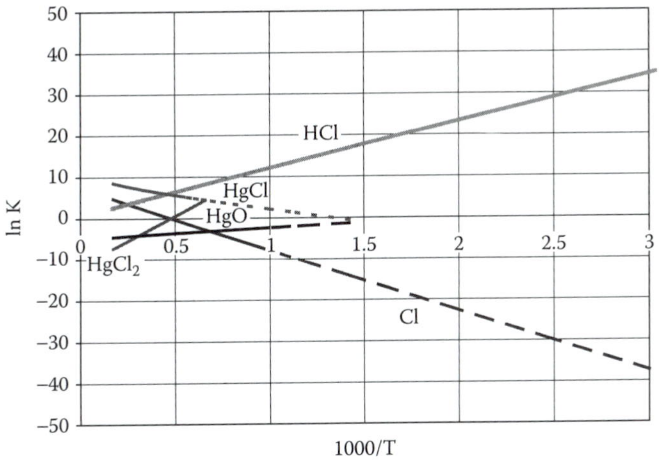

Figure B.7 Atomic equilibrium constants for Hg compounds; $Hg + \frac{1}{2}O_2 \Leftrightarrow HgO$; $Hg + Cl_2 \Leftrightarrow HgCl_2$; $Hg + \frac{1}{2}O_2 \Leftrightarrow HgO$; $\frac{1}{2}H_2 + \frac{1}{2}Cl_2 \Leftrightarrow HCl$;$(1/2)\ Cl_2 \Leftrightarrow Cl$.

References

Abdul-Khalek, I.S., Kittelson, D.B., Graskow, B.R., Wei, Q., and Brear, F., Diesel Exhaust Particles Size: Measurement Issues and Trends, SAE Paper 980525, 1998.

Abraham, J., Bracco, F.V., and Reitz, R.D., Comparisons of computed and measured premixed charge engine combustion, *Combust. Flame*, 60, 309–332, 1985.

Abramovitz, M. and Stegan, I.A., *Handbook of Mathematical Functions*, Dover Publications, New York, 1964.

Afonso, Rui F. 2001, Assessment of Mercury Removal by Existing Air Pollution Control Devices in Full Scale Power Plants, Energy and Environmental Strategies, MA, USA.

Aggarwal, S.K., Fix, G.J., Lee, D.N., and Sirignano, W.A., Numerical optimization studies of axisymmetric unsteady sprays, *J. Computational Phys.*, 50, 1, 101–115, 1983.

Aggarwal, S.K., Fix, G.J., and Sirignano, W.A., Two phase laminar axisymmetric jet flow explicit, implicit, and split-operator approximations, *Numer. Meth. Part. Diff. Equ.*, 1, 279–294, 1985.

Aggarwal, S.K. and Puri, I.K., Flame structure interactions and state relationships in an unsteady partially premixed flame, *AIAA Journal,* 36, 1190–1199, 1998.

Aggarwal S.K. and Sirignano, W.A., Numerical modeling of one-dimensional enclosed homogeneous and heterogeneous deflagrations, *Comput. Fluids,* 12, 2, 145–158, 1984a.

Aggarwal, S.K. and Sirignano, W.A., Ignition of Fuel Sprays Deterministic Calculations for Idealized Droplet Arrays, Twentieth Symposium (International) on Combustion, The Combustion Institute, Pittsburgh, PA, pp. 1773–1780, 1984b.

Aggarwal, S.K. and Sirignano, W.A., Unsteady spray flame propagation in a closed volume, *Combust. Flame,* 62, 69–84, 1985.

Aggarwal, S.K., Tong A., and Sirignano, W.A., A comparison of vaporization models for spray calculations, *AIAA Journal,* 22, 10, 1448–1457, 1984.

Alabanese, V.M., Boyle, J.M., Huhmann, A., and Wallace, A., Evaluation of Hybrid SNCR/SCR for NOx Abatement on a Utility Boiler (TPP-522), Fuel Tech.

Annamalai, K. and Caton, J., Distinctive burning characteristics of carbon particles, *Can. J. Chem. Eng.*, 65, 1027–1032, 1987.

Annamalai, K. and Colaluca, M., Combustion of carbon under free convection, *Chem. Eng. Commun.*, 137, 1–22, 1995.

Annamalai, K. and Durbetaki, P., Characteristics of an Open Diffusion Flame, *Combust. Flame,* 25, 137–139, 1975a.

Annamalai, K. and Durbetaki, P., Extinction of spherical diffusion flames: spalding approach, *Int. J. Heat Mass Transfer,* 170, 1416–1418, 1975b.

Annamalai, K. and Durbetaki, P., Ignition of thermally thin porous pyrolysing solids under normally impinging flames, *Combust. Flame,* 27, 253–266, 1976.

Annamalai, K. and Durbetaki, P., A theory on transition of ignition phase of coal particles, *Combust. Flame*, 29, 193–208, 1977.

Annamalai, K. and Durbetaki, P., Combustion Behavior of Small Char/Graphite Particles, 17th International Symposium on Combustion, University of Leeds, England, Combustion Institute, pp. 169–178, August 20–25, 1979.

Annamalai, K., Freeman, J., Sweeten, M., Mathur, M., O'Dowd, W., Walbert, G., Jones, S., *Fuel,* 82 (10), 1195–1200, 2003.

Annamalai, K., Lau, S.C., and Sekhar, K., A non-integral technique for the approximate solution of transport problems, *Int. Commun. Heat Mass Transfer,* 13, 523–534, 1986.

Annamalai, K., Lau, S.C., Somasundaram, S., and Kim, Y.S., Application of non-integral method to combustion and natural convection problem, *J. Chem. Eng. Commun.*, 65, 231–242, 1988.

Annamalai, K., Madan, A., and Mortada, Y., Ignition of a Cloud of Droplets, ASME/WAM Meeting, Heat Transfer Division, Louisiana, ASME Paper #84-WA/HT-18, December 9–14, 1984.

Annamalai, K. and Morris, T.K, Thermodynamics of Human Fever, Emerging Energy Technology, 8th Annual International Energy week, Book V, pp. 457–461, 1997.

Annamalai, K., Morris, T., and McNicholls, D., A thermodynamic analysis of fever in chemotraphs, AES Vol. 37, Advanced Energy Systems Division, pp. 239–247, 1997.

Annamalai, K. and Puri, I., *Advanced Thermodynamics Engineering,* CRC Press, Boca Raton, FL, 2001.

Annamalai, K. and Ramalingam, S., Group combustion of char particles, *Combust. Flame*, 70, 307–332, 1987.

Annamalai, K., Ramalingam, S., Dahdah, T., and Chi, D., Group combustion of a cylindrical cloud of char/carbon particles, *J. Heat Transfer, Trans. ASME,* 110, 190–200, 1988.

Annamalai, K. and Ryan, W., Evaporation Characteristics for Arrays of Drops Using Point Source Superposition Method, CSS/CI Proceedings, pp. 347–354, 1991.

Annamalai, K. and Ryan, W., Interactive transport processes during gasification and combustion. Part I: drop arrays and clouds, *Prog. Energy Combust. Sci.*, 18(3), 221–295, 1992.

Annamalai, K. and Ryan, W., Interactive process during gasification and combustion; Part II: isolated coal and porous char particles, *Prog. Energy Combust. Sci.*, 19, 383–446, 1993a.

Annamalai, K., Ryan, W., and Chandra, S., Evaporation of multi-component arrays, *J. Heat Transfer, Trans. ASME,* 115, 707–716, 1993b.

Annamalai, K., Ryan, W., and Dhanapalan, S., Interactive processes in gasification and combustion — Part III: coal/char particle arrays, streams, and clouds, *Prog. Energy Combust. Sci.*, 19, 383–446, 1995a.

Annamalai, K., Ryan, W., and Dhanapalan, S., Interactive processes in gasification and combustion — Part III: coal/char particle arrays, streams, and clouds, *Prog. Energy Combust. Sci.*, 20, 487–618, 1995b.

Annamalai, K. and Sibulkin, M., Ignition and flame spread tests of cellular plastics, *J. Fire Flammability,* 9, 445–458, 1978.

Annamalai, K. and Sibulkin, M., Flame spread over combustible surfaces for laminar flow systems. Part I: excess fuel and heat flux, *Combust. Sci. Technol.*, 19, 167–183. 1979a.

Annamalai, K. and Sibulkin, M., Flame spread over combustible surfaces in laminar flow systems. Part II: flame heights and flame spread velocity, *Combust. Sci. Technol.,* 19, 185–193, 1979b.

Annamalai, K. and Stockbridge, S., Combustion of char/carbon in a periodic thermal environment, *Int. J. Heat Mass Transfer,* 27, 1932–1935, 1984.

Annamalai, K., Sweeten, J., and Ramalingam, S.C., Estimation of gross heating values of biomass fuels, *Transactions of the ASAE,* 30, 1205–1208, 1987.

Anthony, D.B., Coal devolatilization and hydro-gasification, *Fuel*, 55, 121, 1976.

Anthony, D.B. and Howard, J.B., Coal devolatilization and hydrogasification, *A.I.Ch.E. J.*, 22, 625, 1976.

Anthony, D.B., Howard, J.B., Hottel, H.C., and Meissner, H.P., Rapid devolatilization of Pulverized Coal, 15th Symposium on Combustion, 1305–1317, 1975.

Anthony, D.B., Howard, J.B., Meissner, H.P., and Hottel, H.C., Apparatus for determining high pressure coal hydrogen reaction kinetics, *Rev. Sci. Instrum.*, 45, 992, 1974.

Arana, C., Pontoni, M., Sen, S., and Puri, I.K., Field measurements of soot volume fractions in laminar partially premixed coflow ethylene/air flames, *Combust. Flame,* 138, 362–372, 2004.

Arrhenius, S., On the influence of carbonic acid in the air upon the temperature of the ground, *Philos. Mag. J. Sci.*, S.5, 41, 237–276, 1896.

Arthur, J.R., Reactions between carbon and oxygen, *Trans. Faraday Soc.*, 47, 164–178, 1951.

ASTM Specifications for Petroleum Products, American Society for Testing Materials, Philadelphia, 1996.

Avedesian, N.M., and Davidson, J.F., Combustion of carbon particles in fluidized bed combustors, *Trans. Inst. Chem. Engineers.*, 51, 121–132, 1973.

Ayyaswamy, P.S., Combustion dynamics of moving droplets, *Encyclopedia of Environmental Technology,* Vol. 1: Thermal Treatment of Hazardous Wastes, Cheremisinoff, P.N., Ed., Gulf Publishing, Houston, 1989, chap. 20.

Bachalo, W.D., Method for measuring the size and velocity of spheres by dual beam light scattering interferometry, *Appl. Opt.*, 19(3), 363–370, 1980.

Badzioch, S. and Hawksley, P.G.W., Kinetics of thermal decomposition of pulverized coal particles, *Ind. Eng. Chem. Process. Des. Dev.*, 9(4), 521–530, 1970.

Bahadori, M.Y. et al., *Modern Developments in Energy Combustion and Spectroscopy,* Williams, F.A., Oppenheim, A.K., Olfe, D.B., and Lapp, M., Eds., Pergamon Press, Oxford, 1993, pp. 49–66.

Ballal, D.R. and Lefebvre, A.H., *Combust. Flame*, 35, 155–168, 1979.

Bartok, W. and Sarofim, A.F., *Fossil Fuel Combustion*, John Wiley & Sons, New York, 1991.

Bar-Ziv, E., Jones, D.B., Spjut, E., Dudek, D.R., Sarofim, A.F., and Longwell, J.P., Measurement of combustion kinetics of a single char particle in an electrodynamic thermogravimetric analyzer, *Combust. and Flame*, 75, 81–106, 1989.

Baukal, C.E., Gershtein, V.Y., and Li, X., Eds., *Industrial Combustion*, CRC Press, Boca Raton, FL, 2000.

Baukal, C.E., Gershtein, V.Y., and Li, X., Eds., *Computational Fluid Dynamics in Industrial Combustion,* CRC Press, Boca Raton, FL, 2001.

Bausch, D.L. and Drysdale, D.D., An evaluation of rate data for CO + OH = CO_2 + H reaction, *Combust. Flame*, 23, 215, 1974.

Beer, J.M. and Chigier, N.A., *Combustion Aerodynamics,* Krieger Publishing, Malabar, FL, 1983.

Bejan, A., *Advanced Engineering Thermodynamics,* John Wiley & Sons, New York, 1988.

Bellan, J. and Cuffel, R., A theory of nondilute spray evaporation based upon multiple drop interactions," *Combust. Flame,* 51, 55–67, 1983.

Bellan, J. and Harstad, K., Analysis of the convective evaporation of nondilute clusters of drops," *Int. J. Heat Mass Tr.,* 30, 125–136, 1987.

Benaissa, A., Gauthier, J.E.D., Bardon, M.F., and Laviolette, M., Modeling evaporation of multicomponent fuel droplets under ambient temperature conditions, *J. Inst. Energy,* 75, 19–26, March 2002.

Ben-Dor, G., Elperin, T., and Krasovitov, B., Numerical analysis of temperature and concentration jumps on transient evaporation of moderately large ($0.01 < Kn < 0.3$) droplets in non-isothermal multicomponent gaseous mixtures, *Heat and Mass Transfer,* 39, 157–166, 2003a.

Ben-Dor, G., Elperin, T., and Krasovitov, B., Effect of thermo- and diffusiophoretic motion of flame-generated particles in the neighbourhood of burning droplets in microgravity conditions, *Proceedings of the Royal Society London A*, 459, 677–703, 2003b.

Berta, P., Puri, I.K., and Aggarwal, S.K., An experimental and numerical investigation of n-heptane/air counterflow partially premixed flame structure and emission, *Combust. Flame,* 145, 740–764, 2006.

Berta, P., Puri, I.K., and Aggarwal, S.K., Structure of partially premixed n-heptane/air counterflow flames, *Proc. Combustion Institute,* 30, 1565–1572, 2005.

Bilger, R.W., *Turbulent Reacting Flow,* Libby, P.A. and Wiliams, F.A., Eds., Springer-Verlag, New York, 1980, pp. 65–114, chap 3.

Bird, R.B., Stewart, W.E., and Lightfoot, E.N., *Transport Phenomena,* John Wiley & Sons, New York, 1960.

Bishop, J.E. and Kinra, V.K., Analysis of elastothermodynamic damping in particle-reinforced metal-matrix composites, *Metall. Mater. Trans.*, 26A(11), 2773–2783, November 1995.

Bishop, J.E. and Kinra, V.K., Equivalence of the mechanical and entropic descriptions of elastothermodynamic damping in composite materials, *Mech. Composite Mater. Struct.*, 3(2), 83–95, 1996.

Blackshear, W., Mechanical Engineering, University of Minnesota, Combustion Notes, 1961.

Blinov, V.I. and Khudiakov, G.N., *Diffusive Burning of Liquids,* Pergamon Press, New York, 1960.

Bodurtha, F.T., *Industrial Explosion Prevention and Protection,* McGraw-Hill, New York, 1980.

Borman, G.L. and Ragland, K.W., *Combustion Engineering,* McGraw-Hill, New York, 1998.

Bose, A.C. and Wendt, J.O.L., Pulverized Coal Combustion: Fuel Nitrogen Mechanisms in the Rich Post-Flame, 22nd Symposium on Combustion, The Combustion Institute, Pittsburgh, PA, 1127–1134, 1988.

Bowman, C.T., Chemistry of gaseous pollutants formation and destruction, *Fossil Fuel Combustion,* Bartok, W. and Sarofim, A.F., Eds., John Wiley & Sons, New York, 1991, chap. 4.

Bowman, C.T., Control of Combustion-Generated Nitrogen Oxide Emissions: Technology Driven by Regulation, 24th International Symposium on Combustion, The Combustion Institute, Sydney, 859–878, 1992.

Bradley, D., Dixon-Lewis, G., Habik, S.E., and Mushi, E.M., The Oxidation of Graphite Powder in Flame Reaction Zones, 20th International Symposium on Combustion, The Combustion Institute, Pittsburgh, PA, 931–940, 1984.

Breen, B.P., Sweterlitsch, J.J., and Gabrielson, J.E., Unites States Patent US 6,357,367, Filed and issued March 19, 2002.

Brokaw, R.S., NASA Technical Report, TR R 81, 1961.

Brzustowski, T.A., Twardus, E.M., Wojcicki, S., and Sobiesiak, A., Interaction of two burning fuel droplets of arbitrary size, *AIAA,* 17, 11, 1234–1242, 1979.

Buckmaster, J.D. and Ludford, G.S.S., Lectures on Mathematical Combustion, Society for Industrial and Applied Mathematics, Philadelphia, 19103, 1983.

Burke, S.B. and Schumann, T.E.W., Diffusion flames, *Ind. Eng. Chem.*, 20, 998, 1928.

Butz, J., Smith, J., Grover, C., Haythornthwaite, S., Fox, M., Hunt, T., Chang, R., and Brown, T.D., in Proceedings of the Mercury in the Environment Specialty Conference, Air and Waste Management Association, Minneapolis, MN, September 15–17, 1999.

Calcote, H.F., Gregory, C.A., Jr., Barnett, C.M., and Gilmer, R.B., Spark ignition, *Ind. Eng. Chem.*, 44, 2656, 1952.

Carey, T.R., Hargrove, O.W., Richardson, C.F., Chang, R., and Meserole, F.B., Factors affecting mercury control in utility flue gas using activated carbon, *J. Air Waste Manage. Assoc.*, 48, 1166–1164, 1998.

Carslaw, H.S. and Jaeger, J.C., *Combustion of Heat Solids,* 2nd edition, Clarendon Press, Oxford, England, 1959.

Caton, J.A., Narney, J.K., II, Cariappa, C., Laster, W.R., The selective non-catalytic reduction of nitric oxide using ammonia at up to 15% oxygen, *Can. J. Chem. Eng.*, 73(3), 345–350, 1995.

Caton, J.A. and Xia, Z., The selective non-catalytic removal (SNCR) of nitric oxides from engine exhaust: comparison of three processes, *ASME Trans. — J. Eng. Gas Turbines Power*, 126, 234–240, April 2004.

Cengel, Y.A. and Boles, M.A., *Thermodynamics*, 4th ed., McGraw-Hill, New York, 2002.

Chambre, P.L., On the solution of the Poisson-Boltzmann equation with application to the theory of thermal explosions, *J. Chem. Phys.*, 20, 1795–1797, 1952.

Channiwala, S.A., HHV Formulae, PhD thesis, On biomass gassification process and technology development, The Indian Institute of Technology, Bombay, 1992; http://www.woodgas.com/proximat.htm.

Chen, S.L., Heap, M.P., Pershing, D.W., Nihart, R.K., Rees, D.P., and Martin, G.B., The fate of coal nitrogen during combustion, *Fuel*, Vol. 61, 1218–1228, 1982.

Chigier, N., *Energy, Combustion, and Environment*, McGraw-Hill, New York, 1981.

Chigier, N., Spray Combustion Processes: A Review, ASME, Winter Annual Meeting, Phoenix, AZ, 1982.

Chigier, N., Group combustion models and laser diagnostic methods in sprays: a review, *Combust. Flame*, 51, 127–139, 1983.

Chin, J.S., Durrett, R., and Lefebvre, A.H., The interdependence of spray characteristics and evaporation history of fuel sprays, *ASME J. Eng. Gas Turbine Power*, 106, 639–644, 1984.

Chiu, H.H. and Liu, F.M., Group combustion of liquid droplets, *Combust. Sci. Technol.*, 17, 127–142, 1977.

Choi, C.W. and Puri, I.K., Flame stretch effects on partially premixed flames, *Combust. Flame*, 123, 119–139, 2000.

Choi, C.W. and Puri, I.K., Contribution of curvature to flame stretch in premixed flames, *Combust. Flame*, 126, 1640–1654, 2001.

Choi, C.W. and Puri, I.K., Response of flame speed to positively and negatively curved premixed flames, *Combust. Theory Modeling*, 7, 205–220, 2003.

Clift, R., Grace, J.R., and Weber, M.E., *Bubbles, Drops, and Particles*, Academic Press, New York, 1978.

Cooper, D.C. and Alley, F.C., *Air Pollution Control: A Design Approach*, Waveland Press, IL, 1990.

Correa, S.M. and Sichel, M., The boundary layer structure of a vaporizing fuel cloud, *Combust. Sci. Technol.*, 121–130, 1982a.

Correa, S.M. and Sichel, M., The Group Combustion of a Spherical Cloud of Monodisperse Droplets, 19th International Symposium on Combustion, 981–991, 1982b.

Coward, H.F. and Jones, G.W., Limits of Flammability of Gases and Vapors, Bulletin 503, Bureau of Mines, 1952.

Davis, M.L. and Cornwell, D.A. *Introduction to Environmental Engineering*, 2nd ed., Chemical Engineering Series, McGraw-Hill, Singapore, 1991.

Daw, C.S., Proceedings of Central States Section of the Combustion Institute, Austin, Texas, March 21–23, 2004.

de Bruyn Kops, S.M., Ph.D. thesis, Numerical simulations of non-premixed turbulent combustion, University of Washington, 1999.

de Bruyn Kops, S.M., Riley, J.J., and Kolsay, G., Direct numerical simulation of reacting scalar mixing layers, *Phys. Fluids*, Vol. 13, No. 5, 1450–1465, 2001.

Deutch, J.M., Felderhof, B.U., and Saxton, M.J., Competitive effects in diffusion controlled reactions, *J. Chem. Phys.*, 11, 4559–4563, 1976.

DeVito, M.S. and Rosenhoover, W.A., in 15th International Pittsburgh Coal Conference, Pittsburgh, PA, September 15–17, 1998.

Dhanapalan, S., Annamalai, K., and Daripa, P., Turbulent Combustion Modeling of Coal: Biomass in a Swirl Burner, Emerging Energy Technology, 8th International Energy Week, Book V, pp. 415–423, 1997a.

Dhanapalan, S., Annamalai, K., and Daripa, P., Turbulent Combustion Modeling of Coal: Biomass Blends in a Swirl Burner I-Preliminary Results, Energy Technology Conference and Exhibition, ASME, pp. 410–418, 1997b.

Dodge, L.G. and Schwalb, J.A., Fuel Spray Evolution: Comparisons of Experiment and CFD Simulation of Non-Evaporating Spray, ASME, 88-FT-27, 1988.

DOE/EIA, *Annual Energy Outlook*, 2002.

DOE/EIA-0383(98).

Drake, M.C. and Blemberg, P.N., *Comb. Sc. Tech.*, 8, 25–37, 1973; 21, 225–258, 1980.

Du, X. and Annamalai, K., Transient ignition of isolated coal particle, *Combust. Flame*, 97, 339–354, 1994.

Duo, W., Dam-Johansen, K., and Ostergaard, K., Kinetics of the gas-phase reaction between nitric oxide, ammonia and oxygen, *Can. J. Chem. Eng.*, 70, 1014–1020, October 1992.

EERC, Enhancing Use of Coal by Gas Reburning, 1988.

Elkotb, M.M., Fuel automization for spray modelling, *Prog. Energy Combust. Sci.*, 8, 61, 1982.

El-Wakil, M.M., *Powerplant Technology,* McGraw-Hill, New York, 1984.

Emmons, H.W., The film combustion of liquid fuel, *Z. Angew. Math. Mech.*, 36, 60, 1956.

Eppich, J.D. and Cosulich, J.D., *Solid Waste and Power,* July–August 1993, p. 27.

Essenhigh, R.H., *Chemistry Coal Utilization*, Elliott, M.A., Ed., 2nd Suppl., John Wiley & Sons, New York, 1981, p. 1153.

Essenhigh, R.H., *Fundamentals of Coal Combustion, Chemistry of Coal Utilization*, Vol II, Elliott, M.A., Ed., Wiley Interscience, New York, 1979, pp. 1153–1312, chap. 19.

Essenhigh, R.H., The influence of coal rank on the burning times of single captive particles, *J. Eng. Power,* 85, 183–190, 1963.

Essenhigh, R.H., NO derivation, private communication, 2004.

Fachini, F.F. and Seshadri, K., Rate-ratio asymptotic analysis of nonpremixed n-heptane flames, *Combust. Sci. Technol.,* 175, 1, 125–155, 2003.

Faeth, G.M., Current status of droplet and liquid combustion, *Prog. Energy Combust. Sci.*, 3, 191–224, 1977.

Faeth, G.M., Laminar jets evaporation and combustion of sprays, *Prog. Energy Combust. Sci.*, 9, 1–76, 1983.

Fay, *J. Aero Sci.*, 681–689, 1954.

Fenimore, C.P., Formation of nitric oxide from fuel nitrogen in ethylene flame, *Combust. Flame,* 19, 289, 1972. [Also see *J. Phys. Chem.* 69, 3593, 1961].

Flagan, R.C. and Seinfeld, J.H., *Fundamentals of Air Pollution Engineering*, Prentice Hall, Englewood Cliffs, NJ, 1988.

Friedman, R., Kinetics of the combustion wave, *American Rocket Soc.*, 24, 349–354, 1953.

Fukutani, S., Kunioshi, N., and Jinno, H., 23rd International Symposium on Combustion, 567–573, 1991.

Gayle, E.E., *Children of the Corn*, Vol. 125, No. 1, 40–43, 2003.

Glarborg, P., Kristensen, P.G., Soren, H.J., and Dam-Johansen, K., A flow reactor study of HNCO oxidation chemistry, *Combust. Flame*, Vol. 98, 241–258, 1994.

Glassman, I., *Combustion*, Academic Press, New York, 1977.

Glassman, I., *Combustion*, 2nd edition, Academic Press, New York, 1987.

Gogos, G., Soh, S., and Pope, D.N., Effects of gravity and ambient pressure on liquid fuel droplet evaporation, *Int. J. Heat Mass Transfer,* 46, 283–296, 2003.

Gollahalli, S.R., Buoyancy effects on the flame structure in the wakes of burning liquid drops, *Combust. Flame*, 29, 21–31, 1977.

Gollahalli, S.R. and Brzustowski, T.A., Experimental Studies on the Flame Structure in the Wake of a Burning Droplet, 14th International Symposium on Combustion, The Combustion Institute, Pittsburgh, PA, 1333–1344, 1973.

Gollahalli, S.R. and Brzustowski, T.A., The Effect of Pressure on the Flame Structure in the Wake of a Burning Hydrocarbon Droplet, 15th International Symposium on Combustion, The Combustion Institute, Pittsburgh, PA, 409–417, 1974.

Goodger, E.M., *Alternate Fuels: Chemical Sources,* Halsted Press, New York, 1980.

Gray, W.A., Kilham, J.K., and Muller, R., *Heat Transfer from Flames,* Elsevier, Amsterdam, 1976.

Haagen-Smit, A.J., *Ind. Eng. Chem.*, 44, 1342–1346, 1952.

Haglund, J.S., McFarland, A.R., Wooldridge, M.S., and Vardiman, S., A Continuous Emission Monitor for Quantitative Measurement of PM_{10} Emissions from Stationary Sources, Air and Waste Management Association's 93rd Annual Conference and Exhibition, Salt Lake City, Utah, June 18–22, 2000.

Hamor, R.J., Smith, I.W., and Tyler, R.J., Kinetics of Combustion of Pulverized Brown Coal Char between 630 and 2200 K, *Combust. Flame*, 21, 153–162, 1973.

Haynes, B.S., Soot and HC in combustion, *Fossil Fuel Combustion*, Bartok, W. and Sarofim, A.F., Eds., John Wiley & Sons, New York, 1991, chap. 5.

Hedley, A.D., Nuruzzaman, A.S.M., and Martin, G.F., Prog. Review No. 62: Combustion of single droplets and simplified spray systems, *J. Inst. Fuel*, 38–54, 1971.

Helble, J.J. and Mamani-Paco, R.S., Bench Scale Examination of Hg Oxidation Under Non-Isothermal Post-Combustion Conditions, presented at the Air and Waste Management Association Annual Meeting, Salt Lake City, Utah, June 18–22, 2000.

Hertzberg, M., Conti, R.S., and Cashdoillar, K.L., Electrical Ignition Energies and Thermal Autoignition Temperatures for Evaluating Explosion Hazards of Dusts, Bureau of Mines Report RI 8988, Pittsburgh, PA, 1985.

Heywood, J.B., *Internal Combustion Engine Fundamentals*, McGraw-Hill, New York, 1988.

Hippinen, I., Lu, Y., Laatikainen, J., Nieminen, M., and Jahkola, A., Gas emissions from the PFB combustion of solid fuels, *J. Inst. Energy*, December, 1992.

Hippo, E. and Walker, P.L., Reactivity of heat treated coals in CO_2 at 900 C, *Fuel*, 54, 245–248, 1975.

Hiroyasu, H. and Kadota, T., SAE Paper 740715, *SAE Trans.*, 83, 1974.

Hirschfelder, J.O., Curtis, C.F., and Bird, R.B., *Molecular Theory of Gases and Liquids*, 4th ed., John Wiley & Sons, New York, 1964.

Hoffman, J.S., Lee, W., Litzinger, T.A., Santavicca, D.A., and Pitz, W.J, The oxidation of propane at elevated pressures: experiments and modelling, *Combust. Sci. Technol.*, 77, 95–125, 1991.

Hottel, H.C. and Hawthorne, W.R., Diffusion in laminar plane jets, 3rd Symposium, 254–266, 1949.

Hottel, H.C. and Sarofim, A.F., *Radiative Heat Transfer*, McGraw-Hill, New York, 1978.

Howard, J.B. and Essenhigh, R.H., Combustion mechanisms in pulverized coal flames, *Combust. Flame*, 10, 92–93, 1966.

Howard, J.B. and Essenhigh, R.H., Pyrolysis of Coal Particles in Pulverized Fuel Flames, I&EC Process Design and Development, 6, 74–79. 1967a.

Howard, J.B. and Essenhigh, R.H., Mechanism of Solid-Particle Combustion with Simultaneous Gas Phase Volatiles Combustion, 11th International Symposium on Combustion, The Combustion Institute, Pittsburgh, PA, pp. 399–408, 1967b.

Hsieh, K.C., Shuen, J.S., and Yang, V., Droplet vaporization in high pressure environments 1: near critical conditions, *Combust. Sci. Technol.*, 76, 111–132, 1991.

http://www.wikipedia.org/wiki/Bunsen_burner.

http://www.zymaxforensics.com/forensicsprimer/images/5_figures/fig_13.htm, Distillation curves.

http://yosemite.epa.gov/oar/globalwarming.nsf/content/EmissionsNationalGlobalWarmingPotentials.html.

Hunter, J.T.B., Wang, H., Litzinger, T.A., and Franklach, M., Oxidation of methane at elevated pressures: experiments and modelling, *Combust. Flame*, 97, 201–224, 1994.

Hurt, R.H., Dudek, D.R., Longwell, J.P., and Sarofim, A.F., The phenomenon of gasification induced carbon densification and its influence on pore structure evolution, *Carbon*, 26, 4, 433–449, 1988.

Incropera, F.P. and DeWitt, D.P., *Heat Transfer*, 5th ed., 2002, pp. 963–967.

Jodal, M., Lauridsen, T.L., and Dam-Johansen, K., NOx removal on a coal-fired utility boiler by selective non-catalytic reduction, *Environ. Prog.*, 11(4), November 1992.

Jones, W.P. and Launder, B.E., The prediction of laminarization with a two-equation turbulence model, *Int. J. Heat Mass Transfer*, 15, 301–314, 1972.
Juntgen, H. and Van Heek, K.H., Gas release from coal as a function of the rate of heating, *Fuel*, 47, 103–117, 1968.
Kanury, M., *Introduction to Combustion Phenomena*, Gordon and Breach, New York, 1975.
Keating, E.L., *Applied Combustion*, Marcel Dekker, New York, 1993.
Kim, J.S., de Ris, J., and Kroesser, W., Laminar Free Convective Burning of Fuel Surfaces, 13th International Symposium on Combustion, The Combustion Institute, Pittsburgh, PA, pp. 949–961, 1971.
Kimber, G.M. and Gray, M.D., Rapid devolatization of small coal particles, *Combust. Flame*, 11, 360–362, 1967.
Kinoshita, C.M., Pagni, P.J., and Beier, R.A., Opposed Flow Diffusion Flame Extensions, 18th International Symposium on Combustion, The Combustion Institute, Pittsburgh, PA, pp. 1853–1860, 1981.
Knight, B.E., *Proc. Inst. of Mech. Eng.*, 104, 1955.
Kobayashi, H., Howard, J.B., and Sarofim, A.F., Kinetics of Rapid Devolatilization of Pulverized Coal, 16th International Symposium on Combustion, 411, 1976.
Kobayashi, H., Ph.D. thesis, Chemical Engineering, MIT, Cambridge, MA, 1972.
Kong, S.C., Han, Z., and Reitz, R.D., The Development and Application of a Diesel Ignition and Combustion Model for Multi-Dimensional Engine Simulation, SAE Paper 950278, 1995.
Kulkarni, A.K., Combustion of Oil and Water-in-Oil Layers Supported on Water, final report on U.S. Department of Commerce NIST Grant No. 60NANB0D0103, 2001.
Kuo, K.K., *Principles of Combustion*, Wiley Interscience, New York, 1986.
Labowsky M., The effects of nearest neighbor interactions on the evaporation rate of cloud particles, *Chem. Eng. Sci.*, 31, 803–813, 1976a.
Labowsky, M., The effect of nearest neighbor interactions on the evaporation rate of cloud particles, *Chem. Eng. Sci.*, 32, 803–813, 1976b.
Labowsky, M., A formalism for calculating the evaporation rates of rapidly evaporating interacting particles, *Combust. Sci. Technol.*, 18, 145–151, 1978.
Labowsky, M., Transfer rate calculations for compositionally dissimilar interacting particles, *Chem. Eng. Sci.*, 35, 1041–1048, 1980.
Labowsky, M. and Rosner, D.E., Group combustion of droplets in fuel clouds. I. Quasi-steady predictions, *Evaporation — Combustion of Fuels*, Advances in Chemistry Series, Zung, J.T., Ed., No. 166, American Chemical Society, Washington, D.C., 63–79. 1978b.
Laidler, K.J., *Reaction Kinetics*, Vol. 1, Pergamon Press, Oxford, New York, 1963.
Langwell, Fuels and combustion, *Fossil Fuel Combustion*, Bartok, W. and Sarofim, A.F., Eds., John Wiley & Sons, New York, 1991, chap. 1.
Laster, R. and Annamalai, K., Ignition delays of liquid drop clouds: results from group combustion theory, *J. Chem. Eng. Commun.*, 105, 201–219, 1991.
Law, C.K., Multi-component droplet combustion with rapid internal mixing, *Combust. Flame*, 26, 219–233, 1976.
Law, C.K., Internal boiling and superheating in vaporizing multi-component droplets, *AIChE*, 24, 626–636, 1978.
Law, C.K., Recent advances in droplet vaporization and combustion, *Prog. Energy Combust. Sci.*, 3, 171–181, 1982.
Law, C.K., Chung, S.H., and Srinivasan, N., Gas phase quasi-steadiness and fuel vapor accumulation effects in droplet burning, *Combust. Flame*, 38, 173–198, 1980.
Law, C.K. and Law, H.K., A d^2-law for multicomponent droplet vaporization, *AIAA J.*, 20(4), 522–527, 1982.
Lawn, C.J., *Principles of Combustion Engineering for Boilers*, Academic Press, Orlando, FL, 1987.
Lee, B.J., Kim, J.S., and Chung, S.H., Effect of Dilution on the Liftoff of Non-Premixed Jet Flames, 25th Symposium on Combustion, The Combustion Institute, Pittsburgh, PA, pp. 1175–1181, 1994.

Lefevbre, A., *Atomization and Sprays*, Hemisphere, New York, 1989.

Lefebvre, A., *Gas Turbine Combustion*, Taylor & Francis, Philadelphia, 1999.

Lewis, B. and Von Elbe, G., *Combustion, Flames, and Explosion of Gases*, 2nd ed., Academic Press, New York, 1961.

Lewis, B. and Von Elbe, G., *Combustion, Flames, and Explosion of Gases*, 3rd ed., Academic Press, New York, 1987.

Lide, D.R., Ed., *CRC Handbook of Chemistry and Physics*, CRC Press, Boca Raton, FL, 1993.

Lin, K.-C. and Faeth, G.M., Shapes of nonbuoyant round laminar-jet diffusion flames in coflowing air, *AIAA*, 37, 6, 759–765, 1999.

Lin, K.-C., Faeth, G.M., Sunderland, P.B., Urban, D.L., and Yuan, Z.-G., Shapes of non-bouyant round luminous hydrocarbon/air laminar jet diffusion flames, *Combust. Flame*, 116, 415–431, 1999.

Linak, W.P., Ryan, J.V., Ghorishi, B.S., and Wendt, J.O.L., *J. Air Waste Manage. Assoc.*, 51, 688–698. 2001.

Lisauskas, R.A., Itse, D.C., Masser, C.C., and Abele, A.R., A prototype burner for NOx/SO2, *Chem. Eng., Prog.*, November 1985.

Lucht, R., Purdue University, private communication.

Lyon, R.K., Method for the Reduction of the Concentration of Combustion Effluents Using Ammonia, U.S. Patent No. 3,900,554, August 19, 1975.

Lyon, R.K., The NH_3-NO-O_2 reaction, *Int. J. Chemical Kinetics*, 8, 315–318, 1976.

Lyon, R.K., Thermal $DeNO_x$, *Environ. Sci. Technol.*, 21(3), 231–236, 1986.

Lyon, R.K. and Benn, D., Kinetics of the NO-NH_3-O_2 Reaction, 17th International Symposium on Combustion, The Combustion Institute, Pittsburgh, PA, pp. 601–610, 1978.

Lyon, R.K. and Hardy, J.E., Discovery and development of the thermal $DeNO_x$ process, *Ind. Eng. Chem. Fundam.*, 25(1), 19–24, 1986.

MacGibbon, B.S., Busnaina, A.A., and Fardi, B., *J. Electrochem. Soc.*, 146(8), 2901–2905, 1999.

Magnussen, B.F., On the Structure of Turbulence and a Generalized Eddy Dissipation Concept for Chemical Reaction in Turbulent Flow. Nineteeth AIAA Meeting, St. Louis, 1981.

Magnussen, B.F. and Hjertager, B.H., On Mathematical Models of Turbulent Combustion with Special Emphasis on Soot Formation and Combustion, in 16th International Symposium on Combustion, The Combustion Institute, Pittsburgh, PA, 1976.

Mahalingam, S., Ferziger, J.H., and Cantwell, B.J., Self similar diffusion flame, *Combust. Flame*, 82, 231–234, 1990.

Makino, A., A theoretical and experimental study of carbon combustion in stagnation flow, *Combust. Flame*, 81, 166–197, 1990.

Makino, A. and Law, C.K., Quasi-Steady and Transient Combustion of a Carbon Particle: Theory and Experimental Comparisons, 21st Symposium on Combustion, The Combustion Institute, Pittsburgh, PA, pp. 183–191, 1986.

Marberry, M., Ray, A.K., and Leung. K., Effect of multiple particle interactions on burning droplets, *Combust. Flame*, 57, 237–245, 1984.

Martin, G.B. and Berkau, E.E., An investigation of the conversion of various fuel N compounds to nitrogen oxides in oil combustion, 70th National AICHE Meeting, 1971.

Marxmann, G.A., 11th Symposium (International) on Combustion, p. 269, 1967.

Matlosz, R.L., Leipziger, S., and Torda, T.P., Investigation of liquid drop evaporation in high temperature and high pressure environment, *Int. J. Heat Mass Transfer*, 15, 831–852, 1972.

Matsui, K., Tsuji, H., and Makino, A., Estimation of the Relative Rates of C-O2 and C-H2O Reactions, *Carbon*, 21, 3, 320–321, 1983a.

Matsui, K., Tsuji, H., and Makino, A., The effect of water vapor concentration on the rate of combustion of an artificial graphite in humid air flow, *Combust. Flame*, 50, 107–118, 1983b.

Matsui, K., Tsuji, H., and Makino, A., A further study of the effects of water vapor concentration on the rate of combustion of an artificial graphite in humid air flow, *Combust. Flame*, 63, 415–427, 1986.

McCreath, C.G. and Chigier, N.A., Liquid Spray Burning in the Wake of a Stabilizer Disc, 14th International Symposium on Combustion, 1355–1363, 1973.

McGowin, C.R. and Hughes, E.E., Efficient and Economical Energy Recovery from Waste by Cofiring with Coal, ACS Symposium Series No. 515, May 1993.

McKinney, D.J. and Dunn-Rankin, J., *Combust. Sci. Technol.*, 161, 18–27, 2000.

Melville, E.K. and Bray, N.C., A model of the two-phase turbulent jet, *Int. J. Heat Mass Transfer*, 22, 647–656, 1979.

Midkiff, K.C., Altenkirch, R.A., and Peck, R.E., Stoichiometry and coal type effects on homogeneous vs. heterogeneous homogeneous combustion in pulverized coal flames, *Combust. Flame*, 16, 139–152, 1986a.

Midkiff, K.C., Altenkirch, R.A., and Peck, R.E., Stoichiometry and coal type effects on homogeneous vs. heterogeneous combustion in pulverized coal flames, *Combust. Flame*, 64, 253–266, 1986b.

Miller, E.S., Dunham, S.J., Olson, and Brown, T.D., Mercury sorbent developments for coal-fired boilers, presented at Conference on Air Quality, McLean, VA, December 1–4, 1998.

Miller, J.A. and Bowman, C.T., Mechanisms and modeling of nitrogen chemistry in combustion, *Prog. Energy Combust. Sci.*, 15, 287–338, 1989.

Miller, J.A., Branch, M.C., and Kee, R.J., A chemical kinetic model for the selective reduction of nitric oxide by ammonia, *Combust. Flame*, 43, 81–98, 1981.

Mitchell, J.W. and Tarbell, J.M., A kinetic model of nitric oxide formation during pulverized coal combustion, *AIChE*, 28(2), 302–311, 1982.

Modell, M. and Reid, R.C., *Thermodynamics and Its Applications*, 2nd ed., Prentice Hall, Englewood Cliffs, NJ, 1983.

Moffat, A.J., The chemistry and mechanism of sulfur release during coal combustion, paper no. CSS/CI-82/08, The Combustion Institute, 1982.

Mukhopadhyay, A. and Puri, I.K., An investigation of stretch effects on rich premixed flames, *Combust. Flame,* 133, 499–502, 2003.

Mukhopadhyay, A., Qin, X., Aggarwal, S.K. and Puri, I.K., On extension of "heatline" and "massline" concepts to reacting flows through use of conserved scalars, *ASME J. Heat Transfer*, 124, 791–799, 2002.

Mukhopadhyay, A., Qin, X., Puri, I.K., and Aggarwal, S.K., Visualization of scalar transport in nonreacting and reacting jets through a unified heatline and massline formulation, *Numer. Heat Transfer, Part A-Applications,* 44, 683–704, 2003.

Mulcahy, M.F.R., Oxygen in Metal and Gaseous Fuel Industries, The Chemical Society, London, p. 175, 1978.

Mulcahy, M.F.R. and Smith, I.W., Kinetics of combustion of pulverized fuel: a review of theory and experiment, *J. Rev. Pure Appl. Chem.*, 19, 81–108, 1969.

Neoh, K.G., Howard, J.B., and Sarofim, A.F., *Particulate Carbon: Formation during Combustion*, Siegla, D.C. and Smith, G.W., Eds., Plenum Press, New York, 1981, pp. 261–282.

Niioka, T. and Williams, F.A., Ignition of reactive solid in a hot stagnation point flow, *Combust. Flame*, 29, 43–54, 1977.

Niksa, S., Helble, J.J., and Fujiwara, N., Kinetice modeling of homogeneous mercury oxidation and the importance of NO and H_2O in predicting oxidation in coal-derived systems, *Environ. Sci. Technol.*, 35, 3701–3706, 2001.

Nishida K., Takagi, T., and Kinoshita, S., Analysis Of Entropy Generation and Exergy Loss during Combustion, Proceedings of the Combustion Institute, 29: 869–874, Part 1, 2003.

NRC, National Research Council of Canada, Combustion Research Lab., M-9, Ottawa, Ontario K1A OR6, Canada.

Nuruzzaman, A.S.M., Hedley, A.B., and Beer, J.M., Combustion rates in self-supporting flames on monosized droplet streams, *J. Inst. Fuel*, 301–310, 1970.

Nuruzzaman, A.S.M., Hedley, A.B., and Beer, J.M., Combustion of Monosized Droplet Streams in Stationary Self-Supporting Flames, 13th International Symposium on Combustion, The Combustion Institute, Pittsburgh, PA, 787–799, 1971.

OECD (Organization for Economic Co-operation and Development) Ed., *Control Technology for Nitrogen Oxide Emissions from Stationary Sources*, Paris, 1983.

Office of Air Quality Planning and Standards, Review of the National Air Quality Standards for Particulate Matter: Policy Assessment of Scientific and Technical Information: OAQPS Staff Paper, United States Environmental Protection Agency, Washington, 1995.

Onuma, Y., Ogasawara, M., and Inoue, T., Further Experiments on the Structure of a Spray Combustion Flame, 16th International Symposium on Combustion, The Combustion Institute, Pittsburgh, PA, 561–567, 1977.

Oran, E.S., Astrophysical Combustion, 30th International Symposium on Combustion, Combustion Institute, Pittsburgh, PA, 2004.

Orloff, L. and de Ris, J., Modeling of Ceiling Fires, 13th Symposium on Combustion, 979–992, 1971.

Orloff, L., de Ris, J., and Markstein, G.H., Upward Turbulent Fire Spread and Burning of Fuel Surface, 15th Symposium on Combustion, The Combustion Institute, Pittsburgh, PA, 183–192, 1974.

Oxtoby, D.W., Freeman, W.A., and Block, T.F., *Chemistry Science of Change*, Saunders College Publishing, 1998.

Pagni, P.J. and Shih, T.M., Excess Pyrolyzate, 16th Symposium, 1329–1343, 1974.

Patankar, S.V., *Numerical Heat Transfer and Fluid Flow*, Hemisphere Publishing, New York, 1980.

PCGC-2, Revised User's Manual, Advanced Combustion Engineering Research Center, Brigham Young University and University of Utah, 1988.

Penner, S.S., *Chemistry Problems in Jet Propulsion*, Pergamon Press, New York, 1957.

Perry, R.H. and Green, D.W., *Perry's Chemical Engineers' Handbook*, 7th ed., McGraw-Hill, New York, 1997.

PETC Review, Pittsburgh, PA, 1990.

Pohl, J.H. and Sarofim, A.F., Proceedings of the Stationary Source Combustion Symposium, EPA report 600-12-76-15a, 1976.

Pohl, J.H. and Sarofim, A.F., Devolatalization and Oxidation of Coal Nitrogen, 16th International Symposium on Combustion, The Combustion Institute, Pittsburgh, PA, 491–501, 1977.

Pope, S.B., PDF methods for turbulent reactive flows, *Prog. Energy Combust. Sci.*, 11, 119, 1985.

Poplow, F., Numerical calculation of the transition from subcritical droplet evaporation to supercritical diffusion, *Int. J. Heat Mass Transfer*, 37, 485–492, 1994.

Priyadarsan, S., Annnamalai, K., Sweeten, J.M., Holtzapple, M., and Mukhtar, S., Co-gasification of Blended Coal with Feedlot and Chicken Litter Biomass, 30th International Symposium on Combustion, 2004.

Qin, X., Choi, C.W., Mukhopadhyay, A., Puri, I.K., Aggarwal, S.K., and Katta, V.R., Triple flame propagation and stabilization in a laminar axisymmetric jet, *Combust. Theory Modeling,* 8, 1–22, 2004.

Qin, X., Puri, I.K., and Aggarwal, S.K., A numerical and experimental investigation of inverse triple flames at normal gravity, *Phys. of Fluids,* 40, 731–740, 2001.

Qin, X., Puri, I.K., and Aggarwal, S.K., Characteristics of lifted triple flames stabilized in the near field of a partially premixed axisymmetric jet, *Proc. Combustion Institute,* 29, 1565–1572, 2003.

Qin, X., Xiao, X., Puri, I.K., and Aggarwal, S.K., Effect of varying composition on temperature reconstructions obtained from refractive index measurements in flames, *Combust. Flame,* 128, 121–132, 2002.

Rahim, F., Elia, M., Ulinski, M., and Metghalchi, M., Burning velocity measurements of methane-oxygen-argon mixtures and an application to extend methane-air burning velocity measurements, *Int. J. Engine Res.*, 3, 81–92, 2002.

Raju, M.S. and Sirignano, W.A., Interaction between two vaporizing droplets in an intermediate Reynolds number flow, A,2(10), *Phys. Fluids*, 1780–1796, 1990.

Rangel, R.H. and Sirignano, W.A., Vaporization, Ignition, and Combustion of Two Parallel Fuel Droplet Streams, presented at ASME/JSME Thermal Engineering Conference, 1, 27–34, 1987.

Reed, R., *Combustion Handbook II*, North American Co., Cleveland, OH, 1997.

Reid, R.C., Prausnitz, J.M., and Poling, B.E., *The Properties of Gases and Liquids*, 4th ed., McGraw-Hill, New York, 1987.

Reitz, R.D. and Kuo, T.W., Modeling of HC Emissions due to Crevice Flows in Premixed Charge Engines, SAE 892085, 1989.

Renksizbulut, M. and Haywood, R.J., Transient droplet evaporation with variable properties and internal circulation at intermediate Reynolds numbers, *Int. J. Multiphase Flow*, 14, 189–202, 1988.

Ronney, P.D., Understanding Combustion Processes through Microgravity Research, Proceedings of the Combustion Institute, Vol. 27, pp. 2485–2506, 1998.

Roper, F.P., Prediction of laminar jet diffusion flame sizes: Part I, *Combust. Flame*, 29, 219–226, 1977a.

Roper, F.P., Prediction of laminar jet diffusion flame sizes: Part II. Experimental verification, *Combust. Flame*, 29, 227–234, 1977b.

Rosenburg, H.S., Curran, L.M., and Slack, A.V., Post combustion models for control of NO_x emissions, *Prog. Energy Combust. Sci.*, 6, 287–302, 1980.

Ryan, W. and Annamalai, K., Group ignition of coal particles, *J. Heat Transfer*, 113, 677–687, 1991.

Ryan, W., Annamalai, K., and Caton, J., Relation between group combustion and drop array studies, *Combust. Flame*, 80, 313–321, 1990.

Ryan, Westinghouse, Boca Raton, private communication, 2005.

Sami, M., Numerical Modeling Of COAL-Feedlot Biomass Blend Combustion and NO_x Emissions in Swirl Burners, Ph.D. dissertation, Texas A&M University, 2000.

Scala, IAS, Paper # 62–154, 1962.

Schlichting, H., *Boundary Layer Theory*, 7th ed., McGraw Hill, London, 1955.

Schneider, T. and Grant, L. (eds.), *Air Pollution by Nitrogen Oxides: proceedings of US-Dutch international symposium,* Elsevier Scientific Publishing, Amsterdam, 1982.

Seiser, R., Pitsch, H., Seshadri, K., Pitz, W.J., and Curran, H.J., Extinction and autoignition of n-heptane in counterflow configuration, *Proc. Combustion Institute,* 28, 2029–2037, 2000.

Senior, C.L., Helble, J.J., and Sarofim, A.F., Conference on Air Quality II: Mercury, Trace Elements, and Particulate Matter, McLean, Virginia, September 19–21, 2000a.

Senior, C.L., Sarofim, A.F., Zeng, T., Helble, J.J., and Mamani-Paco, R., *Fuel. Proc. Technol.*, 63(2–3), 197–214, 2000.

Sergeant, G.D. and Smith, I.W., Combustion rate of bituminous coal char in the temperature 800 to 1700, *Fuel,* 52, 52–57, January, 1973.

Seshadri, K. and Williams, F.A., Laminar flow between parallel plates with injection of a reactant at high Reynolds number *Int. J. Heat Mass Transfer,* 21, 2, 251–253, 1978.

Seshan and Kumar, J.P.L., Caltech. Summer National Meeting AIChE, September 1, 1982.

Shimy, A.A., Calculating flammability characteristics of hydrocarbons and alcohols, *Fire Technol.*, 135–139, 1970.

Sibulkin, M., Kulkarni, A., and Annamalai, K., Effects of Radiation on the Burning of Vertical Fuel Surfaces, 18th International Symposium on Combustion, Waterloo, Canada, pp. 611–617, 1981.

Sibulkin, M., Kulkarni, A., and Annamalai, K., Burning on vertical fuel with finite reaction rate, *Combust. Flame,* 44, 187–199, 1982.

Sirignano, W.A., Fuel droplet vaporization and spray combustion theory, *Prog. Energy Combust. Sci.*, 9, 291–322, 1983.

Sirignano, W.A., An Integrated Approach to Spray Combustion Model Development, ASME, WAM, Dec 7–12, 1986.

Sliger, R.N., Kramlich, J.C., and Marinov, N.M., Towards the development of a chemical kinetic model for the homogeneous oxidation of mercury by chlorine species, *Fuel Proc. Technol.*, 65–66, 423–438, 2000b.

Smith, I.W., 19th International Symposium on Combustion, The Combustion Institute, Pittsburgh, PA, pp. 1045–1065, 1982.

Smith, I.W., Kinetics of combustion of size graded pulverized fuels in the temperature range 1200 to 2270 K, *Combust. Flame,* 17, 303–314, 1971a.

Smith, I.W., The kinetics of combustion of pulverized semi-anthracite in the temperature range 1400 to 2200 K, *Combust. Flame,* 17, 421–428, 1971b.

Smith, J.M. and Van Ness, H.C., *Introduction to Chemical Engineering Thermodynamics,* 4th ed., McGraw-Hill, New York, 1987.

Smith, P.J., Hill, S.C., and Smoot, L.D., Theory for NO Formation in Turbulent Coal Flames, 19th International Symposium on Combustion, The Combustion Institute, Pittsburgh, PA, pp. 1263–1270, 1982.

Smoot, L. and Pratt, D.T., *Pulverized Coal Combustion and Gasification,* Plenum Press, New York, 1979.

Smoot, L.D. and Smith, P.J., *Coal Combustion and Gasification,* Plenum Press, New York, 1985.

Stanmore, B.R., Modeling the combustion behavior of petroleum coke, *Combust. Flame,* 83, 221–227, 1991.

Steward, F.R., Couturier, M., and Morris, K., Firing coal and limestone in a low-NOx burner, to reduce SO2 emissions, *J. Inst. Energy,* December 1992.

Strahle, W.C., *An Introduction to Combustion,* Gordon and Breach Science, Netherlands, 1993, 1998.

Street, J.C. and Thomas A., Carbon formation in premixed flames, *Fuel,* 34, 4–36, 1955.

Street, P.J., Weight, R.P., and Lightman, P., Further investigations of structural changes occurring in pulverized coal particles during rapid heating, *Fuel,* 48, 343–364, 1969.

Strehlow, R.A., *Combustion Fundamentals,* McGraw-Hill, New York, 1984.

Strehlow, R.A., *Fundamentals of Combustion,* International Textbook Co., Scranton, PA, 1968.

Strickland–Constable, R.F., 2nd Conference on Industrial Carbon and Graphite, Society of Chemical Industry, London, 255, 1965.

Sunderland, P.B., Yuan, Z.-G., and Urban, D.L., Shapes of Buoyant and Non-buoyant Laminar Jet Diffusion Flames, Proceedings of 1997 Technical Meeting of Central States Section of Combustion Institute 55-60; also see *Combust. Flame,* 116, 376–386, 1999.

Sutton, D., Kelleher, B., and Ross, J.R.H., *Fuel Process. Technol.*, 73–155, 2001.

Suuberg, E.M., Peters, W.A., and Howard, J.B., Product Compositions and Formation in Rapid Pyrolysis of Pulverized Coal-Implications for Comb., *Ind. Eng. Chem. Process. Des. Develop.*, 17, 37, 1978.

Suzuki, T. and Chiu, H.H., Multi-Droplet Combustion of Liquid Propellants, Proceedings of the 9th International Symposium on Space Technology and Science, Tokyo, Japan, 145–154, 1971.

Sweeten, J.M., Annamalai, K., Thien, B., and McDonald, L., Co-firing of coal and cattle feedlot biomass (FB) fuels, part I: feedlot biomass (cattle manure) fuel quality and characteristics, *Fuel,* 82(10), 1167–1182, 2003.

Tal, R.T., Lee, D.N., and Sirignano, W.A., Heat and momentum transfer around a pair of spheres in viscous fluid, *Int. J. Heat Mass Transfer,* 27, 1953–1962, 1984.

Tennekes, H. and Lumley, J.L., *A First Course in Turbulence,* MIT Press, Cambridge, 1990.

Thien, B. and Annamalai, K., National Combustion Conference, Oakland, CA, March 25–27, 2001.

Thien, B., Annamalai, K., and Bukur, D., Thermogravimetric Analyses of Coal, Feedlot Biomass and Blends in Inert and Oxidizing Atmospheres, Proceedings of IJPGC:IJPGC-2001, JPGC 2001/FACT-19082, New Orleans, LA, June 4–7, 2001.

Tillman, S.T., Annamalai, K., and Caton, J., A lift off and blow off characteristics of laminar non-premixed planar jets, *ASME,* Vol. 352, 101–108, 1997.

Tillman, S.T., Annamalai, K., and Caton, J.A., Investigation of Schmidt Number Dependence on Lift off and Blow off in Laminar Planar Jets, ASME-Energy Sources Technology Conference and Exhibition 1999, ETCE 99-6696, pp. 1–9, 1999.

Tillman, T., Ph.D. thesis, Lift-off and blow-off in gas fired cluster burners, Texas A&M University, College Station, 2000.

Timothy, L.D., Froelich, D., Sarofim, A.F., and Beer, J.M., Soot Formation and Burnout during the Combustion of Dispersed Pulverized Coal Particles, 21st International Symposium on Combustion, The Combustion Institute, Pittsburgh, PA, pp. 1141–1148, 1986.

Timothy, L.D., Sarofim, A.F., and Beer, J.M., Characteristics of Single Particle Coal Combustion, 19th International Symposium on Combustion, 1123–1130, 1982.

Tishkoff, J., A model for the effect of droplet interactions on vaporization, *Int. J. Heat Mass Transfer,* 22, 1407–1415, 1979.

Toqan, M.A., Beer, J.M., Johnson, P., Sun, N., Tsta, A., Shihadeh, A., and Teare, D., 24th International Symposium on Combustion, The Combustion Institute, Pittsburgh, PA, 1391–1397, 1992.

Touloukian, Y.S. and Ho, C.Y., Thermophysical properties of matter, TPRC Data Series, IFI/Plenum, New York, 1970–1972.

Tsuji, H. and Matsui,K., An aerothermo-chemical analysis of combustion of carbon in stagnation flow, the effect of water vapor concentration on the rate of combustion of an artificial graphite in humid air flow, *Combust. Flame,* 26, 283–297, 1976.

Turns, S., *An Introduction to Combustion: Concepts and Applications,* McGraw-Hill, New York, 1996.

Turns, S.R., Myhr, F.H., Bandaru, R.V., and Maund, E.R., Oxides of nitrogen emissions from turbulent flame, Part I: fuel effects and flame radiation, *Combust. Flame,* 87, 319–335, 1991.

Turns, S.R., Myhr, F.H., Bandaru, R.V., and Maund, E.R., Oxides of nitrogen emissions from turbulent flame, Part II: fuel dilution and partial premixing effects, *Combust. Flame,* 43, 255–269, 1993.

Twardus, E.M. and Brzustowski, T.A., The Interaction between Two Burning Fuel Droplets, 5th International Symposium on Combustion Processes, Krakow, Poland, 1977.

Ubhayakar, S.K., Burning characteristics of a spherical particle reacting with ambient oxidizing gas at its surface, *Combust. Flame,* 26, 24–34, 1976.

Ubhayakar, S.K., Stickler, D.B., and Gannon, R.E., Modelling of entrained bed pulverized coal gasifiers, *Fuel,* 56, 281–291, 1977.

Ubhayakar, S.K., Stickler, D.B., Von Rosenberg, C.W., and Gannon, R.E., Rapid Devolatization of Pulverized Coal in Hot Combustion Gases, 16th International Symposium on Combustion, The Combustion Institute, Pittsburgh, PA, pp. 427–436, 1976.

US DOE, www.eere.energy.gov/biopower.

Vargaftik, N.B., *Tables on the Thermophysical Properties of Liquids and Gases,* 2nd ed., Halsted Press, New York, 1975.

Versteeg, H.K. and Malalasekera, W., *An Introduction to Computational Fluid Dynamics — The Finite Volume Method,* John Wiley & Sons, New York, 1995.

Walavalkar, A.Y. and Kulkarni, A.K., Combustion of water-in-oil emulsion layers of diesel supported on water under external heat flux, *Combust. Flame,* 125, 1001–1011, 2001.

Waranatz, R.J. and Miller, J.A., A FO1 Computer Code Package for the Evaluations of Gas Phase Viscosities, Conductivities, and Diffusion Coefficients, SANDIA report, SAND 83-8209 UC-32, March 1983.

Welty, J.R., Wicks, C.E., and Wilson, R.E., *Fundamentals of Momentum, Heat, and Mass Transfer*, 3rd ed., John Wiley & Sons, New York, 1984, 803 pp.

Westbrook, C.K. and Dryer, F.C., PECS, 10, 1–57, 1984.

White, J.R., Detecting pulverizer fires before they start, *Power Eng.*, 42–43, February 1993.

Wicke, E., Contributions to the Combustion Mechanism of Carbon, 5th International Symposium on Combustion, Reinhold Publishing Co., New York, pp. 245–252, 1955. [See Smith, 1983].

Widman, J.F. and Davis, E.J., Evaporation of multi-component droplets, *Aerosol Sci. Technol.*, 27(2), 243–254, August 1997.

Widmer, N.C., West, J., and Cole, J.A., Thermochemical study of mercury oxidation in utility boiler flue gases, presented at Air and Waste Management Association Annual Meeting, Salt Lake City, Utah, June 18–22, 2000.

Wierzba, I. and Karim, G.A., Prediction of Flammability Limits of Fuel Mixtures, AFRC/JFRC International Symposium, October 1998.

Wiesenstein, D.K., Ko, M.K.W., SZe, N.-D., and Rodriguez, J.M., *Geophysical Res. Lett.*, 23, 161–164, 1996.

Wilhelm, S.M., Generation and disposal of petroleum processing waste that contains mercury, *Environ. Prog.*, 18(2), 130–143, 1999.

Wilke, C.R., Diffusional properties of multicomponent gases, *Chem. Eng. Prog.*, 46(2), 95–104, 1950.

Williams, A., *Combustion of Liquid Fuel Sprays*, Butterworths, London, 1990.

Williams, A., Fundamentals of oil combustion, *Prog. Energy Combust. Sci.*, 2, 167–179, 1976.

Williams, F.A., *Combustion Theory*, 2nd ed., Benjamin Cummins Publishing, Menlo Park, CA, 1985.

Williams, F.A., Turbulent combustion, in *The Mathematics of Combustion*, John D. Buckmaster, Ed., Society of Industrial and Applied Mathematics (SIAM), 1985, pp. 97–131, chap. 3.

Winters, J., Carbon underground, *Mech. Eng.*, 125, 46–48, 2003.

Wood, B.J., Wise, H., and Inami, S.H., Heterogeneous combustion of multicomponent fuels, *Combust. Flame*, 26, 17–22, 1976.

Xiao, X. and Puri, I.K., Digital recording and numerical reconstruction of holograms: A new optical diagnostic for combustion, *Appl. Opt.*, 41, 3890–3899, 2002.

Xiao, X., Puri, I. K., and Agrawal, A., Temperature measurements in steady axisymmetric partially premixed flames using rainbow schlieren deflectometry, *Appl. Opt.*, 41, 1922–1928, 2002.

Xin, J. and Megaridis, C.M., Interacting Droplet Dynamics and Vaporization in Dense Spray Systems, CSS/CI 94, pp. 165–170, 1994.

Xiong, T.Y., Law, C.K., and Miyasaka, K., Interactive Vaporization and Combustion of Binary Droplet Systems, 20th International Symposium on Combustion, The Combustion Institute, Pittsburgh, PA, 1781–1787, 1984.

Xue, H. and Aggarwal, S.K., Osborne, R.J., Brown, T.M., and Pitz, R.W., Comparison of predicted and measured counterflow partially premixed flame structures: assessment of reaction mechanisms, *AIAA Journal*, 40, 1236–1239, 2002a.

Xue, H. and Aggarwal, S.K., The structure and extinction of heptane-air partially-premixed flames, *AIAA Journal*, 40, 2289–2297, 2002b.

Xue, H. and Aggarwal, S.K., NOx emissions in n-heptane partially flames, *Combust. Flame*, 132, 723–741, 2003.

Yaverbaum, L.H., *Nitrogen Oxides Control and Removal — Recent Developments*, Noyes Data Corp., New Jersey, 1979.

Yule, A.J., Seng, A.H., Felton, P.G., Ungut, A., and Chigier, N.A., Sprays, drops, dusts, particles: a study of vaporizing fuel sprays by laser techniques, *Combust. Flame*, 44, 71–84, 1982.

Zabetakis, M.G., Flammability Characteristics of Combustible Gases and Vapors, Bulletin 627, Bureau of Mines, 1965a.

Zabetakis, M.G., Fire and Explosion Hazards at Temperature and Pressure Extremes, AIChE Institute of Chemical Engineers Symposium Ser 2, *Chem. Engr. Extreme Cond. Proc. Symp.*, 99–104, 1965b.

Zabetakis, M.G., Lambiris, S., and Scott, G.S., Flame temperatures of Limit Mixtures, 7th Symposium on Combustion, Butterworths, London, p. 484, 1959.

Zelenka, P., Kriegler, W., Herzog, P.L., and Cartellieri, W.P., Ways toward the Clean Heavy-Duty Diesel, SAE Paper 900602, 1990.

Zhou, J., Masutani, S.M., Ishimura, D.M., Turn, S.Q., and Kinoshita, C.M., Release of fuel-bound nitrogen during biomass gasification, *Ind. Eng. Chem. Res.*, 39(3), 626–634. 1997.

Zhuang, F.C., Ynag, B., Zhou, J., and Yang, H., Unsteady Combustion and Vaporization of a Spherical Fuel Droplet Group, 21st International Symposium on Combustion, The Combustion Institute, Pittsburgh, PA, 647–653, 1986.

Zumdahl, S.S., *Chemistry*, DC Heath and Company, Lexington, MA, 1986.

Appendix A

Abraham, M., Annamalai, K., and Claridge, D., Optimization of the storage process for a cool thermal storage system, *J. Energy Resour. Technol.*, 119, 236–241, 1997.

Annamalai, K., Sweeten, J., and Ramalingam, S.C., Estimation of the gross heating values of biomass fuels, *Trans. ASAE*, 30, 1205–1208, 1987.

David, R.L., *CRC Handbook of Chemistry and Physics*, 74th ed., CRC Press, Boca Raton, FL, 1993, pp. 6-203–6-204.

Ebeling, J.M. and Jenkins, B.M., Physical and chemical properties of biomass fuels, *Trans. ASAE*, 28(3), 898–902, 1985.

Incropera, F.P. and DeWitt, D.P., *Heat Transfer*, 5th ed., 2002, pp. 963–967.

Kuo, K.K. and Turns, S., *An Introduction to Combustion: Concepts and Applications*, McGraw-Hill, New York, 1996, pp. 163–168, chap. 3.

Lide, D.R., ed., *CRC Handbook of Chemistry and Physics*, 80th ed., CRC Press, Boca Raton, FL, 1999.

Smith, J. and Van Ness, H.C., *Introduction to Chemical Engineering Thermodynamics*, McGraw-Hill, New York, 1987, chap. 10.

Waranatz, R.J. and Miller, J.A., A FO1 Computer Code Package for the Evaluations of Gas Phase Viscosities, Conductivities, and Diffusion Coefficients, SANDIA report, SAND 83-8209 UC-32, March 1983.

Westbrook, C.R. and Dryer, F.L., Simplified reaction mechanisms for the oxidation of hydrocarbon fuels in flames, *Comb. Sc. Tech.*, 31–43, 1981.

Appendix B

Annamalai, K. and Ryan, W., Interactive process during gasification and combustion; Part II: isolated coal and porous char particles, *Prog. Energy Combust. Sci.*, 19, 383–446, 1993.

Berlad, A.L. and Potter, A.E., The effect of fuel type and pressure on flame quenching, *Combust. Flame*, 1, 1, 127–128, 1957.

Index

A

B

C

D

E

F

G

H

I

M

O

P

Q

R

S

T

U

V

W

Z